Spatial Sound

Spatial sound is an enhanced and immersive set of audio techniques which provides sound in three-dimensional virtual space. This comprehensive handbook sets out the basic principles and methods with a representative group of applications: sound field and spatial hearing; principles and analytic methods of various spatial sound systems, including two-channel stereophonic sound, and multichannel horizontal and spatial surround sound; Ambisonics; wavefield synthesis; binaural playback and virtual auditory display; recording and synthesis, and storage and transmission of spatial sound signals; and objective and subjective evaluation. Applications range from cinemas to small mobile devices.

- The only book to review spatial sound principles and applications extensively
- Covers the whole field of spatial sound

The book suits researchers, graduate students, and specialist engineers in acoustics, audio, and signal processing.

Bosun Xie was born in Guangzhou, China, in 1960. He received a Bachelor's degree in physics and a Master of Science degree in acoustics from the South China University of Technology in 1982 and 1987, respectively. In 1998, he received a Doctor of Science degree in acoustics from Tongji University.

Since 1982, he has been working at the South China University of Technology and is currently the director and a professor at Acoustic Lab., School of Physics and Optoelectronics. He is also a member of the State Key Lab of Subtropical Building Science. His research interests include binaural hearing, spatial sound, acoustic signal processing, room acoustics, the relation between modern physics and classical acoustics. He has published a book entitled "Head-related transfer function and virtual auditory display" and over 300 scientific papers. He owns 20 patents in audio fields. His personal interest is in classical music, particularly classical opera.

He is the vice-president of the Acoustical Society of China (2014–2022), a member of the Audio Engineering Society (AES), and a member of the Acoustical Society of America (ASA).

Spatial Sound

Principles and Applications

Bosun Xie

CRC Press
Taylor & Francis Group
Boca Raton London New York

CRC Press is an imprint of the
Taylor & Francis Group, an **informa** business

First (Chinese) edition published 2019 by the Science Press (Beijing China 2019).
Second (English) edition published 2023
by CRC Press
6000 Broken Sound Parkway NW, Suite 300, Boca Raton, FL 33487-2742

and by CRC Press
4 Park Square, Milton Park, Abingdon, Oxon, OX14 4RN

CRC Press is an imprint of Taylor & Francis Group, LLC

ISBN: 978-0-367-53344-1 (hbk)
ISBN: 978-0-367-53345-8 (pbk)
ISBN: 978-1-003-08150-0 (ebk)

DOI: 10.1201/9781003081500

Typeset in Sabon
by SPi Technologies India Pvt Ltd (Straive)

Contents

Preface

In addition to vision, hearing is a means for humans to acquire external information. Human hearing can perceive not only the loudness, pitch, and timbre of sound but also the spatial attributes of sound. With spatial auditory perception, we can localize a sound source and create spatial auditory sensations of the environment.

Spatial sound or spatial audio aims to record (or simulate), transmit (or store), reproduce the spatial information of a sound field, and recreate the desired spatial auditory events or perceptions. Spatial sound is traditionally applicable to cinema and domestic sound reproductions. Recently, spatial sounds have been increasingly applied to wide fields of scientific research and engineering, such as psychoacoustic and physiological acoustic experiments, room acoustic designs, communication, computers and the internet, multimedia, and virtual reality.

Spatial sound has a long history dating back to more than 100 years ago. Since the 1930s, spatial sound techniques have been developed and used for practical application through the combination of acoustics and electronics. Since the 1990s, computer and digital signal processing techniques have further enabled spatial sound to develop quickly. There have been numerous scientific and technical studies on spatial sound. Various spatial sound techniques based on different physical and auditory principles have developed, and some of these techniques have been widely used.

In China, research on spatial sound began in 1958. Especially, the group at the South China University of Technology has conducted a series of fundamental and application studies on this field. Since 2010, spatial sound has gradually received attention in China, and some other groups have conducted relevant work.

From the point of scientific research, spatial sound is an interdiscipline dealing with acoustics (physics), psychology, and physiology of hearing, electronics and signal processing, computers, and even the art of music. Physical and auditory analysis of a sound field is the foundation of spatial sound. Signal processing, electronics, electroacoustic devices, and instruments are technical means for implementing spatial sound. With wide applications, spatial sound has been an active field in audio and signal processing and is still developing quickly. The development of spatial sound deals with both fundamental and application studies.

Internationally, special topics on spatial sound have been covered in some books, such as *Spatial Hearing* by Prof. Blauert (1997), *Analytic Methods of Sound Field Synthesis* by Dr. Jens Ahrens (2012), *Ambisonics* by Dr. Franz Zotter and Matthias Frank (2019), *Sound Visualization and Manipulation* by Profs. Yang-Hann Kim and Jung-Woo Choi (2013), and *3D Sound for Virtual Reality and Multimedia* by Dr. Duran R. Begault (1994). The author of the present book previously wrote a book entitled *Head-related Transfer Function and Virtual Auditory Display* (Chinese edition in 2008 and English edition in 2013). However, books that cover relatively complete topics on spatial sound are rare. The only book is *Spatial Audio* by Prof.Francis Rumsey (2001), which is one of the series books intended to support college and university courses in music technology, sound recording, multimedia, and their related fields. In fact, writing a book that covers most aspects of the principle and applications of spatial sound is difficult because of the long history, extensive contents, and quick development in this field.

Prof. Xinfu Xie at the South China University of Technology wrote a book entitled *The Principle of Stereo Sound* in 1981. It reviewed and summarized the main international works on spatial sound before the end of the 1970s and contributed to the development of spatial sound in China. However, the book by Prof. XinfuXie was a Chinese edition and published in more than 40 years ago. During the past 40 years, especially since 1990, spatial sound has been developed greatly. Current spatial sound techniques differ considerably from those in 1970 in many aspects of basic physical, auditory principles, and technical means. Therefore, a book on the principles and applications of spatial sound should be rewritten.

The present book systematically states the basic principles and applications of spatial sound and reviews the latest development, especially those from the author's research group. The book focuses on the physical and auditory principles of spatial sound. Another major purpose of the present book is to reveal that various spatial sound techniques are unified under the theoretical framework of spatial function sampling, interpolation, and reconstruction. The original Chinese edition was published by the Science Press (Beijing) in 2019. The present English edition is formed mainly from the Chinese edition with amendments, including the most recent developments from 2019 to 2021.

The book consists of 16 chapters, covering the main issues in the research of spatial sound. Chapter 1 presents the essential principles and concepts of sound field, spatial hearing, and sound reproduction to provide readers with sufficient background information for elaborating the succeeding chapters. Chapter 2 describes the basic principles and some issues related to the applications of two-channel stereophonic sound. Chapters 3–6 discusses the basic principles and traditional analysis of various multichannel horizontal and spatial surround sounds in detail. Chapter 7 presents the methods of microphone and signal simulation techniques for multichannel sounds. Chapter 8 discusses the matrix surround sound and downmixing/upmixing of multichannel sound signals. Chapters 9 and 10 address the principles and methods of physical sound field analysis and reconstruction and discuss the principles of Ambisonics and wave field synthesis in detail. Chapter 11 describes the principle and method of binaural reproduction and virtual auditory display. Chapter 12 presents the method of binaural pressures and auditory model analysis of spatial sound reproduction. Chapters 13–15 discuss some issues related to the application of spatial sounds, including signal storage and transmission, acoustic conditions, requirements and methods for subjective assessment and monitoring. Chapter 16 outlines some representative applications of spatial sound. In addition, two appendices briefly introduce some mathematical tools for the analysis in the main text. The present book lists more than 1000 references at the end, representing the main body of literature in this field.

The present book intends to provide the necessary knowledge and latest results to researchers, graduate students, and engineers who work in the field of spatial sound. Readers can become familiar with the frontier of the field after reading and undertake the corresponding scientific research or technical development work. Because this field is interdisciplinary, reading this book needs some prior understanding of acoustics and signal processing. The References section provides relevant references about previous studies.

The publication of the present book is supported by the National Nature Science Fund of China (12174118) and the National Key Research and Development Program of China (2018YFB1403800). The relevant studies on spatial sound by the author and our group have been supported by a series of grants from the National Nature Science Fund of China (11674105, 19974012, 10374031, 10774049, 11174087, 50938003, 11004064, 11474103, 11574090, and 11104082), the Ministry of Education of China for outstanding young teachers, Guangzhou Science and Technology plan projects (98-J-010-01, 2011DH014, and 2014-Y2-00021), the State Key Lab of Subtropical Building Science, South China University of Technology., South China University of Technology, where the author works, has also provided enormous supports.

With more than 20 years of research experience, Prof. Shanqun Guan, working at the Beijing University of Posts and Telecommunications, has generously provided many guidance and suggestions. The author has also received long-term help and support from Prof. Zuomin Wang, the author's PhD advisor, at Tongji University since the mid-1990s.

The author is especially indebted to Profs. Guangzhen Yu, Xiaoli Zhong, Zhiwen Xie, and Drs. Dan Rao and Qinglin Meng at the South China University of Technology, Dr. Chengyun Zhang at Guangzhou University, and all graduate students who provided support and cooperation. The author also expresses gratitude to Prof. Guangzheng Yu for preparing all figures, Dr. Qinglin Meng for revising the English translation, and PhD students, namely, Haiming Mai, Jianliang Jiang, Kailin Yi, Lulu Liu, Tong Zhao, Jun Zhu, Wenjie Ding, and Shanwen Du, for their help in checking the references and proof of the book.

Many colleagues also provided the author with various kinds of support and help during the author's research work, particularly Prof. Jens Blauert at Ruhr-University Bochum; Prof. Ning Xiang at Rensselaer Polytechnic Institute; Profs. Shuoxian Wu and Yuezhe Zhao at the School of Architecture, South China University of Technology; Profs. Jian Zhong, Hao Shen, Mingkun Cheng, Jun Yang, Xiaodong Li, Yonghong Yan, and Junfeng Li at the Institute of Acoustics at the China Academy of Sciences; Profs. Jianchun Cheng, Boling Xu, Xiaojun Qiu, Yong Shen, Xiaojun Liu, and Jin Lu at Nanjing University; Prof. Dongxing Mao and Dr. Wuzhou Yu at Tongji University; Prof. Changcai Long at the Huazhong University of Science and Technology; Prof. Dean Ta at Fudan University; Prof. Hairong Zheng at the Shenzhen Institute of Advanced Technology, Chinese Academy Science; Profs. Baoyuan Fan and Jingang Yang, Senior Engineers Jincai Wu and Houqiong Zhong at The Third Research Institute of China Electronics Technology Group Company and Senior Engineer Jinyuan Yu at Guoguang Electric Co., Ltd.; Senior Engineer Jiakun Qi at Wuhan Wireless Power Plant Co., Ltd.; and Mr. Heng Wang at Guangzhou DSPPA Audio Co., Ltd. The CRC Press, especially Mr. Tony Moore, Frazer Merritt, Aimee Wragg, Vasudevan Thivya and Anya Hastwell made enormous work on the publication of the present book.

The author would like to thank the abovementioned units and individuals.

The author's parents, Profs. Xingfu Xie and Shujuan Liang, were also acoustical researchers, who cultivated the author's enthusiasm for acoustics. The author's mother also gave great support during the preparation of the present book in 2009. The present book is in memory of the author's parents who have since passed away.

Introduction

This book presents the basic principles, methods, and applications of spatial sound, and summarizes the recent results and progress in this field, including the latest fruits from the author and team. There are 16 chapters in the book. Contents of the book cover sound field and spatial hearing; principles and analytic methods of various spatial sound systems, including two-channel stereophonic sound, multichannel horizontal and spatial surround sound, Ambisonics, wavefield synthesis, binaural playback, and virtual auditory display, recording and synthesis, storage and transmission of spatial sound signals, objective, and subjective evaluation as well as various applications of spatial sound. The book also lists more than 1000 references covering the main original works.

The book is intended for scientific researchers, graduate students, and engineers who deal with the field of spatial sound. Readers could become familiar with the frontier of the field after reading, and undertake corresponding scientific research or technical development work.

Chapter 1

Sound field, spatial hearing, and sound reproduction

In this chapter, the essential concepts, definitions, and principles of sound field, spatial hearing, and spatial sound reproduction are presented to provide readers with sufficient background information for elaborating in the succeeding chapters. In Section 1.1, the coordinate systems used in this book are defined to avoid ambiguity. In Section 1.2, a concise review of sound fields and their physical characteristics, including free fields, reflected sound fields in enclosed spaces, directivity of sound source radiation, and sound receivers is provided. In Section 1.3, the human auditory system and perception are outlined; the hearing threshold, the perception of loudness, and masking are discussed; and the concepts of critical bands and auditory filters are introduced. In Section 1.4, an artificial head model and the principle of binaural sound receivers are presented, and the concept of a head-related transfer function is described. In Section 1.5, the basic definition and scope of spatial hearing are provided. In Section 1.6, various cues and their effective range for the auditory localization of a single sound source are discussed in detail. In Section 1.7, the summing localization and other spatial-related perceptions of multiple sound sources are elaborated. In Section 1.8, the auditory spatial impression related to room reflections is explained. In Section 1.9, the principle, classification, and development of spatial sound are concisely presented.

1.1 SPATIAL COORDINATE SYSTEMS

A specific point in the three-dimensional space is first selected as the origin, and a coordinate system is defined. The position of an arbitrary point in space is specified in terms of its direction and distance with respect to the origin or equivalently specified by the ***position vector*** from the origin to the concerned point. In studies on spatial sound, ***the head center of a listener*** (i.e., ***the midpoint of the straight line connecting the two entrances of the ear canals***) ***is often selected as the origin of the coordinate system.*** Three specific planes are specified in this coordinate system: the ***horizontal plane***, determined by the two vectors from the origin to the front and left (or right) directions; the ***median plane***, determined by the two vectors from the origin to the front and top directions; and the ***lateral plane*** (or the frontal plane), determined by the two vectors from the origin to the top and left (or right) directions. The three planes are perpendicular to one another and intersect at the origin. Various specifications of the coordinate system are presented.

Coordinate system A is an anticlockwise spherical coordinate system with respect to the head center. In Figure 1.1, an arbitrary position in space is described with (r, θ, ϕ), where the distance with respect to the origin is denoted by r with $0 \leq r < +\infty$. The angle between the position vector and the horizontal plane is indicated by the elevation ϕ with $-90° \leq \phi \leq +90°$, where the bottom, top, and horizontal directions are represented by $-90°$, $90°$, and $0°$, respectively. The angle between the horizontal projection of the position vector

DOI: 10.1201/9781003081500-1

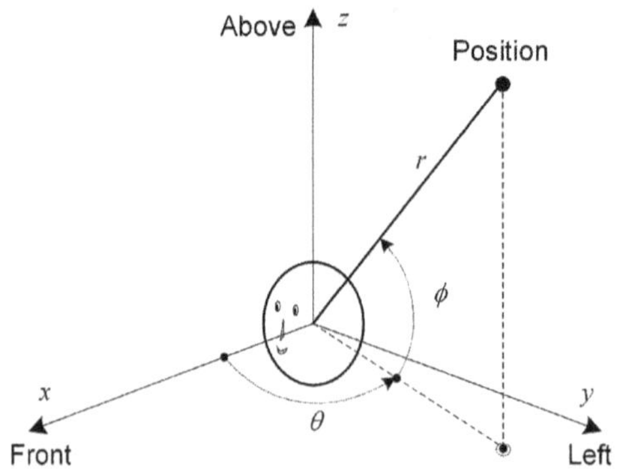

Figure 1.1 Coordinate system A: anticlockwise spherical coordinate system with respect to the head center.

and the front (x-axis) is denoted by the azimuth θ with $-180° < \theta \le 180°$ labeled anticlockwise, where the right, front, left, and back directions in the horizontal plane are indicated by -90°, 0°, 90°, and 180°, respectively.

In some studies, θ varies in the range of $0° \le \theta < 360°$, where 0°, 90°, 180°, and 270° represent the front, left, back, and right directions in the horizontal plane, respectively. The two kinds of θ variation range are equivalent because θ is a periodic variable with $\theta + 360°$ equal to θ.

The complementary angle α of ϕ, i.e., the angle between the position vector and the top direction (z-axis or polar axis), can also be used to denote the elevation with

$$\alpha = 90° - \phi \qquad 0 \le \alpha \le 180°. \tag{1.1.1}$$

In the analysis of the sound field of spatial sound reproduction, always supposing that a listener exists in the sound field is unnecessary. Therefore, coordinate system A can be generally considered as ***a coordinate system with respect to a specific receiver position or a field point*** regardless of the existence or nonexistence of a listener.

For convenience in analyzing the sound field generated by a sound source, the acoustic center of the source is sometimes selected as the origin of the coordinate. Thus, ***coordinate system B, which is a coordinate system with respect to the sound source***, is established. In Figure 1.2, the origin of the coordinate is located at the center of the sound source. The receiver position in space is described with its distance and direction with respect to the

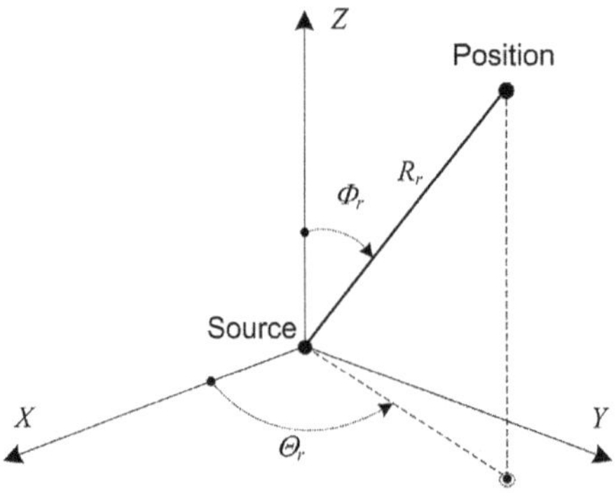

Figure 1.2 Coordinate system B: coordinate system with respect to the sound source.

origin or determined with the position vector $\boldsymbol{R}_r$ from the origin to the receiver position, where the subscript "r" refers to the receiver position. The receiver position can also be determined with Cartesian coordinates (X_r, Y_r, and Z_r) or spherical coordinates (R_r, Θ_r, and Φ_r), where $0 \leq R_r < +\infty$ denotes the distance with respect to the origin, $-180° < \Theta_r \leq 180°$ represents the azimuthal angle (or azimuth), and $0° \leq \Phi_r \leq 180°$ indicates the polar angle, i.e., the angle between the position vector and the polar axis (Z-axis).

The clockwise spherical coordinate system with respect to a listener's head center or a specific receiver position is also used in some literature and books on spatial hearing and sound reproduction. The azimuth in this coordinate system is labeled clockwise. Therefore, in the horizontal plane, azimuths −90°, 0°, 90°, and 180° represent the left, front, right, and back directions, respectively. Overall, the anticlockwise spherical coordinate system with respect to the head center is often used in studies on stereophonic and multichannel sounds, and the clockwise spherical coordinate system with respect to the head center is often utilized in the literature on spatial hearing, head-related transfer function, and virtual auditory display. However, ambiguity may arise if two or more coordinate systems are used in a book without explanation.

The anticlockwise spherical coordinate system with respect to the head center or a specific receiver position, i.e., coordinate system A, is the default in this book to be consistent with most studies on stereophonic and multichannel sounds. The results from the literature have been converted into coordinate system A with specific explanations unless otherwise stated. The default coordinate system here is different from that in the author's other book on head-related transfer functions and virtual auditory display, which utilizes the clockwise spherical coordinate system with respect to the head center (Xie, 2008a, 2013). Moreover, for convenience in analyzing the field generated by a sound source, the coordinate system with respect to sound source, i.e., coordinate system B, is sometimes used in this book with specific explanations. The variables in coordinate systems A and B are labeled with lowercase (r, θ, ϕ) and uppercase (R_r, Θ_r, Φ_r) letters, respectively. $\phi = 0°$ in coordinate system A is the horizontal plane, whereas $\Phi_r = 0°$ in coordinate system B is the direction of the polar axis (Z-axis). Confusion in identifying the two coordinate systems can be eliminated by being careful.

1.2 SOUND FIELDS AND THEIR PHYSICAL CHARACTERISTICS

Some basic physical concepts in sound waves and sound fields are briefly reviewed in this section before they are discussed in the succeeding chapters. The detailed analyses and mathematical derivations can be found in some textbooks on fundamental acoustics (Morse and Ingrad, 1968; Piere 2019; Du et al., 2001).

1.2.1 Free-field and sound waves generated by simple sound sources

Sound waves generated by sound sources propagate in a space with a medium. A sound field is defined as a region in a medium where sound waves are being propagated. Air can be considered as a uniform and isotropic medium. A sound field in air is physically characterized by the temporal and spatial distribution of sound pressure $p(r, t)$ in the time domain or equally characterized by the frequency and spatial distribution of sound pressure $P(r, f)$ in the frequency domain. In this book, vectors are denoted by boldface letters, functions in the time domain are indicated by lowercase letters, and functions in the frequency domain are represented by uppercase letters.

In a source-free region, sound pressure in air satisfies the homogeneous wave equation. In principle, sound pressure can be calculated by solving the wave equation imposed on appropriate initial and boundary conditions. Therefore, a sound field in air is determined by the physical characteristics and position of a sound source and the geometrical and acoustic characteristics of a given boundary.

Free field is an important concept in acoustic analysis. It refers to a special sound field in a uniform and isotropic medium in which the influences of boundaries are completely negligible (i.e., the absence of reflections from the boundary). Cases of the free field under actual conditions are rare. Sound fields can be approximated as a free field only in an anechoic chamber or a local region, which is sufficiently high above the ground. However, the concept of the free field specifies an ideal and standard condition for acoustic analysis. Therefore, many acoustic analyses and measurements are conducted under free-field conditions. In the succeeding chapters, the reproduced sound field are often analyzed and discussed under the assumption of a free field for convenience.

The simplest sound field is the plane wave in an infinite free space. In the coordinate system with respect to a specific receiver position, an arbitrary receiver position is denoted by vector $\boldsymbol{r}$, and the frequency-domain sound pressure of a plane wave is given by

$$P(\boldsymbol{r}, \Omega_S, f) = P_A(f)\exp(-j\boldsymbol{k}\cdot\boldsymbol{r}), \tag{1.2.1}$$

where j is the imaginary unit; f is the frequency; $\boldsymbol{k}$ is the wave vector whose direction is the propagating direction of the plane wave; $|\boldsymbol{k}| = k = 2\pi f/c$ is the wave number; c = 343 m/s is the speed of sound in air; $P_A(f)$ is the complex-valued ***amplitude*** of a harmonic plane wave with frequency f, whose modulus and phase angle are the ***magnitude*** and ***initial phase*** of the plane wave, respectively; and $\boldsymbol{k}\cdot\boldsymbol{r}$ is the scalar product of two vectors. Let $\Omega = (\theta, \phi)$ denote the direction of a receiver position, and $\Omega_S = (\theta_S, \phi_S)$ indicate the ***incident direction*** of the plane wave (at the origin, which is opposite to the propagating direction or the direction of a wave vector), then

$$\boldsymbol{k}\cdot\boldsymbol{r} = -kr\cos\Delta\Omega_S, \tag{1.2.2}$$

where $\Delta\Omega_S = \Omega - \Omega_S$ is the angle between the direction of the receiver position and the incident direction of the plane wave.

The wavefront of a plane wave is an infinite plane, and the amplitude of a plane wave is independent of the receiver position. As such, the overall power of a plane wave in an unbounded space is infinite. Therefore, plane waves can only be approximately generated in a local region, and the plane wave in an unbounded space is unrealized.

The sound field generated by a ***point source*** (***monopole source***) in a free field is relatively simple and can be realized physically. Let vector $\boldsymbol{r}_S$ denote the position of a point source in a coordinate system with respect to a specific receiver position. The frequency-domain sound pressure at an arbitrary receiver position $\boldsymbol{r}$ is expressed in the following equation by appropriately choosing the initial phase of a sound source:

$$\begin{aligned} P(\boldsymbol{r}, \boldsymbol{r}_S, f) &= \frac{Q_p(f)}{4\pi|\boldsymbol{r}-\boldsymbol{r}_S|}\exp\left[-j\boldsymbol{k}\cdot(\boldsymbol{r}-\boldsymbol{r}_S)\right] \\ &= \frac{Q_p(f)}{4\pi|\boldsymbol{r}-\boldsymbol{r}_S|}\exp(-jk|\boldsymbol{r}-\boldsymbol{r}_S|) \\ &= \frac{Q_p(f)}{4\pi R_r}\exp(-jkR_r), \end{aligned} \tag{1.2.3}$$

where $Q_p(f)$ is the complex-valued strength of a point source; the subscript "p" is the point source; $\boldsymbol{R}_r = \boldsymbol{r} - \boldsymbol{r}_S$ is the vector from the point source to an arbitrary receiver position, that is, the vector of the receiver position with respect to the source; and $R_r = |\boldsymbol{R}_r|$ is the distance between the receiver position and the sound source, that is, the radial coordinate of the receiver position in the coordinate system with respect to the sound source.

Equation (1.2.3) indicates that the sound field generated by a point source is a spherical wave in which the wavefront is a spherical surface, and sound pressure is independent of the direction of the receiver position with respect to the sound source. The amplitude of sound pressure is expressed as

$$P_A(f) = \frac{Q_p(f)}{4\pi R_r}. \tag{1.2.4}$$

Therefore, the magnitude $|P_A(f)|$ of sound pressure is inversely proportional to R_r. Every double distance between the receiver position and the sound source causes a −6 dB attenuation in the sound pressure level. This attenuation is the consequence of finite and constant power provided by the point source in the free field.

In Equation (1.2.4), for a spherical wave, the relative variation in pressure amplitude with R_r is $dP_A(f)/P_A(f) = -dR_r/R_r$. $|dP_A(f)/P_A(f)|$ decreases as the distance increases, i.e., $R_r = |\boldsymbol{r} - \boldsymbol{r}_S|$. For a large R_r, $dP_A(f)/P_A(f)$ approaches zero. Therefore, in the local region of a far field where the distance between the receiver position and the sound source is large enough, the amplitude $P_A(f)$ of sound pressure is approximately constant. In this case, the sound field of a point source can be approximated as a plane wave, and Equation (1.2.3) becomes

$$P(\boldsymbol{r}, \boldsymbol{r}_S, f) = P_A(f)\ \exp\left[-j\boldsymbol{k}\cdot(\boldsymbol{r} - \boldsymbol{r}_S)\right]. \tag{1.2.5}$$

The factor $\exp(j\boldsymbol{k} \cdot \boldsymbol{r}_S)$ in Equation (1.2.5) can be omitted by choosing an appropriate initial phase of the sound source, and the incident plane wave at the receiver position near the origin of the coordinate is expressed as

$$P(\boldsymbol{r}, \Omega_S, f) = P_A(f)\exp(-j\boldsymbol{k}\cdot\boldsymbol{r}). \tag{1.2.6}$$

This equation is the plane wave approximation of the far field generated by a point source in which $P_A(f)$ is independent of the sound source distance. Plane wave approximation is usually convenient for the analysis of the reproduced sound field. However, this approximation is only valid within a local region of the far field, and the spherical wave in the whole space can never be approximated as the plane wave.

The sound field generated by a ***straight-line source*** with an infinite length in the free field is relatively complicated. A straight-line source can be regarded as a composition of an infinite number of point sources, which are distributed densely and uniformly in a straight line. The sound field generated by a straight-line source with an infinite length is a cylindrical wave. In the coordinate system with respect to a specific receiver position, if the z-axis is chosen to be parallel to the straight-line source, the sound pressure is independent of the z coordinate of the receiver position. In this case, our analysis can be limited to the horizontal sound field. Here, $\boldsymbol{r}_S$ is the intersect position of the straight-line source and the horizontal plane, and $\boldsymbol{r}$ is the receiver position in the horizontal plane; then, the sound pressure generated by a straight-line source with an infinite length is given by

$$P(\boldsymbol{r}, \boldsymbol{r}_S, f) = -Q_{li}(f)\frac{j}{4}H_0(k|\boldsymbol{r} - \boldsymbol{r}_S|) = -Q_{li}(f)\frac{j}{4}H_0(kR_r), \tag{1.2.7}$$

where $Q_{li}(f)$ is the complex-valued strength of the straight-line source, the subscript "li" is the straight-line source, $R_r = |\boldsymbol{r} - \boldsymbol{r}_S|$ is the distance between the receiver position and the straight-line source, and $H_0(kR_r)$ is the zero-order Hankel function of the second kind. For $kR_r >> 1$, the asymptotic formula of the Hankel function is

$$H_0(kR_r) = \sqrt{\frac{2}{\pi k R_r}}\exp\left(-jkR_r + j\frac{\pi}{4}\right). \tag{1.2.8}$$

It is obtained

$$P(\boldsymbol{r}, \boldsymbol{r}_S, f) = -Q_{li}(f)\frac{j}{4}\sqrt{\frac{2}{\pi k R_r}}\exp\left(-jkR_r + j\frac{\pi}{4}\right). \tag{1.2.9}$$

In this case, the magnitude of sound pressure is inversely proportional to $\sqrt{R_r}$. A double distance between the receiver position and the linear source causes a –3 dB attenuation in the sound pressure level. This feature is observed in the sound field generated by a straight-line source with an infinite length in the free field. Similar to the case of Equation (1.2.6), in the local region of the far field with $kR_r >> 1$, the sound field generated by a straight-line source with an infinite length in the free field can be approximated as a plane wave with a complex-valued amplitude:

$$P_A(f) = -Q_{li}(f)\frac{j}{4}\sqrt{\frac{2}{\pi k R_r}}\exp\left(j\frac{\pi}{4}\right). \tag{1.2.10}$$

The time-domain sound pressures are evaluated by applying the inverse time–frequency Fourier transform to Equations (1.2.3), (1.2.6), and (1.2.7), respectively. For example, the case of a plane wave is expressed as follows:

$$p(\boldsymbol{r}, \Omega_S, t) = \int P(\boldsymbol{r}, \Omega_S, f)\exp(j2\pi f t)df = \frac{1}{2\pi}\int P(\boldsymbol{r}, \Omega_S, \omega)\exp(j\omega t)d\omega, \tag{1.2.11}$$

where $\omega = 2\pi f$ is the angular frequency.

According to spatial Fourier analysis (Williams, 1999), an arbitrary sound field in a source-free region can be decomposed as a linear superposition of the plane wave from various directions. In a region adjacent to the origin (in the coordinate system with respect to a specific receiver position), the frequency-domain sound pressure of the incident wave is given by

$$P(\boldsymbol{r}, f) = \int \tilde{P}_A(\Omega_{in}, f)\exp(-j\boldsymbol{k}\cdot\boldsymbol{r})d\Omega_{in}, \tag{1.2.12}$$

where $\tilde{P}_A(\Omega_{in}, f)$ is the complex-valued amplitude of the plane wave component with f and the incidence from the direction Ω_{in}, i.e., the direction–frequency distribution function of the complex-valued amplitude of the incident plane wave. The integral in Equation (1.2.12) is calculated over all possible incident directions Ω_{in}.

The time-domain sound pressure at the receiver position $\boldsymbol{r}$ is calculated by applying the inverse temporal frequency Fourier transform to Equation (1.2.12):

$$\begin{aligned} p(\boldsymbol{r}, t) &= \iint \tilde{P}_A(\Omega_{in}, f)\exp(-j\boldsymbol{k}\cdot\boldsymbol{r} + j2\pi f t)\, d\Omega_{in}\, df \\ &= \frac{1}{2\pi}\iint \tilde{P}_A(\Omega_{in}, \omega)\exp(-j\boldsymbol{k}\cdot\boldsymbol{r} + j\omega t)\, d\Omega_{in}\, d\omega. \end{aligned} \tag{1.2.13}$$

1.2.2 Reflections from boundaries

Radiations by sound sources in a free field are ideal cases, and reflective boundaries exist in most actual cases. Sound waves in a receiver position are a combination of the direct sound from the sound source and the reflected sounds from boundaries.

The simplest case that deals with reflections from the boundary is the sound field generated by a point source in front of a rigid plane with an infinite extension (such as a wall). In Figure 1.3, the strength of a real point source is $Q_p(f)$. Within the half space in which the real point source is located, sound pressure can be calculated by solving the nonhomogeneous wave equation imposed on the given boundary condition. Alternatively, it can be simply computed by applying the ***acoustic principle of an image source.***

The sound pressure at a receiver position $\boldsymbol{r}$ is the superposition of the free-field sound pressure generated by the real point source and the reflected sound pressure caused by the rigid plane. Reflection is taken as the sound wave originating from an image source S' with the same strength and located at the mirror position against the boundary. The summing sound pressure depends on the wave number (or frequency) and on R_r and R'_r between the receiver position and the two positions of real and image sources. When the path difference between real and image sources to the receiver position is an integer multiple of the wavelength, a constructive interference and a local maximum in sound pressure occur; when this path difference is an odd number multiple of the half wavelength, a destructive interference and a local minimum in the sound pressure are observed.

When the real sound source is exactly located on the surface of the rigid plane, $R_r = R'_r$, the position of the image source coincides with that of the real sound source, and the path difference between two sources to the receiver position vanishes. Accordingly, sound pressure is given by

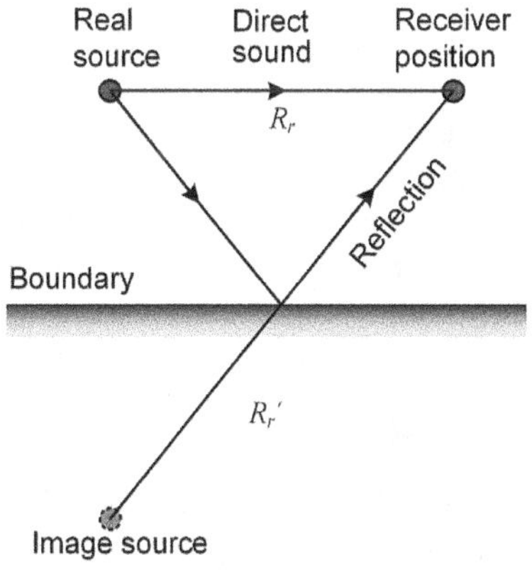

Figure 1.3 Sound field generated by a real point source in front of a rigid plane with an infinite extension.

$$P(R_r, f) = \frac{Q_p(f)}{2\pi R_r}\exp(-jkR_r). \tag{1.2.14}$$

Therefore, sound waves generated by real and image sources are added in phase at the receiver positions, and the frequency-position-dependent interference vanishes. In comparison with Equation (1.2.3), for a given distance between the sound source and the receiver position, the sound pressure magnitude generated by a point source located on the surface of a rigid plane with an infinite extension is twice that generated by a point source with the same strength in the free space. As a result, the sound pressure level increases by 6 dB. A twice sound pressure means a quadruple power density. The overall radiated power is twice the point source with the same strength in the free space because the point source on the surface of a rigid plane with an infinite extension generates sound waves in the half space, resulting in a 3 dB gain in the overall radiated power level.

Similarly, if a real point source is located in the space restricted by two rigid and perpendicular half planes with an infinite extension, the sound pressure at the receiver position can be regarded as the sum of those generated by the real source and three image sources with identical strength. When the real source is exactly located at the intersect line between two planes, the positions of all three image sources coincide with that of the real source, and the summing sound pressure is fourfold that of the real source in the free space. Thus, a 12 dB gain in the sound pressure level is obtained. In this case, the source generates a sound wave in the quarter space, and the overall radiated power is fourfold that of the real source in the free space. As a result, the overall radiated power level increases by 6 dB.

If a real point source is located at the corner formed by three rigid and perpendicular half-planes with an infinite extension, the summing sound pressure at the receiver position is eightfold that of the real source in the free space. Consequently, an 18 dB gain in the sound pressure level occurs. In this case, the source generates a sound wave in the eighth space, and the overall radiated power is eightfold that of the real source in the free space. Thus, a 9 dB gain in the overall radiated power level is achieved.

Therefore, the overall radiated power of a sound source is dependent on the spatial position of the source because of reflective boundaries. This dependency is due to the influence of reflections from boundaries on a sound source, thereby altering the radiation impedance and radiation efficiency of the sound source. The results on this issue are applicable to designing the arrangement of loudspeakers for reproduction in rooms (e.g., listening rooms), as described in Section 14.3.

In enclosed spaces, such as rooms with multiple reflective surfaces, the reflection from each surface is reflected by other surfaces once more, resulting in second-, third-, and higher-order reflections. This process can be modeled by the corresponding order of image sources. The number of image sources rapidly grows as the order of reflection increases, and the distribution of image sources constitutes a three-dimensional array outside the region of an enclosed space. The interference among direct and reflected sounds forms a standing wave in the enclosed space.

The exact solution of a steady-state sound field in a rectangular room with a rigid surface can be obtained via the image source method. Alternatively, the solution can be more conveniently obtained by directly solving the wave equation. The lengths of the sides of the rectangular room are L_x, L_y, and L_z, respectively. The eigenfrequencies of the sound field are evaluated by solving the wave equation imposed on the boundary condition of the zero-normal derivative of sound pressure on a given boundary (Morse, and Ingrad, 1968; Kuttruff, 2009; Du, et al., 2001):

$$f_n = \frac{c}{2}\sqrt{\left(\frac{n_x}{L_x}\right)^2 + \left(\frac{n_y}{L_y}\right)^2 + \left(\frac{n_z}{L_z}\right)^2} \quad n_x, n_y, n_z = 0, 1, 2..... \tag{1.2.15}$$

The sound pressure in a room is a superposition of normal standing waves. If the absorption of boundaries is further considered, the magnitude or power of normal standing waves decays as time increases when the sound source is turned off.

For rooms with complicated shapes, the sound field is more complicated, and the analytical solution of the wave equation cannot be obtained. In the past 40 years, some numerical calculation methods, such as finite-difference time domain and boundary element method, have been developed to analyze sound fields in enclosed spaces (Wu and Zhao, 2003; Svensson et al., 2017a, 2017b).

For a rectangular room, the density of eigenfrequencies on the frequency axis is approximately evaluated with the following equation (Morse and Ingrad, 1968; Kuttruff, 2009; Du et al., 2001):

$$\frac{dN_f}{df} \approx \frac{4\pi V}{c^3} f^2, \tag{1.2.16}$$

where V is the volume of the room. Therefore, the density of eigenfrequenies increases as the volume and square of frequency increase. Equation (1.2.16) is also applied to a high frequency for a room with an arbitrary shape.

The temporal distribution of the reflections can be evaluated. For an impulse sound source, the number of reflections arriving at a receiver position within a small temporal interval increases as the reflection order increases. For a rectangular room with a rigid surface, the temporal density of reflections at the instant t can be evaluated with the image source method (Kuttruff, 2009):

$$\frac{dN_R}{dt} = \frac{4\pi c^3}{V} t^2. \tag{1.2.17}$$

Therefore, the temporal density of the reflections increases with the square of time.

Generally, the transmission from a fixed sound source to a receiver position can be considered a linear time-invariant process. It is physically characterized by the impulse response from the sound source to the receiver position and is called a ***room impulse response*** (***RIR***). The RIR includes the time-domain features of direct and reflected sounds in a room. It can be utilized to evaluate some important room acoustic parameters, such as reverberation time (Section 1.2.4; Schroeder, 1965).

Another basic principle related to sound radiation by a sound source is the ***acoustic principle of reciprocity***. This principle has various formulations. In a simple formulation, pressure is invariant after the positions of the sound source and the receiver are exchanged. The acoustic principle of reciprocity is valid for cases of a free field and a room with reflective boundaries and often applicable to the simplification of the analysis of some acoustic problems.

1.2.3 Directivity of sound source radiation

The point source discussed in Section 1.2.1 is an idealized sound source. A practical sound source possesses certain physical dimensions. For an isotropic spherical sound source in a free space, the sound field outside the sound source is a spherical wave expressed in

Equation (1.2.3). For a small sound source radius R_0 and at low frequencies such that $kR_0 << 1$, an isotropic spherical sound source can be considered a point source, that is, the definition of a point source.

In general, the sound pressure generated by a sound source depends on the direction of the receiver position with respect to the sound source. The directional characteristic of sound source radiation is described with the ***directivity of sound source radiation***, which is defined as the ratio between the far-field ($kR_r >> 1$) sound pressure amplitudes in the (Θ_r, Φ_r) direction and in the polar axis direction ($\Phi_r = 0°$):

$$\Gamma_S(\Theta_r, \Phi_r) = \frac{P_A|_{(\Theta_r, \Phi_r)}}{P_A|_{\Phi_r = 0°}}. \tag{1.2.18}$$

In general, the directivity of sound source radiation depends on frequency, although this dependency is not obviously shown in Equation (1.2.18). The directivity of radiation can also be intuitively demonstrated by the three-dimensional graphs of a directional pattern, which is an enclosed surface in a three-dimensional space. If the directivity of radiation is symmetric around the polar axis and thus azimuth independent, it can be simply demonstrated by the two-dimensional graphs of the polar pattern in a polar coordinate system, which is an enclosed curve in a two-dimensional plane. In this case, the graph of the three-dimensional directional pattern is obtained at a 360° rotation of the two-dimensional polar pattern around the polar axis.

The pressure generated by a ***dipole source*** is a simple example of directional sound source radiation. A dipole source consists of two point sources with an equal strength of magnitude but with an opposite phase. The interval l_d between two point sources satisfies $k\, l_d < 1$. The sound pressure at the receiver position is the sum of those generated by two point sources. In coordinate system B (Figure 1.2), the origin is chosen at the midpoint between two point sources, and the Z-axis (the polar axis) is selected to coincide with the straight line connecting the two point sources. The far-field ($R_r >> l_d$) sound pressure produced by a dipole source is proven as follows (Morse and Ingrad, 1968; Du et al., 2001):

$$P(R_r, \Phi_r, f) = -j\frac{kQ_p(f)l_d}{4\pi R_r}\cos\Phi_r \exp(-jkR_r). \tag{1.2.19}$$

Therefore, sound pressure depends on the wave number k or frequency. According to the definition in Equation (1.2.18), the directivity of radiation for a dipole source is

$$\Gamma_S(\Phi_r) = \cos\Phi_r. \tag{1.2.20}$$

This directivity is symmetric around the polar axis. Figure 1.4 shows the two-dimensional polar pattern $|\Gamma_S(\Phi_r)|$ of a dipole source radiation. A dipolar source exhibits the bidirectional pattern of radiation, and the curve of the polar pattern is similar to the figure of "8." Therefore, a dipole source is sometimes called "a sound source with the directivity of the figure of 8." For example, at low frequencies, a loudspeaker in a free space can be considered a dipole source.

Another example of directional sound source radiation is a circular piston diaphragm installed on an infinitely extended, rigid, and flat baffle. Every point of the diaphragm vibrates with equal magnitudes and phases. In this case, the diaphragm can be divided into a number of infinite small areas. Each small area is regarded as a point source on the baffle, and all

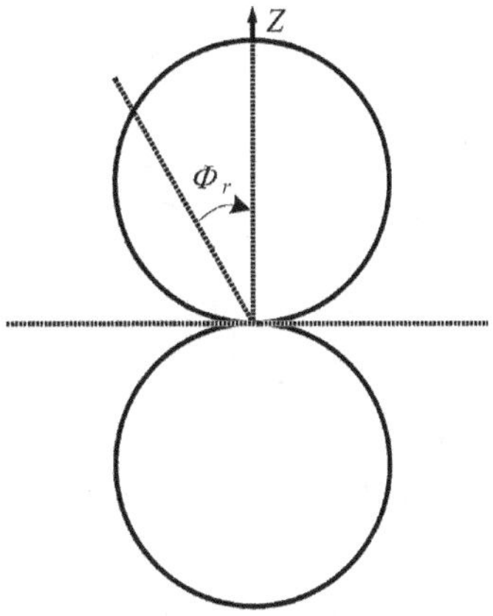

Figure 1.4 Two-dimensional polar pattern of a dipole source radiation.

point sources vibrate with equal magnitudes and phases. The overall pressure generated by the diaphragm is calculated from the superposition or integral over the pressures generated by all point sources. In coordinate system B (Figure 1.2), the origin is chosen at the center of the diaphragm, and the Z-axis (polar axis) is selected at the direction perpendicular to the diaphragm, so the directivity of radiation is given by

$$\Gamma_S(\Phi_r) = \frac{2J_1(ka_P \sin\Phi_r)}{ka_P \sin\Phi_r}, \tag{1.2.21}$$

where a_p is the radius of the diaphragm, and $J_1(ka_p \sin\Phi_r)$ is the first-order Bessel function. At low frequencies with $ka_p \sin\Phi_r << 1, J_1(ka_P \sin\Phi_r) \approx ka_P \sin\Phi_r/2$, so Equation (1.2.21) becomes

$$\Gamma_S(\Phi_r) \approx 1. \tag{1.2.22}$$

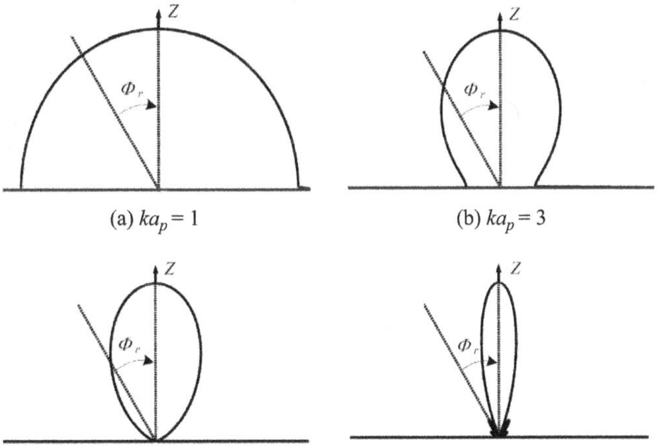

Figure 1.5 Two-dimensional polar patterns of radiation for a circular piston diaphragm in an infinitely extended, rigid, and flat baffle (a) $ka_p = 1$; (b) $ka_p = 3$; (c) $ka_p = 4$; (d) $ka_p = 10$.

In this case, radiation pressure is nearly direction independent within a specific half space.

The directivity pattern expressed in Equation (1.2.21) is symmetric around the polar axis. The two-dimensional polar patterns of radiation for various ka_p are shown in Figure 1.5. As ka_p increases, the directional-dependent characteristic of radiation appears, and the width of the main lobe narrows. If ka_p increases further, some side lobes occur.

The aforementioned piston diaphragms can serve as a model of a loudspeaker installed in large baffles. In addition, a loudspeaker array consists of multiple loudspeakers in boxes (baffles). The directivity of radiation in a loudspeaker array is controlled by the arrangement of loudspeakers and the complex-valued amplitudes of loudspeaker signals. In electroacoustic reproduction, loudspeaker arrays with the desired directivity are often used to concentrate more sound energy in some regions and reduce sound energy in other regions.

1.2.4 Statistical analysis of acoustics in an enclosed space

As mentioned in the preceding sections, in an enclosed space (usually a room) with reflective boundaries, sound waves consist of direct and reflected sounds from boundaries, such as walls, ceiling, and floor. The analysis of the enclosed sound field is one of the main aspects of room acoustics. In this section, an overview of the results of the statistical analysis of acoustics in an enclosed space is briefly presented, but the details of room acoustics are not included. Further details, such as formula derivations, are provided in other studies (Kuttruff, 2009; Beranek, 1996; Wu and Zhao, 2003).

Figure 1.6(a) shows a typical example of sound transmission paths from a sound source to a receiver position (i.e., a listener) in a room. Sound waves reach the receiver position through direct and reflected paths (such as those reflected by walls). Figure 1.6(b) illustrates the corresponding energy impulse response of the room from the sound source to the receiver position in a logarithmic scale. When the direct sound arrives the earliest, early reflections from reflective surfaces (or boundaries) arrive discretely with different time delays. In comparison with the acoustical path of the direct sound, the first reflection with the shortest time delay comes from the reflective surface with a minimum acoustical path difference. After several early reflections, many sounds reflected by more than one surface arrive, and the number of arriving reflections rapidly increases with time. As a result, the temporal density of reflections from various directions to the receiver position dramatically increases. However, the

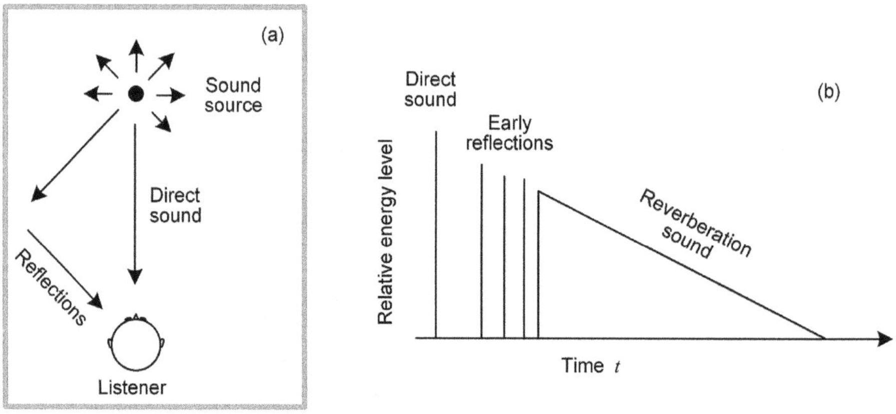

Figure 1.6 (a) Sound transmission in an enclosed space and (b) a typical example of room energy impulse response (energy vs. time curve).

total energy of reflections decreases because of the surface absorption and high-frequency air absorptions, leading to a ***reverberant sound field***. Finally, the reflection level decreases and even reaches below the background noise.

A number of methods for investigating the enclosed sound field have been developed and can be classified into two categories, namely, wave acoustic- and statistical or geometrical acoustic-based methods. Wave acoustic-based methods are used to analyze an enclosed sound field by solving the wave equation imposed by a variety of boundary conditions. However, the corresponding calculation is complicated, and analytic solutions are only possible for some rooms with regular shapes (such as rectangular rooms). Various numerical solutions based on wave acoustics and computer technology are complicated. According to Equation (1.2.16), as the frequency and volume of an enclosed space increase, the density of eigenfrequencies increases. The pressure at the receiver position is the superposition of those contributed by various eigenmodals. In this case, statistical acoustic-based methods are applicable. Such methods are relatively simple but less accurate than wave acoustic-based methods. They are invalid at low frequencies and in small rooms.

In statistical acoustics, the concept of a ***diffuse field*** is important, and it should satisfy the following requirements (Du et al., 2001).

1. In a way, sound propagates as a sound ray with a constant speed c along straight lines, and the energy carried by the sound ray is completely uniform with respect to directional distribution.
2. Sound rays are mutually incoherent; thus, their phases vary randomly.
3. The average sound energy density is constant everywhere in the region under consideration.

Reverberation time, T_{60}, is an important parameter for describing the sound field and the corresponding perceived attributes in a room. It is defined as the duration needed for the sound pressure level to decay by 60 dB after the sound source is suddenly switched off in a diffused field. Therefore, it can be used to measure the decaying speed of reverberation sound. A variety of calculation techniques and formulae have been developed, and the widely used formula is

$$T_{60} = 0.161\frac{V}{-S_{\Sigma}\ln\left(1-\alpha_{abs}\right)}. \tag{1.2.23}$$

Equation (1.2.23) is called ***Eyring's formula***, where V is the room volume, S_{Σ} is the total absorption area, and α_{abs} is the average absorption coefficient. It can be simplified as ***Sabin's formula*** for $\alpha_{abs} < 0.2$:

$$T_{60} = 0.161\frac{V}{S_{\Sigma}\alpha_{abs}}. \tag{1.2.24}$$

Sound energy in a room includes two components: direct sound energy and reverberant (or reflected) sound energy. As for the point source, direct sound pressure is inversely proportional to the source distance with respect to the receiver position ($1/r$ law), or the relationship between the direct energy density and the distance equally obeys an inverse square law. Therefore, the closer distance to the sound source is, the larger the direct sound energy density will be, and vice versa. The ratio of the direct sound energy density to the reverberant sound energy density is called the **direct-to-reverberant energy ratio** K as

$$K = \frac{D_S S_\Sigma}{16\pi R_S^2} \frac{\alpha_{abs}}{1-\alpha_{abs}}. \tag{1.2.25}$$

where R_S is the distance between the sound source and the receiver position, and D_S is the directional factor of the sound source, which is defined as the ratio between the square of the effective value of sound pressure at the receiver position and that caused by an omnidirectional sound source with identical power. D_S is different from that of the directivity of a sound source in Equation (1.2.18). In comparison with the case of $D_S = 1$, the proportion of direct energy increases in the region with $D_S > 1$, and the proportion of direct energy decreases in the region with $D_S < 1$. D_S varies with the sound source position in a room even for an omnidirectional point source. For example, $D_S = 1$ for a point source located at the center of a room; $D_S = 2$, 4, and 8 for a point source located on the surface of a rigid wall (or floor and ceiling; the same in the following discussion), at the intersect line between two rigid and perpendicular walls, and at the corner of three rigid and perpendicular walls, respectively. The direct-to-reverberant energy ratio can be controlled by changing the sound source position in a room. The physical reason underlying this phenomenon is discussed in Section 1.2.2.

Reverberation radius, ***room radius***, or ***critical distance*** R_c is defined as the distance at which the direct sound energy density is equal to that of reverberant sound (i.e., $K = 1$):

$$R_c = \frac{1}{4}\sqrt{\frac{D_S S_\Sigma \alpha_{abs}}{\pi(1-\alpha_{abs})}}. \tag{1.2.26}$$

According to Equation (1.2.26), R_c depends on the total absorption area S_Σ and the average absorption coefficient α_{abs}. R_c also increases as D_S increases. At a position whose distance to the source is less than the reverberation radius R_c (i.e., $K > 1$), a higher direct sound energy than reverberant energy is received, and vice versa.

Strictly speaking, the analysis above is valid only in a diffuse reflected sound field. As for common rooms where the actual reflected sound field may not be ideally diffused, the analysis above can be applied only as a rough approximation. The low-frequency limit for the application scope of statistical acoustics is evaluated on the basis of reverberation time T_{60} and V of the room (Schroeder, 1987):

$$f > f_g = 2000\sqrt{\frac{T_{60}}{V}} \quad (Hz). \tag{1.2.27}$$

Therefore, the validness of statistical acoustics extends to a low-frequency region for a large room.

The normalized cross-correlation coefficient between the sound pressures $p(r_1, t)$ and $p(r_2, t)$ at two receiver positions r_1 and r_2 in reverberant (diffused) sound fields is calculated by

$$\Psi_{12} = \frac{\int p(r_1,t)p(r_2,t)dt}{\left[\int p^2(r_1,t)dt \int p^2(r_2,t)dt\right]^{1/2}}. \tag{1.2.28}$$

For narrow-band signals with the center frequency f_0 or the corresponding wave number k_0, Ψ_{12} is approximated as follows (Cook et al., 1955):

$$\Psi_{12} = \frac{\sin(k_0 \Delta r)}{k_0 \Delta r} \qquad \Delta r = |\mathbf{r}_1 - \mathbf{r}_2|. \tag{1.2.29}$$

Therefore, cross-correlation is oscillating and damping as the product $(k_0\Delta r)$ of the wave number and the distance between two receiver positions increases. When $(k_0\Delta r)$ is large enough, the sound pressures at the two receiver positions are regarded as uncorrelated. In a two-dimensional space (the horizontal plane), Equation (1.2.29) is replaced by

$$\Psi_{12} = J_0(k_0 \Delta r), \tag{1.2.30}$$

where J_0 is the zero-order Bessel function.

1.2.5 Principle of sound receivers

Microphones are electroacoustic devices that convert the mechanical vibration of sound into electrical signals. They are used in acoustic measurement and sound recording to acquire the temporal and spatial information of a sound field. Various types of microphones have different acoustic principles of receivers and exhibit different physical performance.

For simplicity, an ideal microphone model is assumed in this section, i.e., the physical dimension of microphones is much smaller than the wavelength such that the scattering and diffraction of microphones on the incident sound are negligible. Coordinate system A (Figure 1.1; with respect to a specific receiver position in the absence of the head) is conveniently utilized to analyze the principle of the sound received.

The amplitude response of a microphone may depend on the sound source direction with respect to the microphone. This directional response is described with the ***directivity of the microphone***. Directivity is defined as the ratio between the amplitude response for sound incidence from the polar direction $\Delta\alpha$ and that for incidence from the on-axis direction $\Delta\alpha = 0°$ because the directional responses of actual microphones are usually symmetric around the polar axis and azimuth independent:

$$\Gamma_M(\Delta\alpha) = \frac{E|_{\Delta\alpha}}{E|_{\Delta\alpha=0°}}. \tag{1.2.31}$$

The origin of the coordinate is at the center of the diaphragm of the microphone, and the polar axis (z-axis) is consistent with the normal direction of the diaphragm. Therefore, $\Delta\alpha$ is the polar angle of the incident direction of a sound wave, i.e., the complementary angle $\alpha_S = 90° - \phi_S$ in Equation (1.1.1). The subscript "S" denotes the direction of an incident or sound source.

Similar to the cases of the directivity of sound source radiation, the cases of the directivity of polar symmetric microphones can be intuitively demonstrated by the two-dimensional graphs of a polar pattern, and the graph of a three-dimensional directional pattern can be obtained through a 360° rotation of the two-dimensional polar pattern around the polar axis.

Three categories of microphones based on the different principles of the sound receiver are commonly used in spatial sound recording (Blauert and Xiang, 2008; Du et al., 2001; Guan, 1988).

1. Pressure microphone

A pressure microphone responds to the pressure of the sound field alone. For an incident plane wave expressed in Equation (1.2.6), the amplitude response or normalized output signal of a pressure microphone is given by

$$E = P_A A_{mic}\left[\frac{2J_1(ka_M \sin\Delta\alpha)}{ka_M \sin\Delta\alpha}\right], \tag{1.2.32}$$

where a_M is the radius of the circular diaphragm of the microphone, $\Delta\alpha$ is the incident angle of the plane wave with respect to the polar axis (normal direction of the diaphragm), J_1 is the first-order Bessel function, A_{mic} is a coefficient associated with the acoustic-electric efficiency or gain of the microphone, and $P_A = P_A(f)$ is the complex-valued amplitude of the incident sound pressure. Generally, the output signal E of a microphone depends on frequency, i.e., $E = E(f)$, although this frequency dependence is omitted in most of the following equations for conciseness.

According to Equation (1.2.31), the directivity of a pressure microphone is given by

$$\Gamma_M(\Delta\alpha) = \frac{2J_1(ka_M \sin\Delta\alpha)}{ka_M \sin\Delta\alpha}. \tag{1.2.33}$$

The form of Equation (1.2.33) is similar to that of Equation (1.2.21). Equation (1.2.33) can be obtained by substituting Φ_r and a_P in Equation (1.2.21) with $\Delta\alpha$ and a_M, respectively. This substitution is the consequence of the acoustic principle of reciprocity. At low frequencies with $ka_M << 1$, $J_1(ka_M \sin\Delta\alpha) \approx ka_M \sin\Delta\alpha/2$,Equation (1.2.33) becomes

$$\Gamma_M(\Delta\alpha) \approx 1. \tag{1.2.34}$$

Equation (1.2.31) is expressed as

$$E = P_A A_{mic}. \tag{1.2.35}$$

In this case, the magnitude response of pressure microphones is independent of the direction of the incident sound wave. Therefore, pressure microphones are sometimes called ***omnidirectional microphones***. Figure 1.7(a) plots the two-dimensional polar pattern given by Equation (1.2.34), and it is a circle with a unit radius. Usually, a_M of the diaphragm of a microphone is small. Therefore, the condition $ka_M << 1$ is satisfied, and the microphone is omnidirectional within a wide frequency region (even covering the full audible bandwidth). This observation is different from the situation of the sound radiation for a loudspeaker with a large radius of the diaphragm.

2. Pressure-gradient microphone

A pressure-gradient microphone responds to the pressure gradient or pressure difference in the sound field alone. The amplitude response of a pressure-gradient microphone is also directly proportional to the velocity of the medium because the velocity of the medium is directly proportional to the sound pressure gradient. Therefore, a pressure-gradient microphone is sometimes called a ***velocity field microphone***.

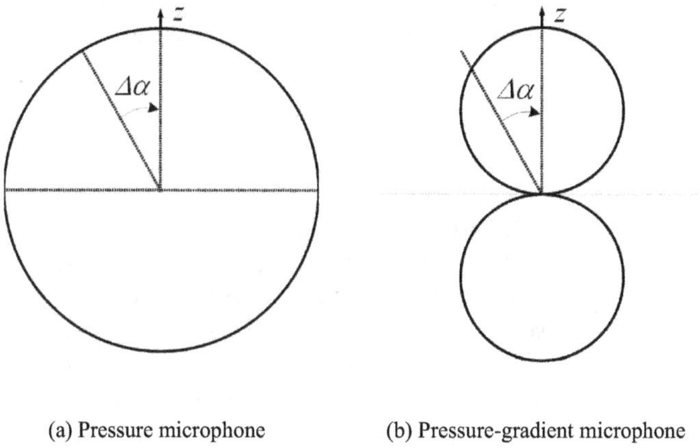

Figure 1.7 Two-dimensional polar patterns of the pressure and pressure-gradient microphones: (a) pressure microphone; (b) pressure-gradient microphone.

For the sound wave generated by a point source, the amplitude response or normalized output signal of a pressure-gradient microphone is

$$E = P_A A_{mic} \cos\Delta\alpha, \tag{1.2.36}$$

where $\Delta\alpha$ is similar to that in Equation (1.2.31), i.e., the sound source direction with respect to the polar axis; P_A is the complex-valued amplitude of the incident sound pressure; and A_{mic} is a coefficient associated with the acoustic-electric efficiency or gain of the microphone.

According to the definition in Equation (1.2.32), the directivity of a pressure-gradient microphone is expressed as

$$\Gamma_M(\Delta\alpha) = \cos\Delta\alpha. \tag{1.2.37}$$

It exhibits a bidirectional pattern. The magnitude response (i.e., the modulus of the amplitude response) maximizes in (and opposite to) the polar axis direction ($\Delta\alpha$ = 0° and 180°) and decreases in the off-axis direction. It becomes zero in the side direction $\Delta\alpha$ = 90°. Within the region of 0° ≤ $\Delta\alpha$ < 90° (facing the sound source), the amplitude response exhibits a positive polarity, i.e., it is in phase with the sound pressure. Within the region of 90° < $\Delta\alpha$ ≤ 180° (facing away from the sound source), the amplitude response exhibits a negative (reversal) polarity, i.e., it is out of phase relative to the sound pressure. Figure 1.7(b) shows the two-dimensional polar pattern of the magnitude response. The curve of the polar pattern is similar to the figure of "8."Therefore, pressure-gradient microphones are also called ***bidirectional microphones*** or ***figure-of-eight microphones***. Usually, the ***main axis*** of a bidirectional microphone is specified as the direction of $\Delta\alpha$ = 0° at which the response of a microphone maximizes with positive polarity.

Even for a constant sound pressure magnitude $|P_A|$, the magnitude response of a pressure-gradient microphone depends on kr_S, i.e., the product of the wave number and the distance of point source

$$|E| \propto A_{mic} \propto \frac{\sqrt{1+k^2r_S^2}}{kr_S}. \tag{1.2.38}$$

The ratio between the magnitude response for a near-field sound source distance with $kr_S << 1$ and that for a far-field sound source distance with $kr_S >> 1$ is given by

$$\frac{|E|_N}{|E|_F} = \frac{c}{2\pi f r_{S,N}} \frac{|P_A|_N}{|P_A|_F}, \tag{1.2.39}$$

where the subscripts "N" and "F" are the cases of near and far fields, respectively, and $r_{S,N}$ is the distance of a near-field sound source. Therefore, for equal sound pressure magnitudes $|P_A|_N = |P_A|_F$, the magnitude response of a near-field sound source is $c/(2\pi f r_{S,N})$ times as much as that of a far-field sound source. In addition, the magnitude response increases as the frequency and sound source distance decrease, or the low-frequency response is relatively increased at the near-field sound source distance. This observation is a feature of pressure-gradient microphones and called the proximity effect. In the practical design of pressure-gradient microphones, their mechanical and electric parameters are appropriately chosen so that A_{mic} in Equation (1.2.36) and the magnitude of response are independent of the frequency of a far-field incidence plane wave (Guan, 1988). This assumption is held for the succeeding discussions on pressure-gradient microphones and a combination of pressure and pressure-gradient microphones unless special notes are instructed.

3. Combination of pressure and pressure-gradient microphones
A combination of pressure and pressure-gradient microphones responds to the pressure and pressure gradient of a sound field. A general formula of the zero- and first-order directional microphone responses is

$$E = P_A A_{mic} \left(B_p + B_v \cos \Delta\alpha \right), \tag{1.2.40}$$

where $\Delta\alpha$ is the direction of a point source with respect to the polar axis, P_A is the complex-valued amplitude of the incident sound pressure, and A_{mic} is a coefficient associated with the acoustic-electric efficiency or gain of microphones. On the right side of Equation (1.2.40), the first and second terms are the pressure and pressure-gradient responses, respectively. Two coefficients, i.e., B_p and $B_v \geq 0$, describe the relative contribution of pressure and pressure-gradient responses.

A variety of ratios between B_p and B_v in Equation (1.2.40)correspond to different directivities or polar patterns of microphones, where $B_v = 0$ and $B_p = 0$ correspond to cases of pressure and pressure-gradient microphones, respectively. For $B_p \neq 0$ and $B_v \neq 0$, Equation (1.2.40) is the output signal or amplitude response of a combination of pressure and pressure-gradient microphones and can be rewritten as

$$E = P_A A'_{mic} \left(1 + b \cos \Delta\alpha \right) \qquad A'_{mic} = B_p A_{mic} \quad b = B_v / B_p, \tag{1.2.41}$$

where A'_{mic} is a new coefficient associated with the acoustic-electric efficiency or gain of the microphones, and b is a parameter that determines the directivity of microphones. In Equation (1.2.32), the directivity of a combination of pressure and pressure-gradient microphones is given by

$$\Gamma_M \left(\Delta\alpha \right) = \frac{B_p + B_v \cos \Delta\alpha}{B_p + B_v} = \frac{1 + b \cos \Delta\alpha}{1 + b}. \tag{1.2.42}$$

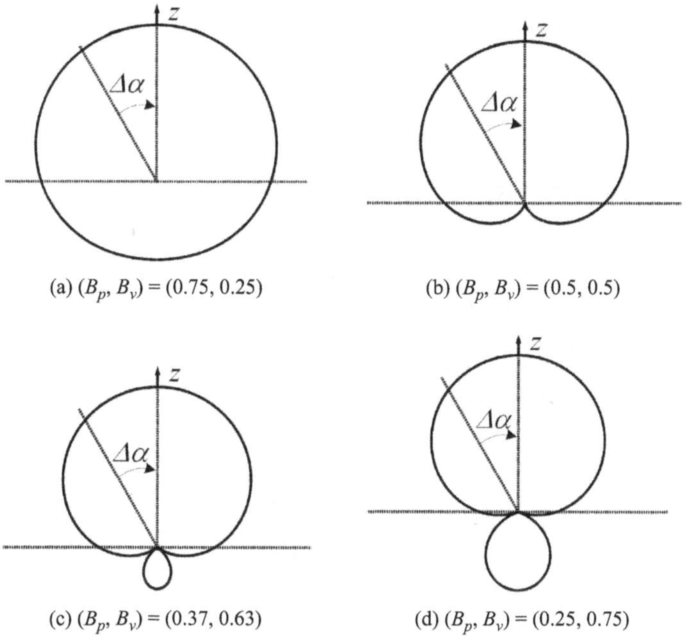

Figure 1.8 Two-dimensional polar patterns $|\Gamma_M(\Delta\alpha)|$ for some combinations of (B_p, B_v) (a) $(B_p, B_v) = (0.75, 0.25)$; (b) $(B_p, B_v) = (0.5, 0.5)$; (c) $(B_p, B_v) = (0.37, 0.63)$; (d) $(B_p, B_v) = (0.25, 0.75)$.

Figure 1.8 shows the two-dimensional polar pattern $|\Gamma_M(\Delta\alpha)|$ for some combinations of B_p and B_v. For convenience, $P_A A_{mic} = 1$ and $B_p + B_v = 1$ are assumed so that the response in the on-axis (main axis) direction $\Delta\alpha = 0°$ is normalized to a unit. As $b = B_v/B_p$ approaches 1, the response exhibits a unidirectional characteristic. In this case, the microphone is sometimes called a ***unidirectional microphone***. In cases of $B_v > B_p$ ($b > 1$), a rear lobe with reverse (negative) polarity occurs in the polar pattern around $\Delta\alpha = 180°$. The magnitude and width of the rear lobe increase as b increases. In practical uses, a variety of the combinations of (B_p, B_v) result in a family of microphones with different polar patterns: ***subcardioid microphone*** for $(B_p, B_v) = (0.75, 0.25)$ or $b = 1/3$; ***cardioid microphone*** for $(B_p, B_v) = (0.5, 0.5)$ or $b = 1$; ***supercardioid microphone*** for $(B_p, B_v) = (0.37, 0.63)$ or $b = 1.7$; and ***hypercardioid microphone*** for $(B_p, B_v) = (0.25, 0.75)$ or $b = 3$. Various combinations of pressure and pressure-gradient microphones also exhibit a proximity effect, but the details are omitted here.

The directional responses or output of the three categories of microphones are not completely independent. For example, the directional output of a combination of pressure and pressure-gradient microphones in Equation (1.2.40) can be derived from the linear sum of the outputs of a separate pressure microphone described in Equation (1.2.35) and a separate pressure-gradient microphone expressed in Equation (1.2.36). In Chapters 2 and 4, the relationships among the directional responses of the three categories of microphones are important for the spatial information recording and transformation in stereophonic and multichannel sounds. If the directional characteristics of a sound field are decomposed with spherical harmonic functions (multipole expansion, Morse and Ingrad, 1968), the output of a pressure microphone corresponds to the monopole (zero-order) component of the sound field; the outputs of three pressure-gradient microphones with their main axes pointing to three orthogonal directions (such as the directions of x-, y-, and z-axes) correspond to the three independent dipole (the first order) components of the sound field. High-order microphones and other microphone arrays have outputs corresponding to the high-order components of the sound field (Chapter 9).

1.3 AUDITORY SYSTEM AND PERCEPTION

1.3.1 Auditory system and its functions

The human peripheral auditory system consists of three main parts: the external, middle, and inner ears (Figure 1.9). The external ear is composed of the pinna and ear canal. The shape and dimension of the pinna with an average length of 65 mm (52–79 mm) vary depending on each individual (Maa and Shen, 2004). The pinna considerably modifies incoming sounds, particularly at high frequencies. In addition, its coupling with the ear canal leads to a series of high-frequency resonance modes. The ear canal is a slightly curved tube with an average diameter of 7 mm and an average length of 27 mm. It acts as a transmission path for sound. The effects of the pinna play an important role in the localization of high-frequency sounds (Section 1.6.4).

The middle ear consists of the eardrum, tympanic cavity, ossicles, associated muscles, ligaments, and Eustachian tube. The eardrum lies at the end of the ear canal and has an average surface area of approximately 66 mm^2 and a thickness of approximately 0.1 mm. The ossicles within the tympanic cavity consist of three small bones: the malleus, incus, and stapes. The malleus is attached to the eardrum, the stapes is terminated at the oval window of the inner ear, and the incus is found in between the malleus and the stapes.

The middle ear acts as an impedance transformer. If sound directly impinges onto the oval window (an interface with low-impedance air outside and high-impedance fluid inside), most of the energy is simply reflected back because of the distinct difference in acoustical impedance between air and fluid. Fortunately, before reaching the oval window, the sound is transduced into the mechanical vibration of the eardrum, ultimately resulting in the motion of the middle ear ossicles. The incoming sound can be effectively coupled into the inner ear

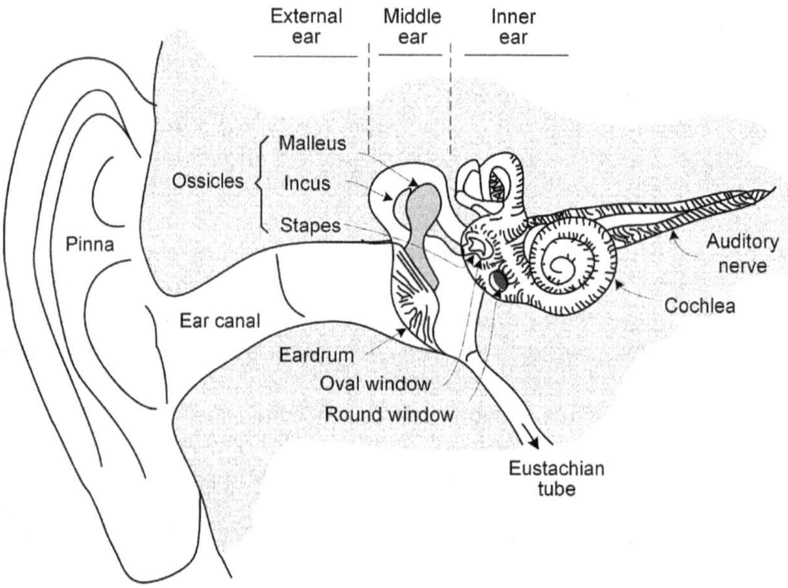

Figure 1.9 Cross-section of the peripheral auditory system of humans (adapted from Geisler, 1998).

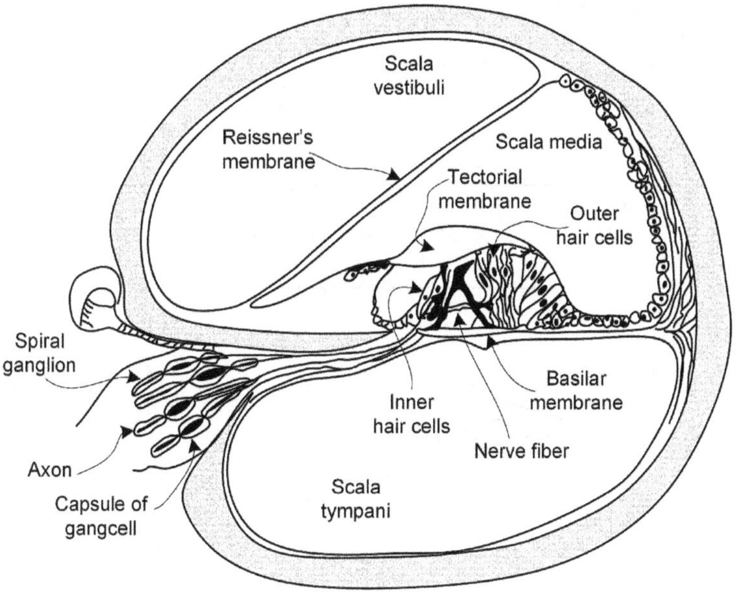

Figure 1.10 Cross-section of the cochlea (adapted from Ando, 1998).

because the lever effect of the ossicles and the differences in the area between the eardrum (about 66 mm^2) and the oval window (about 2 mm^2) facilitate the matching of equivalent acoustical impedances between air and fluid.

The inner ear mainly consists of the cochlea, a bony-wall tube with a coiled (shelled) structure of approximately 2.75 turns. The length of the uncoiled cochlea is approximately 35 mm. The end close to the oval window is called the base, and the other end is called the apex. The cochlea is divided by two membranes, i.e., the basilar membrane and Reissner's membrane, into three chambers: the scala vestibuli, the scala media, and the scala tympani (Figure 1.10). The scala vestibuli and scala tympani, filled with the perilymph, are connected by a small hole at the apex, that is, the helicotrema. Two openings exist between the middle ear and the cochlea: the oval window attached to the stapes and the round window (sealed by a membrane) located at the bony wall of the scala tympani.

The sound pressure in the eardrum is transduced into a piston-like motion of the stapes, which couples sound energy into the scala vestibuli. The inward movement of the stapes increases the fluid pressure in the scala vestibuli, whereas the outward movement decreases fluid pressure. Such pressure variation results in pressure differences across the basilar membrane and ultimately forms a traveling wave along the basilar membrane from the base to the apex. The movement of the oval window causes an antidirectional movement in the membrane covering the round window, leading to a balance in fluid pressure.

In effect, the cochlea acts as a frequency analyzer. The responses of the basilar membrane to the incoming sound are frequency and location on the basilar membrane-dependent because the mechanical properties of the basilar membrane vary continuously and considerably from the base to the apex. As for tones, a displacement envelope (i.e., the maximum magnitude of displacement at each point along the basilar membrane) has a single peak at a location that is dependent on tone frequency. The basilar membrane around the base is relatively narrow and stiff, whereas the one around the apex is wide and soft. As a result, the maximum displacement envelope for high-frequency tones appears near the base and quickly fades in the remaining parts of the basilar membrane. On the contrary, the displacement envelope

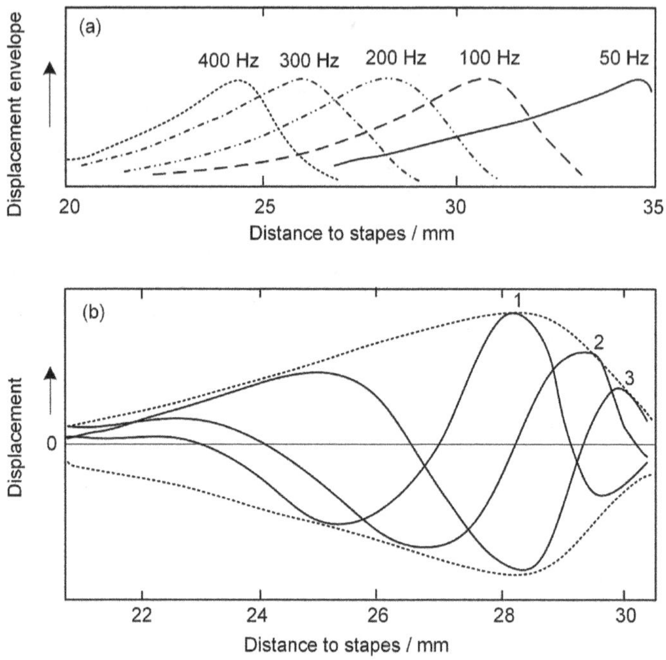

Figure 1.11 Envelopes of the basilar membrane displacement for five low-frequency tones and illustration of a traveling wave for a 200 Hz tone: **(a)** envelope of the basilar membrane displacement; **(b)** traveling wave for a 200 Hz tone, with waveforms at successive times identified with numbers "1" to "3" (adapted from Bekesy, 1960).

for low-frequency tones peaks at the more apical location and then extends along the basilar membrane. Therefore, different tone frequencies correspond to various locations in the basilar membrane, where the displacement envelope of the basilar membrane oscillation is maximized. Figure 1.11 illustrates the envelope of the basilar membrane displacement of five low-frequency tones and the traveling wave of a 200 Hz tone.

The basilar membrane serves as a converter from frequency to location on the basilar membrane. Experimental results show that the linear distance from the apex to the location of the maximum displacement envelope is approximately proportional to the logarithmic frequency above 500 Hz. This phenomenon directly results in relatively fine frequency resolutions at low frequencies and rough frequency resolutions at high frequencies. The case in which two tones are simultaneously presented is considered to illustrate the frequency resolution on the basilar membrane. Two tones can be distinguished from each other if the frequency difference is distinct enough that the corresponding maximum displacement envelopes are located at two distinct positions on the basilar membrane. As frequency difference decreases, the responses of the basilar membrane to the two tones may gradually overlap, resulting in a complicated response pattern. Ultimately, the two tones are indistinguishable if frequencies are close to each other. Almost all frequency-related auditory phenomena, such as the perception of pitch and masking, are related to the frequency selectivity of the inner ear. However, the response width (i.e., resonance width) on the basilar membrane is so wide; as such, it cannot be completely responsible for the sharp frequency resolution of the auditory system. The active process of the auditory system should also be considered.

Another function of the inner ear is to convert mechanical oscillation into neural pulses or activities. Hair cells are found between the basilar membrane and the tectorial membrane.

When the basilar membrane moves up and down, a shearing motion is created between the basilar membrane and the tectorial membrane. Consequently, the stereocilia on the surface of the hair cells bend, causing potassium ions to flow into hair cells and altering the potential difference between the inside and outside of hair cells. In turn, the release of neurotransmitters is induced, and action potentials in the neurons of the auditory nerve are initiated. In this way, hair cells convert the oscillation of the basilar membrane into neural pulses, and the neural system then conveys neural pulses to the high-level system for further processing. However, the conversion mechanism of the inner ear is complex and beyond the scope of this book. More details can be found in textbooks on physiological acoustics (Gelfand, 2010).

1.3.2 Hearing threshold and loudness

Listeners with normal hearing can perceive sound in the frequency range of 20–20 kHz. However, this observation does not mean that sounds with arbitrary strength in this frequency range are detectable. The sound is detectable only when its strength or ***sound pressure level*** (***SPL***) exceeds a certain low threshold. The minimum detectable SPL depends on frequencies and individuals. The ***hearing threshold*** is the minimum SPL of a pure tone that a listener can detect in the absence of other sounds. When the SPL exceeds a certain upper limit, the listener feels uncomfortable or even experiences pain in the ears. The ***threshold of pain*** refers to the SPL beyond which a listener feels uncomfortable.

For a sound above the hearing threshold, ***loudness*** refers to the perception of strength and is one of the important perceptual attributes of sound. It depends on multiple physical factors, including SPL, frequency, and duration of sound. Loudness is measured in terms of loudness level. If a target sound is equal to the loudness of a 1000 Hz reference tone, the SPL of a reference tone is specified as the loudness level of the target sound in the unit of phon. For example, a target sound as loud as a 1000 Hz tone at 70 dB SPL has a loudness level of 70 phons.

Numerous psychoacoustic experiments have been performed to explore the relationship among loudness, SPL, and frequency. ***Normal equal-loudness level contours*** are a measure of SPL over the frequency at which an average human perceives constant loudness when he or she is presented with steady pure tones. These contours are statistically derived from the experimental results from a large number of subjects with normal hearing. Figure 1.12 illustrates the normal equal-loudness level contours published by ISO 226 (2003). It also presents the contour of the hearing threshold (dash line). These contours are measured under the following conditions:

1. Free progressive plane wave directly from the front (in the absence of the subject)
2. Sound signals as pure tones
3. SPL measured at the position of the head center in the absence of the head
4. Binaural listening
5. Subjects in the age range of 18–25 years and with normal hearing

Figure 1.12 shows the following:

1. For a tone at a given frequency, loudness increases as the SPL increases.
2. The contours have a minimum value around the frequency of 3000–4000 Hz. Therefore, hearing is the most sensitive to sound in this frequency range. This finding is the consequence of ear canal resonance.
3. The contours increase at low and high frequencies, implying that loudness for a constant SPL reduces in these frequency ranges.

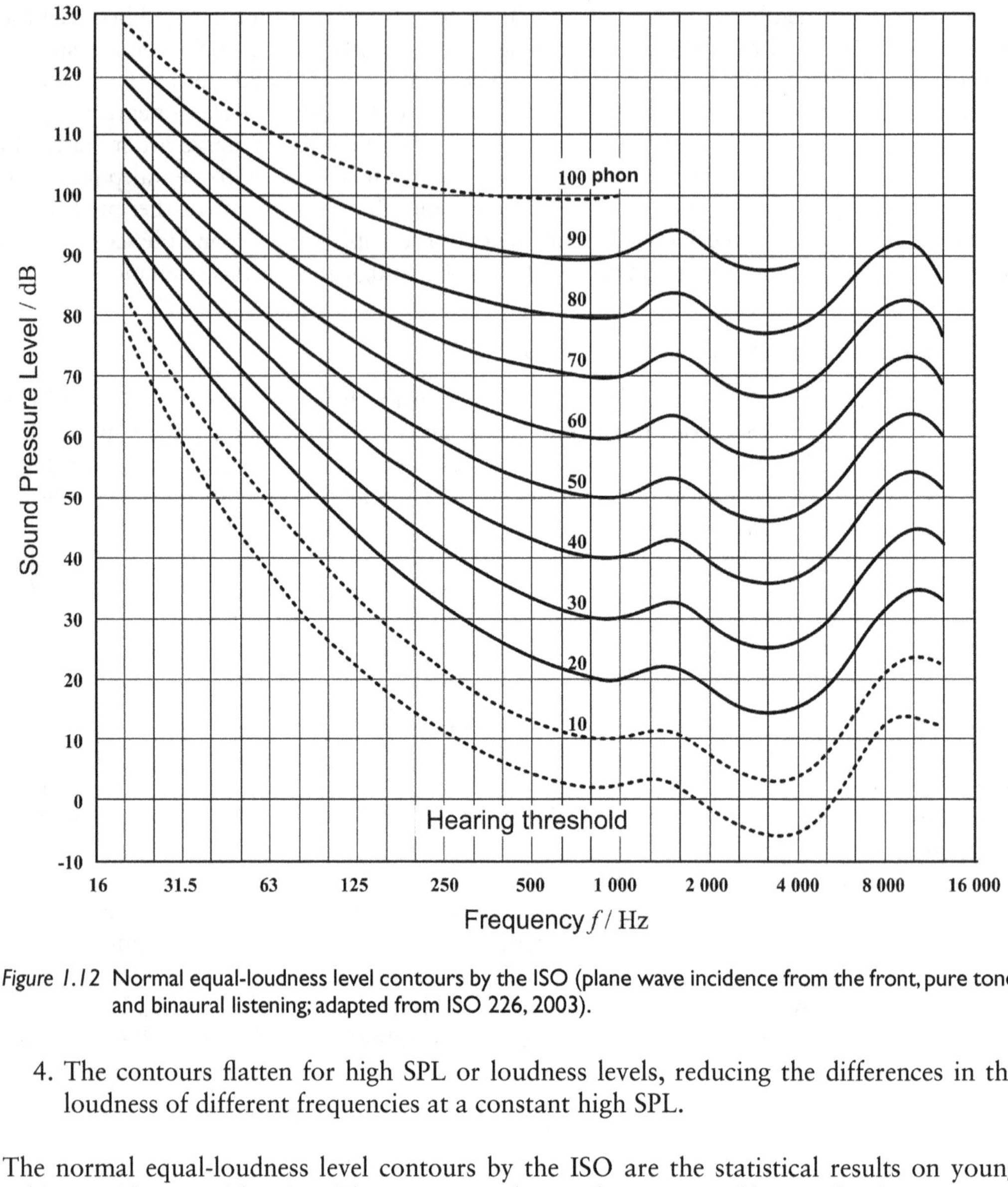

Figure 1.12 Normal equal-loudness level contours by the ISO (plane wave incidence from the front, pure tone, and binaural listening; adapted from ISO 226, 2003).

4. The contours flatten for high SPL or loudness levels, reducing the differences in the loudness of different frequencies at a constant high SPL.

The normal equal-loudness level contours by the ISO are the statistical results on young subjects with normal hearing. They represent the regular pattern of human loudness perception. Some (or even considerable) differences may exist between the ISO contours and those of each individual.

The results of some psychoacoustic experiments indicate that loudness depends on the incident direction of a plane wave (or sound source direction) for a free-field plane wave with a given frequency and SPL. This finding is the ***directional loudness*** in the free field. Directional loudness is an issue related to binaural hearing (Sivonen and Ellermeier, 2008; Moore and Glasberg, 2007; Section 1.6.5).

1.3.3 Masking

Masking refers to the psychoacoustic phenomenon in which the auditory detection threshold of a sound (***target***) may increase in the presence of another sound (***masker***). The ***masking threshold*** is the minimum SPL of a target that is detectable in the presence of a masker. ***The***

amount of masking is the difference in the detectable thresholds of the SPL between the presence and absence of the masker.

The masking threshold and amount of masking vary with multiple factors, including the types, strength, frequency spectrum, temporal relationship, and spatial positions of targets and maskers. Given the types, temporal relationship, and spatial positions of targets and maskers, the masking threshold or amount of masking at various target frequencies and for different SPL and frequencies of a masker can be determined via psychoacoustic measurements. As a result, a series of masking curves or patterns is formed. Masking patterns depend on the type of targets and maskers. The results of a tone masked by another tone and a tone masked by another band-pass noise are common. Two experimental methods are utilized to measure the masking. Accordingly, the SPL is determined at two different reference positions. One is to measure the monaural or binaural masking via headphone presentation, and the SPL is determined at a certain position in the external ear (e.g., the entrance of the ear canal or eardrum). The other is to measure the masking of a free-field target and a masker, and the SPL is identified at the position of head center in the absence of the head. The SPLs measured from two reference positions are different because of the scattering and diffraction effects of the head and pinna when a subject enters the sound field. However, they can be converted to each other by using head-related transfer functions (Section 1.4.2).

Figure 1.13 illustrates the monoaural masking patterns of a tone by another tone (Ehmer, 1959a, 1959b), and these patterns represent the amount of masking as a function of the target frequency at various SPLs of a tone masker. The target sound and masker are presented simultaneously, i.e., the case of ***simultaneous masking***.

Figure 1.13 presents the following:

1. Masking is effective when the frequency spectra of a masker and a target are close to each other.
2. Masking patterns are asymmetric. More masking occurs for the target frequency that is higher than the masker frequency, and less masking occurs for the target frequency that is lower than the masker frequency.
3. As the SPL of the masker increases, the range covered by masking widens, especially toward a high frequency range.

Masking occurs when a masker and a target are presented successively. This phenomenon is called temporal or nonsimultaneous masking. Temporal masking is subdivided into ***backward masking*** (premasking) and ***forward masking*** (postmasking). For backward masking, the target is presented prior to the masker. For forward masking, the target is presented after the masker. The durations of forward and backward masking are different. Generally, forward masking lasts 200 ms, and backward masking only lasts 15–20 ms. However, the mechanism of temporal masking is still unclear.

The masking threshold or amount of masking for a spatially separated masker and target is lower than that for a spatially coincident masker and target. Spatial unmasking is the phenomenon that the spatial separation of the masker and target – in terms of direction and distance – decreases the masking threshold or the amount of masking (Kopčo and Shinn-Cunningham, 2003). It is a binaural phenomenon associated with head-related transfer functions and spatial cues in binaural pressures (Section 1.6.5).

1.3.4 Critical band and auditory filter

Fletcher (1940) investigated the masking of a tone by a band-pass noise and reported that noise in a bandwidth centered at the tone frequency is effective in masking the tone. Conversely, the other noise component outside the bandwidth has no effect on masking.

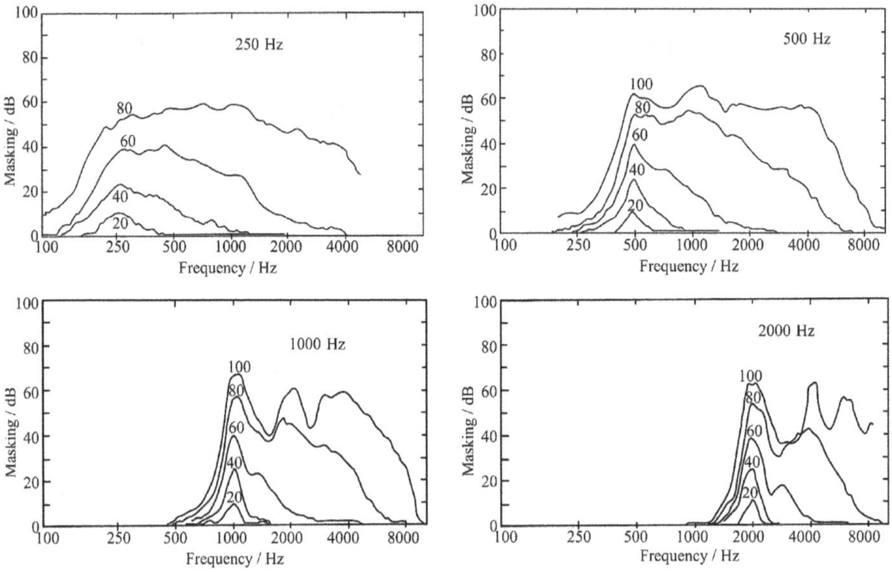

Figure 1.13 Monoaural masking patterns of a tone by another tone. Each panel represents a different frequency of the masker. Each curve in a panel represents the result of the SPL of the masker. The abscissa is the target frequency, and the ordinate is the amount of masking (reproduced from Ehmer 1959a, with the permission of the Acoustical Society of America).

The bandwidth derived in this manner is called the ***critical bandwidth*** at the center frequency.

Fletcher (1940) attributed this phenomenon to the frequency analysis function of the basilar membrane. Each location on the basilar membrane maximally responds to a specific center or characteristic frequency, and the response decreases dramatically if the sound frequency deviates from the characteristic frequency. As such, each location on the basilar membrane acts as a band-pass filter with a specific characteristic (central) frequency. Correspondingly, the entire basilar membrane (strictly the corresponding functions of the auditory system) can be regarded as a bank of overlapping band-pass or ***auditory filters*** with a series of consecutive characteristic frequencies.

The frequency resolution of the auditory system is related to the shape and width of auditory filters. Fletcher (1940) simplified each auditory filter as a rectangular filter. If the bandwidth of the masking noise is within the effective bandwidth of an auditory filter, the noise effectively masks the tone at the characteristic frequency. If the bandwidth of the masking noise is wider than the effective bandwidth, the components of noise outside the effective bandwidth of auditory filters slightly affect masking. The critical bandwidth provides an approximation of the bandwidth of auditory filters. The results of various psychoacoustic experiments indicate that the width of critical bandwidth (Δf_{CB}) in Hz is related to the center f in kHz (Zwicker and Fastl, 1999):

$$\Delta f_{CB} = 25 + 75\left(1 + 1.4f^2\right)^{0.69}. \tag{1.3.1}$$

Then, a new frequency metric related to auditory filters, that is, the ***critical band rate*** (in Bark) can be introduced. One Bark is equal to the width of a critical frequency band and equal to a distance of about 1.3 mm along the basilar membrane. The critical band rate v in Bark is related to f in kHz as

$$v = 13 \arctan(0.76f) + 3.5 \arctan\left(\frac{f}{7.5}\right)^2. \tag{1.3.2}$$

Some recent studies have indicated that an actual auditory filter is not exactly symmetric. However, an equivalent rectangular band-pass filter is a good approximation for convenience, especially at moderate sound levels. The equivalent rectangular band-pass filter has a transmission in its passband equal to the maximum transmission of the specified auditory filter and transmits the same power of white noise as the specified auditory filter. The bandwidth of the equivalent rectangular filter is known as ***equivalent rectangular bandwidth*** (***ERB***). Psychoacoustic experiments demonstrate that the ERB in Hz is related to the center frequency in kHz (Moore, 2012):

$$ERB = 24.7(4.37f + 1). \tag{1.3.3}$$

According to the concept of the equivalent rectangular auditory filter, the ***equivalent rectangular bandwidth number*** (***ERBN***), a new frequency metric, can be introduced, and it is related to frequency in kHz as

$$ERBN = 21.4 \ \log_{10}(4.37f + 1). \tag{1.3.4}$$

According to the aforementioned formulae, the bandwidth in the critical bandwidth model is near 100 Hz within the frequency range of 100 Hz to 500 Hz. Above this frequency, the bandwidth is approximately 20% of the center frequency; conversely, in an ERB model, a bandwidth above 500 Hz is similar to that in the critical bandwidth model and frequency dependent below 500 Hz rather than nearly a constant in the critical bandwidth model.

The bandwidths in critical bandwidth and ERB models increase as frequency increases; hence, the frequency resolution of the auditory system decreases as frequency increases. In this sense, both models can represent the nonuniform frequency resolution of the auditory system. Although both models are widely adopted, recent experiments have proven that the ERB model is more precise than the critical bandwidth model, particularly at frequency below 500 Hz. A variety of auditory phenomena, such as the perception of loudness and masking, are closely related to the properties of auditory filters. An auditory filter model is one of the basic parts of various binaural auditory models. Binaural auditory models should be applied to analyze the perceived attributes of sounds, including those in spatial sound reproduction (Section 12.2). The concepts of auditory filters and maskers are vital to perceptual audio coding (Chapter 13).

1.4 ARTIFICIAL HEAD MODELS AND BINAURAL SIGNALS

1.4.1 Artificial head models

Humans use two ears to receive a sound wave. Binaural sound pressures or signals (strictly binaural pressures in the eardrums) contain the major spatial information of sound. In contrast to the ideal microphone model described in Section 1.2.5, human anatomical structures, such as the head, pinnae, and torso, produce non-negligible scattering and diffraction of sound waves within the audible frequency range. A subject inevitably disturbs the sound field when he/she enters it. The sound waves received by the two ears are modified via scattering and diffraction by anatomical structures.

An ***artificial head*** (or a dummy head, or head and torso simulator) is a kind of a model made of specific materials and used to simulate the anatomical structures, such as the head, torso, and pinnae, of a real human from the perspective of acoustics (Vorlander, 2004; Paul, 2009). Its design is based on the average dimensional properties of certain populations or on the anatomical features of a "standard" or "typical" human subject. Moreover, the acoustic properties of materials used in an artificial head are similar to those of humans. Artificial heads can be used in a variety of acoustic measurements and studies, such as binaural recording, sound quality evaluation, headphone testing, and hearing aid testing, instead of a human subject. In binaural hearing, the effects of diffractions and reflections caused by the head, pinnae, and torso can be simulated using the artificial head. The resultant binaural sound pressures can be recorded by a pair of microphones placed at the ear canal entrance or at a certain point along the ear canal of the artificial head.

Among various artificial head products, Knowles Electronics Manikin for Acoustic Research (KEMAR), originally produced by Knowles Electronics and now acquired by GRAS Sound & Vibration in Denmark, has been used extensively in the field of binaural hearing possibly because of two reasons: (1) KEMAR has a human-like appearance, and (2) the head-related transfer function database from KEMAR is probably the first to become available on the internet (Section 11.2; Gardner and Martin, 1995) and consequently used in a substantial number of relevant studies. KEMAR was originally designed for hearing aid measurement and relevant acoustic research (Burkhard and Sachs, 1975). Its design is based on the average head and torso dimensions from the anatomical measurements of Western male and female adults from the 1950s to the 1960s; this design also satisfies the requirement of IEC 60959 (1990) and ANSI S3.36/ASA58 (1985).

Figure 1.14(a) shows the KEMAR with the detailed head, torso, and pinnae. Basically, two sizes of pinnae are available, but they vary in different applications. As suggested in the specifications of KEMAR, the pair of the small pinnae (DB-060/061 for the right and left ears,

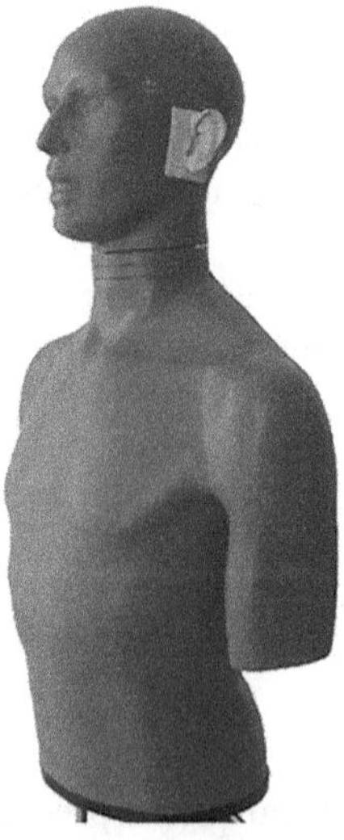

(a)

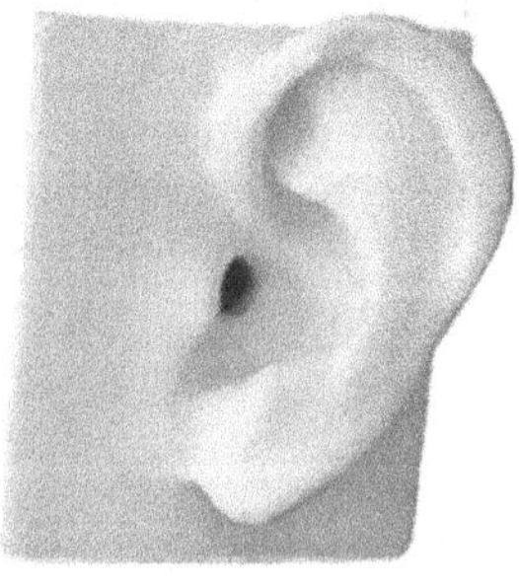

(b)

Figure 1.14 KEMAR and DB-061 pinna: **(a)** KEMAR and **(b)** DB-061 pinna.

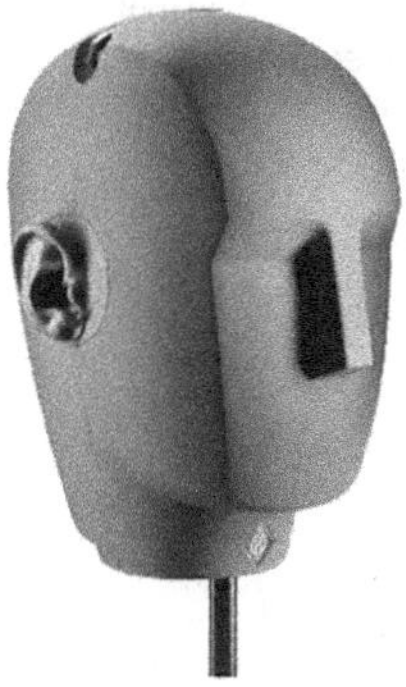

Figure 1.15 Neumann KU-100 artificial head (with the permission of Georg Neumann GmbH).

respectively) is typical of American and European females and Japanese males and females. The pair of the large pinnae (DB-065/066 for the right and left ears, respectively) is typical of American and European males. Figure 1.14(b) illustrates the small pinna DB-061 on KEMAR (note: GRAS Sound &Vibration provides a new type of pinnae now). In addition, KEMAR is equipped with a pair of Zwislocki-occluded ear simulators, satisfying the requirements of ANSI S3.25/ASA80 (1989) for simulating sound transmission through the human ear canal to the eardrum. One end of the occluded ear simulator is connected to the ear canal entrance by a canal extension, and a standard 12.7 mm (1/2 in) pressure field microphone is built-in at the other end (i.e., inside the head). Hence, a microphone diaphragm, analogous to the eardrum, is used to convert sound pressures into electrical signals.

In addition to KEMAR, some other commercial artificial head products are available for different purposes, such as products developed by B&K, Head Acoustics, and 01 dB-Metravib. Figure 1.15 presents the photo of the Neumann KU-100 artificial head (from Georg Neumann GmbH). KU-100 only includes the head and is often employed in binaural recording.

1.4.2 Binaural signals and head-related transfer functions

Binaural signals should be analyzed in studies on spatial hearing and sound reproduction because binaural (sound pressures) signals capture the major information of auditory perception. In nature, transmission from a free-field point sound source to each of the two ears on a fixed head can be regarded as a linear time-invariant (LTI) process (Figure 1.16). ***Head-related transfer functions***, or ***HRTFs***, which describe the overall filtering effect imposed by anatomical structures, are introduced as the acoustic transfer functions of the LTI process. With regard to an arbitrary sound source position, a pair of HRTFs, H_L and H_R, for the left and right ears, respectively, is defined as

$$\begin{aligned} H_L &= H_L(r_S, \theta_S, \phi_S, f, a) = \frac{P_L(r_S, \theta_S, \phi_S, f, a)}{P_{free}(r_S, f)}, \\ H_R &= H_R(r_S, \theta_S, \phi_S, f, a) = \frac{P_R(r_S, \theta_S, \phi_S, f, a)}{P_{free}(r_S, f)}. \end{aligned} \tag{1.4.1}$$

where (r_S, θ_S, ϕ_S) is the sound source position in the coordinate system shown in Figure 1.1; the subscript "S" is the sound source; f is frequency; P_L and P_R are the complex-valued sound pressures in the frequency domain at the left and right ears, respectively; and P_{free} is the

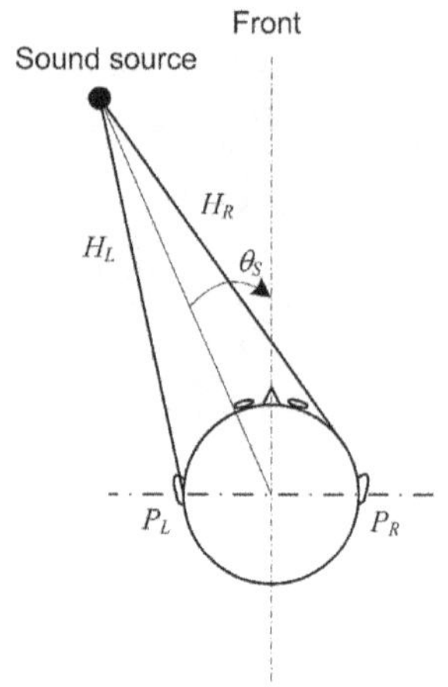

Figure 1.16 Diagram of sound transmission from a point sound source to the two ears.

complex-valued free-field sound pressure in the frequency domain at the center of the head with the head absent.

For a certain spatial position of the sound source, a pair of HRTFs exists, one for each ear. Generally, an HRTF is a function of frequency and sound source position in terms of distance r_S, azimuth θ_S, and elevation ϕ_S. Furthermore, the HRTF is individually dependent because of the unique anatomical structure and dimension of each subject, which is denoted by the variable a. Strictly speaking, a is a set of parameters specifying the dimensions of relevant anatomical structures. Therefore, HRTFs are a pair of multivariable functions. At the far-field distance with $r_S > 1.0$–1.2 m, an HRTF is nearly distance independent and termed the ***far-field HRTF***. At the near-field distance with $r_S < 1.0$ m, the HRTF is distance dependent and termed the ***near-field HRTF***.

The measurement point of P_L and P_R in Equation (1.4.1) varies in different studies, and among them, the eardrum is the most intuitive choice. Mϕller (1992) proved that the entire ear canal can be regarded as a one-dimensional (i.e., direction-independent) transmission line for sound waves whose wavelengths are much larger than the diameter when the ear canal is approximated by a tube with a diameter of 8 mm, which is slightly larger than the average 7 mm (Maa and Shen, 2004). A wavelength of 8 mm corresponds to a frequency of 42.5 kHz, and a quarter of the wavelength approximately refers to a frequency of 10 kHz. Therefore, the approximation of one-dimensional ear canal transmission is held below 10 kHz. Hammershϕi and Mϕller (1996) demonstrated that the entire ear canal can be regarded as a one-dimensional transmission line below 12–14 kHz. Therefore, binaural sound pressures in the definition of HRTFs can be measured at an arbitrary position from the entrance of the ear canal to the eardrum even at the entrance of a blocked ear canal. The ***reference point*** is the position at which binaural sound pressures are measured. Moreover, binaural sound pressures measured at different reference points can be converted to one another. In this book, the sound pressures measured at an arbitrary reference point from the entrance of the ear canal to the eardrum are called binaural pressures unless otherwise stated. This problem is further discussed in Section 11.7.1.

According to Equation (1.4.1), if a pair of HRTFs is known, the binaural sound pressures for a free-field point source at position (r_S, θ_S, ϕ_S) are evaluated by

$$\begin{aligned} P_L(r_S, \theta_S, \phi_S, f, a) &= H_L(r_S, \theta_S, \phi_S, f, a) P_{free}(r_S, f), \\ P_R(r_S, \theta_S, \phi_S, f, a) &= H_R(r_S, \theta_S, \phi_S, f, a) P_{free}(r_S, f). \end{aligned} \tag{1.4.2}$$

In Equation (1.4.1), HRTFs are defined in the frequency domain. Their time-domain counterparts are known as ***head-related impulse responses*** (***HRIRs***) or ***binaural impulse responses***, which are impulse responses from a point sound source to the two ears in the free field. HRTFs and HRIRs are related in terms of temporal frequency Fourier transform:

$$\begin{aligned}
h_L(r_S, \theta_S, \phi_S, t, a) &= \int_{-\infty}^{+\infty} H_L(r_S, \theta_S, \phi_S, f, a) e^{j2\pi ft} df, \\
h_R(r_S, \theta_S, \phi_S, t, a) &= \int_{-\infty}^{+\infty} H_R(r_S, \theta_S, \phi_S, f, a) e^{j2\pi ft} df, \\
H_L(r_S, \theta_S, \phi_S, f, a) &= \int_{-\infty}^{+\infty} h_L(r_S, \theta_S, \phi_S, t, a) e^{-j2\pi ft} dt, \\
H_R(r_S, \theta_S, \phi_S, f, a) &= \int_{-\infty}^{+\infty} h_R(r_S, \theta_S, \phi_S, t, a) e^{-j2\pi ft} dt.
\end{aligned} \tag{1.4.3}$$

As for HRTF, HRIR is a multivariable function that varies with sound source position, time, and anatomical structures and dimensions. Equation (1.4.2) is expressed in the time domain as

$$\begin{aligned}
p_L(r_S, \theta_S, \phi_S, t, a) &= \int_{-\infty}^{+\infty} h_L(r_S, \theta_S, \phi_S, \tau, a) p_{free}(r_S, t-\tau) d\tau = h_L(r_S, \theta_S, \phi_S, t, a) \otimes_t p_{free}(r_S, t), \\
p_R(r_S, \theta_S, \phi_S, t, a) &= \int_{-\infty}^{+\infty} h_R(r_S, \theta_S, \phi_S, \tau, a) p_{free}(r_S, t-\tau) d\tau = h_R(r_S, \theta_S, \phi_S, t, a) \otimes_t p_{free}(r_S, t).
\end{aligned} \tag{1.4.4}$$

where p_L and p_R are sound pressures in the time domain at the left and right ears, respectively, and p_{free} is the sound pressure in the time domain at the original position of the head center in the absence of the head. They are related to the corresponding sound pressures in the frequency domain via the Fourier transform. $\otimes_t$ is convolution.

HRTFs or HRIRs are vital to binaural analysis and virtual auditory display (Chapter 11). This issue is also addressed in detail in another study (Xie, 2008a, 2013). In Section 11.2.2, at low frequencies, horizontal HRTFs can be approximated as

$$H_L(\theta_S, f) \approx 1 + j\frac{3}{2} ka \sin\theta_S \qquad H_R(\theta_S, f) \approx 1 - j\frac{3}{2} ka \sin\theta_S. \tag{1.4.5}$$

where a is the radius of the head. In this case, the HRTF magnitudes of the left and right ears are approximately equal to a unit, but their phases are different.

1.5 OUTLINE OF SPATIAL HEARING

Aside from the perceptions of loudness, pitch, and timbre, spatial perception, which is a subjective perception of the spatial attributes of sound, is implicated in human hearing. The auditory system utilizes the sound coming from a sound source to evaluate the spatial position of the sound source in terms of direction and distance. It is useful in assisting visual attention for seeking objects and warning humans (or animals) to evade potential dangers.

A variety of experimental results have demonstrated that a human's ability to locate a sound source depends on many factors, such as sound source direction and source acoustical properties, and varies among individuals. On average, the human acuity of localization is the highest in the front of the horizontal plane, that is, the ***minimal audible angle*** (***MAA*** or ***localization blur***) $\Delta\theta$ reaches a minimum of 1° to 3°. In other directions, the acuity of localization decreases, and $\Delta\theta$ is about three-fold in lateral directions and twofold in rear directions. In the median plane, the human acuity of localization is relatively low, and MAA varies from $\Delta\phi = 4°$ for white noise to 17° for speech.

A human's ability to evaluate the distance of a single sound source depends on many factors, such as source properties and acoustic environment, and varies among individuals. In general, distance estimation is relatively easy, given the prior knowledge of the sound source.

When two or more sound sources simultaneously radiate sound waves, the combined pressures in the two ears contain the spatial information of multiple sound sources; this information can be used for hearing to form spatial auditory events. The different acoustical attributes of multiple sources lead to different spatial auditory events or perception. For uncorrelated sounds coming from multiple sound sources, hearing may perceive each sound as a separate auditory event and then be able to locate each sound source. For correlated sounds coming from multiple sound sources, under certain situations, hearing may perceive the sounds as a fused auditory event arising from the position of one real sound source while ignoring the other real sources as in the case of the precedence effect (Section 1.7.2). However, if the relative level and arrival time of each sound satisfy certain conditions, hearing perceives a summing ***virtual sound source*** (shortened as a ***virtual source*** and also called ***virtual sound image*** or ***sound image*** in some references) at a spatial position where no real source exists. This phenomenon is the basis of stereophonic and multichannel surround sound reproduction, which is discussed in succeeding chapters. In some other situations, hearing is unable to locate the source position or even likely to perceive an unnatural auditory event, such as lateralization in a headphone presentation.

Here, two terminologies should be clarified: ***localization*** and ***lateralization***. Localization is a process by which the position of an auditory event in a three-dimensional space is determined. By contrast, lateralization is a process by which the lateral displacement of an auditory event in a one-dimensional space is identified, that is, along the straight line connecting the entrances of the two ear canals (Blauert, 1997; Plenge, 1974). Localization and lateralization determinations are commonly used in psychoacoustic experiments. Generally, the former is usually employed in a natural or simulated natural auditory environment, and the latter is often used under unnatural situations, such as a headphone presentation where binaural signals are controlled or changed independently.

In an enclosed space with reflective surfaces, a series of reflections exist with the direct sound through which hearing can acoustically perceive the spatial dimension and form a spatial impression of the environment and sound source. Thus, the spatial information encoded in reflections is vital to the design of room acoustics.

In summary, spatial hearing includes the localization of a single sound source, summing localization, and other auditory events of multiple sound sources and subjective spatial perceptions of environmental reflections. Research on spatial hearing is related to numerous disciplines, such as physics, physiology, psychology, and signal processing. Studies on spatial hearing have been undertaken and reviewed at length in the famous monograph of Blauert (1997).

1.6 LOCALIZATION CUES FOR A SINGLE SOUND SOURCE

Auditory localization refers to ascertaining the apparent or perceived spatial position of a sound source in terms of its direction and distance in relation to a listener. Psychoacoustic studies have demonstrated that directional localization cues for a single sound source include ***interaural time difference*** (*ITD*), ***interaural level difference*** (*ILD*), and ***dynamic*** and ***spectral*** cues (Blauert, 1997). Distance perception, which is another important aspect of auditory localization, is also based on the comprehensive effect of multiple cues. These localization cues are presented as follows.

1.6.1 Interaural time difference

ITD refers to the arrival time difference between the sound waves at the left and right ears and plays an important role in directional localization. In the median plane, ITD is approximately zero because the path lengths from the sound source to both ears are identical. However, when the sound source deviates from the median plane, the path lengths to each ear are different; thus, the ITD becomes nonzero. For example, in the horizontal plane [Figure 1.17(a)], the curved surface of the head is disregarded, and the two ears are approximated by two points in the free space separated by 2 a, where a is the head radius. Ideally, a plane wave is

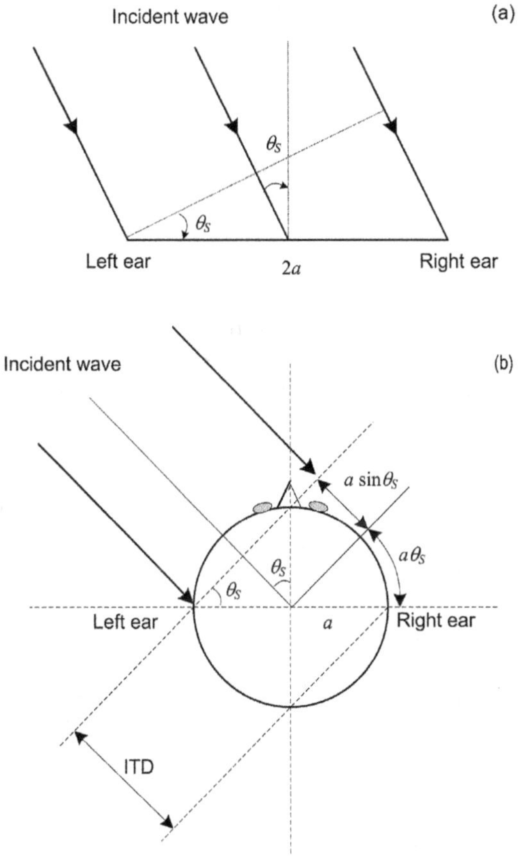

Figure 1.17 Sketch of ITD calculation in the horizontal plane: **(a)** neglecting the effect of the head, and **(b)** considering the curved surface of the head.

generated by a sound source at an infinite distance. It approximately holds for a point source at a distance far larger than a in practice. The ITD for an incident plane wave from the azimuth θ_S can be evaluated by

$$ITD(\theta_S) = \frac{2a}{c}\sin\theta_S, \tag{1.6.1}$$

where c is the speed of sound. Here, the positive ITD indicates that the left ear is relatively closer to the sound source with the received sound leading in time, and vice versa. However, the results here are contrary to those in a clockwise spherical coordinate system because an anticlockwise spherical coordinate system is used in this book.

If the curved surface of the head is considered, as shown in Figure 1.17(b), the head is approximated by a sphere with a, and the ears by two points on opposite sides. As for the curved path of sound around the spherical head, the improved formula of the ITD (Boer, 1940; Woodworth and Schlosberg, 1954) is

$$ITD(\theta_S) = \frac{a}{c}(\sin\theta_S + \theta_S) \qquad 0 \le \theta_S \le \frac{\pi}{2}, \tag{1.6.2}$$

which is usually called Woodworth formula in the literature. In accordance with the spatial symmetry of the spherical head, Equation (1.6.2) can be extended to the incident sounds in other quadrants.

For the frontal incident sound, $\sin\theta \approx \theta$, Equation (1.6.1) is approximately equivalent to Equation (1.6.2), whereas the results of the two equations are different for the incident sound from other directions. Equations (1.6.1) and (1.6.2) show that the ITD varies with azimuth θ_S and consequently acts as a directional localization cue. Moreover, ITD is individual dependent because it is closely related to the anatomical dimensions of the head, which varies among individuals.

Figure 1.18 shows the ITDs calculated using Equation (1.6.2), where a equals 0.0875 m, as it is generally used in the literature. Only the results for the left horizontal plane are given because of the symmetry of the spherical head. In the figure, the ITD is zero at an azimuth of 0° (i.e., the directly frontal direction) and gradually increases as the sound source approaches the lateral region. At an azimuth of 90°, the ITD reaches its maximum of 662 μs.

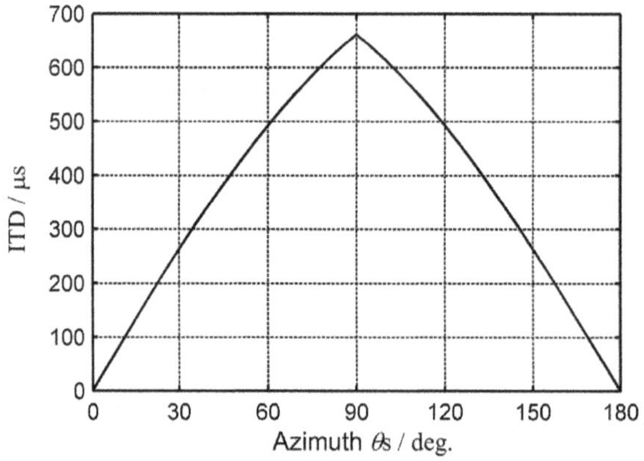

Figure 1.18 Calculated ITD from Equation (1.6.2) as a function of azimuth θ_S.

Subsequently, as the sound source approaches the rear, ITD decreases and returns to zero at an azimuth of 180°.

Further psychoacoustic experiments (Blauert, 1997) have revealed that ***interaural phase delay difference*** is an important cue for localization below approximately 1.5 kHz. In wave acoustics, the phase delay is determined by the phase of the composite sound pressure in the ear caused by the direct and diffracted (by the head etc.) transmission to the ear. Accordingly, the ITD derived from the interaural phase delay is defined as

$$ITD_p(\theta_S, f) = \frac{\Delta\psi}{2\pi f} = \frac{\psi_L - \psi_R}{2\pi f}, \tag{1.6.3}$$

where p is the phase delay; ψ_L and ψ_R are the phases of the sound pressures at the left and right ears, respectively; and $\Delta\psi = \psi_L - \psi_R$ is the interaural phase difference. The time factor of a sinusoidal wave is assumed to be equal to exp $(j2\pi ft)$; hence, $\psi > 0$ means leading in phase, and $\psi < 0$ corresponds to lagging in phase.

Binaural sound pressures and their phases can be evaluated with HRTFs, and the interaural phase delay difference $\mathrm{ITD_p}$ is calculated. Following the rigid spherical head model provided by Kuhn (1977), we calculate the binaural pressures [Equation (11.2.2)] and $\mathrm{ITD_p}$. In this model, the head is approximated as a rigid sphere with a = 0.0875 m, and two ears are located diametrically across the spherical head. Figure 1.19 shows $\mathrm{ITD_p}$ as a function of frequency for the incident plane wave from three different horizontal azimuths θ_S = 30°, 60°, and 90°. ITD_p at a given incident direction is approximately independent of frequency below 0.4 kHz and above 3 kHz. The low-frequency ITD_p is greater than its high-frequency counterpart. For example, ITD_p for θ_S = 90° incidence is 767 µs at 0.1 kHz and 676 µs at 3.0 kHz. Within a mid-frequency of 0.5–3 kHz, ITD_p is frequency dependent and smoothly transitions from an asymptotic result at low frequencies to that at high frequencies.

At low frequencies with $ka << 1$, the HRTFs of a rigid spherical head model is simplified into Equation (1.4.5), and ITD_p is calculated as

$$ITD_p = \frac{2\arctan\left(\frac{3}{2}ka\sin\theta_S\right)}{2\pi f} \approx \frac{3a}{c}\sin\theta_S. \tag{1.6.4}$$

This result is independent of frequency and only direction dependent. The low-frequency ITD_p in Equation (1.6.4) is 1.5 times that given by Equation (1.6.1). Equation (1.6.1) is derived from a shadowless head model, and Equation (1.6.4) is the asymptotic result of wave acoustics. The head shadow effect extends the effective path difference between the two ears. Accordingly, an accurate low-frequency ITD_p can be obtained by substituting a magnified head radius a' = 1.5 a into Equation (1.6.1).

Kuhn (1977) interpreted the asymptotic behavior of $\mathrm{ITD_p}$ at high frequencies by expressing the waves on the surface of a rigid sphere as a series of creeping waves and analyzing their attenuation. Kuhn (1977) also observed a similar behavior in ITD_p of KEMAR.

When the head dimension (path difference between the two ears) equals half of a wavelength, roughly corresponding to a frequency of 0.7 kHz, the pressures at the two ears for lateral incidence are out of phase; thus, interaural phase difference begins to provide ambiguous localization cues. Head or source movement may resolve this ambiguity. However, above 1.5 kHz, the absolute value of the interaural phase difference $\Delta\psi$ may exceed 2π, which increases to a completely ambiguous ITD_p, when the head dimension is larger than the wavelength.

Fortunately, for sounds with a complex wavefront (rather than a sinusoidal sound), psychoacoustic experiments (Henning, 1974; Blauert, 1997) have proven that ***interaural envelope delay difference***, termed ITD_e instead of ITD_p, is another localization cue above 1.5 kHz. Here, the sinusoidal modulation stimulus is presented as an example. Let us assume that a high-frequency sinusoidal carrier has f_c of 3.9 kHz, whose amplitude is modulated by a low-frequency sinusoidal signal with f_m of 0.3 kHz (m is the modulation factor). Then, the binaural sound pressure signals associated with an ITD_e as τ_e are

$$\begin{aligned} p_L(t) &= \left[1 + m\cos(2\pi f_m t)\right]\cos(2\pi f_c t) \\ p_R(t) &= \left\{1 + m\cos\left[2\pi f_m (t - \tau_e)\right]\right\}\cos(2\pi f_c t) \end{aligned}. \tag{1.6.5}$$

In this case, the localization (strictly speaking, the lateralization in headphone presentation) of sound is determined by τ_e, not by the ITD derived from the fine structure of the binaural signals. However, some other studies have pointed out that ITD_e at mid and high frequencies is a relatively weaker localization cue than that of ITD_p at low frequencies (Durlach and Colburn, 1978).

In summary, the auditory system separately uses ITD_p and ITD_e as localization cues in different frequency ranges. Equations (1.6.1) and (1.6.2) are based on geometrical acoustics, so the results are neither ITD_p nor ITD_e. Actually, Equations (1.6.1) and (1.6.2) roughly correspond to the time delay between the wavefront at the positions of two ears. However, the accurate calculations of ITD_p and ITD_e are complex, whereas the calculation of the frequency-independent ITD by using Equations (1.6.1) and (1.6.2) is relatively simple. Therefore, these two equations are often used for approximately evaluating ITD. In fact, ITD_p calculated from a rigid spherical head model is simply related to Equations (1.6.1) and (1.6.2) below 0.4 kHz and above 3 kHz.

The ***interaural group delay difference*** ITD_g, which refers to the interaural difference in the slope of the HRTF (or binaural pressures) phase divided by 2π

$$ITD_g(\theta_S, \phi_S, f) = \frac{1}{2\pi}\left(\frac{d\psi_L}{df} - \frac{d\psi_R}{df}\right). \tag{1.6.6}$$

ITD_g is generally a function of source direction and frequency. In some specific cases (such as $f < 0.4$ kHz and $f > 3$ kHz in Figure 1.19), however, ITD_g is approximately independent of frequency, with $ITD_g \approx ITD_p$. Moreover, for band-pass signals with a bandwidth of much less than its center frequency, the approximation $ITD_g \approx ITD_e$ is valid. This observation can be tested with a sinusoidal amplitude modulation signal in Equation (1.6.5).

The ITD based on the maximal normalized interaural cross-correlation function is used in the analysis of spatial sound reproduction (Section 12.1).

ITDs vary with the evaluated methods (Xie, 2006a). Although the ITD examined via certain methods cannot be directly applied to analyze sound source localization, it is strongly related to ITD_p and ITD_e, so it is meaningful and applicable in practice. When ITDs are compared, these ITDs must be assessed with the same methods. In practice, the analysis of ITDs evaluated via some appropriate methods yields meaningful results.

1.6.2 Interaural level difference

ILD is another important cue for directional localization. When a sound source deviates from the median plane, the sound pressure in the farther ear (contralateral to the sound source) is attenuated (especially at high frequencies) because of the shadowing effect of the head,

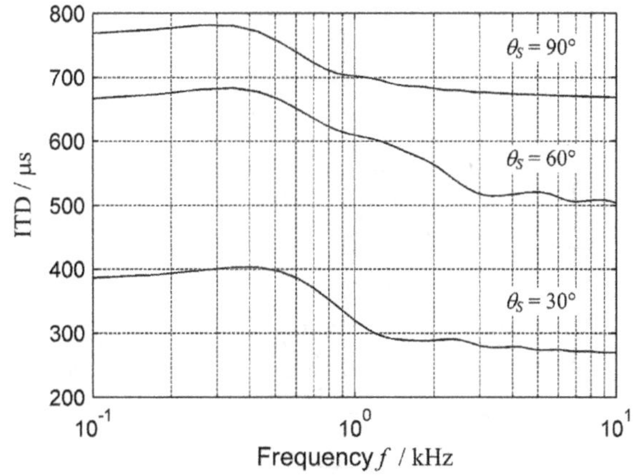

Figure 1.19 ITD_p of the rigid spherical head model varying as a function of the frequency of the incident plane wave at three different horizontal azimuths.

whereas the sound pressure in the nearer ear (ipsilateral to the sound source) increases to some extent. This acoustic phenomenon leads to direction- and frequency-dependent ILD:

$$ILD(r_S, \theta_S, \phi_S, f) = 20 \log_{10} \left| \frac{P_L(r_S, \theta_S, \phi_S, f)}{P_R(r_S, \theta_S, \phi_S, f)} \right| \quad (dB), \tag{1.6.7}$$

where $P_L(r, \theta, \phi, f)$ and $P_R(r, \theta, \phi, f)$ are the frequency-domain sound pressures in the left and right ears, respectively, generated by a sound source at (r, θ, ϕ). The results here are contrary to those in a clockwise spherical coordinate system, considering that an anticlockwise spherical coordinate system is used in this book.

ILD can be evaluated from HRTFs. The binaural pressures for a far-field source distance $r_S >> a$ (incident plane wave) can be calculated on the basis of a rigid spherical head model with a by using the corresponding HRTF formula [Equation (11.2.2)], and the resultant ILD is independent of distance r_S. Figure 1.20 shows the far-field ILD in the horizontal plane as a function of the azimuth of the incident plane wave at different frequencies (ka = 0.5, 1.0, 2.0, 4.0, and 8.0). ka ($k = 2\pi f/c$ is the wave number), instead of f, is adopted as the metric for frequency in the figure.

At low frequencies with small ka, ILD is small and varies smoothly with the azimuth θ_S. For example, the maximum ILD is only 0.5 dB for ka of 0.5, which indicates that the head shadowing effect and the resultant ILD are negligible at low frequencies in the far field. Moreover, for ka larger than 1, ILD tends to increase as the frequency increases, exhibiting a complex relationship with azimuth and frequency. For example, the maximum ILDs are 2.9, 6.7, 12.0, and 17.4 dB for ka = 1.0, 2.0, 4.0, and 8.0, respectively. If a = 0.0875 m is selected, the frequencies corresponding to ka = 0.5, 1.0, 2.0, 4.0, and 8.0 are about 0.3, 0.6, 1.2, 2.5, and 5.0 kHz, respectively. Moreover, psychoacoustic experiments have demonstrated that ILD does not act as an effective directional localization cue until it varies with source direction at frequencies above 1.5 kHz. In addition, ILD is a multivariable function; as such, it is direction and frequency dependent for a given head radius a. From another point of view, for a sound source at certain direction and frequency, ILD depends on anatomical dimensions, that is, ILD is an individual localization cue. The comparison of Figures 1.18 and 1.20 reveals that the ILD for a narrow-band stimulus may be an ambiguous localization cue because the ILD does not vary monotonously with azimuth even within the range of 0° to 90°.

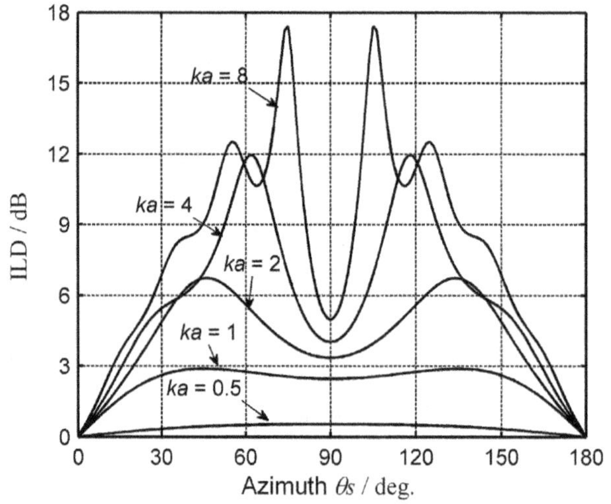

Figure 1.20 Calculated ILD as a function of azimuth at different *ka* with the spherical head model.

ILD varies dramatically with azimuth at high frequencies, such as *ka* of 4.0 and 8.0 (in Figure 1.20). Additionally, the maximum ILD for a sinusoidal sound stimulus (with a single frequency component) does not appear at an azimuth of 90°, where the contralateral ear is exactly opposite the sound source. This finding is due to the enhancement in sound pressure in the contralateral ear by the in-phase interference of multipath diffracted sounds around the spherical head. For a complex sound wave with multiple frequency components, such as octave noise, ILD varies relatively smoothly with azimuth. However, an actual human head is not a perfect sphere, and it is composed of the pinnae and other fine structures. Therefore, the relationship between ILD, sound source direction, and frequency is more complicated than that for a spherical head. Nevertheless, the results from the spherical head model are adequate for qualitatively interpreting some localization phenomena.

1.6.3 Cone of confusion and head movement

ITD and ILD are regarded as two dominant localization cues at low and high frequencies, respectively, which were first stated in classic "duplex theory" proposed by Lord Rayleigh in 1907. However, a set of ITD and ILD are inadequate for determining the unique position of a sound source. In fact, an infinite number of spatial positions possess identical differences in path lengths to the two ears (i.e., identical ITD). When the curved surface of the spherical head is disregarded and when the two ears are approximated by two separated points in a free space, the points with identical ITD form a cone around the interaural axis in a three-dimensional space, which is called "***cone of confusion***" (Figure 1.21). In the cone of confusion, ITD alone is insufficient for determining an exclusive sound source position. Similarly, for a spherical head model and at a far-field distance comparatively longer than the head radius, an infinite point set exists in space within which the ILDs are identical for all points. For an actual human head, even when its nonspherical form and curved surface are considered, the corresponding ITD and ILD are still insufficient for identifying the unique position of a sound source because they do not vary monotonously with the source position. In this case, the cone of confusion persists, but it is no longer a strict cone.

An extreme case of the cone of confusion is the median plane in which the sound pressures received by the two ears are nearly identical; thus, ITD and ILD are zero. In another case, two sound sources are located at the front–back mirror positions at the azimuths of 45° and

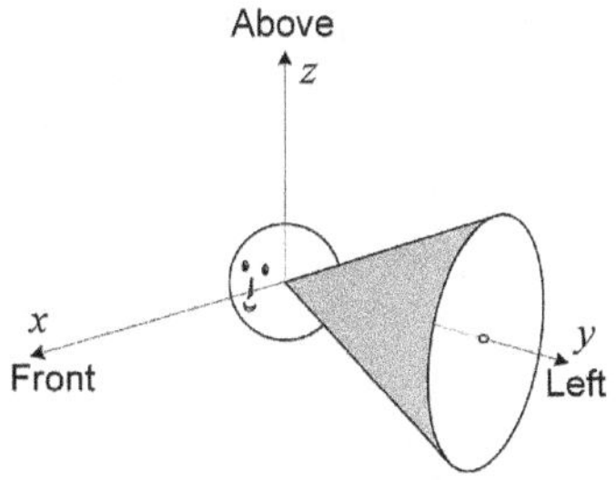

Figure 1.21 Cone of confusion in sound source localization.

135° in the horizontal plane. The resultant ITD and ILD for the two sound source positions are identical as far as a symmetrical spherical head is concerned. ITD and ILD can determine only the cone of confusion in which the sound source is located but not the unique spatial position of the sound source. Therefore, Rayleigh's duplex theory is only effective for lateral localization and ineffective for the front–back and vertical localization.

To address this problem, Wallach (1940) hypothesized that ITD and ILD change introduced by head-turning may be another localization cue (i.e., a ***dynamic cue***). For example, when the head in Figure 1.22 is fixed, ITDs and ILDs for sources at the front (0°) and rear (180°) in the

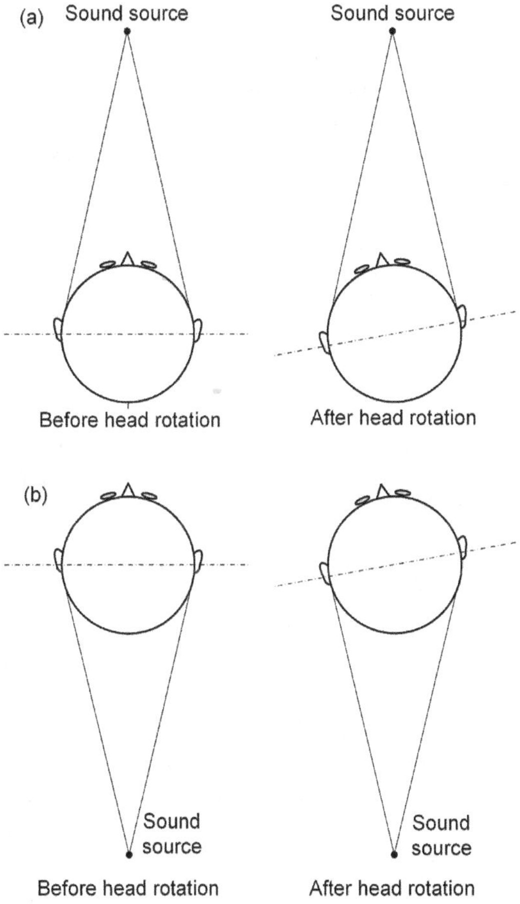

Figure 1.22 Changes in ITD caused by head rotation: sound sources in the (a) front and (b) the rear.

horizontal plane are both zero because of the symmetry of the head. Hence, the two source positions are indistinguishable in terms of ITD and ILD cues. However, head rotation can introduce a change in ITD. If the head is rotated anticlockwise (to the left) around the vertical axis, the right ear comes closer to the front sound source, and the left ear comes closer to the rear sound source. That is, for the same head rotation, the ITD for the front sound source changes from zero to negative; by contrast, the ITD for the rear sound source changes from zero to positive. If the head is rotated clockwise (to the right) around the vertical axis, a completely opposite situation occurs. Head rotation changes not only the ITD but also the ILD and the sound pressure spectra in the ears, although ILD is not a monotonic function of the source azimuth. Therefore, dynamic information aids localization. Previous experiments preliminarily confirmed that head rotation around a vertical axis is necessary to resolve the front–back ambiguity in horizontal localization. This conclusion has been further verified by some recent experiments (Wightman and Kistler, 1999) and applied to virtual auditory displays (Section 11.10.2). In addition, experimental evidence has indicated that the change in ITD provides major dynamic information about front–back localization (Macpherson, 2011).

Wallach also hypothesized that head-turning provides information for vertical localization. Follow-up studies have attempted to verify this hypothesis through experiments. However, completely and experimentally excluding contributions from other vertical localization cues (such as spectral cues; Section 1.6.4) was difficult. Wallach's hypothesis was not widely explored because of the lack of sufficient experimental support. Since the 1990s, nevertheless, the problem of vertical localization has attracted renewed attention to develop a virtual auditory display. Perrett and Noble (1997) first experimentally verified Wallach's hypothesis. Our own work (Rao and Xie, 2005) further demonstrated that the change in ITD introduced by the head movement in two degrees of freedom (turning around the vertical and front–back axes, respectively, i.e., rotating and pivoting or yawing and rolling) provides information for localization in the median plane at low frequencies and allows the quantitative verification of Wallach's hypothesis to be given (Chapter 6). Some recent experiments have also confirmed the contributions of head-turning to vertical localization (Ashby et al., 2013, 2014). In addition, other experiments have investigated the range and pattern of head movement made by listeners (Kim et al., 2013).

1.6.4 Spectral cues

Many studies have suggested that the spectral feature caused by the reflection and diffraction in the pinna and around the head and torso provides helpful information on vertical localization and front-back disambiguity. In contrast to binaural cues (ITD and ILD), the spectral cue is a monaural cue (Wightman and Kistler, 1997).

Batteau (1967) proposed a simplified model to explain the pinna effect. Figure 1.23 shows that direct and reflected sounds arrive at the entrance to the ear canal. The relative delay between the direct and reflected sounds is direction dependent because the incident sounds from different spatial directions are likely to be reflected by the different parts of the pinna. Therefore, peaks and notches in the sound pressure spectra caused by the interference between the direct and reflected sounds are also direction dependent, thereby providing information for directional localization. In Batteau's model, the pinna effect is described as a combination of two reflections with different magnitudes, i.e., A_1 and A_2, and different time delays, i.e., τ_1 and τ_2. Hence, the transfer function of the pinna, including one direct and two reflected sounds, is expressed as

$$H(f) = 1 + A_1 \exp(-j2\pi f \tau_1) + A_2 \exp(-j2\pi f \tau_2). \quad (1.6.8)$$

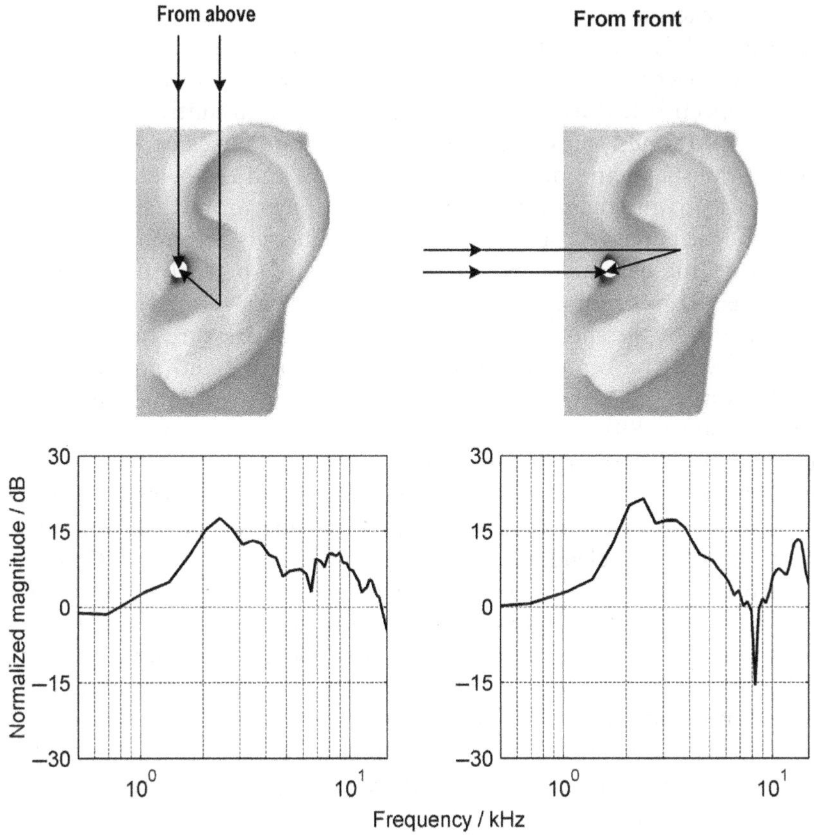

Figure 1.23 Pinna interacting with incident sounds from two typical directions.

Batteau's model achieved limited success because of its considerable simplification. The dimension of the pinna is about 65 mm, so it functions effectively only if the frequency is above 2–3 kHz. At this frequency, the sound wavelength is comparable with the dimension of the pinna. Moreover, the effect of the pinna is prominent at frequencies above 5–6 kHz. The pinna also has a complex and irregular surface, so it cannot be regarded as a reflective plane from the perspective of geometrical acoustics within the entire audible frequency range. This characteristic is the inherent drawback of Batteau's model. Further studies have pointed out that the pinna reflects and diffracts the incident sound in a complex manner (Lopez-Poveda and Meddis, 1996). The interference among direct and multipath reflected/diffracted sounds acts as a filter and then modifies the incident sound spectrum as direction-dependent notches and peaks. In addition, this interference is highly sensitive to the shape and dimension of the pinna, which differs among individuals. Therefore, the spectral information provided by the pinna is an extremely individualized localization cue.

Shaw and Teranishi (1968) and Shaw (1974) investigated the effect of the pinna in terms of wave acoustics and proposed a resonance model, which demonstrates that resonances within pinna cavities and the ear canal form a series of resonance modes at mid and high frequencies of 3, 5, 9, 11, and 13 kHz. This model successfully interprets the peaks in the pressure spectra; among them, the peak at 3 kHz for the first salient resonance is derived from the quarter wavelength resonance of the ear canal, though the existence of the pinna extends the effective length of the ear canal. Hearing is also most sensitive around this frequency (Section 1.3.2). Moreover, the magnitudes of high-order resonance modes vary with the direction of the incident sound except the first one. Hence, the resonance model supports the idea that the spectral cue provided by the pinna is a directional localization cue.

Numerous psychoacoustic experiments have been devoted to exploring the localization cue encoded in spectral features. However, no general quantitative relationship between spectral features and sound source positions has been found because of the complexity and individuality of the shape and dimension of the pinna and the head. Blauert (1997) used narrow-band noise to investigate directional localization in the median plane. Experimental results show that the perceived position of a sound source is determined by the directional frequency band in the ear canal pressures regardless of the real sound source position; that is, the perceived position of a sound source is always located in specific directions, where the frequency of the spectral peak in the ear canal pressure caused by a wide-band sound coincides with the center frequency of the narrow-band noise. Hence, peaks in the ear canal pressure caused by the head and the pinna are important in localization (Middlebrooks et al., 1989).

However, some researchers argued that the spectral notch, especially the lowest frequency notch caused by the pinna (called the ***pinna notch***), is more important for localization in the median plane and even for vertical localization outside the median plane (Hebrank and Wright, 1974; Butler and Belendiuk, 1977; Bloom, 1977; Kulkarni, 1997; Han, 1994). In front of the median plane, the center frequency of the pinna notch varies with elevation in the range of 5 or 6 kHz to about 12 or 13 kHz. This variation is due to the interaction of the incident sound arriving from different elevations to the different parts of the pinna. As a result, different diffraction and reflection delays occur relative to the direct sound. Thus, shifting frequency notch provides vertical localization information. Moore et al. (1989) found that shifting in the central frequency of the exquisitely narrow notch can be easily perceived although hearing is usually more sensitive to the spectral peak.

Other researchers contended that both peaks and notches (Watkins, 1978) or spectral profiles are important in localization (Middlebrooks, 1992). Algazi et al. (2001b) suggested that the change in the ipsilateral spectra below 3 kHz caused by the scattering and reflection of the torso, especially the shoulder, provides vertical localization information for a sound source outside the median plane.

In brief, the spectral feature caused by the diffraction and reflection of an anatomical structure, such as the head and pinna, is an important and individualized localization cue. Although a clarified and quantitative relationship between the spectral feature and the direction of a sound source is far less complete, one's own spectral feature can be used to localize a sound source.

1.6.5 Discussion on directional localization cues

In summary, directional localization cues can be classified as follows:

1. For frequencies approximately below 1.5 kHz, the ITD derived from ITD_p is the dominant cue for lateral localization.
2. Above the frequency of 1.5 kHz, ILD and ITD derived from the interaural envelope delay difference (ITD_e) contribute to lateral localization. As frequency increases (approximately above 4–5 kHz), ILD gradually becomes dominant.
3. A spectral cue is important for localization. In particular, above frequencies of 5–6 kHz, a spectral cue introduced by the pinna is essential for the vertical localization and disambiguation of front–back confusion.
4. The dynamic cue introduced by the slight turning of the head is helpful in resolving front–back ambiguity and vertical localization.

The aforementioned directional localization cues except the dynamic cue can be evaluated from HRTFs. Therefore, HRTFs include the major directional localization cues. These localization cues are individually dependent because of the unique characteristics of anatomical structures and dimensions. The auditory system determines the position of a sound source based on a comparison between the obtained cues and patterns stored from prior experiences. However, even for the same individual, anatomical structures and dimensions vary with time, especially from childhood into adulthood albeit slowly. Therefore, a comparison with prior experiences may be a self-adaptive process, and the high-level neural system can automatically modify stored patterns by using auditory experiences.

Different kinds of localization cues work in different frequency ranges and contribute differently to localization. For sinusoidal or narrow-band stimuli, only the localization cues existing in the frequency range of the stimuli are available; hence, the resultant localization accuracy is likely to be frequency dependent. Mills (1958) investigated localization accuracy in the horizontal plane by using sinusoidal stimuli. He showed that localization accuracy is frequency dependent, and the highest accuracy is $\Delta\theta_S = 1°$ in front of the horizontal plane ($\theta_S = 0°$) at frequencies below 1 kHz. With an average head radius a of 0.0875 m, the corresponding variation in ITD evaluated from Equation (1.6.1) is about 10 μs, or the variation in the low-frequency interaural phase delay difference evaluated from Equation (1.6.4) is about 15 μs. This finding is consistent with the average value of a just noticeable difference in ITD derived from psychoacoustic experiments (Blauert, 1997; Moore, 2012). Conversely, localization accuracy is the poorest around the frequency of 1.5–1.8 kHz, which is the range of difficult or ambiguous localization. This finding may be because the cue of ITD_p becomes invalid within this frequency range; unfortunately, ITD_e is a relatively weak localization cue, and the cue of ILD only begins to work and does not vary significantly with the direction within this frequency range.

In general, when more localization cues are presented in sound signals in the ears, the localization of the sound source position is more accurate because the high-level neural system can simultaneously use multiple cues. This fact is responsible for numerous phenomena. For example, (1) the accuracy of binaural localization is much better than that of monaural localization; (2) the accuracy of localization in a mobile head is usually better than that in an immobile one; and (3) the accuracy of localization for a wide-band stimulus is usually better than that for a narrow-band stimulus, especially when the stimulus contains components above 6 kHz, which can improve accuracy in vertical localization. Despite the absence of some cues, the auditory system can localize the sound source because the information provided by multiple localization cues may be somewhat redundant. For example, high-frequency spectral cues and dynamic cues contribute to vertical localization. When one cue is eliminated, another cue alone still enables vertical localization to some extent (Jiang et al., 2019).

Under some situations, when some cues conflict with others, the auditory system appears to identify the source position according to the more consistent cues. This phenomenon indicates that the high-level neural system can correct errors in localization information. Wightman and Kistler (1992) performed psychoacoustic experiments and proved that ITD is dominant as long as the wide-band stimuli include low frequencies regardless of conflicting ILD. However, if too many conflicts or losses exist in localization cues, accuracy, and quality in localization are likely to be degraded, splitting virtual sources are perceived, or localization is even impossible, as proven by a number of experiments. For example, when a dynamic cue at low frequency conflicts with a spectral cue at high frequencies, the front-back accuracy is degraded. Sometimes, one cue may dominate localization, or two conflicting cues may yield two splitting virtual sources at different frequency ranges (Pöntynen et al., 2016). These cues depend on the characteristics of signals, especially the power spectra of signals (Macpherson,

2011, 2013; Brimijoin and Akeroyd, 2012). These aforementioned results are applicable to spatial sound reproduction. Various practical spatial sound techniques are unable to reproduce the spatial information of a sound field within a full audible frequency range because it is limited by the complexity of the system. Practical spatial sound techniques can create the desired perceived effects to some extent provided that they can reproduce dominant spatial cues.

Aside from acoustic cues, visual cues dramatically affect sound localization. The human auditory system tends to localize sound from a visible source position. For example, in watching television, the sound usually appears to come from the screen, although it actually comes from loudspeakers. However, an unnatural perception may occur when the discrepancy between the visual and auditory location is too large, such as when a loudspeaker is positioned behind a television audience. This result is important for spatial sound reproduction with an accompanying picture (Chapters 3 and 5). This phenomenon further indicates that sound source localization is a consequence of a comprehensive processing of a variety of information received by the high-level neural system.

In Sections 1.3.2 and 1.3.3, directional loudness and spatial unmasking are related to binaural cues. After being scattered and diffracted by anatomical structures, such as the head and pinnae, sound waves are received by the two ears, and sound pressures in the eardrum depend on the source direction. Most variations in subjective loudness with the sound source direction can be analyzed in terms of HRTFs (Sivonen and Ellermeier, 2008).

Spatial unmasking can be partially interpreted with the source position tendency of HRTFs (Kopčo and Shinn-Cunningham, 2003). Other cues also provide information for spatial unmasking. The bandwidth of a masker is assumed to be less than that of an auditory filter, and a target is assumed as a pure tone, whose frequency is within the bandwidth of the masker. At a specific frequency, when the positions of the masker and the target are spatially coincident, the diffractions that anatomical structures, such as the head, cause to the masker and target sounds are the same. Accordingly, a certain target-to-masker sound pressure ratio (target-to-masker ratio) exists for each ear. When the masker and the target are spatially separated, the diffractions imposed on their sounds differ from each other, thereby potentially increasing the target-to-masker ratio of one ear (called the ***better ear***). The auditory system can detect the target with the information provided by the better ear and therefore decrease the masking threshold. Each ear's target-to-masker ratio, which is related to the conditions of the target and the masker (i.e., intensity, frequency, and spatial position), can be evaluated in terms of HRTFs.

1.6.6 Auditory distance perception

Although the ability of the human auditory system to estimate the sound source distance is generally poorer than the ability to locate sound source direction, a preliminary but biased auditory distance perception can still be formed. Experiments have demonstrated that the auditory system tends to significantly underestimate distances to distant sound sources with the physical source distance farther than a rough average of 1.6 m and typically overestimates distances to nearby sound sources with the physical source distance less than a rough average of 1.6 m. This finding suggests that the perceived source distance is not always identical to the physical one. Zahorik (2002a) examined experimental data from a variety of studies and found that the relationship between the perceived distance r_I and the physical distance r_S can be well approximated with a compressive power function by using a linear fit method:

$$r_I = \kappa r_S^{\delta}, \tag{1.6.9}$$

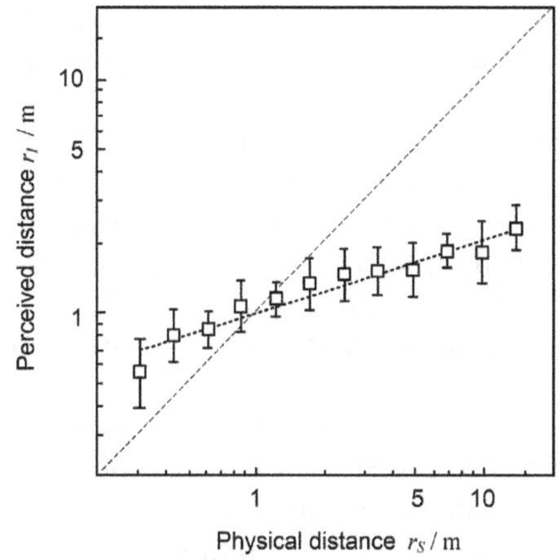

Figure 1.24 Relationship between r_I and r_S obtained using a linear fit for a typical subject, with $\alpha = 0.32$ and $\kappa = 1.00$ (Zahorik, 2002b, with the permission of Zahorik P.).

where κ is a constant whose average is slightly greater than 1 (average of approximately 1.32), and δ is a power-law exponent whose value is influenced by various factors, such as experimental conditions and subjects, so this value varies in a wide range with a rough average of 0.4. In the logarithmic coordinate, the relationship between r_I and r_S is expressed with straight lines having various slopes; among them, a straight line through the origin with a slope of 1 means that r_I is identical to r_S, i.e., the case of unbiased distance estimation. Figure 1.24 shows the relationship between r_I and r_S obtained using a linear fit for a typical subject (Zahorik, 2002b).

Auditory distance perception, which was thoroughly reviewed by Zahorik et al. (2005), is a complex and comprehensive process based on multiple cues. Subjective loudness has been considered an effective cue to distance perception. Generally, loudness is closely related to sound pressure or intensity at a listener's position; usually, strong sound pressure results in high loudness. In a free field, the sound pressure generated by a point sound source with constant power is inversely proportional to the distance between the sound source and the receiver (the $1/r$ law); that is, the SPL decreases by 6 dB for each doubling of the source distance. As a result, a close distance corresponds to a high sound pressure and subsequent high loudness. As such, loudness becomes a cue for distance estimation. However, the $1/r$ law only applies to the free field and deviates in reflective environments. Moreover, the sound pressure and loudness at a listener's position depends on source properties, such as radiated power. Previous knowledge on sound sources or stimuli also influences the performance of distance estimation when loudness-based cues are used. In general, loudness is regarded as a relative distance cue unless the listener is highly familiar with the pressure level of the sound source.

The high-frequency attenuation caused by air absorption may be another cue of auditory distance perception. For a far sound source, air absorption acts as a low-pass filter and thereby modifies the spectra of sound pressures at the receiving position. This effect is important only for an extremely far sound source and negligible in an ordinary-sized room. Moreover, previous knowledge on the sound source may influence the performance of distance estimation when high-frequency attenuation-based cues are used. In general, high-frequency attenuation provides weak information for relative distance perception.

Some studies have demonstrated that the effects of acoustic diffraction and shadowing by the head provide information on evaluating distance for nearby sound sources (Brungart and Rabinowitz, 1999; Brungart et al., 1999; Brungart, 1999). In Section 1.6.2, ILD is nearly independent of the source distance in the far field. However, it varies considerably as the source distance changes within the range of 1.0 m (i.e., in the near field) for a sound source outside the median plane, especially within the range of 0.5 m. ILD is irrelevant to source properties because it is defined as the ratio between the sound pressures in the two ears. Therefore, the near-field ILD is a cue for absolute distance estimation. The pressure spectrum in each ear also changes with the source distance in the near field, which potentially serves as another distance cue. These distance-dependent cues are described by near-field HRTFs. Near-field HRTFs are used to render a virtual source at various distances in a virtual auditory display (Chapter 11). However, this method is reliable only within a target source distance of 1.0 m.

Reflections in an enclosed space are effective cues for distance estimation (Nielsen, 1993). In Equation (1.2.25), the direct-to-reverberant energy ratio is inversely proportional to the square of distance. Therefore, it can be used as a distance cue, although a real reflected sound field may deviate from the diffuse sound field. Equation (1.2.25) is derived on the basis of this ratio. Bronkhorst and Houtgast (1999) indicated that a simple model based on a modified direct-to-reverberant energy ratio can accurately predict the auditory distance perception in rooms. In stereophonic and multichannel sound program production, the perceived distance is often controlled by the direct-to-reverberant energy ratio in program signals. Frequency-dependent boundary absorption modifies the power spectra of reflections, and the proportion of the reflected power increases as the sound source distance increases. Thus, the power spectra of binaural pressures vary with the sound source distance. This finding also provides information for auditory distance perception.

In summary, auditory distance perception is derived from the comprehensive analyses of multiple cues. Although distance estimation has recently received increasing attention, knowledge regarding its detailed mechanism remains incomplete.

1.7 SUMMING LOCALIZATION AND SPATIAL HEARING WITH MULTIPLE SOURCES

The localization of multiple sound sources, as the localization of a single sound source presented in Section 1.6, is another important aspect of spatial hearing (Blauert, 1997). Under a specific situation, the auditory system may perceive a sound coming from a spatial position where no real sound source exists when two or more sound sources simultaneously radiate correlated sounds. Such kind of an illusory or phantom sound source, also called ***virtual sound source*** (shortened as ***virtual source***) or ***virtual sound image*** (shortened as ***sound image***), results from the summing localization of multiple sound sources. In summing localization, the sound pressure in each ear is a linear combination of the pressures generated by multiple sound sources. The auditory system then automatically compares the localization cues, such as ITD and ILD, encoded in binaural sound pressures with the stored patterns derived from prior experiences with a single sound source. If the cues in binaural sound pressures successfully match the pattern of a single sound source at a given spatial position, then a convincing virtual sound source at that position is perceived. However, this case is not always true. Some experimental results of summing localization remain incompletely interpreted. Overall, summing localization with multiple sound sources is a spatial auditory event that should be explained with the psychoacoustic principle (Guan, 1995).

Under some situations, multiple sources may result in other spatial auditory events. For example, in the case of the precedence effect described in Section 1.7.2, a listener perceives sounds as though they come from one of the real sources regardless of the presence of other sound sources. When two or more sound sources simultaneously radiate partially correlated sounds, the auditory system may perceive an extended or even diffusely located spatial auditory event.

All the abovementioned phenomena are related to the summing spatial hearing of multiple sound sources, and they are the subjective consequences of comprehensively processing the spatial information of multiple sources by the auditory system. In addition, the subjective perceptions of environment reflection are closely related to spatial hearing with multiple sources (Section 1.8).

1.7.1 Summing localization with two sound sources

The simplest case of summing localization is the one involving two sound sources. Blumlein (1931) first recognized the application of this psychoacoustic phenomenon to stereophonic reproduction. Since the work of Boer (1940), other researchers have conducted a variety of experiments, i.e., two-channel stereophonic localization experiments, on the summing localization with two sound sources (Leakey, 1959, 1960; Mertens, 1965; Simonson, 1984). Blauert (1997) summarized the results of some early experiments in his monograph.

A typical configuration of summing localization with two sources or two-channel stereophonic loudspeakers is shown in Figure 1.25. A listener locates at a symmetric position with respect to left and right loudspeakers. The azimuths of two loudspeakers are $\pm\theta_0$, or two loudspeakers are separated by a spanned angle of $2\theta_0$.The distance r_0 from the loudspeaker to the head center is much larger than the head radius a. The base line length (distance) between two loudspeakers is $2L_Y$, and the distance between the midpoint of the base line and the head center is L_X. When both loudspeakers are provided identical signals, the listener perceives a single virtual source at the mid-direction between the two loudspeakers, i.e., directly in front of the listener. When the magnitude ratio or ***interchannel level difference*** (***ICLD***) is adjusted between loudspeaker signals, the virtual source moves toward the direction of the loudspeaker with a large signal level. An ICLD larger than approximately 15 dB to 18 dB is sufficient to position the virtual sound source to either of the loudspeakers (full left or full right). Then, the position of the virtual sound source no longer changes even with an increasing level difference.

The above results are qualitatively held for signals that include a low-frequency component below 1.5 kHz. However, the results obtained from various experiments quantitatively differ in terms of signals, experimental conditions, and methods. Some experimental errors

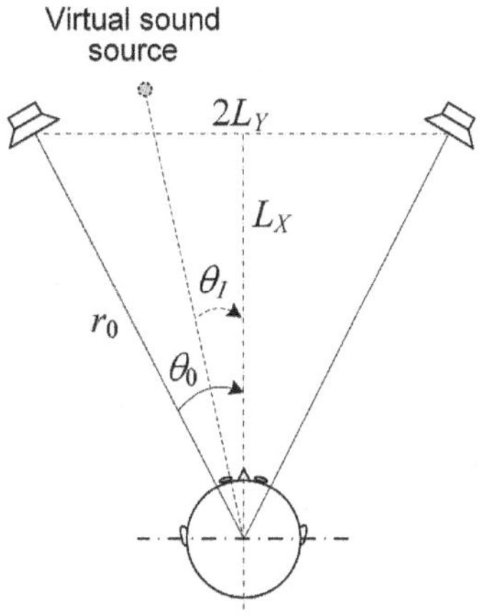

Figure 1.25 Summing localization experiment involving two sound sources (loudspeakers).

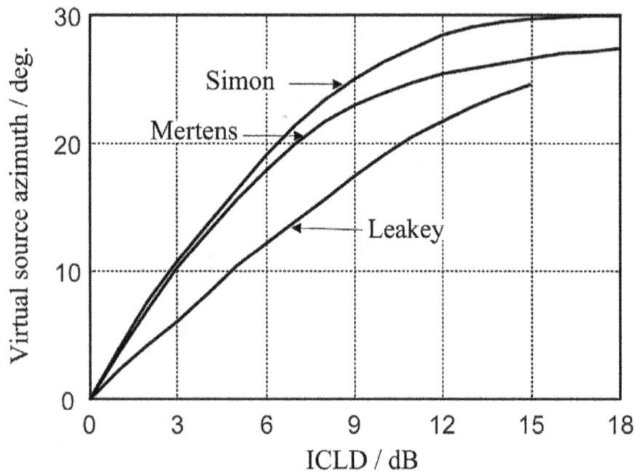

Figure 1.26 Virtual source localization experiments on loudspeaker signals with ICLD. (redrawn on the basis of the results of Leakey 1960, Mertens 1965, and Simonson 1984 and adopted from Wittek and Theile 2002).

may also be included in the results. Figure 1.26 illustrates the experimental results of a virtual source position that varies with the ICLD of loudspeaker signals obtained by Leakey (1960), Mertens (1965), and Simonson (1984). A standard angle span $2\theta_0 = 60°$ (or near 60°) between two loudspeakers was chosen in their experiments. Speech signals were used in the experiments of Leakey and Simonson. The noise signal centered at 1.1 kHz was used by Mertens. In addition, the displacements of the virtual source in the baseline were determined in some experiments. Here, they are converted to the azimuths of the virtual source.

In the same loudspeaker configuration shown in Figure 1.25, if a signal and its delayed version are fed into two loudspeakers, a virtual source moves toward the direction of the loudspeaker with the leading signal. When the ***interchannel time difference*** (**ICTD**) between two loudspeaker signals exceeds a certain upper limit, the virtual source moves to the direction of the loudspeaker. This result is held for impulse-like signals or some other signals with transient characteristics, such as click, speech, and music signals. However, the ICTD cannot be utilized effectively for low-frequency steady signals.

The ICTD required to position the virtual sound source to either of the loudspeakers varies considerably in different experiments and usually depends on the type of signals. It differs from several hundreds of microseconds (μs) to slightly more than 1 millisecond (ms) in most cases. Figure 1.27 illustrates the experimental results of the virtual source position varying with the ICTD obtained by Leakey (1960), Mertens (1965), and Simonson (1984). The conditions of experiments are similar to those mentioned above, but the signal used by Mertens was random noise. ICLD and ICTD are the level and time differences between two loudspeaker signals, respectively. They should not be confused with the ILD and ITD discussed in Section 1.6, which describes the level and time differences between the pressures or signals in the two ears and serve as cues for directional localization.

For some transient signals (rather than all signals), a trading effect exists between ICLD and ICTD. This effect has been experimentally investigated, but results have some differences depending on the type of signals (Leakey, 1959; Mertens, 1965; Blauert, 1997). The general tendency is summarized as follows. For loudspeaker signals with ICLD and ICTD, the movement of a virtual source enhances when the individual effects of ICLD and ICTD are consistent, and the movement of the virtual source becomes cancelled when the individual effects of ICLD and ICTD are opposite. Figure 1.28 illustrates the results of Mertens; that is, trading curves between ICLD and ICTD for the virtual source at $\theta_I = 0°$ and ±30°. The loudspeakers

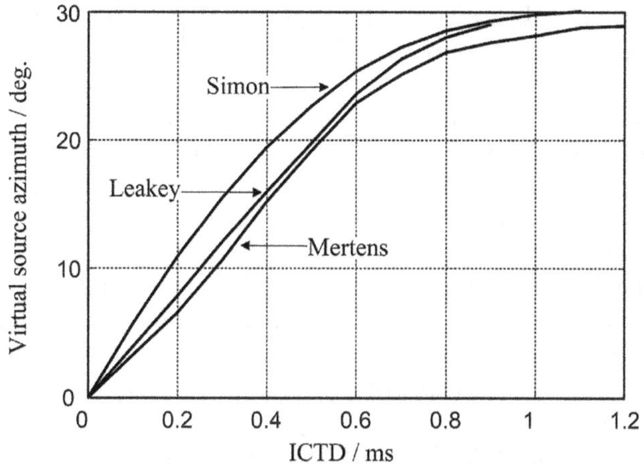

Figure 1.27 Results of virtual source localization experiments for loudspeaker signals with ICTD. (redrawn on the basis of the results of Leakey 1960, Mertens 1965, Simonson 1984 and adopted from Wittek and Theile 2002.)

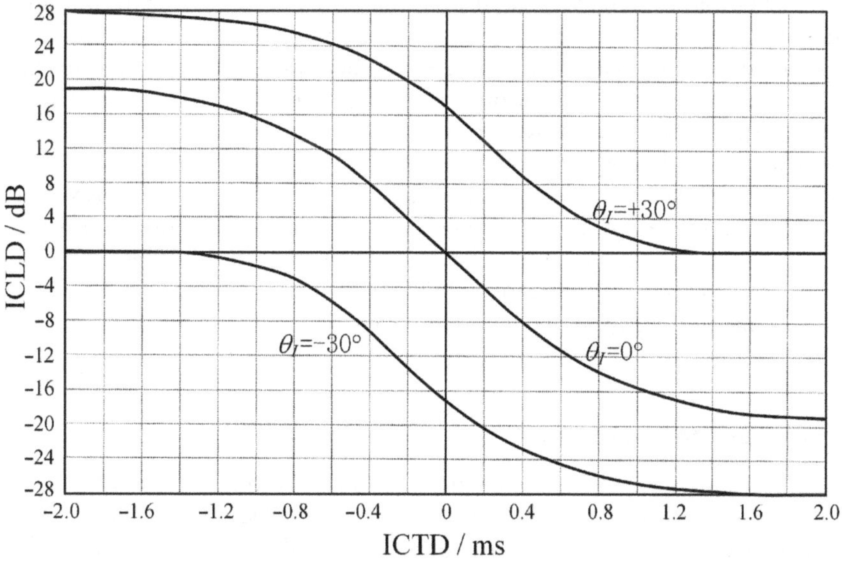

Figure 1.28 Trading curve between ICLD and ICTD. (redrawn on the basis of the data of Mertens, 1965 and adopted from Williams, 2013.)

are arranged at azimuths ±30°, and the signal is a Gaussian burst of white noise. The curves in Figure 1.28 are left-right symmetric. All combinations of ICTD and ICLD in each curve yield the same azimuth perception (i.e., 0°, −30°, or 30°).

As is indicated in Section 2.1 and Chapter 12, stereophonic loudspeaker signals with ICLD only results in appropriate low-frequency ITD_p in the superposed sound pressures in the two ears. In addition, the ILD caused by loudspeaker signals with ICLD only is small below the frequency of 1.5 kHz, which is qualitatively consistent with the case of an actual sound source. The auditory system determines the position of a virtual source based on the comparison between the resultant ITD_p and the patterns stored from previous experiences on real sound sources because ITD_p dominates the lateral localization. The method of recreating a virtual source with ICLD only is effective for signals with a low-frequency component below 1.5 kHz. The resultant virtual source is relatively stable, and it exhibits a relatively low signal-dependent tendency.

As indicated in Section 12.1.4, for low-frequency signals below 1.5 kHz, ICTD only results in a frequency-dependent and inconsistent ITD_p cue. It also produces a conflicting ILD cue. Therefore, for low-pass and steady signals, these inconsistent cues are unable to recreate stable virtual source localization. For transient signals, the auditory system may determine the source position according to the more consistent cues or some unknown psychoacoustic mechanisms, but the direction of a virtual source usually depends on the type and power spectra of signals. In addition, the virtual source may become extended, blurred, unnatural, or even unable to be localized. This property is a drawback of recreating a virtual source with ICTD, as proven by previous experiments. However, this drawback has not been fully explored in other studies.

For loudspeaker signals with ICLD and ICTD, the superposed sound pressures in the two ears contain localization cues in various frequency ranges, and these cues may conflict with one another. For transient stimuli, the auditory system may identify the summing virtual source position based on the more consistent cues or some unknown psychoacoustic mechanisms. The direction of a virtual source is likewise dependent on the type and power spectra of signals. The virtual source may also become extended, blurred, unnatural, or even unable to be localized.

A two-channel stereophonic localization experiment provides important insights into the basic principle of summing localization with two or more sound sources. The principle of summing localization with other two or more loudspeaker (sound source) configurations is similar and will be addressed in the succeeding chapters. The results of summing localization experiments indicate that a virtual source can be recreated at positions where no actual loudspeakers are arranged by appropriately choosing loudspeaker configuration and signals. The position of the virtual source can be controlled with loudspeaker signals (e.g., ICLD and ICTD). This method is important for the reproduction of the directional information of sound. It is one of the psychoacoustic principles of stereophonic and multichannel sound reproduction.

1.7.2 The precedence effect

Two methods are used to create two or more sounds with differences in the arrival time in the receiver position. One is created by the delay among electrical signals fed to each loudspeaker, as described in Section 1.7.1. The other is created by the differences in the transmission paths from each loudspeaker (source) to the receiver position. Summing localization occurs when the difference in the arrival time of two sounds is within some range.

When the difference in arrival time between two sounds exceeds a lower limit, a kind of spatial auditory event that is completely different from summing localization occurs. This effect is known as the ***precedence effect*** or the ***Hass effect.*** The precedence effect refers to a phenomenon in which the auditory system perceives a sound to be coming from the position of the leading sound (regardless of the existence of lagged sounds) for two correlated sounds with transient characteristics (such as click, music, and speech) when the difference in the arrival time falls within a certain boundary, which is defined by a lower-limit τ_L and an upper-limit τ_H. The two sounds are perceived to be a fused spatial auditory event, and localization is dominated by the wavefront that first reaches the two ears; that is, the ability to perceive the second sound as a separate spatial auditory event is suppressed. Thus, the precedence effect is also called the ***law of the first wavefront.*** Conversely, if the difference in arrival time between the sounds exceeds τ_H, then a separate echo is perceived.

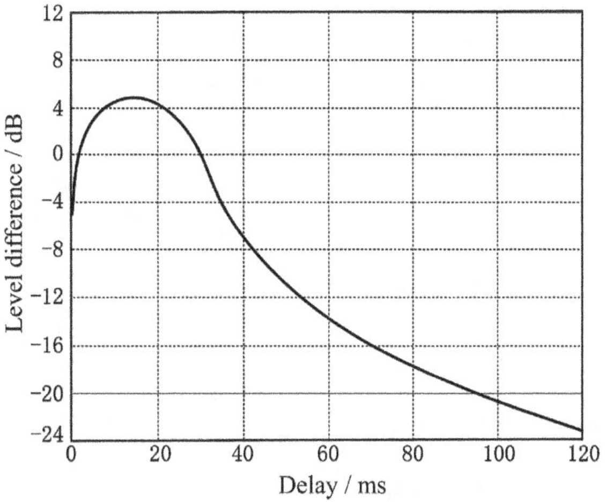

Figure 1.29 Echo threshold for the speech signal obtained from the criterion "echo barely inaudible". (adapted from Meyer and Schodder, 1952.)

τ_L and τ_H depend on some acoustical factors, such as source position, and the properties and relative magnitude of sound signals. Generally, τ_L is approximately 1–3 ms, whereas τ_H is several milliseconds for a single pulse and approximately 50 ms for speech. Some studies have shown that τ_H is relevant to the time duration of the autocorrelation function of the sound signal. Ando (1985) suggested that time duration should be used to evaluate τ_H when the autocorrelation function of the sound signal decays to 10% of its maximum. The precedence effect is robust even when the lagged sound intensity is moderately higher than the leading sound.

Numerous psychoacoustic experiments on the precedence effect have been performed. Figure 1.29 shows the echo threshold of the speech signal obtained from the criterion "echo barely inaudible" (Meyer and Schodder, 1952). The angle between two stereophonic loudspeakers is $2\theta_0 = 80°$. The abscissa and ordinate in the figure are the delay and level difference between the second (lagged) and first (leading) sounds, respectively. The threshold varies among experiments. In addition to the differences in signal type and level, differences in the criterion of the echo threshold are important reasons for the variation in experimental results. In Figure 1.29, the resultant level threshold is lower than those obtained from other criteria because the criterion "echo barely inaudible" is used. The criterion "echo clearly audible" is utilized in other experiments, and the resultant level threshold is higher than that in Figure 1.29. Generally, with an appropriate delay, the echo is clearly audible only when the level of the second sound is 10–15 dB higher than that of the first sound.

In the precedence effect, the lagged sound contributes in a specific way to the overall auditory events, including timbre, loudness, and other spatial attributes, although it is not perceived as a separate auditory event from the leading sound. In this sense, the suppression of the lagged sound in the precedence effect is restricted to the spatial location information and does not include all perceived information. Numerous studies on the mechanism of the precedence effect have been conducted, and some models have been suggested. However, the psychoacoustic phenomena of the precedence effect have not been completely interpreted. Overall, the precedence effect is due to suppression on the nerve responses of the lagged sound by the leading sound in the auditory pathway. Suppression is mostly generated at the level above the monaural pathway. Long-lasting suppression is observed in the inferior colliculus, and the dorsal nucleus of the lateral lemniscus may play an important role in the results occurring in the inferior colliculus. However, the behavioral manifestation of the precedence effect is likely mediated at higher levels in the auditory pathway than inferior colliculus, and the auditory cortex is responsible for the behavioral occurrence of the precedence effect. Details of the precedence effect are described in other studies (Litovsky et al., 1999; Blauert, 1997; Zurek, 1987).

The precedence effect is vital to the directional localization of the direct sound in reflective rooms. In Section 1.2.2, reflections in a room can be regarded as multipath sounds originating from a series of image sound sources. Numerous experiments have indicated that reflections slightly influence the perceived direction of the direct sound as long as the time interval and the relative intensity difference in the direct sound and the reflections satisfy the prerequisites of the precedence effect. However, reflections begin to impair the localization accuracy of the direct sound as the reflected energy increases. In particular, at the position outside the reverberation radius of a room, where the reverberant sound energy density is larger than that of the direct sound, the directional localization of the direct sound becomes difficult and even impossible. In this case, the existence of the reverberation sound decreases the degree of correlation in the binaural sound pressures, thereby leading to the failure of localization.

These results are considerably important in practice. In most daily activities, such as talking in a room, directional localization is not influenced by room reflections because of the precedence effect. In domestic sound reproduction, reasonable loudspeaker arrangements and absorption treatments on room surfaces are required to ensure that the differences in time interval and relative intensity among direct and reflected sounds satisfy the prerequisites of the precedence effect because a listening room is relatively small. As such, the influence of reflections on the summing localization of a stereophonic sound source can be avoided. The acoustical design of halls should be explored to ensure that the differences in time interval and relative intensity among the direct sound and reflections satisfy the prerequisites of the precedence effect to avoid perceived echoes. In a distributing sound reinforcement system with multiple loudspeakers, the perceived sound direction can be controlled by applying appropriate delays to parts of loudspeaker signals according to the rule of the precedence effect. The precedence effect is also applicable to the microphone techniques of stereophonic and multichannel sound recording (Section 2.2.3 and Chapter 7).

1.7.3 Spatial auditory perceptions with partially correlated and uncorrelated source signals

Spatial auditory perceptions are closely related to the interaural correlation of the pressures (signals) in the two ears (Damaske, 1969/1970). This correlation is measured with the ***interaural cross-correlation coefficient*** (*IACC*), which is defined as the value that maximizes the absolute value of the normalized cross-correlation function of binaural pressures in the time domain:

$$\Psi_{LR}(\tau) = \frac{\int_{t_1}^{t_2} p_L(t)\, p_R(t+\tau)\, dt}{\left\{\left[\int_{t_1}^{t_2} p_L^2(t)\, dt\right]\left[\int_{t_1}^{t_2} p_R^2(t)\, dt\right]\right\}^{1/2}}, \tag{1.7.1}$$

$$IACC = \max |\Psi_{LR}(\tau)| \qquad |\tau| \leq 1\ ms, \tag{1.7.2}$$

where $[t_1, t_2]$ is the time window in which the interaural correlation is analyzed, and $|\Psi_{LR}(\tau)|$ is measured over a range of the time parameter τ, typically ±1 ms, which corresponds to the possible maximal ITD for a single sound source. Let τ_{max} denote τ that maximizes $|\Psi_{LR}(\tau)|$. For a sound source in the free field, τ_{max} is the ITD defined through interaural cross-correlation calculation (Section 12.1.1). By definition, $0 \leq \text{IACC} \leq 1$. IACC describes the similarity between the pressures in the two ears, with a large IACC (close to unit) representing a high degree of similarity.

For a single source in a free field and a reflective room, binaural pressures usually exhibit a positive interaural correlation with $\max |\Psi_{LR}(\tau)| = \max [\Psi_{LR}(\tau)]$, and the interaural correlation tends to be zero in an ideal diffused sound field. In these cases, the IACC defined by Equation (1.7.2) does not have ambiguities. However, under some artificially controlled conditions, such as some cases in headphones or two (more) loudspeaker reproductions, binaural pressures may be negatively correlated with $\max |\Psi_{LR}(\tau)| = \max [-\Psi_{LR}(\tau)]$, depending on loudspeaker or headphone signals. However, the cases of positive and negative interaural correlations are distinguishable from the results of Equation (1.7.2) because of the absolute value of $\Psi_{LR}(\tau)$. The unsigned IACC in Equation (1.7.2) is replaced with the following ***sign-IACC*** to resolve this ambiguity:

$$IACC_{sign} = \begin{cases} IACC & \text{if } \Psi_{LR}(\tau_{\max}) > 0 \\ -IACC & \text{if } \Psi_{LR}(\tau_{\max}) < 0 \end{cases}. \tag{1.7.3}$$

The aforementioned calculation can be converted into the frequency domain with Fourier transformation, so Equation (1.7.1) becomes

$$\Psi_{LR}(\tau) = \frac{\int P_L^*(f) P_R(f) \exp(j2\pi f\tau) df}{\left\{\left[\int |P_L(f)|^2 df\right]\left[\int |P_R(f)|^2 df\right]\right\}^{1/2}}, \tag{1.7.4}$$

where $P_L(f)$ and $P_R(f)$ are the frequency-domain counterparts of the binaural signals, and the superscript "*" is the complex conjugation.

The relationships between interaural correlation and spatial auditory perception have been investigated via numerous psychoacoustic experiments by using headphone presentation. Typical results can be described as follows. For wide-band noise signals, the complete interaural correlation causes a fused auditory event inside the head, the auditory event becomes extended and vague because of the reduction of interaural correlation, and completely uncorrelated signals yield two separate auditory events, one at each side of the head (Chernyak and Dubrovsky, 1968). In addition, negative interaural correlation, which seldom occurs for an actual single sound source, results in unnatural auditory events. Blauert (1997) summarized the detailed experimental results. These results indicate that the change in interaural correlation produces different spatial auditory events or perceptions. However, no exact quantitative correspondence exists between interaural correlation and spatial auditory perception.

In Equation (1.7.1), interaural correlation is evaluated in a wide or even full audible bandwidth. However, in Section 1.3.4, human hearing processes sound information based on the frequency resolution of auditory filters. Interaural correlation can be separately evaluated in each auditory filter band to be consistent with the auditory frequency resolution. It can be implemented by properly choosing the frequency range in the integral of Equation (1.7.4). The positive and high interaural correlation in the full audible bandwidth corresponds to the positive and high correlation in all or most individual frequency bands. However, the converse is not always true. In some cases, interaural correlations in all or most individual frequency bands are high, but the overall correlation in the full audible bandwidth is low. In this case, binaural signals provide inconsistent interaural localization cues, such as ITD, in different frequency bands, resulting in extended and vague auditory events. If binaural pressures are uncorrelated within all individual frequency bands, no fused auditory events are created.

Different spatial auditory events or perceptions in two or more loudspeaker reproduction may occur by controlling the correlation between loudspeaker signals. For a pair of front stereophonic loudspeakers with signals $e_L(t)$ and $e_R(t)$ in the time domain, the normalized cross-correlation function between loudspeaker signals is evaluated by

$$\Psi_{chan}(\tau) = \frac{\int e_L(t)e_R(t+\tau)dt}{\left\{\left[\int e_L^2(t)dt\right]\left[\int e_R^2(t)dt\right]\right\}^{1/2}}, \tag{1.7.5}$$

where τ is a time parameter.

Similar to Equation (1.7.3), Equation (1.7.6) can be used to evaluate the degree of correlation between two loudspeaker signals (***interchannel correlation***):

$$ICCC_{sign} = \begin{cases} \max\limits_{\tau} |\Psi_{chan}(\tau)| & \text{if } \Psi_{chan}(\tau_{\max}) > 0 \\ -\max\limits_{\tau} |\Psi_{chan}(\tau)| & \text{if } \Psi_{chan}(\tau_{\max}) < 0 \end{cases}, \tag{1.7.6}$$

where τ_{max} is τ that maximizes $|\Psi_{chan}(\tau)|$. By definition, $-1 \leq ICCC_{sign} \leq 1$.

Alternatively, the following definition of the interchannel correlation between two loudspeaker signals is often used in studies:

$$ICC = \frac{\int e_L(t)e_R(t)dt}{\left\{\left[\int e_L^2(t)dt\right]\left[\int e_R^2(t)dt\right]\right\}^{1/2}}. \tag{1.7.7}$$

Similar to Equation (1.7.4), Equations (1.7.5) and (1.7.7) can be converted into the frequency domain with Fourier transformation.

The relationships between interchannel correlation and spatial auditory perception are also investigated with a series of psychoacoustic experiments. Some differences in the results of different studies are found because of differences in experimental conditions, such as signals, but results show similar patterns. Plenge (1972) conducted a psychoacoustic experiment to investigate the spatial auditory events for a ±30° front stereophonic loudspeaker configuration with identical loudspeaker signal gains but different degrees of interchannel correlation. The signal is a wide-band noise. The results indicate that a positively high interchannel correlation obtained from Equation (1.7.7) causes a fused auditory event (virtual source) with a definite location at the front (Figure 1.30(a)). The auditory event becomes extended and vague because of a reduction in the positive interchannel correlation (Figure 1.30(b)). A partly negative interchannel correlation causes two separate auditory events in the loudspeaker directions. A highly negative interchannel correlation (Figure 1.30(c)), especially completely negative correlation (Figure 1.30(d)), results in some unnatural auditory events, such as a virtual source to the ear or even inside the head (lateralization). Kurozumi and Ohgushi (1983) yielded similar results.

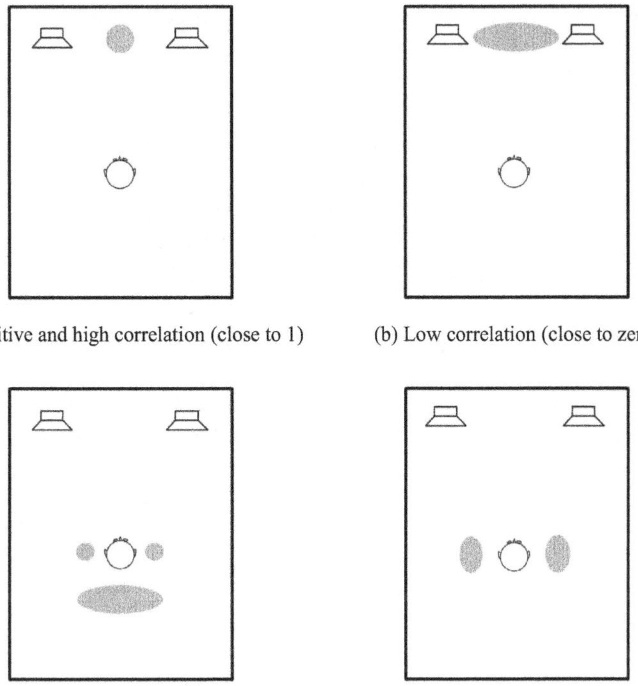

(a) Positive and high correlation (close to 1) (b) Low correlation (close to zero)

(c) Negative correlation (close to −0.8) (d) Completely negative correlation (close to −1)

Figure 1.30 Spatial auditory events for a ±30° front stereophonic loudspeaker configuration with different degrees of interchannel correlation:(a) positive and high correlation (close to 1); (b) low correlation (close to zero); (c) negative correlation (close to −0.8); (d) completely negative correlation (close to − 1). (redrawn with reference to Plenge, 1972.)

Binaural pressures in two loudspeaker reproductions are the superimposition of those caused by each loudspeaker. Changing the interchannel correlation between loudspeaker signals alters the interaural correlation of the resultant binaural signals. For a wide-band noise signal, interaural correlation varies following the interchannel correlation (Section 12.1.6). Therefore, experiments on loudspeaker reproduction yield results similar to those of headphone presentation. Auditory events in loudspeaker reproduction are located outside the head in most cases, and auditory events in headphone presentation are usually located inside the head. The relationship between interchannel correlation and spatial auditory perception is further discussed in Sections 3.3 and 12.1.6.

1.7.4 Auditory scene analysis and spatial hearing

When two or more uncorrelated sound sources exist simultaneously, binaural pressures are a linear combination or mixture of those generated by all sound sources, embodying the mixed information from multiple sources and environmental reflections. In this case, on the one hand, the auditory system, especially the high-level neural system (the same in the following), analyzes and processes information. As a result, the overall auditory scene of multiple sound sources and environment is formed. On the other hand, the auditory system may distinguish a series of auditory objects from the mixed auditory streaming from binaural signals through ***auditory scene analysis*** (Bregman, 1990). Thus, different auditory events in various objects occur.

In psychoacoustics and physiological acoustics, auditory scene analysis deals with the mechanisms and rules of auditory stream segregation and grouping (Yost and Sheft, 1993; Rumsey, 2001). On the basis of temporal, spectral, and spatial information in binaural signals, the high-level neural system analyzes some independent perceptual attributes of sound and their variations, including time order, pitch, timbre, intensity, and directions. Then, this system separates, distributes, and groups the components into different object streams. Objects are identified in terms of pattern recognition from a previous memory. Spatial cues contribute partly to the segregation and grouping of the auditory stream. For example, components with similar spectra are often grouped into an auditory object independent of the spatial cues in these components. However, as stated in Section 1.6.5, even when some spatial cues are absent or conflicting with others, the auditory system may determine the source position based on the more consistent cues.

The theory of auditory scene analysis is being developed continuously. This theory is necessary not only to reveal the mechanisms of human auditory perception in complicated sound fields but also to simulate human hearing, such as speech recognition and machine hearing, with a computer. Future developments in this field will be greatly helpful for analyzing the mechanisms of spatial sound on the high levels of psychoacoustics and physiological acoustics.

1.7.5 Cocktail party effect

If a target speech sound source and one or more interfering sound sources (e.g., competitive speech sources) simultaneously exist, the auditory system can take advantage of the spatial separation of the target and interfering sound sources to more effectively detect the target speech information. This phenomenon is known as the ***cocktail party effect***. In daily life, this effect facilitates the detection of target speech information even in a noisy environment.

The cocktail party effect is a kind of binaural auditory effect associated with the spatial hearing of multiple sources. If the spatial information of sources, such as monoaural presentation via headphones or monophonic reproduction by a loudspeaker, is lost, benefits from the cocktail party effect are subsequently lost. Since the pioneering works of Cherry (1953), a large number of investigations on the mechanism of the cocktail party effect have been carried out, but no final conclusion has thus far been achieved. This effect is generally considered the consequence of the comprehensive processing of binaural sound information by a high-level neural system (e.g., auditory stream segregation and selective attention). Binaural (spatial) information is vital to auditory stream segregation. This issue was thoroughly reviewed by Bronkhorst (2000).

1.8 ROOM REFLECTIONS AND AUDITORY SPATIAL IMPRESSION

1.8.1 Auditory spatial impression

Reflections in enclosed spaces create significant auditory effects and become a key issue in the acoustic design of rooms. In Section 1.6.6, reflections in an enclosed space are effective cues for distance perception. The precedence effect discussed in Section 1.7.2 is also related to spatial hearing in reflective rooms. In this section, another spatial auditory perception caused by room reflections, i.e., auditory spatial impression, is discussed (Shi and Xie, 2008).

Reflections in rooms are not perceived as separate echoes, and they influence slightly on the perceived direction of a direct sound when they satisfy the prerequisites of the precedence effect. However, they contribute to the perceived size of a sound source, along with the

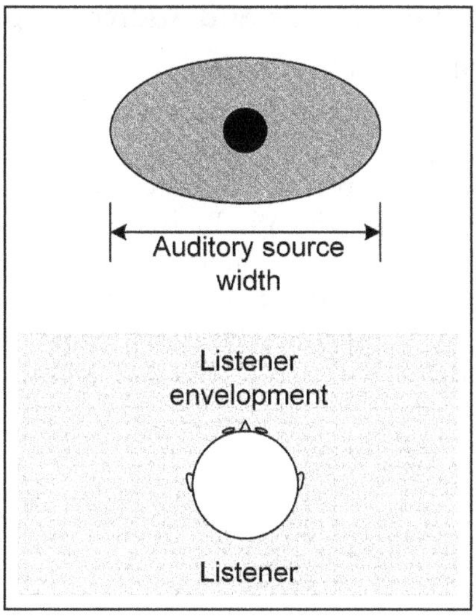

Figure 1.31 Illustration on ASW and LEV (adapted from Morimoto et al., 2001).

perceived spatial and acoustical features of a room. Hence, reflections create comprehensive auditory impressions of a sound source and the environment.

Auditory spatial impression (***ASI***) is an important subjective attribute of sound quality in concert halls. The definitions of ASI still differ. Morimoto et al. (1990) pointed out that ASI consists of at least two distinguishable components, namely, ***auditory source width*** (***ASW***) and ***listener envelopment*** (***LEV***). An ASW is the auditory perception of the broadening of a sound source compared with that of the visual width of an actual source. The LEV, sometimes termed ***spaciousness***, is the sensation of being surrounded by reverberant sounds. Some subsequent studies have also confirmed that the ASI is composed of these two components (Bradley and Soulodre, 1995). Figure 1.31 illustrates the definition of ASW and LEV.

The ASI is a subjective or perceptual attribute, which describes the psychoacoustic responses of human hearing, including a high-level neural system, to the comprehensive spatial information of a sound field. Numerous works have evaluated the ASI with the objective or physical measures of room reflections. Physically, the ASI should depend on the temporal and spatial characteristics of the sound field. After being scattered and diffracted by human anatomical structures, such as the head, pinnae, and torso, sound waves reach the two ears. Binaural pressures embody the auditory information of the sound field. The auditory system, including the high-level neural system, analyzes this information comprehensively, resulting in various subjective auditory perceptions or events, such as ASI. Accordingly, two ways are applied to evaluate the ASI in a hall from the physical measures of sound. One way is to estimate some physical measures, such as arrival time, reflection directions, and direct-to-reverberant energy ratio, of the sound field and then evaluate the ASI with these sound field-related measures (Section 1.8.2). Another way is to estimate some physical measures in binaural pressures and then examine the ASI with these binaural-related measures (Section 1.8.3).

Further analysis should be based on the neurophysiologic mechanisms of comprehensive information processing by human hearing. This analysis is essential not only for fundamental studies on human hearing but also for the development of spatial sound techniques. Although some progresses have been made on this issue (Ando, 1998, 2009; Blauert, 2012; Ahveninen et al., 2014), numerous problems should be further investigated because these mechanisms will be an important and promising field in the future.

1.8.2 Sound field-related measures and auditory spatial impression

Marshall (1967) suggested that early lateral reflections are important for spatial responsiveness. Barron and Marshall (1981) further investigated the ASI caused by a single reflection after a direct sound through a simulated sound field with multiple loudspeakers in an anechoic chamber. Psychoacoustic experimental results indicate that reflections up to 80 ms after the direct sound are essential for the ASI. Reflections from the front slightly contribute to the ASI, and reflections from the lateral direction contribute mostly to the ASI. On the basis of this phenomenon, Barron and Marshall proposed a physical measure of ASI, i.e., the ***early lateral energy fraction*** (LF), which is defined as the ratio of the lateral sound energy to the total sound energy that arrives within the preceding 80 ms after the arrival of the direct sound:

$$LF = \frac{\int_{5ms}^{80ms} p_F^2(t)\,dt}{\int_{0ms}^{80ms} p_{omi}^2(t)\,dt}, \tag{1.8.1}$$

where $t = 0$ is the arrival time of the direct sound, $p_{omi}(t)$ is the impulse response of sound pressure recorded with an omnidirectional microphone, and $p_F(t)$ is the impulse response of sound pressure recorded with a bidirectional microphone (with the directivity of the figure of eight) arranged at the receiver position and with its main axis pointing to the lateral direction. This strategy is equivalent to applying a direction-dependent weight of $\cos^2\gamma$ to the calculation of reflected energy in the numerator in Equation (1.8.1), where γ is the angle between the direction of reflection and the lateral direction. Further studies have confirmed that LF creates an auditory perception of sound source broadening, i.e., the ASW. The International Organization for Standardization suggested LF as a physical measure of ASI and specified the conditions for LF measurement (ISO 3382-1, 2009). Although LF has been widely used in practical measurements, some controversies on its usability remain (Beranek, 1996). Further experiments have also indicated that some problems exist with LF as the only measure of ASW (Marshall and Barron, 2001; Barron, 2000). Nevertheless, some new measures have been proposed.

In addition to ASW, LEV is an important component of ASI (Bradley and Soulodre, 1995, 1996).Similar to the parameter described in Equation (1.8.1), the late lateral energy fraction (LLF), which is defined as the ratio of the lateral sound energy to the total sound energy arriving 80 ms after the arrival of the direct sound, is suggested as a physical measure of LEV:

$$LLF = LF_{80}^{\infty} = \frac{\int_{80ms}^{\infty} p_F^2(t)\,dt}{\int_{80ms}^{\infty} p_{omi}^2(t)\,dt}, \tag{1.8.2}$$

where $p_F(t)$ and $p_{omi}(t)$ are identical to those in Equation (1.8.1). LEV increases as the ratio between the late lateral energy and the direct energy increases or as the ratio between the late lateral energy and the early lateral energy increases. Moreover, the ***late lateral relative level*** (*GLL*), a revised physical measure that is more related to the LEV than to the LLF, is suggested (Bradley and Soulodre; 1995; Evjen et al., 2001):

$$GLL = LG_{80}^{\infty} = 10\log_{10}\left[\frac{\int_{80ms}^{\infty} p_F^2(t)dt}{\int_{0ms}^{\infty} p_A^2(t)dt}\right] \quad (1.8.3)$$

where $p_F(t)$ is identical to that in Equation (1.8.1). $p_A(t)$ is the impulse response of sound pressure for the same source measured at a distance of 10 m in a free field. The results of a subjective LEV experiment are highly correlated with the prediction based on the mean GLL in four octaves from 125 Hz to 1000 Hz.

Previous studies focused on the contribution of late lateral reflections to LEV. Some subsequent studies have suggested that late reflections from the back and other directions, such as upward, also contribute to LEV by simulating the sound field with multiple loudspeakers (Morimoto and Iida, 1993; Morimoto et al., 2001; Furuya et al., 2001, 2005, 2008). However, other studies have proposed that late lateral reflections contribute mostly to LEV (Evjen et al., 2001).

In summary, the ASI in rooms is closely related to the temporal and spatial characteristics of reflections. Early reflections contribute to ASW, and early lateral reflections have the most contribution. Late reflections also contribute to LEV. In addition to the contribution of late lateral reflections, the role of late reflections from other directions to the LEV still has some controversies. From the physical point of view, late reflections are closer to a diffused reverberation field, and LEV should be related to the extent of diffusion of the reverberation sound. However, the energy of the late diffused reverberation decays as time is extended because of boundary absorptions. Therefore, the perceptual contributions of the late diffused reverberation should also decrease with time. In addition, some studies have suggested that early reflections mainly contribute to sound source perceptions, and late reflections mainly contribute to environmental perceptions. However, this classification has some controversies.

1.8.3 Binaural-related measures and auditory spatial impression

The relationships between sound field-related measures and ASI are addressed in Section 1.8.2. This section continues to address the relationships between binaural-related measures and ASI. In Section 1.4.2, HRTFs or HRIRs are introduced to analyze the binaural pressures generated by a point source in a free field. In the cases of reflective spaces, after being scattered and diffracted by human anatomical structures, such as the head, pinnae, and torso, direct and reflected sounds reach the two ears. Binaural pressures embody the main information of direct and reflected sounds. As an extension of HRIRs in the free field defined by Equation (1.4.3), ***binaural room impulse responses*** (***BRIRs***),which are defined as the acoustic impulse responses from a sound source to the two ears and denoted by h_L and h_R, are used to describe the sounds received by the two ears in reflective rooms. In contrast to HRIRs in the free field, BRIRs depend on source and receiver positions rather than the relative position between sources and receivers. Binaural pressures in a reflective room can also be evaluated with Equation (1.4.4), but h_L and h_R in Equation (1.4.4) should be replaced with BRIRs.

In Section 1.7.3, IACC is a binaural-related measure that is closely related to the spatial auditory perception in headphone and loudspeaker reproductions. A low IACC corresponds to broadening and vague auditory events. IACC is also closely related to the physical characteristics of the sound field in a room and the perceptual attribute ASI. In an ideal diffused sound field, IACC $\approx$ 0.

Ando (1985, 1998) investigated the variation in IACC with the direction of a single reflection. The results indicate that the direction for minimizing IACC is frequency dependent. On average, the preferred azimuth of early discrete reflections that minimize IACC is $\theta = 55° \pm 20°$ ($\theta = 0°$ and 90° represent the front and lateral directions, respectively). Ando (1985, 1998) further suggested that IACC is one of the four orthogonal measures strongly related to the preference of listeners in concert halls. Morimoto and Iida (1995) also confirmed that IACC is related to ASW, i.e., ASW decreases as IACC increases.

Some other studies have indicated that the IACC within three octaves centered at 500, 1000, and 2000 Hz are the most effective for the ASI (Okano et al., 1998; Hidaka and Beranek, 2000). Studies have further introduced IACC within different periods of reflections, which are evaluated from the interaural cross-correlation of BRIRs by using formulas similar to Equations (1.7.1) and (1.7.2). Early IACC, denoted by $IACC_E$, is evaluated by choosing the time window of cross-correlation calculation with $t_1 = 0$ and $t_2 = 80$ ms. Late IACC, denoted by $IACC_L$, is examined by selecting the time windows with $t_1 = 80$ ms and $t_2 = 1$ s.

Psychoacoustic experimental results indicate that ASW is closely related to $IACC_E$, and a low $IACC_E$ corresponds to a large ASW. $IACC_E$ describes the similarity between the binaural pressures caused by direct and early reflected sounds within the preceding 80 ms. Lateral reflections cause more difference between binaural pressures because of the diffraction effect of the head, resulting in a low $IACC_E$. The perceived width of the source broadens because of this low interaural correlation. From a physical point of view, $IACC_L$ should be closely related to LEV. However, this hypothesis has not been validated by simulated experiments with multiple loudspeakers (Beranek, 1996).

Some studies have indicated that IACC is usually high below the frequency of 500 Hz at which the wavelength is obviously larger than the dimension of the human head (Hidaka et al., 1995; Martens, 2001). However, a slight change in IACC in this frequency range causes a large variation in the ASI. Therefore, the auditory system is more sensitive to a small differentiation in IACC at low frequency.

In summary, although IACC has been applied to evaluate the ASI in concert halls with appropriate success, many problems remain to be solved. The fused auditory perceptions or scenes caused by direct, early, and late reflected sounds are consequences of the comprehensive and ensemble courses of auditory information processing, which should be analyzed by auditory models, including the complicated processing in the high-level neural system. Analyzing the spatial auditory perceptions caused by direct, early reflected, and late reflected sounds individually is convenient but not always reasonable from the point of view of auditory information processing (from the private discussion between the author of this book and Prof. Ning Xiang). In practical concert halls, the direct and reflected sounds generated by multiple sources form a complex auditory scene, resulting in more complicated psychoacoustic courses of perceptions that need further investigation.

In addition to IACC, the relationships between the ASI and other binaural-related measures have been examined. In Sections 1.6.1 and 1.6.2, ITD and ILD are binaural cues for lateral localization. Blauert and Lindemann (1986) experimentally confirmed that fluctuations in ITD and ILD over time may be responsible for the broadening of auditory events. For periodic fluctuations, when the rates of ITD and ILD fluctuation are less than 2.4 and 3.1 Hz, respectively, a moving auditory event or virtual source is perceived. When the rate of fluctuation increases but does not exceed a certain upper limit (approximately up to 500 Hz for ITD fluctuation; Grantham and Wightman, 1978), a broadening auditory event rather than a moving auditory event is perceived. Griesinger (1992a, 1992b) pointed out that the interaction among direct and single or multiple reflected sounds in rooms causes fluctuations in ITD and ILD over time. Consequently, the perceptions of ASI are obtained. The extent of binaural cue fluctuation increases as the SPL of lateral reflections in a hall increases. This finding is qualitatively consistent with the results of interaural correlation analysis. Mason (2002) analyzed the subjective perceptions caused by binaural cue fluctuations in detail. He suggested that "*when fluctuations are contained within a part of a signal that is perceived to be a sound source, a variation in the magnitude of fluctuation alters the perceived width of that source.*" Furthermore, "*when fluctuations are contained within a part of a signal that is perceived to be reverberation, a variation in the magnitude of fluctuation alters the perceived width of the acoustical environment.*" However, these results are preliminary, so the detailed and quantitative relationships of binaural cue fluctuations, ASI, and underlying mechanisms should be further investigated.

1.9 PRINCIPLE, CLASSIFICATION, AND DEVELOPMENT OF SPATIAL SOUND

1.9.1 Basic principle of spatial sound

Spatial sound or ***spatial audio*** aims to record (or simulate), transmit (or store) and finally reproduce the spatial information of sound and then recreate the desired spatial auditory events or perceptions. The spatial information of sound includes the localization information of sound sources and the comprehensive spatial information of environmental reflections. A complete spatial sound system is composed of three stages, i.e., signal recording (or simulation and synthesis), transmission (or storage), and reproduction. In other words, sound information is initially converted into appropriate electric signals, subsequently transmitted or stored by some media, and finally reproduced with loudspeakers or headphones to recreate the information.

Figure 1.32 depicts a chain of stages from a sound source to auditory perception to elucidate the principle of spatial sound. Generally, a sound wave generated by a source propagates in space through various direct and reflected paths, forming a sound field. When a listener enters the sound field, anatomical structures, such as the head, pinnae, and torso, scatter and diffract the sound wave. The courses of scattering and diffraction of the sound wave encode the temporal and spatial information of the sound field into binaural pressures in the eardrums. After transmission through the middle ears, the mechanical oscillation of sound is converted into neural pulses by the inner ears and then processed by the higher-level neural system. As a result, various auditory perceptions or events occur. The propagation from a sound source to the eardrums is a physical course, which can be analyzed with the knowledge of environmental acoustics and simulated via electroacoustic means. The transmission and processing from the middle ears to the high-level neural system are physiological courses, which should be analyzed with the knowledge of physiological acoustics. However, the knowledge of physiological acoustics, especially information associated

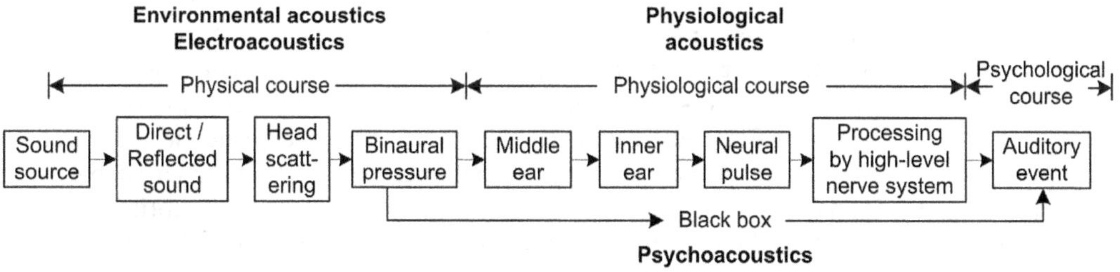

Figure 1.32 Chain of courses from a source to the auditory perception.

with the processing in the higher-level neural system, remains incomplete. Studies on spatial sound take advantage of the psychoacoustic method. In this method, the processing stage in the high-level neural system is often modeled as a "black box," and auditory perceptions or responses (outputs of the black box) are directly linked to the physical attributes of a sound field or binaural pressures (input). However, a one-to-one correspondence between the spatial auditory perceptions and the physical sound field or binaural sound pressures is not always observed. A well-known example is the summing localization with two loudspeakers discussed in Section 1.7.1. In this case, a spatial auditory event (virtual source) created by two stereophonic loudspeakers is similar to that caused by a single source in the front, although the corresponding sound fields are different. Various psychoacoustic methods are useful in practical spatial sound techniques. The physiological acoustic mechanism behind these psychoacoustic methods should be further studied.

The spatial information of sound can be reproduced by one of the following methods and principles. These methods correspond to recreating information in three different stages in the chain shown in Figure 1.32, i.e., in the stages of sound fields, binaural pressures, and auditory perception, respectively. All existing spatial sound techniques are based on these three methods and principles (Xie, 1995, 1999a, 1999b, 2020 Xie and Guan, 2002):

1. Sound field-based methods and principles
   Physically, a sound field in the air can be entirely specified by the spatial and temporal distribution of the sound pressure $p(x, y, z, t)$. For the same listener, identical sound fields should yield the same auditory perceptions. Sound-field-based methods aim to reconstruct an original or target sound field as exactly as possible in an appropriate area, that is, to recreate a sound pressure distribution or wavefront matching with that of the target sound field in this area. Listeners acquire the target spatial information and corresponding auditory perceptions as they enter the reproduced sound fields. High-order Ambisonics and wave field synthesis are two examples of sound-field-based methods. These methods are usually used to create authentic and natural auditory perceptions within a reasonable or even a large listening area, the resultant perceived performances slightly change as the listening position is altered. However, these methods require a large number of signal channels and loudspeakers; consequently, they become complicated.
2. Sound field approximation and psychoacoustic-based methods and principles
   Under certain conditions, two physically different sound fields may produce similar spatial auditory perceptions. Psychoacoustic-based methods aim to recreate target auditory perceptions. The reproduced sound field by these methods is physically different from the original or target sound field. It is a rough approximation of the target sound field at most. With an appropriate number of channels and loudspeakers, psychoacoustic-based methods can be utilized on the basis of appropriate psychoacoustic principles to recreate auditory perceptions similar to those of target sound fields. In other words, these methods aim to "deceive" our hearing. Examples of these methods include two-channel stereophonic sounds, various multichannel surround sounds, and low-order Ambisonics. For example, a virtual source between a pair of stereophonic loudspeakers may be created on the basis of the psychoacoustic principle of summing localization (Section 1.7.1). According to the psychoacoustic principle discussed in Section 1.7.3, sensations similar to those in a concert hall can be created by using two or more loudspeakers with low-correlated signals. Auditory perceptions created by psychoacoustic-based methods may be somewhat different from those of the target sound field, depending on the extent of approximation in the reproduced sound field. The resultant perceived performances may also vary with a listener's position. However,

psychoacoustic-based methods are relatively simple so that they are widely used in practice.

3. Binaural-based methods and principles
   Binaural pressures or signals embody the auditory information in any complicated sound field. Binaural-based methods aim to duplicate and render binaural signals exactly, then to create the desired auditory perceptions. The examples of binaural-based methods include binaural recording and reproduction, and virtual auditory displays. The hardware of these methods is simple because only two signals and reproduction channels are required. Binaural-based methods can create authentic and natural auditory perceptions through a careful design. However, they have some drawbacks (Chapter 11).

As discussed in Chapters 9–11, all three categories of methods deal with the spatial sampling of the sound field. In the sense of the spatial sampling and reconstruction of a sound field, the three categories of methods are related and even consistent. Thus, some techniques used by the three categories of methods can be interchanged. In practice, some combined applications of two or three categories of methods may be utilized to achieve the desired effects.

1.9.2 Classification of spatial sound

Various spatial sound techniques have been introduced and designed on the basis of the three categories of methods and principles (Rumsey, 2001; Blauert and Rabenstein, 2012). From different points of view, there are different classifications and names for various spatial sound techniques. Such differences, together with some commercial declaration, may cause ambiguity.

Spatial sound generally refers to various techniques and systems for recording (or simulating), transmitting (or storage) and reproducing the spatial information of sound. They can be classified according to their acoustic principles and methods discussed in Section 1.9.1. From the point of scientific research, this classification is strict and reasonable. The common spatial sound techniques and systems are classified into three groups based on this classification:

1. Sound field-based techniques and systems
2. Psychoacoustic-based techniques and systems
3. Binaural-based techniques and systems

Spatial sound can be classified on the basis of the dimensionality of the spatial information they recreate:

1. ***Stereophonic sound*** techniques and systems can recreate the spatial information within a certain horizontal–frontal sector (one-dimensional space).
2. ***Horizontal surround sound*** techniques and systems can recreate the spatial information in the horizontal plane (two-dimensional space).
3. ***Three-dimensional spatial surround sound*** techniques and system, shortened as ***spatial surround sound*** (used in this book) or ***3D sound***, can recreate the spatial information in the three-dimensional space with height.

Horizontal and spatial surround sounds are sometimes termed ***surround sound***. Less strictly, the term "***stereophonic sound***" sometimes refers to various spatial sound techniques.

According to the number of reproduction channels, spatial sound techniques and systems can be classified into two-, three-, four-, or five-channel techniques and systems, etc. Techniques and systems with more than two channels are called **multichannel techniques and systems.** Literally, ***two-channel sounds*** include all techniques with two channels (e.g., including virtual auditory display). Multichannel sounds cover all techniques with three or more channels, such as wave field synthesis. However, two-channel sounds usually refer to ***two-channel stereophonic sounds*** based on the psychoacoustic method. ***Multichannel sounds*** refer to techniques with more than two channels based on the psychoacoustic method. Strictly speaking, multichannel sounds involve ***multichannel frontal-stereophonic sound, multichannel horizontal surround sound and multichannel spatial surround sound.*** But in most cases, ***multichannel surround sounds*** or shorten for ***multichannel sound*** refers to the latter two cases.

Spatial sound can classify based on electroacoustic devices used in reproduction, resulting in loudspeaker- and headphone-based spatial sounds. Intuitively, in sound field- and psychoacoustic-based techniques, loudspeakers are used as reproducing devices. In binaural-based techniques, headphones are utilized as reproducing devices. However, these cases do not always occur. With an appropriate signal processing, binaural signals can be converted for loudspeaker reproduction, and multichannel sound signals can be converted for headphone presentation. These strategies are examples of the combined applications of two spatial sound techniques based on different methods and principles (Chapter 11).

Spatial sounds can be classified on the basis of their applications, such as spatial sound techniques and systems for scientific studies (e.g., room acoustic and psychoacoustic experiments) for cinemas, halls, consumers, domestic uses, and mobile devices. Technical requirements on spatial sounds in different applications vary.

1. Physical accuracy is the most essential for scientific applications. Therefore, sound field- and binaural-based techniques are relatively appropriate for this purpose. Psychoacoustic-based techniques are usually inappropriate because any approximation or simplification of the physical information of the sound may cause errors in perceived results before the contributions of the physical information to auditory perceptions are completely understood. Unfortunately, this phenomenon has been disregarded in some studies.
2. For consumer applications, an appropriate compromise between perceived performances and complicity is usually made. As a result, psychoacoustic-based techniques are widely used in this field.
3. For hall and cinema applications, a large listening region is desired. From this point of view, sound-field-based techniques are appropriate. In practice, sound field- and psychoacoustic-based techniques or a combination of two are applicable.
4. For applications on mobile devices, portability and simplicity should be considered first. Therefore, binaural-based techniques are usually preferred.

1.9.3 Developments and applications of spatial sound

Cinema and consumer reproductions are two traditional and important applications of spatial sounds. Numerous commercial spatial sound techniques are originally developed for these two applications. These techniques are developed for practical demands and commercial purposes. From the point of practical uses, these commercial techniques usually do not physically seek an exact reconstruction of the sound field. Instead, on the basis of the rough approximation of sound field and appropriate psychoacoustic principles, they aim to recreate the desired auditory perceptions under the conditions of practical applications.

Spatial sound was first explored more than 100 years ago. The first spatial sound demonstration occurred at the Paris exhibition in 1881. Ader placed a pair of microphones at the front of the stage of Paris Opera and relayed the outputs of microphones to a pair of telephone receivers (Hertz, 1981).

Early works at Bell Labs in the 1930s were originally intended for sound reproduction in cinema and large auditoriums. The proposed "acoustic curtain" consisted of a large number of microphones and loudspeakers to record and exactly reconstruct the acoustic wavefront from the frontal stage. Steinberg and Snow (1934) further investigated the effect of reducing the number of channels for frontal sound reproduction in cinemas. The results revealed that three frontal channels produce a reasonable perceived effect. Sound reproduction techniques with three frontal channels were first used at the end of the 1930s and have been widely used since in 1950s in cinemas. Two- or three-channel-spaced microphone techniques based on the early work of Bell Labs have also been applied to stereophonic recording for consumer reproduction.

For consumer or domestic uses, Blumlein (1931) specified the patent that allows the conversion of signals from two spaced microphones (with interchannel time difference only) to be suitable for stereophonic loudspeaker reproduction (with interchannel level difference). In the early 1940s, Boer (1940) conducted an experiment on summing localization with two loudspeakers. In the late 1950s, Clack et al. (1957) and Leakey (1959) re-analyzed the principle of two-channel stereophonic sounds in accordance with some preliminary theories and models. By the end of the 1950s and at the beginning of the 1960s, two-channel stereophonic sounds became gradually popular in consumer reproduction because of the development of 45°/45° record and FM stereophonic broadcast. Currently, two-channel stereophonic sound remains the most popular spatial sound technique.

A two-channel stereophonic sound can recreate spatial information within a frontal–horizontal sector, which is usually bounded by two frontal loudspeakers. By the end of the 1960s and at the beginning of the 1970s, great efforts were devoted to developing quadraphones (four-channel horizontal surround sound) for consumer music reproduction (Xie X.F., 1981). With four horizontal channels and a square loudspeaker configuration (arranged in left-front, right-front, left-back, and right-back directions, respectively), a discrete quadraphone was intended to reproduce the full 360° horizontal information. However, with analog audio techniques available at that time, transmitting and storing four independent signals were difficult for a discrete quadraphone. Accordingly, a variety of competing 4-2-4 matrix quadraphone techniques were developed. In these matrix techniques, the original four channel signals for the discrete quadraphone were converted into two independent signals by linear matrix encoding and then transmitted or stored by traditional two-channel media. Therefore, the matrix quadraphones were compatible with stereophonic sounds. Two independent signals were converted back to four channel signals by matrix decoding in reproduction. Unfortunately, some fatal defects existed in various quadraphones. Therefore, quadraphones were seldom used in practices, and efforts devoted to the development of quadraphones failed.

Some other horizontal and even spatial surround sound techniques (mainly for consumer reproduction) were introduced in the 1970s. Overall, numerous techniques were developed in this period, but they were seldom used in practice. Nevertheless, studies on quadraphones and other techniques in the 1970s provided good experiences and lessons for the subsequent development of new multichannel surround sounds.

Multichannel surround sounds for cinemas have been successfully developed since the 1950s. In the middle of the 1970s, Dolby Laboratories introduced Dolby Stereo, a four-channel surround sound technique for cinemas (Dolby Laboratories, Surround sound past, present, and future, http://www.dolby.com). On the basis of the lessons from early quadraphones with a square channel configuration, Dolby Stereo includes three frontal channels and loudspeakers, namely, left, center, and right channels and loudspeakers, to recreate frontal-biased localization effects that match with the picture. An additional surround channel is for ambient sound. To store signals in optical soundtracks in a 35 mm film, Dolby Stereo converts the four original signals into two independent signals by matrix encoding. Adaptive decoding is used to convert the two independent signals back to the four channel signals and improve channel separation in reproduction. Since the end of the 1970s, Dolby Stereo has been widely used for film sound in cinemas.

Only two independent signals and only a mono surround channel are present in Dolby Stereo, so it is unable to recreate the ambient effect on the side and rear satisfactorily. Nevertheless, the development of digital audio and processing techniques has enabled the transmission and storage of more channel audio signals for the surround sound. In the 1980s, the new surround techniques were developed for digital film sounds. A group in the Society of Motion Picture and Television Engineers investigated the required independent channels for cinema sound, resulting in the 5.1-channel surround sound in which five independent channels have a full audible bandwidth. Three front channels, including the left, right, and center, are used to recreate the stable virtual source in the front region to match the picture. Two surround channels, including the left surround and the right surround, are attempted to recreate the ambience and occasional localization effect in the side and rear. A low-frequency effect channel (called .1 channel) operates in the frequency range below 120 Hz and drives the subwoofer (usually in the front) to convey some specific low-frequency effects. The signals of five full-bandwidth channels in the 5.1-channel surround sound are completely independent compared with that in Dolby Stereo, freeing from the loss caused by matrix encoding. Two independent surround channels provide more flexibility for the storage of ambient information. Some digital audio coding and compression techniques, such as Dolby Digital (AC-3) and DTS coherent acoustic coding, were developed to transmit and store the 5.1 channel signals effectively (Chapter 13). Combined with these coding techniques, the 5.1-channel surround sound has been widely used for cinemas since the mid-1990s.

In the early age, the multichannel surround sound for cinema and consumer reproductions (mainly for pure music) were developed individually. The situation has changed greatly since the 1980s. Multichannel surround sounds originally intended for cinema were applied to consumer reproduction after some simplification and revision, resulting in the surround sound for "***domestic theaters*** or ***home cinema***." With stereophonic video recorders and laser disks (LDs), a revised version of Dolby Stereo called Dolby Surround was used in consumer reproduction (Dolby Laboratories, Dolby Surround mixing manual, http://www.dolby.com).

Since the mid-1980s, with the development of a high-definition television (HDTV) or a digital television (DTV), international efforts have been devoted to the new generation of surround sound techniques for consumer use. Various possible techniques and formats were investigated by researchers in the USA, Europe, and Japan. In 1994, the 5.1-channel surround sound was recommended by the International Telecommunication Union (ITU) as the standard for multichannel sound systems with and without accompanying pictures (ITU-R BS 775-1, 1994). It has been extensively employed first in LD and then in digital versatile disk (DVDs), domestic theaters, and digital televisions. Moreover, many horizontal and spatial surround sound techniques for cinemas and consumers with more than five channels have been developed since the end of the 1990s, resulting in commercial competition in this field. Object-based spatial sound becomes a new trend of development. Currently, some international standards for spatial sound are specified and continuatively revised.

Some existing spatial sound techniques do not rely on strict acoustic theories. They are designed to meet some practical or commercial requirements. Some practical spatial sound techniques, including those recommended by some international standards, are compromised among various factors rather than the one with optimal physical or perceptual performance.

In addition to various commercial spatial sound techniques, most of them were based on sound field approximation and psychoacoustic principles, another important direction on spatial sound is the principles and techniques for the exact reconstruction of the sound field. For various reasons, these sound-field-based techniques were not widely applied, but they were necessary to elucidate the physical natures of spatial sound. Thus, they provide an important theoretical foundation in this field. This situation has changed since the mid-2000s. The techniques of sound field reconstruction have been increasingly used for public and consumer reproduction.

As a well-known example of sound field reconstruction in a local region, a kind of sound field recording, signal mixing, and sound field reconstruction theory and technique has been extensively developed since the 1970s. This technique is now termed Ambisonics (Gerzon, 1985). In the early time, Ambisonics was developed as a special spatial sound technique based on sound field approximation and psychoacoustic principles. Subsequent studies have indicated that Ambisonics transits gradually from approximation to the exact reconstruction of a target sound field. Ambisonics theory is perfect. Currently, Ambisonics is still a hot topic in spatial sound and being developed. It is also increasingly used in practice.

Wave field synthesis, which has been developed since the end of the 1980s, is another theory and technique of sound field reconstruction in an extended area (Berkhout et al., 1993; Boone et al., 1995). According to the Huygens–Fresnel principle and the Kirchhoff–Helmholtz boundary integral equation, pressure and its derivative on the surface of an arbitrary closed space without sources determine the sound field within a space. Thus, the desired sound field can be reconstructed with a combination of sound waves from an array of monopole and dipole loudspeakers (secondary sources) arranged on boundary surfaces. Under certain conditions, sound field reconstruction can be implemented with a single type of secondary sources. Wave field synthesis usually provides an extending listening area, and it is appropriate for reproduction in auditoriums. Only a few practical examples with wave field synthesis have been presented because of its complicated structures, but applications are increasing.

Since the 1990s, numerous works on sound field reconstruction have been performed to establish the fundamental theory in this direction. Most of these works have focused on Ambisonics and wave field synthesis, which are sometimes referred to the theories and methods originally developed in the fields of active noise control and acoustical holography. Ambisonics and wave field synthesis remain a hot topic in spatial sound.

Parallel to the development of commercial spatial sound techniques, ***binaural recording and reproduction*** techniques and systems have been developed since the end of 1920 and the beginning of the 1930s. Binaural signals are recorded using an artificial head and presented by a pair of headphones. Since then, artificial head techniques, binaural recording, and reproduction have been developed greatly, as reviewed by Paul (2009) in detail. Binaural recording and reproduction had been seldom used in consumer reproduction for a long time because of incompatibilities with loudspeaker reproduction and other problems. However, they have been widely used in studies on psychoacoustics and room acoustics. This situation has changed greatly since the end of the 1980s. With the development of computer and signal processing techniques, virtual auditory displays have been developed quickly (Begault, 1994; Xie, 2008a, 2013). In a virtual auditory display, target binaural signals are synthesized or simulated by HRTF-based signal processing. Since the 1990s, a virtual auditory display has become a hot topic in acoustics and signal processing. It is applicable to vast fields of scientific studies, engineering, and consumer uses, such as psychoacoustic and room acoustic experiments, multimedia and virtual reality, communication, and domestic reproduction

(Xie, 2008a, 2008b, 2008c). Since the end of the 1990s and the beginning of 2000s, sound reproduction in various multimedia and mobile devices has become a new (the third) important application field of spatial sound. Binaural and virtual auditory displays play a significant role in this application.

In China, researchers at the South China University of Technology (SCUT; formerly called the South China Institute of Technology, which was changed to the Guangdong Institute of Technology within a short period in the early 1970s) began the works on stereophonic sounds and set up a laboratory in 1958. This laboratory paused from 1966 to 1976 during the period of "cultural revolution" in China and resumed in 1977. In 1973, Prof. Xingfu Xie at the SCUT explored quadraphones and other multichannel surround sounds for consumer reproduction, proposed, and improved the theory of sound field signal mixing and reproduction. This theory was similar to Ambisonics and called the 4-3-N horizontal surround sound. Limited by the situation in China in those times, these works could not be submitted to international journals and were published in some Chinese journals lately (Xie X.F., 1977, 1978b). In the 1970s and 1980s, Prof. Xingfu Xie also conducted theoretical works on multichannel full- and upper-spatial surround sounds. In those times, Prof. Xingfu Xie published two books on spatial sound in Chinese. In one book entitled *The Principle of Stereophonic Sound* (Xie X.F., 1981), works on spatial sound by international and SCUT groups before that time were comprehensively reviewed. This book also referred to numerous literatures. Another book named *The Researches on Stereophonic Sound* (Xie X.F., 1987), which is a collection of 40 papers by Prof. Xingfu Xie and collaborators, provided a summary of the works by the SCUT group up to that time.

After Prof. Xingfu Xie passed away in 1991, the SCUT group continued studying multichannel surround sound, especially the acoustic and perceptual analysis of 5.1-channel surround sound. In the mid-1990s, especially after 2000, the SCUT group collaborated with Prof. Shanqun Guan at the Beijing University of Posts and Telecommunications and focused on binaural and virtual auditory displays. These works are described in other studies (Xie, 2008a, 2009a, 2013, 2020; Xie et al., 2013b).

Limited by historical conditions, the applications of spatial sound techniques in China began later than those in the USA, Europe, and Japan. In the middle and late 1970s, China began to produce the 45°/45° record. However, the applications of spatial sounds have spread quickly since 1980. In the early 1980s, radios in China began an FM stereophonic broadcast. With the use of a stereophonic tape record, two-channel stereophonic sound has become popular. Since then, two-channel stereophonic sounds have been widely used in consumer reproduction with the popularization of compact disks (CDs). In the middle of the 1980s, cinemas with surround sound emerged in China. Since the development of DVD-video in the middle and late 1990s, domestic theaters with multichannel surround sound (typically 5.1-channel surround sound) have been popularized. Since 2000, spatial sound techniques, including stereophonic, multichannel surround sound, and virtual auditory display, have been applied not only to traditional cinema and consumer reproduction but also to new media, such as computers and mobile devices. Spatial sound applications have also been extended to some new fields, including communication and virtual reality. Overall, China promises a very large space for the applications of spatial sound techniques.

1.10 SUMMARY

An anticlockwise spherical coordinate system with respect to a receiver position is specified in this book. For convenience, the coordinate system with respect to a sound source is also sometimes used.

A sound field is defined as a region in a medium in which sound waves are being propagated. A sound field in air is physically characterized by the temporal and spatial distribution of sound pressure in the time domain or equally characterized by the frequency and spatial distribution of sound pressure in the frequency domain. The free field refers to a special sound field in a uniform and isotropic medium in which the influences of boundary are completely negligible (absence of reflections from boundaries). The sound field generated by a point source in the free field is a spherical wave in which the magnitude of sound pressure is inversely proportional to the distance between the receiver position and the source. The sound field generated by a straight-line source with an infinite length in the free field is a cylindrical wave. In a local region of a far field where the distance between the receiver position and the sound source is large enough, spherical and cylindrical waves can be approximated as a plane wave. In general cases, the sound pressure generated by a sound source depends on the direction of the receiver position with respect to the source. This directional characteristic is described by the directivity of sound source radiation.

When reflective boundaries exist, sound waves in the receiver position are the combination of the direct sound from the sound source and the reflected sounds from boundaries. For a point source in front of a rigid plane with infinite extension, sound pressure can be simply calculated in accordance with the acoustic principle of the image source. In enclosed spaces, such as rooms with multiple reflective surfaces, the interference between direct and reflected sounds forms standing waves. Under certain conditions, statistical acoustic-based methods are applicable to the analysis of a sound field in rooms. Diffused sound fields, reverberation times, and reverberation radius are important concepts in room acoustics.

Three categories of microphones, namely, pressure microphones, pressure-gradient microphones, and a combination of pressure and pressure-gradient microphones, are often used for spatial sound signals recording. These microphones are omnidirectional, bidirectional, and approximately unidirectional (or with some rear lobes).

The human peripheral auditory system consists of three main parts: the external, middle, and inner ears. The external ear is composed of the pinna and ear canal. The shape and dimension of the pinna vary depending on each individual. The pinnae play an important role in the localization of high-frequency sounds. The middle ear acts as an impedance transformer. Consequently, incoming sound can be effectively coupled into the inner ear. The inner ear (cochlea) acts as a frequency analyzer and converts mechanical oscillation into neural pulses or activities. The frequency analysis of the auditory system can be modeled using a bank of overlapping auditory filters, whose bandwidth can be approximated with the critical bandwidth or an ERB. Correspondingly, two new frequency metrics, i.e., Bark and ERB number, are introduced. The psychoacoustic phenomena of subjective loudness and masking are closely related to auditory filters.

The sound waves received by the two ears have been modified via scattering and diffraction from the anatomical structures of listeners. HRTFs describe the overall filtering effect of anatomical structures and contain major directional localization cues. HRIRs are the time-domain counterparts of HRTFs. An artificial head is a kind of model used to simulate the scattering and diffracting effects of anatomical structures, such as the head, torso, and pinnae, on the sound waves. Artificial heads are also applicable to binaural recording and measurements.

Spatial hearing is a subjective perception or sensation of the spatial attributes of sound, which covers the localization of a single sound source, summing localization and auditory event of multiple sound sources, and subjective spatial perception of environmental reflections.

Localization determines two aspects: the direction and distance of sound sources. Directional localization cues for a single sound source include ITD, ILD, dynamic and spectral cues, and others. Different cues take effect within different frequency ranges, and the information provided by these cues may be somewhat redundant. The auditory system determines the direction of a sound source based on a comparison between the obtained cues and patterns stored from previous experiences. Auditory distance perception is also a comprehensive consequence of multiple cues.

Under certain conditions, the cues of summing localization with multiple correlated sound sources are similar to those of a single source. When the combined sound pressures in the two ears caused by multiple sources contain inconsistent or even conflicting localization cues, the auditory system may determine the source position based on more consistent cues or some unknown psychoacoustic mechanisms. However, the perceived direction of a virtual source may depend on the features of signals, such as frequency spectra. Moreover, the virtual source is broadened, blurred, unnatural, or even unable to be localized likely because of inconsistent or conflicting localization cues. Multiple sources with partly correlated signals can create broadening and blur auditory events or even the sensations of envelopment by sound. The precedence effect, cocktail party effect, and auditory scene analysis are closely related to spatial hearing with multiple sound sources.

Room reflections contribute to the perceptions of the overall ASI of sound sources and acoustical environment. The ASI consists of at least two distinguishable components, namely, ASW and LEV. They are determined with the temporal and spatial distributions of early and late reflections, respectively. ASI is closely related to some physical attributes of sound fields or the physical attributes of binaural pressures, such as IACC.

Spatial sound aims to record (or simulate), transmit (or store), and finally reproduce the spatial information of sound and then recreate the desired spatial auditory events or perceptions. Spatial sounds can be implemented with one of the three methods and principles, namely, sound field-based methods and principles, sound field approximation and psychoacoustic-based methods and principles, and binaural-based methods and principles. Spatial sound can be classified on the basis of its acoustic principles and methods. They can also be classified on the basis of the dimensionality of spatial information they recreate, the number of reproduction channels, and their applications.

Since the first exploration of spatial sound more than 100 years ago, numerous spatial sound techniques and systems have been developed.

Chapter 2

Two-channel stereophonic sound

A two-channel stereophonic sound is the simplest and most common spatial sound technique and system. The spatial information within a certain frontal–horizontal sector (one-dimensional space) can be recreated on the basis of the principle of sound field approximation and psychoacoustics by using a pair of frontal loudspeakers. The two-channel stereophonic sound is considered a milestone in the developments and applications of spatial sound techniques and is still the most popular technique in use. This chapter is not intended to review the detailed history and development of the two-channel stereophonic sound. For details, readers can refer to a previous study (Xie X.F., 1981). In this chapter, the basic principles and some issues related to the applications of the two-channel stereophonic sound are presented to provide readers with sufficient background information for further discussion on multichannel surround in succeeding chapters. In Section 2.1, the basic principle of recreating spatial information by the two-channel stereophonic sound is addressed. The corresponding summing localization equations are derived, and some rules in the summing localization of a virtual source are discussed. In Section 2.2, methods for generating two-channel stereophonic signals are introduced, including various microphone recording and signal simulating techniques. In Section 2.3, the compatibility between stereophonic and mono reproduction and the problems of up/downmixing between mono and stereophonic signals are briefly discussed. In Section 2.4, some issues related to practical two-channel stereophonic reproduction, such as loudspeaker arrangement, the compensation for off-central listening position are addressed.

2.1 BASIC PRINCIPLE OF A TWO-CHANNEL STEREOPHONIC SOUND

2.1.1 Interchannel level difference and summing localization equation

The two-channel stereophonic sound is designed on the basis of the results of summing localization with two sound sources (loudspeakers) described in Section 1.7.1. In the late 1950s and beginning of the 1960s, the principle of two-channel stereophonic sound was re-analyzed by some researches (Clack et al., 1957; Leakey, 1959; Bauer, 1961a; Makita, 1962; Mertens, 1965).

In Figure 2.1, a pair of loudspeakers are arranged symmetrically in the front of the listener with azimuths $\theta_L = \theta_0$ and $\theta_R = -\theta_0$. The two loudspeaker signals in the frequency domain are E_L and E_R (stereophonic signals are usually denoted as notation L and R, whereas frequency-domain signals are denoted as notation E in this book). Two identical loudspeaker signals with different amplitudes are written as

$$E_L = E_L(f) = A_L E_A(f) \qquad E_R = E_R(f) = A_R E_A(f), \tag{2.1.1}$$

DOI: 10.1201/9781003081500-2

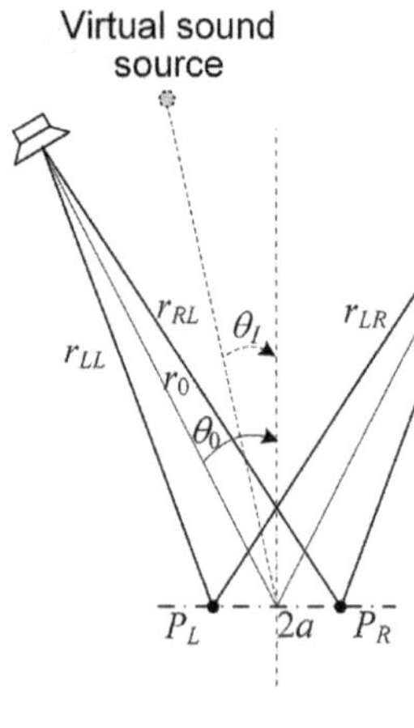

Figure 2.1 Configuration of two-channel stereophonic loudspeakers.

where A_L and A_R are ***normalized amplitudes***, ***relative gains***, or ***panning coefficients*** of the left and right loudspeaker signals, respectively. For in-phase loudspeaker signals with a level difference only, A_L and A_R are real and non-negative numbers. $E_A(f)$ represents the ***signal waveform*** in the frequency domain and determines the overall complex-valued pressure (including magnitude and phase) in reproduction. For harmonic or narrow-band signals, the perceived virtual source direction is independent from $E_A(f)$. Therefore, a unit $E_A(f)$ can be assumed in the analysis. In this case, A_L and A_R can also be regarded as ***normalized loudspeaker signals*** in the frequency domain. If necessary, $E_A(f)$ should be multiplied to the results derived from the normalized loudspeaker signals when the absolute amplitude of the reproduced sound pressures is considered. When A_L and A_R are frequency independent, the two loudspeaker signals in the time domain can be expressed by replacing $E_A(f)$, $E_L(f)$, and $E_R(f)$ in Equation (2.1.1) with their time-domain forms $e_A(t)$, $e_L(t)$, and $e_R(t)$, respectively.

At low frequencies, the head shadow is negligible, and the two ears are approximated as two points in the free space separated by $2a$, where a is the head radius. For simplicity, loudspeakers are approximated as point sources. When the source distance with respect to the head center is much larger than the head radius, i.e., $r_0 >> a$, the incident wave generated by loudspeakers can be further approximated as plane waves. For convenience in analysis, the overall gain of electroacoustic reproduction system is calibrated so that loudspeakers are equivalent to point sources with the strength $Q_p = 4\pi r_0$ for unit input signals. In this case, according to Equations (1.2.4) and (1.2.6), the transfer coefficient from the loudspeaker signal to the pressure amplitude of a free-field plane wave at the origin of the coordinate (the position of head center in the absence of head) is equal to a unit. ***This assumption is held*** for the discussions in the succeeding chapters when plane waves generated by loudspeakers at a far-field distance are considered (here, "loudspeaker signals" are used to refer to the ***input signals of the electroacoustic reproduction system***). Under the above assumption and letting $E_A(f)$ being a unit, the binaural sound pressures in the frequency domain are the superposition of those caused by the incident plane waves from two loudspeakers and can be written as

$$\begin{aligned} P_L' &= A_L \exp(-jkr_{LL}) + A_R \exp(-jkr_{LR}), \\ P_R' &= A_L \exp(-jkr_{RL}) + A_R \exp(-jkr_{RR}). \end{aligned} \tag{2.1.2}$$

where $k = 2\pi f/c$ is the wave number, and c = 343 m/s is the speed of sound, and

$$r_{LL} = r_{RR} \approx r_0 - a\sin\theta_0 \qquad r_{LR} = r_{RL} \approx r_0 + a\sin\theta_0, \tag{2.1.3}$$

denote the distances from the loudspeaker to the ipsilateral (near) and contralateral (far) ears, respectively. At a distance of $r_0 >> a$, if incident waves from loudspeakers are represented as a spherical wave rather than approximated as plane waves, A_L and A_R in Equation (2.1.2) are substituted with $A_L/4\pi r_0$ and $A_R/4\pi r_0$, respectively. Even in this case, the resultant localization equation is identical to that derived by the approximation of a plane wave. In Equations (2.1.2) and (2.1.3), the common phase factor $\exp(-jkr_0)$ represents the linear delay caused by the sound propagation from each loudspeaker to the origin and can be omitted. Omitting this phase factor is equivalent to supplementing an initial linear phase $\exp(jkr_0)$ to the complex-valued strength Q_p of the point source or loudspeakers mentioned above. This manipulation is also equivalent to a normalization that makes the transfer coefficient from the loudspeaker signal to the pressure amplitude of the free-field plane wave at the origin to be a unit. Then, the interaural phase difference is calculated as

$$\Delta\psi_{SUM} = \psi_L - \psi_R = 2\ \arctan\left[\frac{A_L - A_R}{A_L + A_R}\tan\left(ka\sin\theta_0\right)\right], \tag{2.1.4a}$$

or an interaural phase delay difference

$$ITD_{p,SUM} = \frac{\Delta\psi_{SUM}}{2\pi f} = \frac{1}{\pi f}\ \arctan\left[\frac{A_L - A_R}{A_L + A_R}\tan\left(ka\sin\theta_0\right)\right]. \tag{2.1.4b}$$

The subscript "SUM" in the two equations represents the case of summing localization with two loudspeakers. As stated in Section 1.6.5, the interaural phase delay difference is considered a dominant cue for azimuthal localization at low frequencies. A comparison between the combined $\text{ITD}_{p,SUM}$ in Equation (2.1.4b) and the single-source ITD_p derived from prior auditory experiences [Equation (1.6.1)] enables the determination of the azimuthal position θ_I of the summing virtual source as

$$\sin\theta_I = \frac{1}{ka}\arctan\left[\frac{A_L - A_R}{A_L + A_R}\tan\left(ka\sin\theta_0\right)\right], \tag{2.1.5}$$

At low frequencies with $ka << 1$, Equation (2.1.5) can be expanded as a Taylor series of ka (or $ka\ \sin\theta_0$). If the first expansion term is retained, the equation can be simplified as

$$\sin\theta_I = \frac{A_L - A_R}{A_L + A_R}\sin\theta_0 = \frac{A_L/A_R - 1}{A_L/A_R + 1}\sin\theta_0. \tag{2.1.6}$$

This expression is the ***virtual source localization equation for the two-channel stereophonic sound***, i.e., the famous ***stereophonic law of sine***. This law demonstrates that the spatial position θ_I of the summing virtual source is completely determined by the amplitude ratio (A_L/A_R) between the two loudspeaker signals and the half-span angle θ_0 between the two loudspeakers with respect to the listener, but it is irrelevant to frequency and head radius. For an average head radius with a = 0.0875 m, Equation (2.1.6) is quite effective below 0.7 kHz.

Thus, Equation (2.1.6) suggests the following:

1. When A_L and A_R are identical, $\sin\theta_I$ is zero, indicating that the summing virtual source is positioned at the midpoint between two loudspeakers.
2. When A_L is larger than A_R, $\sin\theta_I$ is positive, meaning that the summing virtual source is positioned close to the left loudspeaker.
3. When A_L is far larger than A_R, $\sin\theta_I$ is approximately equal to $\sin\theta_0$, indicating that the summing virtual source is positioned at the left loudspeaker.
4. Similar results are obtained when A_R is larger than A_L because of the left-right symmetry in configuration.

In Equation (2.1.6), $\theta_0 = 30°$ or $2\theta_0 = 60°$(standard stereophonic loudspeaker configuration) is substituted, and the relationship between the position of the virtual source and the interchannel level difference (ICLD) between loudspeaker signals denoted by $d = 20 \log_{10}(A_L/A_R)$ dB is illustrated in Figure 2.2. In Figure 2.2, θ_I varies continuously from 0° to approximately 30°as d increases from 0 dB to +30 dB. This finding is consistent with the results of the virtual source localization experiment with two stereophonic loudspeakers.

Some remarks on summing localization with two stereophonic loudspeakers and the stereophonic law of sine are as follows:

1. The stereophonic law of sine is based on the principle of the summing localization of two sound sources. In stereophonic localization, $\text{ITD}_{p,SUM}$ encoded in the superposed binaural pressures is used by the auditory system to identify the position of the virtual source at low frequencies. $\text{ITD}_{p,SUM}$ is controlled by the ICLD. This finding indicates that transformation occurs from the ICLD at the two loudspeaker signals to $\text{ITD}_{p,SUM}$ at the binaural pressures. The ICLD regarding the loudspeaker signals should not be confused with the interaural level difference (ILD) at the two ears (introduced in Section 1.6.2).
2. The approach of creating localization cues by adjusting the ICLD is invalid above 1.5 kHz for two-channel stereophonic reproduction because the superposed binaural pressures contain only the localization cue of ITD_p, which is an effective cue below 1.5 kHz. For wideband stimuli containing low-frequency components below 1.5 kHz,

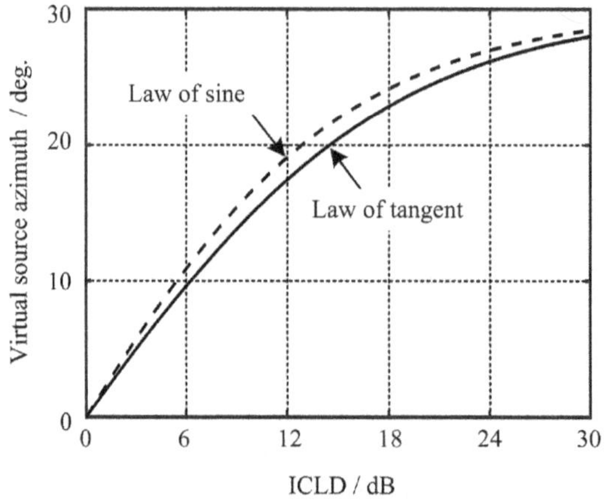

Figure 2.2 Relationship between the position of the virtual source and the interchannel level difference between the loudspeaker signals calculated from the stereophonic laws of sine and tangent, respectively.

creating a virtual source by using ICLD is still valid because of the dominant role of ITD_p in azimuthal localization at low frequencies.

3. An anticlockwise spherical coordinate system with respect to the head center is employed in this book. If the clockwise spherical coordinate system is used, a negative sign should be supplemented to the sine law in Equation (2.1.6).

The law of sine is derived under the assumption that a listener's head is fixed to the front orientation. When the listener's head rotates around the vertical axis with an azimuth $\delta\theta$ ($\delta\theta > 0$ represents an anticlockwise rotation to the left, and $\delta\theta < 0$ denotes a clockwise rotation to the right), the distances from two loudspeakers to two ears in Figure 2.1 become

$$\begin{aligned} r_{LL} &= r_0 - a\sin(\theta_0 - \delta\theta) \quad & r_{RL} &= r_0 + a\sin(\theta_0 - \delta\theta), \\ r_{RR} &= r_0 - a\sin(\theta_0 + \delta\theta) \quad & r_{LR} &= r_0 + a\sin(\theta_0 + \delta\theta). \end{aligned} \tag{2.1.7}$$

Similar to the derivation from Equation (2.1.1) to (2.1.4), the interaural phase delay difference becomes

$$ITD_{p,SUM} = \frac{1}{\pi f}\arctan\frac{A_L \sin\left[ka\sin(\theta_0 - \delta\theta)\right] - A_R \sin\left[ka\sin(\theta_0 + \delta\theta)\right]}{A_L \cos\left[ka\sin(\theta_0 - \delta\theta)\right] + A_R \cos\left[ka\sin(\theta_0 + \delta\theta)\right]}. \tag{2.1.8}$$

The azimuth $\delta\theta$ of the rotation represents the virtual source direction with respect to the fixed coordinate, i.e., $\hat{\theta}_I = \delta\theta$, by choosing the azimuth $\delta\theta$ of rotation so that the listener is oriented to the virtual source direction and the interaural phase delay difference $ITD_{p,SUM}$ given by Equation (2.1.8) consequently vanishes. Here, the notation $\hat{\theta}_I$ is used to denote the azimuth of the virtual source because the result of the head's rotation may be different from that of the fixed head orientation. When $ITD_{p,SUM} = 0$ is substituted in Equation (2.1.8), the following equation is obtained:

$$A_L \sin\left[ka\sin(\theta_0 - \delta\theta)\right] - A_R \sin\left[ka\sin(\theta_0 + \delta\theta)\right] = 0. \tag{2.1.9}$$

At low frequencies with $ka << 1$, Equation (2.1.9) can be expanded as a Taylor series of ka. If only the first expansion term is retained, the virtual source azimuth is determined in accordance with ***the law of tangent***

$$\tan\hat{\theta}_I = \frac{A_L - A_R}{A_L + A_R}\tan\theta_0 = \frac{A_L/A_R - 1}{A_L/A_R + 1}\tan\theta_0. \tag{2.1.10}$$

For an average head radius, Equation (2.1.10) is quite effective below 0.7 kHz. Makita (1962) supposed that the perceived virtual source direction in the superposed sound field is consistent with the inner normal direction (opposite to the direction of the medium velocity) of the superposed wavefront at the receiver position. Equation (2.1.10) can also be derived from Makita's hypothesis (Section 3.2.2). Actually, Makita's hypothesis is equivalent to that of the rotation of the listener's head to the orientation of the virtual source.

The results calculated from Equation (2.1.10) are presented in Figure 2.2. In particular, the span angle between two loudspeakers is also $2\theta_0 = 60°$. The results of Equation (2.1.10) are similar to those of Equation (2.1.6) because $\tan\theta \approx \sin\theta$ for $\theta \le 30°$. Therefore, for loudspeaker configuration with the span angle $2\theta_0 \le 60°$, the perceived virtual source direction is relatively stable during head rotation. In practice, Equations (2.1.6) and (2.1.10) are applied to analyze the virtual source direction in two-channel stereophonic sound reproduction. However, the physical significances of the two equations are different. Equation (2.1.6) represents the case of fixed head orientation to the front, but Equation (2.1.10) denotes the case of the head oriented to the virtual source direction.

In the aforementioned analysis, the summing virtual source with two front stereophonic loudspeakers is located in the frontal–horizontal plane. This assumption is valid for the case of the fixed head orientation to the front and after the head rotation to the orientation of the virtual source. However, a special phenomenon has been observed in some experiments that the perceived direction of the frontal virtual source may be elevated to a higher position in the median plane during head rotation, especially for a large span angle between two loudspeakers. This observation is further addressed in Section 6.1.4.

2.1.2 Effect of frequency

Equation (2.1.6) indicates that the perceived virtual source direction is frequency independent. This principle is consistent with the experimental results only at frequencies below 0.7 kHz. As the frequency of the stimulus increases (for narrow-band stimuli), the perceived virtual source direction deviates from that predicted with Equation (2.1.6). In other words, Equation (2.1.6) is invalid if $f > 0.7$ kHz and therefore should be revised (He et al., 1993).

In fact, interaural phase delay difference is a dominant localization cue below the frequency of 1.5 kHz. Equation (2.1.6) is derived by considering the interaural phase delay difference. In the following, notation ITD_p is used to denote the interaural phase delay difference for a single source and for multiple sources. However, some errors occur at the two stages of approximation made in the derivation of Equation (2.1.6):

1. At the stage of simplifying Equation (2.1.5) to Equation (2.1.6), Equation (2.1.5) is expanded as a Taylor series of $ka \sin\theta_0$, and only the first expansion term is retained. This approximation is appropriate only when $(ka \sin\theta_0) << 1$. For a standard stereophonic loudspeaker configuration with $2\theta_0 = 60°$ and an average head radius of $a = 0.0875$ m, $ka \sin\theta_0 \approx 0.56$ at a frequency of 0.7 kHz. Therefore, retaining the first term in the Taylor expansion causes errors above 0.7 kHz.
2. A simplified (shadowless) head model is used in the aforementioned analysis. However, the effect of the head shadow on ITD_p is non-negligible. In a more actuate calculation of ITD_p, the diffraction and scattering effect of the head should be considered via head-related transfer function (HRTF)-based analysis. As demonstrated in Section 1.6.1, if a is chosen as the actual head radius, the actual (accurate) ITD_p cannot be obtained with Equation (1.6.1). Below the frequency of 3 kHz, the actual ITD_p is larger than that evaluated from Equation (1.6.1), as illustrated in Figure 1.19, because the effect of the head enlarges the effective head radius in the calculation of ITD_p with the simplified head model. Therefore, the accuracy in ITD_p calculation can be moderately improved by using an effective or pre-corrected head radius $a' = \kappa a$ to replace the actual head radius a in the analysis and derivation of Equations (1.6.1) and (2.1.5), where $\kappa \geq 1$ and is generally frequency dependent.

An equivalent head radius $a' = \kappa a$ with an appropriate κ can be used to replace a in Equation (2.1.5) to improve the calculation of the virtual source position above the frequency of 0.7 kHz. The frequency-dependent parameter κ can be evaluated from the rigid spherical head model (Figure 1.19). However, this method is complex and inconvenient for practical calculation. In Figure 1.19, the accurate ITD_p is approximately 1.5 times of ITD_p evaluated from Equation (1.6.1) with a head radius a below 0.4 kHz. Within a mid-frequency region of 0.5–3 kHz, ITD_p is frequency dependent, and κ smoothly transits from an asymptotic result

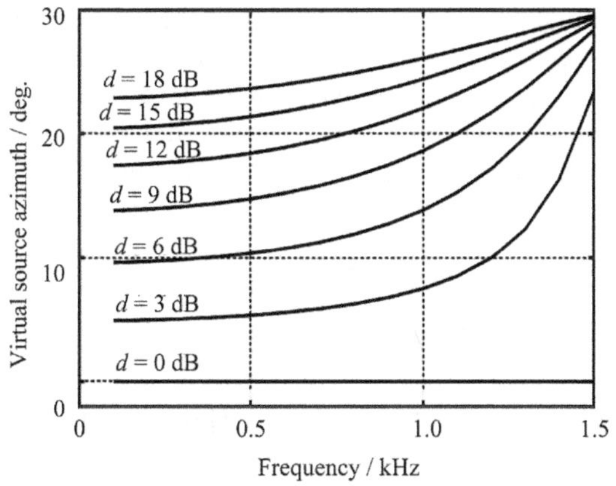

Figure 2.3 Variation in the virtual source direction with frequency for various interchannel level differences of $d = 20\log_{10}(A_L/A_R)$ between loudspeaker signals.

of 1.5 at low frequencies to 1 at high frequencies. However, at low frequencies with $f \le 0.7$ kHz, the higher-order terms in the Taylor expansion of Equation (2.1.5) are negligible, so a is canceled out in the equation, and the final results of Equation (2.1.6) are independent of a. Therefore, below 0.7 kHz, various precorrected head radii lead to identical results. Above the frequency of 1.5 kHz, because ITD_p is no longer an effective localization cue, Equation (2.1.5) becomes invalid. With all these reasons, a should be precorrected in the calculation within the frequency range from 0.7 kHz to 1.5 kHz. In summary, the applicable range of Equation (2.1.5) can be extended to 1.5 kHz by replacing a with $a' = \kappa a$ in the equation. For simplicity, a constant parameter $\kappa = 1.2\sim1.3$ is chosen to approximately match the actual ITD_p within 0.7 kHz to 1.5 kHz. However, the analysis method here is only an approximation. A stricter method should involve the use of HRTFs for calculating ITD_p (Section 12.1.3). The results of the approximate analysis here are basically consistent with those of a strict analysis.

Figure 2.3 illustrates the variation in the virtual source direction with frequency for various ICLDs between loudspeaker signals, as calculated from Equation (2.1.5). The standard stereophonic loudspeaker configuration with $2\theta_0 = 60°$, the average values of $a = 0.0875$ m and $\kappa = 1.25$ are chosen in the calculation. In Figure 2.3, for a given ICLD or $d = 20\log10(A_L/A_R)$ dB except $d = 0$ dB, the virtual source moves toward the direction of the loudspeaker (30°) as frequency exceeds 0.7 kHz, resulting in an unstable virtual source. For example, when $d = 6$ dB, θ_I is 9.6°, 11.2°, and 23.0° for 0.1, 0.7, and 1.4 kHz, respectively. For stimuli with different frequency components, this frequency-dependent characteristic may generate broadened and blurred virtual source. In addition, the frequency dependence of a virtual source becomes more obvious for a large span angle $2\theta_0$ between two loudspeakers.

2.1.3 Effect of interchannel phase difference

Equations (2.1.5) and (2.1.6) are suitable for in-phase loudspeaker signals. Similarly, the analysis can be extended to the summing localization of two loudspeaker signals with a frequency-independent phase difference (Xie, 1998).

In the loudspeaker configuration shown in Figure 2.1, the loudspeakers are provided with signals having a level difference and a phase difference. Then, the complex-valued normalized amplitude of loudspeaker signals is given by

$$A_L = |A_L|\exp(j\eta_L) \qquad A_R = |A_R|\exp(j\eta_R), \tag{2.1.11}$$

where $|A_L|$ and $|A_R|$ are the normalized magnitudes of the left and right loudspeaker signals, respectively, and $-180° < \eta_L$ and $\eta_R \leq 180°$ are the phases of the signals, where $\eta_{L,R} > 0°$ and $\eta_{L,R} < 0°$ represent the leading and lagging phase signals, respectively.

Similar to the derivation of Equation (2.1.5), under the condition of $r_0 >> a$ and with the common phase factor $\exp(-jkr_0)$ omitted, the superposed binaural pressures caused by two loudspeakers are

$$\begin{aligned} P_L' &= |A_L|\exp(j\eta_L + jka\sin\theta_0) + |A_R|\exp(j\eta_R - jka\sin\theta_0), \\ P_R' &= |A_L|\exp(j\eta_L - jka\sin\theta_0) + |A_R|\exp(j\eta_R + jka\sin\theta_0). \end{aligned} \tag{2.1.12}$$

Further omitting the phase factor $\exp[j(\eta_L + \eta_R)/2]$ in Equation (2.1.12) yields

$$\begin{aligned} P_L' &= (|A_L|+|A_R|)\cos\left(ka\sin\theta_0 + \frac{\eta}{2}\right) + j(|A_L|-|A_R|)\sin\left(ka\sin\theta_0 + \frac{\eta}{2}\right), \\ P_R' &= (|A_L|+|A_R|)\cos\left(ka\sin\theta_0 - \frac{\eta}{2}\right) - j(|A_L|-|A_R|)\sin\left(ka\sin\theta_0 - \frac{\eta}{2}\right). \end{aligned} \tag{2.1.13}$$

where $\eta = \eta_L - \eta_R$ represents the interchannel phase difference between the left and right loudspeaker signals. It can be deduced from Equation (2.1.13) that, when

$$ka\sin\theta_0 + \frac{|\eta|}{2} < \frac{\pi}{2}, \tag{2.1.14}$$

we have $\cos(ka\sin\theta_0 \pm \eta/2) > 0$ and the phase angles of binaural pressures lie in the I or IV quadrant. In this case, the interaural phase difference can be evaluated from binaural pressures. If the interaural phase delay difference is a dominant cue for azimuthal localization at low frequencies, when the listener's head is fixed to the front orientation, the $\mathrm{ITD}_{p,SUM}$ evaluated from Equation (2.1.13) can be compared with the single-source ITD_p given by Equation (1.6.1) to determine the azimuthal position θ_I of the summing virtual source as

$$\sin\theta_I = \frac{1}{2ka}\left\{\begin{aligned} &\arctan\left[\frac{|A_L|-|A_R|}{|A_L|+|A_R|}\tan\left(ka\sin\theta_0 - \frac{\eta}{2}\right)\right] \\ &+\arctan\left[\frac{|A_L|-|A_R|}{|A_L|+|A_R|}\tan\left(ka\sin\theta_0 + \frac{\eta}{2}\right)\right] \end{aligned}\right\}. \tag{2.1.15}$$

The direction of the summing virtual source is generally frequency or (ka) dependent. When the two loudspeaker signals are in phase with $\eta = 0°$, Equation (2.1.15) is simplified into Equation (2.1.5). At very low frequencies with $ka << 1$, Equation (2.1.15) can be expanded as a Taylor series of ka. If only the first expansion term is retained, the equation becomes

$$\sin\theta_I = \mathrm{Re}\frac{A_L - A_R}{A_L + A_R}\sin\theta_0 = \frac{|A_L|^2 - |A_R|^2}{|A_L|^2 + |A_R|^2 + 2|A_L||A_R|\cos\eta}\sin\theta_0. \tag{2.1.16}$$

Re denotes the real part of a complex number. In this case, the direction of the virtual source is frequency independent. When $\eta = 0°$, Equation (2.1.16) is simplified into the law of sine given by Equation (2.1.6).

Similar to the derivation of Equation (2.1.10), if the listener's head rotates to the orientation of the virtual source so that $ITD_{p,SUM}$ vanishes, the azimuth of rotation represents the direction of the virtual source with respect to the fixed coordinate. At low frequencies with $ka << 1$, the following equation is obtained:

$$\tan\hat{\theta}_I = \mathrm{Re}\frac{A_L - A_R}{A_L + A_R}\tan\theta_0 = \frac{|A_L|^2 - |A_R|^2}{|A_L|^2 + |A_R|^2 + 2\,|A_L|\,|A_R|\cos\eta}\tan\theta_0. \qquad (2.1.17)$$

Equation (2.1.17) represents the virtual source localization equation for loudspeaker signals with interchannel phase difference and was originally derived from Makita's hypothesis (Makita, 1962).

The applicable frequency limit of Equation (2.1.15) can be estimated from Equation (2.1.14) as

$$f < f_0 = \frac{c(\pi - |\eta|)}{4\pi a\sin\theta_0}. \qquad (2.1.18)$$

Similar to the case in Section 2.1.2, $a' = \kappa a$ should be used to replace a in Equation (2.1.18). Table 2.1 lists the upper-frequency limits (f_0) for various interchannel phase differences. The standard stereophonic loudspeaker configuration with $2\theta_0 = 60°$, $a = 0.0875$ m, and $\kappa = 1.25$ is chosen in the calculation. When two loudspeaker signals are in phase with $\eta = 0°$, it has $f_0 = 1.57$ kHz. Therefore, Equation (2.1.15) or Equation (2.1.5) is valid approximately below 1.5 kHz. As the interchannel phase difference increases, the upper-frequency limit in Equation (2.1.15) decreases. For two out-of-phase loudspeaker signals with $|\eta| \rightarrow \pi$ (180°), Equation (2.1.18) yields $f_0 \rightarrow 0$. Therefore, Equation (2.1.15) is inappropriate for out-of-phase loudspeaker signals.

For out-of-phase loudspeaker signals with $\eta = 180°$ (π), when

$$ka\sin\theta_0 < \frac{\pi}{2}, \qquad (2.1.19)$$

we have $\cos(ka\sin\theta_0 + \eta/2) < 0$ and $\cos(ka\sin\theta_0 - \eta/2) > 0$. In this case, the phase Ψ_L of the left ear pressure lies in the II or III quadrant, and the phase Ψ_R of the right ear pressure lies in the I or IV quadrant. Accordingly, interaural phase difference and $ITD_{p,SUM}$ can be calculated, resulting in the virtual source localization equation for a fixed head oriented to the front direction.

$$\sin\theta_I = \frac{1}{2ka}\left\{\pm\pi - 2\arctan\left[\frac{|A_L| - |A_R|}{|A_L| + |A_R|}\tan\left(\frac{\pi}{2} - ka\sin\theta_0\right)\right]\right\}. \qquad (2.1.20)$$

Table 2.1 Applicable frequency limit of Equation (2.1.15) for various interchannel phase differences

η	0°	30°	60°	90°	120°	150°	180°
f_0 (kHz)	1.57	1.31	1.05	0.78	0.52	0.26	1.57

When $A_R \to 0$, Equation (2.1.20) should lead to $\theta_I \to \theta_0$; when $A_L \to 0$, Equation (2.1.2) should lead to $\theta_I \to -\theta_0$. Therefore, if $A_L > A_R$, a positive sign is chosen before π in the brackets in Equation (2.1.20). If $A_L < A_R$, a negative sign is chosen. At very low frequencies with $ka << 1$, a procedure similar to the derivation of Equation (2.1.16) yields

$$\sin\theta_I = \frac{|A_L| + |A_R|}{|A_L| - |A_R|}\sin\theta_0. \tag{2.1.21}$$

Equation (2.1.21) exactly matches Equation (2.1.16) when $\eta \to 180°$ or equally matches Equation (2.1.6) when the signal amplitude A_R in Equation (2.1.6) is replaced with its opposite phase version $-A_R$. The applicable frequency limit of Equation (2.1.20) can be estimated from Equation (2.1.19) as

$$f < f_0 = \frac{c}{4a\sin\theta_0}. \tag{2.1.22}$$

For a standard stereophonic loudspeaker configuration with $2\theta_0 = 60°$ and $a' = \kappa a$ with $\kappa = 1.25$ and $a = 0.0875$ m, the applicable frequency limit is $f_0 = 1.57$ kHz.

On the basis of Equations (2.1.15) and (2.1.20), the frequency dependence of a virtual source direction is analyzed for various interchannel phase differences (η) and interchannel level differences ($d = 20\log_{10}|A_L/A_R|$). The standard stereophonic loudspeaker configuration with $2\theta_0 = 60°$, $a = 0.0875$ m, and $\kappa = 1.25$ is chosen in the calculation. Figure 2.4 illustrates the following results:

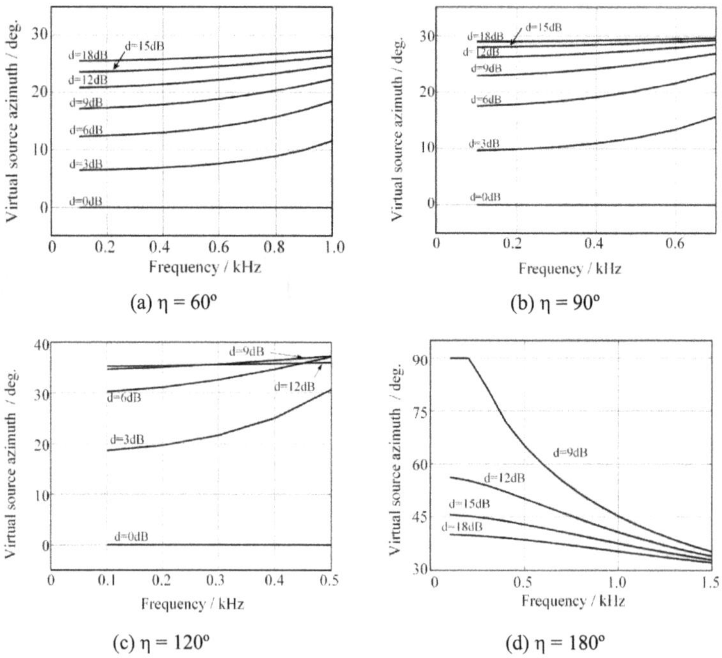

Figure 2.4 Virtual source direction for various interaural phase differences (a) $\eta = 60°$; (b) $\eta = 90°$; (c) $\eta = 120°$; (d) $\eta = 180°$.

1. At very low frequencies (e.g., $f = 0.1$ kHz) and for arbitrary $0° \leq \eta \leq 180°$, the results are consistent with Equation (2.1.16) or Equation (2.1.21). In this case, Equations (2.1.16) and (2.1.21) are good approximations.
2. For $\eta \leq 60°$ and $f \leq 0.6$ kHz, θ_I varies slightly with f, and Equation (2.1.16) is an appropriate approximation. For a given d, the deviation between θ_I for $\eta \neq 0°$ and θ_I for $\eta = 0°$ (in phase) is small or moderate. Within the frequency range of 0.7 kHz < $f < f_0$, as f increases, θ_I moves toward the direction of the loudspeaker with a larger signal magnitude. A larger η causes a larger variation in θ_I with frequency. For example, for $\eta = 60°$, $d = 3$ dB, and $f = 0.1$, 0.6, and 1.0 kHz, θ_I is 6.5°, 7.6°, and 11.6°, respectively.
3. For the interchannel phase difference $\eta > 60°$, θ_I varies with frequency greatly even within the low-frequency range of 0.1 kHz $\leq f \leq$ 0.6 kHz. This variation in θ_I with frequency becomes more obvious as η increases. For $\eta > 90°$, the position of the virtual source may ***exceed the boundary bounded by two loudspeakers***, i.e., $\theta_I > \theta_0$. This virtual source is called "***outside-boundary virtual source.***" In addition, for $\eta = 120°$, even at the low frequency of $f = 0.1$ kHz, when d changes from 0 dB to 3 dB, θ_I increases from 0° to 18.6°. For $d < 6$ dB, θ_I obviously varies with frequency. For example, for $d = 3$ dB, θ_I varies from 18.6° to 30.6° when f increases from 0.1 kHz to 0.5 kHz. Overall, for $\eta > 90°$, the virtual source is very unstable.
4. The case of $\eta = 180°$ is a special one. At very low frequencies, the interchannel level difference with $d < 9$ dB yields $\sin\theta_I > 1$, and the position of an auditory event is uncertain. In fact, when the ICLD of two loudspeaker signals with out-of-phase is small, the interaural phase relationship between binaural signals is different from that of a single source at any spatial position, consequently yielding an unnatural auditory perception. However, when $d > 9$ dB, a virtual source located outside the boundary of the loudspeaker appears (Bauer, 1961a; Xie X.F., 1981). Moreover, within the range of 9 dB $\leq d \leq$ 12 dB, the position of the outside-boundary virtual source varies obviously with frequency even at the low-frequency range of $f \leq 0.5$ kHz. In this case, the outside-boundary virtual source is unstable and blurred. When $d \geq 15$ dB, the position of the outside-boundary virtual source is close to the loudspeakers with a larger signal magnitude and varies slightly with frequency.

The ILD caused by the interchannel phase difference between loudspeaker signals should be analyzed. The ILD in reproduction is evaluated from Equation (2.1.12):

$$ILD = 20\log_{10}\left|\frac{P_L'}{P_R'}\right| = 10\log_{10}\frac{|A_L|^2 + |A_R|^2 + 2\,|A_L|\,|A_R|\cos(2ka\sin\theta_0 + \eta)}{|A_L|^2 + |A_R|^2 + 2\,|A_L|\,|A_R|\cos(2ka\sin\theta_0 - \eta)}\ (dB). \quad (2.1.23)$$

For in-phase loudspeaker signals with $\eta = 0°$ and out-of-phase loudspeaker signals with $\eta = \pm 180°$, when the frequency is below the limit given by Equation (2.1.18) or (2.1.22), Equation (2.1.23) yields $ILD = 0$. This finding is consistent with the low-frequency ILD caused by a single source at the far-field distance. However, $\eta \neq 0°$ and $\pm 180°$ lead to $ILD \neq 0$ in reproduction, even at very low frequencies with $ka << 1$. This value is inconsistent with the result of a single source at the far-field distance. In other words, even if the effect of the head is omitted at low frequencies, the phase difference between loudspeaker signals leads to an additional ILD so that ILD in reproduction deviates from that of a single source at the far-field distance. The localization information provided by this additional ILD usually conflicts with that of the low-frequency ITD_p cue. It is neither consistent with the low-frequency ILD caused by a lateral source at a proximity distance. In the presence of conflicting localization cues, the low-frequency ITD_p may dominate the azimuthal localization, and ILD may slightly contribute to localization, but additional or abnormal ILD in reproduction degrades the naturalness of the perceived virtual source. In particular, a large additional ILD at low frequencies creates a sensation of "oppression". Equation (2.1.23) proves that for a given η, the additional ILD increases with frequency. At a given frequency, the additional ILD initially increases with η and peaks at $\eta = 120°$. For example, for $\eta = 120°$, a standard stereophonic loudspeaker configuration with $2\theta_0 = 60°$ and $a = 0.0875$ m, when $A_L = A_R$ and $f = 0.2$ kHz, the resultant $ILD = -5$ dB and $ITD_p \approx 0$ (precorrecting the head radius for ILD calculation is not required). This combination of ILD and ITD deviates from the results of any single source.

At very low frequencies, Equation (2.1.23) can be expanded as a Taylor series of ka. If only the first expansion term is retained, the equation becomes

$$ILD = -\frac{40}{\ln 10}\frac{2|A_L||A_R|\sin\eta\ \sin\theta_0\ ka}{|A_L|^2 + |A_R|^2 + 2|A_L||A_R|\cos\eta} = -\frac{40}{\ln 10}\,\mathrm{Im}\frac{A_L - A_R}{A_L + A_R}\sin\theta_0(ka). \tag{2.1.24}$$

The notation Im denotes the imaginary part of the complex number. Therefore, the additional ILD is directly proportional to the following "**phasines**":

$$Pha = \mathrm{Im}\frac{A_L - A_R}{A_L + A_R}\sin\theta_0. \tag{2.1.25}$$

The phasines depends on the interchannel phase difference in loudspeaker signals and increases with the half-span angle θ_0 between two loudspeakers. Gerzon (1975b, 1992a) suggested that phasines is a measure of the naturalness of virtual sources in reproduction. However, a strict analysis of naturalness should consider the diffraction and scattering effect of the head. This analysis is presented in Chapter 12.

The results of a virtual source localization experiment validate the aforementioned analysis (Xie, 1998). In summary, interchannel phase difference results in the obvious frequency dependence and instability in a virtual source direction. At the same time, the additional ILD caused by interchannel phase difference degrades the naturalness of the perceived virtual source. In practice, the interchannel phase difference should be reduced within the limit of 60° or higher but not exceeding 90°, except in the case of out-of-phase loudspeaker signals for creating a virtual source outside the boundary of loudspeaker configuration (a virtual source outside the boundary of loudspeaker configuration is unstable). The aforementioned analysis is only suitable for steady stimuli at low frequencies. For other stimuli, the interchannel phase difference may yield different results in summing virtual source localization. In some psychoacoustic experiments, when the transient stimuli with equal magnitudes but with an interchannel phase difference of less than 90° are provided to two stereophonic loudspeakers, the interchannel phase difference causes the summing virtual source moves toward the direction of the loudspeaker with the leading phase signal, and the virtual source becomes extended and blurred (Matsudaira and Fukami, 1973). When the interchannel phase difference exceeds 90°, an extended auditory event that is difficult to be localized is perceived. The effect of the interchannel phase difference on the virtual source of wideband stimuli with transient characteristics is a psychoacoustic phenomenon, which is needed to be interpreted further.

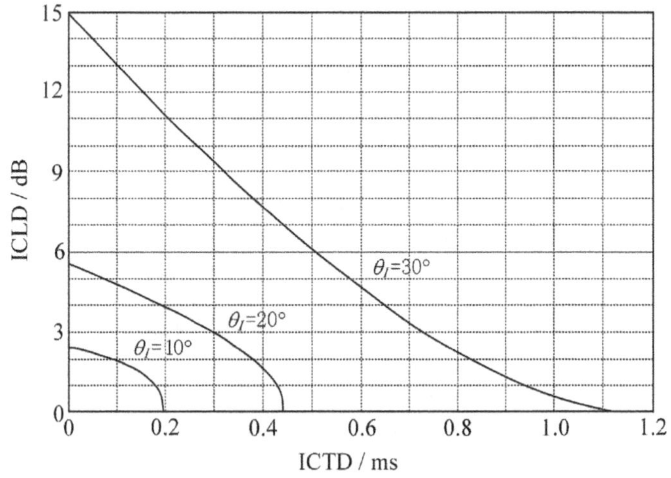

Figure 2.5 Trade-off curves of the combined ICLD and ICTD by Williams. (adapted from Williams 2013).

2.1.4 Virtual source created by interchannel time difference

As stated in Section 1.7.1, for some signals with transient characteristics, the method of interchannel time difference (ICTD) or a combination of ICTD and ICLD can be used to recreate a virtual source. The trade-off curves of the combined ICLD and ICTD are applicable to analyze the direction of the summing virtual source. Figure 2.5 illustrates an example of these curves, which were derived by Williams (1987) through interpolation of the original data obtained by Simonson (1984) and called **Williams curves**. The original data are measured through a localization experiment by using the stimuli of speech and maracas. The trade-off curves for standard loudspeaker configuration with $2\theta_0 = 60°$ and three target azimuths of $\theta_I = 10°, 20°$, and $30°$ are shown in the figure.

The localization results of the combined ICLD and ICTD vary among studies. For example, Figure 1.26 indicates that the mean perceived virtual source azimuth θ_I measured by Simonsen for an ICLD only is larger than those in the two other curves. As stated in Section 1.7.1, the perceived virtual source azimuth depends on stimuli and other experimental conditions. The curves in Figure 2.5 are often used in practice because Simonsen's data are measured from natural stimuli rather than artificial ones. These differences have been observed, and the trade-off curves of the combined ICLD and ICTD have been remeasured using musical stimuli in some studies (Lee, 2010).

ICTD-based summing localization in two or more loudspeakers is a psychoacoustic phenomenon, but physical interpretations or models for this phenomenon have not been developed yet. This situation is different from the case of ICLD-based summing localization. Similar to the precedence effect, a neurophysiologic experiment on cats has demonstrated that the responses of inferior colliculus neurons caused by ICTD signals match those caused by the target source at the summing localization direction (Yin, 1994). Therefore, ICTD-based summing localization may be interpreted at the level of the neurophysiology of hearing.

2.1.5 Limitation of two-channel stereophonic sound

The spatial information of sound includes the localization information of sound sources and the comprehensive spatial information of environment reflections. In two-channel stereophonic sound, the spatial information of sound is represented by the relative relationship between two channels or loudspeaker signals with various manners, resulting in various subjective perceptions or sensations in reproduction.

The directional localization information of the target source can be represented by the ICLD between two-channel signals. This presentation is termed ***amplitude stereophonic sound***, ***level-difference stereophonic sound***, or ***intensity stereophonic sound***. The theory of amplitude stereophonic sound is relatively mature. As stated in previous sections, at low frequencies with $f \leq 0.7$ kHz, the interaural phase delay difference ITD_p created by in-phase loudspeaker signals with ICLD only matches with that of the target source. Within a frequency range of 0.7–1.5 kHz, ITD_p created by ICLD is qualitatively consistent with but quantitatively deviates from that of the target source, resulting in frequency-dependent perceived source direction. At high frequencies (above 1.5 kHz), ICLD may result in interaural localization cues (such as ILD) that are quantitatively inconsistent with those of the target source. However, ICLD does not lead to conflicting interaural localization cues at high frequencies. For wideband stimuli with a low-frequency component below 1.5 kHz, ITD_p dominates azimuthal localization. Therefore, ICLD yields an appropriate localization perception of a virtual source.

Summing localization with two loudspeakers can be further analyzed in terms of the reproduced sound field. Figure 2.6 illustrates the wave front amplitude of the superposed sound pressures created by two stereophonic loudspeakers (approximated as point sources) with identical signal amplitudes $A_L = A_R$. The distance between the loudspeakers and the origin of the coordinate is $r_0 = 2.5$ m. The span angle between two loudspeakers is $2\theta_0 = 60°$. Figure 2.6(a) and 2.6(b) present the results of the harmonic wave at $f = 0.5$ kHz and 1.5 kHz, respectively. In the regions adjacent to either of the loudspeakers, the wavefront is dominated by the spherical wave generated by the loudspeaker. Within a small region adjacent to the central line (bounded by the two dash lines in the figures), the superposed wavefront is approximated to that of a spherical wave incidence from the frontal direction $\theta_I = 0°$. In the far-field distance, the superposed wavefront is further approximated to that of a plane wave incidence from the frontal direction. However, apart from the adjacent region of the central line, the superposed wavefront is no longer a plane or spherical wave. As frequency increases, the width of the region for reconstructing the plane or spherical wavefront narrows. As the receiver position moves toward the back, the span angle between two loudspeakers with respect to the receiver

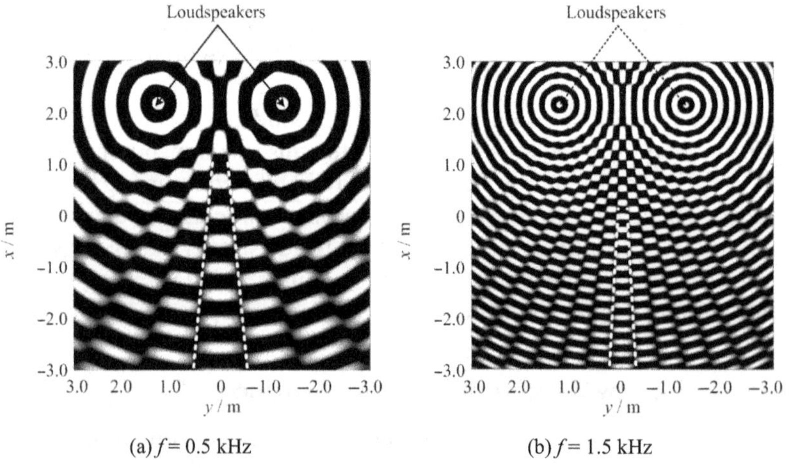

Figure 2.6 Wavefront amplitude of the superposed sound pressures in stereophonic reproduction with identical signal amplitudes (a) $f = 0.5$ kHz; (b) $f = 1.5$ kHz.

position reduces, and the width of the region for reconstructing the plane or spherical wavefront broadens. With an appropriate span angle between two loudspeakers and within the low-frequency range of $f \leq 0.7$ kHz, the amplitude stereophonic sound can reconstruct the target plane or spherical wavefront in a region whose width matches the size of the head (Makita, 1962; Bennett et al., 1985). Therefore, the two-channel amplitude stereophonic sound is a typical example of a spatial sound based on the sound field approximation and psychoacoustics. Overall, the two-channel amplitude stereophonic sound can recreate a relatively authentic and natural virtual source between two loudspeakers.

The frequency-independent interchannel phase difference gives rise to conflicting interaural localization cues and consequently degrades the perceived quality of a virtual source or prevents localization. Two-channel out-of-phase signals may be applicable to recreate an outside-boundary virtual source and then broadens the frontal stereophonic stage. However, the resultant virtual source position is unstable as frequency varies. It is also unstable even when the listening position slightly changes (Section 2.4.2). Moreover, in some cases, two-channel out-of-phase signals may create an unnatural auditory event with an uncertain position.

For some signals with transient characteristics, the method of ICTD or a combination of ICLD and ICTD can be used to recreate a virtual source. This method has been applied to design some microphone techniques for two-channel stereophonic recording. The ICTD-based method is termed ***time (difference) stereophonic sound***. The combination method of ICLD and ICTD is termed a ***combined amplitude and time stereophonic sound*** or ***intensity and time difference stereophonic sound***. In Section 2.1.4, physical models for ICTD-based or a combination of ICLD–ICTD-based summing localization are unavailable. This situation is different from ICLD-based summing localization. Therefore, the time stereophonic sound and the combined amplitude and time stereophonic sound are usually designed on the basis of psychoacoustic experimental results, such as the Williams curves shown in Figure 2.5 (Williams, 1987; Wittek and Theile, 2002). Generally, the perceived quality of the virtual source created via the ICTD-based method is inferior to that created via the ICLD-based method. For practical (wideband) stimuli, the ICTD-based virtual source is blurry, with less naturalness and authenticity. The perceived direction of the virtual source also depends on the spectra and transient characteristics of the stimuli.

Overall, for any loudspeaker signal method, the two-channel stereophonic sound is unable to recreate the spatial information in the full horizontal plane, to say nothing of recreating the spatial information in a fully three-dimensional space. Generally, two-channel stereophonic sound can recreate spatial information within a frontal-horizontal sector bounded by two loudspeakers. Although the case of an outside-boundary virtual source is considered, two-channel stereophonic sound is theoretically able to recreate spatial information extending to the frontal-horizontal quadrants at most. However, the outside-boundary virtual source is usually unstable. These factors are limitations of two-channel stereophonic sound. For many practical applications, such as music reproduction or television sound, if a listener's attention is focused to the frontal direction, two-channel stereophonic sound may meet the requirements to some extent.

The analysis in this section focuses on the methods for representing or encoding the directional information of target virtual sources in two-channel stereophonic signals. The spatial position of an actual sound source or virtual source is specified by its direction and distance. Although auditory distance perception is biased, the relative perceived distance of auditory events in spatial sound reproduction may be controlled by using some appropriate signal simulation and microphone techniques. Various possible cues for auditory distance perception discussed in Section 1.6.6 may be used to control auditory distance perception in reproduction. However, altering the ratio of direct and reflected sound energy in signals is a major means to control the perceived distance in two-channel stereophonic and multichannel sound reproduction. Increasing the relative proportion of reflected sound creates a more distant auditory event or perception.

In Section 1.8, early lateral reflections and late diffuse reverberation are important for the sensations of an auditory source width and listener envelopment in a concert hall. Limited by its ability, two-channel stereophonic sound is unable to recreate the spatial information of these reflections exactly. Appropriate microphone and signal simulation techniques improve the perceived performance of reflected sound in stereophonic sound reproduction to some extent. Some psychoacoustic methods are available for recreating sensations similar to those caused by the reflections in a hall. For example, the perceived virtual source width can be controlled by introducing a small interchannel phase difference between two channel signals. In Section 1.7.3, the auditory event broadens and becomes blurred because of the reduction of positive interchannel correlation.

The aforementioned methods are applicable to represent the spatial information in two-channel stereophonic signals and then recreate various target auditory perceptions or sensations in reproduction. Some methods for a two-channel stereophonic sound are not based on strict acoustic theory. Instead, they are based on psychoacoustic experimental results, relevant experience, and requirements for practical uses. This problem is dealt with in the two-channel stereophonic recording discussed in the next section and the multichannel surround sound discussed in the succeeding chapters. It is also a feature of various spatial sound techniques based on sound field approximation and psychoacoustic principles.

2.2 MICROPHONE AND SIGNAL SIMULATION TECHNIQUES FOR TWO-CHANNEL STEREOPHONIC SOUND

Two-channel stereophonic sound is popular in consumer use. It is usually applied to reproduce music (including classical and pop music), speech, and other program materials. However, the ability of a two-channel stereophonic sound to transmit and reproduce the spatial information of a sound field is limited. The key is the manner by which this limited ability is utilized properly to transmit and reproduce the desired information (including the localization information of direct sound and the comprehensive information of reflections) essential for auditory perceptions as much as possible.

As the first stage in the system chain of a two-channel stereophonic system, signal ***recording*** or ***picking up*** involves using some appropriate microphone techniques to capture the spatial information of an on-site sound field. Signal ***simulation*** or ***synthesis*** is a process by which appropriate signal processing techniques are utilized to artificially create the desired spatial information of sound. In accordance with the basic principle of stereophonic sound discussed in Section 2.1, the spatial information of sound is encoded into two-channel stereophonic signals. Various techniques for stereophonic signals recording and simulation have been developed and can be roughly classified into four categories.

The first category is the ***coincident microphone technique***. It was developed on the basis of Blumlein's patent in the 1930s (Blumlein, 1931). A pair of spatially coincident microphones with appropriate directivity is used to capture stereophonic signals. The directivity of a microphone pair encodes the direction information of a source into two channel signals with direction-dependent ICLD only. The coincident microphone technique can be further divided into two sub-categories, i.e., XY and mid-side (MS) microphone pair techniques.

The second category is the ***spaced microphone technique***. It was first introduced in a demonstration at the Paris exhibition in 1881 (Hertz, 1981) and then developed on the basis of the works at Bell Labs in the 1930s (Keller, 1981). A pair of microphones spaced apart at a larger distance is used to capture stereophonic signals. The path difference between the source and two microphones encodes the direction information of a source into two channel signals with direction-dependent ICTD and some ICLD (ICLD can be enhanced by using a pair of directional microphones if necessary). For an off-central source, the larger path difference creates a larger ICTD that exceeds the lower limit of the precedence effect. As a result, a virtual source forms at the direction of the loudspeaker with the leading signal in reproduction.

The third category is the ***near-coincident microphone technique***. A pair of microphones with appropriate directivity and spaced apart at a smaller distance is used to capture stereophonic signals. The directivity of the microphones and the smaller path difference between a sound source and two microphones encode the direction information of the source into two-channel signals with ICLD and smaller ICTD. According to the principle of summing localization, the combination of ICLD and smaller ICTD yields a virtual source in various directions in reproduction.

The fourth category is the ***spot microphone and pan-pot technique***. Multiple microphones are used to capture a mono signal for each target source (or each set of target sources). Each (mono) source signal is split into two channel signals with a target-direction-dependent amplitude ratio by using a pan-pot or signal processing. Therefore, this technique encodes the direction information of a target source into two channel signals with artificially simulated ICLD (and usually without ICTD). This technique is often used for pop music, video–audio programming, and film recording.

In this section, microphone recording and simulation techniques for a two-channel stereophonic sound are described. Other related studies are used as references for the detailed reviews on this issue (Dooley and Streicher, 1982; Streicher and Dooley, 1985; Lipshitz, 1986; Hibbing, 1989; Julstrom, 1991; Rumsey, 2001).

2.2.1 XY microphone pair

The ***XY microphone pair*** is a sub-category of the coincident microphone technique (Clack et al., 1957; Bauer et al., 1965). A pair of identical microphones with the first-order directivity is used to record stereophonic signals. The main axes of the two microphones point to the left-front and right-front directions symmetrically in the horizontal plane with appropriate azimuths $\pm \theta_m$.

According to the general response of the first-order directional microphone, if the source directions $\Delta\alpha$ with respect to the main axes of two microphones are represented by the elevation ϕ_S and azimuth θ_S of the source, the signal amplitudes of two microphones are

$$\begin{aligned} E_L &= P_A A_{mic}\left[B_p + B_v \cos(\theta_S - \theta_m)\cos\phi_S\right] \\ E_R &= P_A A_{mic}\left[B_p + B_v \cos(\theta_S + \theta_m)\cos\phi_S\right] \\ B_v &\neq 0. \end{aligned} \tag{2.2.1}$$

where $P_A = P_A(f)$ is the amplitude of the incident wave.

For a horizontal source with $\phi_S = 0°$, Equation (2.2.1) becomes

$$E_L = P_A A_{mic}\left[B_p + B_v \cos(\theta_S - \theta_m)\right] \qquad E_R = P_A A_{mic}\left[B_p + B_v \cos(\theta_S + \theta_m)\right]. \tag{2.2.2}$$

The signals in Equation (2.2.2) are amplified and then reproduced by two loudspeakers. For convenience, the normalized amplitudes or relative gains of two channel or loudspeaker signals are adopted in the following discussion:

$$A_L = B_p + B_v \cos(\theta_S - \theta_m) \qquad A_R = B_p + B_v \cos(\theta_S + \theta_m). \tag{2.2.3}$$

The actual loudspeaker signals are obtained by multiplying the amplitude $P_A(f)$ of the incident wavefront and an overall gain factor to Equation (2.2.3).

Equation (2.2.3) indicates that the directional responses of microphones result in source-azimuth-dependent two channel signals and consequently encode the directional information of the source into ICLD. In Equation (2.2.3), B_p and B_v, or particularly $b = B_v/B_p$, specify the directivity of microphones, and the azimuthal parameter θ_m specifies the main axis directions of microphones. Various choices of these parameters result in different directional patterns in the recorded signals (note: for convenience in theoretical analysis, microphones with various first directional responses are assumed to be available, but this assumption is not the case in practice).

The perceived virtual source azimuth at low frequencies is evaluated by substituting Equation (2.2.3) into the law of sine given in Equation (2.1.6) or the law of tangent expressed in Equation (2.1.10). For a head fixed to the front orientation,

$$\sin\theta_I = \frac{B_v \sin\theta_m \sin\theta_S}{B_p + B_v \cos\theta_m \cos\theta_S} \sin\theta_0. \tag{2.2.4a}$$

For a head oriented to the virtual source,

$$\tan\hat{\theta}_I = \frac{B_v \sin\theta_m \sin\theta_S}{B_p + B_v \cos\theta_m \cos\theta_S} \tan\theta_0. \tag{2.2.4b}$$

Therefore, the perceived azimuth depends on the source azimuth θ_S, the directional parameters B_p and B_v of microphones, and the direction θ_m of the main axes of microphones. Usually, the perceived azimuth is inconsistent with that of the target source. In other words, directional distortion exists in reproduction.

From Equations (2.2.3), (2.2.4), and (2.2.4b), some basic features of signals recorded with the XY microphone pair can be observed.

1. The frontal source at $\theta_S = 0°$ always yields two channel signals with an equal amplitude $A_L = A_R$ (null ICLD) and gives rise to a virtual source at the frontal direction of $\theta_I = 0°$. As the source deviates from the frontal direction toward the left or right, the magnitude of the ipsilateral channel signal increases, whereas the magnitude of the opposite channel signal decreases. At the source direction $\theta_S = \pm\theta_m$, the left or right signal amplitude maximizes with a value of $(B_p + B_v)$.
2. For $B_p > B_v$, the microphone response consists of a more omnidirectional component than that of a bidirectional component. The resultant two channel signals are always in phase, and each channel signal never vanishes for any horizontal source azimuth within $-180° < \theta_S \leq 180°$. In this case, the virtual source lies in a region between two loudspeakers but cannot reach either of the loudspeaker (full left or full right) directions.
3. For $B_p \leq B_v$, the microphone response consists of more or equal bidirectional components as compared with that of an omnidirectional component. The left or right

signals vanish at the source azimuth of $\theta_S = \theta_p = -180° + \arccos(B_p/B_v) + \theta_m$ or $180° - \arccos(B_p/B_v) - \theta_m$, respectively, which correspond to the null point of the microphone opposite to source. The null of an opposite channel signal leads to the maximal (infinite) magnitude of ICLD and virtual source at either loudspeaker direction (full left or full right). For $B_p < B_v$, the two channel signals are out of phase when the source azimuth exceeds the above limit. This phenomenon is caused by the reverse polarity of the rear lobe of the opposite microphones. Two-channel out-of-phase signals may yield an unstable "outside-boundary virtual source" or an unnatural auditory event with an uncertain direction.

For recording with a coincident microphone pair, the source azimuth at which the opposite channel signal vanishes such that the virtual source lies at the ipsilateral loudspeaker direction is termed the ***effective recording angle*** and denoted by $\pm\theta_p$. The angular range bounded by the effective left and right recording angles is termed the ***effective recording range***, ***pick up angle***, or ***included angle*** and denoted by $[-\theta_p, +\theta_p]$ or $2\theta_p$. Within the effective recording range, when the target source direction θ_S varies from $-\theta_p$ to $+\theta_p$, the perceived virtual source direction θ_I changes from the $-\theta_0$ direction of the right loudspeaker to the $+\theta_0$ direction of the left loudspeaker. Outside the effective recording range, the virtual source direction is seriously distorted, cases of unstable "outside-boundary virtual sources" or unnatural auditory events with uncertain directions may appear. These cases are undesirable in practice. For the XY microphone pair with $B_p > B_v$, no restrictions from the effective recording angle are imposed, and the effective recording range covers the whole horizontal plane. For $B_p \le B_v$, the effective recording angle is evaluated with the following:

$$\pm\theta_p = \pm 180° \mp \arccos\left(\frac{B_p}{B_v}\right) \mp \theta_m. \tag{2.2.5}$$

The normalized overall power of two channel signals can also be analyzed from Equation (2.2.1):

$$\begin{aligned} Pow(\theta_S, \phi_S) &= A_L^2 + A_R^2 \\ &= 2B_p^2 + 2B_v^2\left(\cos^2\theta_m \cos^2\theta_S + \sin^2\theta_m \sin^2\theta_S\right)\cos^2\phi_S \\ &\quad +4B_pB_v \cos\theta_m \cos\theta_S \cos\phi_S. \end{aligned} \tag{2.2.6}$$

Generally, the overall power depends on source direction and B_p, B_v, and θ_m of the XY microphone pair. The total recorded power for some source regions may be enhanced by appropriately choosing these parameters. Furthermore, the total recorded power for other source regions may be reduced. This method is applicable to controlling the relative signal power between a frontal-direct sound component and sound components reflected from other directions. Thus, distance perception in reproduction can be controlled.

The ***front-to-back half-space pick up ratio*** is defined as the ratio between the recorded signal powers from the frontal-half space and the rear-half space and represented in decibels to evaluate the relative proportion of the recorded power of the front and rear sounds:

$$Pow_{F/B} = 10\log_{10}\frac{\int_{\Omega\in F} Pow(\theta_S, \phi_S)d\Omega}{\int_{\Omega\in B} Pow(\theta_S, \phi_S)d\Omega}(dB), \tag{2.2.7}$$

where the integrals in the numerator and denominator of Equation (2.2.7) are calculated over the solid angle in the frontal-half and rear-half spaces, respectively. The front-to-back half-space pick up ratio of the XY microphone pair is evaluated by substituting Equation (2.2.6) into Equation (2.2.7):

$$Pow_{F/B} = 10\log_{10}\frac{3B_p^2 + B_v^2 + 3B_pB_v\cos\theta_m}{3B_p^2 + B_v^2 - 3B_pB_v\cos\theta_m} \quad (dB). \tag{2.2.8}$$

For coincident microphone pair recording, the ***random energy efficient*** (**REE**) is defined as the ratio between the two-channel means of the total reverberation power from all three-dimensional directions in the resultant stereophonic signals and that of the mono signal recorded with an omnidirectional microphone with the same on-axis sensitivity. For the XY microphone pair recording given by Equation (2.2.1) with the normalized on-axis response $(B_p + B_v) = 1$,the corresponding parameters for the omnidirectional microphone are $B_p = 1$ and $B_v = 0$, the total reverberation power of the omnidirectional microphone signal is calculated with an integral over the solid angle in a full-dimensional space, resulting in 4π. Then,

$$REE = \frac{1}{4\pi}\int_{\Omega}\frac{\left(A_L^2 + A_R^2\right)}{2}d\Omega = B_p^2 + \frac{1}{3}B_v^2 = 1 - 2B_v + \frac{4}{3}B_v^2. \tag{2.2.9}$$

The ***distance factor*** (**DF**) of recording is defined as

$$DF = \frac{1}{\sqrt{REE}}. \tag{2.2.10}$$

DF represents the distance from the microphone at which an on-axis source yields the same recorded reverberation power as that of a closer (mono) omnidirectional microphone. The smaller *REE*, the larger *DF*, i.e., the less reverberation power is recorded. An XY microphone pair with a smaller *REE* should be arranged at a more distant distance from the sound source to keep a constant reverberation power in the recorded signals.

In summary, the virtual source direction in reproduction, effective recording range, and DF are attributes that specify the performance of recording with an XY microphone pair. These attributes are closely related to B_p, B_v, and main axis orientations θ_m of microphones and should be chosen through comprehensive considerations.

1. According to Equation (2.2.4a) or (2.2.4b), for a given (B_v/B_p), when the target source direction lies within the effective recording range, increasing the azimuth θ_m of the main axis orientation of the microphone pair increases the perceived virtual source azimuth θ_I or $\hat{\theta}_I$ and thus broadens the width of the stereophonic stage in reproduction. At the same time, increasing the azimuth θ_m reduces the effective recording angle θ_p, as evaluated with Equation (2.2.5).
2. According to Equations (2.2.4a) and (2.2.4b), for a given θ_m, when the target source direction lies within the effective recording range, increasing B_v/B_p, which is the relative proportion of the bidirectional component in the microphone response, increases the perceived virtual source azimuth and consequently broadens the width of the stereophonic stage in reproduction. At the same time, increasing (B_v/B_p) reduces the effective recording angle θ_p, as evaluated with Equation (2.2.5).

3. According to Equation (2.2.8), $Pow_{F/B}$ can be controlled by changing the directional parameters B_p and B_v and the main axis orientation θ_m of the microphone pair. For B_p, $B_v \neq 0$ (neither omnidirectional nor bidirectional microphones), $\theta_m < 90°$ yields $Pow_{F/B} > 0$ dB. Therefore, the microphone pair with the main axes pointing to the frontal–horizontal plane captures more sound power from the front-half space than that from the rear-half space. By contrast, $\theta_m > 90°$ yields $Pow_{F/B} < 0$ dB. Therefore, the microphone pair with the main axes pointing to the rear-horizontal plane captures more sound power from the rear-half space than that from the front-half space. The microphone pair with the main axes pointing to the two sides of $\theta_m = \pm 90°$ captures equal sound powers from the frontal and rear-half spaces with $Pow_{F/B} = 0$ dB.
4. From Equations (2.2.9) and (2.2.10), the main axis orientation θ_m of the microphone pair do not influence REE and DF. However, if $B_p + B_v = 1$, *REE* and *DF* can be controlled by adjusting B_v.

As the first example of the XY microphone technique, the two channel signals are captured by a pair of bidirectional microphones whose main axes point to the horizontal directions of $\theta_m = \pm 45°$. This example is the case of $B_p < B_v$ in Equation (2.2.3) when the proportion B_p of the omnidirectional component vanishes. If $B_p = 0$ and $B_v = 1$, then the normalized amplitudes of two channel signals are given as

$$A_L = \cos\left(\theta_S - 45^\circ\right) \qquad A_R = \cos\left(\theta_S + 45^\circ\right). \tag{2.2.11}$$

Figure 2.7 (a) illustrates the polar pattern or variation in the normalized amplitude of two channel signals with the source azimuth θ_S. The perceived source azimuth at low frequencies is evaluated with Equations (2.2.4a) and (2.2.4b):

$$\sin\theta_I = \tan\theta_S \sin\theta_0, \tag{2.2.12a}$$

$$\tan\hat{\theta}_I = \tan\theta_S \tan\theta_0. \tag{2.2.12b}$$

From Figure 2.7(a), Equations (2.2.5), (2.2.11), (2.2.12a) and (2.2.12b), and the discussion above, the following results are obtained:

1. A frontal target source at $\theta_S = 0°$ yields $A_L = A_R = 0.707$. In this case, the signal level decreases by −3 dB compared with that of the maximal on-axis output level of the microphones. The effective recording angle is $\theta_p = \pm 45°$, which is consistent with the main axis orientations $\pm\theta_m$ of the microphones. As the target source direction varies from $-45° \leq \theta s \leq 45°$, the virtual source direction changes from the direction of the right loudspeaker (full right) to the direction of the left loudspeaker (full left) and consequently follows the source direction. Figure 2.7 (b) illustrates the virtual source direction at low frequencies evaluated from Equation (2.2.12a) for various half-span angles of loudspeakers. For a standard loudspeaker configuration with $\theta_0 = 30°$ or $2\theta_0 = 60°$, the virtual source direction θ_I is biased toward the frontal direction in comparison with the target source direction θ_S. Therefore, some directional distortions in the virtual source exist. Similar results are obtained from Equation (2.2.12b).
2. For the target source within the rear quadrant between $\theta_S = -135°$ and $\theta_S = 135°$, the amplitude of both signals exhibits a negative (reversal) polarity because of the out-of-phase response in the rear lobe of the microphone. However, the two channel signals

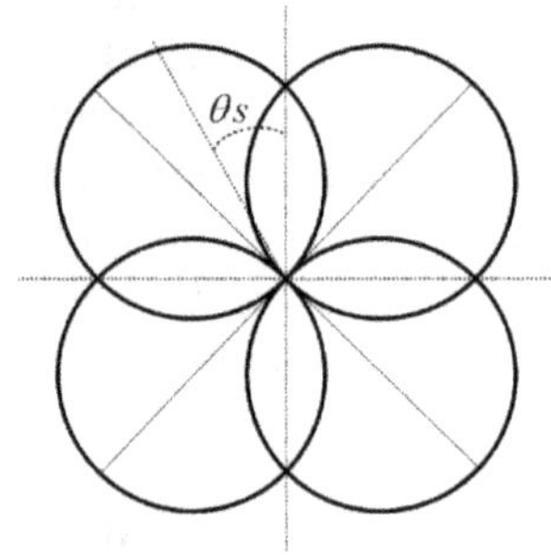

(a) Polar pattern

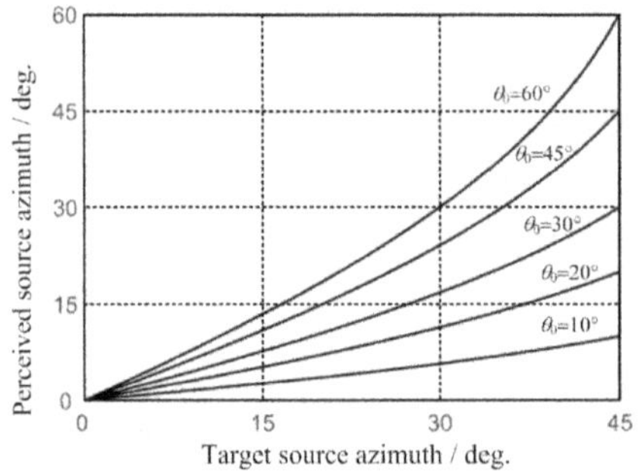

(b) Virtual source direction at low frequencies for various half-span angles θ_0 of loudspeaker arrangement

Figure 2.7 XY recording with a pair of bidirectional microphones whose main axes point to θ_m = ±45°: (a) polar pattern; (b) virtual source direction at low frequencies for various half-span angles θ_0 of loudspeaker arrangement.

are in phase with each other. In this case, the virtual source in reproduction lies in the frontal–horizontal plane with a left–right reversal. The detailed results can be derived in a manner similar to those in (1).

3. For the target source within the left lateral quadrant of 45° < θ_S < 135°, the amplitude of the left and right microphone signals exhibits positive and negative polarities, respectively, resulting in two-channel out-of-phase signals A_L and A_R. According to Equation (2.2.12a), within the left lateral region of 45° < θ_S < 90°, it has $\theta_I > \theta_0$ when $\theta_S \leq$ arctan(1/sinθ_0). In this case, an (unstable) outside-boundary virtual source appears. When $\theta_S >$ arctan(1/sinθ_0), it has sin$\theta_I > 1$; thus, an unnatural auditory event with an uncertain direction occurs. The virtual source position for the target region of 90° < θ_S < 135° can be analyzed similarly. The resultant outside-boundary virtual source (if existing) is mapped into the frontal–horizontal plane with left-right reversal. The analysis of the right lateral quadrant of −135° < θ_S < −45° is similar to that of the left lateral quadrant and omitted here.

In addition, according to Equation (2.2.11), for all horizontal sources with ϕ_S = 0°, the total power of two channel signals remains constant and independent of the target source azimuth:

$$Pow(\theta_S, \phi_S) = A_L^2 + A_R^2 = 1. \tag{2.2.13}$$

If $B_p = 0$ and $B_v = 1$, then Equation (2.2.8) yields $Pow_{F/B} = 0$ dB; Equations (2.2.9) and (2.2.10) yield $DF = \sqrt{3}$.

The XY recording with a pair of bidirectional microphones records the sound signals (waves) from all horizontal directions with constant overall power output and can capture the frontal direct sound and lateral and back reflections uniformly and unbiasedly. When the target source lies within the effective recording range of [−45°, +45°], it can recreate a virtual source in a roughly correct direction in reproduction. Fortunately, lateral and rear sounds in a hall are usually reflections. Accurately localizing these reflections in reproduction is usually unnecessary. These reflections are perceived to be coming from an appropriate spatial region rather than from a point or a narrow region. Actually, reflections in a hall can be modeled as sound waves generated by a series of "image sources" in different directions, with different delays and random phases. When the signals recorded by an XY bidirectional microphone pair are reproduced with a pair of stereophonic loudspeakers, target "image sources" within the rear quadrant bounded by $\theta_S = -135°$ and 135° are mapped into the left–right-reversal virtual sources within the frontal region bounded by two loudspeakers. In addition, the lateral "image sources" create two-channel out-of-phase signals, resulting in the formation of outside-boundary virtual sources or the occurrence of auditory events with an uncertain direction in reproduction. However, human hearing is unable to localize individual "image sources" of reflections separately. Overall, this microphone recording method can recreate an appropriate localization effect of the frontal virtual source and comprehensive perceived effects of reflections to some extent. However, the two channel out-of-phase signals are canceled when they are mixed together. Therefore, they are incompatible with the mono downmixing.

As the second example of the XY microphone technique, the two channel signals are captured by a pair of cardioid microphones. As shown in Figure 2.8(a), the main axes of two microphones point to the horizontal azimuths of $\theta_m = \pm 45°$, respectively (usually $45° \leq \theta_m \leq 90°$). If $B_p = B_v = 0.5$ in Equation (2.2.3), the normalized amplitudes of two channel signals are expressed as

$$A_L = 0.5\left[1+\cos(\theta_S-\theta_m)\right] \qquad A_R = 0.5\left[1+\cos(\theta_S+\theta_m)\right]. \tag{2.2.14}$$

According to Equations (2.2.4a) and (2.2.4b), the virtual source direction at low frequencies is calculated with

$$\sin\theta_I = \frac{\sin\theta_m \sin\theta_S}{1+\cos\theta_m \cos\theta_S}\sin\theta_0, \tag{2.2.15a}$$

$$\tan\hat{\theta}_I = \frac{\sin\theta_m \sin\theta_S}{1+\cos\theta_m \cos\theta_S}\tan\theta_0. \tag{2.2.15b}$$

Equations (2.2.5), (2.2.14), and (2.2.15a) and Figure 2.8 reveal the following:

1. The two channel signals are in phase for all horizontal target source azimuths. Therefore, they are compatible with mono downmixing.
2. The effective recording angle is $\theta_p = 180° - \theta_m$. When the main axes of microphones point to the frontal–horizontal quadrants with $\theta_m < 90°$, the effective recording angle extends to the rear-horizontal quadrant with $\theta_p > 90°$ and is not consistent with θ_m. The main axis direction of each microphone does not match with the null point direction of the opposite microphone. In the left-half horizontal plane, as the target source azimuth increases

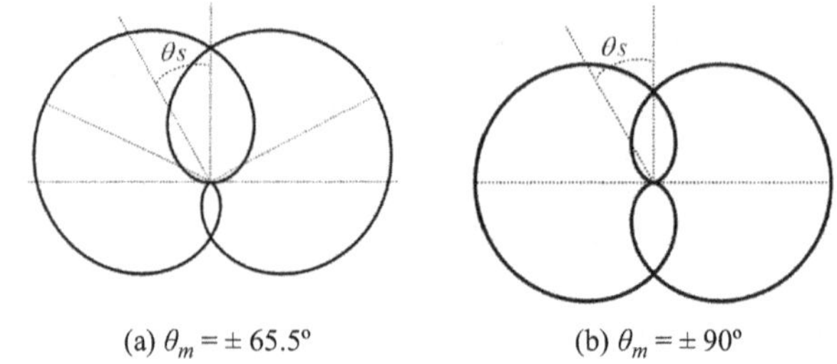

Figure 2.8 Polar pattern of the XY recording with a pair of cardioid microphones (a) $\theta_m = \pm 65.5°$, (b) $\theta_m = \pm 90°$.

from the front of $\theta_S = 0°$, ICLD between two channel signals increases from 0 dB and maximizes (becomes infinity) at $\theta_S = \theta_p$. Accordingly, the virtual source position moves from the front to the direction of the left loudspeaker (full left). As θ_S increases further and exceeds θ_p, ICLD decreases and becomes 0 dB at the back of $\theta_S = 180°$. Accordingly, the virtual source position varies from the direction of the left loudspeaker to the front. The virtual source in the right-half horizontal plane can be analyzed similarly through left–right symmetry. Therefore, the target sources within the effective recording range are mapped or compressed into the virtual sources within the region bounded by two loudspeakers. The target sources within rear ranges of $[-180°, -180° + \theta_p]$ and $[180° - \theta_p, 180°]$ are reflected and mapped into the virtual source (without left–right reversal) in the frontal–horizontal plane within the region bounded by two loudspeakers.

3. For $\theta_m = 90°$, the two microphones are arranged back-to-back, and the main axis direction of each microphone is consistent with the null point direction of the opposite microphone, as shown in Figure 2.8 (b). In reproduction, the target sources in the frontal–horizontal quadrants are mapped or compressed into the virtual sources within the frontal region bounded by two loudspeakers. The target sources in the rear-horizontal quadrants are reflected and compressed into the virtual source within the frontal region bounded by two loudspeakers.

In addition, the overall power of two channel signals is evaluated by Equation (2.2.6):

$$\begin{aligned} Pow(\theta_S, \phi_S) &= A_L^2 + A_R^2 \\ &= 0.5\left[1 + \left(\cos^2\theta_m \cos^2\theta_S + \sin^2\theta_m \sin^2\theta_S\right)\cos^2\phi_S + 2\cos\theta_m \cos\theta_S \cos\phi_S\right]. \end{aligned} \tag{2.2.16}$$

For a horizontal source with $\phi_S = 0°$, the overall power varies with the target source azimuth. The DF calculated from Equations (2.2.9) and (2.2.10) is $DF = \sqrt{3}$. The front-to-back half-space pick up ratio calculated from Equation (2.2.8) is

$$Pow_{F/B} = 10\log_{10}\frac{4 + 3\cos\theta_m}{4 - 3\cos\theta_m}. \tag{2.2.17}$$

Overall, the XY recording with a pair of cardioid microphones leads to some virtual source direction distortions in reproduction. The main axis orientation θ_m of the microphone pair is an important parameter related to the recording performance:

1. The distribution of the virtual source (the width of the stereophonic stage in reproduction) is determined by θ_m. A larger θ_m corresponds to a wider distribution of virtual sources. The distribution of the virtual source for recording with a pair of cardioid microphones is narrower than that for recording with a pair of bidirectional microphones because the proportion of the bidirectional component of the responses of a cardioid microphone pair, e.g., $b = B_v/B_p$ in Equation (2.2.3), is smaller than that of the bidirectional microphone pair. This phenomenon can also be observed from Equation (2.2.4a).
2. According to Equation (2.2.17), when $\theta_m < 90°$, the proportion of the recorded power for the front sound is larger than that for the rear sound. This difference increases as θ_m decreases. For example, θ_m of 45° and 65.5° have $Pow_{F/B}$ of 5.1 and 2.8 dB, respectively. For music recording in a hall, when microphones are arranged far from the sound sources, microphone orientation with $\theta_m < 90°$ is chosen to reduce the captured power of rear reflections. In this case, the captured power ratio between the direct and reflected sounds increases, so an excessive perceived distance in reproduction is avoided. This characteristic also distinguishes the XY recording with a cardioid microphone pair from that with a bidirectional microphone pair.
3. Excessive θ_m reduces the relative recorded power for a frontal target source.

In practical uses, θ_m should be chosen appropriately by comprehensively considering various factors. A larger θ_m for cardioid microphone recording, typically $\theta_m = 65.5°$, is selected to obtain a virtual source distribution in reproduction similar to that of XY recording with a bidirectional microphone pair. In this case, the signal amplitude for a frontal source at $\theta_S = 0°$ decreases by −3 dB compared with the maximal on-axis amplitude of the microphones.

As the third example of the XY microphone technique, two channel signals are recorded by a pair of supercardioid microphones (Figure 2.9). θ_m of recording with the supercardioid microphone pair is usually chosen between 45° and 65.5° because the directional response of a supercardioid microphone lies between those of a cardioid microphone and a bidirectional microphone. The distribution of the virtual source for signals recording with a supercardioid microphone pair is similar to that with a bidirectional microphone pair. Substituting $B_p = 0.37$ and $B_v = 0.63$ of supercardioid microphones into Equation (2.2.8) yields $Pow_{F/B} = 4.7$ dB for $\theta_m = 55°$. A supercardioid microphone pair captures relatively less power of rear reflections than a bidirectional microphone pair does because the magnitude of the rear lobe of a supercardioid microphone is smaller than that of a bidirectional microphone. For a given direct-to-reverberation ratio of the recorded signals, a supercardioid microphone pair allows a more distant arrangement from a target source than in the case of bidirectional microphones.

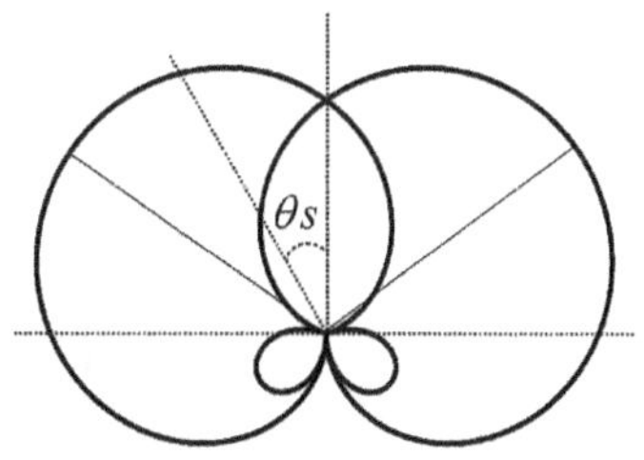

Figure 2.9 Polar pattern of XY recording with a supercardioid microphone pair.

2.2.2 MS transformation and the MS microphone pair

As stated in Section 2.2.1, A_L and A_R are also used to denote the two normalized channel signals. Complete two channel signals are obtained by multiplying the normalized signals with the waveform function in the frequency domain, i.e., $E_L = A_L E_A(f)$, and $E_R = A_R E_A(f)$.

Signals A_L and A_R are independent. Applying a ***sum–difference*** or ***mid-side (MS) transformation*** to A_L and A_R results in the (normalized) sum signal A_M and the difference signal A_S:

$$A_M = \kappa_\mu (A_L + A_R) \qquad A_S = \kappa_\mu (A_L - A_R), \tag{2.2.18}$$

where $\kappa_\mu > 0$ is a coefficient. A_M and A_S are also two independent normalized signals. The complete signals should be $E_M = A_M E_A(f)$ and $E_S = A_S E_A(f)$, which are usually denoted by M and S in other studies. The two original channel signals A_L and A_R can also be derived from A_M and A_S, i.e., from the inverse MS transformation of Equation (2.2.18):

$$A_L = \kappa_\nu (A_M + A_S) \qquad A_R = \kappa_\nu (A_M - A_S) \qquad \kappa_\nu = \frac{1}{2\kappa_\mu}. \tag{2.2.19}$$

From the point of linear algebra, (A_L, A_R) and (A_M, A_S) are two sets of independent variables. MS transformation is a linear transformation from one set of independent variables to another set of independent variables. Equations (2.2.18) and (2.2.19) can be written in a matrix form:

$$\begin{bmatrix} A_M \\ A_S \end{bmatrix} = \kappa_\mu \begin{bmatrix} 1 & 1 \\ 1 & -1 \end{bmatrix} \begin{bmatrix} A_L \\ A_R \end{bmatrix} \quad \begin{bmatrix} A_L \\ A_R \end{bmatrix} = \kappa_\upsilon \begin{bmatrix} 1 & 1 \\ 1 & -1 \end{bmatrix} \begin{bmatrix} A_M \\ A_S \end{bmatrix}. \tag{2.2.20}$$

Therefore, a pair of (A_M, A_S) signals is completely equivalent to a pair of (A_L, A_R) signals. Accordingly, in two-channel stereophonic recording, a pair of (A_M, A_S) signals can be captured by a pair of microphones with appropriate directivities and then converted into a pair of (A_L, A_R) signals through inverse MS transformation. This principle is the basic foundation of ***MS microphone pair recording technique***. This technique, which was derived from Blumlein's patent (Blumlein, 1931), further developed, and widely used (Hibbing, 1989), yields two channel signals with direction-dependent ICLD only. An MS matrix (and inverse matrix) was traditionally implemented via an analog circuit and is now easily implemented through digital signal processing.

Some notations are provided for the choice of κ_μ or κ_ν for MS transformation. If the original captured signals are (A_L, A_R), they are converted into signals (A_M, A_S) by using Equation (2.2.18). If signals (A_L, A_R) are uncorrelated, the superimposed signals obtain a power gain of +3 dB. Therefore, a coefficient of $\kappa_\mu = \kappa_\nu = \sqrt{2}/2$ is selected to supplement a −3 dB attenuation in the power of superimposed signals. In this case, the matrix of inverse MS transformation is exactly identical to that of MS transformation, i.e., MS transformation is a mathematically orthogonal transformation that maintains a constant overall power of signals after transformation:

$$A_M^2 + A_S^2 = A_L^2 + A_R^2. \tag{2.2.21}$$

If signals (A_L, A_R) are completely correlated, the superimposed signals achieve a magnitude gain of +6 dB. $\kappa_\mu = 1/2$ should be chosen to supplement a −6 dB attenuation in the magnitude of superimposed signals to avoid overload. The actual signals (A_L, A_R) may be partly correlated, so $\kappa\mu$ between 1/2 and $\sqrt{2}/2$ or attenuation between –6 and –3 dB is usually chosen.

Similarly, if the original captured signals are (A_M, A_S) and converted into signals (A_L, A_R) through inverse MS transformation in Equation (2.2.19), κ_ν between 1/2 and $\sqrt{2}/2$ is chosen. For example, if κ_ν= 1/2 is chosen, MS and its inverse transformation are given as

$$A_M = A_L + A_R \qquad A_S = \left(A_L - A_R\right), \tag{2.2.22}$$

$$A_L = \frac{1}{2}\left(A_M + A_S\right) \qquad A_R = \frac{1}{2}\left(A_M - A_S\right). \tag{2.2.23}$$

The MS transformation in the form of Equations (2.2.22) and (2.2.23) have been used in some studies. The only difference between various forms of MS transformation is a normalized coefficient that does not alter the characteristics of transformation. The ambiguousness caused by the normalized coefficient can be avoided through a careful observation.

Each MS microphone pair has an equivalent XY microphone pair and vice versa. The general equation of the directional responses of an MS microphone pair can be derived from the general equation of the directional responses of an XY microphone pair expressed in Equation (2.2.1). The following equations are obtained by substituting Equation (2.2.1) into Equation (2.2.18), and normalizing $P_A A_{mic}$ to a unit value for convenience:

$$A_M = 2\kappa_\mu\left(B_p + B_\nu \cos\theta_m \cos\theta_S \cos\phi_S\right) \qquad A_S = 2\kappa_\mu B_\nu \sin\theta_m \sin\theta_S \cos\phi_S, \tag{2.2.24}$$

or

$$A_M = B_p' + B_\nu' \cos\theta_S \cos\phi_S \qquad A_S = B_\nu'' \sin\theta_S \cos\phi_S, \tag{2.2.25}$$

where

$$B_p' = 2\kappa_\mu B_p \qquad B_\nu' = 2\kappa_\mu B_\nu \cos\theta_m \qquad B_\nu'' = 2\kappa_\mu B_\nu \sin\theta_m. \tag{2.2.26}$$

For a horizontal source with ϕ_S = 0°, Equation (2.2.25) becomes

$$A_M = B_p' + B_\nu' \cos\theta_S \qquad A_S = B_\nu'' \sin\theta_S. \tag{2.2.27}$$

Equation (2.2.25) indicates that the sum signal A_M can be captured by arbitrary microphones with zeroth- and first-order directional responses (from omnidirectional to bidirectional). If a directional microphone is used, its main axis should point to the frontal-center direction θ = 0°. The difference signal A_S is always captured by a bidirectional microphone with its main axis pointing to the left side direction with θ = 90°. Various MS directional responses can be derived by choosing different microphone parameters, such as B'_p, B'_ν, and B''_ν in equivalent XY directional responses.

The aforementioned conclusion can be drawn from the left-right symmetry of stereophonic signals. Three independent directional components in zeroth- and first-order microphone responses include an omnidirectional component and two bidirectional components with the main axes of the positive lobe pointing to frontal-center and left-lateral directions, respectively. If the maximal amplitude of these independent components is normalized to a unit, the three independent components can be written as 1, cosθ_S, and sinθ_S. Here, the first and second components are left-right symmetric, i.e., they are invariant against the left-right reflection with θ changing into $-\theta$. The third component is left-right antisymmetric, i.e., it becomes $-\sin\theta_S$ after the left-right reflection. According to the definition in Equation (2.2.18), A_M should be left-right symmetric so that it is a linear combination of the symmetric components 1 and $\cos\theta_S$, yielding the left equation in Equation (2.2.27). A_S should be left–right antisymmetric so that it only includes the component of $\sin\theta_S$. It is the antisymmetry of A_S that results in ICLD and causes the virtual source to depart from the frontal direction.

The direction of a virtual source can be analyzed on the basis of MS signals. Through the MS transformation in Equation (2.2.18), the law of sine in Equation (2.1.6) becomes

$$\sin\theta_I = \frac{A_S}{A_M}\sin\theta_0. \tag{2.2.28a}$$

Furthermore, the law of tangent in Equation (2.2.10) becomes

$$\tan\hat{\theta}_I = \frac{A_S}{A_M}\tan\theta_0. \tag{2.2.28b}$$

Equation (2.2.28a) indicates the following:

1. For a given span angle between loudspeakers, $\sin\theta_I$ is directly proportional to the A_S/A_M ratio. A greater magnitude of A_S/A_M corresponds to a larger deviation of the virtual source from the frontal-center direction.
2. A pair of in-phase left and right signals yields $|A_S| \le |A_M|$ and $|\sin\theta_I| \le |\sin\theta_0|$. In this case, a virtual source lies between two loudspeakers. When A_S/A_M changes from -1 to 0 and $+1$, θ_I changes from the direction of the right loudspeaker (full-right $-\theta_0$) to the central–front (0°) and the left loudspeaker (full-left θ_0).
3. A pair of out-of-phase left and right signals yields $|A_S| > |A_M|$ and $|\sin\theta_I| > |\sin\theta_0|$. If $|A_S/A_M| \le 1/\sin\theta_0$, an unstable outside-boundary virtual source appears (Section 2.1.3). If $|A_S/A_M| > 1/\sin\theta_0$, then $|\sin\theta_I| > 1$, and an unnatural auditory event with an uncertain direction appears.

Equation (2.2.28b) yields similar results. Therefore, the width of virtual source distribution in stereophonic reproduction can be controlled by altering the ratio between A_S and A_M signals. A_S includes more lateral sound components, and A_M covers more frontal sound components. Increasing A_S/A_M broadens the virtual source distribution and vice versa. The virtual source direction in reproduction can also be evaluated by substituting A_M and A_S in Equation (2.2.27) into Equations (2.2.28a) and (2.2.28b):

$$\sin\theta_I = \frac{B_v''\sin\theta_S}{B_p' + B_v'\cos\theta_S}\sin\theta_0, \tag{2.2.29a}$$

$$\tan\hat{\theta}_I = \frac{B_v''\sin\theta_S}{B_p' + B_v'\cos\theta_S}\tan\theta_0. \tag{2.2.29b}$$

These two equations are equivalent to Equations (2.2.4a) and (2.2.4b).

A_R or A_L vanishes when $A_S = \pm A_M$. Substituting this condition into Equation (2.2.27) yields the effective recording angle of an MS pair, which is equivalent to Equation (2.2.5):

$$\pm\theta_p = \pm 180° \mp \arccos\left(\frac{B'_p}{\sqrt{B'^2_v + B''^2_v}}\right) \mp \theta_m \qquad \theta_m = \arctan\left(\frac{B''_v}{B'_v}\right). \tag{2.2.30}$$

Some attributes of recording can also be evaluated in terms of the parameters of MS pairs. The overall power of A_L and A_R is evaluated with the following expression:

$$\begin{aligned} Pow(\theta_S, \varphi_S) &= A_L^2 + A_R^2 \\ &= \frac{1}{2\kappa_\mu^2}\left(B'^2_p + 2B'_p B'_v \cos\theta_S \cos\phi_S + B'^2_v \cos^2\theta_S \cos^2\phi_S + B''^2_v \sin^2\theta_S \cos^2\phi_S\right). \end{aligned} \tag{2.2.31}$$

The front-to-back half-space pick up ratio is examined with the following:

$$Pow_{F/B} = 10\log_{10}\frac{3B'^2_p + B'^2_v + B''^2_v + 3B'_P B'_v}{3B'^2_p + B'^2_v + B''^2_v - 3B'_P B'_v}(dB). \tag{2.2.32}$$

The REE is evaluated with

$$REE = \frac{1}{4\kappa_\mu^2}\left[B'^2_p + \frac{1}{3}\left(B'^2_v + B''^2_v\right)\right]. \tag{2.2.33}$$

According to the aforementioned results, among the parameters of the MS pair, increasing B''_v (relative magnitude of the S microphone output) or decreasing B'_p and B'_v (relative magnitude of the M microphone output) broadens the distribution of the frontal virtual source but narrows the effective recording range in the front. The front-to-back half-space pick-up ratio and DF for an MS pair can be analyzed similarly. The results are identical to those of the XY-equivalent pair and hence omitted here.

The performance of XY recording is controlled by initially changing B_p, B_v, and the main axis orientation θ_m of the microphone pair and then altering A_S/A_M. However, the performance of the equivalent MS recording can be controlled by changing the microphone parameters

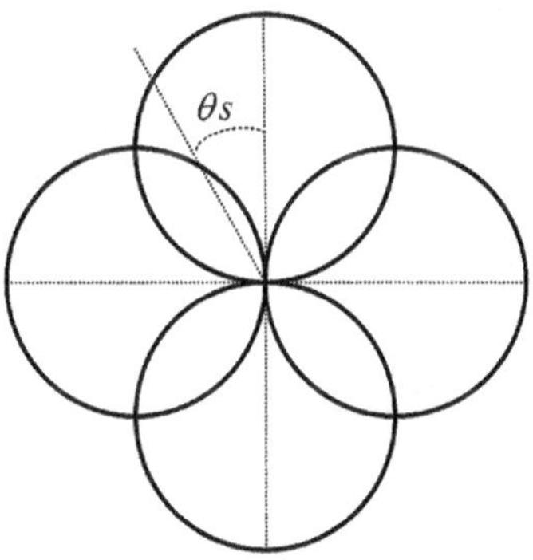

Figure 2.10 Polar patterns of MS recording with a pair of bidirectional microphones whose main axes point to the front and left directions.

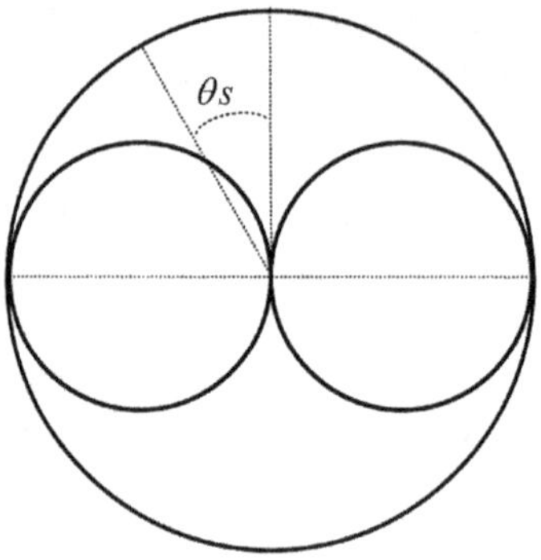

Figure 2.11 Polar pattern of MS recording with an omnidirectional microphone and a bidirectional microphone whose main axis points to the horizontal-left direction.

B'_p, B'_v, and B''_v or by modifying the relative proportion in the mixing of A_M and A_S signals. Changing the main axis orientations of microphone pairs is unnecessary. Therefore, this technique is convenient for practical uses.

As the first example of the MS microphone technique, signals are captured by a pair of bidirectional microphones whose main axes point to the horizontal–front and left direction (Figure 2.10). If $B'_p = 0$ and $B'_v = B''_v = 1$ in Equation (2.2.27), then the normalized amplitude of the MS signals created by a horizontal source at azimuth θ_S is given by

$$A_M = \cos\theta_S \qquad A_S = \sin\theta_S. \tag{2.2.34}$$

After the inverse transformation of Equation (2.2.19) with $\kappa_v = \sqrt{2}/2$, the following equation is obtained:

$$A_L = \cos\left(\theta_S - 45^\circ\right) \qquad A_R = \cos\left(\theta_S + 45^\circ\right). \tag{2.2.35}$$

Equation (2.2.35) is identical to Equation (2.2.11). Therefore, this MS pair is equivalent to an XY pair with bidirectional microphones pointing at ±45° directions, and the resultant virtual source direction is identical to that expressed in Equations (2.2.12a) and (2.2.12b). Here, the MS or its inverse transformation is equivalent to the rotation of a bidirectional microphone pair around the vertical axis with an angle of 45°. The MS transformation with $\kappa_v = \sqrt{2}/2$ is a linear orthogonal transformation geometrically equivalent to spatial rotation.

As the second example of the MS microphone technique, the signals are captured by an omnidirectional microphone and a bidirectional microphone whose main axis points to the horizontal-left direction (Figure 2.11). If $B'_v = 0$, $B'_p = A_{mic,M}$, and $B''_v = A_{mic,S}$ in Equation (2.2.27), the normalized amplitude of the MS signals created by a horizontal source at azimuth θ_S is given by

$$A_M = A_{mic,M} \qquad A_S = A_{mic,S}\sin\theta_S. \tag{2.2.36}$$

Suppose that the on-axis responses of two microphones are identical and normalized as $A_{mic,M} = A_{mic,S} = \sqrt{2}/2$. After the inverse MS transformation of Equation (2.2.19) with $\kappa_v = \sqrt{2}/2$, the following equation is obtained:

$$A_L = 0.5\left[1+\cos\left(\theta_S - 90^\circ\right)\right] \qquad A_R = 0.5\left[1+\cos\left(\theta_S + 90^\circ\right)\right]. \tag{2.2.37}$$

Equation (2.2.37) is identical to Equation (2.2.14) with $\theta_m = 90°$. Therefore, this MS pair is equivalent to an XY pair with cardioid microphones pointing at ±90°.The resultant virtual source direction is also identical to that given by Equations (2.2.15a) and (2.2.15b) with $\theta_m = 90°$:

$$\sin\theta_I = \sin\theta_S \sin\theta_0, \tag{2.2.38a}$$

$$\tan\hat{\theta}_I = \sin\theta_S \tan\theta_0. \tag{2.2.38b}$$

Equation (2.2.38a) yields $\sin\theta_I < \sin\theta_S$. Therefore, for a fixed head orientation to the front, the virtual source distribution narrows compared with that of the target source.

The virtual source distribution can be broadened by increasing the proportion of A_S signals so that the maximum A_S/A_M increases from 1 to $1/\sin\theta_0 > 1$. In this case, A_M and A_S are given as

$$A_M = \frac{\sqrt{2}\sin\theta_0}{1+\sin\theta_0} \qquad A_S = \frac{\sqrt{2}}{1+\sin\theta_0}\sin\theta_S. \tag{2.2.39}$$

The low-frequency virtual source direction for a fixed head oriented to the front is evaluated with Equations (2.2.39) and (2.2.28):

$$\sin\theta_I = \sin\theta_S \tag{2.2.40}$$

Theoretically, Equation (2.2.40) yields $\theta_I = \theta_S$ for a target source in the frontal-half-horizontal plane. Therefore, the virtual source direction exactly matches that of the target source. For a target source in the rear-half-horizontal plane, the virtual source is mapped or reflected to the mirror direction in the frontal-half-horizontal plane. However, this phenomenon is only a theoretical result. In actual situations, in the frontal-half-horizontal plane, the virtual source in reproduction is stable only when the target source lies within the effect recording range of $-\theta_0 \le \theta s \le \theta_0$. For a frontal–horizontal target source outside this range, the resultant left and right channel signals are out of phase, resulting in an unstable outside-boundary virtual source.

For example, when the standard stereophonic loudspeaker configuration has a span angle of $2\theta_0 = 60°$, $1/\sin\theta_0 = 2$ and Equation (2.2.39) becomes

$$A_M = \frac{\sqrt{2}}{3} \qquad A_S = \frac{2\sqrt{2}}{3}\sin\theta_S. \tag{2.2.41}$$

The equivalent A_L and A_R are

$$A_L = \frac{1}{3}\left(1+2\sin\theta_S\right) \qquad A_R = \frac{1}{3}\left(1-2\sin\theta_S\right). \tag{2.2.42}$$

This condition is the case of Equation (2.2.3) with $B_p = 1/3$, $B_v = 2/3$, and $\theta_m = 90°$, i.e., the case of XY recording with a pair of supercardioid-like microphones whose main axes point to ±90° in horizontal planes. It is simply a negative rear lobe in the opposite microphone of

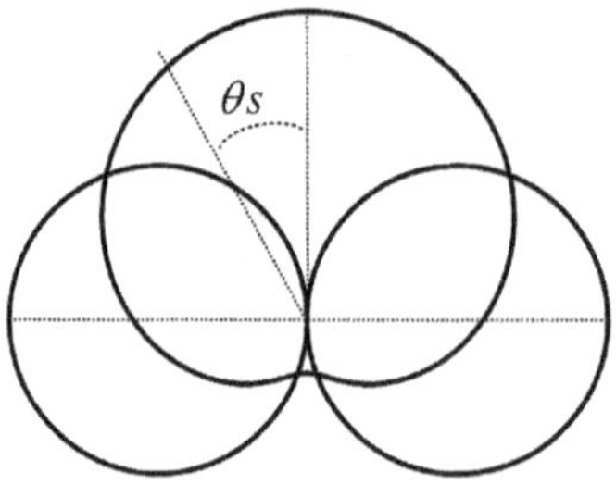

Figure 2.12 Polar pattern of MS recording with a subcardioid-like microphone and a bidirectional microphone whose main axes point to the frontal-center and horizontal-left directions, respectively (the maximal output amplitudes of two microphones are normalized to a unit).

an XY pair that results in two channel out-of-phase signals and an unstable outside-boundary virtual source.

The normalized overall power of two channel signals for a horizontal source with $\phi_S = 0°$ can be evaluated with Equation (2.2.42):

$$Pow(\theta_S, 0^\circ) = A_L^2 + A_R^2 = \frac{2}{9}(1 + 4\sin^2\theta_S). \tag{2.2.43}$$

Therefore, the overall signal power of a lateral source at $\theta_S = \pm 90°$ is five times (a 7.0 dB increase) that of a frontal-center source. For recording music in a hall, the sum signal includes more frontal-direct sound components, and difference signal consists of more lateral reflection components. In addition to broadening the distribution of a virtual source, increasing A_S/A_M increases the relative proportion of the lateral reflection power. If an MS pair specified in Equation (2.2.41) or an equivalent XY pair expressed in Equation (2.2.42) is used to record frontal-direct sounds and reflections, the balance between the power of direct and reflected components in recorded signals is spoiled, creating a distant auditory sensation in reproduction. Although this result is obviously inappropriate, this method may be applied to record lateral reflections and obtain good spaciousness in reproduction.

As the third example of the MS microphone technique, signals are recorded by a subcardioid-like microphone and a bidirectional microphone whose main axes point to the frontal-center and horizontal-left directions, respectively (Figure 2.12). If $B'_p = \sqrt{2}/2$, $B'_v = 1/2$ and $B''_v = 1/2$ in Equation (2.2.27), then the two microphone signals are expressed as follows:

$$A_M = \frac{\sqrt{2}}{2}\left(1 + \frac{\sqrt{2}}{2}\cos\theta_S\right) \qquad A_S = \frac{1}{2}\sin\theta_S. \tag{2.2.44}$$

After the inverse MS transformation is given by Equation (2.2.19) with $\kappa_v = \sqrt{2}/2$, the following equation is obtained:

$$A_L = 0.5\left[1 + \cos(\theta_S - 45^\circ)\right] \qquad A_R = 0.5\left[1 + \cos(\theta_S + 45^\circ)\right]. \tag{2.2.45}$$

This expression is equivalent to the XY recording in Equation (2.2.14) by a pair of cardioid microphones with $\theta_m = 45°$.

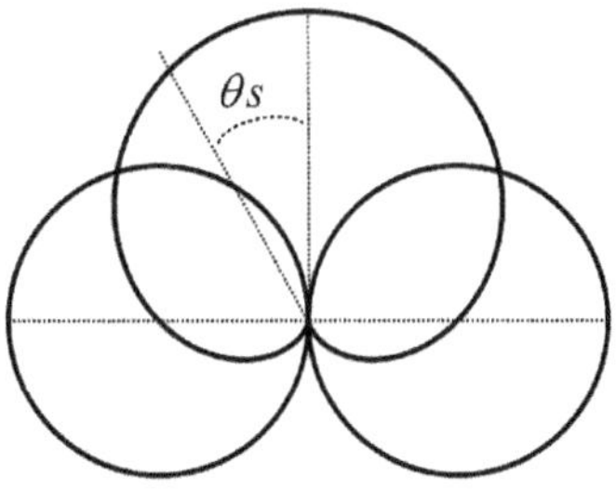

Figure 2.13 Polar pattern of MS recording with a cardioid microphone and a bidirectional microphone whose main axes point to the frontal-center and horizontal-left directions, respectively (the maximal output amplitudes of two microphones are normalized to a unit).

As the fourth example of the MS microphone technique, signals are recorded by a cardioid microphone and a bidirectional microphone whose main axes point to the frontal-center and horizontal-left directions, respectively (Figure 2.13). If $B'_p = B'_v = A_{mic,M}$ and $B''_v = A_{mic,S}$ in Equation (2.2.27), then

$$A_M = A_{mic,M}\left(1+\cos\theta_S\right) \qquad A_S = A_{mic,S}\sin\theta_S. \tag{2.2.46}$$

After the inverse MS transformation expressed in Equation (2.2.19) with $\kappa_v = \sqrt{2}/2$, the following equation is obtained:

$$A_L = B_p + B_v\cos\left(\theta_S - \theta_m\right) \qquad A_R = B_p + B_v\cos\left(\theta_S + \theta_m\right), \tag{2.2.47}$$

where

$$B_p = \frac{\sqrt{2}}{2}A_{mic,M} \quad B_v = \frac{\sqrt{2}}{2}\sqrt{A_{mic,M}^2 + A_{mic,S}^2} \quad \theta_m = \arctan\left(\frac{A_{mic,S}}{A_{mic,M}}\right) \quad 0^\circ \le \theta_m \le 90^\circ. \tag{2.2.48}$$

The condition of $B_p + B_v = 1$ may be supplemented so that the maximum A_L and A_R are normalized to a unit. Equation (2.2.47) is consistent with Equation (2.2.3). Because of $(B_v/B_p) > 1$, Equation (2.2.48) is equivalent to XY recording with a pair of supercardioid-like (or hypercardioid-like) microphones. Changing $A_{mic,S}/A_{mic,M}$ of bidirectional and cardioid components in MS recording is equivalent to altering the directivity B_v/B_p and θ_m in XY recording.

The virtual source direction at low frequencies is evaluated by substituting Equation (2.2.46) in Equation (2.2.28a), or it can also be evaluated with Equation (2.2.28b):

$$\sin\theta_I = \frac{A_{mic,S}}{A_{mic,M}}\frac{\sin\theta_S}{1+\cos\theta_S}\sin\theta_0 = \frac{A_{mic,S}}{A_{mic,M}}\tan\frac{\theta_S}{2}\sin\theta_0 \tag{2.2.49}$$

Therefore, the width of virtual source distribution can be changed by appropriately choosing the ratio between $A_{mic,S}$ and $A_{mic,M}$ to achieve desired effective recording angle.

When $A_{mic,S} = A_{mic,M}$, Equation (2.2.49) becomes

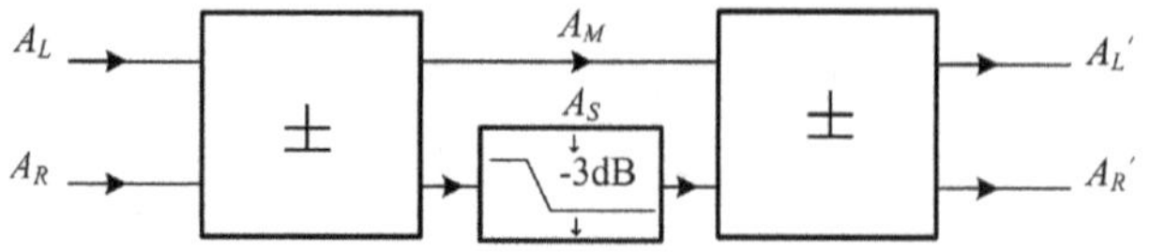

Figure 2.14 Signal processing for correcting the virtual source direction at high frequencies.

$$\sin\theta_I = \frac{\sin\theta_S}{1+\cos\theta_S}\sin\theta_0 = \tan\frac{\theta_S}{2}\sin\theta_0. \tag{2.2.50}$$

According to Equation (2.2.34) and the subsequent discussion, the virtual source direction for an MS pair is identical to that of the equivalent XY pair given in Equation (2.2.12). For a target source within the effective recording range, the virtual source distribution described in Equation (2.2.49) is narrower than that expressed in Equation (2.2.12). Within the azimuthal range of $|\theta_S| < \pi/6$ (30°), $\tan\theta_S \approx 2\tan(\theta_S/2)$. In this case, the width of virtual source distribution is approximately half of that presented in Equation (2.2.12) or the effective recording range is twice that presented in Equation (2.2.12). A cardioid microphone is used in Equation (2.2.46) to record A_M, which increases the relative magnitude of A_M and narrows the virtual source distribution in reproduction compared with that in Equation (2.2.34). In practical recording when a bidirectional XY pair (Figure 2.5) or an equivalent bidirectional MS pair (Figure 2.8) is used, if the frontal stage of target sources (such as orchestra) is excessively wide, a microphone pair should be placed far from target sources so that the effective recording range covers the whole stage of target sources. However, increasing the distance between sources and microphones decreases the direct-to-reverberation ratio in the captured signals. This problem can be avoided by using the MS pair in Figure 2.10 to comprehensively control the effective recording range and perceived distance in reproduction.

As an application of the MS transformation, the problem of correcting the direction of a high-frequency virtual source in stereophonic reproduction is discussed. As stated in Section 2.1.2, for a given ICLD, the virtual source moves toward the direction of either of the loudspeakers as the frequency exceeds 0.7 kHz, leading to an unstable virtual source. The signal processing procedure in Figure 2.14 can be used to correct the virtual source direction at high frequencies. In this procedure, the difference signal at $f > 0.7$ kHz is attenuated by −3 dB to compensate for the frequency-dependent variation in the virtual source direction; other studies have suggested an attenuation of −4 dB to −8 dB (Vanderlyn, 1954; Harwood, 1968; Gerzon, 1986). Similarly, Griesinger (1986) recommended that a difference signal should be appropriately enhanced below 0.3 kHz.

In the above discussion, the directivities or polar patterns of microphones are assumed to be consistent across the recording frequency range. Therefore, for a given target source position, the ICLD between two channel signals is frequency independent. However, the polar patterns of practical microphones vary with frequency. As frequency increases, the main lobe of a directional microphone narrows, resulting in a frequency-dependent ICLD and a frequency-dependent virtual source direction in reproduction. In addition to some methods of high-frequency compensation in microphone directivities, some signal processing procedures similar to that in Figure 2.14 are available to alleviate this problem.

2.2.3 Spaced microphone technique

In the spaced microphone technique, or sometimes called ***the AB recording technique***, a pair of identical microphones spaced apart by a large interval is used to capture stereophonic

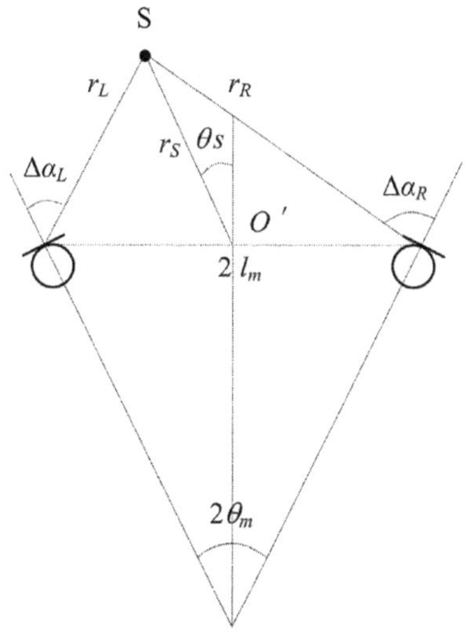

Figure 2.15 Spaced microphone pair.

signals. Usually, the resultant two channel signals include source direction-dependent ICTD and ICLD.

As illustrated in Figure 2.15, a pair of microphones are spaced apart at a distance of $2l_m$. The directivity of microphones is specified by $\Gamma_M(\Delta\alpha)$ as expressed in Equation (1.2.32). The distances from a horizontal point source to two microphones and the midpoint between two microphones are r_L, r_R, and r_S, respectively. The normalized two channel signals are given by

$$A_L = \frac{\Gamma_M(\Delta\alpha_L)}{4\pi r_L}\exp(-jkr_L) \qquad A_R = \frac{\Gamma_M(\Delta\alpha_R)}{4\pi r_R}\exp(-jkr_R), \tag{2.2.51}$$

where $\Delta\alpha_L$ and $\Delta\alpha_R$ are the angles between the source direction and the main axes of two microphones, respectively. In general, the directivity of microphones is introduced to Equation (2.2.51), although an omnidirectional pair with $\Gamma_M(\Delta\alpha_L) = \Gamma_M(\Delta\alpha_R) = 1$ is often used in the spaced microphone technique.

The ICTD and ICLD of two channel signals can be evaluated with Equation (2.2.51):

$$\begin{aligned} \Delta t &= \frac{r_R - r_L}{c}, \\ d &= 20\log_{10}\left|\frac{A_L}{A_R}\right| = 20\log_{10}\left|\frac{\Gamma_M(\Delta\alpha_L)\, r_R}{\Gamma_M(\Delta\alpha_R)\, r_L}\right| \ (dB), \end{aligned} \tag{2.2.52}$$

where $c = 343$ m/s is the speed of sound. In addition to the directivity of microphones, the path difference between a source and two microphones results in an increase in ICLD when the source approaches either of the microphones.

The spaced microphone technique was developed on the basis of the works of wavefront reconstruction at the frontal stage at Bell Labs in the 1930s (Steinberg and Snow, 1934). However, two channel signals are inadequate for accurate reconstruction of a frontal incident wavefront. Therefore, this technique is not based on strict acoustic theorem. On the contrary, for some continuous stimuli, signals from a spaced microphone pair may produce some

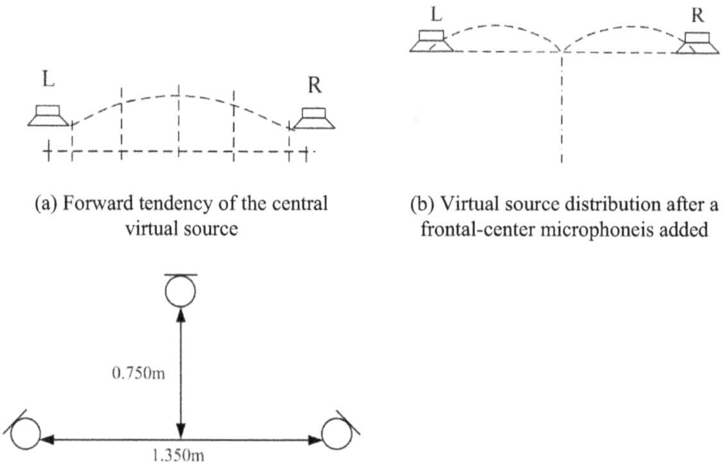

(a) Forward tendency of the central virtual source

(b) Virtual source distribution after a frontal-center microphoneis added

(c) Decca tree

Figure 2.16 Problems associated with a spaced microphone pair: (a) forward tendency of the central virtual source; (b) virtual source distribution after a frontal-center microphone is added; (c) Decca tree.

conflicting interaural localization cues (such as conflicting ITD and ILD). Therefore, evaluating the virtual source direction in reproduction through physical analysis is impossible. Actually, the spaced microphone technique is based on some psychoacoustic experiment for transient stimuli.

For traditional spaced microphone pairs, two microphones are spaced far apart from each other and sometimes called a ***wide-spaced microphone pair***. Moreover, an omnidirectional microphone pair is often used.

For direct sound, the spaced microphone technique applies the precedence effect to recreate the localization perception of virtual sources. When a target source deviates from the central–front direction to an extent such that the ICTD between two channel signals exceeds the lower limit of the precedence effect (the order of 1–3 ms; Section 1.7.2), the virtual source lies at the direction of the loudspeaker with a leading signal in reproduction. However, one problem with a spaced microphone pair is the hole-in-the-middle effect during reproduction, especially for a wide microphone distance. In this case, an off-central source with a certain angle is enough to create the ICTD that reaches the lower limit of the precedence effect and creates the perceived virtual source at the corresponding loudspeaker direction. Another problem with a spaced microphone pair is a forward tendency of the central virtual source. As shown in Figure 2.16(a), when multiple target sources are located in a straight line parallel to the microphone pair, the central source is more distant from either microphone than the off-central source. Therefore, a microphone pair captures a reduced direct-reflection ratio associated with the central source, thereby increasing the perceived virtual source distance. That is, the perceived virtual source lies in an arc between two loudspeakers, and the virtual source at the frontal-center position is slightly forward with respect to those at the positions of the loudspeakers.

A frontal-center microphone is added to overcome this defect, and its output is bridged to the left and right channels. This frontal-center microphone alleviates the problems of hole-in-the-middle effect and the forward tendency of the central virtual source. The perceived virtual source lies in two arcs, as shown in Figure 2.16(b). In this case, the stage of virtual source distribution narrows slightly. Some studies have suggested that a frontal-center microphone should be placed slightly forward with respect to two other outer microphones to further reduce the distance between a frontal-center source and the frontal-center microphone (Grignon, 1949). This three-microphone configuration is usually called a ***Decca tree***, as shown in Figure 2.16(c). Overall, adding a frontal-center microphone improves the perceived effect of the central virtual source. However, comb filtering caused by interference among the signals of the central and outer microphones leads to timbre coloration in reproduction, and this timbre coloration is undesirable.

When a spaced microphone pair is placed beyond the reverberation radius and when the interval between two microphones exceeds the distance specified by the correlated distance of reverberation sound field, the microphone outputs are dominated by low-correlated reflection signals. These phenomena are expressed as Equations (1.2.29) and (1.2.30), and these low-correlated signals result in the good sensation of spaciousness in reproduction.

If virtual source localization alone is considered, the interval between two microphones is chosen so that the maximal ICTD exceeds the lower limit of 1 ms but does not exceed the upper limit of the precedence effect. With the precedence effect, the perceived virtual source is located at the direction of the loudspeaker with a leading signal. However, an excessively wide interval between two microphones results in a decrease in the levels of signals for a frontal-center source because of the wide distance between the source and microphones. In practice, the interval between microphones should be chosen by comprehensively considering multiple factors, such as the width of a source stage (orchestra), and recording performance for direct and reflected sounds.

Overall, the spaced microphone technique usually degrades the perceptual quality of a virtual source but exhibits a better-perceived performance of reflections. Therefore, it is often used to record reflections in a hall.

2.2.4 Near-coincident microphone technique

A pair of directional microphones spaced apart with a small interval (usually 0.17–0.50 m) is often used to capture stereophonic signals to avoid the problems in spaced microphones with a wide interval. This method is called the ***near-coincident microphone technique*** or sometimes called a ***narrow-spaced microphone pair***. The directivities of microphones generate an appropriate ICLD in the captured signals, and the path difference between a source and two microphones introduces additional ICTD.

Generally, ICTD and ICLD between two channel signals can be calculated with Equation (2.2.52). The calculation can be further simplified under certain conditions. As shown in Figure 2.15, the origin O of the coordinate is located at the midpoint between two microphones. The main axes of two microphones point to the left- and right-frontal directions with a span angle of $2\theta_m$. Here, the span angle is specified with respect to the point O′ in the figure rather than to the origin O. Therefore, it is different from the case of a coincident microphone pair. For a horizontal source with distance r_S and azimuth θ_S with respect to the origin, the distances r_L and r_R between the source and two microphones and the angles $\Delta\alpha_L$ and $\Delta\alpha_R$ of the source direction with respect to the main axes of microphones can be calculated from the geometrical relationship. When the source distance with respect to the origin is much greater than the interval between the two microphones, i.e., $r_S >> 2l_m$, the angle satisfies $\Delta\alpha_L \approx \theta_S - \theta_m$, $\Delta\alpha_R \approx \theta_S + \theta_m$. In this case, the approximation $r_L \approx r_R \approx r_S$ is taken in the ICLD calculation. Substituting the above relations into Equation (2.2.52) yields the approximate equation for ICTD and ICLD calculation:

$$\Delta t \approx \frac{2l_m}{c}\sin\theta_S \qquad d \approx 20\log_{10}\left|\frac{\Gamma_M(\theta_S-\theta_m)}{\Gamma_M(\theta_S+\theta_m)}\right| \ (dB). \tag{2.2.53}$$

A near-coincident microphone pair yields two channel signals with ICLD and ICTD. In comparison with the spaced microphone pair with a wide interval, however, the near-coincident microphone pair yields a smaller ICTD that is usually within the range of summing localization illustrated in Figure 1.28; consequently, a virtual source is created through a combination of ICLD and ICTD. At low frequencies, the contribution of small ICTD to summing localization is trivial. The summing localization is dominated by the ICLD caused by the directivities of microphones. This phenomenon is similar to the case of the coincident microphone pair. For high-frequency components with transient characteristics, the ICTD supplements some information for summing localization. Overall, a near-coincident microphone pair can create a virtual source in terms of the principle of summing localization, but the perceived virtual source direction may be stimulus-dependent. In practice, the near-coincident microphone technique is designed on the basis of the results of psychoacoustic experiments, such as the Williams curves in Figure 2.5. Williams analyzed the near-coincident microphone technique in detail and gave a set of curves for design. The details are described in other studies (Williams, 1987, 2013).

Similar to the case of a coincident microphone pair, the effective recording angle $\pm\theta_p$ for a near-coincident microphone pair can be introduced. This parameter is specified as the source azimuth at which the virtual source lies at the ipsilateral loudspeaker direction. The near-coincident microphone pair yields two channel signals with ICLD and ICTD. The rule of summing localization with a combination of ICLD and ICTD does not result in a null signal in the opposite channel for a target source at the direction of the effective recording angle. The effective recording range $[-\theta_p, +\theta_p]$ or $2\theta_p$ of a near-coincident microphone pair can also

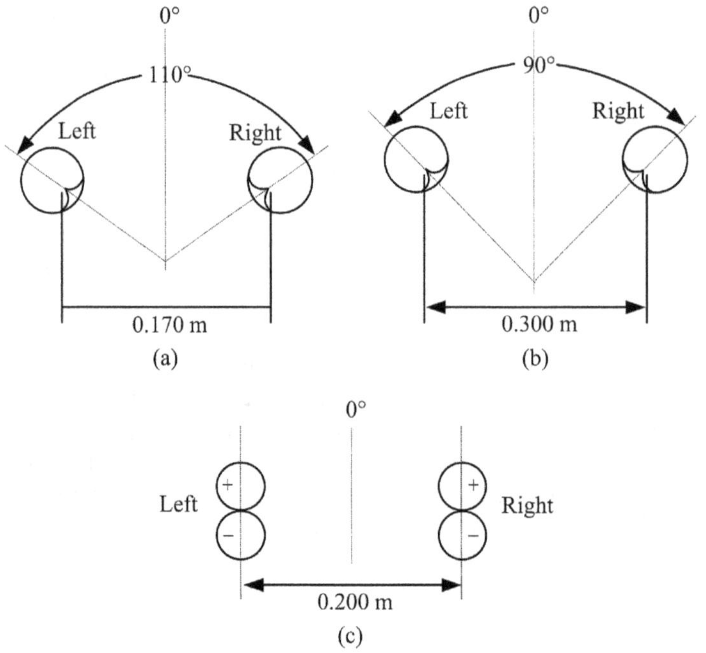

Figure 2.17 Some near-coincident microphone techniques: (a) ORTF; (b) NOS; (c) Faulkner. (adapted from Rumsey, 2001).

be specified by the effect recording angle. Generally, the effective recording range of a near-coincident microphone pair is not necessarily consistent with the span angle between the main axis directions of two microphones. The former may be equal to, larger, or smaller than the latter depending on the directivities and space of microphones or even the characteristics of stimuli. Moreover, the effective recording range of a near-coincident microphone pair can only be evaluated from the results of psychoacoustic experiments, such as the aforementioned Williams curves.

Various near-coincident microphone techniques are often used in practical recording. Figure 2.17 illustrates some examples (Rumsey, 2001).

The ORTF pair was named from the organization of "L'*Office de radiodiffusion-télévision française*." It consists of a pair of cardioid microphones spaced apart by $2l_m = 0.170$ m. The main axes of microphones orientate to the left-frontal and right-frontal direction respectively with a span angle of $2\theta_m = 110°$. The effective recording angle of an ORTF pair is ±47.5°, and the effective recording range is 95°. For example, the resultant ICTD and ICLD are 0.25 ms and 4.9 dB, respectively, by substituting the above microphone interval and main axis directions into Equation (2.2.53), using Equation (1.2.41), and assuming a target source azimuth of $\theta_S = 30°$.

The NOS pair was used by The Nederlandse Omroep Stichting. Similar to an ORTF pair, the NOS pair consists of a pair of cardioid microphones spaced apart by 0.300 m. The span angle between the main axes of two microphones is $2\theta_m = 90°$, the effective recording angle is ±40°, and the effective recording range is 80°. In addition, the Faulkner pair was suggested by Faulkner, a British recording engineer. It consists of a pair of bidirectional microphones spaced apart by 0.200 m. The main axes of microphones point to the frontal directions with respect to each microphone position, respectively.

The near-coincident microphone pair is inferior to the coincident pair if merely the quality of virtual source localization is considered. However, when it is applied to record music signals in a hall, the near-coincident microphone pair usually leads to a better perceived performance. In addition to the quality of a virtual source, the perceived quality related to the reflected sounds, such as spaciousness, is vital to the overall quality of stereophonic reproduction. The near-coincident microphone pair is an appropriate compromise between these two aspects. Furthermore, the near-coincident microphone pair introduces an ICTD with the same order as the ITD, which improves the perceived performance for direct headphone presentation. This phenomenon is addressed in Section 11.9.1. Conversely, the ICTD introduced by a near-coincident microphone pair may cause comb filtering and timbre coloration in mono downmixing.

A baffle can be inserted between two microphones to enhance the high-frequency ICLD between the outputs of a near-coincident microphone pair. An example is an optimal stereo signal (OSS) microphone pair (Jecklin, 1981) in which two omnidirectional microphones are spaced apart by 0.165 m and acoustically separated by a disk with a diameter of 0.28 m. Some studies have suggested arranging a pair of microphones on the two sides of a rigid spherical surface whose size matches the human head (radius of 0.09 m). In addition to the ICTD caused by the path difference between two microphones, the scattering and diffraction by the sphere introduces frequency-dependent ICLD at high frequencies (Theile, 1991b). This spherical microphone is similar to an artificial head for binaural recording (Section 11.1). However, binaural signals from an artificial head can be adapted to loudspeaker reproduction only after appropriate signal processing or conversion (Section 11.8). Therefore, the baffled microphone pairs originated from a practical experience rather than strict theoretical analysis. Their effects have some controversies.

The signals captured by a pair of microphones with a small interval can be approximately converted into two channel signals with ICLD only. This technique was first introduced by Blumlein in his patent in 1931 and called the ***Blumlein difference technique*** or ***Blumlein***

shuffing (Blumlein, 1931; Gerzon, 1994). Although the Blumlein difference technique is not directly used in practical commercial stereophonic recording, a similar technique is applied to record stereophonic signals in mobile devices (Faller, 2010).

A pair of identical omnidirectional microphones are spaced apart by an interval of $2l_m$. Similar to the condition in deriving Equation (2.1.3), when the source distance to the origin is much larger than the interval between two microphones, i.e., $r_S >> 2l_m$, the incident sound to the microphones is approximated as a plane wave from direction θ_S, and the normalized signals of two microphones are given by

$$A_L = \exp\left(jkl_m \sin\theta_S\right) \qquad A_R = \exp\left(-jkl_m \sin\theta_S\right). \tag{2.2.54}$$

In this case, two microphone signals have ICTD but have no ICLD. ICTD can be evaluated from Equation (2.2.53). Applying MS transformation given by Equation (2.2.18) to A_L and A_R and selecting a coefficient of $\kappa_\mu = 1/2$ lead to

$$A_M = \cos\left(kl_m \sin\theta_S\right) \qquad A_S = j\sin\left(kl_m \sin\theta_S\right). \tag{2.2.55}$$

At low frequencies with $kl_m << 1$, Equation (2.2.55) can be expanded as a Taylor series of ka (or ka $\sin\theta_S$). If the first expansion term is retained, the equations become

$$A_M \approx 1 \qquad A_S \approx jkl_m \sin\theta_S. \tag{2.2.56}$$

Therefore, A_M is equivalent to the signal captured by an omnidirectional microphone. A_S is equivalent to the signal captured by a bidirectional microphone with its main axis pointing to the left and with an additional 90° phase shift. Moreover, the magnitude of A_S increases with frequency at a rate of 6 dB/Octave. The difference in A_L and A_R is directly proportional to the velocity of the medium, i.e., directly proportional to the output of a pressure gradient or a bidirectional microphone (Section 1.2.5). Ideally, if the magnitude and phase of A_S are equalized by a filter with a response given by Equation (2.2.57), the resultant signals are equivalent to the MS signals expressed in Equation (2.2.36). The two channel stereophonic signals with ICLD only are derived by applying an inverse MS transformation to the equalized MS signals:

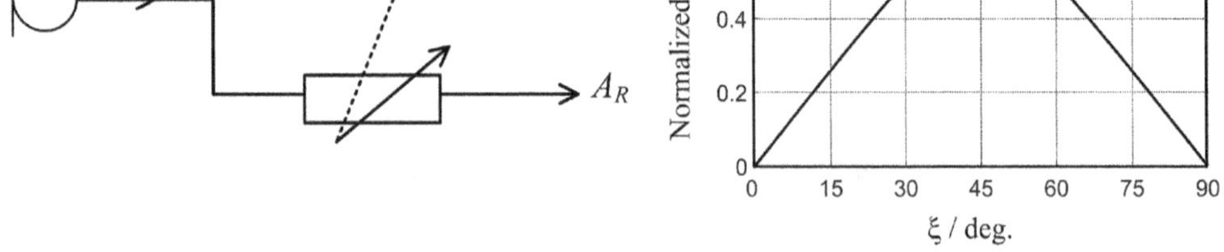

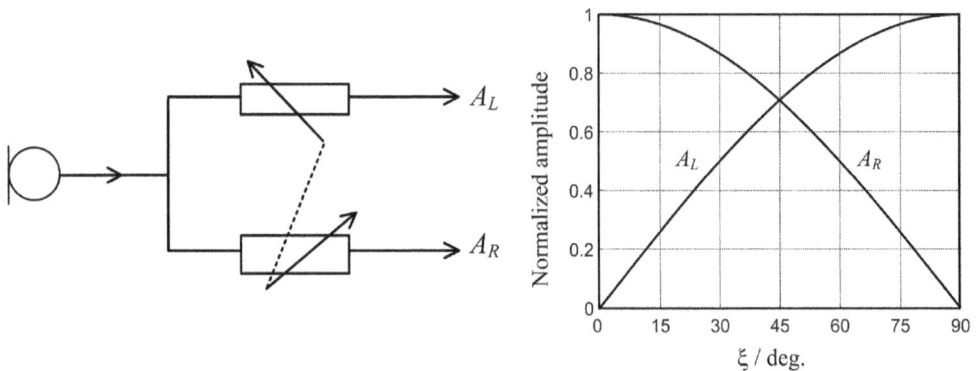

Figure 2.18 Stereophonic signals captured by a spot microphone and simulated by pan-pot (a) Spot microphone and pan-pot; (b)Variation in A_L and A_R with parameter ξ.

$$H_{eq}(f) = \frac{1}{jkl_m} = \frac{c}{j2\pi f l_m}. \tag{2.2.57}$$

In addition, Blumlein originally used an analog circuit with the following response to equalize A_S

$$H_{eq}(f) = 1 + \frac{1}{j2\pi f \tau_m}, \tag{2.2.58}$$

where τ_m is the adjustable parameter. At low frequencies, this equalization and the inverse MS transformation also yielded stereophonic signals with ICLD only.

2.2.5 Spot microphone and pan-pot technique

Multiple spot microphones are used to capture source signals, and each microphone separately captures the signal of each source or each set of sources, resulting in multiple mono-source signals. As illustrated in Figure 2.18(a), the mono signal from each spot microphone is split into two channel signals by a ***pan-pot***. A conventional pan-pot is a dual-ganged variable resistor that controls the relative magnitude or level of the mono signal fed to the left and right channels and then leads to a different ICLD. In this case, the desired directional information of a target source is simulated or synthesized artificially. Currently, the equivalent function of a pan-pot can be easily implemented via digital signal processing.

Usually, the overall power of two channel signals is normalized to a constant (unit) to ensure equal-loudness virtual sources in different directions. Therefore, the normalized amplitudes of two signals from the pan-pot are given by

$$A_L = \sin\xi \qquad A_R = \cos\xi \qquad A_L^2 + A_R^2 = 1, \tag{2.2.59}$$

where $0° \leq \xi \leq 90°$ is a parameter. Figure 2.18(b) illustrates the variation in A_L and A_R with ξ.

From Equation (2.1.6), for a head fixed to the frontal orientation, the direction of a low-frequency virtual source is related to ξ as follows:

$$\sin\theta_I = \frac{\tan\xi - 1}{\tan\xi + 1}\sin\theta_0. \tag{2.2.60}$$

Similarly, according to Equation (2.1.10), for a head oriented to the virtual source, the direction of virtual source is evaluated by

$$\tan\hat{\theta}_I = \frac{\tan\xi - 1}{\tan\xi + 1}\tan\theta_0. \tag{2.2.61}$$

In both cases, when ξ changes continuously from 0° to 90°, signal amplitude A_R decreases continuously from 1 to 0, and A_L increases continuously from 0 to 1. Accordingly, the virtual source moves continuously from $-\theta_0$ (the direction of the right loudspeaker) to θ_0 (the direction of the left loudspeaker). It has $A_L = A_R = 0.707$ for $\xi = 45°$, i.e., a −3 dB drop off compared with the maximal amplitude of a unit value. In this case, the virtual source lies in the directly front direction of $\theta_I = 0°$.

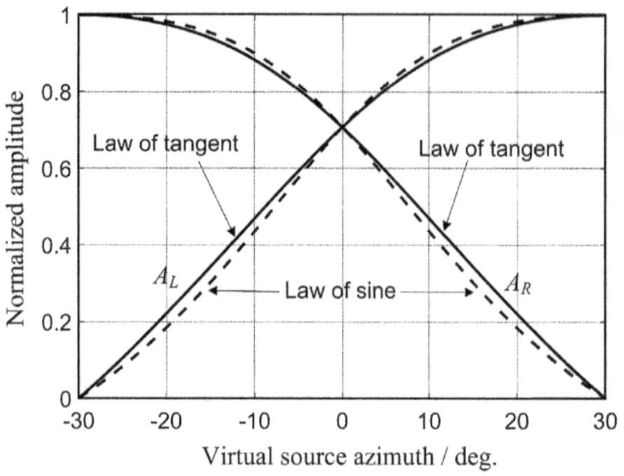

Figure 2.19 Two-channel stereophonic panning curves.

For ***constant-power panning***, the normalized amplitudes of two channel signals can be derived from Equations (2.2.60) or (2.2.61) subjected to the condition of a constant (unit) overall power. For a head fixed to the front orientation, we have

$$A_L = \frac{\sqrt{2}}{2}\frac{\sin\theta_0 + \sin\theta_I}{\sqrt{\sin^2\theta_0 + \sin^2\theta_I}} \qquad A_R = \frac{\sqrt{2}}{2}\frac{\sin\theta_0 - \sin\theta_I}{\sqrt{\sin^2\theta_0 + \sin^2\theta_I}}. \tag{2.2.62}$$

For a head rotating to the orientation of the virtual source, we have

$$A_L = \frac{\sqrt{2}}{2}\frac{\tan\theta_0 + \tan\hat{\theta}_I}{\sqrt{\tan^2\theta_0 + \tan^2\hat{\theta}_I}} \qquad A_R = \frac{\sqrt{2}}{2}\frac{\tan\theta_0 - \tan\hat{\theta}_I}{\sqrt{\tan^2\theta_0 + \tan^2\hat{\theta}_I}}. \tag{2.2.63}$$

Equation (2.2.62) or Equation (2.2.63) gives normalized amplitudes of two channel signals as functions of a virtual source direction, i.e., ***signal panning function***. Figure 2.19 plots the panning functions, i.e., the ***panning curve*** of the left and right channel signals for a stereophonic loudspeaker configuration with a span angle of $2\theta_0 = 60°$. In the front and either loudspeaker directions (0°, ±30°), Equations (2.2.62) and (2.2.63) yield identical results. The two equations yield different results in other directions. However, the difference is trivial if the span angle $2\theta_0$ between two loudspeakers does not exceed 60°.

Applying the transformation $\xi = \theta_S + 45°$ with $-45° \leq \theta_S \leq 45°$, Equation (2.2.59) becomes

$$A_L = \cos\left(\theta_S - 45^\circ\right) \qquad A_R = \cos\left(\theta_S + 45^\circ\right). \tag{2.2.64}$$

This expression is consistent with Equation (2.2.11). Therefore, the nature of synthesizing two-channel stereophonic signals with pan-pot is equivalent to the artificial simulation of the signals from a coincident bidirectional XY microphone pair for a source within the effective recording range. Here, θ_S is the azimuth of a target source in the original sound field to be simulated rather than the direction of the perceived virtual source in reproduction shown in Equation (2.2.12).

In addition to constant-power panning, two-channel signals are sometimes normalized according to the condition of constant unit amplitude, i.e., ***constant-amplitude panning***:

$$A_L + A_R = 1. \tag{2.2.65}$$

The recorded signals in Equation (2.2.37) satisfy the condition of constant amplitude. The constant-amplitude and constant-power panning are relatively appropriate for reproduction in anechoic rooms and rooms with some reverberation, respectively.

In practice, the acoustic characteristics of reproduction rooms are usually frequency dependent. Some studies have introduced a frequency-dependent normalization for two channel signals according to the acoustic characteristics of a reproduction room (Laitinen et al., 2014), i.e.,

$$A_L^{\lambda} + A_R^{\lambda} = 1, \tag{2.2.66}$$

where $1 \leq \lambda = \lambda(f, DTT) \leq 2$ is a parameter depending on frequency (band) and direct-to-total energy ratio (DTT). The DTT can be evaluated using the method in Section 1.2.4. λ =1 and λ =2 corresponds to constant amplitude and constant-power panning, respectively.

Equations (2.2.62) and (2.2.63) are derived from the stereophonic laws of sine and tangent respectively. For practical music stimuli, however, the direction of the perceived virtual source may not exactly match the results of the laws of sine and tangent. Therefore, Lee and Rumsey (2013) derived the relationship between the direction of the perceived virtual source and ICLD based on the fitting of the results of a localization experiment for music stimuli and used this relationship for panning curve design.

2.2.6 Discussion on microphone and signal simulation techniques for two-channel stereophonic sound

Various microphone and signal simulation techniques for two-channel stereophonic sound are presented in the previous sections. These techniques are chosen and used flexibly according to practical requirements.

The MS, XY, and near-coincident microphone techniques are usually chosen for large orchestra recording to achieve a fused sensation in reproduction. The microphone technique and associated parameters, such as the type, directivities, distance to source (orchestra), effective recording angle of coincident microphone pairs, or various parameters of a near-coincident microphone pair, are chosen on the basis of practical conditions. The directivities for some coincident microphone products are adjustable and therefore convenient for practical uses.

The performance of XY and MS microphone pairs is compared in some studies (Hibbing, 1989). Although XY and MS microphone pairs are theoretically equivalent, the MS microphone pair is relatively flexible in practical use. Deriving various XY-equivalent signals from a pair of MS signals is relatively easy, freeing from the restriction on available XY microphone products with the desired directivity. At the same time, a practical directional microphone usually possesses the perfect or desired directivity below a certain frequency. As frequency increases, the main lobe of microphones usually narrows. As a result, the high-frequency output magnitude of the directional microphone decreases for an off-axis source, giving rise to timbre coloration in reproduction. This phenomenon occurs in the direct front source in XY recording, especially in an XY recording with a wide span angle between the main axis orientations of two microphones, because the source lies at the off-axis direction of two microphones. This problem can be avoided in MS recording because the main axis of the M microphone always points in the front direction. Indeed, the effectiveness of the MS pair in reducing timbre coloration depends on the relative importance of the timbre of the front source at the overall stereophonic stage.

From the preceding analysis on XY coincident microphone pairs (or equivalent MS microphone pairs) and near-coincident microphone pairs, the span angle between the main axes of two microphones, effective recording range, and the span angle between two loudspeakers in reproduction are usually not coincident. The span angle between the main axes of two microphones is just a parameter related to microphone configuration. For sound sources within the effective recording range, the virtual sources in reproduction are limited or mapped to a range between two loudspeakers, when the case of outside-boundary virtual source is not considered. For many living recording stereophonic program materials, virtual source positions in reproduction are not exactly consistent with those of the actual source (such as instruments) at the original stage. However, this consistency is not vital because listeners usually do not care about the absolute positions of sound (virtual) sources in reproduction. Recreating the relative position distribution of virtual sources in reproduction is enough.

For on-site stereophonic recording, one important step is to choose the effective recording range. The effective recording range is determined by the width of source stage (span angle of source distribution with respect to microphones). A wide source stage requires a wide effective recording range. In some experiences from on-site recording, for a narrow sound stage (such as quartet), the effective recording range is chosen to be about 10% wider than the total sound stage to leave a side room (Williams and Du, 2001). However, for a wide sound stage (such as an orchestra), the effective recording range is chosen to be about 10% smaller than the sound stage to enable better resolution of the central orchestra. For an excessively wide sound stage, microphones can be placed backward at a more distant position from the sources to reduce the span angle of source distribution with respect to microphones so that all sources at the sound stage can be recorded by a pair of microphones with a smaller effective recording range. However, when the microphone pair is placed at a more distant position from the sources, the captured power of reverberation sound increases compared with that of the direct sound. In this case, a microphone pair with appropriate directivities, such as a cardioid pair in Figure 2.8(a), can be used to reduce the relative proportion of reverberation components in captured signals. Therefore, the choice of the effective recording range, the distance between a source and microphones, and the directivities and orientation of the main axes of microphones are closely related and restrained. An appropriate choice should be on the basis of practical conditions.

In some situations, stereophonic recording with a coincident or near-coincident microphone pair may not satisfy the requirement from the point of acoustics. In these cases, some additional microphones (or microphone sets) are needed. As the first example, for some music program recordings, such as solo in a concerto, the solo should be enhanced from the background of orchestral music. In this case, in addition to a coincident or near-coincident microphone pair for recording orchestral music, an individual microphone is used to capture the source signal of a solo, and the resultant signal is mixed to the two channel signals by pan-pot (for large solo instruments, such as a piano, a pair of additional microphones are needed). As a second example, for an orchestra with excessive width, instruments at two sides of the orchestra are located at a more distant distance from the coincident microphone pair and lead to a low magnitude in microphone outputs. In this case, a pair of outrigger microphones can be supplemented to the two sides to capture the source signals from the instruments at two sides of the orchestra and the outputs of an outrigger microphone pair are mixed with those of the main microphone pairs. The virtual source of the instruments at the two sides lies in the position of two loudspeakers in reproduction because of the precedence effect and the attenuation caused by the large propagating distance between the instruments at the two sides and the coincident microphone pair. As a third example, when a pair of coincident or near-coincident microphones fail to consider both the sound stage of the direct sound and sensation of environmental reflections, the direct and reflected sound can be captured separately by two pairs of microphones and then mixed together. That is, the direct sound is mainly captured by a pair of coincident or near-coincident microphones placed more closely to the orchestra, and the reverberation is mainly captured by other pairs of spaced microphones far from the orchestra. For various stereophonic recordings with two or more sets of microphones, the gain of the signals from each set of microphones should be adjusted to ensure the balance among the intensities of various sound components.

Multiple spot microphones and pan-pots are often used in pop music and video/audio program recording. The signal for each source or each set of sources is individually captured using a microphone with narrow directivity and then split into two channel signals by an individual pan-pot. The signal from an electronic instrument is directly fed to the pan-pot. In practical program production, each source signal can be individually recorded into a synchronous track or be even recorded at separate times provided that the synchronization among all source signals is ensured. Then, each mono signal from synchronous tracks is split into two channel signals and mixed together. The balance among the intensities of the source signal can be controlled by changing the relative gain of each input source signal. In addition, a relative temporal relationship among source signals can be controlled by changing the delay of each source signal.

Multiple microphone and pan-pot recordings are often conducted in a recording room with sound absorption processing. The environmental reflections in a recording room usually do not satisfy the requirements of desired auditory perceptions. As a result, various artificial delay and reverberation techniques are used to supplement the information of reflected sound (Section 7.5). In terms of fused sensation in reproduction, stereophonic recording with various artificial simulation techniques is often inferior to those live recording with coincident or near-coincident microphone techniques. However, an artificial simulation technique can create an exaggerated virtual source effect or spatial auditory effects that do not exist in a natural auditory environment.

Stereophonic recording depends not only on acoustic conditions but also on the program contents to be recorded. Particularly, (classical) music recording involves acoustic or psychoacoustic problems, such as the directions, distance, and depth of sound source (or depth of scene), spatial perception of reflections, timbre, intensity balance, and fused sensation among different sources. It also encounters the problem of reproducing the esthetic content of music performance. Therefore, stereophonic recording deals with scientific (technical) and esthetic problems. In other words, stereophonic recording is a comprehensive application of the knowledge of acoustics, signal processing, and psychology of hearing and arts (Guan, 1988). In accordance with the aforementioned technical principles, stereophonic recording leaves a large space for esthetic creation.

2.3 UPMIXING AND DOWNMIXING BETWEEN TWO-CHANNEL STEREOPHONIC AND MONO SIGNALS

In the early days of stereophonic sound, a number of mono reproduction equipment were still used. To be compatible with mono reproduction, two-channel stereophonic signals are downmixed into mono signal. Downmixing can be simply implemented by adding the left and right channel signals together and then multiplying a gain factor κ_μ:

$$E_M = \kappa_\mu \left(E_L + E_R \right). \tag{2.3.1}$$

For two-channel uncorrelated signals, a gain factor of $\kappa_\mu = \sqrt{2}/2$, or equally a −3 dB attenuation, is chosen. For two-channel correlated signals, a gain factor of $\kappa_\mu = 1/2$, or equally a −6 dB attenuation, is selected. In practice, the gain factor lies between the two aforementioned values.

The spatial information in two-channel stereophonic signals is lost after these signals are downmixed to mono signals. The timbre in mono-downmixing reproduction depends on the microphone and mixing techniques used in the original stereophonic recording. For two-channel uncorrelated or decorrelated signals, mono-downmixing slightly influences the timbre. For two-channel correlated signals that are recorded by coincident microphone techniques or simulated by pan-pot, mono-downmixing slightly influences the timbre because of the absence of ICTD between two channel signals. For two-channel correlated signals with ICTD, such as those captured via a spaced microphone technique, mono-downmixing causes the comb-filtering effect and then degrades the timbre. The comb-filtering effect in mono-downmixing is a defect of the spaced microphone technique. Therefore, two-channel stereophonic signals with ICTD are not fully mono compatible (Harvey and Uecke, 1962).

In the early times of stereophonic sound, numerous mono program materials are available, and some program materials possess an important historical and artistic value. Therefore, mono signals should be upmixed into two-channel stereophonic signals. These upmixing techniques are usually termed ***pseudo-stereophonic sound***. The basic method of pseudo-stereophonic sound is converting mono signals into two channel signals with an appropriate interchannel difference or into two-channel decorrelated signals so as to simulate the spatial information in usual stereophonic signals and then cause some spatial effects. Various methods for upmixing or pseudo-stereophonic sound have been proposed.

Viewed from the audience, bass and treble instruments are usually located on the right and left sides of an orchestra, respectively. The mono signal is split into two channel signals to simulate the directional distribution of instruments in an orchestra. For the left channel signals, a high-frequency component is boosted, and a low-frequency component is attenuated. For the right channel signals, the processing of these components is reversed. Such signal processing can be directly implemented by high-pass and low-pass filters. Alternatively, Xie X.F. (1964a, 1964b, 1964c) suggested a similar method of pseudo-stereophonic sound by using a phase shift circuit. Other early examples include the methods of complementary comb filters, stereophonic reverberator, and signal decorrelation (Schroeder, 1958; Orban, 1970; Eargle, 2006). Some recent methods rely on the algorithms of blind source separation in signal processing. On the basis of the difference in temporal, spectral, and statistical characteristics of different source and ambient components, various sources and ambient components are separated from the mono signal. After some appropriate postprocessing (including decorrelation), various components are re-panned or re-mixed into two-channel pseudo-stereophonic signals (Uhle and Gampp, 2016).

2.4 TWO-CHANNEL STEREOPHONIC REPRODUCTION

2.4.1 Standard loudspeaker configuration of two-channel stereophonic sound

The two loudspeakers for stereophonic reproduction are arranged at the azimuths of $\pm\theta_0$, i.e., at the left- and right-front directions with a span angle of $2\theta_0$. Sections 2.1 and 2.2 indicate that a larger span angle between two loudspeakers widens the distribution of virtual sources (baseline of the sound stage) in reproduction. However, the analysis of Equation (2.1.10) indicates that the head rotation causes a slight variation in the virtual source direction for a loudspeaker configuration with $2\theta_0 \leq 60°$; thus, a virtual source is relatively stable. For loudspeaker configuration with a span angle larger than 60°, the virtual source is unstable under the head rotation. At the end of Section 2.1.2, even for a fixed head, the movement of a virtual source with an increase in frequency becomes more obvious for a large span angle between two loudspeakers.

Numerous experiments and applications have proved that stereophonic loudspeaker configuration with a span angle of $2\theta_0 = 60°$ is an appropriate compromise between the width of virtual source distribution and the stability of virtual source direction. Therefore, the ***span angle in a standard stereophonic loudspeaker configuration is*** $\boldsymbol{2\theta_0 = 60°}$.In this case, two loudspeakers and a listener are located at the three vertex angles of a regular triangle.

In many practical uses, a pair of two-way loudspeaker systems (box) is often used for stereophonic reproduction in which the low- and high-frequency components of signals are reproduced by the woofer and the tweeter, respectively. Aside from the signal processing algorithm shown in Figure 2.14, two-way loudspeaker systems can be laid down with the tweeters being located in the inward position to correct the virtual source direction at high frequencies. The crossover frequency of a two-way loudspeaker system is usually higher than 1.5 kHz. This laying-down arrangement is unable to correct the virtual source direction below the crossover frequency. However, psychoacoustic experiments indicate that the high-frequency virtual source above 1.5 kHz also exhibits a tendency of moving to the directions of loudspeakers, although the direction of a high-frequency virtual source cannot be evaluated with Equation (2.1.5). Therefore, the laying-down arrangement is beneficial to correcting the position of a high-frequency virtual source. However, when one channel signal vanishes (for example, $A_R = 0$), the laying-down arrangement results in azimuthally splitting virtual sources with low- and high-frequency virtual sources being located in different azimuths.

For some practical uses, such as a TV set or a traditional stereophonic-type player, the span angle between two loudspeakers is smaller than the standard of $2\theta_0 = 60°$, which narrows the stereophonic width (width of virtual source distribution) in reproduction. In these cases, the method of an outside-boundary virtual source is applied to correct or expand the stereophonic width in reproduction (Xie X.F., 1981). The reversal phase counterparts of the original left and right channel signals are mixed to the opposite channel with a gain of $0 < \chi < 1$.

$$A'_L = A_L - \chi A_R \qquad A'_R = A_R - \chi A_L. \tag{2.4.1}$$

This reversal phase mixing boosts the difference component in the resultant signals because $(A'_L - A'_R) = (1 + \chi)\,(A_L - A_R) > (A_L - A_R)$. At the same time, it attenuates the sum component in the resultant signals because $(A'_L + A'_R) = (1 - \chi)\,(A_L + A_R) < (A_L + A_R)$. In practical loudspeaker configuration with $2\theta'_0 < 2\theta_0 = 60°$, to match the virtual source direction at low frequencies with that of the standard configuration with $2\theta_0 = 60°$, the law of sine in Equation (2.1.6) yields

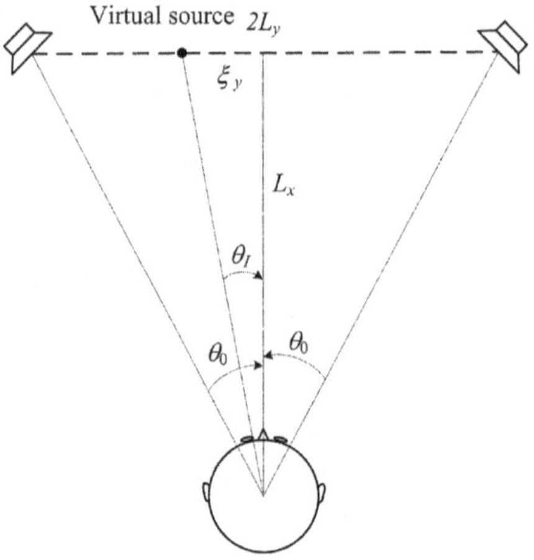

Figure 2.20 Geometrical relationship between the virtual source and the head position.

$$\frac{A'_L - A'_R}{A'_L + A'_R}\sin\theta'_0 = \frac{A_L - A_R}{A_L + A_R}\sin\theta_0. \tag{2.4.2}$$

The gain χ is derived from Equations (2.4.1) and (2.4.2):

$$\chi = \frac{\sin\theta_0 - \sin\theta'_0}{\sin\theta_0 + \sin\theta'_0}. \tag{2.4.3}$$

For a standard loudspeaker configuration with $2\theta_0 = 60°$, a similar method is available to expand the width of virtual source distribution to be outside the region between the two loudspeakers. However, such an excessive expansion produces an unstable virtual source.

The aforementioned method for stereophonic expansion is valid only at low frequencies because the scattering and diffracting effects of the head are ignored. The perceptual quality (such as the definition and naturalness of a virtual source) of this method is not satisfactory. This method was previously applied to some consumer electronic products. With the recent development of digital signal processing, stereophonic expansion based on transaural processing (Section 11.9.2) exhibits a superior performance and has basically taken over the aforementioned method.

2.4.2 Influence of front-back deviation of the head

In previous analysis, a listener's head is located at a default or ideal listening position (sweet point). If the head departs from the default position, the geometrical relationship and propagating paths from two loudspeakers to two ears change. This geometrical change may alter the superposed sound pressure in the two ears and then change the perceived virtual source position.

As shown in Figure 2.20, the two loudspeakers are arranged at the horizontal azimuths of $\pm\,\theta_0$, respectively, with a baseline length of $2L_y$. The distance between the midpoint of the baseline and the head center is L_x. The displacement of the virtual source in the baseline is related to the direction of a virtual source by

$$\xi_y = L_x \tan\theta_I. \tag{2.4.4}$$

The direction of the virtual source is evaluated from Equation (2.1.6). For loudspeaker configuration with $\theta_0 \le \pi/6$ (30°) and virtual source direction $|\theta_I| \le \pi/6$, we have $\sin\theta_I \approx \tan\theta_I$, and $\sin\theta_0 \approx \tan\theta_0 = L_y/L_x$, Equation (2.4.4) yields

$$\xi_y = \frac{A_L - A_R}{A_L + A_R} L_y = \frac{A_L/A_R - 1}{A_L/A_R + 1} L_y. \tag{2.4.5}$$

Equation (2.4.5) indicates that the displacement of the virtual source in the baseline depends on the magnitude ratio A_L/A_R (or ICLD) between two channel signals and the baseline length $2L_y$ for loudspeaker configuration with $2\theta_0 \le 60°$ and the virtual source located between two loudspeakers. However, the displacement is approximately independent from the distance L_x between the midpoint of the baseline and the head. Therefore, a moderate front-back deviation of the head along the central (symmetric) line alters the distance L_x and span angle between two loudspeakers with respect to the head, but this deviation influences slightly on the virtual source position ξ_y. In other words, the virtual source is basically stable against the front-back deviation of the head. However, when the span angle $2\theta_0$ between two loudspeakers exceeds 60°, the virtual source is no longer stable against the front-back deviation of the head. This observation is another reason for choosing a standard span angle of $2\theta_0 = 60°$ between two stereophonic loudspeakers.

In the case of an outside-boundary virtual source with $|\theta_I| > 30°$ created by out-of-phase loudspeaker signals, a front-back deviation of the head along the central line causes variation in a virtual source position. For the case of a lateral virtual source at $\theta_I \approx \pm 90°$ with $\sin\theta_I \approx \pm 1$ in Equation (2.1.21), when the head deviates forward in the central line but when $|A_L/A_R|$ is kept invariant, the span angle $2\theta_0$between two loudspeakers with respect to the head center increases. The result of Equation (2.1.21) becomes $\sin\theta_I > 1$. In this case, the ITD in reproduction exceeds the maximal possible ITD caused by an actual sound source, and an auditory system is unable to identify the virtual source direction in terms of ITD cues. Therefore, in addition to the instability against frequency variation, the outside-boundary virtual source is unstable against the front-back deviation of the head along the central line. This phenomenon is also a defect of the outside-boundary virtual source created by two-channel out-of-phase signals. Some subjective experiments on stereophonic localization with out-of-phase signals also exhibit uncertainty or even confusion.

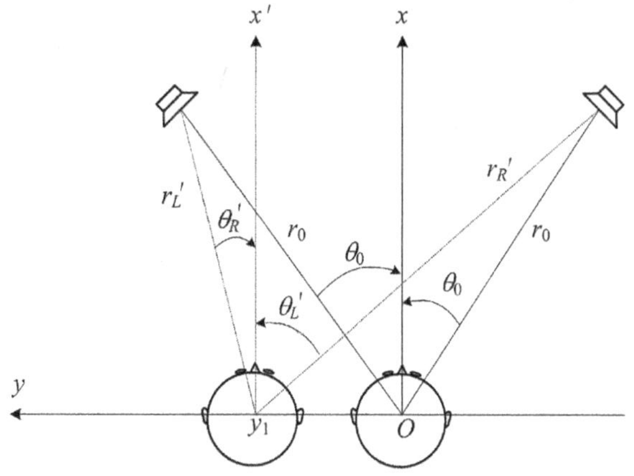

Figure 2.21 Head deviation from the default position to the left.

2.4.3 Influence of lateral translation of the head and off-center compensation

In contrast to the case of the front-back deviation along the central line, a lateral translation of the head away from the central line (off-central line) not only makes the two loudspeakers left-right asymmetric with respect to the head but also (most importantly) leads to the difference in a path length between the two loudspeakers and the head center. As shown in Figure 2.21, two loudspeakers are arranged at the azimuths $\pm\theta_0$, and the distance r_0 with respect to the default position of the head center. When the head center deviates from the default position laterally to the left position at coordinate $(0, y_1)$, the distances and azimuths of the two loudspeakers with respect to the head center become

$$r'_L = \left[(r_0 \sin\theta_0 - y_1)^2 + (r_0 \cos\theta_0)^2\right]^{1/2} \quad r'_R = \left[(r_0 \sin\theta_0 + y_1)^2 + (r_0 \cos\theta_0)^2\right]^{1/2}, \tag{2.4.6}$$

$$\begin{aligned} \theta'_L &= \arccos\left(\frac{r_0}{r'_L}\cos\theta_0\right) = \arctan\left(\frac{r\sin\theta_0 - y_1}{r\cos\theta_0}\right), \\ \theta'_R &= -\arccos\left(\frac{r_0}{r'_R}\cos\theta_0\right) = -\arctan\left(\frac{r\sin\theta_0 + y_1}{r\cos\theta_0}\right). \end{aligned} \tag{2.4.7}$$

Correspondingly, the difference between the path lengths of two loudspeakers results in additional differences in arrival time and sound pressure level:

$$\Delta t = \frac{r'_R - r'_L}{c}, \tag{2.4.8}$$

$$\Delta L = 20\log_{10}\left(\frac{r_R}{r_L}\right) \quad (dB), \tag{2.4.9}$$

where c = 343 m/s is the speed of sound. For example, for loudspeaker configuration with r_0 = 2.0 m and $2\theta_0$ = 60°, a lateral translation of y_1 = 0.05 m leads to a negligible difference in

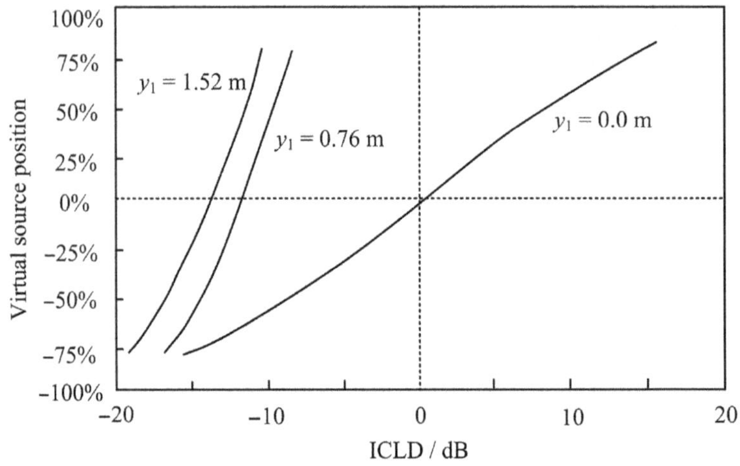

Figure 2.22 Results of a virtual source localization experiment at the off-central position. (adapted from Leakey, 1959).

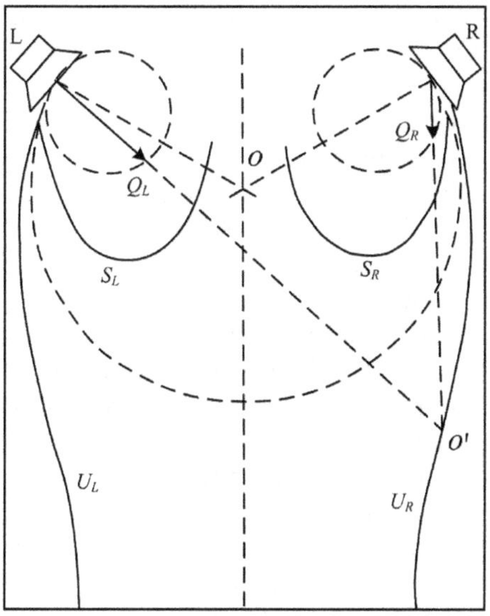

Figure 2.23 Compensation for the influence of the head translation by dipole loudspeakers. (Bauer, 1960, with the permission of the Audio Engineering Society).

ΔL = 0.2 dB. However, this translation leads to a difference in the arrival time of $\Delta t \approx 0.15$ ms, which is enough to cause a perceptible change in a virtual source direction for transient stimuli.

Therefore, auditory events at the off-central position are similar to those created by a combination of ICLD and ICTD described in Section 2.1.5. The analysis in 12.1.5 indicates that the low-frequency localization information in superposed binaural pressures is inconsistent or even conflicting. For some wideband stimuli with transient characteristics, the auditory system may identify the virtual source direction according to the more consistent cues or according to some unknown mechanisms. When Δt and ΔL do not exceed the lower limit of the precedence effect, the perceived virtual source moves in the same direction as the head (toward the direction of the nearer loudspeaker) and becomes vague. The definition and naturalness of a virtual source decrease. When the lateral translation of the head exceeds a certain limit so that Δt and ΔL exceed the lower limit of the precedence effect, the virtual source collapses to the direction of nearer loudspeakers, and the spatial information of stereophonic sound is spoiled.

Some authors investigated the virtual source localization at the off-central position through analysis and experiments. The results of various studies differ, and results depend on stimuli (Clack et al., 1957; Leakey, 1959). Figure 2.22 shows the experimental results from Leakey. A wideband speech stimulus is used. The baseline length between two loudspeakers is $2L_y$ = 3.05 m, and the distance between the midpoint of the baseline and the default position of the head center is L_x = 2.44 m. The results for the translation y_1 = 0.0, 0.76, and 1.52 m are illustrated in the figure. The virtual source position is presented by the percentage of movement in the baseline, i.e., $(\xi_y/L_y) \times 100\%$ (Figure 2.20).

Various methods have been proposed to compensate for the influence caused by head translation and widen the listening region in stereophonic reproduction. A traditional method involves using a pair of loudspeakers with appropriate directivity and axis orientations to compensate for the variation in pressure magnitude at the off-central position (Boer, 1946). Bauer (1960) suggested the use of a pair of dipole loudspeakers. In Figure 2.23, a pair of dipole loudspeakers are arranged in a circle. The main axes of loudspeakers point toward the center of a circle, which is located forward of the default listening position in the central line. At the off-central position O′, the sound pressure caused by the left loudspeaker decreases, and the sound pressure caused by the right loudspeaker increases because of the change in the distances from each loudspeaker to the position O′. However, the directivity of dipole loudspeakers compensates for the pressure variation at the position O′ to some extent. The maximum size of the compensated area is obtained when the angle between the main axes of two loudspeakers lies between 120° and 130°. Within the area bounded by the curves S_L, S_R, U_L, and U_R, the additional level difference between the sound pressures caused by the difference in the path length of two loudspeakers does not exceed 3 dB.

A moving coil loudspeaker installed in an open baffle exhibits a bipolar radiation pattern when the wavelength is larger than the dimension of the baffle. However, the low-frequency power radiated by a dipole source is low. In practice, a pair of bipolar loudspeakers is used to reproduce the stereophonic signals above 250 Hz. An additional omnidirectional woofer arranged in the frontal-center direction is used to reproduce the low-frequency component below 250 Hz of two channel signals.

Dipole loudspeakers compensate for the additional level difference at the off-central position and then balance sound levels from two loudspeakers. However, bipole loudspeakers are unable to compensate for the additional path or time difference at the off-central position. Equations (2.4.8) and (2.4.9) and the example after these two equations indicate that the influence of the additional time difference at the off-central position on a virtual source is greater than that of the additional level difference. Kates (1980) explored the optimal loudspeaker directivity for widening the listening area based on the interaural phase delay difference evaluated from a spherical head model, as expressed in Equation (1.6.4). ILD is fitted from experimental data. The results depend on frequency, loudspeaker configuration, and listening position. Overall, loudspeakers should be more directional than conventional designs, and the main axes should point toward the listening area. On the basis of the experimental results of the trading effect between ICLD and ICTD (similar to those in Section 1.7.1), Aarts (1993) suggested a method that uses the directivity of loudspeakers to compensate for the influence of additional time difference at the off-central position. This trading method is valid for wideband stimuli with transient characteristics.

The required directivity for off-central compensation in a conventional loudspeaker system design is not easily achieved. Some studies have suggested obtaining the required directivity by using a phased loudspeaker array (Davis, 1987), but the resultant loudspeaker system is complicated. Moreover, when a listener deviates away from the central line to a certain distance so that the combination of Δt and ΔL exceeds the lower limit of the precedence effect, the method of compensation by using loudspeaker directivity is invalid.

Conversely, some methods have been proposed to adapt the optimal listening position or a sweet point to an off-central position. These methods are realized by controlling the magnitude and delay of the two loudspeaker signals to compensate for the additional time and level differences at the off-central position. In some domestic reproduction systems, the delays of loudspeaker signals can be controlled manually or controlled automatically in terms of the measurement by a microphone placed in a practical listening position. Kyriakakis (1998) suggested using a head-tracker system to detect the position of a listener's head and then automatically control the delays of loudspeaker signals. The above methods compensate for the additional time difference at the off-central position only. Merchel and Groth (2010) proposed a method that uses an optical face tracker to detect the listener's head position and then adaptively controls the amplitude and delay of loudspeaker signals. However, a free-field transmission model was adopted in this model to evaluate the pressure amplitude and delay in the listening position. Momose et al. (2015) further used an HRTF model to evaluate the ILD in the listening position to improve the accuracy of loudspeaker signal control.

2.5 SUMMARY

The two-channel stereophonic sound is the simplest and most popular spatial sound technique and system. In this system, the spatial information of a sound field is represented by the relative relationship or difference between two-channel signals. It can be used to recreate the spatial information of sound coming from a frontal-horizontal sector and within a local listening region.

For two-channel stereophonic signals with an ICLD only, the directional information represented by the ICLD is converted into an interaural phase delay difference, which is a low-frequency localization cue in superposed binaural pressures. Below the frequency of 0.7 kHz, the direction of a low-frequency virtual source can be evaluated on the basis of the law of sine or tangent. In this case, the virtual source direction is dependent on the ICLD, but it is independent of frequency. As frequency increases, the virtual source direction depends on both ICLD and frequency. For two-channel low-frequency and stationary signals with a constant (frequency-independent) phase difference, the virtual source direction exhibits a complicated dependence on ICLD, interchannel phase difference, and frequency. An interchannel phase difference exceeding 90° (especially 180°) may give rise to an outside-boundary virtual source or an auditory event with an uncertain position. An interchannel phase difference also makes the virtual source blur and degrades the naturalness of the virtual source. The ICTD may lead to some conflicting interaural localization cues. Physical analysis of binaural pressures does not lead to consistent results. For some wideband signals with transient characteristics, however, ICTD may cause the virtual source to move toward the direction of a loudspeaker with a leading signal. This mechanism is a true psychoacoustic phenomenon.

Various principles and methods are applied to encode the spatial information of sound into two channel signals, giving rise to different spatial auditory perceptions in reproduction. Various microphone techniques for two-channel stereophonic recording have been developed. The coincident microphone technique includes an XY and MS microphone pair technique, which is equivalent from the point of MS transformation and can generate two-channel signals with an ICLD only. Various capturing characteristics can be achieved by appropriately choosing the directivity and main axis orientation of the microphone pair. With the spaced microphone technique, two-channel signals with a large ICTD and with ICLD in some instances can be obtained. It can also be used to produce two-channel signals with low correlation. With the near-coincident microphone technique, two channel signals with an appropriate ICLD and a small ICTD are created. Aspot microphone and a pan-pot technique are utilized to simulate the spatial information by artificially controlling the ICLD.

A compatible mono signal can be obtained by a downmixing of two-channel stereophonic signals. In the early times of stereophonic sound, some pseudo-stereophonic sound techniques were proposed to upmix mono signals into two-channel stereophonic signals.

The span angle in a standard stereophonic loudspeaker configuration is $2\theta_0 = 60°$. This span angle is an appropriate compromise between the width of virtual source distribution and the stability of virtual source direction. In some practical uses, the span angle between two loudspeakers is smaller than the standard, and the method of an outside-boundary virtual source was sometimes used to correct or expand the width of virtual source distribution in reproduction.

In stereophonic reproduction, a moderate front-back deviation of the head along the central line slightly influences the virtual source position. However, a lateral translation of the head from the central line influences the virtual source position. Some off-central compensation or adaption methods have been proposed.

Chapter 3

Basic principles and analysis of multichannel surround sound

The two-channel stereophonic sound discussed in Chapter 2 is the simplest spatial sound. However, it is only able to recreate the spatial information of the sound field within a frontal–horizontal sector. Multichannel surround sounds that can recreate the horizontal or even three-dimensional spatial information of the sound field should be developed to enhance the spatial performance in reproduction. Multichannel surround sound has been a major trend in sound reproduction since the end of the 1960s.

Multichannel surround sounds can be regarded as extensions of two-channel stereophonic sound. They are based on the rough approximation of the sound field and psychoacoustic principles of spatial hearing. Some psychoacoustic principles used in multichannel surround sounds are common or similar to those in two-channel stereophonic sound. However, an increase in spatial dimensionality in multichannel surround sound causes new problems. The diversity in the options of the number, configuration, and signals of loudspeakers results in various multichannel surround sound techniques and signal mixing methods. Accordingly, analysis of sound fields and perceived performance in reproduction becomes complicated. Moreover, some misunderstandings of the principle of multichannel surround sounds may cause conceptual and logical confusions.

In this chapter, the basic concepts and principles of multichannel surround sound are discussed as an introduction to multichannel surround sound. Multichannel surround sound is introduced as a consequence of sound field and psychoacoustic-based approximation. On this basis, traditional and simple psychoacoustic-based analysis methods for multichannel surround sound are addressed. The basic considerations and physical and psychoacoustic principles of multichannel surround sound are explained in Section 3.1. The principle of summing localization with multiple sound sources (loudspeakers) is often used in multichannel surround sound reproduction within a small-sized listening region. As such, summing localization with multiple sound sources in a horizontal plane is analyzed, and summing localization equations are derived in Section 3.2. The methods to recreate the subjective sensations, which are similar to those in the environment of a hall, in multichannel surround sound reproduction are outlined in Section 3.3. The discussion provides guidance for a detailed analysis of various multichannel surround sounds in succeeding chapters.

3.1 PHYSICAL AND PSYCHOACOUSTIC PRINCIPLES OF MULTICHANNEL SURROUND SOUND

As stated in Section 1.9.1, spatial sound aims to record (or simulate), transmit (or store) and reproduce the spatial information of a sound field and then recreate the desired spatial auditory events or perceptions. Sound field reconstruction is a method to reproduce spatial information accurately by which a pressure distribution matching with that of the target

DOI: 10.1201/9781003081500-3

sound field is recreated in a certain region. According to the discussion in Section 1.2.1, an arbitrary sound field in a source-free region can be decomposed as a linear superposition of incident plane waves from various directions. Therefore, an intuitive method for sound field reconstruction involves arranging a number of loudspeakers on a spherical surface around a listener or receiver region. If loudspeakers are approximated as point sources and the radius of the spherical surface is large enough, the sound field near the center of the spherical region can be approximated as a superposition of incident plane waves from all loudspeaker directions. The complex-valued amplitude (including magnitude and phase) of each incident plane wave component can be controlled by changing the complex-valued amplitude of the corresponding loudspeaker signal. Various sound fields and associated temporal and spatial information can be theoretically recreated with this method. For example, to reconstruct the sound field of a concert hall within a region (Figure 1.6), the direct sound from the front is recreated by the front loudspeaker; the early lateral reflection is recreated by the lateral loudspeaker, and the late diffused reverberations are recreated by all loudspeakers at various directions.

An arbitrary sound field in the source-free region is generally composed of incident plane waves with a continuous direction distribution. Accordingly, a directional-continual loudspeaker arrangement is required for sound field reconstruction with the aforementioned method. However, this condition is actually infeasible. In practice, a discrete arrangement and finite number of loudspeakers are available to reconstruct the target sound field. Analogous to the temporal sampling of a continual signal in the time domain, a discrete loudspeaker arrangement is equivalent to applying a directional sampling procedure to the directional-continual distribution of loudspeakers. If the directional interval (grid) of the actual loudspeaker arrangement is dense enough so that the physical error caused by directional sampling is negligible, the reconstructed sound field can be regarded as an exact duplication of the target sound field. Since the 1950s, this intuitive method has been used to simulate the sound field and auditory perception in a room (hall). For example, the researchers at the Third Physical Institute of the University of Gottingen used an array with 65 loudspeakers to simulate the sound field in a room (Meyer and Thiele, 1956). Even though many loudspeakers are used, the physical error of the reconstructed sound field at high frequencies is non-negligible. More channels and loudspeakers are required to reconstruct a sound field accurately up to 20 kHz. Accordingly, more signal transmission and storage channels are also needed. Numerous signal channels cause difficulty even for current digital media. The aforementioned intuitive method is mainly for scientific studies rather than for commercial or consumer purposes because of its complicity.

Alternatively, if the directional interval of loudspeakers satisfies the requirement of the Shannon–Nyquist spatial sampling theorem (Chapter 9), the target sound field can still be reconstructed within a certain region and up to a certain upper-frequency limit by appropriate loudspeaker signals. This reconstruction is the ***starting point of an actual spatial sound based on sound field reconstruction***. The physical nature of sound field reconstruction involves the directional sampling and reconstruction of an incident sound field. They are characterized by the appropriate combinations of different loudspeaker arrangements (spatial sampling schemes) and loudspeaker signals (spatial interpolation schemes). As stated in Section 1.9.1, sound field-based methods can create authentic and natural auditory perceptions within a reasonable or even large region, and the resultant perceived performances change slightly as the listener's position varies. However, such methods require a large number of channels and loudspeakers, so they remain complicated.

For practical uses, systems with an appropriate number of channels and loudspeakers are preferred. Practical systems are derived from a simplification of the exact sound field reconstruction in accordance with psychoacoustic principles. These systems are the ***starting point of the multichannel surround sound based on sound field approximation and psychoacoustics*** (Xie, 1999a, 1999b; Xie and Guan, 2002). The related logic and considerations are outlined as follows.

Spatial information is reproduced roughly or even omitted to make a compromise between complexity and perceptual performance in reproduction. When the capacity of transmission or storage media is limited, the spatial information most important for auditory perception is reproduced as exactly as possible at the cost of reducing the accuracy or even omitting the spatial information that is relatively less important for auditory perception. For example, if horizontal information is assumed to be more important than vertical information, the former is retained, and the latter is omitted. In this case, only horizontal channels and loudspeakers are used. This condition is the basic consideration of multichannel horizontal surround sound.

In multichannel surround sound, the spatial information in all directions should be reproduced with an equal accuracy or equal extent of approximation if the spatial information from all directions is assumed to be of equal importance, and the auditory resolutions in all directions are assumed to be identical. By contrast, if the spatial information from some directions (usually from the front) is considered dominant and when auditory resolution is the finest in the front, the spatial information from these directions should be reproduced accurately or with an enhanced approximation. A rough approximation in the reproduction of information from the other directions (such as the rear) is acceptable.

The perceptual importance and appropriate approximation of spatial information from different directions depend on the application. For example, the multichannel horizontal surround sound experiences the following:

1. For music reproduction without an accompanying picture (especially classical music reproduction), the main purpose is to recreate the auditory spatial information of concert halls. In addition to the spatial information of direct sound (usually from the front), the spatial information of early lateral reflections and late diffused reverberations of concert halls should be recreated as exactly as possible to create a good auditory spatial impression (Section 1.8.1). Consequently, in early times (the 1970s), the spatial information from all horizontal directions was subjected to equal approximation, and regular loudspeaker configurations were mostly used in multichannel horizontal surround sound for true music reproduction. However, when the capacity of transmission or storage media is limited, this equal or unbiased manipulation is problematic.
2. For sound reproduction with accompanying pictures, such as reproduction for commercial cinema, televisions, and domestic theaters, the main purpose is to recreate auditory perceptions that match a given picture. The localization information at frontal directions should be recreated accurately and stably because the attention of a listener is induced by the picture and thus focused in frontal directions. Conversely, lateral and rear information includes ambiance or occasional localization information supplementing the effect of an overall auditory scene; as such, it can be reproduced with rough approximation. Therefore, spatial information from different horizontal directions has been manipulated with different approximations in multichannel surround with accompanying picture since the end of the 1980s. Accordingly, irregular (front biased) loudspeaker configurations have been used to enhance accuracy in the reproduction of frontal information at the cost of rough approximation in the reproduction of lateral and rear information. However, they may give rise to some problems in compatible music reproduction without an accompanying picture.

Even for applications with an accompanying picture, requirements for the reproduction of ambient information depend on the content of the programs. For concert programs with pictures or incidental music for movies, ambient sound mainly includes reflections in original halls or simulated reflections by artificial reverberations; in these cases, requirements for the reproduction of ambient sound are similar to those for true music reproduction. Hence, in the development of multichannel surround sound, all ambient sounds can be easily taken for reflections. However, this consideration is incomplete and has caused some controversies (Zacharov, 1998a; Holman, 2000). Reflections are important parts but not all of the ambient sound. In many video/audio and movie programs, ambiance may be direct background sounds, such as raindrops and cheering from viewers in an outdoor sports tournament. In addition, the sound recording director may sometimes add some special localization effects to the side and rear parts, such as the effect of a helicopter flying around the audience. However, the accuracy of these special localization effects is not as crucial as that in front. Overall, lateral and rear information is not always ambient information, let alone always reflected information. Ideally, the content and requirement in ambient sound reproduction should match the given pictures.

For multichannel surround sound reproduction, in addition to recreating sensations similar to reverberation-related listener envelopment in concert halls, the sensations of being enveloped by or immersed in direct sound, such as the sensation of being surrounded by rain, should be recreated. This sensation is also called ***envelopment***. Therefore, the meaning of "envelopment" in multichannel surround sound reproduction is slightly different from that in concert hall acoustics. The former can be recreated by multiple direct sounds or late reverberation. In other words, ***the term "envelopment" has more implications in multichannel surround sound*** (Berg, 2009; George et al., 2010). Nevertheless, some signal processing techniques for recreating the perceived envelopment by multiple direct sounds and reverberation in multichannel surround sound reproduction may be similar.

Overall, the relative importance of spatial information from different directions depends on applications or even program contents. The diversity in applications and program contents leads to different requirements in the approximation of spatial information from various directions. When the information transmission or storage capacity of a system is limited, optimizing the spatial perceptual performances is difficult in all cases. Practical multichannel surround sounds are usually appropriate compromises for most cases.

Simplifying the number of channels and loudspeakers for practical uses is equivalent to applying directional downsampling or rough sampling to the directional distribution of loudspeakers at the far-field distance. Regular and irregular loudspeaker configurations in horizontal plane correspond to uniform and nonuniform downsampling in directions, respectively. After directional downsampling, an arbitrary target sound field cannot be reconstructed exactly by various signal mixing (directional interpolation) methods because of the restriction of the Shannon–Nyquist spatial sampling theorem. The target sound field can be reconstructed roughly within a small region and below a frequency limit rather than the full audible frequency bandwidth at most. In this case, various psychoacoustic-based methods should be used to recreate the desired spatial auditory perceptions, i.e., to "deceive" hearing. For reproduction within a relatively small-sized listening region(such as domestic reproduction), summing localization with interchannel level difference (ICLD) or interchannel time difference (ICTD) is applicable to recreate a virtual source between the directions of loudspeakers (Sections 1.7 and 2.1).As indicated in Section 2.1, for two stereophonic loudspeakers with ICLD only (amplitude panning), the low-frequency sound field adjacent to the central line is a good approximation of the incident plane wave from the target direction. Summing localization with ICLD is a consequence of the dominant role of low-frequency interaural time difference (ITD) in azimuthal localization. However, summing localization with ICTD (time panning) for a transient signal is a psychoacoustic phenomenon. Therefore, the virtual source generated by psychoacoustic-based methods may be perceived not as authentic and natural as that of a real source; moreover, appropriate localization performance may be limited in a narrow listening region or "sweet point."

Spatial auditory perceptions or sensations of reflections (such as those similar to reflections in a concert hall) can be recreated via two psychoacoustic-based methods.

1. ***Indirect method.*** If multiple virtual sources can be recreated simultaneously in arbitrary horizontal or three-dimensional directions, early lateral reflections and late diffused reverberations are simulated as if they are generated by those virtual sources. The overall effect is to generate a low interaural cross-correlation coefficient (IACC) in reproduction, resulting in spatial auditory perceptions similar to those in a concert hall.
2. ***Direct method.*** As stated in Section 1.7.3 and Section 3.3, two or more loudspeakers with low-correlated signals yield low IACC in binaural pressures and directly result in spatial auditory perceptions similar to those in a concert hall. This method is usually simpler than the indirect method. It is effective in a relatively large listening region, but its perceived performance may be sometimes inferior to that of the indirect method.

Some multichannel surround sound techniques are applicable to cinema and domestic or consumer reproduction. Some domestic techniques are simplified from the cinema sound technique. However, psychoacoustic methods used in cinema sound reproduction may be different from those utilized in domestic reproduction because of the difference in the required sizes of a listening region. Accordingly, loudspeaker configuration and signal mixing are different in two cases. For example, for most seats in a cinema, the path difference from the left-frontal and right-frontal loudspeakers to the receiver position is large so that the difference in propagating delay exceeds the upper limit of summing localization with the interchannel time difference (ICTD). In this case, summing localization cannot be applied to recreate a virtual source between two loudspeakers. Instead, in cinema sound, a signal is usually panned to a single loudspeaker to recreate the exact localization perception in the loudspeaker direction to avoid the problem caused by summing localization with two or more loudspeakers. A target plane wave in the loudspeaker direction is reconstructed in this case. A similar method is also applicable to domestic reproduction to recreate directional information with high accuracy and stability. For example, a frontal localization signal (such as a dialog) is often panned to the central loudspeaker only for domestic reproduction with an accompanying picture. Thus, practical multichannel surround sounds can be regarded as a combination of psychoacoustic- and sound field-based techniques (Guan, 1995).

In summary, increasing the number of channels and loudspeakers enhances accuracy in sound field reconstruction. Consequently, the degree of spatial information loss reduces, and the perceived performance improves. However, this process makes the system complicated. The requirements of accuracy and simplicity are usually considered a mutual conflict. Practical techniques and systems are compromised on the basis of psychoacoustic principles through which some spatial information is omitted. They can recreate a rough approximation of the target sound field. Accordingly, practical techniques and systems should be designed in accordance with psychoacoustic principles and in consideration of multiple factors, such as available system resources (transmission or storage capacity of media) and the required size of listening regions. The number, configuration, and signals of loudspeakers should be chosen appropriately to reproduce the important information accurately and create the desired spatial auditory perceptions. These aspects are the main considerations and core problems in the development of multichannel surround (including horizontal and spatial surround) sound.

Various multichannel surround sounds with a different number of channels and loudspeaker configurations have been developed. Various loudspeaker configurations can be regarded as ***different procedures of directional downsampling.*** On the basis of the concepts of directional-downsampling, multichannel surround sound can be subdivided according to the number of channels and the configuration of loudspeakers, such as 5.1 and 7.1 channel techniques (Section 1.9.2). This classification method may be straightforward, but not very strict. It is consistent with the development of the multichannel surround sound and subdivision of the traditional channel-based spatial sound (Section 6.5.1). For each loudspeaker configuration, various loudspeaker signal panning and mixing methods have been developed. Some panning and mixing methods are universal and suitable for different loudspeaker configurations, but some other methods may be only valid for special loudspeaker configurations. In Section 11.5.3, under certain conditions, different panning methods for recreating virtual sources are analogous to ***different directional interpolation procedures in loudspeaker signal amplitudes or gains.*** When the number and configuration of loudspeakers do not satisfy the requirement of the Shannon–Nyquist spatial sampling theorem, the target sound field cannot be reconstructed exactly with these panning methods or directional interpolation procedures. The target sound field can only be reconstructed within a small region and below a certain frequency. However, these signal panning or mixing methods can recreate the desired auditory perceptions in accordance with appropriate psychoacoustic principles.

The final perceived performances in multichannel surround sound reproduction are determined by multiple factors, including the number, configuration, and signals of loudspeakers and room acoustic environments in reproduction. An appropriate design on the number, configuration, and signals of loudspeakers depends on the desired spatial perceived effects and available system resources. This problem is addressed in Chapters 4 to 6. The physical and perceived effects on various numbers, configurations, and signals of loudspeakers are analyzed sufficiently. The cases of ***different numbers of the channels and configurations of loudspeakers*** are discussed separately.

3.2 SUMMING LOCALIZATION IN MULTICHANNEL HORIZONTAL SURROUND SOUND

3.2.1 Summing localization equations for multiple horizontal loudspeakers

One of the basic requirements of spatial sound is the recreation of the localization information of a target sound source. As stated in Section 3.1, for reproduction within a small-sized listening region (as in domestic reproduction), a virtual source may be recreated in accordance with the principle of summing localization with multiple loudspeakers (real sources) to recreate the localization information of target sound sources. This process is an extension of the summing localization of two frontal loudspeakers discussed in Sections 1.7.1 and 2.1 to the cases of multiple loudspeakers. For two completely correlated loudspeaker signals with ICLD only, the perceived direction of a virtual source at low frequencies is evaluated with summing localization Equations (2.1.5), (2.1.6), and (2.1.10). These equations can be extended to the cases of multiple loudspeakers with in-phase or partly out-of-phase signals and ICLD only. For conciseness, the case of horizontal loudspeaker configuration, i.e., the case of multichannel horizontal surround sound, is discussed here (Xie and Liang, 1995; Xie, 2001a). The resultant summing localization equations are applied to analyze various multichannel horizontal surround sounds in Chapters 4 and 5. They are extended to the cases of the three-dimensional spatial loudspeaker configuration in Chapter 6.

Similar to the analysis in Section 2.1, the analysis in this section assumes that the shadow of the head is disregarded, and the two ears are approximated by two points in a free space separated by $2a$, where a is the head radius. For a horizontal sound source at the far-field distance and azimuth θ_S, the incident sound is approximated as a plane wave, and the interaural phase delay difference ITD_p is evaluated with Equation (1.6.1):

$$ITD_p(\theta) = \frac{2a}{c}\sin\theta_S. \quad (3.2.1)$$

If the listener's head rotates around the vertical axis anticlockwise with an azimuth $\delta\theta$, ITD_p becomes

$$ITD_p(\theta_S - \delta\theta) = \frac{2a}{c}\sin(\theta_S - \delta\theta). \quad (3.2.2)$$

Therefore, when the head rotates at an azimuth of $\delta\theta = \theta_S$ to be oriented to the sound source, ITD_p vanishes.

A general note on the symbols of multichannel signals is initially made. Similar to the case of two-channel signals in Section 2.1.1, the complex-valued signal in the frequency domain of the ith loudspeaker is denoted by E_i. The signals of different loudspeakers may be correlated or uncorrelated. When the same signals with different complex-valued amplitudes (including different magnitudes and phase and linear delays) are fed to multiple loudspeakers, the ith loudspeaker signal is given by $E_i = A_i E_A$, where $E_A = E_A(f)$ is the signal waveform in the frequency domain, and A_i is the normalized complex or real-valued amplitude, relative gain, or scalar factor of the ith loudspeaker signal. A_i is dependent on the physical characteristics of the electroacoustic system and independent of the amplitude of input signals. Therefore, using A_i as a set of parameters is convenient to specify the physical characteristics of the system. This parameter is also termed (signal)***panning value*** and expressed as $A_i = A_i(\theta_S)$ as a function of target source direction. It is also termed (signal) ***panning function*** in other studies. A_i is used to represent the normalized loudspeaker signals in the succeeding discussion of summing localization with multiple loudspeakers to emphasize the relative relationship among the complex-valued amplitudes of different loudspeaker signals. A complete loudspeaker signal in the frequency domain should be $E_i = A_i E_A(f)$. If $A_i$is independent of frequency, the complete loudspeaker signals in the time domain are given by $e_i(t) = A_i\, e_A(t)$, where $e_i(t)$ and $e_A(t)$ are the time domain counterparts of E_i and $E_A(f)$, respectively. However, if the signals of two or more loudspeakers are different (such as uncorrelated), loudspeaker signals should be generally represented by $E_i(f)$.

In the case of multiple loudspeaker reproduction in Figure 3.1, M loudspeakers are arranged in a horizontal circle around a listener. The radius r_0 of a circle is much larger than that of the head, i.e., it satisfies the far-field condition $r_0 >> a$. The azimuth and normalized signal (amplitude) of ith loudspeaker are θ_i and A_i ($i = 0, 1, 2\ldots M - 1$), respectively. Similar to the case in Section 2.1.1, the transfer coefficient from the electric input of each loudspeaker to the amplitude of the free-field plane wave at the position of the head center (with the absence of head) is assumed to be a unit value in this section, if no specifications are otherwise provided. Then, the binaural sound pressures in the frequency domain are a linear superposition of the plane wave (far-field) pressures caused by all loudspeakers:

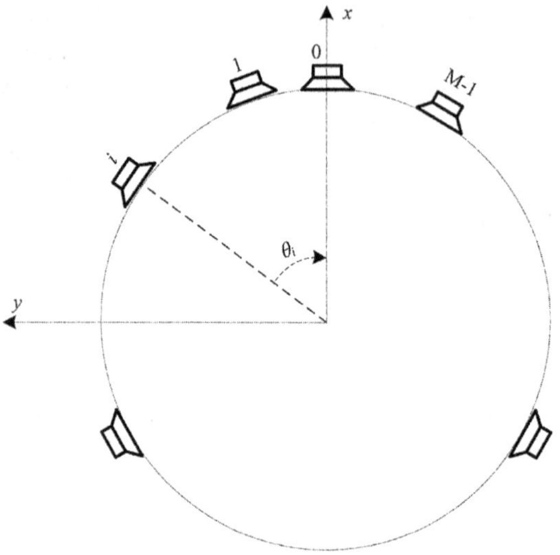

Figure 3.1 Horizontal configuration of loudspeakers.

$$P_L' = \sum_{i=0}^{M-1} A_i \exp(-jkr_{Li}) = \sum_{i=0}^{M-1} A_i \exp\left[-j(kr_0 - ka\sin\theta_i)\right],$$
$$P_R' = \sum_{i=0}^{M-1} A_i \exp(-jkr_{Ri}) = \sum_{i=0}^{M-1} A_i \exp\left[-j(kr_0 + ka\sin\theta_i)\right]. \quad (3.2.3)$$

where r_{Li} and r_{Ri} are the distances between the ith loudspeaker and the left or right ear, respectively.

Similar to the case of the two-channel stereophonic sound in Section 2.1.1, the interaural phase difference in multiple loudspeaker reproduction is calculated from Equation (3.2.3):

$$\Delta\psi_{SUM} = \psi_L - \psi_R = 2\arctan\left[\frac{\sum_{i=0}^{M-1} A_i \sin(ka\sin\theta_i)}{\sum_{i=0}^{M-1} A_i \cos(ka\sin\theta_i)}\right]. \quad (3.2.4)$$

The subscript "SUM" denotes the case of summing localization with multiple loudspeakers, and the interaural phase delay difference is evaluated with

$$ITD_{p,SUM} = \frac{\Delta\psi_{SUM}}{2\pi f} = \frac{1}{\pi f}\arctan\left[\frac{\sum_{i=0}^{M-1} A_i \sin(ka\sin\theta_i)}{\sum_{i=0}^{M-1} A_i \cos(ka\sin\theta_i)}\right]. \quad (3.2.5)$$

As stated in Section 1.6.5, the interaural phase delay difference is a dominant cue for lateral localization at low frequencies. It is assumed that the summing virtual source is located in the horizontal plane (but this assumption is not always true, Section 6.1.4). The perceived virtual source direction can be evaluated by comparing $ITD_{p,SUM}$ in Equation (3.2.5) with that of an actual sound source in Equation (3.2.1), i.e., comparing with patterns stored from prior experiences:

$$\sin\theta_I = \frac{1}{ka}\arctan\left[\frac{\sum_{i=0}^{M-1} A_i \sin(ka\sin\theta_i)}{\sum_{i=0}^{M-1} A_i \cos(ka\sin\theta_i)}\right]. \tag{3.2.6}$$

Equation (3.2.6) is complicated, but it indicates that the direction of the perceived virtual source depends on relative signal amplitudes and frequency (or *ka*).

At low frequencies with $ka << 1$, Equation (3.2.6) can be expanded as a Taylor series of *ka*. If the first expansion term is retained, the equation is simplified to

$$\sin\theta_I = \frac{\sum_{i=0}^{M-1} A_i \sin\theta_i}{\sum_{i=0}^{M-1} A_i}. \tag{3.2.7}$$

Similar to the case of Equation (2.1.8), $ITD_{p,SUM}$ in Equation (3.2.5) becomes Equation (3.2.8)when the listener's head rotates around the vertical axis with an azimuth $\delta\theta$ ($\delta\theta > 0$ represents a anticlockwise rotation):

$$ITD_{p,SUM}(\delta\theta) = \frac{\Delta\psi_{SUM}}{2\pi f} = \frac{1}{\pi f}\arctan\left\{\frac{\sum_{i=0}^{M-1} A_i \sin\left[ka\sin(\theta_i - \delta\theta)\right]}{\sum_{i=0}^{M-1} A_i \cos\left[ka\sin(\theta_i - \delta\theta)\right]}\right\}. \tag{3.2.8}$$

Similar to the derivation of Equation (2.1.10), Equation (3.2.9) describes the perceived virtual source direction at low frequencies with $ka << 1$ by selecting the rotation azimuth $\delta\theta$ so that the interaural phase delay difference given by Equation (3.2.8) vanishes:

$$\tan\hat{\theta}_I = \frac{\sum_{i=0}^{M-1} A_i \sin\theta_i}{\sum_{i=0}^{M-1} A_i \cos\theta_i}. \tag{3.2.9}$$

The notation $\hat{\theta}_I$ denotes the perceived virtual source azimuth with respect to a fixed coordinate when the head is oriented to the virtual source to distinguish the notation θ_I that denotes the perceived virtual source azimuth for a fixed head orientation.

Equations (3.2.7) and (3.2.9), which were derived by Bernfeld (1975), are summing localization equations for multiple horizontal loudspeakers. They indicate that the perceived virtual source direction is independent from the frequency or ka at very low frequencies. In the case of two frontal stereophonic loudspeakers, Equations (3.2.6), (3.2.7), and (3.2.9) are simplified into Equations (2.1.5), (2.1.6), and (2.1.10), respectively. Equations (3.2.6) to (3.2.9) have some issues, such as the effective frequency range and the selection of the equivalent head radius a. They are also similar to the cases of stereophonic sound in Sections 2.1.1 and 2.1.2. As such, they are not repeated here.

3.2.2 Analysis of the velocity and energy localization vectors of the superposed sound field

Virtual source localization equations can be derived by analyzing the superposed sound field of multiple sources or loudspeakers. As mentioned in the discussion of the law of tangent described in Equation (2.1.10) for two-channel stereophonic sound in Section 2.1.1, Makita (1962)supposed that the perceived virtual source direction in the superposed sound field is consistent with the inner normal direction of the superposed wavefront at the receiver position. On the basis of Makita's hypothesis, Gerzon (1992a, 1992b) proposed summing localization theorem by analyzing the velocity and energy (flow) vectors of the superposed sound field.

The coordinate with respect to a special receiver position shown in Figure 1.1 but in the absence of the head is used. The sound pressure at an arbitrary receiver position $\boldsymbol{r} = [x, y, z]^T$ is $p(\boldsymbol{r}, t)$, where superscript "T" denotes matrix transposition. The inner normal direction of the superposed wavefront is opposite to the velocity vector $\boldsymbol{v} = [v_x, v_y, v_z]^T$ of the medium. The components of velocity vector in three perpendicular directions of coordinate axes are given in other studies (Morse and Ingrad, 1968; Du et al., 2001):

$$v_x = -\frac{1}{\rho_0}\int \frac{\partial p}{\partial x}dt \qquad v_y = -\frac{1}{\rho_0}\int \frac{\partial p}{\partial y}dt \qquad v_z = -\frac{1}{\rho_0}\int \frac{\partial p}{\partial z}dt \tag{3.2.10}$$

where ρ_0 is the density of air. Based on Equation (1.2.11), Equation (3.2.10) can be expressed in the frequency domain as:

$$V_x = -\frac{1}{j2\pi f \rho_0}\frac{\partial P}{\partial x} \qquad V_y = -\frac{1}{j2\pi f \rho_0}\frac{\partial P}{\partial y} \qquad V_z = -\frac{1}{j2\pi f \rho_0}\frac{\partial P}{\partial z}. \tag{3.2.11}$$

Let $\boldsymbol{r}_S$ and $\boldsymbol{r}$ be the vector of the source and receiver positions, respectively. According to Equation (1.2.5), the plane wave pressure at a receiver position far from a point source is given by

$$P = P_A \exp\left[-j\boldsymbol{k}\cdot\left(\boldsymbol{r} - \boldsymbol{r}_S\right)\right]. \tag{3.2.12}$$

The amplitude P_A is frequency dependent on a nonharmonic source, i.e., $P_A = P_A(f)$. For conciseness, however, this frequency dependencies omitted in the equations.

Here, summing localization with multiple horizontal loudspeakers or sound sources is analyzed. The analysis is extended to the case of multiple loudspeakers in a three-dimensional space in Section 6.1.2. For the plane wave incident from the horizontal azimuth θ_S, the horizontal wave vector is $\boldsymbol{k} = [k_x, k_y]^T = [-k \cos\theta_S, -k \sin\theta_S]^T$, where $k = |\boldsymbol{k}|$ is the wave number. The horizontal receiver position is specified by vector $\boldsymbol{r} = [x, y]^T = [r \cos\theta, r \sin\theta]^T$. Equation (3.2.12) becomes

$$\begin{aligned} P &= P_A \exp\left[jk(\cos\theta_S x + \sin\theta_S y) - jkr_S\right] \\ &= P_A \exp\left[jkr\cos(\theta - \theta_S) - jkr_S\right]. \end{aligned} \tag{3.2.13}$$

The components of the velocity vector are expressed in Equation (3.2.14) by substituting Equation (3.2.13) into Equation (3.2.11):

$$\begin{aligned} V_x &= -\frac{P_A \cos\theta_S}{\rho_0 c} \exp\left[jk(\cos\theta_S x + \sin\theta_S y) - jkr_S\right], \\ V_y &= -\frac{P_A \sin\theta_S}{\rho_0 c} \exp\left[jk(\cos\theta_S x + \sin\theta_S y) - jkr_S\right]. \end{aligned} \tag{3.2.14}$$

where c is the speed of sound. The complex-valued amplitudes of the medium velocity component in the position adjacent to the origin are given in the following equation by omitting the phase factor $\exp(-jkr_S)$:

$$V_{A,x} = -\frac{P_A \cos\theta_S}{\rho_0 c} \qquad V_{A,y} = -\frac{P_A \sin\theta_S}{\rho_0 c}. \tag{3.2.15}$$

The negative sign on the right side of Equation (3.2.15) indicates that the sound wave is incident from the source $r_S = [x_S, y_S]^T$ to the origin. The normalized amplitude of the medium velocity is defined as

$$V_{norm,x} = -\frac{\rho_0 c V_{A,x}}{P_A} \qquad V_{norm,y} = -\frac{\rho_0 c V_{A,y}}{P_A}. \tag{3.2.16}$$

The division by P_A in Equation (3.2.16) is equivalent to normalizing the amplitude of the plane wave in Equation (3.2.12) to a unit value. The negative sign in Equation (3.2.16) makes up for the opposite directions between the source and medium velocity. The source direction with respect to the origin is determined with the normalized amplitude of medium velocity components:

$$\cos\theta_S = V_{norm,x} \qquad \sin\theta_S = V_{norm,y}. \tag{3.2.17}$$

Therefore, $V_{norm,x}$ and $V_{norm,y}$ constitute a horizontal vector $\boldsymbol{r}_v$, which is also called ***velocity vector*** by Gerzon. $\boldsymbol{r}_v$ is termed "***velocity localization vector***" here to distinguish the actual velocity vector $\boldsymbol{v}$ of the medium expressed in Equation (3.2.10) and related to $\boldsymbol{v}$ by a certain scaling factor as $|\boldsymbol{r}_v| = \rho_0 c |\boldsymbol{v}| / P_A$. Then, it has

$$\boldsymbol{r}_v = \left[V_{norm,x}, V_{norm,y}\right]^T = \left[\cos\theta_S,\ \sin\theta_S\right]^T. \tag{3.2.18}$$

In the case of the incident plane wave, the velocity localization vector is a unit vector with $r_v = |\, r_v \,| = 1$, and it points to the direction of the sound source or incident direction of the plane wave.

For the reproduction with multiple horizontal loudspeakers shown in Figure 3.1, the azimuth of the ith loudspeaker is θ_i ($i = 0, 1, 2...M-1$), whose corresponding position vector is $\boldsymbol{r}_i = [r_0 \cos\theta_i, r_0 \sin\theta_i]^T$. The normalized amplitudes of loudspeaker signals are temporarily assumed to be real values so that these signals are in phase or partly out of phase. After the function $E_A(f)$ for the signal waveform in the frequency domain is omitted, the superposed sound pressure at a receiver position $\boldsymbol{r} = [x, y]^T = [r \cos\theta, r \sin\theta]^T$ adjacent to the origin is given by

$$\begin{aligned} P' &= \sum_{i=0}^{M-1} A_i \exp\left[-j\boldsymbol{k}_i \cdot (\boldsymbol{r} - \boldsymbol{r}_i)\right] \\ &= \sum_{i=0}^{M-1} A_i \exp\left[jk(\cos\theta_i x + \sin\theta_i y) - jkr_0\right] \\ &= \sum_{i=0}^{M-1} A_i \exp\left[jkr(\cos\theta_i - \theta) - jkr_0\right], \end{aligned} \tag{3.2.19}$$

where $\boldsymbol{k}_i$ is the wave vector of the sound wave from the ith loudspeaker, and $|\boldsymbol{k}_i| = k$, $i = 0,1...(M-1)$. At the origin with $r \to 0$, the amplitude of the superposed sound pressure is

$$P'_A = \sum_{i=0}^{M-1} A_i. \tag{3.2.20}$$

The sound pressure of a single plane wave in Equation (3.2.13) is substituted with the superposed sound pressure given in Equation (3.2.19), and the common phase factor $\exp(-jkr_0)$ is omitted in a manner similar to the derivation from Equation (3.2.14) to Equation (3.2.16). For a small kr, the components of the normalized amplitude of the medium velocity of the superposed sound field are evaluated with

$$V'_{norm,x} = \frac{\sum_{i=0}^{M-1} A_i \cos\theta_i}{\sum_{i=0}^{M-1} A_i} \qquad V'_{norm,y} = \frac{\sum_{i=0}^{M-1} A_i \sin\theta_i}{\sum_{i=0}^{M-1} A_i}. \tag{3.2.21}$$

Similar to the case of a single plane wave in Equation (3.2.17), the x and y components of the velocity localization vector $\boldsymbol{r}_v$ are given by

$$r_v \cos\theta_v = V'_{norm,x} = \frac{\sum_{i=0}^{M-1} A_i \cos\theta_i}{\sum_{i=0}^{M-1} A_i} \qquad r_v \sin\theta_v = V'_{norm,y} = \frac{\sum_{i=0}^{M-1} A_i \sin\theta_i}{\sum_{i=1}^{M-1} A_i}, \tag{3.2.22}$$

or

$$r_v = [r_v \cos\theta_v, r_v \sin\theta_v]^T = [V'_{norm,x}, V'_{norm,y}]^T = \left[\frac{\sum_{i=0}^{M-1} A_i \cos\theta_i}{\sum_{i=0}^{M-1} A_i}, \frac{\sum_{i=0}^{M-1} A_i \sin\theta_i}{\sum_{i=0}^{M-1} A_i}\right]^T. \quad (3.2.23)$$

where $r_v = |\, \boldsymbol{r}_v \,|$ is the length (module) of the velocity localization vector and termed ***velocity vector magnitude***; θ_v is the ***velocity localization vector-based azimuth***, i.e., the azimuth of the inner normal direction of the superposed wavefront, or the perceived virtual source azimuth supposed by Makita. Equations (3.2.22) and (3.2.23) are the horizontal summing localization equation by Gerzon based on Makita's theorem and velocity localization vector.

The result of Equation (3.2.23) can be expressed in the form of vector superposition. The direction of the *i*th loudspeaker is represented by a unit directional vector as follows:

$$\hat{\boldsymbol{r}}_i = \frac{\boldsymbol{r}_i}{|\, \boldsymbol{r}_i \,|} = [\cos\theta_i, \sin\theta_i]^T. \quad (3.2.24)$$

The weights are defined as

$$A'_i = \frac{A_i}{\sum_{i'=0}^{M-1} A_{i'}} \qquad i = 0, 1 \ldots (M-1). \quad (3.2.25)$$

Then, the velocity localization vector $\boldsymbol{r}_v$ is the weighted superposition of the unit directional vectors of all loudspeakers:

$$\boldsymbol{r}_v = \sum_{i=0}^{M-1} A'_i \hat{\boldsymbol{r}}_i. \quad (3.2.26)$$

The direction of $\boldsymbol{r}_v$ is the perceived virtual source direction proposed by Makita. Figure 3.2 shows an example of a summing virtual source by two loudspeakers.

Some comments are supplemented to Equation (3.2.22) or (3.2.23):

1. These equations are valid below the frequency of 0.7 kHz.
2. From Equation (3.2.22), the velocity localization vector-based azimuth is evaluated by

$$\tan\theta_v = \frac{\sum_{i=0}^{M-1} A_i \sin\theta_i}{\sum_{i=0}^{M-1} A_i \cos\theta_i}. \quad (3.2.27)$$

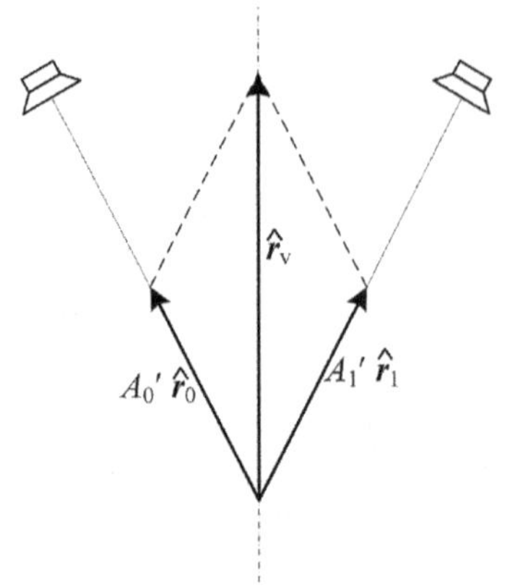

Figure 3.2 Vector representation of a summing virtual source by two loudspeakers.

It is consistent with Equation (3.2.9). Therefore, the perceived azimuth evaluated from the velocity localization vector is equivalent to that examined from the head oriented to the virtual source, i.e., $\theta_v = \hat{\theta}_I$.

3. The comparison of Equations (3.2.22) and (3.2.7) reveals that the perceived azimuth θ_I evaluated from the interaural phase delay difference for the fixed head orientation is related to θ_v as

$$\sin\theta_I = r_v \sin\theta_v = \frac{\sum_{i=0}^{M-1} A_i \sin\theta_i}{\sum_{i=0}^{M-1} A_i}. \tag{3.2.28}$$

θ_I is determined by the y (left-right) component but independent of the x (front-back) component of the velocity localization vector. Generally, θ_I is unequal to θ_v, i.e., the perceived azimuth for the fixed head orientation is different from that for the head oriented to the virtual source. Therefore, the perceived azimuth is unstable under head rotation. When $r_v = 1$, $\theta_I = \theta_v$. In this case, the perceived azimuth is stable under head rotation.

4. In the case of a single source at a far-field distance (incident plane wave), Equation (3.2.18) yields $r_v = 1$, and the perceived azimuth is stable under head rotation. In the case of summing virtual localization with multiple loudspeakers, the velocity factor is calculated from Equation (3.2.23):

$$r_v = \sqrt{V'^2_{norm,x} + V'^2_{norm,y}} = \frac{\sqrt{\left(\sum_{i=0}^{M-1} A_i \cos\theta_i\right)^2 + \left(\sum_{i=0}^{M-1} A_i \sin\theta_i\right)^2}}{\left|\sum_{i=0}^{M-1} A_i\right|}. \tag{3.2.29}$$

r_v is generally unequal to a unit. The closer r_v to a unit is, the more stable the perceived azimuth during head rotation will be. Therefore, r_v is a measure of the stability of the virtual source under head rotation.

5. In practice, the optimized r_v may serve as one of the conditions for designing loudspeaker configuration and signals in a multichannel surround sound. For example, in the case that signals for all $M \geq 2$ loudspeakers are in phase (with all $A_i \geq 0$), Equation (3.2.29) shows that the resultant r_v is less than a unit unless the signal is only fed to one single loudspeaker. Some out-of-phase loudspeaker signals with $A_i < 0$ are required to make r_v equal to or exceed a unit. This phenomenon is addressed in Section 4.1.3.
6. The cases of in-phase or partly out-of-phase loudspeaker signals with real-valued amplitudes are analyzed above. This discussion can be extended to the case of loudspeaker signals with an arbitrary phase (difference) so that A_i is complex valued (Gerzon, 1992a). In this case, Equation (3.2.22) should take the real value of $V'_{norm,x}$ and $V'_{norm,y}$, i.e., $V'_{norm,x}$ and $V'_{norm,y}$ in Equation (3.2.22) should be replaced with $\mathrm{Re}(V'_{norm,x})$ and $\mathrm{Re}(V'_{norm,y})$. For two-channel stereophonic loudspeaker configuration, the results are consistent with Equation (2.1.17). And the analysis of $\mathrm{Im}\,(V'_{norm,y})$, e.g., on the imaginary part of the y (left-right) component of the velocity localization vector, leads to the phasiness in Equation (2.1.25).

As an example of the preceding discussion, the velocity vector magnitude of two-channel stereophonic sound reproduction with loudspeakers arranged at $\theta_i = \pm\,\theta_0$, $A_0 = A_L$, and $A_1 = A_R$ are evaluated with Equation (3.2.29):

$$r_v = \frac{\sqrt{A_L^2 + A_R^2 + 2A_L A_R \cos 2\theta_0}}{|A_L + A_R|}. \tag{3.2.30}$$

When two loudspeaker signals are in phase, $0.707\sqrt{1+\cos 2\theta_0} \leq r_v \leq 1$ is obtained. The minimal r_v occurs at $A_L = A_R$, i.e., for a directly frontal virtual source. However, if $2\theta_0 \leq 60°$, then $0.866 \leq r_v \leq 1$, which is close to the unit. Therefore, the span angle between two stereophonic loudspeakers should not exceed a certain limit to ensure the stability of a virtual source under head rotation. The standard span angle is $2\theta_0 = 60°$. $r_v = 1$ when either A_L or A_R vanishes, i.e., a signal is reproduced with a single loudspeaker. When A_L and A_R are out of phase, r_v exceeds a unit. This phenomenon corresponds to the case of an outside-boundary virtual source or auditory event with an uncertain position. The outside-boundary virtual source is unstable under head rotation (Section 2.1.3). Therefore, some basic characteristics in two-channel stereophonic reproduction are further proved here.

In addition to summing localization theorem based on a velocity localization vector, summing localization theorem based on the energy localization vector of a sound field was proposed by Gerzon. This theorem is applicable to the evaluation of the perceived virtual source direction at mid-high frequencies from 0.7 kHz to 5 kHz.

In the case of a single sound source at the far-field distance (incident plane wave) expressed in Equation (3.2.12) or (3.2.13), the sound intensity vector is

$$\mathbf{I} = \frac{1}{T}\int_0^T \mathrm{Re}(p)\mathrm{Re}(\boldsymbol{v})\,dt = \frac{1}{2}P_A \mathbf{V}_A = -\frac{P_A^2}{2\rho_0 c}\hat{\boldsymbol{r}}. \tag{3.2.31}$$

The notations in Equation (3.2.31) are similar to these in the preceding discussion. $\boldsymbol{v}$ is the velocity vector of the medium, $\mathbf{V}_A = [V_{A,x}, V_{A,y}]^{\mathrm{T}}$ is its amplitude vector, and $\hat{\boldsymbol{r}} = [\cos\theta_S, \sin\theta_S]^T$ is the unit vector from the origin to the source direction. The negative sign on the third equality of Equation (3.2.31) indicates that the direction of sound intensity (energy flow) is just opposite to $\hat{\boldsymbol{r}}$.

For the reproduction with multiple horizontal loudspeakers shown in Figure 3.1, the incoherent superposition of sound intensities caused by M loudspeakers is expressed in the following equation, but the scale of the function $|E_A(f)|^2$ related to the overall sound intensity is omitted:

$$I' = \sum_{i=0}^{M-1} I_i = -\frac{1}{2\rho_0 c}\sum_{i=0}^{M-1} A_i^2 \hat{r}_i. \tag{3.2.32}$$

where $\hat{r}_i$ is the unit directional vector of the *i*th loudspeaker defined in Equation (3.2.24), and A_i^2 is the square of the signal amplitude (assumed to be a real value) of the *i*th loudspeaker. At the origin of the coordinate, the normalized incoherent superposition of power generated by M loudspeakers is given by

$$Pow' = \sum_{i=0}^{M-1} A_i^2. \tag{3.2.33}$$

Similar to the derivation of Equations (3.2.22) and (3.2.23), the perceived virtual source direction at high frequencies is opposite to the direction of sound intensity, then

$$r_E \cos\theta_E = \frac{\sum_{i=0}^{M-1} A_i^2 \cos\theta_i}{\sum_{i=0}^{M-1} A_i^2} \qquad r_E \sin\theta_E = \frac{\sum_{i=0}^{M-1} A_i^2 \sin\theta_i}{\sum_{i=0}^{M-1} A_i^2}. \tag{3.2.34}$$

Similar to Equation (3.2.26), Equation (3.2.34) can be expressed in the form of vector superposition:

$$\boldsymbol{r}_E = \sum_{i=0}^{M-1} A_i''^2 \hat{r}_i \qquad A_i''^2 = \frac{A_i^2}{\sum_{i'=0}^{M-1} A_{i'}^2}. \tag{3.2.35}$$

Equation (3.2.34) is Gerzon's horizontal summing localization equation based on the normalized intensity or energy localization vector. $\boldsymbol{r}_E = [r_E \cos\theta_E, r_E \sin\theta_E]^T$ is the (normalized) ***energy localization vector***, and θ_E is the energy localization vector-based azimuth, which is supposed to be the perceived virtual source azimuth at the high frequency. $r_E = |\boldsymbol{r}_E|$ is the length (module) of the energy localization vector and termed ***energy vector magnitude*** with

$$r_E = \frac{\sqrt{\left(\sum_{i=0}^{M-1} A_i^2 \cos\theta_i\right)^2 + \left(\sum_{i=0}^{M-1} A_i^2 \sin\theta_i\right)^2}}{\sum_{i=0}^{M-1} A_i^2} \tag{3.2.36}$$

By definition, $0 \le r_E \le 1$. For a single source at a far-field distance, $r_E = 1$. r_E describes the concentration of energy flow at azimuth θ_E, so it is a measure of the quality of a virtual source at high frequencies. Gerzon (1992a) suggested that loudspeaker signals at high frequencies should be optimized to obtain a larger r_E (close to unit). This condition is equivalent to concentrating the energy flow to the direction of θ_E and may serve as one of the conditions for designing loudspeaker signals at high frequencies.

Equations (3.2.32) and (3.2.33) are derived from an incoherent superposition of sound intensities caused by M loudspeakers. If all loudspeakers are arranged in a circle with an equal distance to the origin and loudspeaker signals are correlated, then the sound pressure at the origin should be a coherent superposition of those caused by M loudspeakers. However, at the off-origin position, path differences exist among loudspeakers. The phase of the superposed sound pressure varies rapidly with the change in the receiver position because of the path difference and short wavelength at high frequencies. As a result, for signals with a large bandwidth, an incoherent superposition of sound pressures is an appropriate approximation. Therefore, the energy localization vector analysis is suitable for high-frequency signals with a large bandwidth.

3.2.3 Discussion on horizontal summing localization equations

Based on different physical and psychoacoustic hypotheses, several horizontal summing localization equations are derived in the preceding sections:

1. ITD_p-based Equation (3.2.6) for fixed head orientation, which is simplified into Equation (3.2.7) at low frequencies.
2. ITD_p-based equation (3.2.9) for the head oriented to the virtual source.
3. Velocity localization vector-based Equation (3.2.22).
4. Energy localization vector-based Equation (3.2.34).

All these equations yield approximate results. These equations are related to one another but are valid under different conditions. Different equations may reveal different results. These equations may also have different interpretations. Some basic problems associated with these equations should be addressed in detail to avoid confusion.

As stated in Section 1.6, interaural phase delay difference ITD_p is a dominant cue for lateral localization at low frequencies below 1.5 kHz. The dynamic cue caused by head rotation helps to resolve front-back ambiguity and contributes to vertical localization. Equation (3.2.6) or (3.2.7) is based on ITD_p for the fixed head orientation. The perceived azimuth cannot be identified unambiguously from $\sin\theta_I$ in Equation (3.2.6) or (3.2.7) because of the multiple-valued characteristic of the arcsine function. This condition is a consequence of the front-back symmetry in ITD_p. According to Wallach's hypothesis discussed in Section 1.6.3, the dynamic ITD_p variation caused by the head rotation should be supplemented to resolve front-back ambiguity. This phenomenon may yield some new results, as indicated by the discussion in the spatial surround sound in Section 6.1.4. Conversely, $\tan\hat{\theta}_I$ in Equation (3.2.9) is derived from the head oriented to the virtual source so that ITD_p vanishes. It is related but not completely equivalent to the ITD_p variation caused by head rotation. The perceived azimuth may be identified unambiguously by combining $\sin\theta_I$ in Equations (3.2.6) or (3.2.7) and (3.2.9) with $\tan\hat{\theta}_I$ in Equation (3.2.9). In a practical analysis of multichannel sound reproduction, if Equation (3.2.6) or (3.2.7) and (3.2.9) yield consistent results, the perceived azimuth can be identified uniquely, and the perceived azimuth is stable under head rotation around the vertical axis. If the result of Equation (3.2.9) is different from that of Equation (3.2.6) or (3.2.7), the perceived azimuth for the head oriented to the virtual source is different from that for the fixed head orientation. Considering that a listener's head is oriented to the front direction in most cases, Equation (3.2.6) or (3.2.7) can be used to determine the perceived azimuth quantitatively, and Equation (3.2.9) can be applied to resolve front-back ambiguity. Loudspeaker configuration and signals should be designed to reduce the difference between the results of the fixed head orientation and those of the head oriented to a virtual source and consequently enhance the stability of a virtual source under head rotation. The consideration here is consistent with the psychoacoustic principle of auditory localization and followed in the analyses in Chapters 4 and 5. Indeed, the aforementioned equations are valid at low frequencies, and the effective frequency range depends on loudspeaker configuration and signal panning methods. The effective frequency range of Equation (3.2.6) does not exceed 1.5 kHz, and Equations (3.2.7) and (3.2.9) are valid only below 0.7 kHz.

Equation (3.2.22) based on the velocity localization vector is valid below 0.7 kHz and completely equivalent to ITD_p-based Equation (3.2.9) for head rotation. r_v is a measure of the stability of a perceived virtual source. However, identifying the perceived virtual source azimuth based on the direction of the velocity localization vector only is inappropriate (Makita's hypothesis) because the results do not represent the important case of the fixed head orientation. Although the method based on the velocity localization vector has been used in some studies, some examples in Chapters 4 and 5 illustrate the limitation of this method. The additional condition $r_v = 1$ ensures that the perceived azimuth for head rotation is identical to that for the fixed head orientation. In this case, the result of Equation (3.2.22) is consistent with those of Equations (3.2.7) and (3.2.9), and the virtual source is stable under head rotation. In the practical design of loudspeaker configuration and signals, r_v close to a unit is desired.

Equation (3.2.34) based on the energy localization vector is not apparently related to the psychoacoustic cue of mid and high-frequency localization. However, the physical and psychoacoustic mechanisms of high-frequency localization are complicated, so they cannot be represented by some simple equations. In the absence of a simple equation or tool for analyzing the localization at high frequencies, Equation (3.2.34) is a rough and remedial method. In addition, the hypothesis of the incoherent superposition of sound pressures in the derivation of Equation (3.2.34) is not strictly valid. Even so, Equation (3.2.34) has been applied to optimize high-frequency loudspeaker signals from 0.7 kHz to 5 kHz, and appropriate results are yielded. Therefore, Equation (3.2.34) is also applied to the analyses in Chapters 4 and 5.

Traditional methods for analyzing horizontal summing localization are addressed in this section. In these methods, some rough approximations are applied to analyze the reproduced sound field or binaural sound pressures, and some psychoacoustic hypotheses have been introduced. Therefore, the results are valid below a certain frequency limit. A stricter analysis of virtual source localization should be based on a comprehensive consideration of various localization cues discussed in Section 1.6. Above the frequency of 1.5 kHz (or even above 0.7 kHz), the scattering and diffraction of a sound wave by anatomical structures, such as the head and pinnae, should be considered. This consideration can be implemented by evaluating binaural pressures with the head-related transfer function (HRTFs) discussed in Section 1.4.2. Then, the perceived virtual source direction and other perceived attributes are evaluated by comparing binaural pressures in reproduction with those of the target source. However, a comprehensive analysis on various localization cues in binaural pressures is rather complicated. Sophisticated binaural hearing models, including the processing of a high-level system, are required. This analysis is addressed in Chapter 12.

The traditional analysis method discussed in this section is simple although it is mainly valid at low frequencies. As stated in Section 1.6.5, for signals with a low-frequency component, a low-frequency ITD dominates lateral localization. A dynamic cue at low frequencies and a spectral cue at high frequencies help resolve front-back ambiguity. In reproduction with multiple loudspeakers, recreating a high-frequency spectral cue exactly matching that of a target source is difficult, but it usually does not recreate conflicting spectral cues for localization. For most practical signals, such as music and speech, a major component of power spectra is found in the low- and mid-frequency range. Therefore, localization analysis based on a low-frequency ITD and its variation with head rotation reveals some basic features in multichannel horizontal surround sound reproduction. In other words, traditional analysis methods are applicable to the practical design although these methods are not very strict from the views of physics and psychoacoustics. The early works on multichannel surround sound are often based on these methods, which are reviewed in Chapters 4 and 5.

The cases of loudspeaker signals with ICLD only (or amplitude panning) are discussed in this section. For loudspeaker signals with ICTD (time panning), appropriate delays τ_i should be supplemented to each loudspeaker signal in the analysis of the two preceding sections, such as Equation (3.2.3). That is, real-valued A_i in the preceding analyses should be replaced with $A_i \exp(-j2\pi f \tau_i)$. A stricter analysis is based on the calculation of binaural pressures by using HRTFs, the evaluation of localization cues (such as ITD and ILD), and the comparison of the results of multiple loudspeaker reproduction with those of a real sound source (Section 12.1.4). However, the aforementioned methods are mostly valid for narrow-band signals at low frequencies. The summing localization mechanism of transient signals with ICTD is different from that of narrow signals. A comprehensive analysis requires complicated auditory models, but such models are still unavailable because the psychoacoustic mechanism of the summing localization of transient signals with ICTD remains unclear. Available results are only obtained from psychoacoustic experiments. Therefore, the summing localization of signals with ICTD is applied on the basis of psychoacoustic experimental results to the design of two-channel and multichannel microphone techniques (Section 2.2.4 and Chapter 7).

The analyses in Sections 3.2.1 and 3.2.2 focus on the default or central listening position with equal loudspeaker distances. For the off-central position, the distances from each loudspeaker to the center of a listener's head are unequal and denoted by r'_i, and the azimuth of each loudspeaker with respect to the head also changes. Accordingly, in the analysis of superposed sound field and binaural pressures, linear delays and inverse distance factors $\exp(-j2\pi f\, r'_i/c)/r'_I$ should be supplemented to the amplitudes of each loudspeaker signal in the analysis. The azimuths of loudspeakers should be replaced with those with respect to the off-central position. For a slightly off-central position so that the incident wave from each loudspeaker can still be approximated as a plane wave, the virtual source direction can be approximately evaluated from Equation (3.2.22) after the real part in the right side of the equation is taken. Similar to the case of two-channel reproduction with ICTD, a stricter analysis should be based on the superposed binaural pressures evaluated using HRTFs (Section 12.1.5). However, these methods are valid for narrow-band signals at low frequencies. For transient and wideband signals, the rule of summing localization can be only derived from the psychoacoustic experiment. Therefore, the localization analysis methods presented in this section are mainly valid for reproduction in a small-sized listening region with equal loudspeaker distances to the listening position. This condition should be considered.

3.3 MULTIPLE LOUDSPEAKERS WITH PARTLY CORRELATED AND LOW-CORRELATED SIGNALS

As stated in Section 1.7.3, for two stereophonic loudspeakers with partly correlated signals, the perceived auditory events broaden as the cross-correlation of signals reduces. Similar phenomena are observed in multiple loudspeaker reproduction.

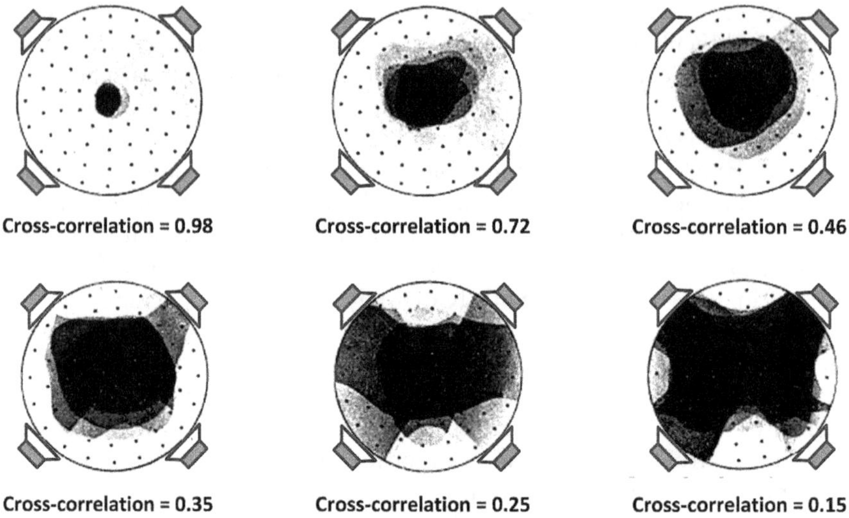

Figure 3.3 "Subjective diffuseness" caused by four loudspeaker configurations with various interchannel correlations. (adapted from Damaske, 1969/1970.)

As shown in Figure 3.3, Damaske (1967/1968) arranged four loudspeakers at azimuths ±45° and ±135° and provided 0.25–2.0 kHz low-pass-filtered white noise with various cross-correlations in loudspeakers. The cross-correlation of signals is shown in Figure 3.3. As the interchannel correlation reduces, auditory events extend to a large region, similar to the sensation of "subjective diffuseness" or listener envelopment in a diffused reverberation sound field. Damaske and Ando (1972) further analyzed the IACC (Section 1.7.3) caused by various loudspeaker configurations and 0.25–2.0 kHz low-pass-filtered white noise with various cross-correlations. The results indicate that for four horizontal loudspeakers at ±54° and ±126° azimuths, the calculated and measured IACC are 0.30 and 0.36 respectively. For five horizontal loudspeakers at ±36°, ±108°, and 180°, the calculated and measured IACC are 0.17 and 0.15, respectively. Therefore, the five-loudspeaker configuration exhibits a lower IACC.

Hiyama et al. (2002) conducted a subjective experiment and demonstrated that six loudspeakers with regular configuration in a horizontal plane are required to recreate the sensation similar to that of the diffused sound field. However, four loudspeakers with irregular configuration at azimuths ±30° and ±90° to ±120°are enough to recreate this sensation.

The aforementioned methods are applicable to the recreation of a subjective sensation in multichannel sound reproduction similar to the listener envelopment caused by the late reverberation in a concert hall. Similar methods are applied to recreate the sensations of being enveloped by or immersed in direct sounds in multichannel surround reproduction. In room acoustics, a low IACC corresponds to a diffused reverberation sound field. However, a diffused reverberation sound field should also create equal power or root-mean-square sound pressures in the two ears. Therefore, in addition to low IACC, nearly equal root-mean-square sound pressures should be applied to the two ears to recreate the listener envelopment related to diffused reverberation in multichannel sound reproduction.

3.4 SUMMARY

The straightforward methods for spatial sound reproduction include physically reconstructing a sound field within a region. However, restricted by the Shannon–Nyquist spatial sampling theorem, these methods require numerous signal and reproduction channels. A practical multichannel surround sound should be simplified on the basis of psychoacoustics. The spatial information that is most important to auditory perception is reproduced as exactly as possible at the cost of reducing accuracy or even omitting the spatial information that is relatively less important to auditory perception. For horizontal surround sound, spatial information from all horizontal directions is equally or unequally important. Different assumptions result in different simplification schemes.

The relative importance of various kinds of spatial information of sound and their appropriate approximation depends on applications, such as sound reproduction with or without accompanying picture. Multichannel surround sound techniques and systems should be designed in accordance with practical requirements. The number, configuration, and signals of loudspeakers should be appropriately chosen by considering the available system resources and adhering to the psychoacoustic principle to retain important information and then recreate the desired perceptual performance. This phenomenon is the main consideration and core problem in multichannel surround sound. Practical techniques and systems are usually appropriate compromises after various factors are considered.

When multichannel surround sound systems are simplified to be adapted to practical applications, the reproduced sound field is different from the target sound field. Psychoacoustic methods are required to recreate various spatial auditory events or perceptions. Common methods include summing localization with multiple loudspeakers, indirect and direct (decorrelated signal) methods for recreating spatial auditory sensations similar to those in a concert hall. The psychoacoustic methods used in cinema sound reproduction may be different from those utilized in domestic reproduction because of the difference in the required sizes of a listening region. Various multichannel surround techniques with different loudspeaker configurations and signals have been developed on the basis of these considerations.

For reproduction within a small-sized listening region, summing localization theorem with multiple loudspeakers provides tools for the analysis and design of multichannel surround sound. In traditional analyses on summing localization, the shadow of the head is neglected, and the interaural phase delay difference ITD_p for the fixed head orientation is used to evaluate the perceived azimuth. Furthermore, the head rotation (related to dynamic cues) is utilized to resolve front-back ambiguity. This simplified analysis is consistent with the psychoacoustic principle of summing localization, but the resultant localization equation is mostly valid below 1.5 kHz. The simplified analysis also indicates that the azimuth of the virtual source of amplitude panning is independent of frequency at low frequencies (up to 0.7 kHz at most). In this case, the localization equation is simple. When the perceived azimuths for the fixed head orientation and the head oriented to a virtual source vary, the virtual source is unstable under head rotation.

Summing localization with multiple loudspeakers can also be analyzed in accordance with theories based on velocity localization vector and energy localization vector. According to the theorem based on the velocity localization vector, the perceived virtual source direction in the superposed sound field is consistent with the inner normal direction of the superposed wavefront at the receiver position. Velocity localization vector-based theorem is valid at low frequencies below 0.7 kHz and equivalent to the case of the head oriented to the virtual source. r_v is a measure of the stability of the virtual source under head rotation. However, identifying the perceived virtual source azimuth based on the direction of the velocity localization vector only is inappropriate because the results do not represent the important case of the fixed head orientation. The additional condition of $r_v = 1$ ensures that the perceived azimuth of the head rotation is identical to that of the fixed head orientation. In this case, the virtual source is stable under head rotation.

The energy localization vector-based theorem is not apparently related to the psychoacoustic cue of high-frequency localization. However, the physical and psychoacoustic mechanisms of high-frequency localization are complicated. In the absence of a simple tool for analyzing localization at high frequencies, the energy localization vector-based theorem is a rough method. Moreover, this theorem may yield meaningful results within the frequency range from 0.7 kHz to 5 kHz in some instances.

The summing localization mechanism of transient signals with an interchannel time difference is different from that of narrow signals. Available results are derived from psychoacoustic experiments.

A subjective sensation that is similar to the listener envelopment caused by the late reverberation in a concert hall may be recreated in multichannel sound reproduction by providing partly or weakly correlated signals to multiple loudspeakers with an appropriate configuration.

Chapter 4

Multichannel horizontal surround sound with a regular loudspeaker configuration

On the basis of the number and configuration of loudspeakers, various multichannel surround sounds are analyzed and discussed in Chapters 4–6.

From the 1970s until the beginning of the 1980s, domestic or consumer horizontal surround sound techniques were mainly intended for music reproduction only within a small-sized listening region. In this period, most techniques were developed to manipulate spatial information from all horizontal directions with equal accuracy or approximation. In other words, systems were attempted to recreate a full 360° horizontal virtual source. Systems were also necessary to recreate the spatial auditory sensations similar to those in a concert hall via the indirect method mentioned in Section 3.1. According to this consideration, regular (uniform) loudspeaker configurations were used in most multichannel horizontal surround sounds in this period. That is, uniform directional sampling was applied to loudspeaker configurations. Various ***signal panning or mixing methods*** for these regular loudspeaker configurations were also developed. Subsequent studies and practices indicated that horizontal surround sounds with regular loudspeaker configurations and a few (such as four) channels were unable to achieve the desired performance. Therefore, the horizontal surround sound with regular loudspeaker configurations was seldom used in practice. However, theoretical and experimental studies in this period provided good lessons and laid down the basis for further developing multichannel surround sound. Derived from a strict theoretical analysis, sound field signal mixing or lower-order Ambisonic signal mixing techniques first developed in the 1970s were essential for the approximate reconstruction of a physical sound field. Higher-order Ambisonics, which is a further development from a lower-order one, should be further explored from the perspective of spatial sound and has been gradually applied to wide areas.

The summing localization theorems presented in Chapter 3 are applied to this chapter to evaluate the performance of some typical early techniques because one attempt of the early multichannel horizontal surround sound is to recreate a full 360°horizontal virtual source. The analyses in Chapters 4–6 are based on traditional analysis methods for multichannel surround sound. In particular, a psychoacoustic cue (ITD) is examined for directional localization at low frequencies, rather than the accuracy of a reconstructed sound field. These analyses are basically consistent with the logic of sound reproduction based on sound field approximation and psychoacoustic principles.

Two typical signal panning or mixing methods in multichannel surround sound, i.e., pair-wise amplitude panning and sound field (Ambisonic) mixing methods, are analyzed in detail. These two signal mixing methods possess a solid theoretical basis and are applicable to regular and irregular horizontal loudspeaker configurations. After an appropriate extension, these signal mixing methods are also applicable to three-dimensional spatial surround sound.

A discrete quadraphone technique and the reasons for its defects are analyzed in Section 4.1. Other horizontal surround sound techniques with regular loudspeaker configurations

DOI: 10.1201/9781003081500-4

are explored, and the effects of increasing loudspeakers in reproduction are investigated in Section 4.2. The transformation of horizontal sound field signals is addressed in Section 4.3, and the basic principle, characteristics, and implementations of the lower-order horizontal Ambisonics are discussed.

4.1 DISCRETE QUADRAPHONE

4.1.1 Outline of the quadraphone

The ***quadraphone***, developed at the beginning of the 1970s, is a four-channel horizontal surround sound technique for consumer reproduction within a small-sized listening region. It is a typical technique and system with a regular loudspeaker configuration. However, efforts in the development of quadraphones failed, and quadraphones were rarely used in practice. As such, the characteristics and defects of quadraphones should be analyzed to understand the physical and perceptual characteristics of a horizontal surround sound field. This analysis also contributes to the succeeding development of various horizontal surround sound techniques.

A quadraphone has four reproduction channels and loudspeakers. Four loudspeakers are arranged in the horizontal left-front (LF), right-front (RF), left-back (LB), and right-back (RB) directions with an equal distance to the head center. The azimuthal spans between two adjacent loudspeakers are usually regular such that the four loudspeakers are located in the four vertex angles of a quadrate. As shown in Figure 4.1, the azimuths of four loudspeakers are given by

$$\theta_{LF} = \theta_0 = 45^\circ \qquad \theta_{LB} = \theta_1 = 135^\circ \qquad \theta_{RB} = \theta_2 = -135^\circ \qquad \theta_{RF} = \theta_3 = -45^\circ. \tag{4.1.1}$$

Therefore, a quadraphonic loudspeaker configuration is an intuitive extension of that of two-channel stereophonic sound.

Quadraphones can be subdivided into two types, the ***discrete quadraphone*** and the ***matrix quadraphone***. The four-channel or loudspeaker signals in a discrete quadraphone are independent (Inoue et al., 1971).Therefore, a discrete quadraphone was previously called the 4-4-4 system, where the first "4" represents the four original recording channels or signals, the second "4" represents the four transmission or storage channels, and the last "4" represents the four reproduction channels or loudspeaker signals. The discrete quadraphone is analyzed in this chapter, and the **matrix quadraphone** is explored in Chapter 8.

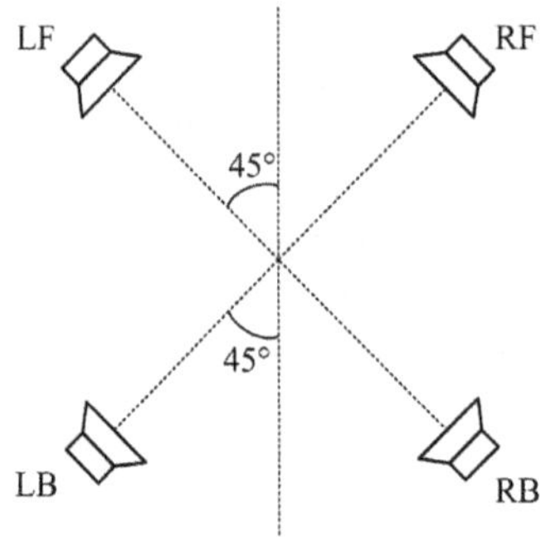

Figure 4.1 The loudspeaker configuration of a quadraphone.

Similar to the case of two-channel stereophonic sound (Section 2.2), various signal recording, panning, and mixing methods have been developed for a discrete quadraphone, causing different perceived effects. Two typical signal panning and mixing methods, i.e., pair-wise amplitude panning and first-order sound field signal mixing, are analyzed in the next two subsections to investigate the summing virtual source localization in discrete quadraphones. These two-signal panning and mixing methods are general and applicable to various multichannel surround sound systems, including systems with irregular loudspeaker configurations (Chapter 5). Moreover, the basic theory of sound field reconstruction in Ambisonics can be developed on the basis of the analysis of sound field signal mixing (Chapter 9).

4.1.2 Discrete quadraphone with pair-wise amplitude panning

Pair-wise amplitude or intensity panning for a discrete quadraphone is similar to a spot microphone and pan-pot technique for two-channel stereophonic sound (Section 2.2.5). This is a local signal panning or mixing method. A signal is fed to a single loudspeaker to recreate a virtual source in its direction or fed to a pair of adjacent loudspeakers with various amplitude ratios or interchannel level difference (ICLD) to recreate a virtual source in the intermediate directions. Therefore, a discrete quadraphone with pair-wise amplitude panning is equivalent to a combination of four pairs of stereophonic channels or loudspeakers at the front, back, and two lateral quadrants. At each loudspeaker direction, the virtual source is created by a single loudspeaker and therefore stable. This observation is a feature of pair-wise amplitude panning for various loudspeaker configurations. However, this situation is different from that of a summing virtual source between two loudspeakers.

The theorem and equations in Section 3.2 are applied to analyze the summing virtual source localization in discrete quadraphones (Xie X.F., 1981). The case of a target virtual source located within the frontal–horizontal quadrant at $-45° \le \theta_S \le 45°$ is analyzed first. In this case, the signal is fed to the LF and RF loudspeakers at $\theta_{LF} = 45°$ and $\theta_{RF} = -45°$, respectively. This case is equivalent to a two-channel stereophonic reproduction with a span angle of $2\theta_0 = 90°$ between two loudspeakers. The perceived virtual source direction at low frequencies is evaluated from Equations (3.2.7) and (3.2.9) with $A_0 = A_{LF}$ and $A_3 = A_{RF}$, respectively, or it is directly examined from Equations (2.1.6) and (2.1.10). For a fixed head,

$$\sin\theta_I = \frac{A_{LF} - A_{RF}}{A_{LF} + A_{RF}}\sin 45° = 0.707\frac{A_{LF}/A_{RF} - 1}{A_{LF}/A_{RF} + 1}. \qquad (4.1.2)$$

For the head oriented to the virtual source,

$$\tan\hat{\theta}_I = \frac{A_{LF} - A_{RF}}{A_{LF} + A_{RF}}\tan 45° = \frac{A_{LF}/A_{RF} - 1}{A_{LF}/A_{RF} + 1}. \qquad (4.1.3)$$

Figure 4.2 illustrates the curves of the perceived virtual source azimuth versus the interchannel level difference $d_1 = 20\log_{10}(A_{LF}/A_{RF})$ evaluated from Equations (4.1.2) and (4.1.3). Similar to two-channel stereophonic reproduction with a standard loudspeaker configuration of $2\theta_0 = 60°$ (Figure 2.2), the virtual source moves from the front (0°) to the direction of the LF loudspeaker (+45°) when d_1 varies from 0 dB to +∞ dB for the fixed head and the head oriented to the virtual source. However, in comparison with Figure 2.2, the results of Equations (4.1.2) and (4.1.3) slightly differ. This difference indicates that head rotation causes some changes in the perceived source direction at low frequencies. The perceived source direction obviously changes in the range of d_1 = 12–18 dB. For example, when d_1 = 12

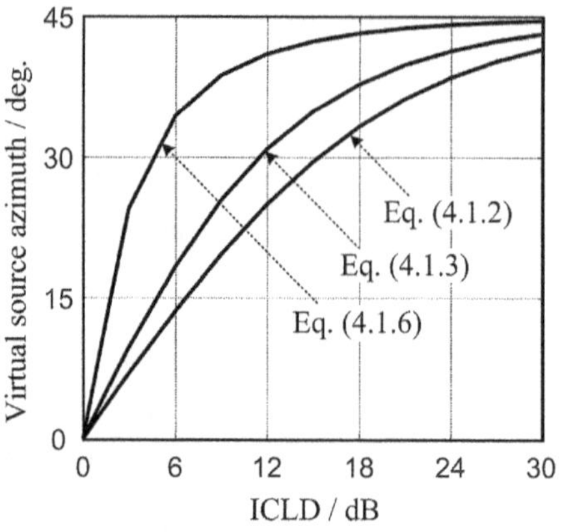

Figure 4.2 Curves of the perceived virtual source azimuth versus the interchannel level difference $d_1 = 20\log_{10}(A_{LF}/A_{RF})$ in the frontal quadrant.

dB or $A_{LF}/A_{RF} = 4$,the results of Equations (4.1.2) and (4.1.3) are $\theta_I = 25.1°$ and $\hat{\theta}_I = 31.0°$, respectively, and the difference is 5.9°.

The stability of a virtual source against head rotation can be evaluated from the velocity vector magnitude r_v given by Equation (3.2.29). Similar to the case of Equation (2.2.59) for a two-channel stereophonic sound, by introducing $0° \le \xi_1 \le 90°$ and normalizing the overall signal power to a unit, the normalized amplitudes of two front channel signals can be written as

$$A_{LF} = \sin\xi_1 \quad A_{RF} = \cos\xi_1 \quad A_{LF}^2 + A_{RF}^2 = 1. \tag{4.1.4}$$

Substituting Equation (4.1.4) into Equation (3.2.29) yields

$$r_v = \frac{\sqrt{2}}{2}\frac{1}{\cos(\xi_1 - 45°)}. \tag{4.1.5}$$

When $\xi_1 = 0°$ or 90°, i.e., either A_{LF} or A_{RF} vanishes, so the virtual source is located at either of the loudspeaker directions, the velocity vector magnitude maximizes to $r_v = 1$. When $\xi_1 = 45°$ or $A_{LF} = A_{RF}$, the velocity vector magnitude minimizes to $r_v = 0.707$. When $A_{LF}/A_{RF} = 4$, Equation (4.1.5) yields $r_v = 0.82$, which is smaller than the ideal value of a unit. Actually, Equation (3.2.30) and the succeeding discussion indicate that a wider span angle of 90° between two front loudspeakers causes an unstable virtual source against head rotation.

As frequency increases, the virtual source direction should be evaluated with Equation (3.2.6) (Xie and Liang, 1995). For a fixed head oriented to the front,

$$\sin\theta_I = \frac{1}{ka}\arctan\left[\frac{A_{LF}/A_{RF} - 1}{A_{LF}/A_{RF} + 1}\tan(ka\sin 45°)\right]. \tag{4.1.6}$$

According to the discussion in Section 2.1.2, the variation in the perceived virtual source direction θ_I with an interchannel level difference d_1 is evaluated by substituting the head radius a in Equation (4.1.6) with a precorrected head radius of $a' = 1.25 \times 0.0875$ m. The result at frequency f = 1 kHz is also illustrated in Figure 4.2. For a given d_1, the virtual source moves toward the direction of the LF loudspeaker (45°) as the frequency increases. For example, when d_1 = 12 dB or A_{LF}/A_{RF} = 4, the virtual source moves from θ_I = 25.0° at low frequencies given by Equation (4.1.2) to θ_I = 41.1° at 1 kHz evaluated from Equation (4.1.6).The difference is 16.1°. As such, a wider span angle of 90° between the left and right loudspeakers also increases the frequency dependence, so the virtual source becomes unstable. Moreover, even for a small ICLD, the virtual source moves obviously as frequency increases above 0.7 kHz, leading to the hole-in-the-middle effect in reproduction. For stimuli with different frequency components, this frequency dependence of a virtual source position may cause broadening and blurry virtual sources. The analysis involving an energy localization vector expressed in Equation (3.2.36) also indicates that the energy vector magnitude minimizes to r_E = 0.707 for $A_{LF} = A_{RF}$. A standard stereophonic loudspeaker configuration of $2\theta_0$ = 60° with equal signal amplitudes yields r_E = 0.866. Therefore, a wider span angle of 90° between the left and right loudspeakers decreases r_E.

The virtual source within a rear quadrant of 135° $\leq \theta_S \leq$ 180 °and −180° < $\theta_S \leq$ −135° can be analyzed similarly. In this case, signals are fed to the LB (θ_{LB} = 135 °) and RB (θ_{RB} = −135°) loudspeakers. The results of the rear quadrant can be obtained from a mirror reflection of those of the front quadrant within −45° $\leq \theta_S \leq$ 45° because of the front-back symmetry. For conciseness, the results of the rear quadrant are omitted here.

The target virtual source within the left lateral quadrant with 45° $\leq \theta_S \leq$ 135 ° is analyzed only because the results of the right lateral quadrant with −135° $\leq \theta_S \leq$ −45 ° can be derived from those of the left lateral quadrant via the left-right symmetry. The signal is fed to the LF (θ_{LF} = 45°) and LB (θ_{LB} = 135°) loudspeakers to create the virtual source within the left lateral quadrant. The perceived virtual source direction at low frequencies is evaluated from Equations (3.2.7) and (3.2.9) with $A_0 = A_{LF}$ and $A_1 = A_{LB}$, respectively. For a fixed head oriented to the front,

$$\sin\theta_I = 0.707. \tag{4.1.7}$$

For the head rotation to orient the virtual source,

$$\tan\hat{\theta}_I = \frac{A_{LF} + A_{LB}}{A_{LF} - A_{LB}} = \frac{1 + A_{LB}/A_{LF}}{1 - A_{LB}/A_{LF}}. \tag{4.1.8}$$

According to Equation (4.1.8),when A_{LB}/A_{LF} varies from 0 to + ∞, or the interchannel level difference $d_2 = 20\log_{10}(A_{LB}/A_{LF})$ varies from −∞ dB to +∞ dB, $\hat{\theta}_I$ changes continuously from 45° to 135°. However, according to Equation (4.1.7), $\sin\theta_I$ is always equal to 0.707, or a virtual source is located at azimuth θ_I = 45° or 135°. Therefore, for the head oriented to the virtual source, changing d_2 enables the virtual source to move continuously from 45° to 135°. For a fixed head oriented to the front, however, the virtual source is located at either the left–front or left–back directions.

The velocity vector magnitude r_v is evaluated from Equation (3.2.29). Similar to the case of two front channels in Equation (4.1.4), the normalized amplitudes of two left-lateral channel signals can be written as Equation (4.1.9) by introducing a parameter 0° $\leq \xi_2 \leq$ 90° and normalizing the overall signal power to unit:

$$A_{LF} = \cos\xi_2 \qquad A_{LB} = \sin\xi_2 \qquad A_{LF}^2 + A_{LB}^2 = 1. \tag{4.1.9}$$

Substituting Equation (4.1.9) into Equation (3.2.29) yields

$$r_v = \frac{\sqrt{2}}{2}\frac{1}{\cos(\xi_2 - 45°)}. \tag{4.1.10}$$

When $\xi_2 = 0°$ or $90°$, i.e., either A_{LB} or A_{LF} vanishes, so the virtual source is located at either loudspeaker direction, the velocity vector magnitude maximizes to $r_v = 1$. When $\xi_2 = 45°$ or $A_{LB} = A_{LF}$, the velocity vector magnitude minimizes to $r_v = 0.707$, although Equation (4.1.8) yields $\hat{\theta}_I = 90°$. Therefore, a pair of lateral loudspeakers symmetrically arranged from front to back is unable to recreate a stable lateral virtual source. This characteristic is a major defect of the discrete quadraphone with pair-wise amplitude panning.

Prior to the theoretical analysis, some authors investigated the summing virtual source localization in a discrete quadraphone with pair-wise amplitude panning via psychoacoustic experiments (Ratliff, 1974; Thiele and Plenge, 1977). The results were consistent with the theoretical analysis. Figure 4.3 illustrates the summing localization results with the LF (45°) and LB (135°) loudspeakers given by Thiele and Plenge (1977). The principle and method of the experiment on virtual source localization is addressed in Section 15.6. The stimulus used in the experiment is band-limited Gaussian impulse noise within 0.6–10 kHz. The median and quartiles of localization results are shown in Figure 4.3. Figure 4.3 indicates that changing the ICLD between the LB and LF loudspeaker signals makes the virtual source approximately jump from the direction of the LF loudspeaker to that of the LB loudspeaker. The quartiles in the localization results also increase.

The aforementioned theoretical and experimental results can be easily interpreted through physical and psychoacoustic analysis. Below the frequency of 1.5 kHz, the interaural phase delay difference is a dominant cue for lateral localization. For quadraphonic reproduction with a fixed head oriented to the front direction, the four loudspeakers are symmetrically arranged from front to back and from left to right with respect to the head. A pair of LF and RF loudspeakers can control interaural difference (such as ITD) between the binaural

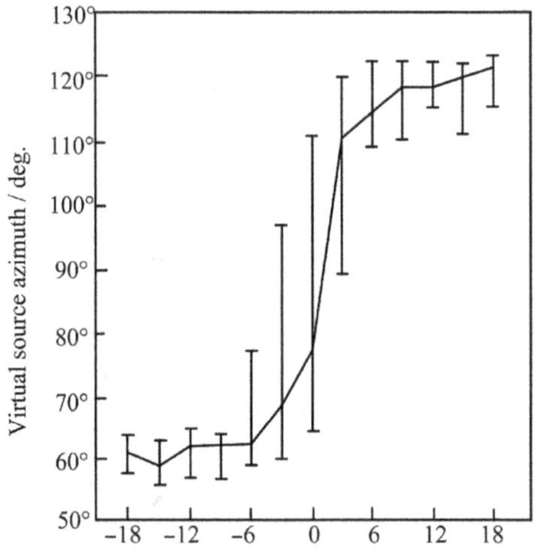

Figure 4.3 Experimental results of summing localization with the left-front (45°)and left-back loudspeakers, where the interchannel level difference is given by $d_2 = 20 \log_{10}(A_{LB}/A_{LF})$ (Thiele and Plenge, 1977, with the permission of the Audio Engineering Society).

pressures and thus recreate a virtual source between loudspeakers by changing the interchannel difference (such as ICLD) between two loudspeaker signals. By contrast, a pair of LF and LB loudspeakers fail to control the interaural difference between the binaural pressures because of the front-back symmetry. The geometrical relationship of the positions of two ears and loudspeakers has been misunderstood in recreating a lateral virtual source with LF and LB loudspeakers in a quadraphone. From the point of psychoacoustics, a stable lateral virtual source cannot be recreated with a pair of lateral loudspeakers symmetrically arranged from front to back (Cooper, 1987). After the head rotates 90° to the left, the original LF and LB loudspeakers are left-right symmetrically located in the RF and left-front directions with respect to the new orientation of the head, respectively. In this case, the LF and LB loudspeakers can recreate a virtual source at the intermediate direction between two loudspeakers. This result is predicted with Equation (4.1.3).

The aforementioned analysis indicates that evaluating the virtual source direction with Equation (3.2.9) for the head oriented to the virtual source only or equally with Equation (3.2.22), derived from the velocity localization vector only, is incomplete or inappropriate. The aforementioned analysis reveals the limitation of Makita's hypothesis. The virtual source direction should be evaluated with a combination of Equations (3.2.7) and (3.2.9), i.e., through a comprehensive analysis of the interaural phase delay difference for a fixed head and the variation in the interaural phase delay difference after head rotation.

4.1.3 Discrete quadraphone with the first-order sound field signal mixing

Similar to the XY microphone technique for two-channel stereophonic sound in Section 2.2.1, the first-order sound field signals for a discrete quadraphone can be recorded by combining the pressure and pressure-gradient microphones discussed in Section 1.2.5 (Yamamoto, 1973). Four identical combinations of pressure and pressure-gradient microphones with their main axes pointing to the horizontal LF(45°), LB(135°), RB (−135°), and RF (−45°) directions are used to capture source signals. The four microphone signals are amplified and then fed to the corresponding loudspeakers. From Equation (1.2.41), when $\Delta\alpha$ between the source direction and main axes of the microphones is represented with the horizontal source azimuth, the normalized amplitudes of the four microphone or loudspeaker signals are given by

$$\begin{aligned} A_0 &= A_{LF} = A_{total}\left[1 + b\cos(\theta_S - 45^\circ)\right] \qquad A_1 = A_{LB} = A_{total}\left[1 + b\cos(\theta_S - 135^\circ)\right], \\ A_2 &= A_{RB} = A_{total}\left[1 + b\cos(\theta_S + 135^\circ)\right] \qquad A_3 = A_{RF} = A_{total}\left[1 + b\cos(\theta_S + 45^\circ)\right]. \end{aligned} \tag{4.1.11}$$

where θ_S is the source azimuth in the original sound field or the target virtual source azimuth in reproduction, and A_{total} is a constant related to the overall gain of the system and determined by the sensitivity A'_{mic} of microphones and the gains of amplifiers. $b > 0$ specifies the directivity of microphones.

Localization theorem (Section 3.2) is applied to analyze the virtual source in reproduction (Xie X.F., 1977). At low frequencies, the virtual source position is evaluated by substituting Equations (4.1.1) and (4.1.11) into Equations (3.2.7) and (3.2.9), respectively. For a fixed head,

$$\sin\theta_I = \frac{b}{2}\sin\theta_S. \tag{4.1.12}$$

For the head oriented to the virtual source,

$$\tan\hat{\theta}_I = \tan\theta_S. \tag{4.1.13}$$

Therefore, the perceived virtual source direction in reproduction depends on parameter b of the microphones but is independent of the overall gain. When

$$b = 2, \tag{4.1.14}$$

Equations (4.1.12) and (4.1.13) yields

$$\sin\theta_I = \sin\theta_S \qquad \tan\hat{\theta}_I = \tan\theta_S. \tag{4.1.15}$$

That is,

$$\theta_I = \hat{\theta}_I = \theta_S. \tag{4.1.16}$$

In this case, the perceived virtual source azimuth matches that of the target source within the full horizontal direction of$-180° \le \theta_S \le 180°$, and the results for the head oriented to the virtual source are consistent with those of a fixed head. Moreover, Equation (3.2.29) verifies that parameter $b = 2$ leads to a unit velocity vector magnitude $r_v = 1$. Therefore, a discrete quadraphone with the first-order sound field signals can theoretically recreate a full 360° horizontal virtual source at low frequencies. This feature is ideal and desirable.

The microphone with $b = 2$ exhibits a supercardioid directional characteristic similar to that in Figure 1.8. Figures 4.4 illustrates the two-dimensional polar pattern of a microphone with $b = 2$. A_{total} is normalized so that the maximal on-axis response of a microphone is a unit. Let θ_i represent the main axis orientation of the ith microphone. The main response (lobe) of each microphone is centered around the main axis direction $\Delta\theta'_i = (\theta s - \theta_i) = 0°$and maximizes at the on-axis direction. The response magnitude decreases as $|\Delta\theta'_i|$ increases and becomes null at $|\Delta\theta'_i| = 120°$. As $|\Delta\theta'_i|$ further increases, the responses exhibit a negative (out-of-phase) rear lobe. The maximal magnitude of the rear lobe drops to −9.5 dB in comparison with that of the main lobe.

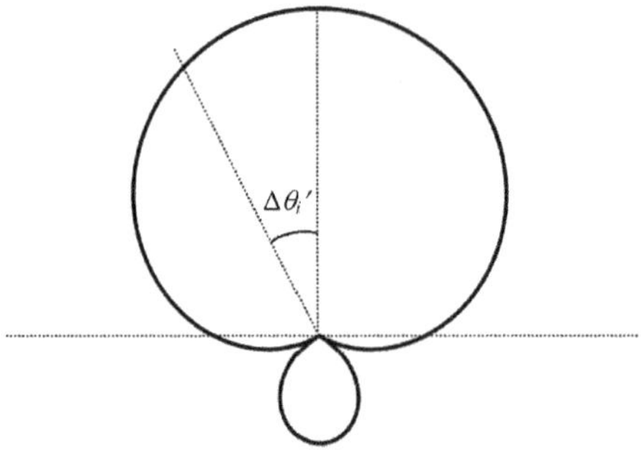

Figure 4.4 Two-dimensional polar pattern of the first-order sound field microphone with $b = 2$ (the maximal on-axis response is normalized to a unit).

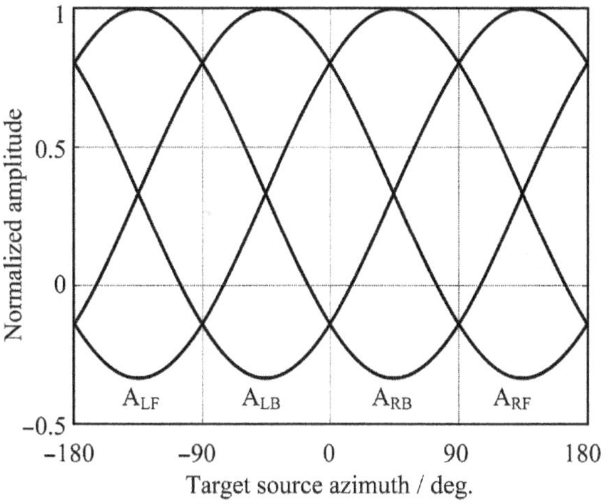

Figure 4.5 Microphone amplitude responses or the panning curve of the first-order sound field signal mixing for a quadraphone.

The pair-wise amplitude panning is a ***local amplitude-based panning or mixing method***. In this method, a signal is mixed or panned to a pair of adjacent loudspeakers to create a virtual source between them. For a virtual source in either of the loudspeaker directions, a signal is panned to one of the loudspeakers only, and the signal of the other loudspeaker vanishes. By contrast, sound field signal mixing is an essentially ***global amplitude-based mixing or panning method***. In this method, a signal is mixed or panned to all loudspeakers to create a virtual source except for a few target source directions. Even for a target source in a loudspeaker direction, the signal may be fed to a given loudspeaker and be spread to other loudspeakers. In other words, loudspeaker signals may encounter crosstalks. This feature distinguishes global amplitude-based mixing from local amplitude-based mixing.

Figure 4.5 illustrates the variations in four microphone amplitude responses versus a target source azimuth or the panning curve of four loudspeaker signals for the first-order sound field signal mixing. One of the four loudspeaker signals vanishes for some special target azimuths at which the angle between a given loudspeaker and the target source is 120°.Thus, the target source is exactly located in the null direction of the corresponding microphone. Even in this case, the signals of the three other loudspeakers remain. Moreover, the signal of the loudspeaker opposite to the target virtual source direction is out of phase. As stated in Section 3.2.2, this out-of-phase loudspeaker signal is necessary to ensure the velocity vector magnitude $r_v = 1$. Therefore, from the point of psychoacoustics, the out-of-phase crosstalk signal from the opposite loudspeaker is beneficial to recreating a stable virtual source at the central listening position. However, at the off-central listening position close to the opposite loudspeakers, excessive crosstalk may cause a virtual source to collapse to the opposite loudspeaker direction.

Equation (4.1.16) is valid at very low frequencies. As frequency increases, Equation (3.2.6) should be used to evaluate the virtual source direction, and Equation (3.2.9) is applied to resolve front-back ambiguity (Xie and Liang, 1995). When Equations (4.1.1) and (4.1.11) are substituted into Equation (3.2.6), the virtual source direction for affixed head is given by

$$\sin\theta_I = \frac{1}{ka}\arctan\left[\sqrt{2}\sin\theta_S \tan\left(\frac{\sqrt{2}}{2}ka\right)\right]. \tag{4.1.17}$$

When

$$\frac{\sqrt{2}}{2}ka < \frac{\pi}{2}, \tag{4.1.18}$$

or

$$f < f_C = \frac{c}{2\sqrt{2}a}, \tag{4.1.19}$$

$\tan\left(\sqrt{2}ka/2\right) > 0$, and $\sin\theta_I$ possesses the same sign or polarity as $\sin\theta_S$. Combined with Equation (3.2.9), θ_I is located in the same quadrant as θ_S. Equation (4.1.19) is the upper frequency limit of Equation (4.1.17). When a precorrected head radius of $a' = 1.25 \times 0.0875$ m to replace a in Equation (4.1.19) and when the speed of sound of $c = 343$ m/s is chosen, Equation (4.1.19) yields $f_C = 1.1$ kHz.

From Equations (4.1.17) and (4.1.18) and under the parameters given above, the variation in the virtual source direction with frequency for various values of θ_S is evaluated. Figure 4.6 illustrates the results of a target source in the left half-horizontal plane with $0° \le \theta_S \le 180°$ for a fixed head. The results of the target source in the right half-horizontal plane can be derived from the left-right symmetry. Figure 4.6 indicates the following:

1. At the special source azimuths of $\theta_S = 0°, 45°, 135°$, and $180°$, θ_I is independent of the frequency.
2. As frequency increases, the virtual source in the left–front quadrant with target azimuth $0° < \theta_S < 45°$ and $45° < \theta_S < 90°$ moves from $\theta_I = \theta_S$ at low frequencies toward the direction of LF loudspeakers (45°). When frequency approaches f_C as expressed in Equation (4.1.19), the virtual source is adjacent to the LF loudspeakers.
3. As frequency increases, the virtual source in the left-back quadrant with a target azimuth of $90° < \theta_S < 135°$and $135° < \theta_S < 180°$ moves from $\theta_I = \theta_S$ at low frequencies toward

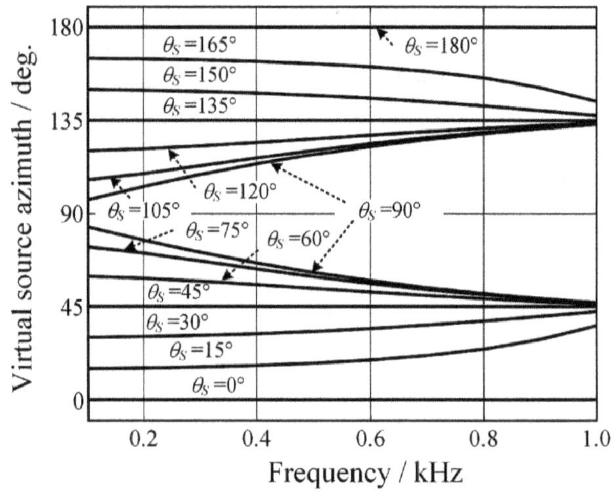

Figure 4.6 Variation in the virtual source direction with frequency for different target azimuths in a discrete quadraphone with the first-order sound field signal mixing.

the direction of LB loudspeakers (135°). When frequency approaches f_C expressed in Equation (4.1.19), the virtual source is adjacent to LB loudspeakers.
4. As frequency increases, the lateral virtual source at the target azimuth $\theta_S = 90°$ splits into two unstable virtual sources. One moves toward the direction of LF loudspeakers (45°), and the other moves toward the direction of LB loudspeakers (135°). When a listener slightly moves the head, the perceived source may jump from the direction of LF loudspeakers to that of LB loudspeakers or vice versa.
5. The direction of a front virtual source within the target region of $0° < \theta_S < 45°$ and a rear virtual source within the target region of $135° < \theta_S < 180°$ varies slightly or moderately as frequency increases. By contrast, the direction of the lateral virtual source varies greatly as frequency increases. For example, when $\theta_S = 15°$, the virtual source direction varies from $\theta_I = \theta_S$ at very low frequencies to $\theta_I = 21.3°$ at $f = 0.7$ kHz, and the difference is $|\theta_I - \theta_S| = 6.3°$. When $\theta_S = 75°$, the virtual source direction varies from $\theta_I = \theta_S$ at very low frequencies to $\theta_I = 53.3°$at $f = 0.7$ kHz, and the difference is $|\theta_I - \theta_S| = 21.7°$.
6. Equation (3.2.7) is valid only up to 0.2 kHz if the deviation between its results and those of Equation (4.1.17) is limited to 15° for all target azimuths within $-180° < \theta_S < 180°$. A large deviation appears in the lateral direction. The deviation reveals the limitation of Equation (3.2.7).

The aforementioned theoretical analysis was proven by a virtual source localization experiment (Xie and Liang, 1995). Therefore, the frequency-dependent direction of a lateral virtual source is a major defect of a discrete quadraphone with the first-order sound field signal mixing. For stimuli with different frequency components (such as music), this frequency dependence may cause broadening and blurry virtual sources.

4.1.4 Some discussions on discrete quadraphones

The virtual source localization at the central listening position of a discrete quadraphone is analyzed in the preceding sections. The results are summarized as follows. For pair-wise amplitude panning, virtual sources at the four loudspeaker directions are recreated by the corresponding single loudspeaker and thus stable. The direction of the virtual sources between the loudspeaker in the front and rear quadrants is unstable under head rotation. Moreover, the direction of the virtual source varies as the frequency increases, leading to a broadening and blurry virtual source and the hole-in-the-middle effect in the front and rear quadrants. A major defect of the quadraphone with pair-wise amplitude panning is that it cannot recreate a stable virtual source in lateral directions. The discrete quadraphone with the first-order sound field signal mixing can theoretically recreate a full 360° horizontal virtual source at very low frequencies, and the virtual source is stable under the head rotation. As frequency increases, virtual sources move toward the nearest loudspeakers and become unstable. The frequency dependence of the lateral virtual source is obvious and begins at frequencies of 0.2–0.3 kHz.

The results indicate that the performance of virtual source localization depends on signal mixing even for the same loudspeaker configuration. Overall, a discrete quadraphone exhibits obvious defects, especially the problem of instability in the lateral virtual source. The localization performance of the first-order sound field signal mixing is better than that of the pair-wise amplitude panning at very low frequencies. However, for sound field signal mixing, binaural pressures are a coherent superposition of the pressures created by the four loudspeakers. When a listener deviates from the central listening position slightly toward an arbitrary direction, the superposed binaural pressures and their corresponding interaural localization cues change. Therefore, the localization performance of a discrete quadraphone with the first-order sound field signal mixing is sensitive to the listening position, resulting in a very small listening region or sweet point. This is another defect of a discrete quadraphone. By contrast, as stated in Section 2.4, the perceived virtual source position in a two-channel stereophonic sound is sensitive to the lateral translation of the head from the central line but insensitive to the front-back deviation of the head along the central line.

As stated in Section 1.9.1, spatial sound aims to reproduce the spatial information of sound, including the localization information of sound sources and the comprehensive spatial information of environment reflections. Although only the localization performance is analyzed in the preceding sections, the results sufficiently explain the problems appearing in a discrete quadraphone. In the early development of discrete quadraphones, an indirect method was used to recreate the spatial auditory sensations of reflections; that is, virtual sources were used in reproduction to simulate the image source of reflections from various directions. For other horizontal surround sounds with regular loudspeaker configurations in which the information of reflection was recreated by the indirect method, the perceived performance of reflection in reproduction could be evaluated preliminarily through the analysis of virtual source localization.

A discrete quadraphone had been seldom applied to practical uses because of the aforementioned defects. Furthermore, efforts devoted to developing discrete quadraphones failed because some important psychoacoustic factors were ignored in the design of discrete quadraphones. A discrete quadraphone with pair-wise amplitude panning was simply designed as four independent two-channel stereophonic sounds. Discrete quadraphones were then utilized to manipulate full 360°horizontal information with equal approximation and to recreate this information. On the contrary, when the capacity for the transmission of spatial information is limited, the unbiased manipulation of directional information leads to a loss of important information. Moreover, a large span angle (90°) between the loudspeakers in quadraphones degrades the accuracy of spatial information reproduction.

The following considerations and methods can be used to improve the surround sound reproduction within a small-sized listening region. In the first method, the number of independent channels and loudspeakers increases to improve the accuracy in spatial information reproduction. Some examples of the first method are given in Section 4.2. In the second method, the limited capacity of signal transmission and storage is fully used when the number of independent signals is given; then, the target sound field is reconstructed with each order approximation by choosing the number of independent signals and loudspeakers. This consideration is for sound field signal transformation and Ambisonics addressed in Section 4.3. In the third method, from practical requirements and importance in auditory perception, spatial information from different directions is manipulated with different approximations, and an irregular loudspeaker configuration is used in reproduction. Current multichannel surround sounds for accompanying picture or general use (Chapters 5 and 6) are usually a combination of the first and third methods.

4.2 OTHER HORIZONTAL SURROUND SOUNDS WITH REGULAR LOUDSPEAKER CONFIGURATIONS

4.2.1 Six-channel reproduction with pair-wise amplitude panning

To recreate a full 360° horizontal virtual source with pair-wise amplitude panning, some authors utilized regular loudspeaker configurations other than that of quadraphones in the 1970s. Figure 4.7 illustrates the results of a virtual source localization experiment with a pair of loudspeakers arranged in the left–front direction at $\theta_{LF} = 30°$ and the left direction at $\theta_L = 90°$, respectively (Thiele and Plenge, 1977). The stimulus used in the experiment was a

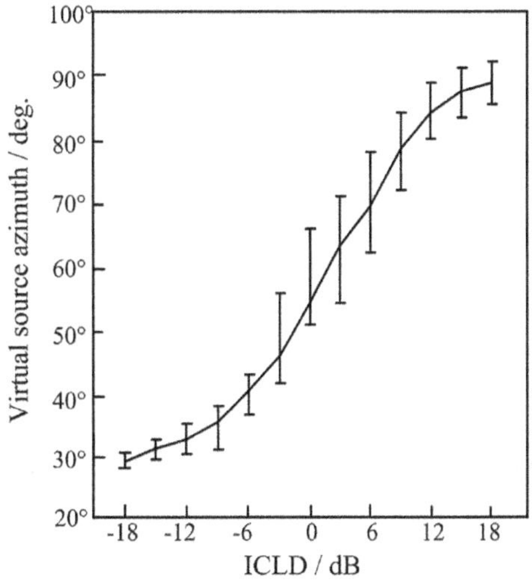

Figure 4.7 Virtual source localization experiment with a pair of loudspeakers arranged in the left–front direction at θ_{LF} = 30° and the left direction at θ_L = 90°, respectively. ICLD is $d_2 = 20 \log_{10}(A_L/A_{LF})$ (Thiele and Plenge, 1977, with the permission of the Audio Engineering Society).

band-limited Gaussian impulse noise within 0.6–10 kHz. The median and quartiles of localization results are shown in Figure 4.7. For this configuration, pair-wise amplitude panning can recreate a lateral virtual source within the range of 30°–90°.

On the basis of the above analysis, Thiele and Plenge (1977) suggested a horizontal six-channel reproduction system. In Figure 4.8,the system involves six horizontal channels, including the left-front (LF), left (L), left-back (LB), right-back (RB), right (R), and right-front (RF) channels. The six loudspeakers are arranged in the horizontal plane with a regular interval of 60°, and the azimuths of loudspeakers are given by

$$\begin{aligned} &\theta_{LF} = \theta_0 = 30^\circ \quad &&\theta_L = \theta_1 = 90^\circ \quad &&\theta_{LB} = \theta_2 = 150^\circ, \\ &\theta_{RB} = \theta_3 = -150^\circ \quad &&\theta_R = \theta_4 = -90^\circ \quad &&\theta_{RF} = \theta_5 = -30^\circ. \end{aligned} \tag{4.2.1}$$

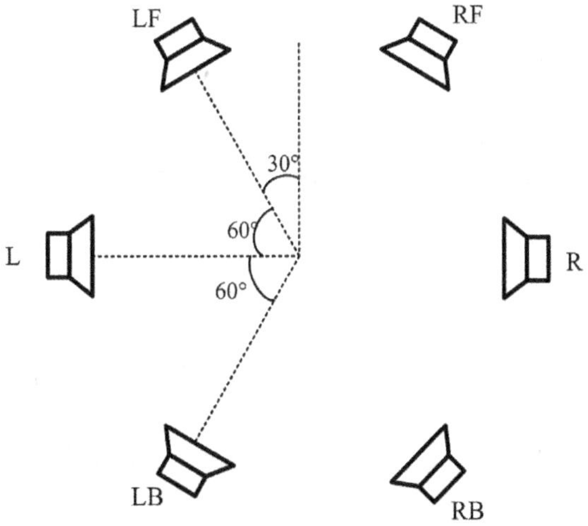

Figure 4.8 Horizontal six-channel reproduction system.

The virtual source localization experiment validates that this system and loudspeaker configuration can recreate a full 360° horizontal virtual source.

The analysis on this six-channel system is similar to that in Section 4.1.2. For a target source within the frontal range of $-30° \le \theta_S \le 30°$, the virtual source is created with the LF ($\theta_{LF} = 30°$) and RF ($\theta_{RF} = -30°$) loudspeakers. In this case, the loudspeaker configuration is identical to that of the standard two-channel stereophonic sound discussed in Section 2.1.1. The virtual source directions at low frequencies of $f \le 0.7$ kHz are also identical to those illustrated in Figure 2.2; the only change is to replace the normalized signals A_L and A_R in Figure 2.2 with the normalized signals A_{LF} and A_{RF} here, respectively. The head rotation only causes a slight variation in the perceived virtual source direction because the span angle between two loudspeakers is appropriate (60°), and the low-frequency virtual source is relatively stable. However, according to the analysis in Section 2.1.2 and Figure 2.3, above the frequency of 0.7 kHz the virtual source moves toward the loudspeaker direction ($\theta_L = 30°$ or $\theta_R = -30°$), resulting in an unstable virtual source.

For a target source within the rear range of $150° \le \theta_S \le 180°$ and $-180° < \theta_S \le -150°$, the virtual source is created by the LB ($\theta_{LB} = 150°$) and RB ($\theta_{RB} = -150°$) loudspeakers. The virtual source positions in this range can be evaluated with a mirror reflection of those in the frontal range with $-30° \le \theta_S \le 30°$because of the front-back symmetry, and the detailed results are omitted here for conciseness.

For a target source within the lateral range of $30° \le \theta_S \le 90°$, the virtual source is recreated by the LF ($\theta_{LF} = 30°$) and L ($\theta_L = 90°$) loudspeakers. The perceived virtual source direction at low frequencies is evaluated with Equations (3.2.7) and (3.2.9). For a fixed head, Equation (3.2.7) yields

$$\sin\theta_I = \frac{A_{LF}\sin 30° + A_L \sin 90°}{A_{LF} + A_L} = \frac{A_L/A_{LF} + 0.5}{A_L/A_{LF} + 1}, \tag{4.2.2}$$

while for the head oriented to the virtual source, Equation (3.2.9) yields

$$\tan\hat{\theta}_I = \frac{A_{LF}\sin 30° + A_L \sin 90°}{A_{LF}\cos 30° + A_L \cos 90°} = \frac{A_L/A_{LF} + 0.5}{0.866}. \tag{4.2.3}$$

According to Equations (4.2.2) and (4.2.3), Figure 4.9 illustrates the variation in the virtual source direction versus the interchannel difference $d_2 = 20\log_{10}(A_L/A_{LF})$ (dB). As d_2 varies from $-\infty$ dB to $+\infty$ dB, the virtual source moves continuously from 30° to 90° for both the cases of the fixed head and the head oriented to the virtual source. The results of the two cases slightly differ; therefore, the virtual source is stable under head rotation.

As frequency increases, Equation (3.2.6) should be used to evaluate the direction of the virtual source. The perceived virtual source direction is given by substituting the signal A_{LF}, A_L, and the azimuths of loudspeakers in Equation (4.2.1) into Equation (3.2.6),

$$\sin\theta_I = \frac{1}{ka}\arctan\left[\frac{\sin(ka/2) + A_L/A_{LF}\ \sin(ka)}{\cos(ka/2) + A_L/A_{LF}\cos(ka)}\right]. \tag{4.2.4}$$

When a in Equation (4.2.4)is replaced with a precorrected head radius of $a' = 1.25 \times 0.0875$ m, the variations in the perceived virtual source direction θ_I versus the interchannel level difference d_2 at various frequencies are evaluated. The result of the frequency of 0.7 kHz

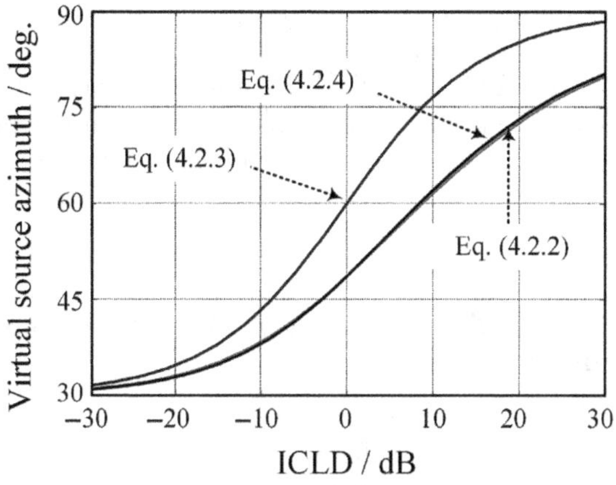

Figure 4.9 Calculated results of the lateral virtual source direction versus ICLD. ICLD is $d_2 = 20\log_{10}(A_L/A_{LF})$. The results of Equation (4.2.4) at a frequency of 0.7 kHz almost coincide with those of Equation (4.2.2).

is also illustrated in Figure 4.9.The direction of a lateral virtual source is basically invariable with a frequency of at least up to 0.7 kHz; thus, a stable lateral virtual source can be recreated at low frequencies. The analysis here is consistent with the experimental results given by Thiele and Plenge (1977).

According to the symmetry, a virtual source direction in the lateral range of $90° \le \theta_S \le 150°$, $-90° \le \theta_S \le -30°$, and $-150° \le \theta_S \le -90°$ can be deduced from the results of $30° \le \theta_S \le 90°$ and are omitted here.

The aforementioned results can also be deduced by analyzing the velocity vector magnitude r_v in reproduction. Similar to the case of Equation (4.1.4) for two frontal channels, the normalized amplitudes of the Land LF channel signals can be expressed in Equation (4.2.5) by introducing $0° \le \xi_2 \le 90°$ and normalizing the overall signal power to a unit:

$$A_L = \sin\xi_2 \qquad A_{LF} = \cos\xi_2 \qquad A_{LF}^2 + A_L^2 = 1. \tag{4.2.5}$$

Substituting Equation (4.2.5) into Equation (3.2.29) yields

$$r_v = \frac{\sqrt{2}}{2}\frac{\sqrt{1+\cos\xi_2\sin\xi_2}}{\cos(\xi_2 - 45°)}. \tag{4.2.6}$$

For $\xi_2 = 0°$ or $90°$, either A_L or A_{LF} vanishes such that the virtual source is located at either of the loudspeaker directions; therefore, the velocity vector magnitude maximizes to $r_v = 1$. For $\xi_2 = 45°$, r_v minimizes with a value of 0.866, which is identical to the result of a frontal stereophonic loudspeaker configuration with a standard span angle of 60°.

Overall, for the six-channel reproduction and loudspeaker configuration shown in Figure 4.8, pair-wise amplitude panning can recreate a full 360° horizontal virtual source at least up to the frequency of 0.7 kHz. The virtual source is relatively stable under head rotation. Therefore, the localization performance of the six-channel system is superior to that of the discrete quadraphone discussed in Section 4.1. This improvement is due to the increasing number of loudspeakers in the six-channel system; consequently, the span angle between any pair of loudspeakers does not exceed 60°. The two loudspeakers at ±90° play an important role in stabilizing or anchoring the latter virtual source. In addition, as stated in Section 3.3, the six-loudspeaker configuration is appropriate for creating spatial auditory sensations similar to those in a diffused sound field. Therefore, increasing the number of channels and loudspeakers improves the accuracy in spatial information reproduction at the cost of increasing the complexity of the system. Indeed, a further increase in the number of channels and loudspeakers further promotes the accuracy in reproduction and makes the system more complicated.

4.2.2 The first-order sound field signal mixing and reproduction with $M \geq 3$ loudspeakers

The first-order sound field signal mixing discussed in Section 4.1.3 can be extended to arbitrarily $M \geq 3$ loudspeaker reproduction (Bernfeld, 1975; Xie X.F., 1981). $M \geq 3$ coincident supercardioid microphones are used to capture source signals. The main axes of microphones point to M uniform directions in the horizontal plane. The main axis orientation of the ith microphones is expressed in Equation (4.2.7) by starting from the direct front and along the anticlockwise direction:

$$\theta_i = \theta_0 + \frac{360° i}{M} \quad i = 0, 1 \ldots (M-1). \tag{4.2.7}$$

If θ_i in Equation (4.2.7) exceeds 180°, it should be replaced with θ_i – 360° because of the periodic variation in θ to keep the azimuthal variable within the range of $-180° < \theta \leq 180°$. This condition is held in the succeeding discussion without explanation. M microphone signals are amplified and reproduced by M loudspeakers with a regular configuration in the corresponding directions. The normalized amplitude of the ith loudspeaker signal is given by

$$A_i(\theta_S) = A_{total}\left[1 + 2\cos(\theta_S - \theta_i)\right], \tag{4.2.8}$$

where θ_S is the azimuth of the sound source in the original sound field or the target source azimuth in reproduction, and A_{total} is a constant related to the overall gain. At very low frequencies, the perceived virtual source azimuth is evaluated by substituting Equations (4.2.7) and (4.2.8) into Equations (3.2.7) and (3.2.9), respectively. For a fixed head and head oriented to the virtual source, the results are

$$\sin\theta_I = \sin\theta_S \qquad \tan\hat{\theta}_I = \tan\theta_S, \tag{4.2.9}$$

or

$$\theta_I = \hat{\theta}_I = \theta_S. \tag{4.2.10}$$

When Equations (4.2.7) and (4.2.8) are substituted into Equation (3.2.29), the velocity vector magnitude satisfies $r_v = 1$. Therefore, a full 360° horizontal virtual source can be recreated at low frequencies, and the virtual source is stable under head rotation. When Equations (4.2.7) and (4.2.8)are substituted into Equation (3.2.6), the virtual source moves as frequency increases and consequently becomes unstable even for reproduction with $M > 4$ loudspeakers.

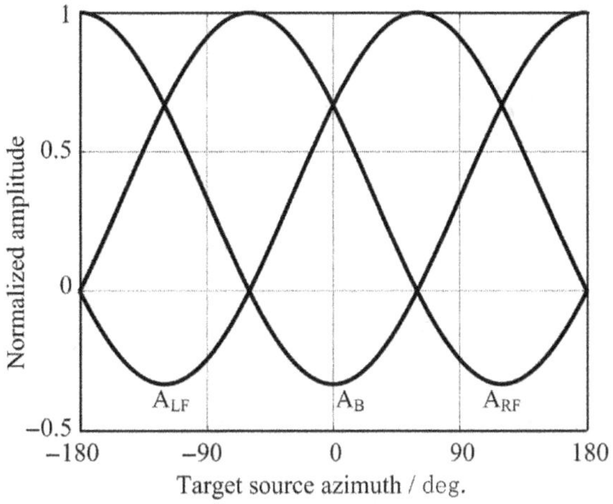

Figure 4.10 Panning curve of the first-order sound field signal mixing for three horizontal loudspeakers (the maximal amplitudes of signals are normalized to a unit).

The case of reproduction with four loudspeakers has been discussed in Section 4.1.3. As a new example, Figure 4.10 illustrates the panning curve of the first-order sound field signal mixing for three horizontal loudspeakers at ±60° and 180°. The subscripts LF, RF, and B represent the left-front, right-front, and back directions, respectively. Notably, when the target source is located at the direction of one loudspeaker, the signals of the two other loudspeakers vanish. For example, when $\theta_S = 60°$, except for the signal of LF loudspeakers, the signals of the RF and B loudspeakers vanish. Therefore, in the three directions of loudspeakers, the crosstalk between channels vanishes completely, and the localization performance is enhanced, which is preferred. However, the situation is different in the cases of the first-order sound field signal mixing for $M \geq 4$ loudspeakers reproduction. Figures 4.11 and 4.12 illustrate the panning curves of the first-order sound field signal mixing for six and eight horizontal loudspeakers, respectively. Figure 4.8 presents the six-loudspeaker configuration. For the eight-loudspeaker configuration, the loudspeakers are arranged at 0°, ±45°, ±90°, ±135°, and 180°. In Figures 4.11 and 4.12, the subscripts F, B, L, and R represent the front, back,

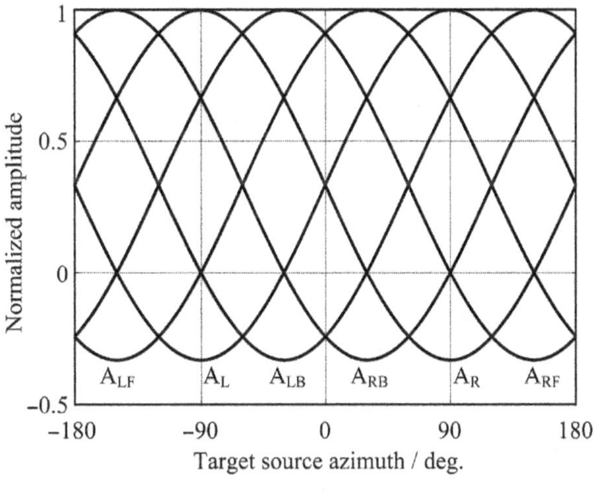

Figure 4.11 Panning curves of the first-order sound field signal mixing for six horizontal loudspeakers (the maximal amplitudes of signals are normalized to a unit).

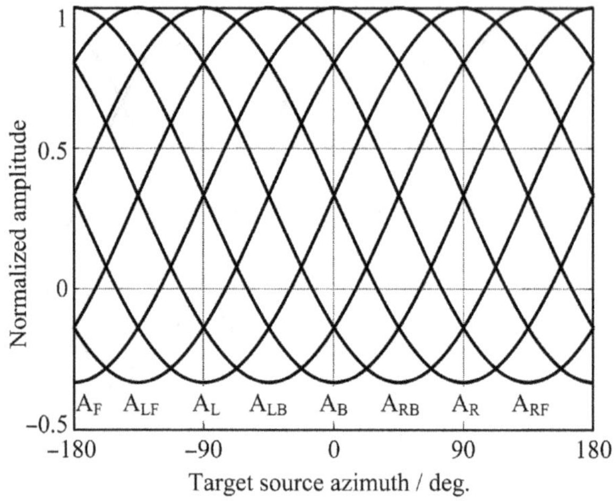

Figure 4.12 Panning curve of the first-order sound field signal mixing for eight horizontal loudspeakers (the maximal amplitudes of signals are normalized to a unit).

left, and right directions, respectively. The LF, LB, RB, and RF denote the left-front, left-back, right-back, and right-front directions, respectively. The figures illustrate the crosstalk among channels and the features of the global amplitude mixing.

Overall, at the central listening position, the first-order sound field signal mixing theoretically leads to a similar virtual source localization performance for arbitrary $M \geq 3$ loudspeaker reproduction. However, Figure 4.5 indicates that the magnitudes and phases of the four loudspeaker signals vary with the target source direction in reproduction with four loudspeakers. The superposed sound fields for a target source in and off loudspeaker directions are obviously different. Increasing the number of loudspeakers in reproduction with the first-order sound field signal mixing makes the sound field more uniform and reduces the difference between the virtual source in and off loudspeaker directions. Thus, the perceived difference between virtual sources at different directions decreases. Increasing the number of loudspeakers may also be beneficial to recreating sensations similar to those in a diffused sound field by using the indirect method described in Section 3.1. However, it may cause some other problems (Sections 9.5.1 and 12.1.3).

4.3 TRANSFORMATION OF HORIZONTAL SOUND FIELD SIGNALS AND AMBISONICS

The first-order sound field signal mixing for the reproduction with four horizontal and arbitrary $M \geq 3$ loudspeakers are addressed in Sections 4.1.3 and 4.2.2. In the preceding discussion, each loudspeaker signal is obtained from the recording by a corresponding supercardioid microphone. The sound field signals for different numbers of microphones or loudspeakers are related. They can be transformed mutually. On the basis of the transformation of a sound field signal, a category of multichannel surround sound techniques, systems, and corresponding signal mixing methods has been developed. The nature of this technique is each order decomposition, approximation, and reconstruction of the sound field. For lower-order reproduction, this technique creates a rough approximation of the target sound field, and psychoacoustic methods are applied to recreate a virtual source. As the order increases, this technique transits continuously from sound field reconstruction with rough approximation to reconstruction with high accuracy. Therefore, this technique and its system are vital in the development of spatial sound theory. The technique began in the 1970s (Cooper and Shiga, 1972; Kohsaka et al., 1972; Gerzon, 1973; Xie X.F., 1977, 1978b). This technique has also been widely studied and become an important part and method of spatial sound (Gerzon, 1985, 1992a; Xie X.F., 1982; Bamford and Vanderkooy, 1995; Xie and Xie, 1996; Daniel et al., 1998, 2003; Daniel 2000; Malham and Myatt, 1995; Poletti, 1996, 2000). This type of system and technique was named with various terms by different authors. Currently, it is customarily termed ***Ambisonics*** or ***Ambisonic-like*** system and technique. Sound field signal transformation and Ambisonics involve cases of horizontal and three-dimensional spatial sound reproduction. The former is addressed in this section, and the latter is discussed in Chapter 6. Ambisonics can be analyzed with various mathematical, physical, and psychoacoustic methods. A traditional psychoacoustic-based method in accordance with the virtual source localization theorem is addressed in this section. This method was used in most studies before 1990. A stricter method based on sound field reconstruction is described in Chapter 9.

4.3.1 Transformation of the first-order horizontal sound field signals

Section 4.1.3 indicates that the first-order sound field signals for four-loudspeaker reproduction can be captured with four coincident supercardioid microphones, as shown in Equation (4.1.11). According to the explanation in Section 3.2.1, $A_0 = A_{LF}$, $A_1 = A_{LB}$, $A_2 = A_{RB}$, and $A_3 = A_{RF}$ can be regarded as four normalized loudspeaker signals, and the complete loudspeaker signals are obtained by multiplying a waveform function E_A to the four normalized signals. When some simple formulas of trigonometric functions are used, and $b = 2$ is set in accordance with Equation (4.1.14), Equation (4.1.11) becomes

$$\begin{aligned} A_{LF} &= A_{total}\left(1+\sqrt{2}\cos\theta_S+\sqrt{2}\sin\theta_S\right) \quad & A_{LB} &= A_{total}\left(1-\sqrt{2}\cos\theta_S+\sqrt{2}\sin\theta_S\right) \\ A_{RB} &= A_{total}\left(1-\sqrt{2}\cos\theta_S-\sqrt{2}\sin\theta_S\right) \quad & A_{RF} &= A_{total}\left(1+\sqrt{2}\cos\theta_S-\sqrt{2}\sin\theta_S\right) \end{aligned}. \tag{4.3.1}$$

A_{total} is a constant related to the overall gain, which is determined by the response magnitude A_{mic} of microphones and the gains of amplification. However, these four signals are not completely independent. Equation (4.3.1) easily proves that the four signals have a linear constraint:

$$A_{LF} + A_{RB} - A_{LB} - A_{RF} = 0 \tag{4.3.2}$$

Therefore, the four signals have only three independent components. Each signal can be derived from a linear combination of three other signals.

Equation (4.3.1) expresses that the four signals A_{LF}, A_{LB}, A_{RB}, and A_{RF} are the linear combinations of the three independent components or signals:

$$W = 1 \quad X = \cos\theta_S \quad Y = \sin\theta_S, \tag{4.3.3}$$

and

$$\begin{aligned} A_{LF} &= A_{total}\left(W+\sqrt{2}X+\sqrt{2}Y\right) \quad & A_{LB} &= A_{total}\left(W-\sqrt{2}X+\sqrt{2}\,Y\right), \\ A_{RB} &= A_{total}\left(W-\sqrt{2}X-\sqrt{2}Y\right) \quad & A_{RF} &= A_{total}\left(W+\sqrt{2}X-\sqrt{2}Y\right). \end{aligned} \tag{4.3.4a}$$

Equation (4.3.4a) can be written as a matrix form:

$$\begin{bmatrix} A_{LF} \\ A_{LB} \\ A_{RB} \\ A_{RF} \end{bmatrix} = A_{total} \begin{bmatrix} 1 & \sqrt{2} & \sqrt{2} \\ 1 & -\sqrt{2} & \sqrt{2} \\ 1 & -\sqrt{2} & -\sqrt{2} \\ 1 & \sqrt{2} & -\sqrt{2} \end{bmatrix} \begin{bmatrix} W \\ X \\ Y \end{bmatrix}. \qquad (4.3.4b)$$

However, the three independent signals W, X, and Y can be derived from the following linear combination of the four signals of A_{LF}, A_{LB}, A_{RB}, and A_{RF}:

$$W = \frac{1}{4A_{total}}\left(A_{LF} + A_{LB} + A_{RB} + A_{RF}\right)$$

$$X = \frac{1}{4\sqrt{2}A_{total}}\left(A_{LF} - A_{LB} - A_{RB} + A_{RF}\right) \qquad (4.3.5a)$$

$$Y = \frac{1}{4\sqrt{2}A_{total}}\left(A_{LF} + A_{LB} - A_{RB} - A_{RF}\right).$$

They can also be written in the matrix form:

$$\begin{bmatrix} W \\ X \\ Y \end{bmatrix} = \frac{1}{4A_{total}} \begin{bmatrix} 1 & 1 & 1 & 1 \\ 1/\sqrt{2} & -1/\sqrt{2} & -1/\sqrt{2} & 1/\sqrt{2} \\ 1/\sqrt{2} & 1/\sqrt{2} & -1/\sqrt{2} & -1/\sqrt{2} \end{bmatrix} \begin{bmatrix} A_{LF} \\ A_{LB} \\ A_{RB} \\ A_{RF} \end{bmatrix}. \qquad (4.3.5b)$$

An ***important conclusion*** can be reached from the above mathematical derivation. The first-order sound field signals in four-loudspeaker reproduction can be recorded with four supercardioid microphones, but the four signals have only three linear independent components, such as W, X, and Y components (signals) in Equation (4.3.3). The four original signals A_{LF}, A_{LB}, A_{RB}, and A_{RF} can be encoded into three independent signals W, X, and Y for transmission or storage by a 3 × 4 matrix in Equation (4.3.5b) and then decoded to recover the four signals A_{LF}, A_{LB}, A_{RB}, and A_{RF} for reproduction by a 4× 3 matrix in Equation (4.3.4b). From the point of linear algebra, matrix encoding and decoding are linear transformation processes with a full rank without losing signal information. In brief, signals can be captured with four microphones, encoded into three independent signals, and decoded into four reproduction signals without any loss of spatial information. This process is the 4-3-4 transformation of first-order horizontal sound field signals (Xie X.F., 1977). The corresponding system is termed the 4-3-4 system, where the first "4" represents the four original recording channels or signals; the second "3" denotes the three transmission or storage channels; and the last "4" corresponds to the four reproduction channels or loudspeaker signals. Figure 4.13 illustrates the block diagram of the 4-3-4 system.

The three independent components given by Equation (4.3.3) are equivalent to the normalized signals (or source signals of a plane wave with the unit amplitude $P_A = E_A = 1$) captured with an omnidirectional microphone and two bidirectional microphones with their main axes pointing to the front and left directions, respectively. Figure 4.14 illustrates the polar

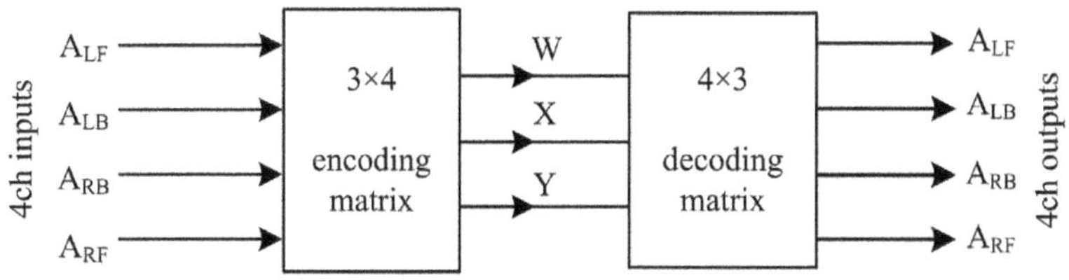

Figure 4.13 Block diagram of the 4-3-4 system.

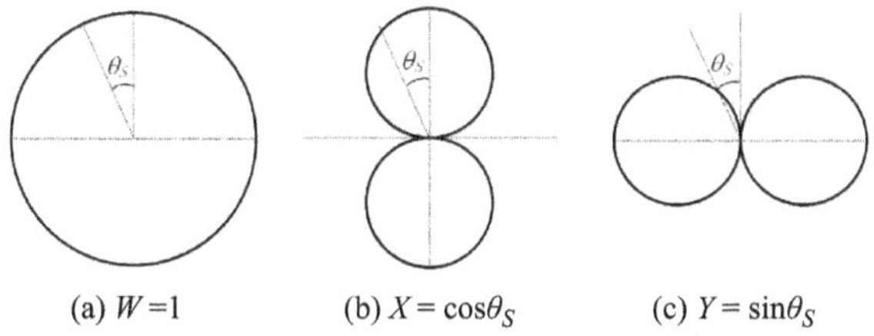

Figure 4.14 Polar patterns of the three independent components W, X, and Y for the first-order horizontal sound field signals (a) $W = 1$; (b) $X = \cos\theta_S$; (c) $Y = \sin\theta_S$.

pattern of these three independent signals. Therefore, the signal W is the sound field pressure; the signals X and $Y$are the velocity components of the sound field at the x (front–back axis) and y (left-right axis) directions. W, X, and Y are also called the first-order horizontal sound field signals. Accordingly, the first-order horizontal sound field signals W, X, and Y can be directly and simply captured by three corresponding microphones. The course of recording with four supercardioid microphones and then encoding the four signals A_{LF}, A_{LB}, A_{RB}, and A_{RF} into the three independent signals W, X, and Y via Equation (4.3.5b) is not needed. The transformation here is analogous to the MS transformation for two-channel stereophonic sound in Section 2.2.2.

According to the theorem of linear algebra, three uncorrelated linear combinations of W, X, and Y leads to a new set of (three) independent first-order sound field signals. That is,

$$\begin{bmatrix} S_1 \\ S_2 \\ S_3 \end{bmatrix} = [T_S] \begin{bmatrix} W \\ X \\ Y \end{bmatrix}. \tag{4.3.6}$$

$[T_S]$ is a 3×3 transformation matrix. If $[T_S]$ is a full rank and thus invertible (i.e., the determinant of matrix satisfies det $[T_S] \neq 0$), W, X, and Y can be derived from S_1, S_2, and S_3.

$$\begin{bmatrix} W \\ X \\ Y \end{bmatrix} = [T_S]^{-1} \begin{bmatrix} S_1 \\ S_2 \\ S_3 \end{bmatrix}. \tag{4.3.7}$$

S_1, S_2, and S_3 are not needed to be converted into W, X, and Y, and then decoded into A_{LF}, A_{LB}, A_{RB}, and A_{RF}. The four loudspeaker signals can be decoded directly from S_1, S_2, and S_3. Equations (4.3.4b) and (4.3.7) yields

$$\begin{bmatrix} A_{LF} \\ A_{LB} \\ A_{RB} \\ A_{RF} \end{bmatrix} = [D_E] \begin{bmatrix} S_1 \\ S_2 \\ S_3 \end{bmatrix}. \tag{4.3.8}$$

The 4 × 3 decoding matrix is given by

$$[D_E] = \begin{bmatrix} 1 & \sqrt{2} & \sqrt{2} \\ 1 & -\sqrt{2} & \sqrt{2} \\ 1 & -\sqrt{2} & -\sqrt{2} \\ 1 & \sqrt{2} & -\sqrt{2} \end{bmatrix} [T_S]^{-1}. \tag{4.3.9}$$

S_1, S_2, and S_3 can be theoretically captured using various microphone techniques. For example, the three normalized microphone signals are evaluated from Equation (1.2.41) by using the three identical combinations of pressure and pressure-gradient microphones with their main axes pointing to three horizontal directions θ_A, θ_B, and θ_C:

$$\begin{aligned} S_1 &= A_{mic}\left[1 + b\cos(\theta_S - \theta_A)\right] = A_{mic}\left(1 + b\cos\theta_A\cos\theta_S + b\sin\theta_A\sin\theta_S\right) \\ S_2 &= A_{mic}\left[1 + b\cos(\theta_S - \theta_B)\right] = A_{mic}\left(1 + b\cos\theta_B\cos\theta_S + b\sin\theta_B\sin\theta_S\right), \\ S_3 &= A_{mic}\left[1 + b\cos(\theta_S - \theta_C)\right] = A_{mic}\left(1 + b\cos\theta_C\cos\theta_S + b\sin\theta_C\sin\theta_S\right) \end{aligned} \tag{4.3.10}$$

where A_{mic} is a constant related to the response magnitude of microphones. A comparison among Equations (4.3.3), (4.3.6), and (4.3.10) yields the 3 × 3 transformation matrix:

$$[T_S] = A_{mic} \begin{bmatrix} 1 & 1 & 1 \\ b\cos\theta_A & b\cos\theta_B & b\cos\theta_C \\ b\sin\theta_A & b\sin\theta_B & b\sin\theta_C \end{bmatrix}. \tag{4.3.11}$$

If the main axis orientations of an arbitrary microphone pair among the three microphones are neither parallel nor opposite, the matrix $[T_S]$ is of full rank and thus invertible; S_1, S_2, and S_3 theoretically serve as independent signals. Moreover, the parameter of the directivity of a microphone is not limited to $b = 2$ (similar to supercardioid). Theoretically, all microphones with $b \neq 0$ are appropriate. In practice, however, b, θ_A, θ_B, and θ_C should be chosen properly to avoid the instability caused by the ill condition of $[T_S]$ and improve the signal–noise ratio in the captured signals. A set of parameters with $b = 1$, $\theta_A = 60°$, $\theta_B = 180°$, and $\theta_C = -60°$ is equivalent to recording with three cardioid microphones with the main axes pointing to the LF, B, and RF directions, respectively. The problem caused by the ill condition of $[T_S]$ can also be avoided by using four microphones to record and then converting into W, X, and Y in accordance with Equation (4.3.5b).

In the preceding discussion, the directional information θ_S of a target source is encoded into the azimuthal-dependent amplitudes of the independent signals X and Y alone with an omnidirectional signal W. The directional information can also be encoded into the azimuthal-dependent phases of independent signals, and the three normalized independent signals can be written as

$$S_1 = 1 \quad S_2 = \exp\left(-j\theta_S\right) \quad S_3 = \exp\left(j\theta_S\right) \tag{4.3.12}$$

The functions $\exp(\pm j\theta_S)$ represent a phase shift of $\pm \theta_S$. According to the Euler formula of $\exp(\pm j\theta_S) = \cos\theta_S \pm j\sin\theta_S$, the independent signals given by Equation (4.3.12) and W, X, and Y expressed in Equation (4.3.3) can be converted mutually. Some coefficients of this conversion are complex values, representing phase shifts in the signals. The four loudspeaker signals can also be decoded from the independent signals given by Equation (4.3.12), and some decoding coefficients should be complex values. This concept corresponds to BMX/TMX/QMX directional encoding series developed in early times (Cooper and Shiga, 1972; Kohsaka et al., 1972; Cooper et al., 1973; Copper, 1974).

In Section 4.2.2, the transformation of the first-order sound field signals can be extended to the cases of reproduction with arbitrary $M \geq 3$ loudspeakers. If $M \geq 3$ loudspeakers are arranged uniformly in the horizontal plane, the azimuth of the ith loudspeaker is expressed in Equation (4.2.7). The loudspeaker signals in Equation (4.2.8) can be written as follows by using Equation (4.3.3):

$$\begin{aligned} A_i\left(\theta_S\right) &= A_{total}\left(1 + 2\cos\theta_i\cos\theta_S + 2\sin\theta_i\sin\theta_S\right) \\ &= A_{total}\left(W + 2\cos\theta_i X + 2\sin\theta_i\ Y\right). \end{aligned} \tag{4.3.13}$$

Therefore, the signals of arbitrary $M \geq 3$ loudspeakers are obtained from a linear combination (matrix decoding) of the three independent signals W, X, and Y. Similar to the case in Equation (4.3.5b), the three independent signals can be derived from A_{LF}, A_{LB}, A_{RB}, and A_{RF} captured with four supercardioid microphones or directly captured with an omnidirectional microphone and two bidirectional microphones. The three independent signals W, X, and Y can also be theoretically derived from a linear combination of three other independent signals. This concept is basic for the ***4-3-N horizontal surround sound system*** (Xie X.F., 1978b, 1982). In this system, source signals are captured with four or three microphones, transmitted, or stored with three channels and then reproduced with arbitrary $N \geq 3$ loudspeakers (note: the number of loudspeakers is denoted by N in original studies, but it is replaced with M in this book).

For the first-order sound field signal mixing, increasing the number of recording microphones and loudspeakers in reproduction (Section 4.2.2) does not increase the number of independent signals and is unable to provide more spatial information of a target sound field. In this case, reproduction with more than four loudspeakers is unable to radically improve the accuracy in spatial information reproduction.

In summary, the analysis in this section indicates that the first-order sound field signals possess three independent components, which are irrelevant to the number and configuration of loudspeakers in reproduction. Independent signals have various forms. Various independent signals can be converted mutually. The reproduced signals of arbitrary $M \geq 3$ loudspeakers are derived from the three independent signals by decoding the matrix. Therefore, the transformation discussed in this section removes the redundant information in the sound field signals to make full use of the capacity of signal transmission and storage and thus simplify the number of transmission and storage channels. It also provides a basis for discussing the horizontal Ambisonics in the next section.

4.3.2 The first-order horizontal Ambisonics

The number and configuration of loudspeakers for Ambisonics are flexible. For convenience in analysis, M loudspeakers are arranged in a horizontal plane, and the azimuth of the ith loudspeaker is expressed in Equation (4.2.7), i.e.,

$$\theta_i = \theta_0 + \frac{360°i}{M} \quad i = 0, 1 \ldots (M-1), \tag{4.3.14a}$$

or it is written in radians:

$$\theta_i = \theta_0 + \frac{2\pi i}{M} \quad i = 0, 1 \ldots (M-1). \tag{4.3.14b}$$

If θ_i in Equation (4.3.14b) exceeds π, it should be replaced with $\theta_i - 2\pi$ to keep the azimuthal variable within the range of $-\pi < \theta \leq \pi$ because of the periodic variation in θ.

Generally, in the first-order Ambisonics, the normalized signals for the ith loudspeakers can be written as the form of sound field signal mixing, i.e., a linear combination or matrix decoding of the independent signals W, X, and Y given in Equation (4.3.3).

$$\begin{aligned} A_i(\theta_S) &= A_{total}\left(W + b\cos\theta_i X + b\sin\theta_i\ Y\right) \\ &= A_{total}\left(1 + b\cos\theta_i \cos\theta_S + b\sin\theta_i \sin\theta_S\right)^{\cdot} \end{aligned} \tag{4.3.15}$$

b, which determines the characteristic of the reproduced sound field, can be derived from various optimized methods. Similar to the cases in the preceding sections, the complete loudspeaker signals are obtained by multiplying a waveform function E_A to the normalized signals.

Some characteristics of the trigonometric function series are used in the succeeding analysis. When θ_i satisfies Equation (4.3.14a) or (4.3.14b), the following is obtained:

$$\sum_{i=0}^{M-1} \cos(q\theta_i) = \sum_{i=0}^{M-1} \sin(q\theta_i) = 0 \quad for\ M \geq q+1, \quad q = 1, 2, 3\ldots.., \tag{4.3.16}$$

and for

$$M \geq q + q' + 1, \tag{4.3.17}$$

the following is derived:

$$\begin{aligned} &\sum_{i=0}^{M-1} \cos(q\theta_i)\cos(q'\theta_i) = \sum_{i=0}^{M-1} \sin(q\theta_i)\sin(q'\theta_i) = \begin{cases} 0 & q \neq q' \\ \dfrac{M}{2} & q = q' \end{cases} \quad q, q' = 1, 2, 3\ldots, \\ &\sum_{i=0}^{M-1} \cos(q\theta_i)\sin(q'\theta_i) = 0. \end{aligned} \tag{4.3.18}$$

Equations (4.3.16) to (4.3.18) are the discrete versions of the following integral orthogonalities of trigonometric functions:

$$\int_{-\pi}^{\pi} \cos(q\theta)d\theta = \int_{-\pi}^{\pi} \sin(q\theta)d\theta = 0 \qquad q = 1, 2, 3..., \tag{4.3.19}$$

and

$$\begin{gathered}\int_{-\pi}^{\pi} \cos(q\theta)\cos(q'\theta)d\theta = \int_{-\pi}^{\pi} \sin(q\theta)\sin(q'\theta)d\theta = \begin{cases} 0 & q \neq q' \\ \pi & q = q' \end{cases} \quad q, q' = 1, 2, 3... \\ \int_{-\pi}^{\pi} \cos(q\theta)\sin(q'\theta)d\theta = 0. \end{gathered} \tag{4.3.20}$$

In the traditional analysis of Ambisonics, b in Equation (4.3.15) is chosen according to the certain optimized physical and psychoacoustic criteria of virtual source localization. Different psychoacoustic criteria lead to different results. At low frequencies, b can be chosen according to the optimization of interaural phase delay difference and its dynamic variation with head rotation or according to the optimization of the velocity localization vector in Section 3.2.2. According to Equations (4.3.16) to (4.3.18), for

$$M \geq 3, \tag{4.3.21}$$

the following equation is obtained:

$$\begin{gathered}\sum_{i=0}^{M-1} A_i(\theta_S)\sin\theta_i = \frac{b}{2} MA_{total} \sin\theta_S \quad \sum_{i=0}^{M-1} A_i(\theta_S)\cos\theta_i = \frac{b}{2} MA_{total} \cos\theta_S \\ \sum_{i=0}^{M-1} A_i(\theta_S) = MA_{total}. \end{gathered} \tag{4.3.22}$$

The perceived direction of a virtual source is evaluated with Equations (3.2.7) and (3.2.9). For a fixed head,

$$\sin\theta_I = \frac{\sum_{i=0}^{M-1} A_i(\theta_S)\sin\theta_i}{\sum_{i=0}^{M-1} A_i(\theta_S)} = \frac{b}{2}\sin\theta_S, \tag{4.3.23}$$

and for the head oriented to the virtual source,

$$\tan\hat{\theta}_I = \frac{\sum_{i=0}^{M-1} A_i(\theta_S)\sin\theta_i}{\sum_{i=0}^{M-1} A_i(\theta_S)\cos\theta_i} = \tan\theta_S. \tag{4.3.24}$$

The condition of the head oriented to the virtual source, or the direction of the velocity localization vector does not restrict b. The condition in Equation (4.3.23) for a fixed head leads to

$$b = 2, \tag{4.3.25}$$

so

$$\sin\theta_I = \sin\theta_S. \tag{4.3.26}$$

The combination of Equations (4.3.24) and (4.3.26) yields

$$\theta_I = \hat{\theta}_I = \theta_S. \tag{4.3.27}$$

Therefore, at very low frequencies, the perceived virtual source direction matches with that of the target source direction within the full horizontal directions of $-180° \le \theta_S \le 180°$. Moreover, the results of a fixed head and those of the head oriented to the virtual source are identical. Equation (3.2.29) proves that the velocity vector magnitude equals to a unit value of $r_v = 1$ when $b = 2$. In other words, $b = 2$ is necessary to create a correct interaural phase delay difference or a unit value of velocity vector magnitude in reproduction. This result is obtained for the special case of four-loudspeaker reproduction in Section 4.1.3 and is extended to the cases of arbitrary $M \ge 3$ loudspeakers. In addition, Equation (4.3.21) indicates that the first-order horizontal Ambisonics requires three loudspeakers at least to ensure the correct interaural phase delay difference and its dynamic variation with head rotation. It also aims to ensure the correct direction of a velocity localization vector and the optimized velocity vector magnitude of $r_v = 1$.

The virtual source localization in horizontal Ambisonics can also be analyzed according to the energy localization vector discussed in Section 3.2.2 although this analysis is incompletely accurate. Equations (4.3.14a) to (4.3.18) yield

$$Pow' = \sum_{i=0}^{M-1} A_i^2(\theta_S) = \left(1 + \frac{b^2}{2}\right) M A_{total}^2 \qquad M \ge 3, \tag{4.3.28}$$

$$\sum_{i=0}^{M-1} A_i^2(\theta_S)\sin\theta_i = MbA_{total}^2 \sin\theta_S \qquad M \ge 4, \tag{4.3.29}$$

$$\sum_{i=0}^{M-1} A_i^2(\theta_S)\cos\theta_i = MbA_{total}^2 \cos\theta_S \qquad M \ge 4. \tag{4.3.30}$$

Equation (4.3.28) is the overall power of the loudspeaker signals, which depend on b. For the first-order Ambisonics and a given b, when the number of loudspeakers is not less than 3, the overall power is a constant and independent from the target source azimuth θ_S. This concept is a feature of horizontal Ambisonics with a regular loudspeaker configuration.

From Equation (3.2.34) and for $M \geq 4$, the direction θ_E of the energy localization vector is evaluated with

$$r_E \cos\theta_E = \frac{\sum_{i=0}^{M-1} A_i^2(\theta_S)\cos\theta_i}{\sum_{i=0}^{M-1} A_i^2(\theta_S)} = \frac{2b\cos\theta_S}{2+b^2},$$
$$r_E \sin\theta_E = \frac{\sum_{i=0}^{M-1} A_i^2(\theta_S)\sin\theta_i}{\sum_{i=1}^{M-1} A_i^2(\theta_S)} = \frac{2b\sin\theta_S}{2+b^2}. \quad (4.3.31)$$

Equation (4.3.31) yields

$$\tan\theta_E = \tan\theta_S. \quad (4.3.32)$$

For a regular loudspeaker configuration, the first-order horizontal Ambisonics with loudspeaker signals expressed in Equation (4.3.15) leads to an energy localization vector with its direction identical to that of a target source. Moreover, the direction of the energy localization vector is independent of b. The above feature of the energy localization vector is desired if the hypothesis that the direction of a mid- and high-frequency virtual source is consistent with that of the energy localization vector is valid. Therefore, Equations (4.3.29) and (4.3.30) require that $M \geq 4$; that is, the theorem of the energy localization vector requires four reproduction channels and loudspeakers at least for the first-order Ambisonics, one more than the requirement from the theorem of velocity localization vector. Therefore, four or more loudspeakers are more appropriate for the first-order Ambisonics to consider the feature of the energy localization vector.

The energy vector magnitude is calculated from Equation (4.3.31) or directly from Equation (3.2.36):

$$r_E = \frac{2b}{2+b^2}. \quad (4.3.33)$$

Therefore, the energy vector magnitude depends on b with $r_E = 0.667$ for $b = 2$.

$b = 2$ is derived from an optimization on the localization performance at low frequencies and is inappropriate for mid and high frequencies. If the theorem of the energy localization vector is valid for mid and high frequencies above 0.7 kHz, the optimized parameter b for mid- and high-frequency localization can be derived by maximizing the energy vector magnitude. In Equation (4.3.33), if

$$\frac{\partial r_E}{\partial b} = 0, \quad (4.3.34)$$

the following equation is obtained:

$$b = \sqrt{2}. \quad (4.3.35)$$

Thus, $r_E = (r_E)_{max} = 0.707$, which is slightly greater than $r_E = 0.667$ for $b = 2$. Figure 4.15 (a) illustrates the polar pattern of the first-order microphone response (i.e., the loudspeaker signal) with $b = \sqrt{2}$.

For b expressed in Equation (4.3.35), the perceived virtual source direction $\hat{\theta}_I$ for the head oriented to the virtual source (i.e., the direction of the velocity localization vector) is still consistent with that of the target direction θ_S. However, the perceived virtual source direction for a fixed head given in Equation (4.3.23) becomes

$$\sin\theta_I = \frac{\sqrt{2}}{2}\sin\theta_S. \tag{4.3.36}$$

The virtual source of a fixed head moves toward the direction of the central front because $|\sin\theta_I| < |\sin\theta_S|$; thus, it deviates from that of the target direction. In addition, Equation (3.2.29) reveals that the velocity vector magnitude in this case is $r_v = 0.707$, which is smaller than the ideal value of $r_v = 1$. Therefore, the first-order signal with b for maximizing the energy vector magnitude does not lead to correct localization at low frequencies. Ideally, b should be separately optimized within different frequency ranges.

For reproduction within a large-sized listening region (such as halls), the out-of-phase crosstalk from opposite channels should be restrained. If the following parameter is chosen (Malham and Myatt, 1995):

$$b = 1, \tag{4.3.37}$$

the signals in Equation (4.3.15) become

$$A_i(\theta_S) = A_{total}(W + \cos\theta_i X + \sin\theta_i\ Y) = A_{total}\left[1 + \cos(\theta_S - \theta_i)\right]. \tag{4.3.38}$$

Equation (4.3.38) is equivalent to the signals recorded with a set of cardioid microphones with their main axes pointing to each direction θ_i. Figure 4.15 (b) presents the polar pattern of the response of the cardioid microphone. This solution of reproduction signals is termed the in-phase solution because the negative rear lobe is eliminated in the polar pattern. A comparison between Figures 4.4 and 4.15 indicates that the magnitude of the out-of-phase rear lobe for $b = \sqrt{2}$ is smaller than that of $b = 2$, which is between the magnitude of $b = 1$ and $b = 2$. For $b = 1$ given in Equation (4.3.37), the virtual source direction for a fixed head is evaluated with Equation (4.3.23):

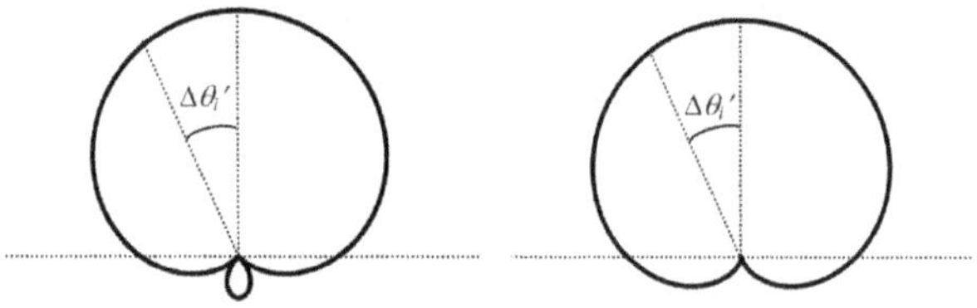

(a) $b = \sqrt{2}$ (b) $b = 1$

Figure 4.15 Polar pattern of the first-order microphone response (i.e., loudspeaker signal) with two different values of b, (a) $b = \sqrt{2}$; (b) $b = 1$

$$\sin\theta_I = \frac{1}{2}\sin\theta_S \tag{4.3.39}$$

In this case, the velocity vector magnitude evaluated from Equation (3.2.29) is $r_v = 0.5$, and the energy vector magnitude evaluated from Equation (4.3.33) is $r_E = 0.667$. Both velocity and energy vector magnitude deviate from the ideal value of the unit. The reduction of a velocity vector magnitude is due to the elimination of an out-of-phase crosstalk in the opposite loudspeaker signal. However, the virtual source localization in the central listening position is insignificant for a reproduction system intended for large-sized listening regions.

The preceding discussion does not deal with the constant A_{total} related to the overall gain. Although the virtual source direction is independent from A_{total},A_{total} determines the overall reproduced pressure or power. The actual loudspeaker (non-normalized) signals in reproduction are

$$\begin{aligned} E_i(\theta_S) &= E_A A_{total}\left(W + b\cos\theta_i X + b\sin\theta_i\ Y\right) \\ &= E_A A_{total}\left(1 + b\cos\theta_i \cos\theta_S + b\sin\theta_i \sin\theta_S\right)^{\cdot} \end{aligned} \tag{4.3.40}$$

E_A is related to the target-reproduced sound pressure. Two methods are used to specify A_{total}. The first method is the constant amplitude normalization. That is to choose A_{total} thus the free-field sound pressure amplitude at the origin in reproduction matches with the target amplitude P_A. Letting $E_A = P_A$ yields

$$\sum_{i=0}^{M-1} A_i(\theta_S) = 1. \tag{4.3.41}$$

According to Equation (4.3.22), the following equation is obtained:

$$A_{total} = \frac{1}{M}. \tag{4.3.42}$$

Another method is the constant power or constant energy normalization. That is, to select A_{total} thus the free-field sound power at the origin in reproduction matches with the target power. Let

$$\sum_{i=0}^{M-1} A_i^2(\theta_S) = 1. \tag{4.3.43}$$

According to Equation (4.3.28), the A_{total} is expressed as follows:

$$A_{total} = \frac{1}{\sqrt{\left(1 + b^2/2\right)M}}. \tag{4.3.44}$$

According to Jot et al. (1999), Table 4.1 lists the parameters and characters of the first-order horizontal Ambisonics with regular loudspeaker configurations and different decoding methods.

Table 4.1 Parameters and characters for the first-order horizontal Ambisonics with regular loudspeaker configurations and different decoding methods.

Criteria for optimization	*Frequency range*	*Listening region*	*Normalization*	A_{total}	*b*
$r_v = 1$	Low	Small	Amplitude	$1/M$	2
$r_v = 1$	Low	Small	Power	$1/\sqrt{3M}$	2
Maximize r_E	Mid and high	Small	Power	$1/\sqrt{2M}$	$\sqrt{2}$
In-phase	Full	Large	Power	$\sqrt{2/(3M)}$	1

4.3.3 The higher-order horizontal Ambisonics

Sound field signals in Ambisonics are extended from the first order to higher orders to improve the accuracy in spatial information reproduction, which are termed ***higher-order Ambisonics*** (**HOA**; Xie X.F., 1978b; Xie and Xie, 1996; Bamford and Vanderkooy, 1995; Daniel et al., 1998, 2003). For the first-order horizontal Ambisonics, the signal of the loudspeaker at azimuth θ_i is given in Equation (4.3.15) and can be written as the linear combination of a target-azimuthal-independent component $W = 1$ and a pair of first-order target-azimuthal harmonics $X = \cos\theta_S$ and $Y = \sin\theta_S$:

$$\begin{aligned} A_i(\theta_S) &= A_{total}\left[W + D_1^{(1)}(\theta_i)X + D_1^{(2)}(\theta_i)Y\right] \\ &= A_{total}\left[1 + D_1^{(1)}(\theta_i)\cos\theta_S + D_1^{(2)}(\theta_i)\sin\theta_S\right], \end{aligned} \tag{4.3.45}$$

where

$$D_1^{(1)}(\theta_i) = b\cos\theta_i \qquad D_1^{(2)}(\theta_i) = b\sin\theta_i. \tag{4.3.46}$$

For the second-order horizontal Ambisonics, two additional second-order target-azimuthal harmonics components are supplemented to the signals expressed in Equation (4.3.45):

$$U = \cos 2\theta_S \quad V = \sin 2\theta_S. \tag{4.3.47}$$

Then, the normalized signal amplitude of the loudspeaker at θ_i becomes a linear combination of five independent components or signals W, X, Y, U, and V:

$$\begin{aligned} A_i(\theta_S) &= A_{total}\left[W + D_1^{(1)}(\theta_i)X + D_1^{(2)}(\theta_i)Y + D_2^{(1)}(\theta_i)U + D_2^{(2)}(\theta_i)V\right] \\ &= A_{total}\left[1 + D_1^{(1)}(\theta_i)\cos\theta_S + D_1^{(2)}(\theta_i)\sin\theta_S + D_2^{(1)}(\theta_i)\cos 2\theta_S + D_2^{(2)}(\theta_i)\sin 2\theta_S\right]. \end{aligned} \tag{4.3.48}$$

Generally, the independent signals of the Q-order horizontal Ambisonics with $Q \geq 1$ consist of the preceding $(2Q + 1)$ azimuthal harmonics up to the Q order:

$$1, \quad \cos q\theta_S, \quad \sin q\theta_S \quad q = 1, 2 \ldots . Q. \tag{4.3.49}$$

The normalized signal amplitude of the loudspeaker at θ_i is a linear combination of $(2Q + 1)$ independent components:

$$A_i(\theta_S) = A_{total}[1 + \sum_{q=1}^{Q}\left[D_q^{(1)}(\theta_i)\cos q\theta_S + D_q^{(2)}(\theta_i)\sin q\theta_S\right]. \tag{4.3.50}$$

Equation (4.3.50) can also be written in the matrix form as

$$\boldsymbol{A} = A_{total}\left[D_{2D}\right]\boldsymbol{S}. \tag{4.3.51}$$

where $\boldsymbol{A} = [A_0(\theta_S), A_1(\theta_S),\ldots A_{M-1}(\theta_S)]^T$ is an $M \times 1$ column matrix or vector composed of M normalized loudspeaker signals; the superscript "T" denotes the matrix transpose; $\boldsymbol{S} = [1, \cos\theta_S, \sin\theta_S, \cos2\theta_S, \sin2\theta_S\ldots, \cos Q\theta_S, \sin Q\theta_S]^T$ is a $(2Q + 1) \times 1$ column matrix composed of $(2Q + 1)$ normalized independent signals; $[D_{2D}]$ is an $M \times (2Q + 1)$ decoding matrix with entries $1, D_q^{(1)}(\theta_i), D_q^{(2)}(\theta_i), q = 1, 2\ldots Q$.

Loudspeaker signals depend on entries $D_q^{(1)}(\theta_i)$ and $D_q^{(2)}(\theta_i)$. When a loudspeaker configuration is regular and θ_i of each loudspeaker is given in Equation (4.3.14a), similar to Equation (4.3.46), $D_q^{(1)}(\theta_i)$ and $D_q^{(2)}(\theta_i)$ take the forms

$$D_q^{(1)}(\theta_i) = 2\kappa_q \cos q\theta_i \qquad D_q^{(2)}(\theta_i) = 2\kappa_q \sin q\theta_i \quad \kappa_q > 0 \quad b = 2\kappa_1. \tag{4.3.52}$$

Then,

$$\begin{aligned} A_i(\theta_S) &= A_{total}\left\{1 + 2\sum_{q=1}^{Q}\left[\kappa_q \cos q\theta_i \cos q\theta_S + \kappa_q \sin q\theta_i \sin q\theta_S\right]\right\} \\ &= A_{total}\left[1 + 2\sum_{q=1}^{Q}\kappa_q \cos q(\theta_S - \theta_i)\right]. \end{aligned} \tag{4.3.53}$$

The decoding parameter κ_q specifies the relative proportion of each azimuthal harmonics in loudspeaker signals. A set of κ_q can be regarded as an azimuthal harmonic window applied to truncate the azimuthal harmonics up to order Q. The loudspeaker signal magnitude maximizes when the target source direction is exactly consistent with the loudspeaker direction, i.e., $\theta_S = \theta_i$, then

$$A_i(\theta_S)_{\max} = A_{total}\left[1 + 2\sum_{q=1}^{Q}\kappa_q\right]. \tag{4.3.54}$$

Similar to the case of the first-order Ambisonics, for a regular loudspeaker configuration with θ_i given in Equations (4.3.14a), (4.3.53), (4.3.16) to (4.3.18) verify that when

$$M \geq Q + 2, \tag{4.3.55}$$

the following equation is obtained:

$$\sum_{i=0}^{M-1} A_i(\theta_S) = MA_{total}$$
$$\sum_{i=0}^{M-1} A_i(\theta_S)\cos\theta_i = \kappa_1 MA_{total}\cos\theta_S \quad \sum_{i=0}^{M-1} A_i(\theta_S)\sin\theta_i = \kappa_1 MA_{total}\sin\theta_S. \tag{4.3.56}$$

Equation (4.3.56) is only related to the parameter or proportion κ_1 of the first-order azimuthal harmonic component and independent from the second or higher-order azimuthal harmonic component. The perceived virtual source direction is evaluated from Equations (3.2.7) and (3.2.9). For a fixed head,

$$\sin\theta_I = \frac{\sum_{i=0}^{M-1} A_i(\theta_S)\sin\theta_i}{\sum_{i=0}^{M-1} A_i(\theta_S)} = \kappa_1 \sin\theta_S, \tag{4.3.57}$$

For the head oriented to the virtual source,

$$\tan\hat{\theta}_I = \frac{\sum_{i=0}^{M-1} A_i(\theta_S)\sin\theta_i}{\sum_{i=0}^{M-1} A_i(\theta_S)\cos\theta_i} = \tan\theta_S. \tag{4.3.58}$$

The condition of the head oriented to the virtual source in Equation (4.3.58), or the direction of the velocity localization vector does not place restrictions on κ_q, and the condition for a fixed head in Equation (4.3.57) only limits κ_1. When

$$\kappa_1 = 1 \quad or \quad b = 2\kappa_1 = 2, \tag{4.3.59}$$

the following equation is obtained:

$$\sin\theta_I = \sin\theta_S. \tag{4.3.60}$$

Equations (4.3.58) and (4.3.60) yield

$$\theta_I = \hat{\theta}_I = \theta_S. \tag{4.3.61}$$

In this case, the perceived virtual source azimuth matches with that of the target source within the full horizontal direction of $-180° \le \theta_S \le 180°$, and the results of the head oriented to the virtual source are consistent with those of a fixed head. Equation (3.2.29) proves that the velocity vector magnitude is $r_v = 1$ when $\kappa_1 = b/2 = 1$. In other words, for $Q > 1$ order horizontal Ambisonics, the optimized velocity localization vector only requires the decoding parameter of the first-order azimuthal harmonics to be $\kappa_1 = b/2 = 1$ without restriction on κ_q for the second or higher-order azimuthal harmonics.

κ_q can also be derived from other physical criteria. In Section 9.1, under the criterion of spatial harmonic decomposition and each order approximation of the target sound field, the parameters should be

$$\kappa_q = b/2 = 1 \qquad q = 1,2\ldots Q. \tag{4.3.62}$$

Loudspeaker signals with the parameter given in Equation (4.3.62)are the conventional or fundamental solution of Ambisonic signals to which a rectangular azimuthal harmonic window is applied to truncate azimuthal harmonic components up to Q.

The two second-order azimuthal harmonic components of the second-order Ambisonics are given in Equation (4.3.47). These two components are equivalent to the signals captured by a pair of the second-order directional microphones, and the polar patterns are illustrated in Figure 4.16. The responses of the second-order azimuthal harmonics vary faster than those of the first-order azimuthal harmonics. Similarly, in comparison with the $(Q - 1)$-order reproduction signals, two additional Q-order target-azimuthal harmonic components are supplemented to the Q-order reproduction signals. These additional components are equivalent to the signals captured by a pair of the Q-order directional microphones. In practice, the realization of the higher-order directional microphones is difficult, but the higher-order azimuthal harmonic components can be derived from the outputs of a spherical microphone array (Section 9.8). In addition, higher-order azimuth harmonic components can be easily simulated by signal processing.

The normalized signals of Q-order horizontal Ambisonics can be obtained by substituting Equation (4.3.62) into Equation (4.3.53):

$$\begin{aligned} A_i(\theta_S) &= A_{total}\left[1 + 2\sum_{q=1}^{Q}(\cos q\theta_i \cos q\theta_S + \sin q\theta_i \sin q\theta_S)\right] \\ &= A_{total}\left[1 + 2\sum_{q=1}^{Q}\cos q(\theta_S - \theta_i)\right] \\ &= A_{total}\frac{\sin\left[\left(Q + \frac{1}{2}\right)(\theta_S - \theta_i)\right]}{\sin\left(\frac{\theta_S - \theta_i}{2}\right)}. \end{aligned} \tag{4.3.63}$$

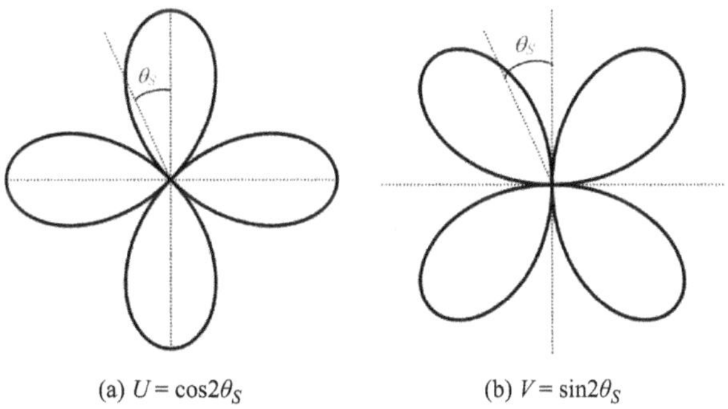

Figure 4.16 Polar patterns of a pair of the second-order directional microphones (a) $U = \cos 2\theta_S$; (b) $V = \sin 2\theta_S$.

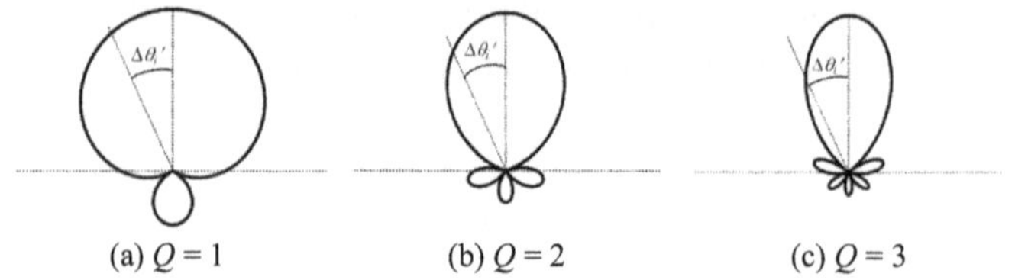

Figure 4.17 Polar patterns of the preceding three orders Ambisonic signals $A_i(\theta_S) = A_i(\theta_S - \theta_i)$ (a) Q = 1; (b) Q = 2; (C) Q = 3

The recording and reproduction of Ambisonic signals are analyzed to obtain insights into the physical nature of Equation (4.3.63). Figure 4.17 illustrates the polar pattern of the preceding three orders Ambisonic signals $A_i(\theta_S) = A_i(\theta_S - \theta_i)$. In Figure 4.17, the maximal magnitude of the signal is normalized to a unit. From the point of signal recording, Equation (4.3.63) is equivalent to a signal captured with a microphone with an appropriate directivity and main axis direction. The main lobe of the signal is centered at $\Delta\theta'_i = (\theta_S - \theta_i) = 0°$. The response $|A_i(\theta_S - \theta_i)|$ maximizes at the on-axis direction $\Delta\theta'_i = 0°$ and then decreases as $|\Delta\theta'_i|$ increases. As $|\Delta\theta'_i|$ further increases, the responses exhibit the side and rear lobes (in-phase or out-of-phase) and null points of the polar patterns at some azimuths. As order Q increases, the width of the main lobe and the responses of the side and rear lobes decrease, sharpening the directivity of the resultant signal.

From the point of reproduction, the reproduced sound field in Ambisonics is a superposition of the pressures caused by M loudspeakers. Equation (4.3.63) represents the signal for the ith loudspeaker with θ_i being the azimuth of the loudspeaker and θ_S being the azimuth of the target source. As Q of reproduction signals increases, the relative signal magnitude of the loudspeaker nearest the target source direction increases, and the relative signal magnitudes (crosstalk) of the other loudspeakers decrease. Consequently, the perceived performance of the virtual source improves, and the listening region widens. However, these improvements occur at the cost of increasing the complexity of the system. For example, Figure 4.18 illustrates the virtual source position for Q = 1, 2, and 3 order horizontal Ambisonic reproduction. The results are evaluated from Equation (3.2.6) for an eight-loudspeaker configuration (Figure 4.12) and a fixed head. The frequency is $f = 0.7$ kHz, and the precorrected head radius is $a' = 1.25 \times 0.0875$ m. Figure 4.18 illustrates the results within $0° \le \theta_S \le 90°$ only because of symmetry. Ideally, the perceived virtual source direction should be consistent with that of the

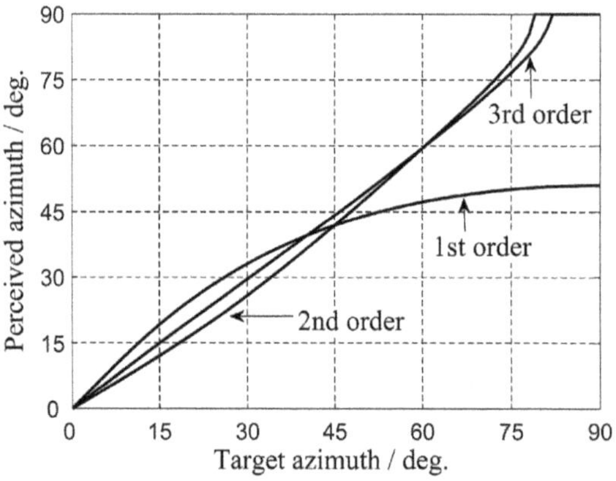

Figure 4.18 Virtual source position for Q = 1, 2, and 3 order horizontal Ambisonic reproduction with an eight-loudspeaker configuration and a fixed head. The frequency is f = 0.7 kHz.

target source direction. The figure also indicates that the movement of the virtual source with frequency decreases as the order increases, and the upper frequency limit for the accurate reproduction of spatial information increases. This problem is addressed in Section 9.3.1.

In the case of the Q-order horizontal Ambisonic signals given in Equation (4.3.14a) for a regular loudspeaker configuration with the number of loudspeakers,

$$M = 2Q + 1 \tag{4.3.64}$$

Equation (4.3.63) verifies that the signals of other $2Q$ loudspeakers vanish when the target source is located at the direction of one loudspeaker except for the signal of this loudspeaker. In other words, the virtual source in loudspeaker directions is recreated by the single loudspeaker, the crosstalks among loudspeakers vanish, and localization performance enhances. The first-order reproduction with three loudspeakers discussed in Section 4.2.2 is a special example of this case. Here, the analysis is extended to the case of arbitrary Q-order reproduction. If $M > (2Q + 1)$ loudspeakers are used in the Q-order reproduction, crosstalks among loudspeaker signals exist even if the target source is located at the direction of one loudspeaker. For horizontal Ambisonics with a regular loudspeaker configuration, the reproduced sound field exhibits symmetry against the rotation around the vertical axis.

Similar to the case of the first-order Ambisonics, the energy localization vector discussed in Section 3.2.2 is applicable to the analysis of the second- and higher-order Ambisonics. The mid-and high-frequency criteria for optimizing the energy localization vector can be used to choose κ_q in Equation (4.3.53) (Daniel et al., 1998).From Equation (4.3.53) and by using Equations (4.3.16) to (4.3.18), when

$$M \geq (2Q + 1), \tag{4.3.65}$$

the following equation is obtained:

$$Pow' = \sum_{i=0}^{M-1} A_i^2(\theta_S) = MA_{total}^2 \left(1 + 2\sum_{q=1}^{Q} \kappa_q^2\right). \tag{4.3.66}$$

Therefore, for regular loudspeaker configurations, the overall free-field sound power at the origin in reproduction is independent from the target source direction θ_S.

Similar to the case of the first-order Ambisonics, by considering Equation (4.3.53), using Equations (4.3.16) to (4.3.18), and applying some simple formulas of trigonometric functions, when

$$M \geq (2Q + 2), \tag{4.3.67}$$

the following equations are obtained:

$$\sum_{i=0}^{M-1} A_i^2(\theta_S)\cos\theta_i = 2MA_{total}^2 \sum_{q=1}^{Q} \kappa_q \kappa_{q-1} \cos\theta_S, \tag{4.3.68}$$

$$\sum_{i=0}^{M-1} A_i^2(\theta_S)\sin\theta_i = 2MA_{total}^2 \sum_{q=1}^{Q} \kappa_q \kappa_{q-1} \sin\theta_S, \tag{4.3.69}$$

where $\kappa_0 = 1$.

According to Equation (3.2.34), the direction θ_E of the energy localization vector satisfies the following equations:

$$
\begin{aligned}
r_E \cos\theta_E &= \frac{\sum_{i=0}^{M-1} A_i^2(\theta_S)\cos\theta_i}{\sum_{i=0}^{M-1} A_i^2(\theta_S)} = \frac{2\sum_{q=1}^{Q} \kappa_q \kappa_{q-1} \cos\theta_S}{1+2\sum_{q=1}^{Q} \kappa_q^2}, \\
r_E \sin\theta_E &= \frac{\sum_{i=0}^{M-1} A_i^2(\theta_S)\sin\theta_i}{\sum_{i=1}^{M-1} A_i^2(\theta_S)} = \frac{2\sum_{q=1}^{Q} \kappa_q \kappa_{q-1} \sin\theta_S}{1+2\sum_{q=1}^{Q} \kappa_q^2}.
\end{aligned}
\tag{4.3.70}
$$

Then,

$$
\tan\theta_E = \tan\theta_S. \tag{4.3.71}
$$

For a regular loudspeaker configuration in the horizontal plane, the direction of the energy localization vector for the Q-order Ambisonic signals given in Equation (4.3.53) matches that of the target source direction and is independent of κ_q (q = 1, 2...Q). This feature is desirable if the hypothesis of the energy localization vector theorem is valid above 0.7 kHz.

For the Q-order horizontal Ambisonics with a regular loudspeaker configuration, the condition that the overall power is given in Equation (4.3.65) is target-direction-independent requires $M \geq (2Q + 1)$ reproduction channels and loudspeakers. The result of the energy localization vector theorem given in Equation (4.3.67)requires one more channel and loudspeaker than that of Equation (4.3.65), i.e., it requires $(2Q + 2)$ channels and loudspeakers at least. Therefore, using $(2Q + 2)$ loudspeakers at least is appropriate for the Q-order horizontal Ambisonics to consider the requirement of the energy localization vector. This number of channels and loudspeakers is minimal for the Q-order horizontal Ambisonic reproduction. The same conclusion is made in Section 4.3.2 for the first-order reproduction, and the conclusion is extended to the arbitrary Q-order reproduction. Therefore, Q = 1, 2, 3, and 4-order horizontal Ambisonics require 4, 6, 8, and 10 loudspeakers, respectively. This conclusion is consistent with the results derived from an evaluation of the width of the directivity of the signals for arbitrary Q-order reproduction (Xie and Xie, 1996). As stated in Section 4.2.2, a further increase in the number of loudspeakers may decrease the perceived difference between a virtual source in and off loudspeaker directions, but it may also cause some other problems.

The energy vector magnitude of the Q-order reproduction is evaluated from Equation (4.3.70) or more directly from Equation (3.2.36):

$$
r_E = \frac{2\sum_{q=1}^{Q} \kappa_q \kappa_{q-1}}{1+2\sum_{q=1}^{Q} \kappa_q^2}. \tag{4.3.72}
$$

For the conventional solution of the Ambisonic signals with $\kappa_q = \kappa_0 = 1$ given in Equation (4.3.62), Equation (4.3.72) yields

$$r_E = \frac{2Q}{1+2Q}. \tag{4.3.73}$$

$Q = 1, 2$, and 3-order Ambisonics have the resultant r_E of 0.667, 0.800, and 0.857, respectively. As Q increases, r_E gradually approaches the unit value, i.e., it approaches the case of ideal reproduction.

If the hypothesis of the energy localization vector theorem is valid above 0.7 kHz, the criterion of maximizing the energy vector magnitude can be applied to choose κ_q in Equation (4.3.53). In Equation (4.3.72). According to the condition:

$$\frac{\partial r_E}{\partial \kappa_q} = 0 \quad q = 1, 2 \ldots Q, \tag{4.3.74}$$

a set of equations for κ_q is obtained

$$\begin{aligned} \kappa_{q-1} - 2r_E\kappa_q + \kappa_{q+1} &= 0 \qquad q = 1, 2\ldots(Q-1) \\ \kappa_{Q-1} - 2r_E\kappa_Q &= 0. \end{aligned} \tag{4.3.75}$$

The solution for these equations is expressed as

$$\kappa_q = \cos\left(\frac{q\pi}{2Q+2}\right) \quad q = 1, 2 \ldots Q. \tag{4.3.76}$$

The maximum energy vector magnitude is

$$\left(r_E\right)_{\max} = \cos\frac{\pi}{2Q+2}. \tag{4.3.77}$$

$Q = 1, 2$, and 3-order horizontal Ambisonics have $(r_E)_{\max}$ of 0.707, 0.866, and 0.924, respectively. As Q increases, $(r_E)_{\max}$ approaches the unit value. However, Equation (4.3.73) indicates that r_E approaches the unit value as Q increases even for the conventional solution of Ambisonic signals. Therefore, the optimization of energy vector magnitude at mid and high frequencies may be unnecessary for choosing κ_q in the HOA.

Similar to the case of the first-order Ambisonics, the in-phase solutions for the second- or higher-order horizontal Ambisonic signals can be derived to restrain the out-of-phase crosstalks from the opposite channels. Various in-phase solutions are provided for the second- and higher-order horizontal Ambisonics. Some additional criteria are applied to derive the in-phase solutions, resulting in maximum r_E, maximum front–back ratio, a maximum integrated front–back ratio, smooth and first-order extended solutions (Monro, 2000). If the following κ_q is chosen for the in-phase solution (Daniel, 2000; Neukom, 2007),

$$\kappa_q = \frac{(Q!)^2}{(Q+q)!(Q-q)!}, \tag{4.3.78}$$

then the loudspeaker signals are given as

$$A_i(\theta_S) = A_{total}\left[1+\cos(\theta_S - \theta_i)\right]^Q. \tag{4.3.79}$$

Similar to Equation (4.3.40) for the first-order Ambisonics, for arbitrary Q-order Ambisonic signals given in Equation (4.3.53), although the virtual source direction is independent from A_{total}, A_{total} determines the overall sound pressure or power in reproduction. For the constant amplitude normalization similar to Equation (4.3.41), A_{total} can be obtained from the first equation in Equation (4.3.56)

$$A_{total} = \frac{1}{M}. \tag{4.3.80a}$$

This equation is identical to the result of the first-order Ambisonics. For the constant power normalization similar to Equation (4.3.43), the following equation can be obtained:

$$A_{total} = \frac{1}{\sqrt{M\left(1+2\sum_{q=1}^{Q}\kappa_q^2\right)}}. \tag{4.3.80b}$$

For $\kappa_q = \kappa_0 = 1$ given in Equation (4.3.62),

$$A_{total} = \frac{1}{\sqrt{M(2Q+1)}}. \tag{4.3.81}$$

For κ_q expressed in Equation (4.3.77),

$$A_{total} = \frac{1}{\sqrt{M(Q+1)}}. \tag{4.3.82}$$

4.3.4 Discussion and implementation of the horizontal Ambisonics

From the point of multichannel loudspeaker signals, Ambisonics is a global amplitude-based mixing or panning method. From the point of information representation, Ambisonic encodes the spatial information of a sound field into a set of independent and universal signals, which are independent from the loudspeaker configuration in reproduction. Diverse and equivalent forms of independent signals are available. The number and configuration of loudspeakers for Ambisonic reproduction are flexible. Loudspeaker signals are derived from independent signals by decoding equations or matrices. Ambisonics can recreate the spatial information of sound field at the central listening position and up to a certain frequency limit. Therefore, Ambisonics is a universal and flexible system from the point of signal transmission and reproduction.

From the point of physics, Ambisonics is a series of systems with various orders that are based on the principle of spatial harmonics decomposition and each order approximation of the sound field. In Section 1.9.1, Ambisonics is a typical example of a gradual transition from the approximate to the exact reconstruction of the target sound field. The lower-order spatial harmonics in Ambisonics represent the rough information of the spatial sound field, and the higher harmonics represent the detailed information of the spatial sound field. When the capacity of signal transmission (and storage) is limited, transmitting the information in the order from rough to detailed is reasonable. Ambisonics reduces the redundancy among the transmitted information and enables efficient use of the finite transmission capacity of the system because the independent information of the spatial sound field is extracted through spatial harmonic decomposition.

The lower-order Ambisonics can reconstruct the target sound field within a very small spatial region and limited frequency range; therefore, appropriate psychoacoustic methods should be applied to create a virtual source and other spatial auditory perceptions. The perception performance of the lower-order (e.g., the first-order) Ambisonics is limited. As the order increases, the spatial region and frequency range of the accurate transmission and reproduction of the information of the spatial sound field are extended gradually. Consequently, the spatial resolution and perceived performance of the reproduced sound field improve, and the listening region widens. However, the required number of independent signals and loudspeakers also increases as the order increases. Therefore, Ambisonics is a series of hierarchical systems in which the performance and complexity increase as the order increases. When the form of independent signals, such as the universal form of spatial harmonics given in Equation (4.3.49), is appropriately chosen, the HOA can be implemented by adding new independent signals to the lower-order Ambisonics and combining into the decoding equation. This characteristic enables the upward and downward compatibilities among different order Ambisonics. Previous studies also suggested some methods for the compatible transmission of mono, stereophonic, and first-order Ambisonic signals (Gerzon, 1985; Xie X.F., 1982).

The aforementioned features are held for horizontal and spatial Ambisonics (Section 6.4). Q-order horizontal Ambisonics involves an odd number of $(2Q + 1)$ independent signals and requires $M_{min} = (2Q + 1)$ or more appropriately $(2Q + 2)$ reproduction channels and loudspeakers; that is, the number of loudspeakers should be equal to or at least slightly larger than the number of independent signals. Further increasing the number of loudspeakers may improve the uniformity of the reproduced sound field but may cause new problems.

Regular horizontal loudspeaker configurations are discussed in the preceding discussion, which is true for most early domestic surround sounds. For irregular loudspeaker configurations, if loudspeakers are arranged front-back and left-right symmetrically so that each pair of loudspeakers is respectively located in two opposite directions alone the diameter of a circle, loudspeaker signals can also be derived easily. For example, the azimuths of four loudspeakers in a rectangular configuration are given as

$$\theta_{LF} = \theta_0 \quad \theta_{LB} = \theta_1 = 180° - \theta_0 \quad \theta_{RB} = \theta_2 = -180° + \theta_0 \quad \theta_{RF} = \theta_3 = -\theta_0. \tag{4.3.83}$$

For the first-order Ambisonics, the signals or decoding equation of the four loudspeakers is derived similarly to that in Section 4.3.2 (Gerzon, 1985):

$$A_i(\theta_S) = A_{total}\left(W + \frac{1}{\cos\theta_i}X + \frac{1}{\sin\theta_i}Y\right). \tag{4.3.84}$$

where $A_i(\theta_S)$ with $i = 0, 1, 2, 3$ denotes A_{LF}, A_{LB}, A_{RB}, and A_{RF}, respectively. When $\theta_0 \neq 45°$, the overall power of the four loudspeaker signals is no longer a constant; instead, it depends on the target source azimuth.

The mathematical derivation of the decoding equation for other irregular loudspeaker configurations is complicated and may deal with the solution of nonlinear equations and numerical calculations. Moreover, some physical and psychoacoustic criteria for optimization may conflict one another; therefore, requirements from these criteria cannot be satisfied, or the solution of mathematical equations is physically inappropriate. This problem is addressed in Section 5.2.3.

In Sections 4.3.2 and 4.3.3, two methods are applied to improve the mid- and high-frequency localization performance in Ambisonic reproduction. The most effective method is to use the second- or higher-order Ambisonics. For example, a virtual source localization experiment on the Q = 1, 2, and 3-order horizontal Ambisonics indicated that the second-order Ambisonic reproduction with six or eight loudspeakers exhibits a stable localization effect of full 360° horizontal virtual sources in the central listening position and for speech and music stimuli (Xie and Xie, 1996). This result is due to the dominant role of interaural phase delay difference and its dynamic variation with head rotation in horizontal localization (Section 1.6.5). Compared with the lower-order Ambisonics, higher-order Ambisonics further improve the perceived performance, but it is more complicated. Another method is to decode the low- and mid-to-high-frequency components with different optimizing criteria, such as the criteria of velocity localization vector and energy localization vector. In practice, a set of ***shelf filters*** is used to decode the low- and mid-to-high-frequency components of independent signals with different optimizing parameters. The crossover frequency of shelf filters is chosen on the basis of psychoacoustic consideration. Gerzon (1985) as well as Gerzon and Barton (1992) suggested a crossover frequency of 0.7 kHz for the first-order Ambisonics. Daniel et al. (1998) suggested a higher crossover frequency for the 2- and higher-order Ambisonics, such as 1.2 kHz for the second-order Ambisonics. As the order increases, the upper frequency limit for the exact reconstruction of the sound field increases. In this case, it may be unnecessary to optimize high-frequency decoding.

In Section 4.3.1, independent signals for Ambisonics have various forms corresponding to various formats of Ambisonics. Choosing W, X, and Y given in Equation (4.3.3) as the independent signals of the first-order horizontal Ambisonics is convenient. In practice, the levels of X and Y are enhanced by 3 dB to ensure a diffused field power response identical to that of M. The independent signals are written as

$$W' = W = 1 \qquad X' = \sqrt{2}X = \sqrt{2}\cos\theta_S \qquad Y' = \sqrt{2}Y = \sqrt{2}\sin\theta_S. \tag{4.3.85}$$

The decoding equation should be appropriately changed to make adjustments for the independent signals in Equation (4.3.85). The Ambisonic system with independent signals given by (4.3.85) is termed the first-order ***B-format*** horizontal Ambisonics (Gerzon, 1985). Figure 4.19 is the block diagram of the B-format first-order horizontal Ambisonics with six

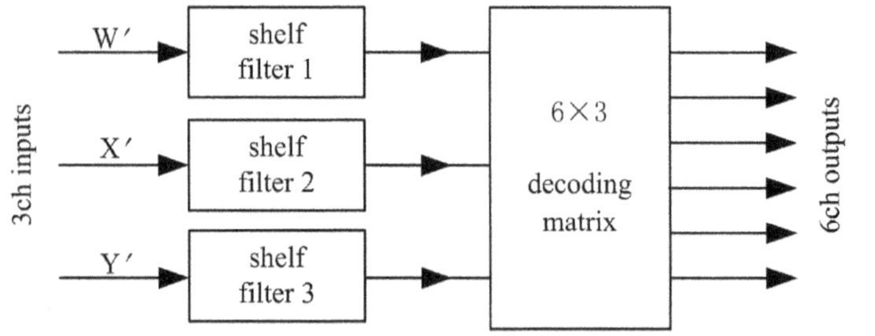

Figure 4.19 Block diagram of the B-format first-order horizontal Ambisonics with six reproduction channels.

reproduction channels. The responses of the shelf filters in the figure vary with frequency so that the gains for X' and Y' signals differ between low- and mid-to-high-frequency ranges. These frequency-dependent responses are equivalent to decoding the low- and mid-to-high-frequency component with different optimized parameters b (Table 4.1). The responses of shelf filters transit smoothly from a low frequency to a mid-to-high frequency. The shelf filter for the W' signal compensates for the phase shift among the signals.

4.4 SUMMARY

In early times, horizontal surround sound techniques for reproduction within a small-sized listening region were mainly intended for music reproduction only. Most early techniques were developed to manipulate spatial information from all horizontal directions with equal accuracy or approximation. Accordingly, regular loudspeaker configurations were used in reproduction. Pair-wise amplitude panning and sound field signal mixing were two common methods for loudspeaker signals. Discrete quadraphones were also widely studied in these periods.

For pair-wise amplitude panning, a discrete quadraphone is unable to recreate a stable virtual source in lateral directions. Moreover, the virtual source in the front and rear directions is unstable against head rotation. For the first-order sound field signal mixing, discrete quadraphones can theoretically recreate a stable 360° virtual source in full horizontal directions at very low frequencies, and a virtual source is stable under head rotation. As frequency increases, however, virtual sources (especially lateral virtual sources) move toward the nearest loudspeakers and become instable. For pair-wise amplitude panning, increasing the number of reproduction channels and loudspeakers (especially increasing lateral loudspeakers) radically improves the virtual source localization. For the first-order sound field signal mixing, increasing the number of reproduction channels and loudspeakers may decrease the perceived difference between virtual sources in and off loudspeaker directions, but it fails to improve the virtual source localization radically.

The nature of sound field signal mixing is spatial harmonic decomposition and each order approximation of the sound field. The first-order horizontal sound field signal mixing involves three independent signals. Various forms of independent signals can be converted mutually.

Ambisonics is developed on the basis of sound field signal mixing and is vital in spatial sound theory. It encodes the spatial information of the sound field into a set of independent and universal signals, which are independent of loudspeaker configuration in reproduction. If the number of loudspeakers exceeds the minimum requirement for each order reproduction, loudspeaker signals can be derived from independent signals via a decoding equation or matrix. Ambisonics can recreate the spatial information of sound in the central listening position and up to a certain frequency limit. In nature, Ambisonics is a series of hierarchical spatial sound systems with different orders. As the order of Ambisonics increases, the spatial region and frequency range for the accurate transmission and reproduction of the spatial information of the sound field are extended gradually. Consequently, its performance improves. However, the number of independent signals and the minimal required number of loudspeakers in reproduction also increase as the order increases. The Q-order horizontal Ambisonics involves an odd number of $(2Q + 1)$ independent signals and at least $(2Q + 1)$ or $(2Q + 2)$ loudspeakers. In addition, further increasing the number of loudspeakers may improve the uniformity of the reproduced sound field, but it may cause new problems.

Chapter 5

Multichannel horizontal surround sound with irregular loudspeaker configuration

Since 1990, domestic or consumer multichannel surround sounds for reproduction within a small-sized listening region have been mainly developed for sound reproduction with accompanying picture and general uses. In these multichannel surround sounds, the spatial information from different horizontal directions is manipulated with different approximations to enhance accuracy in the reproduction of front localization information and recreate the subjective sensation of ambience instead of improving the ability to recreate a full 360° virtual source in a horizontal plane. With this consideration, an irregular horizontal loudspeaker configuration with a front-center channel is usually adopted in these systems, i.e., an irregular or nonuniform azimuthal-sampling procedure is applied. The 5.1-channel surround sound is a typical example of horizontal surround sound with an irregular loudspeaker configuration, which has been widely employed and recommended by the International Telecommunication Union (ITU) as the standard of multichannel sound for domestic uses. Other horizontal surround sounds with more than 5.1-channels have also been developed.

In this chapter, multichannel horizontal surround sounds with irregular loudspeaker configurations, especially 5.1-channel surround sound, are analyzed. The virtual source localization in these systems with pair-wise amplitude panning and Ambisonic-like signal mixing is examined. The basic concepts and requirements of surround sound with accompanying picture and for general uses are outlined in Section 5.1. The signals and localization performance of 5.1-channel surround sounds are explored in Section 5.2. Other multichannel horizontal surround sounds with irregular loudspeaker configurations are briefly introduced in Section 5.3. The problem of low-frequency effect channels in sound reproduction (with accompanying picture) is addressed in Section 5.4.

5.1 OUTLINE OF SURROUND SOUNDS WITH ACCOMPANYING PICTURE AND GENERAL USES

Surround sounds with accompanying picture refers to surround sound techniques for cinema, television, and other video uses. In early times (beginning in the 1950s or earlier times), multichannel surround sounds with accompanying pictures were developed for cinema uses with a large-sized listening region and independent from stereophonic or surround sound techniques for domestic music reproduction with a small-sized listening region. Since the beginning of 1980, this situation has changed radically. Various sound reproduction techniques originally developed for cinema uses were modified and simplified for domestic reproduction with accompanying pictures, thereby producing a surround sound for "domestic theater". Film programs were previously distributed to domestic uses with an analog video recorder. Since the 1990s, the development of high-definition televisions (HDTVs) or digital

DOI: 10.1201/9781003081500-5

televisions (DTVs) and various digital media (such as DVD) techniques have successfully promoted the development of surround sounds with accompanying picture for domestic uses.

In Section 3.1, for sound reproduction with accompanying picture, the major purpose is to recreate auditory perceptions that match with the picture. The localization information at frontal directions should be recreated accurately and stably, and lateral and rear information is usually ambience or occasionally localization information. Therefore, spatial information from different horizontal directions is manipulated with different approximations. The underlying basic consideration is to enhance the accuracy in the frontal information reproduction at the cost of reducing accuracy in lateral and rear information reproduction when system resources are limited.

As stated in Section 1.9.3, sound reproduction technique with three frontal channels was introduced in the early works at Bell Labs in the 1930s. An independent front-central channel has been used for cinema sound to recreate auditory perceptions that match with a given picture, especially for dialogs in films. Since the *Fantasia* by Walt Disney in 1939, a frontal-center channel was used in sound reproduction for cinema, including early cinema surround sounds in the 1950s, matrix surround sound in the 1970s and new discrete surround sounds since the end of the 1980s. The cinema applications of surround sounds and related problems are discussed in Section 16.1.1.

Although the required listening region for domestic sound reproduction is much smaller than that for cinema reproduction, the frontal-center channel is vital to recreating stable frontal virtual sources within an appropriate listening region in domestic reproduction. In studies on sound reproduction techniques for HDTVs, Ohgushi et al. (1987) and Komiyama (1989) conducted psychoacoustic experiments and proved that the acceptable angular deviation between visual and auditory sources is about 11° for acoustic engineers and 20° for a general audience. Adding a frontal-center loudspeaker (usually above or below the screen) to a stereophonic loudspeaker configuration obviously improves the stability of a frontal virtual source. Moreover, the center channel and loudspeaker remove the difference in the timbre between the virtual sources in the front and the left or right loudspeaker directions. Therefore, the frontal-center channel is also adopted in domestic surround sounds with accompanying picture. In comparison with surround sounds for cinemas, surround sounds for domestic uses possess the following features:

1. A group of surround loudspeakers are used for cinema reproduction to recreate the ambient sound in a large-sized listening region. By contrast, restricted by the size of a listening room, a few surround loudspeakers are usually used in domestic reproduction.
2. For domestic reproduction within a small-sized listening region, the principle of summing localization with multiple loudspeakers is applied to recreate a virtual source between the loudspeakers via amplitude or time panning. This application not only provides a large creative space in program production but also presents new problems for further studies on the reproduced sound field, psychoacoustic analysis, signal recording, and simulation of surround sound.

From the point of practical uses, a domestic surround sound with accompanying picture should be compatible with music reproduction. Accordingly, a series of ***general surround sounds with and without accompanying picture*** has been developed. Ideally, these general surround sounds should satisfy the requirements of reproduction with accompanying picture and the requirements of music reproduction only discussed in Section 3.1, although practical systems and techniques are usually a compromise between the two requirements. The general multichannel surround sound techniques and systems for domestic uses developed after the 1990 are based on the consideration of enhancing the accuracy in the frontal spatial information reproduction. Accordingly, irregular loudspeaker configurations with front biases are adopted. With the progress on digital signal processing and media techniques, multichannel audio signals can be transmitted and stored. In most cases, the loudspeaker signals of these multichannel sounds are independent, i.e., these systems are discrete multichannel surround sound systems. In contrast, loudspeaker signals are not completely independent in some matrix surround sound systems, which will be discussed in Chapter 8.

Irregular loudspeaker configurations in a horizontal surround sound are sometimes marked with the notation M_1/M_2 or M_1–M_2, where M_1 is the number of front channels and loudspeakers, and M_2 is the number of lateral/rear channels and loudspeakers.

In the 1980s, researchers in Japan proposed a four-channel 3/1 system and technique for HDTVs (Miyasaka, 1989; Meares, 1991, 1992; Nakabayashi et al., 1991). This system involved four independent channels and signals with full audible bandwidth, i.e., the left *L*, center *C*, and right *R* channels in the front and a surround channel *S* in the rear. The loudspeaker configuration is illustrated in Figure 5.1. The three front channels were used to recreate stable virtual sources in the front region to match the picture. The surround channel was utilized to recreate ambience. The surround signal could also be reproduced with two surround loudspeakers, but the signals for two loudspeakers were identical. Although the four-channel 3/1 technique have been greatly developed (Suzuki et al., 1993; Yoshikawa et al., 1993), it has not been widely used because it only has one independent surround channel and consequently fails to recreate the ambience perfectly.

In Section 1.9.3, the 5.1-channel surround sound was originally developed for a digital film sound. In the early 1990s, international studies on the required number of channels and loudspeaker configurations in a horizontal surround sound for HDTVs and other domestic uses suggested adopting 5.1-channel surround sound (Theile, 1990, 1991a, 1993). In 1994, the 5.1-channel surround sound was recommended by the International Telecommunication Union as the standard of "multichannel stereophonic sound system with and without accompanying picture,"(ITU-R BS 775-1, 1994; and the revised version in 2012 was referred to

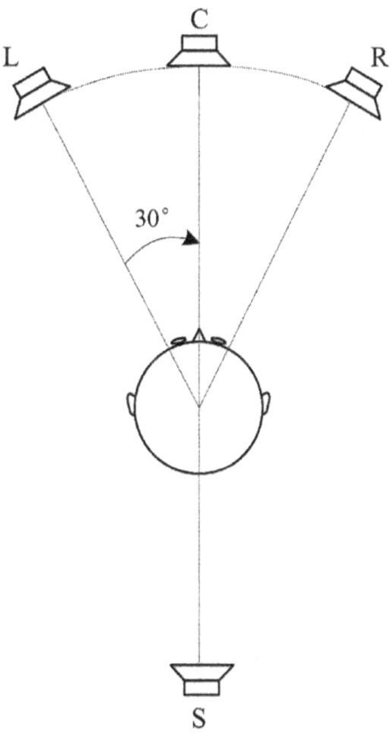

Figure 5.1 Loudspeaker configurations of the four-channel 3/1 surround sound.

ITU-R BS 775-3).Numerous studies on 5.1-channel surround sound have been performed since 1994 (Cohen and Eargle, 1995; Steinke, 1996; Dressler, 1996; Theile and Steinke, 1999). The 5.1-channel surround sound has been widely used in domestic reproduction and sound format for DVD video, DVD audio, and DTV/HDTV. Since the mid of 1990, other multichannel surround sound techniques with accompanying picture and for general uses in domestic reproduction have been developed. These techniques may be modified from cinema sound techniques or may be specially developed for domestic uses. These techniques and systems usually exhibit an improved perceived performance but are more complicated in comparison with 5.1-channel surround sound systems. In the succeeding sections of this chapter, the reproduced sound field, signal mixing, and virtual source localization in 5.1-channel surround sound systems are analyzed in detail. The performance of other multichannel horizontal surround sound systems is also examined briefly.

5.2 5.1-CHANNEL SURROUND SOUND AND ITS SIGNAL MIXING ANALYSIS

5.2.1 Outline of 5.1-channel surround sound

The 5.1-channel surround sound is the first discrete surround sound technique and system that has been successfully used in domestic reproduction. A 5.1-channel surround sound (hereinafter referred to as 5.1-channel sound) involves five independent channels with full audible bandwidth. Three front channels, including left L, right R, and center C, are used to recreate stable virtual sources in the front region to match the picture. Two surround channels, including left surround (LS) and right surround (RS), are used to recreate ambience and occasionally other spatial effects. An optional low-frequency effect channel (LFE, called.1 channel) operates at the frequency range of $f \leq 120$ Hz and drives a subwoofer to convey the specific effects at low frequencies. Figure 5.2 illustrates the loudspeaker configuration of 5.1-channel sound recommended by the ITU. In the figure and for most discussions in this chapter, the LFE channel and subwoofer are omitted. In this case, it may be more appropriate to term the system "the 5 channel 3/2 surround sound." For convenience, it is called

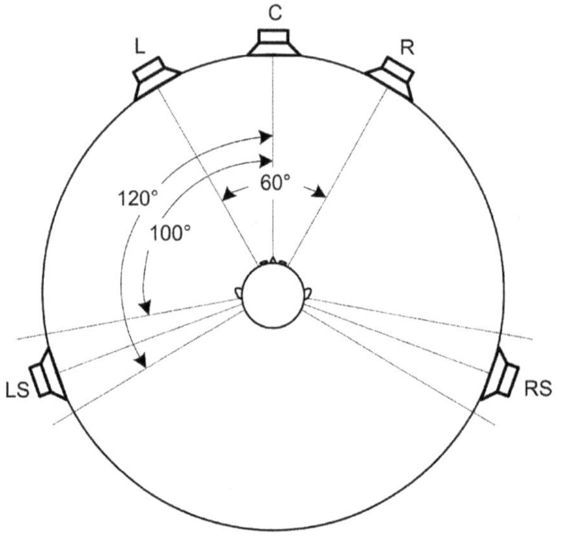

Figure 5.2 Loudspeaker configuration of 5.1-channel sound recommended by the ITU. (ITU-R BS 775-1, 1994).

"5.1-channel sound" and "5.1-channel loudspeaker configuration. "The horizontal azimuths of loudspeakers in 5.1-channel sounds are given by

$$\theta_L = 30^\circ \quad \theta_C = 0^\circ \quad \theta_R = -30^\circ \quad \theta_{LS} = 110^\circ(\pm 10^\circ) \quad \theta_{RS} = -110^\circ(\pm 10^\circ). \qquad (5.2.1)$$

The left and right loudspeakers at ±30° are compatible with those of two-channel stereophonic sound. A pair of surround loudspeakers at the side and slightly backward directions of ±110° are appropriate for ambient sound reproduction. The 5.1-channel loudspeaker configuration is typically an irregular and front-biased configuration that enhances the accuracy in the frontal information reproduction at the cost of the rough reproduction of lateral and rear information. Therefore, a 5.1-channel sound reproduces horizontal sound information with a rough approximation.

The 5.1-channel sound was originally not intended to recreate a full 360° virtual source in the horizontal plane. The two surround channels were originally intended for ambient information reproduction. For example, multiple loudspeakers with decorrelated signals (direct method) described in Sections 3.1 and 3.3 are used to simulate auditory sensations similar to those in a hall. Moreover, as stated in Section 3.3, the configuration of the left, right, and two surround loudspeakers is appropriate for recreating the auditory sensations similar to those in a diffused field.

The ITU standard only specifies the 5.1-channel loudspeaker configuration and does not specify the reproduced signals. Therefore, various signal mixing and microphone techniques are designed for 5.1-channel sound to recreate various spatial auditory effects. For domestic uses with a small-sized listening region, a sound engineer may try to recreate various effects of virtual source localization in 5.1-channel reproduction. Various microphone recording and signal mixing/panning techniques have been developed for 5.1-channel sound. Some of these techniques are derived from the physical analysis of the desired reproduced sound field, and other techniques are developed on the basis of psychoacoustic considerations, practical experiences, and requirements. The pair-wise amplitude panning and Ambisonic-like signal mixing of 5.1-channel sound are analyzed in the next two sections because virtual source localization is one of the basic performances of 5.1-channel sound. The theorems for analyzing the pair-wise amplitude panning and Ambisonic-like signal mixing are relatively mature. This analysis reveals the limitations of 5.1-channel sound and provides a basis for further developing multichannel surround sounds. Practical microphone recording and signal mixing techniques are discussed in Chapter 7, and the other perceived performances of 5.1-channel surround sound are explored in Chapter 12.

5.2.2 Pair-wise amplitude panning for 5.1-channel surround sound

Pair-wise amplitude panning is the most common signal mixing method for 5.1-channel sound, especially various 5.1-channel sound programs with accompanying picture. Similar to the case of quadraphones in Section 4.1.2, the signal is fed to a single loudspeaker to recreate the virtual source at its direction or fed to a pair of adjacent loudspeakers with various amplitude ratios or interchannel level difference (ICLDs) to recreate a virtual source in intermediate directions. The virtual source created by pair-wise amplitude panning can be analyzed on the basis of the summing localization theorem discussed in Section 3.2 (Xie, 1997; Xie, 2001a). Virtual sources in the left-half horizontal plane are analyzed only because of the left-right symmetry.

Two different pair-wise amplitude panning methods are used for the front virtual source at $0^\circ \le \theta \le 30^\circ$ between the center and left loudspeakers. In the first method, signals are fed to the left and right loudspeakers with an appropriate ICLD, and the results are identical to those in two-channel stereophonic sound reproduction with a standard loudspeaker configuration of ±30° (Section 2.1.1). In the second method, a signal is fed to the left and center loudspeakers with an appropriate ICLD. The virtual sources within $-30° \leq \theta \leq 0°$are recreated by the right and center loudspeakers because of symmetry.

Let A_L, A_C, and A_R note the normalized signal amplitudes of the left, center, and right loudspeakers, respectively. Within the azimuthal range of $0° \leq \theta \leq 30°$, the perceived virtual source direction at low frequencies is evaluated from Equations (3.2.7), (3.2.9), and (5.2.1). For a fixed head, the perceived azimuth is

$$\sin\theta_I = \frac{1}{2}\frac{A_L}{A_L + A_C} = \frac{1}{2}\frac{A_L/A_C}{A_L/A_C + 1}, \tag{5.2.2}$$

For the head oriented to the virtual source,

$$\tan\hat{\theta}_I = \frac{A_L}{\sqrt{3}A_L + 2A_C} = \frac{A_L/A_C}{\sqrt{3}\ A_L/A_C + 2}. \tag{5.2.3}$$

For the in-phase signals A_L and A_C, it has $A_L/A_C > 0$, then Equations (5.2.2) and (5.2.3) yield $\sin\theta_I > 0$, $\tan\hat{\theta}_I > 0$. Therefore, the virtual source lies within the left-front quadrant of (0°, 90°). Figure 5.3 illustrates the curves of the perceived virtual source azimuth versus the interchannel level difference $d_1 = 20\log_{10}(A_L/A_C)$. It also shows the results of the virtual source localization experiment, including the means and standard deviations of the perceived azimuth for the orchestral stimulus (Section 15.6.3). When d_1 varies from $-\infty$ dB to $+\infty$ dB, the virtual source moves from the front (0°) to the direction of the L loudspeaker (+30°). Moreover, the difference between the results for a fixed head and those for a head oriented to the virtual source is slight (less than 0.7°); therefore, the virtual source is stable against head rotation.

The stability of the virtual source against head rotation can also be evaluated from the velocity vector magnitude r_v given in Equation (3.2.29). The normalized amplitudes of the

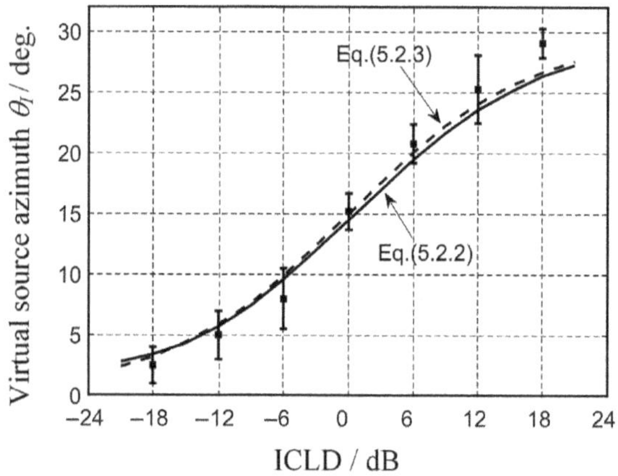

Figure 5.3 Curves of the perceived virtual source azimuth versus the interchannel level difference $d_1 = 20\log_{10}(A_L/A_C)$ of the pair-wise amplitude panning between the L and C loudspeakers in the 5.1-channel loudspeaker configuration. The figure also illustrates the means and standard deviations of the results from a virtual source localization experiment.

left and center channel signals can be written in the following equation by introducing a parameter $0° \le \xi_1 \le 90°$ and normalizing the overall signal power to a unit:

$$A_L = \sin\xi_1 \qquad A_C = \cos\xi_1 \qquad A_L^2 + A_C^2 = 1. \tag{5.2.4}$$

Substituting Equation (5.2.4) into Equation (3.2.29) yields

$$r_v = \frac{\sqrt{2}}{2}\frac{\sqrt{1+0.866\sin 2\xi_1}}{\cos\left(\xi_1 - 45^\circ\right)}. \tag{5.2.5}$$

The velocity vector magnitude minimizes with $r_v = 0.966$ for $\xi_1 = 45°$ or $A_L = A_C$, which is close to the unit.

The perceived virtual source direction for a fixed head at higher frequencies can be evaluated from Equation (3.2.6). A precorrected head radius of $a' = 1.25\times0.0875$ m is used in the evaluation. The results indicate that a virtual source is stable as the frequency increases to 1.5 kHz. This stability is due to the relatively narrow span angle (30°) between the left and center loudspeakers.

For constant-power panning, the panning functions at $0° \le \theta_I \le 30°$, i.e., the normalized amplitudes of the left and center channel signals as functions of the perceived virtual source direction, can be derived from Equation (5.2.2) or (5.2.3). Equation (5.2.2) yields

$$A_L = \frac{2\sin\theta_I}{\sqrt{1-4\sin\theta_I+8\sin^2\theta_I}} \qquad A_C = \frac{1-2\sin\theta_I}{\sqrt{1-4\sin\theta_I+8\sin{}^2\theta_I}}, \tag{5.2.6}$$

and Equation (5.2.3) yields

$$A_L = \frac{2\sin\hat{\theta}_I}{\sqrt{1-1.732\sin 2\hat{\theta}_I+6\sin^2\hat{\theta}_I}} \qquad A_C = \frac{\cos\hat{\theta}_I-1.732\sin\hat{\theta}_I}{\sqrt{1-1.732\sin 2\hat{\theta}_I+6\sin^2\hat{\theta}_I}}. \tag{5.2.7}$$

The panning function at $-30° \le \theta_I \le 0°$ can be derived similarly. Figure 5.4 illustrates the panning curves for the left, center, and right channel signals. It only illustrates the curve of Equation (5.2.7) because the results of Equations (5.2.6) and (5.2.7) differs lightly.

In the lateral region of $30° \le \theta \le 110°$, signals with ICLD are fed to the left and LS loudspeakers to recreate a virtual source between them. Let A_L and A_{LS} denote the normalized signal amplitudes of the L and LS loudspeakers, respectively. The perceived virtual source direction at low frequencies is evaluated from Equations (3.2.7), (3.2.9), and (5.2.1). For a fixed head, the perceived azimuth is

$$\sin\theta_I = \frac{0.940A_{LS}/A_L+0.5}{A_{LS}/A_L+1}. \tag{5.2.8}$$

For the head oriented to the virtual source,

$$\tan\hat{\theta}_I = \frac{0.940A_{LS}/A_L+0.5}{0.866-0.342\ A_{LS}/A_L}. \tag{5.2.9}$$

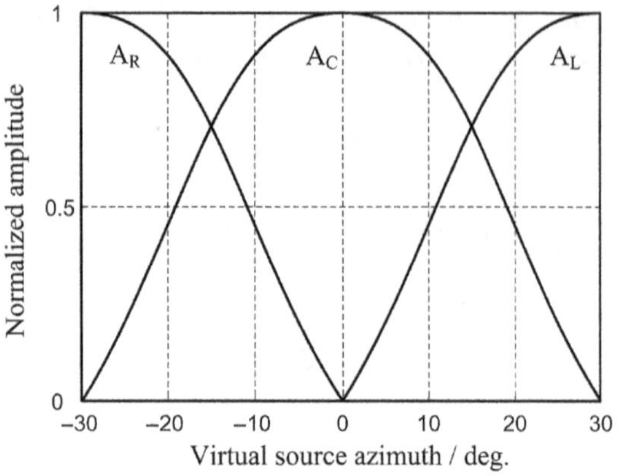

Figure 5.4 Panning curves of the left, center, and right channel signals with pair-wise amplitude panning.

From Equations (5.2.8) and (5.2.9), for the amplitude ratio of $0 \leq A_{LS}/A_L < 2.53$ (or interchannel level difference $-\infty$ dB $\leq d_2 = 20\log_{10}(A_{LS}/A_L) < 8.1$ dB), it has $\sin\theta_I > 0$ and $\tan\hat{\theta}_I > 0$; therefore, the virtual source lies within the left-front quadrant of (0°, 90°). For the amplitude ratio of $2.53 < A_{LS}/A_L < +\infty$(or interchannel level difference 8.1 dB $< d_2 < +\infty$ dB), it has $\sin\theta_I > 0$ and $\tan\hat{\theta}_I < 0$; therefore, the virtual source lies within the left-back quadrant of (90°, 180°). Figure 5.5 illustrates the curves of the perceived virtual source azimuth versus the interchannel level difference evaluated from Equations (5.2.8) and (5.2.9). It also illustrates the results of the virtual source localization experiment, including the means and standard deviations of the perceived azimuth of the orchestral stimulus (Section 15.6.3). For a head oriented to the virtual source, when d_2 varies from $-\infty$ dB to $+\infty$ dB, $\hat{\theta}_I$ moves from 30° to 110°. For a fixed head, when d_2 varies from $-\infty$ dB to 8.1 dB, θ_I moves from 30°to 54.6°; when d_2 varies from 8.1 dB to $+\infty$ dB, θ_I moves from 125.4°to 110°. The results here are abnormal.

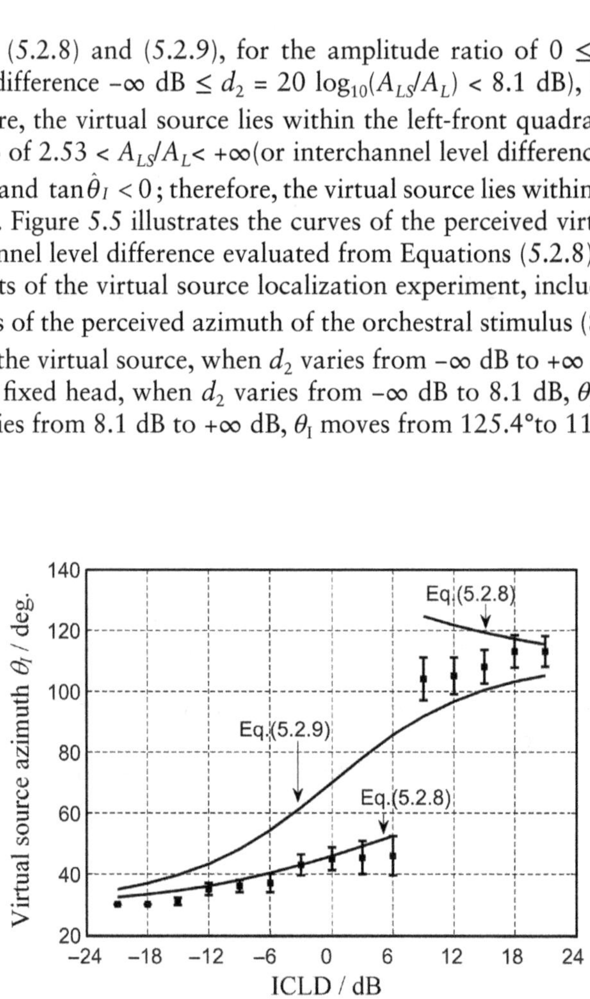

Figure 5.5 Curves of the perceived virtual source azimuth versus the interchannel level difference $d_2 = 20\log_{10}(A_{LS}/A_L)$ for pair-wise amplitude panning between the L and LS loudspeakers in the 5.1-channel loudspeaker configuration. The figure also illustrates the means and standard deviations of the results from a virtual source localization experiment.

The results of Equations (5.2.8) and (5.2.9) are considerably different within the azimuthal region of $54.6° < \theta < 110°$. Therefore, the lateral virtual source is unstable. For a fixed head, the pair-wise amplitude panning is unable to recreate a stable virtual source within the region of $54.6° < \theta < 110°$ even at low frequencies, i.e., a hole of the virtual source exists in this region. In fact, the analysis on quadraphones in Section 4.1.2 indicates that a pair of front-back symmetric lateral loudspeakers is unable to recreate a stable lateral virtual source. A similar problem occurs in the 5.1-channel sound, although the locations of the *L* and *LS* loudspeakers in the 5.1-channel configuration are not completely front-back symmetric.

In the rear region $110° \leq \theta \leq 180°$ (or $-180° < \theta \leq -110°$), signals with ICLD are fed to the *LS* and *RS* loudspeakers to recreate a virtual source between them. Let A_{LS} and A_{RS} denote the normalized signal amplitudes of the LS and RS loudspeakers, respectively. The perceived virtual source direction at low frequencies is evaluated from Equations (3.2.7), (3.2.9), and (5.2.1). For a fixed head, the perceived azimuth is

$$\sin\theta_I = \frac{1 - A_{RS}/A_{LS}}{1 + A_{RS}/A_{LS}} \sin 110°. \tag{5.2.10}$$

For the head oriented to the virtual source, the perceived azimuth is

$$\tan\hat{\theta}_I = \frac{1 - A_{RS}/A_{LS}}{1 + A_{RS}/A_{LS}} \tan 110°. \tag{5.2.11}$$

Figure 5.6 illustrates the curves of the perceived virtual source azimuth versus the inter-channel level difference $d_3 = 20\log_{10}(A_{RS}/A_{LS})$ evaluated from Equations (5.2.10) and (5.2.11). At low frequencies, when d_3 varies from $-\infty$ dB to 0 dB, the virtual source moves from the direction of the LS loudspeaker (110°) to the rear (180°). However, the difference between the results of Equations (5.2.10) and (5.2.11)are obvious. Therefore, the virtual source is unstable against head rotation.

The stability of the virtual source against head rotation can also be evaluated on the basis of the velocity vector magnitude r_v given by Equation (3.2.29). The normalized amplitudes

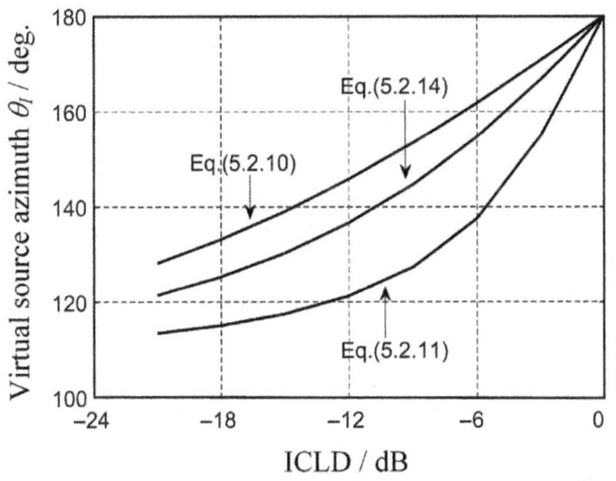

Figure 5.6 Curves of the perceived virtual source azimuth versus the interchannel level difference $d_3 = 20\log_{10}(A_{RS}/A_{LS})$ for the pair-wise amplitude panning between the LS and RS loudspeakers in the 5.1-channel loudspeaker configuration.

of the LS and RS signals can be written in the following equation by introducing a parameter $0° \leq \xi_3 \leq 90°$ and normalizing the overall signal power to a unit:

$$A_{LS} = \cos\xi_3 \quad A_{RS} = \sin\xi_3 \quad A_{LS}^2 + A_{RS}^2 = 1. \tag{5.2.12}$$

Substituting Equation (5.2.12) into Equation (3.2.29) yields

$$r_v = \frac{\sqrt{2}}{2} \frac{\sqrt{1 - 0.766\sin 2\xi_3}}{\cos(\xi_3 - 45°)}. \tag{5.2.13}$$

The velocity vector magnitude minimizes with $r_v = 0.342$ for $\xi_3 = 45°$ or $A_{LS} = A_{RS}$.

As frequency increases, the perceived virtual source direction fora fixed head should be evaluated from Equation (3.2.6),

$$\sin\theta_I = \frac{1}{ka} \arctan\left[\frac{1 - A_{RS}/A_{LS}}{1 + A_{RS}/A_{LS}} \tan(ka\sin 110°)\right]. \tag{5.2.14}$$

Figure 5.6 illustrates the results at frequency $f = 0.5$ kHz. A precorrected head radius of $a' = 1.25 \times 0.0875$ m is used in the evaluation. For a given d_3, the virtual source moves toward the direction of the LS loudspeaker (110°) and thus is unstable as frequency increases. For example, when d_3 =–6 dB, θ_I moves from 161.7° at low frequencies to 154.6° at 0.5 kHz. Similar phenomena occur in two-channel stereophonic sounds and quadraphones (Sections 2.1.2 and 4.1.2), but the phenomenon is more obvious in the rear virtual source in 5.1-channel reproduction. For stimuli with different frequency components, this frequency dependence of the virtual source position may lead to broadening and blurry virtual sources. Another feature is that the perceived azimuth θ_I of the rear virtual source varies quickly with d_3 even at low frequencies. For example, Equation (5.2.10) shows that θ_I varies from 180° to 170.7° when d_3 changes from 0 dB to –3 dB, which may cause a hole in the virtual source in the rear direction. In addition, the perceived elevation of the rear virtual source is elevated when the head rotates around the vertical axis. This problem is addressed in Section 6.1.4. Overall, although the 5.1-channel loudspeaker configuration can theoretically recreate a virtual source in the rear region, the perceived quality of the rear virtual source is poor. The rear virtual source is unstable when the head rotates and as frequency increases. The listening region for the rear virtual source is also narrow. All these problems are caused by the large span angle (140°) between the LS and RS loudspeakers. In Section 4.1.2, a similar problem occurs in a two-channel stereophonic sound and a quadraphone when the span angle between two front loudspeakers exceeds 60°.

According to the aforementioned analysis, the characteristics and limitations of 5.1-channel sound with pair-wise amplitude panning are summarized as follows. It can recreate a stable virtual source in the frontal region. The virtual source in the frontal-center direction is recreated by a single loudspeaker (the crosstalk between loudspeakers vanishes), it is stable and never collapses to the nearest loudspeaker even in the off-central listening position. This feature is vital for sound reproduction with accompanying pictures. However, the virtual source in the rear region is blurry and unstable. Fortunately, the resolution of human hearing in the rear directions is inferior to that in the front directions. Degraded virtual sources in the rear directions are acceptable in sound reproduction with accompanying pictures. A major defect of 5.1-channel sound is that it is unable to recreate a stable virtual source within the lateral region of 54.6° < θ < 110°. This defect influences slightly and remains acceptable in sound reproduction with accompanying pictures. However, as stated in Section 1.8.3, the lateral early reflections from θ = 55° ± 20° are vital to the auditory spatial impression (especially the ASW) in a concert hall. A 5.1-channel sound with pair-wise amplitude panning fails to simulate and reproduce the spatial information of lateral reflections perfectly and therefore displays a defect in music reproduction only. This defect also reveals the limitations of using the indirect method discussed in Section 3.1 to simulate the auditory sensation of a concert hall in 5.1-channel reproduction. Further analysis of the 5.1-channel loudspeaker configuration with LS and RS loudspeakers at ±120° yields similar results (Xie, 1997). A virtual source localization experiment also validates the aforementioned analysis (Xie, 1997; Xie, 2001a; Martin et al., 1999).

The aforementioned characteristics and defects are due to the 5.1-channel loudspeaker configuration. The 5.1-channel sound was originally designed for sound reproduction with accompanying pictures; therefore, an irregular and front-biased loudspeaker configuration is used. It was not originally intended to recreate the full 360° virtual source in the horizontal plane. Using 5.1-channel sound for general reproduction (including music reproduction only) is a compromise. The analysis in this section reveals the limitation of 5.1-channel sound with pair-wise amplitude panning. Other signal panning and mixing methods are also available to recreate a virtual source in the 5.1-channel sound, and the decorrelated signal method is applicable to simulate the subjective sensation in reflective environments.

5.2.3 Global Ambisonic-like signal mixing for 5.1-channel sound

In addition to pair-wise amplitude panning, other signal panning and mixing methods have been explored for the 5.1-channel sound. Sound field or Ambisonic-like signal mixing methods similar to those discussed in Section 4.3 are suggested for recreating the virtual source in various horizontal directions, although the 5.1-channel loudspeaker configuration is inherently inappropriate for recreating a full 360° horizontal virtual source. Deriving Ambisonic-like decoding matrices or signals for irregular loudspeaker configurations based on physical and psychoacoustic criteria is complicated and sometimes difficult. Despite their complexity, some authors derived Ambisonic-like signals and mixing methods for 5.1-channel loudspeaker configurations. These methods are theoretically attractive but rarely used in practical program production.

As a general case, if M loudspeakers are arranged in a horizontal circle with regular or irregular azimuthal intervals and the azimuth of the ith loudspeaker is θ_i, the normalized signal amplitude of the ith loudspeaker is a linear combination of azimuthal harmonics up to the order Q,

$$A_i(\theta_S) = A_{total}\left\{D_0^{(1)}(\theta_i) + \sum_{q=1}^{Q}\left[D_q^{(1)}(\theta_i)\cos q\theta_S + D_q^{(2)}(\theta_i)\sin q\theta_S\right]\right\} \quad i = 0, 1\ldots(M-1), \quad (5.2.15)$$

where θ_S is the azimuth of the target source. In contrast to the case of a regular loudspeaker configuration given by Equation (4.3.50), the coefficients $D_0^{(1)}(\theta_i)$ for different loudspeakers are different in the case of irregular loudspeaker configurations; as such, they cannot be identically normalized to a unit. Given the loudspeaker configuration and the order Q, the decoding matrix or loudspeaker signals are derived by searching for a set of decoding coefficients $\left\{D_0^{(1)}(\theta_i), D_q^{(1)}(\theta_i), D_q^{(2)}(\theta_i), \quad q = 1, 2\ldots Q, \quad i = 0, 1..(M-1)\right\}$ so that the reproduced sound field satisfies certain optimized criteria. Equation (5.2.15) involves $[M(2Q + 1)]$ decoding coefficients to be determined. Before the optimized criteria are applied, the decoding coefficients are simplified by considering symmetry. Even for irregular loudspeaker configurations, loudspeaker arrangements are usually left-right symmetric. For an arbitrary pair of left-right symmetric loudspeakers i and i', their azimuths satisfy $\theta_i = -\theta_{i'}$ and $\theta_i \neq 0°$or 180°. The coefficients in Equation (5.2.15) for a pair of left-right symmetric loudspeakers satisfy the following equation because $\cos q\theta_S$ is an even function with $\cos(-q\theta_S) = \cos q\theta_S$ and thsin$q\theta_S$ is an odd function with $\sin(-q\theta_S) = -\sin q\theta_S$:

$$D_0^{(1)}(\theta_{i'}) = D_0^{(1)}(\theta_i) \qquad D_q^{(1)}(\theta_{i'}) = D_q^{(1)}(\theta_i) \quad D_q^{(2)}(\theta_{i'}) = -D_q^{(2)}(\theta_i) \quad q = 1, 2 \ldots Q. \qquad (5.2.16)$$

For loudspeakers at $\theta_i = 0°$ and 180°, the coefficients satisfy

$$D_q^{(2)}(\theta_i) = 0 \qquad q = 1, 2 \ldots Q. \qquad (5.2.17)$$

In addition, if a pair of loudspeakers i and i'' is arranged front-back symmetrically, their azimuths satisfy $\theta_{i''} = (180° - \theta_i)$ in the left-half horizontal plane or $\theta_{i''} = (-180° - \theta_i)$ in the right-half horizontal plane. Since $\cos[q(\pm 180° - \theta_S)] = (-1)^q \cos q\theta_S$, $\sin[q(\pm 180° - \theta_S)] = (-1)^{q+1} \sin q\theta_S$, then

$$\begin{aligned} &D_0^{(1)}(\theta_{i''}) = D_0^{(1)}(\theta_i) \\ &D_q^{(1)}(\theta_{i''}) = (-1)^q D_q^{(1)}(\theta_i) \quad D_q^{(2)}(\theta_{i''}) = (-1)^{q+1} D_q^{(2)}(\theta_i) \quad q = 1, 2 \ldots Q. \end{aligned} \qquad (5.2.18)$$

Symmetry reduces the number of decoding coefficients to be determined and simplifies the procedures of optimization. For the ITU 5.1-channel loudspeaker configuration, the number of loudspeakers is $M = 5$. For $Q = 1$ to 4 order signals, Equation (5.2.13) involves $[M(2Q + 1)] = 15, 25, 35$, and 45 unknown coefficients, respectively. By considering the left-right symmetry, the number of coefficients is reduced to $(5Q + 3) = 8, 13, 18$, and 23, respectively. Therefore, using symmetry is a mathematical skill for deriving Ambisonic-decoding equations and signals.

Similar to cases of regular loudspeaker configurations, the criteria of the optimized velocity localization vector (equivalent to optimized interaural phase delay difference and its dynamic variation with head rotation) and optimized energy localization vector are used to derive the decoding coefficients. From Equation (5.2.15), the following quantities are evaluated as the functions of the coefficients to be determined:

$$\begin{aligned} &P'_A = \sum_{i=0}^{M-1} A_i \qquad V'_x = \sum_{i=0}^{M-1} A_i \cos\theta_i \qquad V'_y = \sum_{i=0}^{M-1} A_i \sin\theta_i \\ &Pow' = \sum_{i=0}^{M-1} A_i^2 \qquad I'_x = \sum_{i=0}^{M-1} A_i^2 \cos\theta_i \qquad I'_y = \sum_{i=0}^{M-1} A_i^2 \sin\theta_i. \end{aligned} \qquad (5.2.19)$$

Ideally, the decoding coefficients are derived with the following optimized criteria (Gerzon and Barton, 1992):

1. Criterion 1. The virtual source direction θ_v evaluated from the velocity localization vector given by Equation (3.2.22) should be equal to θ_E evaluated from the energy localization vector given by Equation (3.2.34), and θ_v and θ_E should be as close to the target source direction as possible, i.e.,

$$\theta_v = \theta_E \approx \theta_S. \tag{5.2.20}$$

According to Equation (5.2.19), the first equality in Equation (5.2.20) yields

$$V'_y I'_x = V'_x I'_y. \tag{5.2.21}$$

2. Criterion 2. For all target source azimuths θ_S or part of target source azimuths for which the accuracy of localization is important, the velocity vector magnitude r_v given by Equation (3.2.29) is optimized as close to the unit as possible at low frequencies below 0.4–0.7 kHz. The energy vector magnitude r_E expressed in Equation (3.2.36) is optimized as close to the unit as possible within the mid- and high-frequency range of 0.7–4.0 kHz.
3. Criterion 3. The overall sound pressures P'_A and overall power Pow' in the origin should be a constant and independent from the target source azimuth θ_S.

For the first-order signals and if the velocity localization vector is considered only, the optimized criterion yields a set of linear equations of the decoding coefficients, which can be easily solved. In Equation (5.2.15), the amplitudes of the first-order loudspeaker signals are written as

$$A_i(\theta_S) = A_{total}\left[D_0^{(1)}(\theta_i) + D_1^{(1)}(\theta_i)\cos\theta_S + D_1^{(2)}(\theta_i)\sin\theta_S\right] \qquad i = 0, 1\ldots(M-1) \tag{5.2.22}$$

Letting $\theta_v = \theta_S$, $r_v = 1$, and applying the criterion of $P'_A = P_A = 1$ (i.e., the overall reproduced sound pressure in the origin is equal to the unit target sound pressure), Equations (3.2.20) and (3.2.22) yield

$$\sum_{i=0}^{M-1} A_i(\theta_S) = 1 \quad \sum_{i=0}^{M-1} A_i(\theta_S)\cos\theta_i = \cos\theta_S \quad \sum_{i=0}^{M-1} A_i(\theta_S)\sin\theta_i = \sin\theta_S. \tag{5.2.23}$$

The left sides of the three equations above represent the normalized pressure, x, and y components of the velocity localization vector of the reproduced sound field in the origin, respectively. The right sides of the above equations represent the three corresponding physical quantities in the target sound field, respectively. Therefore, Equation (5.2.23) shows that the three physical quantities in reproduction match with those in the target sound field. Substituting Equation (5.2.22) into Equation (5.2.23) yields a linear matrix equation for the unknown coefficients:

$$\mathbf{S}_{2D} = A_{total}\left[Y_{2D}\right]\left[D_{2D}\right]\mathbf{S}_{2D}, \tag{5.2.24}$$

where $S_{2D} = [W, X, Y]^T$ is a 3 ×1 column matrix or vector composed of the first-order independent signals given by Equation (4.3.3), the subscript "2D" denotes the case of two-dimensional (horizontal) reproduction. $[D_{2D}]$ is an M×3 decoding matrix to be determined:

$$[D_{2D}] = \begin{bmatrix} D_0^{(1)}(\theta_0) & D_1^{(1)}(\theta_0) & D_1^{(2)}(\theta_0) \\ D_0^{(1)}(\theta_1) & D_1^{(1)}(\theta_1) & D_1^{(2)}(\theta_1) \\ \vdots & \vdots & \vdots \\ D_0^{(1)}(\theta_{M-1}) & D_1^{(1)}(\theta_{M-1}) & D_1^{(2)}(\theta_{M-1}) \end{bmatrix}. \quad (5.2.25)$$

$[Y_{2D}]$ is a 3 × M matrix with its entries being the cosine and sine of each loudspeaker azimuth:

$$[Y_{2D}] = \begin{bmatrix} 1 & 1 & \dots & 1 \\ \cos\theta_0 & \cos\theta_1 & \dots & \cos\theta_{M-1} \\ \sin\theta_0 & \sin\theta_1 & \dots & \sin\theta_{M-1} \end{bmatrix}. \quad (5.2.26)$$

For $M > 3$, the unknown coefficients in Equation (5.2.24) can be solved from the following pseudo-inverse methods:

$$A_{total}[D_{2D}] = pinv[Y_{2D}] = [Y_{2D}]^T \left\{[Y_{2D}][Y_{2D}]^T\right\}^{-1}. \quad (5.2.27)$$

The solution given by Equation (5.2.27) is valid for regular and irregular loudspeaker configurations and satisfies the condition of constant-amplitude normalization given by Equation (4.3.80a). When M= 3, if the matrix $[Y_{2D}]$ is invertible, the solution of Equation (5.2.24) is directly obtained from the inverse matrix of $[Y_{2D}]$. For regular loudspeaker configuration, the solution or loudspeaker signals expressed in Equation (5.2.27) are equivalent to Equation (4.3.15) with b = 2 and constant-amplitude normalization. For irregular loudspeaker configurations such as 5.1-channel configuration, the solution given by Equation (5.2.27) is often close to singular and thus unstable (Neukom, 2006). In this case, a slight change in loudspeaker positions or responses may influence the performance in reproduction. In addition, the following undesirable situation may occur. That is, the signal amplitude for a loudspeaker near the target source direction is large; the signal amplitude for some other loudspeakers is also large but out of phase. The destructive interference among the sound wave from all loudspeakers makes the superposed sound pressure in the origin match with that of the target source, but the overall power of all loudspeaker signals increases dramatically. Moreover, the superposed sound pressure at the off-central (off-origin) position deviates from that of the target plane wave obviously.

If the energy localization vector is considered, the optimized criteria yield a set of nonlinear equations of the unknown coefficients. Solving these equations deals with the problem of nonlinear optimization and thus is complicated. The equations are usually solved by numerical methods. The following problems may occur in solving the equations:

1. For some practical irregular loudspeaker configurations, it may only be able to satisfy the aforementioned optimized criteria partly or approximately rather than completely and exactly. In addition, it may only be able to satisfy the optimized criteria in some

target source directions rather than in all target source directions. In this case, the error in target source directions that are important for auditory perception (such as frontal directions) is considered preferentially in the optimization procedure.
2. The optimized solution depends on the optimized criteria and the measures of errors. A comprehensive measure of the overall error should be a weighted combination of the errors evaluated from the aforementioned Criteria 1 to 3. The weights are chosen in accordance with the relative importance of different errors. Different error weights lead to various results.
3. The final results of some nonlinear optimization procedures depend on the initial parameters. Inappropriate initial parameters may lead to local rather than global optimized solutions.
4. Under certain optimized criteria, two or more sets of the solution of the decoding coefficients may be available. A set of appropriate solutions should be finally identified from physical and auditory analysis or even from experience.
5. Some solutions may be nearly singular and unstable so that slight variations in parameters (such as loudspeaker azimuths) remarkably change the resultant coefficients. Such a solution is obviously inappropriate.

With these problems, deriving the decoding coefficients for irregular loudspeaker configurations becomes mathematically difficult. Despite these difficulties, some studies have derived Ambisonic-decoding coefficients and signals for irregular loudspeaker configurations, especially for the 5.1-channel configuration by using various nonlinear optimization methods.

Gerzon and Barton (1992) first investigated the solution of the decoding coefficients of second-order Ambisonic-like signals for irregular configurations with five loudspeakers. They presented their results at the 92nd Convention of the Audio Engineering Society held in Vienna and termed Vienna decoders. The azimuths of loudspeakers in Gerzon's configuration slightly differ from those recommended by the ITU. In Gerzon's configuration, the span angle between the left and right loudspeakers in the front was wider than that recommended by the ITU, and the span angle between a pair of surround loudspeakers was narrower than that recommended by the ITU. The ITU did not publish the standard of 5.1-channel sound when Gerzon presented his work in 1992. Gerzon outlined a method for solving decoding coefficients but did not give a detailed mathematical derivation. Many set solutions are available for the decoding coefficients. These solutions depend on some predetermined parameters that should be chosen on the basis of certain auditory localization criteria. The optimization of a velocity localization vector below 0.4 kHz and an energy localization vector at above 0.7 kHz yields two sets of decoding coefficients at low and mid-high frequencies, respectively. Accordingly, the low- and mid-high-frequency components of independent signals are individually decoded by two decoding matrices rather than a common decoding matrix and shelf filters similar to those in Figure 4.19. The overall gains of two decoding matrices are chosen to balance the timbre between low and mid-high frequencies so that the root-mean-square value of decoding coefficients at mid-high frequencies matches with that at low frequencies. This signal mixing method improves the horizontal localization performance and the stability of the front virtual source.

Craven (2003) gave the frequency-independent decoding equation for an ITU 5.1-channel configuration with the 4th-order Ambisonic-like signal mixing. Decoding coefficients were derived from the cost function based on the aforementioned optimized criteria. The conjugate-gradient method was used to solve the nonlinear optimization problem. The final convergence was accelerated using the second derivatives in a Newton iteration. Craven presented the final results of (un-normalized) loudspeaker signal amplitudes or gains as shown in the following equations but did not provide the detailed mathematical derivation:

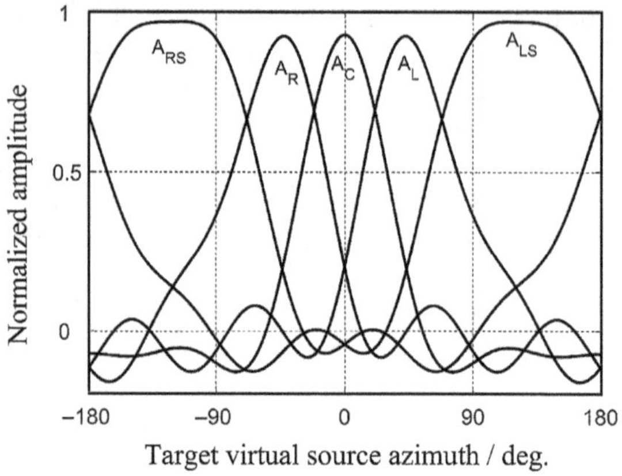

Figure 5.7 The 4th-order Ambisonic-like signal panning curves for an ITU 5.1-channel loudspeaker configuration.

$$
\begin{aligned}
A_L &= 0.167 + 0.242\cos\theta_S + 0.272\sin\theta_S - 0.053\cos 2\theta_S + 0.222\sin 2\theta_S \\
&\quad -0.084\cos 3\theta_S + 0.059\sin 3\theta_S - 0.070\cos 4\theta_S + 0.084\sin 4\theta_S \\
A_C &= 0.105 + 0.332\cos\theta_S + 0.265\cos 2\theta_S + 0.169\cos 3\theta_S + 0.060\cos 4\theta_S \\
A_R &= 0.167 + 0.242\cos\theta_S - 0.272\sin\theta_S - 0.053\cos 2\theta_S - 0.222\sin 2\theta_S \\
&\quad -0.084\cos 3\theta_S - 0.059\sin 3\theta_S - 0.070\cos 4\theta_S - 0.084\sin 4\theta_S \\
A_{LS} &= 0.356 - 0.360\cos\theta_S + 0.425\sin\theta_S - 0.064\cos 2\theta_S - 0.118\sin 2\theta_S \\
&\quad -0.047\sin 3\theta_S + 0.027\cos 4\theta_S - 0.061\sin 4\theta_S \\
A_{RS} &= 0.356 - 0.360\cos\theta_S - 0.425\sin\theta_S - 0.064\cos 2\theta_S + 0.118\sin 2\theta_S \\
&\quad +0.047\sin 3\theta_S + 0.027\cos 4\theta_S + 0.061\sin 4\theta_S
\end{aligned} \quad . \tag{5.2.28}
$$

Figure 5.7 illustrates the signal panning curves of Equation (5.2.28). The amplitudes of A_L, A_R, A_{LS}, and A_{RS} signals, especially the two surround signals, are asymmetric about the axis direction of the corresponding loudspeakers. This asymmetry is adapted to an irregular loudspeaker configuration. By contrast, for regular loudspeaker configurations, the amplitudes of loudspeaker signals given by Equation (4.3.53) are even functions of the variable $(\theta_S - \theta_i)$ and therefore symmetric about the axis directions of the corresponding loudspeakers. The symmetry can also be observed in the example illustrated in Figure 4.17. In addition, for a regular loudspeaker configuration, the Q-order Ambisonic reproduction requires at least $(2Q + 1)$ or $(2Q + 2)$ loudspeakers (Section 4.3.3). Therefore, a regular five-loudspeaker configuration is appropriate for the first- and second-order Ambisonic reproduction. Conversely, for an irregular loudspeaker configuration, higher-order azimuthal harmonic components are necessary to fit the required asymmetric polar pattern of loudspeaker signals.

Craven (2003) presented some evaluated results for loudspeaker signals given in Equation (5.2.28). Overall, the velocity vector magnitude $r_v$is improved at the lateral direction $\theta_S = 90°$ and the rear direction $\theta_S = 180°$ in comparison with those of the pair-wise amplitude panning. For example, at the rear direction $\theta_S = 180°$, r_v is 0.693for loudspeaker signals expressed in Equations (5.2.13) and 0.342 for pair-wise amplitude panning. This finding is due to the out-of-phase and small-amplitude A_L and A_R signals in Equation (5.2.28). However, the accuracy of the virtual source near $\theta_S = 50°$ should be further improved.

Poletti (2007)used the criterion of minimizing the overall square error between the reproduced sound pressures and target plane wave pressures to design Ambisonic signals.

A weighted combination of the power of each loudspeaker signal is used as a penalty function and added to the cost function of the square error to enhance the stability and avoid the excessive overall power of loudspeaker signals caused by an irregular loudspeaker configuration. The weights in the penalty function depend on the difference between the direction of each loudspeaker and the direction of the target source. On the basis of this method, Poletti designed a decoding equation for an ITU 5.1-channel loudspeaker configuration with the 4th-order Ambisonic-like signal mixing.

Wiggins (2007) used a Tabu search method to derive the decoding coefficients for an ITU 5.1-channel loudspeaker configuration with the 4th-order Ambisonic-like signal mixing. L (uniform) target source directions $\theta_{S,l}$ with $l = 0, 1, 2\ldots (L-1)$ are chosen within the azimuthal region of $-180° < \theta_S \leq 180°$ (or considering the left–right symmetry within the azimuthal region of $-180\ ° < \theta_S \leq 180°$). Six root-mean-square (RMS) errors are evaluated:

$$
\begin{aligned}
Err_1 &= \sqrt{\frac{1}{L}\sum_{l=0}^{L-1}\left[1-P'_A(\theta_{S,l})\right]^2} \quad & Err_2 &= \sqrt{\frac{1}{L}\sum_{l=0}^{L-1}\left[1-Pow'(\theta_{S,l})\right]^2},\\
Err_3 &= \sqrt{\frac{1}{L}\sum_{l=0}^{L-1}\left[1-r_v(\theta_{S,l})\right]^2} \quad & Err_4 &= \sqrt{\frac{1}{L}\sum_{l=0}^{L-1}\left[1-r_E(\theta_{S,l})\right]^2},\\
Err_5 &= \sqrt{\frac{1}{L}\sum_{l=0}^{L-1}\left[\theta_{v,l}-\theta_{S,l}\right]^2} \quad & Err_6 &= \sqrt{\frac{1}{L}\sum_{l=0}^{L-1}\left[\theta_{E,l}-\theta_{S,l}\right]^2}.
\end{aligned}
\tag{5.2.29}
$$

1. Err_1 is the RMS error of the reproduced sound pressure $P'_A(\theta_{S,,l})$ at the origin over the L target directions. $P'_A(\theta_{S,,l})$ is calculated from Equation (5.2.19), and the target sound pressure at the origin is normalized to a unit.
2. Err_2 is the RMS error of the reproduced power $Pow'(\theta_{S,,l})$ at the origin over the L target directions. $Pow'(\theta_{S,,l})$ is calculated from Equation (5.2.19), and the target power is normalized to a unit.
3. Err_3 is the RMS error of the velocity vector magnitude $r_v(\theta_{S,,l})$ over the L target directions. $r_v(\theta_{S,,l})$ is calculated from Equation (3.2.29), and the ideal value of r_v is a unit.
4. Err_4 is the RMS error of the energy vector magnitude $r_E(\theta_{S,,l})$ over the L target directions. $r_E(\theta_{S,,l})$ is calculated from Equation (3.2.36), and the ideal value of r_E is a unit.
5. Err_5 is the RMS error of the velocity localization vector-based azimuth $\theta_{v,,l}$ over L target directions. $\theta_{v,,l}$ is calculated from Equation (3.2.27), and the ideal value is $\theta_{v,,l} = \theta_{S,,l}$.
6. Err_6 is the RMS error of the energy localization vector-based azimuth $\theta_{E,,l}$ over the L target directions. $\theta_{E,,l}$ is calculated from Equation (3.2.34), and the ideal value is $\theta_{E,,l} = \theta_{S,,l}$.

The cost function for solving the decoding coefficients is a weighted combination of the six RMS errors:

$$
Err = w_1 Err_1 + w_2 Err_2 + w_3 Err_3 + w_4 Err_4 + w_5 Err_5 + w_6 Err_6. \tag{5.2.30}
$$

The weights w_1–w_6 determine the relative contributions of each term in Equation (5.2.30). Increasing the weight of a certain term reduces the error caused by this term but at the cost of increasing the error caused by the other terms. Therefore, the weights should be chosen in accordance with certain psychoacoustic rules and the desired performance in reproduction.

Various weights and combinations of errors lead to different optimized decoding coefficients. For example, at low frequencies, the reproduced sound pressure, the velocity localization vector-based azimuth, the velocity vector magnitude, and the energy localization vector-based azimuth should be optimized preferentially. Accordingly, the combination of Err_1, Err_3, Err_5, and Err_6 is preferentially chosen as the cost function. At mid-high frequencies, the reproduced power, energy vector magnitude, velocity localization vector-based azimuth, and energy localization vector-based azimuth should be optimized preferentially. Accordingly, the combination of Err_2, Err_4, Err_5, and Err_6 are preferentially chosen as the cost function. This choice of a frequency-dependent cost function leads to two sets of the decoding equations for low- and mid-high frequencies.

In the Tabu search, given the form of the overall cost function, decoding coefficients are initialized at some random values (or some predetermined values that have been derived from other methods). Each coefficient is increased or decreased at a predetermined step size in a given order but is restricted within a predetermined bound. Then, the variation in the overall cost (error) function is evaluated. Decoding coefficients change toward the direction of a decreasing overall cost function, and the best results are kept. A convergent result is obtained by a recursive search. On the basis of the initial coefficients expressed in Equation (5.2.28), Wiggins derived a set of decoding coefficients by using a Tabu search. In comparison with the decoding coefficients given in Equation (5.2.28), decoding with Tabu-search-based coefficients improves the accuracies in a virtual source direction (especially for the front source) and power level, although the maximum r_v and r_E decrease slightly.

New optimized criteria, such as uniformity and standard deviation in localization, have also been supplemented to the cost function for solving Ambisonic-decoding coefficients (Moore and Wakefield, 2008). Accordingly, the cost function is constructed by supplementing the weighted combination of the standard deviations of Err_3, Err_4, Err_5, and Err_6 over the L target directions into Equation (5.2.30). The optimized procedure for the second-order Ambisonic-like signal mixing has been written as software (Heller et al., 2010). Some other mathematical algorithms, such as artificial neuron networks and genetic algorithms, have also been used for nonlinear optimization to derive the Ambisonic-like decoding equations and coefficients of irregular loudspeaker configurations (Tsang and Cheung, 2009; Tsang et al., 2009).

Some studies have indicated that all aforementioned optimized criteria are difficult to satisfy for the second-order Ambisonic-like signal mixing via an ITU5.1-channel loudspeaker configuration. Four loudspeaker configurations are relatively appropriate for the second-order Ambisonic-like signal mixing. Adding a front-center loudspeaker is not always beneficial to the lateral virtual source for the second-order Ambisonic-like signal mixing. Therefore, region-dependent signal mixing methods can be used. Three front loudspeakers with pair-wise amplitude panning (or local Ambisonic-like mixing discussed in Section 5.2.4) can be used to recreate the virtual source within the front region of $-30° \leq \theta_S \leq 30°$. The four loudspeakers of L, R, LS, and LS with Ambisonic-like signal mixing are used to recreate lateral and rear virtual sources. The design of this signal mixing method is relatively simple. For example, a frequency-dependent gain is designed to reduce the variation in the virtual source position with frequency (Xie, 2001a). Heller et al. (2010) illustrated more examples for four horizontal loudspeakers with irregular configurations.

5.2.4 Optimization of three frontal loudspeaker signals and local Ambisonic-like signal mixing

In multichannel sound reproduction, a horizontal plane can be divided into subregions, and the virtual source in each subregion is recreated by loudspeakers in the same subregion. The loudspeaker signals in each subregion can also be optimized. For ***local sound field signal***

mixing or ***local Ambisonic-like signal mixing***, the normalized loudspeaker signals in each subregion are a linear combination of azimuthal harmonics. This mixing differs from global Ambisonic (-like) signal mixing. In the latter, the normalized signals of all loudspeakers are a linear combination of azimuthal harmonics.

For 5.1-channel sound, the virtual source in the frontal region is recreated by the three frontal loudspeakers. On the basis of the virtual sound localization theorem, Gerzon (1990) first derived the local Ambisonic-like signal mixing for three frontal loudspeakers. Generally, similar to Equation (5.2.22), Equation (5.2.31) expresses the normalized signals for three frontal loudspeakers with the first-order local Ambisonic-like signal mixing:

$$\begin{aligned} A_L(\theta_S) &= A_{total}\left(D_{0,L}^{(1)} + D_{1,L}^{(1)}\cos\theta_S + D_{1,L}^{(2)}\sin\theta_S\right) \\ A_R(\theta_S) &= A_{total}\left[D_{0,R}^{(1)} + D_{1,R}^{(1)}\cos\theta_S + D_{1,R}^{(2)}\sin\theta_S\right], \\ A_C(\theta_S) &= A_{total}\left[D_{0,C}^{(1)} + D_{1,C}^{(1)}\cos\theta_S + D_{1,C}^{(2)}\sin\theta_S\right] \end{aligned} \tag{5.2.31}$$

Form the left–right symmetry, decoding coefficients satisfy $D_{0,L}^{(1)} = D_{0,R}^{(1)}$, $D_{1,L}^{(1)} = D_{1,R}^{(1)}$, $D_{1,L}^{(2)} = -D_{1,R}^{(2)}$, $D_{1,C}^{(2)} = 0$. If the optimization at low frequencies is considered only, the criteria of the optimized interaural phase delay difference and its variation with the head rotation or the equivalent criteria of the optimized velocity localization vector in Equation (5.2.31) are used to derive the decoding coefficients. Gerzon (1990) presented the results for three frontal loudspeakers arranged at 0° and ±45°. Here, the results for the frontal loudspeakers with an ITU configuration are given. The decoding coefficients are derived by substituting Equation (5.2.31) into Equation (3.2.22) and letting $\theta_v = \theta_S$ and $r_v = 1$. Then, the normalized loudspeaker signals in Equation (5.2.31) become

$$\begin{aligned} A_L &= A_{total}\left[1 - \cos\theta_S + \left(2-\sqrt{3}\right)\sin\theta_S\right] \\ A_R &= A_{total}\left[1 - \cos\theta_S - \left(2-\sqrt{3}\right)\sin\theta_S\right] \\ A_C &= A_{total}\left(-\sqrt{3} + 2\cos\theta_S\right) \end{aligned} \tag{5.2.32}$$

Similar to Equation (4.3.41), Equation (5.2.33) shows A_{total}(overall gain) for constant-amplitude normalization:

$$A_{total} = \frac{1}{2-\sqrt{3}}. \tag{5.2.33}$$

A_{total} for constant-power normalization is slightly complicated. Equation (5.2.32) proves that a target-azimuth-independent A_{total} results in an overall power of $Pow' = A_L^2 + A_R^2 + A_C^2$ that varies with the target azimuth θ_S. The virtual source direction depends on the relative magnitude (and phase or time) relationships among the loudspeaker signals. As such, a target-azimuth-dependent A_{total} can be chosen so that the overall power of loudspeaker signals is normalized to a unit, then

$$A_{total} = \frac{1}{\left[11 + 8\left(1-\sqrt{3}\right)\sin^2\theta_S - 4\left(1+\sqrt{3}\right)\cos\theta_S\right]^{1/2}}. \tag{5.2.34}$$

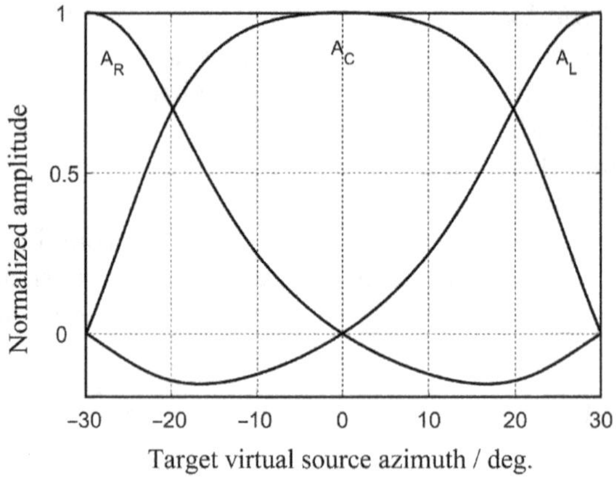

Figure 5.8 The first-order local Ambisonic-like panning curves for three frontal loudspeakers.

This set of signals is difficult to be created with a practical microphone recording technique, but it can be easily simulated via signal processing.

Figure 5.8 illustrates the first-order local Ambisonic-like panning curves for the three frontal loudspeakers. The signals are normalized according to Equation (5.2.34). For a target source at each loudspeaker direction ($\theta_S = 0°$ or $\pm 30°$), the signals for two other loudspeakers vanish. In this case, the virtual source is recreated by a single loudspeaker without a cross-talk (with infinite interchannel separation). This ideal feature is similar to that in pair-wise amplitude panning. However, for the first-order local Ambisonic-like panning, the virtual source between two loudspeakers is recreated by three loudspeakers, and the signal for the third loudspeaker is out of phase. This feature is found in Ambisonic-like panning. As stated in Sections 3.2.2 and 4.1.3, an out-of-phase signal is essential to ensure that the velocity vector magnitude is $r_v = 1$ and to stabilize the virtual source. Equations (3.2.7) and (3.2.9) prove that the perceived virtual source direction matches with that of the target source at low frequencies and within the region of $-30° \leq \theta \leq 30°$.

Equation (3.2.34) verifies that the energy localization vector-based azimuth θ_E is inconsistent with the velocity localization vector-based azimuth θ_v for signal mixing given by Equation (5.2.32). Gerzon (1990) also pointed out that the mismatch of θ_E and θ_v may lead to the directional distortion of a virtual source at high frequencies, such as the movement of a high-frequency virtual source near the front toward the direction of central–frontal loudspeakers. Gerzon (1992c) further proposed an optimized method through which the velocity localization vector at low frequencies and the energy localization vector at mid-high frequencies are considered. That is, the decoding coefficients are chosen so that θ_E matches with θ_v.

The aforementioned optimized signal mixing for three frontal loudspeakers is theoretically perfect, but it is rarely used in practical program production.

5.2.5 Time panning for 5.1-channel surround sound

Time panning for 5.1-channel surround sound is similar to the case of a two-channel stereophonic sound. Some studies have investigated the possibility of using pair-wise time panning to recreate a virtual source in 5.1-channel reproduction with transient stimuli. Using female speech as a stimulus, Martin et al. (1999) explored the summing localization with ICTD only in the 5.1-channel loudspeaker configuration. They arranged the three frontal loudspeakers identical to those in Figure 5.2, but they arranged two surround loudspeakers in azimuths of±120°.

The virtual source localization experiment revealed that ICTD alone could recreate the virtual source between left and center loudspeaker pair (within 0° and 30°) l. An ICTD of about 1.2 ms was enough to move the virtual source to the direction of the loudspeaker with a leading signal. Overall, the virtual source created by ICTD only was blurry, and the dispersion in localization increased, especially in the case of center channel signal leading in time. For the pair of the left and right surround loudspeakers, ICTD alone could recreate the rear virtual source between the loudspeakers. An ICTD of about 0.6 ms was sufficient to move the virtual source to the direction of the loudspeaker with the leading signal. This result was smaller than that of the pair of left and center loudspeakers or the traditional pair of stereophonic loudspeakers (about 1.1–1.2 ms; Figure 1.26). This observation was also due to the wide span angle between the left and right surround loudspeakers. For the pair of the left and LS loudspeakers, ICTD alone failed to recreate a stable virtual source between two loudspeakers. The perceived virtual source jumped from one loudspeaker to another, and the perceived direction exhibited a larger dispersion.

The aforementioned results were obtained from psychoacoustic experiments rather than theoretical analysis, and the detailed results were dependent on the type of stimuli. Moreover, multiple loudspeakers with ICTD not only lead to confused localization but also cause timbre coloration because of the interference among loudspeaker signals. This situation should be avoided in practice.

For transient stimuli, a combination of ICLD and ICTD can recreate a virtual source between the center and left (or right) loudspeakers and two surround loudspeakers. Similar to the case of two-channel stereophonic sound discussed in Sections 1.7.1 and 2.1.4, the effect of ICLD and ICTD enhances or cancels depending on whether the individual effects of ICLD and ICTD are consistent or different. The quantitative relationships among the summing virtual source direction and ICTD/ICLD in the 5.1-channel loudspeaker configuration are different from those of two-channel stereophonic sound because of the difference in the span angles between the adjacent loudspeaker pairs.

The experimental results discussed in this section exhibit the regular patterns and limitation of summing localization in 5.1-channel loudspeaker configuration with ICTD only or a combination of ICLD and ICTD. These patterns provide some guidance for a 5.1-channel microphone recording design.

5.3 OTHER MULTICHANNEL HORIZONTAL SURROUND SOUNDS

Some multichannel horizontal surround sound systems and techniques with more than 5.1-channels have also been developed since the 1990s to improve the reproduction performance and compete commercially. Some of these techniques were originally developed for cinema uses and then simplified for consumer uses, and others were directly developed for consumer uses. Irregular loudspeaker configurations were usually used in these systems.

A rear surround (BS) channel is added to the 5.1-channel system to improve the rear virtual source localization and ambient sound reproduction, leading to 6.1-channel horizontal surround sound. An example is DTS-ES 6.1 developed by DTS Inc. DTS-ES 6.1 has two working modes, namely, matrix and discrete modes (http://dts.com/). In the matrix mode, the signal of the rear surround channel is derived from the left and right surround channel signals by a matrix decoding, which is discussed in Section 8.1.4. In the discrete mode, the signal of the rear surround channel is independent. The discrete mode is discussed here. In addition, the rear surround signal of the DTS-ES 6.1 discrete mode is mixed into the left and right surround signals and then separated with a decoding matrix in discrete 6.1-channel reproduction to be compatible with the 5.1-channel system.

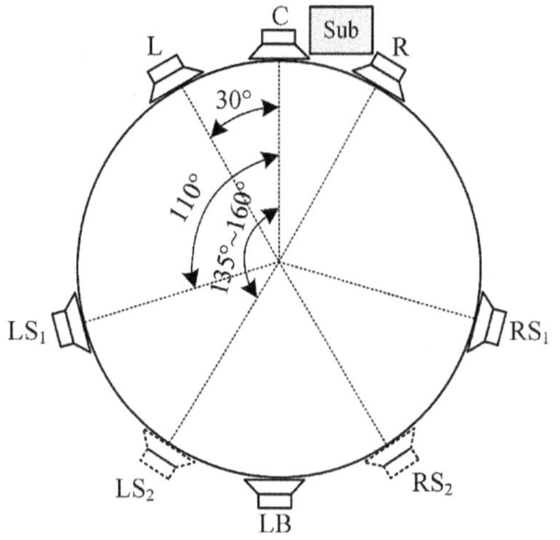

Figure 5.9 Typical 6.1-channel loudspeaker configuration.

Figure 5.9 illustrates the typical 6.1-channel loudspeaker configuration for domestic uses. The rear channel signal is reproduced by a single loudspeaker LB in the rear of 180°. It can also be reproduced by a pair of rear surround loudspeakers LS_2 and RS_2 arranged at ±135° to ±160°. Moreover, moving the left and right surround loudspeakers, LS1 and LS2, to the two side directions of ± 90° improves the reproduction of the lateral virtual source.

In the 1990s, the *Sony Cinema Products Corporation* in the USA developed a 7.1-channel surround sound system, namely, the Sony Dynamic Digital Sound (SDDS), for cinema uses (Steinke, 1996). The 7.1-channel system SDDS comprises five front channels, two surround channels, and an LFE channel, providing auditory perceptions matching with the picture and appropriate ambience.

The ITU standard also supports the 7.1-channel reproduction as an optional function (ITU-R BS 775-1, 1994), which contains seven channels with a full audible bandwidth and an optional LFE channel. The three front loudspeakers are arranged at 0° and ±30°. Dolby Labs and DTS Inc. further developed the 7.1-channel systems for cinema and domestic uses and corresponding multichannel digital transmission and storage techniques (Sections 13.6 and 13.7). The 7.1-channel system is an optional sound format for Blue-ray disk. For domestic uses, the loudspeaker configuration of the Dolby 7.1-channel system is consistent with that of the ITU recommendation. The DTS-HD by DTS Inc. supports seven different 7.1-channel loudspeaker configurations for domestic uses (called configurations 1 to 7, DTS Inc., 2006), which improve the reproduction performance in different aspects. Configuration 1 is the most common (Figure 5.10). A pair of the lateral (left and right) surround loudspeaker LS_1 and RS_1 at ±90° improve the reproduction of lateral information. A pair of the rear surround loudspeakers LS_2 and RS_2 at ±150° improves the reproduction of rear information. Configuration 5 is similar to that in Figure 5.10, but the four surround loudspeakers are arranged at ±110° and ±150°. Configurations 1 and 5 are also consistent with the ITU recommendation. The other loudspeaker configurations in DTS-HD are for three-dimensional-spatial surround sound (Chapter 6). In addition to the aforementioned systems, more multichannel spatial surround sound systems have been developed (Section 6.4).

Similar to the case of the 5.1-channel system, the loudspeaker configurations for the aforementioned systems are specified, but the loudspeaker signals are flexible. Various signal mixing methods for these systems may be applied to recreate different localization and other spatial auditory effects. For example, for a 6.1-channel loudspeaker configuration with a pair of left and right surround loudspeakers at ± 90° and a rear surround loudspeaker at 180°, the pair-wise amplitude panning, global, and local Ambisonic-like mixing are available to recreate a low-frequency full-360° virtual source in the horizontal plane. The analysis is similar to those for the 5.1-channel loudspeaker configuration. For the second-order local Ambisonic-like signal mixing (Xie, 2001b), a virtual source within ±90° in the frontal-half-horizontal plane is recreated by the five loudspeakers of the left, center, right, and left surround and the right surround. The loudspeaker signals (gains) with constant-amplitude normalization are expressed in the following equation by using the method similar to the derivation of Equation (5.2.27):

$$
\begin{aligned}
A_L &= -2.15 + 4.31\cos\theta_S - 2.15\cos 2\theta_S + 0.58\sin 2\theta_S \\
A_C &= 3.73 - 6.46\cos\theta_S + 3.73\cos 2\theta_S \\
A_R &= -2.15 + 4.31\cos\theta_S - 2.15\cos 2\theta_S - 0.58\sin 2\theta_S \\
A_{LS} &= 0.79 - 1.08\cos\theta_S + 0.5\sin\theta_S + 0.29\cos 2\theta_S - 0.29\sin 2\theta_S \\
A_{RS} &= 0.79 - 1.08\cos\theta_S - 0.5\sin\theta_S + 0.29\cos 2\theta_S + 0.29\sin 2\theta_S.
\end{aligned}
\tag{5.3.1}
$$

The loudspeaker signals (gains) with constant-power normalization can be derived by dividing each signal in Equation (5.3.1) by the mean-square-root amplitude over all five signals (gains).

The virtual source in the rear-half-horizontal plane is recreated by the LS, rear surround (BS), and RS loudspeakers with the first-order local Ambisonic-like signal mixing. Through the method similar to the derivation of Equation (5.2.27), the loudspeaker signals (gains) with constant-amplitude normalization are presented as

$$
\begin{aligned}
A_{LS} &= 0.5 + 0.5\cos\theta_S + 0.5\sin\theta_S \\
A_{BS} &= -\cos\theta_S \\
A_{RS} &= 0.5 + 0.5\cos\theta_S - 0.5\sin\theta_S.
\end{aligned}
\tag{5.3.2}
$$

The signal mixing expressed in Equations (5.3.1) and (5.3.2) satisfies the optimized criteria of the velocity localization vector at low frequencies, i.e., $\theta_v = \theta_S$ and $r_v = 1$. For the target source at each loudspeaker direction, the signals for other loudspeakers vanish; thus, channel separation is infinite, which is desired. For the target source between a pair of loudspeakers, the signals for some loudspeakers are out of phase.

The analysis of signal mixing for 7.1-channel or more loudspeaker configurations is similar. Equation (3.2.6) is also applicable to analyze the frequency dependence or stability of the virtual source (Chapter 4 and Section 5.2.2). Moreover, the theorem of the energy localization vector is applied to examine the localization at mid-high frequencies. For brevity, these analyses are omitted here.

Overall, increasing the independent channels and loudspeakers in multichannel sound improves the accuracy in spatial information reproduction and enlarges the listening region. In particular, for pair-wise amplitude panning, increasing a pair of lateral loudspeakers at ±90° obviously improves the lateral virtual source localization, which has been proved in Section 4.2.1. For the 7.1-channel loudspeaker configuration in Figure 5.10, if the center loudspeaker is ignored, the configuration of six other loudspeakers is arranged regularly. As stated in Chapter 4, especially in Section 4.2, this regular six-loudspeaker configuration is appropriate for recreating a full 360° virtual source in the horizontal plane. However, when independent channels and loudspeakers increase, the hardware and signal processing of the

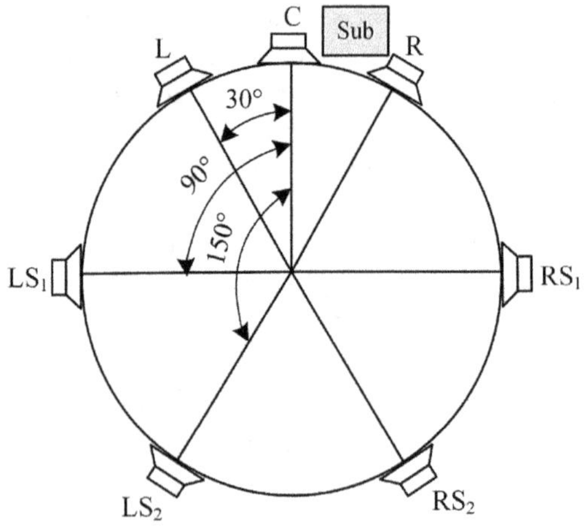

Figure 5.10 Configuration I of 7.1-channel loudspeakers in DTS-HD.

system becomes complicated. As such, the development of multichannel surround sound is considered (Section 3.1). When the available system resources are sufficient, increasing the number of independent channels and manipulating all spatial information with an equal accuracy improve the overall perceived performance. When the system resources are limited, spatial information should be manipulated with different approximations according to their importance in auditory perception. The key is how to make a compromise between the performance and complexity. The newly developed digital signal processing, transmission, and storage techniques, including object-based coding technique discussed in Section 13.4.8, remarkably alleviates the limitation of independent channels and signals. However, using an excessive number of channels in most consumer reproduction is still difficult. In fact, the 7.1-channel sound has been used as the optional sound format in a Blue-ray disk, other multichannel sounds with more than 5.1-channels (including some spatial surround sounds) are still being developed quickly up to now (2021). Their prospects depend on the requirements from their practical uses.

5.4 LOW-FREQUENCY EFFECT CHANNEL

A low-frequency effect (LFE) channel, such as the ".1" channel in 5.1-channel sound, is an independent channel operating below the frequency limit of 120 Hz. It is originally intended to convey the special low-frequency content with a greater pressure level in the cinema sound, such as an explosion sound. These special low-frequency contents are inappropriate to be transmitted with the main channel signals because of their greater levels. In cinema sound, the in-band gain of the LFE channel is usually 10 dB higher than that of the individual main channel. This result is obtained by increasing the gain in the reproduction channel rather than increasing the transmitted or recorded level in the LFE channel. Moreover, it does not mean that the overall level of the LFE channel is 10 dB higher than that of an individual main channel because of the narrow bandwidth of the LFE channel. The LFE channel should not be confused with the ***bass management*** in multichannel sound reproduction. The former refers to the independent channel for a low-frequency content. The latter refers to a mixing of the low-frequency components of the main channel signals (and possibly together with the LFE channel signal), so they are reproduced by a (or more) subwoofer. Bass management can be used for domestic and cinema reproduction, which is addressed in Section 14.3.3.

For domestic uses, an LFE channel is mainly utilized for reproduction with accompanying picture. In music reproduction only, programs usually do not contain contents with a much greater level at low frequencies except for some special programs such as the cannon sound in Tchaikovsky's 1812 Overture or the explosion sound in J. Strauss's Explosion Polka. Therefore, the low-frequency content of music signals can be transmitted by the main channels, and the independent LFE channel is unnecessary (Dolby Laboratories, 2000). Moreover, the transmission and reproduction of the low-frequency components of music signals with a single channel may degrade spaciousness. The LFE channel is optional in domestic reproduction. Therefore, the low-frequency components of signals are also mixed into the main channels to ensure the overall timbre balance even if the LFE channel and subwoofer are unavailable. Accordingly, when cinema programs are transferred into consumer media (such as DVD video), the content of the LFE channel should be re-mixed into the main channels. Moreover, the low-frequency components that are important to integrity should not be placed in LFE channels (SMPTE 320M, 1999).

The arrangement of the subwoofer in the 5.1-channel system is not specified in the ITU standard. In addition, some authors suggested using two independent LFE channels (i.e., the " .2" channels). The optimization of the number of LFE channels and the arrangement of subwoofers are closely related to the acoustic normal modes of the listening room (Section 1.2.2). They are also related to the low-frequency characteristics of program contents. This problem is addressed in Section 14.3.3.

5.5 SUMMARY

Multichannel surround sounds with accompanying picture were originally developed for cinema uses and then modified and simplified for domestic reproduction. Domestic surround sound with accompanying picture should be compatible with music reproduction, leading to a series of general surround sounds with and without accompanying pictures. Practical systems and techniques are usually a compromise between the two requirements of reproduction with accompanying picture and music reproduction only.

Since the early 1990s, most of the multichannel horizontal surround sound techniques have been based on the consideration that enhances the accuracy in the reproduction of frontal information at the cost of a rough approximation in the reproduction of lateral and rear information. Irregular loudspeaker configurations with a front-center loudspeaker are adopted to recreate auditory perceptions that match with the picture.

The5.1-channel sound has been recommended by the ITU as the standard of "multichannel stereophonic sound system with and without accompanying picture" and widely used in the domestic reproduction. Various signal mixing and recording methods are available for 5.1-channel sound. The common pair-wise amplitude panning can recreate a stable virtual source in the frontal region, but the virtual source in the rear region is blurry and unstable. A major problem with pair-wise amplitude panning is that it fails to recreate a stable virtual source within the lateral region, i.e., a "hole of virtual source" is found in the lateral region. Therefore, when used for music reproduction, a 5.1-channel sound with pair-wise amplitude panning is unable to simulate and reproduce the spatial information of lateral reflections perfectly and exhibit a defect. Some studies have explored the global Ambisonic-like signal mixing for 5.1-channel sound to recreate the virtual source in various horizontal directions. Deriving the Ambisonic-decoding coefficients deals with the problem of nonlinear optimization and is complicated because of the irregular loudspeaker configuration of 5.1-channel sound. The local Ambisonic-like signal mixing is also applicable to recreate the virtual source in the frontal region. Overall, the 5.1-channel sound is not originally intended and is inappropriate for recreating a full 360° virtual source in the horizontal plane; it is a compromise when it is used for general reproduction purposes.

Some other multichannel horizontal surround sounds with more than 5.1-channels and irregular loudspeaker configurations, such as 6.1 and 7.1-channel sounds, have also been developed. Increasing independent channels and loudspeakers in multichannel sound improves the accuracy of spatial information reproduction and enlarges the listening region. At the same time, it makes the system complicated. The prospects of various multichannel sounds depend on the requirements from practical uses.

The signal of the LFE channel is independent. The LFE channel is originally intended for conveying a special low-frequency content with a greater pressure level in the cinema sound. For domestic uses, the LFE channel is optional. The ITU standard has not specified the arrangement of the subwoofer in the 5.1-channel system, although some arrangements improve the perception of a sound field at low frequencies.

Chapter 6

Multichannel spatial surround sound

Multichannel horizontal surround sounds are discussed in Chapters 3 to 5. A real space possesses three dimensionalities. Multichannel three-dimensional spatial surround sounds, shorten for multichannel spatial surround sound or multichannel 3D (surround) sound, should be developed to recreate the three-dimensional spatial information of sound. As an extension of multichannel horizontal surround sounds, multichannel spatial surround sounds are considered the new generation of spatial sound techniques and are addressed in this chapter. In Section 6.1, the summing localization theorems of a horizontal virtual source are extended to a three-dimensional space to provide a basis for the succeeding analyses. The principle of summing localization with two loudspeakers in the median and sagittal planes is analyzed in Section 6.2. The vector base amplitude panning, a typical signal mixing method for spatial surround sound, is examined in Section 6.3. The principle of spatial Ambisonics, another typical spatial surround sound technique and signal mixing method, is discussed in Section 6.4. Some examples of spatial Ambisonic reproduction are also given. Some advanced spatial surround sound systems and related problems are addressed in Section 6.5.

6.1 SUMMING LOCALIZATION IN MULTICHANNEL SPATIAL SURROUND SOUND

6.1.1 Summing localization equations for spatial multiple loudspeaker configurations

Similar to the cases of horizontal surround sound, virtual source localization is an important aspect of spatial surround sound. Summing localization equations for multiple horizontal loudspeakers in Section 3.2.1 should be extended to spatial multiple loudspeaker configurations to analyze the virtual source localization in a spatial surround sound (Xie X.F., 1988; Rao and Xie, 2005; Xie et al., 2019). The foundation of this extension is Wallach's (1940) hypothesis, discussed in Section 1.6.3.

The coordinate shown in Figure 1.1 is used. At low frequencies, the head shadow is neglected, and the two ears are approximated by two points in a free space and separated by a distance of $2a$ (head diameter). For a point source in the direction of (r_S, θ_S, ϕ_S) and a source distance of $r_S >> a$, the incident wave can be approximated as a plane wave, and the pressures in the two ears are given by

$$P_L = P_A \exp\left[-jk\left(r_S - a\sin\theta_S\cos\phi_S\right)\right] \qquad P_R = P_A \exp\left[-jk\left(r_S + a\sin\theta_S\cos\phi_S\right)\right], \quad (6.1.1)$$

DOI: 10.1201/9781003081500-6

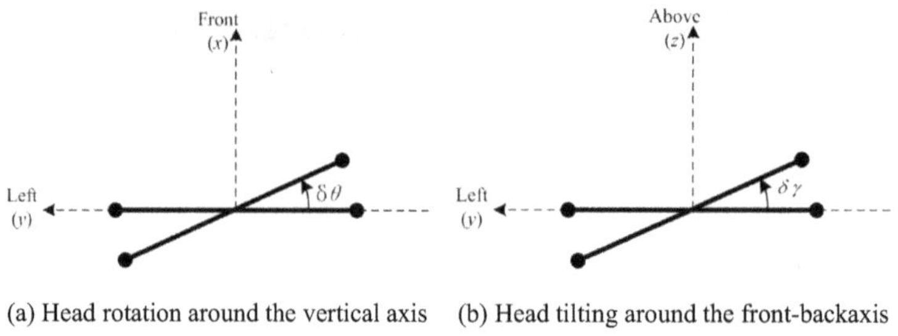

Figure 6.1 Head-turning around the vertical and front-back axes: (a) head rotation around the vertical axis; (b) head tilting around the front-back axis.

where P_A is the amplitude, and k is the wave number. The interaural phase difference in pressures is calculated from Equation (6.1.1) as

$$\Delta\psi = \psi_L - \psi_R = 2ka\sin\theta_S\cos\phi_S, \tag{6.1.2}$$

or the interaural phase delay difference ITD_p is given by

$$ITD_p(\theta_S, \phi_S) = \frac{\Delta\psi}{2\pi f} = \frac{2a}{c}\sin\theta_S\cos\phi_S. \tag{6.1.3}$$

When the sound source is located in the cone of confusion in Figure 1.21, $\sin\theta_S \cos\phi_S$ and ITD_p are constant. Therefore, ITD_p only is inadequate for determining the unique position of a sound source.

In Figure 6.1(a), if the head rotates around the vertical (z) axis anticlockwise with a small azimuth $\delta\theta$ in the horizontal (x–y) plane, ITD_p becomes

$$ITD_p(\theta_S - \delta\theta, \phi_S) = \frac{2a}{c}\sin(\theta_S - \delta\theta)\cos\phi_S. \tag{6.1.4}$$

Letting $\delta\theta \to 0$, the variation rate or derivative of ITD_p with respect to $\delta\theta$ is expressed as

$$\frac{dITD_p(\theta_S, \phi_S)}{d(\delta\theta)} = -\frac{2a}{c}\cos\theta_S\cos\phi_S. \tag{6.1.5}$$

Equation (6.1.5) indicates that the variation in ITD_p with head rotation is relevant to the elevation ϕ_S. For a given $\theta_S \neq 90°$, the magnitude of ITD_p variation maximizes in the horizontal plane with $\phi_S = 0°$. As the source elevation departs from the horizontal plane to a high or low elevation, the magnitude of variation decreases. At the top or bottom with $\phi_S = \pm 90°$, ITD_p is invariant against head rotation. Therefore, ITD_p variation caused by head rotation provides information on vertical displacement from the horizontal plane. However, the variation in ITD_p with head rotation alone does not provide enough information on up-down discrimination because of the even function characteristic of $\cos\phi_S$.

As shown in Figure 6.1 (b), if the head turns around the front-back (x) axis to the left with a small angle $\delta\gamma$ (tilting), the positions of the left and right ears become $(0, a\cos\delta\gamma, -a\sin\delta\gamma)$ and $(0, -a\cos\delta\gamma, a\sin\delta\gamma)$ in Cartesian coordinates, respectively. Similar to the above derivation, ITD_p becomes

$$ITD_p(\theta_S, \phi_S, \delta\gamma) = \frac{2a}{c}(\sin\theta_S\cos\phi_S\cos\delta\gamma - \sin\phi_S\sin\delta\gamma). \tag{6.1.6}$$

Letting $\delta\gamma \to 0$, the variation rate or derivative of ITD_p with respect to $\delta\gamma$ is given by

$$\frac{dITD_p(\theta_S, \phi_S)}{d(\delta\gamma)} = -\frac{2a}{c}\sin\phi_S. \tag{6.1.7}$$

In a horizontal plane with $\phi_S = 0°$, ITD_p is invariant against head tilting. As the source departs from the horizontal plane to a high or low elevation, the magnitude of ITD_p variation with head tilting increases. Head tilting provides supplementary information on up-down discrimination because of the odd function characteristic of $\sin\phi_S$. This analysis is the mathematical expression of Wallach's hypothesis and was experimentally validated by Perrett and Noble (1997).

In the case of summing localization with multiple loudspeakers, if M loudspeakers are arranged on a spherical surface with a large radius $r_0 >> a$, then the incident wave near the origin can be approximated as a plane wave. Let (θ_i, ϕ_i) be the direction of the ith loudspeaker and A_i be the normalized amplitude or gain of the corresponding loudspeaker signal. According to Section 3.2.1, for a unit signal waveform in the frequency domain, binaural sound pressures are a linear superposition of the plane wave (far-field) pressures caused by all loudspeakers; for the arbitrary signal waveform $E_A(f)$, the following equation should be multiplied by $E_A(f)$:

$$\begin{aligned} P'_L &= \sum_{i=0}^{M-1} A_i \exp\left[-jk\left(r_0 - a\sin\theta_i\cos\phi_i\right)\right] \\ P'_R &= \sum_{i=0}^{M-1} A_i \exp\left[-jk\left(r_0 + a\sin\theta_i\cos\phi_i\right)\right] \end{aligned}. \tag{6.1.8}$$

Similar to the case of two-channel stereophonic sound in Section 2.1.1, the interaural phase delay difference is evaluated using

$$ITD_{p,SUM} = \frac{\Delta\psi_{SUM}}{2\pi f} = \frac{1}{\pi f}\arctan\left[\frac{\sum_{i=0}^{M-1} A_i \sin(ka\sin\theta_i\cos\phi_i)}{\sum_{i=0}^{M-1} A_i \cos(ka\sin\theta_i\cos\phi_i)}\right]. \tag{6.1.9}$$

ITD_p is the dominant cue for lateral localization at low frequencies; as such, the lateral position of the virtual source for a fixed head is found by comparing Equations (6.1.9) and (6.1.3):

$$\sin\theta_I\cos\phi_I = \frac{1}{ka}\arctan\left[\frac{\sum_{i=0}^{M-1} A_i \sin(ka\sin\theta_i\cos\phi_i)}{\sum_{i=0}^{M-1} A_i \cos(ka\sin\theta_i\cos\phi_i)}\right]. \tag{6.1.10}$$

Equation (6.1.10) indicates that the virtual source direction generally depends on ka or frequency. At very low frequencies with $ka << 1$, Equation (6.1.10) can be expanded as a Taylor series of ka. If the first expansion term is retained, the equation is simplified as

$$\sin\theta_I \cos\phi_I = \frac{\sum_{i=0}^{M-1} A_i \sin\theta_i \cos\phi_i}{\sum_{i=0}^{M-1} A_i}. \tag{6.1.11}$$

In this case, the virtual source direction is independent of ka or frequency.

If the head rotates around the vertical (z) axis anticlockwise with a small azimuth $\delta\theta$, the variation rate $ITD_{p,SUM}$ with $\delta\theta$ can also be evaluated. At very low frequencies with $ka << 1$, the result is

$$\frac{dITD_{p,SUM}}{d(\delta\theta)} = -\frac{2a}{c}\frac{\sum_{i=0}^{M-1} A_i \cos\theta_i \cos\phi_i}{\sum_{i=0}^{M-1} A_i}. \tag{6.1.12}$$

If ITD_p variation caused by head rotation provides information on vertical displacement from a horizontal plane, then comparing Equation (6.1.12) with Equation (6.1.5) yields

$$\cos\theta_I' \cos\phi_I' = \frac{\sum_{i=0}^{M-1} A_i \cos\theta_i \cos\phi_i}{\sum_{i=0}^{M-1} A_i}. \tag{6.1.13}$$

Similarly, if the head tilts around the front-back axis with a small angle $\delta\gamma$, the variation rate $ITD_{p,SUM}$ with $\delta\gamma$ can also be evaluated. If head tilting provides supplementary information on up-down discrimination, then comparing the result of multiple loudspeakers with Equation (6.1.7) at very low frequencies with $ka << 1$ yields

$$\sin\phi_I'' = \frac{\sum_{i=0}^{M-1} A_i \sin\phi_i}{\sum_{i=0}^{M-1} A_i}. \tag{6.1.14}$$

Equations (6.1.11), (6.1.13), and (6.1.14) are a set of summing localization equations for multiple loudspeakers in a three-dimensional space.

In addition, if the head rotates around the vertical axis with an angle $\delta\theta$, $ITD_{p,SUM}$ in Equation (6.1.9) becomes

$$ITD_{p,SUM} = \frac{1}{\pi f}\arctan\left\{\frac{\sum_{i=0}^{M-1} A_i \sin\left[ka\sin(\theta_i - \delta\theta)\cos\phi_i\right]}{\sum_{i=0}^{M-1} A_i \cos\left[ka\sin(\theta_i - \delta\theta)\cos\phi_i\right]}\right\}. \quad (6.1.15)$$

The perceived virtual source direction at low frequencies of $ka << 1$ is expressed in the following equation by selecting the rotation azimuth $\delta\theta$, so that $ITD_{p,\,SUM}$ given in Equation (6.1.15) vanishes:

$$\tan\hat{\theta}_I = \frac{\sum_{i=0}^{M-1} A_i \sin\theta_i \cos\phi_i}{\sum_{i=0}^{M-1} A_i \cos\theta_i \cos\phi_i}. \quad (6.1.16)$$

Equation (6.1.16) only determines that the virtual source is located in the left-right symmetrical vertical plane with respect to the new orientation of the head and then identifies the azimuth $\hat{\theta}_I$ of the virtual source with respect to the fixed coordinate. The virtual source is not certainly located in the horizontal plane with $\phi_I = 0°$, though it may be located at an arbitrary elevation in the left-right symmetrical vertical plane with respect to the new orientation of the head. If Equations (6.1.11), (6.1.13), and (6.1.14) yield values are nearly consistent with $\theta_I \approx \theta'_I$, $\phi_I = \phi'_I = \phi''_{Im}$ and the result of Equation (6.1.13) is nonzero, then Equation (6.1.16) can also be derived by dividing Equation (6.1.11) with Equation (6.1.13).

6.1.2 Velocity and energy localization vector analysis for multichannel spatial surround sound

The velocity and energy localization vector analysis of horizontal sound reproduction in Section 3.2.2 can be extended to the cases of spatial sound reproduction (Gerzon, 1992a). The unit vector of an arbitrary three-dimensional source direction (θ_S,ϕ_S) with respect to the origin of a coordinate can be written as $\hat{r} = [\cos\theta_S\cos\phi_S,\ \sin\theta_S\cos\phi_S,\ \sin\phi_S]^T$. If M loudspeakers are arranged on a spherical surface, the direction of the ith loudspeaker is represented by the unit vector $\hat{r}_i = [\cos\theta_i\cos\phi_i,\ \sin\theta_i\cos\phi_i,\ \sin\phi_i]^T$, and the normalized signal amplitude of the ith loudspeaker is A_i, then the velocity localization vector is evaluated with the following equation as a three-dimensional extension of Equation (3.2.26):

$$r_v = \sum_{i=0}^{M-1} A'_i \hat{r}_i \qquad A'_i = \frac{A_i}{\sum_{i'=0}^{M-1} A_{i'}}. \quad (6.1.17)$$

If the perceived virtual source direction is consistent with the direction of a velocity localization vector (the inner normal direction of the superposed wavefront), then a set of summing localization equations for multiple loudspeakers in a three-dimensional space is derived by decomposing Equation (6.1.17) into three Cartesian components:

$$r_v \cos\theta_v \cos\phi_v = \frac{\sum_{i=0}^{M-1} A_i \cos\theta_i \cos\phi_i}{\sum_{i=0}^{M-1} A_i} \qquad r_v \sin\theta_v \cos\phi_v = \frac{\sum_{i=0}^{M-1} A_i \sin\theta_i \cos\phi_i}{\sum_{i=0}^{M-1} A_i}$$

$$r_v \sin\phi_v = \frac{\sum_{i=0}^{M-1} A_i \sin\phi_i}{\sum_{i=0}^{M-1} A_i} \quad . \tag{6.1.18}$$

In the case of the unit velocity vector magnitude $r_v = 1$, Equation (6.1.18) is equivalent to Equations (6.1.13), (6.1.11), and (6.1.14). A correct interaural phase delay difference in reproduction requires that $r_v = 1$. The closer r_v to a unit, the more stable the virtual source against head rotation. Generally, the velocity vector magnitude is evaluated with the following equation:

$$r_v = \frac{\sqrt{\left(\sum_{i=0}^{M-1} A_i \cos\theta_i \cos\phi_i\right)^2 + \left(\sum_{i=0}^{M-1} A_i \sin\theta_i \cos\phi_i\right)^2 + \left(\sum_{i=0}^{M-1} A_i \sin\phi_i\right)^2}}{\left|\sum_{i=0}^{M-1} A_i\right|}. \tag{6.1.19}$$

Similarly, the perceived virtual source direction at high frequencies is assumed to be just opposite to the direction of sound intensity. Equation (3.2.24) for horizontal summing localization can be extended to the case of a three-dimensional space:

$$r_E \cos\theta_E \cos\phi_E = \frac{\sum_{i=0}^{M-1} A_i^2 \cos\theta_i \cos\phi_i}{\sum_{i=0}^{M-1} A_i^2} \qquad r_E \sin\theta_E \cos\phi_E = \frac{\sum_{i=0}^{M-1} A_i^2 \sin\theta_i \cos\phi_i}{\sum_{i=0}^{M-1} A_i^2}$$

$$r_E \sin\phi_E = \frac{\sum_{i=0}^{M-1} A_i^2 \sin\phi_i}{\sum_{i=0}^{M-1} A_i^2} \quad . \tag{6.1.20}$$

The energy velocity magnitude is evaluated as follows:

$$r_E = \frac{\sqrt{\left(\sum_{i=0}^{M-1} A_i^2 \cos\theta_i \cos\phi_i\right)^2 + \left(\sum_{i=0}^{M-1} A_i^2 \sin\theta_i \cos\phi_i\right)^2 + \left(\sum_{i=0}^{M-1} A_i^2 \sin\phi_i\right)^2}}{\sum_{i=0}^{M-1} A_i^2}. \tag{6.1.21}$$

A single source always yields $r_E = 1$. In the design of mid- and high-frequency signal mixing for a spatial surround sound, r_E may be optimized to as close to a unit as possible.

6.1.3 Discussion on spatial summing localization equations

The summing localization equations in a three-dimensional space are derived in Sections 6.1.2 and 6.1.3. Similar to the case of horizontal summing localization in Section 3.2.3, these equations are based on different physical and psychoacoustic hypotheses and restrictions. They are related to but different from each other; therefore, they are valid under different conditions.

As indicated in the succeeding sections, experimental results proved that appropriate three-dimensional loudspeaker configurations and signal mixing can recreate a summing virtual source in vertical directions. Previous experiments indicated that the high-frequency spectral cue caused by the diffraction of the pinna is important to front-back and vertical localization. However, the analyses in Section 12.2.2 reveal that a high-frequency spectral cue in multiple loudspeaker reproduction does not match with that of the target source. Therefore, the high-frequency spectral cue does not account for vertical summing localization. As stated in Section 1.6.5, auditory localization is a comprehensive consequence of multiple localization cues. The cooperative effects of multiple localization cues enhance accuracy in localization. However, the information provided by various localization cues may be somewhat redundant. When some cues are unavailable, the auditory system may still be able to localize sound to some extent. As stated in Section 1.6.3, Wallach hypothesized that the variations in ITD_p caused by head rotation and tilting provide information on vertical localization, and this hypothesis has been experimentally validated. Equations (6.1.11), (6.1.13), and (6.1.14) are derived on the basis of Wallach's hypothesis; therefore, the dynamic ITD_p variation caused by head turning accounts for the vertical summing localization at low frequencies in multiple loudspeaker reproduction. The examples in Sections 6.2 and 6.4 can be regarded as the experimental validation of Equations (6.1.11), (6.1.13), and (6.1.14). Some experiments have demonstrated that subjects can discriminate the sound source in up-and-down directions even for a fixed head (Perrett and Noble, 1997). The scattering and diffraction effects of the torso and shoulders may provide supplementary information on up-down discrimination at low and mid frequencies. The same experiments have also shown that increasing the dynamic cues further improves the up-down discrimination.

Similar to the cases of horizontal surround sound in Section 3.2.3, if Equations (6.1.11), (6.1.13), and (6.1.14) yield consistent results, the perceived virtual source direction can be identified uniquely, and the perceived direction is stable as the head turns. If the three equations reveal inconsistent results, Equation (6.1.11) based on ITD_p can be reasonably used to identify the lateral displacement of the virtual source (cone of confusion).Equation (6.1.13) based on head rotation and Equation (6.1.14) based on head tilting can be comprehensively utilized to identify the virtual source position in the cone of confusion. If the difference in the results of the three equations is obvious, the virtual source is unstable during head turning. This situation should be avoided in a practical signal mixing design. The analyses in Sections6.2 and 6.4 are based on the logic and consideration presented here.

Equations (6.1.11), (6.1.13), and (6.1.14) are only valid at low frequencies. For wideband stimuli, vertical localization is the comprehensive consequence of dynamic cues at low frequencies, spectral cues caused by the head and pinna at high frequencies, and even spectral cues caused by the torso and shoulders at mid and low frequencies. In practical multichannel spatial sound reproduction, if dynamic and spectral cues provide consistent information, a summing virtual source is accurate and stable. If two cues or a cue in different frequency bands provide inconsistent or even conflicting information, the auditory system may utilize dominant or some partly consistent cues to localization, but the virtual source becomes blurry. Moreover, excessive conflicting information may lead to splitting virtual sources in different directions at different frequencies or auditory events with an uncertain position, depending on the power spectra of the stimuli. Strict analysis should comprehensively consider various localization cues and should be based on the mechanisms of auditory information processing by the high-level neural system. Unfortunately, this analysis is currently not feasible. Therefore, the aforementioned localization theorem and equations for spatial surround sound are incomplete. In some instances, they may qualitatively interpret some experimental results. In other instances, they fail to do so. However, the succeeding discussions in this chapter indicate that the results of these localization equations are approximately or qualitatively consistent with the experimental findings in most cases. Therefore, in the absence of other strict means of analysis, the aforementioned localization theorem and equations are still helpful to the practical analysis and design of multichannel spatial surround sound.

Equation (6.1.16) is derived by considering head rotation to an orientation so that $ITD_{p,SUM}$ vanishes. It is related but not completely equivalent to the ITD_p variation caused by head rotation.

The velocity localization vector-based Equation (6.1.18) and energy localization vector-based Equation (6.1.20) are similar to the cases of horizontal reproduction in Section 3.2.3. In the case of $r_v = 1$, Equation (6.1.18) is equivalent to Equations (6.1.11), (6.1.13), and (6.1.14). In general cases, the velocity localization vector-based direction (Makita hypothesis) is not always the practical perceived direction. The mid- and high-frequency virtual source direction given by Equation (6.1.20) is also a rough approximation. Other features of the aforementioned virtual source localization theorems for spatial surround sound (such as the applicable frequency range) are similar to those for horizontal surround sound in Section 3.2.3 and are omitted here.

6.1.4 Relationship with the horizontal summing localization equations

The relationship between the spatial and horizontal summing localization equations is analyzed in this section. A special summing spatial auditory phenomenon, i.e., the virtual source-elevated effect in horizontal loudspeaker arrangement, is addressed.

If all loudspeakers are arranged in the horizontal plane with $\phi_i = 0°$, Equations (6.1.11), (6.1.13), and (6.1.14) yield the following results.

In terms of ITD_p for a fixed head orientation,

$$\sin\theta_I \cos\phi_I = \frac{\sum_{i=0}^{M-1} A_i \sin\theta_i}{\sum_{i=0}^{M-1} A_i}. \tag{6.1.22}$$

In terms of the variation rate of ITD_p caused by head rotation around the vertical axis,

$$\cos\theta_I' \cos\phi_I' = \frac{\sum_{i=0}^{M-1} A_i \cos\theta_i}{\sum_{i=0}^{M-1} A_i}. \tag{6.1.23}$$

In terms of the variation rate of ITD_p caused by head tilting around the front-back axis,

$$\sin\phi_I'' = 0°. \tag{6.1.24}$$

If the three equations above yield consistent results, i.e., a fixed head orientation, head rotation, and head tilting provide consistent low-frequency localization information, the solutions of the three equations satisfy $\phi_I = \phi'_I = \phi''_I = 0°and\theta_I = \theta'_I$. In this case, Equation (6.1.22) is identical to summing localization Equation (3.2.7) in the horizontal plane. In addition, dividing Equation (6.1.22) by Equation (6.1.23) yields Equation (3.2.9) and leads to $\hat{\theta}_I = \theta_I$. In this case, as stated in Section 3.2.3, combining Equations (3.2.7) and (3.2.9) is appropriate to examine the horizontal summing localization. The analysis of horizontal Ambisonics in Chapter 4 is based on this consideration.

If Equations (6.1.22), (6.1.23), and (6.1.24) yield inconsistent results, i.e., a fixed head orientation, head rotation, and head tilting provide inconsistent low-frequency localization information, an appropriate solution that is approximately consistent with experimental results should be obtained by considering the psychoacoustic principle of auditory localization. Low-frequency ITD_P is a dominant cue for lateral localization; as such, Equation (6.1.22) should be initially used to identify the cone of confusion in which the virtual source is located. However, the two situations or hypotheses may occur, depending on which vertical or elevation localization information among the inconsistent information provided by Equations (6.1.23) and (6.1.24) is dominant.

The first situation or hypothesis considers the result of Equation (6.1.24). According to this equation, the dynamic ITD_P variation caused by head tilting around the front-back axis provides information that the virtual source is located in the horizontal plane. If this information is basically consistent with other vertical localization information (such as spectral information), the auditory system may identify the vertical position based on the more consistent information. Therefore, the perceived virtual source is located in the horizontal plane. In this case, $\phi_I = \phi'_I = \phi''_I = 0°$, and Equations (6.1.22) and (6.1.23) become

$$\sin\theta_I = \frac{\sum_{i=0}^{M-1} A_i \sin\theta_i}{\sum_{i=0}^{M-1} A_i}, \tag{6.1.25}$$

and

$$\cos\theta_I' = \frac{\sum_{i=0}^{M-1} A_i \cos\theta_i}{\sum_{i=0}^{M-1} A_i}. \tag{6.1.26}$$

If the qualitative results of Equations (6.1.25) and (6.1.26) are consistent, the horizontal azimuth of the virtual source is determined using Equation (6.1.25), which is based on ITD_P. The front-back ambiguity is resolved with Equation (6.1.26), which is based on ITD_P variation caused by the head rotation around the vertical axis.

In the second situation or hypothesis, if the comprehensive information of the dynamic ITD_P variation caused by the head rotation around the vertical axis and other cues (such as spectral cues) is considered, the perceived virtual source is not always located in the horizontal plane. In this case, the result of Equation (6.1.24) should be discarded, and the perceived virtual direction is determined via a combination of Equations (6.1.22) and (6.1.23).

The first hypothesis is used in the analysis of the pair-wise amplitude panning for a horizontal loudspeaker configuration in Section 3.2.3 and in Chapters 4 and 5. Actual horizontal loudspeaker configurations and signal mixing are unable to recreate the exact spectral cue of a horizontal target source at high frequencies, but they may not obviously conflict with that of the horizontal target source. As such, Equation (3.2.9), based on the head oriented to the virtual source, is used to replace Equation (6.1.26) for the analysis of summing localization with a horizontal loudspeaker configuration in Chapters 4 and 5 and to evade the second situation. Consequently, the results basically match with experimental findings in most cases, but they may exhibit some limitations.

In practical horizontal loudspeaker reproduction, the two aforementioned situations may occur depending on the competition and coordination among different localization cues. The final results may depend on various related conditions, such as the head-turning mechanism and stimulus characteristics (such as spectra).

The virtual source-elevated effect in a horizontal loudspeaker arrangement is analyzed to illustrate the second situation. For a two-channel stereophonic loudspeaker arrangement with a large span angle and an identical signal amplitude, the perceived virtual source is located at a high elevation, especially as the head rotates around the vertical axis. A similar phenomenon occurs in reproduction with two rear loudspeakers with a large span angle (such as two surround loudspeakers in a 5.1-channel configuration). This virtual source-elevated effect was observed in early studies on stereophonic sound (Boer, 1947). By the end of the 1950s, Leakey (1959)interpreted this effect via ITD variation caused by head rotation. Some studies have provided experimental results with additional details (Lee, 2017).

A pair of stereophonic loudspeakers are arranged in a horizontal plane with elevations of $\phi_L = \phi_R = 0°$ and azimuths of $\theta_L = \theta_0$ and $\theta_R = -\theta_0$. The amplitudes of two loudspeaker signals are identical to $A_L = A_R$. For a fixed head, the law of sine in Equation (2.1.6) predicts that the low-frequency virtual source is located at the horizontal front direction of $\theta_I = 0°$ and $\phi_I = 0°$. However, Equations (6.1.22), (6.1.23), and (6.1.24) yield

$$\sin\theta_I \cos\phi_I = 0, \tag{6.1.27}$$

$$\cos\theta_I' \cos\phi_I' = \cos\theta_0, \tag{6.1.28}$$

and

$$\sin\phi_I'' = 0. \tag{6.1.29}$$

The three equations indicate the following results:

1. Equation (6.1.27) yields $\sin\theta_I = 0$ or $\cos\phi_I = 0$, $\theta_I = 0°$ or $180°$, or $\phi_I = \pm 90°$. Therefore, for a fixed head oriented to the front, $ITD_p = 0$, and the virtual is located in the median plane.
2. If the virtual source is located at the frontal-median plane, then $\cos\theta'_I = 1$, Equation (6.1.28) yields $\cos\phi'_I = \cos\theta_0$. Therefore, during the head rotation around the vertical

axis, the dynamic ITD_p variation provides information that the virtual source deviates from the direct front to a high or low elevation in the median plane. This deviation increases as the half-span angle θ_0 of the loudspeaker pair increases. For a pair of horizontal loudspeakers arranged at the two sides with $\pm\theta_0 = \pm 90°$, Equation (6.1.28) yields $\cos\phi'_I = 0$, and the perceived virtual source is located at the top (or bottom) direction.

3. The result of Equation (6.1.29) is inconsistent with that of Equation (6.1.28), indicating that the head tilting around the front-back axis fails to provide consistent information for up-down discrimination.

The virtual source-elevated effect also occurs in the case of regular horizontal loudspeaker configurations with more loudspeakers (such as four) and identical signal amplitudes. In this case, ITD_p and its dynamic variation with head rotation on a vertical axis vanish, and this observation only matches with a real source in the top or bottom direction. However, Equation (6.1.24) provides an inconsistent or conflicting information on $\sin\phi''_I = 0$ and fails to yield consistent information on up-down discrimination. This inconsistency reveals the limitation of the analysis in this section. On the other hand, the virtual source-elevated effect is applicable to recreate the perception of "flying over" with a horizontal loudspeaker configuration (Jot et al., 1999). A similar method is also applicable to an irregular loudspeaker configuration in a horizontal plane.

The above example indicates that the experimental results of a virtual source elevated in a horizontal loudspeaker arrangement can be partly and qualitatively interpreted by the dynamic cue caused by head turning. Some quantitative differences may be observed between the theoretical and experimental results, but the above analysis is at least qualitatively consistent with Wallach's hypothesis, that is, the head rotation around the vertical axis provides information on the vertical displacement of a sound source from the horizontal plane. The virtual source elevated in a horizontal loudspeaker arrangement may be a comprehensive consequence of multiple cues, especially for wideband stimuli. For example, the scattering and diffraction of the torso do not provide information that the virtual source is located below the horizontal plane. The auditory system has adapted to a natural environment with sound sources on the top rather than below directions. However, the effect of multiple localization cues may cause a quantitative difference between the practical perceived directions and those predicted by Equations (6.1.27), (6.1.28), and (6.1.29). Moreover, other studies have interpreted the virtual source-elevated effect with a spectral cue or an interchannel crosstalk (Lee, 2017). Therefore, the analysis in this section is partly based on some hypotheses and thus incomplete; more psychoacoustic validations and corrections are required.

6.2 SIGNAL MIXING METHODS FOR A PAIR OF VERTICAL LOUDSPEAKERS IN THE MEDIAN AND SAGITTAL PLANE

The summing localization and other spatial auditory perceptions with a pair of loudspeakers in the horizontal plane are analyzed in Section 1.7.1 and in Chapters 2 to 5. They include recreating a virtual source between loudspeakers with pair-wise amplitude and time panning and recreating spatial auditory perceptions with decorrelated loudspeaker signals. These methods are applicable to two-channel stereophonic and horizontal surround sound reproduction. In multichannel spatial surround sound, a virtual source should also be recreated in the median plane and other sagittal planes with a pair of adjacent loudspeakers in different elevations. The feasibility of this method is analyzed in the following section. A comparison between the analysis and experiment also validates or reveals the limitations of the localization theorem in Section 6.1.

The virtual source localization equations in Section 6.1.1 are used to analyze the summing localization in the median plane at low frequencies (Rao and Xie, 2005; Xie and Rao, 2015; Xie et al., 2019). If all loudspeakers are arranged in the median plane with $\theta_i = 0°$ or $180°$, then $\sin\theta_i = 0$ and $i = 0,1...(M-1)$. Equation (6.1.9) or (6.1.11) yields the interaural phase delay difference $ITD_{p,SUM} = 0$, and the summing virtual source (if it exists) is located in the median plane with its perceived azimuth satisfying $\theta_I = 0°$ or $180°$. Substituting this result into Equation (6.1.13) and letting $\theta_I = \theta'_I$ yield

$$\cos\phi'_I = \pm\frac{\sum_{i=0}^{M-1} A_i \cos\theta_i \cos\phi_i}{\sum_{i=0}^{M-1} A_i}. \tag{6.2.1}$$

When the virtual source is located in the frontal-median plane with $\theta_I = 0°$ or when the variation rate of $ITD_{p,SUM}$ with head rotation calculated from Equation (6.1.12) is negative, the positive sign should be chosen in Equation (6.2.1). Otherwise, the negative sign should be selected in Equation (6.2.1). According to Wallach's hypothesis, Equation (6.2.1)is used to evaluate the elevation angle of the summing virtual source in the median plane quantitatively, and Equation (6.1.14) is applied to resolve the up-down ambiguity.

Equation (6.2.1) is used to analyze the examples of summing localization with two loudspeakers in the median plane. As shown in Figure 6.2, five loudspeakers numbered 0, 1, 2, 3, and 4 are arranged in the median plane. Loudspeakers 0, 1, and 2 are arranged in the frontal-median plane with an azimuth of $\theta_i = 0°$ and elevations of $\phi_0 = -\phi_2$, $\phi_1 = 0°$, and $0° < \phi_2 < 90°$, respectively. Loudspeaker 3 is arranged on the top with an azimuth of $\theta_3 = 0°$ and an elevation of $\phi_3 = 90°$. Loudspeaker 4 is arranged in the rear-median plane with an azimuth of $\theta_4 = 180°$and an elevation of $\phi_4 = \phi_2$. The normalized amplitudes of loudspeaker signals are denoted by A_0, A_1, A_2, A_3, and A_4. For pair-wise amplitude panning between loudspeakers 1 and 2, A_0, A_3, and A_4 vanish. In this case, Equation (6.1.14) yields $\sin\phi''_I \geq 0$, and the virtual source is located in the upper half of the median plane. The variation rate of $ITD_{p,SUM}$ with

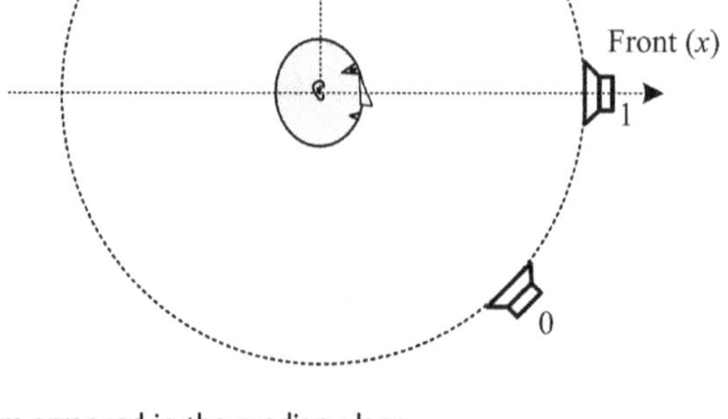

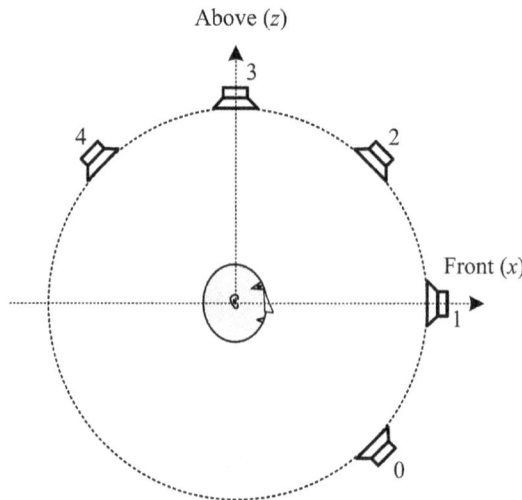

Figure 6.2 Five loudspeakers arranged in the median plane.

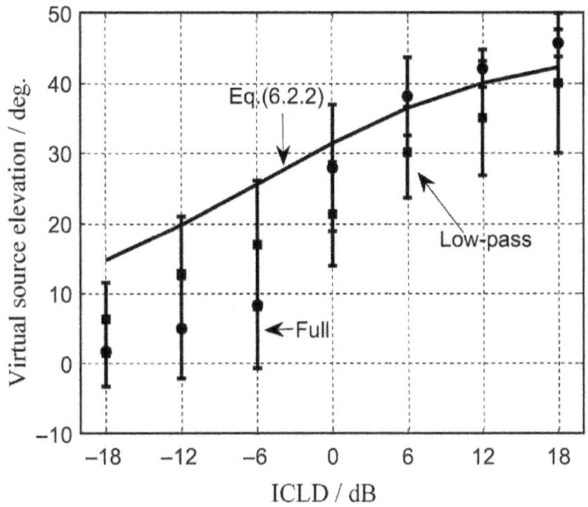

Figure 6.3 Elevation of a virtual source from the calculation and experiment for pair-wise amplitude panning between loudspeakers 1 and 2.

the head rotation calculated from Equation (6.1.12) is negative. Accordingly, the elevation of the virtual source in the frontal-median plane is evaluated from Equation (6.2.1) as

$$\cos\phi_I' = \frac{A_1 + A_2\cos\phi_2}{A_1 + A_2} = \frac{1 + A_2/A_1\cos\phi_2}{1 + A_2/A_1}. \tag{6.2.2}$$

According to Equation (6.2.2), when the interchannel level difference (ICLD) $d_{21} = 20\log_{10}(A_2/A_1)$ changes from $-\infty$ dB to $+\infty$ dB, the elevation of the virtual source varies from 0° to ϕ_2. In this case, the panning can recreate a virtual source between the two adjacent loudspeakers. Figure 6.3 illustrates the result calculated from Equation (6.2.2) with $\phi_2 = 45°$.

Similar analyses indicate that pair-wise amplitude panning is also able to recreate a virtual source between two adjacent loudspeakers 0 and 1 or 2 and 3. By contrast, for pair-wise amplitude panning between loudspeakers 0 and 2, Equation (6.2.1) yields

$$\cos\phi_I' = \cos\phi_2 = \cos\phi_0. \tag{6.2.3}$$

Therefore, for a pair of up–down symmetrical loudspeakers, the virtual source is located in the direction of either loudspeaker 0 or 2 in spite of the interchannel level difference $d_{20} = 20\log_{10}(A_2/A_0)$, i.e., pair-wise amplitude panning fails to recreate a virtual source between two loudspeakers. This result is similar to the case when a pair of lateral loudspeakers with a front-back symmetrical arrangement fails to recreate a stable lateral virtual source (Section 4.1.2).

To analyze the pair-wise amplitude panning between loudspeakers 2 and 4, which have a front-back symmetrical arrangement, selecting a new elevation angle of $-180° < \varphi \leq 180°$ in the median plane is convenient. As shown in Figure 6.4, the new elevation angle is related to the default angle when $\theta = 0°$ and $\varphi = 90° - \phi$ and when $\theta = 180°$ and $\varphi = -90° + \phi$. Therefore, the top, front, and back directions are denoted by $\varphi = 0°$, 90°, and −90°, respectively. As shown in Figure 6.4, the span angle between two loudspeakers is $2\varphi_2 = 2(90° - \phi_2)$, where the elevation angles of the front and back loudspeakers are φ_2 and $\varphi_4 = -\varphi_2$, respectively. For the normalized loudspeaker signal amplitudes A_2 and A_4, Equation (6.2.1) yields

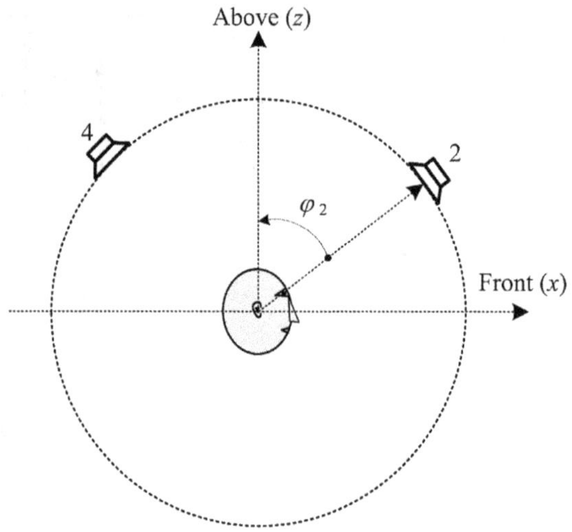

Figure 6.4 Pair of loudspeakers with a front-back symmetrical arrangement in the median plane.

$$\sin\varphi_I' = \frac{A_2 - A_4}{A_2 + A_4}\sin\varphi_2 = \frac{A_2/A_4 - 1}{A_2/A_4 + 1}\sin\varphi_2. \qquad (6.2.4)$$

The form of Equation (6.2.4) is similar to the law of sine for stereophonic reproduction in the front horizontal plane described in Equation (2.1.6), resulting in a similar variation pattern of a virtual source angle with an interchannel level difference. Therefore, pair-wise amplitude panning can recreate a virtual source between a pair of loudspeakers with a front-back symmetrical arrangement in the median plane. For a half-span angle of $\varphi_2 = 45°$ between two loudspeakers, Figure 6.5 illustrates the variation in the perceived elevation angle φ'_I with ICLD = d_{24}= 20 $\log_{10}(A_2 / A_4)$ (dB). For $d_{24} \to +\infty$, $d_{24} = 0$ and $d_{24} \to -\infty$, φ'_I are 45°, 0°, and −45°, respectively.

The physical origins of Equation (6.2.4) are different from those of law of sine for stereophonic reproduction. The law of sine for stereophonic reproduction is derived from ITD_p,

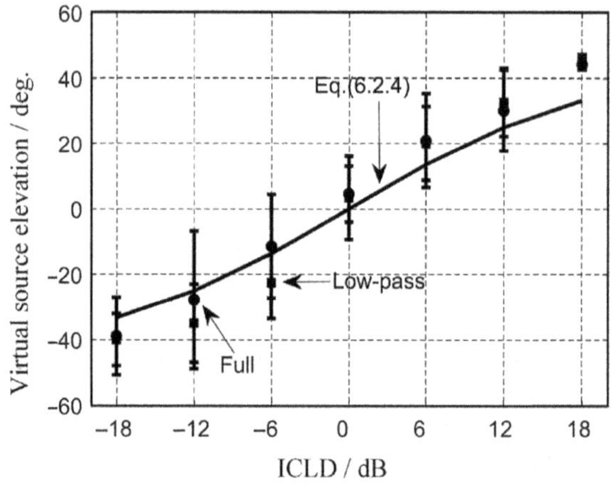

Figure 6.5 Elevation of a virtual source from the calculation and experiment for pair-wise amplitude panning between a pair of front-back symmetrical loudspeakers 2 and 4. In the figure, solid circle and square represent the results of full-bandwidth and low-pass filtered pink noise, respectively.

which is a dominant cue for azimuthal localization at low frequencies. Equation (6.2.4) is derived from the variation in ITD_p caused by head rotation, which is supposed to be a cue of elevation localization at low frequencies.

Figures 6.3 and 6.5 also illustrate the results of virtual source localization experiments, including the means and standard deviations from eight subjects. Pink noise with a full-audible bandwidth and a 500 Hz low-pass filtered bandwidth are used as stimuli. The experimental results exhibit a tendency similar to that of the analysis. However, in Figure 6.3, for pink noise with a full-audible bandwidth and ICLD $d_{21} = -6$ dB, the elevation angle from the experimental result is closer to the direction of the front loudspeaker 1 than that from the analysis. This inconsistency may be attributed to the following reasons:

1. The accuracy of elevation localization is limited compared with that of azimuthal localization.
2. The approximate (shadowless) head model is used in the analysis.
3. The spectral cue in amplitude panning deviates from that of the target (real) source. The mismatched spectral cue may influence the localizations.
4. The localization information provided by head rotation is inconsistent with that given by tilting for pair-wise amplitude panning.

These reasons should be further examined experimentally.

For a pair of loudspeakers 0 and 2 with an up-down symmetrical arrangement in the median plane, the virtual source jumps from the direction of one loudspeaker to another when the ICLD changes. This experimental result is also consistent with that of theoretical analysis. In addition, the standard deviation of the experimental results in Figures 6.3 and 6.5 is larger than those of usual localization experiments in the horizontal plane. These standard deviations are larger probably because pair-wise amplitude panning with two loudspeakers in the median plane only provides information on dynamic ITD_p variation for vertical localization. The perceived virtual source becomes blurry (for stimuli with a full-audible bandwidth) because of incorrect spectral information at high frequencies.

Some studies have explored the feasibility of recreating a virtual source between two loudspeakers in the median plane to develop spatial surround sound. The results vary across works. Pulkki (2001a) experimentally demonstrated that pair-wise amplitude panning fails to recreate virtual sources between two loudspeakers at $\phi = -15°$ and $30°$ in the median plane for pink-noise or narrow-band stimuli. This finding is similar to the case of pair-wise amplitude panning between loudspeakers 0 and 2 shown in Figure 6.2, although the loudspeaker configuration in Pulkki's experiment was not completely symmetrical in the up-and-down direction. An analysis similar to that in Equation (6.2.3) can interpret Pulkki's experimental results.

Wendt et al. (2014) conducted a virtual source localization experiment with a pair of loudspeakers at an elevation of $\phi = \pm 20°$ in the median plane. They used pink noise as a stimulus and set ICLD to 0, ±3, and ±6 dB. During the experiment, they instructed their subjects to keep their head immobile to the frontal orientation. The results indicated that the perceived elevation of a virtual source can be controlled by ICLD, but the quantitative results are obviously subject-dependent. Therefore, reproduction fails to create consistent localization cues.

In the development of advanced multichannel spatial sound, a series of virtual source localization experiments is also conducted to investigate the summing localization in the median plane (ITU-R Report, BS 2159-7, 2015). The experimental results from the ITU-R Report indicate that pair-wise amplitude panning for white noise stimuli can recreate a virtual source between two loudspeakers at $\phi = 0°$ and $30°$ or $0°$ and $-30°$ in the median plane; however, it fails to recreate a virtual source between two loudspeakers at $\phi = -30°$ and $30°$. The results of the aforementioned analysis are also consistent with those in the ITU-R Report.

Lee (2014) investigated the perceived virtual source movement caused by ICLD in the median plane. They arranged two loudspeakers at elevations of 0° and 30° in the frontal-median plane and used the signals of a cello and a bongo as stimuli. The results indicated that an ICLD of 6–7 dB is enough to make the virtual source fully move in the direction of either loudspeaker. An ICLD of 9–10 dB is sufficient to completely mask the crosstalk from the loudspeaker with a weaker signal so that the crosstalk is inaudible. The thresholds of the full movement and masking are lower than those of the horizontal stereophonic loudspeaker configuration (Section 1.7.1).

However, Barbour (2003) experimentally revealed that pair-wise amplitude panning for pink noise and speech stimuli fails to recreate a virtual source between two loudspeakers with a front-back symmetrical arrangement in the median plane with ($\theta = 0°$, $\phi_2 = 60°$) and ($\theta = 180°$, $\phi_4 = 60°$), or $\varphi_{2,4} = \pm 30°$ for the new elevation angle in Figure 6.4. It is also unable to recreate a virtual source between two loudspeakers arranged in the median plane with one at $\phi = 0°$ and another at $\phi = 45°$, $60°$, or $90°$. Therefore, Barbour's results differ from those of the analysis of Equation (6.2.4) and those of the aforementioned experiments possibly because the dynamic cue caused by head-turning might not be fully utilized in Barbour's experiment. Actually, as stated in Section 1.6, for wideband stimuli, dynamic and spectral cues contribute to vertical localization. However, a pair of loudspeakers in the median plane fails to synthesize the correct or desired spectral cues above the frequency of 5–6 kHz (see the discussion in Section 12.2.2). Therefore, summing localization is impossible if a dynamic cue is not fully utilized. In other words, summing localization in the median plane is caused by the dynamic cue rather than the spectral cue. However, a mismatched spectral cue may degrade the perceived quality of a virtual source, making the virtual source blurry and unstable.

Equation (6.2.4) is based on Wallach's hypothesis that head rotation provides dynamic information on vertical localization. As stated in Section 1.6.3, since Wallach (1940) proposed the hypotheses on front-back and vertical localization, the contribution of head rotation to front-back discrimination has been verified via numerous experiments. However, few appropriate experiments have been performed to validate the contribution of head-turning to vertical localization because completely and experimentally excluding contributions from other vertical localization cues is difficult. Therefore, the analysis and experiments presented in this section can be regarded as a quantitative validation of Wallach's hypothesis on vertical localization (Rao and Xie, 2005; Xie et al., 2019), e.g., head rotation provides information on source vertical displacement from the horizontal plane, and head tilting yields additional information on up-down discrimination. The analysis in this section aims to validate Wallach's hypothesis on vertical localization. The analysis and experiment in this section are also a validation of the summing localization equations derived in Section 6.1.1.

In summary, the equations derived in Section 6.1.1 are applicable to the analysis of the summing localization in the median plane. The analysis and experiments indicate that a pair-wise amplitude panning for appropriate two-loudspeaker configurations in the median plane can recreate a virtual source between loudspeakers, but the position of a virtual source may be blurry. However, for some other up-down symmetrical loudspeaker configuration, a pair-wise amplitude panning fails to recreate a virtual source between loudspeakers. These results are applicable to designs of loudspeaker configurations in the median plane. Notably, the results are appropriate for stimuli with dominant low-frequency components. For stimuli with dominant high-frequency components, the results are different.

This analysis can be extended to the vertical summing localization in sagittal planes (cone of confusion) other than the median plane (Xie et al., 2017b). For example, by using pair-wise amplitude panning, a pair of loudspeakers with an up-down symmetrical arrangement at (θ, ϕ) and (θ, $-\phi$) fails to recreate a virtual source between them. By contrast, a pair of loudspeakers arranged asymmetrically at (θ, ϕ) and (θ, 0°) can recreate a virtual source between them.

The vertical summing localization by interchannel time difference (ICTD) or pair-wise time panning has also been explored in some studies. Lee (2014) conducted an experiment with cello and bongo stimuli and indicated that pair-wise time panning fails to recreate a virtual source between two vertical loudspeakers, especially in the median plane. Tregonning and Martin (2015) experimentally demonstrated that ICTD between the signals of horizontal and elevated loudspeakers changes the perceived elevation, but it never makes the virtual source fully move in the direction of either loudspeaker. For speech and conga stimuli, an ICTD of 5 ms leads to a maximal vertical movement of the virtual source. As ICTD further increases, the width and vertical spread of the virtual source increase.

Litovsky et al. (1999) experimentally investigated the precedence effect of two sound sources (loudspeakers) in the median plane and obtained results that vary from other studies. They indicated that the delay for localization dominance in the median plane is similar to that in the horizontal plane, but the precedence effect is weaker in the median plane than in the horizontal plane. Conversely, Wallis and Lee (2014)observed that the precedence effect does not operate in the median plane. In addition, decorrelated signals are less effective for creating an auditory event with a vertical spread than for creating an auditory event with a horizontal extension (Gribben and Lee, 2014).

Overall, cues for horizontal and vertical auditory perceptions differ; as such, rules for summing spatial auditory events and perceptions in horizontal and vertical directions also differ. Moreover, different studies may yield various results, which may depend on the characteristics of stimuli. Hence, further studies are needed.

6.3 VECTOR BASE AMPLITUDE PANNING

The reproduced sound field and perceived effect of a multichannel sound depend on loudspeaker configuration and signal mixing. Pair-wise amplitude panning is a common signal mixing method and is applicable to a pair of adjacent loudspeakers in the horizontal, median, and sagittal planes to recreate a virtual source between loudspeakers. Other signal mixing methods are used to recreate a virtual source in a three-dimensional space. Pulkki (1997) proposed a ***vector base amplitude panning*** (***VBAP***) method. It is another typical and general signal mixing method that can be regarded as an extension of pair-wise amplitude panning in the case of three-dimensional loudspeaker configurations.

VBAP can be conveniently analyzed in terms of the velocity localization vector discussed in Section 6.1.2. The origin of coordinates is located at a spherical center and coincident with the head center. A spherical surface is divided by spherical–triangular grids. M loudspeakers are arranged in the vertices of the grid. The direction (θ_i,ϕ_i) of the ith loudspeaker is represented by a unit vector $\hat{r}_i = \left[\cos\theta_i \cos\phi_i,\ \sin\theta_i \cos\phi_i,\ \sin\phi_i\right]^T$ pointing from the origin to the loudspeaker. The target source direction Ω_S = (θ_S,ϕ_S) is represented by a unit vector $\hat{r}_S = \left[\cos\theta_S \cos\phi_S,\ \sin\theta_S \cos\phi_S,\ \sin\phi_S\right]^T$. The virtual source within a spherical triangle is recreated by the loudspeakers forming the triangle. Therefore, a virtual source at different directions may be recreated by loudspeakers in different spherical triangles. This feature is similar to that of pair-wise amplitude panning in a horizontal plane.

Figure 6.6 illustrates the principle of VBAP. The unit vectors of three active loudspeakers are $\hat{r}_1$, $\hat{r}_2$, and $\hat{r}_3$. The amplitudes A_1, A_2, and A_3 of three loudspeaker signals are real and non-negative values (in phase). According to the discussion in Sections 3.2.2 and 6.1.2 and Equation (6.1.17), the linear combination of $\hat{r}_1$, $\hat{r}_2$, and $\hat{r}_3$ with weights of A_1, A_2, and A_3 is

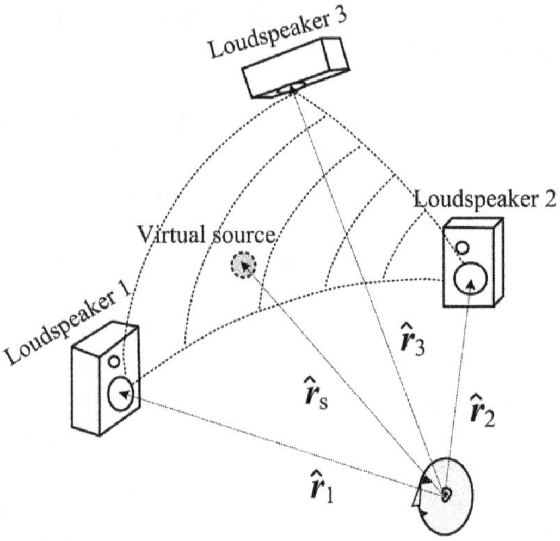

Figure 6.6 Principle of VBAP. (Pulkki, 1997, with permission of the Audio Engineering Society.)

the velocity localization vector r_v in reproduction, which should be matched with the target virtual source direction (Makita's hypothesis), then

$$\hat{r}_S = A_1 \hat{r}_1 + A_2 \hat{r}_2 + A_3 \hat{r}_3. \tag{6.3.1}$$

Equation (6.3.1) can be written in a matrix form as follows:

$$\hat{r}_S = \left[\hat{r}_1, \hat{r}_2, \hat{r}_3\right] \boldsymbol{A}, \tag{6.3.2}$$

where

$$\hat{r}_S = \left[\cos\theta_S \cos\phi_S,\ \sin\theta_S \cos\phi_S,\ \sin\phi_S\right]^T \qquad \boldsymbol{A} = \left[A_1,\ A_2,\ A_3\right]^T$$

$$\left[\hat{r}_1, \hat{r}_2, \hat{r}_3\right] = \begin{bmatrix} \cos\theta_1 \cos\phi_1 & \cos\theta_2 \cos\phi_2 & \cos\theta_3 \cos\phi_3 \\ \sin\theta_1 \cos\phi_1 & \sin\theta_2 \cos\phi_2 & \sin\theta_3 \cos\phi_3 \\ \sin\phi_1 & \sin\phi_2 & \sin\phi_3 \end{bmatrix}. \tag{6.3.3}$$

The loudspeaker signal amplitudes can be solved from Equation (6.3.2) as

$$\boldsymbol{A} = \left[\hat{r}_1, \hat{r}_2, \hat{r}_3\right]^{-1} \hat{r}_S. \tag{6.3.4}$$

The directions of the three loudspeakers are unparallel; as such, $\hat{r}_1$, $\hat{r}_2$, and $\hat{r}_3$ are not linearly correlated, and the inverse matrix in Equation (6.3.4) certainly exists. As expected, Equation (6.3.4)proves that a virtual source is recreated only by the corresponding loudspeakers if a target source is located in the same direction as any of the loudspeakers. If the target source is located in an arc between two adjacent loudspeakers, the virtual source is recreated by the two loudspeakers, and the signal for the other loudspeaker vanishes. If the target source is located between two adjacent loudspeakers in a horizontal plane, VBAP is simplified into horizontal pair-wise amplitude panning.

The resultant signal amplitudes A_1, A_2, and A_3 should be normalized further. The normalized amplitudes of actual loudspeaker signals are A_1, A_2, and A_3 multiplying A_{total}. For constant-power normalization, the following equation is obtained:

$$A_{total} = \frac{1}{\sqrt{A_1^2 + A_2^2 + A_3^2}}, \tag{6.3.5}$$

The aforementioned discussion is the basic principle of VBAP.

VBAP satisfies the low-frequency optimized criterion of the direction of the velocity localization vector. However, it does not satisfy the optimized criterion of the unit velocity vector magnitude $r_v = 1$ except the target virtual source located in the loudspeaker direction. If the head rotates to an orientation so that ITD_p vanishes, the lateral displacement of the perceived virtual source with respect to the fixed coordinate is identical to that of the target source. However, for a fixed head with a frontal orientation, the lateral displacement of the perceived virtual source may differ from that of the target source because of the mismatched ITD_p in reproduction. Pulkki (2001a) also validated this result by using a binaural auditory model. In addition, the vertical displacement of the perceived virtual source may differ from that of the target source because of the mismatched variation in ITD_p caused by head rotation. The accuracy of the perceived virtual source direction in reproduction depends on the positions of three active loudspeakers on a spherical surface and can be analyzed using the localization equations in Section 6.1.1. The situation here is similar to the cases of pair-wise amplitude panning in a horizontal plane (Section 4.1.2) or in a median plane (Section 6.2).

Four active loudspeakers with at least one out-of-phase loudspeaker signal are needed to further satisfy the optimized criterion of $r_v = 1$. The function of an out-of-phase loudspeaker signal is similar to the cases of horizontal reproduction described in Sections 3.2.2 and 4.1.3. Out-of-phase loudspeaker signals are used in global or local Ambisonic signal mixing for horizontal and spatial reproduction, as discussed in Sections 4.3, 5.2.3, 5.2.4, and Section 6.4. The analysis here also reveals the difference between VBAP and Ambisonic signal mixing.

VBAP employs three active loudspeakers to recreate a virtual source within a spherical triangle formed by loudspeakers. Therefore, the error of the perceived virtual source direction is within the region bounded by active loudspeakers despite the position of a listener. The crosstalk from loudspeakers in opposite directions and a serious localization error in off-central listening position are averted. As the total number of loudspeakers increases, the active loudspeaker triangle grid becomes dense, and the localization error is further reduced. VBAP is simple and easily implemented in program production. Therefore, it possesses some advantages in practical uses. Some experiments on VBAP with various loudspeaker configurations have validated the aforementioned analysis (Pulkki, 2001a; Wendt et al., 2014).

6.4 SPATIAL AMBISONIC SIGNAL MIXING AND REPRODUCTION

6.4.1 Principle of spatial Ambisonics

Spatial Ambisonics is another typical multichannel spatial sound and signal mixing method (Gerzon, 1973). It was developed in the 1970s and originally termed Periphony. Currently, it is a hot topic in spatial sound. As an extension of horizontal Ambisonics, spatial Ambisonics involves the decomposition and reconstruction of a sound field by a series of directional harmonics (spherical harmonic functions). Similar to horizontal Ambisonics, spatial Ambisonics can be analyzed with various mathematical and physical methods. A traditional analysis based on a virtual source localization theorem is addressed in this section, and a stricter analysis based on sound field reconstruction is presented in Chapter 9.

The first-order spatial Ambisonics involves four independent signals. If the maximal amplitude of independent signals is normalized to a unit, a set of normalized independent signals (amplitudes) can be chosen as

$$W = 1 \qquad X = \cos\theta_S \cos\phi_S \qquad Y = \sin\theta_S \cos\phi_S \qquad Z = \sin\phi_S, \tag{6.4.1}$$

where (θ_S, ϕ_S) represent the azimuth and elevation of the target source in the original sound field. Equation (6.4.1) represents the signals captured by four coincident microphones in the original sound field. W is the signal from an omnidirectional microphone (spherical symmetrical directivity); X, Y, and Z are signals from three bidirectional microphones (symmetrical directivity around the polar axis) with their main axes pointing to the front, left, and top directions, respectively. The directional patterns of these four microphones are illustrated in Figure A.1 in Appendix A. W, X, Y, and Z also represent the pressure and the x (front-back), y (left-right), and z (up-down) components of the velocity of the medium in the original sound field, respectively.

In reproduction, M loudspeakers are arranged on a spherical surface around a listener. The distance from each loudspeaker to the origin (center of the sphere) satisfies the condition of a far-field distance. The direction of the ith loudspeaker is (θ_i, ϕ_i) with $i = 0,1,2 \ldots M-1$, and loudspeaker signals can be written as a form of sound field signal mixing, i.e., as a linear combination of the independent signals W,X,Y, and Z given in Equation (6.4.1):

$$\begin{aligned} A_i(\theta_S, \phi_S) &= A_{total}\left[D'^{(1)}_{00}(\theta_i, \phi_i)W + D'^{(1)}_{11}(\theta_i, \phi_i)X + D'^{(2)}_{11}(\theta_i, \phi_i)Y + D'^{(1)}_{10}(\theta_i, \phi_i)Z\right] \\ &= A_{total}\left[\begin{aligned} &D'^{(1)}_{00}(\theta_i, \phi_i) + D'^{(1)}_{11}(\theta_i, \phi_i)\cos\theta_S\cos\phi_S + D'^{(2)}_{11}(\theta_i, \phi_i)\sin\theta_S\cos\phi_S \\ &+D'^{(1)}_{10}(\theta_i, \phi_i)\sin\phi_S \end{aligned}\right] \\ i &= 0, 1 \ldots (M-1), \end{aligned} \tag{6.4.2}$$

where the decoding coefficients $D'^{(1)}_{00}(\theta_i, \phi_i)$, $D'^{(1)}_{11}(\theta_i, \phi_i)$, $D'^{(2)}_{11}(\theta_i, \phi_i)$, and $D'^{(1)}_{10}(\theta_i, \phi_i)$ can be chosen by using various methods depending on the loudspeaker configuration and optimized criteria for a reproduced sound field.

Generally, for left-right symmetrical loudspeaker configurations, decoding coefficients satisfy symmetrical relationships similar to those in Equations (5.2.16)–(5.2.18). If loudspeaker configurations are also up-down symmetric, for any pair of up-down symmetrical loudspeakers I and i', their azimuths and elevations satisfy $\theta_i = \theta_{i'}$ and $\phi_i = -\phi_{i'}$, and

$$\begin{aligned} &D'^{(1)}_{00}(\theta_i, \phi_i) = D'^{(1)}_{00}(\theta_{i'}, \phi_{i'}) \quad D'^{(1)}_{11}(\theta_i, \phi_i) = D'^{(1)}_{11}(\theta_{i'}, \phi_{i'}) \quad D'^{(2)}_{11}(\theta_i, \phi_i) = D'^{(2)}_{11}(\theta_{i'}, \phi_{i'}) \\ &D'^{(1)}_{10}(\theta_i, \phi_i) = -D'^{(1)}_{10}(\theta_{i'}, \phi_{i'}). \end{aligned} \tag{6.4.3}$$

Symmetry greatly simplifies the decoding coefficients.

The decoding coefficients in Equation (6.4.2) are solved in accordance with various optimized criteria. Similar to the case of horizontal Ambisonics with an irregular loudspeaker configuration in Section 5.2.3, the optimized criterion of the perceived virtual source direction matching that of a target source can be used when the head is fixed and when the head is turning at low frequencies; as such, $\theta_I = \theta'_I = \theta_S$ and $\phi_I = \phi'_I = \phi''_I = \phi_S$ in Equations (6.1.11), (6.1.13), and (6.1.14). With these parameters, a set of linear equations for decoding coefficients can be obtained. Furthermore, $\theta_v = \theta_S$, $\phi_v = \phi_S$, and $r_v = 1$ can be set in Equation (6.1.18) by using the optimized criterion of the velocity localization vector; the normalized pressure amplitude in the origin is set to a unit, so the following equation is obtained:

$$\begin{array}{ll}\sum_{i=0}^{M-1} A_i(\theta_S, \phi_S) = 1 & \sum_{i=0}^{M-1} A_i(\theta_S, \phi_S)\cos\theta_i \cos\phi_i = \cos\theta_S \cos\phi_S \\ \sum_{i=0}^{M-1} A_i(\theta_S, \phi_S)\sin\theta_i \cos\phi_i = \sin\theta_S \cos\phi_S & \sum_{i=0}^{M-1} A_i(\theta_S, \phi_S)\sin\phi_i = \sin\phi_S \end{array}. \quad (6.4.4)$$

Substituting Equation (6.4.2) into Equation (6.4.4) leads to a set of linear equations or a matrix equation for decoding coefficients:

$$S'_{3D} = A_{total}\left[Y'_{3D}\right]\left[D'_{3D}\right] S'_{3D}, \quad (6.4.5)$$

where $S'_{3D} = [W, X, Y, Z]^T$ is a 4 × 1 column matrix or vector composed of independent signals in Equation (6.4.1), the subscript "3D" denotes the case of three-dimensional spatial Ambisonics; and $[D'_{3D}]$ is an M ×4 decoding matrix to be solved:

$$\left[D'_{3D}\right] = \begin{bmatrix} D'^{(1)}_{00}(\theta_0, \phi_0) & D'^{(1)}_{11}(\theta_0, \phi_0) & D'^{(2)}_{11}(\theta_0, \phi_0) & D'^{(1)}_{10}(\theta_0, \phi_0) \\ D'^{(1)}_{00}(\theta_1, \phi_1) & D'^{(1)}_{11}(\theta_1, \phi_1) & D'^{(2)}_{11}(\theta_1, \phi_1) & D'^{(1)}_{10}(\theta_1, \phi_1) \\ \vdots & & & \\ D'^{(1)}_{00}(\theta_{M-1}, \phi_{M-1}) & D'^{(1)}_{11}(\theta_{M-1}, \phi_{M-1}) & D'^{(2)}_{11}(\theta_{M-1}, \phi_{M-1}) & D'^{(1)}_{10}(\theta_{M-1}, \phi_{M-1}) \end{bmatrix}. \quad (6.4.6)$$

$[Y'_{3D}]$ is a 4 × M matrix with its entries related to the directions of loudspeakers:

$$\left[Y'_{3D}\right] = \begin{bmatrix} 1 & 1 & \dots & 1 \\ \cos\theta_0 \cos\phi_0 & \cos\theta_1 \cos\phi_1 & \dots & \cos\theta_{M-1} \cos\phi_{M-1} \\ \sin\theta_0 \cos\phi_0 & \sin\theta_1 \cos\phi_1 & \dots & \sin\theta_{M-1} \cos\phi_{M-1} \\ \sin\phi_0 & \sin\phi_1 & \dots & \sin\phi_{M-1} \end{bmatrix}. \quad (6.4.7)$$

Similar to the case of Equation (5.2.24), for $M > 4$, the decoding coefficients in Equation (6.4.5) can be solved from the following pseudo-inverse methods if the loudspeaker configuration is chosen appropriately so that the matrix $[Y'_{3D}]$ is well conditioned (Section 9.4.1):

$$A_{total}\left[D'_{3D}\right] = pinv\left[Y'_{3D}\right] = \left[Y'_{3D}\right]^T \left\{\left[Y'_{3D}\right]\left[Y'_{3D}\right]^T\right\}^{-1}. \quad (6.4.8)$$

The solution given in Equation (6.4.8) satisfies the condition of constant-amplitude normalization. For $M = 4$, the decoding coefficients can be directly solved from the inverse of the matrix $[Y'_{3D}]$. Except for some regular loudspeaker configurations, the solution shown in Equation (6.4.8) usually does not satisfy the criterion of constant power in reproduction:

$$Pow' = \sum_{i=0}^{M-1} A_i^2 \neq const. \tag{6.4.9}$$

For the solution expressed in Equation (6.4.8), the direction of the energy localization vector evaluated using Equation (6.1.20) generally does not coincide with that of the velocity localization vector, i.e., $(\theta_E, \phi_E) \neq (\theta_v, \phi_v)$. However, as indicated by Equation (6.4.4), the first-order spatial Ambisonics with a decoding matrix given by Equation (6.4.8) is an example in which Equations (6.1.11), (6.1.13), and (6.1.14) are consistent. In other words, the first-order spatial Ambisonics creates ITD_p, and its dynamic variation with head rotation and tilting that match with those of the target source at very low frequencies.

Similar to the case of horizontal Ambisonics (Section 4.3.1), spatial Ambisonics has various forms of independent signals. In principle, four independent linear combinations of W, X, Y, and Z can serve as a set of independent signals of spatial Ambisonics. Different sets of independent signals can be converted mutually. The decoding equations for different sets of independent signals are related via linear matrix transformation. In practice, four independent signals can be captured by four cardioid or subcardioid microphones with symmetrical directivity around the polar axis. The main axes of four microphones are pointed to the four non-parallel directions: left-front-up (LFU), left-back-down (LBD), right-front-down (RFD), and right-back-up (RBU) (Figure 6.7). The normalized amplitudes of microphone signals are

$$A'_{i'}(\Delta\Omega'_{i'}) = A_{mic}\left[1 + b\cos(\Delta\Omega'_{i'})\right] \qquad i' = 0, 1, 2, 3. \tag{6.4.10}$$

where $0.5 \leq b \leq 1$, $\Delta\Omega'_i{}'$ is the angle between the target source direction and the main axis of the ith microphone, and A_{mic} is a coefficient associated with the acoustic-electric efficiency or gain of microphones. The independent signals expressed in Equation (6.4.10) are termed A-format spatial Ambisonic signals.

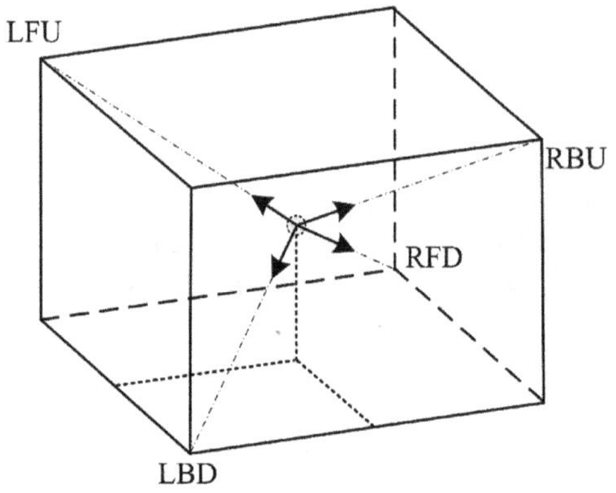

Figure 6.7 Main axis directions of the four microphones for recording the A-format first-order spatial Ambisonic signals.

W, X, Y, and Z given in Equation (6.4.1) can be conveniently chosen as the independent signals of the first-order spatial Ambisonics. W, X, Y, and Z can be derived from the signals presented in Equation (6.4.10) through linear transformation. Let

$$A'_{LFU} = A'_0(\Delta\Omega'_0) \quad A'_{LBD} = A'_1(\Delta\Omega'_1) \quad A'_{RFD} = A'_2(\Delta\Omega'_2) \quad A'_{RBU} = A'_3(\Delta\Omega'_3). \tag{6.4.11}$$

The linear combinations of the four signals in Equation (6.4.1) yield

$$\begin{aligned} W'' &= A'_{LFU} + A'_{LBD} + A'_{RFD} + A'_{RBU} = 4A_{mic} \\ X'' &= A'_{LFU} - A'_{LBD} + A'_{RFD} - A'_{RBU} = \frac{4b}{\sqrt{3}} A_{mic} \cos\theta_S \cos\phi_S \\ Y'' &= A'_{LFU} + A'_{LBD} - A'_{RFD} - A'_{RBU} = \frac{4b}{\sqrt{3}} A_{mic} \sin\theta_S \cos\phi_S \\ Z'' &= A'_{LFU} - A'_{LBD} - A'_{RFD} + A'_{RBU} = \frac{4b}{\sqrt{3}} A_{mic} \sin\phi_S. \end{aligned} \tag{6.4.12}$$

The normalized independent signals W, X, Y, and Z are obtained by equalizing the signals in Equation (6.4.12) with an appropriate gain. In practice, the levels of X, Y, and Z are increased by 3 dB to ensure a diffused-field power response identical to that of the M signal. The independent signals are written as

$$\begin{aligned} &W' = W = 1 \quad X' = \sqrt{2}X = \sqrt{2}\cos\theta_S \cos\phi_S \quad Y' = \sqrt{2}Y = \sqrt{2}\sin\theta_S \cos\phi_S \\ &Z' = \sqrt{2}Z = \sqrt{2}\sin\phi_S. \end{aligned} \tag{6.4.13}$$

The decoding equation should be appropriately changed to accommodate the independent signals in Equation (6.4.13). The spatial Ambisonic system with independent signals given by Equation (6.4.13) is termed the first-order ***B-format spatial Ambisonics*** (Gerzon, 1985).

The second- and higher-order directional harmonics (appendix A) are supplemented as new independent signals to improve the performance in reproduction, and the first-order spatial Ambisonics can be extended to the higher-order spatial Ambisonics. Under the optimized criterion of matching each order approximation of reproduced sound pressures with those of the target source within a given listening region and below a certain frequency limit, the decoding coefficients or matrix and loudspeaker signals are solved via a pseudo-inverse method similar to Equation (6.4.8). This problem is addressed in Section 9.3.2. For the second- and higher-order spatial Ambisonics with decoding coefficients given in Equation (6.4.4), localization Equations (6.1.11), (6.1.13), and (6.1.14) yield consistent results. Equation (6.4.4) only specifies the zero and first-order decoding coefficients, flexibly leaving room for choosing the second- and higher-decoding coefficients.

Moreover, the equivalence between A-format and B-format spatial Ambisonics allows a post-equalization of recorded source signal at a given direction (Favrot and Faller, 2020). By using an appropriate matrix, the recorded B-format signals can be converted into a set of A-format signals with the main axis of polar pattern of one of A-format signal $A'_0(\Delta\Omega'_0)$ points to the concerned source direction. The resultant signal $A'_0(\Delta\Omega'_0)$ is equalized and, and the equalized signal along with the other un-equalized A-format signals are converted back into B-format signals by an invert matrix.

Similar to the case of horizontal Ambisonics, the decoding coefficients at high frequencies can be derived according to various optimized physical and psychoacoustic criteria (Arteaga, 2013), such as optimized direction of the energy localization vector in Equation (6.1.20), maximized energy vector magnitude in Equation (6.1.21), and constant overall power; thus,

$$\theta_E = \theta_v \quad \phi_E = \phi_v \quad \max(r_E) \qquad Pow' = const. \tag{6.4.14}$$

However, if the energy localization vector is considered, the optimized criteria yield a set of nonlinear equations of decoding coefficients, and nonlinear optimization methods are required. For higher-order spatial Ambisonics with irregular loudspeaker configurations, nonlinear optimization is complicated and can be usually solved with numerical methods (Scaini and Arteaga, 2014). Moreover, it may satisfy various optimized criteria partly or approximately, rather than completely and exactly. For example, Zotter et al. (2012) provided a constant power solution and Epain et al. (2014) provided a constant angular spread solution for irregular loudspeaker configuration. As the order of spatial Ambisonics increases, the upper-frequency limit of the exact reconstruction of a sound field increases. In this case, optimizing high-frequency decoding may be unnecessary. All these aspects are similar to the horizontal Ambisonics with an irregular loudspeaker configuration in Section 5.2.3, so they are no longer described here. For reproduction within a large-sized listening region, the in-phase solution of decoding coefficients can be derived in a manner similar to the horizontal case in Section 4.3.2 (Malham and Myatt, 1995; Daniel, 2000; Neukom, 2007).

The features and limitations of spatial Ambisonics are similar to those of horizontal Ambisonics (discussed in Section 4.3.4, so they are no longer presented in this section).

Virtual source localization experiments for spatial Ambisonics with different orders and loudspeaker configurations have been conducted (Capra et al., 2007; Morrell and Reiss, 2009; Power et al., 2012; Xie et al., 2017a, 2019). Overall, the results indicate that spatial Ambisonics can recreate a virtual source at three-dimensional directions (including vertical directions). The localization performance improves as the order increases. The third-order spatial Ambisonics usually exhibits a satisfactory localization performance. In Section 12.2.2, the upper frequency limit for Ambisonics to reconstruct target binaural pressures accurately increases as the order increases. The third-order spatial Ambisonics can reconstruct target binaural pressures approximately up to 1.87 kHz and can reconstruct ITD_p and its dynamic variation with head turning below 1.5 kHz. In other words, the third-order spatial Ambisonics can recreate correct localization information below 1.5 kHz. Therefore, the results of localization experiments on spatial Ambisonics can be regarded as experimental validations of the localization hypothesis and theorem presented in Section 6.1.

6.4.2 Some examples of the first-order spatial Ambisonics

The first-order spatial Ambisonics can be implemented via various loudspeaker configurations. Some examples are given here. These examples are used to illustrate the principle of first-order spatial Ambisonics. They can theoretically optimize the localization performance at low frequencies (velocity localization vector) but may fail to optimize the overall perceived performance. As a consequence, they may not completely satisfy the practical requirements.

A classic example is reproduction with a loudspeaker configuration, as shown in Figure 6.8 (Xie X.F., 1988; Gerzon, 1992a). Eight loudspeakers are arranged in the vertices of a hexahedron (cuboid), and the listener's head is located at the center of the hexahedron with an equal distance to all loudspeakers. This loudspeaker configuration is left-right and up-down symmetric. The position of the ith loudspeaker is denoted by (θ_i, ϕ_i) with $i = 0, 1,\ldots 7$. From Equation (6.4.8), the normalized loudspeaker signals (amplitudes) for low-frequency optimization are given by

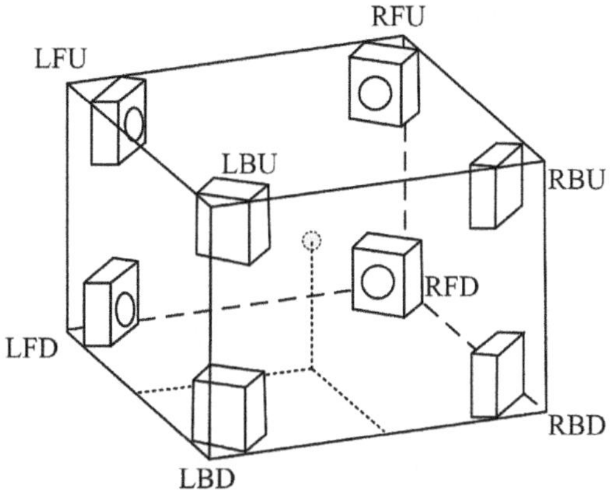

Figure 6.8 Hexahedral configuration of eight loudspeakers.

$$A_i(\theta_S,\phi_S)=\frac{1}{8}\left[W+\frac{1}{\cos\theta_i\cos\phi_i}X+\frac{1}{\sin\theta_i\cos\phi_i}Y+\frac{1}{\sin\phi_i}Z\right]$$
$$=\frac{1}{8}\left[W+\frac{1}{\cos\gamma_{x,i}}X+\frac{1}{\cos\gamma_{y,i}}Y+\frac{1}{\cos\gamma_{z,i}}Z\right]\qquad i=0,1\ldots 7, \tag{6.4.15}$$

where $\gamma_{x,i}$, $\gamma_{y,i}$, and $\gamma_{z,I}$ are the angles between the directional vector of the ith loudspeaker (with respect to the center of the cuboid or origin) and the x, y, and z axes of the coordinate; $\cos\gamma_{x,i}$, $\cos\gamma_{y,i}$, and $\cos\gamma_{zIi}$ are the direction cosines of the ith loudspeakers. This hexahedral configuration is accommodated for most practical room shapes in reproduction.

In a special case of a regular hexahedral configuration, the positions of eight loudspeakers are

$$\begin{array}{llllll} LFU: & \theta_{LFU}=45^\circ & \phi_{LFU}=35.3^\circ & LFD: & \theta_{LFD}=45^\circ & \phi_{LFD}=-35.3^\circ \\ LBU: & \theta_{LBU}=135^\circ & \phi_{LBU}=35.3^\circ & LBD: & \theta_{LBD}=135^\circ & \phi_{LBD}=-35.3^\circ \\ RFU: & \theta_{RFU}=-45^\circ & \phi_{RFU}=35.3^\circ & RFD: & \theta_{RFD}=-45^\circ & \phi_{RFD}=-35.3^\circ \\ RBU: & \theta_{RBU}=-135^\circ & \phi_{RBU}=35.3^\circ & RBD: & \theta_{RBD}=-135^\circ & \phi_{RBD}=-35.3^\circ \end{array}, \tag{6.4.16}$$

where the notations L, R, F, B, U, and D in the subscripts denote left, right, front, back, up, and down, respectively. Substituting Equation (6.4.16) into Equation (6.4.15) yields the (matrix) decoding equation for loudspeaker signals:

$$\begin{bmatrix} A_{LFU} \\ A_{LFD} \\ A_{LBU} \\ A_{LBD} \\ A_{RFU} \\ A_{RFD} \\ A_{RBU} \\ A_{RBD} \end{bmatrix} = \frac{1}{8}\begin{bmatrix} 1 & 1 & 1 & 1 \\ 1 & 1 & 1 & -1 \\ 1 & -1 & 1 & 1 \\ 1 & -1 & 1 & -1 \\ 1 & 1 & -1 & 1 \\ 1 & 1 & -1 & -1 \\ 1 & -1 & -1 & 1 \\ 1 & -1 & -1 & -1 \end{bmatrix}\begin{bmatrix} W \\ \sqrt{3}X \\ \sqrt{3}Y \\ \sqrt{3}Z \end{bmatrix}. \tag{6.4.17}$$

Equation (6.4.17) is equivalent to the normalized signals captured by eight coincident hypercardioid microphones with a symmetrical directivity around the polar axis, and the main axes of eight microphones are pointed to the directions of eight vertices of the regular hexahedron:

$$A_i\left(\Delta\Omega'_{i'}\right) = \frac{1}{8}\left[1 + 3\cos\left(\Delta\Omega'_{i'}\right)\right], \tag{6.4.18}$$

where $\Delta\Omega'_{i'}$ is the angle between the target source direction and the main axis of the *i*th microphone (which coincides with the direction of the *i*th loudspeaker). For a loudspeaker configuration in Equation (6.4.16) and the signals in Equation (6.4.17), the velocity localization vector in reproduction satisfies the optimized criteria of $\theta_v = \theta_S$, $\phi_v = \phi_S$, and $r_v = 1$. At the same time, the overall power of loudspeaker signals is a constant. The energy localization vector satisfies the optimized criteria of $\theta_E = \theta_S$ and $\phi_E = \phi_S$, but the energy vector magnitude is $r_E = 0.5$. The regular hexahedral configuration is one of a few cases in which the direction of a virtual source satisfies the optimized criteria of velocity localization vector and energy localization.

Another example is the first-order spatial Ambisonics with a regular tetrahedral loudspeaker configuration, as shown in Figure 6.9 (Xie X.F., 1988). The listener's head is located at the center of the regular hexahedron with an equal distance to all loudspeakers. Four loudspeakers are arranged in the four vertices of the regular tetrahedron, i.e., in left-front-down (LFD), right-front-down (RFD), back–down (BD), and up (U) directions with respect to the listener. This loudspeaker configuration is the simplest among configurations for first-order spatial Ambisonics because it involves four independent signals and requires at least four loudspeakers in reproduction. The positions of the eight loudspeakers are:

$$\begin{array}{llllll} LFD: & \theta_{LFD} = 60^\circ & \phi_{LFD} = -19.47^\circ & RFD: & \theta_{RFD} = -60^\circ & \phi_{RFD} = -19.47^\circ \\ BD: & \theta_{BD} = 180^\circ & \phi_{BD} = -19.47^\circ & U: & \theta_U = 0^\circ & \phi_U = 90^\circ \end{array}. \tag{6.4.19}$$

Substituting Equation (6.4.19) into Equation (6.4.8) yields the (matrix) decoding equation for loudspeaker signals:

$$\begin{bmatrix} A_{LFD} \\ A_{RFD} \\ A_{BD} \\ A_U \end{bmatrix} = \frac{1}{4}\begin{bmatrix} 1 & 1.414 & 2.449 & -1 \\ 1 & 1.414 & -2.449 & -1 \\ 1 & -2.828 & 0 & -1 \\ 1 & 0 & 0 & 3 \end{bmatrix}\begin{bmatrix} W \\ X \\ Y \\ Z \end{bmatrix}. \tag{6.4.20}$$

For this loudspeaker configuration and signals, the overall power of loudspeaker signals is a constant, but the direction of a virtual source does not satisfy the optimized criterion of the energy localization vector.

Similar to the case of horizontal Ambisonics, the first-order spatial Ambisonic reproduction has more loudspeaker configurations. The decoding equation and loudspeaker signals for these configurations can also be derived. For example, for the conical configuration with M loudspeakers shown in Figure 6.10 (a), $(M - 1)$ loudspeakers are arranged in the circle of the bottom of the cone, and a loudspeaker is arranged in the vertex of the cone (up direction). The listener's head center is located at the center of the cone with an equal distance to

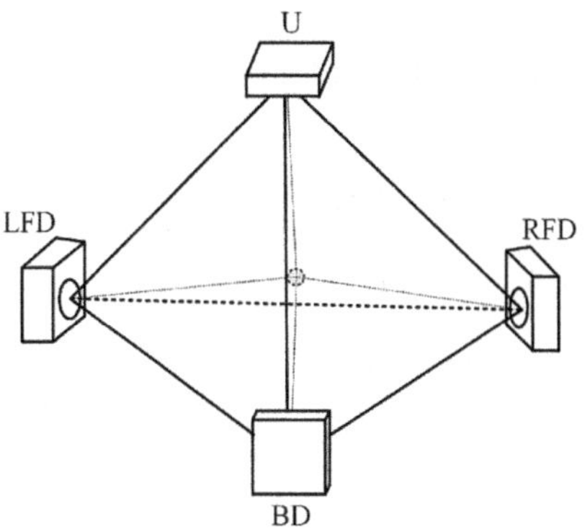

Figure 6.9 Regular tetrahedral configuration of four loudspeakers.

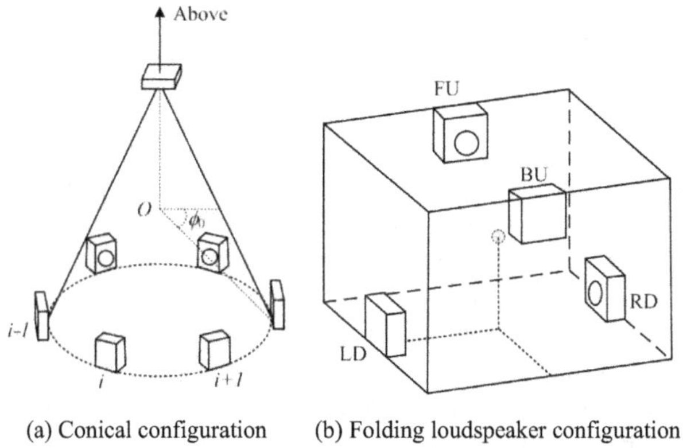

Figure 6.10 Two other loudspeaker configurations: (a) conical configuration; (b) folding loudspeaker configuration.

all loudspeakers. The folding loudspeaker configuration with four loudspeakers shown in Figure 6.10 (b) is also theoretically available for the first-order spatial Ambisonic reproduction (Xie and Xie, 1992, 1992b). However, the details are not presented here because of page limitations.

6.4.3 Local Ambisonic-like signal mixing for vertical loudspeaker configuration

The spatial Ambisonic signal mixing described in the preceding sections is a global signal mixing method through which signals are fed to all spatial loudspeakers to recreate a virtual source. Similar to the case of horizontal reproduction, local Ambisonic-like signal mixing – by which signals are fed to loudspeakers arranged in the median or sagittal plane – is available to recreate a vertical virtual source in a given plane (Yi and Xie, 2020).

In the first-order local Ambisonic-like signal mixing, three loudspeakers, i.e., loudspeakers 0, 1, and 2 in Figure 6.2, are used to recreate a virtual source in the frontal median plane. Similar to the derivation of Equation (5.2.32) for three frontal horizontal loudspeakers, the normalized signal amplitudes for three vertical loudspeakers are a linear combination of elevation harmonics up to the first order and given by

$$A_0 = A_{total}\left(1-\cos\phi_S - \frac{1-\cos\phi_2}{\sin\phi_2}\sin\phi_S\right)$$
$$A_1 = A_{total}\left(-2\cos\phi_2 + 2\cos\phi_S\right) \qquad (6.4.21)$$
$$A_2 = A_{total}\left(1-\cos\phi_S + \frac{1-\cos\phi_2}{\sin\phi_2}\sin\phi_S\right),$$

where ϕ_S is the target source elevation in the median plane, and ϕ_2 is the elevation of loudspeaker 2. For constant-power normalization, the normalized factor A_{total} is expressed as

$$A_{total} = \frac{1}{\sqrt{2\left(1-\cos\phi_S\right)^2 + 2\left(\frac{1-\cos\phi_2}{\sin\phi_2}\sin\phi_S\right)^2 + 4\left(\cos\phi_S - \cos\phi_2\right)^2}}. \qquad (6.4.22)$$

Figure 6.11 presents the local Ambisonic-like panning curves of Equations (6.4.21) and (6.4.22) with $\phi_2 = 45°$. For target source elevation in each loudspeaker direction ($\phi_S = 0°$ or $\pm 45°$), the signals for two other loudspeakers vanish, and a virtual source is recreated by a single loudspeaker. For target source elevation between two loudspeakers, the virtual source is recreated by three loudspeakers, and the signal for the third loudspeaker is out of phase. The panning curves in Figure 6.11 are extended to create a target source outside the boundary of a three-loudspeaker array.

The three localization equations can be proved to yield consistent results by substituting Equation (6.4.21) into Equations (6.1.11), (6.1.13), and (6.1.14):

$$\sin\theta_I \cos\phi_I = 0 \qquad \cos\theta_I' \cos\phi_I' = \cos\phi_S \qquad \sin\phi_I'' = \sin\phi_S \qquad (6.4.23)$$

or

$$\theta_I = \theta_I' = 0° \qquad \phi_I = \phi_I' = \phi_I'' = \phi_S. \qquad (6.4.24)$$

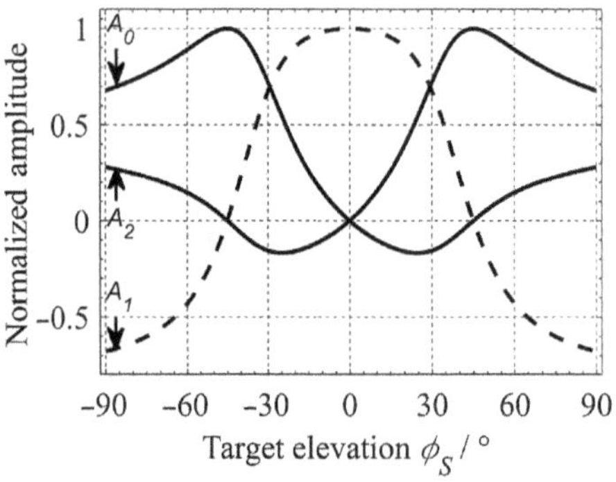

Figure 6.11 The first-order local Ambisonic-like panning curves for three vertical loudspeakers in a median plane.

In this case, ITD_p and its dynamic variation with head rotation and tilting in reproduction are consistent with those of the target source, and the perceived virtual source direction matches that of the target source at low frequencies. This feature is superior to that of vertical pair-wise amplitude panning discussed in Section 6.2. Further analysis using head-related transfer functions (HRTFs) (Section 12.1.3) yields similar results.

Another feature of local Ambisonic-like signal mixing given by Equation (6.4.21) is that it can theoretically recreate a vertical virtual source at the elevation outside the boundary of a three-loudspeaker array. A virtual source localization experiment for a three-loudspeaker configuration with elevations of $\phi_0 = -45°$, $\phi_1 = 0°$, and $\phi_2 = 45°$ indicates that the local Ambisonic-like signal mixing can recreate a stable virtual source within the elevation region of ±60°. The local Ambisonic-like signal mixing can also be extended to vertical loudspeaker configurations in other sagittal planes.

6.4.4 Recreating a top virtual source with a horizontal loudspeaker arrangement and Ambisonic signal mixing

The first-order spatial Ambisonics encodes spatial information into four independent signals, namely, W, X, Y, and Z, and requires at least four loudspeakers arranged in space (such as a regular tetrahedral configuration) in reproduction. The first-order horizontal Ambisonics excludes Z and loses the vertical localization information. However, if the three independent signals for horizontal Ambisonics are revised as per (Jot et al., 1999), we have:

$$W_1 = \sqrt{1 + \sin^2 \phi_S} \qquad X_1 = \cos\theta_S \cos\phi_S \qquad Y_1 = \sin\theta_S \cos\phi_S. \tag{6.4.25}$$

Then, a perceived virtual source can be recreated in the top direction by using horizontal loudspeaker configurations. The nature of this method is the virtual source-elevated effect discussed in Section 6.1.4.

For a regular configuration with $M \geq 3$ loudspeakers in the horizontal plane, the independent signals W, X, and Y of the first-order horizontal Ambisonics are replaced with W_1, X_1, and Y_1 expressed in Equation (6.4.25). For a target source in a horizontal plane with $\phi_S = 0°$, the independent signals in Equation (6.4.25) are identical to those of the first-order horizontal Ambisonics in Equation (4.3.3), thereby causing an identical horizontal localization effect. For a target source in the top direction $\phi_S = 90°$, $X_1 = Y_1 = 0$ and the amplitudes of all loudspeaker signals are equal:

$$A_i\left(\phi_S = 90°\right) = \sqrt{2} A_{total} \qquad i = 0, 1 \ldots (M-1). \tag{6.4.26}$$

According to the analysis in Section 6.1.4, the perceived virtual source is located in the top direction.

6.5 ADVANCED MULTICHANNEL SPATIAL SURROUND SOUNDS AND PROBLEMS

6.5.1 Some advanced multichannel spatial surround sound techniques and systems

For practical uses, especially for ultra-high-definition video/audio and new cinema sound requirements, multichannel spatial surround sounds have emerged as new-generation sound reproduction techniques after multichannel horizontal surround sounds with and without accompanying pictures. Since 2000, multichannel spatial surround sound has been considered an important trend in the audio field (Rumsey, 2013).

Multichannel spatial surround sound aims to improve the capability of reproducing the three-dimensional spatial information of sounds in all or some of the following aspects depending on the applications and contents of programs:

1. Improve the sound effects that match with pictures, i.e., recreate frontal virtual sources in left-right and vertical dimensions in the entire screen area because an ultra-high-definition video with a large screen provides larger horizontal and vertical view angles.
2. Enhance the three-dimensional virtual source effect, including lateral, rear, and top virtual sources, and improve the resolution in the reproduction of the spatial information of sound.
3. Increase the reproduction of the three-dimensional spatial information of discrete early reflections and late diffused reverberation to produce a good auditory spatial impression.
4. Improve the reproduction of non-reflective ambience components to obtain a good ***immersive sense.***
5. Enlarge the listening region.
6. Ensure compatibility with existing and common techniques and systems.

Similar to the case of horizontal surround sound, the main considerations and core problems in a multichannel spatial surround sound are the appropriate selection of the number of channels, loudspeaker configurations, and signal mixing based on the psychoacoustic principle and available system resources to achieve the desired spatial auditory perceptions. However, for multichannel spatial surround sounds, more choices are available in the number of channels and loudspeaker configurations. For practical uses and commercial competition, various multichannel spatial surround sound techniques with different loudspeaker configurations have been developed since 2000. These techniques are also termed ***immersive sound or audio.*** Some techniques and systems were originally developed for cinema sound reproduction with a large-sized listening region. Subsequently, they were simplified for domestic or consumer uses. Other techniques and systems were directly established for domestic reproduction with a small-sized listening region. For domestic uses, these techniques involve an irregular loudspeaker configuration in a three-dimensional space and can be applied to reproduction with and without accompanying picture.

In 2002, Dabringhaus published the first music program by using the 2 + 2 + 2 channel technique (www.mde.de/frame2.htm). This technique is based on the ITU-5.1-channel configuration, but the center loudspeaker and subwoofer are omitted. Instead, a pair of loudspeakers arranged above the L and R loudspeakers are supplemented (Ehret et al., 2007).

As stated in Section 5.3, some other loudspeaker configurations in DTS-HD are for 7.1-channel spatial surround sound (DTS Inc., 2006), which can partly recreate vertical sound information.

1. In configuration 2, a pair of loudspeakers on the two sides of $\theta = \pm 90°$ above the horizontal plane are supplemented to the standard 5.1-channel configuration. This configuration improves the reproduction of vertical information at the lateral-up and top directions; it is also compatible with 5.1-channel reproduction. However, it does not improve the reproduction of lateral and rear information in the horizontal plane.
2. In configuration 3, a loudspeaker in the horizontal rear direction and another loudspeaker in the top direction are supplemented to the standard 5.1-channel configuration to improve the reproduction of rear and top information.

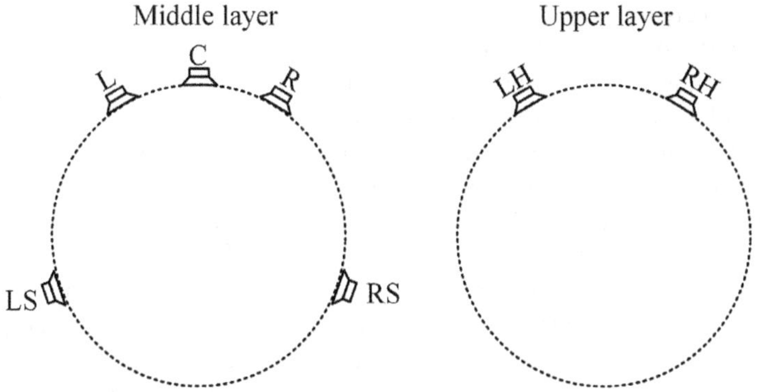

Figure 6.12 Loudspeaker configuration 4 in a DTS-HD 7.1-channel system.

3. In configuration 4, a pair of loudspeakers is supplemented to the positions above the left and right loudspeakers in a standard 5.1-channel configuration (Figure 6.12, where the subwoofer is omitted).
4. In configuration 7, a loudspeaker is supplemented to the position above the center loudspeaker, and another loudspeaker is supplemented to the rear in standard 5.1-channel configuration.

Configurations 4 and 7 are beneficial to recreating a vertical localization effect that matches with a large-screen video.

Layer-wise loudspeaker configurations are often used in multichannel spatial surround sound reproduction. For instance, Auro Technologies introduced a series of multichannel spatial surround sound systems and loudspeaker configurations (termed Auro-3D). These loudspeaker configurations are constructed with supplemented upper-layer loudspeakers into a 5.1 or 7.1-channel horizontal configuration and are compatible with horizontal reproduction. The Auro 9.1 channel system, which was introduced in 2016, is intended for domestic reproduction and is composed of a two-layer loudspeaker configuration (Figure 6.13, where the subwoofer is omitted; Theile and Wittek, 2011). The middle or horizontal layer at an elevation of $\phi = 0°$ involves five loudspeakers identical to that of an ITU-5.1-channel configuration, i.e., located at azimuths of $\theta = 0°$, $\pm 30°$, and $\pm 110°$ respectively. The upper layer at an elevation of $\phi = 30°$ involves four loudspeakers located above the horizontal left, right, left-surround, and right-surround loudspeakers, i.e., located at azimuths $\pm 30°$ and $\pm 110°$, respectively. This loudspeaker configuration may fail to recreate a vertical virtual source in the frontal-median

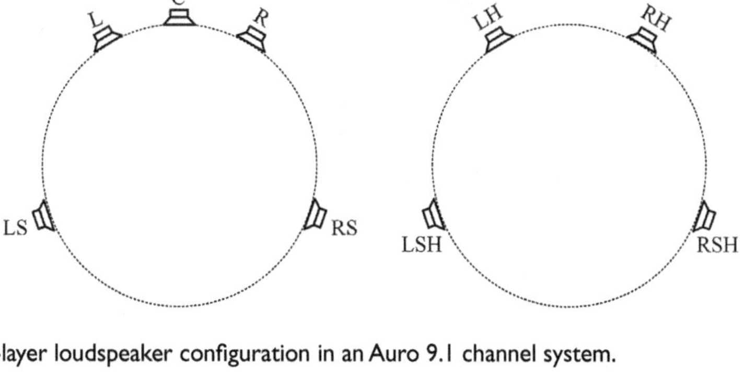

Figure 6.13 Two-layer loudspeaker configuration in an Auro 9.1 channel system.

plane, but it can recreate vertical localization information in the directions of upper-layer loudspeakers. On the basis of the Auro 9.1 channel system, the Auro 10.1-channel system is constructed by adding a top channel; the Auro 11.1-channel system is developed by adding a top channel and an upper-center channel in the upper layer; and the Auro 13.1-channel system is created by adding a top channel, an upper-center channel in the upper layer, and left-back and right-back surround channels in the middle layer (i.e., 7.1-channels are present in the middle layer). The Auro 11.1 and 13.1-channel systems were originally developed for cinema sound, but they are applied to domestic reproduction after simplification.

The 10.2-channel sound, which was proposed by Holman in cooperation with the Integrated Media System Center at the University of California, was developed for domestic and cinema reproduction; it is shortened for USC 10.2-channel system (Holman, 1996, 2001; ITU-R Report, BS 2159-7, 2015). For domestic reproduction, it consists of a two-layer loudspeaker configuration (Figure 6.14). The middle layer at an elevation of $\phi = 0°$ involves eight loudspeakers at azimuths of $\theta = 0°, \pm30°, \pm60°, \pm110°$, and 180°. In comparison with the ITU-5.1-channel sound, the USC 10.2-channel sound can be used to extend the virtual source distribution in the frontal horizontal plane and improve the reproduction of early lateral reflections by adding a pair of left/right-wide loudspeakers at larger azimuths of $\theta = \pm60°$. Adding a back (rear) surround loudspeaker at $\theta = 180°$ enhances the reproduction of a virtual source and reflection from the rear. The upper layer at an elevation of $\phi = 45°$ consists of a pair of left/right-height loudspeakers at an azimuth of $\phi = 45°$ to improve the reproduction of a virtual source and reflection from vertical directions. In practice, two types of loudspeakers with different radiation patterns are arranged at an azimuth of ±110° in the middle layer. One type is a loudspeaker with a traditional direct radiation pattern for recreating localization effects. Another type is a loudspeaker with a dipolar radiation pattern, which is elevated above the direct radiation loudspeaker. Combined with the reflections from the lateral and rear walls of a listening room, dipolar-type loudspeakers enhance ambience reproduction. Therefore, the USC 10.2-channel system has 12 loudspeakers with a full-audible bandwidth. In addition, two low-frequency effect channels are available for decorrelated low-frequency contents. The system employs bass management for all full-audible channels (Section 14.3.3). All low-frequency components below 120 Hz are reproduced by a pair of subwoofers arranged on two sides to enhance the auditory spatial impression at low frequencies.

Another 10.2-channel sound was introduced by Samsung Electronics Co., Ltd. in the Republic of Korea for ultra-high-definition digital television (Kim et al., 2010; ITU-R Report, BS 2159-7, 2015). It is shortened for the Samsung 10.2-channel system. The two-layer configuration of the main loudspeakers is illustrated in Figure 6.15. The loudspeaker configuration

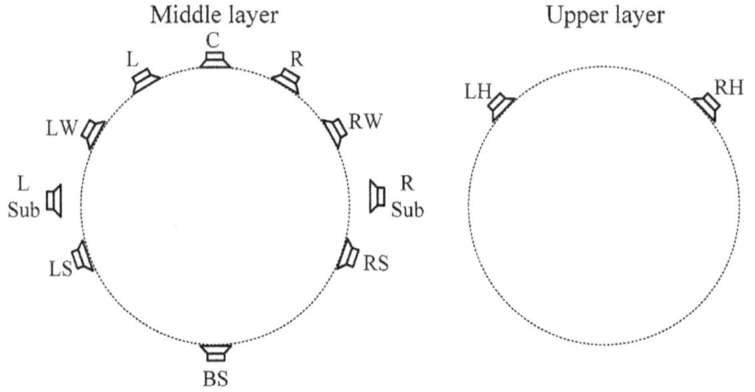

Figure 6.14 Two-layer loudspeaker configuration for a USC 10.2-channel sound.

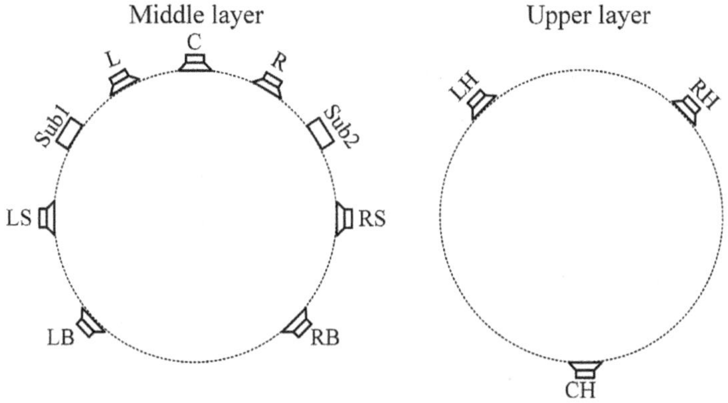

Figure 6.15 Two-layer loudspeaker configuration for Samsung 10.2-channel sound.

in the middle layer at an elevation of ϕ = 0° is similar to that of the optional 7.1-channel configuration in the ITU standard (ITU-R BS 775-1, 1994), which involves seven loudspeakers located at azimuths of θ = 0°, ±30°, ±90°, and ±135°, respectively. The four horizontal surround loudspeakers can also be arranged within the azimuthal region of ±60°–±150°. The upper (top) layer at ϕ = 45° (or 30°–45°) involves three loudspeakers arranged in the Left-front-up (LFU), right-front-up (RFU), and back-up (BU) directions with azimuths θ = ± 45° (or ±30° –± 45°) and 180°, respectively. The back-up loudspeaker can also be arranged within the range of θ = 180° and ϕ = 45°–90°. Two low-frequency effect channels are also present, and two subwoofers are arranged below the middle layer. The loudspeaker configuration in the Samsung 10.2-channel system improves the localization and reflection reproduction in rear and vertical directions.

The 22.2-channel sound by NHK (Japan Broadcasting Corporation) was developed for sound reproduction of ultra-high-definition video system called Super High Vision (Hamasaki et al., 2004, 2007; ITU-R Report, BS 2159-7, 2015). The Super High Vision comprises 100° horizontal–frontal viewing angle and about 4,000 scanning lines (7680 × 4320-pixel video image), the definition is 16 times that of HDTV and twice that of a 70 mm film. The 22.2-channel sound improves the ability to recreate three-dimensional spatial information, including localization in a three-dimensional space, envelopment, and ambience in consumer reproduction. It consists of a three-layer loudspeaker configuration (Figure 6.16). Some international

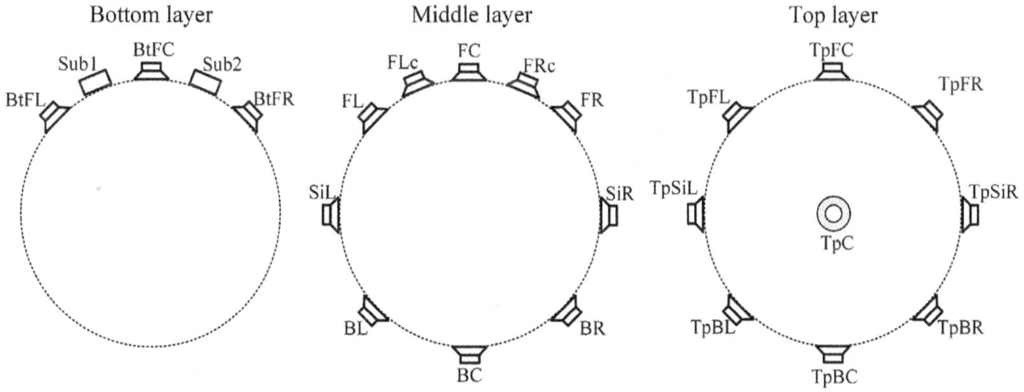

Figure 6.16 Three-layer loudspeaker configuration for the 22.2-channel sound.

standards specify the channel labeling and loudspeaker positions in the 22.2-channel system. However, channel labeling may vary among these standards. The labeling specified by ITU is illustrated in Figure 6.16 [ITU-R, BS 2051-2 (2018)], which is identical to those of the Society of Motion Picture and Television Engineers (SMPTE ST 2036-2, 2008) but different from those of ISO/IEC (ISO/IEC 23001-8, 2015). In addition, the loudspeaker positions of the ITU and ISO/IEC standards slightly differ, but both standards allow variations in loudspeaker positions within a certain region. Loudspeaker positions may also be slightly different in other studies and even in studies from the same group in different periods. The loudspeaker positions in the ITU standard are specified as follows:

1. Middle (or ear) layer
   Five frontal loudspeakers at an elevation of $\phi = 0° - 5°$ and azimuths of $\theta = 0°$, $\pm 22.5° - \pm 30°$, and $\pm 45° - \pm 60°$
   Five lateral and back (rear) surround loudspeakers at an elevation $\phi = 0° - 15°$ and azimuths of $\theta = \pm\ 90°$, $\pm 110° - \pm 135°$, and 180°
2. The top (or upper) layer
   Eight loudspeakers in elevation $\phi = 30° - 45°$ and azimuths $\theta = 0°$, $\pm 45° - \pm 60°$, $\pm 90°$, $\pm 110° - \pm 135°$, and 180°
   A loudspeaker directly in the top direction with an elevation of $\phi = 90°$
3. Bottom layer
   Three loudspeakers at an elevation of $\phi = -15°$ to $-30°$ and azimuths of $\theta = 0°$ and $\pm 45° - \pm 60°$
4. Subwoofers
   Two subwoofers at a low elevation of $\phi = -15°$ to $-30°$ and an azimuth of $\theta = \pm 30° - \pm 90°$

Here, all loudspeakers are assumed to be arranged on a spherical surface with an equal distance to the center. If loudspeakers are arranged at different distances to the center, appropriate delay and magnitude correction for loudspeaker signals are needed to compensate for the unequal loudspeaker distances.

6.5.2 Object-based spatial sound

The preceding discussions focus on ***channel-based*** spatial sounds. The spatial information of sound is represented by channel (loudspeaker) signals. In a channel-based method, the configuration of loudspeakers in reproduction is predefined, and loudspeakers signals are prepared at the stage of program production according to the desired perceived effects and the physical or psychoacoustic rules of signal mixing. The resultant signals are inappropriate for reproduction with different loudspeaker configurations unless some signal conversion procedures are supplemented. As the number and options of loudspeakers increase, especially for multichannel spatial surround sound, this problem becomes serious. Ambisonics is a ***scene-based*** spatial sound. The spatial information of sound is encoded into a set of independent signals. Independent signals, which are independent of the number and configuration of loudspeakers, are decoded into loudspeaker signals in reproduction according to the practical number and configuration of loudspeakers. In comparison with channel-based methods, the scene-based method is relatively flexible.

Since 2000, ***object-based*** spatial sound has been considered a remarkable development. In an object-based spatial sound, audio contents with identical spatial and other properties are grouped into an ***audio object*** to be transmitted. A set of ***metadata***, which describe the temporary spatial and other properties of audio objects (such as the temporary position of a target source), are transmitted with audio objects. The metadata can be regarded as the parameters and ***side information*** of audio objects or ***data about data***. Audio objects and metadata are independent from the number and configuration of loudspeakers or even the manners or techniques in reproduction. During reproduction, audio objects are distributed to loudspeakers according to the information of metadata, practical loudspeaker configuration, and certain signal mixing rules. Object-based methods can recreate virtual sources and other spatial auditory perceptions (such as envelopment similar to that caused by diffused reverberation). In an object-based method, the signal mixing or ***signal rendering*** is moved from the stage of program production to the stage of reproduction. This method is flexible and able to cater to different numbers and configurations of loudspeakers. By using different signal rendering methods, object-based methods can also be used for spatial sound reproductions with different principles and techniques, such as wave field synthesis (Chapter 10) and binaural technique (Chapter 11).

Various object-based spatial sound techniques or combinations of channel- and object-based techniques have been developed. Examples include the ISO/IEC MPEG-H 3D Audio standard (ISO/IEC 23008-3, 2015; Herre et al., 2014,2015), the Dolby Atmos by Dolby Laboratories (2012, 2015, 2016), the AuroMax by Auro Technologies and Bacro Audio Technologies (2015), and the Multi-Dimensional Audio (MDA) by DTS Inc. (ETSI TS 103 223, V1.1.1, 2015). The basic considerations of these examples are similar.

Dolby Atmos was first introduced by Dolby Laboratories in 2012 as a spatial sound technique for cinema and then modified for domestic (consumers) use. Currently, it is widely used in many domestic theaters and cinemas. In addition to audio objects, Dolby Atmos supports the transmission of parts of audio contents called "***beds***" with traditional channel-based methods. In other words, the Dolby Atmos involves two types of audio contents, namely, audio objects and beds. Audio objects are appropriate for directional localization contents, and beds are designed for ambient contents. In reproduction, loudspeaker signals are rendered by an appropriate mixing of audio objects and beds according to practical loudspeaker configurations and by using a special processor. Dolby Atmos for cinema uses supports up to 118 audio objects, 5.1, 7.1, or even 9.1-channel beds and reproduction of up to 64 loudspeakers. The 5.1 and 7.1-channel beds are similar to previous Dolby 5.1- and 7.1-channel systems. The 9.1-channel beds involve three frontal channels, four horizontal surround channels on the sides and rear, and two (left and right) top channels. Figure 6.17 illustrates the block diagram of the Dolby Atmos for cinema uses (the B-chain processing in the figure is discussed in Section 14.7).

In 2014, Dolby Laboratories introduced the Dolby Atmos for domestic use, which is similar to but is modified or simplified from the Dolby Atmos for cinema use (Dolby Laboratories, 2016). For domestic use, the number and configuration of loudspeakers are flexible, and Dolby Atmos supports the reproduction of up to 24 horizontal loudspeakers and 10 top loudspeakers. Typical configurations are constructed by adding four top (overhead) loudspeakers to the horizontal 5.1- or 7.1-channel configurations, leading to the 9.1- or 11.1-channel configuration (Figure 6.18). The horizontal left and right loudspeakers are arranged at azimuths

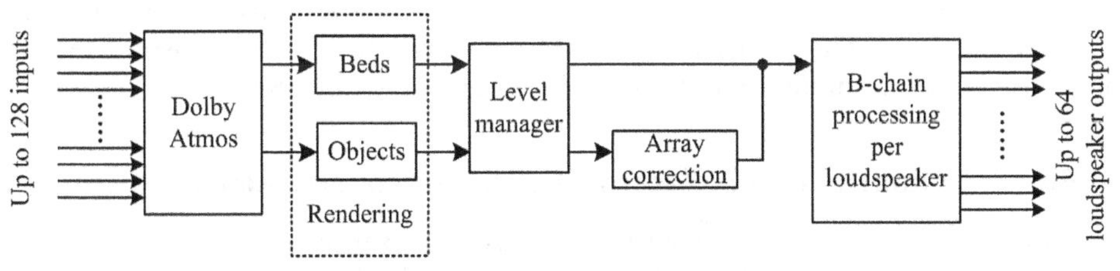

Figure 6.17 Block diagram of Dolby Atmos for cinema uses. (adapted from Dolby Laboratories, 2012.)

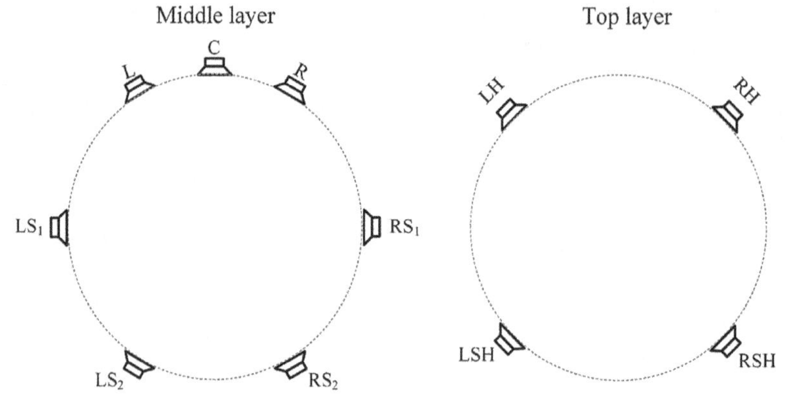

Figure 6.18 Typical loudspeaker configuration of Dolby Atmos for domestic use.

of ±22°–±30°, and the four horizontal surround loudspeakers are arranged at azimuths of ±90°–±110° and ±135°–±150°, respectively. A pair of top loudspeakers are arranged in front of the listening position, and another pair of top loudspeakers are arranged behind the listening position. The subwoofer is arranged on the floor between the horizontal center and left loudspeakers. If two top loudspeakers are used in reproduction, they are arranged slightly in front of the listening position. This loudspeaker configuration favors the reproduction of beds in Dolby Atmos. However, the details of the signal rendering in Dolby Atmos have yet to be published until 2021.

The MPEG-H 3D Audio is a standard for new-generation spatial sound and signal coding techniques by the International Organization for Standardization (ISO) and the Moving Pictures Expert Group (MPEG) under the International Electrotechnical Commission (IEC). It supports many existing spatial sound techniques and formats, including object- and channel-based methods, Ambisonics, VBAP, binaural reproduction, and audio coding techniques. The details are addressed in Section 13.5.6.

The principle and structures of other object-based techniques and methods are similar to MPEG-H 3D Audio. The Advanced Television Systems Committee (ATSC) in the USA specified the standard of ATSC Audio 3.0, an immersive audio standard for television (ATSC Standard Doc A/342-1, 2017). The ATSC Audio 3.0 involves an 11.1-channel loudspeaker configuration similar to that in Figure 6.18 and supports channel-based, object-based, and Ambisonic reproduction. AuroMax was developed for cinema reproduction, but it can be modified for domestic uses. AuroMax involves 11.1- or 13.1-channel beds, so it is compatible with Auro 11.1- or 13.1-channel sound, as described in Section 6.5.1. It also has more audio objects. AuroTechnologies recommended examples of 20.1-, 22.1-, and 26.1-channel reproduction. Another feature of AuroMax is that it supports the wave field synthesis technique to recreate virtual sources at distances beyond the boundaries of a loudspeaker array (Chapter 10). The Multi-Dimensional Audio (MDA) group, which consists of five corporations, including DTS, also developed an object-based spatial sound technique. The principle is similar to the above. The rendering stage also involves VBAP and Ambisonic techniques. Channel-based beds are modeled as a special MDA group of objects.

6.5.3 Some problems related to multichannel spatial surround sound

Commercial spatial sound is developed from horizontal to three-dimensional reproduction and from channel-based methods to object-based methods. A combination of channel- and object-based spatial sound is a new-generation technique. Since 2010, the Audio Engineering

Society (AES) has organized many conferences and sessions on this special issue (Rumsey, 2013; and The AES 40th Conference; The AES 138th Convention). The perceived performance of multichannel spatial surround sound is greatly prior to traditional multichannel horizontal surround sound but at the cost of increasing the complexity. Multichannel spatial surround sounds are being developed quickly, and various techniques have been established, but they may cause some confusion.

To standardize multichannel spatial surround sound, some international organizations and societies have formulated some standards and specifications, which are being constantly revised and updated. Early in 2006, the Committee on Digital Cinema Technology under SMPTE established the standard of distribution master audio channel mapping and channel labeling for 16-channel digital cinema sound (SMPTE 428-3, 2006). The SMPTE ST 2036-2 (2008) specified the audio channel mapping and labeling for program production in ultra-high-definition television. The SMPTE ST 2098-5 (2018) further defines names and abbreviations for immersive audio channels and immersive sound field groups associated with digital cinema immersive audio presentation. This standard also provides informative guidance on typical locations of cinema loudspeakers used for immersive audio reproduction.

Among the works and standards of ITU, ITU-R BS 1909 (2012) specified the performance requirements for an advanced multichannel stereophonic (and spatial surround) sound system for use with or without an accompanying picture, including a virtual source in all directions around a listener, the sensation of three-dimensional spatial impression, a stable virtual source in the entire area of high-resolution large-screen digital imagery, and excellent sound quality over a wide region. The ITU-R Report BS 2159-7 (2015) compared multichannel sounds for domestic and broadcasting uses and gave the results of the subjective assessment of various loudspeaker configurations. The ITU-R BS 2051-2 (2018) recommended possible loudspeaker configurations, channel labeling, and metadata requirements for multichannel spatial surround sound. Among the standards of ISO/IEC, in addition to the MPEG-H 3D Audio mentioned in Section 6.5.2 (ISO/IEC 23008-3, 2015), the IEC 62574 (2011) and ISO/IEC 23001-8 (2015) specified various loudspeaker configurations, channel labels, and program reproduction levels for multichannel spatial surround sound. Notably, loudspeaker positioning and channel labeling in ISO/IEC standards are different from those in ITU standards.

Similar to the case of 5.1-channel sound, various standards of multichannel spatial surround sound for domestic reproduction only recommend the loudspeaker configurations, but these standards do not restrict signal mixing for loudspeakers. Various signal mixing methods, such as traditional pair-wise amplitude panning, VBAP, Ambisonics, and some spatial microphone techniques (Section 7.3), are applicable to multichannel spatial surround sound with different loudspeaker configurations. Theoretically, the virtual source localization theorems in Section 6.1 can be applied to optimize loudspeaker configuration and signal mixing, but the results are usually valid in the central listening position rather than the off-central listening position even for reproduction within a small-sized region. However, these analyses are beneficial to the design of multichannel spatial surround sound for a small-sized listening region. They also help reveal and validate the limitation of systems and techniques. Object-based methods allow more flexible and diverse loudspeaker configurations and signal mixing, which should also be optimized in accordance with the psychoacoustic principles of spatial hearing.

As an example of flexible signal mixing, a signal mixing for loudspeaker configuration similar to the DTS 7.1-channel configuration 2 described in Section 6.5.1 is outlined (Rao and Xie, 2004). In this configuration, a pair of left-high and right-high loudspeakers are arranged at $\theta = \pm 90°$ and $\phi = 30°–45°$. A pair of horizontal surround loudspeakers is moved backward to $\theta = \pm 135°$. Only the case of a virtual source in the left-half space is outlined because of the left-right symmetry.

1. In the left-front space ($0° \leq \theta \leq 30°$, $0° \leq \phi \leq 90°$), a virtual source is recreated by four loudspeakers, i.e., the left-front, center, left-high, and right-high loudspeakers. For a target source in the median plane with $\theta = 0°$, the signal for the left-front loudspeaker vanishes, and a virtual source is recreated by the center, left-high, and right-high loudspeakers.
2. In the front-lateral space ($30° \leq \theta \leq 90°$, $0° \leq \phi \leq 90°$), a virtual source is recreated by three loudspeakers, i.e., the left-front, left-high, and right-high loudspeakers. In the lateral plane, the signals of left-front loudspeakers vanish, and the virtual source is recreated by the left-high and right-high loudspeakers.
3. In the rear-lateral space ($90° < \theta \leq 135°$, $0° \leq \phi \leq 90°$), a virtual source is recreated by three loudspeakers, i.e., the left-surround, left-high, and right-high loudspeakers.
4. In the rear space ($135° \leq \theta \leq 180°$, $0° \leq \phi \leq 90°$), a virtual source is recreated by four loudspeakers, i.e., the left-surround, right-surround, left-high, and right-high loudspeakers. In the horizontal plane, the signals of left-high and right-high loudspeakers vanish, and the virtual source is recreated by the left-surround and right-surround loudspeakers.

The analysis via the localization equations in Section 6.1.1 and experiments indicated that this loudspeaker configuration and signal mixing can recreate a virtual source in the upper-half space. In addition to recreating a stable virtual source in the front and rear region, the stability of a lateral virtual source is improved in comparison with 5.1-channel reproduction.

The differences in the distances from each loudspeaker to an off-central listening position lead to variations in the propagating time. In this case, the analysis of summing localization and other spatial auditory perception becomes complicated and may deal with the mechanisms of the high-level nervous system to processing binaural information comprehensively (Sections 1.7.1 and 1.7.2). This issue has been explored in some studies. However, complete models for analysis and optimized reproduction performance at the off-central listening position are still unavailable. As such, future studies should develop such models. In the current stage, the analysis, optimization, and evaluation of off-central performance are mainly based on psychoacoustic experiments.

The results of psychoacoustic or subjective assessment experiments on multichannel spatial surround sound generally depend on loudspeaker configurations, signal types, and mixing methods. As the number of reproduction channels increases, virtual sources in many directions can be recreated by the corresponding single loudspeaker, thereby reducing the dependence of the summing virtual source with two or more loudspeakers. In this case, the listening region is enlarged, and the perceived localization quality at the off-central position is improved. This conclusion is valid for both channel- and object-based methods. For example, subjective assessments (by using a method similar to that in Section 15.4) on two-channel stereophonic sound, 5.1-channel sound, and 22.2-channel sound indicate that 22.2-channel sound is obviously superior to the two other types in terms of various perceived attributes (ITU-R Report, BS 2159-7, 2015). Subjective experiments also reveal that the 22.2-channel sound in domestic reproduction demonstrates a good subjective effect within a larger listening region, e.g., off-central to the left or right about 1–2 m and to the front or back about 1 m (Hamasaki et al., 2007). Conversely, using the NHK 22.2-channel sound as a reference, Kim et al. (2010) assessed the subjective directional quality and overall perceived quality of multichannel spatial sound reproduction with various numbers of loudspeakers in the upper (top) layer. They used the stimuli with VBAP signal mixing, including a moving virtual source soundtrack. The results indicated that the Samsung 10.2-channel configuration with three loudspeakers in the upper layer yields grading scales similar to those of the 22.2-channel sound. Howie et al. (2017) conducted a double-blind listening test to evaluate a listener's discrimination among four common channel-based spatial surround sound reproduction with music stimuli, namely, the NHK 22.2-channel sound, ATSC 11.1-channel sound, Samsung 10.2-channel sound, and Auro 9.1-channel sound. The results revealed that listeners can easily discriminate between NHK 22.2-channel sound and the three other reproductions. Listeners can also discriminate between the three other reproductions with a significantly lower success rate.

In channel- and object-based methods, multichannel spatial surround sound requires more independent signals, which demand an increase in the bandwidth of transmission or capacity of storage. The current digital transmission and storage techniques meet these demands (Chapter 13).

Overall, compared with multichannel horizontal surround sound, multichannel spatial surround sound improves the ability to recreate spatial information and perceive performance. However, it requires more reproduction channels and is more complicated. Therefore, the problem of how to make a compromise between complexity and perceptual performance in reproduction remains. In addition, multichannel spatial surround sound opens up a new dimensionality, and new problems should be further investigated, including loudspeaker configuration and signal mixing/rendering, microphone technique, up/down mixing, signal transmission, and storage, related psychoacoustics, and subjective assessments. The prospect of multichannel spatial surround sound depends on practical demands.

6.6 SUMMARY

Multichannel spatial surround sound aims to recreate the three-dimensional spatial information of sound. It is an extension of multichannel horizontal surround sound and a new-generation sound reproduction technique.

Horizontal summing localization theorems are extended to a three-dimensional space. However, the mechanisms of summing localization and other auditory events in a vertical direction are different from those in a horizontal plane. In addition to spectral cues at high frequencies, the dynamic variation in ITD_p caused by head-turning is a cue for vertical localization, as proposed by Wallach. The summing localization equations for a multichannel spatial surround sound at low frequencies are derived by considering ITD_p and its dynamic variation caused by head rotation and tilting. The analysis of velocity and energy localization vectors in horizontal reproduction is also extended to three-dimensional space reproduction.

The summing localization with pair-wise amplitude panning in the median plane can be interpreted and predicted by the summing localization theorems at low frequencies. Conversely, corresponding summing localization experiments are regarded as experimental validation of the theorems. For some appropriate loudspeaker configurations in the median or sagittal plane, the pair-wise amplitude panning can recreate a virtual source between a pair of adjacent loudspeakers but the virtual source may be blurry. However, for up-down symmetrical loudspeaker configurations, the pair-wise amplitude panning fails to recreate the virtual source between the loudspeakers.

Some optimized loudspeaker configuration and signal mixing methods can be derived from the summing localization theorem, such as the VBAP and Ambisonics methods, although the derivations are suitable for central listening positions rather than the off-central listening positions. Even so, the related analyses and experiments are beneficial to the design of multichannel spatial surround sound because these analyses and experiments reveal and validate the limitation of the systems and techniques at least.

Multichannel spatial surround sounds are currently a frontier field in sound reproduction, especially the object-based spatial sound is now a trend of development. Various advanced multichannel spatial surround sounds have been developed. As compared with multichannel horizontal surround sound, multichannel spatial surround sound improves abilities to recreate spatial information and improves perceived performances greatly. However, it requires more reproduction channels and is more complicated. Some international organizations and societies have formulated some standards and specifications for multichannel spatial surround sounds. These standards are constantly revised and updated. Numerous new problems related to multichannel spatial surround sound need to be further investigated. The prospect of multichannel spatial surround sound depends on practical demands.

Chapter 7

Microphone and signal simulation techniques for multichannel sound

Similar to the case of a two-channel stereophonic sound, one stage of a multichannel sound technique involves converting or encoding the spatial information of sound into multichannel signals by using various microphone or signal simulation techniques based on the principle of spatial hearing. Thus, the desired spatial auditory perception can be recreated in reproduction. For channel- and scene-based spatial sounds, signal recording or simulation is an important stage in program production. For an object-based spatial sound, simulation or synthesis is a key step in signal rendering. For such a purpose, various microphone and signal simulation techniques for multichannel sounds have been developed.

In this chapter, multichannel microphone and signal simulation techniques are discussed. Basic considerations on signal recording and simulation for multichannel sounds are presented in Section 7.1. Various microphone techniques of 5.1-channel and other multichannel sounds are discussed in Sections 7.2 and 7.3. The techniques for simulating or synthesizing multichannel localization signals are shown in Section 7.4. The simulation of the reflections for stereophonic and multichannel sounds is explored in Section 7.5. The directional audio coding technique for simulating or synthesizing directional localization signals and diffused reverberation is described in Section 7.6.

7.1 BASIC CONSIDERATIONS ON THE MICROPHONE AND SIGNAL SIMULATION TECHNIQUES FOR MULTICHANNEL SOUNDS

As stated in Section 1.9.1, a multichannel sound or a spatial sound aims to record (or simulate), transmit (or store), and reproduce the spatial information of sounds. It subsequently recreates the desired spatial auditory events or perceptions. The spatial information of sound includes the localization information of sound sources and the comprehensive spatial information of environmental reflections. In Section 3.1, a practical multichannel sound is usually unable to reconstruct the target sound field exactly within the full audible frequency range. Therefore, multichannel sound signals should be created in accordance with appropriate psychoacoustic principles and methods to recreate the desired auditory events or perceptions in reproduction (Theile, 2001; Rumsey, 2001). Related psychoacoustic principles involve the principle of summing localization and recreating the comprehensive auditory perceptions of various ambient sounds (especially reflections). For summing localization with two or more loudspeakers, the interchannel level differences (ICLD)-based method can be interpreted by or derived from physical analysis. By contrast, the interchannel time difference (ICTD)-based method or a combination of ICTD- and ICLD-based methods, which are applicable to signals with transient characteristics, is based on psychoacoustic results, or even from experiences rather than from physical analysis. The comprehensive information of environmental reflections can be recreated via direct and indirect methods (Section 3.1), which are mainly

DOI: 10.1201/9781003081500-7

based on the results of psychoacoustic experiments or experiences. Some signal recording and simulation techniques may be interpreted at the neurophysiological level of hearing (Section 2.1.4).

Two ways can be applied to obtain multichannel sound signals: (1) onsite recording via a set of microphones with appropriate configuration and directivities and (2) simulation or synthesis via signal processing. Various microphone and signal simulation techniques have been designed and used in practice on the basis of the aforementioned psychoacoustic principles. Two typical examples are microphone and signal simulation techniques based on the principles of pair-wise amplitude panning (or its three-dimensional extension to VBAP) and Ambisonics. Nevertheless, practical microphone and signal simulation techniques for multichannel sounds are not limited to these two principles. The principles of pair-wise amplitude panning and Ambisonic signal mixing are analyzed in detail in the preceding chapters because on the one hand, these analyses reveal the abilities and limitations of various multichannel sounds. On the other hands, pair-wise amplitude panning and Ambisonic signal mixing are two common methods of multichannel sound recording.

Two-channel stereophonic signals can be obtained through various microphone and signal simulation techniques (Section 2.2), such as coincident, near-coincident, and spaced microphone techniques, spot microphone and pan-pot technique, and various simulation techniques for reflections. These techniques can be extended to the cases of multichannel sound. However, microphone and signal simulation techniques for multichannel sounds are more complicated than those for two-channel stereophonic sounds. Some special problems should be considered.

First, more than two loudspeakers may contribute to summing auditory events in multichannel sound reproduction, thereby complicating the relationship between loudspeaker signals and resultant auditory events. Consequently, signal recording and simulation also become complicated. For directional localization signal recording, coincident microphone techniques recreate signals with ICLD only. They can be designed in terms of the summing localization theorems presented in Section 3.2. Spaced and near-coincident microphone techniques recreate signals with ICTD or a combination of ICTD and ICLD. However, no physical models are available for ICTD-based summing localization or a combination of ICTD- and ICLD-based summing localization. Moreover, the detailed results of localization experiments of three or more channel signals with ICTD or a combination of ICTD and ICLD are unavailable. Even for two-channel signals with ICTD or a combination of ICTD and ICLD, the results of the localization experiment on various two-loudspeaker configurations are incomplete except for those on two frontal stereophonic loudspeakers. In addition, the comb filtering effect and resultant timbre coloration caused by the interference among multichannel signals with ICTD is a complicated problem. Therefore, the design of multichannel-spaced and near-coincident microphone techniques can only refer to those of two-channel microphone techniques. Nevertheless, various microphone techniques are not only for directional localization signals but also for ambient signal recording. Many practical multichannel (especially 5.1-channel) microphone techniques are designed on the basis of qualitative considerations or even experiences. The design of microphone techniques for a multichannel spatial surround sound is more complicated because the mechanism of auditory vertical perception is different from that of horizontal perception.

Second, multichannel sound increases the dimensionality and improves the ability in spatial information reproduction compared with that of a two-channel stereophonic sound. Therefore, spatial information may be recreated from various directions not limited to the region between two frontal stereophonic loudspeakers. Multichannel sounds should recreate more natural and richer auditory scenes than those of a two-channel stereophonic sound. However, the abilities of a multichannel sound to recreate spatial information are still limited because of loudspeaker configuration and signal mixing. For example, an ITU 5.1-channel loudspeaker configuration was originally not intended for recreating a full 360° horizontal virtual source (Section 5.2). Pair-wise amplitude panning fails to recreate a stable lateral virtual source. The virtual source in the rear region is also unstable. Furthermore, pair-wise time panning leads to a similar problem. Global Ambisonic-like signal mixing may improve the localization performance of a lateral virtual source in a narrow listening region. Therefore, the limitations of the reproduction system and loudspeaker configuration should be considered in the design of microphone and signal simulation techniques for multichannel sounds.

Third, in many cases, the compatibility of stereophonic downmixing and mono downmixing should be considered in the design of microphone and signal simulation techniques for multichannel sounds and stereophonic sounds, respectively.

The design of recording techniques aims to capture the target spatial information effectively, represent information with appropriate signal relationships, and recreate information correctly in reproduction. In addition to the aforementioned problems, the physical characteristics of original or target sound fields should be considered in the design and selection of a microphone technique. The physical characteristics and spatial information of a target sound field vary considerably across different acoustic environments, which require different designs of microphone techniques for recording. For classical music performance in a concert hall, in addition to direct sounds from the orchestra, ambient sounds are mainly reflections of the hall. Ideally, microphone techniques for classical music recording should meet the following requirements (although practical techniques usually only meet some of these requirements):

1. Ability to capture the localization information of frontal target sources (orchestra) to recreate stable, definite, and natural frontal virtual sources in reproduction.
2. Ability to capture the spatial information of early lateral reflections and late diffused reverberation to recreate good auditory spatial impression in reproduction.
3. Ability to balance the perceived loudness of different sources (instruments) and to balance the proportion of direct and reflected sounds; ability to recreate a fused auditory scene in reproduction.

Microphone techniques for the recording of outdoor sports tournaments, where the ambience involves cheering from viewers, are different from those for music recording in a concert hall. Various microphone techniques are also required for different types of music recording because of differences in the acoustic characteristics of different concert halls, the directivities of instruments, the distribution and width of an orchestra on a stage. Moreover, an appropriate choice of microphone techniques for recording depends on the physical characteristics of reproduction systems and environments. Ideally, signals recorded through a microphone technique should be appropriate for as many reproduction systems and environments as possible. However, for a channel-based spatial sound, microphone techniques with such a universal characteristic are unavailable.

The appropriate selection of a microphone technique depends on the desired effects in reproduction. These effects are related to psychoacoustics in reproduction, such as direction, distance, depth of sources (or depth of scene), timbre, intensity balance, and fused sensation among different sources, subjective spaciousness, and envelopment. It is also associated with the esthetic content of the musical performance. Therefore, similar to stereophonic recording, multichannel sound recording deals with scientific (technical) and esthetic problems (Guan, 1988). In practice, an appropriate choice is based on a comprehensive consideration of all aspects. An appropriate compromise should be made if fulfilling all the requirements is impossible.

Overall, various microphone and signal simulation techniques are available for multichannel sounds, but "optimal" or "standard" techniques are unavailable. Appropriate techniques are designed and chosen on the basis of practical conditions.

7.2 MICROPHONE TECHNIQUES FOR 5.1-CHANNEL SOUND RECORDING

7.2.1 Outline of microphone techniques for 5.1-channel sound recording

A 5.1-channel sound is a typical multichannel horizontal surround sound with an irregular loudspeaker configuration. Microphone techniques for a 5.1-channel sound have been widely explored and developed. Various microphone techniques for recording different sources and environments have also been established. In this section, some typical microphone techniques for classical music recording in a hall are mainly discussed. Microphone techniques for other sources and environments, such as natural sources in outdoor environments and sports tournaments in indoor or outdoor environments, may greatly differ from the techniques discussed in this section (Kirby et al., 1998).

Microphone techniques for 5.1-channel sound recording can be classified in different ways. Similar to the cases of two-channel stereophonic microphone techniques, microphone techniques for 5.1-channel sound recording can be classified into coincident, near-coincident, and spaced microphone techniques in terms of their principle and configuration. A combination of microphone techniques based on different principles may be used to record localization and ambient information because the principles of recording are different. Therefore, a classification of microphone techniques for 5.1-channel sound recording in terms of their principles may be inconvenient. Rumsey (2001) suggested classifying microphone techniques into two categories. The first category is the main microphone technique through which a single array of microphones close to one another is used to record information on frontal and surround channels. The second category is the combination of frontal and surround microphone techniques through which two separate microphone arrays are used to record information on frontal and surround channels. In the aforementioned techniques, the central channel signal may be recorded with a separate microphone in some instances. Moreover, some additional microphones (such as spot microphones) may be supplemented with various microphone techniques, and their outputs are mixed to the main channels to enhance source information.

LFE channel signals in a 5.1-channel sound are usually derived from the main channel signals rather than recorded using a separate microphone. LEF channels are even omitted in classical music program production. Therefore, the technique presented in this section may be appropriately called "microphone techniques for 5-channel (3/2) surround sound recording." However, these techniques are still termed "microphone techniques for 5.1-channel sound recording" to be consistent with preceding discussions.

7.2.2 Main microphone techniques for 5.1-channel sound recording

In the main microphone techniques, a single microphone array is used to capture localization and ambient information. The microphone array is usually placed close to the orchestra (within the distance of the reverberation radius of a hall). This category of microphone techniques is based on the hypothesis that system and loudspeaker configuration can recreate a full 360° horizontal virtual source in reproduction (but this assumption is inappropriate); in this way, similar auditory sensations caused by reflections in a hall can be recreated via indirect methods (Section 3.1). Therefore, the main microphone technique aims to capture information on full 360° horizontal localization.

In accordance with the principle of summing localization with ICLD only, the main microphones for 5.1-channel sound recording can be theoretically constructed with an array of coincident microphones with appropriate directivities. However, the design of such an array is complicated, and the resultant performance may not satisfy the practical requirements (Section 7.2.3). Therefore, many practical main microphone techniques for 5.1 channel sound recording are based on the principle of summing localization with a combination of pair-wise ICLD and ICTD. A horizontal plane is divided into five sectors in accordance with a 5.1-channel loudspeaker configuration, and a pair of adjacent loudspeakers are used to recreate a virtual source within the corresponding sector. Accordingly, an array of five near-coincident microphones with appropriate directivities is utilized to capture signals. Figure 7.1 illustrates the configuration of this array. Five microphones are arranged in the left L (or left-front LF), center C (or central-front CF), right R (or right-front RF), left-surround LS (or left-back LB), and right-surround RS (or right-back RB) directions and spaced to one another with appropriate distances. The C microphone is placed slightly forward with respect to L and R microphones; therefore, the three frontal microphones form a ***front triplet*** or ***LF*** and ***RF pairs***. The LS and RS microphones form the ***surround*** or ***back pair***. The L and LS microphones form the ***left side pair***, and the R and RS microphones form the ***right-side pair***. Therefore, the five microphones form five adjacent pairs. Cardioid microphones are often used for the main microphone array. The main axes of microphones are pointed to the left-front (or left), central-front, right-front (or right), left-back, and right-back directions. The consistency of the directions of the main axes of microphones with those of loudspeakers is unnecessary. Microphones with other directivities (such as supercardioid or hypercardioid microphones) or an array of microphones with different directivities (but the left and right microphones should be symmetric) may also be used. Distances between microphones are typically in the order of 0.1–1.0 m.

The main microphone array and corresponding parameters (such as directivities, directions of the main axis, and positions of microphones) are designed in terms of the rule and hypothesis on the summing localization between two adjacent channels with a combination of ICLD and ICTD. Although the principle of the main microphone techniques for 5.1-channel sound recording is similar to that of the near-coincident microphone techniques for stereophonic recording, some special problems of the main microphone techniques should be considered.

Inter-pair crosstalk is a problem. Generally, the sound wave caused by a target source is captured with all five microphones. However, the main microphone techniques rely on the principle of summing localization with adjacent channels. For a given target source direction,

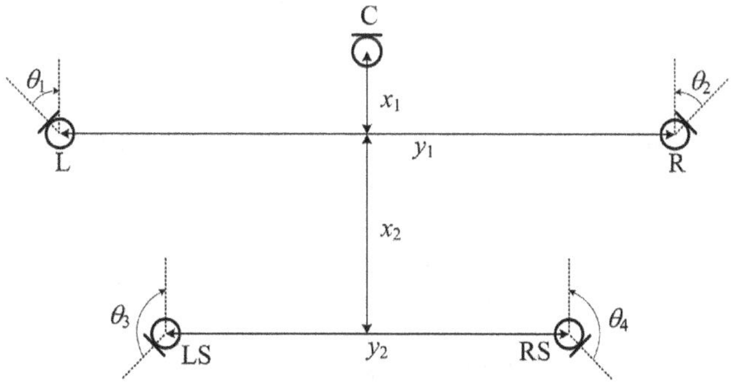

Figure 7.1 Typical configuration of the main microphone array for 5.1-channel sound recording.

except the signals from the concerned microphone pair, crosstalks from other three microphones interrupt summing localization and result in complicated or uncertain auditory events. The influence of crosstalks should be restrained as far as possible. According to the psychoacoustic principle of a summing spatial auditory event with a multiple sound source or loudspeakers, three methods are available to reduce the influence of crosstalks.

1. Gain difference method. Gain differences between channels are introduced so that crosstalks from other channels are attenuated. Psychoacoustic experiments indicate that the influence of crosstalks on virtual source localization can be omitted when the level of crosstalks is attenuated at least −18 dB with respect to that of the desired signals.
2. Time difference method. According to the precedence effect, time differences between channels are introduced so that crosstalks from other channels are delayed, and their influence on virtual source localization can be omitted. The delay required by precedence effect depends on the type of stimuli. It usually lies between 1.5 ms to 40 ms.
3. Combination method of gain and time differences. Gain and time differences between channels are introduced simultaneously. The required amounts of attenuations and delays, which may depend on the type of stimuli, are determined from the results of psychoacoustic experiments.

The gain or magnitude of a crosstalk can be reduced by appropriately choosing the directivities and main axis directions of microphones. However, calculation reveals that cardioid microphones usually reduce the gains of crosstalks by an order of several decibels, depending on the angle between the source and main axes of microphones. Therefore, in most cases, the directivities of microphones alone are not enough to restrain the influence of crosstalks. The distances between microphones introduce time differences between channel signals. For a plane wave incident from the direction parallel to the straight line between two microphones, a 0.51 m interval between microphones introduces a time difference of 1.5 ms between channel signals. Such a microphone configuration can be achieved.

The combination method is often used in practical main microphone arrays, e.g., microphones with appropriate directivities and main axis directions are spaced apart with some distances. It imposes less critical restrictions on the directivities and interval of microphones in comparison with the gain or time difference method alone because of the cooperative effects of gain and time differences. A near-coincident configuration of cardioid microphones is usually sufficient. In designing this microphone array, other factors, in addition to localization, should be considered. The main microphone array in Figure 7.1 is designed on the basis of the combination method. For example, when a target source is located at the direction between the C and L microphones, the cardioid directivity and main axis directions of the microphones introduces some attenuation in the R, LS, and RS microphone outputs. Intervals between microphones introduce some delays in the three aforementioned microphone outputs. Therefore, attenuations and delays restrain the influence of crosstalks from these microphone outputs. For a target virtual source located at other directions, the analyses are similar to above case.

Another problem is the effective recording range or ***sector*** covered by each adjacent microphone pair and the critical link between the effective recording ranges. As stated in Sections 2.2.4 and 2.2.6, the effective recording range of a near-coincident stereophonic microphone pair covers a frontal sector with the left-right symmetry. The range is generally determined by the directivity, the directions of the main axes, and the interval of the microphone pair and specifically identified by the summing localization rule of a combination of ICTD and ICLD.

This range is not necessarily consistent with the span angle between the main axis directions of two microphones. The former may be equal to, larger, or smaller than the latter.

A 5.1-channel main microphone array, which involves five adjacent near-coincident microphone pairs, is different from a stereophonic microphone pair. The effective recording ranges of five adjacent microphone pairs in the 5.1-channel main microphone array should cover the full 360° horizontal azimuth continuously without an overlap, which requires a comprehensive and careful design of the effective recording range of each adjacent microphone pair.

Each adjacent pair with the corresponding effective recording range in the 5.1-channel main microphone array can be obtained by manipulating the rotation to a stereophonic microphone pair. Figure 7.2 illustrates an example of a manipulation of anticlockwise rotation to a stereophonic microphone pair. Moreover, ***recording range offset*** may be required to link the effective recording range of each adjacent pair critically. The recording range offset can be realized by ***time offset***, ***level or intensity offset***, or ***combined time and level offset***. Time offset refers to introducing some additional time differences to microphone outputs. Level offset corresponds to introducing some additional level differences to microphone outputs. The combined time and level offset refers to introducing both additional time and level differences to microphone outputs. For example, for the L and C microphone pair, when the output of the C is advance in time, or the gain of C microphone is enhanced with respect to that of the L, the virtual sources in reproduction are offset toward the front (clockwise) and the effective recording range rotates anticlockwise.

The additional time and level difference in the recording range offset can be implemented by changing the microphone positions in a process called a ***microphone position offset***. For example, in the microphone array (Figure 7.1), the C microphone is placed slightly forward with respect to the L and R microphones, and its main axis is oriented to the central-front direction. The output of the C microphone is enhanced and advanced in time with respect to that of the L microphone. Therefore, the effective recording range rotates anticlockwise as desired. In other words, a lightly forward position of the C microphone causes a time and level offset of the effective recording range. The analyses on other adjacent microphone pairs are similar. The additional time and level difference in the recording range offset can also be implemented by introducing appropriate electronic delay and attenuation (or boost) to microphone outputs in a process called ***electronic offset***.

Overall, the design of the main microphone array is to choose the configuration, directivities, main axis directions, and intervals of microphones appropriately to meet the requirement of crosstalk restraint and critical link. The design is based on a conversion of the

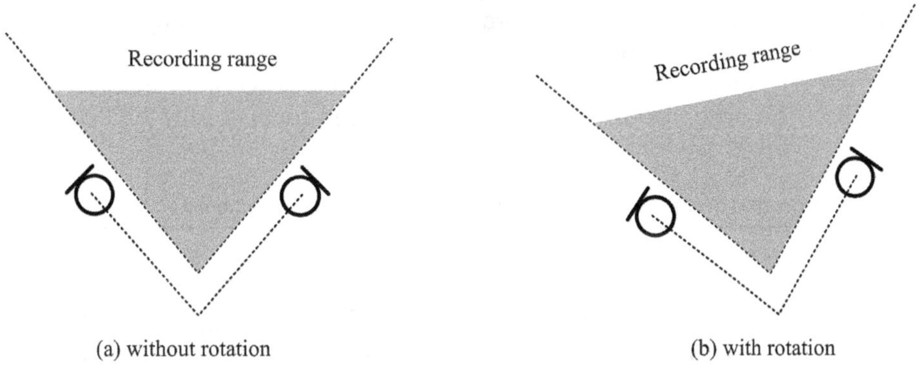

Figure 7.2 Manipulation of anticlockwise rotation to a stereophonic microphone pair.

psychoacoustic experiment results of stereophonic localization with a combination of ICLD and ICTD. Williams and collaborators (Williams and Du, 1999, 2000, 2001, 2004, Williams, 2013) analyzed the main microphone arrays and presented the methods and software for array parameter design. Many practical main microphone arrays are designed on the basis of these analyses and methods. The steps of the design of the main microphone array are outlined as follows:

1. Choose the microphones with appropriate directivities.
2. Choose the interval and main axis directions of the three frontal microphones according to the required effective recording range in the two frontal sectors.
3. Choose the interval and main axis directions of the two surround (back) microphones according to the required effective recording range in rear sector.
4. Choose the interval between three frontal microphones and two surround microphones according to the requirements of the critical link in the side sectors. If necessary, some additional electric offsets of the time and level between the outputs of three frontal microphones and two surround microphones are introduced. However, the level offset changes the front-to-back half-space pick up ratio of power.

The effective recording ranges of the three frontal microphones and the two surround microphones are selected according to practical requirements. Williams (2003) provided some results for reference. The effective recording range for the three frontal microphones usually lies between ±50° and ±90°, and effective recording range for the two surround microphones usually lies between 30° and 100°. The accuracy on the effective recording range of three frontal microphones are not so crucial as that of a stereophonic microphone pair because the effective recording ranges of all adjacent microphone pairs have been linked critically (Section 2.2.6). Moreover, the effective recording range of the three front microphones can be designed to be consistent with that of the stereophonic pair composed of L and R microphones to be compatible with two-channel stereophonic microphone recording (Williams, 2007). The aforementioned analysis can be extended to the cases of the sound incidence above and below the horizontal plane (Williams, 2002).

Three examples of the main microphone array with cardioid microphones for 5.1-channel recording are listed in Table 7.1. The notations of the microphone positions in the table are illustrated in Figure 7.1. The first example in the table comes from many designs by Williams (Williams and Du, 2001). No electronic offsets are introduced in this example. The effective recording ranges of the three frontal microphones and two surround microphone pairs are ±72° and 72°, respectively. A critical link requires that the effective recording range of a side microphone pair is also 72°. In other words, the effective recording ranges of all adjacent microphone pairs in this example are equal.

The second example in Table 7.1 is the main microphone array of a true space recording system (TSRS). This array has also been designed and used commercially (Williams and Du,

Table 7.1 Three examples of the main microphone array with cardioid microphones

				Recording range		
Examples	$x_1/y_1/x_2/y_2$ *(m)*	$\theta_{1,2}$	$\theta_{3,4}$	*Front*	*Back*	*Electronic offset*
Williams	0.17/0.61/0.415/0.48	±90°	±160°	±72°	72°	No
TSRS	0.23/0.88/0.23/0.56	±70°	±156°	±60°	60°	−2.4 dB to the front output
INA-5	0.179/0.35/0.515/0.6	±90°	±150°	±90°	60°	No

1999). The positions, main axis directions of the microphones, and effective recording ranges of each pair in the TSRS array are different from those of the above example. An electronic offset, e.g., a −2.4 dB attenuation to the outputs of three frontal microphones with respect to those of two surround microphones, is introduced to make a critical link.

The third example in Table 7.1 is the main microphone array of INA-5 (Herrmann et al., 1998). According to the original literature, the interval between the L and C microphones is 0.25 m; the interval between the L and R microphones is 0.35 m; and the interval between the L and LS (or between the R and RS) microphones is 0.60 m. The position parameters of microphones in the original literature have been converted into the parameters remarked in Figure 7.1.

The principle and configuration of the main microphone array of the optimum cardioid triangle (OCT) is also similar to those in Table 7.1 (Theile, 2001; Wittek and Theile, 2002). The array of the OCT was originally developed for the recording of three frontal channels and then for 5.1-channel recording. As shown in Figure 7.3, the main microphone array of the OCT involves a cardioid microphone with its main axis pointing to the frontal direction of 0°, a pair of L and R hypercardioid microphones with their main axis pointing to the lateral directions ±90°, and a pair of surround cardioid microphones with their main axis pointing to the back directions. Hypercardioid microphones reduce the crosstalks between channels. The intervals between microphones are $x_1 = 0.08$ m, $y_1 = 0.40$–0.90 m, $x_2 = 0.40$ m, and $y_2 = 0.6$–1.10 m.

Other examples of 5.1-channel main microphone arrays with different configurations, directivities, main axis directions of microphones and different effective recording ranges in each sector are presented. Williams and collaborators gave numerous examples of array designs with cardioid, supercardioid, and hypercardioid microphones. The details are omitted here.

The aforementioned main microphone technique for 5.1-channel sound recording is moderately successful in practice, but it is still **problematic from the perspective of the psychoacoustic principle of multichannel sounds.**

1. The aforementioned main microphone technique is based on the assumption that the 5.1-channel sound with pair-wise amplitude and time panning can recreate a full 360° horizontal virtual source in reproduction. However, the 5.1-channel sound is inappropriate for a full 360° virtual source reproduction, i.e., it fails to recreate a lateral virtual source, and the rear virtual source is unstable (Sections 5.2.2 and 5.2.5). Even with this

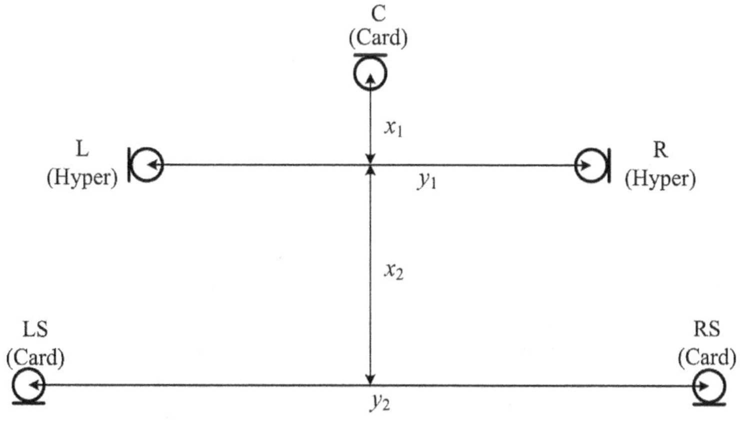

Figure 7.3 Main microphone array of the OCT for 5.1-channel recording.

defect, if the lateral and rear directions are dominated by the information of environmental reflections, the intervals among microphones lead to the decorrelated outputs of environmental reflections and thus recreate some desired sensations of envelopment in reproduction. In this case, hearing is more tolerant to the defect of lateral and rear virtual source in reproduction.

2. Ideally, crosstalks in the main microphone array should be restrained at all. However, a practical main microphone array is unable to restrain the crosstalks completely. In particular, a combination method of gain and time differences may restrain the influence of crosstalk on localization, but cardioid microphones can reduce the gains of crosstalks by an order of several decibels. The interference and timbre coloration caused by a crosstalk in reproduction are non-negligible, but studies on this issue are inadequate.
3. The critical link in the main microphone array places some restrictions on the distance, relative delay, and gain between the three frontal microphones and the two surround microphones; consequently, the ratio of the captured direct and reflected sounds is limited to some extent.
4. A 5.1-channel main microphone array is designed on the basis of the localization experiment results of traditional two-channel stereophonic sounds with a combination of ICTD and ICLD, e.g., on the basis of the results derived from the Williams curves (or similar curves) in Section 2.1.4. However, the summing localization of two adjacent channel signals with ICTD and ICLD depends on loudspeaker configurations. **Simon and Mason** (2010) **conducted** a virtual source localization experiment **and preliminarily confirmed this result although their experiment was based on a regular configuration with eight loudspeakers rather than the 5.1-channel loudspeaker configuration. Therefore, the direct conversion of the localization results of two-channel stereophonic sounds to that of a 5.1-channel loudspeaker configuration is problematic, especially in the lateral region. However, the complete localization experiment results of a 5.1-channel loudspeaker configuration with a combination of ICTD and ICLD are still unavailable.**

7.2.3 Microphone techniques for the recording of three frontal channels

In a combination of frontal and surround microphone techniques, two separate microphone arrays are used to capture frontal localization and ambient information, the outputs of microphones are mixed to form 5.1-channel signals. This category of microphone techniques aims to capture the localization information of frontal sources (sound stage) and the ambient information of the environment rather than the full 360° horizontal localization information. Various microphone arrays are used for the recording of three frontal channels because the 5.1-channel sound is originally intended to reproduce the frontal localization information by using three frontal channels.

Microphone techniques for the three frontal channels were developed in the early times of spatial sound. The outputs of three microphones with a configuration similar to this of the Decca tree in Figure 2.16 (c) can be used as the signals of three frontal channels in a method called a spaced microphone technique. In this technique, omnidirectional microphones are used, and the precedence effect is utilized to recreate the localization perception of virtual sources. Similar to the case of two-channel stereophonic techniques, this method is simple. The delays introduced by the intervals among microphones are enough to restrain the influence of crosstalks among microphones to localization in terms of the precedence effect. However, this method exhibits some problems and defects when it is used to capture frontal localization information. The first defect is the degradation of the perceptual quality of a virtual source in reproduction. The second defect is that omnidirectional microphones capture too many reflected components from the rear directions. The third defect is that the virtual source is anchored to the position of three frontal loudspeakers, thereby creating the hole between loudspeakers in reproduction.

The second defect can be overcome by using microphones with appropriate directivities and main axis directions. For traditional two-channel stereophonic recording, the Decca tree involves three omnidirectional microphones. Localization sounds from frontal directions and reflections from rear directions are captured with these microphones and then reproduced together with a pair of frontal stereophonic loudspeakers. However, for the 5.1-channel sound, frontal information and back information are separately reproduced by three frontal loudspeakers and two surround loudspeakers. The frontal microphones with appropriate directivities and main axis directions restrain rear reflections in the three frontal channels, and the rear reflections are captured with another microphone array.

The third defect mentioned above becomes serious when recording is performed in a very wide source stage (such as a large orchestra), where an excessively wide-spaced microphone array is required. Similar to the case in Section 2.2.3, two additional microphones are supplemented into the three microphone arrays to form a five-microphone array distributed in the line across the stage width to address this problem. The outputs of two additional microphones are mixed into the left, center, and right channels with an appropriate gain. Figure 7.4 illustrates an example of a five-microphone array for three frontal channels recorded by Theile (2001). This method eliminates the defect of the hole between loudspeakers in reproduction, but the interference among microphone outputs may create comb filtering and timbre coloration.

To overcome the third defect mentioned above, Edwin (2002) suggested using a wide-spaced omnidirectional microphone pair and an ORTF triple at the intermediate position to record three frontal channels. The ORTF triple involves three near-coincident cardioid microphones arranged similarly to that of three frontal microphones in Figure 7.1. The interval and main axis directions of the left and right microphones in the ORTF triple are similar to those in an ORTF pair for two-channel stereophonic recording (Section 2.2.4). The left and right outputs of the wide-spaced microphone pair are fed to the left and right channels, respectively. The output of the C microphone in the ORTF triple is directly fed to the center channel. The left output of the ORTF triple is panned to the position between the frontal-center and the full left, and the right output of the ORTF triple is panned to the position between the frontal-center and the full right.

Two pairs of stereophonic microphones can also be used to capture three frontal channel signals in a very wide source stage. Each stereophonic microphone pair is an XY (or their

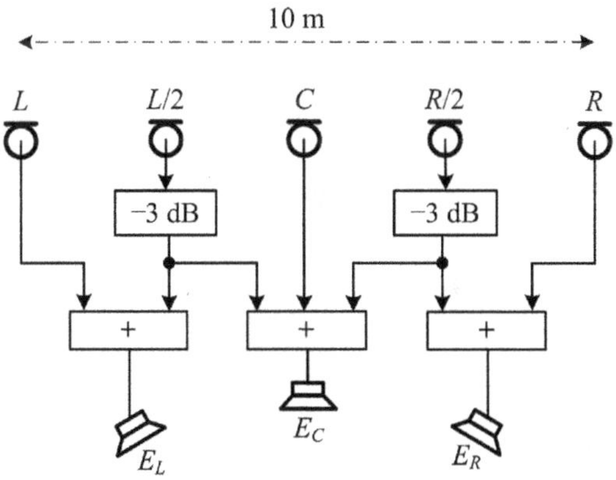

Figure 7.4 Five-microphone array for the recording of three frontal channels. (adapted from Theile, 2001).

equivalent MS) or near-coincident pairs. Two microphone pairs are widely spaced and used to record the source signals in the LF and RF regions. The output of the left microphone in the left pair is fed to the left channel, the output of the right microphone in the right pair is fed to the right channel, and the outputs of the two other microphones are mixed to the central channel. Figure 7.5 illustrates an example given by Germanenn (1998). The delays in the output of the left microphone in the left pair and the right microphone in the right pair compensate for the directional shift of virtual sources because of the propagating attenuation in the central channel signal.

In practice, a triangular microphone array shown in Figure 7.6 is often used to capture three frontal channel signals. In contrast to the aforementioned array of three omnidirectional microphones with a wide space, the array shown in Figure 7.6 involves three directional microphones with a close space. This microphone array creates three frontal channel signals with appropriate ICLD and ICTD and thus recreates a frontal virtual source within the range of three frontal loudspeakers in reproduction. The microphone array in Figure 7.6 is equivalent to the three frontal microphones in the main microphone array for 5.1-channel recording (Section 7.2.2). More exactly, the main microphone array for 5.1-channel recording is an extension of the triangular microphone array in Figure 7.6. The analysis and design of this triangular microphone array, including effective recording range, directivities and main axis directions of microphones, and spaces between microphones, are similar to those in Section 7.2.2. Williams (2004) provided numerous design examples.

Table 7.1 illustrates the example of an INA-5 main microphone array. The microphone configuration of the three frontal channels in INA-5 comes from the INA-3 array for the recording of three frontal channels. INA-3, which was designed by Herrmann et al. (1998),

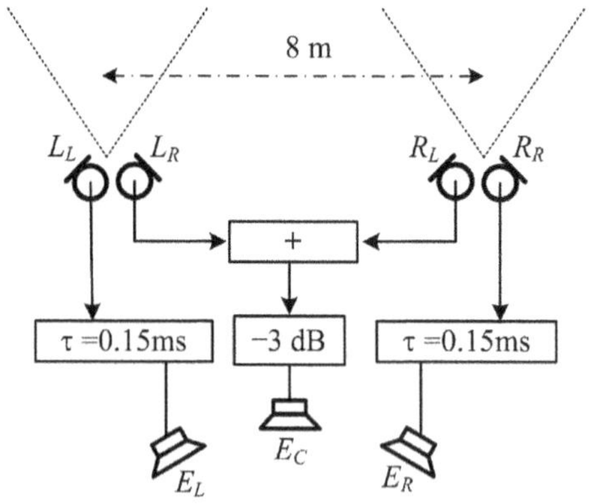

Figure 7.5 Three frontal channel recording with two pairs of stereophonic microphones. (adapted from Germanenn, 1998).

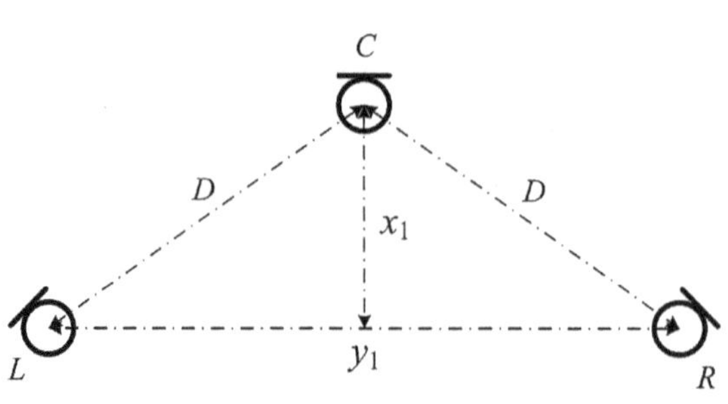

Figure 7.6 Triangular microphone array with a close space for three frontal channel recording.

Table 7.2 Positional parameters and effective recording range of the INA-3 microphone array

$x_l/y_l/D$ (m)	Effective recording range	Directions of main axes of the left and right microphones
0.28/1.26/0.69	±50° (100°)	±50°
0.26/0.92/0.53	±60° (120°)	±60°
0.23/0.68/0.41	±70° (140°)	±70°
0.21/0.49/0.32	±80° (160°)	±80°
0.18/0.35/0.25	±90° (180°)	±90°

involves three cardioid microphones. The main axes of three microphones point to the LF (or left), C, and RF (or right) directions. Table 7.2 lists the effective recording range of the INA-3 array with different parameters (Theile, 2001).

Theile also proposed the OCT array with three frontal channel microphones shown in Figure 7.7. (Theile, 2001; Wittek and Theile, 2002). The OCT frontal microphone array is equivalent to the three frontal microphones in the OCT main microphone array for 5.1-channel recording (Figure 7.3). The OCT main microphone array is constructed by adding two surround (back) microphones to the OCT frontal microphone array. A pair of omnidirectional microphones can be supplemented to the OCT frontal microphone array to enhance the low-frequency response of array. The high-pass filtered output of the left (right) supercardioid and low-pass filtered output of the left (right) omnidirectional microphone are mixed to the left (right) channel. The crossover frequency of the filter is 100 Hz because the low-frequency component below 100 Hz slightly contributes to localization. The effective recording ranges of this array are ±55° and ±45° for spaces of 0.6 and 0.8 m between the left and right microphones, respectively.

A baffle can also be inserted between the left and right microphones of the three frontal microphones to reduce crosstalks. Figure 7.8 illustrates an example suggested by Hamasaki (Rumsey, 2001). Three near-coincident cardioid microphones are spaced at 0.3 m with one another. A pair of omnidirectional outrigger microphones spaced about 2–3 m is supplemented. The outputs of the outrigger microphones are low-passed filtered at 250 Hz and mixed to the left and right channels to improve the low-frequency quality.

Some studies have suggested using a line array of three directional microphones to capture the three frontal channel signals. For example, Klepko (1997) proposed a line array with three microphones spaced at 0.175 m to one another. A cardioid microphone with its main axis pointing to the front is used for the center channel, and two supercardioid microphones with

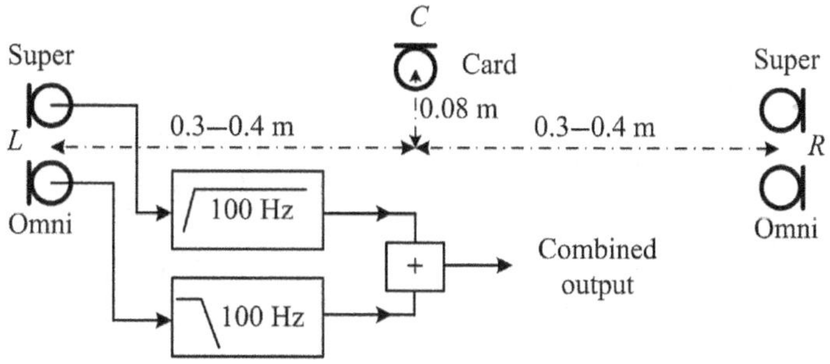

Figure 7.7 OCT frontal microphone array by Theile.

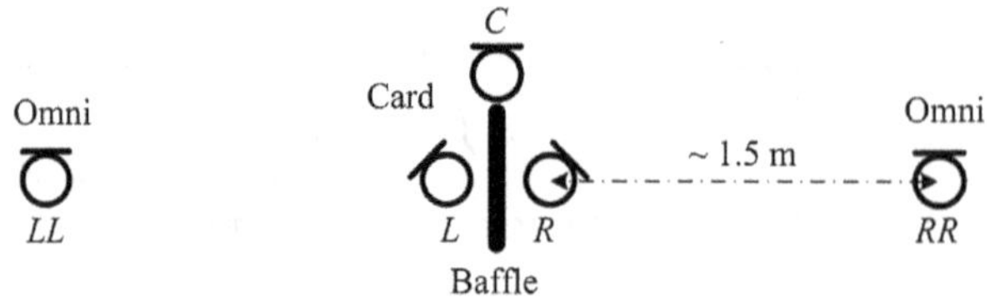

Figure 7.8 Baffled microphone array of three frontal channels recorded by Hamasaki (adapted from Rumsey, 2001).

their main axis pointing to the LF and RF directions are utilized for the left and right channels, respectively. Supercardioid microphones aim to restrain the crosstalk from an opposite channel. However, as stated in Section 7.2.2, a line configuration of microphones is not good for the control of the effective recording range of an adjacent microphone pair and the critical link between pairs. As stated by Theile (2001), crosstalks in the outputs of a line microphone array influence the quality of reproduction even if supercardioid microphones are used in the array.

Theoretically, three frontal channel signals can be captured by three coincident microphones with appropriate directivities (McKinnie and Rumsey, 1997). The directivities and main axis directions of coincident microphones can be designed in accordance with local Ambisonic-like signal mixing in Section 5.2.4. The resultant microphone outputs or loudspeaker signals are given in Equation (5.2.32) and shown in Figure 5.8. However, microphone products with directivities expressed in Equation (5.2.32) are unavailable. Alternatively, the signals in Equation (5.2.32) can be linearly decoded from the outputs of an omnidirectional microphone and two bidirectional microphones with their main axes pointing to the front and left directions, respectively. Moreover, the signals in Equation (5.2.32) do not satisfy the condition of constant power in outputs.

McKinnie and Rumsey (1997) proposed to use a MS pair (with inverse MS transformation) to capture the L and R channel signals and another coincident microphone to capture the C channel signal. The microphone for the S signal is the bidirectional microphone with its main axis pointing to the left directions. The microphones for M and C signals can be chosen from one of the three methods (Rumsey, 2001):

1. A supercardioid microphone and a bidirectional microphone are used for M and C signals, respectively, and the main axes of two microphones point to the frontal direction.
2. A supercardioid microphone and a hypercardioid microphone are used for M and C signals, respectively, and the main axes of two microphones point to the frontal direction. This method leads to the lowest rear recording among the three methods.
3. An omnidirectional microphone and a supercardioid microphone are used for M and C signals, respectively, and the main axes of supercardioid microphone points to the frontal direction.

Cohen and Eargle (1995) used three coincident microphones with a second-order directivity to capture the three frontal channel signals. The main axes of microphones point to azimuths 0° and ±74°. The normalized amplitudes of microphone outputs are

$$\begin{aligned} A_L &= \left[0.5 + 0.5\cos(\theta_S - 74°)\right]\cos(\theta_S - 74°) \\ A_C &= (0.5 + 0.5\cos\theta_S)\cos\theta_S \\ A_R &= \left[0.5 + 0.5\cos(\theta_S + 74°)\right]\cos(\theta_S + 74°), \end{aligned} \qquad (7.2.1)$$

where θ_S is the target source azimuth in the original sound field. The normalized magnitude of C, L, and R microphone outputs maximize to a unit at $\theta_S = 0°$ and ±74°, respectively. When a target source is located midway between the main axis directions of two adjacent microphones, the normalized magnitudes of these two microphone outputs decrease by −3 dB with respect to the maximal on-axis output of a unit. A direct realization of the microphones with a second-order directivity may be difficult, but the signals given in Equation (7.2.1) can be derived for the outputs of an appropriate microphone array.

The virtual source localization performance in the reproduction of signals captured by the aforementioned coincident microphone array can be analyzed on the basis of the theorems presented in Section 3.2. Similar to the case of two-channel stereophonic sound, virtual source positions in the reproduction of three frontal channels may not be exactly consistent with those of the actual source at the original stage, but recreating the relative position distribution of virtual sources in reproduction is enough for live recording.

7.2.4 Microphone techniques for ambience recording and combination with frontal localization information recording

As stated in Section 7.2.3, two separate microphone arrays can be used to capture the frontal localization and ambient information in 5.1-channel recording. For live recording in a concert hall, ambiences are mainly reflections. In this case, ambient information is usually recorded with a wide-spaced microphone array arranged relatively far from the sources. The resultant decorrelated reflected signals recreate subjective sensations similar to those in the concert hall by using the direct method stated in Section 3.1. For 5.1-channel recording, the outputs of the ambient microphone array may be fed to the two surround channels only; accordingly, the three frontal channel microphones should be involved in the recording of ambient information. Alternatively, the outputs of ambient microphone array may be fed to frontal and surround channels to recreate the sensations of envelopment in reproduction. Many techniques for ambience recording have been developed, but some of them are based on experience rather than strict acoustic theory. The combinations of ambient microphone and frontal channel microphone arrays in Section 7.2.3 result in various practical 5.1-channel microphone techniques. Furthermore, 5.1-channel recording with two separate microphone arrays is flexible. The performance of frontal localization information recording and ambience recording can be optimized separately with the less restrictive relation, and the direct-to-reverberation ratio in the recording is easily controlled. Various combinations of the frontal channel and ambient microphone arrays are available for a practical choice. An appropriate electronic delay may be supplemented to ambient signals to reduce their influence on frontal localization according to the precedence effect.

In the direct method of ambience recording, a pair of wide-spaced microphones is used to capture the decorrelated reflected signals. The theoretical basis of this method is expressed in Equations (1.2.29) and (1.2.30). An example of the combination of the frontal channel and ambient microphone arrays is the ***Fukada tree*** shown in Figure 7.9 (Fukada et al., 1997; Fukada, 2001). The configuration of three frontal microphones, including left (L), center (C), and right (R) microphones, is similar to that of the Decca tree. The Decca tree for two-channel stereophonic recording involves three omnidirectional microphones. The captured signals include frontal localization information and rear reflected information, and they are then reproduced by a pair of frontal stereophonic loudspeakers. In 5.1-channel reproduction, the frontal and rear information is reproduced by frontal and surround loudspeakers, respectively. Accordingly, three cardioid microphones with their main axes pointing to ± 55° to ± 65° in the LF and RF directions and to 0° in the C direction are used in the Fukada tree to capture the frontal source and frontal reflected signals. The directivity of the three frontal

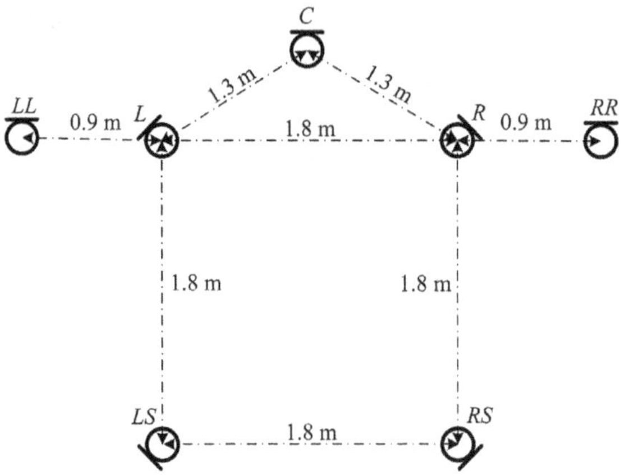

Figure 7.9 Fukada tree.

microphones reduces the captured power of rear reflections. Two omnidirectional outrigger microphones, namely, LL and RR, are sometimes added outside the left and right microphones. The outputs of the outrigger microphones are usually panned between the left and left surround (or right and right surround) channels to increase the recording width of the frontal stage. A pair of left-back (surround) and right-back (surround) cardioid microphones, denoted by LS and RS, are used to record surround channels. They are located at the reverberation radius of the hall and spaced at a distance not less than the reverberation radius. Their main axes point to ±135° to ±150°. The outputs of LS and RS microphones are dominated by decorrelated reverberation from the rear. As stated in Section 7.2.3, the three frontal channel signals captured with wide-spaced microphone arrays such as Fukada tree result in the degraded quality of a virtual source. However, the three frontal microphones capture the ambience from the frontal at the same time. Wide-spaced microphone arrays reduce the cross-correlation among the outputs and then improve the auditory spatial impression in reproduction.

In addition to the wide-spaced microphone array, a near-coincident pair whose main axes pointing to the left-back and right-back directions or even a XY coincident pair (or its equivalent MS pair) can also be used to capture rear reflection. The combination of this near-coincident pair and three appropriate frontal channel microphones results in a complete 5.1-channel microphone technique. In contrast to the main microphone array in Section 7.2.2, the two back (surround) microphones in this technique are located far from the three frontal microphones (e.g., at a distance of 2–3 m or more). Accordingly, the outputs of two back microphones are mainly rear reflections, and they possess a low correlation with the three frontal channel outputs so that the summing localization between the frontal and rear channels is ignored. A pair of coincident or near-coincident rear microphones is not enough to record the decorrelated reverberation signals. However, when the outputs of these two microphones are fed to a pair of (rear) surround loudspeakers, a subjective sensation similar to those caused by reflections in a hall may be recreated by using the indirect method in Section 3.1. For example, the 5.1-channel recording technique suggested by DPA involves an array similar to the Decca tree to capture the frontal channel signals and a near-coincident ORTF pair to capture the surround channel signals (Nymand, 2003). The distance between the adjacent frontal microphones varies from 0.6 m to 1.2 m. The rear ORTF pair is located 8–10 m from the frontal array. Berg and Rumsey (2002) also used a near-coincident cardioid pair to capture rear reflections, but they utilized three coincident microphones for the recording of three frontal channels.

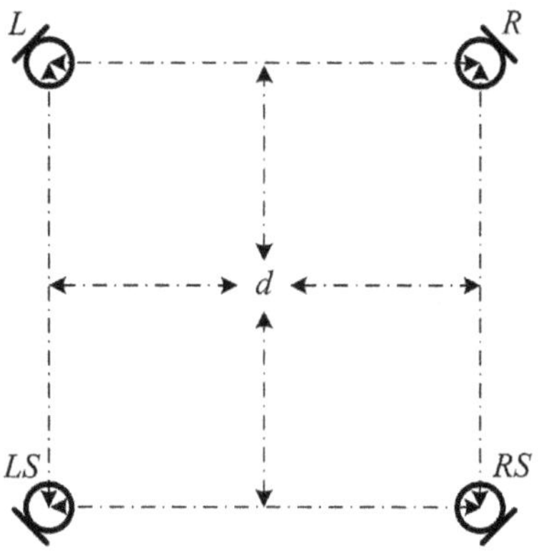

Figure 7.10 IRT cross.

Ambience can also be captured by four cardioid or omnidirectional microphones with a square arrangement. The four outputs are fed to the L, R, LS and the RS channels (Theile, 2001). The configuration of microphones is shown in Figure 7.10 and is called IRT-cross. When four cardioid microphones are used, the main axes of the microphones point to the LF, RF, LB and RB directions. The distance between the adjacent microphones varies from 0.25 m to 0.4 m. The distance of omnidirectional microphones is usually larger than that of cardioid microphones because the directivity of cardioid microphones also contributes to the decorrelation of the recorded signals in a reflected sound field. Theile combined the OCT frontal microphone array in Figure 7.7 with the IRT-cross in Figure 7.10 to construct a complete 5.1-channel microphone technique. The IRT-cross is located some distance behind the three frontal channel microphone array.

Hamasaki and Hiyama (2003) of NHK also proposed to use an array of four directional microphones with a square arrangement to capture the reflections in halls. This array is termed the Hamasaki square. The outputs of four microphones are fed to the L, R, and LS and the RS channels. Various configurations of the directivities of microphones are found in the Hamasaki square. Configuration 1 in Figure 7.11 (a) involves four bidirectional microphones with their main axes pointing to the lateral directions. It aims to capture lateral reflections and restrain the frontal direct sound and rear reflections to reduce their influence on frontal localization in reproduction. Configuration 2 in Figure 7.11 (b) involves two bidirectional microphones and two cardioid microphones. The main axes of cardioid microphones point to the rear directions. This configuration aims to capture rear reflections. Configuration 3 in Figure 7.11 (c) involves four bidirectional microphones with their main axes pointing to the lateral directions and two cardioid microphones with their main axes pointing to the rear. This configuration aims to capture lateral and rear reflections. The spaces between microphones in Figure 7.11 are chosen in terms of the correlation of the microphone outputs in the reverberation field, usually within the range of 2–3 m.

The Hamasaki square can be combined with other arrays to construct a complete 5.1-channel microphone technique. Hamasaki's original scheme was a combination of the frontal array in Figure 7.8 with a wide-spaced cardioid pair to capture rear/surround channel signals. The cardioid pair is located 2–3 m behind the frontal array and spaced apart at a distance of about 3 m. The main axes of the cardioid pair point to the left-back and right-back directions to capture the rear reflections. A Hamasaki square is shown in Figure 7.11(a) can be added to the above array to capture the ambient signals, and the outputs are mixed to the L, R, and the LS and the RS channels. The microphones in the Hamasaki square are spaced apart at a distance of 1 m (smaller than the latter choice of 2–3 m). In another scheme, Hamasaki also

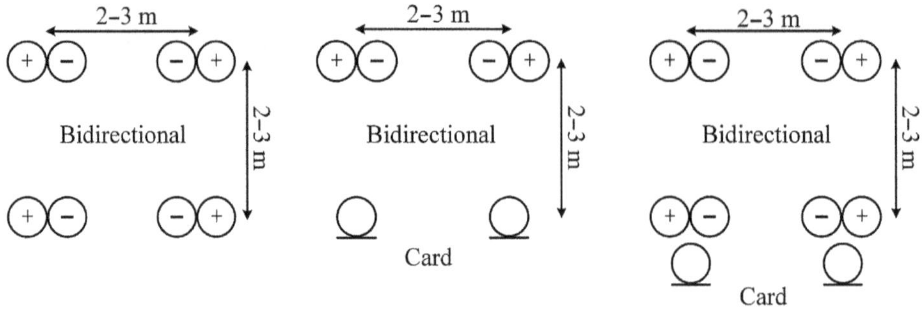

Figure 7.11 The Hamasaki square: (a) configuration 1; (b) configuration 2; (c) configuration 3.

suggested using a five-microphone array similar to Figure 7.4 to capture the three frontal channel signals, but supercardioid microphones instead of cardioid microphones are used. The microphones are spaced apart at a distance of 1.5 m. An omnidirectional pair spaced apart by 4 m is also added. Their outputs are low-pass filtered with a crossover frequency of 250 Hz and mixed to the left and right channels to enhance the low-frequency recording. A Hamasaka square is placed 2–10 m behind the frontal array, which is determined on the basis of the required ratio of direct and reflected sounds in the captured signals.

Klepko (1997) proposed to use an omnidirectional pair placed in the two ears of an artificial head to capture ambient signals and combine them with a lined array of three microphones (Section 7.2.3) for 5.1-channel recording. The artificial head is placed 1.24 m behind the line array. As stated in Section 1.4, an artificial head simulates the anatomical structures of a real human from the perspective of acoustics. Binaural signals from artificial head recording are originally appropriate for headphone presentation. As stated in Section 11.8, a crosstalk cancellation processing should be supplemented when binaural signals are reproduced through loudspeakers. The effect of the head shadow partly plays a natural role in crosstalk cancellation because surround loudspeakers in 5.1-channel reproduction are arranged at the azimuths (±110°). Therefore, crosstalk cancellation processing is omitted in Klepko's scheme. However, the final binaural signals in reproduction undergo the scattering and diffraction of the head/pinna twice (i.e., one is in the course of recording with the artificial head, and the other is in the course of reproduction to a listener, resulting in variation in the spectra in final binaural signals (pressures). Thus, timbre coloration occurs. Consequently, binaural signals from artificial head recording should be equalized.

The original purpose of artificial head recording is to make up for the deficiency of other methods, e.g., to recreate a virtual source within the rear region of ±90° with a pair of surround loudspeakers in 5.1-channel reproduction and to recreate the sensations similar to that in a hall by the indirect methods in Section 3.1. However, as stated in Section 11.8, even if crosstalk cancellation is included, the listening region of binaural signal reproduction via loudspeakers is narrowed. For a pair of surround loudspeakers with a wide span angle of 140°, a slight lateral translation of the head position spoils virtual source localization. On the other hand, the binaural signals captured by an artificial head in a nearly diffused reverberation field possess approximately equal power spectra and random phases. The scattering and diffracting effects of the artificial head enhance the randomness of binaural signals so that they are decorrelated. When reproduced by a pair of surround loudspeakers, these decorrelated signals lead to the sensation of envelopment in reproduction, and the perceived effect is less sensitive to the listening position. Therefore, using the artificial head for ambience recording is effective.

Some representative examples of microphone techniques for ambience recording and their combinations with microphone techniques for frontal channel recording are presented in the aforementioned discussion. Many other examples with similar designs and principles are described. The aforementioned microphone techniques for ambience recording are limited. A set of ambient microphones may sometimes not satisfy the practical requirement. Two or even more subsets of ambient microphones may be used to record the reflections. These subsets of ambient microphones are complementary and work together. The principles of these subsets are similar to the discussion above. The outputs of these subsets are mixed to channel signals. These subset techniques are also used for ambience recording in 5.1-channel sound. Hamasaki's original scheme mentioned above involves two subsets of microphones for ambience recording. Many examples of using two or even more subsets of ambient microphones are not discussed here.

7.2.5 Stereophonic plus center channel recording

In three frontal channel microphone techniques and 5.1-channel main microphone techniques, designing microphone arrays becomes complicated and difficult because of the introduction of the C channel. Consequently, the C can be captured with a separate microphone.

Three frontal channel signals can be captured through stereophonic plus center channel methods (shorten for stereo plus C) by which a stereophonic microphone pair is used to capture localization information in the frontal range, and the outputs are fed to the L and R channels. A separate cardioid or supercardioid microphone with its main axis pointing to the front direction is used to capture the C channel signal to enhance the stability of a center virtual source in reproduction. Various traditional stereophonic microphone techniques, such as a XY coincident pair or its equivalent MS pair, and a near-coincident pair, are suitable for this purpose. Stereo plus C is compatible with two-channel stereophonic recording.

For example, three channel signals are recorded with stereo plus C in the 5.1-channel microphone technique by Danmarks Radio (Sawaguchi, editor, 2001). This stereo plus C involves a stereophonic MS pair (Figure 2.13), e.g., a cardioid microphone with its main axis pointing to the frontal direction and a bidirectional microphone with its main axis pointing to the left direction. The C channel signal is captured with a separate cardioid microphone arranged in front and has its main axis pointing to the front direction. Back reflections can be captured with another similar MS pair but has the main axis of the cardioid microphone pointing to the back direction or alternatively captured with a wide-spaced cardioid microphone pair. The back microphone pair is placed 8–10 m behind the frontal MS pair.

A four-channel main microphone array can be used to capture localization and ambient information. Four-channel outputs are fed to the L, R, and LS the RS channels. A separate cardioid or supercardioid microphone with its main axis pointing to the front direction is used to capture the C channel signal to enhance the stability of the frontal-center virtual source in reproduction. This type of technique is termed "quadraphone plus C" (Williams, 2007; Martin, 2005). Various quadraphonic microphone techniques, including coincident and near-coincident microphone techniques, are suitable for this purpose.

An example is the main double MS array plus C configuration. The main double MS array involves three coincident microphones, e.g., two cardioid microphones with their main axes pointing to the front and back directions and one bidirectional microphone with its main axis pointing to the left direction. The frontal-pointing cardioid microphone and the bidirectional microphone form a front MS pair, and their outputs serve as the L and R channel signals after an inverse MS transformation. The back-pointing microphone and the bidirectional microphone form a back MS pair, and their outputs serve as the LS and RS channel signals after an inverse

MS transformation. The main double MS array only requires three coincident microphones because the front and back MS pairs share the same bidirectional microphone. The center channel signal is captured with a separate microphone. In another configuration, the main array with four coincident cardioid microphones is used to replace the main double MS array. The main axes of the four cardioid microphones point to the front, back, left, and right directions. Four signals that are equivalent to the outputs of the main double MS array can be derived from the outputs of four cardioid microphones by using the method in Section 4.3.1. The main array with four coincident cardioid microphones, together with a separate microphone for capturing the C channel signal, constructs a complete microphone configuration for 5.1-channel recording.

7.3 MICROPHONE TECHNIQUES FOR OTHER MULTICHANNEL SOUNDS

7.3.1 Microphone techniques for other discrete multichannel sounds

In Section 5.3, some more than 5.1-channel horizontal surround sound systems and techniques are introduced. Microphone techniques for these discrete multichannel sounds have also been developed. The basic considerations and design methods for these microphone techniques are similar to those for 5.1-channel sound. For example, Williams (2008) designed a main microphone array (Figure 7.12) for 7.1-channel recording, which can be regarded as an extension of the main microphone array for 5.1-channel recording. The main microphone array in Figure 7.12 involves seven cardioid microphones spaced at appropriate distances. The main axes of these microphones point to 0°, ±40°, ±110°, and ±160°. The outputs of back microphones are attenuated by −10 dB. The effective recording range is ± 40° for the three front microphones, 70° for a frontal–lateral microphone pair, and 60° for a back–lateral microphone pair. In addition to main arrays with cardioid microphones, main arrays with hypercardioid microphones are available for 7.1-channel recording.

Similar to the case of 5.1-channel recording, two separate microphone arrays can be used to record frontal localization information and ambient information in 7.1-channel recording. Various three frontal channel microphone techniques in Section 7.2.3 are also applicable to 7.1 channel recording.

The quadraphone plus C described in Section 7.2.5 can be extended to more channel recording (Williams, 2008). For example, the microphone configuration in Figure 7.13 yields eight channel signals. It involves an inner quad array with four cardioid microphones spaced apart by 0.245 m (near coincident). The main axes of inner microphones point to the

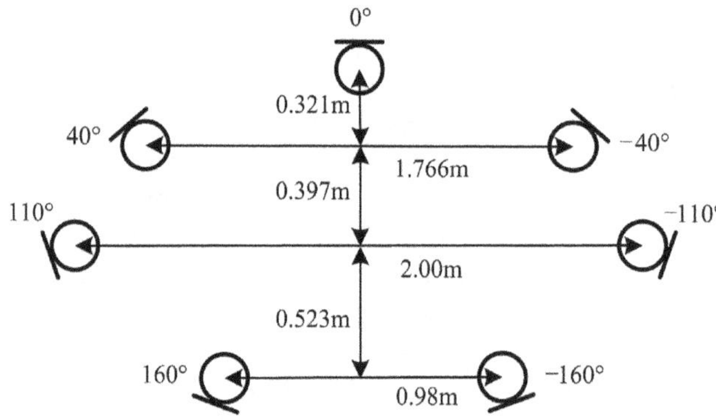

Figure 7.12 7.1-Channel main microphone array by Williams.

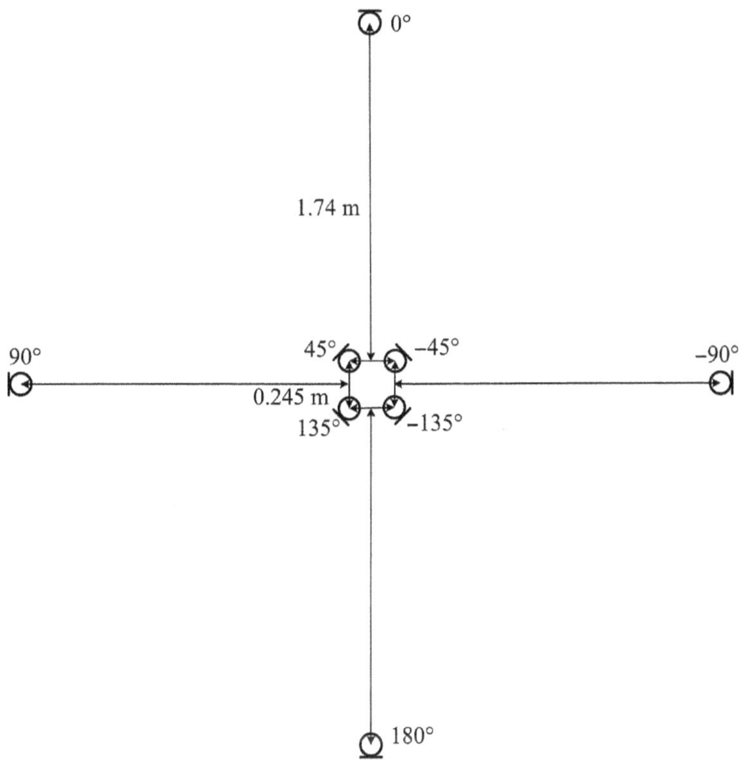

Figure 7.13 8-Microphone configuration.

LF (45°), RF (−45°), LB (135°), and RB (−135°) directions. This configuration also involves an outer array with four cardioid microphones. All outer microphones are 1.74 m from the center of the inner array. The main axes of the outer microphones point to the front (0°), left (90°), right (−90°), and back directions. The outputs of all outer microphones are delayed by about 4 ms. This configuration is adapted to recording for different channels. For 7.1-channel recording, the outer back microphone is turned off. For 6.1-channel recording, the outer left and outer right microphones are turned off. For 5.1-channel recording, the outer left, back, and right microphones are turned off. A configuration similar to that in Figure 7.13 with subcardioid microphones can be designed.

Lee (2011) proposed a 7.1-channel microphone technique for an effective perspective control. As shown in Figure 7.14, five pairs of coincident microphones are spaced apart at appropriate distances. The LF, C, and RF pairs consist of cardioid microphones. The main axes of the microphones in the LF pair point to the LF and LB directions. The main axes of the microphones in the RF pair point to the RF and RB directions. The main axes of the microphones in the C pair point in the F and B directions. These three front pairs capture frontal localization and ambient signals to recreate an overall perspective in reproduction. The linear mixing of the outputs of each front pair serves as the corresponding frontal channel signal. Changing the proportion of mixing is equivalent to microphone recording with different directivities and main axis directions. The two back microphone pairs are spaced apart with a wide distance. Each back pair consists of a hypercardioid microphone with its main axis pointing to the lateral direction and a cardioid microphone with its main axis pointing to the rear direction. The two back pairs capture the lateral and rear reflections and lead to four surround channel signals.

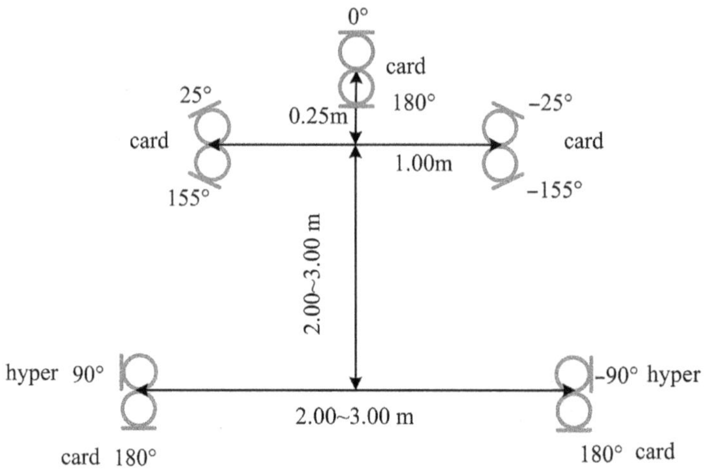

Figure 7.14 7.1-Channel microphone technique for an effective perspective control.

Spatial surround sound is a tendency in multichannel sounds. Some multichannel spatial surround sounds are introduced in Section 6.5.1. Microphone techniques for these multichannel spatial surround sounds have also been developed. Similar to the case of horizontal surround sound, spatial surround sound involves practical microphone and signal mixing techniques that are not always based on the virtual source localization theorem from Sections 6.1 to 6.4 and the corresponding optimized results. They may be based on a comprehensive consideration of multiple factors, such as virtual source localization, envelopment, ambient effect, and size of the listening region in reproduction. They are designed in terms of the results of psychoacoustic experiments or even experiences. Although some considerations in the design are similar to those for horizontal surround sound, the increased number and the vertical arrangement of loudspeakers complicate microphone techniques in spatial surround sound.

Layer-wise microphone configurations and arrays are often used in recording because these configurations are often used in multichannel spatial surround sound reproduction. They are constructed by adding an array of upper-layer microphones to various horizontal arrays to capture reflections and occasional localization information from a high elevation. Similarly, if necessary, an array of bottom-layer microphones may be added to collect information from a low elevation. The considerations and methods of designing the microphone array in each layer are similar to those of the horizontal microphone array. Coincident, near-coincident, and spaced arrays are applicable to each layer. Appropriate ICLD and ICTD between channel signals in each layer and different layers are obtained by choosing directivities, main axis directions of microphones, distances between microphones in the same layer, and distance between different layers. As stated in Section 6.2, the mechanism and rule of summing auditory events in vertical directions differ from those in the horizontal direction. For low-frequency stimuli, ICLD may recreate a virtual source between two vertical loudspeakers with appropriate configuration. However, for wideband stimuli, vertical summing localization with ICLD may even be impossible, or quantitative results may quite differ from those in the horizontal plane. ICTD fails to recreate a stable virtual source between two vertical loudspeakers. Moreover, in the vertical direction, the results of a precedence effect and an auditory event or sensation created by decorrelated signals are quite different from those in the horizontal plane. Therefore, some considerations and methods for the design of horizontal microphone arrays cannot be directly extended to these cases, including vertical directions. As pointed out by Theile and Wittek (2011), the two-layer (middle and upper layers) loudspeaker configurations are originally used to recreate virtual sources in the elevations of different layers and enhance the reproduction of reflections from high elevations rather than to recreate the vertical virtual sources between two layers. One requirement is restraining the disturbance from the signals of the upper layer to localization for a horizontal target source.

For a channel-based spatial surround sound, microphone technique and array are usually designed in accordance with the loudspeaker configuration in reproduction. Various microphone techniques for different loudspeaker configurations have been designed. According to the space between microphones, Lee (2021) classified these techniques into three categories, e.g., horizontally spaced and vertically spaced technique, horizontally spaced and vertically coincident technique, horizontally coincident and vertically coincident technique. A few examples are outlined here. Detailed reviews and analysis of various three-dimensional microphone techniques are referred to Lee's paper.

Theile and Wittek (2011) proposed an OCT-3D microphone array for Auro 9.1 channel sound (Figure 6.13). They added an upper-layer configuration of microphones to the OCT 5.1-channel main microphone array in Figure 7.3. The four supercardioid microphones in the upper layer are located above the left, right, LS, and the RS microphones in the horizontal layer. The main axes of the four supercardioid microphones point to the above directions.

Wittek and Theile (2017) proposed an ORTF-3D microphone array for multichannel spatial surround sound (such as Auro and Dolby Atmos). The ORTF-3D array involves eight supercardioid microphones arranged in two layers (horizontal/middle and upper layers), i.e., four microphones are placed in each layer, forming a cubic configuration. The adjacent microphones are spaced apart by 0.1–0.2 m. The arbitrary two adjacent microphones can be regarded as a near-coincident ORTF pair similar to that in Section 2.2.4. Each ORTF pair creates two adjacent channel signals with ICLD and ICTD. In addition, four microphones in each layer can be regarded as a microphone array for horizontal surround sound recording, i.e., the ORTF surround array.

Williams (2012) designed a microphone configuration for 11 (or 11.1)-channel recording. The configuration involves a 7-channel middle microphone layer and a 4-channel upper microphone layer. The 7-channel middle microphone layer consists of seven hypercardioid microphones arranged similarly to those in Figure 7.13 (in the absence of an outer back microphone). The upper layer is located at an appropriate distance above the horizontal layer. The microphones in the upper layer are arranged in a square and spaced apart by 0.55 m. Bidirectional or supercardioid microphones are used in the upper layer, and the main axes of the microphones point to the top directions. When supercardioid microphones are used in the upper layer, the zero in the directional responses of microphones restrains the disturbance of upper-layer outputs to the horizontal virtual source localization.

Geluso (2012) suggested adding "Z microphones" to a horizontal array to capture the elevation information. That is, a set of bidirectional microphones with their main axes pointing to top directions are added to an original horizontal array. The position of each bidirectional microphone coincides with that of a corresponding microphone in the original horizontal array. Consequently, each new added bidirectional microphone and the corresponding microphone in the original horizontal array form a vertical MS pair, and the corresponding horizontal and high-elevation channel signals can be derived from the outputs of this vertical MS pair. For example, the output of the LF microphone in the horizontal array serves as the signal of the horizontal LF channel. The sum of the outputs of the LF microphone in the horizontal array and the coincident bidirectional microphone serves as the signal of the left-front-up channel. Various horizontal microphone arrays can be extended for recording the elevation information by using this method. This method yields signals with ICLD only between adjacent vertical channels.

Some microphone techniques for 22.2-channel recording have been developed (Hamasaki et al., 2004; ITU-R Report BS 2159-7, 2015; Howie et al., 2016). These techniques are basically extensions of 5.1-channel microphone techniques. Moreover, Ono et al. (2013) of NHK proposed a spherical microphone array with baffles for 22.2-channel recording. A spherical region with a diameter of 0.45 m is divided into angular segments, including eight parts horizontally and three parts vertically. Omnidirectional microphones are placed in each angular segment. With these baffles, the microphone in each angular segment has a narrow directivity to reduce the crosstalk between channels. Measurement indicates that the microphone in each angular segment exhibits a narrow and nearly constant (frequency-independent) directivity above 6 kHz. The directivity widens as frequency decreases and becomes almost omnidirectional below 500 Hz. A signal processing method similar to beamforming (Section 9.8) is applicable to the improvement of the directivity below 800 Hz.

Lee and Gribben (2014) investigated the effect of the space between upper and middle microphone layers on the perceived spatial impression and overall preference. The OCT-3D microphone array for Auro 9.1-channel sound is analyzed through psychoacoustic experiments and interaural correlation calculation (Section 12.1.6). These results indicate that a variation in the space between two layers from 0.5 m to 1.5 m does not cause significant differences in the perceived spatial impression. A 0 m (coincident) space between two layers yields slightly better results. Lee and Gribben pointed out that the perceived results are associated with a vertical interchannel crosstalk from the signals of the upper layer and the pattern of the spectral change in binaural signals. The directivities of microphones restrain the influence of the crosstalk of upper-layer signals to the horizontal localization. The threshold of level difference for full masking in vertical directions is lower than that in horizontal directions; therefore, the space between two layers slightly influences the perceived performance.

Overall, with the development of multichannel sound, especially multichannel spatial surround sound, various multichannel microphone techniques are developed. Various microphone techniques are established, even for the same multichannel sound and loudspeaker configuration. The diversity in multichannel sound and loudspeaker configuration complicates the problem. Some typical examples of microphone techniques for more than 5.1-channel sounds are outlined in this section. The psychoacoustic rules for summing vertical spatial auditory events are still incomplete, causing some difficulties in the design of microphone techniques for multichannel spatial surround sound. Therefore, the multichannel microphone technique is still being developed.

7.3.2 Microphone techniques for Ambisonic recording

As stated in the preceding chapters, Ambisonics is a special signal recording and mixing technique for multichannel sound. Various microphone techniques for Ambisonic recording may be designed. The advantage of Ambisonic recording is its flexibility. The recorded signals are independent of the loudspeaker configuration in reproduction and therefore suitable for various configurations.

As stated in Section 4.3, regular loudspeaker configurations in the horizontal plane with global Ambisonic signal mixing can recreate a full 360° virtual source and auditory sensations similar to those caused by reflections in a hall via indirect methods. The three independent signals of the first-order horizontal Ambisonics can be captured by coincident microphones, such as an omnidirectional microphone and two bidirectional microphones with their main axes pointing to the front and left directions, respectively. Three independent signals can also be captured by other coincident microphones.

Similarly, spatial Ambisonic signal mixing is applicable to regular (uniform) and nearly regular (near-uniform) loudspeaker configurations in a three-dimensional space (or loudspeaker configurations satisfying the discrete orthogonalities of spherical harmonic functions in Section 9.4.1). The four independent signals of the first-order spatial Ambisonics can be captured by coincident microphones. For example, independent signals in Equation (6.4.1) can be captured by an omnidirectional microphone and three bidirectional microphones with their main axes pointing to the front, left, and top directions, respectively. The A-format independent signals in Equation (6.4.11) can be captured by four cardioid-like microphones with symmetric directivity around the polar axis. Figure 7.15 illustrates an example of the microphone set for A-format Ambisonic signal recording. This tetrahedral microphone set is made by Sennheiser (Core Sound has similar product). Four microphones are located on the four surfaces of a tetrahedron with their main axes pointing to the normal directions of the surfaces.

Independent signals for higher-order horizontal and spatial Ambisonics can be theoretically captured by a set of coincident microphones, including microphones with higher-order directivities. However, making microphones with higher-order directivities is difficult. In early times, this difficulty hindered higher-order Ambisonic recording. The microphone array for higher-order Ambisonic recording has been developed (Section 9.8).

According to Sections 5.2.3 and 6.4.1, global Ambisonic signal mixing may be theoretically applied to irregular loudspeaker configurations in the horizontal plane and three-dimensional loudspeaker configurations that do not satisfy the discrete orthogonalities of spherical harmonic functions. The Ambisonic independent signals recorded by various methods can be decoded into loudspeaker signals via an appropriate decoding matrix. However, the decoding matrix of some loudspeaker configurations is unstable (Section 9.4.1). The Ambisonic signal mixing methods for ITU 5.1-channel loudspeaker configuration are theoretically derived in Section 5.2.3, but these signal mixing methods usually exhibit some defects, such as a narrow

Figure 7.15 Tetrahedral microphone set for A-format Ambisonic signal recording by Sennheiser.

listening region and instability in virtual sources. A slight deviation from the central listening position may spoil the virtual source localization and causes the perceived timbre coloration. Therefore, Ambisonic signal recording and mixing were rarely used in practical 5.1-channel program production. In fact, an irregular 5.1-channel loudspeaker configuration is originally inappropriate for global Ambisonic signal mixing. Conversely, local Ambisonic signal mixing similar to that discussed in Section 5.2.4 may be applied to improve the frontal virtual source localization at the central listening position. The method of three coincident microphones with appropriate directivities for three-frontal-channel recording in Section 7.2.3 is similar to the local Ambisonic mixing, although this method is not often used in practical signal recording.

The number of loudspeakers increases in some irregular loudspeaker configurations. Therefore, spans between adjacent loudspeakers are reduced, and the uniformity of loudspeaker configurations enhances. Accordingly, the performance of reproduction with global Ambisonic signal mixing improves. An example is a configuration 1 of 7.1-channel loudspeakers in Figure 5.10. The signal recording and mixing methods for Ambisonic reproduction with irregular loudspeaker configuration should be further explored.

7.4 SIMULATION OF LOCALIZATION SIGNALS FOR MULTICHANNEL SOUNDS

In addition to recording with a set of onsite microphones, simulation can be used to obtain multichannel sound signals. Signal simulation is not only a common method in the stage of program production in channel-based spatial sound but also an important element in the stage of signal rendering in object-based spatial sound. The signal simulation of multichannel sounds involves two parts: (1) to simulate the localization information of a virtual source and (2) to simulate the information of reflections. The details of these two parts are addressed in this and next sections.

7.4.1 Methods of the simulation of directional localization signals

Similar to the case of two-channel stereophonic sound, directional localization signals can be created through simulation. That is, the mono signal from a spot microphone or an electronic instrument is rendered to channels based on certain signal panning or mixing rules to create multichannel signals with desired amplitude and time relationship and recreate a virtual source in reproduction. In principle, various signal panning methods discussed in Chapters 4–6 are applicable to simulate multichannel signals for directional localization. However, amplitude panning or mixing methods are used in most practice. Amplitude panning or mixing methods create multichannel in-phase (or out-of-phase at most) signals with an interchannel level difference. Time panning is rarely used to simulate multichannel sound signals because it is relatively complicated, and the perceived quality of its virtual source is inferior to that of amplitude panning. For simple pair-wise amplitude panning, signals can be rendered via a pan-pot. In current program production, signals can be easily rendered to channels by using digital signal processing software according to various rules.

In Chapters 4–6, pair-wise amplitude panning is the simplest and common method to recreate a virtual source in multichannel horizontal surround sound. It is often used for pop-music and video/audio program production. The pair-wise amplitude panning is also applicable to the recreation of a virtual source in the median and other sagittal planes (Section

6.2). Similar to the case of two-channel stereophonic sound in Equation (2.2.59), for a pair of adjacent channels i and $i + 1$, the normalized amplitudes of channel signals under constant power condition are written in the form of sine/cosine function:

$$A_{i+1} = \sin\xi_i \qquad A_i = \cos\xi_i \qquad A_{i+1}^2 + A_i^2 = 1 \qquad 0° \le \xi_i \le 90°, \tag{7.4.1}$$

where ξ_i is a parameter. For $\xi_i = 0°$, A_i maximizes to a unit, and $A_{i+1} = 0$, so a virtual source is located at the direction of the ith loudspeaker. For $\xi_i = 90°$, A_{i+1} maximize to a unit, and $A_i = 0$, so a virtual source is located at the direction of the $(i + 1)$th loudspeaker. For $\xi_i = 45°$, the amplitudes of two signals are equal to $A_i = A_{i+1} = 0.707$, i.e., a −3 dB drop off compared with the maximal amplitude at a unit.

In Sections 4.1.2 and 5.2.2, for an arbitrary pair of adjacent loudspeakers in the horizontal plane and a fixed head oriented to the front, pair-wise amplitude panning is not always able to recreate the virtual source between two loudspeakers. Therefore, the signal panning function cannot always be derived from Equation (3.2.7) for the arbitrary target virtual source azimuth θ_I. This feature is different from the case of two-channel stereophonic sound in Equation (2.2.62). In practice, the signal panning function is derived with Equation (3.2.9), i.e., the conditions of the head oriented to the virtual source and constant power, yielding the normalized amplitudes of two-channel signals as a function of the target azimuth $\hat{\theta}_I$:

$$\begin{aligned}
A_i &= \frac{\cos\theta_{i+1}\sin\hat{\theta}_I - \sin\theta_{i+1}\cos\hat{\theta}_I}{\sqrt{\left(\cos\theta_{i+1}\sin\hat{\theta}_I - \sin\theta_{i+1}\cos\hat{\theta}_I\right)^2 + \left(\sin\theta_i\cos\hat{\theta}_I - \cos\theta_i\sin\hat{\theta}_I\right)^2}}. \\
A_{i+1} &= \frac{\sin\theta_i\cos\hat{\theta}_I - \cos\theta_i\sin\hat{\theta}_I}{\sqrt{\left(\cos\theta_{i+1}\sin\hat{\theta}_I - \sin\theta_{i+1}\cos\hat{\theta}_I\right)^2 + \left(\sin\theta_i\cos\hat{\theta}_I - \cos\theta_i\sin\hat{\theta}_I\right)^2}}.
\end{aligned} \tag{7.4.2}$$

Even for pair-wise amplitude panning, the panning function does not always follow Equation (7.4.1) or (7.4.2). Figure 5.4 illustrates the panning curves of Equation (5.2.7) or (7.4.2) for the three front channels of 5.1-channel sound. The signal mixing of three frontal channels was investigated in Bell Labs in the 1930s (Snow, 1953), and the results have been widely used for cinema sound production. Figure 7.16 illustrates the panning curves of three frontal channel signals given by Bell Labs. These curves are different from those in Figure 5.4. In the direction neighboring the left, center, and right loudspeakers, the amplitude of the corresponding channel signal remains in the unit. In fact, the large-sized listening region of cinema sound makes the principle of summing localization cannot be applied to recreate the virtual source between two loudspeakers. Signal panning aims to recreate the localization effect on the directions of three loudspeakers. However, some programs for cinema sound may be used for domestic reproduction.

The VBAP discussed in Section 6.3 is a three-dimensional extension of pair-wise amplitude panning and has been used for an object-based virtual source rendering in the MPEG-H 3D Audio. Localization signals in multichannel sound can also be simulated using other mixing methods or rules, such as horizontal or spatial Ambisonic methods.

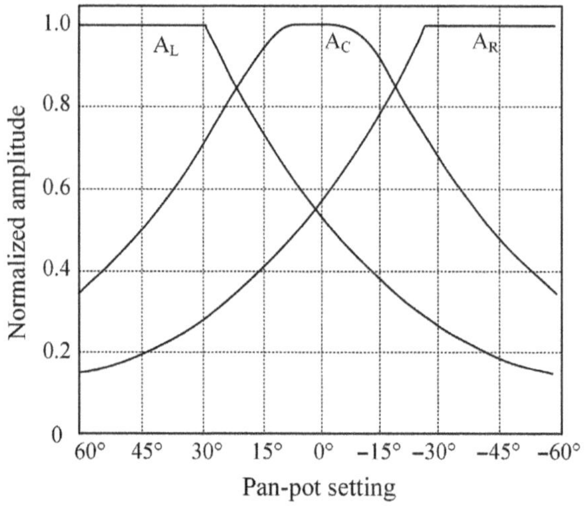

Figure 7.16 Three frontal channel panning curves by Bell Labs.

7.4.2 Simulation of virtual source distance and extension

The preceding sections focus on the recording and simulation of the directional information of a sound source. In addition to direction, distance is the spatial information of a sound source. Accordingly, the perceived distance is another important auditory attribute of the sound source. The perceived distance refers to the apparent distance of a sound source that a listener perceives. Although the ability of the human auditory system to estimate sound source distance is generally inferior to the ability to locate the sound source direction, a preliminary but biased auditory distance perception can still be formed. Therefore, virtual sources or auditory events should be recreated at various perceived distances in spatial sound reproduction. As stated in Section 1.6.6, auditory distance perception is based on the comprehensive information provided by multiple cues, but the conclusion is not as obvious as that of auditory directional localization. As such, techniques for controlling the perceived distance of virtual sources or auditory events in spatial sound reproduction remain limited and immature (Xie and Yu, 2021).

In principle, controlling the perceived distance in reproduction may be realized by controlling various auditory distance cues (Section 1.6.6) in reproduction. In the case of a free field, distance-dependent sound pressure and subjective loudness are cues for relative auditory distance perception, but they are ineffective for absolute auditory distance perception. The high-frequency attenuation caused by air absorption is a weak cue for relative auditory distance perception. In an enclosed space, the direct-to-reflected energy ratio depends on the source distance. In the case of diffused reflections, Equation (1.2.25) indicates that the direct-to-reverberant energy ratio is inversely proportional to the square of distance. In two-channel stereophonic and multichannel sound program production, a common and relatively effective method for controlling the perceived distance in reproduction is changing the direct-to-reflected energy ratio of signals. For onsite recording with various microphone techniques, the direct-to-reflected energy ratio of the recorded signals can be changed by choosing the distance of a microphone set to the sound source, the directivities and orientation of microphones, and the relative gains between the microphones for direct sound recording and those for reflected sound recording. For simulated multichannel sound signals, the direct-to-reflected energy ratio of signals can be controlled by changing the gain ratio of direct sound signals and artificial reverberation signals. The method of artificially simulating reverberation signals is addressed in Section 7.5. Some studies have also suggested controlling the perceived distance in reproduction by changing the relative delay and gain of simulated early reflections (Gerzon, 1992e).

As mentioned in Section 1.6.6, for the proximal source distance within 1.0 m, the scattering and shadow effects caused by the head and pinnae also provided information on auditory distance perception. For sound field-based spatial sound reproduction, the perceived distance may also be controlled by reconstructing the curve wavefront of a proximal source (Sections 9.3.4 and 10.2.4).

The simulation of a point source in reproduction is addressed. In practice, the perception of a spatially extended source should be recreated. The perceived dimensionality of a spatially extended source involves an auditory source width (ASW), the perceived vertical spread of the source, and the perceived depth of the source. As stated in Section1.8.1, the ASW in a concert hall is closely related to the intensity and directions of early reflections. An appropriate ASW is a desired perceived attribute of a concert hall. For classical music reproduction, the appropriate ASW may also be similar to that in a concert hall. For pop music or some other program reproductions, sharp virtual sources with definite directions and spatially extended source may be preferred depending on the perceived or even esthetic requirements in reproduction. However, the ASW should be controlled in program production.

For onsite recording in a concert hall, the original reflection information, especially that of the early lateral reflections, may be captured and reproduced using appropriate methods, consequently resulting in the desired ASW. Ambisonic recording and reproduction are an example. For simulating the information of the early later reflections of a concert hall, various delay algorithms presented in Section 7.5.1 may be used, and the delayed signals may be fed to the corresponding channels in terms of certain signal mixing rules.

The perceived width of an auditory event is closely related to the IACC of binaural pressures. As stated in Sections 1.7.3 and 3.3, for reproduction with two or more loudspeakers, the perceived width of auditory events can be controlled by changing the interchannel correlation of signals and thus changing the resultant IACC. This method is commonly used in practice. The algorithm of audio signal decorrelation is addressed in Section 7.5.4. Vilkamo and Pulkki (2014) suggested a method of the adaptive optimization of interchannel correlation to control the perceived width of auditory events in stereophonic and surround sound reproduction. The width of perceived auditory events may be extended by creating multiple decorrelated versions of a stimulus, and these decorrelated stimuli may then be reproduced as multiple "virtual sources" according to certain signal mixing rules (Potard and Burnett, 2004). However, this method is more complex than the aforementioned method of controlling the interchannel correlation.

Other methods are utilized to control the perceived width of auditory events in reproduction. The method of fluctuations in interaural time difference (ITD) and interaural level difference (ILD) over time (Section 1.8.3) may be applicable to sound reproduction. That is, signal mixing changes over time to create fluctuations in the ITD and ILD of binaural superposed pressures in reproduction, thereby broadening auditory events. For pair-wise amplitude panning, variation in ξ_i in Equation (7.4.1) over time periodically within a certain range leads to the fluctuation of interchannel level difference (Griesinger, 1992b). However, the applicability of this method in practice should be further proved.

Some studies have suggested decomposing a mono input stimulus into time-frequency or subband (such as equivalent rectangular bandwidth (ERB)) components and distributing them to different directions randomly, thus recreating spatially extended virtual sources (Pihlajamaki et al., 2014). The time–frequency or subband signals can be distributed to loudspeakers in different directions or virtual sources created by certain mixing rules. Various signal panning or mixing methods, such as Ambisonics and pair-wise amplitude panning, are applicable to the latter case (Zotter et al., 2014). The spatial extension of virtual sources can be improved by distributing the neighboring band contents into far away directions. Therefore, some constraints should be applied to the randomly spatial distribution of subband contents.

The aforementioned methods may be inappropriate for controlling the vertical spread of the perceived virtual source because the mechanisms of spatial hearing in the vertical direction are different from those in the horizontal directions. For example, changing the interchannel correlation between the signals of two loudspeakers in the median plane or other cones of confusion does not effectively alter IACC. The psychoacoustic experiment by Gribben and Lee (2018) indicated that for a pair of loudspeakers arranged at elevation $\phi = 0°$ and $30°$ in the vertical planes of $\theta = 0°$, $\pm30°$ and $\pm110°$ respectively, the perceived virtual source spreads vertically with the decrease of interchannel correlation at frequencies around 0.5 kHz and above. The loudspeaker configuration in this experiment is basically consistent with the 9.1-channel system in Figure 6.13. Pulkki et al. (2019) experimentally examined the summing spatial auditory perception caused by a pair of loudspeakers in the median plane with uncorrelated pink noise stimuli and without head movement. The result indicated that two sources (loudspeakers) can be perceived individually when they are separated in elevation by 60° or more. When the two sources are separated by less than 60°, they are perceived inaccurately, biased, and spatially compressed but nevertheless not as point-like virtual sources. Overall, dynamic cues and spectral cues for vertical localization may also contribute to the vertical spread of the perceived virtual source, and the underlying mechanism should be further studied.

The perceived depth of an auditory event (source or scene) refers to the overall perceived front-back dimensionality or distance of this particular event. It is closely related to but different from the perceived distance of the auditory event. In practice, the perceived depth is controlled by changing the reflections in reproduction. Further studies on the mechanism of controlling the perceived depth are also needed.

7.4.3 Simulation of a moving virtual source

In the preceding discussions, virtual sources are assumed to be static or immobile. In practical program production, a moving virtual source should be stimulated on a given trajectory in some instances. Chowning (1971) first simulated a moving virtual source in a quadraphone with pair-wise amplitude panning. Similar methods are applicable to stereophonic and other multichannel loudspeaker configurations with different signal mixing methods.

Generally, a moving sound source leads to the following time-dependent changes:

1. Change in the source direction with respect to a listener
2. Change in the source distance with respect to a listener
3. Change in reflected sounds in a reflective environment
4. Doppler frequency shift of a fast-moving sound source

Accordingly, these changes should be simulated for a moving virtual source in sound reproduction.

The instantaneous position of a moving sound source is specified by its time-varying distance $r_S(t)$, azimuth $\theta_S(t)$, and elevation $\phi_S(t)$ with respect to a listener. The functions $r_S(t)$, $\theta_S(t)$, and $\phi_S(t)$ are a set of parametric equations for the trajectory of a moving sound source. Given the signal panning or mixing method in a stereophonic or multichannel sound, the functions or mapping between a target source direction and channel signals are known. The normalized amplitudes of channel signals are changed according to time-varying $\theta_S(t)$ and $\phi_S(t)$ to simulate a moving virtual source in a freefield at a constant distance and instantaneous direction of $\theta_S(t)$ and $\phi_S(t)$. For example, the time-varying normalized amplitudes of three frontal channel signals are given by letting $\phi_S(t) = 0°$, $\hat{\theta}_I = \theta_S(t)$ and substituting these time-dependent parameters into Equation (7.4.2) to simulate a moving virtual source across the three frontal loudspeakers in a 5.1-channel configuration with pair-wise amplitude panning. As stated in Section 5.2.2, for a 5.1-channel loudspeaker configuration and a fixed head oriented to the front, pair-wise amplitude panning fails to recreate a stable lateral virtual source. The virtual source in the rear region is also blurry and unstable. However, in practical program production, pair-wise amplitude panning is often used to recreate a moving virtual source in lateral and rear directions. For a fast-moving virtual source, human hearing is less sensitive and thus more tolerant to jumping in a virtual source direction.

In addition to pair-wise amplitude panning, other signal mixing methods can be used to recreate a moving virtual source in various multichannel sound reproductions. For pair-wise amplitude panning, the signals of a moving virtual source are previously recreated by manually manipulating a pan-pot in a console. Currently, signals of a moving virtual source can be easily created with the software of digital signal processing in accordance with the rules of signal mixing.

Direct sound magnitude should vary with the source distance to simulate a time-varying-distance virtual source. The magnitude of direct sound from a point source is inversely proportional to the source distance with respect to a listener, and the normalized amplitude or gain of all channel signals should be scaled with a factor of $1/r_S(t)$. For spatial sound based on sound field reconstruction, such as near-field-compensated higher-order Ambisonics discussed in Section 9.3.4, a distance-dependent curve wavefront may be simulated by manipulating the time-varying parameter $r_S(t)$ of a target distance in channel signals.

In a reflective environment, the change of a source position alters the directions, delays (distances), and magnitudes of all reflections with respect to the listener. The magnitude spectra of reflections also depend on the source position. Strictly, all these changes should be considered in the simulation of a moving virtual source. However, exactly simulating all these changes is actually difficult. For some applications requiring high accuracy, such as scientific studies on a virtual auditory environment, changes in several proceeding early discrete reflections caused by a moving sound source are simulated. That is, the instantaneous delays, directions, and power spectra of early discrete reflections are evaluated and simulated according to the geometrical acoustic model of a room (Section 7.5.5). The resultant reflected signals are fed to channels in terms of certain signal mixing rules.

In a diffused reverberation field, numerous reflections reach a listener from various directions at every instant. Human hearing is unable to detect the change in each individual reflection. In practical program production, reflections are usually approximated as a diffused field, and the change in the statistical characteristics of the sound field is simulated. For example, the direct-to-reverberant energy ratio of various source distances is evaluated with Equation (1.2.25), and the relative gain of direct sound with respect to that of artificial reverberation is changed to simulate the change in the direct-to-reverberant energy ratio caused by the variation in the source distance. This method simulates the relative change in direct and reverberant sound, providing important perceptual information on the variation in the distance of a moving virtual source.

The Doppler frequency shift should be considered in the simulation of a fast-moving virtual source in multichannel sounds. If a listener is immobile and a source moves at velocity v_S, then the projections of v_S to the line that connects the source and the listener are v_{S1}. For a sound radiation with the static frequency f_0, the movement of the source creates the received frequency for the listener as follows (Krebber et al., 2000):

$$f = \frac{c}{c - v_{S1}} f_0 \qquad v_{S1} < c, \tag{7.4.3}$$

where c is the speed of sound. The Doppler frequency shift is given by

$$\Delta f = f - f_0 = \frac{v_{S1}}{c - v_{S1}} f_0 \approx \frac{v_{S1}}{c} f_0. \tag{7.4.4}$$

The approximation on the right side of Equation (7.4.4) is held at $v_{S1} << c$. When the source approaches the listener with $v_{S1} > 0$, the received frequency increases with $\Delta f > 0$. When the source moves away from the listener, the received frequency decreases. In digital signal processing, the Doppler frequency shift can be implemented by changing (up/down) the sampling frequency of input stimuli, a task equivalent to applying instantaneous linear frequency modulation to input stimuli.

7.5 SIMULATION OF REFLECTIONS FOR STEREOPHONIC AND MULTICHANNEL SOUNDS

The reproduction of the comprehensive information of environmental reflections is also the aim of stereophonic and multichannel sound. In addition to live recording onsite (Sections 7.2 and 7.3), the signals of environmental reflections can be artificially simulated. For recording with a spot microphone and a pan-pot technique, the resultant signals are "dry" with fewer reflection components because of the closer distance of a spot microphone to the sound source. In this case, some simulated reflections should be supplemented with the signals. For object-based spatial sound reproduction, environmental reflections should be simulated artificially on the basis of the metadata parameters of the target reflected sound fields. Two basic methods are commonly available for simulating room or environmental reflections. ***Perception-based methods*** simulate reflections according to some predetermined statistical (general) room acoustic parameters, such as reverberation time, to recreate the desired auditory perception. Various artificial delay and reverberation algorithms are based on perception-based methods. ***Physics-based methods*** simulate reflections by modeling or measuring the acoustic transmission characteristics of a room.

Artificial delay and reverberation have been used in audio program productions, including mono, stereophonic, and multichannel sound program production. In pop music and video/audio program productions, artificial delay and reverberation signals are usually added to stereophonic or multichannel sound signals created by spot microphones and pan-pots to simulate environmental reflections. Spring reverberators, plate reverberators, echo chambers, loop magnetic tape, and BBD (*bucket brigade* device) analog delay lines were previously used to simulate delay and reverberation signals (Xie X.F., 1981). Since 1990, various digital delay lines and reverberators have been widely used to replace previous devices, but some traditional concepts and algorithms are still used in digital delay lines and reverberators. Delay and reverberation algorithms often used in stereophonic and multichannel sounds are first discussed in this section. Then, the methods of simulating reflection signals based on the spatial room impulse responses are discussed. The simulation of reflections and reverberation is described in detail in other studies (Gardner, 2002; Dattorro, 1997; Välimäki et al., 2012).

The algorithms in this section are discussed on the basis of digital signal processing because artificial delay and reverberation algorithms are currently implemented by digital signal processing. Digital signal processing is explained in detail in another book (Oppenheim et al., 1999).

7.5.1 Delay algorithms and discrete reflection simulation

As stated in Section 1.2.4, room reflections consist of early discrete reflections with different delays and late reverberation. Artificial delay algorithms are designed to simulate early discrete reflections.

In digital signal processing, a signal is represented by a discrete time sequence sampled from a continuous-time analog signal. The discrete time t_n takes the form of

$$t_n = nT = \frac{n}{f_s} \qquad -\infty < n < +\infty, \tag{7.5.1}$$

where T and f_s are the sampling period and the sampling frequency, respectively. Usually, discrete time is expressed by the integer n for simplification. An original analog signal can be completely recovered from the discrete signal as long as the Shannon–Nyquist sampling theorem is fulfilled.

In a digital delay line, the output $e_y(n)$ is an m-sample delay version of the input $e_x(n)$ in the time domain

$$e_y(n) = e_x(n-m). \tag{7.5.2}$$

The length of delay in seconds is calculated as

$$\tau_D = \frac{m}{f_s}. \tag{7.5.3}$$

The Z transforms of the input and output are given by

$$E_x(z) = \sum_{n=-\infty}^{+\infty} e_x(n) z^{-n} \qquad E_y(z) = \sum_{n=-\infty}^{+\infty} e_y(n) z^{-n}. \tag{7.5.4}$$

The representation of Equation (7.5.2) in the Z-domain is given by

$$E_y(z) = z^{-m} E_x(z). \tag{7.5.5}$$

Therefore, multiplying a factor of z^{-m} to the signal in the Z-domain is equivalent to the delay of the corresponding signal in the time domain by m samples.

A digital delay line is a linear-time-invariant (LTI) system. The output in the Z-domain can be expressed as the input multiplied with the system function $H_D(z)$ of the digital delay line:

$$E_y(z) = H_D(z) E_x(z). \tag{7.5.6}$$

In the comparison of Equation (7.5.5) with Equation (7.5.6), the system function is given by

$$H_D(z) = z^{-m}. \tag{7.5.7}$$

This equation is equivalent to the response of an all-pass filter with a linear phase. Equation (7.5.6) can be written as the input/output equation in the discrete time domain

$$e_y(n) = \sum_{q=-\infty}^{+\infty} h(q)e_x(n-q) = h_D(n) \otimes_t e_x(n) \qquad -\infty < n < +\infty, \tag{7.5.8}$$

where the notation "$\otimes_t$" is the convolution in the discrete time domain, and $h_D(n)$ is the impulse response of the discrete time system and can be obtained by applying an inverse Z-transformation to Equation (7.5.7). The resultant $h_D(n)$ is a unit sampling sequence $\delta(n)$ with

$$h_D(n) = \delta(n-m) \quad \delta(n) = \begin{cases} 1 & n = 0 \\ 0 & n \neq 0 \end{cases}. \tag{7.5.9}$$

Discrete reflections in reproduction are simulated by mixing the delayed signal $e_y(n)$ to different channels based on certain panning or mixing rules.

If the weighted delayed signals $e_x(n - m)$ with gain g is mixed with the original signal $e_x(n)$, the resultant signal is given by

$$e_y(n) = e_x(n) + ge_x(n-m). \tag{7.5.10}$$

Equation (7.5.10) simulates the case of a direct sound, along with single reflection. A gain (attenuation) with $|g| < 1$ simulates surface material absorption. g can be regarded as the pressure reflection coefficient. The impulse response and system function that correspond to Equation (7.5.10) are respectively provided by

$$h_D(n) = \delta(n) + g\delta(n-m), \tag{7.5.11}$$

$$H_D(z) = 1 + gz^{-m}. \tag{7.5.12}$$

The algorithm in Equation (7.5.10) is implemented with the finite impulse response (FIR) filter structure shown in Figure 7.17(a).

Let $z = \exp(j\omega)$ in a system equation such as Equation (7.5.6) and ω denotes the digital angular frequency. The relationship among ω, conventional analogy frequency f, and the sampling frequency f_s of the digital system is given by

$$\omega = \frac{2\pi f}{f_s}, \tag{7.5.13}$$

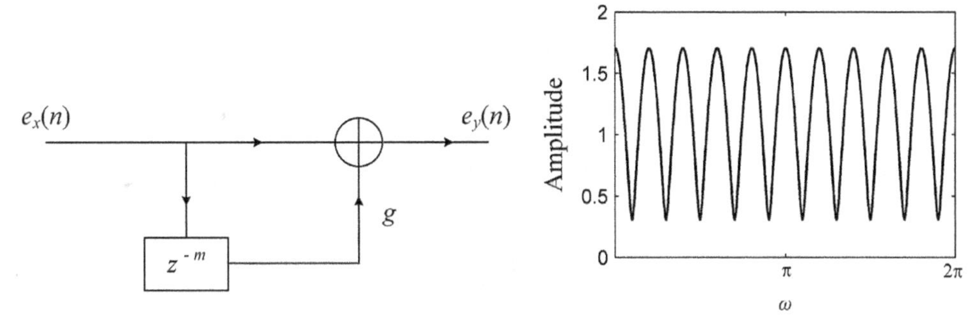

Figure 7.17 FIR comb filter structure of a direct sound and a single reflection simulation: (a) FIR filter structure; (b) comb filtering response.

$E_x[\exp(j\omega)]$ and $E_y[\exp(j\omega)]$ are the Fourier transformation of discrete-time signals (sequence) $e_x(n)$ and $e_y(n)$, respectively. $H_D[\exp(j\omega)]$ is the Fourier transformation of the impulse response $h_D(n)$, i.e., the transfer function of the digital system. It is the value of $H_D(z)$ in the unit circle in the Z-plane with $|z| = 1$ and a periodic function of ω with a period of 2π. In the following equation, $H_D[\exp(j\omega)]$ is denoted by $H_D(\omega)$ for convenience. After the aforementioned transformation, Equation (7.5.12) becomes

$$H_D(\omega) = 1 + g\exp(-jm\omega) \qquad |H_D(\omega)| = \sqrt{1 + 2g\cos(m\omega) + g^2}. \tag{7.5.14}$$

Figure 7.17 (b) shows the variations in the magnitude response $|H_D(\omega)|$ with ω within the range of $0 \le \omega < 2\pi$ for $m = 10$ and $g = 0.7$. The magnitude $|H_D(\omega)|$ exhibits a comb filtering characteristic, with notch $(1 - g)$ at $\omega = (2q + 1)\pi/m$ and peak $(1 + g)$ at $\omega = 2q\pi/m$ for $q = 0,1,2\ldots(m - 1)$. This characteristic stems from the interference caused by the simulated direct sound and reflection. Therefore, Figure 7.17(a) shows an FIR comb filter structure.

The outputs of Q delay lines with different lengths m_q are mixed to various channels in accordance with certain signal mixing rules to simulate multiple early discrete reflections with various delays caused by the different surfaces in a room. If these delayed signals are mixed with different gains g_q and then combined with the original signal $e_x(n)$,

$$e_y(n) = e_x(n) + \sum_{q=1}^{Q} g_q e_x(n - m_q), \tag{7.5.15}$$

and the impulse response and system function corresponding to Equation (7.5.15) are given by

$$h_D(n) = \delta(n) + \sum_{q=1}^{Q} g_q \delta(n - m_q), \tag{7.5.16}$$

$$H_D(z) = 1 + \sum_{q=1}^{Q} g_q z^{-m_q}. \tag{7.5.17}$$

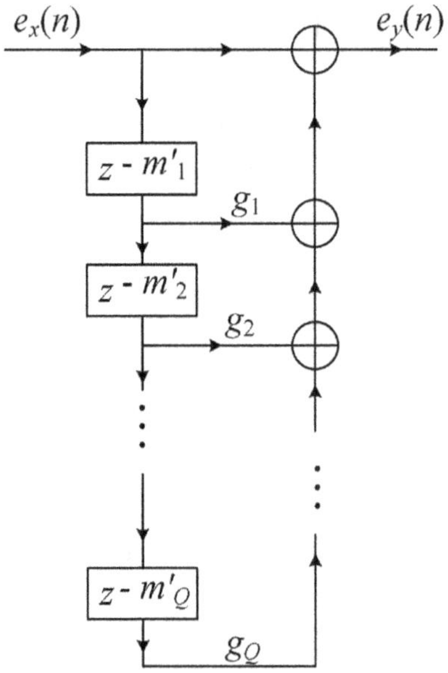

Figure 7.18 FIR filter structure for simulating multiple early discrete reflections.

The algorithms for simulating multiple early discrete reflections are implemented with the FIR filter structure shown in Figure 7.18, with $m_1 = m'_1$, $m_2 = m'_1 + m'_2$, ... $m_Q = m'_1 + m'_2 + ... + m'_Q$. The lengths of the delay lines in algorithms are chosen in accordance with the arrival times of early reflections in a modeled room or environment, such as the position of image sources for reflections (Section 1.2.2). In some practical cases, the lengths of delay lines can be selected according to the subjective preferred delay times for early reflections (Ando, 1985), provided that the requirement of the precedence effect is satisfied.

g_q in the delay lines of Equation (7.5.16)is independent of frequency. Accordingly, the spectrum of each simulated reflection is identical to that of the input signal (direct sound), but the former is attenuated by a constant factor. The gain in each delay line can be replaced with a low-pass filter unit with the system function $G_{LOW,q}(z)$ to simulate the frequency-dependent surface and air absorption. In this case, the overall impulse response and system function become

$$h_D(n) = \delta(n) + \sum_{q=1}^{Q} g_{LOW,q}(n - m_q), \tag{7.5.18}$$

$$H_D(z) = 1 + \sum_{q=1}^{Q} G_{LOW,q}(z) z^{-m_q}, \tag{7.5.19}$$

where $g_{LOW}(n)$ is the impulse response of $G_{LOW,q}(z)$.

The low-pass filter unit can be implemented with an FIR or infinite impulse response (IIR) structure. Well-designed filters can simulate various surfaces and air absorptions. Figure 7.19 depicts an example of a first-order IIR low-pass filter, whose system function is provided by

$$G_{LOW}(z) = \frac{b_0 + b_1 z^{-1}}{1 + a_1 z^{-1}}. \tag{7.5.20}$$

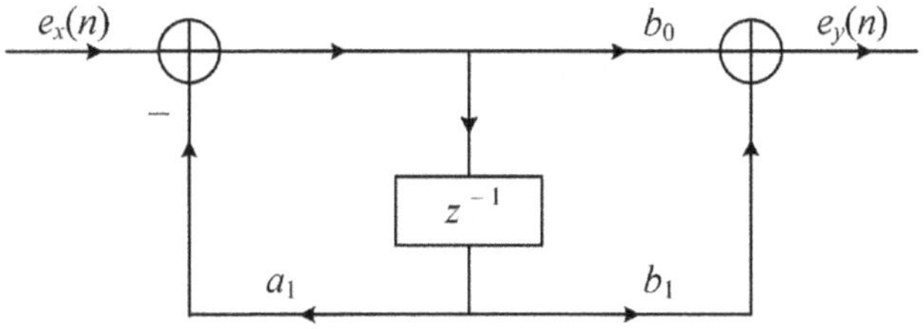

Figure 7.19 First-order IIR low-pass filters.

7.5.2 IIR filter algorithm of late reverberation

For the late diffused reverberation of a room, the temporal density of reflections from various directions increases as time is extended because of the increasing number of reflections from the boundary surfaces, as expressed in Equation (1.2.17). The overall energy of reverberations exponentially decays as time is prolonged because of the absorption of boundary surfaces. Late reverberation signals can be simulated with various perceived-based reverberation algorithms based on the statistical acoustic parameters of reverberation sound fields. Reverberation signals with a certain temporal density of reflections are enough to enhance auditory perceptions because of the limited resolution of human hearing. Schroeder (1962) suggested that a temporal density of not less than 1000/s is enough. On the basis of the results of psychoacoustic experiments, Kuttruff (2009) suggested a limit of 2000/s. Some other studies have also suggested a temporal density of ≥4000/s (Rubak and Johansen, 1998).

A plain reverberation algorithm simulates the successive reflections and decay in a room. It is implemented by combining the outputs of an infinite number of delay lines with lengths of $m, 2m, 3m$... samples and gains of g, g^2, g^3 .... ($g < 1$) relative to the original signal $e_x(n)$

$$e_y(n) = e_x(n) + \sum_{q=1}^{\infty} g^q e_x(n - qm). \tag{7.5.21}$$

Equation (7.5.21) can be written in a recursive form; thus,

$$e_y(n) = e_x(n) + g e_y(n - m). \tag{7.5.22}$$

Correspondingly, the impulse response and system function are

$$h_{REV}(n) = \delta(n) + \sum_{q=1}^{\infty} g^q \delta(n - qm), \tag{7.5.23}$$

$$H_{REV}(z) = 1 + \sum_{q=1}^{\infty} g^q z^{-qm} = \frac{1}{1 - g z^{-m}}. \tag{7.5.24}$$

The impulse response $h_{REV}(n)$ in Equation (7.5.23) consists of an infinite series of unit impulses weighted with an infinite power series of g. The algorithm given by Equation (7.5.23) is called the ***plain reverberation algorithm***, which is implemented by the IIR filter structure shown in Figure 7.20. The structure comprises a feedback loop with an m-sample

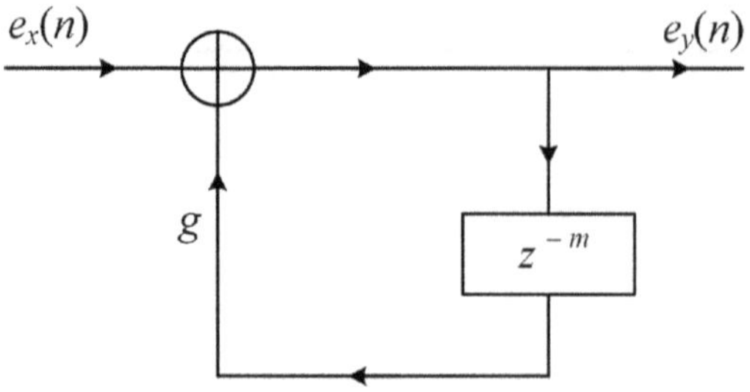

Figure 7.20 IIR filter structure for the plain reverberation algorithm.

delay line and g. The input and output of Figure 7.20 satisfy Equation (7.5.21), and this conclusion is easy to prove.

Equation (7.5.23) shows that each simulated reflection is delayed by m samples and attenuated g times compared with the preceding reflection. The magnitude of the qth reflection is g^q times that of direct sound. As stated in Section 1.2.4, the reverberation time of the plain reverberation algorithm is evaluated by assuming $20\log_{10}g^q = -60$ (dB):

$$T_{60} = \frac{-3m}{f_s \log_{10} g}. \tag{7.5.25}$$

For a given sampling frequency and delay line length m, T_{60} increases with the feedback g. The plain reverberation algorithm is simple with adjustable reverberation time and accommodates the exponential energy decay of natural reverberation. However, it suffers from the following drawbacks:

1. Reverberation time is frequency independent, preventing it from simulating frequency-dependent reverberation time caused by surface and air absorptions, which usually decrease as frequency increases in real rooms.
2. The equal time interval between two successive reflections tends to create fluttering. Moreover, the simulated reflection density f_s/m is usually small and invariable with time, which contradicts the phenomenon of increasing reflection density with time in real rooms.
3. The system function $H_{REV}(z)$ in Equation (7.5.24) includes m poles located at an equal interval in a circle with the radius $g^{1/m}$ in the Z-plane:

$$z_p = g^{1/m} \exp\left(j\frac{2\pi p}{m}\right) \qquad p = 0, 1, 2\ldots(m-1). \tag{7.5.26}$$

These poles cause the response magnitude $|H_{REV}(\omega)|$ to vary with frequency, resulting in comb filtering characteristics similar to those shown in Figure 7.17(b). These characteristics cause subjective coloration in timbre. The peaks in $|H_{REV}(\omega)|$ correspond to the poles of $H_{REV}(z)$. The digital angular frequencies of peaks are evaluated by substituting $z = \exp(j\omega)$ in Equation (7.5.26) as

$$\omega_p = \frac{2\pi p}{m} \qquad p = 0, 1\ldots(m-1). \tag{7.5.27}$$

A delay less than $m = 44$ in the plain reverberation algorithm is needed to obtain the temporal density of reflections larger than 1000/s at a sampling frequency of 44.1 kHz.

Accordingly, in Equation (7.5.25), $g = 0.9966$ is needed to simulate the reverberation time of $T_{60} = 2$ s. In this case, the poles in Equation (7.5.26) are located close to the unit circle $|z| = 1$ in the Z-plane and lead to instability in the IIR filter. For a given reverberation time T_{60}, the delay m is increased, and g is reduced to improve the stability of the IIR filter. However, this process in turn reduces the temporal density of reflections. As such, using a single plain reverberation unit for late reflection simulation often causes perceivable artifacts. This problem prompts the improvement of the plain reverberation algorithm.

g in Equation (7.5.25) controls reverberation time. The ***low-pass reverberation algorithm*** shown in Figure 7.21 is implemented to simulate the surface and air absorption-induced decrease in reverberation time at high frequencies, where g in Figure 7.20 is replaced with a low-pass filter unit $G_{LOW}(z)$ (Moorer, 1979). The impulse response and system function of the low-pass reverberation algorithm are given by

$$\begin{aligned} h_{REV}(n) = \delta(n) + g_{LOW}(n-m) + g_{LOW}(n) \otimes_t g_{LOW}(n-2m) \\ + g_{LOW}(n) \otimes_t g_{LOW}(n) \otimes_t g_{LOW}(n-3m) + \ldots\ldots, \end{aligned} \tag{7.5.28}$$

$$H_{REV}(z) = \frac{1}{1 - G_{LOW}(z) z^{-m}}, \tag{7.5.29}$$

where $g_{LOW}(n)$ is the impulse response related to the low-pass filter $G_{LOW}(z)$, and "$\otimes_t$" denotes convolution. Similar to the case in Equation (7.5.19), the low-pass filter unit can be implemented by an FIR or IIR structure and not repeated here.

In addition to the decrease in reverberation time at high frequencies, the reflection density in the low-pass reverberation algorithm increases with time. This behavior is consistent with the characteristics of late reflections in real rooms. High-order reflections in the low-pass reverberation algorithm are simulated through multiple convolutions with $g_{LOW}(n)$, which increases the temporal density of reflections.

To produce a flat static magnitude response and reduce the perceived timbre coloration, Schroeder (1962) proposed the well-known ***all-pass reverberation algorithm***. The impulse response and system function of this algorithm are given by

$$\begin{aligned} H_{REV}(z) &= \frac{-g + z^{-m}}{1 - gz^{-m}} = B_1 + \frac{B_2}{1 - gz^{-m}} = (B_1 + B_2) + B_2 \sum_{q=1}^{\infty} g^q z^{-qm} \\ B_1 &= -\frac{1}{g} \qquad B_2 = \frac{(1-g^2)}{g}, \end{aligned} \tag{7.5.30}$$

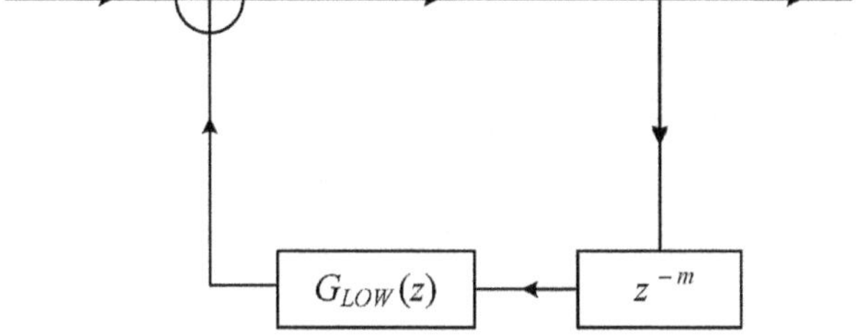

Figure 7.21 Low-pass reverberation algorithm.

$$h_{REV}(n) = (B_1 + B_2)\delta(n) + B_2\sum_{q=1}^{\infty} g^q \delta(n - qm). \tag{7.5.31}$$

$h_{REV}(n)$ comprises an infinite series of unit impulses with time-decaying gains. The magnitude of the $(q + 1)$th reflected impulse is g times that of the qth reflected impulse. The magnitude response of the system satisfies the following equation:

$$|H_{REV}(\omega)| = 1. \tag{7.5.32}$$

Therefore, the magnitude response is frequency independent. The all-pass reverberation algorithm can be implemented by the all-pass IIR filter structure shown in Figure 7.22. Equation (7.5.30) corresponds to the following input–output equation:

$$e_y(n) = -ge_x(n) + e_x(n - m) + ge_y(n - m). \tag{7.5.33}$$

The all-pass reverberation algorithm cannot completely eliminate timbre coloration because of the short-term frequency analysis of human hearing. The flat static magnitude response of the algorithm is the consequence of long-term Fourier analysis.

Several all-pass reverberation units can be connected in series to increase reflection density. The ratios of delay in the all-pass reverberation unit are an irrational number so that reflections are inconsistent at the same instant. In practice, the ***Schroeder reverberation algorithm*** or the structure shown in Figure 7.23 is often adopted. The algorithm consists of several parallel plain reverberation units and a series connection of all-pass reverberation units. Reducing the coloration caused by comb filters necessitates using plain reverberation units with incommensurate delays, and the ratio between the largest and smallest delays is about 1.5. The temporal density of reflections caused by Q parallel plain reverberation units is given by

$$\frac{dN_R}{dt} = \sum_{i=1}^{Q} \frac{1}{\tau_i} = \sum_{i=1}^{Q} \frac{f_s}{m_i}. \tag{7.5.34}$$

The density of the frequency model (eigenfrequenies) is given by

$$\frac{dN_f}{df} = \sum_{i=1}^{Q} \tau_i = \frac{1}{f_s}\sum_{i=1}^{Q} m_i. \tag{7.5.35}$$

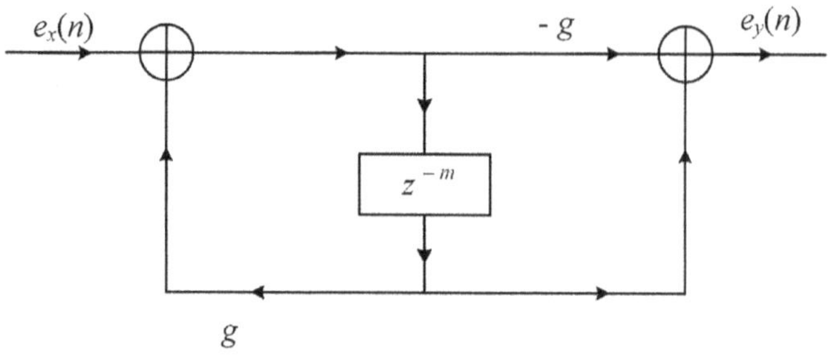

Figure 7.22 IIR filter structure of the all-pass reverberation algorithm.

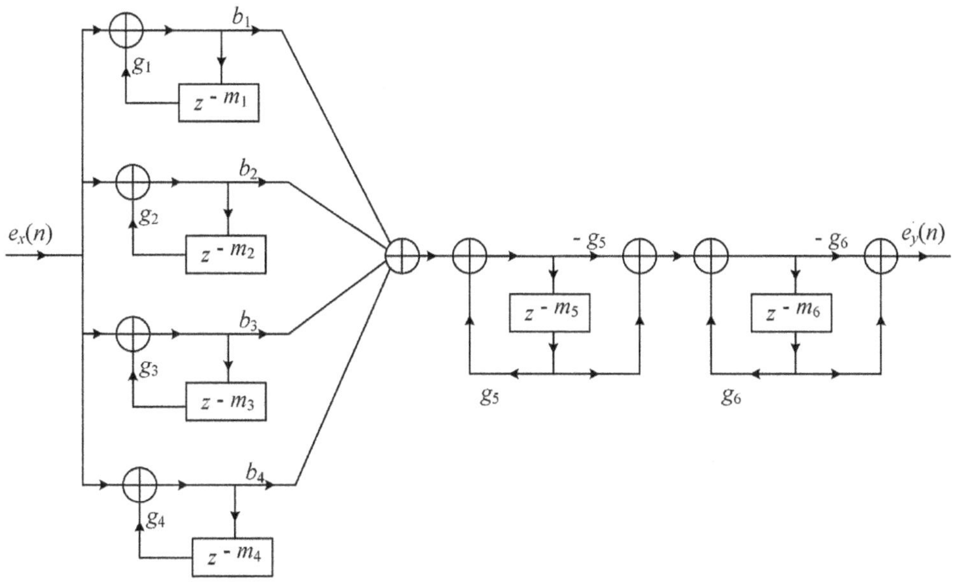

Figure 7.23 Schroeder reverberation algorithm.

where $\tau_i = m_i / f_s$ is the delay of the *i*th plain reverberation unit, m_i is the corresponding delay in the measured sample, and f_s is the sampling frequency. The series connection of all-pass reverberation units is intended to increase the reflection density. Other elements in Figure 7.23 are similar to those in the aforementioned discussion. However, the simulated temporal density of reflections in Figure 7.23 is constant over time. The low-pass reverberation units in Figure 7.21 can be used to replace the plain reverberation units in Figure 7.23, thereby enabling the simulation of increasing reflection density with time in real rooms (Gardner, 2002). The parameters of the reverberation algorithm can be chosen in terms of statistical acoustic characteristics of a target hall. In practice, they can be chosen on the basis of various optimal perception methods (Bai and Bai, 2005).

On the basis of the work of Stanuter and Puckette (1982), Jot and Chaigne (1991) proposed a ***feedback delay network*** (***FDN***) structure of reverberation algorithms. As shown in Figure 7.24, the outputs of N parallel delay lines with different lengths are connected to all N inputs by an $N \times N$ feedback matrix $[A]$. The feedback matrix and delay in each delay line may be designed so that the outputs of delay lines are uncorrelated. Generally, the matrix $[A]$ is chosen as a unitarity multiplying with a gain $|g| < 1$ to ensure the stability of the algorithm. The FDN structure is applicable to creating multichannel uncorrelated reverberations, and the inputs of N parallel delay lines can serve as multichannel signal inputs. Stanuter and Puckette illustrated an example of a four-input–four-output structure. Appropriate filters (such as low-pass filters) can be inserted before/after each delay line to simulate frequency-dependent reverberation time, but they also cause reflection density to increase with time. The FDN is the general form of an IIR reverberation structure. In the FDN structure with N inputs and N outputs, if the feedback matrix $[A]$ is a diagonal matrix, channels are decoupled, and their structure is simplified into the case of N independent plain or low-pass reverberation units

Various reverberation algorithms in the time domain are discussed above. Some reverberation algorithms in the time-frequency domain have also been proposed to simulate the frequency-dependent decay of late reverberation in rooms. Nikolic (2002) used quadrature mirror filters (QMF) or wavelet decomposition to decompose the input signal into 2–16

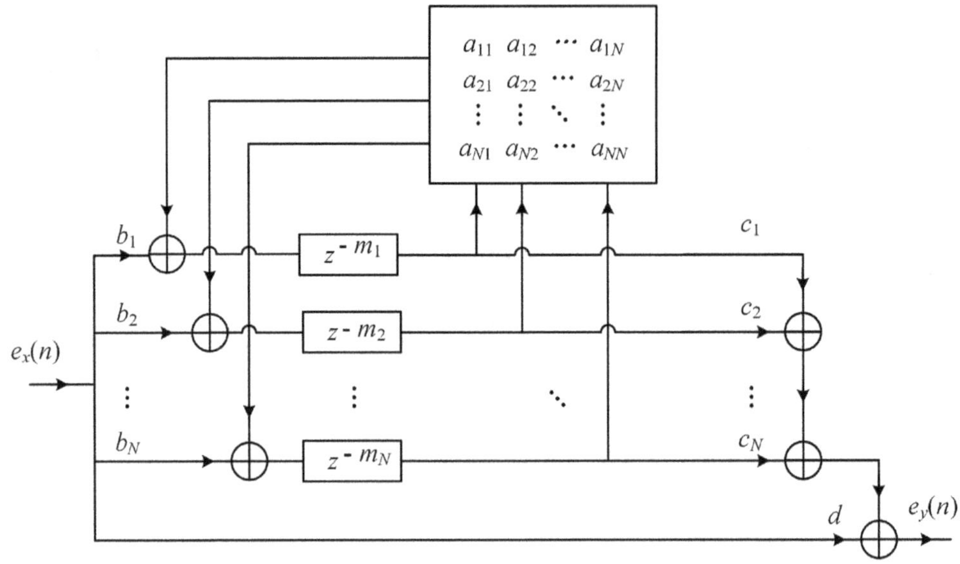

Figure 7.24 Feedback delay network structure of reverberation algorithms.

subband signals. Each subband signal is processed with different FDN units similar to that in Figure 7.20 to simulate the exponential decay of late reverberation. The feedback gain in each subband is controlled independently to accommodate the frequency-dependent decay of late reverberation. Vilkamo et al. (2011) proposed a similar reverberation algorithm through which the input signal is decomposed into 64 subbands by QMF. Each subband signal undergoes two series processing units. The first unit is an FDN similar to that in Figure 7.20, and the second unit is a decorrelator used to increase the temporal density of reflections and create the decorrelated subband reverberations. Output reverberation signals are the combination of all subband signals.

7.5.3 FIR, hybrid FIR, and recursive filter algorithms of late reverberation

Various feedback delay loops or networks described in Section 7.5.2 are IIR or recursive filter algorithms of reverberation simulation. Given the target room impulse response, a reverberation signal can be simulated with a ***finite impulse response*** (***FIR***) filter or equally by convoluting the input signal with the room impulse response. This principle is called a ***convolving reverberator***. A target room impulse response can be approximately obtained with perceived-based methods. In a simple method, a segment of Gauss white noise with a uniform power spectrum and a random phase is created. The length of the segment matches the target reverberation time. The room impulse response is simulated by applying an exponential decay envelope to the segment of white noise (Moorer, 1979). Frequency-dependent reverberation time is simulated by the bass-pass filtering of white noise and applying different exponential decay factors in various bands.

However, the length of a room's impulse response is in the order of the reverberation time of the room. For example, the length of a room impulse response is 88,200 points at a sampling frequency of 44.1 kHz to simulate the reflections of a concert hall with a reverberation time of 2.0 s. Implementing signal convolution with such a long impulse response by using a direct-form FIR filter in digital signal processing is computationally complex. A block transform method based on fast Fourier transform (FFT) improves the efficiency of calculation but leads to an additional delay in the output signal. To solve this problem, Gardner (1995) proposed a hybrid convolution method without an input-output delay. In this method, a filter response with a length of 16 N is divided into seven blocks with lengths of 2 N, N, N, 2 N, 2 N, 4 N, and 4 N, respectively. Convolution with the first block is implemented with the direct-form FIR filter, and convolutions with remanent blocks are individually implemented with FFT algorithms.

Even with hybrid convolution, the computational cost of FIR-filter-based reverberation algorithms is still high. Some hybrid FIR filters and recursive loop algorithms have been proposed to reduce the computational cost. These hybrid algorithms are constructed by inserting some FIR filters into recursive structures. Figure 7.25 illustrates the hybrid algorithm proposed by Rubak and Johansen (1998, 1999). The impulse response of the FIR filter is constructed by applying an exponential decay envelope to a segment of a relatively short pseudo-random sequence (noise) (100–200 ms) to simulate the random phase characteristic of a room impulse response. A recursive filter is constructed with feedback of the output of the FIR filter to the input with a delay and attenuation. A pseudo-random sequence with an infinite length is created by appropriately choosing the length of the FIR filter and the delay m and g of the feedback. This pseudo-random sequence simulates the exponential decay characteristic of reverberation with a sufficient time density. The recursive structure enables an FIR filter that is shorter than the target reverberation time and consequently reduces the computational cost. Similar to the low-pass reverberation algorithm in Section 7.5.2, the shortening in reverberation time at high frequencies can be simulated by inserting a low-pass filter in the feedback loop in Figure 7.25. In the subband-based reverberation algorithm proposed by Vilkamo at al. (Section 7.5.2), the signal processing in each subband is similar to that in Figure 7.25, but the random-sequence-based FIR filter is series connected after the recursive loop rather than inserted into the recursive loop.

One problem with the hybrid FIR filter and recursive loop reverberation algorithm in Figure 7.25 is that the unwanted periodicity in the decay reverberation signal may lead to audible artifacts. To solve this problem, Lee et al. (2009) suggested updating or switching a random sequence-based FIR filter. However, updating the random sequence-based FIR filter increases the computational cost and may lead to audible artifacts.

Some studies have suggested deriving an FIR filter based on a sparse random sequence to further simplify a hybrid FIR and recursive loop reverberation algorithm because reverberation with a certain temporal density is enough for auditory perception (Karjalainen and Järveläinen, 2007).

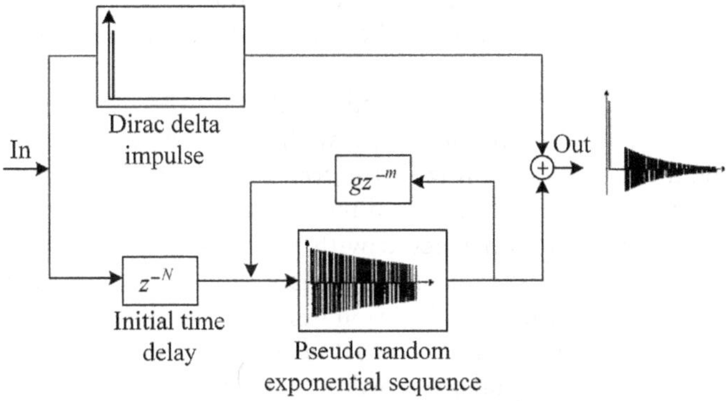

Figure 7.25 Hybrid FIR and recursive filter algorithms of reverberation (adapted from Rubak and Johansen, 1999).

7.5.4 Algorithms of audio signal decorrelation

As stated in Section 3.3, the subjective sensation similar to those caused by reflections in a hall or the sensations of immersion in direct sounds may be recreated by providing partly or low correlated signals to two or more loudspeakers with an appropriate configuration. Section 7.4.2 indicates that the perceived width of a virtual source or an auditory event can be controlled through the interchannel correlation between signals. Section 7.2 indicates that various multichannel microphone techniques are available for recording decorrelated reflected signals to recreate the sensation of envelopment in reproduction. Decorrelated signals can also be created artificially. The FDN reverberation algorithm in Section 7.5.2 is an example.

Two or more decorrelated signals from a mono input signal can be created by using a pair or a set of decorrelating filters. For example, in digital signal processing, two-channel decorrelated signals can be created by filtering the input signal in the frequency domain with a pair of decorrelating filters, namely, $H_1(\omega)$ and $H_2(\omega)$, or equally by convoluting the input signal $e_x(n)$ in the time domain with a pair of impulse responses, namely, $h_1(n)$ and $h_2(n)$:

$$e_{y1}(n) = h_1(n) \otimes_t e_x(n) \qquad e_{y2}(n) = h_2(n) \otimes_t e_x(n). \tag{7.5.36}$$

Various algorithms are available for decorrelating filters (Kendall, 1995). The simplest algorithm is to delay the input signal. That is, an input signal is delayed m samples by using a linear phase filter given by Equation (7.5.9). The delayed version and original signal are a pair of decorrelated signals, i.e., $e_{y1}(n) = e_x(n-m)$, and $e_{y2}(n) = e_x(n)$. However, as stated in the discussion of the precedence effect in Section 1.7.2, if the delay exceeds the duration of the autocorrelation function of the sound signal, a perceived echo may occur. Accordingly, the maximal delay of decorrelation evaluated from Equation (7.5.3) should not exceed about 40 ms. However, a less delay may be insufficient to decorrelate the signals. Therefore, decorrelation with a signal delay is simple but has a limited effect. This method is also unable to control the correlation between two output signals exactly. Moreover, the superposition of two channel signals with a relative delay in the listening position leads to a comb filtering effect and thus timbre coloration.

Decorrelated signals can be created by applying a random delay to each critical band (Bouéri and Kyirakakis, 2004). An input signal is decomposed into 23 subband signals with an ERB. Random delays are applied to each of the subbands, but the maximum allowable delay in each subband decreases as the central frequency of subbands increases. The decorrelated version of the original signal is obtained from a mixing of the delayed subband signals. This algorithm improves the perceived effect, but it may still exhibit timbre artifacts.

Another decorrelation algorithm is realized by filtering the input signal with a pair of random-phase filters. If filters are implemented with an FIR structure, they can be designed on the basis of the method of sampling in the frequency domain. That is, the magnitude responses of two filters are units (all-pass filters), and the phase responses of two filters at each discrete frequency are constructed with two sets of random number sequences within the range of $\pm\pi$ ($\pm 180°$). Ideally, all-pass filters do not change the power spectra of signals and then do not cause timbre coloration in reproduction. The frequency-dependent random phases of filters alter the interchannel correlation of two output signals, resulting in a pair of decorrelated signals. Accordingly, $h_1(n)$ and $h_2(n)$ in Equation (7.5.36) represent the impulse responses of a pair of all-pass filters with frequency-dependent random phases. However, this method only ensures that the magnitude responses of resultant filters are units and phase responses are random at discrete frequencies because of the limited frequency resolution of the FIR filter. The magnitude responses between two adjacent discrete frequencies fluctuate across frequencies and result in timbre coloration. As the length (order) of FIR filters increases, magnitude fluctuation decreases, but signal processing becomes complicated. Optimizing filter phases in each discrete frequency may also reduce magnitude fluctuation, but it should not influence the decorrelation characteristics of filters.

All-pass filters based on a reciprocal-maximal length sequence (MLS) are available for audio signal decorrelation (Xie et al., 2012). Binary MLS is a kind of a deterministic and periodic sequence, but it possesses some characteristics similar to random noise (Vanderkooy, 1994). A reciprocal MLS pair consists of a binary MLS and its sequence in a reversed order. Xiang and Schroeder (2003) proved that a reciprocal MLS pair exhibits the characters similar to the responses of a pair of all-pass filters with a random phase, i.e., nearly uniform (constant) power spectra and low-valued cross-correlation. Therefore, the impulse responses (FIR coefficients) of a pair of decorrelated filters can be designed in accordance with a reciprocal MLS pair. Taking advantage of the deterministic and periodic characteristics of MLS, the design of MLS-based decorrelation filters is controllable and repeatable. The perceived low-frequency performance of decorrelation filters at a given length can be optimized or improved by the properly circle time shift of the inverse-order MLS filter coefficient by using the circle time shift characteristics of MLS. This process is similar to adjusting the relative phase between filters. The results of subjective experiments indicate that optimizing a scheme improves the relatively perceived performance of 511-point reciprocal MLS-based filters (at a sampling of 44.1 kHz) so that it is better than the original 1023-point filters.

Similar to the aforementioned algorithm, in the reverberation algorithm with the hybrid FIR and recursive loop described in Section 7.5.3, two or more channel decorrelated reverberation signals can be obtained by using two or more uncorrelated white noise sequences or reciprocal MLS with an exponential decay envelope for FIR filter responses. For example, Xiang et al. (2019) suggested using reciprocal MLS in each subband to obtain two-channel decorrelated reverberation signals. This method can be extended to multichannel decorrelated reverberation signals.

In sound reproduction, two signals should be created with various interchannel correlations through the weighted mixing of two uncorrelated signals $e_{y1}(n)$ and $e_{y2}(n)$, which are expressed as follows:

$$
\begin{aligned}
e'_{y1}(n) &= \cos\gamma\, e_{y1}(n) + \sin\gamma\, e_{y2}(n) \\
e'_{y2}(n) &= \sin\gamma\, e_{y1}(n) + \cos\gamma\, e_{y2}(n),
\end{aligned}
\qquad (7.5.37)
$$

where γ is a parameter. The interchannel correlation between two signals is evaluated from Equation (1.7.6) as

$$
corre = \sin 2\gamma. \qquad (7.5.38)
$$

Therefore, the interchannel correlation of the two signals is controlled with γ. When the parameter varies within the range of $0 \le \gamma \le \pi/4$, the interchannel correlation changes from zero (uncorrelated) to a unit (completely correlated) as expected.

7.5.5 Simulation of room reflections based on physical measurement and calculation

The reflection simulations discussed in Sections 7.5.1–7.5.4 simulate the subjective perceptions of reflections according to some statistical acoustic parameters of a hall. In multichannel sound signal production, the room impulse responses of convolution-based reflection simulation can be derived through physical-based measurement or simulation.

In various microphone techniques described in Sections 7.2 and 7.3 for recording in halls, when the positions of a sound source and microphones are fixed, the transmission from the sound source to microphone outputs can be represented with an LTI system with one input and multiple outputs. The impulse responses of the LTI system describe the acoustic (including direct and reflected) transmissions from the source to microphones, directivities, and acoustic-electric conversion of microphones. Therefore, given the spatial room impulse responses of a hall, signals recorded with onsite microphones can be physically simulated by convoluting an anechoic stimulus with spatial room impulse responses. Generally, simulated signals involve the spatial information of direct and reflected sounds, depending on the microphone technique to be simulated. If the microphone technique is chosen appropriately so that the reflected components are dominant in impulse responses, and the outputs of convolution are an approximate simulation of the reflected signals of the hall. This principle is the basis for simulating reflected sound signals with the convolution of spatial room impulse responses. In practice, the direct, early reflection, and late reverberation parts of spatial room impulse responses can be separated by using an appropriate time window, and the corresponding signal components can be individually simulated. Kleczkowski et al. (2015) conducted a subjective experiment and indicated that the perceived quality is improved by the angular separation of direct and reflected sound components in 5.0- or 7.0-channel sound reproduction involving spatial room impulse responses measured with a coincident microphone technique.

Sound sources and a set of microphones are arranged in a hall to measure spatial room impulse responses. Various stimuli and techniques commonly used for acoustic impulse response measurement, such as MLS (Section 7.5.4) and sweep signal-based techniques, are applied to measure spatial room impulse responses (Stan et al., 2002). The spatial room impulse responses of various source positions should be measured to simulate the spatial information caused by sound sources at different positions (such as different instruments in various positions of an orchestra). The final perceived effects in reproduction are greatly influenced by the physical characteristics (such as directivity) of the sound source used for measurement. Ideally, the physical characteristics of sound source used for measurement should be similar to those of the practical sound source (instrument), but this process may be difficult to be implemented in practice. The omnidirectional characteristic and upper-frequency limit (usually 8 kHz) of dodecahedral loudspeaker systems commonly used in room acoustic measurement cannot meet this requirement. In practice, anechoic or "dry" stimuli are usually captured by spot microphones close to the sound sources in a recording room with a short reverberation time.

Various microphone techniques in Sections 7.2 and 7.3 are possibly applicable to the measurement of spatial room impulse responses and the simulation of reverberation (Farina and Ayalon, 2003). Gerzon (1975a) first suggested using the spatial room impulse responses of Ambisonics to simulate the reflections in a concert hall. With the development of digital signal processing, this method can be implemented easily. The spatial room impulse responses of the first-order A-format Ambisonics can be measured and derived by using microphones similar to those in Figure 7.15. The spatial room impulse responses of higher-order Ambisonics can be measured and derived by using the microphone array described in Section 9.8.

Fernando (2014) proposed another method for simulating diffused reverberation in higher-order spatial Ambisonics. The "dry" source signal is allocated to K virtual sources uniformly located in space. These virtual source signals are convolved with K-independent and decorrelated room impulse responses. The outputs of K convolutors are encoded back into Ambisonic signals and mixed with original Ambisonic dry signals. However, Fernando did not mention how to acquire the decorrelated room impulse responses. Fernando's method can be regarded as an indirect method of reflection simulation in multichannel sound reproduction described in Section 3.1.

Some special microphone techniques have also been designed for the measurement of spatial room impulse responses (Kessler, 2005; Woszczyk et al., 2009). For example, Woszczyk et al. (2010) designed an array with eight microphones, including six horizontal microphones and two high-elevation microphones, to measure spatial room impulse responses. The resultant impulse responses are applicable to recreating the ambient signals of 22.2 channel reproduction.

In addition to measurement, physical simulation can be performed to obtain spatial room impulse responses. Complete physical simulation involves source simulation (such as radiation pattern simulation) and transmission or room acoustic simulation (e.g., surface reflection, scattering and absorption, and air absorption simulation), resulting in sound field distribution in a room; consequently, sound field distribution is converted into information in multichannel sound signals. Computational simulation has become the most important method for analyzing room acoustics because of the rapid development of computer techniques.

In principle, the sound pressure inside a room can be calculated by solving the wave equation subject to certain boundary conditions. However, the analytical solution of the wave equation can be obtained only in rare cases, such as the one achieved for a rectangular room. In general, the wave equation of an acoustic field in a room is solved through various numeral methods, such as the boundary element method and the finite difference time-domain method. Limited to extensive computational workloads, these ***wave acoustic-based numerical methods*** are suitable only for low-frequency and small-room simulation.

For high frequencies and smooth boundary surfaces, an acoustic field in a room can be approximately simulated through ***geometrical acoustic-based methods***. Geometrical acoustic-based methods basically disregard the wave nature of sound, so they are insufficiently accurate as wave acoustic-based numerical methods. However, the computational workloads of geometrical acoustic-based methods are relatively low. ***Image-source*** and ***ray-tracing methods*** are two common examples of geometrical acoustic-based methods. Image-source methods decompose the reflected sound field into the radiations of multiple image sources in a free space. Ray-tracing methods approximate the sound radiation from a point sound source as a number of rays. Each ray carries a certain amount of energy and propagates at sound speed. When a ray comes in contact with the boundary surface, some of its energy is reflected, and the rest is absorbed in accordance with some rules. Geometrical acoustic-based methods yield the arrival time, direction, energy, and spectrum of direct sound and each reflection. Some other work decomposed the measured spatial room impulse responses into a set of image sources, which are used for room acoustic analysis and multichannel convolution-based reverberation algorithm (Tervo et al., 2013).

Simulated room acoustic field information can be converted into spatial room impulse responses in accordance with appropriate signal mixing rules. The details of room acoustic simulation are beyond the scope of this book, but they are described in related books and literature (Lehnert and Blauert, 1992; Kleiner et al., 1993; Svensson and Kristiansen, 2002; Wu and Zhao, 2003; Vorländer, 2008).

Overall, reflections simulated with spatial room impulse responses may yield relative authentic and natural auditory perceptions. In this method, signals are recorded and produced in a "virtual environment," i.e., creating the signals of a target environment by using the "virtual recording" algorithm. It is relatively simple and low cost because onsite recording is not required. Therefore, "virtual recording" is a promising method for the simulation of multichannel sound signals. However, deriving spatial room impulse responses through physical-based simulation is difficult, and the computational cost of convolution with spatial room impulse responses is large. "Virtual recording" is also a preliminary method, but further studies and practices are required.

7.6 DIRECTIONAL AUDIO CODING AND MULTICHANNEL SOUND SIGNAL SYNTHESIS

In Sections 7.5.1–7.5.3, environmental reflections are simulated in terms of some general (statistical) results about room acoustic parameters (such as reverberation time) and psychoacoustics of reflections. The directional and diffused information of a sound field may be estimated to improve the performance in recreating the perceived effect of a target sound field, and directional and diffused signals may be simulated in terms of the estimated information. According to this consideration, the method of ***spatial impulse response rendering*** (***SIRR***) was first proposed to simulate multichannel sound signals (Merimaa and Pulkki, 2005; Pulkki and Merimaa, 2006). On the basis of SIRR, ***directional audio coding*** (***DiRAC***, Pulkki, 2007), a general method for enhancing the spatial information of continuous sound signals, was further developed. The application of DiRAC to the recording and synthesis of multichannel sound signals is discussed in this section. DiRAC can also serve as a coding method of spatial audio signals (Section 13.4.5).

DiRAC involves two stages: sound field analysis stage and signal synthesis or simulation stage. In DiRAC, spatial and timbre perceptions in a sound field are determined by physical information such as the direction of energy propagation, extent of diffusion, and power spectra. The desired perceived effects can be obtained by correctly recreating physical information in reproduction.

In the stage of sound field analysis, directional and diffused information is estimated from the analysis of the energy density and energy propagation (power flow) of an original sound field. Instantaneous energy density is defined as the sound energy (power) in a unit volume at an instant t:

$$\varepsilon(t) = \frac{1}{2}\rho_0\left[\left|\boldsymbol{v}(t)\right|^2 + \frac{1}{\rho_0^2 c^2}\left|p(t)\right|^2\right], \tag{7.6.1}$$

where $p(t)$ is the sound pressure at the receiver position, $\boldsymbol{v}$(t) is the velocity vector of the medium (air), and c is the speed of sound.

The energy propagation of the sound field is described using the sound intensity vector, which is defined as the power carried by *sound* waves per unit area in a direction perpendicular to that area:

$$\boldsymbol{I} = \frac{1}{T}\int_0^T p(t)\boldsymbol{v}(t)dt = \overline{p(t)\boldsymbol{v}(t)}. \tag{7.6.2}$$

$p(t)$ and $\boldsymbol{v}(t)$ are real-valued functions. If $p(t)$ and $\boldsymbol{v}(t)$ are complex-valued functions, the real part is taken in the calculation of Equation (7.6.2). The second equality on the right side of Equation (7.6.2) denotes the average over time. For a point source in the free field, the direction of the sound intensity vector is opposite the source direction. The instantaneous sound intensity vector is defined as

$$I(t) = p(t)v(t). \tag{7.6.3}$$

The instantaneous sound intensity vector involves two parts. The direction of first-part oscillates with time so that it does not contribute to the sound intensity vector (mean power flow) in Equation (7.6.2). The instantaneous sound intensity with local oscillation is used to estimate the proportion of the diffused component in an original sound field because of the isotropic characteristic of a diffused field. The second part in the instantaneous sound intensity vector represents the energy propagation along a special direction and contributes to the sound intensity vector in Equation (7.6.2). The power flow in the second part is used to estimate the proportion of a directional sound field. Accordingly, the proportion of time-oscillating instantaneous sound intensity is estimated as

$$\Psi = 1 - \frac{|\overline{I(t)}/c|}{\overline{\varepsilon(t)}} = 1 - \frac{2\rho_0 c\,|\overline{p(t)v(t)}|}{\overline{p^2(t) + (\rho_0 c)^2 |v(t)|^2}}. \tag{7.6.4}$$

By definition, $0 \le \Psi \le 1$, which is called ***diffuseness***, describes the extent of the diffusion of a sound field. $\Psi = 1$ indicates complete diffusion, and $\Psi = 0$ represents energy propagation without a time-oscillating component. Therefore, the directional information of the energy propagation of direct and discrete reflections and the proportion of diffused reflections can be estimated from the analysis of the sound intensity vector and Ψ.

The spatial information of sound field varies with time and frequency and should be analyzed in the time-frequency domain. For example, the discrete time samples of sound pressure signal $p(t)$ are denoted by $p(n)$. Applying a time-frequency transform to $p(n)$ yields $P(n', k)$, where n is the discrete time, k is the discrete frequency or index of the frequency band, and n' is the index of the time frame. Similarly, other physical quantities, such as the velocity of a medium, can be transformed into the time-frequency domain. Various subband filters are available to transform the physical quantity or signals to the time-frequency domain. Subband filters with the bandwidth of auditory filters usually enhance the perceived effects, but they are complicated. The method of ***short-time Fourier transform*** (***STFT***) is used to simplify signal processing. The STFT of sound pressure $p(n)$ is calculated by

$$P(n', k) = \sum_{n=NL}^{NH} W(n)p(n'+n)\exp\left(-j\frac{2\pi}{N}kn\right) \qquad k = 0, 1 \ldots (K-1), \tag{7.6.5}$$

where $NL \le 0$ and $NH > 0$ are the initial and end times for STFT calculation; $N = NH - NL + 1$ is the length of a frame or a block for STFT calculation; and $W(n)$ is an appropriate time window function. The shortcoming of STFT is its uniform frequency resolution, which is inconsistent with the frequency resolution of human hearing. A short-time frame in STFT may lead to insufficient frequency resolution at low frequencies. Conversely, a long-time frame in STFT may lead to a transient error at high frequencies.

In the implementation of DiRAC in the time-frequency domain, the diffuseness $\Psi(n', k)$ and sound intensity vector $\boldsymbol{I}(n', k)$ should be calculated for each frequency or frequency band and each time frame:

$$\Psi(n', k) = 1 - \frac{2\rho_0 c \left| \sum_{n=NL1}^{NH1} P(n'+n, k)\,\boldsymbol{V}(n'+n, k) W_1(n, k) \right|}{\sum_{n=NL1}^{NH1} \left[\left|P(n'+n, k)\right|^2 + (\rho_0 c)^2 \left|\boldsymbol{V}(n'+n, k)\right|^2 \right] W_1(n, k)}, \tag{7.6.6}$$

$$\boldsymbol{I}(n', k) = \frac{1}{NH_2 - NL_2} \sum_{n=NL_2}^{NH_2} P(n'+n, k)\,\boldsymbol{V}(n'+n, k) W_2(n, k), \tag{7.6.7}$$

where $\boldsymbol{V}(n'+n, k)$ is the STFT of the velocity vector of the medium; $NL_1 \leq 0$ and $NH_1 > 0$ or NL_2 and NH_2 are the initial and end times for short-time average; $W_1(n, k)$ and $W_2(n, k)$ are the appropriate time window functions. For the complex-valued functions P and $\boldsymbol{V}$, the real parts of $\Psi(n', k)$ and $\boldsymbol{I}(n', k)$ are taken in the two equations. Hanning windows are used in the literature (Pulkki, 2007). The time lengths of windows are chosen on the basis of an informal listening test. The time length of $W_1(n, k)$ is 10–50 times the period of the central frequency of the corresponding frequency band but limited to a range of 3 ms to 50 ms. The time length of $W_2(n, k)$ is only three times the period of the central frequency of the corresponding frequency band but limited to a minimum of 1 ms.

In practice, the sound pressure and velocity vector of a medium in sound field analysis can be measured through appropriate microphone techniques. A convenient method involves using the first-order Ambisonic microphone technique. The first-order Ambisonic microphone technique yields four independent signals, e.g., a signal from an omnidirectional microphone (spherical symmetrical directivity) and signals from three bidirectional microphones (symmetrical directivity around the polar axis) with their main axes pointing to the front, left, and top directions. For an incident plane wave in a free field, the normalized amplitudes of four independent signals are given by Equation (6.4.1). These four signals are directly proportional to the sound pressure and three orthogonal components of the velocity vector of a medium. In the time-frequency domain, the four independent signals are

$$\begin{array}{ll} E_W(n', k) \propto P(n', k) & E_X(n', k) \propto \rho_0 c V_X(n', k) \\ E_Y(n', k) \propto \rho_0 c V_Y(n', k) & E_Z(n', k) \propto \rho_0 c V_Z(n', k) \end{array}. \tag{7.6.8}$$

The B-format independent signals of the first-order Ambisonics in Equation (6.4.13) differ from Equation (7.6.8) by a scale of gain. The diffuseness $\Psi(n', k)$ and sound intensity vector $\boldsymbol{I}(n', k)$ in the time-frequency domain are calculated by substituting Equation (7.6.8) into Equations (7.6.6) and (7.6.7). If the direction of the sound intensity vector is explored only, the denominator in Equation (7.6.7) can be ignored in the calculation.

In the stage of signal simulation or synthesis in DiRAC, multichannel sound signals (including directional signals and diffused signals) are simulated or synthesized from the recorded sound pressure (signal) waveform of an original sound field in terms of the information from the aforementioned sound field analysis and practical loudspeaker configuration. In signal synthesis, because human hearing decodes only single (dominant) spatial cues for each critical band at one time, the spatial information of multiple sources can be separated from signals in the time-frequency domain. As a simple example, the mono sound pressure (signal) waveform in the time-frequency domain is $E_A(n', k)$. It can be derived from the output $E_W(n', k)$ of the omnidirectional microphone in the aforementioned Ambisonic recording or recorded with a separate microphone.

According to the definition of diffuseness in Equation (7.6.4), the directional component (direct and early reflected components) in signal $E_A(n', k)$ is directly proportional to $\sqrt{1-\Psi(n', k)}$. Based on the analysis of the sound intensity vector in Equation (7.6.7), the direction of the instantaneous power flow and the target virtual source direction $\Omega_S(n', k) = [\theta_S(n', k), \phi_S(n', k)]$ in each frequency band are estimated. Given the target virtual source direction and loudspeaker configuration in reproduction, the directional signals of loudspeakers are derived in accordance with an appropriate signal panning or mixing rule. For example, if the VBAP in Section 6.3 is used, the normalized gain $A_i(n', k)$ of the ith loudspeaker signal is obtained, then the signal of that loudspeaker is given by

$$E_i(n', k) = \sqrt{1-\Psi(n', k)}A_i(n', k)E_A(n', k). \tag{7.6.9}$$

The normalized gain $A_i(n', k)$ varies with frequency (band) and time frame. The weighted temporal average of $A_i(n', k)$ is needed to avoid the audible artifacts caused by fast temporal variations in $A_i(n', k)$, i.e., the following $A'_i(n', k)$ is used to replace $A_i(n', k)$ in Equation (7.6.9):

$$A'_i(n', k) = \frac{\sum_{n''=NL3}^{NH3} A_i(n'+n'', k)\left[1-\Psi(n'+n'', k)\right]W_3(n'', k)}{\sum_{n''=NL3}^{NH3}\left[1-\Psi(n'+n'', k)\right]W_3(n'', k)}, \tag{7.6.10}$$

where $W_3(n'',k)$ is a time window function. Hanning windows are used in the literature (Pulkki, 2007). The time length of $W_3(n'',k)$ is 100 times the period of the central frequency of the corresponding frequency band but limited to a maximum of 1000 ms.

The diffused component in the signal $E_A(n', k)$ is directly proportional to $\sqrt{\Psi(n', k)}$. Accordingly, the diffused signals of loudspeakers are derived by scaling the mono signal $E_A(n', k)$ in the time-frequency domain with a factor $\sqrt{\Psi(n', k)}$ and decorrelating processing. Decorrelation is implemented by convoluting the signal with random sequences having an exponential decay envelope. However, for some transient signals, such as applause-like signals, decorrelation processing with low temporal resolution may cause audible artifacts. A high temporal resolution in DiRAC encoding improves the perceived quality (Laitinen et al., 2011).

The aforementioned method of scaling mono waveform signals with factors $\sqrt{1-\Psi(n', k)}$ and $\sqrt{\Psi(n', k)}$ is insufficient to separate directional and diffused components effectively. Directional components should be simulated from multiple pressure waveform signals to enhance their synthesis. Multiple pressure waveform signals, which are equivalent to the outputs of a set of virtual cardioid microphones with their main axis directions matching the loudspeaker directions in reproduction, can be derived from the linear combinations of the recorded first-order Ambisonic signals in Equation (7.6.8). The directional components of loudspeaker signals are simulated from the outputs of virtual microphones by using a method similar to Equation (7.6.9) (Vilkamo et al., 2009). In this case, input pressure waveform signals for the simulation of each loudspeaker signal are different. This method results in the attenuation of diffused sound by −4.8 dB in the simulated directional component. A similar method is available for the simulation of a diffused component.

DiRAC and SIRR are for the simulation or synthesis of multichannel sound signals. They are based on the same principle and closely related. They differ in terms of their design to manipulate different types of sound signals or responses. The implementations of DiRAC and SIRR are also partly different. In DiRAC, continuous Ambisonic signals are analyzed, and continuous sound waveform signals are manipulated and transformed into the time-frequency domain in accordance with Equation (7.6.5). Multichannel sound signals are directly synthesized using the aforementioned method. In other words, DiRAC focuses on the analysis and manipulation of continuous sound signals from onsite recording.

SIRR focuses on the analysis and manipulation of the spatial impulse responses of an acoustic environment. This process is called "spatial impulse response rendering." The Ambisonic spatial impulse responses and mono impulse responses of an acoustic environment are measured. These measured impulse responses are transformed into the time-frequency domain and used as the Ambisonic signals $E_W(n', k)$, $E_X(n', k)$, $E_Y(n', k)$, and $E_Z(n', k)$ in Equation (7.6.8) for sound field analysis and mono wavefront signal $E_A(n', k)$ for synthesis. The time-frequency domain representations of the impulse responses for the simulation of multichannel sound signals are synthesized from $E_A(n', k)$ and the analysis of $E_W(n', k)$, $E_X(n', k)$, $E_Y(n', k)$, and $E_Z(n', k)$. They are then transformed into the time domain. However, short-time average processing may be unnecessary for impulse signals. The signals of multichannel sound reproduction are obtained by convoluting the continuous mono signal recorded onsite with the synthesized time-domain impulse responses for the simulation of multichannel sound signals. Therefore, SIRR synthesizes the impulse responses for the simulation of multichannel sound signals from the mono impulse responses of an acoustic environment by analyzing the Ambisonic spatial impulse response of an acoustic environment and then simulating multichannel sound signals through convolution. From this point, SIRR is similar to a convolving reverberator. Pulkki and Merimaa (2006) conducted a psychoacoustic experiment and indicated that SIRR can recreate a natural spatial auditory effect.

DiRAC/SIRR is a general method for rendering the spatial information of a sound field. Recording and analysis in DiRAC are independent of loudspeaker configuration, signal mixing, and reproduction. The captured signals and information from the analysis stage in DiRAC are applied to synthesize signals for various multichannel sounds and other spatial sounds, such as signals for wave-field synthesis (Chapter 10) and binaural reproduction (Chapter 11). The simulation of localization signals in multichannel sound is not limited to VBAP, and other methods, such as higher-order Ambisonic method, are applicable. The recorded signals for sound field analysis in SIRR/DiRAC are not limited to B-format first-order Ambisonic signals. Pulkki et al. (2013) demonstrated that sound field is divided into sectors and signals from a higher-order Ambisonic microphone array is used for sound field analysis to enhance the perceived quality in DiRAC reproduction. In other studies, signals from near-coincident and spaced microphone arrays in Section 7.2 and 7.3 are used for sound field analysis (Politis et al., 2015). In traditional multichannel sound, recording with near-coincident and spaced microphone techniques usually exhibits a good perceptual envelopment because of a low interchannel correlation in captured reverberation signals but at the cost of degrading the localization quality of direct sound. Through DiRAC, signals are analyzed on the basis of near-coincident and spaced microphone recording. Signals are enhanced and re-synthesized to improve the perceived effects of direct and reverberation sound. For example, in 5.1-channel sound recording, the 5.1-channel main microphone array is similar to that in Figure 7.1, and the horizontal direction of direct sound and diffuseness of sound field can be estimated from the outputs of two pairs of opposite microphones (LF and LB pair and RF and LB pair). According to the spatial information of a sound field in the time-frequency domain, the directional and reverberation components of recorded signals are manipulated, and multichannel sound signals are re-synthesized. Coherent and incoherent parts in reverberation components can be manipulated separately in signal re-synthesis because of the inherent partly decorrelating characteristic (especially at high frequencies) of the reverberation components captured by near-coincident and spaced microphone arrays.

7.7 SUMMARY

Signal recording and simulation is one of the stages of practical multichannel sounds. In the program production of channel-based multichannel sounds or the signal rendering of object-based multichannel sounds, multichannel sound signals should be created in accordance with the psychoacoustic principle of spatial hearing to recreate the desired spatial auditory events or perceptions in reproduction. Multichannel sound signals can be recreated through onsite recording or artificial simulation.

The 5.1-channel sound is a typical multichannel horizontal surround sound with an irregular loudspeaker configuration. Its microphone techniques are relatively matured and can be roughly classified into two categories. The first category is the main microphone technique through which a single array of microphones close to one another is used to capture information on frontal and surround channels. The second category is a combination of frontal and surround microphone techniques through which two separate microphone arrays are used to capture information on frontal and surround channels. These 5.1-channel microphone techniques are based on the principle of summing localization with ICLD/ICTD and the principle of recreating auditory sensations similar to those caused by reflections with decorrelated signals. Various 5.1-channel microphone techniques and some microphone techniques for other discrete multichannel sounds, including microphone techniques for spatial surround sound, have been developed. They are similar to those for 5.1-channel sound, but they can capture more spatial information and are more complicated.

Directional localization signals for multichannel sound can be created through simulation. That is, a mono signal is rendered to channels in accordance with certain signal panning or mixing rules to create multichannel signals with the desired amplitude and time relationship. Pair-wise amplitude panning is commonly used in multichannel horizontal surround sounds. Other signal panning or mixing methods, such as the Ambisonic method, may also be utilized. In addition to the simulation of virtual sources in various directions, the control of the perceived width (ASW), distance, and depth of virtual sources can be achieved by changing the ratio of reflected sound, interchannel correlation, and interaural cue fluctuations. Time-varying signal panning or mixing is applicable to the simulation of a moving virtual source.

Two basic methods, i.e., perception- and physics-based methods, are commonly available for simulating room or environmental reflections. In perception-based methods, reflections are synthesized on the basis of some predetermined statistical (general) room acoustic parameters. These methods are the basis of various artificial delay, reverberation, and signal decorrelation algorithms. In physics-based methods, reflections are simulated by convoluting signals with the spatial impulse responses of a room. Generally, perception-based methods are simple, but physics-based methods are accurate.

In terms of the information of a sound field obtained through recording and analysis, spatial impulse response rendering and directional audio coding simulate re-synthesize multichannel sound signals to improve the perceived effect of direct and reflected sounds in reproduction.

Chapter 8

Matrix surround sound and downmixing/upmixing of multichannel sound signals

Matrix surround sound is a special category of multichannel sound techniques and systems. The basic consideration of matrix surround sound is that original multichannel sound signals are linearly encoded or matrixed into fewer independent signals for transmission and then decoded into more channel signals for reproduction when the number of transmission (or storage) channels is limited. Conversion or downmixing/upmixing between two-channel stereophonic signals and mono signal is addressed in Section 2.3. Similar situations occur in multichannel sound, that is, a larger number of channel signals are converted into a smaller number of channel signals (including two-channel signals) for reproduction or vice versa. The former is for compatible reproduction with systems of a smaller number of channels. Similar to the pseudo-stereophonic sound in early times, the latter is used to synthesize or simulate the effects similar to those of more channel reproduction by post-processing on fewer numbers of independent signals. Although the purposes of matrix surround sound and the downmixing/upmixing of multichannel sound signals are different, they are closely related to one another.

This chapter focuses on the matrix surround sound and the downmixing/upmixing of multichannel sound signals. In Section 8.1, various matrix surround sounds, including matrix quadraphone in early times, four-channel matrix surround sound by Dolby Labs and some recent multichannel matrix surround sounds are introduced. In Section 8.2, the problem of downmixing multichannel sound signals into two-channel stereophonic signals is discussed. In Section 8.3, the problem of upmixing multichannel sound signals, especially the upmixing of two-channel stereophonic signals into multichannel sound signals, is discussed.

8.1 MATRIX SURROUND SOUND

8.1.1 Matrix quadraphone

The discrete quadraphone in Section 4.1 involves four independent signals. In the end of the 1960s and the beginning of the 1970s when analog media were dominant, transmission and storage of four-channel signals are practically difficult. The number of independent signals should be reduced to adapt to the traditional two-channel media (such as 45°/45° disk recording system or analog stereophonic broadcast). Therefore, a series of ***matrix quadraphones*** was developed at that time. In a matrix quadraphone, the four original signals of a discrete quadraphone are encoded or linearly combined into two independent signals for transmission. Then, the two independent signals are decoded or linearly combined back into four signals for reproduction. Figure 8.1 shows the block diagram of a matrix quadraphone (sometimes called 4-2-4 matrix quadraphone). Four loudspeakers in a matrix quadraphone

DOI: 10.1201/9781003081500-8

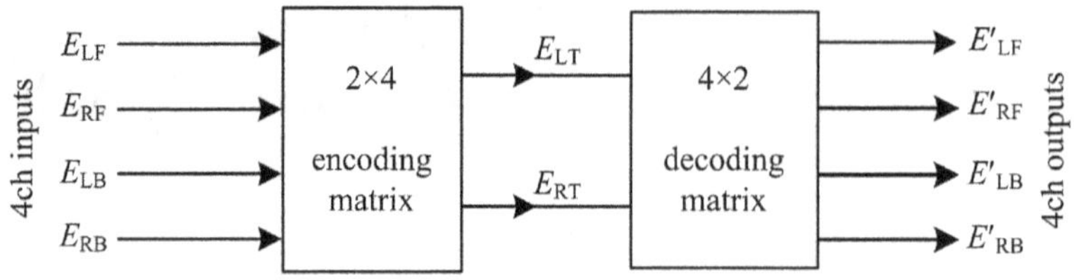

Figure 8.1 Block diagram of a matrix quadraphone.

are arranged identically to those in Figure 4.1. Matrix quadraphones are described in detail in other studies (Xie X.F., 1981; Eargle, 1971b; Woodward, 1977).

If E_{LF}, E_{RF}, E_{LB}, and E_{RB} are assumed to be the original left-front, right-front, left-back, and right-back-channel signals, respectively, they may be correlated or uncorrelated. Differentiating from Equation (4.3.4b), the order of four original signals is rearranged to demonstrate the symmetry between the left and right signals. The four original signals are converted into two independent signals via a 2×4 ***encoding matrix***,

$$\begin{bmatrix} E_{LT} \\ E_{RT} \end{bmatrix} = \begin{bmatrix} 2\times 4 \\ Encording \\ Matrix \end{bmatrix} \begin{bmatrix} E_{LF} \\ E_{RF} \\ E_{LB} \\ E_{RB} \end{bmatrix}. \tag{8.1.1}$$

The two independent signals are then converted back into four loudspeaker signals or reproduced signals by a ***decoding matrix***.

$$\begin{bmatrix} E'_{LF} \\ E'_{RF} \\ E'_{LB} \\ E'_{RB} \end{bmatrix} = \begin{bmatrix} 4\times 2 \\ Decording \\ Matrix \end{bmatrix} \begin{bmatrix} E_{LT} \\ E_{RT} \end{bmatrix}. \tag{8.1.2}$$

Given the 2 × 4 encoding matrix and the 4 × 2 decoding matrix, four loudspeaker signals can be obtained by substituting Equation (8.1.1) into Equation (8.1.2):

$$\begin{bmatrix} E'_{LF} \\ E'_{RF} \\ E'_{LB} \\ E'_{RB} \end{bmatrix} = \begin{bmatrix} 4\times 2 \\ Decording \\ Matrix \end{bmatrix} \begin{bmatrix} 2\times 4 \\ Encording \\ Matrix \end{bmatrix} \begin{bmatrix} E_{LF} \\ E_{RF} \\ E_{LB} \\ E_{RB} \end{bmatrix} = \begin{bmatrix} 4\times 4 \\ Reproducing \\ Matrix \end{bmatrix} \begin{bmatrix} E_{LF} \\ E_{RF} \\ E_{LB} \\ E_{RB} \end{bmatrix}. \tag{8.1.3}$$

Loudspeaker signals depend on encoding and decoding matrices. Based on different considerations (including technical and commercial considerations), various encoding and decoding matrices, e.g., various matrix quadraphones, were developed in the 1970s. Table 8.1 lists some typical encoding and decoding matrices (Xie X.F., 1981).

Loudspeaker signals vary in different matrix quadraphones, but they exhibit some similar characteristics. The encoding and decoding matrices by Scheiber (1971) are taken as example. The four loudspeaker signals are expressed as follows:

Table 8.1 Encoding/decoding matrices for some representative matrix quadraphones (the coefficients in the matrices are accurate to two decimal places)

System and authors	*Encoding matrix*	*Decoding matrix*	*Loudspeaker signals*
Dynaco	$\begin{bmatrix} 1 & 0.25 & 1 & -0.5 \\ 0.25 & 1 & -0.5 & 1 \end{bmatrix}$	$\begin{bmatrix} 1 & 0 \\ 0 & 1 \\ 0.64 & -0.36 \\ -0.36 & 0.64 \end{bmatrix}$	$\begin{bmatrix} 1 & 0.25 & 1 & -0.5 \\ 0.25 & 1 & -0.5 & 1 \\ 0.55 & -0.2 & 0.82 & -0.68 \\ -0.2 & 0.55 & -0.68 & 0.82 \end{bmatrix}$
EV (Electro Voice, EVX-4, Durbin, 1972)	$\begin{bmatrix} 1 & 0.3 & 1 & -0.5 \\ 0.3 & 1 & -0.5 & 1 \end{bmatrix}$	$\begin{bmatrix} 1 & 0.2 \\ 0.2 & 1 \\ 0.76 & -0.61 \\ -0.61 & 0.76 \end{bmatrix}$	$\begin{bmatrix} 1.06 & 0.5 & 0.9 & -0.3 \\ 0.5 & 1.06 & -0.3 & 0.9 \\ 0.59 & -0.38 & 1.06 & -0.99 \\ -0.38 & -0.59 & -0.99 & 1.06 \end{bmatrix}$
Zenith decoder		$\begin{bmatrix} 1 & 0 \\ 0 & 1 \\ 0.68 & -0.53 \\ -0.53 & 0.68 \end{bmatrix}$	$\begin{bmatrix} 1 & 0.3 & 1 & -0.5 \\ 0.3 & 1 & -0.5 & 1 \\ 0.52 & -0.33 & 0.95 & -0.81 \\ -0.33 & 0.52 & 0.81 & 0.95 \end{bmatrix}$
Scheiber (1971)	$\begin{bmatrix} 0.92 & 0.38 & 0.92 & -0.38 \\ 0.38 & 0.92 & -0.38 & 0.92 \end{bmatrix}$	$\begin{bmatrix} 0.92 & 0.38 \\ 0.38 & 0.92 \\ 0.92 & -0.38 \\ -0.38 & 0.92 \end{bmatrix}$	$\begin{bmatrix} 1 & 0.71 & 0.71 & 0 \\ 0.71 & 1 & 0 & 0.71 \\ 0.71 & 0 & 1 & -0.71 \\ 0 & 0.71 & -0.71 & 1 \end{bmatrix}$
Tappan	$\begin{bmatrix} 0.71 & 0 & 1 & -0.71 \\ 0 & 0.71 & -0.71 & 1 \end{bmatrix}$	$\begin{bmatrix} 1.41 & 1 \\ 1 & 1.41 \\ 1 & 0 \\ 0 & 1 \end{bmatrix}$	$\begin{bmatrix} 1 & 0.71 & 0.71 & 0 \\ 0.71 & 1 & 0 & 0.71 \\ 0.71 & 0 & 1 & -0.71 \\ 0 & 0.71 & -0.71 & 1 \end{bmatrix}$
Tria	$\begin{bmatrix} 1 & 0.71 & 0.71 & 0 \\ 0.71 & 1 & 0 & 0.71 \end{bmatrix}$	$\begin{bmatrix} 1 & 0 \\ 0 & 1 \\ 1.41 & 1 \\ -1 & 1.41 \end{bmatrix}$	$\begin{bmatrix} 1 & 0.71 & 0.71 & 0 \\ 0.71 & 1 & 0 & 0.71 \\ 0.71 & 0 & 1 & 0.71 \\ 0 & 0.71 & 0.71 & 1 \end{bmatrix}$
CBS-SQ (Bauer et al., 1971, 1973a, 1973b)	$\begin{bmatrix} 1 & 0 & -0.71j & 0.71 \\ 0 & 1 & -0.71 & 0.71j \end{bmatrix}$	$\begin{bmatrix} 1 & 0 \\ 0 & 1 \\ 0.71j & -0.71 \\ 0.71 & -0.71j \end{bmatrix}$	$\begin{bmatrix} 1 & 0 & -0.71j & 0.71 \\ 0 & 1 & -0.71 & 0.71j \\ 0.71j & -0.71 & 1 & 0 \\ 0.71 & -0.71j & 0 & 1 \end{bmatrix}$
New Orleans (Bauer et al., 1973b)	$\begin{bmatrix} 0.92 & 0.38j & 0.92 & -0.38j \\ -0.38j & 0.92 & 0.38j & 0.92 \end{bmatrix}$		
Sansui QS (Itho, 1972; Bauer et al., 1973b)	$\begin{bmatrix} 0.92 & 0.38 & 0.92j & 0.38j \\ 0.38 & 0.92 & -0.38j & -0.92j \end{bmatrix}$	$\begin{bmatrix} 0.92 & 0.38 \\ 0.38 & 0.92 \\ -0.92j & 0.38j \\ -0.38j & 0.92j \end{bmatrix}$	$\begin{bmatrix} 1 & 0.71 & 0.71j & 0 \\ 0.71 & 1 & 0 & -0.71j \\ -0.71j & 0 & 1 & 0.71 \\ 0 & 0.71j & 0.71 & 1 \end{bmatrix}$

$$\begin{array}{ll} E'_{LF} = E_{LF} + 0.71E_{RF} + 0.71E_{LB} & E'_{RF} = 0.71E_{LF} + E_{RF} + 0.71E_{RB} \\ E'_{LB} = 0.71E_{LF} + E_{LB} - 0.71E_{RB} & E_{RB} = 0.71E_{RF} - 0.71E_{LB} + E_{RB} \end{array}. \quad (8.1.4)$$

Ideally, each loudspeaker signal should be exactly identical to the corresponding original channel signal after encoding and decoding. However, Equation (8.1.4) indicates that each loudspeaker signal is a weighted mixing of the desired original channel signal and other channel signals. In other words, crosstalk occurs in practical loudspeaker signals. For example, four original signals are created by pair-wise amplitude panning (Section 4.1.2). When three original signals vanish except one original signal, e.g., except $E_{LF} \neq 0$, the loudspeaker signals in Equation (8.1.4) become

$$E'_{LF} = E_{LF} \quad E'_{RF} = 0.71E_{LF} \quad E'_{LB} = 0.71E_{LF} \quad E'_{RB} = 0. \quad (8.1.5)$$

In this case, the crosstalk of the opposite channel vanishes, but crosstalks of adjacent channels occur. The magnitude ratio between the crosstalks of an adjacent channel and the desired signal is 0.71/1 = 0.71 (–3 dB), i.e., the separation between adjacent channels is only 3 dB.

When two original front-channel signals are identical so that $E_{LF} = E_{RF} \neq 0$ and the two back-channel signals vanish with $E_{LB} = E_{RB} = 0$, the loudspeaker signals in Equation (8.1.4) become

$$E'_{LF} = E'_{RF} = 1.71E_{LF} \qquad E'_{LB} = E'_{RB} = 0.71E_{LF}. \quad (8.1.6)$$

In this case, the magnitude ratio between the crosstalks of the back channels and the desired signal in the front channels is 0.71/1.71 = 0.41 (–7.7 dB), i.e., the separation between the front and back channels is only 7.7 dB. The crosstalk and channel separation in other cases can be analyzed similarly.

For the CBS-SQ (stereo-quadraphonic) matrix in Table 8.1, a ±90° phase shift is applied to the original back (surround) channel signals during encoding. Another ±90° phase shift is also introduced to the decoding for the back-channel signals (Bauer et al., 1971, 1973a). The four final loudspeaker signals are given as

$$\begin{array}{ll} E'_{LF} = E_{LF} - 0.71jE_{LB} + 0.71E_{RB} & E'_{RF} = E_{RF} - 0.71E_{LB} + 0.71jE_{RB} \\ E'_{LB} = 0.71jE_{LF} - 0.71E_{RF} + E_{LB} & E'_{RB} = 0.71E_{LF} - 0.71jE_{RF} + E_{RB} \end{array}. \quad (8.1.7)$$

An analysis similar to the above indicates that crosstalks between the left and right channels vanish, and separation is +∞dB. However, the separation between front and back channels is only 3 dB. Moreover, if two original front-channel signals are identical so that $E_{LF} = E_{RF} \neq 0$ and two original back-channel signals vanish with $E_{LB} = E_{RB} = 0$, the separation between the front and back channels is 0 dB, and the two back-channel signals for reproduction are out of phase.

The encoding/decoding matrices of the Tappan, Tria, and Scheiber systems in Table 8.1 are different. However, the final loudspeaker signals for these three systems are identical. This result indicates that encoding/decoding matrices are not unique. In addition, some encoding/decoding matrices in Table 8.1 are related to one another. This relation is equivalent to the two inserted 2×2 full-rank matrices with $[T_2][T_1] = [I]$ between the encoding and decoding matrices on the right side of the second equality in Equation (8.1.3), where $[I]$ is a

2×2 identity matrix. The multiplication of the 2×2 matrix $[T_1]$ with the given 2×4 encoding matrix in Equation (8.1.3) yields a new 2×4 encoding matrix. The multiplication of the given 4×2 decoding matrix in Equation (8.1.3) with the 2×2 matrix $[T_2]$ yields a new 4×2 decoding matrix. The encoding matrix of the New Orleans system can be obtained by adding ±90° phase shift to part the original signals in the encoding of the Scheiber system. Applying some different phase transformations to the encoding/decoding matrix of the New Orleans system yields the Sansui QS and BMS systems (Cooper and Shiga, 1972), which are constructed by omitting the S_3 signal and leaving the S_1 and S_2 signals in Equation (4.3.2), respectively. The encoding/decoding matrix of the regular matrix (RM) system, which was developed from the Sansui QS system, was recommended by the Japan Phonograph Record Association as the engineering standard of the 4-2-4 matrix quadraphone. The British Broadcast Corporation (BBC) also developed a BBC-H encoding/decoding matrix and system, the magnitudes of matrix coefficients are similar to those in the Scheiber or Sansui QS matrix, but phase shifts other than ±90° are introduced in some coefficients (Meares and Ratliff, 1976; Juhasz and Piret, 1980).

In addition to the above examples, other examples of a 4-2-4 matrix quadraphone can be analyzed similarly. A major problem with various 4-2-4 matrix quadraphones is the crosstalk among channels. For the four original channel signals with pair-wise amplitude panning, the crosstalk in matrix surround sound reproduction results in obvious distortion in the perceived virtual source direction. This result can be proved by using Equations (3.2.7) and (3.2.9) of summing localization. Similarly, Equations (3.2.7) and (3.2.9) verify that various matrix quadraphones fail to recreate a full–360° horizontal virtual source even at a low frequency for the four original channel signals with the sound field signal mixing described in Section 4.1.3 (Xie and Xie, 1992b), or they are equally unable to satisfy the condition of the optimal direction and magnitude ($r_v = 1$) of the velocity localization vector.

In the 1970s, great efforts were devoted to developing 4-2-4 matrix quadraphones and investigating the related techniques, including the basic theory and design method of matrix encoding (Gerzon, 1975b; White, 1976), the conversion among various quadraphonic signals, the compatibility between matrix and discrete quadraphones, the standard of matrix quadraphones(Eargle, 1972; Bauer, 1979; Juhasz and Piret, 1980), and subjective experimental comparison among 4-2-4 (BMX) matrix quadraphones and 4–3-4 or 4-4-4 channel reproduction (Woodward, 1975a,1975b). Overall, various 4-2-4 matrix quadraphones exhibit some fatal defects, including directional distortion in a perceived virtual source and a narrow listening region. Therefore, as stated in Section 1.9.3, 4-2-4 matrix quadraphones are not the ideal horizontal surround sound system, and efforts devoted to the development of 4-2-4 matrix quadraphones failed. In fact, even discrete quadraphones exhibit some serious defects. At the encoding stage of 4-2-4 matrix quadraphones, the four original signals are linearly combined into two independent signals, so spatial information is lost and cannot be recovered in decoding. According to linear algebra theory, 2×4 matrix encoding is linear transformation with a reduced rank; as such, recovering the four original signals exactly via linear transformation or matrix decoding is impossible. With this characteristic, the 4-2-4 matrix quadraphone differs from the transformation of the first-order horizontal sound field signals. In the transformation of the first-order horizontal sound field signals given as Equation (4.3.5b), the 3×4 matrix encoding does not reduce the rank of the four original channel signals, so these signals can be recovered with 4×3 matrix decoding in Equation (4.3.4b). Matrix surround sound often refers to systems in which the ***rank of original signals is reduced in matrix encoding***. 4-2-4 matrix quadraphones have rarely been used for practical purposes, but the development of 4-2-4 matrix quadraphones in the 1970s provided good experiences and lessons for the subsequent production of other matrix surround sound.

8.1.2 Dolby Surround system

Dolby Surround originated from a type of matrix surround sound technique for film sound reproduction in cinemas by Dolby Laboratories. Drawn a lesson from the 4-2-4 matrix quadraphone, this type of technique was designed on the basis of new consideration; that is, under the condition of the limited capacity of transmission or storage media, a front-biased reproduction of the spatial information of sound is adopted to ensure that localization effects match with the picture. The surround channel is mainly intended for ambient sound and some special effects. In the mid-1970s, Dolby Labs introduced the ***Dolby Stereo***, a four-channel matrix surround sound technique for cinemas. Dolby Stereo involves three original front channels and a surround channel. Dolby Stereo converts the four original signals into two independent signals by matrix encoding to store signals in optical soundtracks in a 35 mm film on the basis of previous 4-2-4 matrix quadraphones. Since the end of the 1970s, Dolby Stereo has been widely used for film sound reproduction in cinema. ***Dolby Surround** technique, a consumer version of* Dolby Stereo, was subsequently used for domestic reproduction by using a stereophonic video-type recorder and a *laser disk (LD) as signal storage media (*Dolby Laboratories, 1998; Julstrom, 1987).

The principles of matrix encoding in Dolby Surround and Dolby Stereo are basically identical. The four original signals, including left, center, and right channel signals in the front and a surround channel signal, are denoted by E_L, E_R, E_C, and E_S. They may be correlated or uncorrelated. The encoding equation is given as

$$E_{LT} = E_L + 0.71E_C - 0.71jE_S \qquad E_{RT} = E_R + 0.71E_C + 0.71jE_S \tag{8.1.8}$$

They may also be written in a matrix form:

$$\begin{bmatrix} E_{LT} \\ E_{RT} \end{bmatrix} = \begin{bmatrix} 1 & 0.71 & 0 & -0.71j \\ 0 & 0.71 & 1 & 0.71j \end{bmatrix} \begin{bmatrix} E_L \\ E_C \\ E_R \\ E_S \end{bmatrix}. \tag{8.1.9}$$

The original left channel signal E_L is directly mixed to the E_{LT} channel; the original right channel signal E_R is directly mixed to the E_{RT} channel; the center channel signal E_C is mixed to E_{LT} andE_{RT} channels after a −3 dB attenuation; and the surround channel signal E_S is mixed to E_{LT} and E_{RT} channels after ±90° phase shift and a −3 dB attenuation. The subscript *LT* and *RT* denote the left total and right total signals, respectively.

Linear (matrix) decoding was used in early Dolby Surround, resulting in four-channel-reproduced signals E'_L, E'_C, E'_R, and E'_S. The decoding equation is given as

$$\begin{bmatrix} E'_L \\ E'_C \\ E'_R \\ E'_S \end{bmatrix} = \begin{bmatrix} 1 & 0 \\ 0.71 & 0.71 \\ 0 & 1 \\ 0.71 & -0.71 \end{bmatrix} \begin{bmatrix} E_{LT} \\ E_{RT} \end{bmatrix}. \tag{8.1.10}$$

Substituting Equation (8.1.9) into Equation (8.1.10) leads to

$$E'_L = E_L + 0.71E_C - 0.71jE_S \qquad E'_C = 0.71E_L + E_C + 0.71E_R$$
$$E'_R = E_R + 0.71E_C + 0.71jE_S \qquad E'_S = 0.71E_L - 0.71E_R - jE_S. \qquad (8.1.11)$$

After decoding is completed, the crosstalk between opposite channels vanishes, i.e., the separation between opposite channels is infinite. In particular, the original center channel signal does not appear in the surround channel after decoding and vice versa, which is an advantage of Dolby Surround encoding/decoding. However, the magnitude ratio between the crosstalk of an adjacent channel and the desired signal is 0.71/1 = 0.71 (−3 dB), and the separation between adjacent channels is only 3 dB. In fact, crosstalk in traditional 4-2-4 matrix systems is inevitable.

Figure 8.2 shows the block diagram of the Dolby Surround encoder. A band-pass filter within 100 Hz–7 kHz and noise reduction processing are added to the surround channel. Figure 8.3 illustrates the block diagram of the basic Dolby Surround decoder used in early times. The surround channel signal is subjected to an anti-aliasing filter, a low-pass (more strictly a band-pass filter within 100 Hz–7 kHz), and modified Dolby B noise reduction. A low-pass filter is used to reduce the side effects of matrix encoding/decoding. Moreover, an adjustable delay of the order of 20–30 ms is added to the surround channel. This delay aims to take advantage of the precedence effect (Section 1.7.2) to reduce the influence of surround channel signals on front localization.

In domestic use, the loudspeaker configuration for Dolby Surround is similar to that of the 5.1 channel surround sound. It involves the left, center, and right loudspeakers in the front and a pair of surround loudspeakers in the sides and slightly back or slightly above the horizontal plane. The simplest method is to feed the same surround signal to two surround loudspeakers because only a single surround signal is derived after decoding. Some post-processing techniques, such as signal decorrelation, are also applied to derive two surround

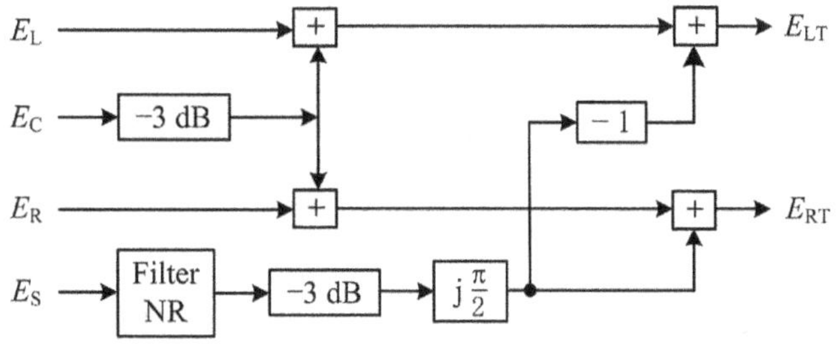

Figure 8.2 Block diagram of a Dolby Surround encoder.

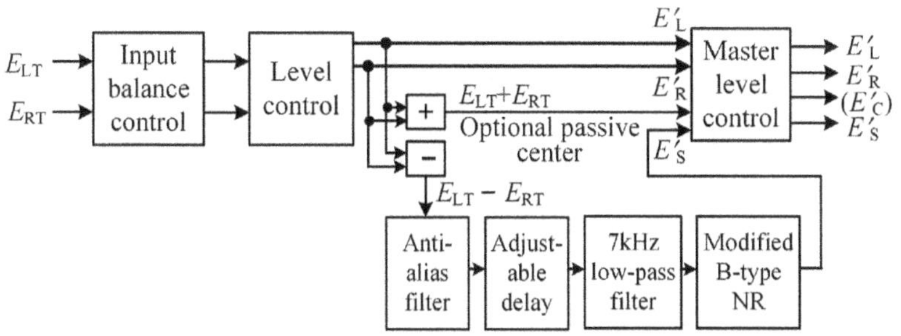

Figure 8.3 Block diagram of a basic Dolby Surround decoder.

channel signals and improve the perceived effect of ambient sound in reproduction. A three-loudspeaker configuration is also used in the early Dolby Surround reproduction. The left and right channel signals are reproduced by a pair of left and right loudspeakers in the front; the center channel signal is simultaneously reproduced by the left and right loudspeakers with −3 dB attenuation (phantom center); and the surround channel signal is reproduced by a back loudspeaker.

8.1.3 Dolby Pro-Logic decoding technique

Passive decoding is used in early Dolby Surround. In this technique, the coefficients in a decoding matrix are constants. A major problem of passive decoding in Dolby Surround is the crosstalk between adjacent channels that spoil virtual source localization in reproduction. In 1987, ***logic decoding*** was used in ***Dolby Pro-Logic***, which was also modified from film sound for domestic or consumer reproduction (Dolby Laboratories, 1998; Hull, 1999), to enhance the separation between channels. Logic decoding is a ***time-variant and adaptive or active decoding technique*** through which decoding coefficients vary with instantaneous characteristics or relationship among the encoded signals. The relative magnitude and phase between two encoded signals, namely, E_{LT} and E_{RT}, are detected to evaluate the dominant component of four original signals at each instant. The obtained information enables the adaptive decoder to steer the signals to appropriate channels with a smooth control of gain so that the gains of output channels with dominant component are enhanced, and the gains of other output channels are attenuated. The adaptive decoder enhances the separation between channels up to an order of 30 dB.

To illustrate the principle of Dolby Pro-Logic decoder, a case is considered in which three original front-channel signals are created via pair-wise amplitude panning and the original surround channel signal is an ambient signal. Starting from the direction of the right channel to the direction of the left channel anticlockwise, the original signals of three front channels are created. The second column in Table 8.2 illustrates the variation in the three original front-channel signals and the case of surround channel signal only. The notation "↑"in the table denotes the signal magnitude increases smoothly, and the notation "↓" corresponds to the signal magnitude reduces smoothly. Ideally, the allocation of four-channel signals after adaptive decoding should be identical to that of the original signals.

Table 8.2 Basic information of the adaptive decoding provided by the encoding signals E_{LT} and E_{RT}

Target direction	*Original signal*	*Encoding signals*	*Relationship between E_{LT} and E_{RT}*	*Relationship between $(E_{LT} + E_{RT})$ and $(E_{LT} - E_{RT})$*
Right	$E_R = 1$, Other = 0	$E_{LT} = 0$ $E_{RT} = 1$	$E_{LT} = 0$	Out of phase, $\lvert E_{LT} + E_{RT}\rvert = \lvert E_{LT} - E_{RT}\rvert$
Right→ Center	$\lvert E_R\rvert\downarrow$, $\lvert E_C\rvert\uparrow$, Other = 0	$E_{LT} = 0.71 E_C$ $E_{RT} = E_R + 0.71 E_C$	In phase	Out of phase, $\lvert E_{LT} + E_{RT}\rvert > \lvert E_{LT} - E_{RT}\rvert$
Center	$E_C = 1$, Other = 0	$E_{LT} = 0.71 E_C$ $E_{RT} = 0.71 E_C$	In phase, $E_{LT} = E_{RT}$	$\lvert E_{LT} - E_{RT}\rvert = 0$
Center→Left	$\lvert E_C\rvert\downarrow$, $\lvert E_L\rvert\uparrow$, Other = 0	$E_{LT} = E_L + 0.71 E_C$ $E_{RT} = 0.71 E_C$	In phase	In phase, $\lvert E_{LT} + E_{RT}\rvert > \lvert E_{LT} - E_{RT}\rvert$
Left	$E_L = 1$, Other = 0	$E_{LT} = 1$ $E_{RT} = 0$	$E_{RT} = 0$	In phase, $\lvert E_{LT} + E_{RT}\rvert = \lvert E_{LT} - E_{RT}\rvert$
Surround(back)	$E_S = 1$ Other = 0	$E_{LT} = -0.71 jE_S$ $E_{RT} = 0.71 jE_S$	Out of phase	$\lvert E_{LT} + E_{RT}\rvert = 0$

The third column in Table 8.2 illustrates the two encoded signals E_{LT} and E_{RT} given as Equation (8.1.8). The fourth column in Table 8.1 shows the relative magnitudes and phases between two encoded signals. The relative magnitude and phase change with the variation in the panning of original signals, which provide information on the separation of the front and back dominant components. Thus, the crosstalk between the center and surround channels decreases. When the original signals are panned from the right channel through the center channel to the left channel, E_{LT} and E_{RT} signals are in phase. When the original signal is panned to the full right or full left, either E_{LT} or E_{RT} vanishes. When the original signal is fed to the surround channel only, E_{LT} and E_{RT} are out of phase. Therefore, signal steering in adaptive decoding can be controlled by the detected information from E_{LT} and E_{RT}. When E_{LT} and E_{RT} are in phase, the gain of the surround channel in the decoding outputs is restrained. When either of E_{LT} or E_{RT} vanishes, the gains of surround and center channels in the decoding outputs are restrained. When E_{LT} and E_{RT} are out of phase, the gains of the three frontal channels in the decoding outputs are restrained.

Similarly, the relative magnitude and phase between the sum $(E_{LT} + E_{RT})$ and difference $(E_{LT} - E_{RT})$ of the encoded signals in the fifth column of Table 8.2 provide the left–right information of the dominant component. This information is applied to control the signal steering and smooth transition of the gain of the left and right channel signals in the decoding outputs. When the original signal is panned to center or surround channel only, either $|E_{LT} - E_{RT}|$ or $|E_{LT} + E_{RT}|$ vanishes, the gains of the left and right channels in the adaptive decoding outputs are restrained. When the original signal is panned between the right and center channels, signals $(E_{LT} + E_{RT})$ and $(E_{LT} - E_{RT})$ are out of phase, the gains of the left and surround channels in the adaptive decoding outputs are restrained, and the gains of the right and center channels in the adaptive decoding outputs are controlled on the basis of the relative magnitude between $|E_{LT} + E_{RT}|$ and $|E_{LT} - E_{RT}|$. When the original signal is panned between the left and center channels, the signals $(E_{LT} + E_{RT})$ and $(E_{LT} - E_{RT})$ are in phase, and the gains of the left and center channels are controlled on the basis of the relative magnitude between $|E_{LT} + E_{RT}|$ and $|E_{LT} - E_{RT}|$.

The overall power of all channels outputs in a decoder should be a constant when the coefficients of decoding change adaptively. Therefore, the gain of a channel in the decoding output should be increased smoothly when the gain of an adjacent channel output is reduced smoothly. Moreover, the response time of an adaptive decoder should be chosen carefully. A short response time is beneficial to recreating the virtual source with a rapid change in the direction but is inclined to creating an audible discontinuous artifact.

According to the aforementioned principle, adaptive decoders can be implemented with various methods (Gundry, 2001). The early Dolby Pro-Logic decoder is implemented with an analog circuit. Controlling signals are derived from the relative magnitude and phase between signals E_{LT} and E_{RT} and between $(E_{LT} + E_{RT})$ and $(E_{LT} - E_{RT})$. The gain of each channel in the decoding outputs is controlled by voltage-controlled amplifiers. Dolby Pro-Logic decoding can also be implemented via digital signal processing. This function is found in some products of Dolby digital surround sound processor. Figure 8.4 shows the block diagram of a Dolby Pro-Logic decoder.

8.1.4 Some developments on matrix surround sound and logic decoding techniques

Since the end of the 1980s, the development of digital techniques has solved the problems of the transmission and storage of multichannel signals. However, matrix surround sound techniques are still developed because two-channel transmission and storage media still occupy a large proportion in practice even in the age of digital techniques. The transmission and

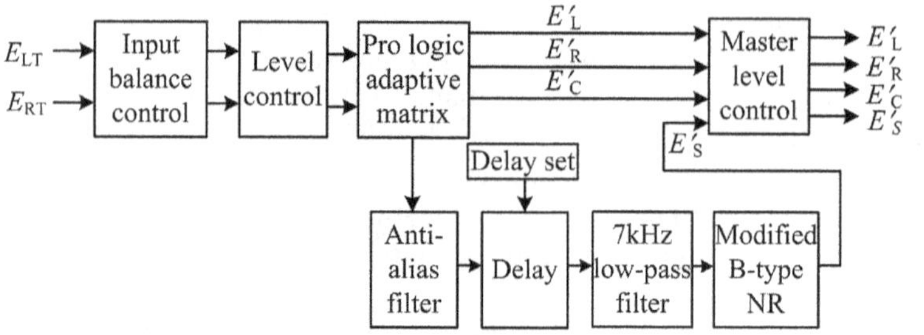

Figure 8.4 Block diagram of Dolby Pro-Logic.

storage of two-channel signals reduce the bit rate of data and thus compress the data of multichannel sound.

After the development of Dolby Pro-Logic, various encoding and adaptive decoding techniques for matrix surround sound have also been established. One feature of these new matrix surround sound techniques is that they can manipulate more than four-channel spatial information. For example, in the encoding of five-channel sound signals into two-channel independent signals, the general equation is given as

$$\begin{aligned} E_{LT} &= \kappa_L E_L + \kappa_C E_C + \kappa_{LS} E_{LS} + \kappa_{RS} E_{RS} \\ E_{RT} &= \chi_R E_R + \chi_C E_C + \chi_{LS} E_{LS} + \chi_{RS} E_{RS}. \end{aligned} \tag{8.1.12}$$

Equation (8.1.12) can be written as a matrix form

$$\begin{bmatrix} E_{LT} \\ E_{RT} \end{bmatrix} = \begin{bmatrix} \kappa_L & \kappa_C & 0 & \kappa_{LS} & \kappa_{RS} \\ 0 & \chi_C & \chi_R & \chi_{LS} & \chi_{RS} \end{bmatrix} \begin{bmatrix} E_L \\ E_C \\ E_R \\ E_{LS} \\ E_{RS} \end{bmatrix} \tag{8.1.13}$$

Generally, the encoding coefficients κ_L, χ_R, κ_C, and χ_C are real values, and κ_{LS}, κ_{RS}, χ_{LS}, and χ_{RS} are real or complex values. The coefficients should satisfy $|\kappa_{LS}|^2 + |\chi_{LS}|^2 = 1$ and $|\kappa_{RS}|^2 + |\chi_{RS}|^2 = 1$ to maintain the constant overall power of signals after encoding. If $\kappa_L = \chi_R = 1$ and $\kappa_C = \chi_C = 0.71$ are chosen, the original left, center, and right channel signals are encoded identical to those in Dolby Stereo or Dolby Surround. The original left- and right-surround signals are encoded into the independent signals E_{LT} and E_{RT} with appropriate gains and phase shifts.

Different encoding coefficients are used in various five-channel matrix surround sound techniques and systems. Constant coefficients are used in passive encoding. For more sophisticated ***active or adaptive encoding***, encoding coefficients are changed adaptively according to the instantaneous characteristics of input signals to optimize the encoding performance of different signals. The two-channel encoded signals are converted back into five or more channels reproduced signals. Various decoding matrices and methods are used in different matrix surround sound techniques. In addition to decoding the two-channel signals encoded from original five-channel signals, some newly developed decoding methods can decode the encoded signals from Dolby Stereo or Dolby Surround and manipulate the usual two-channel stereophonic signals to obtain five or more channel reproduced signals.

A simple choice of encoding coefficients is described as follows (Faller and Schillebeeckx, 2011):

$$\kappa_L = \chi_R = 1 \quad \kappa_C = \chi_C = \frac{\sqrt{2}}{2}$$
$$\kappa_{LS} = \frac{\sqrt{3}}{2}j \quad \kappa_{RS} = \frac{1}{2}j \quad \chi_{LS} = -\frac{1}{2}j \quad \chi_{RS} = -\frac{\sqrt{3}}{2}j. \tag{8.1.14}$$

Lexicon Logic 7 is another matrix surround sound technique with adaptive decoding developed in the mid of 1990 (Griesinger, 1996, 1997a). It preserves the left–right and front–back separation of the reproduced signals and enhances the front–back balance for different types of program materials. Lexicon Logic 7 is intended for high-quality domestic or consumer reproduction. In addition to be used for reproduction with an accompanying picture, it improves performance for music reproduction. Lexicon Logic 7 encodes the original 5.1-channel signals into two-channel signals and then decodes into five- or seven-channel signals. The loudspeaker configuration for seven-channel reproduction is similar to that of the discrete 7.1 channel system in Figure 5.10. A pair of side surround loudspeakers improves the lateral localization in reproduction with an accompanying picture and auditory spatial impression in music reproduction.

Two major considerations of Lexicon Logic 7 encoding are as follows:

1. It can effectively encode the original 5.1-channel signals so that the encoded signals can be decoded with minimal loss.
2. The encoded signals should be stereophonically compatible.

An adaptive encoding algorithm is used on the basis of this consideration in Lexicon Logic 7. According to the information detected from the relative magnitudes and phases between the original five-channel inputs, encoding coefficients are changed adaptively. The encoding of the original left, center, and right channel signals is similar to that of Dolby Stereo or Dolby Surround, e.g., taking $\kappa_L = \chi_R = 1$, $\kappa_C = \chi_C = 0.71$ in Equation (8.1.12).

The encoding of original surround channel signals is complicated. In the adaptive encoding algorithm, the encoding coefficients for surround signals in Equation (8.1.12) are written as follows:

$$\begin{aligned} &\kappa_{LS} = 0.91\left[w_1(E_L, E_{LS}) - w_2(E_L, E_{LS})j\right] \quad \kappa_{RS} = 0.38\left[-w_1(E_R, E_{RS}) - w_2(E_R, E_{RS})j\right] \\ &\chi_{LS} = 0.38\left[-w_1(E_L, E_{LS}) + w_2(E_L, E_{LS})j\right] \quad \chi_{RS} = 0.91\left[w_1(E_R, E_{RS}) + w_2(E_R, E_{RS})j\right]. \end{aligned} \tag{8.1.15}$$

Where two functions w_1 and w_2 vary with the relative magnitude and phase of the original input signals. In the basic operation of the encoder, two functions become $w_1 = 0$, $w_2 = 1$, then $\kappa_{LS} = -0.91j$, $\chi_{RS} = 0.91j$, $\kappa_{RS} = -0.38j$, $\chi_{LS} = 0.38j$. The original surround signals are mixed to the encoded signals with a ±90° phase shift because all the encoding coefficients are imaginary values. For a single original surround signal only (e.g., E_{LS}) or two decorrelated surround signals, the overall power of two encoded signals is identical to that of original signals. In this case, the basic operation of the encoder is appropriate. For two identical original surround signals, the basic operation of the encoder leads to an undesired boost of 1.29 times (2.2 dB) in each encoded output signal. In this case, adaptive encoding algorithm should reduce w_2 by a factor up to 1/1.29 or –2.2 dB. Moreover, when the original left- and right-surround signals are similar in terms of magnitude but are out of phase, the basic operation of the encoder generates two almost identical encoded signals $E_{LT} \approx E_{RT}$. These encoded signals are mistakenly decoded as a center channel signal in reproduction. Through adaptive encoding, w_1 increases to create a 90° phase difference between the encoded signals E_{LT} and E_{RT} and thus avoid this error.

For classical music recording, surround channels often record reverberation. To be compatible with stereophonic reproduction, original surround signals are mixed to the two encoded signals with a –3 dB attenuation in accordance with the standard in Europe. In the adaptive encoding in Lexicon Logic 7, the relative signal levels of three original front-channel signals are compared with two original surround channel signals. If the maximum of two surround channel levels is lower than the maximum of the three front-channel levels with –3 dB, the surround channel signals are processed as reverberation and attenuated by changing w_2 in Equation (8.1.15). A maximal attenuation of –3 dB is reached when the surround channel level is lower than that of the front-channel level by at least –8 dB.

Lexicon Logic 7 can decode the two encoded signals E_{LT} and E_{RT} into five- or seven-channel outputs. The principle of adaptive decoding in Lexicon Logic 7 is similar to that of Dolby Pro-Logic. Different decoding modes are used for music and programs with an accompanying picture. The practical decoder of Lexicon Logic 7 is complicated and described in Griesinger's studies.

Dolby Labs developed new generations of matrix surround sound and adaptive encoding/decoding technique. ***Dolby Pro-Logic II*** encodes the original 5.1 channel signals into two independent signals by using a matrix method (Dressler, 2000). Encoding is also expressed in Equation (8.1.12), and the encoding coefficients are

$$\begin{gathered}\kappa_L = \chi_R = 1 \quad \kappa_C = \chi_C = 0.71 \\ \kappa_{LS} = -\sqrt{\frac{19}{25}}j \quad \kappa_{RS} = -\sqrt{\frac{6}{25}}j \quad \chi_{LS} = \sqrt{\frac{19}{25}}j \quad \chi_{RS} = \sqrt{\frac{6}{25}}j.\end{gathered} \tag{8.1.16}$$

The 5.1-channel reproduced signals are derived from the two encoded signals via adaptive decoding (Gundry, 2001). In comparison with traditional Dolby Surround and Dolby Pro-Logic, Dolby Pro-Logic II has two surround channels. All five main channels are full audible bandwidths. Pro-Logic II also enables bass management. In accordance with different modes (such as movie, Pro-Logic, and music modes), Pro-Logic II can decode two-channel encoded signals or upmix stereophonic signals into 5.1-channel outputs. The music mode is mainly for upmixing the two-channel stereophonic inputs into 5.1-channel outputs. A Dolby Pro-Logic II decoder provides some user-adjustable parameters for optional controls, including dimension (front and back sound field) control and center width control, because universal or optimal methods for upmixing have yet to be developed. A high-frequency shelf filter is also provided in the surround channel to model the high-frequency roll-off of ambience caused by room reflection and absorption. In addition, Dolby Pro-Logic II utilizes feedback control to improve the dynamic characteristic in adaptive decoding.

The ***Dolby Pro-Logic IIx*** introduced in 2002 can further decode or upmix the two-channel encoded signals, two-channel stereophonic signals, and 5.1-channel signals into horizontal 6.1- and 7.1-channel signals.

Introduced in 2009, ***Dolby Pro-Logic IIz*** embeds two front-height channels into 5.1- or 7.1-channel horizontal surround sound to improve the reproduction of vertical information in the front, leading to 5.1+2 or 7.1+2 channel reproduction (Tsingos et al., 2010). The signals of two front-height channels are mixed with original 5.1- or 7.1-channel signals by matrix encoding and delivered with 5.1- or 7.1-channel media. For example, in 7.1+2

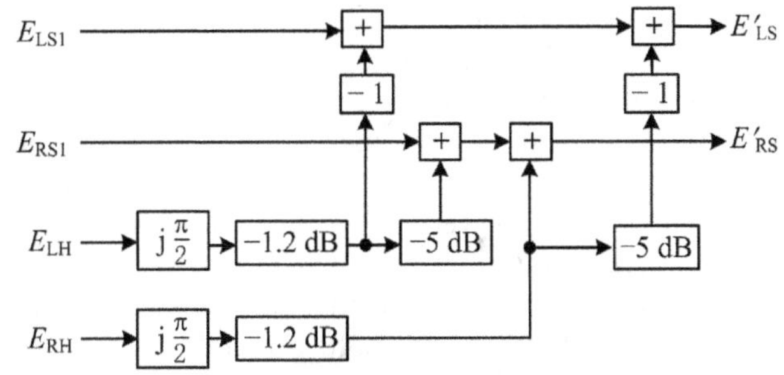

Figure 8.5 Block diagram of a Dolby Pro-Logic IIz encoder.

channel reproduction, original signals involve seven horizontal channel signals, e.g., left-front E_L, center E_C, right-front E_R, left-side surround E_{LS1}, right-side surround E_{RS1}, left-back surround E_{LS2}, right-back surround E_{RS2}, and two height signals, e.g., the left-front height E_{LH} and right-front height E_{RH}. It also involves a signal E_{LFE} of a low-frequency effect channel. The signals E_L, E_C, E_R, E_{LS2}, E_{RS2}, and E_{LFE} are delivered separately; the signals E_{LS1}, E_{RS1}, E_{LH}, and E_{RH} are mixed into two signals E'_{LS} and E'_{RS} by encoding matrix and then delivered. Therefore, all the signals are delivered via 7.1-channel media. Prior to be encoded, the signals E_{LS1} and E_{RS1} may be delayed appropriately to reduce their influence on localization. Figure 8.5 illustrates the block diagram of the Dolby Pro-Logic Iizencoder. The encoding equation is given as

$$
\begin{aligned}
E'_{LS} &= E_{LS1} - \sqrt{\frac{19}{25}} jE_{LH} - \sqrt{\frac{6}{25}} jE_{RH} \\
E'_{RS} &= E_{RS1} + \sqrt{\frac{6}{25}} jE_{LH} + \sqrt{\frac{19}{25}} jE_{RH}.
\end{aligned}
\tag{8.1.17}
$$

The adaptive decoding outputs of Pro-Logic Iiz are reproduced by 5.1+2 or 7.1+2 loudspeaker configuration. The horizontal loudspeaker configuration is similar to that of 5.1 or 7.1-channel configuration in Figures 5.1 and 5.10. Two additional height loudspeakers are arranged above the horizontal left and right loudspeakers approximately at azimuths θ = ±30° and elevation ϕ = 45° or at wider azimuths of θ = ±45° and elevation of ϕ = 45°.The 5.1+2 loudspeaker configuration in Pro-Logic IIz is similar to the loudspeaker configuration 4 for 7.1-channel sound in Figure 6.12.

Pro-Logic IIz adaptive decoding generates optimal results of Pro-Logic IIz-encoded signal inputs. Moreover, Pro-Logic IIz decoder can generates up to 7.1+2 channel outputs for any inputs from stereophonic to 5.1- or 7.1-channel to improve the performance in reproduction.

Dolby Digital Surround EX for cinema use, which was introduced by Dolby Laboratories and Lucasfilm THX in 1998, also uses matrix encoding and adaptive decoding (Dolby Laboratories, 2002). Dolby Digital Surround EX is a 6.1-channel system. It is constructed by adding a rear-surround channel to a 5.1-channel system to improve the localization in the rear. The original signals involve the left, center, right, left-surround, rear-surround, and right-surround signals, as well as a signal for a low-frequency effect channel. The left-surround, rear-surround, and right-surround signals are encoded into two signals by matrix encoding and delivered, together with the three front-channel signals through 5.1-channel media. Adaptive decoding generates three surround channel outputs from the two encoded surround signals and then realizes 6.1-channel reproduction.

Other domestic matrix surround sound techniques have been developed, and their principles are similar to previously established techniques. The Circle Surround 5-2-5 (CS 5.1) is also a five-channel matrix surround sound technique and work in either a music mode or a video mode. CS 5.1 is claimed to be more appropriate for music reproduction in comparison with Dolby Surround. The CS 5.1 decoder is also suitable for two-channel stereophonic signal inputs without being encoded. However, the results of a formal subjective experiment on CS 5.1 are not found in previous studies.

DTS NEO:6 is also a matrix surround sound that can decode two-channel signals into a maximum of six-channel surround sound signals. Its principle is similar to that of Dolby Pro-Logic IIx. In Section 5.3, DTS-ES 6.1 has two working modes, namely, matrix and discrete modes. The principle of the matrix mode is similar to that of Dolby Digital Surround EX. The details of DTS NEO:6 and DTS-ES 6.1 can be found in a related webpage (http://dts.com/).

Various adaptive (logic) decoding techniques are discussed in Section 8.1.3 and this Section. Adaptive decoding reduces the crosstalk and then enhances the separation between channels. However, adaptive decoding is effective for only one dominant directional component at each instant and unable to manipulate two or more directional components at the same time. In the case of only one target source at each instant, such as a dialog and some special effects in video programs, matrix surround sound with adaptive decoding can create a good perceived effect by using the psychological features of human hearing and vision. However, in the case of multiple simultaneous target sources, such as an orchestra, adaptive decoding is basically invalid and may cause a variation in the perceived virtual source position with time. After the Dolby Pro-Logic, the performance of music reproduction has been improved in various adaptive decoding techniques discussed in this section. However, traditional adaptive decoding techniques are in principle more appropriate for sound reproduction with an accompanying picture. They exhibit some limitations for music reproduction. More elaborate decoding algorithms in a time-frequency domain are used in some recent matrix surround sound techniques. As discussed in Section 8.3.7, these techniques can simultaneously decode two or more dominant directional components and improve performance in reproduction.

8.2 DOWNMIXING OF MULTICHANNEL SOUND SIGNALS

Similar to the case of stereophonic sound, ***downmixing*** refers to the conversion of signals for reproduction with a larger number of channels to signals for reproduction with a smaller number of channels. To be compatible with stereophonic reproduction, the problem of downmixing was considered in the early time of multichannel sound. The matrix surround sounds in Section 8.1 deal with encoding multichannel sound signals into a smaller number of signals. The two-channel encoded signals in matrix surround sound are usually stereo compatible. However, the encoded signals in matrix surround sound aim to optimize reproduction after decoding. They may not be directly optimal for stereophonic reproduction because the perceived balance between the sounds from different directions is not always preserved after downmixing. Therefore, the problem of the downmixing of multichannel sound signals should be considered separately. Moreover, the spatial audio signal coding discussed in Section 13.4.5 deals with the problem of downmixing.

For example, the left-front, left-back, right-front, and right-back signals of discrete quadraphone in Section 4.1 are denoted by E_{LF}, E_{LB}, E_{RF}, and E_{RB}. The left-front and left-back signals are mixed. Likewise, the right-front and right-back signals are mixed. The two-channel downmixed signals are reproduced by a pair of stereophonic loudspeakers:

$$E_{L0} = E_{LF} + E_{LB} \qquad E_{R0} = E_{RF} + E_{RB}. \tag{8.2.1}$$

For quadraphone signals with pair-wise amplitude panning in Section 4.1.2, the downmixing reproduction with a standard stereophonic loudspeaker configuration leads to the appearance of the front-target virtual source within the region bounded by two stereophonic loudspeakers. The front stereophonic width in downmixing reproduction is limited to the ±30° for standard stereophonic loudspeaker configuration and narrower than that in quadraphone reproduction with a ±45° front loudspeaker pair. Moreover, the rear-target virtual source is mapped or reflected to the mirror direction between the ±30° front loudspeakers in downmixing reproduction.

For the first-order sound field signal mixing in Section 4.1.3, the stereophonic downmixing signals are given as substituting Equation (4.1.11) into Equation (8.2.1) and letting $b = 2$,

$$E_{L0} = 2E_A A_{total}\left(1+\sqrt{2}\sin\theta_S\right) \qquad E_{R0} = 2E_A A_{total}\left(1-\sqrt{2}\sin\theta_S\right), \tag{8.2.2}$$

where $E_A = E_A(f)$ is the signal waveform in the frequency domain. Equation (8.2.2) is similar to the stereophonic signals derived from an MS pair with an omnidirectional microphone and a bidirectional microphone whose main axis points to the left. The analysis of virtual source localization for signals given as Equation (8.2.2) is similar to that in Section 2.2.2.

Various discrete multichannel sounds with an irregular loudspeaker configuration developed after 1990 also deal with the problem of downmixing. For example, the five signals with a full audible bandwidth in 5.1-channel sound are denoted by E_L, E_C, E_R, E_{LS}, and E_{RS}. Generally, the 5.1-channel sound signals are downmixed into two-channel stereophonic signals by using the following equation:

$$E_{L0} = E_L + \kappa_C E_C + \kappa_S E_{LS} \qquad E_{R0} = E_R + \kappa_C E_C + \kappa_S E_{RS}. \tag{8.2.3}$$

where κ_C and κ_S are the gains for mixing the center and surround channel signals to the two-channel stereophonic signals. The signal of the low-frequency effect channel is downmixed similarly to the center channel signals and omitted here.

Similar to the case of quadraphone, when 5.1-channel signals are downmixed for stereophonic reproduction, auditory events are usually limited within the region bounded by two stereophonic loudspeakers. This phenomenon is inevitable in the stereophonic downmixing of multichannel sound signals, which disrupts the balance of spatial components or information in original multichannel sound signals. For example, for music reproduction, three front-channel signals in 5.1-channel sound are usually the direct sound components of target sound sources, and two surround channel signals are usually reverberation. When 5.1-channel signals are downmixed and reproduced by a pair of stereophonic loudspeakers, reverberation comes from the front directions, resulting in a change in the directional masking between direct and reverberation sounds. Therefore, mixing the surround channel signals to two-channel stereophonic signals with a unit gain destroys the perceived balance between direct and reverberation sounds. Moreover, mixing the center channel signals to two-channel stereophonic signals directly narrows the front stereophonic width unless $E_C = 0$.

Ideally, the optimal gains for center and surround channel signals downmixing depend on the program materials. In the ITU standard (ITU-R BS 775-1, 1994), the gains of downmixing are $\kappa_C = \kappa_S = 0.707$, e.g., −3 dB. In this case, downmixing maintains the constant overall power of the center channel signal. However, the recommendation by ITU is not always optimal for all program materials. The practical gains for center and surround channel signals downmixing vary from 0 dB to −6 dB. The attenuation of surround channel signals in downmixing is to avoid the excessive reverberation or ambient component in the downmixed signals. For example, Dolby Laboratories (1997) suggested the gains $\kappa_C = 0.707, 0.596, 0.500$

and $\kappa_S = 0.707, 0.596, 0.500$, corresponding to –3, –4.5, and –6 dB for downmixing. In the DVD-Audio (Section 13.9.3), the gains of downmixing are stored on the disk, allowing for the control of downmixing in reproduction. In some studies, signal processing in a time-frequency domain is applied to separate direct and ambient components in surround channel signals and downmix them with adaptive gains (Faller and Schillebeeckx, 2011).

In Equation (8.2.3), the original left (right) channel signal of 5.1-channel sound is only downmixed to the stereophonic channel at the same side. In some other downmixing methods, the reversal phase counterpart of the original left (right) channel signal is also mixed to the opposite stereophonic channel with an appropriate gain. For example, the downmixing equation by Gerzon (1992f) is given as

$$\begin{aligned}
E_{L0} &= 0.8536E_L + 0.5000E_C - 0.1464E_R + 0.3536\kappa_{S1}\left(E_{LS} + E_{RS}\right) \\
&\quad +0.3536\kappa_{S2}\left(E_{LS} - E_{RS}\right) \\
E_{R0} &= -0.1464E_L + 0.5000E_C + 0.8536E_R + 0.3536\kappa_{S1}\left(E_{LS} + E_{RS}\right) \\
&\quad -0.3536\kappa_{S2}\left(E_{LS} - E_{RS}\right).
\end{aligned} \tag{8.2.4}$$

where κ_{S1} ranges from 0.500 to 0.707, and κ_{S2} varies from 1.414 κ_{S1} to 1.414. For the original left (right) channel signals, mixing the reversal phase component to an opposite channel helps preserve the front stereophonic width in reproduction. The principle is similar to the method of creating an outside-boundary virtual source or stereophonic width expansion in stereophonic reproduction (Sections 2.1.3 and 2.4.1). The analysis on the frontal virtual source localization of downmixed signals in Equation (8.2.4) is similar to that in Section 3.2. Indeed, the performance of stereophonic width expansion depends on signal frequency and is effective at the central listening position. Similarly, for original surround channel signals, mixing the reversal phase component to an opposite channel aims to expand the distribution of spatial auditory events in stereophonic reproduction although spatial auditory events are mapped to the front region.

Lexicon Logic 7 encoding in Section 8.1.4 can also be regarded as a downmixing method. In a basic operation of an encoder, the downmixing equation is given as

$$\begin{aligned}
E_{L0} &= E_L + 0.707E_C - 0.91jE_{LS} - 0.38jE_{RS} \\
E_{R0} &= E_R + 0.707E_C + 0.38jE_{LS} + 0.91jE_{RS}.
\end{aligned} \tag{8.2.5}$$

In addition to the downmixing to two-channel stereophonic signals, 5.1-channel signals should be downmixed to three-channel signals for 3/0 or 2/1 loudspeaker configuration and four-channel signals for 3/1 or 2/2 loudspeaker configuration. These downmixing methods are specified in accordance with the ITU standard (ITU-R BS 775-1, 1994).

Similar to the downmixing of 5.1-channel signals, more than 5.1-channel signals should be downmixed to two-channel stereophonic signals or a smaller number of channel signals to adapt to delivery media and reproduction configuration. This problem was considered in the early time of discrete multichannel sound with irregular loudspeaker configurations (Theile, 1991a). With the development of various channel-based multichannel sounds, the problem of downmixing should be further considered. For example, Auro 9.1-channel signals can be easily downmixed for 5.1-channel reproduction. Hamasaki et al. (2007) suggested a method for downmixing 22.2-channel signals to 5.1-channel and 2-channel stereophonic signals. Sugimoto et al. (2015) analyzed the results of subjective experiments in detail. Kim et al. (2010) also suggested a method for downmixing 22.2-channel signals to USC 10.2-channel signals, Samsung 10.2-channel signals, and optional 7.1-channel signals in the ITU standard (Section 5.3, and ITU-R BS 775-1, 1994).The MPEG-H 3D Audio standard also includes the method for the downmixing of channel-based signals (Section 13.5.6). Object-based spatial sounds (Section 6.5.2) have no problem in downmixing because of their flexibility.

In traditional downmixing methods, original signals are mixed with different gains and the same or reversal phases. Similar to the case of the mono downmixing of stereophonic signals, for multichannel decorrelated signals or correlated signals with interchannel level difference (ICLD) only, downmixing slightly influences the timbre. By contrast, for multichannel correlated signals with interchannel time difference (ICTD), downmixing causes the comb-filtering effect and then degrades the timbre.

According to the discussion in Chapter 7, coincident microphone techniques and pan-pot technique generates multichannel sound signals with ICLD only. Spaced microphone techniques and near-coincident microphone techniques generate multichannel sound signals with ICTD. The aforementioned downmixing methods are inappropriate for multichannel sound signals with ICDT. In accordance with the ITU standard (ITU-R BS 775-1, 1994), Zielinski et al. (2003) evaluated eight downmixed algorithms by using 5.1-channel program materials recorded with different methods via subjective experiments and confirmed that a comb-filtering effect degrades the timbre in downmixing reproduction.

Some signal processing algorithms are used to reduce timbre coloration in downmixing. However, conditions and methods for multichannel sound recording and program production vary considerably. The frequency spectra of dominant components and the relative phase in multichannel signals also vary with time. These variations are unknown in advance. Therefore, adaptive signal processing algorithms are needed. Some studies have suggested enhancing or attenuating components in some frequency bands in the downmixed signals by using an adaptive equalization algorithm (Faller and Baumgarte, 2003). Some other studies have suggested adaptively aligning the phases of signals in the time-frequency domain before downmixing (Breebaart et al., 2005; Samsudin et al., 2006; *Gnann and Spiertz*, 2008; Hoang et al., 2010) or adaptively equalize the magnitude and phase spectra of signals to decrease spatial and timbre distortion in reproduction (Thompson et al., 2009). Some of these algorithms were originally designed for the parametric coding of stereophonic signals and then applied to the downmixing of 5.1-channel signals. Vilkamo et al. (2014) extended these algorithms to the downmixing of more (such as 22.2) channel signals. The correlations of input signals in the time-frequency domain are evaluated, and the phases of downmixing matrix coefficients are adaptively adjusted to control the relative phase between highly correlated input signals. Subjective experiments validated the performance of this algorithm.

To reduce the timbre coloration caused by coherent mixing in downmixing, Adami et al. (2014) also suggested converting two-channel signals to the time-frequency domain via short-time Fourier transform. They also identified the coherent signal part and suppressed it in one channel prior to downmixing.

8.3 UPMIXING OF MULTICHANNEL SOUND SIGNALS

8.3.1 Some considerations in upmixing

Upmixing refers to the conversion of signals for reproduction with a smaller number of channels to signals for reproduction with a larger number of channels. Since 1990, discrete multichannel sounds have been popular in domestic reproduction. However, numerous program materials, especially some program materials with historic and artistic values, were recorded with two-channel stereophonic sound. Moreover, the number of channels in reproduction increases with the development of multichannel sound techniques. Therefore, the upmixing of multichannel sound signals is practically significant. In addition, upmixing is significant to multichannel audio signal coding (Section 13.4.5). Signal upmixing can be implemented in either the program production stage or the reproduction stage. Decoding in matrix surround sound (Section 8.1) focuses on the conversion of a small number of special encoded signals to a larger number of signals to recover the spatial information in reproduction as far as possible. Using a decoder originally intended for matrix surround sound to upmix the unencoded stereophonic signals may not lead to optimal effects on reproduction. Conducting a subjective experiment, Rumsey (1999) assessed the perceived performance of some stereophonic upmixing algorithms originally for matrix surround sound decoding. The result indicated that these algorithms slightly improve spatial impression in reproduction but generally reduce the quality of a front virtual source. Therefore, the problem of upmixing stereophonic/multichannel sound signals should be considered separately. This problem has been considered in the designed of some matrix surround sound decoders. For different input signals, these decoders operate in various decoding and upmixing modes. The performance of stereophonic upmixing has been improved in some newly designed matrix surround sound decoders (Section 8.1.4).

Original stereophonic or multichannel signals involve directional and ambient sound components. Two considerations and methods are developed for upmixing. In one method, original signals are upmixed for reproduction with more channels, but the original spatial (especially localization) information is preserved. Reproduction with more channels is performed to enlarge the listening region. In another method, directional and ambient sound components are blindly separated from original signals by analyzing the directional and ambient information and extracting related parameters, especially correlation characteristic, relative magnitude, and phase among original signals (Merimaa et al., 2007). For example, Härmä (2010) proposed a method for classifying the original stereophonic signals according to their statistical characteristics. The separated directional and ambient components are processed, enhanced, re-panned, or re-mixed into more channels according to the extracted acoustic parameters. Through this method, spatial information can be reproduced more appropriately. Furthermore, it may enlarge the listening region and improve the perceived effect.

A combination of the two aforementioned methods may be used in practice. Generally, in various discrete multichannel surround sounds with irregular loudspeaker configuration (such as 5.1-channel sound), front channels are often used for directional component reproduction, and surround channels are often utilized for ambient component reproduction. In this case, the ambient component is not limited to the diffused reverberation, it may include non-reflected ambient sound (Section 3.1). Therefore, in upmixing, directional and ambient components are separated in original signals, processed, and enhanced. Then, front and surround channel signals are synthesized. Traditional upmixing algorithms are usually based on passive processing. These algorithms can even be implemented through analog signal processing, but their ability to separate information is limited. More recently, various adaptive signal processing algorithms in the time-frequency domain are used to improve the performance of upmixing. These algorithms are usually implemented via digital signal processing and relatively complicated. Various upmixing algorithms have been proposed and evaluated by subjective experiments. Overall, the resultant performance depends on multiple factors, including upmixing algorithms, the properties of program materials, methods, and attributes for evaluation and subject's experience (Sporer et al., 2006; Bai and Shih, 2007; Chétry et al., 2007; Barry and Kearney, 2009; Marston, 2011; Schoeffler et al., 2014).

8.3.2 Simple upmixing methods for front-channel signals

Various discrete multichannel sounds with irregular loudspeaker configuration are usually developed for sound reproduction with accompanying picture. As stated in Section 5.1, configurations with a front-center loudspeaker are often used to recreate a stable front virtual source matched with a picture, e.g., three or more front loudspeakers are used for front sound reproduction. Early in the 1950s, some simple methods for upmixing stereophonic signals to three front-channel signals were proposed (Klipsch, 1958). The center channel signal is derived through a combination of the original left and right channel signals E_L and E_R with a –3 dB attenuation. The three front-channel signals in the frequency domain created by upmixing are expressed in Equation (8.3.1):

$$E'_L = E_L \qquad E'_C = 0.707\left(E_L + E_R\right) \qquad E'_R = E_R. \tag{8.3.1}$$

Upmixing via Equation (8.3.1) and reproducing with three front loudspeakers improve the stability of front virtual sources and enlarge the listening region. However, this method narrows the front stereophonic width even at the central listening position. This result can be easily confirmed by substituting Equation (8.3.1) into the virtual source localization Equations (3.2.7) and (3.2.9). A simple method to remove this defect is to arrange the left and right loudspeakers with a span angle larger than the standard of ±30°, such as ±45°.

On the basis of velocity localization and energy localization vector theorems (Section 3.2.2), Gerzon (1992b) proposed an upmixing method in accordance with optimal psychoacoustic criteria. This method is appropriate for upmixing to an arbitrary number of front-channel signals. For example, in the upmixing of stereophonic signals to three front-channel signals, if the left and right loudspeakers are arranged at ±45° azimuths, the three resulting front-channel signals are

$$\begin{aligned} &E'_L = 0.885E_L - 0.115E_R \quad E'_C = 0.451E_L + 0.451E_R \\ &E'_R = -0.115E_L + 0.885E_R. \end{aligned} \tag{8.3.2}$$

Equation (8.3.2) can be written in a matrix form

$$\begin{bmatrix} E'_L \\ E'_C \\ E'_R \end{bmatrix} = \begin{bmatrix} 0.885 & -0.115 \\ 0.451 & 0.451 \\ -0.115 & 0.885 \end{bmatrix} \begin{bmatrix} E_L \\ E_R \end{bmatrix}. \tag{8.3.3}$$

Original stereophonic signals are mixed to the opposite channels with a reversal phase and an appropriate gain. Similar to the case in Section 2.4.1, the reversal phase in opposite channel expands the front stereophonic width and compensates for the effect caused by the center channel (loudspeaker). In addition, for two original channel decorrelated signals, the upmixing in Equation (8.3.2) maintains the overall power of signals as follows:

$$E_L'^2 + E_C'^2 + E_R'^2 = E_L^2 + E_R^2. \tag{8.3.4}$$

Therefore, upmixing expressed in Equation (8.3.2) preserves the perceived balance among ambient components.

Gerzon (1992d) also extended the psychoacoustic optimal method to the upmixing/downmixing of the arbitrary number of front-channel signals. Indeed, this psychoacoustic optimum is mainly valid for the central listening position.

In simple upmixing methods, more channel signals are derived from the linear mixing of original channel signals. The basic consideration is to preserve the features of virtual source localization after upmixing and reproducing by more loudspeakers. This type of passive upmixing is simple and can be implemented via analog signal processing, but it cannot separate different front directional components. It also exhibits a limited effect in enlarging the listening region. It fails to separate the front directional and ambient components. Moreover, for original signals with ICTD, such as stereophonic signals captured by spaced microphones in Section 2.2.3, a simple linear mixing causes a comb-filtering effect and then degrades the timbre.

8.3.3 Simple methods for Ambient component separation

Side and rear-surround channels in practical multichannel sound are often used for ambience reproduction, including reverberation in music reproduction and other ambience with less requirements on directional accuracy (such as applause). Indeed, surround channels may occasionally be used to reproduce some special directional effects. Therefore, a common practice of upmixing stereophonic signals to multichannel sound signals is separating or extracting ambience components and then mixing to surround channels.

For original stereophonic music signals, especially those recorded via coincident microphone techniques described in Section 2.2, differences in left and right channel signals include more reflected components from the two sides of a hall:

$$E_S = E_L - E_R. \tag{8.3.5}$$

The reflected component in original stereophonic signals can be extracted with signal E_S and mixed to surround channels. This passive processing method can be implemented via analog signal processing. However, through this simple method, directional and ambient components cannot be separated completely. For example, directional components not coming from the direct front directions leads to $E_L \neq E_R$ and then to $E_S \neq 0$. These directional components leak to the surround channels after upmixing. With the precedence effect, E_S can be delayed appropriately prior to upmixing to reduce the influence of surround channel signals on front localization.

Various post-processing methods can be applied to the delayed version of E_S to convert the mono surround signal to more surround channel signals to enhance the envelopment and ambient effects on reproduction. A common post-processing method is signal decorrelation, which has been used to convert mono surround signals in Dolby Pro-Logic decoding output to two or more surround channel signals. Decorrelation can also be used to convert the two surround signals in 5.1-channel sound to four surround signals for 7.1-channel reproduction. As stated in Section 7.5.4, decorrelation by delaying signals is simple, but it may lead to a comb-filtering effect. Random-phase filters or reciprocal MLS-based filters can be used to improve the performance of surround signal decorrelation (Xie et al., 2012).

Artificial delay and reverberation algorithms can be applied to the extracted signals in Equation (8.3.5) to generate two or more surround channel signals and enhance the reflection information in reproduction. Various delay and reverberation algorithms in Section 7.5 are applicable for this purpose. Different perceptions in various halls can be recreated by choosing different parameters (such as reverberation time, high-frequency attenuation, and direct-to-reverberant energy ratio) in the simulation of reflection. As post-processing methods, delay and reverberation–based algorithms for surround signal upmixing were proposed in the early times of quadraphones (Eargle, 1971a) and appropriate for various loudspeaker configurations. Since 1980, this type of algorithms has been applied to some consumer products (including products with 5.1-channel loudspeaker configuration). This type of algorithms may create some desired perceived effect, but the reflected information is obtained via artificial simulation and different from that of actual halls. Therefore, this type of algorithms is a "pseudo-surround sound" method at most.

8.3.4 Model and statistical characteristics of two-channel stereophonic signals

One important stage in upmixing is the separation of directional and ambient components in original stereophonic signals. The algorithms in Sections 8.3.2 and 8.3.3 cannot effectively separate the directional and ambient components. The statistical processing of signals in a time domain or a time-frequency domain can be further used to separate directional and ambient components in original signals because of differences in the statistical characteristics of directional and ambient components. A priori knowledge of statistical characteristics and model of input signals are needed for such a separation, and algorithms are usually implemented via digital signal processing. For example, in two-channel stereophonic signals, the statistical signal model is first established in the following.

Original two-channel stereophonic signals in the frequency domain are denoted by E_L and E_R, and the corresponding signals in the time domain are given as $e_L(t)$ and $e_R(t)$. The discrete time signals or samples of $e_L(t)$ and $e_R(t)$ are expressed as $e_L(n)$ and $e_R(n)$, where the integer n is the discrete time. Original two-channel stereophonic signals can be decomposed into directional and ambient components and can be represented by the following signal model:

$$e_L(n) = e_{L,dir}(n) + e_{L,amb}(n) \qquad e_R(n) = e_{R,dir}(n) + e_{R,amb}(n). \tag{8.3.6}$$

where $e_{L,dir}(n)$ and $e_{R,dir}(n)$ are directional components of left and right channel signals; $e_{L,amb}(n)$ and $e_{R,amb}(n)$ are ambient components of left and right channel signals. Signal processing aims to separate the four components above from the two signals $e_L(n)$ and $e_R(n)$. This problem is underdetermined, and some additional conditions are required to obtain appropriate solutions.

Suppose that the temporal mean of the directional component in stereophonic signals is zero. For a directional component with a nonzero temporal mean, the mean or direct current component should be extracted. Directional components in two stereophonic signals are highly correlated, and the mean of their product is nonzero:

$$\overline{e_{L,dir}(n)e_{R,dir}(n)} \neq 0. \tag{8.3.7}$$

The ambient components in two stereophonic signals are uncorrelated (strictly low correlated), and the temporal mean of their product is zero:

$$\overline{e_{L,amb}(n)e_{R,amb}(n)} = 0. \tag{8.3.8}$$

Ambient and directional components are also uncorrelated with

$$\overline{e_{L,amb}(n)e_{L,dir}(n)} = \overline{e_{L,amb}(n)e_{R,dir}(n)} = 0$$
$$\overline{e_{R,amb}(n)e_{R,dir}(n)} = \overline{e_{R,amb}(n)e_{L,dir}(n)} = 0. \quad (8.3.9)$$

Equations (8.3.7) to (8.3.9) reveal the differences in the statistical characteristics of directional and ambient components in two-channel stereophonic signals and constitute the basic hypotheses of a stereophonic signal model. Moreover, the following hypothesis is sometimes supplemented because the ambient powers in the left and right channel signals are approximately equal in many cases:

$$\overline{e_{L,amb}^2(n)} = \overline{e_{R,amb}^2(n)} = \sigma_{amb}^2 \quad (8.3.10)$$

If the directional components in two-channel stereophonic signals are created via amplitude panning or coincident microphone techniques so that they involve ICLD only (Sections 2.1.1 and 2.2.5), these directional components can be written as

$$e_{L,dir}(n) = A_L e_{A,dir}(n) \qquad e_{R,dir}(n) = A_R e_{A,dir}(n). \quad (8.3.11)$$

where A_L and A_R are the normalized amplitudes or gains (panning coefficients) of the directional components in left and right channel signals. For a static target virtual source with a fixed position, A_L and A_R are independent from time and frequency; $e_{A,dir}(n)$ is the signal waveform of the directional component in the time domain and related to $E_A(f)$ in Equation (2.1.1) via inverse discrete Fourier transform. For constant power normalization, A_L and A_R satisfy Equation (2.2.59), that is,

$$A_L^2 + A_R^2 = 1. \quad (8.3.12)$$

For constant amplitude normalization, A_L and A_R satisfy Equation (2.2.65), that is,

$$A_L + A_R = 1 \quad (8.3.13)$$

For stereophonic signals captured by using a spaced microphone pair, the directional components $e_{L,dir}(n)$ and $e_{R,dir}(n)$ involve ICTD and ICLD and can be written as

$$e_{L,dir}(n) = a_L(n) \otimes_t e_{A,dir}(n) \qquad e_{R,dir}(n) = a_R(n) \otimes_t e_{A,dir}(n), \quad (8.3.14)$$

where $a_L(n)$ and $a_R(n)$ are obtained through the inverse discrete Fourier transform of A_L and A_R in the frequency domain expressed in Equation (2.2.51). The notation "$\otimes_t$," denotes convolution in the time domain.

On the basis of the aforementioned model and hypotheses, directional and ambient components in stereophonic signals can be decomposed and separated via statistical signal processing in the time domain. However, signal processing in the time domain is valid in the case of only one dominant directional component at each instant and unable to manipulate two or more directional components at the same time. This limitation is similar to the one that occurs in Dolby Pro-Logic decoding.

Usually, original stereophonic signals simultaneously involve two or more directional components. Two cases are considered. In the first case, the frequency spectra of different directional components do not overlap. In the second case, the frequency spectra of various directional components overlap, but the frequency spectra of dominant directional components do not overlap. The second case is similar to the situation of auditory streaming separation and grouping in auditory scene analysis (Section 1.7.4). If human hearing decodes only a single (dominant) directional cue per each critical band at one time, then different directional components are separated in various critical bands. Under above conditions and hypotheses, the temporal statistical characteristics and related parameters of stereophonic signals can be individually estimated in each frequency band, then the (dominant) directional component of each target source and the ambient components are separated. The separated directional and ambient components are re-mixed to more channels in upmixing.

Signal decomposition in the time-frequency domain can be realized by various band-pass filters. Filters with an ERB or CB bandwidth are most appropriate from the point of auditory perception but are complicated. From the point of simplifying signal processing, the short-time Fourier transform (STFT) similar to that in Equation (7.6.5) is applicable to the time-frequency decomposition of signals. The STFT of original stereophonic signals are represented by $E_L(n', k)$ and $E_R(n', k)$ with $k = 0, 1\ldots(N-1)$, where n is the discrete time, n' is the discrete temporal index of the STFT (such as the initial time of the frame), k is the frequency index of the STFT, and N is the length of frame in STFT. $E_L(n', k)$ and $E_R(n', k)$ are related to the time domain signals by

$$
\begin{aligned}
E_L(n', k) &= \sum_{n=NL}^{NH} W(n) e_L(n'+n) \exp\left(-j\frac{2\pi}{N}kn\right) \\
E_R(n', k) &= \sum_{n=NL}^{NH} W(n) e_R(n'+n) \exp\left(-j\frac{2\pi}{N}kn\right).
\end{aligned}
\qquad (8.3.15)
$$

where $W(n)$ is an appropriate time window function, $NL \leq 0$ and $NH > 0$ are the initial and end times for STFT calculation, and $N = NH - NL + 1$ is the length of the frame or block for STFT calculation. In addition, band-pass filters with an ERB or CB bandwidth can be implemented through the combination of the STFT coefficients, and the corresponding signals in the time-frequency domain are also represented by $E_L(n', k)$ and $E_R(n', k)$. However, n' is the temporal variable in filter outputs, and k is the index of frequency bands.

When two-channel stereophonic signals are transformed into the time-frequency domain, the signal model and hypotheses given as Equations (8.3.6) to (8.3.13) are still valid, a requirement is to substitute the time domain signals in these equations with the corresponding time-frequency domain signals. In the time-frequency domain processing, the temporal statistics of stereophonic signals are individually estimated in each frequency band, and the extracted directional and ambient components are re-mixed to more channels in accordance with the estimated information.

Various algorithms for upmixing stereophonic or multichannel sound signals in the time domain or time-frequency domain have been developed. Some of these algorithms are appropriate for time domain and time-frequency domain processing. The principle of these algorithms is similar to that of the blind separation and enhancement of microphone array signals in the field of acoustic signal processing (Goodwin, 2008a). The concepts and models of these algorithms are also similar to those of auditory scene analysis in Section 1.7.4. These algorithms were compared in some works (Bai and Shih, 2007; Merimaa et al., 2007; Goodwin, 2008b). Some typical algorithms are addressed in Sections 8.3.5–8.3.9.

8.3.5 A scale-signal-based algorithm for upmixing

Avendano and Jot (2004) proposed a scale-signal-based algorithm in time-frequency domain for upmixing stereophonic signals to multichannel signals. The cross-correlation between original stereophonic signals in the time-frequency domain is first evaluated:

$$\Phi_{uv}(n', k) = \overline{E_u(n'+n, k)E_v^*(n'+n, k)} \qquad u, v = L, R. \tag{8.3.16}$$

where the notation "*" denotes complex conjugation. Equation (8.3.16) represents the temporal mean or expectation over $NL1 \leq n \leq NH1$. $NL_1 \leq 0$ and $NH_1 > 0$ are the initial and end times for short-time mean, and $N_1 = NH_1 - NL_1 + 1$ is the time length of the frame for mean calculation. Statistical analysis in Equation (8.3.16) varies with the time frame index n' because signals are non-stationary. In practice, Equation (8.3.16) is approximately calculated by using the following iterative method:

$$\Phi_{uv}(n', k) = \mu\Phi_{uv}(n'-1, k) + (1-\mu)E_u(n', k)E_v^*(n', k), \tag{8.3.17}$$

where $0 \leq \mu \leq 1$ is a forgetting factor. The short-time normalized cross-correlation function of two-channel signals is calculated by using the following:

$$\Psi(n', k) = \frac{|\Phi_{LR}(n', k)|}{\sqrt{|\Phi_{LL}(n', k)\Phi_{RR}(n', k)|}}. \tag{8.3.18}$$

By definition, $0 \leq \Psi(n', k) \leq 1$. The closer $\Psi(n', k)$ to a unit value is, the more coherent the two signals will be.

The similarity function is defined as

$$\Lambda(n', k) = \left.\frac{2|\Phi_{LR}(n', k)|}{|\Phi_{LL}(n', k) + \Phi_{RR}(n', k)|}\right|_{1-\mu=1}. \tag{8.3.19}$$

The partial similarity function is defined as

$$\Lambda_u(n', k) = \left.\frac{|\Phi_{uv}(n', k)|}{\Phi_{uu}(n', k)}\right|_{1-\mu=1} \qquad u = L, R \quad v \neq u. \tag{8.3.20}$$

For the cross-correlation function in Equations (8.3.19) and (8.3.20), a forgetting factor $\mu = 0$ or $(1 - \mu) = 1$ is chosen in the iteration calculation of Equation (8.3.17). The panning coefficients or power ratio and the target source direction in each frequency band and each instant of stereophonic signals can be estimated from the two functions given as Equations (8.3.19) and (8.3.20).

The aforementioned derived functions and parameters are applicable to estimate and re-mix the directional and ambient components in stereophonic signals. According to the definition and characteristics of $\Psi(n', k)$, the closer$[1 - \Psi(n', k)]$ to a unit value is, the more numerous the uncorrelated component in stereophonic signals will be. Therefore, $[1 - \Psi(n', k)]$ serves as an ambience index. The two-channel ambient components can be extracted from stereophonic signals by scaling them with an appropriate nonlinear mapping function $\Gamma_{amb}[1 - \Psi(n', k)]$and then re-mixed or fed to the surround channels:

$$\hat{E}_{L,amb}(n', k) = \Gamma_{amb}\left[1 - \Psi(n', k)\right]E_L(n', k)$$
$$\hat{E}_{R,amb}(n', k) = \Gamma_{amb}\left[1 - \Psi(n', k)\right]E_R(n', k). \tag{8.3.21}$$

The continuous and smooth mapping function $\Gamma_{amb}[1 - \Psi(n', k)]$ should be designed so that the outputs of Equation (8.3.21) are not modified for a large ambience index and greatly attenuated for a small ambience index.

Another mapping function $\Gamma_{dir}[\Lambda(n', k), \Lambda_u(n', k)]$ can be designed on the basis of $\Lambda(n', k)$ and $\Lambda_u(n', k)$ to extract the directional components in original stereophonic signals:

$$\hat{E}_{dir}(n', k) = \Gamma_{dir}\left[\Lambda(n', k), \Lambda_u(n', k)\right]\left[E_L(n', k) + E_R(n', k)\right]. \tag{8.3.22}$$

The extracted directional components in the time-frequency domain are re-mixed to more channels according to a certain panning rule and the estimated target source direction from the original stereophonic signals. The final upmixing outputs in the time domain are obtained by applying inverse STFT to those in the time-frequency domain. The design of the aforementioned mapping functions was described in detail in a previous study (Avendano and Jot, 2004). In addition, the above mapping method is applicable to stereophonic expansion (Cobos and Lopez, 2010).

The upmixing given as Equations (8.3.21) and (8.3.22) is only an example of the scale-signal-based decomposition algorithm. Generally, in this algorithm, stereophonic inputs are scaled by appropriate gain functions and then used as the estimations of directional and ambient components. Gain functions are derived in terms of the correlation between original stereophonic signals:

$$\hat{E}_{L,dir}(n', k) = \Gamma_{L,dir}\left[\Phi_{uv}(n', k)\right]E_L(n', k) \quad \hat{E}_{R,dir}(n', k) = \Gamma_{R,dir}\left[\Phi_{uv}(n', k)\right]E_R(n', k)$$
$$\hat{E}_{L,amb}(n', k) = \Gamma_{L,amb}\left[\Phi_{uv}(n', k)\right]E_L(n', k) \quad \hat{E}_{R,amb}(n', k) = \Gamma_{R,amb}\left[\Phi_{uv}(n', k)\right]E_R(n', k). \tag{8.3.23}$$

In a special case, the directional components in left and right channels are estimated by subtracting ambient components from stereophonic inputs. Then, the gain functions of directional and ambient components are related via the following equations:

$$\Gamma_{L,dir}(n', k) = 1 - \Gamma_{L,amb}\left[\Phi_{uv}(n', k)\right] \qquad \Gamma_{R,dir}(n', k) = 1 - \Gamma_{R,amb}\left[\Phi_{uv}(n', k)\right]. \tag{8.3.24}$$

The scale-signal-based decomposition algorithms individually manipulate left and right stereophonic inputs and extract directional and ambient components with different gain functions derived from the correlation of input signals. As a result, the algorithm does not completely separate directional and ambient components. Generally, the two-channel estimated ambient components are correlated with each other. Likewise, the estimated directional and ambient components have a correlation. Therefore, the estimated directional and ambient components do not satisfy Equations (8.3.8) and (8.3.9).

In this algorithm, directional and ambient components are separated in terms of the correlation between left and right stereophonic inputs. When the directional component in original stereophonic signals is panned to a single channel, the correlation between stereophonic inputs vanishes. In this case, the directional component is mistaken for an ambient component. To solve this problem, Merimaa et al. (2007) suggested separating directional and ambient components in terms of the estimated power spectra of ambience. According to the model and hypothesis presented in Equations (8.3.6) to (8.3.10), the square norm of cross-correlation in the time-frequency domain calculated from Equation (8.3.16) is given as

$$\begin{aligned} \left|\Phi_{LR}(n',k)\right|^2 &= \sigma_{amb}^4(n',k) - \sigma_{amb}^2(n',k)\left[\Phi_{LL}(n',k)+\Phi_{RR}(n',k)\right] \\ &\quad +\Phi_{LL}^2(n',k)\Phi_{RR}^2(n',k). \end{aligned} \tag{8.3.25}$$

The power of ambient component at each frequency band is evaluated by

$$\sigma_{amb}^2(n',k) = \frac{1}{2}\left\{\Phi_{LL}(n',k)+\Phi_{RR}(n',k)-\sqrt{\left[\Phi_{LL}(n',k)-\Phi_{RR}(n',k)\right]^2+4\left|\Phi_{LR}(n',k)\right|^2}\right\}. \tag{8.3.26}$$

The ratio between the ambient power and total power in left and right stereophonic inputs is expressed as

$$\Gamma_{L,amb}^2(n',k) = \frac{\sigma_{amb}^2(n',k)}{\left|E_L(n',k)\right|^2} = \frac{\sigma_{amb}^2(n',k)}{\Phi_{LL}(n',k)} \quad \Gamma_{R,amb}^2(n',k) = \frac{\sigma_{amb}^2(n',k)}{\left|E_R(n',k)\right|^2} = \frac{\sigma_{amb}^2(n',k)}{\Phi_{RR}(n',k)}. \tag{8.3.27}$$

Stereophonic inputs are re-mixed to two ambient (surround) channels with scale functions as given in Equation (8.3.28):

$$\begin{aligned} \hat{E}_{L,amb}(n',k) &= \Gamma_{L,amb}(n',k)E_L(n',k) = \frac{\sigma_{amb}(n',k)}{\sqrt{\Phi_{LL}(n',k)}}E_L(n',k) \\ \hat{E}_{R,amb}(n',k) &= \Gamma_{R,amb}(n',k)E_R(n',k) = \frac{\sigma_{amb}(n',k)}{\sqrt{\Phi_{RR}(n',k)}}E_R(n',k). \end{aligned} \tag{8.3.28}$$

The scale-signal-based decomposition algorithms are extended to separate directional and ambient components in more than two input signals (Goodwin, 2008a). They are implemented by calculating the cross-correlation matrix of each pair of input signals. After being normalized by the overall power of input signals, the determinant of a cross-correlation matrix serves as the cross-correlation index of input signals. In addition, the consideration of the algorithm in this section is similar to that of DiRAC in Section 7.6. The method of DiRAC is also applicable to upmixing (Pulkki, 2007).

8.3.6 Upmixing algorithm based on principal component analysis

An ***algorithm based on principal component analysis*** (***PCA***) utilizes the correlation between stereophonic input signals to separate directional and ambient components. It is applicable to input signals of time and time-frequency domains (Briand et al., 2006; Goodwin and Jot, 2007). Applying PCA to signals in the time domain yields a common directional component and a common ambient component within a full audible bandwidth. Conversely, applying

PCA to signals in the time-frequency domain yields individual directional and ambient components in each frequency band. For simplicity, the PCA-based algorithm in the time domain is discussed in this section. For the PCA-based algorithm in the time-frequency domain, time domain signals in calculation should be replaced with the corresponding signals in the time-frequency domain. In addition, PCA-based algorithm is applicable to two-channel and multi-channel inputs (Goodwin, 2008a). For simplicity, the case of two-channel inputs is analyzed.

In conventional PCA-based algorithms for upmixing, the common directional component $\hat{e}_{A,dir}(n)$ and the left and right ambient components $\hat{e}_{L,amb}(n)$ and $\hat{e}_{R,amb}(n)$ are estimated and then re-mixed to more channels. The two-channel stereophonic signal model in Equations (8.3.6) to (8.3.11) is used, and the condition of constant power in Equation (8.3.12) is supplemented. The statistical characteristics of input signals are calculated over each time frame or block. In each time frame with the length of N samples, the 2×2 covariance matrix of stereophonic input signals $e_L(n)$ and $e_R(n)$ are given as

$$[COV] = \begin{bmatrix} \mathrm{cov}(e_L, e_L) & \mathrm{cov}(e_L, e_R) \\ \mathrm{cov}(e_R, e_L) & \mathrm{cov}(e_R, e_R) \end{bmatrix}. \qquad (8.3.29)$$

The entries of the covariance matrix are given as

$$\mathrm{cov}(e_u, e_v) = \frac{1}{N-1}\sum_{n=NL}^{NH}\left\{\left[e_u(n'+n) - \bar{e}_u\right]\left[e_v(n'+n) - \bar{e}_v\right]\right\} \qquad u, v = L, R, \qquad (8.3.30)$$

where $\bar{e}_u$ and $\bar{e}_v$ are the temporal means of the input signals over a time frame; $NL \le 0$ and $NH > 0$ are the initial and end times of the frame, respectively; $N = NH - NL + 1$ is the length of the frame. The entries of the covariance matrix are calculated from Equations (8.3.6) and (8.3.11):

$$\begin{aligned} &\mathrm{cov}(e_L, e_L) = A_L^2\sigma_{A,div}^2 + \sigma_{amb}^2 \quad \mathrm{cov}(e_R, e_R) = A_R^2\sigma_{A,div}^2 + \sigma_{amb}^2 \\ &\mathrm{cov}(e_L, e_R) = \mathrm{cov}(e_R, e_L) = A_L A_R \sigma_{A,dir}^2. \end{aligned} \qquad (8.3.31)$$

A pair of eigenvalues of the matrix [COV] can be found by solving the following eigen equation:

$$[COV]\hat{a} = \sigma^2\hat{a}. \qquad (8.3.32)$$

The larger eigenvalues σ_1^2 and the smaller eigenvalues σ_2^2 are expressed as

$$\sigma_1^2 = \sigma_{A,dir}^2 + \sigma_{amb}^2 \qquad \sigma_2^2 = \sigma_{amb}^2, \qquad (8.3.33)$$

where $\sigma_{A,dir}^2$ and σ_{amb}^2 are the expected or mean power of directional and ambient components in the input signals. A pair of 2×1orthogonal and normalized eigenvectors of matrix [COV] is also obtained. The eigenvector with respect to σ_1^2 is denoted by $\left[\hat{a}_L, \hat{a}_R\right]^T$, with

$$\hat{a}_L = A_L \qquad \hat{a}_R = A_R. \tag{8.3.34}$$

The common directional or correlated component in stereophonic inputs is estimated from the following weighted combination:

$$\hat{e}_{A,dir}(n) = \hat{a}_L e_L(n) + \hat{a}_R e_R(n). \tag{8.3.35}$$

The ambient or uncorrelated components in stereophonic inputs are estimated by subtracting the directional component from the inputs:

$$\hat{e}_{L,amb}(n) = e_L(n) - \hat{a}_L \hat{e}_{A,dir}(n) \qquad \hat{e}_{R,amb}(n) = e_R(n) - \hat{a}_R \hat{e}_{A,dir}(n). \tag{8.3.36}$$

Substituting Equation (8.3.34) into Equations (8.3.35) and (8.3.36) and using Equation (8.3.6) yield

$$\begin{aligned} \hat{e}_{A,dir}(n) &= e_{A,dir}(n) + A_L e_{L,amb}(n) + A_R e_{R,amb}(n) \\ \hat{e}_{L,amb}(n) &= A_R^2 e_{L,amb}(n) - A_L A_R e_{R,amb}(n) \\ \hat{e}_{R,amb}(n) &= A_L^2 e_{R,amb}(n) - A_L A_R e_{L,amb}(n). \end{aligned} \tag{8.3.37}$$

According to the hypotheses given as Equations (8.3.7) to (8.3.12), the expected or mean power of the estimated directional components expressed in Equation (8.3.37) is calculated with

$$\overline{\hat{e}^2_{A,dir}(n)} = \overline{e^2_{A,dir}(n)} + \sigma^2_{amb} = \sigma^2_{A,dir} + \sigma^2_{amb} = \sigma^2_1. \tag{8.3.38}$$

Equation (8.3.38) involves two parts. One is the mean overall power of directional components in stereophonic inputs. Another is the mean power of ambient components in either input.

In Equation (8.3.37), the conventional PCA-based algorithm has the following problems:

1. In conventional PCA-based algorithms, the power of directional or correlated component is assumed to be dominant in stereophonic input signals. If this assumption is not true, the conventional PCA-based algorithm is invalid. If the directional component in stereophonic input signals vanishes, the solution in Equation (8.3.34) cannot be obtained for the eigen Equation (8.3.32). In this case, the estimation given as Equation (8.3.37) is invalid.
2. The estimated power of the left ambient component in the signal $\hat{e}_{L,amb}(n)$ is smaller than that in the actual left input signal except the case of $A_R = 1$ because of $0 \le A_R \le 1$. A similar problem occurs in the estimated signal $\hat{e}_{R,amb}(n)$, that is, the ambient component in each input signal is underestimated.
3. Ambient components in inputs are leaked to the estimated directional component $\hat{e}_{A,dir}(n)$; therefore, directional components in inputs cannot be separated completely.

The estimated directional component does not satisfy Equation (8.3.9). This result can also be observed in Equation (8.3.38).

4. The right ambient component in input is leaked to the estimated left ambient component and vice versa. The estimated ambient component does not satisfy Equation (8.3.8). Therefore, left and right ambient components in inputs cannot be separated completely, and the estimated left and right ambient components are dependent or correlated. Further decorrelation processing is needed.
5. The extent of directional and ambient component separation depends on the panning coefficients A_L and A_R of the directional component in stereophonic input signals. When the directional component in original stereophonic signals is panned to a single channel so that $A_L = 1$ and $A_R = 0$ or $A_L = 0$ and $A_R = 1$, the directional and ambient component cannot be separated. For example, $A_L = 1$ and $A_R = 0$, the left ambient component in input is completely mixed to the estimated directional component, and the estimated left ambient component vanishes.

Therefore, the validity in conventional PCA-based algorithm for upmixing depends on the relative proportion of directional and ambient components in stereophonic inputs. The algorithm is invalid when the directional component is not dominant, especially when it vanishes. Moreover, the validity of the algorithm depends on A_L and A_R for the directional component in stereophonic input signals. The algorithm is invalid when the directional component in original stereophonic signals is panned to a single channel.

To solve the problem in conventional PCA-based algorithm for upmixing caused by the relative proportion of directional and ambient components in stereophonic inputs, Goodwin (2008b) proposed a revised PCA-based algorithm. The normalized cross-correlation function Ψ of stereophonic input signals similar to that expressed in Equation (8.3.18) is calculated, and directional and ambient components are estimated by the following equations:

$$\begin{aligned}
\hat{e}'_{A,dir}(n) &= \Psi \ \hat{e}_{A,dir}(n) \\
\hat{e}'_{L,amb}(n) &= \Psi \left[e_L(n) - A_L \hat{e}_{A,dir}(n) \right] + (1-\psi) e_L(n) \\
\hat{e}'_{R,amb}(n) &= \Psi \left[e_R(n) - A_R \hat{e}_{A,dir}(n) \right] + (1-\psi) e_R(n).
\end{aligned} \tag{8.3.39}$$

When the directional component in stereophonic inputs vanishes, the calculation yields $\Psi = 0$, and the directional component estimated from Equation (8.3.39) vanishes. In this case, original stereophonic inputs directly serve as the estimated left and right ambient components. When stereophonic inputs contain the directional component only and the ambient component vanishes, the calculation yields $\Psi = 1$. With Equation (8.3.39), the directional component can be accurately estimated, and the estimated ambient component vanishes. The aforementioned characteristics are desired. However, Goodwin's revised PCA-based algorithm does not solve the problem related to the accuracy of the separation of directional and ambient components and the problem related to A_L and A_R. When either of A_L or A_R in stereophonic inputs vanishes, the calculation yields $\Psi = 0$. In this case, the full directional component is mixed to the estimated ambient output according to Equation (8.3.39).

Baek et al. (2012) proposed another revised PCA-based algorithm for upmixing. The directional and ambient components are estimated by the following equations:

$$\begin{aligned}
\hat{e}''_{A,dir}(n) &= \sqrt{\frac{\sigma^2_{A,dir}}{\sigma^2_{A,dir}+\sigma^2_{amb}}}\,\hat{e}_{A,dir}(n) \\
\hat{e}''_{L,amb}(n) &= e_L(n) - \left(1-\sqrt{\frac{\sigma^2_{amb}}{\sigma^2_{A,dir}+\sigma^2_{amb}}}\right) A_L\,\hat{e}_{A,dir}(n) \\
\hat{e}''_{R,amb}(n) &= e_R(n) - \left(1-\sqrt{\frac{\sigma^2_{amb}}{\sigma^2_{A,dir}+\sigma^2_{amb}}}\right) A_R\,\hat{e}_{A,dir}(n).
\end{aligned} \tag{8.3.40}$$

$\sigma^2_{A,dir}$ and σ^2_{amb} in Equation (8.3.40) are eigenvalues expressed in Equation (8.3.33).

The powers of the directional component and the left and right ambient components estimated from Equation (8.3.40) are equal to those of the corresponding components in stereophonic inputs. When the directional component in stereophonic inputs vanishes, the calculation yields $\sigma^2_{A,dir}=0$, and the directional component estimated from Equation (8.3.40) vanishes. In this case, original stereophonic inputs directly serve as the estimated left and right ambient components. When stereophonic inputs contain the directional component only and the ambient component vanishes, the calculation yields $\sigma^2_{amb}=0$. Furthermore, Equation (8.3.40) leads to an accurate estimation of the directional component, and the estimated ambient component vanishes. When either A_L or A_R is equal to a unit and another vanishes, e.g., $A_L=1$ and $A_R=0$, Equations (8.3.37) and (8.3.40) yield

$$\begin{aligned}
\hat{e}''_{A,dir}(n) &= \sqrt{\frac{\sigma^2_{A,dir}}{\sigma^2_{A,dir}+\sigma^2_{amb}}}\left[e_{A,dir}(n)+e_{L,amb}(n)\right] \\
\hat{e}''_{L,amb}(n) &= \sqrt{\frac{\sigma^2_{amb}}{\sigma^2_{A,dir}+\sigma^2_{amb}}}\left[e_{A,dir}(n)+e_{L,amb}(n)\right] \\
\hat{e}''_{R,amb}(n) &= e_{R,amb}(n).
\end{aligned} \tag{8.3.41}$$

The estimated right ambient component is equal to the right ambient component in stereophonic inputs. However, the directional component in stereophonic inputs is leaked to the estimated left ambient output, and the left ambient component in stereophonic inputs is leaked to the estimated directional output.

Before the development of conventional PCA-based algorithms, Irwan and Aarts (2002) proposed a method to separate directional and ambient components by maximizing the expectation of the estimated power of directional components. This method is similar to or even approximately equivalent to conventional PCA-based algorithms. If the directional component in original stereophonic signals is created by amplitude panning and satisfies Equation (8.3.11), directional and ambient components are estimated from the linear combination of stereophonic inputs $e_L(n)$ and $e_R(n)$:

$$\begin{aligned}
\hat{e}_{A,dir}(n) &= \hat{a}_L(n)e_L(n)+\hat{a}_R(n)e_R(n) \\
\hat{e}_{amb}(n) &= \hat{a}_R(n)e_L(n)-\hat{a}_L(n)e_R(n).
\end{aligned} \tag{8.3.42}$$

$\hat{a}_L(n)$ 和 $\hat{a}_R(n)$ is a pair of weights that vary with time adaptively and satisfy

$$\hat{a}_L^2(n)+\hat{a}_R^2(n)=1. \tag{8.3.43}$$

Therefore, Equation (8.3.42) is a linear-orthogonal transformation, which maintains the constant overall power of signals.

If the directional component is dominant in original stereophonic signals, the weights $\hat{a}_L(n)$ and $\hat{a}_R(n)$ in Equation (8.3.42) are chosen to maximize the expectation of the estimated power of directional components $\hat{e}_{A,dir}(n)$:

$$\max\left[\overline{\hat{e}_{A,dir}^2(n)}\right]=\max\left\{\overline{\left[\hat{a}_L(n)e_L(n)+\hat{a}_R(n)e_R(n)\right]^2}\right\}. \tag{8.3.44}$$

The resultant $\hat{e}_{A,dir}(n)$ represents the maximal correlated component of stereophonic inputs and serves as the estimated directional component. According to Equation (8.3.44), the weights satisfy the following iterative equation:

$$\begin{aligned}\hat{a}_L(n)&=\hat{a}_L(n-1)+\mu\hat{e}_{A,dir}(n-1)\left[e_L(n-1)-\hat{a}_L(n-1)\hat{e}_{A,dir}(n-1)\right]\\ \hat{a}_R(n)&=\hat{a}_R(n-1)+\mu\hat{e}_{A,dir}(n-1)\left[e_R(n-1)-\hat{a}_R(n-1)\hat{e}_{A,dir}(n-1)\right].\end{aligned} \tag{8.3.45}$$

The step length μ of iteration is chosen as

$$0<\mu<\frac{2}{e_L^2(n)+e_R^2(n)}. \tag{8.3.46}$$

The ambient component and directional component associated with the panning coefficients are calculated with Equations (8.3.42) and (8.3.45).

The ratio of the panning coefficients of the directional components between left and right stereophonic inputs can be estimated with $\hat{a}_L(n)$ and $\hat{a}_R(n)$. According to Equation (8.3.43) and by introducing the parameter $0 \le \xi(n) \le 90°$, the weights $\hat{a}_L(n)$ and $\hat{a}_R(n)$ can be written as

$$\hat{a}_L(n)=\sin\xi(n) \qquad \hat{a}_R(n)=\cos\xi(n). \tag{8.3.47}$$

If Equation (8.3.11), e.g., the directional component in stereophonic inputs is considered only, the signal amplitude with constant power normalization can be written in the form of Equation (2.2.59), then

$$\begin{aligned}e_{L,dir}(n)&=A_Le_{A,dir}(n)=\sin\xi e_{A,dir}(n)\\ e_{R,dir}(n)&=A_Re_{A,dir}(n)=\cos\xi e_{A,dir}(n),\end{aligned} \tag{8.3.48}$$

where $e_{A,dir}(n)$ is the signal waveform of the directional component in the discrete time domain. A_L and A_R are the panning coefficients of directional components. ξ is related to the amplitude ratio of the directional component between the left and right inputs or the target virtual source direction. Substituting Equations (8.3.47) and (8.3.48) into the first formula of Equation (8.3.42) and using the statistical characteristics of signals expressed in Equations (8.3.6) to (8.3.12), the expectation of the power in the estimated directional component $\hat{e}_{A,dir}(n)$ is calculated by

$$\overline{\hat{e}^2_{A,dir}(n)} = \overline{\cos^2\left[\xi - \varsigma(n)\right] e^2_{A,dir}(n)} + \sigma^2_{amb}. \tag{8.3.49}$$

If the target virtual source direction in stereophonic input signals does not change and thus the parameter ξ in Equation (8.3.49) is invariable within the time period to be considered, maximizing the power of the signal $\hat{e}_{A,dir}(n)$ yields

$$\varsigma(n) = \xi = \arctan\left(\frac{\hat{a}_L(n)}{\hat{a}_R(n)}\right). \tag{8.3.50}$$

In this case, $\hat{a}_L(n)$ and $\hat{a}_R(n)$ are an estimation of the panning coefficients of the directional component. The power of directional components is estimated by

$$\overline{\hat{e}^2_{A,dir}(n)} = \overline{e^2_{A,dir}(n)} + \sigma^2_{amb} = \sigma^2_{A,dir} + \sigma^2_{amb}. \tag{8.3.51}$$

$\sigma^2_{A,dir}$ and σ^2_{amb} are the expectations of the overall power of the directional component in the stereophonic input and the expectation of the power of ambient component in each input, respectively. The ambient component in inputs is leaked to the estimated directional output, so the directional component cannot be separated completely. The algorithm of maximizing the expectation of the estimated power of directional components in this section is basically equivalent to conventional PCA-based algorithms. The only difference is that the panning coefficients of directional components are estimated by solving the eigen equation in conventional PCA-based algorithms. In this algorithm, panning coefficients are estimated via adaptive iteration based on the short-time statistical characteristics of each frame of input signals. If the direction of the target virtual source in stereophonic input signals is invariable within the time period to be considered, the results of the algorithm here are consistent with those of conventional PCA-based algorithms. Crossover processing between frames is needed to avoid the audible artifact caused by the sudden change between time frames in conventional PCA-based algorithms.

In upmixing the two-channel stereophonic input signals to 5.1-channel surround sound signals, the estimated ambient component is mixed to two surround channel outputs, and the estimated directional component $\hat{e}_{A,dir}(n)$ is re-panned to the three frontal channel outputs in accordance with a certain panning rule and the estimated $\hat{a}_L$ and $\hat{a}_R$ (the ICLD or target virtual source direction in stereophonic inputs).

On the basis of the estimated directional component $\hat{e}_{A,dir}(n)$ and the ambient component $\hat{e}_{amb}(n)$, Irwan and Aarts (2002) further used an adaptive panning algorithm for upmixing.

For this purpose, the normalized cross-correlation coefficient between stereophonic inputs should be analyzed. The time samples of inputs are divided into a frame or a block, the length of each frame is N_1 sample, and the cross-correlation coefficient in each frame is calculated with

$$\Psi(n') = \frac{\sum_{n=NL1}^{NH1} \left\{ \left[e_L(n'+n) - \bar{e}_L \right] \left[e_R(n'+n) - \bar{e}_R \right] \right\}}{\sqrt{\left\{ \sum_{n=NL1}^{NH1} \left[e_L(n'+n) - \bar{e}_L \right]^2 \right\} \left\{ \sum_{n=NL1}^{NH1} \left[e_R(n'+n) - \bar{e}_R \right]^2 \right\}}}, \tag{8.3.52}$$

where $\bar{e}_L$ and $\bar{e}_R$ are the temporal means of the input signals $e_L(n)$ and $e_R(n)$ over a time frame; $NL \leq 0$ and $NH > 0$ are the initial and end times of the frame, respectively; and $N1 = NH1 - NL1 + 1$ is the length of the frame. The cross-correlation coefficient can be calculated through an iterative method. By definition, $-1 \leq \Psi(n) \leq 1$. A $\Psi(n)$ close to a unit represents highly correlated inputs; a $\Psi(n)$ close to zero represents almost uncorrelated inputs; a $\Psi(n) < 0$ represents negative-correlated inputs in which the out-of-phase component between two inputs is dominant. If the case of negative-correlated inputs is merged into the case of uncorrelated inputs, the normalized cross-correlation coefficient is re-defined as

$$\Psi_0(n) = \begin{cases} \Psi(n) & 0 \leq \Psi(n) \leq 1 \\ 0 & \text{otherwise} \end{cases}. \tag{8.3.53}$$

Then following parameter is defined as

$$\rho(n) = \arcsin\left[1 - \Psi_0(n)\right]. \tag{8.3.54}$$

By definition, $0 \leq \rho(n) \leq \pi/2$. $P(n)$ increases as the uncorrelated or ambient component increases and decreases as the directional component in stereophonic inputs increases. Therefore, $\rho(n)$ is applicable to the control of the proportions or gains of directional and ambient components re-mixed to different channels.

On the basis of the estimated $\hat{e}_{A,dir}(n)$, $\hat{e}_{amb}(n)$, and the aforementioned parameter, four-channel upmixed signals, including the left signal $e'_L(n)$, the center signal $e'_C(n)$, the right signal $e'_R(n)$, and the surround signal $e'_S(n)$ can be created in accordance with the following matrix method:

$$\begin{bmatrix} e'_L(n) \\ e'_C(n) \\ e'_R(n) \\ e'_S(n) \end{bmatrix} = \begin{bmatrix} \kappa_L(n) & g\hat{a}_L(n) \\ \kappa_C(n) & 0 \\ \kappa_R(n) & g\hat{a}_R(n) \\ 0 & \kappa_S(n) \end{bmatrix} \begin{bmatrix} \hat{e}_{A,dir}(n) \\ \hat{e}_{amb}(n) \end{bmatrix}. \tag{8.3.55}$$

with

$$\kappa_L(n) = \begin{cases} \left[\hat{a}_L^2(n) - \hat{a}_R^2(n)\right] & \text{if } \left[\hat{a}_L^2(n) - \hat{a}_R^2(n)\right] > 0 \\ 0 & \text{otherwise} \end{cases}$$

$$\kappa_R(n) = \begin{cases} \left[\hat{a}_R^2(n) - \hat{a}_L^2(n)\right] & \text{if } \left[\hat{a}_R^2(n) - \hat{a}_L^2(n)\right] > 0 \\ 0 & \text{otherwise} \end{cases} \tag{8.3.56}$$

$$\kappa_C(n) = 2\hat{a}_L(n)\hat{a}_R(n)$$
$$\kappa_S(n) = \sin\rho(n) = 1 - \Psi_0(n).$$

where g is a parameter that maintains the constant overall power of the signal. When $g = \cos^2\rho\,(n)$ is chosen, the overall power of a signal is unchanged after upmixing via Equation (8.3.55). Equation (8.3.55) only yields a mono surround signal. Two surround signals can be further created by decorrelating methods.

The aforementioned algorithm is appropriate for stereophonic inputs with ICLD only (amplitude panning). For stereophonic inputs with ICTD (time panning), He (2015) suggested that a time alignment processing should be applied prior to PCA.

8.3.7 Algorithm based on the least mean square error for upmixing

Faller (2006) proposed an algorithm based on the *least mean square error* (*LMSE*) for the separation of directional and ambient components in the time-frequency domain. The mathematical derivation in this section is different from but completely equivalent to that in Faller's original paper. Stereophonic input signals are converted to the time-frequency domain with a CB bandwidth by combining the coefficients of STFT. Then, the autocorrelation and cross-correlation coefficients of input signals are calculated similar to Equation (8.3.16). According to the signal model expressed in Equations (8.3.6) to (8.3.11), the following results are obtained:

$$\Phi_{LL}(n',k) = A_L^2\sigma_{A,dir}^2(n',k) + \sigma_{amb}^2(n',k) \quad \Phi_{RR}(n',k) = A_R^2\sigma_{A,dir}^2(n',k) + \sigma_{amb}^2(n',k)$$
$$\Phi_{LR}(n',k) = A_L A_R \sigma_{A,dir}^2(n',k). \tag{8.3.57}$$

where

$$\sigma_{A,dir}^2(n',k) = \overline{E_{A,dir}(n',k)E_{A,dir}^*(n',k)}. \tag{8.3.58}$$

Equation (8.3.58) represents the expectation of the power of the directional component in the time-frequency domain. $A_L(n', k)$ and $A_R(n', k)$ are the panning coefficients of the directional component in the left and right inputs. According to Equation (8.3.57) and using either the constant power normalization in Equation (8.3.12) or constant amplitude normalization in Equation (8.3.13), the temporal mean power $\hat{\sigma}_{A,dir}^2(n',k)$ of the directional component, the temporal mean power $\hat{\sigma}_{amb}^2(n,k)$ of ambient component, and the left and right panning coefficients $\hat{A}_L(n',k)$ and $\hat{A}_R(n',k)$ of the directional component are estimated. They are all the functions of the correlation coefficients $\Phi_{LL}(n', k)$, $\Phi_{RR}(n', k)$ and $\Phi_{LR}(n', k)$.

If the directional and ambient components can be estimated from the linear weighted combinations of two stereophonic inputs, then

$$\begin{aligned}\hat{E}_{A,dir}(n',k) &= \hat{a}_{L,dir}(n',k)E_L(n',k)+\hat{a}_{R,dir}(n',k)E_R(n',k)\\ \hat{E}_{L,amb}(n',k) &= \hat{a}_{LL,amb}(n',k)E_L(n',k)+\hat{a}_{LR,amb}(n',k)E_R(n',k)\\ \hat{E}_{R,amb}(n',k) &= \hat{a}_{RL,amb}(n',k)E_L(n',k)+\hat{a}_{RR,amb}(n',k)E_R(n',k),\end{aligned} \quad (8.3.59)$$

where $\hat{a}_{L,dir}(n',k)$ and $\hat{a}_{R,dir}(n',k)$ et al. are weights to be determined. Equation (8.3.6) and the signal model in Equation (8.3.11) are substituted into Equation (8.3.59):

$$\begin{aligned}E_L(n',k) &= A_L E_{A,dir}(n',k)+E_{L,amb}(n',k)\\ E_R(n',k) &= A_R E_{A,dir}(n',k)+E_{R,amb}(n',k).\end{aligned} \quad (8.3.60)$$

The weights for combination are chosen to minimize the square errors between the estimated directional and ambient components in Equation (8.3.5) and the actual ones:

$$\min\left\{\overline{\left|E_{A,dir}(n',k)-\hat{E}_{A,dir}(n',k)\right|^2}\right\} \quad \min\left\{\overline{\left|E_{L,amb}(n',k)-\hat{E}_{L,amb}(n',k)\right|^2}\right\}$$
$$\min\left\{\overline{\left|E_{R,amb}(n',k)-\hat{E}_{R,amb}(n',k)\right|^2}\right\}. \quad (8.3.61)$$

Then, the weights for combination are solved as functions of $\hat{\sigma}^2_{A,dir}(n',k)$, $\hat{\sigma}^2_{amb}(n',k)$, $\hat{A}_L$, and $\hat{A}_R$ obtained via estimation, or more exactly as the functions of the correlation coefficients $\Phi_{LL}(n',k)$, $\Phi_{RR}(n',k)$, and $\Phi_{LR}(n',k)$. The estimated directional and ambient components are obtained by substituting the resultant weights into Equation (8.3.59). In the case of multiple directional components at the same time, the estimated panning coefficients $\hat{A}_L$ and $\hat{A}_R$ vary with subband and time, yielding the directional information in stereophonic input signals. According to this information and the conventional summing localization theorem, the separated directional and ambient components are re-mixed or upmixed to more channel outputs. In addition, He, et al. (2014) compared the LMSE- and PCA-based algorithms for

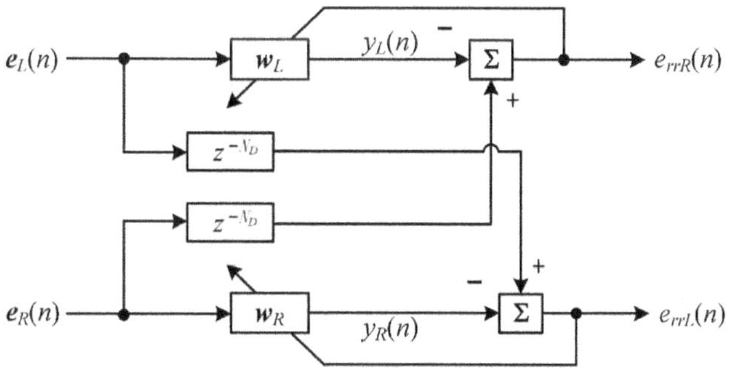

Figure 8.6 Adaptive NMLS-based algorithm in the time domain(redrawn according to Usher and Benesty, 2007).

the separation of directional and ambient components and introduced a method for error analysis.

The aforementioned algorithm is appropriate not only for upmixing stereophonic signals but also for two-channel matrix-encoded signals (Faller, 2007). Matrix-encoded signals are decomposed into subband signals, and directional and ambient components in each subband are separated and then re-mixed to more channel outputs (such as 5.1-channel). In contrast to conventional logic decoding described in Sections 8.1.3 to 8.1.4, the algorithm discussed here can simultaneously separate more than one dominant directional components if these components do not overlap in the time-frequency domain. Therefore, the algorithm discussed here is also an advanced matrix decoding algorithm. This type of algorithm is involved in the new generation of matrix surround sound by Dolby Laboratories (Vinton et al., 2015). It can also upmix the unmatrixed stereophonic signals into more channel signals.

8.3.8 Adaptive normalized algorithm based on the least mean square for upmixing

Figure 8.6 is the block diagram of an ***adaptive normalized algorithm based on the least mean square*** in the time domain (**adaptive NLMS-based algorithm**; Usher and Benesty, 2007). In this algorithm, the correlated component in each channel is cancelled by using the correlation between stereophonic inputs, and directional and ambient components are separated.

The left signal $e_L(n)$ in the stereophonic input is fed to an adaptive N-point finite impulse response (FIR) filter, and the output of filter is given as

$$y_L(n) = \sum_{i=0}^{N-1} w_L(n, i) e_L(n - i), \tag{8.3.62}$$

where $w_L(n, i)$ with i =0,1,...(N–– 1) is the N coefficient or N-point impulse response of the filter at instant n, and these coefficients vary with time. Let

$$\mathbf{w}_L(n) = \left[w_L(n, 0), w_L(n, 1) \cdots w_L(n, N-1)\right]^T. \tag{8.3.63}$$

Equation (8.3.63) represents a N×1 vector or column matrix of coefficients, and superscript "T" denotes matrix transpose. Let

$$\mathbf{e}_L(n) = \left[e_L(n), e_L(n-1) \cdots e_L(n - N + 1)\right]^T. \tag{8.3.64}$$

Equation (8.3.64) represents a N×1 vector or column matrix for input signal. Then, Equation (8.3.62) is written as

$$y_L(n) = w_L^T(n) e_L(n). \tag{8.3.65}$$

Some ICTDs may exist between left and right stereophonic input signals. Therefore, the right channel signal $e_R(n)$ is delayed or advanced by N_D samples, e.g., $e_R(N–N_D)$ is used as a reference signal. The delay is within the range of ± 10 ms, corresponding to the maximal ICTD in the signals of a pair spaced microphones with an interval of about 3.4 m. For simplicity, however, the delay is omitted in the following discussion. In Equation (8.3.6), the error between the output of the adaptive filter and the reference signal is calculated with the following:

$$err_L(n) = e_R(n) - y_L(n) = e_{R,dir}(n) + e_{R,amb}(n) - w_L^T(n)e_L(n). \quad (8.3.66)$$

The coefficient $w(n)$ of the adaptive FIR filter is chosen to minimize the expectation or temporal mean of the square error between the filter output and the reference signal:

$$\min\left[\overline{err_L^2(n)}\right] = \min\left[\overline{e_R(n) - y_L(n)}\right]^2 = \min\left[\overline{e_{R,dir}(n) + e_{R,amb}(n) - w_L^T(n)e_L(n)}\right]^2. \quad (8.3.67)$$

Under an ideal condition, Equation (8.3.67) leads to the following orthogonality between input and error signals and thus removes the correlated components between stereophonic signals:

$$\overline{e_L(n)err_L(n)} = 0. \quad (8.3.68)$$

According to the LMS algorithm in adaptive signal processing (Uncini, 2015) and Equation (8.3.67), a set of iterative equations for the filter output and coefficients are obtained:

$$\begin{aligned} y_L(n) &= \mathbf{w}_L^T(n)\mathbf{e}_L(n) \\ err_L(n) &= e_R(n) - y_L(n) = e_R(n) - \mathbf{w}_L^T(n)\mathbf{e}_L(n), \\ \mathbf{w}_L(n+1) &= \mathbf{w}_L(n) + 2\mu err_L(n)\mathbf{e}_L(n) \end{aligned} \quad (8.3.69)$$

where μ is the step length that controls the stability and the convergence of the adaptive algorithm. A method of the ***normalized least mean square*** (***NLMS***) is also applicable. The step length is normalized with the power of the input signal $e_L(n)$, and the third formula in Equation (8.3.69) becomes

$$\mathbf{w}_L(n+1) = \mathbf{w}_L(n) + \frac{2\mu'}{\mathbf{e}_L^+(n)\mathbf{e}_L(n) + \lambda} err_L(n)\mathbf{e}_L(n),. \quad (8.3.70)$$

where μ' and λ are positive constants.

Similarly, the right channel signal $e_R(n)$ in original stereophonic inputs is fed to another adaptive N-point FIR filter w_R, and the left channel signal $e_L(n)$ is used as the reference signal. Minimizing the expectation or temporal mean of the square error between the output of the filter w_R and reference signal leads to a set of iterative equations similar to Equation (8.3.69).

According to the hypothesis on the statistical characteristics of stereophonic signals expressed in Equations (8.3.7) and (8.3.8), the output $y_L(n)$ of the filter w_L mainly contains the correlated or directional components of the right stereophonic input and serves as an estimation of $e_{R,dir}(n)$. Error signal $err_L(n)$ mainly comprises the ambient component of the right stereophonic signal and denotes the estimation of $e_{R,amb}(n)$. Similarly, the output $y_R(n)$ of the filter w_R serves as an estimation of the directional component of the left stereophonic signal $e_{L,dir}(n)$; the error signal $err_R(n)$ represents as an estimation of the ambient component $e_{L,amb}(n)$ of the left stereophonic input. Then, directional and ambient components are separated via the following equation:

$$\begin{array}{ll} \hat{e}_{L,dir}(n) = y_R(n) & \hat{e}_{L,amb}(n) = err_R(n) \\ \hat{e}_{R,dir}(n) = y_L(n) & \hat{e}_{R,amb}(n) = err_L(n) \end{array}. \tag{8.3.71}$$

The aforementioned algorithm is valid in the case that directional components are involved in both channels of inputs (such as time-difference stereophonic signals). The estimated left and right ambient components directly serve as the left- and right-surround channel signals for stereophonic to 5.1-channel sound upmixing. The estimated left and right directional components are converted into three frontal channel signals by other appropriate methods. However, for stereophonic inputs with ICLD or amplitude panning only, when the directional component is panned to the right or left loudspeaker direction, either A_L or A_R in Equation (8.3.11) vanishes. In this case, the algorithm erroneously identifies the directional component for ambient component and outputs it for an error signal. Then, the outputs of adaptive filters vanish. Therefore, the algorithm is only valid when A_L and A_R do not vanish, especially target virtual source close to the front direction with $a_L \approx a_R$.

8.3.9 Some advanced upmixing algorithms

Some advanced signal processing techniques such as deep neural networks are also applicable for two-to-five-channel upmixing (Choi and Chang, 2020).

Various algorithms for the separation of directional and ambient components in the time domain or time-frequency domain can be extended to the upmixing of more channel signals, including upmixing two-channel stereophonic inputs to more than 5.1-channel outputs or upmixing more than two-channel inputs to a larger number of outputs. The principle is similar to that of upmixing stereophonic inputs to 5.1-channel outputs. Generally, directional components from each source and ambient component are separated or estimated by analyzing the signal characteristics in the time-frequency domain and the relationship among input signals. The estimated components are subjected to some appropriate post-processing and then re-mixing to different output channels. Indeed, an increase in the number of input and output channels complicates the separation and re-mixing of components.

Thompson et al. (2012) proposed an algorithm to separate the directional and ambient components by analyzing the correlation between pairs of input signals. Vilkamo et al. (2013) proposed a framework of covariance matrix analysis and applied to multichannel signal upmixing through which the upmixing matrix is derived in terms of the covariance matrix of input signals. Moreover, the power of ambient component in each input is equal, but this case is not always observed in practical multichannel sound signals. Faller et al. (2013) proposed another upmixing algorithm for multiple outputs in which cascading and paralleling units for upmixing two-channel inputs are used to generate more channel outputs. The more channel outputs are reproduced through a circular loudspeaker configuration. Faller also illustrated an example of upmixing five-channel input signals to 13-channel output signals.

Kraft and Zölzer (2016) used an algorithm similar to that in Section 8.3.6 to upmix the stereophonic inputs to multichannel spatial surround sound outputs. The new generation matrix surround sound by Dolby Laboratories (Section 8.3.7) can also upmix stereophonic inputs to 5.1-channel outputs or upmix 5.1-channel inputs to more channel outputs (Vinton et al., 2015). The DTS Neo:X technique can also upmix two to 7.1 channel inputs to outputs up to 11.1 channels (www.dts.com).

8.4 SUMMARY

Matrix surround sound and signals downmixing/upmixing are designed for different purposes, but they are closely related. When the transmission (or storage) capacity of media is limited, matrix surround sound aims to reduce the number of independent signals by matrix encoding and retrieve multichannel signals in reproduction by decoding. Various 4-2-4 matrix quadraphones were developed in the early times of multichannel sounds. Among them, four original channel signals, e.g., the left-front, right-front, left-back, and right-back-channel signals, are encoded into two independent signals and then decoded back to four-channel signals for reproduction. The ability of spatial information reproduction in an original discrete quadraphone is limited. Information is further lost in matrix encoding. Therefore, various 4-2-4 matrix quadraphones are not ideal systems and have rarely been used in practical applications.

Dolby Stereo and Dolby Surround were developed for sound reproduction with accompanying picture. Under the condition of the limited capacity of transmission or storage media, a frontal-biased reproduction of the spatial information of sound is adopted to ensure accurate localization that matches with the picture within a larger listening region. The surround channel is mainly intended for ambient information and some special effects. Dolby Stereo/Surround encodes four original signals, including left, center, right, and surround signals, into two independent signals. Independent signals are decoded to four signals in reproduction. Linear or passive decoding was applied to early Dolby Surround technique. Adaptive or logic decoding used in the subsequent Dolby Pro-Logic technique obviously improves channel separation in reproduction. However, Dolby Pro-Logic decoding is effective in only one dominant directional component at each instant and unable to manipulate two or more directional components at the same time. It is more appropriate for sound reproduction with accompanying picture and exhibits some limitations for music reproduction. Some new matrix encoding/decoding techniques after Dolby Pro-Logic can manipulate more than four-channel information and improve the performance in music reproduction.

Downmixing aims to convert signals for reproduction with a larger number of channels into signals for reproduction with a smaller number of channels. The problem of downmixing to stereophonic reproduction has been considered in the early time of multichannel sound. Two-channel encoded signals in matrix surround sound are usually stereo compatible, but they do not always have an optimal effect on stereophonic reproduction. A downmixing method for 5.1-channel signals to stereophonic signals is also recommended in accordance with the ITU standard. Ideal downmixing coefficients or gains should be adapted to the characteristics of program materials. Various methods have been proposed on the basis of psychoacoustic principles to improve downmixing.

Upmixing aims to convert signals for reproduction with a smaller number of channels into signals for reproduction with a larger number of channels. Decoding algorithms in matrix surround sound may be applicable to upmixing, but they do not cause an optimal effect on unmatrixed stereophonic signals. Two considerations and methods for upmixing are available. In one method, original signals are upmixed for reproduction with more channels, but the original spatial (especially localization) information is preserved. In the other method, directional and ambient components are separated and extracted from original signals. The separated components are processed, enhanced, and re-mixed into more channels. Traditional upmixing algorithms are usually based on passive processing, but their ability to separate information is limited. Nevertheless, the overall performance of upmixing is improved with recent methods based on the statistical processing of random signals in the time domain or time-frequency domain.

Chapter 9

Physical analysis of multichannel sound field recording and reconstruction

Multichannel sounds based on sound field approximation and psychoacoustics are addressed in Chapters 3 to 6, where lower-order Ambisonics is introduced as a signal mixing technique for multichannel sound. The analyses in the aforementioned chapters are based on traditional methods and tools, i.e., the psychoacoustic principles of spatial hearing, especially on the principles of summing localization with multiple sound sources. Some basic rules and features of multichannel sound reproduction can be preliminarily obtained by analyzing localization-related binaural cues, such as interaural time difference (ITD) and its dynamic variation at low frequencies.

The method of physical sound field analysis is further discussed in this chapter. It is a useful tool for examining the sound field recording and reconstruction of various multichannel sounds, especially Ambisonics. The purposes considerations of this chapter are as follows:

1. Physical sound field analysis is a universal tool that provides a strictly theoretical basis and a design method for spatial sound. It is an essential tool for spatial sound based on physical sound field reconstruction. Some new techniques in sound field reconstruction can be developed with the help of this tool.
2. Although many practical multichannel sound systems and techniques are designed on the basis of psychoacoustic principle, they can be regarded as methods of approximate or rough reconstruction of a target sound field under certain conditions. Therefore, from the points of spatial sampling and reconstruction, the analysis of a physical sound field yields insights into the nature and content of approximation in multichannel sound reproduction and provides important guidance for the design of loudspeaker configuration, signal recording, and mixing.
3. Evolution from lower-order Ambisonics to higher-order Ambisonics is a typical example of a transition from spatial sound based on sound field approximation and psychoacoustics to the one based on the accurate reconstruction of a physical sound field. The analysis of Ambisonic-reconstructed sound field is an important example of (1) and (2) and provides insight into the nature of Ambisonics. It also reveals some new results about recording and reproduction.

This chapter, together with the analysis of wave field synthesis in Chapter 10, forms a relatively complete discussion on spatial sounds based on physical sound field reconstruction. Some details on sound field analysis and reconstruction are also described in other studies (Ahrens, 2012; Kim and Choi, 2013; Zotter and Frank, 2019). In Section 9.1, Ambisonic recording and reproduction are analyzed from the point of beamforming. Ambisonic signals with different orders are derived from various order approximations of ideal recorded and reproduced signals. In Section 9.2, the general formulations of multichannel sound field reconstruction are introduced, and the formulation in a spatial domain

DOI: 10.1201/9781003081500-9

and two examples of formulation in a spatial-spectral domain are presented. In Section 9.3, the reconstructed sound field of Ambisonics in the spatial-spectral domain is analyzed in detail; the decoding equation and deriving signals for arbitrary-order Ambisonics with various secondary source arrays or loudspeaker configurations are derived; the theorem of the spatial sampling and reconstruction of sound field are discussed; near-field compensated higher-order Ambisonics is addressed; and the applications of spatial-spectral analysis in some Ambisonic-like techniques are outlined. In Section 9.4, the secondary sources arrays and stability of Ambisonic sound field are analyzed, and some spatial transformations of Ambisonic sound field are discussed. In Section 9.5, errors in Ambisonic sound field are evaluated, and the problems of spatial aliasing caused by discrete secondary source arrays in a horizontal circle or on a spatial spherical surface are analyzed. In Section 9.6, the basic method of spatial domain analysis of multichannel sound field is introduced, and method of multiple matching receiver positions and their relations with the mode-matching method are outlined. In Section 9.7, the problem of active compensation for reflections in a listening room for multichannel sound reproduction is addressed. In Section 9.8, microphone array techniques for sound field recording, especially Ambisonic recording, are discussed on the basis of the spatial sampling and reconstruction theorem of a sound field.

9.1 EACH ORDER APPROXIMATION OF IDEAL REPRODUCTION AND AMBISONICS

9.1.1 Each order approximation of ideal horizontal reproduction

In this section, the recorded and reproduced signals of Ambisonics are derived from each order approximation of an ideal reproduction (Xie and Xie, 1996). The case of horizontal reproduction is discussed first. Similar to the discussion in Section 3.1, the discussion here starts with an ideal horizontal reproduction by arranging an infinite number of loudspeakers on a circle uniformly and continuously around a listener or receiver region. If the radius r_0 of the circle is large enough, then the incident wave in a region near the center of the circle can be approximated as the superposition of plane waves from each loudspeaker. According to Equation (1.2.12), for an original or target plane wave with unit amplitude and incident from a horizontal azimuth θ_S, the azimuthal distribution function of the complex-valued amplitude of the incident wave is taken in the form of Dirac delta function $\tilde{P}_A(\theta_{in}) = \delta(\theta_S - \theta_{in})$, where θ_{in} is the azimuthal coordinate of a sound field. For an ideal reproduction, the azimuthal distribution function $A(\theta', \theta_S)$ of the normalized amplitude of loudspeaker signals should match with that of the target sound field, where θ' is the azimuth of continuous loudspeaker arrangement. Letting $\theta' = \theta_{in}$, the normalized signal amplitude for loudspeakers at an arbitrary azimuth θ' is given as

$$A(\theta', \theta_S) = \tilde{P}_A(\theta_{in}) = \delta(\theta' - \theta_S) = \delta(\theta_S - \theta'). \tag{9.1.1}$$

As in the case of the preceding chapters, a unit transfer coefficient from the loudspeaker signal to the pressure amplitude of the free-field plane wave at the origin is assumed in Equation (9.1.1), e.g., $E_A = P_A$. Therefore, in an ideal reproduction, only the loudspeaker at azimuth $\theta' = \theta_S$ is active, and the other loudspeakers are inactive.

$A(\theta', \theta_S)$ is a periodic function of azimuth θ_S or θ' with a period of 2π (360°), so it can be expanded into a complex- or real-valued Fourier series within the azimuthal region of $(-\pi, \pi]$:

$$A(\theta', \theta_S) = \frac{1}{2\pi} \sum_{q=-\infty}^{\infty} \exp(-jq\theta') \exp(jq\theta_S)$$
$$= \frac{1}{2\pi} \left[1 + 2\sum_{q=1}^{\infty} (\cos q\theta' \cos q\theta_S + \sin q\theta' \sin q\theta_S) \right]. \tag{9.1.2}$$

Therefore, the normalized amplitudes of loudspeaker signals in an ideal reproduction can be decomposed into a linear combination of azimuthal harmonics $\{\exp(jq\theta_S)\}$ or $\{\cos(q\theta_S), \sin(q\theta_S)\}$ with infinite orders. Complex- and real-valued Fourier expansions are mathematically equivalent. The case of real-valued Fourier expansion is discussed in the following.

Equation (9.1.2) is analyzed from the points of multichannel sound recording and reproduction to obtain insights into physical significance. From the point of multichannel sound reproduction, Equation (9.1.2) represents the azimuthal distribution of the normalized amplitude of loudspeaker signals as a function of the azimuth θ'. It also represents the azimuthal distribution function of the normalized amplitude of the free-field plane wave incident to the center of the circle. As a zero-order approximation, the expansion in Equation (9.1.2) is truncated only to the term of $q = 0$, and other terms are omitted. Then, the normalized signal amplitude of the loudspeaker at arbitrary azimuth θ' is given as

$$A(\theta', \theta_S) = \frac{1}{2\pi}. \tag{9.1.3}$$

Equation (9.1.3) is equivalent to presenting the mono signal captured by an omnidirectional microphone to all loudspeakers in reproduction. The sound pressure at the center of the circle is a superposition of plane waves with equal amplitude and phase from the continuous azimuthal directions of all loudspeakers. Therefore, zero-order reproduction cannot recreate the spatial information of a target plane wave, or can create a perceived virtual source at the top direction similar to the case in Section 6.4.3.

As the first-order approximation, the expansion in Equation (9.1.2) is truncated up to the term of $q = 1$, and higher terms are omitted. Then, the normalized signal amplitude of loudspeakers at arbitrary azimuth θ' is expressed as

$$A(\theta', \theta_S) = \frac{1}{2\pi}\left[1 + 2\cos\theta' \cos\theta_S + 2\sin\theta' \sin\theta_S\right]. \tag{9.1.4}$$

Except for the difference in overall gain, Equation (9.1.4) is directly proportional to the conventional solution of the normalized amplitude of loudspeaker signals for first-order horizontal Ambisonics in Equations (4.3.15) and (4.3.25). Variations in the azimuthal distribution function or horizontal polar pattern of the normalized amplitude with the difference in $\theta_S - \theta' = \theta_S - \theta_i$ between the target and loudspeaker azimuths are illustrated in Section 4.1.3 and Figures 4.4, 4.5, and 4.17. For an arbitrary loudspeaker, the normalized signal magnitude maximizes when the loudspeaker azimuth coincides with the target source azimuth at $\theta' = \theta_S$. When the loudspeaker azimuth deviates from the target source azimuth, the normalized signal magnitude reduces and gradually vanishes. However, as the loudspeaker azimuth further deviates from the target source azimuth, a weak crosstalk with a reversal phase occurs in the loudspeakers close to the direction opposite to the target source. As proven in

Equation (4.3.27), at the central listening position and low frequencies, the perceived virtual source direction in the first-order Ambisonic reproduction matches with that of the target source for the fixed head and the head oriented to the virtual source.

As the second-order approximation, the expansion in Equation (9.1.2) is truncated up to the term of $q = 2$, and the higher terms are omitted. Then, the normalized signal amplitude of loudspeakers at the arbitrary azimuth θ' is expressed as

$$A(\theta', \theta_S) = \frac{1}{2\pi}\left[1 + 2\cos\theta'\cos\theta_S + 2\sin\theta'\sin\theta_S + 2\cos 2\theta'\cos 2\theta_S + 2\sin 2\theta'\sin 2\theta_S\right]. \quad (9.1.5)$$

Except the difference in the overall gain, Equation (9.1.5) is directly proportional to the conventional solution of the normalized amplitude of loudspeaker signals for the second-order horizontal Ambisonics in Equations (4.3.53) and (4.3.62). The variation in azimuthal distribution function or horizontal polar pattern of normalized amplitude with the difference $\theta_S - \theta' = \theta_S - \theta_i$ between target and loudspeaker azimuths are illustrated in Figure 4.17. In comparison with the case of the first-order reproduction, the relative signal magnitude of the loudspeaker nearest the target source azimuth increases, and the relative signal magnitudes (crosstalk) of the other loudspeakers decrease. Therefore, the incident power in reproduction is more focused on the target azimuth of $\theta' = \theta_S$, and the performance of directional information reproduction is improved.

When the expansion in Equation (9.1.2) is truncated up to the term of $q = 3$ or higher, the conventional solution of loudspeaker signals for the third- or higher-order horizontal Ambisonics is achieved. As illustrated in Section 4.3.3 and Figure 4.17, as the order increases, the relative signal magnitude of the loudspeaker consistent with the target source direction increases, and the relative signal magnitudes (crosstalk) of the other loudspeakers decrease. Then, the incident power in reproduction is gradually focused on the target azimuth of $\theta' = \theta_S$. In other words, as the order increases, the approximated reproduction approaches ideal reproduction, and the reproduction of spatial information gradually improves. When the order in Equation (9.1.2) tends to infinity, the approximated reproduction achieves the limitation of ideal reproduction. Generally, when the expansion in Equation (9.1.2) is truncated to an arbitrary-order Q, the normalized amplitude of loudspeaker signals is expressed as

$$\begin{aligned} A(\theta', \theta_S) &= \frac{1}{2\pi}\left[1 + 2\sum_{q=1}^{Q}\left(\cos q\theta'\cos q\theta_S + \sin q\theta'\sin q\theta_S\right)\right] \quad Q \geq 1 \\ &= \frac{1}{2\pi}\left[1 + 2\sum_{q=1}^{Q}\cos q\left(\theta' - \theta_S\right)\right]. \end{aligned} \quad (9.1.6)$$

From the points of multichannel sound recording, arbitrary $Q \geq 1$ order Ambisonic signals $A(\theta', \theta_S)$ in Equation (9.1.6) are the linear combinations of $(2Q + 1)$ independent signals or azimuthal harmonics. These independent signals can be theoretically recorded using $(2Q + 1)$ coincident directional microphones. As stated in Sections 4.3.2 and 4.3.3, 1, $\cos\theta_S$, and $\sin\theta_S$ are three normalized signals recorded with an omnidirectional microphone and two bidirectional microphones with their main axes pointing to the front and left directions, respectively; $\cos q\theta_S$ and $\sin q\theta_S$ ($q \geq 2$) are normalized signals recorded with higher-directional microphones. In the polar patterns of the preceding three-order normalized signal amplitude

$A(\theta', \theta_S)$ in Figure 4.17, horizontal Ambisonic signal recording can be regarded as a horizontal ***beamforming*** method. Beamforming enhances the recorded outputs at the target azimuth of $\theta_{in} = \theta' = \theta_S$ and restrains the outputs at other azimuths, where θ' is a parameter of beam direction. As the order Q increases, the beam becomes sharp, thereby improving the azimuthal resolution of recording. When the order Q tends to infinity, Equation (9.1.1) or (9.1.2) represents the case of recording with the method of ideal beamforming. Equation (9.1.6) indicates that the horizontal beam direction can be steered to an arbitrary azimuth without altering the beam shape by changing θ'. This feature is common for horizontal Ambisonics.

The reconstruction of a plane wave with unit amplitude and incident from a horizontal direction is discussed above. For a plane wave with arbitrary amplitude $P_A(f)$, Equation (9.1.1) is the azimuthal distribution function of incident plane wave amplitudes after being normalized by a factor of $P_A(f)$. Therefore, actual loudspeaker signals are obtained by multiplying the normalized amplitude $A(\theta', \theta_S)$ with a signal waveform $E_A(f) = P_A(f)$ in the frequency domain. According to Equation (1.2.12), an arbitrary sound field in a source-free region can be decomposed as a linear superposition of the plane wave from various directions. In this case, the azimuthal distribution function $\tilde{P}_A(\theta_{in}, f)$ of incident plane wave amplitudes with respect to the origin is no longer a Dirac delta function. If a set of the aforementioned coincident directional microphones are used to capture the sound field signals, the resultant microphone outputs are the superposition of the contribution of plane waves from all directions. For example, for ideal microphones in which the transfer functions from the incident plane wave amplitude to microphone outputs are a unit, the normalized amplitudes of the omnidirectional microphone and two bidirectional microphones are given as

$$
\begin{aligned}
W_\Sigma &= \int_{-\pi}^{\pi} \tilde{P}_A(\theta_{in}, f)\, d\theta_{in} \quad X_\Sigma = \int_{-\pi}^{\pi} \tilde{P}_A(\theta_{in}, f) \cos\theta_{in}\, d\theta_{in}, \\
Y_\Sigma &= \int_{-\pi}^{\pi} \tilde{P}_A(\theta_{in}, f) \sin\theta_{in}\, d\theta_{in},
\end{aligned}
\tag{9.1.7}
$$

where the subscript "Σ" denotes the outputs caused by the superposition of plane waves.

The outputs from microphones are decoded using Equation (9.1.6), the unnormalized signal amplitude of the loudspeaker at θ' is expressed as

$$
\begin{aligned}
E_\Sigma(\theta', f) &= \frac{1}{2\pi} \left\{ \int_{-\pi}^{\pi} \tilde{P}_A(\theta_{in}, f)\, d\theta_{in} + 2\sum_{q=1}^{Q} \left[\cos q\theta' \int_{-\pi}^{\pi} \tilde{P}_A(\theta_{in}, f) \cos q\theta_{in}\, d\theta_{in} \right.\right. \\
&\quad \left.\left. + \sin q\theta' \int_{-\pi}^{\pi} \tilde{P}_A(\theta_{in}, f) \sin q\theta_{in}\, d\theta_{in} \right] \right\} \\
&= \tilde{P}_{A,0}^{(1)}(f) + \sum_{q=1}^{Q} \left[\tilde{P}_{A,q}^{(1)}(f) \cos q\theta' + \tilde{P}_{A,q}^{(2)}(f) \sin q\theta' \right],
\end{aligned}
\tag{9.1.8}
$$

where θ_{in} is substituted by θ' in the second equality of Equation (9.1.8). Equation (9.1.8) represents a $Q \geq 1$ order truncation of the azimuthal Fourier expansion of the azimuthal distribution function $\tilde{P}_A(\theta_{in}, f)$ of the unnormalized amplitude of the incident plane wave. The coefficients of the azimuthal Fourier expansion are given as

$$\tilde{P}_{A,0}^{(1)}(f) = \frac{1}{2\pi}\int_{-\pi}^{\pi} \tilde{P}_A(\theta_{in}, f)\, d\theta_{in} \quad q = 1, 2 \ldots Q$$
$$\tilde{P}_{A,q}^{(1)}(f) = \frac{1}{\pi}\int_{-\pi}^{\pi} \tilde{P}_A(\theta_{in}, f)\cos q\theta_{in} d\theta_{in} \quad \tilde{P}_{A,q}^{(2)}(f) = \frac{1}{\pi}\int_{-\pi}^{\pi} \tilde{P}_A(\theta_{in}, f)\sin q\theta_{in} d\theta_{in}. \tag{9.1.9}$$

In Equation (9.1.8), for an arbitrary target sound field, the Q-order Ambisonic-independent or encoding signals can be theoretically recorded by (2Q + 1) coincident microphones with different order directional characteristics. The preceding Q-order azimuthal harmonic components of the target sound field can be recovered from these microphone outputs after decoding. If the target or incident sound field is spatially bandlimited, i.e., all the $q > Q$-order azimuthal harmonics in the azimuthal Fourier expansion of the azimuthal distribution function $\tilde{P}_A(\theta_{in}, f)$ of the incident plane wave amplitude vanishes, the target sound field can be recovered exactly by using the Q-order Ambisonic signals from decoding outputs. The corresponding equations in the time domain can be obtained by applying an inverse Fourier transform similar to that in Equation (1.2.13) and the above equations in the frequency domain.

The Q-order ***circular sinc function*** is defined as

$$c\sin(\theta', \theta_{in}, Q) = \left[1 + 2\sum_{q=1}^{Q}(\cos q\theta' \cos q\theta_{in} + \sin q\theta' \sin q\theta_{in}]\right]$$
$$= \frac{\sin\left[\left(Q + \frac{1}{2}\right)(\theta_{in} - \theta')\right]}{\sin\left(\frac{\theta_{in} - \theta'}{2}\right)}. \tag{9.1.10}$$

The magnitude of the circular sinc function maximizes at $\theta' = \theta_{in}$ and spreads to two sides around the center of $\theta' = \theta_{in}$. Except a constant gain, the polar patterns of $Q = 1$, 2, and 3 order circular sinc function are identical to those in Figure 4.17. The first equality on the right side of Equation (9.1.8) can be written as

$$E_\Sigma(\theta', f) = \frac{1}{2\pi}\int_{-\pi}^{\pi} \tilde{P}_A(\theta_{in}, f)\, c\sin(\theta', \theta_{in}, Q)\, d\theta_{in}. \tag{9.1.11}$$

Equation (9.1.11) indicates that Q-order-reproduced signals are obtained by multiplying a Q-order circular sinc function (Q-order azimuthal sampling function) to $\tilde{P}_A(\theta_{in}, f)$ of the incident plane wave amplitude and then superposing (taking an integral) over all the azimuths. When Q tends to infinite, the circular sinc function reaches the Dirac delta function

$$\lim_{Q\to\infty} c\sin(\theta', \theta_{in}, Q) = \delta(\theta' - \theta_{in}). \tag{9.1.12}$$

In this case, Equation (9.1.11) approaches the limitation of an ideal sampling

$$E_{\Sigma}(\theta', f) = \tilde{P}_A(\theta', f) = \int_{-\pi}^{\pi} \tilde{P}_A(\theta_{in}, f)\delta(\theta' - \theta_{in})d\theta_{in}. \tag{9.1.13}$$

In the aforementioned discussion, a uniform and continuous configuration with infinite numbers of loudspeakers around a listener is supposed in reproduction. However, a finite number of loudspeakers are used in practical reproduction. For simplicity, the case of a uniform configuration with M loudspeakers arranged on a circle is discussed here. Let θ_i, $i = 0, 1\ldots(M - 1)$ denote the azimuth of the ith loudspeaker. For a target plane wave with unit amplitude and incident from azimuthal θ_S, M loudspeaker signals are equivalent to M azimuthal samples of $A(\theta', \theta_S)$ in Equation (9.1.6) for continuous loudspeaker configuration. The maximal allowing the azimuthal interval of loudspeakers or equally the minimal number of loudspeakers required for the Q-order reproduction cannot be evaluated with Equation (9.1.6), but this parameter should be derived from the analysis of the reproduced sound field in Section 9.3. However, the overall signal gain for reproduction with M loudspeakers can be derived by analyzing the sound pressure at the origin (center of the circle). For a continuous and uniform configuration with an infinite number of loudspeakers, the normalized amplitudes of the loudspeaker signals for arbitrary Q-order Ambisonic reproduction are expressed in Equation (9.1.6); the reproduced sound pressure in the frequency domain and at the origin is given as

$$P'_A = \int_{-\pi}^{\pi} A(\theta', \theta_S)d\theta'. \tag{9.1.14}$$

For the finite Q-order and infinite-order (ideal) reproduction, the result of integral is shown as

$$P'_A = P_A = 1. \tag{9.1.15}$$

In this case, the amplitude of the reproduced sound pressure at the origin is exactly equal to that of the target sound field.

For the Q-order Ambisonic reproduction with M loudspeakers, the normalized amplitudes of loudspeaker signals are the samples of $A(\theta', \theta_S)$ in Equation (9.1.6) at $\theta' = \theta_i$ of M loudspeaker directions. Accordingly, the integral over the continuous azimuth θ' is replaced by the summation over discrete azimuths θ_i. In the case of uniform configuration with M loudspeakers, the interval between adjacent loudspeakers is $2\pi / M \approx d\theta'$, then Equation (9.1.14) becomes

$$P'_A = \sum_{i=0}^{M-1} A(\theta_i, \theta_S)\frac{2\pi}{M} = \left\{\sum_{i=0}^{M-1} \frac{1}{M}\left[1 + 2\sum_{q=1}^{Q}\left(\cos q\theta_i \cos q\theta_S + \sin q\theta_i \sin q\theta_S\right)\right]\right\}. \tag{9.1.16}$$

The normalized amplitude of the actual signal of the ith loudspeaker is $A_i(\theta_S)$. For a target plane wave with a unit amplitude, the reproduced sound pressure at the origin should satisfy the following equation:

$$P'_A = \sum_{i=0}^{M-1} A_i(\theta_S) = 1. \tag{9.1.17}$$

Comparing Equation (9.1.16) and (9.1.17) yields

$$\begin{aligned} A_i(\theta_S) &= \frac{1}{M}\left[1 + 2\sum_{q=1}^{Q}(\cos q\theta_i \cos q\theta_S + \sin q\theta_i \sin q\theta_S)\right] \\ &= \frac{1}{M}\left[1 + 2\sum_{q=1}^{Q}\cos q(\theta_S - \theta_i)\right] \\ &= \frac{1}{M}\frac{\sin\left[\left(Q + \frac{1}{2}\right)(\theta_S - \theta_i)\right]}{\sin\left(\frac{\theta_S - \theta_i}{2}\right)}. \end{aligned} \tag{9.1.18}$$

Equation (9.1.18) is the amplitude of the reproduced signals for the Q-order Ambisonics with constant-amplitude normalization given by Equations (4.3.63) and (4.3.80a). Therefore, for reproduction involving a finite number of horizontal loudspeakers with uniform configuration, the Q-order approximation of ideal reproduction leads to the Ambisonic decoding equation, the amplitude of the reproduced signals, and the normalized factor of the overall amplitude. The normalized signal amplitude of the ith loudspeaker in discrete configuration cannot be obtained directly by letting $\theta' = \theta_i$ in $A(\theta', \theta_S)$ for continuous configuration expressed in Equation (9.1.16). An overall gain and normalized factor should be supplemented.

In conclusion, the following statements have been mathematically proven:

1. An ideal horizontal reproduction requires an infinite number of loudspeakers arranged continuously and uniformly on a circle with a far-field radius.
2. The loudspeaker signals for an ideal reproduction can be expanded into an azimuthal Fourier series. The Q-order approximation of the azimuthal Fourier expansion is equivalent to the Q-order horizontal Ambisonic reproduced signals, which are the linear combination of $(2Q + 1)$ independent or encoded signals. The decoding equation and reproduced signals of the conventional solution of Ambisonics is a natural consequence of each order approximation of ideal reproduced signals.
3. As the order Q of Ambisonics increases, the approximated reproduction gradually approaches ideal reproduction, but higher-order reproduction requires more independent signals and becomes complicated.
4. The independent signals of the Q-order Ambisonics, which involve the preceding Q-order azimuthal harmonic components of the target sound field, can be theoretically recorded by $(2Q + 1)$ coincident microphones with appropriate directivities. Decoding can be regarded as beamforming processing.

9.1.2 Each order approximation of ideal three-dimensional reproduction

The analysis in Section 9.1.1 can be extended to the case of reproduction in a three-dimensional space. According to the statement in Section 1.1, for convenience in mathematical expression, two new angles with respect to the head center are introduced. They are related to the angles (θ, ϕ) in Figure 1.1:

$$\alpha = 90^\circ - \phi \quad \beta = \theta. \tag{9.1.19}$$

For simplicity, three-dimensional direction is denoted by $\Omega = (\alpha, \beta)$ or $\Omega = (\theta, \phi)$.

Similar to the case in Section 9.1.1, ideal three-dimensional reproduction is designed by arranging an infinite number of loudspeakers on a spherical surface uniformly and continuously around a listener or receiver region. If radius r_0 of the sphere is large enough, the incident wave in a region close to the center of the sphere can be approximated as the superposition of plane waves from each loudspeaker. The original or target sound field is a plane wave with a unit amplitude and incident from the direction Ω_S. Let Ω_{in} and Ω' denote the direction in the original sound field and the direction of loudspeakers in reproduction, respectively. For ideal reproduction, the directional distribution function $A(\Omega', \Omega_S)$ of the normalized amplitude of the loudspeaker signals should match that of the target sound field, i.e., the Dirac delta function. If $\Omega' = \Omega_{in}$, the normalized amplitude of the signal for loudspeakers at the arbitrary direction Ω' is given as

$$A(\Omega', \Omega_S) = \tilde{P}_A(\Omega_{in}) = \delta(\Omega' - \Omega_S). \tag{9.1.20}$$

In driving Equation (9.1.20), it is assumed that the transfer coefficient from the loudspeaker signal to the pressure amplitude of the free-field plane wave at the origin is a unit.

According to Appendix A, $A(\Omega', \Omega_S)$ can be decomposed by the real- or complex-valued spherical harmonic functions of Ω_S or Ω'. Decomposition by real- or complex-valued spherical harmonic functions is mathematically equivalent and can be converted mutually. The real-valued spherical harmonic functions are denoted by $\left\{Y_{lm}^{(\sigma)}(\Omega), l = 0, 1, \ldots, m = 0, 1 \ldots l, \sigma = 1, 2\right\}$, and the complex-valued spherical harmonic functions are denoted by $\{Y_{lm}(\Omega), l = 0, 1, \ldots, m = 0, \pm 1 \ldots \pm l\}$. Equation (9.1.20) is decomposed as

$$A(\Omega', \Omega_S) = \sum_{l=0}^{\infty}\sum_{m=0}^{l}\sum_{\sigma=1}^{2} Y_{lm}^{(\sigma)}(\Omega') Y_{lm}^{(\sigma)}(\Omega_S) = \sum_{l=0}^{\infty}\sum_{m=-l}^{l} Y_{lm}^{*}(\Omega') Y_{lm}(\Omega_S), \tag{9.1.21}$$

where the notation "*" denotes complex conjugation. $Y_{l0}^{(2)}(\Omega) = 0$, which is preserved in Equation (9.1.21) for convenience in writing. The case of decomposition by the real-valued spherical harmonic function is discussed in the following.

As the zero-order approximation, the decomposition in Equation (9.1.21) is truncated up to the term of $l = 0$. When the expression of $Y_{00}^{(1)}(\Omega)$ given in Appendix A is used, the normalized signal amplitude of the loudspeaker at Ω' is expressed as

$$A(\Omega', \Omega_S) = \frac{1}{4\pi}. \tag{9.1.22}$$

Equation (9.1.22) is equivalent to the feed of the mono signal captured by an omnidirectional microphone to all loudspeakers in reproduction. Therefore, zero-order reproduction cannot recreate the spatial information of the target sound field.

As the first-order approximation, the decomposition in Equation (9.1.21) is truncated up to the term of $l = 1$. When the expression of the real-valued spherical harmonic functions in Equation (A.5) in Appendix A is applied, the normalized signal amplitude of the loudspeaker at $\Omega' = (\alpha', \beta')$ is given as

$$\begin{aligned} &A(\Omega', \Omega_S) \\ &= \frac{1}{4\pi}\left[1 + 3\sin\alpha'\cos\beta'\sin\alpha_S\cos\beta_S + 3\sin\alpha'\sin\beta'\sin\alpha_S\sin\beta_S + 3\cos\alpha'\cos\alpha_S\right] \\ &= \frac{1}{4\pi}\left[1 + 3\cos(\Delta\Omega')\right], \end{aligned} \tag{9.1.23}$$

where $\Delta\Omega'$ is the angle between Ω' and Ω_S. In the case of hexahedral loudspeaker configuration, except for the difference in overall gain, Equation (9.1.23) is consistent with the normalized amplitude of loudspeaker signals for the first-order spatial Ambisonic reproduction presented in Equation (6.4.18). In other words, the first-order spatial Ambisonics is equivalent to the first-order approximation of ideal spatial reproduction.

After Equation (9.1.19) is converted into the default coordinate (θ, ϕ) of this book, Equation (9.1.23) becomes

$$A(\Omega', \Omega_S) = \frac{1}{4\pi}\left[W + 3\cos\theta'\cos\phi' X + 3\sin\theta'\cos\phi' Y + 3\sin\phi' Z\right], \tag{9.1.24}$$

where W, X, Y, and Z are the four independent (normalized) signals for the first-order spatial Ambisonics given in Equation (6.4.1), which can be recorded by four coincident microphones in the original sound field, i.e., by an omnidirectional microphone and three bidirectional microphones with their main axes pointing to the front, left, and top directions, respectively.

The higher-order approximation of the decomposition in Equation (9.1.21) leads to the loudspeaker signals (amplitudes) for higher-order spatial Ambisonics. Generally, for arbitrary $(L - 1) \geq 1$ order approximation, the normalized signal amplitude of loudspeakers at Ω' is expressed as

$$A(\Omega', \Omega_S) = \sum_{l=0}^{L-1}\sum_{m=0}^{l}\sum_{\sigma=1}^{2} Y_{lm}^{(\sigma)}(\Omega') Y_{lm}^{(\sigma)}(\Omega_S), \tag{9.1.25}$$

where $A(\Omega', \Omega_S)$ is a linear combination of L^2 independent signals or spherical harmonic components $\left\{Y_{lm}^{(\sigma)}(\Omega_S), l = 0, 1, \ldots (L-1), m = 0, 1 \ldots l, \sigma = 1, 2\right\}$. In other words, the $(L - 1)$-order spatial Ambisonics involves L^2 independent or encoded signals. These independent signals can be theoretically recorded by L^2 coincident microphones whose directivities match with those of spherical harmonic functions with various orders. This phenomenon is equivalent to the sound field recording by a multipole method. The directional patterns of $l = 0, 1, 2$ order spherical harmonic functions are illustrated in Appendix A. Loudspeaker signals in reproduction are decoded from the independent signals by using Equation (9.1.25). According to Equation (A.10) in Appendix A, the spherical harmonic functions $Y_{lm}^{(\sigma)}(\Omega_S)$ used here are normalized so that the integral of their square norm over direction is a unit. This normalization is for mathematical convenience, but it is different from the normalization of signals W, X, Y, and Z in Equation (6.4.1). In Equation (6.4.1), the maximal magnitudes of W, X, Y, and Z are normalized to a unit to avoid signal overload in practice. Independent signals or spherical harmonic functions can be normalized by various methods. Independent signals with different normalizations are mathematically equivalent except those for gain factors. The methods of normalization should be noticed to avoid confusion (Charpentier, 2017).

Similar to the case of horizontal recording, Equation (9.1.25) can be regarded as a recording method on the basis of directional beamforming. According to the summation formula of spherical harmonic functions given in Equation (A.17) in Appendix A, Equation (9.1.25) can be written as

$$A(\Omega', \Omega_S) = \frac{1}{4\pi} \sum_{l=0}^{L-1} \left\{ (2l+1) P_l \left[\cos(\Delta\Omega') \right] \right\}, \tag{9.1.26}$$

where $P_l[\cos(\Delta\Omega')]$ is the l-order Legendre polynomials, $\Delta\Omega'$ is the angle between Ω' and Ω_S. As the order $L - 1$ increases, the directivity or directional pattern of beamforming becomes sharp; accordingly, Equation (9.1.26) approaches the loudspeaker signals for ideal reproduction. However, the number of independent or encoded signals also increases with the order and system become complicated. When the order $(L - 1)$ tends to infinity, Equation (9.1.25) or (9.1.26) reaches the case of recording with ideal beamforming given in Equation (9.1.20). Equations (9.1.25) and (9.1.26) indicate that the directional beam can be steered to arbitrary direction without altering the beam shape by changing Ω'. This feature is common in spatial Ambisonics.

Similar to the case in a horizontal sound field, an arbitrary spatial sound field in a source-free region can be decomposed as a linear superposition of plane waves from various directions. In this case, the directional distribution function $\tilde{P}_A(\Omega_{in}, f)$ of incident plane wave amplitudes with respect to the origin is no longer a Dirac delta function. Spatial Ambisonic recording can also be regarded as a directional sampling on $\tilde{P}_A(\Omega_{in}, f)$ of incident plane wave amplitudes. The outputs of various order directional microphones are the spherical harmonic components of $\tilde{P}_A(\Omega_{in}, f)$. When the order $(L - 1)$ tends to infinite, loudspeaker signals in reproduction approach the limitation of the Dirac delta function.

The discrete configuration with a finite number of loudspeakers is used in practical reproduction. The loudspeaker configurations of spatial Ambisonics are more complex than those of horizontal Ambisonics. In many practical cases, the decoding equations and reproduced signals of higher-order spatial Ambisonics cannot be derived directly from Equation (9.1.25). Instead, they should be derived through sound field analysis in Sections 9.2 and 9.3.

9.2 GENERAL FORMULATION OF MULTICHANNEL SOUND FIELD RECONSTRUCTION

9.2.1 General formulation of multichannel sound field reconstruction in the spatial domain

As stated in Sections 1.9.1 and 3.1, ideal spatial sound reproduction can be achieved through the physical reconstruction of the target sound field within a region. Sound field reconstruction is an important aspect of spatial sound. It is closely related not only to conventional multichannel sound techniques but also to some methods in other fields, such as active noise control and acoustical holography (Fazi and Nelson, 2010, 2013). Therefore, analysis of the reconstructed sound field is important to the evaluation of the accuracy or extent of approximation in practical multichannel sound reproduction and development of new reproduction techniques. Moreover, the mathematical expression and analysis methods of sound field reconstruction vary in different literature, although they are actually equivalent (Ahrens and Spors, 2008b; Poletti, 2005b). For unification, a ***general formulation*** of ***multichannel sound field reconstruction*** is addressed in this section and Sections 9.2.2 and 9.2.3. The formulation of sound field reconstruction in a ***spatial domain***, or more strictly in the ***frequency and spatial domain***, is first discussed here because a sound field can be represented by sound pressure as a function of receiver position and frequency.

Some terms are first introduced. In the analysis of sound field reconstruction, including the analyses in this chapter and those on acoustical holography and wave field synthesis in Chapter 10, the terms "***secondary sources***" and "***driving signals***" are used to represent the ideal sound sources of reproduction and the signals of these ideal sound sources, respectively. Accordingly, the term "***secondary source array***" is used to represent the configuration of secondary sources. In practice, the ideal sound sources are approximately realized by loudspeakers. Therefore, the term "loudspeaker" is usually referred to a practical secondary source, the corresponding configuration is called "***loudspeaker configuration***" and the corresponding driving signals are termed "***loudspeaker signals.***" In the following discussion, these terms are used flexibly and alternately.

Suppose that secondary sources are arranged continuously and uniformly in space; the frequency domain deriving signals for secondary source at position $\boldsymbol{r}'$ is $E(\boldsymbol{r}', f)$; the frequency domain transfer function from the secondary source to arbitrary receiver position $\boldsymbol{r}$ is $G(\boldsymbol{r}, \boldsymbol{r}', f)$. The sound pressure at the receiver point is a superposition of those caused by all secondary sources:

$$P'(\boldsymbol{r}, f) = \int G(\boldsymbol{r}, \boldsymbol{r}', f) E(\boldsymbol{r}', f) d\boldsymbol{r}'. \tag{9.2.1}$$

The integral in Equation (9.2.1) is calculated over the whole region of the secondary source distribution.

Under the free-field condition, the transfer function $G(\boldsymbol{r}, \boldsymbol{r}', f)$ in Equation (9.2.1) depends on the physical characteristic or radiation pattern of secondary sources. If secondary sources are point sources with unit strength in the free field, $G(\boldsymbol{r}, \boldsymbol{r}', f)$ is the frequency domain sound pressure at the receiver position $\boldsymbol{r}$ caused by the source at $\boldsymbol{r}'$ and termed ***free-field Green's function in a three-dimensional space (and frequency domain)***. Letting $Q_p(f) = 1$ and substituting $\boldsymbol{r}_S$ with $\boldsymbol{r}'$ in Equation (1.2.3) yield

$$\begin{aligned} G^{3D}_{free}(\boldsymbol{r}, \boldsymbol{r}', f) &= G^{3D}_{free}(|\boldsymbol{r} - \boldsymbol{r}'|, f) \\ &= \frac{1}{4\pi |\boldsymbol{r} - \boldsymbol{r}'|} \exp\left[-j\boldsymbol{k} \cdot (\boldsymbol{r} - \boldsymbol{r}')\right] \\ &= \frac{1}{4\pi |\boldsymbol{r} - \boldsymbol{r}'|} \exp(-jk|\boldsymbol{r} - \boldsymbol{r}'|), \end{aligned} \tag{9.2.2}$$

where the superscript "$3D$" denotes the case of a point source in a three-dimensional space, and the subscript "*free*" denotes the free field. In the following discussion, these superscript and subscript are preserved or omitted depending on a situation. Equation (9.2.2) indicates that $G^{3D}_{free}(\boldsymbol{r}, \boldsymbol{r}', f)$ only depends on the relative distance $|\boldsymbol{r} - \boldsymbol{r}'|$ between the sound source and the receiver position. $G_{free}^{3D}(\boldsymbol{r}, \boldsymbol{r}', f)$ is also invariant after an exchanging of $\boldsymbol{r}$ and $\boldsymbol{r}'$. This invariance is the consequence of the acoustic principle of reciprocity.

In the local region of a far field where the distance between the receiver position and the source is large enough, the spherical wave caused by a point source can be approximated as a plane wave. In this case, the secondary source can be modeled by a plane wave source in the free field. If the strength of a point source is chosen to be $Q_p = 4\pi r'$ or if $P_A(f) = 1$ in Equation (1.2.3), the transfer function from a free-field plane wave source to a region close to the origin is given as

$$G_{free}^{pl}(\boldsymbol{r}, \boldsymbol{r}', f) = \exp\left[-j\boldsymbol{k}\cdot(\boldsymbol{r} - \boldsymbol{r}')\right]. \tag{9.2.3}$$

where the superscript "*pl*" denotes the case of a plane wave source. By choosing an appropriate initial phase of the point source in Equation (1.2.3) to cancel the factor $\exp(j\boldsymbol{kr}')$ in Equation (9.2.3), or directly letting $P_A(f) = 1$ in Equation (1.2.6), Equation (9.2.3) becomes

$$G_{free}^{pl}(\boldsymbol{r}, \boldsymbol{r}', f) = \exp(-j\boldsymbol{k}\cdot\boldsymbol{r}). \tag{9.2.4}$$

where the wave vector $\boldsymbol{k}$ in Equation (9.2.4) depends on the direction of a secondary source.

Similarly, if secondary sources are straight-line sources with unit strength and an infinite length arranged perpendicular to the horizontal plane (two-dimensional space) in the free field, $G(\boldsymbol{r}, \boldsymbol{r}', f)$ is the frequency domain sound pressure at the horizontal receiver position $\boldsymbol{r}$ caused by straight-line sources intersecting the horizontal plane at $\boldsymbol{r}'$. For convenience, $\boldsymbol{r}'$ is called the "***horizontal position of the straight-line source.***" $G(\boldsymbol{r}, \boldsymbol{r}', f)$ is termed the ***free-field Green's function in a two-dimensional space or horizontal plane (and frequency domain)***. Letting $Q_{li}(f) = 1$ in Equation (1.2.7) yields

$$G_{free}^{2D}(\boldsymbol{r}, \boldsymbol{r}', f) = -\frac{j}{4}H_0\left(k\left|\boldsymbol{r} - \boldsymbol{r}'\right|\right). \tag{9.2.5}$$

where the superscript "2*D*" denotes the case of a straight-line source in a two-dimensional space (horizontal plane).

Similar to the case of a point source in a three-dimensional space, in the local region of a far field where the distance between the horizontal receiver position and the straight-line source is large enough, the cylindrical wave caused by a straight-line source can be approximated by a plane wave according to the discussion in Equations (1.2.7) to (1.2.10). When an appropriate initial phase and the frequency-dependent strength of the straight-line source are chosen, the horizontal transfer function from a straight-line source to a region close to the origin is also expressed as the plane wave approximation of Equation (9.2.4).

In addition to three ideal and relatively simple secondary sources, e.g., point source, plane wave source, and straight-line source, other practical secondary sources (such as loudspeakers) exhibit more complicated radiation patterns (such as frequency-dependent directivity). In these cases, reconstructed sound pressures can be calculated using Equation (9.2.1), but $G(\boldsymbol{r}, \boldsymbol{r}', f)$ from the secondary source to the receiver position is more complicated and does not even have an analytical expression.

Equation (9.2.1) can be extended to the case of reproduction in reflective environments, such as in a listening room with reflections. In this case, the free-field transfer function in Equation (9.2.1) should be replaced with the transfer function in reflective environments, e.g., the frequency domain counterparts of room impulse responses in Section 1.2.2.

Practical sound field reconstruction or sound reproduction is implemented through a discrete array with a finite number of secondary sources. If M secondary sources are arranged at the position $\boldsymbol{r}_i$ ($i = 0, 1...M - 1$), the integral over the continuous region of secondary source distribution in Equation (9.2.1) is replaced with the summation of all the discrete secondary sources:

$$P'(\boldsymbol{r}, f) = \sum_{i=0}^{M-1} G(\boldsymbol{r}, \boldsymbol{r}_i, f) E_i(\boldsymbol{r}_i, f). \tag{9.2.6}$$

Equation (9.2.1) or (9.2.6) is the ***general formulation of multichannel sound field reconstruction in the spatial domain.*** The driving signals $E_i(\boldsymbol{r}_i, f)$ of the discrete secondary source array cannot be obtained directly by letting $\boldsymbol{r}' = \boldsymbol{r}_i$ in $E(\boldsymbol{r}', f)$ for continuous secondary source distribution. For a uniform array in a horizontal circle, the overall gain in the driving signals of continuous secondary source distribution and discrete array differs.

Given the characteristics, the array and driving signals of secondary sources, Equation (9.2.1) or (9.2.6) can be used to analyze the reconstructed sound field. Conversely, given the target sound field, Equations (9.2.1) and (9.2.6) can be used to search for the type, array, and driving signals of secondary sources and design the reproduction system. This discussed content is the main point of reconstructed sound field analysis in the following sections.

9.2.2 Formulation of spatial-spectral domain analysis of circular secondary source array

Equations (9.2.1) and (9.2.6) are formulated in a spatial domain or more strictly in frequency and spatial domains, where the reconstructed sound pressure is a function of frequency f and receiver position $\boldsymbol{r}$. The reconstructed sound field can also be analyzed in a ***spatial-spectral domain*** or more strictly a ***frequency and spatial-spectral domain.*** For some regular secondary source array and appropriate coordinate systems, transforming Equations (9.2.1) and (9.2.6) to the spatial-spectral domain for analysis is convenient and may lead to significant results. This phenomenon is similar to the transformation of the signals in the time domain to the frequency domain for analysis in signal processing. However, the appropriate spatial spectrum representation of the reconstructed sound field depends on the secondary source array and chosen coordinate system.

A usual array involves arranging secondary sources uniformly on a horizontal circle, which is often used for horizontal Ambisonic reproduction. Spatial-spectral domain analysis of circular secondary source array is discussed in this section (Bamford and Vanderkooy, 1995; Daniel, 2000; Ward and Abhayapala, 2001; Poletti, 1996, 2000). In this case, a polar coordinate system is convenient for analysis.

The secondary sources are arranged uniformly and continuously on a horizontal circle at $r' = r_0$. The position of secondary sources is denoted by the polar coordinate (r_0, θ'). An arbitrary receiver position inside the circle is denoted by (r, θ). The line element on the circle is $r_0 d\theta'$. The line integral in Equation (9.2.1) is calculated over the circle $r' = r_0$. If r_0 is merged into the driving signals $E(\boldsymbol{r}', f)$ as an overall gain, Equation (9.2.1) can be expressed as a one-dimensional convolution over the azimuth θ:

$$P'(r, \theta, r_0, f) = \int_{-\pi}^{\pi} G(r, \theta - \theta', r_0, f) E(\theta', f) d\theta'. \tag{9.2.7}$$

The arguments indicate the dependence of each function on physical variables.

In Equation (9.2.7), the reconstructed sound pressure is a periodic function of θ with a period of 2π and therefore can be expanded into a real- or complex-valued Fourier series of θ. Real- and complex-valued Fourier expansions are mathematically equivalent. The real-valued expansion is consistent with the preceding discussion of horizontal Ambisonics. The complex-valued expansion is convenient for mathematical expression and analysis. Here, both forms of Fourier expansions are given as

$$\begin{aligned} P'(r,\theta,r_0,f) &= \sum_{q=0}^{\infty}\left[P_q'^{(1)}(r,r_0,f)\cos q\theta + P_q'^{(2)}(r,r_0,f)\sin q\theta\right] \\ &= \sum_{q=-\infty}^{+\infty} P_q'(r,r_0,f)\exp(jq\theta). \end{aligned} \tag{9.2.8}$$

The real-valued Fourier coefficients of expansion are calculated with the following:

$$P_0'^{(1)}(r,r_0,f) = \frac{1}{2\pi}\int_{-\pi}^{\pi} P'(r,\theta,r_0,f)\,d\theta,$$

$$P_q'^{(1)}(r,r_0,f) = \frac{1}{\pi}\int_{-\pi}^{\pi} P'(r,\theta,r_0,f)\cos q\theta\, d\theta \quad P_q'^{(2)}(r,r_0,f) = \frac{1}{\pi}\int_{-\pi}^{\pi} P'(r,\theta,r_0,f)\sin q\theta\, d\theta. \tag{9.2.9}$$

$$q = 1, 2, 3\ldots$$

where $P_0'^{(2)}(r,r_0,f) = 0$, which is preserved in Equation (9.2.8) for convenience in writing. The complex-valued Fourier coefficients of expansion are calculated as follows:

$$P_q'(r,r_0,f) = \frac{1}{2\pi}\int_{-\pi}^{\pi} P'(r,\theta,r_0,f)\exp(-jq\theta)\,d\theta \quad q = 0, \pm 1, \pm 2,\ldots. \tag{9.2.10}$$

The relationship between real- and complex-valued Fourier coefficients is expressed in the following equations:

$$\begin{aligned} P_0'(r,r_0,f) &= P_0'^{(1)}(r,r_0,f), \\ P_q'(r,r_0,f) &= \frac{1}{2}\left[P_q'^{(1)}(r,r_0,f) - jP_q'^{(2)}(r,r_0,f)\right] \quad q > 0, \\ P_q'(r,r_0,f) &= \frac{1}{2}\left[P_{-q}'^{(1)}(r,r_0,f) + jP_{-q}'^{(2)}(r,r_0,f)\right] \quad q < 0. \end{aligned} \tag{9.2.11}$$

Equation (9.2.8) shows that reconstructed sound field can be represented by the azimuthal Fourier coefficients of sound pressure expressed in Equation (9.2.9) or (9.2.10). These coefficients are the ***azimuthal spectrum*** of sound pressure, which is a special and appropriate form of the spatial spectrum representation of the horizontal sound field created through the circular array of secondary sources.

Similarly, transfer functions from a secondary source to receiver positions can be expanded into real- or complex-valued azimuthal Fourier series as

$$G(r,\theta-\theta',r_0,f)=\sum_{q=0}^{\infty}\left\{G_q^{(1)}(r,r_0,f)\cos\left[q(\theta-\theta')\right]+G_q^{(2)}(r,r_0,f)\sin\left[q(\theta-\theta')\right]\right\}$$
$$=\sum_{q=-\infty}^{\infty}G_q(r,r_0,f)\exp\left[jq(\theta-\theta')\right]. \quad (9.2.12)$$

The real-valued Fourier coefficients of expansion are calculated as

$$G_0^{(1)}(r,r_0,f)=\frac{1}{2\pi}\int_{-\pi}^{\pi}G(r,\theta,r_0,f)d\theta,$$
$$G_q^{(1)}(r,r_0,f)=\frac{1}{\pi}\int_{-\pi}^{\pi}G(r,\theta,r_0,f)\cos q\theta d\theta \quad G_q^{(2)}(r,r_0,f)=\frac{1}{\pi}\int_{-\pi}^{\pi}G(r,\theta,r_0,f)\sin q\theta d\theta \quad (9.2.13)$$
$$q=1,2,3\ldots$$

The complex-valued Fourier coefficients of expansion are related to those of real-valued expansion by an equation similar to Equation (9.2.11). For a secondary source with its main axis pointing to the origin and symmetric against the main axis, it has $G_q^{(2)}(r,r_0,f)=0$. The real- or complex-valued Fourier coefficients of expansion in Equation (9.2.13) represent the spatial or azimuthal spectrum of transfer function from secondary sources to receiver positions.

Given the type, array, and orientations of the main axis of the secondary sources, the azimuthal spectrum of the transfer functions from secondary sources to receiver positions is calculated using Equation (9.2.13). For example, for a plane wave source, the transfer function is expressed as Equation (9.2.4). The azimuthal spectrum is calculated by substituting Equation (9.2.4) into Equation (9.2.13) or directly by expanding a plane wave into Bessel–Fourier series,

$$G_{free}^{pl}(r,\theta-\theta',f)=\exp\left[jkr\cos(\theta-\theta')\right]$$
$$=J_0(kr)+2\sum_{q=1}^{\infty}j^qJ_q(kr)(\cos q\theta\cos q\theta'+\sin q\theta\sin q\theta')$$
$$=J_0(kr)+2\sum_{q=1}^{\infty}j^qJ_q(kr)\cos\left[q(\theta-\theta')\right] \quad (9.2.14)$$
$$=\sum_{q=-\infty}^{+\infty}j^qJ_q(kr)\exp\left[jq(\theta-\theta')\right],$$

where $J_q(kr), q=0,1,2\ldots$ are the q-order Bessel functions with $J_{-q}(kr)=(-1)^qJ_q(kr)$. Comparing Equation (9.2.12) with Equation (9.2.14) yields

$$G_0^{(1)}(r,f)=J_0(kr) \quad G_q^{(1)}(r,f)=2j^qJ_q(kr) \quad G_q^{(2)}(r,f)=0 \quad q=1,2,3\ldots$$
$$G_q(r,f)=j^qJ_q(kr) \quad q=0,\pm1,\pm2\ldots \quad (9.2.15)$$

For simplicity, the subscript "*free*" for a free field and superscript "*pl*" for a plane wave source are omitted in Equation (9.2.15).

Similarly, for a straight-line secondary source perpendicular to a horizontal plane, the transfer function from a source to a receiver position is expressed in Equation (9.2.5), e.g., by the free-field Green's function in a horizontal plane. The distance between a source and a receiver position is

$$|\boldsymbol{r}-\boldsymbol{r}'| = \sqrt{r^2 + r_0^2 - 2rr_0\cos(\theta-\theta')}. \tag{9.2.16}$$

Equation (9.2.5) can be written as

$$\begin{aligned} G_{free}^{2D}(\boldsymbol{r}, \boldsymbol{r}', f) &= G_{free}^{2D}(r, \theta-\theta', r_0, f) \\ &= -\frac{j}{4}H_0\left[k\sqrt{r^2 + r_0^2 - 2rr_0\cos(\theta-\theta')}\right]. \end{aligned} \tag{9.2.17}$$

The Hankel function of the second kind can be expanded as

$$\begin{aligned} H_0(k|\boldsymbol{r}-\boldsymbol{r}'|) &= J_0(kr)H_0(kr_0) + 2\sum_{q=1}^{+\infty} J_q(kr)H_q(kr_0)\cos\left[q(\theta-\theta')\right] \\ &= \sum_{q=-\infty}^{+\infty} J_q(kr)H_q(kr_0)\exp\left[jq(\theta-\theta')\right] \qquad r < r_0. \end{aligned} \tag{9.2.18}$$

Substituting Equation (9.2.18) into Equation (9.2.17) and comparing with Equation (9.2.12) yield

$$\begin{aligned} &G_0^{(1)}(r, r_0, f) = -\frac{j}{4}J_0(kr)H_0(kr_0), \\ &G_q^{(1)}(r, r_0, f) = -\frac{j}{2}J_q(kr)H_q(kr_0) \quad G_q^{(2)}(r, r_0, f) = 0 \quad q = 1, 2, 3.... \end{aligned} \tag{9.2.19}$$

and

$$G_q(r, r_0, f) = -\frac{j}{4}J_q(kr)H_q(kr_0) \quad q = 0, \pm 1, \pm 2\ldots . \tag{9.2.20}$$

The equation for driving signals can be obtained from the spatial-spectral or azimuthal-spectral domain representation of the transfer function from a secondary source to a receiver position. Substituting Equation (9.2.8) and the Fourier expansion in Equation (9.2.12) into Equation (9.2.7) and using the following equations of trigonometric functions,

$$\begin{aligned} \cos\left[q(\theta-\theta')\right] &= \cos q\theta\cos q\theta' + \sin q\theta\sin q\theta', \\ \sin\left[q(\theta-\theta')\right] &= \sin q\theta\cos q\theta' - \cos q\theta\sin q\theta'. \end{aligned} \tag{9.2.21}$$

the following equation is obtained:

$$\begin{aligned}&\sum_{q=0}^{\infty} P_q'^{(1)}(r, r_0, f)\cos q\theta + \sum_{q=1}^{\infty} P_q'^{(2)}(r, r_0, f)\sin q\theta \\ &= \sum_{q=0}^{\infty}\left\{\int_{-\pi}^{\pi}\left[G_q^{(1)}(r, r_0, f)\cos q\theta' - G_q^{(2)}(r, r_0, f)\sin q\theta'\right]E(\theta', f)d\theta'\right\}\cos q\theta \\ &+\sum_{q=1}^{\infty}\left\{\int_{-\pi}^{\pi}[G_q^{(1)}(r, r_0, f)\sin q\theta' + G_q^{(2)}(r, r_0, f)\cos q\theta']E(\theta', f)d\theta'\right\}\sin q\theta.\end{aligned} \tag{9.2.22}$$

The left side of Equation (9.2.22) characterizes the azimuthal variation in the reconstructed sound pressure, and each term of $\cos q\theta$ or $\sin q\theta$ represents a mode of azimuthal variation.

Given the target sound field, the reconstructed sound pressure in Equation (9.2.8) should match with the target sound pressure $P(r, \theta, f)$. Accordingly, the azimuthal spectrum representation of the constructed sound pressure on the left side of Equation (9.2.22) should be substituted with those of the target sound pressure, e.g., substituted with $P_q^{(1)}(r, f)$ and $P_q^{(2)}(r, f)$. The coefficients for each azimuthal mode on the two sides of Equation (9.2.22) should be equal because each azimuthal mode is independent. This equality leads to a set of equations for driving signals $E(\theta', f)$:

$$\begin{aligned}&\int_{-\pi}^{\pi} G_0^{(1)}(r, r_0, f)E(\theta', f)d\theta' = P_0^{(1)}(r, f), \\ &\int_{-\pi}^{\pi}\left[G_q^{(1)}(r, r_0, f)\cos q\theta' - G_q^{(2)}(r, r_0, f)\sin q\theta'\right]E(\theta', f)d\theta' = P_q^{(1)}(r, f), \\ &\int_{-\pi}^{\pi}\left[G_q^{(1)}(r, r_0, f)\sin q\theta' + G_q^{(2)}(r, r_0, f)\cos q\theta'\right]E(\theta', f)d\theta' = P_q^{(2)}(r, f).\end{aligned} \tag{9.2.23}$$

A finite number of secondary sources are used in practical reproduction. M secondary sources are arranged in a horizontal circle with radius r_0, then the azimuth of the ith secondary source is θ_i ($i = 0, 1\ldots M - 1$). The integral over the azimuth in Equation (9.2.23) is replaced with the summation of discrete secondary source azimuths:

$$\begin{aligned}&\sum_{i=0}^{M-1} G_0^{(1)}(r, r_0, f)E_i(\theta_i, f) = P_0^{(1)}(r, f), \\ &\sum_{i=0}^{M-1}\left[G_q^{(1)}(r, r_0, f)\cos q\theta_i - G_q^{(2)}(r, r_0, f)\sin q\theta_i\right]E_i(\theta_i, f) = P_q^{(1)}(r, f), \\ &\sum_{i=0}^{M-1}\left[G_q^{(1)}(r, r_0, f)\sin q\theta_i + G_q^{(2)}(r, r_0, f)\cos q\theta_i\right]E_i(\theta_i, f) = P_q^{(2)}(r, f).\end{aligned} \tag{9.2.24}$$

The physical significance of Equations (9.2.23) and (9.2.24) is that a matching of the horizontal reconstructed and target sound field requires the matching of their corresponding azimuthal Fourier or harmonic components or vice versa. The method of solving driving signals via Equation (9.2.23) or (9.2.24) is a ***mode-matching method.*** Equation (9.2.23) or (9.2.24) is valid when the two sides of the equality in these equations do not vanish. In addition, $E_i(\theta_i, f)$ of discrete secondary sources cannot be directly obtained by letting $\theta' = \theta_i$ in $E(\theta', f)$ for continuous secondary sources in Equation (9.2.7). An overall gain or normalized factor should be supplemented.

Through azimuthal spectrum representation, Equation (9.2.7) can be converted to a form different from Equation (9.2.23). The driving signal $E(\theta', f)$ is a periodic function of azimuth θ' with a period 2π, so it can be expanded as a real- or complex-valued Fourier series:

$$\begin{aligned} E(\theta', f) &= \sum_{q=0}^{\infty}\left[E_q^{(1)}(f)\cos q\theta' + E_q^{(2)}(f)\sin q\theta'\right] \\ &= \sum_{q=-\infty}^{+\infty} E_q(f)\exp(jq\theta'). \end{aligned} \tag{9.2.25}$$

The real- or complex-valued azimuthal Fourier coefficients can be calculated similarly to Equation (9.2.9) or (9.2.10). In the complex-valued Fourier expansion, when Equations (9.2.8), (9.2.12), and (9.2.25) are substituted into Equation (9.2.7), a convolution between two functions in the spatial domain becomes a multiplication between two corresponding functions in the azimuthal–spectral domain:

$$P'_q(r, r_0, f) = 2\pi G_q(r, r_0, f)E_q(f) \quad q = 0, \pm 1, \pm 2 \ldots; \tag{9.2.26}$$

This equation is the formulation of multichannel sound field reconstruction in the azimuthal-spectral domain.

Equation (9.2.26) can be expressed in the form of real-valued azimuthal Fourier coefficients, but it is relatively complicated. For a secondary source with its main axis pointing to the origin and symmetric against the main axis, it has $G_q^{(2)}(r, r_0, f) = 0$. In this case, Equation (9.2.26) is expressed by real-valued azimuthal Fourier coefficients as

$$\begin{aligned} P_0'^{(1)}(r, r_0, f) &= 2\pi G_0^{(1)}(r, r_0, f)E_0^{(1)}(f), \\ P_q'^{(\sigma)}(r, r_0, f) &= \pi G_q^{(1)}(r, r_0, f)E_q^{(\sigma)}(f) \quad q = 1, 2, 3 \ldots, \sigma = 1, 2. \end{aligned} \tag{9.2.27}$$

According to Equation (9.2.26), given the driving signals and transfer functions from secondary sources to receiver positions in the azimuthal-spectral domain, the reconstructed sound pressure in the azimuthal–spectral domain can be evaluated. Or, given the azimuthal spectrum representation $P_q^{(\sigma)}(r, f)$ or $P_q(r,f)$ of the target sound field, the driving signals of secondary sources can be found by substituting the azimuthal spectrum representation of reconstructed sound pressures in Equation (9.2.26) with $P_q^{(\sigma)}(r, f)$ or $P_q(r, f)$:

$$E_q(f) = \frac{P_q(r, f)}{2\pi G_q(r, r_0, f)} \quad q = 0, \pm 1, \pm 2 \ldots; \ldots \tag{9.2.28}$$

For a secondary source with its main axis pointing to the origin and symmetric against the main axis, Equation (9.2.27) also yields

$$E_0^{(1)}(f) = \frac{P_0^{(1)}(r,f)}{2\pi G_0^{(1)}(r,r_0,f)} \quad E_q^{(\sigma)}(f) = \frac{P_q^{(\sigma)}(r,f)}{\pi G_q^{(1)}(r,r_0,f)}$$
$$q = 0, 1, 2\ldots, \quad \sigma = 1,2. \tag{9.2.29}$$

In a spatial domain, driving signals can be found by substituting the coefficients in Equation (9.2.28) or (9.2.29) into the azimuthal Fourier expansion. In addition, Equation (9.2.28) or (9.2.29) is valid only when $G_q(r,r_0,f) \neq 0$ or $G_q^{(1)}(r,r_0,f) \neq 0$. The aforementioned method for deriving the driving signals is a ***spatial-spectral division method*** (***SDM***; Ahrens and Spors, 2008b, 2010). Under certain conditions, spatial SDM is equivalent to the mode-matching method, but the former is more general than the latter. It is applicable to the uniform secondary source arrays in a horizontal circle and a straight line (Section 10.3.3), as well as on a spatial spherical surface. For the arbitrary target sound field and characteristics of secondary sources, Equation (9.2.28) or (9.2.29) does not always lead to an appropriate solution for driving signals. An example of this phenomenon is discussed in Section 10.3.3.

9.2.3 Formulation of spatial-spectral domain analysis for a secondary source array on spherical surface

The analysis in Section 9.2.2 can be extended to the secondary source array on a spherical surface (Daniel 2000; Poletti, 2005b). In this case, using a spherical coordinate for analysis is convenient. Similar to the case in Section 9.1.2, the receiver position is specified by (r, Ω). Secondary sources are arranged uniformly and continuously on a spherical surface with radius $r' = r_0$, and the position of secondary sources is specified by (r_0, Ω'). The area element on the spherical surface is $r_0^2 d\Omega'$, and the integral for the position of all secondary sources in Equation (9.2.1) is calculated over the spherical surface $r' = r_0$. If r_0^2 is merged into the driving signals of $E(r',f)$ as an overall gain, for a secondary source with rotational-symmetric directivity around the main axis pointing to the origin, Equation (9.2.1) can be expressed as a two-dimensional angular convolution of Ω over the spherical surface (Rafaely, 2004; Ahrens and Spors, 2008b):

$$P'(r,\Omega,r_0,f) = \int G(r,\Delta\Omega,r_0,f)E(\Omega',f)d\Omega', \tag{9.2.30}$$

where $\Delta\Omega = (\Omega - \Omega')$ is the angle between the vectors of the receiver position and the secondary source.

Sound pressure is decomposed by real- or complex-valued spherical harmonic functions in accordance with the method described in Appendix A to convert Equation (9.2.30) into spatial spectrum representation:

$$\begin{aligned} P'(r,\Omega,r_0,f) &= P_{00}'^{(1)}(r,r_0,f) + \sum_{l=1}^{\infty}\sum_{m=0}^{l}\left[P_{lm}'^{(1)}(r,r_0,f)Y_{lm}^{(1)}(\Omega) + P_{lm}'^{(2)}(r,r_0,f)Y_{lm}^{(2)}(\Omega)\right] \\ &= \sum_{l=0}^{\infty}\sum_{m=0}^{l}\sum_{\sigma=1}^{2}P_{lm}'^{(\sigma)}(r,r_0,f)Y_{lm}^{(\sigma)}(\Omega) \\ &= \sum_{l=0}^{\infty}\sum_{m=-l}^{+l}P_{lm}'(r,r_0,f)Y_{lm}(\Omega), \end{aligned} \tag{9.2.31}$$

where $Y_{l0}^{(2)}(\Omega) = 0$, which is preserved in the secondary equality of Equation (9.2.31) for convenience in writing. The coefficients of spherical harmonic decomposition are the ***spherical harmonic spectrum*** of the reconstructed sound pressure, which is a special and appropriate form of the spatial spectrum representation of the sound field created by a spherical surface array of secondary sources. Spherical harmonic coefficients can be calculated using Equation (A.12) in Appendix A:

$$\begin{aligned} P'^{(\sigma)}_{lm}(r, r_0, f) &= \int P'(r, \Omega, r_0, f) Y^{(\sigma)}_{lm}(\Omega) d\Omega \quad l = 0, 1, 2\ldots, \quad m = 0, 1\ldots l, \sigma = 1, 2, \\ P'_{lm}(r, r_0, f) &= \int P'(r, \Omega, r_0, f) Y^{*}_{lm}(\Omega) d\Omega \quad l = 0, 1, 2\ldots, \quad m = 0, \pm 1, \ldots \pm l, \end{aligned} \tag{9.2.32}$$

where the superscript "*" denotes a complex conjugation. The relationship between real- and complex spherical harmonic coefficients is presented in Equation (A.13) in Appendix A.

The transfer function from a secondary source to receiver positions can also be decomposed by real- or complex-valued spherical harmonic functions:

$$G(r, \Delta\Omega, r_0, f) = \sum_{l=0}^{\infty}\sum_{m=0}^{l}\sum_{\sigma=1}^{2} G^{(\sigma)}_{lm}(r, r_0, f) Y^{(\sigma)}_{lm}(\Delta\Omega) = \sum_{l=0}^{\infty}\sum_{m=-l}^{l} G_{lm}(r, r_0, f) Y_{lm}(\Delta\Omega). \tag{9.2.33}$$

Spherical harmonic coefficients or spherical harmonic spectra are obtained using an equation similar to Equation (9.2.32).

For a secondary source with rotational–symmetric directivity around the main axis pointing to the origin, the spherical harmonic coefficients vanish except at $m = 0$, where the decomposition can be simplified as

$$G(r, \Delta\Omega, r_0, f) = \sum_{l=0}^{\infty} G^{(1)}_{l0}(r, r_0, f) Y^{(1)}_{l0}(\Delta\Omega) = \sum_{l=0}^{\infty} G_{l0}(r, r_0, f) Y_{l0}(\Delta\Omega). \tag{9.2.34}$$

Equations (A.1) and (A.2) for spherical harmonic functions in Appendix A yield

$$Y^{(1)}_{l0}(\Delta\Omega) = Y_{l0}(\Delta\Omega) = \sqrt{\frac{2l+1}{4\pi}} P_l\left[\cos(\Delta\Omega)\right], \tag{9.2.35}$$

where $P_l[\cos(\Delta\Omega)]$ is the l-order Legendre polynomials. Then, the spherical harmonic coefficients in Equation (9.2.34) are calculated as

$$G^{(1)}_{l0}(r, r_0, f) = G_{l0}(r, r_0, f) = \int G(r, \Omega, r_0, f) Y^{(1)}_{l0}(\Omega) d\Omega = \int G(r, \Omega, r_0, f) Y^{*}_{l0}(\Omega) d\Omega. \tag{9.2.36}$$

Through the summation formula of spherical harmonic functions given by Equation (A.17) in Appendix A, Equation (9.2.34) can be written as

$$\begin{aligned} G(r, \Delta\Omega, r_0, f) &= \sum_{l=0}^{\infty}\sum_{m=0}^{l}\sum_{\sigma=1}^{2} \sqrt{\frac{4\pi}{2l+1}} G^{(1)}_{l0}(r, r_0, f) Y^{(\sigma)}_{lm}(\Omega') Y^{(\sigma)}_{lm}(\Omega) \\ &= \sum_{l=0}^{\infty}\sum_{m=-l}^{l} \sqrt{\frac{4\pi}{2l+1}} G_{l0}(r, r_0, f) Y^{*}_{lm}(\Omega') Y_{lm}(\Omega). \end{aligned} \tag{9.2.37}$$

The spherical harmonic coefficients in Equation (9.2.37) can be calculated similar to those in Equation (9.2.32).

For a secondary plane wave source, the transfer function from the source to the receiver position is given in Equation (9.2.4). The spherical harmonic spectrum representation of the transfer function can be calculated using Equation (9.2.36) or directly determined with the following formula of the spherical harmonic decomposition of a plane wave:

$$\begin{aligned}\exp(-j\boldsymbol{k}\cdot\boldsymbol{r}) &= \exp\left[jkr\cos(\Delta\Omega)\right]\\ &= \sum_{l=0}^{\infty}(2l+1)j^{l}j_{l}(kr)P_{l}\left[\cos(\Delta\Omega)\right]\\ &= \sum_{l=0}^{\infty}\sqrt{4\pi(2l+1)}j^{l}j_{l}(kr)Y_{l0}^{(1)}(\Delta\Omega),\end{aligned} \tag{9.2.38}$$

where $j_l(kr)$ is the l-order spherical Bessel function. Through the comparison of Equations (9.2.38) and (9.2.34), the spherical harmonic spectrum representation of the transfer function is given as

$$G_{l0}^{(1)}(r,f) = G_{l0}(r,f) = \sqrt{4\pi(2l+1)}j^{l}j_{l}(kr) \quad l = 0,1,2.... \tag{9.2.39}$$

For the secondary point source described in Equation (9.2.2), when $|\boldsymbol{r}'| = r_0 > r$, the transfer function can be decomposed as

$$G(\boldsymbol{r},\boldsymbol{r}',f) = \frac{1}{4\pi\,|\boldsymbol{r}-\boldsymbol{r}'|}\exp\left[-j\boldsymbol{k}\cdot(\boldsymbol{r}-\boldsymbol{r}')\right] = -jk\sum_{l=0}^{\infty}\sqrt{\frac{2l+1}{4\pi}}j_{l}(kr)h_{l}(kr_0)Y_{l0}^{(1)}(\Delta\Omega), \tag{9.2.40}$$

where $h_l(kr_0)$ is the l-order spherical Hanker function of the second kind. Similar to the derivation of Equation (9.2.39), the spherical harmonic spectrum representation of the transfer function is expressed as

$$G_{l0}^{(1)}(r,r_0,f) = G_{l0}(r,r_0,f) = -jk\sqrt{\frac{2l+1}{4\pi}}j_{l}(kr)h_{l}(kr_0) \quad l = 0,1,2.... \tag{9.2.41}$$

A given target sound pressure $P(r, \Omega, f)$ can also be decomposed similar to Equation (9.2.31), yielding the spherical harmonic spectrum $P_{lm}^{(\sigma)}(r,f)$. In a mode-matching method, equations for driving signals can be derived by matching the constructed sound pressure in Equation (9.2.31) with the target sound pressure and substituting the results and Equation (9.2.37) into Equation (9.2.30):

$$\sqrt{\frac{4\pi}{2l+1}}G_{l0}^{(1)}(r,r_0,f)\int E(\Omega',f)Y_{lm}^{(\sigma)}(\Omega')d\Omega' = P_{lm}^{(\sigma)}(r,f). \tag{9.2.42}$$

A finite number of secondary sources are used in practice. M secondary sources are arranged on a spherical surface with radius r_0, and the direction of the ith secondary sources is Ω_i ($i = 0,1...M - 1$). The integral in Equation (9.2.42) is replaced with the following summation of the discrete directions of secondary sources:

$$\sqrt{\frac{4\pi}{2l+1}}G_{l0}^{(1)}(r, r_0, f)\sum_{i=0}^{M-1}E_i(\Omega_i, f)Y_{lm}^{(\sigma)}(\Omega_i) = P_{lm}^{(\sigma)}(r, f). \tag{9.2.43}$$

$E_i(\Omega_i, f)$ of discrete array cannot be obtained directly by letting $\Omega' = \Omega_i$ in $E(\Omega', f)$ of continuous array. For a uniform secondary source array, an overall gain factor should be supplemented.

Moreover, for continuous and uniform secondary source array on a spherical surface, the driving signals $E(\Omega', f)$ can also be decomposed by real- or complex-valued spherical harmonic functions:

$$E(\Omega', f) = \sum_{l=0}^{\infty}\sum_{m=0}^{l}\sum_{\sigma=1}^{2}E_{lm}^{(\sigma)}(f)Y_{lm}^{(\sigma)}(\Omega') = \sum_{l=0}^{\infty}\sum_{m=-l}^{l}E_{lm}(f)Y_{lm}(\Omega'). \tag{9.2.44}$$

The integral on the left side of Equation (9.2.42) is the real-valued spherical harmonic coefficient $E_{lm}^{(\sigma)}(f)$. The relationship of the spherical harmonic spectrum representation of the reconstructed sound pressure, the transfer functions from the secondary sources to the receiver positions, and the driving signals are obtained by replacing $P_{lm}^{(\sigma)}(r, f)$ on the right side of Equation (9.2.42) with that of the reconstructed sound pressure:

$$P_{lm}^{\prime(\sigma)}(r, r_0, f) = \sqrt{\frac{4\pi}{2l+1}}G_{l0}^{(1)}(r, r_0, f)E_{lm}^{(\sigma)}(f). \tag{9.2.45}$$

Equation (9.2.45) can also be expressed in terms of complex-valued spherical harmonic coefficients

$$P'_{lm}(r, r_0, f) = \sqrt{\frac{4\pi}{2l+1}}G_{l0}(r, r_0, f)E_{lm}(f). \tag{9.2.46}$$

Equations (9.2.45) and (9.2.46) are the formulations of multichannel sound field reconstruction in a spherical harmonic spectral domain.

According to Equation (9.2.46), the driving signals for a secondary source on a spherical surface can be determined with the spatial-spectral division method (Ahrens and Spors, 2008b):

$$E_{lm}^{(\sigma)}(f) = \sqrt{\frac{2l+1}{4\pi}}\frac{P_{lm}^{\prime(\sigma)}(r, r_0, f)}{G_{l0}^{(1)}(r, r_0, f)}. \tag{9.2.47}$$

$$E_{lm}(f) = \sqrt{\frac{2l+1}{4\pi}}\frac{P'_{lm}(r, r_0, f)}{G_{l0}(r, r_0, f)}. \tag{9.2.48}$$

9.3 SPATIAL-SPECTRAL DOMAIN ANALYSIS AND DRIVING SIGNALS OF AMBISONICS

Ambisonics is a typical example of the physical reconstruction of a sound field within a local region with each order approximation. In this section, the Ambisonic sound field is analyzed in the spatial-spectral domain, and the driving signals of Ambisonics under various conditions are derived on the basis of the mode-matching method discussed in Sections 9.2.2 and 9.2.3.

9.3.1 Reconstructed sound field of horizontal Ambisonics

The case of far-field horizontal Ambisonics is first analyzed (Bamford and Vanderkooy, 1995; Poletti, 1996, 2000). As illustrated on the left side of Figure 9.1, a target sound field is created by a point source at r_S and θ_S. For a far-field source distance, the sound field close to the origin can be approximated as an incident plane wave with a unit amplitude. A circular region around the origin with radius $r \ll r_S$ is considered. According to Equation (1.2.6), the frequency domain sound pressure at an arbitrary position with the polar coordinate (r, θ) on the circular region is given as

$$P(r, \theta, \theta_S, f) = \exp\left[jkr\cos(\theta - \theta_S)\right]. \tag{9.3.1}$$

Similar to Equation (9.2.14), Equation (9.3.1) can be expanded as a Bessel–Fourier series,

$$\begin{aligned} P(r, \theta, \theta_S, f) &= J_0(kr) + 2\sum_{q=1}^{\infty} j^q J_q(kr)(\cos q\theta_S \cos q\theta + \sin q\theta_S \sin q\theta) \\ &= \sum_{q=0}^{\infty} J_q(kr)\left(B_q^{(1)} \cos q\theta + B_q^{(2)} \sin q\theta\right), \end{aligned} \tag{9.3.2}$$

and

$$B_q^{(1)} = B_q^{(1)}(\theta_S) = \begin{cases} 1 & q = 0 \\ 2j^q \cos q\theta_S & q > 0 \end{cases} \quad B_q^{(2)} = B_q^{(2)}(\theta_S) = \begin{cases} 0 & q = 0 \\ 2j^q \sin q\theta_S & q > 0 \end{cases}, \tag{9.3.3}$$

where $B_0^{(2)} \sin q\theta_S = 0$, which is preserved in Equation (9.3.2) for convenience in writing. The azimuthal Bessel–Fourier coefficients presented in Equation (9.3.3) are directly proportional to the amplitudes of each order independent signal of Ambisonics.

The comparison of Equation (9.3.2) with Equation (9.2.8) yields the azimuthal spectrum representation of target sound pressure:

$$\begin{aligned} &P_0^{(1)}(r, f) = B_0^{(1)} J_0(kr) = J_0(kr) \quad P_q^{(1)}(r, f) = B_q^{(1)} J_q(kr) = 2j^q J_q(kr)\cos q\theta_S \\ &P_q^{(2)}(r, f) = B_q^{(2)} J_q(kr) = 2j^q J_q(kr)\sin q\theta_S \quad q = 1, 2, 3\ldots \end{aligned} \tag{9.3.4}$$

As illustrated on the right side of Figure 9.1, M secondary sources are arranged on a circle with radius r_0. r_0 is large enough so that in receiver positions close to the origin, each

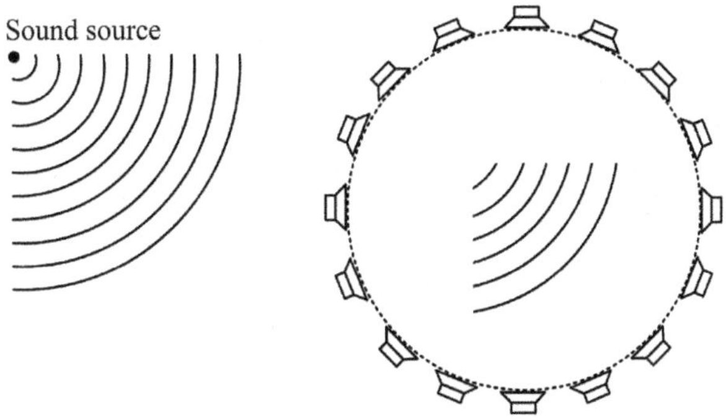

Figure 9.1 Target incident sound field and reconstructed by M secondary sources arranged on a horizontal circle.

secondary source can be approximated as a plane wave source. The azimuth of the ith secondary source is θ_i, and the driving signal is $E_i(\theta_i, \theta_S, f) = A_i(\theta_S)\, E_A(f), i = 0,1\ldots(M-1)$, where the normalized amplitude $A_i(\theta_S)$ depends on θ_S. For a target plane wave with a unit amplitude, the signal waveform is $E_A(f) = 1$. When Equations (9.3.4) and (9.2.15) are substituted into Equation (9.2.24), the normalized amplitude of driving signals satisfies the following equations:

$$\sum_{i=0}^{M-1} A_i(\theta_S) = 1 \quad \sum_{i=0}^{M-1} A_i(\theta_S)\cos q\theta_i = \cos q\theta_S \quad \sum_{i=0}^{M-1} A_i(\theta_S)\sin q\theta_i = \sin q\theta_S \qquad (9.3.5)$$
$$q = 1,2,3\ldots$$

Equation (9.3.5) is derived by a mode-matching method, i.e., by matching each order term of azimuthal harmonic representation or azimuthal Bessel–Fourier series of the reconstructed sound field with that of the target sound field. Equation (9.3.5) is valid only when $J_q(kr) \neq 0$. If kr is equal to the zeros of a Bessel function, Equation (9.3.5) may be invalid. In this case, the problem on the uniqueness of driving signals occurs.

Similar to Equation (9.2.8), Equation (9.3.2) presents the target sound pressure as a weighted combination of an infinite number of spatial or azimuthal harmonics (1, $\cos q\theta_S$, and $\sin q\theta_S$), and the weights of azimuthal harmonics constitute the discrete spatial spectrum or azimuthal spectrum $P_q^{(\sigma)}(r, f)$ of the target sound pressure. The normalized amplitudes of driving signals are also expanded into a weighted combination of azimuthal harmonics (1, $\cos q\theta_S$, and $\sin q\theta_S$) to reconstruct the target sound field:

$$A_i(\theta_S) = A_{\text{total}} \left\{ D_0^{(1)}(\theta_i) + \sum_{q=1}^{\infty} \left[D_q^{(1)}(\theta_i)\cos q\theta_S + D_q^{(2)}(\theta_i)\sin q\theta_S \right] \right\}. \qquad (9.3.6)$$

Equation (9.3.6) can be regarded as an azimuthal Fourier expansion of $A_i(\theta_S)$ within the region of $-\pi < \theta_S \leq \pi$ because $A_i(\theta_S)$ is a periodic function of θ_S with a period of 2π. In the extreme case of the uniform and continuous secondary source array on the circle, $D_0^{(1)}(\theta')$ is a constant; $D_q^{(1)}(\theta')$ and $D_q^{(2)}(\theta')$ are directly proportional to $\cos q\theta'$ and $\sin q\theta'$, respectively, where q = 1, 2.3..., and θ' is the azimuth of a secondary source in continuous array. Multiplying the azimuthal Fourier expansion of the normalized amplitudes with the signal wavefront $E_A(f)$, the resultant $E_A(f)\cos q\theta_S$ and $E_A(f)\sin q\theta_S$ are directly proportional to $E_q^{(1)}(f)$ and $E_q^{(2)}(f)$ in Equation (9.2.25), respectively.

For sound field reconstruction with a finite number of loudspeakers arranged on a circle, similar to the derivation from Equations (5.2.22) to (5.2.26), substituting Equation (9.3.6) into Equation (9.3.5) yields a matrix equation with respect to the decoding coefficients:

$$\mathbf{S}_{2D} = A_{\text{total}}\left[Y_{2D}\right]\left[D_{2D}\right]\mathbf{S}_{2D}, \tag{9.3.7}$$

where $S_{2D} = [1, \cos\theta_S, \sin\theta_S, \cos 2\theta_S, \sin 2\theta_S, \ldots]^T$ is a $\infty \times 1$ column matrix or vector composed of the normalized amplitude of the independent signals of horizontal Ambisonics, and $[D_{2D}]$ is an $M \times \infty$ decoding matrix:

$$\left[D_{2D}\right] = \begin{bmatrix} D_0^{(1)}(\theta_0) & D_1^{(1)}(\theta_0) & D_1^{(2)}(\theta_0) & D_2^{(1)}(\theta_0) & D_2^{(2)}(\theta_0) & \cdots \\ D_0^{(1)}(\theta_1) & D_1^{(1)}(\theta_1) & D_1^{(2)}(\theta_1) & D_2^{(1)}(\theta_1) & D_2^{(2)}(\theta_1) & \cdots \\ \vdots & \vdots & \vdots & \vdots & \vdots & \cdots \\ D_0^{(1)}(\theta_{M-1}) & D_1^{(1)}(\theta_{M-1}) & D_1^{(2)}(\theta_{M-1}) & D_2^{(1)}(\theta_{M-1}) & D_2^{(2)}(\theta_{M-1}) & \cdots \end{bmatrix}. \tag{9.3.8}$$

$[Y_{2D}]$ is a $\infty \times M$ matrix whose entries are various orders of cosine and sine functions of the azimuths of the secondary sources:

$$\left[Y_{2D}\right] = \begin{bmatrix} 1 & 1 & \ldots & 1 \\ \cos\theta_0 & \cos\theta_1 & \ldots & \cos\theta_{M-1} \\ \sin\theta_0 & \sin\theta_1 & \ldots & \sin\theta_{M-1} \\ \cos 2\theta_0 & \cos 2\theta_1 & \ldots & \cos 2\theta_{M-1} \\ \sin 2\theta_0 & \sin 2\theta_1 & \ldots & \sin 2\theta_{M-1} \\ \vdots & \vdots & \ldots & \vdots \end{bmatrix}. \tag{9.3.9}$$

For arbitrary finite $Q \geq 1$ order sound field reconstruction, the summation in Equation (9.3.6) is truncated up to the order Q. In this case, S_{2D} becomes a $(2Q + 1) \times 1$ column vector, $[D_{2D}]$ becomes a $M \times (2Q + 1)$ decoding matrix, and $[Y_{2D}]$ becomes a $(2Q + 1) \times M$ matrix. When $M > (2Q + 1)$ and matrix $[Y_{2D}][Y_{2D}]^T$is well-conditioned, the decoding matrix or coefficients in Equation (9.3.8) can be solved using the pseudoinverse method:

$$A_{\text{total}}\left[D_{2D}\right] = \text{pinv}\left[Y_{2D}\right] = \left[Y_{2D}\right]^T \left\{\left[Y_{2D}\right]\left[Y_{2D}\right]^T\right\}^{-1}. \tag{9.3.10}$$

Equation (9.3.10) is valid for both uniform (regular) and non-uniform (irregular) secondary source arrays and satisfies the condition of constant-amplitude normalization expressed in Equation (4.3.80a). Equation (9.3.10) is an extension of the decoding matrix in Equation (5.2.27) for the first-order horizontal Ambisonics to that for the arbitrary higher-order horizontal Ambisonics. However, the mode-matching method (azimuthal harmonics) rather than the traditional virtual source localization theorem is used here to derive the decoding matrix.

For uniform secondary source arrays, the azimuths of secondary sources are expressed in Equation (4.3.14). When $M = (2Q + 1)$, the matrix $[Y_{2D}]$ in Equation (9.3.9) is of full rank and thus invertible. In this case, the decoding matrix $A_{total}[D_{2D}]$ is directly calculated with the inverse matrix $[Y_{2D}]^{-1}$. Generally, for uniform array with $M \geq (2Q + 1)$ secondary sources, the decoding coefficients are expressed as

$$D_0^{(1)}(\theta_i) = 1 \qquad D_q^{(1)}(\theta_i) = 2\cos q\theta_i \qquad D_q^{(2)}(\theta_i) = 2\sin q\theta_i,$$
$$A_{total} = \frac{1}{M}. \tag{9.3.11}$$

For uniform secondary source array in a horizontal circle, the trigonometric function series in each row of the matrix in Equation (9.3.9) satisfies discrete orthogonality. The decoding coefficients in Equation (9.3.11) can also be derived from Equation (9.3.6) by using the discrete orthogonality of trigonometric function presented in Equations (4.3.16), (4.3.17), and (4.3.18).

The driving signals are obtained by substituting the decoding coefficients in Equation (9.3.11) into Equation (9.3.6)

$$A_i(\theta_S) = \frac{1}{M}\left[1 + 2\sum_{q=1}^{Q}\left(\cos q\theta_i \cos q\theta_S + \sin q\theta_i \sin q\theta_S\right)\right]. \tag{9.3.12}$$

Equation (9.3.12) is the conventional solution of the reproduced signals for the Q-order horizontal Ambisonics expressed in Equation (4.3.63) and with the amplitude normalization of Equation (4.3.80a).

Substituting Equation (9.3.12) back into Equation (9.3.5) and using the discrete orthogonality of trigonometric function in Equations (4.3.16), (4.3.17), and (4.3.18) can prove that Equation (9.3.5) is satisfactory for all $q \leq Q$ provided that the number of horizontal secondary sources with uniform configuration exceeds the following lower limit:

$$M \geq (2Q + 1). \tag{9.3.13}$$

Therefore, for the Q-order horizontal Ambisonic reproduction, the reconstructed sound field matches with that of the target sound field up to the Q-order Bessel–Fourier components; that is, the constructed sound field is accurate up to the Q-order azimuthal harmonic components. Equation (9.3.13) is the minimal number of secondary sources required for the Q-order horizontal Ambisonic reproduction, which has been discussed in Section 4.3.3. A higher number of secondary source than that given in Equation (9.1.13) are needed to satisfy the requirement of energy localization vector theorem, and the result is expressed in Equation (4.3.67).

Similar to the case in Section 9.1.1, an arbitrary incident sound field in a source-free region can be decomposed into a superposition of a plane wave incident from various directions. The azimuthal distribution function of the incident plane wave amplitude is $\tilde{P}_A(\theta_{in}, f)$. In this case, the incident sound field can also be expanded into the Bessel–Fourier series similar to the second equality in Equation (9.3.2), but the coefficients $B_q^{(1)}$ and $B_q^{(2)}$ of expansion are no longer expressed as Equation (9.3.3). The amplitudes of independent signals from the Q-order horizontal Ambisonic recording are proportional to $B_q^{(1)}$ or $B_q^{(2)}$. When these independent signals are decoded by Equation (9.3.10) and then reproduced, the reconstructed sound field matches with that of the target sound field for all $q \leq Q$-order azimuthal harmonic components. If the azimuthal spectrum of the target sound field is spatially or azimuthally bandlimited, then all $q > Q$-order azimuthal harmonic components in the expansion of

$\tilde{P}_A(\theta_{in}, f)$ vanish. In this case, if the number of secondary sources satisfies the requirement of Equation (9.3.13), the Q-order horizontal Ambisonics can construct the azimuthal harmonic components of the target sound field up to the order Q and within a circular region with the desired radius r.

Based on the discussion in this section and that in Section 9.1.1, the following conclusion can be made. For a target horizontal sound field with an azimuthal band limit of maximal order Q, $(2Q + 1)$ coincident microphones with various order directivities are required for recording independent signals. These independent signals are decoded and reproduced by $M \geq (2Q + 1)$ secondary sources to reconstruct the target sound field below a certain frequency limit and within a certain region. In other words, the azimuthal sampling rate of a sound field should be at least twice the bandwidth of the azimuthal harmonic spectrum of the target sound field to reconstruct the target sound field accurately. This phenomenon is the spatial or azimuthal sampling and recovery theorem of a sound field and is analogous to the Shannon–Nyquist theorem for time-domain signal sampling and recovery. The spatial sampling and recovery theorem reveals the relationship of the number of independent signals in sound field recording, the minimal required number of secondary sources (loudspeakers), and the spatial or azimuthal bandwidth of the sound field.

Strictly speaking, a practical plane wave is not spatially bandlimited. However, Equations (9.2.15) and (9.2.24) indicate that the q-order azimuthal harmonic components of an incident plane wave are proportional to the q-order Bessel function $J_q(kr)$ for a given kr. Figure 9.2 illustrates the curves of the $q = 0$ to 3 order Bessel functions $J_q(kr)$. Generally, $J_q(kr)$ oscillates and decays when its order q is not less than $[exp(1)kr/2]$. Therefore, for a given region with r and k, only the azimuthal harmonic components with orders $q \leq Q$ contribute significantly to the target or reconstructed sound field. The Bessel functions $J_q(kr)$ serves as low-pass spatial filters so that the sound field in the concerned region is approximately spatially bandlimited. In this case, the expansion in Equation (9.3.2) can be truncated up to a certain order, and the order Q of truncation is chosen as

$$Q = \text{integer}\left[\frac{\exp(1)kr}{2}\right], \tag{9.3.14}$$

where exp(1) = 2.7183 and function "integer" represent a roundup to an integer.

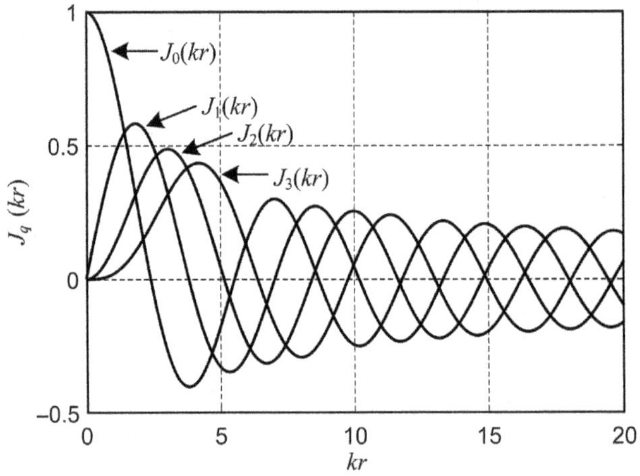

Figure 9.2 Curves of the $q = 0$ to 3 order Bessel functions $J_q(kr)$.

The truncation order Q can also be evaluated with a slightly different method. The minimal number of secondary sources for Q-order reproduction is $(2Q + 1)$, which corresponds to an azimuthal interval of $2\pi / (2Q + 1)$ between adjacent secondary sources for a uniform array. For $(2Q + 1)$ receiver positions located in a horizontal circle with r, the arc length between two adjacent receiver positions is $2\pi r / (2Q + 1)$. If the arc length between two adjacent receiver positions is required to be smaller than the half of a wavelength, the truncation order is evaluated as

$$Q = \text{integer}\left(kr + \frac{1}{2}\right) \approx \text{integer}(kr). \tag{9.3.15}$$

Equations (9.3.14) and (9.3.15) indicate that Q increases as kr increases; that is, the required order Q of the horizontal Ambisonics increases as r and frequency increase to reconstruct a target plane wave within a circular region with radius r around the origin. Conversely, as Q increases, the upper frequency limit and radius of a region for the accurate reconstruction of a sound field increases, thereby improving the performance of sound field reconstruction. This feature of Ambisonics is important. For example, as shown in Equation (9.3.14), within a circular region $r = a = 0.0875$ m (approximately the mean radius of a human head), the upper frequency limits of accurate sound field reconstruction are 0.46, 0.92, and 1.38 kHz at $Q = 1$, 2, and 3 order reproduction, respectively. Alternatively, the results evaluated from Equation (9.3.15) are 0.62, 1.25, and 1.87 kHz for $Q = 1$, 2, and 3 order reproduction, respectively. The results of Equations (9.3.14) and (9.3.15) differ. Equation (9.3.14) or (9.3.15) also indicates that $Q = 44$ or 32 order reproduction is needed to reconstruct the sound field accurately within a region with $r = a = 0.0875$ m and up to the frequency of 20 kHz. However, implementing this high-order reproduction is difficult.

For example, Figure 9.3 illustrates the simulated amplitude of the reconstructed sound field of $Q = 8$ order horizontal Ambisonics with $M = 18$ secondary sources. The secondary sources are arranged in a circle with $r_0 = 5.0$ m. The target sound field is a harmonic plane wave at $f = 1.0$ kHz and incident from $\theta_S = 30°$. Figure 9.3 only shows the sound field within a circular region of $r \leq 3.0$ m. Within a circular region of $r \leq 0.45$ m (bounded by a dashed line in Figure 9.3), the wavefront of the target plane wave is accurately reconstructed. However, outside this region, larger errors occur in the reconstructed sound field. This result

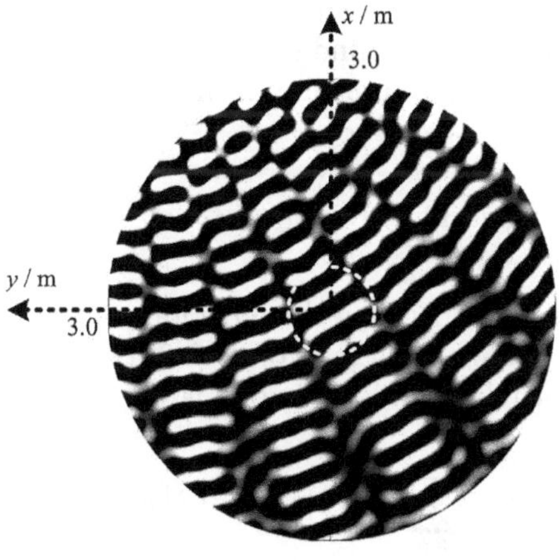

Figure 9.3 Simulated amplitude of the reconstructed sound field of $Q = 8$ order horizontal Ambisonics with $M = 18$ secondary sources and at $f = 1.0$ kHz.

is consistent with that estimated from Equation (9.3.15). These errors are attributed to the following: spatial aliasing in the reconstructed sound field and secondary sources arranged in a circle with a finite radius of $r_0 = 5.0$ m. The sound waves created by secondary sources can be approximated as plane waves only in a small region close to the origin.

The decoding Equation (9.3.12) is derived by matching each order of the azimuthal harmonic component of the reconstructed sound field with that of the target sound field, and the resultant decoding coefficients are independent of frequency. Although the problem of high-frequency error in the reconstructed sound field cannot be solved on the basis of a physical principle, the perceived performance of high-frequency reproduction can be improved by using some psychoacoustic methods. In particular, the perceived performance at a high frequency can be enhanced by replacing the constant coefficients κ_q in Equation (9.3.12) or Equation (4.3.53) with frequency-dependent coefficient $\kappa_q = \kappa_q(f)$, which is equivalent to applying frequency windows to different azimuthal harmonic components in the Bessel–Fourier expansion (Poletti, 2000). Various optimal psychoacoustic methods for the frequency windows are available. These optimal methods preserve the decoding coefficients derived from the matching of azimuthal harmonic components at low frequencies and revising the decoding coefficients (e.g., reducing the gain of decoding coefficients for higher-order azimuthal harmonic components) at high frequencies. The shelf filters and optimization of decoding for low and high frequencies are separately illustrated as examples in Sections 4.3.3 and 4.3.4. However, according to Equation (4.3.66), frequency windows may alter the overall power spectra in reproduction and cause timbre coloration.

In summary, the decoding equation and driving signals of arbitrary Q-order horizontal Ambisonics with M uniform or non-uniform secondary source configurations are derived from the azimuthal spectrum representation of a sound field. For a uniform secondary source array, the spatial sampling and recovery theorem of a sound field is validated, and the minimal number of secondary sources of Q-order horizontal Ambisonics is proven to be $(2Q + 1)$. The upper frequency and region of accurate sound field reconstruction increase as the order of Ambisonics increases.

9.3.2 Reconstructed sound field of spatial Ambisonics

The analysis in Section 9.3.1 can be extended to the case of spatial Ambisonics (Jot et al., 1999; Daniel, 2000; Ward and Abhayapala, 2001; Poletti, 2005b). The target or original sound field is created by a point source at Ω_S and r_S. It can be approximated as an incident plane wave with a unit amplitude within a region close to the origin. The sound field on a spherical surface with radius $r \ll r_S$ around the origin is considered. According to Equation (1.2.6), the frequency domain sound pressure at the arbitrary receiver position (r, Ω) on a spherical surface is given as

$$P(r, \Omega, \Omega_S, f) = \exp\left[jkr\cos(\Delta\Omega_S)\right], \tag{9.3.16}$$

where $\Delta\Omega_S = (\Omega_S - \Omega)$ is the angle between the directions of an incident plane wave and the vector of receiver position. Similar to Equation (9.2.38), Equation (9.3.16) can be decomposed by real-valued spherical harmonic functions as

$$\begin{aligned} P(r, \Omega, \Omega_S, f) &= \sum_{l=0}^{\infty}(2l+1)j^l j_l(kr)P_l\left[\cos(\Delta\Omega_S)\right] \\ &= 4\pi\sum_{l=0}^{\infty}\sum_{m=0}^{l}\sum_{\sigma=1}^{2} j^l j_l(kr)Y_{lm}^{(\sigma)}(\Omega_S)Y_{lm}^{(\sigma)}(\Omega) \end{aligned}$$

$$= \sum_{l=0}^{\infty} \sum_{m=0}^{l} \sum_{\sigma=1}^{2} B_{lm}^{(\sigma)} j_l(kr) Y_{lm}^{(\sigma)}(\Omega), \tag{9.3.17}$$

where $j_l(kr)$ and $P_l[\cos(\Delta\Omega_S)]$ are the l-order spherical Bessel function and l-order Legendre polynomials, respectively. The summation formula of a spherical harmonic function in Equation (A.17) in Appendix A is used to derive the second equality on the right side of Equation (9.3.17). Furthermore,

$$B_{lm}^{(\sigma)} = 4\pi j^l Y_{lm}^{(\sigma)}(\Omega_S), \tag{9.3.18}$$

where $B_{lm}^{(\sigma)}$ are a set of spherical harmonic coefficient of decomposition and proportional to various order independent signals of Ambisonics for a target plane wave source. In comparison with Equation (9.2.31), the spherical harmonic spectrum representation of the target sound pressure is evaluated using the following:

$$P_{lm}^{(\sigma)}(r,f) = B_{lm}^{(\sigma)} j_l(kr) = 4\pi j^l j_l(kr) Y_{lm}^{(\sigma)}(\Omega_S) \quad l = 0,1,2\ldots; \quad m = 0,1,2\ldots l, \quad \sigma = 1,2. \tag{9.3.19}$$

In reproduction, M secondary sources are arranged on a spherical surface with radius r_0. r_0 is large enough so that the incident wave from each secondary source can be approximated as a plane wave at a receiver region close to the origin. The direction of the ith secondary source is Ω_i. The driving signal is $E_i(\Omega_i, \Omega_S, f) = A_i(\Omega_S)\, E_A(f)$, $i = 0, 1\ldots(M-1)$, and the normalized amplitude $A_i(\Omega_S)$ depends on Ω_S. For a target plane wave with a unit amplitude, $E_A(f) = 1$, substituting Equation (9.2.39) and (9.3.19) into Equation (9.2.43) yields a set of equations for the normalized amplitudes of driving signals:

$$\sum_{i=0}^{M-1} A_i(\Omega_S) = 1 \quad \sum_{i=0}^{M-1} A_i(\Omega_S) Y_{lm}^{(\sigma)}(\Omega_i) = Y_{lm}^{(\sigma)}(\Omega_S) \tag{9.3.20}$$
$$l = 1,2,3\ldots \quad m = 0,1\ldots l. \quad \text{if } m \neq 0, \quad \sigma = 1,2; \quad \text{if } m = 0, \quad \sigma = 1.$$

Equation (9.3.20) is valid for $j_l(kr) \neq 0$ similar to the case of horizontal reproduction.

$A_i(\Omega_S)$ is also decomposed by real-valued spherical harmonic functions to solve the driving signals:

$$A_i(\Omega_S) = A_{\text{total}} \sum_{l=0}^{\infty} \sum_{m=0}^{l} \sum_{\sigma=1}^{2} D_{lm}^{(\sigma)}(\Omega_i) Y_{lm}^{(\sigma)}(\Omega_S). \tag{9.3.21}$$

In extreme cases of uniform and continuous secondary source array on a spherical surface, $D_{lm}^{(\sigma)}(\Omega')$ is proportional to $Y_{lm}^{(\sigma)}(\Omega')$, where Ω' is the continuous direction of secondary source distribution. After multiplication with the signal waveform $E_A(f)$ in the frequency domain, $E_A(f) Y_{lm}^{(\sigma)}(\Omega')$ is proportional to $E_{lm}^{(\sigma)}(\Omega, f)$ in Equation (9.2.44).

Substituting Equation (9.3.21) into Equation (9.3.20) yields a matrix equation for unknown decoding coefficients:

$$\mathrm{S}_{3D} = A_{\text{total}} [Y_{3D}][D_{3D}]\, \mathrm{S}_{3D},. \tag{9.3.22}$$

where $S_{3D} = \left[Y_{00}^{(1)}(\Omega_S), Y_{11}^{(1)}(\Omega_S), Y_{11}^{(2)}(\Omega_S), Y_{10}^{(1)}(\Omega_S)....\right]^T$ is a $\infty \times 1$ column matrix or vector composed of the normalized amplitude of independent signals of spatial Ambisonics, and $[D_{3D}]$ is an $M \times \infty$ decoding matrix whose entries are $D_{lm}^{(\sigma)}(\Omega_i)$. The rows of this matrix are arranged in the order of $\Omega_0, \Omega_1... \Omega_{M-1}$, and its columns are arranged in the order of (l, m, σ) = (0,0,1), (1,1,1), (1,1,2), (1,0,1)... etc. $[Y_{3D}]$ is an $\infty \times M$ matrix whose entries are $Y_{lm}^{(\sigma)}(\Omega_i)$. The rows of this matrix are arranged in the order of $Y_{00}^{(1)}$, $Y_{11}^{(1)}$, $Y_{11}^{(2)}, Y_{10}^{(1)}$.... etc., and its columns are arranged in the order of $\Omega_0, \Omega_1... \Omega_{M-1}$. Here, each entry in the signal S_{3D} is normalized so that the integral of their square amplitude in all directions is a unit. This normalization is different from that expressed in Equation (6.4.1), where the maximal magnitudes of W, X, Y, and Z are normalized to a unit. Accordingly, $[Y_{3D}]$ and $[D_{3D}]$ here are different from those in Equations (6.4.5) to (6.4.8). However, the results of different normalizations are equivalent.

For arbitrary $(L - 1) \geq 1$ order reproduction, the summation in Equation (9.3.21) is truncated up to the order $l = (L - 1)$, and S_{3D}, $[D_{3D}]$, and $[Y_{3D}]$ become an $L^2 \times 1$ column matrix, an $M \times L^2$ decoding matrix, and an $L^2 \times M$ matrix, respectively. When $M > L^2$ and when $[Y_{3D}][Y_{3D}]^T$ is well-conditions, the decoding coefficients or matrix in Equation (9.3.22) can be solved using the following pseudoinverse method:

$$A_{total}\left[D_{3D}\right] = pinv\left[Y_{3D}\right] = \left[Y_{3D}\right]^T \left\{\left[Y_{3D}\right]\left[Y_{3D}\right]^T\right\}^{-1}. \tag{9.3.23}$$

Equation (9.3.23) is an extension of Equation (6.4.8) for the first-order spatial Ambisonics to an arbitrary higher-order spatial Ambisonics, and the results satisfy the condition of constant-amplitude normalization.

Driving signals are obtained by substituting decoding coefficients into Equation (9.3.21). The decoding coefficients and normalized amplitude of driving signals depend on secondary source configuration. The problem of secondary source configuration on a spherical surface is generally complicated. However, for a special secondary source array or discrete directional sampling on a spherical surface, the spherical harmonic functions of secondary source directions satisfy the discrete orthogonality given in Equation (A.20) in Appendix A up to the order $(L - 1)$. For real-valued spherical harmonic functions, discrete orthogonality yields

$$\sum_{i=0}^{M-1} \lambda_i Y_{l'm'}^{(\sigma')}(\Omega_i) Y_{lm}^{(\sigma)}(\Omega_i) = \delta_{ll'}\delta_{mm'}\delta_{\sigma\sigma'} \quad l, l' \leq (L-1), \quad 0 \leq m, m' \leq l \quad \sigma = 1,2, \tag{9.3.24}$$

where $\delta_{ll'}$ is the Kronecker delta function, the weight λ_i depends on the configuration or directional sampling scheme of secondary sources. Equation (9.3.24) can be written in a matrix form:

$$[Y_{3D}][\Lambda][Y_{3D}]^T = [I], \tag{9.3.25}$$

where $[\Lambda]$ = diag $[\lambda_0, \lambda_1, ...\lambda_{M-1}]$ is an $M \times M$ diagonal matrix, and $[I]$ is an $L^2 \times L^2$ identify matrix. When the secondary source configuration satisfies the aforementioned discrete orthogonality of spherical harmonic functions, the exact solution of the decoding equation and the driving signals in Equation (9.3.20) for arbitrary (L – 1) order reproduction are obtained as

$$A_i(\Omega_S) = \lambda_i A(\Omega_i, \Omega_S) = \lambda_i \sum_{l=0}^{L-1}\sum_{m=0}^{l}\sum_{\sigma=1}^{2} Y_{lm}^{(\sigma)}(\Omega_i) Y_{lm}^{(\sigma)}(\Omega_S) = \frac{\lambda_i}{4\pi}\sum_{l=0}^{L-1}(2l+1)P_l\left[\cos\Delta\Omega_i'\right], \tag{9.3.26}$$

where $A(\Omega_i, \Omega_S)$ is directly obtained by letting $\Omega' = \Omega_i$ in the driving signals of Equation (9.1.25) for uniform and continuous secondary array; $\Delta\Omega'_i = \Omega_S - \Omega_i$ is the angle between the directions of the ith secondary source and the target source; and $P_l[\cos(\Delta\Omega'_i)]$ is the l-order Legendre polynomials. The summation formula of spherical harmonic functions in Equation (A.17) in Appendix A is used to obtain the second equality on the right side of Equation (9.3.26). From Equations (9.3.22) and (9.3.25), the pseudoinverse solution of the decoding matrix becomes

$$A_{\text{total}}\left[D_{3D}\right] = \left[\Lambda\right]\left[Y_{3D}\right]^T. \tag{9.3.27a}$$

For nearly uniform secondary source arrays, all weights are approximately equal to $\lambda_i = \lambda = 4\pi/M$, Equation (9.3.27a) becomes

$$A_{\text{total}}\left[D_{3D}\right] = \lambda\left[Y_{3D}\right]^T. \tag{9.3.27b}$$

Similar to the case of horizontal sound field, an arbitrary spatial incident sound field can be decomposed by spherical harmonic functions in the form of the third equality on the right side of Equation (9.3.17), but the coefficients $B_{lm}^{(\sigma)}$ of decomposition are no longer the form given in Equation (9.3.18). Moreover, for a non-harmonic incident sound field, $B_{lm}^{(\sigma)}$ depends on frequency. The amplitudes of independent signals from Ambisonic recording are proportional to $B_{lm}^{(\sigma)}$. These independent signals are decoded by Equation (9.3.21), where $B_{lm}^{(\sigma)}$ is used to replace $Y_{lm}^{(\sigma)}(\Omega_S)$ in Equation (9.3.21). Then, they are reproduced by secondary sources. If the array of M secondary sources satisfies the condition of discrete orthogonality, the spherical harmonic decomposition of reconstructed sound field matches with that of the target sound field up to the order (L – 1). Generally, if a target incident sound field is spatially bandlimited so that all the l > (L – 1)-order components in the spherical harmonic decomposition of the directional distribution function $\tilde{P}_A(\Omega_{in}, f)$ of the amplitude of the incident plane wave vanish, the preceding (L – 1) order spherical harmonic components of the reconstructed sound field match with those of the target sound field within the desired spherical region with radius r. The discrete orthogonality of spherical harmonic functions depends on M and configuration of secondary sources. The Shannon–Nyquist spatial sampling theorem requires the minimal number of secondary sources:

$$M \geq L^2. \tag{9.3.28}$$

For most practical arrays (Section 9.4.1), the required number of secondary sources exceeds the lower limit given in Equation (9.3.28) and may be much higher than the lower limit in some instances.

The spatial sampling and recovery theorem are extended to the case of spatial Ambisonic sound field in the above discussion. In summary, L^2 coincident microphones with different order directivities are needed to record the independent signals of (L – 1) order spatially bandlimited sound field. Independent signals are decoded and reproduced by M secondary sources arranged on a spherical surface with a far-field radius to reconstruct the target sound field. The minimal required number of secondary sources depends on the order and configuration with the lower limit given in Equation (9.3.28).

A practical target plane wave is not spatially bandlimited. Similar to the horizontal case in Equations (9.3.14) and (9.3.15), the l-order spherical harmonic component of an incident plane wave is proportional to the l-order spherical Bessel function $j_l(kr)$. For a given kr, the truncation order $(L - 1)$ in Equations (9.3.21) and (9.3.26) can be chosen according to the feature that $j_l(kr)$ oscillates and decays when its order l is not less than [exp(1)kr/2]:

$$(L-1) = \text{integer}\left[\frac{\exp(1)kr}{2}\right]. \tag{9.3.29}$$

Alternatively, $(L - 1)$ is chosen similar to that in Equation (9.3.15):

$$(L-1) = \text{integer}(kr). \tag{9.3.30}$$

As the order $(L - 1)$ increases, the upper frequency limit and radius of the spherical region for the accurate reconstruction of a sound field increase, which improves the performance of sound field reconstruction. The quantitative analysis of some examples also yields results similar to those of horizontal cases in Section 9.3.1. Appropriate frequency windows of decoding coefficients are also applicable to the improvement of the perceived performance at high frequencies.

9.3.3 Mixed-order Ambisonics

According to the Shannon–Nyquist spatial sampling theorem of sound field, the directional resolution of Ambisonic-reconstructed sound field increases as the order of Ambisonics increases. Higher-order Ambisonics is preferred to reconstruct the target sound field accurately. However, the number of independent signals and the minimal number M_{min} of secondary sources for Ambisonic reproduction also increase as the order increases. In particular, M_{min} for horizontal Ambisonics increases linearly with Q, i.e., $M_{min} = (2Q + 1)$, and M_{min} for spatial Ambisonics increases with the square of the order, i.e., $M_{min} = L^2$ for $(L - 1)$ order reproduction. Considering that the horizontal auditory resolution of humans is higher than the vertical resolution, ***mixed-order Ambisonics*** (**MOA**) reconstructs a horizontal sound field with higher-order spatial harmonics and reconstructs a vertical sound field with lower-order spatial harmonics; thus, a compromise between perceptual performance and system complexity is obtained (Daniel, 2000; Favrot et al., 2011; Márschall et al., 2012). The accuracy or resolution of a reconstructed sound field in horizontal and vertical directions can be controlled by adjusting the combined orders of horizontal and vertical harmonics. Therefore, MOA is based on both principles of sound field reconstruction and psychoacoustics.

The independent signals of MOA involve vertical harmonics up to the $(L - 1)$ orders and horizontal harmonics up to the Q orders with $Q > (L - 1)$. M secondary sources are divided into several sets. Each set of secondary sources is arranged in an elevation plane. The number of elevation planes should not be fewer than L to adapt $(L - 1)$ order vertical reproduction. Similarly, the number of horizontal secondary sources should not be fewer than $(2Q + 1)$ to adapt the Q-order horizontal reproduction. However, if no secondary sources are arranged in the horizontal plane, the number of secondary sources in each elevation adjacent to the horizontal plane should not be fewer than $(2Q + 1)$. If secondary sources are arranged on a spherical surface with a far-field radius, the normalized amplitudes of the driving signals for the ith secondary source at Ω_i is written as Equation (9.3.31) to recreate a virtual source at the far-field distance and Ω_S:

$$A_i(\Omega_S) = A_{\text{total}}\left[\sum_{l=0}^{L-1}\sum_{m=0}^{l}\sum_{\sigma=1}^{2} D_{lm}^{(\sigma)}(\Omega_i) Y_{lm}^{(\sigma)}(\Omega_S) + \sum_{l=L}^{Q}\sum_{\sigma=1}^{2} D_{ll}^{(\sigma)}(\Omega_i) Y_{ll}^{(\sigma)}(\Omega_S)\right]. \tag{9.3.31}$$

The notations in Equation (9.3.31) are similar to those in Equation (9.3.21). The first term in the square brackets on the right side of Equation (9.3.31) is equivalent to the driving signals of the (L – 1)-order spatial Ambisonics, which is obtained by truncating the summation in Equation (9.3.21) up to (L – 1) order. The second terms in the square brackets on the right side of Equation (9.3.31) are the additional driving signals for the implementation of horizontal Q-order reproduction, which involves a linear combination of $2[Q - (L - 1)] = 2(Q - L + 1)$ (normalized) independent signals. Each independent signal takes the form of $Y_{ll}^{(\sigma)}(\Omega_S)$. Figure A.1 in Appendix A indicates that the magnitude of $Y_{ll}^{(\sigma)}(\Omega_S)$ peaks in the horizontal plane and vanishes at the top and bottom directions. $Y_{ll}^{(\sigma)}(\Omega_S)$ is the horizontal projection or component of the l-order spatial Ambisonic signal. The combination of these horizontal components improves the horizontal resolution of the reconstructed sound field. Similar to the case in Section 9.3.2, the decoding coefficient $D_{\text{lm}}^{(\sigma)}(\Omega_i)$ in Equation (9.3.31) can be obtained by matching each spherical harmonic component in the reconstructed sound field with that of the target sound field.

As the target source deviates from the horizontal plane, the azimuthal pattern in the signal magnitude given in Equation (9.3.31) transits gradually from the pattern of Q-order horizontal Ambisonics to that of (L – 1) order spatial Ambisonics. Transition is determined by the characteristics of $Y_{ll}^{(\sigma)}(\Omega_S)$ in the second terms in the square brackets on the right side of Equation (9.3.31). From Equation (A.1) in Appendix A,

$$Y_{lm}^{(\sigma)}(\Omega_S) \propto P_l^m(\cos\alpha_S)\begin{cases}\cos m\beta_S & \sigma = 1\\ \sin m\beta_S & \sigma = 2\end{cases}. \tag{9.3.32}$$

The elevation dependence of spherical harmonic functions is determined by the associated Legendre polynomials $P_l^m(\cos\alpha_S)$. Higher-order associated Legendre polynomials represent the rapid variation with elevation. To obtain smooth transition characteristics, an elevation-bandlimited truncation up to the (L – 1) order can be applied to Q-order horizontal signals. That is, the function in the following equation is used to replace $Y_{ll}^{(\sigma)}(\Omega_S)$ on the right side of Equation (9.3.31):

$$Y_{ll}^{(\sigma)}(\Omega_S) \rightarrow P_{L-1}^m(\alpha_S)\begin{cases}\cos l\beta_S & \sigma = 1\\ \sin l\beta_S & \sigma = 2\end{cases}. \tag{9.3.33}$$

As an experimental validation, (L – 1) = 3 and Q = 5 MOA are implemented by the four-layer loudspeaker configuration with 28 + 1 secondary sources shown in Figure 9.6 (Mai et al., 2018). A virtual source localization experiment reveals that MOA improves horizontal localization and preserves the localization performance in other directions.

In addition, Grandjean et al. (2021a, 2021b) derived the drive signals of MOA with different method. The drive signals for horizontal loudspeakers (secondary point sources) are derived by a stationary phase approximation to the derive signals of secondary sources distribution on a cylindrical surface. The derivation is similar to that of 2.5-dimensional wave field synthesis in Section 10.2.3.

9.3.4 Near-field compensated higher-order Ambisonics

Far-field Ambisonics is addressed in preceding sections, where secondary source array is used to reconstruct the target plane wave with each order approximation. In addition to reconstruct plane wave incident from different directions, a sound field for different source distances should be reconstructed to create auditory distance perception. A psychoacoustic-based method to control the perceived distance by changing the direct-to-reflected energy ratio of the signals is discussed in Section 7.4.2. In Section 1.6.6, auditory distance perception is a consequence of the comprehensive processing of multiple cues. In the free-field case, acoustic diffraction and shadowing by the head provide information on evaluating distance for nearby sound sources. Accordingly, if the sound field of a near-field target source can be approximately reconstructed within a region and when a listener enters the reconstructed sound field, diffraction and shadowing by the head provide distance perception cues and then lead to virtual sources or auditory events in a proximal distance. Generally, a ***near-field compensated higher-order Ambisonics*** (*NFC–HOA*) reconstructs the sound field caused by a proximal target source within a region close to the head through the spatial harmonic decomposition of the sound field; at the same time, it compensates for the effect of the finite distance of secondary source array (Daniel, 2003; Daniel and Moreau, 2004).

The case of three-dimensional sound field reconstruction is first discussed. The position of a target source with respect to the origin is specified by $\boldsymbol{r}_S$ or the distance and direction (r_S, Ω_S), and Ω_S is identical to that in Section 9.1.2. The arbitrary receiver position is specified by $\boldsymbol{r}$ or the distance and direction (r, Ω). According to Equation (1.2.3), for a point source with the unit strength $Q_p = 1$, the sound pressure at the receiver position is given as

$$P(\boldsymbol{r}, \boldsymbol{r}_S, f) = \frac{1}{4\pi\,|\boldsymbol{r} - \boldsymbol{r}_S|}\exp(-jk|\boldsymbol{r} - \boldsymbol{r}_S|). \tag{9.3.34}$$

For $r_S > r$, Equation (9.3.34) can be decomposed by real- or complex-valued spherical harmonic functions similar to Equations (9.2.37) and (9.2.40):

$$\begin{aligned} P(\boldsymbol{r}, \boldsymbol{r}_S, f) &= -jk\sum_{l=0}^{\infty}\sum_{m=-l}^{l} j_l(kr)h_l(kr_S)Y_{lm}^{*}(\Omega_S)Y_{lm}(\Omega) \\ &= -jk\sum_{l=0}^{\infty}\sum_{m=0}^{l}\sum_{\sigma=1}^{2} j_l(kr)h_l(kr_S)Y_{lm}^{(\sigma)}(\Omega_S)Y_{lm}^{(\sigma)}(\Omega) \\ &= \sum_{l=0}^{\infty}\sum_{m=0}^{l}\sum_{\sigma=1}^{2} B_{lm}^{(\sigma)}(r_S, k)j_l(kr)Y_{lm}^{(\sigma)}(\Omega), \end{aligned} \tag{9.3.35}$$

where $j_l(kr)$ and $h_l(kr_S)$ are the l-order spherical Bessel function and l-order spherical Hankel function of the second kind, respectively. The coefficients $B_{lm}^{(\sigma)}(\mathrm{r}_S, k)$ depend on source distance and wavenumber:

$$B_{lm}^{(\sigma)}(r_S, k) = B_{lm}^{(\sigma)}(r_S, \Omega_S, f) = -jkh_l(kr_S)Y_{lm}^{(\sigma)}(\Omega_S). \tag{9.3.36}$$

The spherical harmonic coefficients or spherical harmonic spectrum representations of the target sound field are obtained by comparing Equations (9.3.36) and (9.2.31):

$$P_{lm}^{(\sigma)}(r_S, \Omega_S, f) = -jkj_l(kr)h_l(kr_S)Y_{lm}^{(\sigma)}(\Omega_S). \quad (9.3.37)$$

In NFC–HOA reproduction, M secondary sources are arranged on a spherical surface with radius $r' = r_0 > r_S$, the direction of the ith secondary source is Ω_i. Generally, secondary sources are not always localized in far-field distances, so they should be modeled as point sources. Driving signals, which are denoted by $E_i(r_S, \Omega_S, r_0, \Omega_i, f) = A_i(r_S, \Omega_S, r_0, \Omega_i, f)\, E_A(f)$, $i = 0,1\ldots(M - 1)$, depend on r_S and Ω_S, r_0 and Ω_i, and frequency f. The signal waveform in the frequency domain is assumed to be a unit with $E_A(f) = 1$. For a unit driving signal, a secondary source is equivalent to a point source with strength of $Q_p = 1$ in Equation (1.2.3). When Equations (9.3.37) and (9.2.41) are substituted into Equation (9.2.43) and each spherical harmonic component in the reconstructed sound field is matched with that in the target sound field, the normalized amplitude of driving signals satisfies the following equations:

$$\begin{gathered}
h_0(kr_0)\sum_{i=0}^{M-1} A_i(r_S, \Omega_S, r_0, \Omega_i, f) = h_0(kr_S), \\
h_l(kr_0)\sum_{i=0}^{M-1} A_i(r_S, \Omega_S, r_0, \Omega_i, f)Y_{lm}^{(\sigma)}(\Omega_i) = h_l(kr_S)Y_{lm}^{(\sigma)}(\Omega_S), \\
l = 1,2,3\ldots \qquad m = 0,1,\ldots l. \ \text{ if } m \neq 0, \ \ \sigma = 1,2; \ \text{ if } m = 0, \ \ \sigma = 1.
\end{gathered} \quad (9.3.38)$$

Equation (9.3.38) is valid when $j_q(kr) \neq 0$.

$A_i(r_S, \Omega_S, r_0, \Omega_i, f)$ is also decomposed by spherical harmonic functions to solve driving signals:

$$A_i(r_S, \Omega_S, r_0, \Omega_i, f) = A_{\text{total}}\sum_{l=0}^{\infty}\sum_{m=0}^{l}\sum_{\sigma=1}^{2} D_{lm}^{''(\sigma)}(r_S, r_0, \Omega_i, f)Y_{lm}^{(\sigma)}(\Omega_S), \quad (9.3.39)$$

where $D_{lm}^{''(\sigma)}(r_S, {}_0, \Omega_i, f)$ is a set of coefficients to be determined. Substituting Equation (9.3.39) into Equation (9.3.38) and truncating to $(L - 1)$ order yield a matrix equation for unknown coefficients:

$$[\Xi]S_{3D} = A_{\text{total}}[Y_{3D}][D''_{3D}]\ S_{3D}, \quad (9.3.40)$$

where $[Y_{3D}]$ and S_{3D} are identical to those in Equation (9.3.22). $[D''_{3D}]$ is similar to $[D_{3D}]$ in Equation (9.3.22), but the entries of the former depend on r_S, r_0, Ω_i, and f. $[\Xi]$ is an $L^2 \times L^2$ diagonal matrix for distance encoding, and its entries related to the l-order spherical harmonic component are given as

$$\Xi_l(kr_S, kr_0) = \frac{h_l(kr_S)}{h_l(kr_0)}. \quad (9.3.41)$$

Similar to Equation (9.3.22), for the $(L - 1)$-order reproduction, S_{3D} is an $L^2 \times 1$ column matrix or vector, $[D''_{3D}]$ is an $M \times L^2$ decoding matrix, and $[Y_{3D}]$ is an $L^2 \times M$ matrix. When $M > L^2$ and $[Y_{3D}][Y_{3D}]^T$is well-condition, the decoding matrix can be solved with the pseudoinverse method:

$$A_{\text{total}}\left[D''_{3D}\right] = \left\{\text{pinv}\left[Y_{3D}\right]\right\}\left[\Xi\right] = \left[Y_{3D}\right]^T \left\{\left[Y_{3D}\right]\left[Y_{3D}\right]^T\right\}^{-1}\left[\Xi\right]. \tag{9.3.42}$$

In comparison with the case of far-field Ambisonics in Equation (9.3.23), a wave number- or frequency-dependent matrix $[\Xi]$ is supplemented to the above equation. It is a filtering matrix for near-field distance encoding, and its entries are proportional to the ratio between the l-order spherical Hankel functions of the target and secondary source distances. Therefore, $[\Xi]$ changes the curve wavefronts of different spherical waves and converts the information of a secondary source distance to a target source distance. The distance-dependent factor can be merged into the NFC–HOA-encoded signals, i.e., the target direction-dependent but frequency-independent normalized signal amplitude vector S_{3D} in Equation (9.3.22) is filtered with $[\Xi]$:

$$\begin{aligned} S''_{3D} &= \left[\Xi\right]S_{3D} \\ &= [\Xi_0(kr_S, kr_0)Y_{00}^{(1)}(\Omega_S),\ \Xi_1(kr_S, kr_0)Y_{11}^{(1)}(\Omega_S), \\ &\quad \Xi_1(kr_S, kr_0)Y_{11}^{(2)}(\Omega_S),\ \Xi_1(kr_S, kr_0)Y_{10}^{(1)}(\Omega_S),\ldots]^T. \end{aligned} \tag{9.3.43}$$

The comparison of Equations (9.3.42) and (9.3.43) shows that $[D_{3D}]$ in Equations (9.3.21) and (9.3.22) is independent of frequency. Therefore, the decoding matrix of NFC–HOA is identical to that of far-field Ambisonics. The target source distance information is encoded in the filtered signal S''_{3D}, and the perceived virtual source distance is controlled by adjusting the distance parameter of $[\Xi]$ in encoding. S''_{3D} is a normalized complex-valued amplitude vector that involves magnitude and phase information.

When secondary sources are arranged at a far-field distance, the asymptotic formula of the spherical Hanker function of the secondary kind for a large argument yields

$$\lim_{\varsigma\to\infty} h_l(\varsigma) = \frac{j^{l+1}}{\varsigma}\exp(-j\varsigma). \tag{9.3.44}$$

Therefore, when $kr_0 \gg 1$,

$$\Xi_l(kr_S) = kr_0(-j)^{l+1}h_l(kr_S)\exp(jkr_0). \tag{9.3.45}$$

When $kr_S \ll 1$, the asymptotic formula of the spherical Hanker function of the secondary kind for small argument yields

$$\lim_{\varsigma\to 0} h_l(\varsigma) = \frac{j(2l-1)!!}{\varsigma^{l+1}}. \tag{9.3.46}$$

When r_S is a finite target source distance in the near field, the entries of the distance-encoding matrix becomes

$$\lim_{k\to 0} \Xi_l\left(kr_S\right) \propto \frac{1}{k^l}. \tag{9.3.47}$$

Therefore, a low-frequency boost for $l \geq 1$ order signal components is needed, and the quality of boost increases as the order increases and the frequency decreases. The low frequency of the l-order signal component is boosted by $l \times (-6)$ dB / Oct. Accordingly, the entries of filtering $[\Xi]$ in Equation (9.3.43) are divergent at low frequencies. This divergence should be avoided in practical encoding processing.

However, if M secondary sources are arranged at a finite distance r_0, $[\Xi]$ of the finite secondary source distance given in Equation (9.3.41) is used for filtering. From Equation (9.3.46), the entries of $[\Xi]$ at low frequencies tend to

$$\lim_{k\to 0} \left| \Xi_l\left(kr_S, kr_0\right) \right| = \left(\frac{r_0}{r_S}\right)^{l+1}. \tag{9.3.48}$$

According to Equation (9.3.44), the entries of $[\Xi]$ at high frequencies tend to

$$\lim_{k\to \infty} \left| \Xi_l\left(kr_S, kr_0\right) \right| = \frac{r_0}{r_S}. \tag{9.3.49}$$

The low-frequency boost of the entries of $[\Xi]$ with respect to high frequencies is

$$\text{boost} = \left(\frac{r_0}{r_S}\right)^{l}. \tag{9.3.50}$$

The boost is independent of frequency, so the problem of divergence at low frequencies is avoided. The function of $[\Xi]$ is to boost the low-frequency component of a signal for a target source in the region inside the secondary source array. NFC–HOA aims to reconstruct the curve wavefront of a proximal target source in reproduction. The information of curve wavefront caused by a target source is encoded in $B_{lm}^{(\sigma)}\left(\mathbf{r}_S, k\right)$ of the target sound field in Equation (9.3.36), and $B_{lm}^{(\sigma)}\left(r_S, k\right)$ is proportional to $h_l(kr_S)$. Each secondary source arranged at a finite distance r_0 also creates a curve wavefront. The spherical harmonic coefficients of the reconstructed sound field caused by all secondary sources are proportional to $h_l(kr_0)$ rather than $h_l(kr_S)$. When the distance of secondary sources is different from that of the target source, the curve wavefront of the reconstructed sound field does not match with that of the target source. The compensation for the secondary source distance by the denominator of Equation (9.3.41) corrects such a mismatched curve wavefront and results in the desired reproduced sound field. For target source direction outside the median plane, when a listener enters the reconstructed sound field, the scattering and shadow effects caused by anatomical structures, such as the head, lead to target distance-dependent interaural level difference (ILD) at low frequencies and thus provide information for distance perception. Given r_S and $r_0 > r_S$, NFC–HOA can reconstruct the target sound field inside the region $r < r_S$ at most.

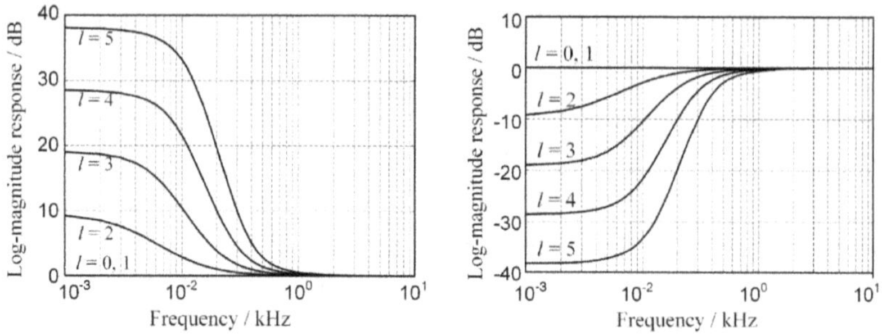

(a) $r_0 = 1.5$ m, $r_S = 0.5$ m (b) $r_0 = 0.5$ m, $r_S = 1.5$ m

Figure 9.4 Normalized logarithmic magnitude of $(r_S / r_0)\ \Xi_l(kr_S, kr_0)$ for order $l = 0$ to 5

In practical use, a problem is associated with the driving signal given in Equation (9.3.42). Equation (9.3.50) indicates that the divergence of the function $\Xi_l(kr_S, kr_0)$ at low frequencies can be avoided when finite target and secondary source distances are considered at the same time. However, for $r_S < r_0$, the low-frequency magnitude of $\Xi_l(kr, kr_0)$ still increases as l increases. For example, for $r_0 = 1.5$ m and $r_S = 0.25$ m, Equation (9.3.50) leads to an increase in 6^l time at a low-frequency magnitude. When $l = 6$ order, the increase is 93 dB and even exceeds the dynamic range of a practical electroacoustic system. Figure 9.4 (a) illustrates the variation in the normalized magnitude $(r_S / r_0)\ \Xi_l(kr_S, kr_0)$ of the distance-encoding function given in Equation (9.3.41) with a frequency for $l = 0$ to 5, $r_0 = 1.5$ m, and $r_S = 0.5$ m. As frequency increases, the normalized logarithmic magnitude approaches 0 dB. Therefore, normalization is convenient for analysis. Otherwise, a $20\log_{10}(r_0/r_S)$ offset to the logarithmic magnitude should be supplemented. An appropriate choice of the NFC–HOA order is more complicated than that of far-field Ambisonics. The problems on the accurate reconstruction of sound field and excessive boost at a low-frequency response of higher-order filter $\Xi_l(kr_S, kr_0)$ should be considered comprehensively. Orders are usually chosen to vary with frequency. If the accuracy of the reconstructed sound field is considered only, according to Equations (9.3.35) and (9.3.40) to (9.3.43), the contribution of l-order spherical harmonic component to the reconstructed sound field is proportional to $j_l(kr)\ \Xi_l(kr_S, kr_0)$. Given wavenumber k and distance $r < r_S < r_0$, the order $(L - 1)$ of truncation should be chosen according to the variation in $j_l(kr)\ \Xi_l(kr_S, kr_0)$ with l so that all the $l \geq L$ order spherical harmonic components are negligible within the region with radius r. At high frequencies, $\Xi_l(kr_S, kr_0)$ tends to be a constant (r_0 / r_S), the order of truncation can be chosen according to the oscillation and decay characteristics of $j_l(kr)$, and the result is consistent with that of far-field Ambisonics given in Equation (9.3.29), i.e., the required order increases as kr increases. Therefore, high-order NFC–HOA is needed to reconstruct the target near field at high frequencies and within an appropriate region. For example, according to Equation (9.3.29), $(L - 1) = 11$ order NFC–HOA at least is needed to reconstruct the target sound field up to 5 kHz and within a region matching the head radius of $r = a = 0.0875$ m. Accordingly, $L^2 = 144$ secondary sources (loudspeakers) is needed. If the size of a region should comprise the pinnae, a larger $r = 0.12$ m should be chosen, then $(L - 1) = 15$ order NFC–HOA with $L^2 = 256$ secondary sources at least is required. Therefore, NFC–HOA usually works in certain low- and middle-frequency ranges, and using NFC–HOA to reconstruct a target sound field within a full audible bandwidth is impractical. Fortunately, some psychoacoustic experiments indicate that the third-order NFC–HOA can recreate auditory distance perception to some contents due to the major contribution of low- and middle-frequency (below 3 kHz) ILD on auditory distance perception in a free field (Favrot and Buchholz, 2012).

At low frequencies (usually below some hundreds or 1 kHz, depending on r_S and r_0), NFC–HOA with the order chosen according to Equation (9.3.29) is not enough to reconstruct the target sound field accurately. A slightly higher order than that given in Equation (9.3.29) should be chosen according to the variation in function $j_l(kr)\ \Xi_l(kr_S, kr_0)$ with l. For example, Daniel and Moreau (2004) suggested choosing a frequency-dependent truncation order so that the magnitude of reconstructed sound pressure maximizes at the receiver position coincident with the target source position. This choice is consistent with the physical characteristic of the target sound field caused by a point source and reduces the power at $r > r_S$. To avoid the excessive boost on the low-frequency response of distance-encoding filter $\Xi_l(kr_S, kr_0)$, Favrot and Buchholz (2012) suggested applying some regularization to the filter response $\Xi_l(kr_S, kr_0)$ to restrict its gain at low frequencies. However, regularization may influence the accuracy of sound field reconstruction.

Overall, for $r_S < r_0$, NFC–HOA can recreate a virtual source at various near-field distances to some extent. However, a much higher order is needed for high-frequency reproduction, and some difficulties occur at low-frequency processing. Therefore, some difficulties are associated with the practical application of NFC–HOA.

The case of using NFC–HOA to recreate a target virtual source inside the secondary source array is discussed above. NFC–HOA is also applicable to recreating a virtual source outside the secondary source array with $r_S > r_0$. The results of the analysis are identical to those given in Equations (9.3.34) to (9.3.43). However, in the case of $r_S > r_0$, the filtering matrix expressed in Equation (9.3.41) attenuates the low-frequency components of driving signals. This result can also be observed in the low-frequency approximation shown in Equation (9.3.48). For example, Figure 9.4(b) illustrates the variation in the logarithmic magnitude of $(r_S/r_0)\ \Xi_l(kr_S, kr_0)$ expressed in Equation (9.3.41) with the frequency at $l = 0$ to 5, $r_0 = 0.5$ m, and $r_S = 1.5$ m. In the case of $r_S > r_0$, the problem of excessive boost at the low-frequency response of higher-order filter $\Xi_l(kr_S, kr_0)$ is avoided. From this point of view, recreating the virtual source outside the secondary array is relatively easier than inside the array. However, arranging multiple secondary sources in a spherical surface with a smaller radius is difficult because of the practical size of secondary sources. Therefore, this method is limited.

Strictly speaking, the influence of a secondary source distance should be compensated when the secondary source array with a finite source distance is used to reconstruct a target plane wave. Similar to the derivation of Equations (9.3.37) to (9.3.41), the filter response for the compensation of the secondary source distance is obtained by replacing the target sound field of a point source in Equation (9.3.37) with that of a plane wave in Equation (9.3.19):

$$\Xi_l\left(kr_S, kr_0\right) = \frac{4\pi j^{l+1}}{kh_l\left(kr_0\right)}. \tag{9.3.51}$$

When the secondary sources are arranged at a far-field distance with $r_0 \gg 1$, they can be approximated as plane wave sources. Similar to the case in Section 2.1.1, a factor of $4\pi r_0 \exp(jkr_0)$ should be supplemented to the denominator of Equation (9.3.51) so that the pressure amplitude of a free-field plane wave at the origin is normalized to a unit for a secondary source with a unit driving signal. According to the asymptotic formula given in Equation (9.3.44), Equation (9.3.51) becomes a unit. This example is a case of reconstructing the target plane wave with secondary source array in the far-field distance in Section 9.3.2 in which the compensation of the secondary source distance is not required.

Although the reference distance of secondary sources is included in the encoded signals given in Equation (9.3.43), the practical distance of secondary sources can be adjusted in reproduction. If the practical distance of secondary sources is r_1, the practical distance of secondary sources can be adapted by multiplying the following factor to each l-order independent signal in Equation (9.3.43):

$$F_l(kr_1, kr_0) = \frac{h_l(kr_0)}{h_l(kr_1)}. \tag{9.3.52}$$

If the spherical Bessel and spherical harmonic decomposition of a spatial sound field is substituted with the Bessel–Fourier expansion of horizontal sound field, a similar method is theoretically applicable to the horizontal NFC–HOA (Daniel et al., 2003). The Bessel–Fourier expansion of a horizontal sound field is valid when the sound field is independent from a vertical coordinate. Therefore, secondary straight-line sources with an infinite length arranged perpendicular to the horizontal plane should be used to reconstruct the cylindrical wave caused by a proximal straight-line source with infinite length and perpendicular to the horizontal plane. If the target sound field is created by a straight-line source with unit strength of $Q_{li} = 1$ and at the horizontal position (r_S, θ_S), the sound pressure at the horizontal receiver position is calculated with Equation (1.2.7). In reproduction, M secondary straight-line sources are arranged uniformly in a horizontal circle with $r' = r_0$. The azimuth of the ith secondary source is θ_i. The strength of the secondary straight-line source is a unit for a unit driving signal. The normalized amplitudes of the driving signals for Q-order reproduction are

$$\begin{aligned} A_i(r_S, \theta_S, r_0, \theta_i, f) &= \frac{1}{M}\left\{\frac{H_0(kr_S)}{H_0(kr_0)} + 2\sum_{q=1}^{Q}\frac{H_q(kr_S)}{H_q(kr_0)}(\cos q\theta_i \cos q\theta_S + \sin q\theta_i \sin q\theta_S)\right\} \\ &= \frac{1}{M}\left\{\frac{H_0(kr_S)}{H_0(kr_0)} + 2\sum_{q=1}^{Q}\frac{H_q(kr_S)}{H_q(kr_0)}\cos\left[q(\theta_S - \theta_i)\right]\right\}. \end{aligned} \tag{9.3.53}$$

Equation (9.3.53) is consistent with the result obtained from the spatial SDM in Equation (9.2.28) (Wu and Abhayapala, 2009).

When the target sound field is a far-field plane wave, $r_S \gg 1$, and the following asymptotic formula of the Hanker function of the secondary kind for a large argument $\xi \gg 1$ can be applied:

$$H_q(\xi) \approx \sqrt{\frac{2}{\pi\xi}} \exp\left(-j\xi + j\frac{q\pi}{2} + j\frac{\pi}{4}\right). \tag{9.3.54}$$

The driving signal amplitudes of secondary sources can be obtained by substituting Equation (9.3.54) into Equation (9.3.53). According to Equation (1.2.8), the factor $Q_{li} = 4j\sqrt{(\pi k r_S)/2}\exp(jkr_S - j\pi/4)$ should be supplemented to normalize the amplitude of reconstructed plane wave at the origin to be a unit:

$$A_i(r_S, \theta_S, r_0, \theta_i, f) = \frac{4}{M}\left\{\frac{j}{H_0(kr_0)} + 2\sum_{q=1}^{\infty}\frac{j^{q+1}}{H_q(kr_0)}\cos\left[q(\theta_S - \theta_i)\right]\right\}. \tag{9.3.55}$$

The Hankel functions in the denominator of Equation (9.3.55) are used to compensate for the influence of the finite distance of secondary sources.

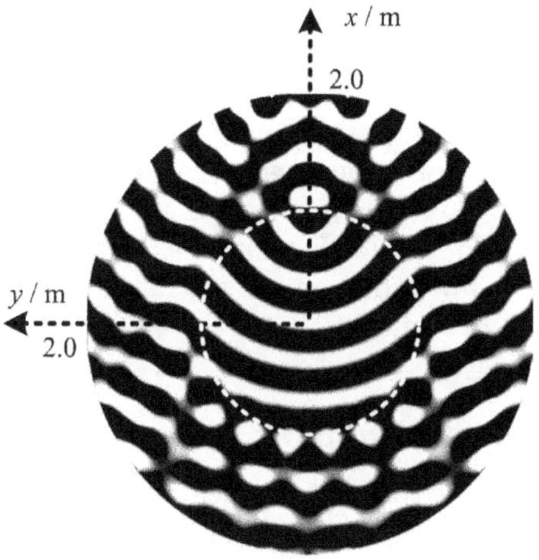

Figure 9.5 Amplitude distribution of the horizontal NFC–HOA reconstructed sound pressures.

For example, Figure 9.5 illustrates the amplitude distribution of the horizontal NFC–HOA reconstructed sound pressures within a circular region with radius $r \leq 2.0$ m. $M = 36$ secondary sources are arranged uniformly in a circle with $r = 2.0$ m. $Q = 17$ order horizontal NFC–HOA is used. The target sound field is a cylindrical wave created by a harmonic straight-line source with unit strength and $f = 1.0$ kHz at $r_S = 1.0$ m and $\theta_S = 0°$. Horizontal NFC–HOA can reconstruct the target cylindrical wavefront within a circular region around the origin (bounded by the dash circle). However, a larger error occurs in the reconstructed sound field outside the circular region.

The radiation pattern of a practical loudspeaker is close to a point rather than a straight-line source. If horizontal secondary point sources at a finite distance were used to reconstruct the sound field created by a near-field target source, the required driving signals would depend on receiver position (Ahrens and Spors, 2008b). This problem is addressed in Section 10.3.3. When secondary sources are arranged in a far-field distance and thus can be approximated as plane wave sources, the reconstructed sound field in a local region close to the origin is approximately independent from vertical coordinates. In this case, the sound field of a target straight-line source at a proximal distance may be reconstructed. However, the problem of low-frequency divergence in high-order harmonic components similar to that in Equation (9.3.47) occurs.

When target or secondary sources are located at a far-field distance and whether they are point or line sources, the local sound field close to origin can be approximated as a plane wave. In this case, the target or reconstructed sound field can be expanded into a Bessel–Fourier series, and the array of secondary point sources can be used to reconstruct the target plane wave field approximately. The discussion in Section 9.3.1 is based on this hypothesis. For $kr_0 \gg 1$, the asymptotic formula of the Hanker function of the second-ary kind in Equation (9.3.54) is substituted into Equation (9.3.55) again, and a factor of $Q_{li} = 4j\sqrt{(\pi k r_0)/2}\exp(jkr_0 - j\pi/4)$ is supplemented to the denominator of Equation (9.3.55); thus, the plane wave amplitude in the origin created by a secondary source with a unit driving signal is normalized to a unit, and Equation (9.3.55) becomes

$$\begin{aligned} A_i(\theta_S) &= \frac{1}{M}\left\{1 + 2\sum_{q=1}^{Q}\cos\left[q(\theta_S - \theta_i)\right]\right\} \\ &= \frac{1}{M}\left[1 + 2\sum_{q=1}^{Q}\left(\cos q\theta_i \cos q\theta_S + \sin q\theta_i \sin q\theta_S\right)\right]. \end{aligned} \tag{9.3.56}$$

Equation (9.3.56) is consistent with the driving signals of Q-order horizontal far-field Ambisonics given in Equation (9.3.12).

Ambisonics can also be used to recreate focusing virtual sources at different distances. Ahrens and Spors (2008c) illustrated an example of recreating a two-dimensional focusing virtual source by using an array of secondary straight-line sources arranged on a circle and perpendicular to a horizontal plane. Driving signals are derived from the azimuthal Fourier spectrum of target focusing sound field via spatial SDM, and focusing virtual sources with various main axis orientations of radiation are recreated. The problem of focusing virtual sources is addressed in detail in Section 10.2.4.

9.3.5 Ambisonic encoding of complex source information

The Ambisonic signals and reconstructed sound field of a target point source and plane wave source are discussed from Sections 9.3.1 to 9.3.4. Most practical sound sources possess a certain size and radiation pattern. Radiation patterns or directivities are usually frequency dependent. When Ambisonic signals are recorded by a set of coincident microphones with appropriate directivities or by a spherical microphone array discussed in Section 9.8, the resultant signals involve the spatial information related to the target source characteristics. In the synthesis of Ambisonic signals by signal processing, the spatial information of target sources can also be simulated. This concept is known as ***O-format Ambisonics*** (Menzies, 2002; Menzies and Al-Akaidi, 2007).

In the presence of a sound source with a certain shape and size, its position and orientation are given. A coordinate system with respect to the sound source in Figure 1.2 is convenient for analyzing the source radiation in a free field. The origin of coordinates is located at the center of the source, and the receiver position outside the source is specified by spherical coordinates (R_r, Θ_r, Φ_r) or simply denoted by (R_r, Ω_r). The position and geometrical characteristics of sources are represented by a set of variables or a set of the vector $\boldsymbol{R}_S$. In the region outside the source, the radiated sound pressure can be decomposed by real-valued spherical harmonic functions as

$$P(R_r, \Omega_r, \boldsymbol{R}_S, f) = \sum_{l'=0}^{\infty}\sum_{m'=0}^{l'}\sum_{\sigma'=1}^{2} O_{l'm'}^{(\sigma')}(\boldsymbol{R}_S, f)\, h_{l'}(kR_r)\, Y_{l'm'}^{(\sigma')}(\Omega_r). \tag{9.3.57}$$

where $h_{l'}(kR_r)$ is the l'-order spherical Hankel function of the second kind. Therefore, the characteristics of radiation are determined by the coefficients $O_{l'm'}^{(\sigma')}(\boldsymbol{R}_S, f)$ of decomposition. These coefficients depend on the frequency, shape, size, position, and directivity of a source. A practical source radiation pattern is usually spatially bandlimited so that the summation in Equation (9.3.57) can be truncated up to some order. In this case, the finite number of $O_{l'm'}^{(\sigma')}(\boldsymbol{R}_S, f)$ can be obtained through calculation or measurement. For a receiver position at the far-field distance with $kR_r \gg 1$, when the asymptotic formula of the spherical Hanker function of the secondary kind given in Equation (9.3.44) is used, Equation (9.3.57) becomes

$$P(R_r, \Omega_r, \boldsymbol{R}_S, f) = \frac{\exp(-jkR_r)}{kR_r} \sum_{l'=0}^{\infty}\sum_{m'=0}^{l'}\sum_{\sigma'=1}^{2} j^{l+1} O_{l'm'}^{(\sigma')}(\boldsymbol{R}_S, f)\, Y_{l'm'}^{(\sigma')}(\Omega_r). \tag{9.3.58}$$

The summation in Equation (9.3.58) determines the directivity of sound source radiation defined in Equation (1.2.18).

A coordinate system with respect to a specific receiver position Figure 1.1 is convenient for analyzing the Ambisonic encoding and sound field reconstruction. In this coordinate system, an arbitrary receiver position is denoted by (r, Ω). The position and geometrical characteristics of sources are represented by a set of variables or a set of vector r_S. In a spherical region centered at the origin, the incident sound field can also be decomposed by real-valued spherical harmonic functions:

$$\begin{aligned} P(r,\Omega,r_S,f) &= \sum_{l=0}^{\infty}\sum_{m=0}^{l}\sum_{\sigma=1}^{2} P_{lm}^{(\sigma)}(r,r_S,f)Y_{lm}^{(\sigma)}(\Omega) \\ &= \sum_{l=0}^{\infty}\sum_{m=0}^{l}\sum_{\sigma=1}^{2} B_{lm}^{(\sigma)}(r_S,f)j_l(kr)Y_{lm}^{(\sigma)}(\Omega), \end{aligned} \quad (9.3.59)$$

where

$$P_{\mathrm{lm}}^{(\sigma)}(r,r_S,f) = B_{\mathrm{lm}}^{(\sigma)}(r_S,f)j_l(kr). \quad (9.3.60)$$

$B_{lm}^{(\sigma)}(\mathbf{r}_S, f)$ represent Ambisonic-encoding information. For a target plane wave incident from Ω_S, $B_{lm}^{(\sigma)}(\mathbf{r}_S, f)$ expressed in Equation (9.3.18) is proportional to the spherical harmonic function $Y_{lm}^{(\sigma)}(\Omega_S)$, i.e., proportional to the amplitudes of independent signals of far-field Ambisonics. For a complex sound source, however, $B_{lm}^{(\sigma)}(\mathbf{r}_S, f)$ is more complicated. To solve Ambisonic driving signals via the method in Section 9.2.3, the relationship between the spherical harmonic spectrum $P_{lm}^{(\sigma)}(r, \mathbf{r}_S, f)$ [or equivalent coefficients $B_{lm}^{(\sigma)}(\mathbf{r}_S, f)$] of a target sound field and the decomposition coefficients of the characteristics of source radiation should be determined. Equations (9.3.57) and (9.3.59) are representations of the same target sound field in two different coordinates and thus should be matched with

$$\sum_{l=0}^{\infty}\sum_{m=0}^{l}\sum_{\sigma=1}^{2} B_{lm}^{(\sigma)}(r_S,f)j_l(kr)Y_{lm}^{(\sigma)}(\Omega) = \sum_{l'=0}^{\infty}\sum_{m'=0}^{l'}\sum_{\sigma'=1}^{2} O_{l'm'}^{(\sigma')}(\mathbf{R}_S,f)h_{l'}(kR_r)Y_{l'm'}^{(\sigma')}(\Omega_r). \quad (9.3.61)$$

Equation (9.3.61) connects the spherical harmonic decomposition of the radiated sound field in two different coordinates. The orthogonality of spherical harmonic functions given in Equation (A.10) in Appendix A yields

$$B_{lm}^{(\sigma)}(r_S,f) = \frac{1}{j_l(kr)}\sum_{l'=0}^{\infty}\sum_{m'=0}^{l'}\sum_{\sigma'=1}^{2} A_{lm,l'm'}^{(\sigma,\sigma')}(f)O_{l'm'}^{(\sigma')}(\mathbf{R}_S,f), \quad (9.3.62)$$

where

$$A_{lm,l'm'}^{(\sigma,\sigma')}(f) = \int Y_{lm}^{(\sigma)}(\Omega)h_{l'}(kR_r)Y_{l'm'}^{(\sigma')}(\Omega_r)d\Omega. \quad (9.3.63)$$

If the geometrical relationship between two coordinates (r, Ω) and (R_r, Ω_r) is given, the integral in Equation (9.3.63) can be calculated using previously described methods (Menzies and Al-Akaidi, 2007). Calculation is simplified by choosing the polar axes of two coordinate systems to be consistent with the straight line that connects the origins of two coordinate systems. According to Equation (9.3.62) and for $j_l(k, r) \neq 0$, $B_{lm}^{(\sigma)}(r_S, f)$ or the spherical harmonic spectrum $P_{lm}^{(\sigma)}(r, r_S, f)$ of the target sound field in the coordinate system (r, Ω) can be obtained when $O_{l'm'}^{(\sigma')}(\mathbf{R}_S, f)$ of sound source radiation is given.

Similar to the case of NFC–HOA in Section 9.3.4, in reproduction, M secondary sources are arranged on a spherical surface with r_0 and position of the ith secondary source is (r_0, Ω_i). Each secondary source is a point source with unit strength when the amplitude of a driving signal is a unit. The normalized amplitude of an actual driving signal is $A_i(r_S, r_0, \Omega_i, f)$. Substituting Equations (9.3.60), (9.3.62), and (9.2.41) into Equation (9.2.43) and matching each spherical harmonic component of the reconstructed sound field with that of target sound field yield

$$\begin{gathered} h_0(kr_0)\sum_{i=0}^{M-1} A_i(r_S, r_0, \Omega_i, f) Y_{00}^{(\sigma)}(\Omega_i) = \frac{j}{k} B_{00}^{(\sigma)}(r_S, f) \\ h_l(kr_0)\sum_{i=0}^{M-1} A_i(r_S, r_0, \Omega_i, f) Y_{lm}^{(\sigma)}(\Omega_i) = \frac{j}{k} B_{lm}^{(\sigma)}(r_S, f) \\ l = 1, 2, 3\ldots \quad m = 0, 1\ldots l. \quad \text{if } m \neq 0, \quad \sigma = 1, 2 \quad \text{if } m = 0, \ \sigma = 1. \end{gathered} \tag{9.3.64}$$

To solve driving signals, $A_i(r_S, r_0, \Omega_i, f)$ is represented as the linear combination of the encoding signals $B_{lm}^{(\sigma)}(r_S, f)$

$$A_i(r_S, r_0, \Omega_i, f) = A_{\text{total}} \sum_{l=0}^{\infty}\sum_{m=0}^{l}\sum_{\sigma=1}^{2} D_{lm}^{\prime\prime(\sigma)}(\Omega_i, r_0, f) B_{lm}^{(\sigma)}(r_S, f), \tag{9.3.65}$$

where $D_{lm}^{\prime\prime(\sigma)}(\Omega_i, r_0, f)$ is a set of decoding coefficients to be determined. Similar to the NFC–HOA in Section 9.3.4, a matrix equation for decoding coefficients can be obtained by substituting Equation (9.3.65) into Equation (9.3.64) and truncating up to order $(L - 1)$. The normalized amplitudes of driving signals are derived by substituting the resultant decoding coefficients $D_{lm}^{\prime\prime(\sigma)}(\Omega_i, r_0, f)$ into Equation (9.3.65). If the distance information of target and secondary sources are merged into Ambisonic-independent signals, the $L^2 \times 1$ vector of independent signal for $(L - 1)$ order Ambisonics is redefined as

$$S_{3D}'' = \frac{j}{k}\left[h_0^{-1}(kr_0) B_{00}^{(1)}(r_S, f), h_1^{-1}(kr_0) B_{11}^{(1)}(r_S, f), h_1^{-1}(kr_0) B_{11}^{(2)}(r_S, f), h_1^{-1}(kr_0) B_{10}^{(1)}(r_S, f)\ldots\right]^T. \tag{9.3.66}$$

The $M \times 1$ vector of driving signal amplitudes is defined as

$$\mathbf{A} = \left[A_0(r_S, r_0, \Omega_0, f), A_1(r_S, r_0, \Omega_1, f), \ldots A_{M-1}(r_S, r_0, \Omega_{M-1}, f)\right]^T. \tag{9.3.67}$$

Then, driving signal amplitudes are obtained using the following decoding equations:

$$\mathbf{A} = A_{\text{total}}\left[D_{3D}\right] S_{3D}''. \tag{9.3.68}$$

Similar to the reconstruction of a plane wave field, $M \times L^2$ decoding matrix $[D_{3D}]$ is obtained via the pseudoinverse method in Equation (9.3.23)

$$A_{\text{total}}\left[D_{3D}\right] = \text{pinv}\left[Y_{3D}\right] = \left[Y_{3D}\right]^T \left\{\left[Y_{3D}\right]\left[Y_{3D}\right]^T\right\}^{-1}. \tag{9.3.69}$$

The decoding matrix of Ambisonics with a complex source is identical to that of far-field Ambisonics for the reconstruction of a target plane wave. However, Ambisonics with a complex source differs from far-field Ambisonics because the information of target and secondary sources is encoded into the vector S''_{3D} of independent signals in Equation (9.3.60).

The above analysis has two stages that deal with the truncation order of spherical harmonic decomposition. The first stage is the truncation order in the summation of Equation (9.3.62). At this stage, the spherical harmonic coefficients of source radiation characteristics are converted to the spherical harmonic spectrum of the target sound field. This truncation order is determined by the radiation characteristics of the target source and the relative position between the source and the receiver position. The second stage is the truncation order in the summation of driving signals of Equations (9.3.64) and (9.3.65), i.e., the order $(L - 1)$ of Ambisonic reproduction, which is determined by the reconstructed sound field. Similar to the case of NFC–HOA in Section 9.3.4, the contribution of the l-order spherical harmonic component to the reconstructed sound field is proportional to

$$l \text{ order term} \propto \frac{j_l(kr)}{kh_l(kr_0)} B_{lm}^{(\sigma)}(r_S,k). \tag{9.3.70}$$

Therefore, the order of truncation should be determined by the contribution of Equation (9.3.70).

In the aforementioned discussion, secondary sources are ideal point sources or plane wave sources. A method similar to the aforementioned discussion can be used to analyze the reconstructed sound field by secondary sources with various directivities or radiation patterns, but this analysis is complicated and omitted here (Poletti et al., 2010a, 2010b; Poletti and Abhayapala 2011).

9.3.6 Some special applications of spatial-spectral domain analysis of Ambisonics

From Sections 9.3.1 to 9.3.5, Ambisonic-reconstructed sound field and driving signals are analyzed using the mode-matching method in a spatial-spectral domain. Ahrens and Spors (2008b) used the spatial SDM in Sections 9.2.2 and 9.2.3 to analyze the Ambisonic-reconstructed sound field and driving signals for secondary source arrays in a horizontal circle and on a spatial spherical surface, yielding results consistent with those of the mode-matching method. Therefore, two methods for analyses are basically equivalent.

In exhibition, public offices, and other open environments, a target sound should be reproduced within a local zone, and the interference of the reproduced sound in other zones should be reduced. Furthermore, different audio extents should be reproduced in different zones without mutual interferences. This type of technique focuses on the problem of ***spatial multizone sound field reconstruction or reproduction*** and is commercially termed "personal audio space." The nature of this technique involves the use of a secondary source array to reconstruct the target sound field in a local zone and reduce the reproduced sound pressure in other zones as far as possible. Reproducing different audio extents in different zones can be implemented through the local reproduction of each audio extent in each zone because of the linear superposition principle of a sound field. Using the mode-matching method in the spatial-spectral domain in Section 9.2.2, Wu and Abhayapala (2011) analyzed multizone sound field reconstruction with an array of secondary straight-line sources in a horizontal plane. Two or more circular subzones are involved in a global zone bounded by a circle of secondary sources. The sound field in each subzone can be expanded into the Fourier series of a local azimuth in a local polar coordinate centered at that subzone. The azimuthal Fourier coefficients are the azimuthal or spatial spectrum representation of the local sound field at that subzone. The spatial spectrum representation of the overall sound field is obtained by converting that of a local sound field of each subzone to the global polar coordinate centered in the global zone and superposing the contribution of all local sound fields. Driving signals are derived using the method in Section 9.2.2.

A similar method is applicable to the reconstruction of a single target sound field in multiple zones (for multiple listeners; Poletti and Betlehem, 2014). For example, according to the analysis in Section 9.3.1, especially Equation (9.3.15), given the upper frequency limit, reconstructing a target sound field in a large zone with radius r requires a much higher-order Ambisonics and a large number of secondary sources in horizontal sound field reconstruction. K small zones with radius $r_1 \ll r$ are chosen within a large zone with radius r. The target sound field is reconstructed in each small zone for each listener. Q_1-order Ambisonic sound field reconstruction in each small zone requires $(2Q_1 + 1)$ secondary sources. In large zones, Q-order sound field reconstruction requires $(2Q + 1)$ secondary sources. Q and Q_1 are determined by a formula similar to Equation (9.3.15). The number of secondary sources required for reconstructing a sound field in each small zone is much fewer than that for reconstructing a sound field in the whole large zone. If

$$K(2Q_1 + 1) \le (2Q + 1), \tag{9.3.71}$$

the total number of secondary sources required for reconstructing a sound field in K small zones is fewer than that for reconstructing a sound field in the whole large zone. Actually, creating a single sound field in multiple zones is analogous to multiband sampling in signal processing. Generally, the sampling frequency of a continuous time-domain signal should satisfy the requirement of Shannon–Nyquist sampling theorem. However, if the signal is sparse in the frequency domain, i.e., if the signal only involves some sub-band components, each sub-band component can be sampled at a much lower sampling frequency. The continuous time-domain signal can be recovered from the samples in all sub-bands. Sampling the signal with a frequency domain sparse characteristic in each sub-band reduces the overall data rate.

Below the upper frequency limit imposed by the Shannon–Nyquist spatial sampling theorem, target sound fields in each subzone can be reconstructed through multizone sound field reconstruction. However, a potential problem is that a large error of the reconstructed sound field may occur in each subzone above the upper frequency limit. These errors may lead to obvious audible interference or distortion. Moreover, most existing analyses are conducted under the condition of an ideal free-field reproduction environment, but practical reproduction may differ from ideal reproduction. For example, some errors in the position and characteristics of secondary sources may exist, or multiple scattering occurs when two or more listeners enter the sound field. Consequently, all these factors lead to errors in the final reconstructed sound field. Therefore, the influence of various errors on the stability of multizone sound field reconstruction should be further explored. Overall, multizone sound field reconstruction is still an academic topic. Numerous studies are needed for practical uses.

9.4 SOME PROBLEMS RELATED TO AMBISONICS

9.4.1 Secondary source array and stability of Ambisonics

Secondary source array or loudspeaker configuration of Ambisonics is relatively flexible. Various secondary source arrays for reproduction have been developed. In Section 9.3, the minimal number of secondary sources required for different order reproductions is analyzed, and the decoding matrix or driving signals are derived. In Section 9.3.1, for horizontal Ambisonics with uniform secondary source array, each row in the matrix described in Equation (9.3.9) satisfies the discrete orthogonality of trigonometric functions. Therefore, uniform (regular) secondary source arrays are often used in horizontal Ambisonics. For non-uniform (irregular) secondary source array, the decoding matrix and driving signals can be solved with the pseudoinverse method in Equation (9.3.10).

For spatial Ambisonics, in Section 9.3.2, if the array of M secondary sources satisfies the discrete orthogonality of spherical harmonic functions given in Equation (9.3.24), an exact solution of the decoding equation and driving signals of the $(L - 1)$-order spatial Ambisonics can be found. However, only a few secondary source arrays satisfy the discrete orthogonality of spherical harmonic functions. Some examples are presented as follows:

1. ***Equiangle array***. The elevation $\alpha = 90° - \phi$ and the azimuth $\beta = \theta$ are uniformly sampled at $2L$ angles, respectively; $M = 4L^2$ secondary sources are arranged in the directions of sampling.
2. ***Gauss–Legendre node array***. Elevation is first sampled at L angles that are chosen according to the Gauss–Legendre nodes. Then, the azimuth in each elevation is sampled at $2L$ angles. $M = 2L^2$ secondary sources are arranged in the directions of sampling.
3. ***Uniform or nearly uniform array***. Directional samples are uniformly or nearly uniformly distributed on the spherical surface so that the distance between neighboring samples is constant or nearly constant. Secondary sources are arranged in the directions of sampling. The number of secondary sources is usually larger than or at least equal to the lower limit given in Shannon–Nyquist spatial sampling theorem, i.e., $M \geq L^2$.

The equiangle array is intuitive, but it requires a fourfold number of secondary sources than the lower limit given in the Shannon–Nyquist spatial sampling theorem. In fact, the actual angular interval between two adjacent azimuthal sampling decreases as the direction deviates from the horizontal plane to the high and low elevations. Accordingly, an elevation-dependent weight λ_i is introduced into the driving signals in Equation (9.3.26) to reduce or avoid the overemphasis of the contribution of secondary sources in high and low elevations to the reconstructed sound field. Therefore, the efficiency of equiangle arrays is not high from the point of directional sampling and the reconstruction of sound field. As such, this array is seldom used in practical reproduction.

Gauss–Legendre node array requires twice the number of secondary sources in comparison with the lower limit given by the Shannon–Nyquist spatial sampling theorem and thus more efficient than the equiangle array. However, this array imposes more restrictions on the directions of secondary sources. In particular, no secondary sources may exist in a horizontal plane (equator), which is inconsistent with the requirement of enhancing the stability of a horizontal virtual source in reproduction. Therefore, Gauss–Legendre node array is usually inappropriate for practical uses.

A regular secondary source array usually exhibits a high efficiency and leads to a stable reconstructed sound field. A horizontal array is ***regular*** if each row of matrix $[Y_{2D}]$ in Equation (9.3.9) satisfies the discrete orthogonality given in Equation (4.3.18), i.e., it satisfies $[Y_{2D}]$

$[Y_{2D}]^T$ = const. Therefore, a horizontal uniform array with M secondary sources is regular up to the following order of Ambisonic reproduction: $Q = (M - 1) / 2$ if M is an odd number, or $(M - 2)/2$ if M is an even number. Similarly, a spatial array is regular if $[Y_{3D}]$ in Equation (9.3.22) satisfies the discrete orthogonality expressed in Equation (9.3.24) or (9.3.25) with a constant weight of $\lambda_0 = \lambda_1 = \ldots = \lambda_{M-1}$. However, only five spatial arrays are strictly regular (Daniel, 2000). These regular arrays can be realized by arranging the secondary sources in the vertices or centers of faces of some polyhedrons, including tetrahedrons, hexahedrons (cube), octahedrons, icosahedrons, and dodecahedrons (Hollerweger, 2006). Tetrahedral, hexahedral, and octahedral arrays are regular for the first-order spatial Ambisonic reproduction. Some examples of the first-order reproduction with tetrahedral and hexahedral arrays are illustrated in Section 6.4.2. Icosahedral and dodecahedral arrays are regular up to the second-order reproduction although dodecahedral arrays provide 20 secondary sources that exceed the minimal number of $(3 + 1)^2 = 16$ for the third-order reproduction. Moreover, secondary sources are arranged in the horizontal plane only for some polyhedral arrays, such as face-center array in a hexahedron. Various nearly uniform arrays provide more secondary sources (Appendix A) for higher-order Ambisonic reproduction. Usually, the number M of secondary sources in a nearly uniform array is 1.3–1.5 times of the lower limit given in Equation (9.3.28). If M exceeds $1.5L^2$, a constant weight $\lambda_i = \lambda$ can be approximately chosen in the calculation of Equations (9.3.24) to (9.3.26). Therefore, the efficiency of nearly uniform array is relatively high.

In addition to the aforementioned arrays, some other arrays satisfy the discrete orthogonality expressed in Equation (9.3.24; Lecomte et al., 2015). These arrays may be mathematically perfect, but they are restricted in practical uses. In particular, arranging secondary sources in the bottom is usually inconvenient. Irregular or non-uniform arrays are often used in accordance with the practical reproduction space, and the pseudoinverse method in Equation (9.3.23) is used to solve the driving signals. However, non-uniform arrays may cause instability in the reconstructed sound field. Therefore, feasibility and stability should be considered comprehensively in the design of practical spatial arrays. Instability may also occur in horizontal Ambisonic reproduction with irregular arrays.

Some perturbations on reproduction systems, such as slight errors in secondary source positions and slight differences in the characteristics of secondary sources, are inevitable. Instability means that the reconstructed sound field is sensitive to these small perturbations. Stability depends on the number and configuration of secondary sources in reproduction and is closely related to the pseudoinverse calculation of $[Y_{2D}]$ in Equation (9.3.10) or $[Y_{3D}]$ in Equation (9.3.22) for deriving the decoding matrix. According to numerical analysis theory, stability can be evaluated on the basis of the condition number of $[Y_{2D}]$ or $[Y_{3D}]$ (Sontacchi, 2003). For example, in spatial Ambisonics,

$$\mathrm{cond}[Y_{3D}] = \frac{\gamma_{\max}[Y_{3D}]}{\gamma_{\min}[Y_{3D}]}, \tag{9.4.1}$$

where $\gamma_{\max}[Y_{3D}]$ and $\gamma_{\min}[Y_{3D}]$ are the largest and smallest singular values of $[Y_{3D}]$ in Equation (9.3.22). Because $[Y_{3D}]$ is an $L^2 \times M$ matrix, $[Y_{3D}][Y_{3D}]^T$ is an $L^2 \times L^2$ real symmetric matrix with K-positive eigenvalues of $\gamma_0^2 \geq \gamma_1^2 \geq \ldots \geq \gamma_{K-1}^2 > 0$. K singular values of $[Y_{3D}]$ are $\gamma_0 \geq \gamma_1 \geq \ldots \gamma_{K-1} > 0$, and the largest and smallest singular values are given by $\gamma_{\max}[Y_{3D}] = \gamma_0$ and $\gamma_{\min}[Y_{3D}] = \gamma_{K-1}$. By definition, the condition number is not less than a unit. The smaller (closer to unit) the condition number is, the more stable the system will be.

As simple cases, the hexahedral array of eight loudspeakers and the tetrahedral array of four loudspeakers in Section 6.4.2 are first analyzed. For the $(L - 1) = 1$ order spatial

Ambisonics, the condition numbers of $[Y_{3D}]$ for two arrays are 1.00 and 1.01; therefore, reproduction is stable. However, for the $(L - 1) = 2$ order spatial Ambisonics, the condition number for both arrays are infinite. In fact, the number of secondary sources (four and eight) in both arrays does not reach the lower limit of the second-order reproduction in Equation (9.3.28).

Three kinds of layer-wise arrays are further analyzed (Liu and Xie, 2013a):

1. Three-layer array of 28 secondary sources (three-layer 28)
   Upper, middle (horizontal), and bottom layers are located at $\phi = 45°, 0°$, and $-45°$, respectively. Eight secondary sources with a uniform azimuth of $\theta = 0°, 45°, \ldots, 315°$ are arranged in each of the upper and bottom layers. Twelve secondary sources with a uniform azimuth of $\theta = 0°, 30°, \ldots, 330°$ are arranged in the middle layer.
2. Three-layer array of 32 secondary sources (three-layer 32)
   Upper, middle (horizontal), and bottom layers are located at $\phi = 45°, 0°$, and $-45°$, respectively. Eight secondary sources with a uniform azimuth of $\theta = 0°, 45°, \ldots, 315°$ are arranged in each of the upper and bottom layers. Sixteen secondary sources with a uniform azimuth of $\theta = 0°, 22.5°, \ldots, 337.5°$ are arranged in the middle layer.
3. Five-layer array of 36 secondary sources (five-layer 36)
   Twelve secondary sources with a uniform azimuth of $\theta = 0°, 30°, \ldots, 330°$ are arranged in the horizontal plane at $\phi = 0°$; eight secondary sources with a uniform azimuth of $\theta = 0°, 45°, \ldots, 315°$ are arranged in each elevation plane at $\phi = \pm 30°$; four secondary sources with a uniform azimuth of $\theta = 0°, 90°, 180°$, and $270°$ are arranged in each of the elevation plane at $\phi = \pm 60°$.

The three aforementioned arrays are easy to implement and often used for research on Ambisonics. For comparison, a nearly uniform array of 36 secondary sources (nearly uniform 36), equiangle array of 36 secondary sources (equiangle 36), and Gauss–Legendre node array of 32 secondary sources (Gauss–Legendre 32) are also analyzed. The condition numbers of $[Y_{3D}]$ of various arrays for the preceding four-order Ambisonics are listed in Table 9.1 (Liu, 2014).

Nearly uniform 36 is the best among various arrays in Table 9.1. It is available up to the fourth-order reproduction with appropriate condition numbers of $[Y_{3D}]$. Five-layer 36 exhibits similar results, but its condition numbers are higher than those of the nearly uniform 36. Gauss–Legendre 32 is available up to the third-order reproduction. Equiangle 36, three-layer 32, or three-layer 28 are available up to the second-order reproduction. Noticeably, after secondary sources in the horizontal plane at $\phi = 0°$ increase, the condition numbers of

Table 9.1 Condition numbers of various secondary source arrays

Cond$[Y_{3D}]$	*1-order*	*2-order*	*3-order*	*4-order*
Nearly uniform 36	1.11	1.23	1.41	1.60
Equiangle 36	2.00	2.91	∞	∞
Gauss–Legendre 32	1.50	1.87	1.88	∞
Three-layer 28	1.25	1.67	∞	∞
Three-layer 32	1.51	2.03	∞	∞
Five-layer 36	1.30	1.53	1.90	3.07
Four-layer 28 + 1	1.15	1.43	6.31	∞
Five-layer 28 + 2	1	1	2.25	∞

three-layer 32 for the first- and second-order reproduction increase although they are still within a reasonable range. Therefore, the stability of three-layer 32 is inferior to that of three-layer 28. For first- and second-order reproduction, the condition number of three-layer 28 is smaller than Gauss–Legendre 32. The stability of equiangle 36 is worse. As such, increasing the number of secondary sources does not always improve stability. An appropriate array with fewer secondary sources can also reconstruct a stable sound field.

No top and bottom secondary sources are used in the two kinds of three-layer arrays, which cause holes in arrays and result in instability in higher-order reproduction. Adding secondary sources in top and bottom directions improves stability in reproduction. For example, a four-layer array of 28 + 1 secondary sources (four-layer 28 + 1) is constituted by adding a top secondary source at $(\theta, \phi) = (0°, 90°)$ to the three-layer 28. The condition numbers of four-layer 28 + 1 for first-, second-, and third-order reproduction are 1.15, 1.43, and 6.32, respectively. The four-layer 28 + 1 is appropriate for practical uses, thereby improving the stability of 1- and 2-order reproduction. It is applicable to the third-order reproduction although the condition number of third-order reproduction is high. Figure 9.6 (a) illustrates the position of secondary sources in the four-layer 28 + 1 array, and Figure 9.6 (b) presents a photo of a practical array in Acoustic Lab, South China University of Technology. Moreover, a five-layer array of 28 + 2 secondary sources (five-layer 28 + 2) is constituted by adding a bottom secondary source at $(\theta, \phi) = (0°, -90°)$ to the four-layer 28 + 1. The condition numbers of five-layer 28 + 2 for first, second, and third-order reproduction are 1.00, 1.00, and 2.25, respectively. Therefore, five-layer 28 + 2 further enhances stability, but arranging a secondary source in the bottom layer is inconvenient in practice.

The absence of the top and bottom secondary sources is equivalent to a sharp truncation of the spatial distribution of secondary sources on the upper and low boundaries of space. This sharp discontinuity in spatial distribution leads to ripple oscillations in the spatial spectrum and consequently leads to errors in the reconstructed sound field. This phenomenon is known as the spatial Gibbs effect and analogous to the Gibbs effect in the time or frequency domain signal processing. For example, truncating a time domain signal with a rectangular time window leads to rippling oscillations of a signal magnitude spectrum in the pass-band and the stop-band. Increasing the secondary sources in top and bottom directions eliminates spatial discontinuity and consequently removes the spatial Gibbs effect. When the top and bottom secondary sources are unavailable, applying an appropriate spatial window to driving signals to smoothen spatial discontinuity can reduce the influence of spatial Gibbs effects.

When the condition number of $[Y_{2D}]$ or $[Y_{3D}]$ is large, the pseudoinverse solution of Equation (9.3.10) or (9.3.23) is unstable. For example, this phenomenon occurs in Ambisonic reproduction with irregular 5.1-channel configuration. In this case, the magnitudes or gains of some decoding coefficients are large and even exceed the dynamic range of an electroacoustic

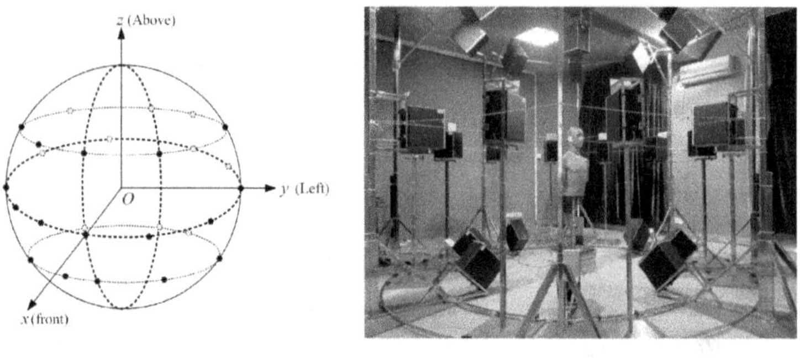

(a) Positions of secondary sources

(b) Photo of a practical array

Figure 9.6 Four-layer array with 28 + 1 secondary sources.

system. In this case, some regularization methods are applicable to solving the decoding matrix. For example, the solution in Equation (9.3.23) can be replaced with the following solution:

$$A_{\text{total}}\left[D_{3D}\right]=\left[Y_{3D}\right]^{T}\left\{\left[Y_{3D}\right]\left[Y_{3D}\right]^{T}+\varepsilon\left[I\right]\right\}^{-1}, \tag{9.4.2}$$

where $[I]$ is an $L^2 \times L^2$ identity matrix, and ε is a regularization parameter. Regularization restricts the excessive gain of some decoding coefficients and thus improves stability, but it reduces accuracy in sound field reconstruction. ε is chosen according to a compromise between stability and accuracy. In Section 5.2.3, a penalty function, which is composed of the weighted overall power of the driving signals of each secondary source, can be added to the cost function for a solution of decoding coefficients based on the least square error (LMS) to further improve stability (Poletti, 2007).

For uniform or nearly uniform secondary source arrays on a spherical surface, the Ambisonic decoding equation is relatively simple. Driving signals satisfy the condition of constant overall power, which is invariant with the target source direction In this case, the reconstructed sound field is stable. However, non-uniform secondary source arrays may not possess these features. Some studies have suggested decoding independent signals of $(L-1)$-order Ambisonics for reproduction with $M_{\text{vir}} \geq L^2$ virtual secondary sources (virtual loudspeakers) that are nearly uniformly arranged on a spherical surface. M_{vir} virtual secondary sources are created with the vector base amplitude panning (VBAP) method in Section 6.3 by M actual secondary sources that are arranged non-uniformly on the spherical surface (Batke and Keiler, 2010; Boehm, 2011; Zotter and Frank, 2012). The maximal allowable order of Ambisonic reproduction depends on angular intervals between actual secondary sources. The overall effect is to decode or convert the independent signals of $(L-1)$-order spatial Ambisonics to be reproduced with a non-uniform array with M actual secondary sources. The aforementioned combination of Ambisonics and VBAP method improves the stability of decoding and satisfies the condition of the constant overall power of driving signals. However, some defects of Ambisonics and VBAP may occur in reproduction.

Conversely, M_{vir} virtual secondary sources are the first to be created by Ambisonics with M actual secondary sources, and VBAP signals of target sources are reproduced by M_{vir} secondary sources. However, this method can create stable virtual secondary sources only when M actual secondary sources are arranged uniformly or nearly uniformly. Therefore, this method fails to take advantage of a combination of VBAP and Ambisonics.

9.4.2 Spatial transformation of Ambisonic sound field

According to Equations (9.3.7) to (9.3.10) and Equations (9.3.21) to (9.3.23), for horizontal or spatial Ambisonics, the normalized amplitudes of the driving signals of M secondary sources are represented by an $M \times 1$ column vector (matrix) $\boldsymbol{A} = [A_1, A_2 \ldots A_{M-1}]^T$. The amplitudes of $(2Q+1)$ or L^2 normalized independent signals are represented by a $(2Q+1) \times 1$ or $L^2 \times 1$ column vector or matrix $\boldsymbol{S}$. Then, driving signals are related to independent signals by using the following decoding equation:

$$\boldsymbol{A}=\left[D\right]\boldsymbol{S}, \tag{9.4.3}$$

where $[D]$ is an $M \times (2Q+1)$ or $M \times L^2$ decoding matrix, and the constant A_{total} is merged into $[D]$ or assumed to be unit. The subscript "2D" or "3D" are temporarily dropped from $[D]$ and $\boldsymbol{S}$. The vector $\boldsymbol{S}$ of normalized independent signals includes the spatial information of the target sound field. If the target sound field undergoes some special spatial transformations, the original normalized independent signal $\boldsymbol{S}$ becomes a set of $(2Q + 1) \times 1$ or $L^2 \times 1$ new normalized independent signals $\boldsymbol{S}'$. Variations in independent signals alter driving signals. If the decoding matrix $[D]$ is unchanged, new independent signals and driving signals can be written as

$$\boldsymbol{S}' = [T]\boldsymbol{S} \qquad \boldsymbol{A}' = [D]\boldsymbol{S}' = [D][T]\boldsymbol{S}, \tag{9.4.4}$$

where $[T]$ is a $(2Q + 1) \times (2Q + 1)$ or $L^2 \times L^2$ matrix for spatial transformation. Variations in driving signals lead to a change in the reconstructed sound field. Therefore, Equation (9.4.4) represents the spatial transformation of the target and corresponding reconstructed sound fields. The situation here is different from independent signal transformation described in Section 4.3.1. In Section 4.3.1, both the forms of independent signals and decoding matrix vary, but the target and reconstructed sound fields are invariant.

Equation (9.4.4) can also be written as:

$$[D'] = [D][T] \qquad \boldsymbol{A}' = [D']\boldsymbol{S} \tag{9.4.5}$$

Equation (9.4.5) indicates that spatial transformation can also be equivalently implemented by changing the decoding matrix from $[D]$ to $[D']$ while keeping the independent signals $\boldsymbol{S}$ unchanged.

Some interesting special spatial transformations are described. The first is the sound field rotation in which the whole sound field is rotated around a certain axis with an angle. In a sound field, a listener's head turning around a certain axis with an angle of $-\gamma$ is equivalent to the whole sound field rotating around the same axis with an opposite angle γ with respect to the listener's head. The rotation of the head or sound field alters binaural pressures. In dynamic and real-time virtual auditory environments with headphones in Section 11.10, binaural pressures should be simulated according to the temporary orientation of the listener's head. This phenomenon is an application of sound field rotation.

In Q-order horizontal Ambisonics and for a target plane wave with a unit amplitude and incident from θ_S, the normalized (amplitudes of) independent signals are 1, $\cos q\theta_S$, and $\sin q\theta_S$, with $q = 1,2\ldots Q$. When the target sound field is rotated anticlockwise around the z-axis (vertical axis) with an azimuth γ_1, independent signal 1 is invariant, and the other independent signals become

$$\begin{aligned} \cos q\theta_S &\rightarrow \cos\left[q(\theta_S + \gamma_1)\right] = \cos q\gamma_1 \cos q\theta_S - \sin q\gamma_1 \sin q\theta_S, \\ \sin q\theta_S &\rightarrow \sin\left[q(\theta_S + \gamma_1)\right] = \sin q\gamma_1 \cos q\theta_S + \cos q\gamma_1 \sin q\theta_S. \end{aligned} \tag{9.4.6}$$

If independent signals are represented by a $(2Q + 1) \times 1$ column vector $\boldsymbol{S} = [1, \cos\theta_S, \sin\theta_S, \ldots\cos Q\theta_S, \sin Q\theta_S]$, the $(2Q + 1) \times (2Q + 1)$ matrix of two-dimensional rotation transformation in Equation (9.4.4) is denoted by $[T_R(\gamma_1)]_{2D}$, then

$$\left[T_R(\gamma_1)\right]_{2D} = \begin{bmatrix} 1 & 0 & 0 & \cdots & 0 & 0 \\ 0 & \cos\gamma_1 & -\sin\gamma_1 & & 0 & 0 \\ 0 & \sin\gamma_1 & \cos\gamma_1 & & 0 & 0 \\ \vdots & & & & & \vdots \\ 0 & 0 & 0 & & \cos Q\gamma_1 & -\sin Q\gamma_1 \\ 0 & 0 & 0 & \cdots & \sin Q\gamma_1 & \cos Q\gamma_1 \end{bmatrix}. \tag{9.4.7}$$

Equation (9.4.7) is a block diagonal matrix. Therefore, rotation around the z-axis leads to mixing between each pair of independent signals in the same order.

In $(L-1)$-order spatial Ambisonics and for a target plane wave with a unit amplitude and incident from direction Ω_S, (normalized) independent signals are a set of spherical harmonic functions $Y_{lm}^{(\sigma)}(\Omega_S)$, $l = 0, 1 \ldots (L-1)$, $m = 0, 1 \ldots l$, $\sigma = 1, 2$. According to the representation of real-valued spherical harmonic functions in Equation (A.1) in Appendix A, when the target sound field is rotated anticlockwise around the z-axis (vertical axis) with γ_1, independent signals $Y_{l0}^{(\sigma)}(\Omega_S)$ are invariant, and the other independent signals with $m \neq 0$ become

$$
\begin{aligned}
Y_{lm}^{(1)}(\Omega_S) &\rightarrow \cos m\gamma_1 Y_{lm}^{(1)}(\Omega_S) - \sin m\gamma_1 Y_{lm}^{(2)}(\Omega_S), \\
Y_{lm}^{(2)}(\Omega_S) &\rightarrow \sin m\gamma_1 Y_{lm}^{(1)}(\Omega_S) + \cos m\gamma_1 Y_{lm}^{(2)}(\Omega_S).
\end{aligned} \tag{9.4.8}
$$

Therefore, rotation around the z-axis leads to mixing between each pair of independent signals with the same (l, m) but with different σ. The matrix of rotation transformation can also be found in Equation (9.4.8).

Generally, when the target sound field is rotated around an axis at arbitrary directions with an angle γ, (normalized) independent signals $Y_{lm}^{(\sigma)}(\Omega_S)$ become

$$
Y_{lm}^{(\sigma)}(\Omega_S) \rightarrow \sum_{m'=0}^{l} \sum_{\sigma=1}^{2} T_{lmm'}^{(\sigma,\sigma')} Y_{lm'}^{(\sigma')}(\Omega_S). \tag{9.4.9}
$$

Rotation leads to a mix among the independent signals of the same order l but with different m and σ. The matrix $[T_R(\gamma)]_{3D}$ of three-dimensional rotation transformation can be obtained if the coefficients $T_{lm\,m'}^{(\sigma,\sigma')}$ are given. The result is also a block diagonal matrix. However, the derivation of the coefficients or entries of a three-dimensional rotation transformation is complicated. In general, rotation around an arbitrary axis can be decomposed into three successive rotations: (1) rotation around the original z-axis with an angle γ_1; (2) rotation around the next y' axis with an angle γ_2; and (3) rotation around the new z-axis with an angle γ_3. Angles $(\gamma_1, \gamma_2, \gamma_3)$ are termed Euler angles (Goldstein, 1980). Therefore, the matrix of three-dimensional rotation transformation can be written as the product of the three rotation matrices with three Euler angles, i.e., $[T_R(\gamma)]_{3D} = [T_R(\gamma_3)]\ [T_R(\gamma_2)]\ [T_R(\gamma_1)]$. The detailed solution of matrix $[T_R(\gamma)]_{3D}$ of three-dimensional rotation transformation can be found in the analysis of angular momentum and rotation operator in a textbook of quantum mechanics (Zeng, 2007; Joshi, 1977).

The aforementioned analysis is for the plane wave field incident from a single direction. For an arbitrary target sound field, independent signals cannot be expressed in simple forms, but the above matrix for three-dimensional rotation transformation is still valid. Another important feature of rotation transformation is that it keeps the overall power of independent signals unchanged:

$$
|S'|^2 = S'^{+} S' = S^{+} S = |S|^2. \tag{9.4.10}
$$

The superscript "+" denotes the transpose and conjugation of the matrix. According to Equation (9.4.4), the matrix for rotation transformation is a unitary matrix that satisfies $[T_R]^{+}[T_R] = [I]$, where $[I]$ is an identity matrix. For real-valued independent signals, the matrix for rotation transformation is simplified into an orthogonal matrix, i.e., rotation transformation is a linear and orthogonal transformation.

The second spatial transformation of interest is spatial reflection in which a target source is reflected in a radially opposite direction. Spatial reflection changes a plane wave incident from direction $\Omega_S = (\alpha_S, \beta_S)$ or (θ_S, ϕ_S) to that incident from $\Omega_S' = (180° - \alpha_S, 180° + \beta_S)$ or $(180° + \theta_S, -\phi_S)$. According to the real-valued representation of the spherical harmonic functions of Equation (A.1) in Appendix A, the spatial reflection changes the independent signal $Y_{lm}^{(\sigma)}(\Omega_S)$ into

$$Y_{lm}^{(\sigma)}(\Omega_S) \to (-1)^l Y_{lm}^{(\sigma)}(\Omega_S). \tag{9.4.11}$$

The matrix $[T_F]$ of spatial reflection transformation for $(L - 1)$-order spatial Ambisonics can be found in Equation (9.4.11). $[T_F]$ is an $L^2 \times L^2$ diagonal matrix, whose entries are +1 or −1. Similar to the case of rotation, spatial reflection keeps the overall power of independent signals unchanged.

The third spatial transformation of interest is dominance or zoom (Gerzon and Barton, 1992). In the first-order horizontal Ambisonics and for a target plane wave with unit amplitude and incident from θ_S, the three independent signals in Equation (4.3.3) satisfy the following relationship:

$$W^2 - X^2 - Y^2 = 0\ . \tag{9.4.12}$$

Dominance at a special direction is a transformation of signals $\boldsymbol{S} = [W, X, Y]^T$ to signals $\boldsymbol{S}' = [W', X', Y']^T$ in Equation (9.4.4) so that the following relationship is held:

$$W'^2 - X'^2 - Y'^2 = 0. \tag{9.4.13}$$

For example, the following transformation is considered:

$$\begin{aligned} W \to W' &= \frac{1}{2}\left(\chi + \chi^{-1}\right)W + \frac{1}{2}\left(\chi - \chi^{-1}\right)X, \\ X \to X' &= \frac{1}{2}\left(\chi - \chi^{-1}\right)W + \frac{1}{2}\left(\chi + \chi^{-1}\right)X \qquad Y \to Y' = Y, \end{aligned} \tag{9.4.14}$$

where $\chi > 0$ is a parameter of transformation. Substituting Equation (4.3.3) into Equation (9.4.14) yields

$$\begin{aligned} &W' = \frac{1}{2}\left(\chi + \chi^{-1}\right) + \frac{1}{2}\left(\chi - \chi^{-1}\right)\cos\theta_S \quad X' = \frac{1}{2}\left(\chi - \chi^{-1}\right) + \frac{1}{2}\left(\chi + \chi^{-1}\right)\cos\theta_S \\ &Y' = \sin\theta_S. \end{aligned} \tag{9.4.15}$$

New driving signals are obtained by substituting Equation (9.4.15) into the decoding Equation (9.4.3) for the first-order horizontal Ambisonics with uniform secondary source array. For a target plane wave incident from the front $\theta_S = 0°$, the normalized amplitudes of new driving signals are χ times of that of original driving signals, i.e., $A'_i(0°) = \chi A_i(0°)$; for target plane wave incident from the back $\theta_S = 180°$, the normalized amplitudes of new driving signals are χ^{-1} times of that of original driving signals, i.e., $A'_i(180°) = \chi^{-1} A_i(180°)$. The overall effect is that the driving signal amplitudes for a target front incident plane wave are χ^2 times of that for a target back incident plane wave. $\chi > 1$ corresponds to forward dominance in which the driving signals for the front incidence are enhanced, and driving signals for the back incidence are restrained. Conversely, $0 < \chi < 1$ corresponds to backward dominance in which driving signals for the back incidence are enhanced, and driving signals for front incidence are restrained.

Dominance changes the perceived virtual source direction in reproduction. After dominance, the azimuth of the velocity localization vector is evaluated by substituting Equation (9.4.15) into Equation (3.2.22):

$$\theta_v = \arccos\frac{\mu + \cos\theta_S}{1 + \mu\cos\theta_S} \quad \mu = \frac{\chi^2 - 1}{\chi^2 + 1}. \tag{9.4.16}$$

For $\chi > 1$, forward dominance moves and focuses all the perceived virtual sources to the front direction except the target direction at $\theta_S = 0°$ and 180°. Conversely, for $0 < \chi < 1$, backward dominance moves and focuses all the perceived virtual sources to the back direction except the target direction at $\theta_S = 0°$ and 180°. The physical nature of dominance is similar to that of changing the front-to-back half-space pick up ratio and perceived virtual source direction in a stereophonic MS microphone technique in Section 2.2.2.

Dominance at a special direction can be extended to the case of the first-order spatial Ambisonics, which includes four independent signals. For a target plane wave incident from Ω_S, the four independent signals are represented by $Y_{00}^{(1)}(\Omega_S)$, $Y_{11}^{(1)}(\Omega_S)$, $Y_{11}^{(2)}(\Omega_S)$, $Y_{10}^{(1)}(\Omega_S)$. According to Equation (A.5) in Appendix A, the four independent signals satisfy the following relationship:

$$3\left[Y_{00}^{(1)}(\Omega_S)\right]^2 - \left[Y_{11}^{(1)}(\Omega_S)\right]^2 - \left[Y_{11}^{(2)}(\Omega_S)\right]^2 - \left[Y_{10}^{(1)}(\Omega_S)\right]^2 = 0. \tag{9.4.17}$$

Independent signals become the form of Equation (6.4.1) by using a different normalization:

$$\begin{aligned} W &= \sqrt{4\pi}Y_{00}^{(1)}(\Omega_S) = 1 \quad X = \sqrt{\frac{4\pi}{3}}Y_{11}^{(1)}(\Omega_S) = \cos\theta_S\cos\phi_S \\ Y &= \sqrt{\frac{4\pi}{3}}Y_{11}^{(2)}(\Omega_S) = \sin\theta_S\cos\phi_S \quad Z = \sqrt{\frac{4\pi}{3}}Y_{10}^{(1)}(\Omega_S) = \sin\phi_S. \end{aligned} \tag{9.4.18}$$

The four independent signals in Equation (9.4.18) satisfy the following relationship:

$$W^2 - X^2 - Y^2 - Z^2 = 0. \tag{9.4.19}$$

Dominance at a special direction is the transformation of $\boldsymbol{S} = [W, X, Y, Z]^T$ to $\boldsymbol{S}' = [W', X', Y', Z']^T$ in Equation (9.4.4) so that the following relationship is held:

$$W'^2 - X'^2 - Y'^2 - Z'^2 = 0. \tag{9.4.20}$$

The aforementioned transformation of dominance at a special direction is the famous ***Lorentz transformation*** in relativity theory (Jackson, 1999).

The forth spatial transformation of interest is the directional emphasis transformation, in which the reproduced sound pressure is emphasized in some special target directions without change of perceived virtual source direction (Kleijn, 2018; Zotter and Frank, 2019). For simplicity, the case of reconstructing target plane wave with directional emphasis is analyzed. Because an arbitrary target sound field in a source-free region can be decomposed as a linear superposition of the plane waves from various directions, the following result is valid for arbitrary target sound field. A target plane wave incident from arbitrary direction Ω_S with magnitude emphasis at a special direction Ω_e is expressed as:

$$P(r, \Omega, \Omega_S, \Omega_e, f) = F(\Omega_S, \Omega_e)\exp\left[jkr\cos(\Delta\Omega_S)\right], \tag{9.4.21}$$

where $\Delta\Omega_S = (\Omega_S - \Omega)$ is the angle between the direction of an incident plane wave and the vector of receiver position. The magnitude of function $F(\Omega_S,\Omega_e)$ maximizes at a target incident direction $\Omega_S = \Omega_e$, which represents the characteristic of directional emphasis.

The plane wave can be decomposed by real-valued spatial harmonic functions according to Equation (9.3.17)

$$\exp\left[jkr\cos(\Delta\Omega_S)\right] = 4\pi\sum_{l'=0}^{\infty}\sum_{m'=0}^{l'}\sum_{\sigma'=1}^{2} j^{l'} j_{l'}(kr) Y_{l'm'}^{(\sigma')}(\Omega_S) Y_{l'm'}^{(\sigma')}(\Omega). \tag{9.4.22}$$

Equation (9.4.22) can be expressed as the form of scale product of two matrix or vectors:

$$\exp\left[jkr\cos(\Delta\Omega_S)\right] = \boldsymbol{H}^T \boldsymbol{S}, \tag{9.4.23}$$

where $\boldsymbol{S} = \boldsymbol{S}_{3D} = \left[Y_{00}^{(1)}(\Omega_S), Y_{11}^{(1)}(\Omega_S), Y_{11}^{(2)}(\Omega_S), Y_{10}^{(1)}(\Omega_S)\ldots\right]^T$ is a $\infty \times 1$ column matrix or vector composed of the normalized amplitude of independent signals of spatial Ambisonics. It is identical to that given in Equation (9.3.22). $\boldsymbol{H}^T$ is a $1 \times \infty$ row matrix or vector with its entries being $4\pi j^{l'} j_{l'}(kr) Y_{l'm'}^{(\sigma')}(\Omega)$.

$F(\Omega_S,\Omega_e)$ can also be decomposed by real-valued spherical harmonic functions as:

$$F(\Omega_S, \Omega_e) = \sum_{l''=0}^{\infty}\sum_{m''=0}^{l''}\sum_{\sigma''=1}^{2} F_{l''m''}^{(\sigma'')}(\Omega_e) Y_{l''m''}^{(\sigma'')}(\Omega_s). \tag{9.4.24}$$

The coefficients $F_{l''m''}^{(\sigma'')}(\Omega_e)$ can be evaluated by Equation (A.12) in Appendix A.

The product of two spherical harmonic functions should be calculated to derive the matrix of directional emphasis transformation. The product of two complex-valued spherical harmonic functions can be expressed as

$$Y_{l'm'}(\Omega_S) Y_{l''m''}(\Omega_S) = \sum_{\substack{l=|l'-l''| \\ m=m'+m''}}^{l=l'+l''} C(l'm', l''m'', lm) Y_{lm}(\Omega_S), \tag{9.4.25}$$

where $C(l'm', l''m'', lm)$ are Clebsch–Gordan coefficients. They are real-valued and can be found in textbooks of quantum mechanics (Joshi, 1977; Zeng, 2007). From Equation (9.4.25) and the relation between real- and complex-valued spherical harmonic functions in Equation (A.9) in Appendix A, the product of two real-valued spherical harmonic functions can also be expressed as a linear combination of real-valued spherical harmonic functions of different orders:

$$Y_{l'0}^{(1)}(\Omega_S)Y_{l''0}^{(1)}(\Omega_S) = \sum_{l=|l'-l''|}^{l'+l''} C(l'0, l''0, l0)Y_{l0}^{(1)}(\Omega_S). \tag{9.4.26}$$

For $m' > 0$,

$$\begin{aligned} Y_{l'm'}^{(1)}(\Omega_S)Y_{l''0}^{(1)}(\Omega_S) &= \sum_{\substack{l=|l'-l''| \\ m=m'}}^{l'+l''} C(l'm', l''0, lm)Y_{lm}^{(1)}(\Omega_S) \\ Y_{l'm'}^{(2)}(\Omega_S)Y_{l''0}^{(1)}(\Omega_S) &= \sum_{\substack{l=|l'-l''| \\ m=m'}}^{l'+l''} C(l'm', l''0, lm)Y_{lm}^{(2)}(\Omega_S). \end{aligned} \tag{9.4.27}$$

And for $m', m'' > 0$,

$$\begin{aligned} Y_{l'm'}^{(1)}(\Omega_S)Y_{l''m''}^{(1)}(\Omega_S) &= \frac{\sqrt{2}}{2}\left[\sum_{\substack{l=|l'-l''| \\ m=m'+m''}}^{l'+l''} C(l'm', l''m'', lm)Y_{lm}^{(1)}(\Omega_S) + \sum_{\substack{l=|l'-l''| \\ m=m'-m''}}^{l'+l''} C(l'm', l''m'', lm)Y_{lm}^{(1)}(\Omega_S)\right] \\ Y_{l'm'}^{(2)}(\Omega_S)Y_{l''m''}^{(2)}(\Omega_S) &= \frac{\sqrt{2}}{2}\left[\sum_{\substack{l=|l'-l''| \\ m=m'-m''}}^{l'+l''} C(l'm', l''m'', lm)Y_{lm}^{(1)}(\Omega_S) - \sum_{\substack{l=|l'-l''| \\ m=m'+m''}}^{l'+l''} C(l'm', l''m'', lm)Y_{lm}^{(1)}(\Omega_S)\right] \\ Y_{l'm'}^{(1)}(\Omega_S)Y_{l''m''}^{(2)}(\Omega_S) &= \frac{\sqrt{2}}{2}\left[\sum_{\substack{l=|l'-l''| \\ m=m'+m''}}^{l'+l''} C(l'm', l''m'', lm)Y_{lm}^{(2)}(\Omega_S) - \sum_{\substack{l=|l'-l''| \\ m=m'-m''}}^{l'+l''} C(l'm', l''m'', lm)Y_{lm}^{(2)}(\Omega_S)\right] \\ Y_{l'm'}^{(2)}(\Omega_S)Y_{l''m''}^{(1)}(\Omega_S) &= \frac{\sqrt{2}}{2}\left[\sum_{\substack{l=|l'-l''| \\ m=m'+m''}}^{l'+l''} C(l'm', l''m'', lm)Y_{lm}^{(2)}(\Omega_S) + \sum_{\substack{l=|l'-l''| \\ m=m'-m''}}^{l'+l''} C(l'm', l''m'', lm)Y_{lm}^{(2)}(\Omega_S)\right]. \end{aligned} \tag{9.4.28}$$

Substituting Equations (9.4.23) and (9.4.24) into Equation (9.4.21) and using Equations (9.4.26) to (9.4.28) yield

$$P(r, \Omega, \Omega_S, \Omega_e, f) = \boldsymbol{H}^T[T_F]\boldsymbol{S}, \tag{9.4.29}$$

where ***HT*** and ***S*** are identical to those in Equation (9.4.23). $[T_F]$ is a sparse $\infty \times \infty$ transformation matrix with its entries being related to the $F_{l''m''}^{(\sigma)}(\Omega_e)$ in Equation (9.4.24) and Clebsch-Gordan coefficients in Equation (9.4.25).

Within a local reproduction region centered at origin, the plane wave field is spatial bandlimited and the spherical decomposition of plane wave in Equation (9.4.22) can be truncated to order $L' - 1$. If function $F(\Omega_S, \Omega_e)$ is also spatial bandlimited, Equation (9.4.24) can be truncated to order $L'' - 1$. Accordingly, Equation (9.4.30) can be truncated to a spherical harmonic order of $(L - 1) = (L' - 1) + (L'' - 1) = L' + L'' - 2$, and ***HT***, ***S*** and $[T_F]$ become $1 \times (L' + L'' - 2)$, $(L' + L'' - 2) \times 1$ and $(L' + L'' - 2) \times (L' + L'' - 2)$ matrix, respectively. A comparison of Equations (9.4.29) and (9.4.23) indicates that the independent signals of a target plane wave field with directional emphasis are related to those of original plane wave field by following transformation:

$$S' = [T_F]S. \tag{9.4.30}$$

Therefore, directional emphasis in spatial Ambisonic reproduction can be implemented by applying a transformation of Equation (9.4.30) to the Ambisonic-independent signals, or equally, by changing the decoding matrix from $[D]$ to $[D']$ with $[T] = [T_F]$ according to Equation (9.4.5) while keeping the independent signals S unchanged.

The aforementioned analysis of directional emphasis transformation is analogous to the theorem of addition of two angular moments in quantum mechanics (Zeng, 2007; Joshi, 1977). In fact, various symmetric transformations are important issues in modern physics. They are applicable to quantum mechanics, quantum field and particle theory, condensed matter physics, and even acoustics (Schroeder, 1989). Group theory is a useful mathematical tool for symmetric analysis. Gerzon (1973) applied group theory to analyze spatial Ambisonics in his early work. According to group theory (Joshi, 1977), all rotations around z-axis constitute the axial rotation group SO(2); all rotations in a three-dimensional space constitute the three-dimensional rotation group SO(3); spatial reflection (inversion) and identity constitute a group $S(2)$ of order 2; SO(3) and S(2) groups constitute the transformation group O(3) in a three-dimensional space. The analysis of secondary source arrays for spatial Ambisonics in Section 9.4.1 is similar to that of crystallographic point groups. Short et al. (2007) applied the theory of the special unitary *group* $SU(n)$ of the degree n to the transformation of multichannel sound signals. Moreover, the directional (beamforming) pattern of driving signals for the first-order Ambisonics with the tetrahedral array of four secondary sources in Figure 6.9 is analogous to electron distribution in a tetrahedral solid (such as silicon); furthermore, the directional pattern of driving signals for the second-order Ambisonics with the array of more secondary sources is analogous to electron distribution in transition metals (Economou, 2006). The discussion in this section shows that the methods in different branches of physics are interchangeable.

9.5 ERROR ANALYSIS OF AMBISONIC-RECONSTRUCTED SOUND FIELD

9.5.1 Integral error of Ambisonic-reconstructed wavefront

Sections 9.3.1 and 9.3.2 indicate that Ambisonics can reconstruct a target plane wave field up to a certain frequency limit and within a circular region centered at the origin. Errors in Ambisonic-reconstructed sound fields involve two parts, i.e., errors caused by approximation in truncating spatial harmonic decomposition of a sound field up to a finite order, and spatial-spectral aliasing errors caused by reproduction with discrete and finite secondary source array. The relationship of the errors in Ambisonic-reconstructed sound field, frequency, and size of a region is analyzed in this section.

Bamford and Vanderkooy (1995) suggested using the following (normalized) mean (integral) complex amplitude error of wavefront to evaluate the error in the reconstructed sound field:

$$Err_1(r,f) = Err_1(kr) = \frac{\int_{-\pi}^{\pi} |P'(r,\theta,f) - P(r,\theta,f)|\,d\theta}{\int_{-\pi}^{\pi} |P(r,\theta,f)|\,d\theta}. \tag{9.5.1}$$

Equation (9.5.1) is the mean normalized absolute value of errors between the reconstructed pressure amplitude $P'(r, \theta, f)$ and the target pressure amplitude $P(r, \theta, f)$ over a circle of the receiver position with radius r. For a target incident plane wave with a unit amplitude, the integral in the denominator of Equation (9.5.1) is 2π.

The error in reconstructed sound field can also be evaluated with the following mean square complex amplitude error of wavefront:

$$Err_2(r, f) = Err_2(kr) = \frac{\int_{-\pi}^{\pi} \left|P'(r, \theta, f) - P(r, \theta, f)\right|^2 d\theta}{\int_{-\pi}^{\pi} \left|P(r, \theta, f)\right|^2 d\theta}. \tag{9.5.2}$$

For a target incident plane wave with a unit amplitude, the integral in the denominator of Equation (9.5.2) is also 2π. The error criteria in Equations (9.5.1) and (9.5.2) are appropriate for arbitrary secondary source arrays and driving signals and not limited to Ambisonics.

For Q-order horizontal far-field Ambisonics, if $P(r, \theta, f)$ is a plane wave with a unit amplitude given in Equation (9.3.1), and the reconstructed sound pressure $P'(r, \theta, f)$ is a truncation of Equation (9.3.2) up to the Q-order, and the mean square error caused by truncation is calculated from Equation (9.5.2) as

$$Err_2(kr) = 1 - \left|J_0(kr)\right|^2 - 2\sum_{q=1}^{Q}\left[J_q(kr)\right]^2. \tag{9.5.3}$$

Errors are independent of the target plane wave direction. When the radius of a region and a wave number satisfies the condition of $kr \leq Q$ in Equation (9.3.15) for ideal reconstruction, the mean square error expressed in Equation (9.5.3) is less than 0.1 or −10 dB. In this case, a target sound field can be reconstructed accurately (Ward and Abhayapala, 2001).

Discrete arrays with a finite number of secondary sources are used in practical horizontal Ambisonics, which cause mirror spatial spectra and aliasing in driving signals. The overall error in the reconstructed sound field is a mix of truncation errors and spatial-spectral aliasing errors. In Sections 9.3.1 and 9.3.2, driving signals are derived by matching each azimuthal harmonic component of the reconstructed sound field with that of the target sound field up to the Q-order. This operation is equivalent to minimizing the mean square error of the reconstructed sound field in Equation (9.5.2). For example, M secondary sources are arranged uniformly in a horizontal circle and satisfy the condition of $M \geq (2Q + 1)$, the target sound field is a plane wave with a unit amplitude expressed in Equation (9.3.1), and driving signals are given in Equation (9.3.12). Ambisonic-reconstructed sound pressure is calculated from Equations (9.3.12) and (9.2.14). Then, the mean square error of the reconstructed sound field can be calculated from Equation (9.5.2). However, analytic results are relatively complicated (Poletti, 2000). Instead, some numerical results have been described (Ward and Abhayapala, 2001; Poletti, 2005b).

Figure 9.7 illustrates the mean square errors of the reconstructed sound field for Q = 1-, 2-, and 3-order horizontal Ambisonics with M = 8 secondary sources and the target plane wave from $\theta_S = 22.5°$. Errors are expressed in decibels. They increase as kr increases at least for $kr \leq 4$. That is, the higher the frequency is and the larger the distance from the origin is, the larger the error will be. However, for a given kr, errors decrease as the Ambisonic order increases, or given the errors, the maximal allowable kr increases with the order of Ambisonics. For example, given the error of $Err_2(kr) \leq -14$ dB, Q = 1-, 2-, and 3-order reproduction have the

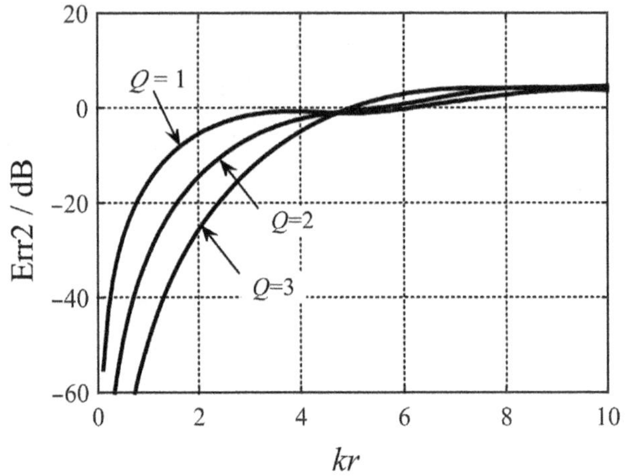

Figure 9.7 Mean square errors of the reconstructed sound field for Q = 1-, 2-, and 3-order horizontal Ambisonics with M = 8 secondary sources arranged at 0°, ±45°, ±90°, ±135°, and 180° and the target plane wave from θ_S = 22.5°.

maximal allowable kr of 1.1, 2.0, and 2.9, respectively. For a region with $r = a = 0.0875$ m (average head radius), the corresponding upper frequency limits are 0.7, 1.2, and 1.9 kHz. This example indicates that Ambisonics can reconstruct a target sound field within a region centered at the origin, and the radius of the region and upper frequency limit for accurate reconstruction increases with the order. This phenomenon is a basic feature of Ambisonics.

Figure 9.8 illustrates the mean square error of complex amplitude of the wavefront for Q = 2-order horizontal Ambisonics with M = 6, 8, and 12 secondary sources to explore the influence of the number of secondary sources on the error of the reconstructed sound field for a given order reproduction. For M = 6 array, which satisfies the number given in Equation (4.3.67), secondary sources are arranged in ±30°, ±90°, and ±150°. For M = 8 array, secondary sources are arranged identical to the example in Figure 9.7. For M = 12 array, secondary sources are arranged from the azimuth of 0° with a uniform azimuthal interval of 30°. For a target incident plane wave, the mean error in Equation (9.5.2) can be equivalently

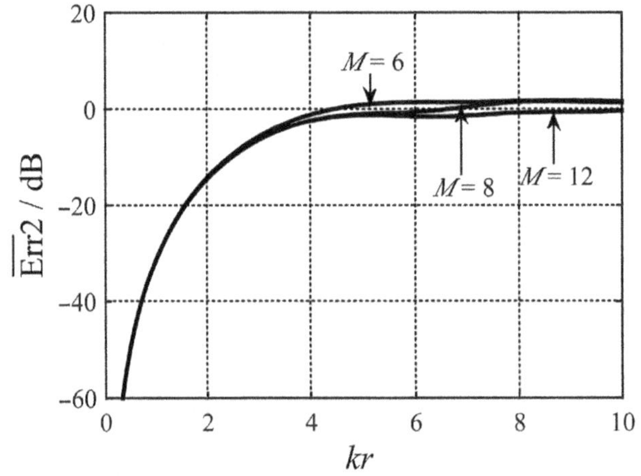

Figure 9.8 Mean square error of complex amplitude of the wavefront for Q = 2-order horizontal Ambisonics with M = 6, 8, and 12 secondary sources.

calculated on the basis of the mean error in a fixed receiver position over a variation in the incident plane wave direction $-180° < \theta_S \leq 180°$. The results are illustrated in Figure 9.8. For $kr \leq 2$ that does not exceed the limit given by Shannon–Nyquist spatial sampling theorem in Equation (9.3.15), errors in different numbers of secondary sources are almost identical. They increase as kr increases, but they are still less than −14 dB. For $kr > 2$, they are obvious; for $kr = 4$, they reach 0 dB. When $kr > 3$, errors depend on the number M of secondary sources. Moreover, they reduce slightly or at least do not increase when M increases from 6 to 8. They further reduce when M increases from 8 to 12.

Below the limit of $kr < Q$ of Shannon–Nyquist spatial sampling theorem, Ambisonics can theoretically reconstruct a target sound field with slight errors. If the number of secondary sources is larger than the lower limit of $M = (2Q + 1)$, errors are basically independent from the number of secondary sources. Above the limit of Shannon–Nyquist spatial sampling theorem, i.e., for $kr > Q$, the error of the reconstructed sound field increases obviously. Errors generally depend on the order and number of secondary sources and target source (or plane wave) directions because the reconstructed sound field is the coherent superposition of those caused by multiple secondary sources. In the above case, increasing the number of secondary sources slightly reduces the error in the reconstructed sound field for $Q > kr$. In other cases, increasing the number of secondary sources may increase the error. The analysis of binaural pressures in Section 12.1.3 yields similar results.

Solvang (2008) analyzed the relationship between ***spectral distortion*** (***SD***) in a reconstructed plane wave and the number of secondary sources in horizontal Ambisonics, where $M \geq (2Q + 1)$ secondary sources are arranged uniformly in a circle for Q-order reproduction. The conclusion for $kr < Q$ is similar to that observed in Figure 9.8, i.e., the SD in the reconstructed sound pressure is small and basically independent of the number of secondary sources. SD is obvious for $kr > Q$. In this case, further increasing the number of secondary sources to $M > (2Q + 1)$ increases SD. However, using $M > (2Q + 1)$ secondary sources reduces the SD around the region of $kr = Q$. Therefore, Solvang concluded that the number of secondary sources should be chosen as a compromise of the SD at $kr > Q$ and around $kr = Q$. Solvang's conclusion is different from the observation in Figure 9.8 at $kr > Q$ and around the region of $kr = Q$.

However, Solvang used a mean SD over receiver positions as the error criterion. SD is defined as the ratio between the power spectra of the reconstructed sound pressure and the target sound pressure. A deviation of SD from a unit (0 dB) indicates an error in reconstructed sound pressures. When the mean SD over receiver positions is calculated, errors in different receiver positions may be wiped out. For example, if the SDs at two receiver positions are 1.5 and 0.5, respectively, the mean SD is a unit (0 dB), which is obviously inappropriate. Therefore, the mean SD in Solvang's analysis is low, and the conclusion should be revised. By contrast, the mean square complex amplitude error of wavefront in Equation (9.5.2) is used for the calculation in Figure 9.8, and the problem is avoided.

The aforementioned error analysis can be extended to spatial reproduction. For $P(r, \Omega, f)$ and $P'(r, \Omega, f)$, Equation (9.5.2) can be extended to the integral overall a spherical surface with radius r:

$$Err_2(kr) = \frac{\int \left| P'(r, \Omega, f) - P(r, \Omega, f) \right|^2 d\Omega}{\int \left| P(r, \Omega, f) \right|^2 d\Omega}. \tag{9.5.4}$$

For a target incident plane wave with unit amplitude, the integral in the denominator of Equation (9.5.4) is 4π.

For the $(L-1)$-order far-field spatial Ambisonics, if $P(r, \Omega, f)$ is a plane wave with a unit amplitude given in Equation (9.3.16) and $P'(r, \Omega, f)$ is a truncation of Equation (9.3.17) up to the $(L-1)$ order, the mean square error caused by truncation is calculated from Equation (9.5.4) as

$$Err_2(kr) = 1 - \sum_{l=0}^{L-1} (2l+1)\left[j_l(kr)\right]^2. \tag{9.5.5}$$

Errors are independent from the target plane wave direction. When the radius of a region and the wave number satisfy the condition of Equation (9.3.30) for ideal reconstruction, the mean square error given in Equation (9.5.5) is less than 0.04 or −14 dB (Ward and Abhayapala, 2001). Similar to the case of horizontal reproduction, discrete arrays with a finite number of secondary sources are used in practical spatial Ambisonics. The overall error in the reconstructed sound field is a mix of truncation and mirror spatial-spectral errors. This analysis is similar to that of horizontal Ambisonics (Poletti, 2005b) and thus omitted here.

9.5.2 Discrete secondary source array and spatial-spectral aliasing error in Ambisonics

Errors caused by the truncation of the spatial harmonic decomposition of Ambisonic sound field are discussed in Section 9.5.1, and a mix of truncation and mirror spatial-spectral errors is analyzed. In this section, mirror spatial-spectral error in Ambisonic-reconstructed sound field caused by discrete array of secondary sources is further analyzed (Spors and Rabenstain, 2006). In the case of digital signal processing, converting the time domain signal to the frequency domain signal is convenient for analyzing the mirror frequency spectra and frequency aliasing caused by the discrete sampling of a continuous time signal. Similarly, converting the sound pressure in the spatial domain to that in the spatial-spectral domain is convenient for the analysis in this section. For convenience in mathematical expression, complex-valued Fourier series is expanded here. In Section 9.2.2, the expansions of complex- and real-valued Fourier series are equivalent.

For continuous and uniform secondary source arrays on a horizontal circle with radius r_0, Equation (9.2.26) gives the relationship of the reconstructed sound pressures, the transfer function from secondary sources to receiver positions, and the driving signals in the spatial-spectral (azimuthal spectrum) domain:

$$P_q'(r, r_0, f) = 2\pi G_q(r, r_0, f) E_q(f) \quad q = 0, \pm 1, \pm 2\ldots \tag{9.5.6}$$

A discrete and uniform secondary source array on a horizontal circle is equivalent to applying spatial or azimuthal sampling to the continuous azimuthal function of the driving signal $E(\theta', f)$ of the secondary source along θ'. For a discrete and uniform array of M secondary sources with an azimuthal interval of $2\pi/M$, the azimuthal-sampled driving signal is

$$E_{\text{samp}}(\theta', f) = E(\theta', f) \sum_{i=0}^{M-1} \delta\left(\theta' - \frac{2\pi i}{M}\right) \frac{2\pi}{M}. \tag{9.5.7}$$

The spatial-spectral domain representation of driving signals is given by the azimuthal Fourier coefficients in Equation (9.5.7):

$$E_{\text{samp},q}(f) = \sum_{v=-\infty}^{+\infty} E_{q+vM}(f). \tag{9.5.8}$$

The spatial or azimuthal spectrum of the sampled driving signal is an infinite repetition of the original azimuthal spectrum with displacements of the integral multiplication of M along q axis, or the q-order azimuthal spectrum component becomes a mix of all original $q + vM$ order azimuthal spectrum components. The azimuthal–spectral domain representation of the reconstructed sound field by discrete secondary source array can be obtained by substituting $E_q(f)$ in Equation (9.5.6) with $E_{samp,q}(f)$ in Equation (9.5.8):

$$P'_{\text{samp},q}(r, r_0, f) = 2\pi \sum_{v=-\infty}^{+\infty} G_q(r, r_0, f) E_{q+vM}(f) \quad q = 0, \pm 1, \pm 2 \ldots \tag{9.5.9}$$

With the azimuthal Fourier representation given in Equation (9.5.9), the reconstructed sound pressure in spatial domain is expressed as

$$\begin{aligned} P'_{\text{samp}}(r, \theta, r_0, f) &= \sum_{q=-\infty}^{\infty} P'_{\text{samp},q}(r, r_0, f) \exp(jq\theta) \\ &= 2\pi \sum_{v=-\infty}^{+\infty} \sum_{q=-\infty}^{+\infty} G_q(r, r_0, f) E_{q+vM}(f) \exp(jq\theta) \\ &= 2\pi \sum_{v=-\infty}^{+\infty} \sum_{q=-\infty}^{\infty} G_{q+vM}(r, r_0, f) E_q(f) \exp\left[j(q+vM)\theta\right]. \end{aligned} \tag{9.5.10}$$

In Equation (9.5.10), the term of $v = 0$ represents the ideal or target reconstructed sound pressure in Equation (9.5.6), and all terms of $v \neq 0$ represent the contributions of mirror spatial or azimuthal spectrum caused by azimuthal sampling. If the azimuthal spectrum of driving signals is azimuthally bandlimited, i.e., for an odd number of discrete secondary sources, the azimuthal spectrum satisfies

$$E_q(f) = \begin{cases} E_q(f) & -\dfrac{M-1}{2} \leq q \leq \dfrac{M-1}{2} \\ 0 & \text{other} \end{cases}, \tag{9.5.11a}$$

for an even number of discrete secondary sources, the azimuthal spectrum satisfies

$$E_q(f) = \begin{cases} E_q(f) & (-M/2+1) \leq q \leq M/2 \\ 0 & \text{other} \end{cases}. \tag{9.5.11b}$$

No overlap exists between target and mirror azimuthal spectra, i.e., no azimuthal spectrum aliasing.

Even if the target driving signal is azimuthally bandlimited and thus satisfies the anti-aliasing condition given in Equation (9.5.11a) or (7.5.11b), the mirror azimuthal spectra caused by azimuthal sampling still appear in the reconstructed sound field and consequently lead to error. This observation is because Green's or transfer functions from secondary sources to receiver positions are not always azimuthally bandlimited, so they may not satisfy the condition of anti-azimuthal-aliasing in some receiver positions. For example, the spatial- or azimuthal-spectral domain representation of the transfer function from a secondary plane wave source to the receiver position is expressed in Equation (9.2.15). The reconstructed sound pressure in the spatial domain is given by substituting Equation (9.2.15) into Equation (9.5.10):

$$P'_{\text{samp}}(r,\theta,r_0,f)=2\pi\sum_{v=-\infty}^{+\infty}\sum_{q=-\infty}^{\infty}E_q(f)\,j^{q+vM}J_{q+vM}(kr)\exp\left[j(q+vM)\theta\right]. \tag{9.5.12}$$

If a driving signal is azimuthally bandlimited with azimuthal harmonics up to order $Q = \max(|q|)$, then similar to the case in Equation (9.3.14), $J_q(kr)$ oscillates and decays when its order q is not less than $[\exp(1)kr / 2]$. Under the condition given by Equation (9.5.13),

$$kr<\frac{2Q}{\exp(1)}, \tag{9.5.13}$$

the function $J_{q+vM}(kr)$ serves as anti-azimuthal-aliasing filtering to remove the influence of mirror azimuthal spectra effectively. Conversely, when the frequency and the distance of the receiver position increase so that Equation (9.5.13) is not satisfied, the mirror azimuthal spectra cause error in reconstructed sound pressure.

Similarly, the azimuthal spectrum representation of the transfer function from a straight-line secondary source to the receiver position is given in Equation (9.2.5). The reconstructed sound pressure in the spatial domain is given by substituting Equation (9.2.20) into Equation (9.5.10):

$$P'_{\text{samp}}(r,\theta,r_0,f)=-\frac{j\pi}{2}\sum_{v=-\infty}^{+\infty}\sum_{q=-\infty}^{+\infty}E_q(f)J_{q+vM}(kr)H_{q+vM}(kr_0)\exp\left[j(q+vM)\theta\right]. \tag{9.5.14}$$

The aforementioned discussion is suitable for reconstruction with straight-line secondary sources.

In Equation (9.5.10), the contribution of $v = 0$ term is denoted by $P'_{ta}(r, \theta, r_0, f)$, the contributions of all $v \neq 0$ terms are denoted by $P'_{er}(r, \theta, r_0, f)$. The relative energy error in the reconstructed sound field caused by azimuthal sampling is evaluated by

$$Err_3(r,\theta)=\frac{\int P'^2_{er}(r,\theta,r_0,f)\,df}{\int P'^2_{ta}(r,\theta,r_0,f)\,df}. \tag{9.5.15}$$

The above analysis can be extended to spatial Ambisonics, but the spatial spectrum aliasing in spatial Ambisonics is more complicated than horizontal Ambisonics. In addition, the analysis of azimuthal spectrum aliasing in this section is suitable not only for horizontal Ambisonics but also for arbitrary sound field reconstruction with a uniform secondary source array in a horizontal circle. This analysis can also be extended to arbitrary three-dimensional sound field reconstruction technique with a uniform secondary source array on a spherical surface. Therefore, the analysis in this section is general.

9.6 MULTICHANNEL RECONSTRUCTED SOUND FIELD ANALYSIS IN THE SPATIAL DOMAIN

9.6.1 Basic method for analysis in the spatial domain

The general formulation for multichannel sound field reconstruction in the spatial domain, including the formulations for continuous secondary source array in Equation (9.2.1) and discrete and finite secondary source array in Equation (9.2.6), is presented in Section 9.2.1. The methods for analyzing a multichannel reconstructed sound field in two special spatial-spectral domains are discussed in Sections 9.2.2 and 9.2.3, and the Ambisonic sound field is analyzed in Section 9.3. The analysis in the spatial-spectral domain is convenient and appropriate for some regular second source arrays, such as uniform or nearly uniform array in a circle or spherical surfaces.

Equation (9.2.1) or (9.2.6) can be directly analyzed and solved in the spatial domain. For continuous and uniform secondary source arrays in a horizontal circle or on a spherical surface, analyses in the spatial domain and the spatial-spectral domain are basically equivalent. However, for arbitrary discrete arrays with a finite number of secondary sources, analyses in the spatial domain often lead to significant results. Given the secondary source array and driving signals, reconstructed sound pressures can be directly evaluated from Equations (9.2.1) and (9.2.6) in the spatial domain. Conversely, under certain conditions, given secondary source array and target sound field, driving signals can be derived directly in the spatial domain.

In the case of discrete array with a finite number of secondary sources, M secondary sources are arranged at positions $\boldsymbol{r}_i$ $(i = 0, 1 \ldots M - 1)$. The receiver position or ***controlled point*** in the sound field is specified by $\boldsymbol{r}_{contro}$, then the sound pressures at a controlled point are calculated with Equation (9.2.6):

$$P'(\boldsymbol{r}_{\text{contro}}, f) = \sum_{i=0}^{M-1} G(\boldsymbol{r}_{\text{contro}}, \boldsymbol{r}_i, f) E_i(\boldsymbol{r}_i, f). \tag{9.6.1}$$

According to Equation (9.6.1) and under certain conditions, driving signals $E_i(\boldsymbol{r}_i, f)$ can be derived by minimizing the error of the reconstructed sound pressures at a set of receiver positions or controlled points within a special spatial region. Indeed, the resultant driving signals depend on the chosen error criteria and controlled points. This phenomenon is the basic consideration and method of multichannel reconstructed sound field analysis in the spatial domain.

9.6.2 Minimizing error in reconstructed sound field and summing localization equation

M secondary sources are arranged in a horizontal circle with radius r_0, the azimuth of the ith secondary source is θ_i, and the normalized amplitude of driving signals is A_i. For a unit signal waveform $E_A(f) = 1$ in the frequency domain, the driving signals of secondary sources are $E_i = A_i$. Two controlled points $\boldsymbol{r}_{\text{contro},L}$ and $\boldsymbol{r}_{\text{contro},R}$ are located at a distance of $r = a$ (radius of head) and azimuth of ±90°, respectively. This simplified head model is used for binaural pressure analysis in Sections 2.1.1 and 3.2.1 in which the effect of head shadow is neglected. Secondary sources are located at a far-field distance with $r_0 \gg a$; according to Equation (9.6.1), pressures at two controlled points are the superposition of pressures caused by incident plane waves from secondary sources:

$$\begin{aligned} P_L' &= P'(r_{\text{contro},L}, f) = \sum_{i=0}^{M-1} A_i \exp\left[jka\cos\left(90^\circ - \theta_i\right)\right], \\ P_R' &= P'(r_{\text{contro},R}, f) = \sum_{i=0}^{M-1} A_i \exp\left[jka\cos\left(-90^\circ - \theta_i\right)\right]. \end{aligned} \tag{9.6.2}$$

Given the target pressures P_L and P_R at two controlled points (two ears), two linear Equations for the normalized amplitudes A_i of driving signals can be obtained by matching the left sides of Equation (9.6.2) with the target pressures. However, in the case of more than two secondary sources ($M > 2$), the equations are underdetermined with more unknown A_i than equations exist. Therefore, a unique solution of driving signal amplitudes cannot be obtained. Conversely, given the driving signal amplitudes, the pressures at two controlled points can be calculated from Equation (9.6.2), and the direction of summing virtual sources can be evaluated.

If a target (virtual) source is located at the far-field distance $r_I \gg a$ and at the azimuth of θ_I, according to Equation (1.2.6), the pressures of a plane wave at two controlled points created by the target sources are given as

$$\begin{aligned} P_L &= P(r_{\text{contro},L}, \theta_I, f) = P_A \exp\left[jka\cos\left(90^\circ - \theta_I\right)\right], \\ P_R &= P(r_{\text{contro},R}, \theta_I, f) = P_A \exp\left[jka\cos\left(-90^\circ - \theta_I\right)\right]. \end{aligned} \tag{9.6.3}$$

The overall square error of complex-valued pressures at two controlled points is evaluated using

$$Err_4 = \left|P'(r_{\text{contro},L}, f) - P(r_{\text{contro},L}, \theta_I, f)\right|^2 + \left|P'(r_{\text{contro},R}) - P(r_{\text{contro},R}, \theta_I, f)\right|^2. \tag{9.6.4}$$

Equations (9.6.2) and (9.6.3) are substituted into Equation (9.6.4) to evaluate the virtual source direction, and the amplitude P_A and the incident azimuth θ_I of a target plane wave are chosen to minimize the error in Equation (9.6.4) or equally

$$\frac{\partial Err_4}{\partial P_A} = 0 \quad \frac{\partial Err_4}{\partial \theta_I} = 0. \tag{9.6.5}$$

The optimal matched target or virtual source direction is found by using the following equation:

$$\sin\theta_I = \frac{1}{ka}\arctan\left[\frac{\sum_{i=0}^{M-1} A_i \sin(ka\sin\theta_i)}{\sum_{i=0}^{M-1} A_i \cos(ka\sin\theta_i)}\right]. \tag{9.6.6}$$

Equation (9.6.6) is the summing localization equation of multiple horizontal secondary sources (loudspeakers) for a fixed head in Equation (3.2.6). At low frequencies with $ka \ll 1$, Equation (9.6.6) is simplified into Equation (3.2.7).

The optimal matched target or virtual plane wave amplitude and the corresponding minimal square error $Err_{4,\min}$ can be evaluated from Equation (9.6.5). The general results are complicated and omitted here. At low frequencies with $ka \ll 1$, the best-matched plane wave amplitude is given as

$$P_A = \sum_{i=0}^{M-1} A_i. \tag{9.6.7}$$

It is the sum of the normalized amplitudes of the driving signals of secondary sources, if the normalized amplitudes of driving signals satisfy

$$\sum_{i=0}^{M-1} A_i = 1. \tag{9.6.8}$$

The optimal matched target sound field is a plane wave with a unit amplitude and incident from azimuth θ_I,

$$P_A = 1. \tag{9.6.9}$$

If the controlled points are continuously and uniformly distributed in a circle centered at the origin and $r = a$, where a can be the head radius, but this condition is not limited to this phenomenon. Under the far-far-field condition, the superposed pressure at the controlled point (a, θ) caused by M secondary sources is given as

$$P'(r_{\text{contro}}, f) = P'(a, \theta, f) = \sum_{i=0}^{M-1} A_i \exp\left[jka\cos(\theta - \theta_i)\right]. \tag{9.6.10}$$

If a target (virtual) source is located at a far-field distance and at an azimuth of $\hat{\theta}_I$, the pressure of the plane wave at the controlled point (a, θ) created by the target sources is given as

$$P(r_{\text{contro}}, \hat{\theta}_I, f) = P(a, \theta, \hat{\theta}_I, f) = P_A \exp\left[jka\cos(\theta - \hat{\theta}_I)\right]. \tag{9.6.11}$$

The integral square error of complex-valued pressure (reconstructed wavefront) over the circle of controlled points is evaluated by

$$Err_5 = \int_{-\pi}^{\pi} \left| P'(a, \theta, f) - P(a, \theta, \hat{\theta}_I, f) \right|^2 d\theta. \tag{9.6.12}$$

Equations (9.6.10) and (9.6.11) are substituted into Equation (9.6.12) to evaluate the virtual source direction, and P_A and $\hat{\theta}_I$ of the target plane wave are chosen to minimize the error in Equation (9.6.12) or equally

$$\frac{\partial Err_5}{\partial P_A} = 0 \quad \frac{\partial Err_5}{\partial \hat{\theta}_I} = 0. \tag{9.6.13}$$

The calculation in Equation (9.6.13) is complicated. However, at low frequencies with $ka \ll 1$, the optimal matched target or virtual source direction is found using the following:

$$\tan\hat{\theta}_I = \frac{\sum_{i=0}^{M-1} A_i \sin\theta_i}{\sum_{i=0}^{M-1} A_i \cos\theta_i}. \tag{9.6.14}$$

The optimally matched amplitude of the target plane wave is determined with the following:

$$P_A = \sum_{i=0}^{M-1} A_i. \tag{9.6.15}$$

Equation (9.6.14) is the summing localization equation in Equation (3.2.9) for the head oriented to the virtual source.

Here, the summing localization equations of multichannel sound reproduction are derived from the criteria of minimizing reconstructed pressure errors in the controlled points. Unlike the derivation in Section 3.2, the psychoacoustic cue (ITD) of low-frequency localization is not considered here. Choosing different controlled points and error criteria and minimizing pressure error lead to different summing localization equations. Minimizing the square error of complex-valued pressures at two ears results in the localization equation for a fixed head; minimizing the integral square error of complex-valued pressures over a circle leads to the localization equation for the head oriented to the virtual source. For horizontal Ambisonics with conventional driving signal mixing, Equations (4.3.60) and (4.3.61) indicate that the two optimal matched conditions can be satisfied at the same time at low frequencies. However, for some other signal panning or mixing methods, such as pair-wise amplitude panning (Section 4.1.2), the two optimal matched conditions cannot be satisfied at the same time. In these cases, the perceived virtual source direction for a fixed head and head oriented to the virtual source is different, especially for a pair of stereophonic loudspeakers (secondary sources) with a large span angle and a pair of side loudspeakers. Therefore, analyzing the reconstructed sound field helps provide insights into the physical nature of summing localization equations.

This method can be extended to the analysis of the reconstructed sound field of multichannel spatial surround sound, but it is omitted here because of the limitation of length.

9.6.3 Multiple receiver position matching method and its relation to the mode-matching method

Under a certain condition, given the discrete and finite secondary source array and controlled points, driving signals can be solved from Equation (9.6.1). Without the loss of generality, the sound pressures at O controlled points specified by the position vector $\boldsymbol{r}_{\text{contro}, o}$, $o = 0, 1 \ldots (O - 1)$ are given as

$$P'\left(\boldsymbol{r}_{\text{contro},o}, f\right) = \sum_{i=0}^{M-1} G\left(\boldsymbol{r}_{\text{contro},o}, \boldsymbol{r}_i, f\right) E_i\left(\boldsymbol{r}_i, f\right) \quad o = 0,1 \ldots (O-1). \tag{9.6.16}$$

Equation (9.6.16) can be written as a matrix form:

$$\boldsymbol{P}'' = [G]\boldsymbol{E}, \tag{9.6.17}$$

where $\boldsymbol{P}' = [P'(\boldsymbol{r}_{\text{contro},0}, f), P'(\boldsymbol{r}_{\text{contro},1}, f), P'(\boldsymbol{r}_{\text{contro},O-1}, f)]^T$ is an $O \times 1$ column vector or matrix composed of the sound pressures at O controlled points; $\boldsymbol{E}$ is an $M \times 1$ column vector or matrix composed of the driving signals of M secondary sources; and $[G]$ is an $O \times M$ matrix composed of the complex-valued transfer functions from M secondary sources to O controlled points, whose entries are $G_{oi} = G(\boldsymbol{r}_{\text{contro},o}, \boldsymbol{r}_i, f)$, $o = 0, 1 \ldots (O - 1)$, $i = 0, 1 \ldots (M - 1)$.

Equation (9.6.17) is the general formulation for controlling the sound pressures at multiple receiver positions by multiple secondary sources. This formulation is suitable for various sound field reconstruction systems and secondary source arrays. From the point of signal processing, this problem occurs in a ***multi-input and multi-output*** (***MIMO***) system. If driving signals in Equation (9.6.17) are chosen so that the reconstructed sound pressures at the O controlled points match with the target sound pressures, then the $O \times 1$ vector on the left side of Equation (9.6.17) becomes $\boldsymbol{P}' = \boldsymbol{P}$. In this case, Equation (9.6.17) is a matrix equation or a set of linear equations with respect to vector $\boldsymbol{E}$ or M driving signals $E_i(\boldsymbol{r}_i, f)$. Solving the vector $\boldsymbol{E}$ of driving signals is a ***multichannel inverse filtering*** problem (Nelson et al., 1996).

When the number of controlled points is equal to the number of secondary sources and matrix $[G]$ is a full rank, i.e., rank $[G] = O = M$, a unique solution of Equation (9.6.17) for driving signals is given as

$$\boldsymbol{E} = [G]^{-1} \boldsymbol{P}. \tag{9.6.18}$$

In this case, the errors in the reconstructed sound pressures at the O controlled points vanish.

Generally, the rank of matrix $[G]$ is rank$[G] = K \le \min(O,M)$. When the number of controlled points is fewer than that of secondary sources, i.e., $K \le O < M$, Equation (9.6.17) is underdetermined, and infinite sets of the solution for driving signals exist. The pseudoinverse solution that minimizes the overall power of driving signals is given as

$$\boldsymbol{E} = [G]^{+} \left\{[G][G]^{+}\right\}^{-1} \boldsymbol{P}. \tag{9.6.19}$$

where superscript "+" denotes the transpose and conjugation of the matrix. Regularization can be applied to the solution to avoid the ill condition or instability in the pseudoinverse of matrix $\{[G][G]^{+}\}$ at some frequencies:

$$\boldsymbol{E} = [G]^{+} \left\{[G][G]^{+} + \varepsilon[I]\right\}^{-1} \boldsymbol{P}, \tag{9.6.20}$$

where $[I]$ is an $O \times O$ identity matrix, and ε is a regularization parameter that balances the stability and accuracy of the solution.

When the number of controlled points is larger than that of secondary sources, i.e., $O > M \geq K$, Equation (9.6.17) is overdetermined and thus without the exact solution. However, an approximate or pseudoinverse solution of driving signals can be found by minimizing the square norm of the error (cost function) between the complex-valued amplitude vectors of the reconstructed and target sound pressures:

$$\min\left(Err_6\right) = \min\left[\left(\boldsymbol{P} - \boldsymbol{P}'\right)^{+}\left(\boldsymbol{P} - \boldsymbol{P}'\right)\right] = \min\left\{\sum_{o=0}^{O-1}\left|P\left(\boldsymbol{r}_{\text{contro},o}, f\right) - P'(\boldsymbol{r}_{\text{contro},o}, f)\right|^2\right. . \quad (9.6.21)$$

The result is given as

$$\boldsymbol{E} = \left\{[G]^{+}[G]\right\}^{-1}[G]^{+}\boldsymbol{P}. \quad (9.6.22)$$

Regularization can be applied to the solution to avoid the ill condition or instability in the pseudoinverse of the matrix $\{[G]^{+}[G]\}$ at some frequencies:

$$\boldsymbol{E} = \left\{[G]^{+}[G] + \varepsilon[I]\right\}^{-1}[G]^{+}\boldsymbol{P}. \quad (9.6.23)$$

where $[I]$ is an $M \times M$ identity matrix.

The method discussed above is the ***least square error method*** for controlling the sound pressures at multiple receiver positions (Kirkeby and Nelson, 1993). Indeed, driving signals obtained by the aforementioned method may not satisfy the causality and thus may be unrealizable. Kirkeby et al. (1996) further proposed a method to obtain causal driving signals in the time domain.

Equation (9.6.17) can also be solved by the method of ***singular value decomposition*** (***SVD***). If the rank of $O \times M$ transfer matrix $[G]$ in Equation (9.6.17) is $K = \text{rank}\,[G] \leq \min(O, M)$, $\{[G][G]^{+}\}$ and $\{[G]^{+}[G]\}$ are $O \times O$ and $M \times M$ Hermitian matrices, respectively; they share K real and positive eigenvalues $\delta_0^2 \geq \delta_1^2 \geq \ldots \geq \delta_{K-1}^2 \geq 0$, and other eigenvalues are zeros:

$$\left\{[G][G]^{+}\right\}\boldsymbol{u}_{\kappa} = \delta_{\kappa}^2\boldsymbol{u}_{\kappa} \quad \left\{[G]^{+}[G]\right\}\boldsymbol{v}_{\kappa} = \delta_{\kappa}^2\boldsymbol{v}_{\kappa} \quad \kappa = 0,1\ldots(K-1), \quad (9.6.24)$$

where the eigenvectors $\boldsymbol{u}_{\kappa}$ and $\boldsymbol{v}_{\kappa}$ are $O \times 1$ left singular value vector and $M \times 1$ right singular value vector of matrix $[G]$, respectively, and they satisfy the following orthogonality and normalization:

$$\boldsymbol{u}_{\kappa'}^{+}\boldsymbol{u}_{\kappa} = \boldsymbol{v}_{\kappa'}^{+}\boldsymbol{v}_{\kappa} = \begin{cases} 1 & \kappa' = \kappa \\ 0 & \kappa' \neq \kappa \end{cases}. \quad (9.6.25)$$

The SVD of $[G]$ is given as

$$[G] = [U][\Delta][V]^{+}. \quad (9.6.26)$$

where $[\Delta]$ is an $O \times M$ singular value matrix, whose K non-zero left-diagonal entries are the singular values of $[G]$ in descending order, i.e., $\delta_0 \geq \delta_1 \geq \ldots \geq \delta_{K-1} \geq 0$, then

$$[\Delta] = \begin{bmatrix} \delta_0 & 0 & 0 & \ldots & & & 0 \\ 0 & \delta_1 & 0 & \ldots & & & 0 \\ & & \ldots & & & & \\ 0 & 0 & \ldots & \delta_{K'-1} & & \ldots & 0 \\ & & & \vdots & & & \\ & & & & & \ldots & 0 \end{bmatrix}. \tag{9.6.27}$$

$[U]$ and $[V]$ are $O \times O$ and $M \times M$ unitarity matrices, respectively, with $[U]^{-1} = [U]^{+}$ and $[V]^{-1} = [V]^{+}$. The preceding K columns of matrix $[U]$ are constructed from the K orthonormal and normalized eigenvectors $\boldsymbol{u}_\kappa$, whereas the preceding K columns of matrix $[V]$ are constructed from the K orthonormal and normalized eigenvectors $\boldsymbol{v}_\kappa$. Substituting Equation (9.6.26) into Equation (9.6.17) yields

$$\boldsymbol{P}' = [U][\Delta][V]^{+}\boldsymbol{E}. \tag{9.6.28}$$

If $\boldsymbol{P}' = \boldsymbol{P}$ at each given frequency, driving signals are solved from Equation (9.6.28):

$$\boldsymbol{E} = [V][1/\Delta][U]^{+}\boldsymbol{P}, \tag{9.6.29}$$

where $[1/\Delta]$ is an $M \times O$ diagonal matrix with K non-zero left-diagonal entries:

$$[1/\Delta] = \begin{bmatrix} \delta_0^{-1} & 0 & 0 & \ldots & & & 0 \\ 0 & \delta_1^{-1} & 0 & \ldots & & & 0 \\ & & \ldots & & & & \\ 0 & 0 & \ldots & \delta_{K-1}^{-1} & & \ldots & 0 \\ & & & \vdots & & & \\ & & & & & \ldots & 0 \end{bmatrix}. \tag{9.6.30}$$

When δ_K is small, the corresponding matrix entry δ_κ^{-1} in Equation (9.6.30) is large, leading to the instability of driving signals in Equation (9.6.29). In this case, the δ_K in Equation (9.6.27) that is larger than a certain threshold is retained, and other small δ_K are discarded. Thus, the solution of driving signals given in Equations (9.6.29) and (9.6.30) becomes stable.

The aforementioned method is essentially a ***multiple receiver position matching*** method through which sound fields are sampled spatially, and sound pressure at O receiver positions is controlled to match with those of target sound pressures as far as possible. If the target sound field is spatially bandlimited, the receiver region can be sampled by a grid of controlled points and the distance between adjacent controlled points does not exceed the minimal half wavelength. According to Shannon–Nyquist spatial sampling theorem, a match of the reconstructed sound pressures at all the controlled points means an accurate reconstruction of target sound field in the concerned region (i.e., the Gibbs effect on the boundary is neglected). Otherwise, spatial aliasing errors occur in the reconstructed sound field. Similar to the case of Ambisonics in Section 9.4.1, the controlled points and secondary source array should be chosen appropriately to obtain stable driving signals by solving Equation (9.6.17) so that the transfer matrix $[G]$ is well-conditioned within the concerned frequency range.

To increase the frequency limit of anti-spatial aliasing, Kolundžija et al. (2011) suggested using secondary sources only that contribute mostly to the reconstructed sound field to control the sound pressures at receiver positions. Active secondary sources in array are selected according to the positions of the target source and reconstructed region on the basis of appropriate geometrical acoustic criteria. In addition, equalization can be introduced to the filters for driving signals; in this way, the overall sound power at the controlled points can be constant.

As an example of multiple receiver positions matching method, horizontal far-field Ambisonics is considered. The target sound field is a plane wave with a unit amplitude and incident from θ_S. O controlled points are located uniformly in a circle with radius r, and the azimuth of the oth controlled point is θ_o, $o = 0, 1 \ldots (O - 1)$. Target sound pressures at controlled points are calculated with Equations (9.3.1) and (9.3.2). M secondary sources are arranged in a circle with radius r_0, azimuth of the ith secondary source is θ_i, the corresponding normalized amplitude of driving signal is A_i, $i = 0, 1 \ldots (M - 1)$. For secondary sources at a far-field distance so that they can be approximated as plane wave sources, the reconstructed sound pressures at O controlled points are calculated using Equation (9.2.14). Matching the reconstructed sound pressures with the target sound pressures at the O controlled points yields a set of O equations:

$$J_0(kr) + 2\sum_{q=1}^{\infty} j^q J_q(kr)(\cos q\theta_S \cos q\theta_o + \sin q\theta_S \sin q\theta_o)$$
$$= \sum_{i=0}^{M-1} A_i \left[J_0(kr) + 2\sum_{q=1}^{\infty} j^q J_q(kr)(\cos q\theta_i \cos q\theta_o + \sin q\theta_i \sin q\theta_o) \right] \tag{9.6.31}$$
$$o = 0, 1 \ldots (o - 1)$$

According to the discussion in Equations (9.3.14) and (9.3.15), the summation of azimuthal harmonics in Equation (9.6.31) can be truncated up to order Q = integer (kr), which is equivalent to the sampling of the sound field along a circle with radius r at an interval of half wavelength. Moreover, driving signals satisfy Equation (9.3.5) if the number of the controlled points or azimuthal sampling points satisfies the condition of $O = M \geq (2Q + 1)$. This result can be proven by multiplying $\cos q\theta_o$ or $\sin q\theta_o$ to both sides of Equation (9.6.31), thereby summing over θ_o and using the discrete orthogonality of trigonometric functions given in Equations (4.3.16) to (4.3.18). The above example indicates that a match of sound pressures at discrete azimuthal sampling points yields results identical to those obtained by a match of sound pressure in a whole continuous circle if the condition of Shannon–Nyquist spatial sampling theorem is satisfied.

Multiple receiver position-matching methods are closely related to the mode-matching method in Section 9.2.2 (Nelson and Kahana, 2001). By substituting $\boldsymbol{P}$ of target sound pressures with $\boldsymbol{P}'$ of arbitrary reconstructed sound pressure and using $[U]^+ = [U]^{-1}$, Equation (9.6.28) becomes

$$[U]^+ \boldsymbol{P}' = [\Delta][V]^+ \boldsymbol{E}, \tag{9.6.32}$$

Or

$$P'_U = [\Delta] E_V, \tag{9.6.33}$$

where

$$P'_U = [U]^+ P' \quad E_V = [V]^+ E. \tag{9.6.34}$$

Therefore, $O \times O$ unitarity matrices $[U]^+$ transform the sound pressure vector $\boldsymbol{P}'$ of the controlled points to a new vector $\boldsymbol{P}'_U$, and $M \times M$ unitarity matrices $[V]^+$ transform the driving signal vector $\boldsymbol{E}$ to a new vector $\boldsymbol{E}_V$. Vector $\boldsymbol{P}'_U$ and $\boldsymbol{E}_V$ are the equivalent representations of $\boldsymbol{P}'$ and $\boldsymbol{E}$, which represent the spatial modes of sound field (pressure) and driving signals, respectively. By using Equations (9.6.24) to (9.6.27), Equation (9.6.32) can be written as

$$\boldsymbol{u}_\kappa^+ \boldsymbol{P}' = \delta_\kappa \boldsymbol{v}_\kappa^+ \boldsymbol{E} \quad \kappa = 0, 1 \ldots (K-1). \tag{9.6.35}$$

Equation (9.6.35) indicates that a special spatial mode component $\mathbf{u}_\kappa^+ \mathbf{P}'$ of the sound field is created by the corresponding spatial mode component $\mathbf{v}_\kappa^+ \mathbf{E}$ of driving signals in the SVD representation. Therefore, this method is applied to control the K independent modes of the reconstructed sound field.

Spatial Ambisonics is analyzed as an example to obtain insights into the relationship between multiple receiver position matching and mode-matching methods. Ambisonics, or more strictly, spatial harmonics decomposition and reconstruction, can be regarded as a method of controlling the independent modes of a reconstructed sound field. M secondary sources are arranged on a spherical surface with radius r_0 and at Ω_i, $i = 0, 1 \ldots (M-1)$. O controlled points are located on a spherical surface with $r < r_0$ and at Ω_o, $o = 0, 1 \ldots (O-1)$. Similar to Equation (9.3.22), an $L^2 \times M$ matrix $[Y_{3D}(\Omega_i)]$ associated with the secondary source array is introduced, and its entries are the real-valued spherical harmonic functions $Y_{lm}^{(\sigma)}(\Omega_i)$ of the secondary source direction. Each row of the matrix corresponds to a given (l, m, σ) with $l = 0, 1 \ldots (L-1)$, $m = 0, 1 \ldots l$, $\sigma = 1, 2$; and each column of the matrix corresponds to a special secondary source direction. Similarly, an $L^2 \times O$ matrix $[Y_{3D}(\Omega_o)]$ associated with the locations of controlled points is introduced, and its entries are the real-valued spherical harmonic functions $Y_{lm}^{(\sigma)}(\Omega_o)$ of the controlled point directions. Each row of the matrix corresponds to a given (l, m, σ) with $l = 0, 1 \ldots (L-1)$, $m = 0, 1 \ldots l$, $\sigma = 1, 2$, and each column of the matrix corresponds to a special controlled point direction.

If both secondary sources and controlled points are uniformly or nearly uniformly distributed on spherical surfaces and the number of secondary sources and the number of controlled points satisfy the requirement of Shannon–Nyquist spatial sampling theorem, and if the directional sampling of spherical harmonic functions satisfies the discrete orthogonality given in Equation (A.25) in Appendix A, the matrices $[Y_{3D}(\Omega_i)]$ and $[Y_{3D}(\Omega_o)]$ satisfy

$$\frac{4\pi}{M}[Y_{3D}(\Omega_i)][Y_{3D}(\Omega_i)]^T = [I] \quad \frac{4\pi}{O}[Y_{3D}(\Omega_o)][Y_{3D}(\Omega_o)]^T = [I], \tag{9.6.36}$$

where $[I]$ is an $L^2 \times L^2$ identity matrix. In Equation (9.6.17), the entries of $[G]$ of the transfer function can be decomposed by real-valued spherical harmonic functions. For secondary point sources, the following equation is obtained from Equations (9.2.37) and (9.2.41) or directly from Equation (9.3.35):

$$G_{oi} = G(r_{\text{contro},o}, r_i, f) = \sum_{l=0}^{\infty}\sum_{m=0}^{l}\sum_{\sigma=1}^{2} g_l Y_{lm}^{(\sigma)}(\Omega_o) Y_{lm}^{(\sigma)}(\Omega_i) \quad g_l = -jkh_l(kr_0)j_l(kr). \tag{9.6.37}$$

Truncating Equation (9.6.37) up to the order $(L - 1)$, matrix $[G]$ can be written as

$$[G] = [Y_{3D}(\Omega_o)]^T [g][Y_{3D}(\Omega_i)], \tag{9.6.38}$$

where $[g]$ is an $L^2 \times L^2$ diagonal matrix, whose diagonal entries associated with the l-order spherical harmonic functions are denoted by g_l. Substituting Equation (9.6.38) into Equation (9.6.17) yields

$$\boldsymbol{P}' = [Y_{3D}(\Omega_o)]^T [g][Y_{3D}(\Omega_i)]\boldsymbol{E}. \tag{9.6.39}$$

Multiplying $[Y_{3D}(\Omega_o)]$ to both sides of Equation (9.6.39) and using Equation (9.6.36) yields

$$\frac{4\pi}{O}[Y_{3D}(\Omega_o)]\boldsymbol{P}' = [g][Y_{3D}(\Omega_i)]\boldsymbol{E}. \tag{9.6.40}$$

The left side of Equation (9.6.40) represents an $L^2 \times 1$ column vector $\boldsymbol{P}_{lm}'^{(\sigma)}$ of the preceding $(L - 1)$ order spherical harmonic coefficients (spectrum) of the sound pressures at O discrete controlled points. This result can be derived by using the discrete orthogonality of the spherical harmonic function given in Equation (9.3.68). The component of vector $\boldsymbol{P}_{lm}'^{(\sigma)}$ is given in $P_{lm}'^{(\sigma)}(r, r_0, f)$ in Equation (9.2.31). Similarly, the right side of Equation (9.6.40) represents an $L^2 \times 1$ column vector $\boldsymbol{E}_{lm}^{(\sigma)}$ of the preceding $(L - 1)$ order spherical harmonic coefficients (spectrum) of the driving signals of M secondary sources. The component of vector $\boldsymbol{E}_{lm}^{(\sigma)}$ is expressed in $E_{lm}^{(\sigma)}(f)$ in Equation (9.2.44). Then, Equation (9.6.40) becomes

$$\boldsymbol{P}_{lm}'^{(\sigma)} = \frac{M}{4\pi}[g]\boldsymbol{E}_{lm}^{(\sigma)}. \tag{9.6.41}$$

That is,

$$P_{lm}'^{(\sigma)}(r, r_0, f) = \frac{M}{4\pi} g_l E_{lm}^{(\sigma)}(f). \tag{9.6.42}$$

Equation (9.6.42) indicates that a special spatial mode component of the sound field is also created by the corresponding spatial mode component of driving signals in spherical harmonic representation. Equation (9.6.42) is equivalent to Equation (9.2.45) except for a scale caused by the discrete secondary source array.

Two matrices, namely, an $L^2 \times O$ matrix $[T_U]$ and an $L^2 \times M$ matrix $[T_V]$ are introduced, and they satisfy $[T_U]\,[T_U]^+ = [I]$ and $[T_V]\,[T_V]^+ = [I]$, where $[I]$ is an $L^2 \times L^2$ identity matrix. Inserting these two matrices into Equation (9.6.39) yields

$$\boldsymbol{P}' = [Y_{3D}(\Omega_o)]^T [T_U][T_U]^+ [g][T_V][T_V]^+ [Y_{3D}(\Omega_i)]\boldsymbol{E}. \tag{9.6.43}$$

Comparing Equations (9.6.43) and (9.6.28) yields

$$[U]=\left[Y_{3D}(\Omega_o)\right]^T[T_U] \quad [V]^+=[T_V]^+\left[Y_{3D}(\Omega_i)\right] \quad [\Delta]=[T_U]^+[g][T_v]. \tag{9.6.44}$$

Spherical harmonic decomposition and SVD are two different spatial mode decomposition methods, and they are related to matrix transformation in Equation (9.6.44). Spherical harmonic decomposition is more suitable for secondary source arrays that satisfy the discrete orthogonality of spherical harmonic functions, and SVD is more suitable for various irregular secondary source arrays. Hamdan and Fazi (2021) analyzed the mode of multichannel sound fields and their relationship to amplitude panning by SVD in detail. In addition to spherical harmonic decomposition, spatial wavelet decomposition (which is closely related to Ambisonics and VBAP), is also applicable to multichannel sound field reconstruction (Scaini and Arteaga, 2020).

Multiple receiver position matching is a general method for sound field reconstruction. It is appropriate for various secondary source arrays, which are not limited to Ambisonics with regular circular arrays and nearly uniform spherical surface arrays. Indeed, the above example reveals the relationship between multiple receiver position matching and mode-matching methods. In addition, multiple receiver position matching and mode-matching methods are common in active noise control (Nelson and Elliott, 1992; Kuo and Morgan, 1999). A similarity exists between sound reproduction and active noise control (Yang and Gan, 2008). The former is used to reconstruct a target sound field within a certain region, and the latter is applied to create a sound field to cancel the primary (original) sound field within a certain region.

Kirkeby and Nelson (1993); Kirkeby et al. (1996) investigated the reconstruction of a plane wave in a local region by using two or four secondary sources in the front as an application of multiple receiver position matching. Moreover, the original signals for multichannel sound can be first upmixed to more channel signals and fed to an appropriate secondary source array to reconstruct a target sound field and enlarge the listening region in discrete multichannel (such as 5.1 channel) sound reproduction. On the basis of multiple receiver position matching, Bai et al. (2014) proposed a method for multichannel surround sound signal upmixing and sound field reconstruction. The multiple receiver position matching method is also applicable to spatial multizone sound field reconstruction in Section 9.3.6 (Poletti, 2008; Park et al., 2010). One practical application of this method is personal audio reproduction in cars (Cheer et al., 2013).

9.7 LISTENING ROOM REFLECTION COMPENSATION IN MULTICHANNEL SOUND REPRODUCTION

In the preceding sections, multichannel sound field reconstruction is implemented in a free-field environment (such as an anechoic chamber). However, most practical sound reproductions are conducted in reflective environments. Reflections from a listening room cause errors in a reconstructed sound field. In this case, appropriate preprocessing can be applied to driving signals to compensate for the influence of reflections from listening rooms. The physical essence of this compensation is that secondary sources create a sound field with a reversal phase with respect to the reflected sound field and then cancel the reflective sound field in a certain region through destructive interference and recreate the target sound field. Therefore, this technique is an active sound field control method and is applicable to Ambisonics and other spatial sounds based on sound field reconstruction, such as wavefield synthesis in

Chapter 10. However, the active method of reflection control is usually valid at certain controlled points or a certain region. The sound field error or deviation may be larger outside the region.

A direct method involves using a multiple receiver position matching method to cancel reflections from a listening room (Mourjopoulos, 1994; Asano and Swanson, 1995; Gauthier et al., 2005). The principle of this method is similar to that in Section 9.6.3 as long as $G(r_{\text{contro, o}}, r_i, f)$ in Equation (9.6.16) and the subsequent equation are taken as the transfer functions from secondary sources to the controlled points with the existence of reflections from the listening room. These transfer functions can be obtained through measurement.

Reflections from the listening room can also be cancelled by the mode-matching method. Sontacchi and Hoeldrich (2000) proposed a relatively simple method. For example, in spatial Ambisonics, similar to the measurement of the spatial room impulse responses of Ambisonics in Section 7.5.5, for a given secondary array in a listening room, a set of spatial room impulse responses from each secondary source to the origin (center of secondary source array) are first measured using Ambisonic recording techniques (such as a set of coincident microphones or microphone arrays in Section 9.8). Spatial room impulse responses of the ith secondary source are denoted by $g_{i,lm}^{(\sigma)}(t)$, $i = 0,1,\ldots(M-1)$; $l = 0,1,\ldots(L-1)$; $m = 0,1\ldots l$; $\sigma = 1, 2$, corresponding to the (l, m, σ) order spherical harmonic components or independent signals $Y_{lm}^{(\sigma)}$ of Ambisonics. For example, spatial impulse responses corresponding to the B-format first-order Ambisonic-independent signals W, X, Y, and Z or spherical harmonic components $Y_{00}^{(1)}$, $Y_{11}^{(1)}$, $Y_{11}^{(2)}$, $Y_{10}^{(1)}$ (Sections 6.4.1 and 9.3.2) can be measured with an omnidirectional microphone and three bidirectional microphones with their main axes pointing to the front, left, and top directions, respectively. Each impulse response $g_{i,lm}^{(\sigma)}(t)$ involves two parts: the first part is the response of direct sound from the secondary source to the receiver position, which is necessary for sound field reconstruction, and the second part is the response of listening room reflections, which should be cancelled. The listening room reflection part $g_{i,lm}^{\prime(\sigma)}(t)$ of each impulse response $g_{i,lm}^{(\sigma)}(t)$ is extracted by an appropriate time window and then multiplied by −1 to inverse its phase to serve as the impulse response to cancel listening room reflections. The frequency domain response of $-g_{i,lm}^{\prime(\sigma)}(t)$ is denoted by $-G_{i,lm}^{\prime(\sigma)}(f)$.

In a free-field reproduction environment, the normalized driving signal amplitudes of $(L - 1)$ order Ambisonics are given by the following decoding equation:

$$A = A_{\text{total}}\left[D_{3D}\right]S_{\text{3D}}, \tag{9.7.1}$$

where $A = [A_0, A_1 \ldots A_{M-1}]^T$ is an $M \times 1$ column vector or matrix composed of the normalized amplitudes of M driving signals, and A_{total} is constant for normalization. $[D_{3D}]$ and S_{3D} are $M \times L^2$ decoding matrix and $L^2 \times 1$ column vector of independent signals, respectively, as described in Equations (9.3.22) and (9.3.23).

Each driving signal of each secondary source in a free-field reproduction environment is filtered with $\left[-G_{i,lm}^{\prime(\sigma)}(f)\right]$ to cancel listening room reflection, and the filtered signals from all secondary sources are summed from Ambisonic-independent signals with listening room reflection cancellation:

$$S_{\text{can},lm}^{(\sigma)}(f) = -\sum_{i=0}^{M-1} G_{i,lm}^{\prime(\sigma)}(f)A_i. \tag{9.7.2}$$

The signals in Equation (9.7.2) can be written in the form of an $L^2 \times 1$ column vector or matrix:

$$S_{can} = \left[S^{(1)}_{can,00}(f), S^{(1)}_{can,11}(f), S^{(2)}_{can,11}(f), S^{(1)}_{can,10}(f) \ldots S^{(1)}_{can,(L-1)0}(f) \right]^T. \qquad (9.7.3)$$

S_{can} is decoded and added to the normalized driving signal amplitudes for a free-field reproduction environment, leading to normalized driving signal amplitudes with listening room reflection cancellation:

$$A' = A_{total}[D_{3D}]S_{3D} + A_{total}[D_{3D}]S_{can} = A_{total}[D_{3D}]\{[I] + A_{total}[G][D_{3D}]\}S_{3D}, \qquad (9.7.4)$$

where $[I]$ is an $L^2 \times L^2$ identity matrix, and $[G]$ is an $L^2 \times M$ matrix, whose entries are $-G'^{(\sigma)}_{i,lm}(f)$ in Equation (9.7.2). Each row of the matrix $[G]$ corresponds to a given spherical harmonic component of (l, m, σ), and each column of the matrix corresponds to a special secondary source direction.

Betlehem and Abhayapala (2005) used the mode-matching method in the spatial-spectral domain (Section 9.2.2) to analyze listening room reflection cancellation. For horizontal Ambisonic reproduction with an array of straight-line secondary sources, given the transfer functions from secondary sources to receiver positions (the frequency domain counterparts of spatial room impulse responses), the azimuthal Fourier coefficients of these transfer functions can be calculated. Betlehem and Abhayapala analyzed the azimuthal Fourier coefficients of the transfer functions from secondary sources to discrete receiver positions located uniformly in two concentric circles with different radii (similar to the dual-radius open spherical microphone array mentioned in Section 9.8.3). They derived the driving signals by matching the pressure within a circular region of a certain radius with that of the target pressure in the sense of the least square pressure error. The simulation indicates that this method exhibits good performance for reproduction in reflective rooms and reduces the error of the reconstructed sound field up to 5 dB, which is better than that of the conventional multiple receiver point matching method with the least square error. This simulation is equivalent to canceling the reflections in each order according to their azimuthal spectrum. Indeed, effective reflection cancellation is realized only when secondary source array satisfies the requirement of Shannon–Nyquist spatial sampling theorem.

Lecomte et al. (2018) extended the aforementioned mode-matching method to spatial Ambisonics. Moreover, the room responses are corrected in the Ambisonic-independent signals rather than after decoding to simplify the number of required filters.

9.8 MICROPHONE ARRAY FOR MULTICHANNEL SOUND FIELD SIGNAL RECORDING

Various microphone techniques for stereophonic and multichannel sounds are discussed in Chapters 2, 4, 6, and 7. In particular, the independent signals of Ambisonics can be theoretically recorded with a coincident configuration of microphones with various orders of directivities. However, coincident microphone configurations can usually record the first-order Ambisonic signals only because of difficulty in realizing microphones with higher-order directivities (although microphones with two-order directivities can be achieved; Cengarle et al., 2011). Moreover, arranging multiple coincident microphones is difficult. Therefore, conventional microphone techniques are not enough to record higher sound field signals. Some new techniques are required.

The analyses in Sections 9.1 and 9.3 indicate that Ambisonics is a physical scheme of the spatial sampling and reconstruction of a sound field. A microphone array is an effective method for the spatial sampling of a sound field, where the spatial information of sound field is recorded through the special configuration of multiple microphones. Under certain conditions, the spatial harmonic components of the sound field or independent signals of Ambisonics can be obtained by applying appropriate frequency and spatial filtering processing to microphone array outputs. Moreover, signals with other different directivities can be derived from the weighted combinations of microphone array outputs by appropriately choosing frequency-dependent weights. As stated in Sections 9.1 and 9.2, Ambisonic signal recording can be regarded as a beamforming method; therefore, Ambisonic driving signals can be obtained directly by applying beamforming processing to microphone array outputs. Beamforming aims to enhance the recorded outputs at the target direction and restrain the outputs in other directions. Signals recorded by a microphone array are also applicable to other multichannel sound reproduction and even binaural reproduction (Section 11.6.1). Therefore, a microphone array is a flexible method for sound field recording, as discussed in this section.

9.8.1 Circular microphone array for horizontal Ambisonic recording

The independent signals of horizontal Ambisonics are represented by azimuthal harmonics of different orders. They can be recorded with a horizontal circular microphone array (Poletti, 2000, 2005a; Zotkin et al., 2010; Hulsebos et al., 2002).

M' (omnidirectional) microphones are arranged uniformly in a horizontal circle with radius r_M, which is equivalent to an M'-point sampling of the sound field along the circle. The circle is open or located on the surface of a rigid cylinder with infinite length. Similar to Equation (4.2.7), the azimuth of the i'th microphone is given as

$$\theta_{i'} = \theta_0 + \frac{360^\circ i'}{M'} \quad i' = 0,1\ldots(M'-1). \tag{9.8.1}$$

For a target plane wave with a unit amplitude and incident from the horizontal azimuth θ_S, the pressure received by the i'th microphone (or the microphone output) can be written as

$$E_{i'}(\theta_{i'}, kr_M, \theta_S) = P_{i'}(\theta_{i'}, kr_M, \theta_S) = \sum_{q=0}^{+\infty} R_q(kr_M)(\cos q\theta_{i'} \cos q\theta_S + \sin q\theta_{i'} \sin q\theta_S),. \tag{9.8.2}$$

where $\sin q\theta_S = 0$ for $q = 0$, which is preserved in Equation (9.8.2) for convenience in writing. In the first equality in Equation (9.8.2), the transfer coefficient from the pressure at the receiver position to the microphone output is assumed to be a unit, i.e., $E_A = P_A$. The radial function $R_q(kr_M)$ depends on frequency and r_M. For an open circular array, $R_q(kr_M)$ is found by comparing Equations (9.3.2) and (9.8.2) and letting $r = r_M$:

$$R_q(kr_M) = \begin{cases} J_0(kr_M) & q = 0 \\ 2j^q J_q(kr_M) & q \geq 1 \end{cases}. \tag{9.8.3}$$

For a rigid cylindrical microphone array, the solution of scattering the plane wave with a rigid cylinder leads to the following (Morse and Ingrad, 1968):

$$R_0(kr_M) = J_0(kr_M) - \frac{dJ_0(kr_M)/d(kr_M)}{dH_0(kr_M)/d(kr_M)} H_0(kr_M) = -\frac{2j}{\pi kr_M} \frac{1}{dH_0(kr_M)/d(kr_M)},$$

$$R_q(kr_M) = 2j^q\left[J_q(kr_M) - \frac{dJ_q(kr_M)/d(kr_M)}{dH_q(kr_M)/d(kr_M)} H_q(kr_M)\right] = -\frac{4j^{q+1}}{\pi kr_M} \frac{1}{dH_q(kr_M)/d(kr_M)}. \tag{9.8.4}$$

$$q = 1, 2, 3\ldots,$$

where $H_q(kr_M)$ is the Hankel function of a secondary kind. If a sound field is spatially bandlimited, the contribution of all $l \geq L$ order azimuthal harmonic components in Equation (9.8.2) is negligible. When Equation (9.8.2) and the discrete orthogonality of trigonometric functions are used and when

$$M' \geq (2Q+1), \tag{9.8.5}$$

the following equation is obtained:

$$S_0^{(1)} = \frac{1}{M'R_0(kr_M)} \sum_{i'=0}^{M'-1} E_{i'}(\theta_{i'}, kr_M, \theta_S) = 1,$$

$$S_q^{(1)} = \frac{2}{M'R_q(kr_M)} \sum_{i'=0}^{M'-1} E_{i'}(\theta_{i'}, kr_M, \theta_S)\cos q\theta_{i'} = \cos q\theta_S, \tag{9.8.6}$$

$$S_q^{(2)} = \frac{2}{M'R_q(kr_M)} \sum_{i'=0}^{M'-1} E_{i'}(\theta_{i'}, kr_M, \theta_S)\sin q\theta_{i'} = \sin q\theta_S \qquad q = 1, 2\ldots Q.$$

These equations are the normalized amplitudes of the $(2Q + 1)$ independent signals of Q-order horizontal Ambisonics. Therefore, Equation (9.8.6) indicates that the preceding Q-order independent signals of horizontal Ambisonics are obtained by the weighted summations of M' microphone outputs and equalizing (filtering) with $[1/R_q(kr_M)]$. Its processing is valid only when $R_q(kr_M) \neq 0$. This is the basic principle of horizontal Ambisonic recording by a circular microphone array.

In the case of a spatially bandlimited incident sound field that only involves the azimuthal harmonics up to order Q., all $q > Q$ components can be omitted in the summation of Equation (9.8.2), and a Q-order recording is enough to capture the spatial information of a sound field. Equation (9.8.5) gives the minimal number of microphones for Q-order recording, which is the consequence of Shannon–Nyquist spatial sampling theorem. Spatial aliasing occurs when the condition of Equation (9.8.5) is not satisfied, i.e., when the incident sound field involves $Q' > Q$ azimuthal harmonic components, but it is recorded with an array with $M' = (2Q + 1) < (2Q' + 1)$ microphones. In this case, $q > Q$-order azimuthal harmonic components are mixed to $q \leq Q$ azimuthal harmonic components given in Equation (9.8.6), thereby causing errors in recorded signals. When Q' tends to infinite, all $q + vM'$ $(v = 1, 2, 3\ldots)$ order azimuthal harmonic components are mixed to $q \leq Q$-order azimuthal harmonic signals. The upper limit of Q in Equation (9.8.2) can be evaluated by a formula similar to

Equation (9.3.14) or (9.3.15) by substituting r of the reconstructed region with r_M of a microphone array because the Bessel function $J_q(kr)$ oscillates and decays when its order q is not less than $[\exp(1)kr / 2]$. In Equations (9.3.14) and (9.8.5), Q and the minimal required number of microphones M' increases with kr_M. Given M' and r_M, spatial aliasing occurs above the certain upper frequency limit.

When the independent signals of Equation (9.8.6) are obtained and the uniform secondary source arrays in a horizontal plane are given, driving signals can be derived from the decoding equation expressed in Equation (4.3.63) or (9.3.12). Indeed, driving signals can be directly derived from microphone array outputs. For horizontal array with $M \geq (2Q + 1)$ secondary sources, the amplitudes of driving signals for Q-order reproduction and secondary sources at θ_i are given in the following [with amplitude normalized, see Equation (9.3.12) or (4.3.80a)]:

$$A_i(\theta_S) = \frac{1}{M}\left[1 + 2\sum_{q=1}^{Q}\left(\cos q\theta_i \cos q\theta_S + \sin q\theta_i \sin q\theta_S\right)\right]$$

$$= \frac{1}{MM'}\left\{\frac{1}{R_0(kr_M)}\sum_{i'=0}^{M'-1} E_{i'}(\theta_{i'}, kr_M, \theta_S)\right. \tag{9.8.7}$$

$$\left. + \sum_{q=1}^{Q}\frac{4}{R_q(kr_M)}\sum_{i'=0}^{M'-1}\left[E_{i'}(\theta_{i'}, kr_M, \theta_S)(\cos q\theta_{i'} \cos q\theta_i + \sin q\theta_{i'} \sin q\theta_i\right]\right\}.$$

This equation corresponds to the driving signals derived from the beamforming processing of microphone array outputs.

The case of a target plane wave with a unit amplitude and incident from a given direction is analyzed above. Similar to the case in Sections 9.1.1 and 9.3.1, for a general incident sound field in the horizontal plane, the sound pressure at a receiver position close to the origin is the superposition of plane wave incident from various azimuths. Accordingly, the contributions of these plane waves are included in each order independent signal given in Equation (9.8.6). The desired driving signals are decoded from these independent signals. Microphone array recording is an azimuthal sampling procedure on the azimuthal distribution function of incident plane wave amplitude. The corresponding time domain signals are obtained by applying an inverse Fourier transform similar to Equation (1.2.13) to the frequency domain signals.

Two azimuthal sampling procedures are used in recording and reproduction. The first azimuthal sampling occurs at the stage of recording through which M' microphones are used to sample the sound pressure in receiver positions along a circle. The second azimuthal sampling occurs in the stage of Ambisonic reproduction by which M secondary sources are arranged in a circle. The schemes of azimuthal sampling for recording and reproduction may be identical or different, but they should satisfy the requirement of Shannon–Nyquist azimuthal sampling theorem, i.e., $M' \geq (2Q + 1)$ and $M \geq (2Q + 1)$.

9.8.2 Spherical microphone array for spatial Ambisonic recording

The circular microphone array in Section 9.8.1 can be extended to a spherical microphone array for spatial Ambisonic recording (Rafaely, 2004, 2005; Meyer and Elko, 2004; Poletti 2005b; Moreau et al., 2006; Zotkin et al., 2010), and the details of spherical microphone arrays are also referred to the book of Rafaely (2015). A spherical microphone array involves

M' microphones arranged on an open or rigid spherical surface with radius r_M. The direction of the i'th microphone is $\Omega_{i'}$, $i' = 0, 1 \ldots (M' - 1)$. This array is equivalent to sampling the sound field on a spherical surface. For a plane wave with a unit amplitude and incident from Ω_S, the pressure received by the i'th microphone (or microphone output) can be decomposed by real- or complex-valued spherical harmonic functions. The real-valued spherical harmonic function decomposition is given as

$$E_{i'}(kr_M, \Omega_{i'}, \Omega_S) = P_{i'}(kr_M, \Omega_{i'}, \Omega_S) = \sum_{l=0}^{\infty}\sum_{m=0}^{l}\sum_{\sigma=1}^{2} R_l(kr_M) Y_{lm}^{(\sigma)}(\Omega_{i'}) Y_{lm}^{(\sigma)}(\Omega_S), \qquad (9.8.8)$$

where $Y_{l0}^{(2)}(\Omega) = 0$, which is preserved in Equation (9.8.8) for convenience in writing, and $R_l(kr_M)$ is a radial function. In the first equality in Equation (9.8.8), the transfer coefficient from the pressure at the receiver position to the microphone output is assumed to be a unit, i.e., $E_A = P_A$.

For an open spherical microphone array, similar to Equations (9.3.16) and (9.3.17), the plane wave pressure received by a microphone can be decomposed with spherical harmonic functions:

$$P_{i'}(kr_M, \Omega_{i'}, \Omega_S) = \exp\left[jkr_M \cos(\Delta\Omega_{S,i'})\right] = 4\pi \sum_{l=0}^{\infty}\sum_{m=0}^{l}\sum_{\sigma=1}^{2} j^l j_l(kr_M) Y_{lm}^{(\sigma)}(\Omega_{i'}) Y_{lm}^{(\sigma)}(\Omega_S), \qquad (9.8.9)$$

where $j_l(kr_M)$ is the l-order spherical Bessel function, $\Delta\Omega_{S,i'} = \Omega_S - \Omega_{i'}$ is the angle between the (source) direction of the incident plane wave and the direction of the i'th microphone. Comparing Equations (9.8.8) and (9.8.9), $R_l(kr_M)$ can be written as

$$R_l(kr_M) = 4\pi j^l j_l(kr_M). \qquad (9.8.10)$$

For a rigid spherical microphone array, the solution of plane wave scattering by a rigid sphere leads to the following (Du et al., 2001; Morse and Ingrad, 1968):

$$\begin{aligned} R_l(kr_M) &= 4\pi j^l \left[j_l(kr_M) - \frac{dj_l(kr_M)/d(kr_M)}{dh_l(kr_M)/d(kr_M)} h_l(kr_M) \right] \\ &= -\frac{4\pi j^{l+1}}{(kr_M)^2} \frac{1}{dh_l(kr_M)/d(kr_M)} \end{aligned} \qquad (9.8.11)$$

where $h_l(kr_M)$ is the l-order spherical Hankel function of the second kind.

M' microphones are arranged on the spherical surface according to the directional sampling scheme given in Appendix A. If the sound field is spatially bandlimited, the contribution of all $l \geq L$ order spherical harmonic components in Equation (9.8.9) is negligible. The normalized amplitudes of preceding $(L - 1)$ order independent signals of spatial Ambisonics are expressed in Equation (9.8.12) by using the discrete orthogonality of spherical harmonic functions given in Equation (A.20) in Appendix A:

$$S_{lm}^{(\sigma)}(\Omega_S) = \frac{1}{R_l(kr_M)} \sum_{i'=0}^{M'-1} \lambda'_{i'} E_{i'}(kr_M, \Omega_{i'}, \Omega_S) Y_{lm}^{(\sigma)}(\Omega_{i'}) = Y_{lm}^{(\sigma)}(\Omega_S) \tag{9.8.12}$$
$$l = 0, 1, 2 \ldots (L-1) \quad m = 0, 1 \ldots l \qquad \sigma = 1, 2.$$

Similar to Equation (9.3.26), the weight $\lambda'_{i'}$ depends on the configuration or directional sample scheme of the microphones. Equation (9.8.12) indicates that the preceding $(L - 1)$ order independent signals of spatial Ambisonics are obtained by the weighted summations of M' microphone outputs and equalizing with $1/R_l(kr_M)$. The number M' of microphones and the truncation order $(L - 1)$ depends on a directional sampling scheme, but it should at least satisfy the requirement of $M' \geq L^2$ by Shannon–Nyquist spatial sampling theorem. For example, for nearly uniform configuration, a spherical microphone array with $M' = 64$ microphones can theoretically record the spherical harmonic signals up to $(L - 1) = 7$ order, but it cannot actually reach this order.

Similar to the case of a circular microphone array, a spatial aliasing error occurs in the microphone array outputs if the number M' of microphones does not satisfy the requirement of Shannon–Nyquist spatial sampling theorem. The upper frequency limit of spherical microphone array recording can also be evaluated with a formula similar to Equation (9.3.29) or (9.3.30) provided that r of reconstructed regions is substituted by r_M of spherical microphone array. For $r_M = 0.1$ m, it can be evaluated from the condition of $kr_M \leq (L - 1)$ in Equation (9.3.30) that the upper frequency limit for a spherical array with $M' = 64$ nearly uniform microphone configuration is 3.8 kHz.

When the independent signals of Equation (9.8.12) are obtained and the secondary source arrays are given, the decoding equation and driving signals can be derived in accordance with the method given in Section 9.3.2. Indeed, driving signals can also be directly derived from microphone array outputs. This derivation is equivalent to the beamforming processing of microphone array outputs.

If M secondary sources are arranged on a spherical surface according to the schemes in Appendix A, the secondary source array satisfies the discrete orthogonality. According to Equation (9.3.26), for the $(L - 1)$-order spatial Ambisonics, the normalized amplitude of the driving signal of the ith secondary source can be written as the following beamforming form:

$$A_i(\Omega_S) = \lambda_i \sum_{l=0}^{L-1} \sum_{m=0}^{l} \sum_{\sigma=1}^{2} Y_{lm}^{(\sigma)}(\Omega_i) Y_{lm}^{(\sigma)}(\Omega_S) \quad i = 0, 1 \ldots (M-1). \tag{9.8.13}$$

Substituting Equation (9.8.12) into Equation (9.8.13) yields

$$A_i(\Omega_S) = \lambda_i \sum_{i'=0}^{M'-1} \lambda'_{i'} \xi_i(kr_M, \Omega_i, \Omega_{i'}) E_{i'}(kr_M, \Omega_{i'}, \Omega_S) \quad i = 0, 1 \ldots (M-1), \tag{9.8.14}$$

where λ_i and $\lambda'_{i'}$ depend on the configurations of secondary source and microphones in the arrays, respectively, and

$$\xi_i(kr_M, \Omega_i, \Omega_{i'}) = \sum_{l=0}^{L-1} \sum_{m=0}^{l} \sum_{\sigma=1}^{2} \frac{Y_{lm}^{(\sigma)}(\Omega_i) Y_{lm}^{(\sigma)}(\Omega_{i'})}{R_l(kr_M)}. \tag{9.8.15}$$

Therefore, driving signals can be obtained through a frequency-dependent weighted summation (beamforming) of microphone array outputs.

The case of a target plane wave with unit amplitude and incident from a given direction is analyzed above. Similar to the case in Sections 9.1.2 and 9.3.2, for a general incident sound field, the sound pressure at a receiver position close to the origin is the superposition of plane waves incident from various directions. Accordingly, the contributions of these plane waves are included in each order independent signal given in Equation (9.8.12). In this case, the unnormalized amplitude of the driving signal of the ith secondary source in Equation (9.8.14) is replaced with the following equation:

$$E_i = A_i E_A = \lambda_i \sum_{l=0}^{L-1} \sum_{m=0}^{l} \sum_{\sigma=1}^{2} \tilde{P}_{A,lm}^{(\sigma)} Y_{lm}^{(\sigma)}(\Omega_i) \quad i = 0, 1 \ldots (M-1),. \tag{9.8.16}$$

where $\tilde{P}_{A,lm}^{(\sigma)}$ is the spherical harmonic coefficient of the direction-frequency distribution function $\tilde{P}_A(\Omega_{in}, f)$ of the complex-valued amplitude of the incident plane wave, and $E_A = E_A(f)$ is the signal waveform in the frequency domain. Equation (9.8.16) corresponds to the $(L - 1)$-order truncation of the spherical harmonic decomposition of $\tilde{P}_A(\Omega_{in}, f)$. The target sound field can be reconstructed accurately from the recorded signals if $\tilde{P}_A(\Omega_{in}, f)$ is spatially bandlimited. In this case, all components higher than $(L - 1)$ order are negligible as a consequence of Shannon–Nyquist spatial sampling theorem. The corresponding time domain signals are obtained by applying an inverse Fourier transform similar to Equation (9.8.16).

Two directional sampling procedures are used in recoding and reproduction. The first directional sampling occurs in the stage of recording through which M' microphones are used to sample the sound pressure on a spherical surface. The second directional sampling occurs in the stage of Ambisonic reproduction. At this stage, M secondary sources are arranged in a spherical surface. The schemes of directional sampling for recording and reproduction may be identical or different, but they should satisfy the requirement of Shannon–Nyquist spatial sampling theorem, i.e., $M' \geq L^2$, and $M \geq L^2$.

Figure 9.9 is a photo of a spherical microphone array (ordered from B&K) in the author's laboratory. It involves 64 microphones located nearly uniformly on the surface of a rigid sphere with $r_M = 0.0975$ m (the same order as the head radius).

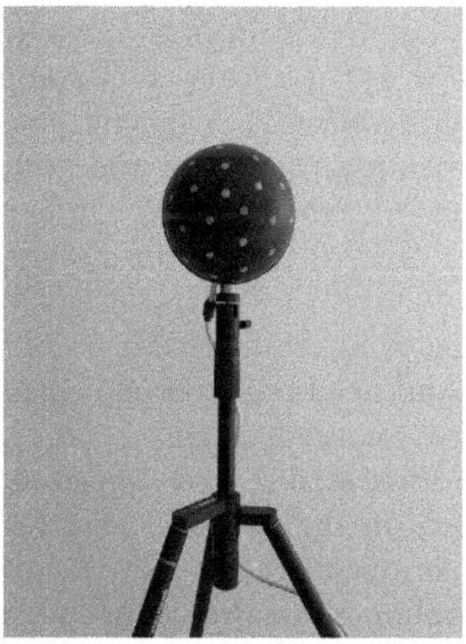

Figure 9.9 Photo of a spherical microphone array (order from B&K).

9.8.3 Discussion on microphone array recording

Some problems associated with circular and spherical microphone arrays should be addressed. In Equations (9.8.6) and (9.8.7) or Equations (9.8.12) and (9.8.15), equalization with the inverse of the radial function $R_l(kr_0)$ is included. Here, the spherical microphone array is taken as an example. For an open spherical array, $R_l(kr_M)$ is given in Equation (9.8.10) and proportional to the spherical Bessel function $j_l(kr_M)$. It exhibits zeros at $kr_M > 0$ because $j_l(kr_M)$ oscillates and decays as kr_M increases. Moreover, for all $l \geq 1$ order, $j_l(kr_M)$ tends zeros when kr_M appears to be zeros. The zeros of spherical Bessel functions lead to divergence in the inverse of $R_l(kr_M)$.

For a rigid spherical microphone array, $R_l(kr_0)$ is given in Equation (9.8.11), and the zeros of $R_l(kr_M)$ at $kr_M > 0$ can be averted. However, when $kr_M < l$, $R_l(kr_M)$ in Equation (9.8.11) decreases quickly and tends to zero, leading to a boost of $l \times (-6)$ dB / Oct in the magnitude response of the equalization filter of $1/R_l(kr_M)$ at low frequencies. An excessive boost may cause a response to exceed the dynamic range of the system. One solution to this problem is applying some regularization to restrict the gain of an equalization filter at low frequencies, and the response of equalization filter becomes

$$F(kr_M) = \frac{R_l^*(kr_M)}{\left|R_l(kr_M)\right|^2 + \varepsilon^2}, \tag{9.8.17}$$

where ε is a regularization parameter.

Similarly, a rigid circular microphone array averts the divergence in the inverse of a radial function caused by the zeros of Bessel functions. In addition, Koyama et al. (2016) used a circular microphone array arranged on the equator of a rigid spherical surface to record horizontal sound field signals. Array outputs are converted for reproduction with a horizontal secondary source array.

For circular and spherical microphone arrays and for open and rigid arrays, the magnitude response of an equalization filter enlarges at low frequencies when r_M of arrays is small, especially for a higher order. Equalization becomes difficult because of a large magnitude response at low frequencies. An array with a large radius is preferred to improve the signal–noise ratio and directional resolution at low frequencies. However, for a given number and configuration of microphones, increasing the radius of arrays reduces the upper frequency limit of recording and inclines to cause spatial aliasing at high frequencies. In practice, the radius of arrays is chosen as a compromise between the low- and high-frequency performances of recording. For this purpose, a dual-radius spherical array is designed with microphones being arranged on and at a distance from the surface (Jin et al., 2014). Moreover, frequency-dependent order can be chosen in Ambisonic recording and reconstruction, i.e., a lower order is chosen for a low frequency and a higher order is chosen for a high frequency.

Rigid array averts the divergence in the inverse of radial function, but they may create multi-scattering between array and neighbor environmental boundaries (such as walls and floor) and then cause an error in recorded signals (Yu et al., 2012a). Open arrays are free from multi-scattering problems. Some studies have suggested using a dual-radius open spherical array to avert the divergence in the inverse of the radial function of an open array (Balmages and Rafaely, 2007). An open circular array composed of a combination of pressure and pressure gradient microphones can also be used for recording (Poletti, 2000, 2005b; Melchior et al., 2009). According to Equation (1.2.40), the output of the i'th microphone in the array is proportional to the linear combination of sound pressure and its radial gradient at the microphone position:

$$E_{i'} = A_1 P_{i'} - (1 - A_1)\rho_0 c V_{r,i'} = A_1 P_{i'} - j(1 - A_1)\frac{\partial P_{i'}}{\partial (kr)}\bigg|_{r=r_M}, \tag{9.8.18}$$

where ρ_0 and c are the density of medium and speed of sound, respectively; $V_{r,i'}$ is the radial speed of the medium at the position of the i'th microphone; $0 < A_1 < 1$ is a parameter that determines the directivity of the microphone; $P_{i'}$ is given in Equation (9.8.9). Equation (3.2.11) is used to obtain the second equality on the right side of Equation (9.8.18). For an incident plane wave with a unit amplitude, the sound pressures at microphone positions are given in Equation (9.8.9). Substituting these pressures into Equation (9.8.18) yields

$$E_{i'}(\Omega_{i'}, kr_M, \Omega_S) = 4\pi \sum_{l=0}^{\infty}\sum_{m=0}^{l}\sum_{\sigma=1}^{2} j^l \left\{A_1 j_l(kr_M) - j(1 - A_1)\frac{d[j_l(kr_M)]}{d(kr_M)}\right\} Y_{lm}^{(\sigma)}(\Omega_{i'}) Y_{lm}^{(\sigma)}(\Omega_S)]. \tag{9.8.19}$$

Accordingly, radial functions become

$$R_l(kr_M) = 4\pi j^l \left\{A_1 j_l(kr_M) - j(1 - A_1)\frac{d[j_l(kr_M)]}{d(kr_M)}\right\}. \tag{9.8.20}$$

Radial functions behave similarly to those of a rigid spherical array and avert the zeros of radial functions at $kr_M > 0$.

In addition to circular and spherical arrays, some other arrays with different shapes are used to record the information of the sound field, and the outputs of arrays are converted to Ambisonic signals. For example, a hemispherical microphone array placed on the surface of an infinitely extended, rigid, and flat baffle is available for recording (Li and Duraiswami, 2006). Generally, a three-dimensional open array with an arbitrary shape and composed of M' microphones is considered (Poletti, 2005b). The position of the i'th microphone is $(r_{i'}, \Omega_{i'})$. For an incident plane wave with unit amplitude and from Ω_S, the sound pressure at the position of a microphone (microphone output) can be decomposed by real-valued spherical harmonic functions according to Equations (9.8.8) and (9.8.10):

$$E_{i'}(kr_{i'}, \Omega_{i'}, \Omega_S) = P_{i'}(kr_{i'}, \Omega_{i'}, \Omega_S) = 4\pi \sum_{l=0}^{\infty}\sum_{m=0}^{l}\sum_{\sigma=1}^{2} j^l j_l(kr_{i'}) Y_{lm}^{(\sigma)}(\Omega_{i'}) Y_{lm}^{(\sigma)}(\Omega_S). \tag{9.8.21}$$

If the size of array is of the order r_M, Equation (9.8.21) can be truncated up to order $(L - 1) \approx kr_M$, yielding

$$\boldsymbol{E} = [K]\boldsymbol{S}_{3D}, \tag{9.8.22}$$

where $\boldsymbol{E}$ is an $M' \times 1$ column matrix or vector composed of the outputs of M' microphones, as described in the following:

$$\boldsymbol{E} = \left[E_0(kr_0, \Omega_0, \Omega_S), E_1(kr_1, \Omega_1, \Omega_S)\ldots, E_{M'-1}(kr_{M'-1}, \Omega_{M'-1}, \Omega_S)\right]^T. \tag{9.8.23}$$

S_{3D} is an $L^2 \times 1$ column matrix or vector composed of the preceding $(L - 1)$ order independent signals or spherical harmonic components:

$$S_{3D} = \left[Y_{00}^{(1)}(\Omega_S), Y_{11}^{(1)}(\Omega_S), Y_{11}^{(2)}(\Omega_S), Y_{10}^{(1)}(\Omega_S) \ldots . Y_{(L-1)0}^{(1)}(\Omega_S) \right]^T . \quad (9.8.24)$$

$[K]$ is an $M' \times L^2$ matrix, whose entries are the forms $4\pi j^l j_l(kr_{i'}) Y_{lm}^{(\sigma)}(\Omega_{i'})$. For $M' \geq L^2$, the pseudoinverse solution of Equation (9.8.18) with regularization is given as

$$S_{3D} = \left\{ [K]^+ [K] + \varepsilon [I] \right\}^{-1} [K]^+ E. \quad (9.8.25)$$

where $[I]$ is an $L^2 \times L^2$ identity matrix, and ε is a regularization parameter. Therefore, the preceding $(L - 1)$ order-independent signals of Ambisonics can be derived from M' microphone outputs.

The above pseudoinverse with regularization is general and suitable for various microphone arrays. For circular or spherical arrays with non-uniform or irregular microphone configurations, Ambisonic-independent signals cannot be derived by using the discrete orthogonality of trigonometric or spherical harmonic functions in Sections 9.8.1 and 9.8.2. In this case, the pseudoinverse method similar to Equation (9.8.25) is applicable. Similar to the discussion on the secondary source array for Ambisonics in Section 9.4.1, the configuration of microphones should be chosen appropriately so that the solution via the pseudoinverse method is stable. Ambisonic-independent signals can be derived through another method. That is, $I > L^2$ regular directions are first chosen; then, independent signals are derived by minimizing the square error between the independent signals and spatial harmonic components of plane wave incident from I directions (Sun and Svensson, 2011; Weller et al., 2014).

In addition to recording higher-order Ambisonic signals, microphone array and beamforming techniques are applicable to creating or simulating microphone outputs with various directivities for different multichannel sound reproduction. For example, circular microphone arrays are applicable to creating multichannel signals for reproduction with an irregular horizontal secondary source (loudspeaker) array (such as 5.1 or 7.1 channel configuration; Hulsebos et al., 2003). Spherical microphone arrays are applicable to creating spatial Ambisonic signals for non-uniform secondary source arrays or mixed-order Ambisonic signals (Weller et al., 2014).

9.9 SUMMARY

The accurate reconstruction of a physical sound field is an ideal method of spatial sound reproduction. Under certain conditions, many practical multichannel sounds can be regarded as an approximation of ideal sound field reconstruction. Analyzing a multichannel sound field provides insights into the physical nature and extent of approximation in reproduction. It also reveals new methods and techniques for multichannel sound.

Generally, a multichannel reconstructed sound field is the linear superposition of those created by all secondary sources. The relationship of the reconstructed sound pressure, transfer functions of a secondary source to receiver positions, and driving signals can be described with a general equation. For secondary source arrays in a horizontal circle or on a spherical surface, analyzing the frequency-spatial spectra is convenient through azimuthal Fourier expansion or spherical harmonic decomposition. Given the target sound field and the transfer functions from secondary sources to receiver positions, driving signals can be derived by mode-matching or spatial SDM.

Ambisonics provides a typical example of the transition from spatial sound based on sound field approximation and psychoacoustics to that based on the accurate reconstruction of a physical sound field. On the one hand, Ambisonic decoding can be regarded as a beamforming processing. The conventional solution of Ambisonic signals is the consequence of spatial harmonic decomposition and each order approximation of the target sound field. As the order increases, approximated reproduction gradually approaches the ideal reproduction. However, higher-order reproduction requires more independent signals and becomes complicated. Ambisonics can also be regarded as a scheme of the spatial sampling and reconstruction of target sound fields. Therefore, it should satisfy the condition of Shannon–Nyquist spatial sampling theorem. The Q-order horizontal Ambisonics requires $(2Q + 1)$ independent signals and $M \geq (2Q + 1)$ secondary sources. It can reconstruct a target sound field up to the Q-order azimuthal harmonic components within a certain region and below a certain frequency. The $(L - 1)$-order spatial Ambisonics requires L^2-independent signals. The required number of secondary sources depends on the secondary source configuration, but at least $M \geq L^2$ secondary sources are required. The practical secondary source array is chosen according to the stability of the reconstructed sound field. The $(L - 1)$ order spatial Ambisonics can accurately reconstruct target sound fields up to the $(L - 1)$ order spherical harmonic components within a certain region and frequency.

In addition to conventional higher-order far-field Ambisonics, some special methods and applications of Ambisonics are developed. On the basis of the psychoacoustic principle of spatial hearing, a MOA reconstructs a horizontal sound field with higher-order spatial harmonics and a vertical sound field with lower-order spatial harmonics, thereby reaching a compromise between perceptual performance and system complexity. A near-field compensated higher-order Ambisonics (NFC–HOA) reconstructs the sound field caused by a proximal target source within a region close to the head and compensates for the effect of the finite distance of the secondary source array. The Ambisonic method is also applicable to encoding and reproducing the complex target source information. Spatial multizone reproduction is a special application of Ambisonics.

The secondary source array of Ambisonics is relatively flexible. Under certain conditions, reproduction can be implemented through various appropriate arrays. Uniform or nearly uniform arrays improve the stability of a reconstructed sound field, but some non-uniform arrays are feasible for practical uses.

The independent signals and reconstructed sound field of Ambisonics exhibit some interesting characteristics under the spatial transformations of rotation, reflection, and dominance and directional emphasis. These transformations are closely related to some important concepts in modern physics.

A multichannel reconstructed sound field can be analyzed in the spatial domain. Summing localization equations of multichannel sound reproduction can be derived by minimizing the error in a reconstructed sound field under certain error criteria. The multiple receiver position matching method, which is related to a mode-matching method, can be used to derive the driving signals of arbitrary secondary source arrays. The method of active sound field control is also applied to compensate for the influence of the reflections from a listening room on a multichannel reproduced sound field.

A microphone array is an effective method for the spatial sampling and recording of a sound field. Circular and spherical microphone arrays are applied to record horizontal and spatial Ambisonic signals, respectively.

Chapter 10

Spatial sound reproduction by wave field synthesis

An ideal spatial sound reproduction technique and system should be able to reconstruct a target sound field within a large region. In this process, a listener receives correct spatial information when he/she enters the reconstructed sound field. Therefore, this ideal reproduction technique shows potential for application in a large listening region and favors sound reproduction in halls.

In Chapter 9, basic methods and concepts of sound field reconstruction are introduced, and the reconstructed sound field of Ambisonics is analyzed in detail. A moderate-order Ambisonics can reconstruct a target sound field only within a small region and at low frequencies unless a higher-order Ambisonics is used. This reconstruction is the consequence of Shannon–Nyquist spatial sampling theorem. Even for conventional multichannel sounds in which psychoacoustic principles have been incorporated to simplify the system, the listening region is usually small unless a system with a number of channels and loudspeakers are used (Section 3.1).

Wave field synthesis (WFS) is another sound field-based technique and system that aims to reconstruct a target sound field in an extended region. In this chapter, WFS is described in detail and analyzed within the framework of the general formulations of sound field reconstruction with multiple secondary sources. In Section 10.1, the basic principle and method, i.e., the traditional analyses of WFS, are presented. In Section 10.2, the general theory of WFS is discussed from the point of mathematical and physical analysis. In Section 10.3, the characteristics of WFS, especially spatial aliasing caused by a discrete secondary source array, are analyzed in the spatial spectrum domain. In Section 10.4, the relationship among acoustical holography, WFS, and Ambisonics is described. In Section 10.5, WFS equalization under nonideal conditions is discussed.

10.1 BASIC PRINCIPLE AND IMPLEMENTATION OF WAVE FIELD SYNTHESIS

10.1.1 Kirchhoff–Helmholtz boundary integral and WFS

WFS is physically based on the ***Huygens–Fresnel principle***. According to this principle, every point (area element) on a wavefront is a secondary source of wavelets. These wavelets spread out in a forward direction at the same speed and frequency as the source wave. The new wavefront is a line tangent to all the wavelets. According to the Huygens–Fresnel principle, the original sound field can be reconstructed if signals of wavelets are captured by a microphone array arranged on the wavefront of the original sound field and then reproduced by a secondary source (loudspeaker) array with the same distribution as that of the microphone array. However, the wavefront of the original sound field is usually unknown in advance. In practice, signals in the original sound field are captured by a microphone array arranged on a fixed

DOI: 10.1201/9781003081500-10

boundary surface (or curve) and then reproduced by the corresponding secondary source array. This principle is the basic concept of ***wavefront reproduction technique and system.***

At the early stage of spatial sound, a primitive technique based on the Huygens–Fresnel principle was proposed for sound field recording and reproduction. For instance, in the work at Bell Labs in the 1930s, an "acoustic curtain" (array in vertical plane) of pressure (omnidirectional) microphones in front of a stage was suggested for recording wavefronts; the signals were then reproduced through a loudspeaker array with the same configuration as that of the microphone array on the receiver side (Steinberg and Snow, 1934; Snow, 1953). Through the simplification of the "acoustic curtain," two- and three-channel-spaced microphone techniques have been developed, as described in Section 2.2.3. After simplification, the reproduced sound field is greatly different from the sound recorded and reproduced by an "acoustic curtain." Olson (1969) also recommended using 15 microphones arranged in a close horizontal curve to record the sound field and then reproduce signals with the corresponding loudspeaker configuration. Of the total number of microphones, seven are frontal microphones for recording the direct sound from the stage, and eight are lateral and rear microphones for recording the reflections of halls.

These previous studies only suggested the conceptual method of recording and reconstructing the wavefront of a sound field without strict mathematical and physical justification. They also did not derive the required radiation characteristics and driving signals of secondary sources. Since 1988, Berkhout (1988), Berkhout et al. (1993) and the group at Delft University of Technology have conducted a series of pioneering works on wavefront reconstruction (Boone et al., 1995; Vries, 1996, 2009). The technique and principle of ***WFS*** were developed on the basis of the principle of ***acoustical holography*** (Williams, 1999), and the radiation characteristics and driving signals of secondary sources were derived. Since the 1990s, WFS has been recognized as an interesting topic in spatial sound technique. As a part of the research on creating interactive audiovisual environments, numerous works on WFS under the framework project of EC IST CARROUSO have been conducted by European groups (Brix et al., 2001). Since then, theory of WFS has been greatly improved (Spors et al., 2008; Ahrens 2012).

Mathematically, the Huygens–Fresnel principle is described by ***Kirchhoff–Helmholtz boundary integral equation.*** As illustrated in Figure 10.1, the frequency-dependent sound pressure $P(r, f)$ in an arbitrary source-free and closed space V' is determined by the pressure and its normal derivative on the boundary surface S' of V':

$$P(\boldsymbol{r}, f) = -\iint_{S'} \left[\frac{\partial P(\boldsymbol{r}', f)}{\partial n'} G_{free}^{3D}(\boldsymbol{r}, \boldsymbol{r}', f) - P(\boldsymbol{r}', f) \frac{\partial G_{free}^{3D}(\boldsymbol{r}, \boldsymbol{r}', f)}{\partial n'} \right] dS' \quad \boldsymbol{r} \in V', \tag{10.1.1}$$

where f is the frequency; $\boldsymbol{r}$ and $\boldsymbol{r}'$ are the vector of the receiver position inside V' and the vector of a point on the boundary surface S', respectively; and $\partial/\partial n'$ is an inward-normal derivative on the surface of S'. The integral is calculated over the entire boundary surface S'. $G_{free}^{3D}(\boldsymbol{r}, \boldsymbol{r}', f)$ is free-field Green's function in a three-dimensional space (and frequency domain) expressed in Equation (9.2.2), which represents the sound pressure at the receiver position $\boldsymbol{r}$ caused by a monopole point source at position $\boldsymbol{r}'$ with the unit strength

$$\begin{aligned} G_{free}^{3D}(\boldsymbol{r}, \boldsymbol{r}', f) &= G_{free}^{3D}(|\boldsymbol{r} - \boldsymbol{r}'|, f) \\ &= \frac{1}{4\pi\,|\boldsymbol{r} - \boldsymbol{r}'|} \exp[-j\boldsymbol{k} \cdot (\boldsymbol{r} - \boldsymbol{r}')] \\ &= \frac{1}{4\pi\,|\boldsymbol{r} - \boldsymbol{r}'|} \exp(-jk|\boldsymbol{r} - \boldsymbol{r}'|). \end{aligned} \tag{10.1.2}$$

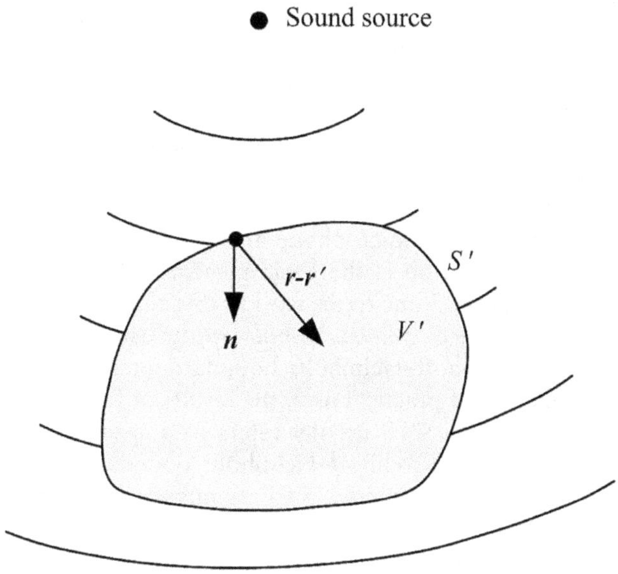

Figure 10.1 Sketch of Kirchhoff–Helmholtz integral.

In Equation (10.1.1), the normal derivative of Green's function is calculated by

$$\begin{aligned}\frac{\partial G_{free}^{3D}(r, r', f)}{\partial n'} &= \frac{1 + jk\,|r - r'|}{4\pi |r - r'|^2} \frac{(r - r') \cdot n'}{|r - r'|} \exp\left[-jk \cdot (r - r')\right] \\ &= \frac{1 + jk\,|r - r'|}{4\pi |r - r'|^2} \cos\theta_{rn} \exp\left(-jk\,|r - r'|\right),\end{aligned} \tag{10.1.3}$$

where $\boldsymbol{n}'$ is a unit vector in the inward-normal direction at $\boldsymbol{r}'$ on S'; and $(\boldsymbol{r} - \boldsymbol{r}')/|\,\boldsymbol{r} - \boldsymbol{r}'\,|$ is a unit vector pointing from $\boldsymbol{r}'$ to $\boldsymbol{r}$. Equation (10.1.3) describes the sound pressure at $\boldsymbol{r}$ caused by a dipole source with unit strength and at $\boldsymbol{r}'$, with the main axis of the dipole source pointing to the $\boldsymbol{n}'$ direction. θ_{rn} is the angle between vectors $(\boldsymbol{r} - \boldsymbol{r}')$ and $\boldsymbol{n}'$.

According to Equation (3.2.11), $\partial P(\mathrm{r}', f)/\partial n'$ is directly proportional to the medium velocity component in the inward-normal direction of surface S':

$$\frac{\partial P(r', f)}{\partial n'} = -j2\pi f \rho_0 V_n(r', f). \tag{10.1.4}$$

Equation (10.1.1) and the discussion above indicate that the closed boundary surface can be equivalent to the continuous distribution of two types of secondary sources, i.e., monopole (point) and dipole secondary sources on the surface. The strength of monopole secondary sources is directly proportional to the derivative of pressure in the outward-normal direction (or an equally medium velocity component in the inward-normal direction) of S'. The strength of the dipole secondary source is directly proportional to the pressure on the surface S'. Therefore, the pressures and medium velocity on a closed surface S' in the original sound fields can be captured by uniform and continuous arrays of pressure (omnidirectional) and velocity field (bidirectional) microphones, respectively. The outputs of these two microphone arrays are used as driving signals of the corresponding dipole and monopole secondary source arrays arranged on the closed boundary surface of the receiver space to reconstruct the target sound field. As stated in Section 1.2.5, however, practical velocity field microphones are designed so that their magnitude responses are independent of the frequency of a far-field incident plane wave. These velocity field microphones cannot be used directly to capture the medium velocity on the boundary surface. In practice, the pressure and its inward-normal derivative on a boundary surface in the original sound field can also be evaluated through calculation and simulation, and the microphone array for recording in the original sound field is unnecessary. This description is the basic principle of sound reproduction via an acoustical holographic technique. Some terms used in this chapter should be explained. The acoustical holographic technique or ***acoustical holography*** usually refers to techniques and systems based on accurate Kirchhoff–Helmholtz boundary integral and accurate sound field reconstruction. From the point of practical uses, the results of Kirchhoff–Helmholtz boundary integral should be simplified. WFS usually refers to a specified technique and system based on an approximation of the Kirchhoff–Helmholtz boundary integral. As in the case of Chapter 9, when acoustical holography and WFS are analyzed, the terms secondary source and driving signals are used.

The following problems should be considered to transform from ideal acoustical holography to practical WFS.

1. Simplification of the types of secondary sources
   A complete acoustical holography requires two types of secondary sources with different radiation characteristics, e.g., monopole and dipole sources. In practice, one type of secondary sources alone is preferred. Therefore, the types of secondary sources should be simplified.
2. Simplification of spatial dimensionality in reproduction
   Sound information reproduction in a three-dimensional space requires a three-dimensional secondary source array arranged on the closed boundary surface S' of volume V' and thus requires a large number of secondary sources. Numerous secondary sources are usually impractical. Moreover, secondary sources arranged on a closed surface are often in conflict with visual requirements of many practical applications. In this case, spatial dimensionality in reproduction should be reduced to simplify the secondary source array.
3. Discrete and finite secondary source array
   An ideal WFS requires a continuous array with an infinite number of secondary sources. However, a discrete and finite array of secondary sources is available in practical WFS. This discrete array leads to a spatial aliasing error, and a finite array causes an edge diffraction effect in the reconstructed sound field.

These problems are addressed in Sections 10.1.2 to 10.1.5 from the point of the practical implementation of WFS.

10.1.2 Simplification of the types of secondary sources

According to Equation (10.1.1), acoustical holography requires secondary source arrays of monopole and dipole types. The system can be simplified when an array of either monopole or dipole secondary sources is enough to reconstruct the target sound field. As illustrated in Figure 10.2, a target (primary) source is located on one (the left) side of an infinite (vertical) plane S'_1. A source-free half-space V' of the receiver is located on another side (right side) of the plane S'_1. The plane S'_1 and the hemispherical surface S'_2 with an infinite radius on the

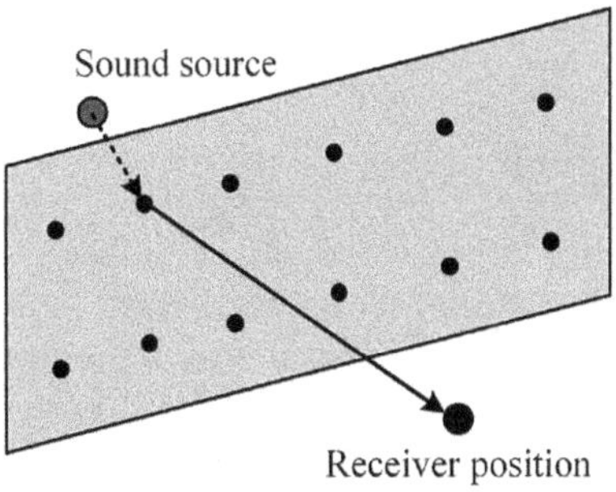

Figure 10.2 Sketch of a secondary source array arranged in an infinite vertical plane.

receiver side constitute a closed boundary surface S'. The integral over S' in Equation (10.1.1) involves the integral over S'_1 and S'_2. Equation (10.1.1) is simplified as the integral over plane S'_1 only because the integral over S'_2 vanishes:

$$P(r,f) = -\iint_{S'_1} \left[\frac{\partial P(r',f)}{\partial n'} G^{3D}_{free}(r,r',f) - P(r',f) \frac{\partial G^{3D}_{free}(r,r',f)}{\partial n'} \right] dS'. \quad (10.1.5)$$

Therefore, the infinite plane S'_1 divides the space into two parts. When the target source is located on one side of S'_1, continuous and uniform secondary source arrays arranged in S'_1 can exactly reconstruct the target sound field on an entire half-space of the receiver side of the array.

In the infinite plane S'_1, Equation (10.1.5) can be calculated using ***Rayleigh integrals*** (Williams, 1999; Ahrens, 2012). The ***Rayleigh integral of the first type*** yields

$$P(r,f) = -2\iint_{S'_1} \frac{\partial P(r',f)}{\partial n'} G^{3D}_{free}(r,r',f) dS'. \quad (10.1.6)$$

In Equation (10.1.6), the integral associated with the normal derivative of Green's function vanishes, and the integral associated with Green's function is twice of that in Equation (10.1.5).

If the ***Rayleigh integral of the second type*** is used, Equation (10.1.5) becomes

$$P(r,f) = 2\iint_{S'_1} P(r',f) \frac{\partial G^{3D}_{free}(r,r',f)}{\partial n'} dS'. \quad (10.1.7)$$

In Equation (10.1.7), the integral associated with Green's function vanishes, and the integral associated with the normal derivative of Green's function is twice of that in Equation (10.1.5).

Equations (10.1.6) and (10.1.7) indicate that an array of either monopole or dipole secondary sources arranged in an infinite vertical plane is enough to reconstruct the target sound field in the entire half-space of the receiver side of the array. In this case, the type of secondary sources for WFS is simplified. In practice, monopole secondary sources are relatively simple and approximately realized by usual loudspeaker systems. Accurate sound field can be reconstructed with an array of a single type of secondary sources only in a few secondary source configurations.

10.1.3 WFS in a horizontal plane with a linear array of secondary sources

If the target source and receiver positions are restricted in the horizontal plane, the secondary source array in Figure 10.2 can be further simplified. Figure 10.3 is a horizontal projection drawing of the WFS system. In particular, a coordinate system with respect to a specific receiver position is used. The horizontal plane is specified by the x-y axes. The z-axis points to the top direction (perpendicular to the horizontal plane in the figure and pointing to the reader). The target monopole (point) source is located at position $\boldsymbol{r}_S$ or $(x_S, y_S, z_S) = (r_S \cos\theta_S, r_S \sin\theta_S, 0)$, where $r_S = |\boldsymbol{r}_S|$ and θ_S are the distance and azimuth of the target source with respect to the origin, respectively. An arbitrary receiver position is specified by vector $\boldsymbol{r}$ or $(x, y, z) = (r \cos\theta, r \sin\theta, 0)$, where $r = |\boldsymbol{r}|$ and θ are the distance and azimuth of the receiver position with respect to the origin, respectively. In accordance with Figure 10.2, secondary sources are uniformly and continuously arranged in the infinite vertical plane $x = x'$. Monopole (point) secondary sources are used in reproduction. The position of an arbitrary secondary source is specified by vector $\boldsymbol{r}'$ or (x', y', z'). The area element in Equation (10.1.6) is $dS' = dy'dz'$ and the integral is calculated over the plane of $x = x'$, and the pressure in the half-space of the receiver is given by

$$P'(\boldsymbol{r}, f) = -2\iint_{x=x'} \frac{\partial P(\boldsymbol{r}', \boldsymbol{r}_S, f)}{\partial n'} G_{free}^{3D}(\boldsymbol{r}, \boldsymbol{r}', f)dy'dz', \tag{10.1.8}$$

where the reconstructed pressure at the receiver position $\boldsymbol{r}$ is denoted by $P'(\boldsymbol{r}, f)$ to be distinguished from the pressure $P(\boldsymbol{r}, \boldsymbol{r}_S, f)$ caused by the target source. Ideally, the reconstructed pressure should be equal to that caused by a target source.

According to Equation (1.2.3), the pressure at $\boldsymbol{r}'$ in the plane $x = x'$ caused by a target point source with strength $Q_P(f)$ is given as

$$P(\boldsymbol{r}', \boldsymbol{r}_S, f) = \frac{Q_P(f)}{4\pi |\boldsymbol{r}' - \boldsymbol{r}_S|}\exp[-j\boldsymbol{k}\cdot(\boldsymbol{r}' - \boldsymbol{r}_S)] = \frac{Q_P(f)}{4\pi |\boldsymbol{r}' - \boldsymbol{r}_S|}\exp(-jk|\boldsymbol{r}' - \boldsymbol{r}_S|)|. \tag{10.1.9}$$

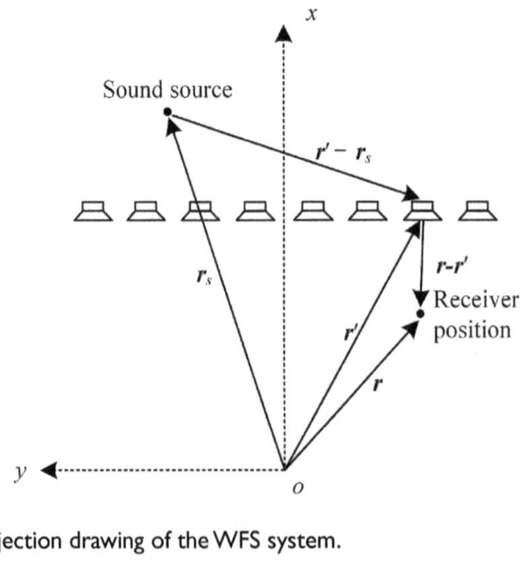

Figure 10.3 Horizontal projection drawing of the WFS system.

The inward-normal derivative of pressure is calculated as follows:

$$\begin{aligned}\frac{\partial P(r', r_S, f)}{\partial n'} &= -\frac{\partial P(r', r_S, f)}{\partial x'}\\ &= \frac{Q_p(f)}{4\pi}\frac{1+jk\,|r'-r_S|}{|r'-r_S|^2}\frac{x'-x_S}{|r'-r_S|}\exp(-jk\,|r'-r_S|),\end{aligned} \tag{10.1.10}$$

where

$$|r'-r_S| = \sqrt{(x'-x_S)^2+(y'-y_S)^2+z'^2} \quad |r-r'| = \sqrt{(x-x')^2+(y-y')^2+z'^2}. \tag{10.1.11}$$

Substituting Equations (10.1.10) and (10.1.2) into Equation (10.1.8) yields

$$\begin{aligned}&P'(r', f)\\ &= \frac{2Q_p(f)}{(4\pi)^2}\iint\limits_{x=x'}\left\{\frac{1+jk\,|r'-r_S|}{|r'-r_S|^2}\frac{x_S-x'}{|r'-r_S|}\frac{1}{|r-r'|}\exp\left[-jk\left(|r'-r_S|+|r-r'|\right)\right]\right\}dy'dz'\\ &= \frac{2Q_p(f)}{(4\pi)^2}\int\limits_{x=x'}dy'\int_{-\infty}^{+\infty}F(z')\exp\left[j\eta(z')\right]dz',\end{aligned} \tag{10.1.12}$$

where

$$\begin{aligned}F(z') &= \frac{1+jk\,|r'-r_S|}{|r'-r_S|^2}\frac{x_S-x'}{|r'-r_S|}\frac{1}{|r-r'|},\\ \eta(z') &= -k\left(|r'-r_S|+|r-r'|\right).\end{aligned} \tag{10.1.13}$$

Under a far-field condition, we have $|\eta(z')| >> 1$. In this case, the integral over z' in Equation (10.1.12) can be approximately calculated using the ***stationary phase method*** (Ahrens, 2012):

$$I = \int_{-\infty}^{+\infty}F(z')\exp\left[j\eta(z')\right]dz' = \sqrt{\frac{2\pi j}{d^2\eta(z')/dz'^2\big|_{z=z_a}}}F(z'_a)\exp\left[j\eta(z'_a)\right], \tag{10.1.14}$$

where z'_a is the stationary phase point and can be evaluated using the following equation:

$$\left.\frac{d\eta(z')}{dz'}\right|_{z'=z'_a} = 0. \tag{10.1.15}$$

According to Equations (10.1.13) and (10.1.11), $z'_a = 0$. Of all the secondary sources arranged in the vertical plane $x = x'$, the secondary sources at the intersect line between the vertical and horizontal planes ($x = x'$, $z = z' = 0$) mainly contribute to the integral in

Equation (10.1.14). Substituting $z'_a = 0$ into Equation (10.1.14), then substituting the result of integral into Equation (10.1.12), and using the far-field approximation $k\,|\,r' - r_S| >> 1$ yield

$$P'(r, f) = Q_P(f) \int_{-\infty}^{+\infty} \left[\sqrt{\frac{jk}{2\pi}} \times \sqrt{\frac{|r' - r_S|}{|r' - r_S| + |r - r'|}} \times \frac{x_S - x'}{|r' - r_S|} \frac{\exp(-jk|r' - r_S|)}{|r' - r_S|} \right] \times \left[\frac{1}{4\pi} \frac{\exp(-jk|r - r'|)}{|r - r'|^{1/2}} \right] dy'. \quad (10.1.16)$$

For the integral on the right side of Equation (10.1.16), the second square brackets can be regarded as the transfer function from the secondary source at r' to the receiver position at r. After $Q_p(f)$ is multiplied, the first square brackets can be regarded as the driving signal of the secondary source. However, the transfer function here is inversely proportional to the square root of the distance $|r - r'|$ between the secondary source and the receiver position rather than inversely proportional to $|r - r'|$. A secondary source with such a radiation characteristic is difficult to be realized. If this mismatched distance dependence of secondary source radiation is merged with driving signals, Equation (10.1.16) is consistent with the general formulation of multichannel sound field reconstruction given by Equation (9.2.1):

$$P'(r, f) = \int_{-\infty}^{+\infty} G_{free}^{3D}(r, r', f) E(r', r_S, f) dy', \quad (10.1.17)$$

where $G_{free}^{3D}(r, r', f)$ is the free-field Green's function in a three-dimensional space expressed in Equation (10.1.2), i.e., the transfer function from a secondary point source to the receiver position; and

$$E(r', r_S, f) = Q_P(f) \sqrt{\frac{jk}{2\pi}} \times \sqrt{\frac{|r' - r_S||r - r'|}{|r' - r_S| + |r - r'|}} \frac{x_S - x'}{|r' - r_S|} \frac{\exp(-jk|r' - r_S|)}{|r' - r_S|}. \quad (10.1.18)$$

Equation (10.1.17) indicates that the sound field of a target point source in a horizontal plane can be approximately reconstructed by an infinite linear array of secondary monopole (point) sources. The $E(r', r_S, f)$ in Equation (10.1.18) is the driving signal of the secondary source located at r'. Therefore, when both the target source and the receiver are restricted in the horizontal plane, WFS can be approximately implemented by an infinite linear array of secondary monopole point sources. Thus, this process is a remarkable simplification in comparison with WFS involving an infinite array of secondary sources in a vertical plane.

Equation (10.1.18) can also be written as

$$E(r', r_S, f) = 4\pi \sqrt{\frac{jk}{2\pi}} \times \sqrt{\frac{|r' - r_S||r - r'|}{|r' - r_S| + |r - r'|}} \times \cos\theta_{sn}\ P(r', r_S, f). \quad (10.1.19)$$

The right side of Equation (10.1.19) involves a multiplication of three terms. $P(\boldsymbol{r}', \boldsymbol{r}_S, f)$ is the pressure at $\boldsymbol{r}'$ of the secondary source caused by the target point source at $\boldsymbol{r}_S$, and θ_{sn} is the angle of the target point source with respect to the horizontal-outward-normal direction at $\boldsymbol{r}'$ of the secondary source array. Therefore, the third term $\cos\theta_{sn}\, P(\boldsymbol{r}', \boldsymbol{r}_S, f)$ can be regarded as the output of a bidirectional (velocity field) microphone at $\boldsymbol{r}'$, with the main axis of the microphone pointing to the horizontal-outward-normal direction of the secondary source array. Here, the magnitude responses of bidirectional microphones are assumed independent of the frequency for a far-field incidence plane wave.

The first term $4\pi\sqrt{jk/2\pi}$ in Equation (10.1.19) is the response of a high-pass filter with an appropriate gain. This high-pass filter can be easily obtained because it is independent from the positions of the target source, the secondary source, and the receiver. If the driving signals of secondary sources are created via simulation, all secondary sources can share a common high-pass filter.

The second term in Equation (10.1.19) aims to equalize the distance-dependent magnitude of secondary source radiation. However, this term depends on the target source position $\boldsymbol{r}_S$, the secondary source position $\boldsymbol{r}'$, and the receiver position $\boldsymbol{r}$. In other words, a given magnitude equalization is valid only at a special receiver position. However, this result is unreasonable. In practice, the magnitude of the reconstructed sound pressure is equalized at a given reference position $\boldsymbol{r}_{ref}$, so Equation (10.1.19) becomes

$$E(\boldsymbol{r}', \boldsymbol{r}_S, f) = 4\pi\sqrt{\frac{jk}{2\pi}} \times \sqrt{\frac{|\boldsymbol{r}' - \boldsymbol{r}_S||\boldsymbol{r}_{ref} - \boldsymbol{r}'|}{|\boldsymbol{r}' - \boldsymbol{r}_S| + |\boldsymbol{r}_{ref} - \boldsymbol{r}'|}} \cos\theta_{sn}\; P(\boldsymbol{r}', \boldsymbol{r}_S, f). \tag{10.1.20}$$

In a receiver position deviating from the reference position, the phase of reconstructed pressure is correct, but magnitude errors occur (Sonke et al., 1998). In the approximation of a target plane wave, a double source distance of the receiver position causes a –3 dB attenuation in the level of the reconstructed sound pressure. For a target point source, a double source distance of the receiver position causes an attenuation between –3 dB and –6 dB in the level of the reconstructed sound pressure. This magnitude error in a reconstructed sound field is due to the ***mismatched distance dependence of secondary source radiation***, i.e., using secondary point sources to replace secondary straight-line sources with an infinite length.

When the target source is distant from the secondary source array with $|\boldsymbol{r}' - \boldsymbol{r}_S| >> 1$, Equation (10.1.20) is simplified as

$$E(\boldsymbol{r}', \boldsymbol{r}_S, f) = 4\pi\sqrt{\frac{jk}{2\pi}} \times \sqrt{|\boldsymbol{r}_{\text{ref}} - \boldsymbol{r}'|} \cos\theta_{sn}\; P(\boldsymbol{r}', \boldsymbol{r}_S, f). \tag{10.1.21}$$

Equation (10.1.21) is also valid for a target plane wave provided that $P(\boldsymbol{r}', \boldsymbol{r}_S, f)$ on the right side is replaced by the pressure $P(\mathrm{r}', f) = P_A(f) \exp(-j\boldsymbol{k} \cdot \boldsymbol{r}')$ of the target plane wave. This condition is due to the plane wave approximation of the far-field spherical wave caused by a point source.

For a receiver position distant from the secondary source array with $|x - x'| >> 1$, the driving signal in Equation (10.1.19) is simplified as

$$E(\boldsymbol{r}', \boldsymbol{r}_S, f) = 4\pi\sqrt{\frac{jk}{2\pi}} \times \sqrt{\frac{|\boldsymbol{r}' - \boldsymbol{r}_S||x - x'|}{|x' - x_S| + |x - x'|}} \cos\theta_{sn}\; P(\boldsymbol{r}', \boldsymbol{r}_S, f). \tag{10.1.22}$$

When the integral over y' in Equation (10.1.16) is also approximately calculated using the stationary phase method, the reconstructed pressure caused by the driving signals in Equation (10.1.22) is identical to that caused by the driving signals in Equation (10.1.19). In this case, the magnitude of the reconstructed sound pressure can be equalized on a reference straight line ($x = x_{ref}$, $z = 0$) parallel to the y axis. Therefore, the magnitude of the reconstructed sound pressure is correct in the reference straight line, and a magnitude error occurs for a receiver position deviating from the reference straight line. However, calculation indicates that the magnitude error of the reconstructed pressure caused by the driving signals in Equation (10.1.22) is small within an appropriate region. For example, when $|r'-r_S| \approx |x'-x_S|$ = 2.0 m and $|x_{ref}-x'|$ = 2.0 m, the magnitude errors at the receiver positions $|x-x'|$ = 3.0 and 4.0 m are 0.8 and 1.2 dB, respectively. In addition, some studies have suggested equalizing the magnitude of the reconstructed sound pressure on a reference contour in a region so that the mean pressure magnitude error within the region is minimized (Sonke et al., 1998).

The signal processing of WFS is conveniently implemented in the time domain. For example, the time-domain counterpart of Equation (10.1.20) is given as

$$e(r', r_S, t) = 4\pi \sqrt{\frac{|r' - r_S||r_{ref} - r'|}{|r' - r_S| + |r_{ref} - r'|}} \cos\theta_{sn} \times h_{hp}(t) \otimes_t p(r', r_S, t), \tag{10.1.23}$$

where the notation "$\otimes_t$" is the convolution manipulation in the time domain; $e(r', r_S, t)$ and $p(r', r_S, t)$ are the driving signals and pressure in the time domain, respectively; and $h_{hp}(t)$ is the impulse response of the high-pass filter expressed as follows:

$$h_{hp}(t) = \int \sqrt{\frac{jk}{2\pi}} \exp(j2\pi ft) df \tag{10.1.24}$$

The linear array of secondary sources can reconstruct the target sound field in a half-horizontal plane of the receiver, but more complicated secondary source arrays are needed to reconstruct a three-dimensional target sound field. In practice, the number and configuration of secondary sources should be simplified. For example, Rohr et al. (2013) suggested using two frontal rows of secondary sources at heights of 0 and 1.20 m relative to the position of a listener's head and two top rows of secondary sources for three-dimensional WFS.

10.1.4 Finite secondary source array and effect of spatial truncation

In Section 10.1.3, an infinite array of secondary sources is used in WFS. In practice, a finite array of secondary sources is utilized. This process is equivalent to truncating the driving signals (10.1.18) for an infinite array with a rectangular spatial window:

$$E'(r', r_S, f) = w(r') E(r', r_S, f). \tag{10.1.25}$$

The function of a rectangular spatial window is given as

$$w(r') = \begin{cases} 1 & -D_L/2 \le y' \le D_L/2 \\ 0 & other \end{cases}, \tag{10.1.26}$$

where D_L is the practical length of an array.

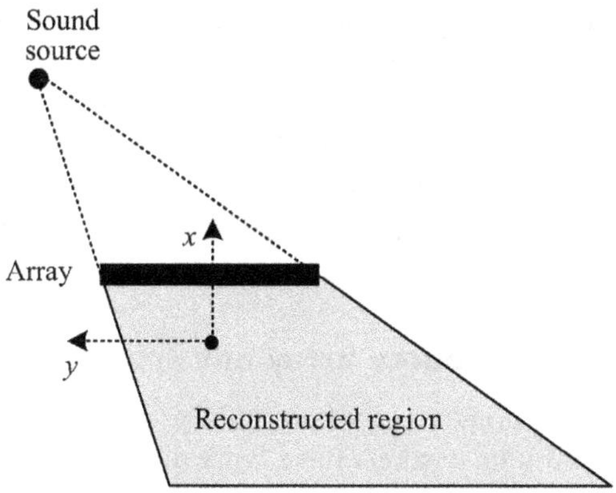

Figure 10.4 Reconstructed region of a finite linear array of secondary sources.

The first problem associated with a finite array of secondary sources is that the reproduced region is narrowed. From the point of geometrical acoustics at least, when a target source or receiver deviates from the central line of the array so that the source-to-receiver connection line does not intersect with the array, the array cannot produce the correct pressure at the receiver position. In other words, sound field reconstruction is restricted within the region in Figure 10.4.

Another problem associated with a finite array of secondary sources is the edge effect (Boone et al., 1995). Similar to the case of truncating the time-domain signal by using a rectangular temporal window, truncating the driving signals for an infinite array with a rectangular spatial window results in a spatial Gibbs effect. In this process, the two edges of a finite array generate diffracted waves. Figure 10.5 illustrates the reconstructed wavefront of a target impulsive point source by using a finite linear array of secondary monopole sources. The target wavefront is supplemented with two additional wavefronts that can be equivalent to the wavefronts created by a pair of point sources with opposite phase to each other. The two additional wavefronts are lagged to the target wavefront, and their magnitudes are much weaker than that of the target wavefront. The arrival time difference between the additional and target wavefronts increases as the receiver position approaches the central line of the secondary source array.

When the arrival time difference between the additional and target wavefronts does not exceed the upper limit of the precedence effect (Section 1.7.2), the interference between the target and edge diffracted waves only causes the perceived timbre coloration. Otherwise, a perceivable echo may occur (especially at the receiver position close to the central line of the

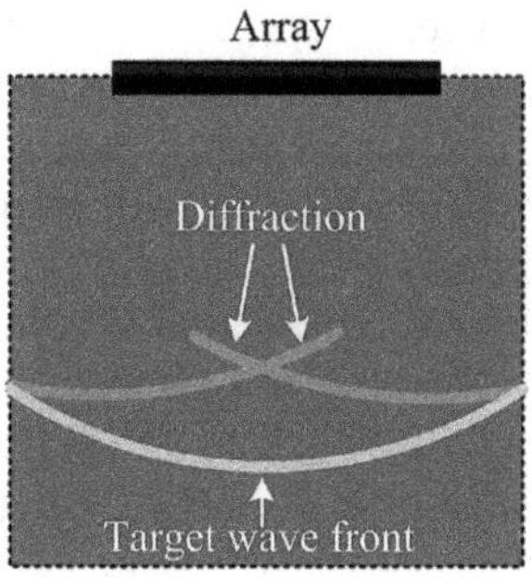

Figure 10.5 Reconstructed wavefront of a target impulsive point source by using a finite linear array of secondary monopole sources.(adapted from Vries, 2009).

array). Therefore, the additional edge diffracted waves caused by the finite array should be suppressed as far as possible. Similar to the case of signal processing in the time domain, a direct method involves weighting the driving signals at the two ends of the array via spatial windows with a smooth transition (such as a cosine window) to avoid the spatial Gibbs effect caused by an abrupt truncation in the rectangular spatial window. However, spatial windows with smooth transition reduce the effective length of the array and consequently further narrow the reconstructed region.

10.1.5 Discrete secondary source array and spatial aliasing

The case of a spatially continuous array of secondary sources is discussed in the preceding sections. However, actual loudspeakers have finite dimensions, which creates an interval between the centers of adjacent loudspeakers. Accordingly, a discrete secondary source array is used in practical WFS, and this process is equivalent to sampling the continuous array and the corresponding driving signal in Equation (10.1.18) or (10.1.20) at discrete spatial positions. For a discrete, linear, and uniform array of secondary sources, the position of the ith secondary source is specified by the vector $r_i = (x', y_i, 0)$, where $i = 0,12...$, and the integral in Equation (10.1.17) is replaced by the following summarization:

$$P'(r, f) = \sum_i G_{free}^{3D}(r, r_i, f) E(r_i, r_S, f) \Delta y', \tag{10.1.27}$$

where $\Delta y'$ is the interval between two adjacent secondary sources. Therefore, the driving signal of the ith secondary source should be $E_i(r_i, r_S, f) = E_{samp}(r_i, r_S, f) = E(r_i, r_S, f)\,\Delta y'$, where $E_{samp}(r_i, r_S, f)$ denotes the driving signals after spatial sampling.

According to the Shannon–Nyquist spatial sampling theorem, the interval between two adjacent secondary sources should not exceed the half of the wavelength. Then, the upper frequency limit of anti-spatial-aliasing reproduction is related to the interval between two adjacent secondary sources as:

$$f_{\max} = \frac{c}{2\Delta y'}, \tag{10.1.28}$$

where c is the speed of sound. Usually, the practical interval between two adjacent secondary sources lies between 0.15 and 0.3 m. Accordingly, the upper frequency limit evaluated from Equation (10.1.28) varies between 1.1 and 0.57 kHz. An interval of 0.008 m between two adjacent secondary sources is needed for an upper frequency limit of 20 kHz. However, such a small interval is infeasible in practice.

Equation (10.1.28) is the result of the worst case in which the target plane wave is incident from the direction parallel to the linear array. In a general case shown in Figure 10.6, the target sound wave is incident from an angle with respect to the linear array or equally incident from an angle θ_{sn} with respect to the outward-normal direction of the array. In this case, the phase difference between the driving signals of two adjacent secondary sources is $2\pi f\,\Delta y' \sin\theta_{sn} / c$ for a harmonic target source at a far distance. However, the phase difference should not exceed π to avoid spatial aliasing. Then, the upper frequency limit of anti-spatial-aliasing reproduction is evaluated using the following equation (Start et al., 1995):

$$f_{\max} = \frac{c}{2\Delta y' \,|\sin\theta_{sn}|}. \tag{10.1.29}$$

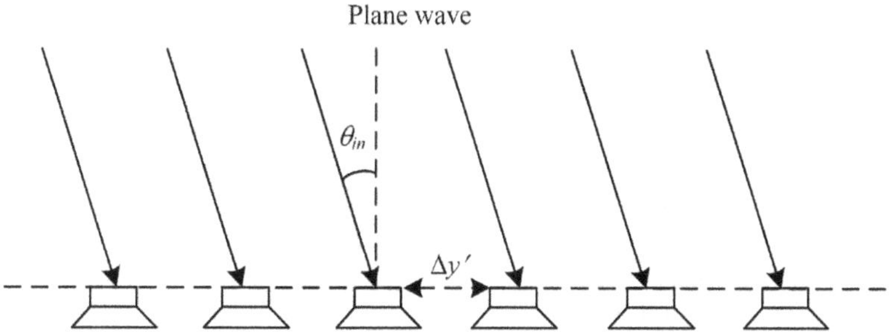

Figure 10.6 Target plane wave incident from an angle θ_{sn} with respect to the outward-normal direction of the array.

For $|\theta_{sn}| < 90°$, the upper frequency limit given in Equation (10.1.29) is larger than that given in Equation (10.1.28), especially for a small incident angle $|\theta_{sn}|$ because $|\sin\theta_{sn}| < 1$. In practice, if the target source direction is restricted within the front region, the upper frequency limit for a linear array with the specific interval of adjacent secondary sources can be increased. For example, when the target source direction is restricted within $|\theta_{sn}| < 60°$, the upper frequency limit is 1.3 kHz for an interval of 0.15 m of adjacent secondary sources.

An infinite linear array of secondary monopole point sources is located at $x' = 3.0$ m to illustrate the sound field reconstructed by a discrete array, and the interval between adjacent secondary sources is 0.15 m. The target sound field is a harmonic plane wave incident from the horizontal azimuth of $\theta_S = 45°$. In this case, the upper frequency limit of anti-spatial-aliasing reproduction is about 1.6 kHz. Figure 10.7 illustrates the simulated wavefront amplitude of the reconstructed pressures within the region of $-3.0 \text{ m} \le x \le 3.0$ m and $-3.0 \le y \le 3.0$ m. Figure 10.7 (a) is the result of $f = 1.0$ kHz. Except for some magnitude errors caused by the mismatched radiation characteristic of secondary sources, the wavefront of the target plane wave can be reconstructed without spatial aliasing within the concerned region. In this simulation, far-field approximation is used, but it is invalid at the receiver positions close to the secondary source array, so a reconstructed error occurs in these positions. Figure 10.7 (b) is the result of $f = 2.0$ kHz, that is, spatial aliasing leads to a confusing reconstructed wavefront within the concerned region.

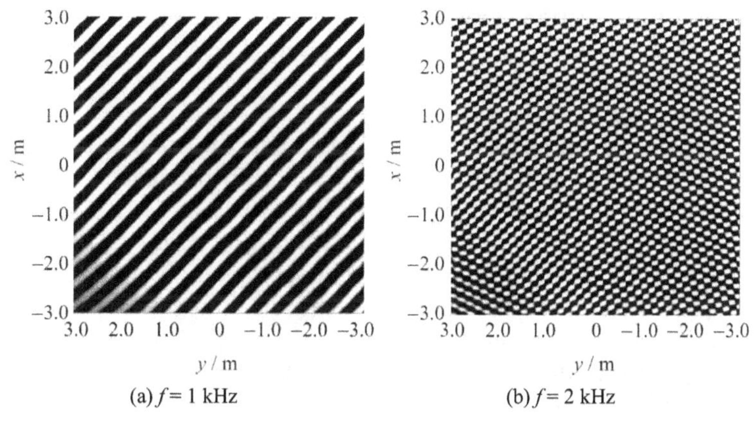

Figure 10.7 Wavefront amplitude reconstructed by an infinite linear array of secondary monopole point sources for the target plane wave incident from a horizontal azimuth of $\theta_S = 45°$ (a) $f = 1$ kHz; (b) $f = 2$ kHz.

The aforementioned analysis indicates that spatial aliasing occurs when the frequency exceeds the upper limit of anti-spatial-aliasing reconstruction given by the interval of adjacent secondary sources. For the practical interval of adjacent secondary sources, the upper frequency limit is far from the desired value of 20 kHz. Fortunately, as stated in Section 1.6.5, the ITD below 1.5 kHz is a dominant lateral localization cue as long as the wideband stimuli include low–frequency components. Therefore, reproduction yields appropriate localization perception providing that target sound field can be reconstructed within a frequency range up to 1.5 kHz. However, spatial aliasing at high frequency cause timbre coloration in reproduction (Wittek, 2007; Wierstorf et al., 2014; Xie et al., 2015b). Specifically, driving signals are frequency dependent because high-pass filtering is included in Equation (10.1.19) to equalize them. Below the upper frequency limit for anti-spatial-aliasing reproduction, the spectra of the superposed sound pressure caused by all secondary sources are similar to those caused by target sources. Above the upper frequency limit, the spectra of the superposed sound pressure change, and timbre coloration occurs. In practice, a filter can be used to equalize the driving signals of secondary sources below the upper frequency limit for anti-spatial-aliasing reproduction, and a flat response of the filter is chosen above the upper frequency limit. Wittek et al. (2007) also suggested a hybrid reproduction method through which sound components below the frequency of anti-spatial aliasing are reproduced by WFS, and components above that frequency limit are reproduced by conventional stereophonic sound. Psychoacoustic experimental results reveal that the hybrid reproduction method yields a localization performance similar to that of reproduction by WFS only but reduces timbre coloration. The influence of spatial aliasing on WFS is further addressed in Section 10.3.2.

10.1.6 Some issues and related problems on WFS implementation

For an infinite linear array of secondary sources in Section 10.1.3, the target source and receiver region are restricted in half-horizontal planes on the two sides of the array. For a finite linear array of secondary sources in Section 10.1.4, the positions of the target source and the receiver are further restricted. A closed polygonal array of secondary sources composed of multiple finite sub-linear arrays can be used to reconstruct the sound field of the target source at various horizontal azimuths. Figure 10.8 illustrates the example of a rectangular array containing four finite sub-linear arrays. A finite sub-linear array is chosen to reconstruct the sound field according to the position of the target source. When the target source is located at the direction close to a vertex angle of the rectangular array, two adjacent sub-linear arrays may be used to reconstruct a sound field (Section 10.2.2). The curved array of secondary sources may also be used for WFS (Start, 1996), which is addressed in Sections 10.2.2 and 10.2.3.

In practical uses, the driving signals of secondary sources are often created with the ***model-based method***. For example, in practical music recording, instruments on a stage are divided into some groups according to their positions. The signals of each group of instruments are captured by a spot microphone at a close distance, and the driving signals of secondary sources are synthesized from the output signal of the spot microphones according to the position of the group of instruments. The overall driving signals are obtained from a mixture of these driving signals from each group.

In addition to reconstructing the direct sound caused by target sources, reconstructing a reflected sound field in a target room or hall, not a receiver or listening room, can be performed through WFS. As stated in Section 7.5.5, the information of reflections can be obtained via physical simulation. Under the approximation of geometrical acoustics, reflections in a hall can be modeled by a number of image sources. Therefore, a reflected sound

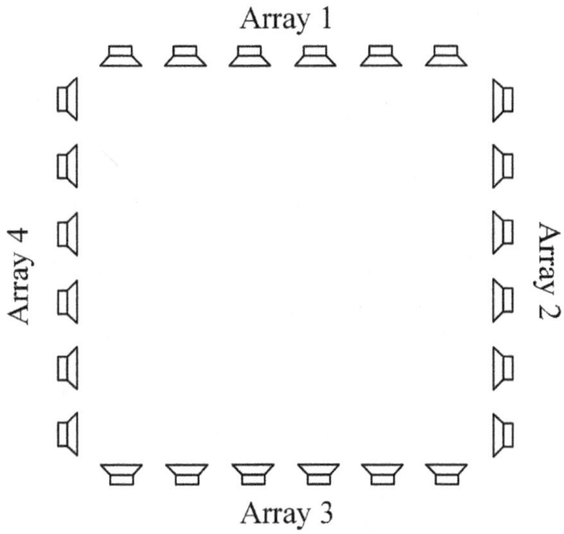

Figure 10.8 Rectangular array composed of four finite sub-linear arrays.

field can be theoretically reconstructed by simulating all image sources in WFS. However, as the order of reflection increases, the number of image sources increases quickly, thereby causing difficulty in simulation. Nevertheless, this problem can be solved by simulating discrete early reflections through an image source method and simulating the late diffused reverberation via artificial reverberation algorithms (Vries et al., 1994b). Sonke and Vries (1997) proposed succeeding WFS processing. Actually, some considerations and methods in sound field approximation and psychoacoustics have been incorporated into practical WFS.

The driving signals of secondary sources in WFS can also be recorded with an appropriate microphone array, and the original sound field, including direct and reflected sound fields, can be physically reconstructed. This is a ***data-based method*** to derive driving signals. According to the basic principle of WFS, a microphone array whose configuration is identical to that of a secondary source array can be used to capture the signals of pressure or medium velocity in the original sound field. A microphone array with a configuration that differs from that of a secondary source array can also be used. In this case, the outputs of the microphone array should be converted to the driving signals of the secondary source array by a signal processing matrix to simulate acoustical transmissions from the positions of microphones to the positions of secondary sources (Berkhout et al., 1993). Or alternatively, the driving signals of secondary sources can be directly derived from the outputs of microphone arrays by signal processing because of the flexibility of WFS. For example, an arbitrary horizontal sound field in a source-free region can be decomposed as a linear superposition of the plane wave from various azimuths. Therefore, the azimuthal–frequency distribution function of the complex-valued amplitude of the incident plane wave can be first recorded and analyzed by a microphone array, and driving signals in WFS are derived from the resultant azimuthal–frequency distribution function to reconstruct incident plane waves from various directions. Hulsebos and Vries (2002) and Hulsebos et al. (2002) analyzed the reconstruction sound field in WFS with driving signals derived from three different microphone arrays, namely, linear, cross, and circular arrays. They found that a linear array with omnidirectional or bidirectional microphones cannot discriminate front and rear incident sound waves. Only the linear array with hypercardioid microphones can discriminate the front and suppress the rear incident sound waves. A linear array is invalid when a sound wave is incident from the direction parallel to the array, but this problem can be solved by a cross array with hypercardioid microphones. In addition, finite linear and cross microphone arrays cause an edge effect. In comparison with a linear and cross array, a circular microphone array discussed in Section 9.8.1 is more appropriate. If three-dimensional information is recorded, the spherical microphone array described in Section 9.8.2 is appropriate.

In addition to onsite recording, reflections of a target sound field can be simulated by convoluting with spatial room impulse responses similar to that in Section 7.5.5. The spatial room impulse responses of a target hall or room are initially measured by a microphone array or obtained through calculation and subsequently converted to impulse responses for secondary sources. The driving signals of secondary sources are obtained by convoluting the input stimulus with the impulse responses for secondary sources. A method similar to DiRAC in Section 7.6 is also applicable to WFS (Gauthier et al., 2014a, 2014b). The spatial information of the original sound field is recorded and analyzed by an appropriate microphone array. Then, the driving signals in WFS are simulated according to the parameters of the original sound field obtained from the analysis.

Similar to the case of a multichannel sound in Section 7.4.3, WFS is applicable to recreating a moving virtual source. As a direct method, piecewise static simulation is applied through which a moving virtual source is simulated as a series of static virtual sources in each short period. The driving signals in different short periods change according to the temporary position of the target source. However, piecewise static simulation causes some problems (Franck et al., 2007). One of the problems is time-variant coloration. According to Equation (10.1.29), the upper frequency limit of anti-spatial aliasing depends on the target source direction. Variations in the target source direction lead to a time-variant upper frequency limit and thus time-variant spatial aliasing. Nevertheless, this problem is solved by reducing the interval between adjacent secondary sources. Other problems include the need for a fractional delay in signals to simulate moving virtual sources, errors in the simulation of a Doppler frequency shift, and spectral broadening in a source signal, which is difficult to be overcome. Some methods, such as those focusing on the Doppler frequency shift and simulating a target source with complex radiation characteristics (Ahrens and Spors, 2008a, 2011), deriving signals via the spatial spectral division method, and applying the stationary phase method in the time domain (Firtha and Fiala, 2015a, 2015b), have been proposed to improve the simulation of a moving virtual source in WFS.

Similar to the upmixing of multichannel sound signals in Section 8.3, signal blind separation and extraction have been suggested separating source signals in stereophonic signals and create the driving signals of WFS (Cobos and Lopez, 2009).

In foregoing discussions, secondary sources are supposed to be ideal monopole point sources. Although practical loudspeaker systems possess a certain directivity, the influence of directivity can be compensated by the inverse filtering $1/\Gamma_S(\Phi, f)$ to the driving signals in Equation (10.1.22), where $\Gamma_S(\Phi, f)$ is the frequency-dependent directivity of the loudspeaker system, and Φ is the angle between the secondary source-to-receiver connection line and the inward-normal direction of the array (Vries, 1996). However, such a compensation is valid for receiver positions at a special direction. Another study has recommended using multiactuator panels (MAPs) as the secondary source array of WFS (Boone, 2004). The advantage of MAPs is that they can create uniform sound radiation within wide frequency range and spatial region. They also satisfy the visual requirement (Pueo et al., 2010).

The group at the Delft University of Technology explored possible WFS applications (Boone and Verheijen, 1998), including commercial cinema, virtual reality theaters, and teleconference systems. WFS can also be applied to sound reinforcement (Vries et al., 1994a). Some applications of WFS are described in Chapter 16.

10.2 GENERAL THEORY OF WFS

The basic principle of WFS is discussed in Section 10.1 in terms of conventional analysis. From the point of implementation and application, the analysis is focused on the simplification of WFS so that it can be implemented by a finite and discrete linear array of secondary sources. In this section, the general and strict theory of WFS is presented on the basis of Green's function of Helmholtz equation (Spors et al., 2008; Ahrens, 2012). Although the analysis in this section does not always lead to practical WFS techniques and systems, insights into the physical nature of WFS and its relation with other spatial sound techniques should be obtained.

10.2.1 Green's function of Helmholtz equation

For convenience in the follow-up discussion, some issues related to Green's function of Helmholtz equation in acoustics and the method of mathematical physics are briefly outlined (Morse and Ingrad, 1968). Here, Green's function of a three-dimensional Helmholtz equation is discussed first. In a uniform and isotropic medium, the frequency domain pressure at an arbitrary receiver position $\boldsymbol{r}$ caused by a point source with unit strength and at position $\boldsymbol{r}_S$ satisfies the following inhomogeneous Helmholtz equation:

$$\nabla^2 G^{3D}(\boldsymbol{r}, \boldsymbol{r}_S, f) + k^2 G^{3D}(\boldsymbol{r}, \boldsymbol{r}_S, f) = -\delta^3(\boldsymbol{r} - \boldsymbol{r}_S), \tag{10.2.1}$$

where k is the wave number, ∇^2 is the three-dimensional Laplace operator, and $\delta^3(\boldsymbol{r} - \boldsymbol{r}_S)$ is the Dirac delta function in a three-dimensional space:

$$\delta^3(\boldsymbol{r} - \boldsymbol{r}_S) = \delta(x - x_S)\delta(y - y_S)\delta(z - z_S). \tag{10.2.2}$$

The solution $G^{3D}(\boldsymbol{r}, \boldsymbol{r}_S, f)$ of Equation (10.2.1) subjected to a certain boundary condition is ***Green's function in a three-dimensional space (and frequency domain)***, which represents the pressure (including the pressures of direct sound from a source and reflections from a boundary) at a receiver position caused by a point source. Generally, Green's function is invariant after the source and receiver positions are exchanged:

$$G^{3D}(\boldsymbol{r}, \boldsymbol{r}_S, f) = G^{3D}(\boldsymbol{r}_S, \boldsymbol{r}, f), \tag{10.2.3}$$

which is a ***mathematical formulation of the acoustic principle of reciprocity***.

An infinite free space is a special boundary in which no reflections from boundaries occur. In this case, Green's function is subjected to the Sommerfeld radiation condition:

$$\lim_{r \to \infty} r\left[\frac{\partial G^{3D}(\boldsymbol{r}, \boldsymbol{r}_S, f)}{\partial r} + jkG^{3D}(\boldsymbol{r}, \boldsymbol{r}_S, f)\right] = 0, \tag{10.2.4}$$

where $r = |\boldsymbol{r}|$. Accordingly, $G^{3D}_{free}(\boldsymbol{r}, \boldsymbol{r}_S, f)$ of Equation (10.2.1) is the free-field Green's function in a three-dimensional space in Equation (10.1.2). Here, $\boldsymbol{r}'$ is replaced with $\boldsymbol{r}_S$ because the source does not necessitate locating on the surface S' of region V'.

Equation (10.2.1) is a linear and inhomogeneous partial differential equation. The solution of Equation (10.2.1) can be expressed as the sum of a particular solution of the inhomogeneous equation and a general solution of the homogeneous equation. If the free-field Green's function in a three-dimensional space is chosen as the particular solution of inhomogeneous equations, the general form of Green's function in a three-dimensional space can be written as

$$G^{3D}(r, r_S, f) = G^{3D}_{free}(r, r_S, f) + \chi(r), \tag{10.2.5}$$

where the $\chi(r)$ satisfies following homogeneous Helmholtz equation:

$$\nabla^2 \chi(r) + k^2 \chi(r) = 0. \tag{10.2.6}$$

The physical significance of Equation (10.2.6) is that the pressure caused by a point source includes the pressure of direct sound and the pressure of boundary reflections. Various $\chi(r)$ or equally different boundary reflections can be chosen so that Green's function in Equation (10.2.5) satisfies different boundary conditions. If a rigid boundary is selected so that the normal component of the medium velocity vanishes on the boundary, Green's function satisfies the following ***Neumann boundary condition***:

$$\left. \frac{\partial G^{3D}(r, r_S, f)}{\partial n'} \right|_{S'} = 0. \tag{10.2.7}$$

In this case, the solution of Equation (10.2.1) is ***Neumann Green's function in a three-dimensional space*** and denoted by $G^{3D}_{Neu}(r, r_S, f)$.

For a general source distribution in a three-dimensional space, strength distribution is no longer the Dirac delta function and represented by $\rho^{3D}(r, f)$. The frequency domain pressure satisfies the following inhomogeneous Helmholtz equation:

$$\nabla^2 P(r, f) + k^2 P(r, f) = -\rho^{3D}(r, f). \tag{10.2.8}$$

$P(r, f)$ is the solution of Equation (10.2.8) subjected to a certain boundary condition.

If the pressure and its inward-normal derivative on a closed boundary surface are known, the pressure at an arbitrary receiver position in region V' inside the surface is calculated as

$$\begin{aligned} P(r, f) = &\iiint_{V'} G^{3D}(r, r'', f)\rho^{3D}(r'', f)dV' \\ &- \iint_{S'} \left[\frac{\partial P(r', f)}{\partial n'} G^{3D}(r, r', f) - P(r', f)\frac{\partial G^{3D}(r, r', f)}{\partial n'} \right] dS', \end{aligned} \tag{10.2.9}$$

where $G^{3D}(r, r'', f)$ is a three-dimensional Green's function.

In the case that all sources are located at the V' inside the boundary, using a different Green's function (under different boundary condition) in Equation (10.2.9) yields the same results. Especially, if the free-field Green's function in a three-dimensional space is used in Equation (10.2.9), the first integral on the right side of the equation represents the contribution of direct sound caused by source distribution inside the boundary surface S'. The second integral on the right side of the equation denotes the contribution of reflections from the boundary surface. Alternatively, if an appropriate Green's function is chosen to match a given boundary condition, the second integral on the right side of Equation (10.2.9) vanishes. Accordingly, because the contribution of boundary reflection is included in the Green's function, the overall pressure in receiver position is a weighted integral of Green's function over V' with a weight of source strength distribution $\rho^{3D}(\boldsymbol{r}, f)$. This result is a consequence of the principle of linear superposition for the solutions of Helmholtz equation.

In the case that V' is a source-free region, such as the cases of acoustical holography and WFS, the first integral on the right side of Equation (10.2.9) vanishes and the secondary integral on the right side of Equation (10.2.9) represents the pressure caused by the source outside the boundary. In this case, free-field Green's function $G^{3D}_{free}(\boldsymbol{r}, \boldsymbol{r}', f)$ in a three-dimensional space should be used in calculated, the Equation (10.2.9) is simplified into the Kirchhoff–Helmholtz integral in Equation (10.1.1).

In Green's function of a two-dimensional Helmholtz equation, a uniform and infinite straight-line source is perpendicular to the horizontal (x-y) plane and intersects the horizontal plane at $\boldsymbol{r}_S$ or $(x_S, y_S, z_S) = (r_S\cos\theta_S, r_S\sin\theta_S, 0)$, as shown in Figure 10.9. The position of an arbitrary point in the straight-line source is represented by the vector $\boldsymbol{r}_l$ or the coordinate $(x_l, y_l, z_l) = (x_S, y_S, z_l) = (r_S\cos\theta_S, r_S\sin\theta_S, z_l)$. The straight-line source creates pressure independent from the vertical coordinate z, resulting in identical pressure distribution in each plane parallel to the horizontal plane. Therefore, the pressure in the horizontal plane is analyzed only. An arbitrary receiver position in the horizontal plane is specified by a vector $\boldsymbol{r}$ or coordinate $(x, y, z) = (r\cos\theta, r\sin\theta, 0)$. In this case, $\boldsymbol{r}_S$ and $\boldsymbol{r}$ can be regarded as vectors in a horizontal plane (two-dimensional space). For a straight-line source with unit strength, the frequency domain pressure at an arbitrary receiver position satisfies the following two-dimensional inhomogeneous Helmholtz equation:

$$\nabla^2 G^{2D}(\boldsymbol{r}, \boldsymbol{r}_S, f) + k^2 G^{2D}(\boldsymbol{r}, \boldsymbol{r}_S, f) = -\delta^2(\boldsymbol{r} - \boldsymbol{r}_S), \tag{10.2.10}$$

where k is the wave number, ∇^2 is the two-dimensional Laplace operator, and $\delta^2(\boldsymbol{r} - \boldsymbol{r}_S)$ is the Dirac delta function in a two-dimensional space.

$$\delta^2(\boldsymbol{r} - \boldsymbol{r}_S) = \delta(x - x_S)\delta(y - y_S). \tag{10.2.11}$$

$G^{2D}(\boldsymbol{r}, \boldsymbol{r}_S, f)$ of Equation (10.2.10) subjected to a certain boundary condition is ***Green's function in a two-dimensional space*** (***and frequency domain***), which represents the pressure

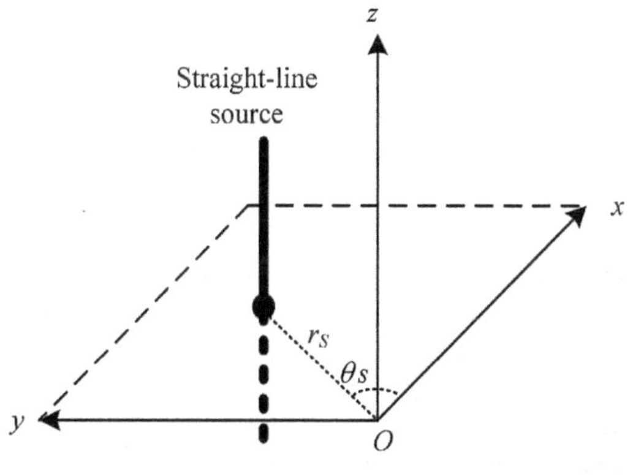

Figure 10.9 Infinite straight-line source perpendicular to the horizontal plane.

(including pressures of direct sound from a source and reflections from a boundary) at the receiver position caused by a straight-line source.

A uniform straight-line source can be regarded as a continuous distribution of point sources in a straight line. According to the principle of linear superposition for the solutions of Helmholtz equation, Green's function in a two-dimensional space can be derived by an integral of Green's function in a three-dimensional space over z_l:

$$G^{2D}(r, r_S, f) = \int_{-\infty}^{+\infty} G^{3D}(r, r_l, f)\, dz_l. \tag{10.2.12}$$

Substituting Equation (10.1.2) into Equation (10.2.12) yields the ***free-field Green's function in a two-dimensional space*** (***and frequency domain***) in Equation (9.2.5):

$$G^{2D}_{free}(r, r_S, f) = -\frac{j}{4} H_0(k|r - r_S|), \tag{10.2.13}$$

where $H_0(k|r - r_S|)$ is the zero-order Hankel function of the second kind. At a far-field distance with $k|r - r_S| >> 1$, Equation (10.2.13) represents the cylindrical wave caused by a straight-line source with unit strength and perpendicularly intersecting with horizontal plane at position r_S.

For a general straight-line source distribution in a two-dimensional space, strength distribution is represented by $\rho^{2D}(r, f)$, where 2D denotes the two-dimensional space. The frequency domain pressure satisfies the following two-dimensional inhomogeneous Helmholtz equation:

$$\nabla^2 P(r, f) + k^2 P(r, f) = -\rho^{2D}(r, f). \tag{10.2.14}$$

The pressure $P(r, f)$ is the solution of Equation (10.2.14) subjected to a certain boundary condition, and the resultant pressure is independent of z.

Suppose that the closed boundary surface is a cylindrical surface and two planes at $z = \pm\infty$. L'_Σ is the intersecting curve line between the cylindrical surface and horizontal plane. In the horizontal region S'_Σ closed by L'_Σ, pressure at an arbitrary receiver position can be calculated using Green's function in a two-dimensional space:

$$\begin{aligned} P(r, f) = &\int_{S'_\Sigma} G^{2D}(r, r'', f)\rho^{2D}(r'', f)\, dS'_\Sigma \\ &- \oint \left[\frac{\partial P(r', f)}{\partial n'} G^{2D}(r, r', f) - P(r', f)\frac{\partial G^{2D}(r, r', f)}{\partial n'} \right] dL'_\Sigma, \end{aligned} \tag{10.2.15}$$

where $P(r', f)$ is the pressure at the curve line L'_Σ, and $\partial/\partial n'$ is the derivative in the inward-normal direction of L'_Σ. The first surface integral on the right side of Equation (10.2.15) is calculated over the horizontal region S'_Σ. The secondary curvilinear integral is calculated over the horizontal curve L'_Σ.

The physical significance and discussion of Equation (10.2.15) are similar to those of Equation (10.2.9). Especially, if S'_Σ is a source-free region, the first integral on the right side of Equation (10.2.15) vanishes and the secondary integral on the right side of Equation

(10.2.15) represents the pressure caused by the source outside the boundary. In this case, free-field Green's function $G_{free}^{2D}(\mathbf{r}, \mathbf{r}', f)$ should be used, Equation (10.2.15) is simplified into the two-dimensional (horizontal) Kirchhoff–Helmholtz integral:

$$P(r,f) = -\int_{L'_\Sigma} \left[\frac{\partial P(r',f)}{\partial n'} G_{free}^{2D}(r,r',f) - P(r',f) \frac{\partial G_{free}^{2D}(r,r',f)}{\partial n'} \right] dL'_\Sigma. \qquad (10.2.16)$$

10.2.2 General theory of three-dimensional WFS

The three-dimensional Kirchhoff–Helmholtz integral in Equation (10.1.1) indicates that an ideal three-dimensional acoustical holography generally requires arrays of secondary monopole and dipole sources. However, the radiated fields of secondary monopole and dipole sources are closely related to each other. In some cases, an array with a single type of secondary sources is enough to reconstruct the target sound field. The secondary source array in an infinite vertical plane in Section 10.1.2 is an example.

Generally, choosing an array of secondary monopole sources for three-dimensional WFS is convenient. The radiation characteristics of a secondary monopole source are determined by the free-field Green's function in a three-dimensional space. Secondary monopole sources are arranged on a closed surface S', and the driving signal of the secondary source at $\boldsymbol{r}'$ is denoted by $E^{3D}(\boldsymbol{r}', f)$. With the exception of a difference in the overall gain at most, the pressure at an arbitrary receiver position $\boldsymbol{r}$ in V' inside S' can be expressed as the general formulation in Equation (9.2.1):

$$P'(r,f) = \iint_{S'} G_{free}^{3D}(r,r',f) E^{3D}(r',f) dS'. \qquad (10.2.17)$$

The key of a WFS design is appropriately choosing the driving signals $E^{3D}(\boldsymbol{r}', f)$ so that the reconstructed pressure $P'(\boldsymbol{r}, f)$ at receiver region V' exactly or approximately match with the target pressure $P(\boldsymbol{r}, f)$.

To derive the driving signals, Neumann Green's function, which satisfies the rigid boundary condition in Equation (10.2.7), is used to replace the free-field Green's function in the Kirchhoff–Helmholtz integral in Equation (10.1.1). In this case, the integral associated with the normal derivative of Green's function vanishes, and Equation (10.1.1) becomes

$$P'(r,f) = -\iint_{S'} \frac{\partial P(r',f)}{\partial n'} G_{Neu}^{3D}(r,r',f) dS'. \qquad (10.2.18)$$

Equation (10.2.18) indicates that sound field is created by a closed array of secondary sources whose radiation characteristics are specified by Neumann Green's function. The driving signals of secondary sources are the reversal phase versions of the inward-normal derivative of pressures in the boundary surface S' or simply the outward-normal derivative of pressures in the boundary surface. However, for an arbitrary secondary source array, Equation (10.2.18) does not ensure that the reconstructed pressure in V'' is equal to the target pressure.

Neumann Green's function can be derived by various methods. However, the exact and analytical solution can be obtained only in a few boundaries with regular geometric shapes. Deriving Neumann Green's function via the image source method is relatively simple and intuitive (Spors et al., 2008). For a boundary of an infinite plane, the image source method yields exact Neumann Green's function. However, for an arbitrary boundary, this method usually results in approximate Neumann Green's function. If a point source is located at $\boldsymbol{r}'$ on the boundary surface S', Neumann Green's function is expressed as Equation (10.2.19) by replacing $\boldsymbol{r}_S$ in Equation (10.2.5) with $\boldsymbol{r}'$:

$$G_{Neu}^{3D}(\boldsymbol{r},\boldsymbol{r}',f) = G_{free}^{3D}(\boldsymbol{r},\boldsymbol{r}',f) + \chi(\boldsymbol{r}). \tag{10.2.19}$$

$\chi(\boldsymbol{r})$, which satisfies the homogeneous Helmholtz equation in the receiver region V', represents the contribution of the reflections from the rigid boundary surface. According to the acoustic principle of reciprocity, Green's function is invariant after the source and receiver positions are exchanged. As illustrated in Figure 10.10, if the source is at $\boldsymbol{r}$ in V' and the receiver is at $\boldsymbol{r}'$ on the boundary surface S', the direct pressure at the receiver position is expressed as $G_{free}^{3D}(\boldsymbol{r}',\boldsymbol{r},f)$. For a rigid boundary surface, the reflected pressure $\chi(\mathbf{r}')$, or more strictly $\chi(\boldsymbol{r}',\boldsymbol{r})$, at $\boldsymbol{r}'$ can be approximated by the free-field pressure caused by an image source with unit strength and outside the boundary surface. The image source is located at a mirror position of $\boldsymbol{r}_{min} = \boldsymbol{r}_{min}(\boldsymbol{r})$ of the source at $\boldsymbol{r}$ against the tangent plane of S' at $\boldsymbol{r}'$. Then, the overall pressure at $\boldsymbol{r}'$ is evaluated as

$$\begin{aligned} &G_{free}^{3D}(\boldsymbol{r}',\boldsymbol{r},f) + \chi(\boldsymbol{r}',\boldsymbol{r}) \\ &= \frac{1}{4\pi\,|\boldsymbol{r}'-\boldsymbol{r}\,|}\exp(-jk|\boldsymbol{r}'-\boldsymbol{r}|) + \frac{1}{4\pi\,|\boldsymbol{r}'-\boldsymbol{r}_{mir}(\boldsymbol{r})\,|}\exp\left[-jk|\boldsymbol{r}'-\boldsymbol{r}_{mir}(\boldsymbol{r})|\right]. \end{aligned} \tag{10.2.20}$$

After the source and receiver positions are exchanged again, the resultant Neumann Green's function is given as

$$G_{Neu}^{3D}(\boldsymbol{r},\boldsymbol{r}',f) = \frac{1}{4\pi\,|\boldsymbol{r}-\boldsymbol{r}'\,|}\exp(-jk|\boldsymbol{r}-\boldsymbol{r}'|) + \frac{1}{4\pi\,|\boldsymbol{r}_{mir}(\boldsymbol{r})-\boldsymbol{r}'\,|}\exp\left[-jk|\boldsymbol{r}_{mir}(\boldsymbol{r})-\boldsymbol{r}'|\right]. \tag{10.2.21}$$

Thus, Equation (10.2.21) satisfies the Neumann boundary condition in Equation (10.2.7).

On the boundary surface S', $|\boldsymbol{r}'-\boldsymbol{r}| = |\boldsymbol{r}'-\boldsymbol{r}_{mir}(\boldsymbol{r})|$, then the two terms in Equation (10.2.21) are equal, and

$$G_{Neu}^{3D}(\boldsymbol{r},\boldsymbol{r}',f) = 2G_{free}^{3D}(\boldsymbol{r},\boldsymbol{r}',f) = \frac{1}{2\pi\,|\boldsymbol{r}-\boldsymbol{r}'\,|}\exp(-jk|\boldsymbol{r}-\boldsymbol{r}'|). \tag{10.2.22}$$

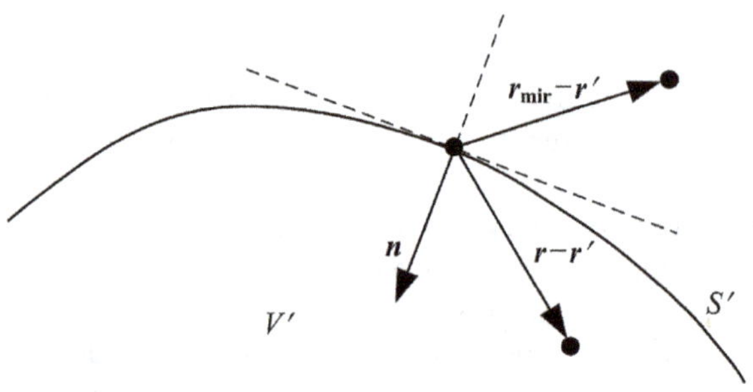

Figure 10.10 Solving Neumann Green's function by an image source method.

In this case, Neumann Green's function is twice that of free-field Green's function, which is the consequence of reflections from a rigid boundary surface. Substituting Equation (10.2.22) into Equation (10.2.18) yields

$$P'(r,f) = -2\iint_{S'} \frac{\partial P(r',f)}{\partial n'} G^{3D}_{free}(r,r',f)\,dS'. \tag{10.2.23}$$

In comparison with Equation (10.2.17), Equation (10.2.23) indicates that sound field is created by an array of secondary monopole sources, and special secondary sources with radiation characteristics specified by Neumann Green's function are not required. The driving signals for secondary monopole sources are given as

$$E^{3D}(r',f) = -2\frac{\partial P(r',f)}{\partial n'}. \tag{10.2.24}$$

For a uniform and continuous array of secondary sources in an infinite plane, the result of Equation (10.2.24) is identical to that of Equation (10.1.6).

When secondary dipole sources are eliminated by using Neumann Green's function in Kirchhoff–Helmholtz integral, the reconstructed sound field does not vanish outside the receiver region V'. For a uniform and continuous array of secondary sources in an infinite vertical plane, the constructed sound field outside the half-space V' of the receiver is a mirror of that in V'. This mirror sound field does not influence the reconstructed sound field in the half-space V' of the receiver. However, for an arbitrary closed boundary surface, the constructed pressure $P'(r,f)$ in Equation (10.2.23) is usually inconsistent with the target pressure $P(r,f)$. This inconsistency is due to the disturbance on the reconstructed sound field by the additional boundary reflections of Neumann Green's function.

The situation can be explained by a simple example illustrated in Figure 10.11 (Nicol and Emerit, 1999). The receiver region is a cuboid. Secondary sources are arranged in four

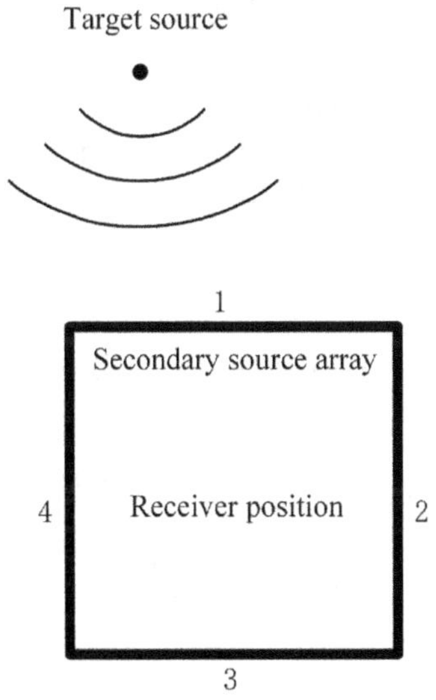

Figure 10.11 Acoustical holographic reproduction in a cuboidal region (adapted from Nicol and Emerit, 1999).

vertical planes labeled with planes 1, 2, 3, and 4. These four planes, together with two other planes at z = ± ∞, constitute a closed boundary surface. The ideal acoustical holography in Equation (10.1.1) is considered. For a target source located at the position illustrated in the figure and within the receiver region, the following characteristics are observed:

1. Secondary source arrays in plane 1 create radiations that possess the same directional propagating components as those of target sources. These secondary sources contribute mostly to the reconstruction of a target sound field in the receiver region. Moreover, the contributions of monopole and dipole source arrays are identical.
2. Secondary arrays in planes 2 and 4 contribute to cancelling the edge diffractions caused by the finite width of secondary source arrays in plane 1.
3 Secondary arrays in plane 3 create radiations that possess the directional propagating components opposite to those of target sources, but this characteristic is not desired. However, monopole and dipole arrays should create radiations with opposite phases so that they cancel each other.

Therefore, array of secondary monopole sources in plane 1 is enough to reconstruct the sound field of the target source behind the array if the array is sufficiently wide. Similar to the monopole secondary sources discussed in Section 10.1.6, arrays of monopole secondary sources in four planes are enough to reconstruct the sound field of the target sources behind four planes. The sound field is reconstructed by an active secondary source array in one of the four planes, and the active array is chosen according to the target source position. This process is equivalent to applying a spatial window to the driving signals of secondary sources in Equation (10.2.24), and the spatial window depends on the target source direction.

A spatial window can be designed for an array of secondary monopole sources on an arbitrary closed boundary surface (Spors et al., 2008). According to Section 3.2.2, the propagating direction of a sound wave is consistent with the medium velocity. The relationship of the frequency domain pressure and medium velocity in a sound field is expressed in Equation (3.2.11) and can be written as the following vector form.

$$\mathrm{V}(\boldsymbol{r}) = -\frac{1}{j2\pi f \rho_0}\nabla P(\boldsymbol{r}) \tag{10.2.25}$$

The spatial window is chosen as follows. At $\boldsymbol{r}'$ on the boundary surface, the weight of the spatial window is a unit if the velocity vector of the medium has a positive component on the inward-normal direction $\boldsymbol{n}'$; otherwise, the weight of the spatial window is zero. Therefore, the spatial window is given as

$$w(\boldsymbol{r}') = \begin{cases} 1 & \mathrm{V}(\boldsymbol{r}')\cdot \boldsymbol{n}' > 0 \\ 0 & other \end{cases}. \tag{10.2.26}$$

After a spatial window is applied, the driving signal in Equation (10.2.24) is revised as

$$E^{3D}(\boldsymbol{r}', f) = -2w(\boldsymbol{r}')\frac{\partial P(\boldsymbol{r}', f)}{\partial n'} \tag{10.2.27}$$

In most cases, the driving signals in Equation (10.2.27) create the target sound field in the receiver region approximately rather than exactly. Similar to the case of the finite linear array of a secondary source in Section 10.1.4, the spatial window in Equation (10.2.26) causes an edge effect, but spatial windows with a smooth transition reduce the edge effect.

For a target incident plane wave, the pressure at an arbitrary receiver position is expressed in Equation (1.2.6). Substituting Equation (1.2.6) into Equation (10.2.27) yields

$$\begin{aligned} E_{pl}^{3D}(\boldsymbol{r}', f) &= 2w(\boldsymbol{r}')P_A(f)\ j(\boldsymbol{k}\cdot\boldsymbol{n}')\exp(-j\boldsymbol{k}\cdot\boldsymbol{r}') \\ &= 2w(\boldsymbol{r}')P_A(f)jk\cos\theta_{sn}\exp(-j\boldsymbol{k}\cdot\boldsymbol{r}'), \end{aligned} \tag{10.2.28}$$

where the subscript "pl" denotes the target plane wave, and $\theta_{sn} = \theta_S$ is the angle between the direction of the incident plane wave and the outward-normal direction at $\boldsymbol{r}'$ of the boundary surface.

For a target point (monopole) source at $\boldsymbol{r}_S$, the pressure at an arbitrary receiver position is presented in Equation (1.2.3). Substituting Equation (1.2.3) into Equation (10.2.27) yields

$$\begin{aligned} E_{p}^{3D}(\boldsymbol{r}', f) &= 2w(\boldsymbol{r}')Q_p(f)\frac{1+jk\,|\boldsymbol{r}'-\boldsymbol{r}_S|}{4\pi|\boldsymbol{r}'-\boldsymbol{r}_S|^2}\,\frac{(\boldsymbol{r}'-\boldsymbol{r}_S)\cdot\boldsymbol{n}'}{|\boldsymbol{r}'-\boldsymbol{r}_S|}\,\exp\left[-j\boldsymbol{k}\cdot(\boldsymbol{r}'-\boldsymbol{r}_S)\right] \\ &= 2w(\boldsymbol{r}')Q_p(f)\frac{1+jk\,|\boldsymbol{r}'-\boldsymbol{r}_S|}{4\pi|\boldsymbol{r}'-\boldsymbol{r}_S|^2}\cos\theta_{sn}\exp(-jk\,|\boldsymbol{r}'-\boldsymbol{r}_S|), \end{aligned} \tag{10.2.29}$$

Where subscript "p" denotes the target point source, and θ_{sn} is the angle between the target source direction and the outward-normal direction at $\boldsymbol{r}'$ of the boundary surface.

10.2.3 General theory of two-dimensional WFS

Secondary sources are infinite straight-line sources perpendicular to the horizontal plane. They are arranged uniformly and continuously to form a vertical cylindrical surface. L'_Σ is the intersecting curve line between the cylindrical surface and the horizontal plane $z = 0$. In the subsequent discussion, this case is simply termed "secondary straight-line sources arranged in a horizontal curve L'_Σ". Thus, the target source is outside the horizontal region S'_Σ closed by L'_Σ and the receiver position is in S'_Σ.

For an ideal acoustical holography, the reconstructed pressure in the receiver region is calculated with the two-dimensional Kirchhoff–Helmholtz integral in Equation (10.2.16). In this case, the reconstructed pressure distribution is independent from the vertical (z) coordinate and thus identical in each plane parallel to the horizontal plane. Equation (10.2.16) also indicates that arrays of secondary monopole and dipole straight-line sources are generally needed for an ideal acoustical holography. If an array of secondary monopole straight-line sources only is used, the reconstructed pressure at a horizontal receiver position $\boldsymbol{r}$ is calculated as

$$P'(\boldsymbol{r}, f) = \int_{L'_\Sigma} G_{free}^{2D}(\boldsymbol{r}, \boldsymbol{r}', f)E^{2D}(\boldsymbol{r}', f)dL'_\Sigma, \tag{10.2.30}$$

where

$$E^{2D}(\boldsymbol{r}', f) = -2w(\boldsymbol{r}')\frac{\partial P(\boldsymbol{r}', f)}{\partial n'}. \tag{10.2.31}$$

Equation (10.2.31) describes the driving signals of secondary monopole straight-line source. The spatial window similar to that in Equation (10.2.26) is chosen, but in this case, n' is the unit vector at the inward-normal direction of L'_{Σ}. Equations (10.2.30) and (10.2.31) indicate that an array of secondary monopole straight-line sources can reconstruct the target sound field approximately in a horizontal receiver region, or more exactly, can reconstruct the vertical direction-independent sound field in a vertical cylindrical region inside the array of secondary sources, by appropriately choosing the driving signals.

For a target incident plane wave, the driving signals of secondary straight-line sources are identical to those expressed in Equation (10.2.28)

$$\begin{aligned} E_{pl}^{2D}(\boldsymbol{r}', f) &= 2w(\boldsymbol{r}')P_A(f)\ j(\boldsymbol{k}\cdot\boldsymbol{n}')\exp(-j\boldsymbol{k}\cdot\boldsymbol{r}') \\ &= 2w(\boldsymbol{r}')P_A(f)\,jk\cos\theta_{sn}\exp(-j\boldsymbol{k}\cdot\boldsymbol{r}')\ . \end{aligned} \tag{10.2.32}$$

For a target straight-line source, the radiated pressure in Equation (1.2.7) is described as

$$P(\boldsymbol{r}', \boldsymbol{r}_S, f) = -Q_{li}(f)\frac{j}{4}H_0(k|\boldsymbol{r}' - \boldsymbol{r}_S|). \tag{10.2.33}$$

Driving signals are derived by substituting Equation (10.2.33) into Equation (10.2.31) and using the derivative formula $dH_0(\xi)/d\xi = -H_1(\xi)$ of the zero-order Hankel function of the second kind as

$$\begin{aligned} E_{li}^{2D}(\boldsymbol{r}',f) &= -w(\boldsymbol{r}')Q_{li}(f)\frac{jk}{2}\frac{(\boldsymbol{r}' - \boldsymbol{r}_S)\cdot\boldsymbol{n}'}{|\boldsymbol{r}' - \boldsymbol{r}_S|}H_1(k|\boldsymbol{r}' - \boldsymbol{r}_S|) \\ &= -w(\boldsymbol{r}')Q_{li}(f)\frac{jk}{2}\cos\theta_{sn}H_1(k|\boldsymbol{r}' - \boldsymbol{r}_S|), \end{aligned} \tag{10.2.34}$$

where $H_1(k\,|\boldsymbol{r}'-\boldsymbol{r}_S|)$ is the first-order Hankel function of the second kind, and θ_{sn} is the angle between the target source direction and the outward-normal direction at $\boldsymbol{r}'$ of the boundary surface.

According to the asymptotic formula of the Hankel function of the second kind, when $k\,|\boldsymbol{r} - \boldsymbol{r}'| >> 1$, the following equation is obtained:

$$H_0(k|\boldsymbol{r} - \boldsymbol{r}'|) = \sqrt{\frac{2}{\pi k\,|\boldsymbol{r} - \boldsymbol{r}'|}}\exp\left[-jk|\boldsymbol{r} - \boldsymbol{r}'| + j\frac{\pi}{4}\right]. \tag{10.2.35}$$

As in Section 10.1.3, at the far-field, the magnitude of cylindrical wave created by a straight-line secondary source is inversely proportional to the square root of the distance between the secondary source and receiver position (–3 dB law rather than –6 dB law for a point source). In practice, WFS can be conveniently implemented using secondary point sources. Substituting Equation (9.2.35) into (9.2.5) and using Equation (10.1.2) yield

$$G_{free}^{2D}(\boldsymbol{r}, \boldsymbol{r}', f) = \sqrt{\frac{2\pi\,|\boldsymbol{r} - \boldsymbol{r}'|}{jk}}\frac{1}{4\pi}\frac{\exp(-jk|\boldsymbol{r} - \boldsymbol{r}'|)}{|\boldsymbol{r} - \boldsymbol{r}'|} = \sqrt{\frac{2\pi\,|\boldsymbol{r} - \boldsymbol{r}'|}{jk}}\,G_{free}^{3D}(\boldsymbol{r}, \boldsymbol{r}', f). \tag{10.2.36}$$

The reconstructed pressure at $\boldsymbol{r}$ is approximated by substituting Equation (10.2.36) into Equation (10.2.30) as follows:

$$P'(\boldsymbol{r}, f) = \int_{L'_\Sigma} \sqrt{\frac{2\pi\,|\boldsymbol{r}-\boldsymbol{r}'|}{jk}}\; E^{2D}(\boldsymbol{r}', f) G^{3D}_{free}(\boldsymbol{r}, \boldsymbol{r}', f) dL'_\Sigma. \tag{10.2.37}$$

In Equation (10.2.37), when the target source and receiver position are restricted in the horizontal plane, a sound field can be reconstructed by a uniform and continuous array of secondary monopole point sources arranged on L'_Σ instead of a straight-line source. In this case, driving signals should be equalized; that is,

$$P'(\boldsymbol{r}, f) = \int_{L'_\Sigma} G^{3D}_{free}(\boldsymbol{r}, \boldsymbol{r}', f) E^{2.5D}(\boldsymbol{r}', f) dL'_\Sigma. \tag{10.2.38}$$

The equalized driving signals for secondary monopole point sources are related to those for secondary monopole straight-line sources by the following equation:

$$E^{2.5D}(\boldsymbol{r}', f) = \sqrt{\frac{2\pi\,|\boldsymbol{r}-\boldsymbol{r}'|}{jk}}\; E^{2D}(\boldsymbol{r}', f). \tag{10.2.39}$$

Equation (10.2.38) specifies a 2.5-dimensional WFS or reproduction. Equation (10.2.39) describes 2.5-dimensional driving signals, which are obtained through frequency- and distance-dependent equalization of two-dimensional driving signals. Equalization depends on the receiver position. In 2.5-dimensional reproduction, the reconstructed sound field deviates from the target sound field when either the source or receiver position deviates from the horizontal plane.

The driving signals in Equation (10.2.39) depend on the receiver position. As indicated in Section 10.1.3, the magnitude of driving signals is equalized at a given reference position $\boldsymbol{r}_{ref}$ in practice. Then, 2.5-dimensional driving signals in Equation (10.2.39) become

$$E^{2.5D}(\boldsymbol{r}', f) = \sqrt{\frac{2\pi\,|\boldsymbol{r}_{ref}-\boldsymbol{r}'|}{jk}}\; E^{2D}(\boldsymbol{r}', f). \tag{10.2.40}$$

For a target plane wave, substituting Equation (10.2.32) into Equation (10.2.40) yields

$$\begin{aligned} E^{2.5D}_{pl}(\boldsymbol{r}', f) &= 2w(\boldsymbol{r}')\sqrt{jk} \times \sqrt{2\pi\,|\boldsymbol{r}_{ref}-\boldsymbol{r}'|} \times P_A(f)\cos\theta_{sn}\exp(-j\boldsymbol{k}\cdot\boldsymbol{r}') \\ &= 4\pi w(\boldsymbol{r}')\sqrt{\frac{jk}{2\pi}}\sqrt{|\boldsymbol{r}_{ref}-\boldsymbol{r}'|}\cos\theta_{sn} P(\boldsymbol{r}', f), \end{aligned} \tag{10.2.41}$$

where $P(\mathbf{r}', f) = P_A(f)\exp(-j\boldsymbol{k}\cdot\boldsymbol{r}')$ is the pressure of the target plane wave at the position of secondary sources. The physical significance of Equation (10.2.41) is similar to that of Equations (10.1.20) and (10.1.21) except for the spatial window $w(\boldsymbol{r}')$ of secondary sources arranged in L'_Σ.

Spors et al. (2008) suggested deriving 2.5-dimensional driving signals of secondary sources from the pressure of a three-dimensional target point source rather than a target straight-line source for a target source at the horizontal position r_S because of the uneven frequency-spectral characteristics of the radiation of a target straight source in Equation (10.2.33). Substituting $E^{2D}(r', f)$ in Equation (10.2.40) with $E_p^{3D}(r', f)$ in Equation (10.2.29) yields

$$E_p^{2.5D}(r', f) = 2w(r')Q_p(f)\sqrt{\frac{2\pi\,|r_{ref} - r'|}{jk}}\,\frac{1 + jk\,|r' - r_S|}{4\pi|r' - r_S|^2}\,\frac{(r' - r_S)\cdot n'}{|r' - r_S|}\exp\left[-jk\cdot(r' - r_S)\right]. \quad (10.2.42)$$

For $k\,|r' - r_S| >> 1$, Equation (10.2.42) becomes

$$\begin{aligned} E_p^{2.5D}(r', f) &= 2w(r')Q_p(f)\sqrt{jk} \times \sqrt{2\pi\,|r_{ref} - r'|}\,\frac{(r' - r_S)\cdot n'}{|r' - r_S|}\,\frac{\exp\left[-jk\cdot(r' - r_S)\right]}{4\pi\,|r' - r_S|} \\ &= 2w(r')\sqrt{jk} \times \sqrt{2\pi\,|r_{ref} - r'|}\cos\theta_{sn}P(r', r_S, f) \\ &= w(r')4\pi\sqrt{\frac{jk}{2\pi}} \times \sqrt{|r_{ref} - r'|}\cos\theta_{sn}P(r', r_S, f), \end{aligned} \quad (10.2.43)$$

where θ_{sn} is the angle of the target source with respect to the outward-normal direction at r' of the secondary source array. For a horizontal linear array of secondary monopole point sources in Section 10.1.3, $w(r') = 1$. Equation (10.2.43) is equivalent to Equation (10.1.21) when $|r' - r_S| >> 1$. Therefore, a horizontal linear array of secondary monopole point sources is a special case of the discussion in this section.

If driving signals are directly derived from the pressure of a target straight-line source, the two-dimensional driving signals in Equation (10.2.34) should be equalized or multiplied with the factor $\sqrt{jk}$ so that the radiation of the target straight-line source has a frequency-spectral characteristic identical to that of a point source. Then, the equalized two-dimensional driving signals are converted into 2.5-dimensional driving signals by using Equation (10.2.40):

$$E_{li}^{2.5D}(r', f) = -w(r')Q_{li}(f)\sqrt{2\pi\,|r_{ref} - r'|}\,\frac{jk}{2}\,\frac{(r' - r_S)\cdot n'}{|r' - r_S|}H_1(k|r' - r_S|). \quad (10.2.44)$$

When $k\,|r' - r_S| >> 1$, the asymptotic formula of the first-order Hankel function of the second kind yields

$$H_1(k|r' - r_S|) = \sqrt{\frac{2}{\pi k\,|r' - r_S|}}\exp\left(-jk|r' - r_S| + j\frac{\pi}{2} + j\frac{\pi}{4}\right). \quad (10.2.45)$$

In the case of $Q_{li}(f) = Q_p(f)$, Equation (10.2.44) becomes

$$\begin{aligned} E_{li}^{2.5}(r', f) &= \sqrt{2\pi\,|r' - r_S|}\,E_p^{2.5D} \\ &= w(r')Q_{li}(f)\sqrt{\frac{jk}{2\pi}} \times \sqrt{2\pi\,|r_{ref} - r'|}\,\frac{(r' - r_S)\cdot n'}{|r' - r_S|}\,\frac{\exp(-jk|r' - r_S|)}{\sqrt{|r' - r_S|}}. \end{aligned} \quad (10.2.46)$$

The 2.5-dimensional driving signals for synthesizing a target straight-line source are identical to those for synthesizing a target point source except for a distance-dependent factor $\sqrt{2\pi \mid r' - r_S \mid}$.

Far-field approximation is assumed in the aforementioned derivation of driving signals. This assumption is appropriate for a target source far from the secondary source array in comparison with wavelength. Lee et al. (2013) derived 2.5-dimensional driving signals for a target source close to the secondary source array. They combined the result with the far-field solution with weights to form driving signals for an arbitrary source distance in relation to the secondary source array.

As stated in Section 10.1.6, the driving signals of WFS can be created by using model- and data-based methods. Given the target sound field, the model-based method can be used to create three-dimensional or 2.5-dimensional driving signals. However, the output of a microphone array cannot be directly used as driving signals in the data-based method. For an arbitrary secondary source array arranged in a three-dimensional curved surface or a two-dimensional curve, a spatial window should be applied to driving signals, but this process is difficult to be implemented directly on the outputs of a microphone. Alternatively, an arbitrary incident sound field in a source-free region can be decomposed as a superposition of incident plane waves from various directions. The azimuthal Fourier or spatial spherical harmonic components of a sound field can be evaluated on the basis of the outputs of a circular or spherical microphone array and then converted to the complex amplitude distribution of incident plane waves from various directions by a beamforming algorithm. Driving signals are created in terms of the complex amplitude distribution of incident plane waves and the spatial window in Equation (10.2.26). When the outputs of a spherical microphone array are used to create 2.5-dimensional driving signals, the three-dimensional complex amplitude distribution of incident plane wave should be projected into the horizontal plane through various methods. For instance, the direction of beamforming can be restricted in the horizontal plane. Other projection methods are also used (Ahrens and Spors, 2012).

10.2.4 Focused source in WFS

In the aforementioned discussion, a target source is assumed to be outside the secondary source array, and the region inside the boundary bounded by the array is source free. WFS can also reconstruct the sound field of a target virtual source inside the boundary. For example, a horizontal linear array of secondary sources can create the sound field of a target virtual source between the array and the receiver position. Such a target virtual source is created on the basis of the principle of acoustic focusing (Spors et al., 2009). When the delay and magnitude of the driving signal of each secondary source are chosen appropriately, the sound waves created by all secondary sources arrive simultaneously (in phase) and thus converge at the focal point. Then, they diverge from the focal point toward the receiver positions as if they are created by a target virtual source at the focal point. The virtual source created by this method is termed ***focused virtual source*** or ***focused source***. Through this method, the target sound field can be reconstructed in a region between the focal point and receiver position, but it is invalid in the region between active secondary sources and the focal point.

Various methods are used to create a focused source in sound reproduction. For instance, the ***time-reversal technique*** is applied to produce a focused source in WFS. Consider a case that the sound waves created by a target source propagate to the position of each secondary source with different delays and magnitudes. In the corresponding time-reversal course, the sound waves created by all secondary sources converge at the position of the target source and then diverge forward. Therefore, the driving signals of secondary sources can be obtained by applying time-reversal manipulation to the driving signal of conventional WFS to create a focused source. According to the definition of Fourier transform, if the Fourier transform of a time-domain signal $e(t)$ is $E(f)$, the Fourier transform of the time-reversal signal $e(-t)$ is $E(-f) = E^*(f)$, e.g., the complex conjugation of $E(f)$.

For example, in horizontal WFS, secondary point (monopole) sources are arranged in a horizontal curve L'_Σ. Driving signals in conventional 2.5-dimensional WFS are expressed in Equation (10.2.43) to create a virtual point source at $\boldsymbol{r}_S$. The driving signals of secondary sources are obtained by replacing $\boldsymbol{r}_S$ with $\boldsymbol{r}_{fs}$ and applying complex conjugation manipulation to Equation (10.2.43) to recreate a focused virtual source at position $\boldsymbol{r}_{fs}$:

$$\begin{aligned} E_{fs}^{2.5D}(\boldsymbol{r}', \boldsymbol{r}_{fs}, f) &= 2w_{fs}(\boldsymbol{r}')Q_p(f)\sqrt{-jk} \times \sqrt{2\pi\,|\boldsymbol{r}_{ref} - \boldsymbol{r}'|}\,\frac{(\boldsymbol{r}' - \boldsymbol{r}_{fs})\cdot\boldsymbol{n}_{fs}}{|\boldsymbol{r}' - \boldsymbol{r}_{fs}|}\,\frac{\exp\left[jk\cdot(\boldsymbol{r}' - \boldsymbol{r}_{fs})\right]}{4\pi\,|\boldsymbol{r}' - \boldsymbol{r}_{fs}|} \\ &= w_{fs}(\boldsymbol{r}')Q_p(f)\sqrt{\frac{k}{2\pi j}}\sqrt{|\boldsymbol{r}_{ref} - \boldsymbol{r}'|}\,\frac{(\boldsymbol{r}' - \boldsymbol{r}_{fs})\cdot\boldsymbol{n}_{fs}}{|\boldsymbol{r}' - \boldsymbol{r}_{fs}|}\,\frac{\exp\left[jk\cdot(\boldsymbol{r}' - \boldsymbol{r}_{fs})\right]}{|\boldsymbol{r}' - \boldsymbol{r}_{fs}|}. \end{aligned} \tag{10.2.47}$$

The spatial window is given as

$$w_{fs}(\boldsymbol{r}') = \begin{cases} 1 & (\boldsymbol{r}_{fs} - \boldsymbol{r}')\cdot\boldsymbol{n}_{fs} > 0 \\ 0 & other \end{cases}, \tag{10.2.48}$$

where $\boldsymbol{n}_{fs}$ is a unit vector at the main axis direction of the radiation of the focused sources. Therefore, the position and main axis direction of the focused source can be controlled by driving signals.

The driving signals for creating a focused source can also be obtained by replacing $\boldsymbol{r}_S$ with $\boldsymbol{r}_{fs}$ and applying complex conjugation manipulation to Equation (10.2.46), e.g.,

$$E_{fs}^{2.5}(\boldsymbol{r}', \boldsymbol{r}_{fs}, f) = w_{fs}(\boldsymbol{r}')Q_{li}(f)\sqrt{\frac{k}{2\pi j}}\sqrt{2\pi\,|\boldsymbol{r}_{ref} - \boldsymbol{r}'|}\,\frac{(\boldsymbol{r}' - \boldsymbol{r}_{fs})\cdot\boldsymbol{n}_{fs}}{|\boldsymbol{r}' - \boldsymbol{r}_{fs}|}\,\frac{\exp(jk|\boldsymbol{r}' - \boldsymbol{r}_{fs}|)}{\sqrt{|\boldsymbol{r}' - \boldsymbol{r}_{fs}|}}. \tag{10.2.49}$$

Driving signals in the frequency domain are given by Equation (10.2.47) or (10.2.49). Signal processing can be conveniently implemented in the time domain. They should be supplemented with a pre-delay to ensure the causality in time-reversal signals. Thus, driving signals in the time domain are given as

$$e_{fs}^{2.5}(\boldsymbol{r}', \boldsymbol{r}_{fs}, t) = q(t) \otimes_t h_{high}(t) \otimes_t \delta(t - \tau_0), \tag{10.2.50}$$

where $q(t)$ is the signal waveform in the time domain and related to the strength of the target focused source, and $\tau_0 = |\boldsymbol{r}' - \boldsymbol{r}_S|/c$ is the maximal propagating delay from the secondary source to the position of the focused source. $h_{hp}(t)$ is the impulse response of a high-pass filter expressed as

$$h_{high}(t) = \int \sqrt{\frac{k}{2\pi j}} \exp(j2\pi ft)df. \tag{10.2.51}$$

Similar to the case of conventional 2.5-dimensional WFS, Equation (10.2.51) represents the high-pass filter for equalizing the frequency characteristics of driving signals. In practice, a filter can be used to equalize the driving signals below the upper frequency limit for anti-spatial-aliasing reproduction, and a flat response of the filter is chosen above the upper frequency limit.

The reconstructed sound field of a focused source exhibits some special physical characteristics. Variations in pressure magnitude with the distance between a receiver position and a focused source deviate from the $1/r$ law because of the mismatched distance dependence of secondary source radiation in 2.5-dimensional reproduction. If the distance between the focused source and the linear array is Δr_1, the distance between a horizontal receiver position and the focused source is Δr_2. According to the condition of the constant radiated power of a linear array, the radiated sound pressure caused by the array is inversely proportional to $\sqrt{(\Delta r_1 + \Delta r_2)\Delta r_2}$. For a receiver position distant from the focused source with $\Delta r_2 >> \Delta r_1$, pressure is inversely proportional to Δr_2, e.g., double distance to the focused source causes a −6 dB attenuation at the pressure level. This observation is consistent with the characteristic of a target point source. For a receiver position close to the focused source with $\Delta r_2 << \Delta r_1$, pressure is inversely proportional to $\sqrt{\Delta r_2}$, e.g., double distance to the focused source causes a −3 dB attenuation at the pressure level. This observation is consistent with the characteristic of a target straight-line source. In general, double distance to the focused source causes attenuation ranging from −3 dB to −6 dB at the pressure level.

For example, in a horizontal focused source created by an infinite linear array of secondary (monopole) point sources, secondary sources are arranged in the horizontal straight line of $x' =$ 3.0 m, and the interval between two adjacent secondary sources is 0.15 m. The target (focused) harmonic source is at a horizontal azimuth of $\theta_S = 0°$ and a distance of $r_S = 2.0$ m from the origin. The frequency is $f = 1$ kHz. Figure 10.12 illustrates the simulated wavefront amplitude of the reconstructed sound pressures within the region of $-3.0 \text{ m} \leq x \leq 3.0$ m and $-3.0 \text{ m} \leq y \leq 3.0$ m. The sound waves from all secondary sources converge at the focal point and then diverge from the focal point similar to the wavefront created by a harmonic point source.

Similar to the case of conventional WFS, a discrete secondary source array is used to create a focused source in practice, but it may cause spatial aliasing in a reconstructed sound field. However, in a receiver region close to the focused source, spatial aliasing almost vanishes, or the upper frequency limit of anti-spatial-aliasing is much higher than (about several times as) that of conventional WFS because of the in-phase superposition of sound waves caused by all secondary sources in the position of the focused source. Therefore, the upper frequency

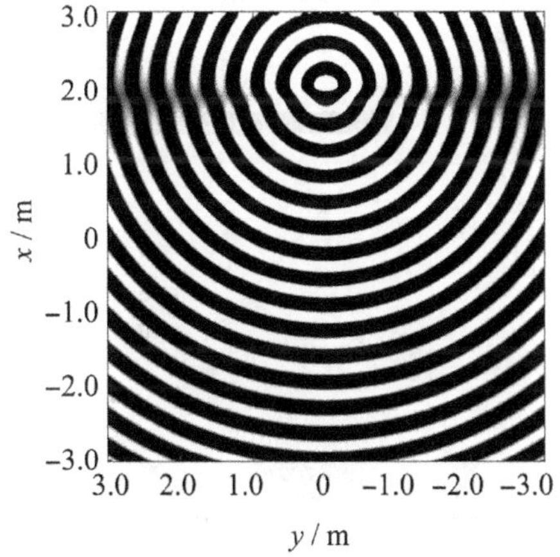

Figure 10.12 Simulated wavefront amplitude of a focused source in WFS.

limit of anti-spatial aliasing depends on the receiver position. However, for wideband stimuli, spatial aliasing creates an additional sound wave to those caused by the focused source. The additional sound wave arrives at the receiver position before the sound wave from the focused source does and thus leads to a ***pre-echo***. The pre-echo is a combined consequence of creating a focused source through time reversal and spatial aliasing. The direction of the pre-echo is different from that of the focused source. According to the precedence effect in Section 1.7.2, a pre-echo may dominate auditory localization and cause a localization error. A psychoacoustic experiment has proven the problem of pre-echoes (Spors et al., 2009). Some studies have suggested selecting an active secondary source (spatial window) to reduce the influence of convergent waves (Song et al., 2012).

- Similar to the case of conventional WFS in Section 10.1.4, a finite array of secondary sources in practice may cause some problems in the creation of a focused source. First, the finite array of secondary sources narrows the region of reproduction. According to the principle of geometrical acoustics, Figure 10.13 illustrates the region of sound field reconstruction for a focused source with a finite array. This region is determined by the extensions of straight lines connecting two edges of the array and the focused source. Second, the finite array causes an edge effect, e.g., the two edges of an array generate diffracted waves. The edge effect can be suppressed by a spatial window with smooth transition. Lastly, finite array increases the size of a focal point because of the diffraction in a finite array. These findings have been validated by a subjective experiment on a focused source in dynamic binaural (virtual) WFS (Wierstorf et al., 2013).

Equations (10.2.9) and (10.2.15) indicate that the Kirchhoff–Helmholtz integral requires a source-free receiver region inside a given boundary. For sources inside the boundary, the sound field does not satisfy the condition of the Kirchhoff–Helmholtz integral. The contribution of these sources is calculated by an additional term of the integral of Green's function weighted with source strength distribution inside the boundary. Choi and Kim (2012) pointed out that the equivalent radiated sound field of a focused source is divided into two components. One component is the desired sound field of a target focused source, which represents the diverging waves from the focal point to the receiver region and satisfies the inhomogeneous Helmholtz equation in Equation (10.2.8) or (10.2.14). Another component is the time-reversal radiated waves, which correspond to the converging waves from secondary sources to the focal point and satisfy Equation (10.2.8) or (10.2.14). The difference in the two components satisfies the homogeneous Helmholtz equation. The physical nature of the

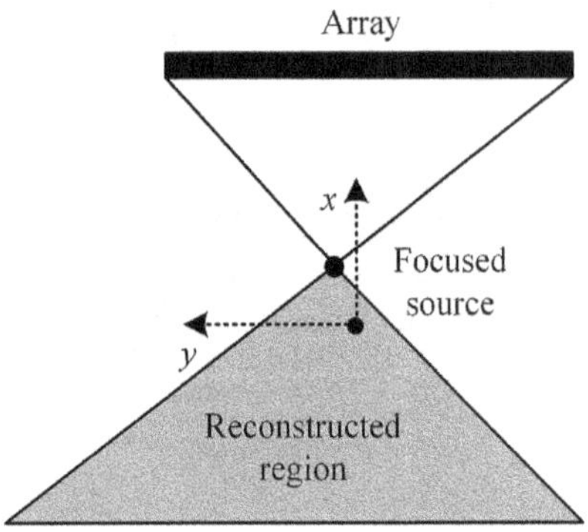

Figure 10.13 Sound field reconstruction for a focused source with a finite array. (adapted from Spors et al., 2009).

focused source method is to design the directivities of the two components so that the major energy of a target radiated component propagates from the focused source to the receiver region; at the same time, the major energy of another component propagates from the focal point to the secondary source array. After time reversal, the latter is equivalent to the major energy radiated by the secondary source array converging to the focal point.

WFS aims to reconstruct a target sound field within an extended region and up to a certain frequency limit. Spatial aliasing occurs at a lower frequency if a smaller number of secondary sources are used. However, for some practical uses, the target sound field should be reconstructed at a higher frequency limit within a smaller region by using a given number of secondary sources. As such, local WFS is developed for this purpose (Spors and Ahrens, 2010a). The basic idea of local WFS is to create a series of focused sources with a small interval; then, target sound fields are reconstructed in a small region closed by these "high-density" focused sources, and the "driving signals" of these focused sources are controlled. In brief, local WFS improves the accuracy in sound field reconstruction at the cost of reducing the reproduction region, but errors in the reconstructed sound field may increase outside the reproduction region.

Similarly, a hybrid method of WFS and higher-order Ambisonics can be used to recreate a virtual source at different distances and directions (Sontacchi, and Holdrich, 2002). A series of virtual secondary sources (virtual loudspeakers) in a circle with a constant radius is created via WFS, and higher-order Ambisonics is implemented through these virtual secondary sources. The perceived virtual source direction and distance are controlled by changing the Ambisonic signal mixing for virtual secondary sources.

10.3 ANALYSIS OF WFS IN THE SPATIAL-SPECTRAL DOMAIN

10.3.1 General formulation and analysis of WFS in the spatial-spectral domain

WFS can be implemented by using an array of a single type of secondary sources after some simplifications are applied to an ideal acoustical holography under certain conditions. In this case, the reconstructed sound pressure is calculated with Equations (10.2.17), (10.2.30), and (10.2.38). These three equations are consistent with the general formulation of multichannel sound field reconstruction in Equation (9.2.1). Accordingly, methods of spatial and spatial spectral domain analysis similar to those in Sections 9.2 to 9.6 are applied to examine WFS. However, these methods are general and not limited to the cases of WFS and Ambisonics.

For a continuous array of secondary sources arranged on a three-dimensional curved surface or a horizontal curve, the reconstructed pressure is expressed in Equation (9.2.1):

$$P'(r,f) = \int G(r,r',f)E(r',f)dr'. \tag{10.3.1}$$

The integral in Equation (10.3.1) is calculated over the curved surface or a curve where the secondary sources are arranged. For a secondary point or straight-line sources in a free field, $G(r, r', f)$ denotes the free-field Green's function $G_{free}^{3D}(r, r', f)$ in a three-dimensional space in Equation (10.1.2) or the free-field Green's function $G_{free}^{2D}(r, r', f)$ in a two-dimensional space in Equation (10.2.13), respectively. Equation (10.3.1) is also applicable to secondary sources of other types.

For some types of secondary sources (such as monopole point and straight-line sources) and some regular arrays, the right side of Equation (10.3.1) can be written as a spatial convolution of $G(\mathbf{r}, \mathbf{r}', f)$ and $E(\mathbf{r}', f)$ over the coordinate by choosing an appropriate coordinate. In this case, the formulation in the spatial domain can be converted into that in the spatial spectral domain by spatial Fourier transform or spherical harmonic decomposition. Accordingly, the convolution manipulation in the spatial domain becomes multiplication in the spatial spectral domain. This is the mathematical basis of Fourier acoustics and acoustical holography (Williams, 1999). The cases of a uniform and continuous secondary source array on a horizontal circle and a spherical surface are discussed in Sections 9.2.2 and 9.2.3. The cases of uniform and continuous secondary source arrays on an infinite plane or a line are further addressed in the following.

For an array of secondary monopole (point) sources arranged in an infinite vertical plane in Figure 10.2, a Cartesian coordinate system is chosen so that the vertical plane is located at x'. Then, the coordinates of an arbitrary secondary source and the receiver position are denoted by (x', y', z') and (x, y, z), respectively. According to Equation (10.1.2), the free-field transfer or Green's function from a secondary source to the receiver position is given as

$$G_{free}^{3D}(\mathbf{r}, \mathbf{r}', f) = G_{free}^{3D}(x - x', y - y', z - z', f)$$
$$= \frac{1}{4\pi\sqrt{(x - x')^2 + (y - y')^2 + (z - z')^2}}$$
$$\exp\left[-jk\sqrt{(x - x')^2 + (y - y')^2 + (z - z')^2}\right]. \tag{10.3.2}$$

Accordingly, Equation (10.3.1) can be written as a form of spatial convolution over the infinite vertical plane where secondary sources are arranged:

$$P'(x - x', y, z, f) = \int_{-\infty}^{+\infty}\int_{-\infty}^{+\infty} G_{free}^{3D}(x - x', y - y', z - z')E^{3D}(x', y', z', f)dy'dz'$$
$$= G_{free}^{3D} \otimes_{yz} E^{3D}, \tag{10.3.3}$$

where the notation "$\otimes_{yz}$" denotes the two-dimensional spatial convolution manipulation over the y and z coordinates.

The two-dimensional spatial Fourier transform of three-dimensional Green's function over the y and z coordinates is given as

$$G_{free,k}^{3D}(x - x', k_y, k_z, f) = \int_{-\infty}^{+\infty}\int_{-\infty}^{+\infty} G_{free}^{3D}(x - x', y, z, f)\exp(jk_y y)\exp(jk_z z)dydz \tag{10.3.4}$$
$$= \int_{-\infty}^{+\infty}\int_{-\infty}^{+\infty} \frac{\exp\left[-jk\sqrt{(x - x')^2 + y^2 + z^2}\right]}{4\pi\sqrt{(x - x')^2 + y^2 + z^2}}\exp(jk_y y)\exp(jk_z z)dydz,$$

where (k_y, k_z) in the spatial spectral domain are the y and z components of the wave vector, respectively. Similarly, the two-dimensional spatial Fourier transforms of pressure and driving signals are denoted by $P_k'(x, k_y, k_z, f)$ and $E_k^{3D}(x', k_y, k_z, f)$, respectively, and calculated similarly to Equation (10.3.4). The subscript "k" is supplemented to all functions in the spatial spectral domain to distinguish from these in the spatial domain. Converted into the spatial spectral domain, Equation (10.3.3) becomes

$$P_k'\left(x-x', k_y, k_z, f\right) = G_{free,k}^{3D}\left(x-x', k_y, k_z, f\right) E_k^{3D}\left(x', k_y, k_z, f\right). \tag{10.3.5}$$

For two-dimensional (horizontal) reproduction in Section 10.2.3, a coordinate system shown in Figure 10.3 is chosen. An infinite array of secondary monopole straight-line sources is arranged in the line $x = x'$ parallel to the y axis. The horizontal coordinates of an arbitrary secondary source and the receiver position are denoted by (x', y') and (x, y), respectively. The free-field Green's function from an arbitrary secondary source to the receiver position is given as Equation (10.2.13) as

$$G_{free}^{2D}\left(r, r', f\right) = G_{free}^{2D}\left(x-x', y-y', f\right) = -\frac{j}{4} H_0\left(k\sqrt{\left(x-x'\right)^2 + \left(y-y'\right)^2}\right). \tag{10.3.6}$$

Equation (10.3.1) is written as one-dimensional spatial convolution over the y' coordinate:

$$P'\left(x-x', y, f\right) = \int_{-\infty}^{+\infty} G_{free}^{2D}\left(x-x', y-y', f\right) E^{2D}\left(y', f\right) dy' = G_{free}^{2D} \otimes_y E^{2D}. \tag{10.3.7}$$

The one-dimensional spatial Fourier transform of two-dimensional Green's function with respect to variable y is written as

$$\begin{aligned} G_{free,\,k}^{2D}\left(x-x', k_y, f\right) &= \int_{\infty}^{+\infty} G_{free}^{2D}\left(x-x', y, f\right) \exp\left(jk_y y\right) dy \\ &= -\frac{j}{4} \int_{-\infty}^{+\infty} H_0\left(k\sqrt{\left(x-x'\right)^2 + y^2}\right) \exp\left(jk_y y\right) dy, \end{aligned} \tag{10.3.8}$$

where k_y in the spatial spectral domain is the y component of the wave vector. Similarly, the one-dimensional spatial Fourier transforms of pressure and driving signals are denoted by $P_k'(x,k_y,f)$ and $E_k^{2D}\left(x', k_y, f\right)$, respectively, and calculated similarly to Equation (10.3.8). Converted into the spatial spectral domain, Equation (10.3.7) becomes

$$P_k'\left(x-x', k_y, f\right) = G_{free,k}^{2D}\left(x-x', k_y, f\right) E_k^{2D}\left(x', k_y, f\right). \tag{10.3.9}$$

Practical horizontal WFS is usually implemented with an array of secondary monopole point sources arranged on a line $x = x'$ parallel to the y axis, e.g., implemented by 2.5-dimensional WFS in Equation (10.2.38). Secondary sources and receiver positions are restricted in the horizontal plane, so $z = z' = 0$. In this case, Equation (10.3.1) is written as one-dimensional spatial convolution over an infinite line in which secondary sources are arranged:

$$P'\left(x-x', y, f\right) = \int_{-\infty}^{\infty} G_{free}^{3D}\left(x-x', y-y', f\right) E^{2.5D}\left(y', f\right) dy' = G_{free}^{3D} \otimes_y E^{2.5D}. \tag{10.3.10}$$

The one-dimensional spatial Fourier transform of three-dimensional Green's function with respect to variable y is written as

$$G_{free,k}^{3D}\left(x-x', k_y, f\right) = \int_{-\infty}^{+\infty} G_{free}^{3D}\left(x-x', y, f\right)\exp\left(jk_y y\right)dy$$
$$= \int_{-\infty}^{+\infty} \frac{\exp\left[-jk\sqrt{\left(x-x'\right)^2+y^2}\right]}{4\pi\sqrt{\left(x-x'\right)^2+y^2}}\exp\left(jk_y y\right)dy. \qquad (10.3.11)$$

Similarly, the one-dimensional spatial Fourier transforms of pressure and driving signals are denoted by $P_k'(x,k_y,f)$ and $E_k^{2.5D}\left(x', k_y, f\right)$, respectively, and calculated similarly to Equation (10.3.11). Converted into the spatial spectral domain, Equation (10.3.10) becomes

$$P_k'\left(x-x', k_y, f\right) = G_{free,k}^{3D}\left(x-x', k_y, f\right)E_k^{2.5D}\left(x', k_y, f\right). \qquad (10.3.12)$$

The analysis method in the spatial spectral domain in this section is not limited to the cases of secondary sources in an infinite vertical plane or a linear array. Similar methods are available for some other regular arrays, such as circular or spherical arrays. Some basic methods are presented in Sections 9.2.2 and 9.2.3. Although the analyses in Sections 9.2.2 and 9.2.3 are focused on Ambisonics, the results are available for WFS.

10.3.2 Analysis of the spatial aliasing in WFS

As an example of analysis in the spatial-spectral domain, spatial aliasing in WFS with a discrete secondary source array is analyzed in this section (Spors and Rabenstain, 2006; Spors and Ahrens, 2009; Ahrens and Spors, 2010). The two-dimensional sound field created by an array of secondary straight-line sources is analyzed to individually evaluate the error caused by the discrete array. Secondary straight-line sources avoid the error caused by the mismatched distance dependence of secondary source radiation in 2.5-dimensional reproduction although this type of secondary source is difficult to be realized in practice.

For two-dimensional (horizontal) reproduction in Section 10.2.3 and the coordinate system shown in Figure 10.3, the reconstructed sound pressure and Green's function in the spatial-spectral domain are expressed in Equations (10.3.9) and (10.3.8), respectively. Using the Fourier transform formula of the zero-order Hanker function of the second kind yields

$$\int_{-\infty}^{+\infty} H_0\left(k\sqrt{\left(x-x'\right)^2+\left(y-y'\right)^2}\right)\exp\left(jk_y y\right)dy$$
$$= \exp\left(jk_y y'\right)\times\begin{cases}\dfrac{2}{\sqrt{k^2-k_y^2}}\exp\left(-j\sqrt{k^2-k_y^2}\,|x-x'|\right) & |k_y|<k=\dfrac{2\pi f}{c} \\ \dfrac{2j}{\sqrt{k_y^2-k^2}}\exp\left(-\sqrt{k_y^2-k^2}\,|x-x'|\right) & |k_y|>k=\dfrac{2\pi f}{c}\end{cases}, \qquad (10.3.13)$$

Equation (10.3.8) becomes

$$G_{free,k}^{2D}(x-x',k_y,f)=\begin{cases}-\frac{j}{2}\frac{1}{\sqrt{k^2-k_y^2}}\exp\left(-j\sqrt{k^2-k_y^2}\,|x-x'|\right) & |k_y|<k=\frac{2\pi f}{c}\\ \frac{1}{2}\frac{1}{\sqrt{k_y^2-k^2}}\exp\left(-\sqrt{k_y^2-k^2}\,|x-x'|\right) & |k_y|>k=\frac{2\pi f}{c}\end{cases}, \tag{10.3.14}$$

where f is the frequency, and c is the speed of sound.

Equation (10.3.14) indicates that the spatial spectrum representation of the free-field Green's function in a two-dimensional space involves two parts. The first part is the contribution of a ***propagating*** (***traveling***) ***wave*** for $|k_y| < 2\pi f/c$, whose magnitude is independent from the distance $|x - x'|$ between a receiver position and an array. The second part is the contribution of an ***evanescent wave*** for $|k_y| > 2\pi f / c$, whose magnitude decays exponentially as $|x - x'|$ increases. The contribution of evanescent waves is significant only for the receiver position close to a secondary source array and at a low frequency. However, Equation (10.3.14) indicates that the two-dimensional Green's function is not spatially bandlimited due to the contribution of evanescent waves.

The driving signals in Equation (10.3.7) depend on a target source. If the target source is an infinite monopole straight-line source and perpendicular to the horizontal plane, then it intersects the horizontal plane at r_S or (x_S, y_S). For a target source with unit strength, the radiated pressure can be calculated using $Q_{li}(f) = 1$ in Equation (10.2.33), and driving signals are derived by choosing the spatial window $w(r') = 1$ in Equation (10.2.34). The following spatial spectrum representation of the driving signals is derived by replacing the (x', y') with (x_S, y_S) and (x, y) with (x', y') in Equation (10.3.13), calculating the derivative with respect to x', using the derivative formula $H_1(\xi) = - dH_0(\xi)/d\xi$ of the Hankel function, and comparing with Equation (10.2.34):

$$\begin{aligned}E_k^{2D}(k_y,f)&=\int_{-\infty}^{+\infty}E^{2D}(y',f)\exp(jk_yy')dy'\\ &=\exp(jk_yy_S)\begin{cases}\exp\left(-j\sqrt{k^2-k_y^2}\,|x_S-x'|\right) & |k_y|<k=2\pi f/c\\ \exp\left(-\sqrt{k_y^2-k^2}\,|x_S-x'|\right) & |k_y|>k=2\pi f/c\end{cases}.\end{aligned} \tag{10.3.15}$$

Driving signals also involve propagating and evanescent components. The later decay exponentially as $|x_S - x'|$ increases. Evanescent components are significant only for the target source position close to a secondary source array and at a low frequency. Similarly, driving signals are not spatially bandlimited because of the contribution of evanescent components.

The reconstructed pressure in the spatial spectral domain is evaluated by substituting Equations (10.3.14) and (10.3.15) into Equation (10.3.9):

$$\begin{aligned}&P_k'(x-x',k_y,f)\\ &=\begin{cases}-\frac{j}{2}\frac{1}{\sqrt{k^2-k_y^2}}\exp(jk_yy_S)\exp\left(-j\sqrt{k^2-k_y^2}\,|x-x_S|\right) & |k_y|<k=\frac{2\pi f}{c}\\ \frac{1}{2}\frac{1}{\sqrt{k_y^2-k^2}}\exp(jk_yy_S)\exp\left(-\sqrt{k_y^2-k^2}\,|x-x_S|\right) & |k_y|>k=\frac{2\pi f}{c}\end{cases}.\end{aligned} \tag{10.3.16}$$

Compared with Equations (10.2.33) and (10.3.14), Equation (10.3.16) is the spatial spectrum representation of the pressure caused by a monopole straight-line source with unit strength and at position (x_S, y_S). Therefore, the target sound field can be exactly reconstructed in the receiver region by using a continuous array of a secondary straight-line source.

When a discrete array of monopole straight-line sources is used in reproduction, the continuous driving signal $E^{2D}(y', f)$ is spatially sampled along y' with an interval of $\Delta y'$. The sampled driving signals are given as

$$E^{2D}_{samp}(y', f) = E^{2D}(y', f)\sum_{v=-\infty}^{+\infty}\delta(y' - v\Delta y')\Delta y'. \tag{10.3.17}$$

The subscript "*Samp*" denotes the driving signals after being spatially sampled. Applying a spatial Fourier transform to Equation (10.3.17) yields the spatial spectrum representation of driving signals:

$$E^{2D}_{samp,k}(k_y, f) = \sum_{v=-\infty}^{+\infty} E^{2D}_{k}\left(k_y - \frac{2\pi v}{\Delta y'}, f\right). \tag{10.3.18}$$

The spatial spectrum of the sampled driving signal is an infinite repetition of the original spatial spectrum with displacements of the integral multiplication of $2\pi/\Delta y'$ along the k_y axis, or the spatial spectrum at k_y becomes a mix of all original values at $k_y - 2\pi v/\Delta y'$. The reconstructed pressure caused by a discrete array is calculated by substituting Equations (10.3.18) and (10.3.14) into Equation (10.3.9):

$$P'_{samp,k}(x - x', k_y, f) = \sum_{v=-\infty}^{+\infty} G^{2D}_{free,k}(x - x', k_y, f) E^{2D}_{k}\left(k_y - \frac{2\pi v}{\Delta y'}, f\right). \tag{10.3.19}$$

The mix of repeated or mirror spatial spectra with the original spatial spectrum leads to spatial aliasing because driving signals are not spatially bandlimited. At the same time, two-dimensional Green's function cannot act as an effective spatial low-pass filter to driving signals because it is also not spatially bandlimited. As a result, the mirror spatial spectra caused by the spatial sampling of the driving signals occur in the reconstructed sound fields. Driving signals and Green's function include the contributions of propagating and evanescent components, and the reconstructed sound fields involves the following parts of contributions from the combination of driving signals and Green's function.

The first part is the contribution of the overlapping of the propagating components of Green's function and driving signals. This contribution is evaluated by letting $|k_y| < 2\pi f/c$ and summing v that satisfies $|k_y - 2\pi v/\Delta y'| < 2\pi f/c$ in Equation (10.3.19). Therein, the term $v = 0$ creates the ideal or target reconstructed pressure in Equation (10.3.16). All the terms of $v \neq 0$ create spatial aliasing. Spatial aliasing vanishes when the frequency spectrum of the target sound field is bandlimited and thus satisfies

$$f < f_{\max} = \frac{c}{2\Delta y'}. \tag{10.3.20}$$

Equation (10.3.20) is the condition of anti-spatial aliasing when the propagating components of Green's function and driving signals are considered. The result is consistent with Equation (10.1.28).

The second part is the contribution of the overlapping of the propagating component of Green's function and the evanescent component of driving signals. This contribution is evaluated by letting $|k_y| < 2\pi f/c$ and summing v that satisfies $|k_y - 2\pi v/\Delta y'| > 2\pi f/c$ in Equation (10.3.19). For the target source far from the secondary source array and at a high frequency, the contribution of this part is minimal.

The third part is the contribution of the overlapping of the evanescent component of Green's function and the propagating component of driving signals. This contribution is evaluated by letting $|k_y| > 2\pi f/c$ and summing v that satisfies $|k_y - 2\pi v/\Delta y'| < 2\pi f/c$ in Equation (10.3.19).

The fourth part is the contribution of the overlapping of the evanescent components of Green's function and driving signals. This contribution is evaluated by letting $|k_y| > 2\pi f/c$ and summing v that satisfies $|k_y - 2\pi v/\Delta y'| > 2\pi f/c$ in Equation (10.3.19).

The contributions of the third and fourth parts to the reconstructed sound field depend on the distance between the receiver position and the secondary source array. For the receiver position far from the secondary source array and at a high frequency, the contributions of the third and fourth parts decrease quickly. No strict anti-spatial-aliasing condition can be derived for the second, third, and fourth parts, except for the first part, because Green's function and driving signals are not spatially bandlimited.

This analysis is for the case of the reconstructed sound field of a target infinite straight-line source. A similar analysis can be applied to the case of the reconstructed sound field of a target plane wave source (Spors and Rabenstain, 2006). The result is an approximation of the above case under the condition that the target source is distant from the secondary source array; thus, the contribution of the evanescent component of driving signals is insignificant.

The case of an infinite secondary source array is analyzed in the preceding discussion. As stated in Section 10.1.4, a finite secondary source array is used in practice, and it is equivalent to truncating the driving signals for an infinite array with a rectangular spatial window in Equations (10.1.25) and (10.1.26). In the spatial spectral domain, this truncation is equivalent to convoluting the driving signals with the spatial window function, and driving signals become

$$\frac{1}{2\pi} E_k^{2D}(k_y, f) \otimes_{k_y} w_k(k_y) = \frac{1}{2\pi} \int_{-\infty}^{+\infty} E_k^{2D}(k'_y, f) w_k(k_y - k'_y) dk'_y, \tag{10.3.21}$$

where $w_k(k_y)$ is the spatial Fourier transform of the spatial window function. For a rectangular spatial window expressed in Equation (10.1.26), the result is given as

$$w_k(k_y) = \int_{-\infty}^{+\infty} w(\mathrm{r}) \exp(jk_y y) dy = D_L \frac{\sin\left(\frac{k_y}{2} D_L\right)}{\frac{k_y}{2} D_L}. \tag{10.3.22}$$

The reconstructed sound pressure in Equation (10.3.9) becomes

$$P'_k(x - x', k_y, f) = \frac{1}{2\pi} \left[E_k^{2D}(k_y, f) \otimes_{k_y} w_k(k_y)\right] G_{free,k}^{2D}(x - x', k_y, f). \tag{10.3.23}$$

The reconstructed sound field of a finite linear array can be analyzed on the basis of Equation (10.3.23). The basic results are presented in Section 10.1.4, including a narrowed region of reproduction and edge effect. For a finite and discrete array of secondary sources, the errors caused by spatial aliasing depend on the positions of the target source and the receiver with respect to the array. Overall, errors reduce as the distance between the receiver position and the array increases. At the same time, the anti-spatial-aliasing condition for the overlapping of the propagating components of Green's function and driving signals is no longer the simple form of Equation (10.3.20). Instead, the condition depends on the receiver position. This phenomenon is described in detail in another study (Spors and Ahrens, 2009).

The case of two-dimensional WFS with a linear array of secondary monopole straight-line sources is analyzed above. A 2.5-dimensional WFS with a linear array of secondary monopole point sources is often used in practice. The analysis of the reconstructed sound field and spatial aliasing can be extended to the case of 2.5-dimensional reproduction provided that three-dimensional Green's function and 2.5-dimensional driving signals are used in the analysis according to Equations (10.3.10) and (10.3.11). Moreover, this analysis can be applied to the case of WFS with a vertical plane array of secondary point sources. The results are similar to those of above analysis (Ahrens and Spors, 2010).

Spatial aliasing occurs in WFS with other discrete arrays of secondary sources. For example, spatial aliasing in horizontal Ambisonics with a uniform and discrete array of secondary sources arranged in a circle is analyzed in Section 9.5.2. Similar analysis is applicable to WFS with a uniform and discrete array of secondary sources in a circle. In contrast to the case of Ambisonics, the azimuthal spectrum of driving signals in WFS is not spatially bandlimited and therefore does not satisfy Equation (9.5.11a) or (9.5.11b). In this case, spatial aliasing occurs. For a target plane wave source with unit strength, driving signals of two-dimensional WFS are expressed as Equation (10.2.32) and can be written as

$$E_{pl}^{2D}(\theta', f) = 2w(r')jk\cos\theta_{sn}\exp\left[jk\cos(\theta_S - \theta')\right], \qquad (10.3.24)$$

where θ_S is the horizontal azimuth of an incident plane wave, and θ' is the azimuth of secondary sources in the circle. The azimuthal spectrum of driving signals is no longer spatially bandlimited because of the influence of the spatial window $w(r')$. This result can be proven by expanding Equation (10.3.24) into an azimuthal Fourier series of θ' although the analytical expression of azimuthal Fourier coefficients cannot be always found. Generally, the driving signals of secondary sources are not spatially bandlimited for target straight-line sources, point sources, and plane wave sources and for the horizontal-circular array of secondary straight-line sources and point sources. In all these cases, spatial aliasing occurs. Moreover, driving signals of three-dimensional WFS with a secondary source array on a spherical surface is not spatially bandlimited. The relative energy error in the reconstructed sound field caused by spatial aliasing is evaluated using Equation (9.5.15).

The method presented in this section is also applicable to the analysis of the spatial aliasing of a focused source in WFS (Spors et al., 2009). In this method, an infinite linear array of secondary monopole point sources is used, and Green's function of secondary sources in the spatial spectral domain is given in Equation (10.3.11). The 2.5-dimensional driving signals are expressed in Equation (10.2.47) or (10.2.49) and can be converted to a spatial spectral domain via a spatial Fourier transform. Similar to the case of Equations (10.3.14) to (10.3.16), Green's function of a point source and driving signals are not spatially bandlimited, and both of them involve the propagating and evanescent components. When a discrete secondary source array is used, the reconstructed sound fields involve four parts of contributions from the combination of propagating and evanescent components of driving signals and

Green's function. The anti-spatial-aliasing condition for the overlapping of the propagating components of Green's function and driving signals is also presented in Equation (10.3.20). Evanescent component contribution and spatial aliasing almost vanish in the region close to the focal point (Section 10.2.4). In creating a focused source, a finite secondary source array narrows the reproduction region and causes an edge effect (Section 10.2.4).

10.3.3 Spatial-spectral division method of WFS

In Sections 9.2.2 and 9.2.3, the spatial-spectral division method is used to derive the Ambisonic driving signals for a uniform and continuous secondary source array on a circular or spherical surface. This method is also applicable to WFS with a line or plane array of secondary sources and may yield more general results than the conventional method. When the spatial spectrum representations of target pressure and transfer function from a secondary source to the receiver position are given, the driving signals can be derived by the spatial-spectral division method. However, this method is not always valid.

1. The spatial-spectral division method is valid only when the transfer or Green's function in the spatial spectral-domain does not vanish. The zeroes of the transfer function in the spatial spectral-domain correspond to the inner models of a closed boundary. This condition has been presented in the discussion on the uniqueness of driving signals in Section 9.3.1.
2. For an arbitrary secondary source array, the driving signals derived from the spatial-spectral division method may depend on the receiver position with respect to the array. The dependence of target pressure and transfer function on the receiver position can be canceled by the spatial-spectral division only in some special cases. In these cases, the resultant driving signals are independent of the receiver positions. Moreover, the spatial-spectral division method does not always lead to the appropriate driving signals for an arbitrary target sound field. In other words, a given secondary source array cannot always reconstruct an arbitrary target sound field.

Horizontal reproduction is considered as the first example (Ahrens and Spors, 2010). An array of secondary point sources is arranged in an infinite straight line $x = x'$ parallel to the y axis. Driving signals are derived by using Equation (10.3.12) and matching the reconstructed pressure with the target pressure as follows:

$$E_k\left(x', k_y, f\right) = \frac{P_k\left(x, y, f\right)}{G^{3D}_{free,k}\left(x - x', k_y, f\right)}, \tag{10.3.25}$$

where the superscript "2.5D" in the driving signals is temporarily dropped to distinguish them from the 2.5-dimensional driving signals in Section 10.2.3.

If the target sound field is a horizontal plane wave, the pressure is expressed in Equation (1.2.6):

$$P(x, y, f) = P_A(f)\exp(-j\boldsymbol{k}_{pw} \cdot \boldsymbol{r}) = P_A(f)\exp(-jk_{pw,x}x)\exp(-jk_{pw,y}y), \tag{10.3.26}$$

where $\boldsymbol{k}_{pw}$ is the wave vector of a plane wave; the subscript "pw" denotes the plane wave; and $k_{pw,x}$ and $k_{pw,y}$ are the x and y components of the wave vector, respectively. The z component of the wave vector vanishes, then $k^2_{pw} = k^2_{pw,x} + k^2_{pw,y}$. In the coordinate illustrated in

Figure 10.3, $k_{pw,x}$ and $k_{pw,y}$ are negative because the target plane wave propagates along the direction with the projection in the $-x$ and $-y$ axes. Equation (10.3.26) can be converted to the spatial spectral domain by a spatial Fourier transform with respect to y. The result is given as

$$\begin{aligned} P_k(x, k_y, f) &= \int_{-\infty}^{+\infty} P(x, y, f)\exp(jk_y y)dy \\ &= P_A(f)\exp(-jk_{pw,x}x)2\pi\delta(k_y - k_{pw,y}). \end{aligned} \tag{10.3.27}$$

The spatial spectrum representation of free-field Green's function in a three-dimensional space is expressed in Equation (10.3.11). If the target source and receiver position are restricted in the horizontal plane, the result is given as

$$G_{free,k}^{3D}(x - x', k_y, f) = \begin{cases} -\dfrac{j}{4}H_0\left(\sqrt{k^2 - k_y^2}\,|x - x'|\right) & |k_y| < k = 2\pi f/c \\ \dfrac{1}{2\pi}K_0\left(\sqrt{k_y^2 - k^2}\,|x - x'|\right) & |k_y| > k = 2\pi f/c \end{cases}, \tag{10.3.28}$$

where H_0 and K_0 are the zero-order Hankel function and the zero-order modified Bessel function of the second kind, respectively. Similar to Equation (10.3.14), Equation (10.3.28) involves the contributions of propagating and evanescent waves. At a receiver position distant from the array, only the contribution of propagating waves is considered. The spatial spectrum representation of driving signals can be obtained by substituting Equations (10.3.27) and (10.3.28) into Equation (10.3.25) as

$$E_k(x - x', k_y, f) = \frac{P_k(x, k_y, f)}{G_{free,k}^{3D}(x - x', k_y, f)} = 4jP_A(f)\frac{2\pi\delta(k_y - k_{pw,y})}{H_0\left(\sqrt{k^2 - k_y^2}\,|x - x'|\right)}\exp(-jk_{pw,x}x). \tag{10.3.29}$$

Driving signals in the frequency-spatial domain are obtained by applying an inverse spatial Fourier transform to Equation (10.3.29):

$$\begin{aligned} E(x - x', y', f) &= \frac{1}{2\pi}\int_{-\infty}^{+\infty} E_k(x - x', k_y, f)\exp(-jk_y y')dk_y \\ &= 4jP_A(f)\frac{\exp(-jk_{pw,x}x - jk_{pw,y}y')}{H_0\left(\sqrt{k_{pw}^2 - k_{pw,y}^2}\,|x - x'|\right)} \\ &= 4j\frac{\exp[-jk_{pw,x}(x - x')]P(r',f)}{H_0[k_{pw,x}(x - x')]}, \end{aligned} \tag{10.3.30}$$

where $P(r',f) = P_A(f)\exp(-jk_{pw}\cdot r')$ is the pressure of the target plane wave at the position of secondary sources. The driving signal in Equation (10.3.30) depends on the receiver position x. Similar to the case in Section 10.1.3, when the magnitude of reconstructed pressure is equalized on a reference straight line ($x = x_{ref}, z = 0$) parallel to the y axis, Equation (10.3.30) becomes

$$E(x-x', y', f) = 4j\frac{\exp\left[-jk_{pw,x}\left(x_{ref}-x'\right)\right]P(r', f)}{H_0\left[k_{pw,x}\left(x_{ref}-x'\right)\right]}, \tag{10.3.31}$$

In the case of $|\, k_{pw,x}(x_{ref} - x')\,| >> 1$, the driving signal in Equation (10.3.31) can be simplified using the asymptotic formula of the Hankel function of the second kind in Equation (10.2.35) and the relation $[k_{pw,x}(x_{ref} - x')] = k_{pw} \cos\theta_{sn} \,|\, x_{ref} - x'|$, where $\theta_{sn} = \theta_S$ is the incident direction of the plane wave:

$$E(x-x', y', f) = 4\pi\sqrt{\frac{jk_{pw}}{2\pi}}\sqrt{|\,x_{ref}-x'\,|}\sqrt{\cos\theta_{sn}}P(r', f). \tag{10.3.32}$$

The expression in Equation (10.3.32) slightly differs from the driving signals of an infinite linear array of secondary point sources in Equation (10.1.21) or the 2.5-dimensional driving signals in Equation (10.2.41). Equation (10.3.32) is obtained through an equalization at a reference straight line ($x = x_{ref}, z = 0$) parallel to the y axis. Equation (10.1.21) is obtained through an equalization of the pressure magnitude caused by each secondary source in a reference circle centered at $\mathbf{r}'$ ($|\, \mathbf{r}_{ref} - \mathbf{r}'\,|$ = const.). As stated in Section 10.1.3, the driving signals in Equation (10.1.21) or Equation (10.2.41) is equivalent to Equation (10.3.32) only for the reference position distant from the secondary source and in the sense of stationary phase approximation. Moreover, Equation (10.1.21) or (10.2.41) differs from Equation (10.3.32)by a factor of $\sqrt{\cos\theta_{sn}}$. For the target plane wave incident near the frontal direction, the difference is small because of $\cos\theta_{sn} \approx 1$. However, the difference is significant for the target plane wave deviating from the frontal direction. This difference is also related to the reference position for magnitude equalization in the derivation of Equation (10.1.21) or Equation (10.2.41). Ahrens and Spors (2010) suggested revising the driving signals for an infinite linear array of secondary point sources with Equation (10.3.32). Furthermore, the spatial spectral division method is applicable to the derivation of driving signals for creating a focused source in WFS, and the results are similar to those in Section 10.2.4(Spors and Ahrens, 2010b).

The second example of the spatial spectral division method is similar to the first example, but the target source is a point source outside the array rather than a plane wave source (Spors and Ahrens, 2010c). When the receiver position (reference line) and the target point source are located at positions distant from the array in comparison with the wavelength, the spatial spectral division method yields the same driving signals as those in Sections 10.1.3 and 10.2.3. Otherwise, this method produces different driving signals.

The third example of the spatial spectral division method is the case of half three-dimensional space reproduction with an array of secondary point sources arranged in an infinite vertical plane (Figure 10.2; Ahrens and Spors, 2010), and the target source is a point source or a plane wave source. The free-field Green's function in a three-dimensional space is expressed in Equation (10.3.2). Driving signals are obtained by converting the calculation to a spatial spectral domain via Equations (10.3.4) and (10.3.5). The resultant driving signals are independent from the receiver position and consistent with those in Section 10.1.2.

One advantage of the spatial spectral division method is that it enables designing secondary source arrays and driving signals based on a target reconstructed sound field. It is applicable to some special cases of sound field reconstruction, such as compensating for the directivity of secondary sources. Some examples in Ambisonics are illustrated in Sections 9.2 and 9.3. Similar methods are applicable to WFS (Ahrens and Spors, 2009).

In some cases, the spatial spectral division method produces results that are difficult to be obtained through conventional methods (Sections 10.1 and 10.2). In some special cases, this method may yield driving signals identical to those obtained through conventional methods. In other cases, it may provide different results. Such variations are due to different mathematical and physical approximations in various methods.

10.4 FURTHER DISCUSSION ON SOUND FIELD RECONSTRUCTION

10.4.1 Comparison among various methods of sound field reconstruction

Various methods of sound field reconstruction and control are addressed in Chapter 9 and the preceding sections of this chapter. These methods can be classified into three types:

1. Direct control of sound pressure in a receiver region
2. Control of sound pressure and its normal derivative (velocity of medium) on a boundary surface
3. Control of the spatial spectrum or mode in the receiver region

Multiple receiver positions matching method in Section 9.6.3 belong to the first type of method. Through this method, the target sound field (pressure) can be reconstructed accurately or approximately within a continuous region specified by the grid of controlled points if the interval between adjacent points satisfies the condition of Shannon–Nyquist spatial sampling theorem. Otherwise, only the target pressures at discrete controlled points can be reconstructed, and errors in pressures at receiver positions between controlled points occur. This method is theoretically appropriate for various secondary source configurations. However, only the minimum-square-error solution of driving signals can be derived in most cases. Moreover, an appropriate minimum-square-error solution of driving signals requires that the number of secondary sources is not fewer than the number of controlled points. Therefore, a smaller number of secondary sources can only control the pressure at a smaller number of points and thus reconstruct the target sound field within a small or local region. A large number of secondary sources are necessary to reconstruct the target sound field in an extended region by using the multiple receiver positions matching method

The second type of method is based on Kirchhoff–Helmholtz boundary integral. This type of method controls the sound field on the boundary surface rather than in multiple receiver positions. This method includes ideal acoustical holography and WFS. A complete acoustical holography requires arrays of two types of secondary sources (secondary monopole and dipole sources) on the boundary surface to control sound field. If the interval between two adjacent secondary sources of each type satisfies the condition of Shannon–Nyquist spatial sampling theorem, acoustical holography can theoretically control the sound fields in two global regions inside and outside the boundary at the same time rather than in a small or local region. This feature distinguishes the acoustical holography from the first type of sound field control. Moreover, the dimensionality of a closed boundary surface is lower than that of the region inside the boundary. The required number of samples on the boundary surface is smaller by an order of magnitude than that in the entire region inside the boundary to satisfy the condition of Shannon–Nyquist spatial sampling theorem. Therefore, this requirement is usually smaller than that in the first type of method. Under certain conditions, acoustical holography can be approximately implemented by a single type (monopole or dipole) of secondary sources, as in the case of WFS. When the condition of Shannon–Nyquist spatial sampling theorem is satisfied, WFS can reconstruct the target sound field in the region inside the boundary at most. However, the mathematical and physical approximation in WFS may cause some errors in the reconstructed sound field.

Ambisonics belongs to the third type of method. On the basis of spatial harmonic decomposition or multiple expansion, Ambisonics controls and matches the spatial harmonic spectra of a sound field with each order approximation. It can reconstruct the target sound field within a receiver region and below a certain frequency limit imposed by Shannon–Nyquist spatial sampling theorem. By representing the sound field inside the boundary surface with the weighted superposition of orthogonal basis functions (spatial harmonics), Ambisonics removes the redundant information caused by the correlation between pressures of different receiver positions to some extent. Therefore, it can efficiently control the sound field with a finite number of secondary sources.

The three types of methods of sound field reconstruction are illustrated. However, these methods and examples are closely related. The reconstructed sound fields of these methods are the simplified or approximate solutions of Helmholtz equation subjected to a certain boundary condition. These solutions are closely associated or even equivalent in some cases because of the uniqueness of the solution of Helmholtz equation subjected to a certain boundary condition. Differences in the reconstructed sound field of different methods are due to various mathematical and physical approximations used in different methods.

Previous studies neglected on the relation among the different methods of sound field control and reconstruction. For example, WFS and Ambisonics were regarded as two different methods and usually investigated separately. The mathematical and physical methods for WFS and Ambisonics analyses were different. Practical WFS was conventionally derived using the Rayleigh integral and stationary phase method on Kirchhoff–Helmholtz boundary integral equation or acoustical holography. First-order Ambisonics was conventionally derived on the basis of virtual source localization theory and psychoacoustic hypothesis (Chapters 3, 4, and 6). Furthermore, high-order Ambisonics was derived through spatial harmonic decomposition and matching of a sound field (Chapter 9).

Some preliminary relations among different methods of sound field reconstruction are discussed in Chapter 9 and the preceding sections of this chapter. For example, the spatial spectrum and reconstructed sound field of Ambisonics and WFS are analyzed in Sections 9.3 and 10.3.3, and driving signals are derived. In Section 9.6.3, the driving signals of Ambisonics are derived from multiple receiver positions matching method and the relation between this method and sound-field-mode control is analyzed. In addition, the condition of the multiple receiver positions matching method in Section 9.6.3 can be modified so that pressure and its normal derivative on a closed boundary surface match with those in the target sound field. In this case, according to Kirchhoff–Helmholtz boundary integral equation and under the condition of Shannon–Nyquist spatial sampling theorem, the target sound field in a source-free region inside the boundary can be exactly reconstructed (Ise, 1999).

Therefore, similarities and differences in various methods of sound field reconstruction can be explored by investigating the relationship among them to obtain insights into the nature of sound field reconstruction and develop practical techniques. As examples, the relationship among acoustical holography, WFS, and Ambisonics is analyzed in Sections 10.4.2, 10.4.3, and 10.4.4 (Nicol and Emerit, 1999; Daniel et al., 2003; Spors and Wierstorf, 2008; Poletti and Abhayapala, 2011).

10.4.2 Further analysis of the relationship between acoustical holography and sound field reconstruction

The reconstructed sound field in the spatial spectral domain for an ideal acoustical holography is first analyzed. For simplicity, two-dimensional (horizontal) reconstruction is considered. In this case, the horizontal receiver position is specified by polar coordinates (r, θ), and the sound field is independent from the vertical (z) coordinate. A target infinite straight-line source is arranged within a circular region $r_{S1}<r_S< r_{S2}$. In the region of $r<r_{S1}$, the ***interior radiated sound field*** created by the target source can be expanded as a Bessel–Fourier series similar to Equation (9.3.2):

$$P(r,f)=P(r,\theta,f)=\sum_{q=0}^{\infty}J_q(kr)\left[B_q^{(1)}(f)\cos q\theta+B_q^{(2)}(f)\sin q\theta\right] \quad r<r_{S1}, \qquad (10.4.1)$$

where $J_q(kr)$ is the q-order Bessel function; $B_q^{(1)}(f)$ and $B_q^{(2)}(f)$ are the coefficients of expansion; and $B_0^{(2)}(f)=0$, which is preserved in Equation (10.4.1) for convenience in writing. In the approximation of a target plane wave with a unit amplitude, these coefficients are simplified into the form of Equation (9.3.3) and with $B_0^{(1)}(f)=1$.

In the region of $r>r_{S2}$, the ***exterior radiated sound field*** created by a target source can also be expanded as a Bessel–Fourier series:

$$P(r,f)=P(r,\theta,f)=\sum_{q=0}^{\infty}H_q(kr)\left[C_q^{(1)}(f)\cos q\theta+C_q^{(2)}(f)\sin q\theta\right] \quad r>r_{S2}, \qquad (10.4.2)$$

where $H_q(kr)$ is the q-order Hankel function of the secondary kind, $C_q^{(1)}(f)$ and $C_q^{(2)}(f)$ are the coefficients of expansion, and $C_0^{(2)}(f)=0$.

In sound field reconstruction, two arrays of secondary monopole and dipole straight-line sources are arranged uniformly and continuously in a horizontal circle with the radius r_0. The main axes of secondary dipole sources point to the center of the circle. The position of the arbitrary source is denoted by $\mathbf{r}' = (r_0, \theta')$. The driving signals of monopole and dipole straight-line sources are $E_{mon}(\theta', f)$ and $E_{dip}(\theta', f)$, respectively. The reconstructed sound pressure is expressed as

$$P'(r,f)=P'(r,\theta,f)=\int_{-\pi}^{\pi}\left[E_{mon}(\theta',f)G_{free}^{2D}(\mathbf{r},\mathbf{r}',f)+E_{dip}(\theta',f)\frac{\partial G_{free}^{2D}(\mathbf{r},\mathbf{r}',f)}{\partial n'}\right]d\theta', \qquad (10.4.3)$$

where $G_{free}^{2D}(\mathbf{r},\mathbf{r}',f)$ is the free-field Green's function in a two-dimensional space expressed in Equations (9.2.17) and (10.2.13):

$$G_{free}^{2D}(\mathbf{r},\mathbf{r}',f)=G_{free}^{2D}(r,\theta-\theta',r_0,f)=-\frac{j}{4}H_0\left[k\sqrt{r^2+r_0^2-2rr_0\cos(\theta-\theta')}\right]. \qquad (10.4.4)$$

n' is the inward-normal direction (pointing to the center of circle). The integral in Equation (10.4.3) is originally calculated over the circle $r'= r_0$, and the curvilinear element is $r_0 d\theta'$. r_0 is merged into the driving signals $E_{mon}(\theta', f)$ and $E_{dip}(\theta', f)$ as an overall gain, so the integral in

Equation (10.4.3) is calculated over the azimuth θ. Equation (10.4.3) is valid for the interior region $r < r_0$ of the circle and the exterior region $r > r_0$ of the circle.

Similar to the case in Section 9.2.2, Green's function in Equation (10.4.4) is expanded as Bessel–Fourier series to analyze the reconstructed sound field in the spatial spectral domain. The expansion differs in the regions of $r < r_0$ and $r > r_0$:

$$G_{free}^{2D}(r, r', f) = -\frac{j}{4}\begin{cases} [J_0(kr)H_0(kr_0) + 2\sum_{q=1}^{\infty} J_q(kr)H_q(kr_0)(\cos q\theta' \cos q\theta + \sin q\theta' \sin q\theta) & r < r_0 \\ [J_0(kr_0)H_0(kr) + 2\sum_{q=1}^{\infty} J_q(kr_0)H_q(kr)(\cos q\theta' \cos q\theta + \sin q\theta' \sin q\theta) & r > r_0 \end{cases}. \tag{10.4.5}$$

The driving signals $E_{mon}(\theta', f)$ and $E_{dip}(\theta', f)$ can also be expanded as a Fourier series of θ':

$$\begin{aligned} E_{mon}(\theta', f) &= \sum_{q=0}^{\infty}\left[E_{mon,q}^{(1)}(f)\cos q\theta' + E_{mon,q}^{(2)}(f)\sin q\theta'\right] \\ E_{dip}(\theta', f) &= \sum_{q=0}^{\infty}\left[E_{dip,q}^{(1)}(f)\cos q\theta' + E_{dip,q}^{(2)}(f)\sin q\theta'\right]. \end{aligned} \tag{10.4.6}$$

The coefficients $E_{mon,q}^{(1)}(f)$, $E_{mon,q}^{(2)}(f)$, $E_{dip,q}^{(1)}(f)$, $E_{dip,q}^{(2)}(f)$ of Fourier expansion are calculated similarly to Equation (9.2.9), and $E_{mon,0}^{(2)} = E_{dip,0}^{(2)} = 0$. The following equations can be obtained by (1) substituting Equations (10.4.5) and (10.4.6) into Equation (10.4.3); (2) matching the reconstructed pressures in the regions of $r< \min(r_{S1}, r_0)$ and $r> \max(r_{S2}, r_0)$ with the target pressures in Equations (10.4.1) and (10.4.2) respectively; and (3) using the integral orthogonality of trigonometric functions in Equations (4.3.19) and (4.3.20) or using the mode-matched method in Section 9.2.2:

$$H_q(kr_0)E_{mon,q}^{(1)}(f) - \frac{\partial H_q(kr_0)}{\partial r_0}E_{dip,q}^{(1)} = \frac{2j}{\pi}B_q^{(1)}(f) \qquad q = 0, 1, 2\ldots, \tag{10.4.7}$$

$$J_q(kr_0)E_{mon,q}^{(1)}(f) - \frac{\partial J_q(kr_0)}{\partial r_0}E_{dip,q}^{(1)} = \frac{2j}{\pi}C_q^{(1)}(f) \qquad q = 0, 1, 2,\ldots, \tag{10.4.8}$$

and

$$H_q(kr_0)E_{mon,q}^{(2)}(f) - \frac{\partial H_q(kr_0)}{\partial r_0}E_{dip,q}^{(2)} = \frac{2j}{\pi}B_q^{(2)}(f) \qquad q = 1, 2, 3\ldots, \tag{10.4.9}$$

$$J_q(kr_0)E_{mon,q}^{(2)}(f) - \frac{\partial J_q(kr_0)}{\partial r_0}E_{dip,q}^{(2)} = \frac{2j}{\pi}C_q^{(2)}(f) \qquad q = 1, 2, 3\ldots. \tag{10.4.10}$$

The relation of $\partial/\partial n' = -\partial/\partial r_0$ has been used in the calculation of inward-normal derivative.

Equations (10.4.7) to (10.4.10) are a set of equations that connect the driving signals and target sound field in the spatial spectral domain. The target interior and exterior sound field can be independent and determined by the coefficients $\left[B_q^{(1)}(f), B_q^{(2)}(f)\right]$ and $\left[C_q^{(1)}(f), C_q^{(2)}(f)\right]$, respectively. Given these coefficients and for each order q, Equations (10.4.7) and (10.4.8) are a pair of linear equations with two unknowns, and the unique solution of equations is given as

$$E_{mon,q}^{(1)}(f) = r_0\left[B_q^{(1)}(f)\frac{\partial J_q(kr_0)}{\partial r_0} - C_q^{(1)}(f)\frac{\partial H_q(kr_0)}{\partial r_0}\right]$$
$$E_{dip,q}^{(1)}(f) = r_0\left[B_q^{(1)}(f)J_q(kr_0) - C_q^{(1)}(f)H_q(kr_0)\right]. \qquad (10.4.11)$$

Similarly, the solution of Equations (10.4.9) and (10.4.10) is expressed as

$$E_{mon,q}^{(2)}(f) = r_0\left[B_q^{(2)}(f)\frac{\partial J_q(kr_0)}{\partial r_0} - C_q^{(2)}(f)\frac{\partial H_q(kr_0)}{\partial r_0}\right]$$
$$E_{dip,q}^{(2)}(f) = r_0\left[B_q^{(2)}(f)J_q(kr_0) - C_q^{(2)}(f)H_q(kr_0)\right]. \qquad (10.4.12)$$

The relationship $H_q(\xi) = J_q(\xi) - jY_q(\xi)$ among the Hankel function $H_q(\xi)$ of the second kind, the Bessel function $J_q(\xi)$, and the Neumann function $Y_q(\xi)$ and the following Wronskian formula are used to derive Equations (10.4.11) and (10.4.12):

$$J_q(\xi)\frac{\partial Y_q(\xi)}{\partial \xi} - Y_q(\xi)\frac{\partial J_q(\xi)}{\partial \xi} = \frac{2}{\pi\xi}. \qquad (10.4.13)$$

Driving signals are obtained by substituting Equations (10.4.11) and (10.4.12) into Equation (10.4.6).

When the target interior radiated sound field is given only, the coefficients satisfy Equations (10.4.7) and (10.4.9). In this case, the equations are underdetermined, that is, the number of unknown coefficients is more than that of equations, resulting in an infinite set of the exact solutions $E_{mon,q}^{(1)}(f)$, $E_{mon,q}^{(2)}(f)$, $E_{dip,q}^{(1)}(f)$, $E_{dip,q}^{(2)}(f)$ of coefficients. In other words, an infinite set of the combinations of the driving signals of secondary monopole and dipole sources exist, which can reconstruct target interior sound field. Similarly, when the target exterior radiated sound field is given only, Equations (9.4.8) and (9.4.10) yield an infinite set of the combinations of the driving signals of secondary monopole and dipole sources.

When only an array of one type of secondary sources is used, for example, if secondary monopole straight-line sources are only used, $E_{dip,q}^{(1)}(f) = E_{dip,q}^{(2)}(f) = 0$ in Equations (10.4.7) to (10.4.10). Therefore, the equations are overdetermined, i.e., more equations than unknown coefficients exist. In this case, no exact solutions of $E_{mon,q}^{(1)}(f)$ and $E_{mon,q}^{(2)}(f)$ exist, so simultaneously controlling the interior and exterior sound fields is impossible. If the interior sound field is controlled while the exterior sound field is ignored, the exact solutions of $E_{mon,q}^{(1)}(f)$

and $E_{mon,q}^{(2)}(f)$ are obtained, resulting in the driving signals of near-field-compensated higher-order Ambisonics in Equation (9.3.53). Similarly, an array of one type of secondary sources can control the exterior sound field but cannot simultaneously control the interior and exterior sound fields.

In ideal acoustical holography, the driving signals of secondary sources are chosen according to the two-dimensional Kirchhoff–Helmholtz boundary integral equation in Equation (10.2.16):

$$\begin{aligned} E_{mon}(\theta', f) &= -r_0 \frac{\partial P(r_0, \theta', f)}{\partial n'} = r_0 \frac{\partial P(r_0, \theta', f)}{\partial r_0} \\ E_{dip}(\theta', f) &= r_0 P(r_0, \theta', f), \end{aligned} \tag{10.4.14}$$

where $P(r_0, \theta', f)$ is the pressure of the target interior radiation at the boundary (r_0, θ') specified by the secondary source array, as expressed in Equation (10.4.1). Driving signals in the spatial spectral domain are obtained by substituting Equation (10.4.1) into Equation (10.4.14) and comparing with Equation (10.4.6):

$$\begin{aligned} E_{mon,q}^{(1)}(f) &= r_0 \frac{\partial J_q(kr_0)}{\partial r_0} B_q^{(1)}(f) \quad E_{dip,q}^{(1)}(f) = r_0 J_q(kr_0) B_q^{(1)}(f) \quad q = 0, 1, 2\ldots \\ E_{mon,q}^{(2)}(f) &= r_0 \frac{\partial J_q(kr_0)}{\partial r_0} B_q^{(2)}(f) \quad E_{dip,q}^{(2)}(f) = r_0 J_q(kr_0) B_q^{(2)}(f) \quad q = 1, 2, 3\ldots \end{aligned} \; . \tag{10.4.15}$$

Substituting Equation (10.4.15) into the left sides of Equations (10.4.7) to (10.4.10) and substituting the coefficients on the right sides of Equations (10.4.7) and (10.4.10) with the Bessel–Fourier coefficients $B_q'^{(1)}(f)$, $B_q'^{(2)}(f)$, $C_q'^{(1)}(f)$, and $C_q'^{(2)}(f)$ of the reconstructed sound field yield

$$B_q'^{(1)}(f) = B_q^{(1)}(f) \quad B_q'^{(2)}(f) = B_q^{(2)}(f) \qquad C_q'^{(1)}(f) = C_q'^{(2)}(f) = 0. \tag{10.4.16}$$

Therefore, in the region inside the array with $r<r_0$, the Bessel–Fourier coefficients of the reconstructed pressure match with those of the target interior radiated pressure in Equation (10.4.1), e.g., $P'(r, \theta, f) = P(r, \theta, f)$; thus, the reconstructed sound field matches with the target interior radiated sound field. In the region outside the array with $r > r_0$, the Bessel–Fourier coefficients of the reconstructed pressure vanish, so the target exterior radiated sound pressure satisfies $P'(r, \theta, f) = 0$.

In conclusion, for two arrays of secondary monopole and dipole straight-line sources arranged uniformly and continuously in a horizontal circle with r_0, driving signals and the reconstructed sound field possess the following features:

1. When arrays of two types of secondary sources are used simultaneously, the unique solution of driving signals can be derived from the given target interior and exterior radiated sound fields. Conversely, the interior and exterior sound fields can be controlled simultaneously if the driving signals of two types of secondary sources are independent.
2. When arrays of two types of secondary sources are used simultaneously, infinite sets of driving signals can be derived from either the given target interior or exterior radiated

sound fields only. In particular, when the interior radiated sound field is given only, the resultant driving signals for two types of secondary sources are correlated and not unique.

3. An array of a single type of secondary sources (such as monopole straight-line sources) can control either the interior radiated sound field of the array or exterior radiated sound field of the array. However, it cannot control both fields at the same time.
4. As a special example of case (1), driving signals of secondary monopole and dipole sources are derived from Kirchhoff–Helmholtz boundary integral equation. Then, the reconstructed sound field inside the array matches with that of the target sound field of interior radiation, and the reconstructed sound field of the exterior radiation of the array vanishes.

Although the aforementioned analysis focuses on the arrays of secondary straight-line sources arranged on a horizontal circle, similar analysis and conclusion can be extended to the arrays of secondary sources arranged on a spherical surface (Daniel et al., 2003). Moreover, discrete and finite arrays in practical reproduction cause a spatial aliasing error, which is analyzed in Sections 9.5.2 and 10.3.2.

This analysis reveals the nature of acoustical holography and sound field reconstruction. Ideal acoustical holography pertains to Case (4). Dipole straight-line sources cancel the exterior radiation of monopole straight sources so that the overall exterior radiation vanishes in the region outside the arrays. The general formulation of sound field reconstruction with a single type of secondary source in Sections 9.2 and 10.3, including Ambisonics and practical WFS, pertains to Case (3). Therefore, for Ambisonics and practical WFS, the sound field outside the secondary source array does not vanish.

In practice, a target sound field should be reconstructed within a secondary source array, and the exterior radiated sound pressures outside the array should be reduced. This reduction helps minimize the influence of listening room reflections on reproduction. In contrast to the case in Section 9.7 in which the influence of listening room reflections is controlled by active cancellation, here, the influence of listening room reflections is controlled by reducing the exterior radiated sound pressure or energy. Therefore, some differences exist between the physical principles of two methods. The signal processing of active cancellation is complicated, and the performance of cancellation is sensitive to the variation in the acoustic environment. Secondary sources with appropriate directivity can exactly control the interior radiated sound field and reduce the exterior radiated sound pressures (although they cannot control the exterior sound field exactly at the same time). Poletti et al. (2010a) proved that an array of secondary sources with hypercardioid directivity reduces the exterior radiated sound pressures and the influence of listening room reflections on the reconstructed sound field. Therefore, this secondary source array obviously increases the direct-to-reverberant energy ratio within the reconstructed region inside the array compared with the case of an array of monopole secondary sources.

Certainly, ideal acoustical holography is the most effective and direct method for eliminating or reducing the exterior radiation outside the array. However, ideal acoustical holography requires two types of secondary source array with independent driving signals. For arrays of secondary sources on a spherical surface, Poletti et al. (2010b) proved that the combination of the arrays of secondary monopole and dipole sources is equivalent to an array of secondary sources with variable first-order directivity. Poletti and Abhayapala (2011) further suggested that a tangential dipole component should be included in secondary straight-line sources with first-order directivity in a horizontal-circular array. The additional tangential dipole component improves the accuracy of interior sound reconstruction and reduces the exterior radiation near and above the upper frequency limit of Shannon–Nyquist spatial sampling theorem. Using secondary sources with higher-order directional components further reduces the exterior radiation and thereby minimizes the influence of listening room reflections on interior sound field (Betlehem and Poletti, 2014).

Chang and Jacobsen (2012) suggested using a circular double-layer array of secondary sources to control sound fields. Secondary sources with the first-order directivity (a combination of monopole and dipole sources) are arranged in two concentric circular layers. The main axes of secondary sources in the outer and inner layers point to the outward- and inward-normal directions of the circles, respectively. The driving signals of secondary sources are derived by using the multiple receiver positions matching and least square error methods similar to those in Section 9.6.3. From the point of the multipole expansion of a sound field, this array of secondary sources is closely related to the array of secondary monopole and dipole sources on a circle and able to control interior and exterior sound fields independently.

Actually, the discussion in this section can be regarded as a kind spatial multizone sound field reconstruction. Here, a two-dimensional space is divided into two sub-regions. One sub-region is inside the circular array, and the other is outside. By contrast, in Section 9.3.6, the region inside a circular array is divided into some sub-regions. The discussions in this section and Section 9.3.6 differ in the division of sub-regions.

10.4.3 Further analysis of the relationship between acoustical holography and Ambisonics

The relation between ideal acoustical holography and Ambisonics can be observed preliminarily from the discussion in Section 10.4.2. If the interior radiated sound field is controlled by an array of secondary monopole straight-line sources only, the driving signals of secondary dipole straight-line sources vanish, e.g., $E_{dip}(\theta', f) = 0$. In this case, Equation (10.4.3) is simplified into the general formulation of multichannel sound field reconstruction in Equation (9.2.1) or (9.2.7), and Equations (10.4.7) to (10.4.10) are equivalent to Equation (9.2.27).

A two-dimensional acoustical holography in a circular region with radius r_0 is considered to further explore the relation between ideal acoustical holography and Ambisonics, and the reconstructed sound field is expressed in Equation (10.2.16). After the line integral along the circle is converted to an integral over the azimuth, Equation (10.2.16) becomes

$$P'(r,f) = -\int_{-\pi}^{\pi}\left[\frac{\partial P(r',f)}{\partial n'}G_{free}^{2D}(r,r',f) - P(r',f)\frac{\partial G_{free}^{2D}(r,r',f)}{\partial n'}\right]r_0 d\theta'. \tag{10.4.17}$$

The corresponding driving signals of secondary monopole and dipole sources are given in Equation (10.4.14).

If the target sound field is created by a monopole straight-line source with unit strength and located at $r_S = (r_S, \theta_S)$ outside the circular array of secondary source ($r_S > r_0$), similar to the case in Section 9.2.2, then converting Equation (10.4.17) to the spatial spectral domain is convenient for analysis. For this purpose, the target pressure $P(r', f)$ in the boundary and Green's function $G_{free}^{2D}(\mathbf{r}, \mathbf{r}', f)$ are expanded as Bessel–Fourier series according to Equation (9.2.18):

$$\begin{aligned} P(r',f) &= -\frac{j}{4}H_0(k|r'-r_S|) \\ &= -\frac{j}{4}[J_0(kr_0)H_0(kr_S) + 2\sum_{q=1}^{\infty}J_q(kr_0)H_q(kr_S)\cos q(\theta'-\theta_S). \end{aligned} \tag{10.4.18}$$

$$G_{free}^{2D}(r, r', f) = -\frac{j}{4} H_0(k \mid r - r' \mid)$$
$$= -\frac{j}{4}\left[J_0(kr)H_0(kr_0) + 2\sum_{q=1}^{\infty} J_q(kr)H_q(kr_0)\cos q(\theta - \theta')\right]. \tag{10.4.19}$$

Equation (10.4.20) can be derived through the following steps: (1) substituting Equations (10.4.18) and (10.4.19) into Equation (10.4.17); (2) using the integral orthogonalities of trigonometric functions in Equations (4.3.19) and (4.3.20); (3) applying the relationship $H_q(\xi) = J_q(\xi) - jY_q(\xi)$ among the Hankel function of the secondary kind, the Bessel function, and the Neumann function; and (4) using the Wronskian formula in Equation (10.4.13)

$$P'(r, f) = \int_{-\pi}^{\pi} G_{free}^{2D}(r, r', f) E(r', f) d\theta', \tag{10.4.20}$$

where

$$E(\theta_S, r_S, r_0, \theta', f) = \frac{1}{2\pi}\left\{\frac{H_0(kr_S)}{H_0(kr_0)} + 2\sum_{q=1}^{\infty}\frac{H_q(kr_S)}{H_q(kr_0)}(\cos q\theta' \cos q\theta_S + \sin q\theta' \sin q\theta_S)\right\}$$
$$= \frac{1}{2\pi}\left\{\frac{H_0(kr_S)}{H_0(kr_0)} + 2\sum_{q=1}^{\infty}\frac{H_q(kr_S)}{H_q(kr_0)}\cos\left[q(\theta_S - \theta')\right]\right\}. \tag{10.4.21}$$

Equation (10.4.20) indicates that the interior pressure can be equivalently created by an array of secondary monopole straight-line sources only, and driving signals are expressed in Equation (10.4.21). For a target straight-line source with unit strength, driving signals of secondary sources are equal to their normalized amplitudes, e.g., $E(\theta_S, r_S, r_0, \theta', f) = A(\theta_S, r_S, r_0, \theta', f)$. Equation (10.4.21) is the driving signal of the horizontal near-field-compensated Ambisonics with an infinite order. Equation (10.4.21) is consistent with Equation (9.3.53) except for a normalized gain. The difference in normalized gain is due to the variation in continuous and discrete secondary source arrays.

In conclusion, if only the target interior sound field is controlled, secondary monopole and dipole sources in acoustical holography or Kirchhoff–Helmholtz boundary integral equation have closely related radiation. Therefore, the target interior sound field can be equivalently reconstructed by an array of secondary monopole sources only. In other words, transition from acoustical holography to Ambisonics can occur without forcing the driving signals of secondary dipole sources to vanish. This analysis can be extended to the case of spatial Ambisonics (Daniel et al., 2003; Poletti, 2005b).

10.4.4 Comparison between WFS and Ambisonics

WFS and higher-order Ambisonics, which use an array of a single type of secondary sources, can be derived from the simplification of acoustical holography or Kirchhoff–Helmholtz boundary integral equation. However, the conditions and methods for simplification differ in two cases.

In WFS, Kirchhoff–Helmholtz boundary integral equation is approximately calculated by Rayleigh integrals or appropriate (Neumann) Green's function to simplify the types of secondary sources. WFS can be theoretically achieved by an arbitrary array of secondary sources. When a curved array is used, a target source direction-dependent spatial window should be applied to driving signals to reconstruct the target sound field correctly. For example, a spatial window enables secondary sources in half of the horizontal-circular array to participate in the reconstruction of a target plane wave. Therefore, this process can be regarded as a local signal mixing method and is similar to local Ambisonic signal mixing in Section 5.2.4. Moreover, driving signals in WFS are not spatially bandlimited. In case of horizontal WFS, a stationary phase method enables the substitution of secondary straight-line sources with point sources. However, mismatched secondary sources lead to errors in the reconstructed spectrum and the overall magnitude of pressure. The former can be pre-equalized by applying a special filter to the driving signals, but the latter can only be equalized at a special reference position or line.

Ideally, a horizontal Ambisonics requires an array of secondary straight-line sources arranged in a circle. For far-field approximation, secondary monopole straight-line sources can be substituted by point sources. Spatial Ambisonics requires an array of secondary monopole point sources. For secondary source arrays in a horizontal circle and a spherical surface, the driving signals of secondary monopole and dipole sources for controlling an interior sound field are dependent. Therefore, an array of secondary monopole sources is enough to control the interior sound field. Corresponding driving signals can be obtained from a combination of the pressure and normal velocity of a medium on the boundary of a circle or spherical surface (Section 9.8.3). When the target sound field is decomposed by spatial harmonics, driving signals are represented by a weighted combination of these spatial harmonics. In actual Ambisonics, spatial harmonic decomposition is truncated up to a certain order; thus, driving signals are spatially bandlimited. Ambisonic driving signals pertain to global signal mixing. All secondary sources in the circular or spherical array take part in the reconstruction of a target sound field, and the spatial window for these driving signals is usually not required. Given an upper frequency limit, Ambisonics can reconstruct the target sound field within a local region centered around the origin rather than an extended region within an array.

In practical WFS, M secondary sources are utilized to control the pressure or normal velocity of a medium on the boundary and reconstruct the target sound field in the entire region inside the boundary. As indicated in Section 10.1.5, Shannon–Nyquist spatial sampling theorem requires that the arc length between adjacent secondary sources on the boundary should not exceed half of the wavelength in the worst case. Therefore, more secondary sources are needed for reproduction in a larger region. By contrast, as indicated in Section 9.6.3, horizontal Ambisonics can be equivalent to a scheme of controlling the pressures at O uniform receiver positions in a circle with radius r through a uniform array of M secondary sources arranged in a circle with radius $r_0 > r$. In Equation (9.3.15), Shannon–Nyquist spatial sampling theorem requires that the number of secondary sources should satisfy $M \geq O$, and the arc length between adjacent receiver positions should not exceed half of the wavelength. The required spatial samples on a circle with radius r is smaller than that on a circle with radius $r_0 > r$ to satisfy the spatial sampling theorem. In other words, Ambisonics reconstructs the target sound field in a smaller region (rather than the entire region inside the array) through fewer secondary sources than WFS. A similar method is used in local WFS in Section 10.2.4. That is, when a smaller number of secondary sources are used, a local WFS improves the accuracy in sound field reconstruction at the cost of reducing the reproduction region.

The reconstructed sound field of WFS and Ambisonics exhibits different physical and perceptual characteristics because of the aforementioned differences between them (Spors and Wierstorf, 2008). However, WFS and Ambisonics can be analyzed using similar methods because their reconstructed sound field and driving signals are closely related to each other.

A horizontal-circular array of secondary monopole straight-line sources is considered, and its radius is r_0. Spatial spectral analysis of a circular array is discussed in Section 9.2.2, and the problems of spatial aliasing and mirror spatial spectra are addressed in Section 9.5.2. The discussions in Chapter 9 focus on Ambisonics, but some general methods and results are applicable to WFS.

The reconstructed sound field of WFS and Ambisonics can be evaluated by substituting the driving signals into Equation (9.2.1) or (9.2.6). In the case of a target plane wave, the driving signals of WFS and Ambisonics are expressed in Equations (10.2.32) and (9.3.53), respectively. Analyses on the horizontal-circular array of secondary monopole straight-line sources, including the calculation of a relative energy error in Equation (9.5.15), lead to the following results.

For WFS, the following conditions are observed:

1. The active secondary sources do not constitute a close curve or curved surface because a spatial window is applied to the driving signals of secondary sources. Therefore, the problem of interior eigen modes in an enclosed space does not occur. A unique solution of driving signals is available to all frequencies or more strictly for all (kr_0). The problem of instability in the solution of driving signals does not take place. However, the spatial window causes an edge effect.
2. Even if the spatial aliasing error is ignored, an error occurs in the reconstructed sound field inside an array. This result is also observed in other arrays of secondary sources.
3. For a target plane wave, driving signals are not spatially bandlimited. The analysis in Section 9.5.2 indicates that the discrete array of secondary sources leads to a spatial aliasing error in the reconstructed sound field. Above a certain frequency limit, spatial aliasing leads to the obvious interference of a sound field.
4. The spatial distribution of errors caused by a discrete array is irregular. Above a certain frequency limit, spatial aliasing occurs in the entire receiver region. However, when the receiver position is far from the secondary source array, the spatial aliasing error is reduced.
5. Spatial aliasing in the reconstructed sound field may lead to perceivable timbre coloration.

For higher-order Ambisonics, the following conditions are presented:

1. Active secondary sources constitute a close curve or curved surface. Interior eigen modes occur at some frequencies or more strictly at some (kr_0). At these frequencies, the solutions of driving signals are not unique [see the discussion after Equation (9.3.5)]. In other words, the interior sound field cannot be controlled at some frequencies, and the problem of instability in the solution of driving signals occurs.
2. The driving signals of a Q-order Ambisonics are spatially bandlimited. If the number of secondary sources satisfies $M \geq (2Q + 1)$, driving signals do not cause spatial aliasing. However, the mirror spatial spectra of the driving signals caused by a discrete array may cause an error in the reconstructed sound field.
3. The spatial distribution of an error caused by mirror spatial spectra is regular. When the number of secondary sources satisfies the condition of $M \geq (2Q + 1)$, all $v \neq 0$ terms in the summarization of Equation (9.5.12) can be omitted if kr is smaller than a certain value [Equation (9.3.14)] because the Bessel function $J_q(kr)$ oscillates and decays when its order q is not less than $[exp(1)kr/2]$. Therefore, Ambisonics can reconstruct the target sound field in a circular region centered at the origin and up to a certain frequency. The radius of this region and the upper frequency limit, or more strictly the

upper limit of kr, is evaluated from Equation (9.3.14) or (9.3.15). The upper limit of kr increases as the order of Ambisonics increases. When kr exceeds the upper limit, the mirror spectra lead to an obvious error in the reconstructed pressures. A much higher-order Ambisonics can theoretically reconstruct the target sound field accurately inside the array and within a full audible frequency range.
4. Mirror spatial spectra cause timbre coloration.

For a target plane wave, the driving signals of WFS are not spatially bandlimited. However, WFS can reconstruct the target plane wave field in the entire receiver region inside the array below an upper frequency limit (Section 10.1.5) if the error in the receiver region close to the secondary source array is ignored. By contrast, the driving signals of Ambisonics are spatially bandlimited. However, a target plane wave is also spatially bandlimited below a certain frequency limit and within a local region centered at the origin. Therefore, Ambisonics can reconstruct the target plane wave below the frequency limit and within the local region. According to the Shannon–Nyquist spatial sampling theorem, the upper frequency limit and radius of the region are related to the order of Ambisonics. For horizontal Ambisonics, the result is given in Equation (9.3.14) or (9.3.15). These features distinguish WFS from Ambisonics.

A spatially bandlimited driving signal of WFS leads to local WFS similar to that in Section 10.2.4, where a target plane wave is reconstructed in a local region (Hahn et al., 2016). Equations (10.2.27), (10.2.31), and (10.2.40) indicate that the driving signals of secondary monopole sources in WFS are directly proportional to the normal derivative of the target pressure at the positions of secondary sources. For a horizontal target plane wave and a circular array of secondary sources, the target pressure can be expanded as azimuthal Fourier series and truncated to a certain order. Spatially bandlimited driving signals are obtained by substituting the truncated series into Equations (10.2.31) and (10.2.40). Similarly, spatially bandlimited driving signals of WFS with a spherical array of secondary sources can be derived by decomposing the pressure or its normal derivative by spherical harmonic functions and truncating to a certain order. The analysis here further reveals the relationship between WFS and Ambisonics.

10.5 EQUALIZATION OF WFS UNDER NONIDEAL CONDITIONS

The analyses in the preceding sections are for ideal secondary sources and in an ideal reproduction environment. Under nonideal conditions, errors occur in the reconstructed sound field of WFS. In this case, signal processing for equalization or compensation for the reconstructed sound field is needed.

In the preceding discussions on 2.5-dimensional WFS, secondary sources are assumed to be monopole point sources with an ideal frequency response. Practical secondary sources usually possess directivity and nonideal frequency responses. These nonideal characteristics, together with the physical restrictions of WFS, lead to errors in the reconstructed sound field and perceivable timbre coloration. Below the upper frequency limit imposed by Shannon–Nyquist spatial sampling theorem, equalization can be introduced to the signal processing of WFS to reduce timbre coloration. In addition to the method suggested by Vries (1996) in Section 10.1.6, the proposed methods involve the equalization of the nonideal frequency response and directivity of secondary sources in WFS by incorporating a multiple receiver positions matching method similar to that in Section 9.6.3. This method is applicable to a secondary source array composed of conventional loudspeakers and multiactuator panels (Corteel,

2006). The responses of equalizing filters are derived by minimizing the error between the reconstructed and target pressures in multiple receiver positions.

Similar to the case of Ambisonics, reflections in a listening room cause error in the reconstructed sound field of WFS. A multiple receiver positions matching method similar to that in Section 9.6.3 can also be incorporated into WFS to compensate for the influence of reflections in a listening room (Fuster et al., 2005). However, as stated in Section 10.4.1, this method is valid in a local receiver region, and WFS aims to reconstruct the target sound field in an extended receiver region. Therefore, the effect of multiple receiver positions matching method on controlling listening room reflections in WFS is limited. A similar problem occurs in the aforementioned equalization of the nonideal frequency response and directivity of secondary sources in WFS. As such, some other methods for compensating the listening room reflections in WFS should be developed (Corteel and Nicol, 2003).

Gauthier and Berry (2006, 2007, 2008) proposed the ***adaptive* WFS(*AWFS*)** to compensate for the influence of listening room reflections and other nonideal conditions on the reconstructed sound field of WFS. Similar to the multiple receiver positions matching method, AWFS considers M secondary sources and O controlled points with $M > O$. Two $O \times 1$ column vector $\boldsymbol{P}$ and $\boldsymbol{P}'$ denote the target and reconstructed pressures at the controlled points, respectively. Equation (9.6.17) yields

$$\boldsymbol{P}' = [G]\boldsymbol{E}, \tag{10.5.1}$$

where $[G]$ is an $O \times M$ matrix composed of the transfer functions from M secondary sources to O controlled points. These transfer functions involve the contribution of the direct transmission and reflections of a listening room. $\boldsymbol{E}$ is a $M \times 1$ column vector of practical driving signals for M secondary sources. Under ideal reproduction conditions, the driving signals of conventional WFS are denoted by the $M \times 1$ column vector $\boldsymbol{E}_{WFS}$. The cost function of the error of the reconstructed sound field is defined as

$$\begin{aligned} Err_6 &= (\boldsymbol{P} - \boldsymbol{P}')^+ (\boldsymbol{P} - \boldsymbol{P}') + \varepsilon (\boldsymbol{E} - \boldsymbol{E}_{WFS})^+ (\boldsymbol{E} - \boldsymbol{E}_{WFS}) \\ &= \{\boldsymbol{P} - [G]\boldsymbol{E}\}^+ \{\boldsymbol{P} - [G]\boldsymbol{E}\} + \varepsilon (\boldsymbol{E}_{WFS} - \boldsymbol{E})^+ (\boldsymbol{E}_{WFS} - \boldsymbol{E}). \end{aligned} \tag{10.5.2}$$

The first term on the right side of Equation (10.5.2) represents the square error between the reconstructed and target pressures. The second term is a penalty function consisting of the power of the difference between the practical driving signals and those of conventional WFS. With the penalty function, practical driving signals approach those of conventional WFS and are derived by minimizing the cost function as

$$\begin{aligned} \boldsymbol{E} &= \left\{[G]^+ [G] + \varepsilon [I]\right\}^{-1} \left\{[G]^+ \boldsymbol{P} + \varepsilon \boldsymbol{E}_{WFS}\right\} \\ &= \boldsymbol{E}_{WFS} + \left\{[G]^+ [G] + \varepsilon [I]\right\}^{-1} [G]^+ \left\{\boldsymbol{P} - [G]\boldsymbol{E}_{WFS}\right\}. \end{aligned} \tag{10.5.3}$$

where $[I]$ is an $M \times M$ identity matrix. The secondary equality in Equation (10.5.3) indicates that practical driving signals have two parts. The first part includes the driving signal $\boldsymbol{E}_{WFS}$ of conventional WFS. The second part functions in the compensation of nonideal reproduction conditions, such as listening room reflections. The real and positive regularization parameter ε determines the proportion of penalty function in cost functions. $\varepsilon = 0$ corresponds to the case of the multiple receiver positions matching method in Section 9.6.3, and a large ε refers to the case of conventional WFS. In addition to the method of minimizing the cost function in Equation (10.5.3), the method of singular value decomposition (SVD) of the transfer matrix and independent mode control in Section 9.6.3 can be used for AWFS. In Section 9.3.6, the relationship between the multiple receiver positions matching method and Ambisonics is analyzed for a case of a horizontal-circular array of secondary sources and controlled points in a circle. Similarly, AWFS can be implemented by reconstructing the target sound field with conventional WFS and simultaneously compensating for the reflections from a listening room with higher-order Ambisonics.

In some works, the method in the spatial spectral domain (or wave number domain) is used to compensate for the reflections of a listening room (Spors et al., 2004, 2007). If the condition of Shannon–Nyquist spatial sampling theorem is satisfied, WFS can reconstruct incident target plane waves from various directions. In practice, the spatial transfer functions or impulse responses of a listening room are initially measured. Then, the reflections included in impulse responses are decomposed into an incident plane wave from various directions. These reflected plane wave components are compensated in the listening region by the plane waves created by WFS. For three-dimensional WFS, compensating for the reflections within the entire receiver region is theoretically possible. For practical 2.5-dimensional WFS, the results are different. For example, Spors et al. (2003) investigated the compensation of reflections in horizontal WFS by using three linear arrays of secondary sources. The spatial impulse responses of a room are measured with a circular microphone array. They found that compensation reduces the influence of reflections within a local receiver region rather than the entire receiver region below the upper frequency limit imposed by the Shannon–Nyquist spatial sampling theorem. Local compensation is due to the magnitude error caused by mismatched characteristics of secondary sources in 2.5-dimensional WFS. Moreover, only the horizontal reflections of a listening room can be compensated by 2.5-dimensional WFS.

A stationary control may not meet the requirement because of the time-varying characteristics of room acoustics (such as temperature that influences the speed of sound). Alternatively, error sensors (microphones) are used to detect pressures at controlled positions, and adaptive method is utilized for compensation processing (Sports et al., 2005). Plane wave decomposition in a spatial spectral domain is also applicable to the adaptive compensation of reflections.

10.6 SUMMARY

Based on Kirchhoff–Helmholtz boundary integral equation, ideal acoustical holography reconstructs a sound field in the source-free region inside a close boundary through uniform and continuous arrays of secondary monopole and dipole sources on the boundary surface. WFS is a spatial sound technique based on the principle of sound field reconstruction. It is simplified from Kirchhoff–Helmholtz boundary integral equation and ideal acoustical holography. When a target source is located on one side of an infinite (vertical) plane, a continuous and uniform array of single-type secondary sources arranged on the infinite plane can reconstruct the target sound field in half of the space of the receiver side of the array. Driving signals are calculated from Rayleigh integrals. When the target source and receiver position are restricted in the horizontal plane, an infinite linear array can reconstruct the target sound field in the half plane of the receiver side of the array. In this case, driving signals are calculated using the stationary phase method.

WFS can be analyzed strictly on the basis of Green's function of Helmholtz equation. Using an array of secondary monopole sources to reconstruct the sound field is mathematically equivalent to the calculation of Kirchhoff–Helmholtz boundary integral by using Neumann

Green's function with a rigid boundary condition. Except for an infinite planar (or linear) array, Neumann Green's introduces boundary reflections and thus disturbs the reconstructed sound field. Accordingly, a spatial window on driving signals is required to eliminate boundary reflections. In this case, the target sound field can be approximately reconstructed within the receiver region. If the target source and the receiver position are restricted in the horizontal plane, an array of secondary straight-line sources is theoretically required for two-dimensional WFS. When secondary straight-line sources are replaced with secondary point sources in 2.5-dimensional WFS, high-pass filtering should be applied to equalize the frequency characteristic of driving signals. The distance dependence of the magnitude of the reconstructed sound pressure should also be equalized. However, the magnitude of reconstructed pressure can only be equalized at a reference position or in a line (or a contour). A magnitude error in the reconstructed sound field occurs when the receiver position deviates from the reference position or reference line. The time-reversal technique is applicable to creating a focused source between an array and a receiver position.

The general formulation of multichannel sound field reconstruction is applicable to the analysis of WFS. For some regular arrays of secondary sources, analysis can be conveniently performed in a spatial spectral domain. The array and driving signals of secondary sources can be derived by using the spatial spectral division method. The discrete and finite array of secondary sources in practical WFS results in spatial aliasing and edge effect. Spatial spectrum analysis is a useful tool for evaluating the error caused by spatial aliasing and edge effect.

Various methods for sound field reconstruction and control can be classified into three types, e.g., control of the pressure in a receiver region, control of the pressure and normal velocity of a medium on the boundary surface, and control of the spatial spectrum or mode in the receiver region. The multiple receiver positions matching method, acoustical holography (and WFS), and Ambisonics are typical examples of these three types of methods. These methods and examples are closely related to one another, which is the consequence of the uniqueness of the solution of Helmholtz equation subjected to a certain boundary condition. The reconstructed sound field and the perceived characteristics of various methods vary because different mathematical and physical approximations are used. Similarities and differences in various methods should be investigated to obtain insights into the nature of sound field reconstruction and develop practical techniques. For ideal acoustical holography, arrays of secondary monopole and dipole sources with independent driving signals can simultaneously control the interior and exterior sound fields. The target sound field can be reconstructed inside the boundary, while the exterior sound field outside the boundary is eliminated by appropriately choosing driving signals. WFS and Ambisonics can be simplified from ideal acoustical holography. Conversely, an array with a single type of secondary sources can control the interior or exterior sound field one at a time.

Equalization or compensation can be applied to WFS to reduce the influence of a nonideal reproduction environment on the reconstructed sound field, such as the influence of reflections from a listening room. Various methods, including multiple receiver positions matching and the spatial spectral control method, are applicable to equalization in WFS. These applications are examples of interdependency and complementarity of different methods of sound field control and reconstruction.

Chapter 11

Binaural reproduction and virtual auditory display

Binaural reproduction and virtual auditory display is a category of spatial sound techniques. In contrast to the multichannel sound and wave field synthesis discussed in the preceding chapters, binaural reproduction and virtual auditory display aim to reconstruct binaural sound pressures (sound pressures at two receiver positions) exactly rather than reconstruct the physical sound field exactly or approximately within a certain region. Binaural reproduction and virtual auditory display are based on the hypothesis that a target auditory event or experience can be recreated if the pressures generated by the target sound field at the eardrums can be replicated or synthesized exactly. Binaural reproduction and virtual auditory display deal with binaural signals, which can be obtained by binaural recording or signal processing (binaural synthesis). The HRTFs defined in Section 1.4.2 are essential for binaural synthesis.

HRTFs, binaural reproduction, and virtual auditory display are not discussed comprehensively in this book because they are a wide range of topics and have been addressed in detail in another book (Xie, 2008a, 2013). Instead, the basic principles of HRTFs, binaural reproduction, and virtual auditory display are concisely reviewed in the current chapter for the completeness of this book. The details are referred to the author's other book and some review articles (Xie, 2009a; Zhong and Xie, 2004; Yu and Xie, 2007; Xie and Guan, 2012; Xie et al., 2013b).

In Section 11.1, the basic principles of binaural reproduction and virtual auditory display are outlined. In Section 11.2, three methods for acquiring HRTFs, including measurement, calculation, and customization, are addressed. In Section 11.3, the primary physical features of HRTFs in the time and frequency domains are presented. In Section 11.4, the filters of HRTF-based signal processing or binaural synthesis are outlined. In Section 11.5, the methods of HRTF analysis in spatial and spatial spectrum domains, including the spatial interpolation and basic function decomposition of HRTFs, are presented. In particular, the concepts of the spatial interpolation of HRTFs, virtual auditory display, and multichannel sound are related and united within the framework of the spatial sampling and reconstruction theorem of the sound field. In Section 11.6, the results of Section 11.5 are used to simplify the signal processing of binaural synthesis. In Section 11.7, the problems of binaural reproduction through headphones are discussed. In Section 11.8, the problems of binaural reproduction through loudspeakers (transaural reproduction) are discussed. As a special application of virtual auditory display, the virtual reproduction of stereophonic and multichannel sounds is presented in Section 11.9. The rendering system of a dynamic and real-time virtual auditory environment, where the dynamic localization cue caused by head turning is introduced in binaural synthesis, is addressed in Section 11.10.

DOI: 10.1201/9781003081500-11

11.1 BASIC PRINCIPLES OF BINAURAL REPRODUCTION AND VIRTUAL AUDITORY DISPLAY

11.1.1 Binaural recording and reproduction

According to the discussion in Section 1.4, for an arbitrary sound field, binaural sound pressures or signals (strictly binaural pressures in the eardrums) contain the major spatial information of a sound event. Therefore, a pair of microphones can be placed at the positions of the two ears of an artificial head or a human subject to capture binaural signals. Binaural signals are amplified, transmitted, stored, and reproduced by a pair of headphones. Binaural reproduction creates signals or information at the eardrums matched with those in the original sound field, realizing the reproduction of the spatial information of sound (Mϕller, 1992). This phenomenon is the basic principle of the ***binaural recording and playback system*** (or ***technique***).Binaural recording and playback systems are also called ***artificial head or dummy head stereophonic systems*** because an artificial head is often used for binaural recording. In other words, an artificial head is used instead of a potential listener to "listen" in the original sound field to recreate the localization information of the sound source and the comprehensive spatial information of acoustic environment.

Figure 11.1 shows the block diagram of binaural recording and playback system. In Section 1.4.2, binaural sound pressures can be captured at an arbitrary position from the entrance of the ear canal to the eardrum even at the entrance of a blocked ear canal. A pair of filters F_L and F_R in the figure is used to equalize the binaural signals captured at different reference points and compensate for the linear distortion caused by the nonideal transfer characteristics of the electroacoustic system (such as microphones and headphones) to ensure an exact replication of target binaural pressures at the eardrums (Section 11.7.1).

The scattering and diffraction of anatomical structures on sound waves and thus the pressures received by the two ears are individually dependent because of the difference in the anatomical shapes and dimensions of different individuals. If binaural signals are recorded from a listener's own ears, they match the individual characteristics of the listener and result in an authentic subjective performance of reproduction. Otherwise, if binaural signals are recorded

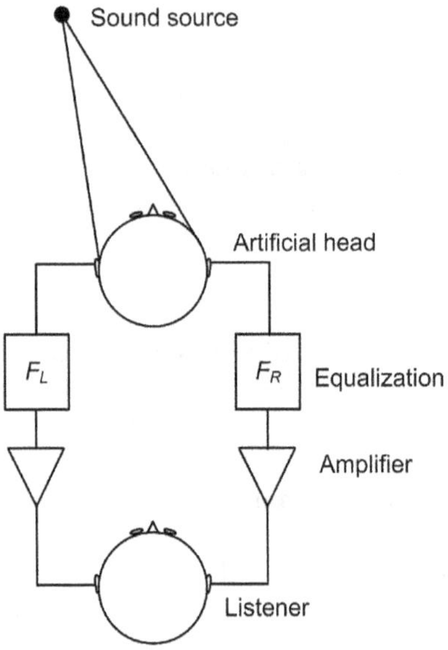

Figure 11.1 Block diagram of the binaural recording and playback system.

from an artificial head (or other human subjects), the subjective performance of reproduction depends on the extent of similarity in anatomical shapes and dimensions between the recording artificial head (or a human subject) and the listener. Generally, the more similar they are, the more authentic the subjective performance will be. Hence, the subjective performance of nonindividualized binaural recording and playback ***varies across listeners***. The localization performance of binaural recording from an artificial head is expected to be generally inferior to that of individual binaural recordings because an artificial head is merely an approximate model of a real human head, as revealed by many subjective listening tests (Møller et al., 1996, 1999).

Binaural recording and playback were introduced at early stages of spatial sound research. However, compared with conventional stereophonic or multichannel sound techniques, this technique had been rarely used in the field of domestic sound reproduction for at least two reasons. First, the artificial head technique was immature at that time. Second, recorded binaural signals, originally intended for headphone reproduction, were not compatible with loudspeaker reproduction without further signal processing. However, it has been a useful tool for evaluating sound quality in halls and for conducting some psychoacoustic experiments because it can capture and recreate the major spatial information of a sound event. In the last two or three decades, binaural recording and playback techniques have been increasingly used in scientific research.

11.1.2 Virtual auditory display

In addition to being recorded from an artificial head or human subject, binaural signals can be synthesized by HRTF-based signal processing. As discussed in Section 1.4.2, free-field binaural pressures, created by a point source at the (r_S, θ_S, ϕ_S) position, are determined by a pair of HRTFs, $H_L(r_S, \theta_S, \phi_S, f)$ and $H_R(r_S, \theta_S, \phi_S, f)$, or head-related impulse responses (HRIRs), $h_L(r_S, \theta_S, \phi_S, t)$, and $h_R(r_S, \theta_S, \phi_S, t)$. According to Equation (1.4.2), given HRTFs or HRIRs, free-field binaural signals can be synthesized by filtering the mono stimulus $E_A(f)$ with HRTFs in the frequency domain:

$$E_{L,ear}(f) = H_L(r_S, \theta_S, \phi_S, f)E_A(f) \qquad E_{R,ear}(f) = H_R(r_S, \theta_S, \phi_S, f)E_A(f), \qquad (11.1.1)$$

or according to Equation (1.4.4), it can be equally synthesized by convoluting the mono stimulus $e_A(t)$ with HRIR in the time domain:

$$e_{L,ear}(t) = h_L(r_S, \theta_S, \phi_S, t) \otimes_t e_A(t) \qquad e_{R,ear}(t) = h_R(r_S, \theta_S, \phi_S, t) \otimes_t e_A(t). \qquad (11.1.2)$$

When the resultant binaural signals are presented over a pair of headphones, the pressures at the listener's two ears are directly proportional to those generated by a real source at the spatial position (r_S, θ_S, ϕ_S), leading to an authentic virtual sound source. In the two above equations, the subscript "ear" denotes binaural signals and distinguishes them from the left and right channel stereophonic signals for loudspeaker reproduction.

In comparison with the binaural sound pressures generated by a real source, as shown in Equation (1.4.2), the synthesized binaural pressures (or signals) presented in Equation (11.1.1) lack a scaling factor and a frequency-independent linear delay because the magnitude and phase variations in $P_{free}(r_S, f)$ with a source distance (in detail, the magnitude attenuation with the inverse of the source distance and the linear delay introduced by free propagation) are omitted in synthesis. The variation in the magnitude of $P_{free}(r_S, f)$ with the source distance can be implemented by scaling the gain of the input stimulus $E_A(f)$ with $1/r$ to render a single virtual source, whereas omitting the linear delay in $P_{free}(r_S, f)$ has no influence on auditory effects. If multiple virtual sources at different spatial positions are rendered simultaneously, however, the gain should be scaled, and different linear delays, $\tau_S = r_S/c$ should be added for each input stimulus. For simplicity, the steps of scaling input gains and adding linear delay are disregarded in the subsequent discussion unless otherwise stated.

In reflective environments, according to the discussion in Section 1.8.3, the corresponding binaural signals $e_L(t)$ and $e_R(t)$ can be synthesized by convoluting the mono stimulus $e_A(t)$ with a pair of binaural room impulse responses (h_L and h_R):

$$e_{L,ear}(t) = h_L(t) \otimes_t e_A(t) \qquad e_{R,ear}(t) = h_R(t) \otimes_t e_A(t). \tag{11.1.3}$$

When the resultant binaural signals are presented over a pair of headphones, the pressures at the two ears of a listener are directly proportional to those generated by the real source in the reflective environment, leading to the authentic spatial auditory perception of the sound event caused by direct and reflected sounds.

The aforementioned technique of replicating the spatial auditory perception of sound events by HRTF- or HRIR-based binaural synthesis is known as ***a virtual auditory display*** (***VAD***). Occasionally, it is called a virtual auditory space, virtual sound or audio, 3D sound or 3D audio, ***binaural technique, or reproduction***. Binaural technique or reproduction is generally used to include not only VAD but also binaural recording and playback techniques. A VAD aims to authentically replicate real spatial hearing, including the localization of a single sound source and the overall spatial auditory perception of multiple sound sources and environmental reflections. In this sense, VADs are sometimes called ***virtual auditory environment displays*** or virtual acoustic environment displays. However, the terms virtual auditory environment display and virtual acoustic environment display are employed in cases where primary emphasis is placed on recreating or synthesizing the overall spatial information of acoustical environments.

Figure 11.2 shows a block diagram of frequency domain signal processing for synthesizing a free-field virtual source. Theoretically, equalization (or compensation) to nonideal transfer characteristics in a playback chain is necessary to recreate the binaural pressures at the eardrums accurately. However, equalization is often omitted in some applications in which accuracy is noncritical because the transfer characteristics in the playback chain are highly dependent on the headphone type and listener; consequently, signal processing may be infeasible. Given that the processing of equalization is independent of the sound source position, it is disregarded in most of the succeeding discussions for simplicity except in Section 11.7, where the method of equalization processing is addressed extensively.

The perceived performance of VAD varies across listeners because HRTFs are individual dependent. Subjective experiments have indicated that the perceived performance of

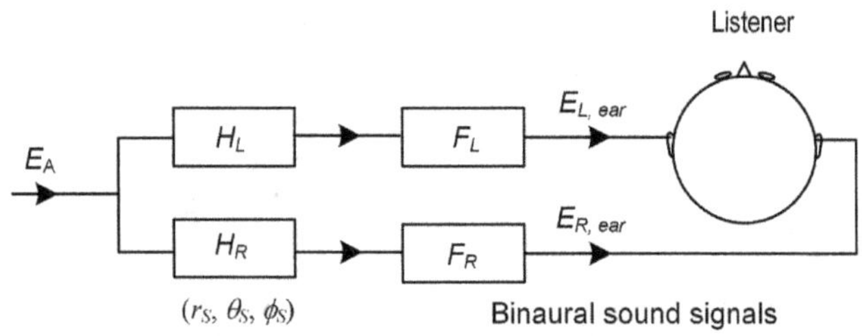

Figure 11.2 Block diagram of frequency domain signal processing in synthesizing a free-field virtual sound source.

individualized HRTFs is relatively ideal (Wenzel et al., 1993). Otherwise, if nonindividualized HRTFs are used for binaural synthesis, the perceived performance depends on the similarity between the listener's own HRTFs and those used for binaural synthesis.

Since 1989, when Wightman and Kistler (1989a, 1989b) simulated a free-field virtual source in a three-dimensional space by synthesizing binaural signals for headphone presentation, the VAD technique has been developed and applied to extensive fields (Begault, 1994; Wightman and Kistler, 2005; Xie, 2008a, 2013).

11.2 ACQUISITION OF HRTFS

HRTFs or HRIRs are key data for binaural synthesis. The three common methods for acquiring HRTFs, that is, measurement, calculation, and customization, are discussed in this section.

11.2.1 HRTF measurement

HRTFs can be obtained through measurement, which is the most common and accurate method. They can be measured from artificial heads and human subjects. Furthermore, individualized HRTFs can be obtained through measurement from human subjects. In measurement, however, human subjects are inclined to make small movements of the head and body. Human subjects may unconsciously generate noise. All these factors are harmful to measurements. However, statistical averages across certain populations are often needed in many studies related to human hearing. Such statistics require samples from many human subjects and are complicated. Therefore, HRTFs measured from artificial heads, which represent the average characteristics of a certain population, are also needed.

The principle of HRTF measurement is similar to that of transfer function measurement in acoustic systems. Analog measurement methods are frequently employed at early stages of research (Shaw, 1974), and digital measurement methods are currently used. Figure 11.3 shows a typical block diagram of the HRTF measurement. The measuring signal generated by a computer is rendered to a loudspeaker after it passes through a digital-to-analog (D/A) converter and a power amplifier. The resultant signals are recorded by a pair of microphones at the subject's two ears and then delivered to the computer after amplification and analog-to-digital (A/D) conversion. HRTFs or HRIRs can be obtained after the necessary signal processing is implemented using a computer. This method is appropriate for both human subjects and artificial head measurements.

As stated in Section 1.4.2, the entire ear canal can be regarded as a one-dimensional transmission line below 12–14 kHz. Therefore, binaural pressures can be measured at an arbitrary reference point from the entrance of the ear canal to the eardrum. For a human subject, measuring the binaural pressures at the entrances of blocked ear canals by using a pair of

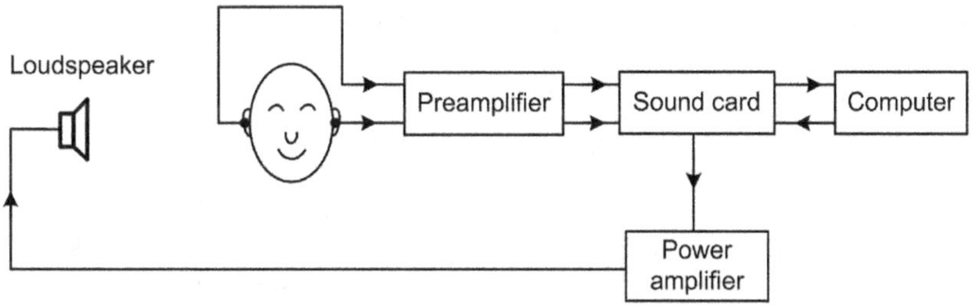

Figure 11.3 Block diagram of HRTF measurement.

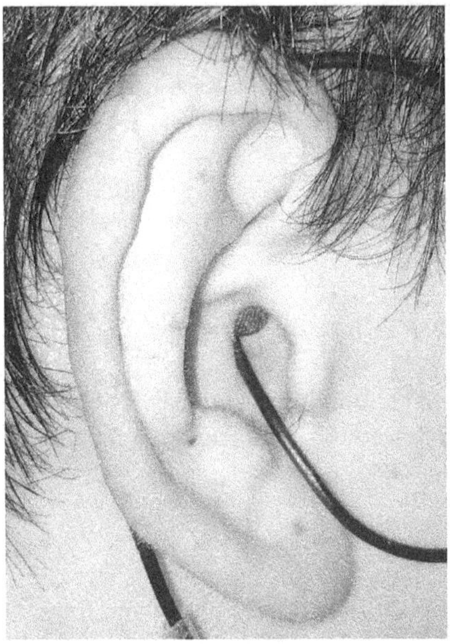

Figure 11.4 Blocked ear canal configuration with the positioned microphone.

miniature microphones is convenient and safe. This method is called the blocked ear canal technique (Mϕller, 1992). Figure 11.4 shows a photograph of the blocked ear canal configuration with a positioned microphone.

Various measuring signals and methods, such as maximal-length sequences (MLSs) mentioned in Section 7.5.4 and sweep signals, used for acoustic transfer function measurement are applicable to HRTF measurements. MLSs are pseudo-random sequences with deterministic and periodic structures, but they possess autocorrelation and power spectral characteristics similar to white noise. The MLS and cross-correlation-based methods for HRTF measurements have the advantage of noise immunity. The signal-to-noise ratio can be greatly improved by the averaging over multiple measurements. A disadvantage of the MLS-based method is its sensitivity to the nonlinear property of the system under evaluation (Vanderkooy, 1994). In addition, the nonideal transfer characteristics of the measurement system, such as the responses of the microphone and sound source (loudspeaker), should be compensated by an appropriate equalization scheme. However, equalization is invalid outside the working frequency range of the measurement system, such as below the low-frequency limit of the sound source. In this case, the asymptotic characteristic of HRTFs can be used to correct the errors in the measured HRTFs at low frequencies (Xie, 2009b).

In accordance with the definition of HRTF, measurements should be performed in an anechoic chamber. In practice, some measurements are performed in a nonechoic room. An appropriate time window should be applied to raw binaural impulse responses to obtain actual HRIRs and remove unwanted reflections from the environment. The length of the time window is determined by the time difference between the direct sound and the first reflection from the environment. A short time window causes errors in measurement, especially at low frequencies (uncertainty principle).

For example, Figure 11.5 is a photo of the second-generation HRTF measurement system (in a non-anechoic room) at the author's lab (Xie et al., 2013b; Yu et al., 2018a). This system is applicable to far- and near-field HRTF measurements for artificial heads and human subjects. Multiple sound sources (loudspeaker systems) are arranged at different elevations. The

Figure 11.5 Photo of the second generation of the HRTF measurement system at the author's lab.

sound source distance with respect to the head center can be adjusted by changing the length of the rod supporting the source with a maximal distance of 1.2 m. A horizontal turntable is adopted on which a rod is installed to support the artificial head or seat for a human subject. The direction of the sound source with respect to the subject can be altered by choosing the sound source at different elevations and rotating the turntable. The measurement system is controlled using a computer.

At the source distance of $r_S > 1.0$–1.2 m with respect to the head center, far-field HRTFs are approximately independent from a distance. Far-field HRTF measurement is relatively simple, and the technique for measurement is mature. Far-field HRTF is a continuous function of the source direction (θ_S, ϕ_S) and frequency, but the data in discrete and finite directions are usually measured. If a digital technique is used in measurement, the resultant data are discrete sequences in the frequency or time domain.

The far-field HRTFs/HRIRs of artificial heads and human subjects have been measured by some groups, and HRTF databases have been set up (Wightman and Kistler, 1989a; Gardner and Martin, 1995; Mφller et al., 1995b; Genuit and Xiang, 1995; Blauert et al., 1998; Riederer, 1998; Bovbjerg et al., 2000; Algazi et al., 2001a; Takane et al., 2002; IRCAM Lab, 2003; Grassi et al., 2003; Xie et al., 2007; Begault et al., 2010; Majdak et al., 2007).

The HRTFs of a KEMAR artificial head are measured by the MIT Media Lab and are available on the Internet at http://sound.media.mit.edu/resources/KEMAR.html). The database contains far-field ($r = 1.4$ m) HRIRs of 710 spatial directions at 14 elevations varying from −40° to 90°. The length of each HRIR is 512 points at a sampling frequency of 44.1 kHz. Binaural pressures are measured using a pair of microphones fixed at the end of the occluded ear simulator DB-100. The resultant HRTFs include information of ear canal resonance. The left ear of KEMAR is mounted with a small pinna (DB-061) and the right ear with a large pinna (DB-065). Thus, the data for the left and right ears are asymmetric or different. In practice, the unmeasured data for both pinnae can be conveniently derived using the left–right symmetrical characteristics of KEMAR. The MIT-KEMAR HRTF data have been widely used in research on binaural hearing and VAD.

Another commonly used HRTF database is the CIPIC database. It is also available on the Internet at https://www.ece.ucdavis.edu/cipic/spatial-sound/hrtf-data/. The CIPIC database contains the HRIRs of 43 human subjects (27 men and 16 women, mainly Western people). HRIRs are measured at a source distance of $r_S = 1.0$ m and 1250 directions by using the blocked ear canal technique. The length of each HRIR is 200 points at a sampling frequency of 44.1 kHz. Some other HRTF databases are available from the SOFA website (https://www.sofaconventions.org/).

Considering the statistical differences in the anatomical shapes and dimensions of various populations, the author's lab set a far-field HRTF database of Chinese subjects in 2005 (Xie et al., 2007). The database contains the HRIRs of 52 human subjects (26 male and 26 female). HRIRs were measured at a source distance of r_S = 1.5 m and 493 directions by using the blocked ear canal technique. The length of each HRIR is 512 points at a sampling frequency of 44.1 kHz.

Near-field HRTF measurements are relatively difficult to perform (Yu and Xie, 2007). In the proximal source distance, the influence of the size and directivity of sound source and multiple scattering between a source and a subject should be considered; therefore, conventional small loudspeaker systems cannot serve as sound sources for near-field HRTF measurements. A special point sound source in the near field is required. In addition, the distance-dependent characteristics of near-field HRTFs require time-consuming measurements at various source directions and distances. As a result, only a few groups have measured the near-field HRTFs of artificial heads over the years (Brungart and Rabinowitz 1999; Hosoe et al., 2005; Gong et al., 2007).

Using a special spherical dodecahedron sound source, Yu et al. measured the near-field HRTFs for KEMAR with DB 60/61 small pinnae with reference points at the ends of a pair of Zwislocki-occluded ear simulators. The resultant database included HRIRs at 10 source distances of 0.20, 0.25, 0.30, 0.40, 0.50, 0.60, 0.70, 0.80, 0.90, and 1.00 m, and 493 directions at each source distance. The length of each HRIR is 512 points at a sampling frequency of 44.1 kHz. Using the second-generation measurement system shown in Figure 11.5, Yu et al. (2018b) further measured the near-field HRTFs of 56 human subjects at seven source distances and 685 directions at each distance, establishing the near-field HRTF database of human subjects.

11.2.2 HRTF calculation

Calculation is another way to acquire HRTFs through which the wave or Helmholtz equation is solved, subject to the boundary conditions presented by the anatomical geometric model, including the head, pinnae, and torso. A rigid spherical head model is the simplest for HRTF calculation by which the head is approximated as a rigid sphere with the radius a, and the pinnae and torso are omitted. For a point source at a far-field distance $r_S >> a$, the incident wave can be approximated as a plane wave. The pressure at the sphere surface and the distance-independent far-field HRTFs can be calculated using analytical solutions for the scattering problem of a sphere (Kuhn, 1977; Cooper, 1982; Du et al., 2001; Morse and Ingrad, 1968). Given the angle $\Delta\Omega_S$ between the incident direction of the plane wave and that of the concerned ear, the far-field HRTFs are calculated using the following equation:

$$H(\Omega_S, f) = -\frac{1}{(ka)^2} \sum_{l=0}^{\infty} \frac{(2l+1) j^{l+1} P_l(\cos\Delta\Omega_S)}{dh_l(ka)/d(ka)}, \tag{11.2.1}$$

where $P_l(\cos\Delta\Omega_S)$ is the Legendre polynomial of degree l, and $h_l(ka)$ is the l-order spherical Hankel function of the second kind.

When two ears are located at the opposite positions on the head surface (±90°), the far-field HRTFs for horizontal source azimuth of θ_S are calculated as

$$H_L(\theta_S, f) = -\frac{1}{(ka)^2} \sum_{l=0}^{\infty} \frac{(2l+1)j^{l+1} P_l(\sin\theta_S)}{dh_l(ka)/d(ka)}.$$
$$H_R(\theta_S, f) = -\frac{1}{(ka)^2} \sum_{l=0}^{\infty} \frac{(2l+1)j^{l+1}(-1)^l P_l(\sin\theta_S)}{dh_l(ka)/d(ka)}. \quad (11.2.2)$$

Equation (11.2.2) is derived in an anticlockwise spherical coordinate system, which is different from the equation in a clockwise spherical coordinate system (Xie, 2008a, 2013). At a low frequency of $ka << 1$, only the terms $l = 0$ and 1 are retained, and higher-order terms can be omitted in Equation (11.2.2). The HRTFs of the rigid spherical head model can then be approximated to Equation (1.4.5).

The distance-dependent near-field HRTFs for a finite source distance r_S can be calculated using a rigid spherical head model (Duda and Martens, 1998). Given the angle $\Delta\Omega_S$ between the direction of the source and that of a concerned ear, near-field HRTFs are calculated by the following equation:

$$H(\rho, \Omega_S, ka) = -\frac{\rho \exp(jka\rho)}{ka} \sum_{l=0}^{\infty} \frac{(2l+1)h_l(ka\rho)P_l(\cos\Delta\Omega_S)}{dh_l(ka)/d(ka)} \qquad \rho = \frac{r_S}{a}. \quad (11.2.3)$$

When $\rho >> 1$, Equation (11.2.3) is simplified to Equation (11.2.1).

A snowman model can be used for HRTF calculations to consider the influence of the torso (Algazi et al., 2002). In this model, the head and torso are simplified into two spheres with different radii. HRTFs can be solved through multiple scattering or expansion, but mathematical calculation is complicated.

Analytical solutions of HRTFs can only be obtained for rare simplified head/torso models. In these models, the influence of the head shape and pinnae on HRTFs is ignored. The results represent the mid- and low-frequency features and do not reflect the fine high-frequency features of HRTFs. The spherical head model is valid at frequencies below 3 kHz (Cooper, 1987). The snowman model is also valid in these mid- and low-frequency ranges.

Some numerical methods have been developed for HRTF calculation to account for the influence of the head shape and pinnae and improve the accuracy of calculation (especially at high frequencies). Among these methods, the ***boundary element method*** (***BEM***) is often used (Katz, 2001; Kahana and Nelson, 2007; Otani and Ise, 2006; Rui et al., 2013). In the BEM calculation, the problem of acoustic scattering by anatomical structures is expressed as a Kirchhoff–Helmholtz integral equation similar to that in Equation (10.1.1). A laser three-dimensional (3D) scanner or other scanning device is first used to acquire the images or data of human (or artificial head) geometrical surfaces. Geometrical surfaces are then discretized into a mesh of triangular elements, and the longest elements should not exceed the length of 1/4 to 1/6 of the shortest wavelength. After discretization, the Kirchhoff–Helmholtz integral equation becomes a set of linear algebraic equations. The acoustic principle of reciprocity can be used to improve the calculation efficiency in which the source is located at the entrance of the ear canal, and a receiver is located at an arbitrary spatial position. In this case, the sound pressure at the boundary (head surface) is calculated only once (or twice, each one for the source located at each ear position) because the source position is fixed. The fast multipole-accelerated boundary element method (FMM BEM) is applicable to speeding up the HRTF calculation (Gumerov et al., 2010).

HRTFs of up to a frequency of 20 kHz are calculated using the BEM. However, the BEM calculation is time consuming. In current personal computer platforms, it often takes dozens or even hundreds of hours to calculate a set of HRTFs in full spatial directions depending on the frequency range and interval, directional resolution of calculation, and the hardware and software of computers.

11.2.3 HRTF customization

Measurement or calculation yields relatively accurate individualized HRTFs. In practice, however, acquiring individualized HRTFs for each user of the VAD is difficult through measurement or calculation. Therefore, some simple methods for estimating or customizing individualized HRTFs have been developed. These methods are usually classified into two categories: anthropometry-based customization and subjective selection-based customization (Zhong and Xie, 2012).

HRTFs are a consequence of the scattering and diffraction of incident sound waves by anatomical structures. Anthropometry-based customization assumes that individualized HRTFs and certain individualized anatomical features are highly correlated; therefore, individualized HRTFs can be approximately estimated or matched from anthropometric measurements. The methods of anthropometry-based customizations include the following:

1. In anthropometry matching methods (Zotkin et al., 2004), the anthropometric parameters of the new individual are measured. The individualized HRTFs of the new individual are chosen by matching their anthropometric parameters with those in the baseline database.
2. In frequency scaling method (Middlebrooks, 1999a, 1999b), the logarithmic HRTF magnitudes of a new individual are estimated by scaling those of an existing subject in frequency. The frequency scaling factor is determined by the anthropometric parameters of the new and existing subjects.
3. In anthropometry-based linear regression methods (Jin et al., 2000; Nishino et al., 2007), on the basis of baseline HRTFs and anthropometric parameters, the statistical model between HRTFs and anthropometric parameters is derived through linear regression. The HRTFs of a new individual are estimated by substituting their anthropometric parameters into the statistical model.

In subjective selection-based customization, it is supposed that the matched HRTFs promise a relatively good perceived performance. Therefore, the matched HRTFs can be customized by appropriate subjective tests, such as localization tests.

The major problem with anthropometry-based customization is the selection of a minimal set of complete and independent anthropometric parameters. In addition, subjective selection-based customization may yield uncertain results. The design of a subjective test to reduce uncertainty should be further studied. In addition, a baseline HRTF database with a sufficient number of subjects is required for HRTF customization because of the diversity of HRTFs. However, customizing individualized HRTFs from the databases with numerous subjects is difficult, especially for subjective selection-based customization. Fortunately, the clustering analysis of individualized HRTFs indicates that HRTFs of most subjects can be classified into a few clusters and represented by the corresponding cluster center (So et al., 2010; Xie and Zhong, 2012; Xie et al., 2013b; Xie and Tian, 2014, Xie et al., 2015a). HRTF customization from a few cluster centers is relatively simple.

Overall, HRTF customization is simpler than measurement or calculation and usually yields a modest perceived performance. However, the accuracy of customization is inferior to that of measurement or calculation. HRTF customization should be improved in many aspects.

11.3 BASIC PHYSICAL FEATURES OF HRTFS

As the core data for binaural hearing research and binaural synthesis in VADs, HRTFs possess some important physical features. These features are discussed in this section. As indicated in Section 1.6, directional localization cues, such as interaural time difference (ITD) and interaural level difference (ILD), can be evaluated using HRTFs. This problem is discussed further in Chapter 12.

11.3.1 Time-domain features of far-field HRIRs

HRIRs are the time domain counterparts of the HRTFs. Figure 11.6 illustrates the far-field HRIRs of the KEMAR artificial head with the DB-061 small pinnae and occluded ear simulator DB-100 measured by the MIT Media Lab (Section 11.2.1) at azimuths of 0°, 30°, 60°, 90°, 120°, 150°, and 180° in the horizontal plane. In the figure, the clockwise spherical coordinate system used in the original literature (Gardner and Martin, 1995) is converted into the (default) anticlockwise spherical coordinate system in this book (Figure 1.1). The symmetric version of HRIRs with the small pinnae in the right ear is obtained by applying a left–right mirror reflection to the corresponding HRIRs in the left ear because two different sizes of pinna are mounted on KEMAR (a DB-061 small pinna on the left ear and a DB-065 large pinna on the right ear) during measurement. For θ_S= 0° and 180°, only the left ear HRIR is shown in this figure.

In Figure 11.6, the HRIR amplitude for the preceding 30–58 samples is approximately zero. These zero-amplitude samples correspond to the propagation delay from the sound source to the ear and are caused by the neglect of linear delay in the free-field pressure P_{free} in Equation (1.4.1). In practice, a time window is usually applied to the measured HRIRs; thus, the initial delay has only a relative significance. The main body of HRIRs, which reflect the complicated interactions between incident sound waves and the head, torso, and pinna, persists for about 50 or 60 samples (i.e., about 1.1–1.4 ms at a sampling frequency of 44.1 kHz). Subsequently, HRIR amplitudes return to nearly zero. When the sound source deviates from the front and back directions, the initial delay difference in HRIRs in the left and right ears reflects the propagation time difference from the sound source to the left and right ears. This difference is known as ITD. At an azimuth of θ_S = 90°, for instance, the HRIR in the right ear lags behind the one in the left ear with a relative delay of 28 samples (approximately 635 μs at a sampling frequency of 44.1 kHz). Moreover, when the sound source is located contralateral to the concerned ear, such as at θ_S = 90° for the right ear, the HRIR amplitude is visibly attenuated because of the shadow effect of the head.

11.3.2 Frequency domain features of far-field HRTFs

Figure 11.7 presents the normalized (logarithmic) HRTF magnitudes of KEMAR with the small pinnae at azimuths of 0°, 30°, 60°, 90°, 120°, 150°, and 180° in the horizontal plane. Similar to the case of HRIRs, left ear data and their left-right mirror reflections are used in the figure. At frequencies below 0.4–0.5 kHz, the normalized magnitudes of HRTFs $20\log_{10}|H|$ approach 0 dB and are roughly independent of frequency because the scattering and shadowing effects of the head are negligible. The decrease in magnitude below 150 Hz is caused by the low-frequency limit of the loudspeaker response used in the HRTF measurement rather than by the HRTF itself. A 2–4 dB difference between the left and right ear HRTF magnitudes is observed at a lateral azimuth of 90° because of the finite source distance relative to the head center (r_S = 1.4 m) even at low frequencies. This difference is larger than that observed in an infinitely distant source (plane-wave incidence; Section 1.6.2).

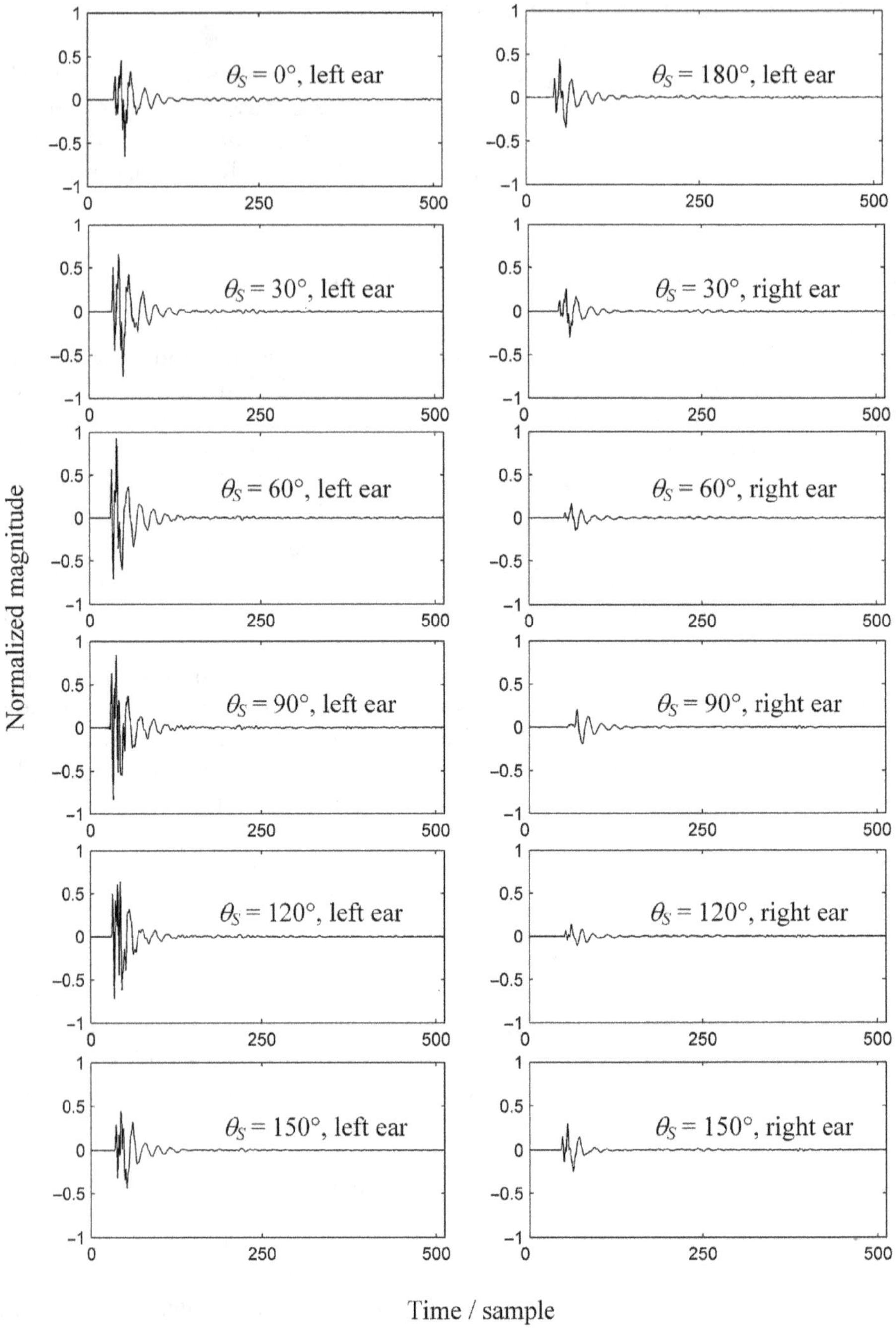

Figure 11.6 KEMAR-HRIRs at various horizontal azimuths.

As frequency increases, the normalized magnitudes of HRTFs vary with the frequency and azimuth in a complex manner. This complexity is attributed to the overall filtering effects of the head, pinna, torso, and ear canal. The apparent peak in the HRTF magnitude at 2–3 kHz results from the resonance of the occluded ear simulator of KEMAR. Above 4 kHz, the contralateral HRTF magnitudes (such as the magnitude of the right ear at an azimuth of 90°) are visibly attenuated because of the low-pass filtering properties of the head shadow. The ipsilateral HRTF magnitudes (such as the magnitude of the left ear at an azimuth of 90°) increase to a certain extent although some notches occur. This phenomenon is partially attributed to the approximate mirror-reflection effect of the head on the ipsilateral incidence at high frequencies, thereby leading to an increase in pressure of ipsilateral sound sources. In fact, a 6 dB increase in pressure is obtained on the surface of a rigid plane with infinite extension compared with that of the case of a free field.

The difference between the high-frequency HRTF magnitudes at azimuths of 0° and 180° is also observed in Figure 11.7. This difference is caused by the front–back asymmetry in the head shape, ear location, and diffraction effect of the pinna. In Section 1.6, the aforementioned features of HRTFs provide important cues for localization.

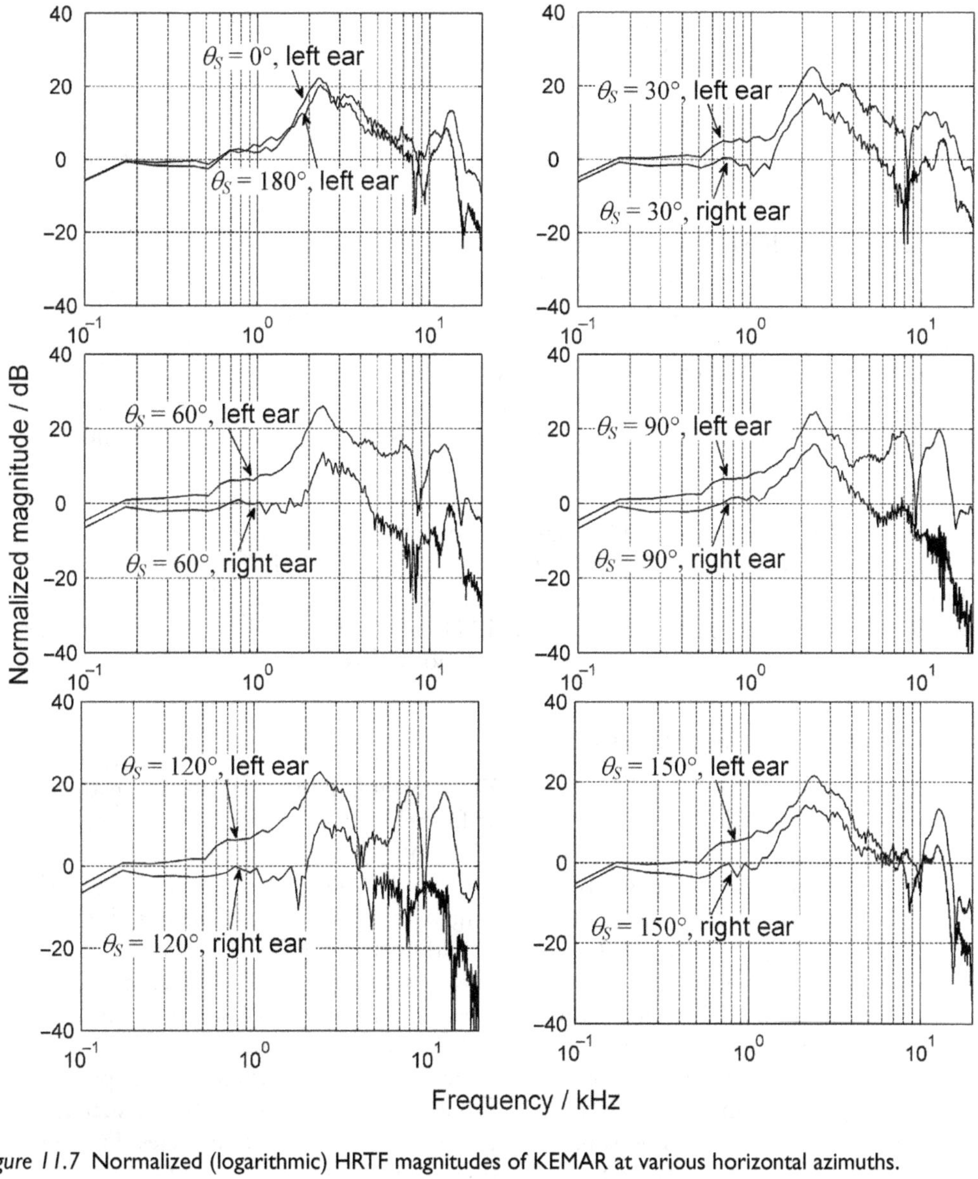

Figure 11.7 Normalized (logarithmic) HRTF magnitudes of KEMAR at various horizontal azimuths.

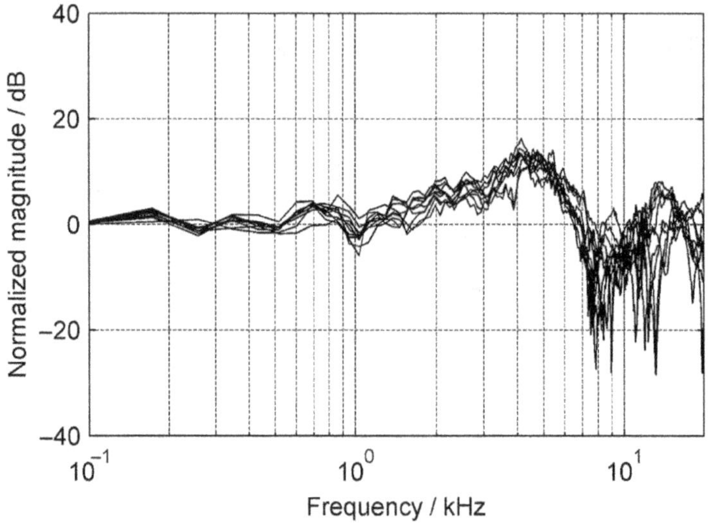

Figure 11.8 HRTF magnitudes of the left ear of 10 subjects at an azimuth of 0° in the horizontal plane.

In Section 1.6.4, above the frequencies of 5–6 kHz, a spectral cue introduced by the pinnae is important for front-back and vertical localization. The elevation-dependent pinna notch may be an important vertical localization cue in the median plane. An inspection of the KEMAR–HRTFs indicates that the center frequency of the pinna notch increases with an increase in elevation in the frontal–median plane. When the elevation exceeds 60°, the pinna notch disappears. Figure 1.23 is plotted according to KEMAR–HRTFs. In addition, considerable differences exist between the high-frequency HRTF magnitudes of different individuals. For some individuals, the pinna notch at high-frequency HRTF magnitudes may not be obvious.

Figure 11.8 plots the normalized magnitudes of left ear HRTFs at $\theta = 0°$ and $\phi = 0°$ for 10 subjects to demonstrate the individuality of HRTFs. These HRTFs are measured at the entrance of the blocked ear canal and randomly selected from the Chinese subject HRTF database. The intersubject differences in the HRTF magnitudes are clearly visible above 6–7 kHz. Similar results are obtained for other source locations.

11.3.3 Features of near-field HRTFs

At a source distance of $r_S < 1.0$ m, near-field HRTFs depend on the source distance and exhibit some physical features different from those of far-field HRTFs (Brungart and Rabinowitz, 1999; Yu et al., 2012b, 2018b). Figure 11.9 illustrates the HRTF magnitude spectra of a human subject at a horizontal azimuth of $\theta_S = 90°$ and two source distances of $r_S = 0.2$ and 1.0 m. HRTFs are measured at the entrance of the blocked ear canals. The HRTF magnitude spectra modestly vary with the source distance within the range of 0.5 m $< r_S <$ 1.0 m and obviously within the range of $r_S \leq 0.5$ m. Moreover, within the near-field distance of $r_S < 1.0$ m, the ipsilateral (left) HRTF magnitude increases as r_S decreases when a direct propagation path from the source to the concerned ear exists; the contralateral HRTF magnitude decreases as r_S decreases because of the enhancement of the head shadow when a direct propagation path is missing. Variations in the HRTF magnitude with r_S increase the ILD associated with a decrease in r_S. This phenomenon is particularly prominent at low frequencies. For example, the ILD below 0.3 kHz exceeds 20 dB for a lateral source distance close to the head surface.

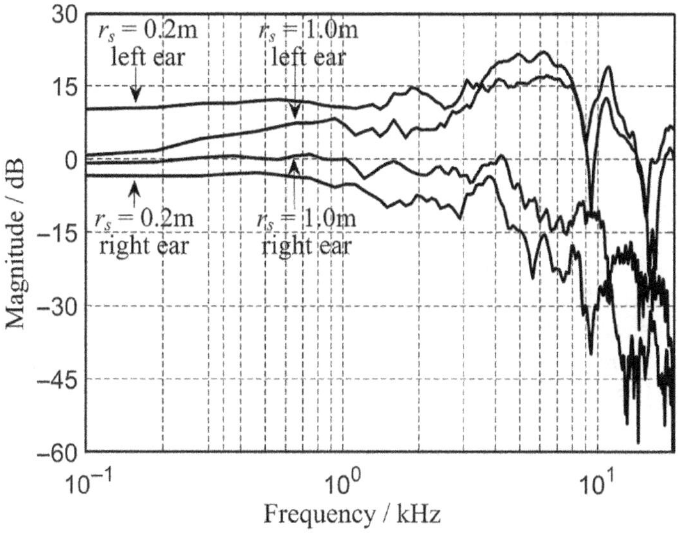

Figure 11.9 HRTF magnitude spectra of a human subject at the horizontal azimuth of θ_S = 90° and two source distances.

11.4 HRTF-BASED FILTERS FOR BINAURAL SYNTHESIS

A binaural thesis is needed for VAD. According to the discussion in Section 11.1.2, a straightforward scheme for binaural synthesis involves filtering the mono input with a pair of HRTFs in the frequency domain or equally convoluting the mono input with a pair of HRIRs in the time domain. Direct schemes usually suffer from low computational efficiency. Therefore, practical binaural synthesis is often implemented using various digital filters whose responses match with or approximate those of HRTFs.

Prior to designing filters, some physical redundancy or perceptual insignificance in the detail of HRTFs can be removed. Through this manipulation, HRTFs and resultant filters can be simplified. For example, the length of the measured HRIRs usually ranges from 128 points to 4096 points at a sampling frequency of 44.1–50 kHz, that is, 2.5–80 ms, and a short HRIR corresponds to a simple filter. According to Section 11.3.1, the main energy of HRIRs is concentrated at 1.1–1.4 ms. Hence, truncating HRIRs with a time window shortens the duration of HRIRs while preserving their essence.

Minimum-phase approximation helps reduce the length of HRIRs. At a upper frequency limit of 10 kHz, an HRTF can be approximated as a product of the minimum-phase function $H_{min}(\theta_S, \phi_S, f)$ and the linear phase function $\exp[-j2\pi f\, T(\theta_S, \phi_S)]$ (Kulkarni et al., 1999):

$$H(\theta_S, \phi_S, f) \approx H_{\min}(\theta_S, \phi_S, f)\, \exp\left[-j2\pi f T(\theta_S, \phi_S)\right]. \tag{11.4.1}$$

The minimum-phase is related to the logarithmic HRTF magnitude by using the Hilbert transform:

$$\psi_{\min}(\theta_S, \phi_S, f) = -\frac{1}{\pi}\int_{-\infty}^{+\infty} \frac{\ln|H(\theta_S, \phi_S, x)|}{f - x} dx. \tag{11.4.2}$$

Thus, the minimum-phase function $H_{min}(\theta_S, \phi_S, f)$ can be accurately reconstructed by its magnitude. The minimum-phase HRIRs in the time domain can be obtained by applying an inverse Fourier transform to the frequency-domain counterpart $H_{min}(\theta, \phi, f)$:

$$h_{\min}(\theta_S, \phi_S, t) = \int_{-\infty}^{+\infty} H_{\min}(\theta_S, \phi_S, f) e^{j2\pi ft} df. \tag{11.4.3}$$

A minimum-phase HRIR can be truncated by an appropriate time window because its effective length is shorter than that of the original HRIR. Then, binaural synthesis is implemented using minimum-phase HRTF-based filters, along with appropriate linear delays.

Truncating HRIRs with a time window results in the frequency domain smoothing of HRTFs. By directly smoothing these HRTFs in the frequency domain, HRTFs can also be simplified. Kulkarni and Colburn (1998) indicated that fine structures in HRTF magnitudes are insignificant to auditory perception because of the limited frequency resolution of human hearing (Section 1.3.4); therefore, they can be smoothed to some content. A psychoacoustic experiment has demonstrated that ipsilateral and contralateral HRTF magnitudes above 5 kHz can be smoothed only with a bandwidth of 2.0 and 3.5 ERB in horizontal and lateral planes, respectively (Xie and Zhang, 2010), without the introduction of audible artifacts. Binaural HRTF magnitudes above 5 kHz can also be simultaneously smoothed with bandwidths of 2.0 and 3.5 ERB for ipsilateral and contralateral ears, respectively. In fact, the shadow effect of a head attenuates the high-frequency pressure at the lateral ear greatly, thus the fine structures in the high-frequency HRTF of lateral ear is not always detectable. In a median plane, binaural HRTFs above 5 kHz can be simultaneously smoothed with a bandwidth of 2.0 ERB without the introduction of audible artifacts.

Common HRTF-based filter models are classified into two categories: ***moving average*** (***MA***) ***model*** and ***autoregressive moving average*** (***ARMA***) model. An MA model is a ***finite impulse response*** (***FIR***) filter model because the length of its impulse response is finite. An ARMA model is an ***infinite impulse response*** (***IIR***) filter model because the length of its impulse response is infinite.

In a complex Z domain, the system function of a digital filter is represented by $H(z)$. For a causal system, $H(z)$ is related to the impulse response of the system by the following Z transformation:

$$H(z) = \sum_{n=0}^{\infty} h(n) z^{-n}, \tag{11.4.4}$$

where n is the index of discrete time.

For a Q-order or $N = Q + 1$ point MA or FIR filter model, the system function is

$$H(z) = b_0 + b_1 z^{-1} + \ldots + b_Q z^{-Q} = \sum_{q=0}^{Q} b_q z^{-q}, \tag{11.4.5}$$

where $b_0, b_1, \ldots b_Q$ are the set of $(Q+1)$ filter coefficients.

For a (Q, P)-order ARMA or IIR filter model, the system function is

$$H(z) = \frac{b_0 + b_1 z^{-1} + \ldots + b_Q z^{-Q}}{1 + a_1 z^{-1} + \ldots + a_P z^{-P}} = \frac{\sum_{q=0}^{Q} b_q z^{-q}}{1 + \sum_{p=1}^{P} a_p z^{-p}}, \tag{11.4.6}$$

where (a_p, b_q) are a set of (Q+P+1) filter coefficients.

With HRTF-based filter designs, the coefficients in Equation (11.4.5) or Equation (11.4.6) can be selected appropriately so that filter responses exactly or approximately match with the target HRTF or HRIR in some mathematical or perceptual criteria. An HRTF-based filter design has been developed with various methods, including conventional time windowing or frequency sampling methods for FIR filters and Prony or the Yule–Walker method for IIR filters. A ***balanced model truncation (BMT)*** method is used to design the IIR filter from the original HRIR with a finite length (Mackenzie et al., 1997). To improve the computational efficiency of binaural synthesis for various target virtual source directions, Haneda et al. (1999) proposed a ***common acoustical pole and zero*** (***CAPZ***) model for HRTF-based filters. In this model, HRTFs in M directions are represented by an ARMA model with direction-independent poles and direction-dependent zeros. Poles represent the direction-independent peaks in HRTF magnitudes caused by ear canal resonance, and zeros denote the variation in the directions of HRTFs. For a group of HRTFs in M directions, the CAPZ model involves fewer parameters and is simpler than conventional ARMA models.

The performance of HRTF-based filters can be evaluated using various error criteria. HRTF-based filters are usually designed to minimize certain physical errors, such as the square error between filter and target responses. Some methods based on auditory error criteria, such as the logarithmic error criterion method (Blommer and Wakefield, 1997) and the frequency-warped method (Härmä et al., 2000), have also been suggested for HRTF-based filter designs. Therefore, HRTF-based filter designs are an important topic in VAD (Xie, 2008a, 2013; Huopaniemi et al., 1999).

11.5 SPATIAL INTERPOLATION AND DECOMPOSITION OF HRTFS

11.5.1 Directional interpolation of HRTFs

Far-field HRTFs are continuous functions of the source direction (θ_S, ϕ_S). As stated in Section 11.2.1, HRTFs are usually measured in discrete and finite directions, that is, sampled in directions around a spatial spherical surface or a horizontal circle. Under certain conditions, HRTFs in unmeasured directions can be reconstructed or estimated from the measured data by using various interpolation schemes.

For example, at a constant source distance $r = r_S$, HRTFs at the arbitrary unmeasured azimuth θ_S can be estimated from the HRTFs measured at M horizontal azimuths [that is, $H(\theta_i, f)$ with $i = 0, 1, \ldots, M - 1$] by using the ***linear interpolation scheme***:

$$\hat{H}(\theta_S, f) \approx \sum_{i=0}^{M-1} A_i H(\theta_i, f), \tag{11.5.1}$$

where the subscript for the left or right ear is omitted, $\hat{H}(\theta_S, f)$ is the interpolated HRTF, and $A_i = A_i(\theta_S)$ is a set of weights related to the target azimuth. Various interpolation schemes can be developed using different methods for selecting the measured azimuths and weights. In each direction, digital measurement leads to HRTF at N discrete frequencies; therefore, Equation (11.5.1) is the directional interpolation equation for N discrete frequencies of $f = f_k$ $(k = 0, 1 \ldots N - 1)$.

Equation (11.5.1) can be extended to three-dimensional spatial directions as

$$\hat{H}(\theta_S, \phi_S, f) \approx \sum_{i=0}^{M-1} A_i H(\theta_i, \phi_i, f), \tag{11.5.2}$$

where $\hat{H}(\theta_S, \phi_S, f)$ is the interpolated HRTF at the arbitrary unmeasured direction (θ_S, ϕ_S) and the $H(\theta_i, \phi_i, f)$ with $i = 0,1,2 \ldots M-1$ are HRTFs at M measured directions. Equations (11.5.1) and (11.5.2) are HRTF linear interpolation equation in the frequency domain and applicable to both complex-valued HRTFs and HRTF magnitudes. Interpolation of HRTF magnitudes alone improved performance.

Because of the linear characteristic of temporal-frequency Fourier transformation, Equations (11.5.1) and (11.5.2) are also applicable to the HRIRs in the time domain. For example, the time domain version of Equation (11.5.2) is given by

$$\hat{h}(\theta_S, \phi_S, t) \approx \sum_{i=0}^{M-1} A_i h(\theta_i, \phi_i, t). \tag{11.5.3}$$

If HRTFs satisfy the minimum-phase approximation given by Equation (11.4.1), Equations (11.5.1) to (11.5.3) are also applicable to the minimum-phase HRTFs or HRIRs. Interpolation of minimum-phase HRTFs improved performance. However, the resultant HRTF is not always a minimum-phase function because a weighted sum of minimum-phase functions does not always result in a minimum-phase function. In addition, some work suggested imposing arrival time correction on HRIR interpolation (Matsumoto et al., 2004). That is, prior to interpolation, the arrival time of the HRIRs for each source direction is made synchronous by shifting the onset time of each HRIR. Time correction also improves the performance of interpolation.

A simple example for directional interpolation is ***adjacent linear interpolation.*** Within the azimuthal region $\theta_i < \theta_S < \theta_{i+1}$ in the horizontal plane, a HRTF at azimuth θ_S is approximated by the first-order term of its Taylor expansion of θ_S

$$\begin{aligned} H(\theta_S, f) &\approx H(\theta_i, f) + \left.\frac{\partial H(\theta_S, f)}{\partial \theta_S}\right|_{\theta_S = \theta_i} (\theta_S - \theta_i) \\ &\approx H(\theta_i, f) + \frac{H(\theta_{i+1}, f) - H(\theta_i, f)}{\theta_{i+1} - \theta_i}(\theta_S - \theta_i), \end{aligned} \tag{11.5.4}$$

or

$$\hat{H}(\theta_S, f) \approx A_{i+1} H(\theta_{i+1}, f) + A_i H(\theta_i, f). \tag{11.5.5}$$

The weights are given by

$$A_{i+1} = \frac{\theta_S - \theta_i}{\theta_{i+1} - \theta_i} \quad A_i = 1 - \frac{\theta_S - \theta_i}{\theta_{i+1} - \theta_i}. \tag{11.5.6}$$

Therefore, the unmeasured HRTF at θ_S is approximated as the weighted sum of a pair adjacent HRTFs, and the weights A_i and A_{i+1} are independent from frequency. Equation (11.5.5) is the equation of conventional adjacent linear interpolation, a special case of Equation (11.5.1).

Bilinear interpolation is a three-dimension extension of adjacent linear interpolation (Wightman et al., 1992). Given that HRTFs are measured at a constant source distance $r = r_S$, the spherical surface (upon which the source is located) is sampled along both azimuthal and elevation directions, resulting in a measurement grid, with its vertices representing the source directions for measurement. The HRTF at an unmeasured direction within the grid are approximated as a weighted sum or average of the HRTFs associated with the four nearest directions:

A ***spherical triangular interpolation*** scheme is established (Freeland et al., 2004). The measured positions consist of a triangular grid on a spherical surface. The HRTF in an unmeasured direction within a grid is approximated as a weighted sum of the measured HRTFs at the three adjacent vertices of the grid.

Similar to the case of the HRTF-based filter in Section 11.4, the performance of interpolation can be evaluated using various error criteria. For example, the relative energy error is defined as

$$Err_R(\theta_S, \phi_S, f) = \frac{\left|H(\theta_S, \phi_S, f) - \hat{H}(\theta_S, \phi_S, f)\right|^2}{\left|H(\theta_S, \phi_S, f)\right|^2} \quad (\times 100\%), \tag{11.5.7}$$

where $H(\theta_S, \phi_S, f)$ and $\hat{H}(\theta_S, \phi_S, f)$ are the target and interpolated HRTFs, respectively.

11.5.2 Spatial basis function decomposition and spatial sampling theorem of HRTFs

HRTFs are multivariable functions. Even for a specified individual and in a far field, HRTFs are complex-valued functions of the source direction (θ_S, ϕ_S) and frequency f. This multivariable-dependent characteristic yields the substantial dimensionality of the entire HRTF data, so the analysis and representation of HRTFs are complicated. Alternatively, an efficient or low-dimensional representation of HRTFs can be achieved by decomposing HRTFs into a weighted sum of appropriate basis functions, where the dependencies of HRTFs on different variables are separately represented by variations in basis functions and weights. The basis function decompositions of HRTFs are applicable to the binaural synthesis of multiple virtual sources in VAD.

HRTF linear decomposition is categorized into two basic types: ***spatial basis function decomposition*** and ***spectral shape basis function decomposition***. The former is addressed here, and the latter is discussed in Section 11.5.4. Several spatial basis function decomposition schemes exist. Among them, the spatial harmonic decomposition scheme, which is closely related to Ambisonics, leads to the spatial sampling theorem of HRTFs and is applicable to simplifying HRTF representation and binaural synthesis in VAD.

The azimuthal harmonic decomposition of horizontal HRTFs is discussed first (Zhong and Xie, 2005, 2009). At each given elevation $\phi_S = \phi_0$, such as the horizontal plane $\phi_S = 0°$, a far-field HRTF for a specified individual and ear is a continuous function of the azimuth with a period of 2π. Therefore, it can be expanded as a real- or complex-valued azimuthal Fourier series as

$$\begin{aligned} H(\theta_S, f) &= H_0^{(1)}(f) + \sum_{q=1}^{+\infty}\left[H_q^{(1)}(f)\cos q\theta_S + H_q^{(2)}(f)\sin q\theta_S\right] \\ &= \sum_{q=-\infty}^{+\infty} H_q(f)\exp(jq\theta_S). \end{aligned} \tag{11.5.8}$$

Therefore, $H(\theta_S, f)$ is decomposed into a weighted sum of infinite orders of azimuthal harmonics. The azimuthal harmonics $\{\cos q\theta_S, \sin q\theta_S\}$ or $\{\exp(jq\theta_S)\}$ are an infinite set of orthogonal basis functions. They depend only on θ_S. The frequency-dependent coefficients or weights $\left\{H_q^{(1)}(f), H_q^{(2)}(f)\right\}$ or $\{H_q(f)\}$ represent the azimuthal spectrum of HRTFs. They can be evaluated from the continuous $H(\theta_S, f)$ as follows:

$$\begin{aligned} &H_0^{(1)}(f) = \frac{1}{2\pi}\int_{-\pi}^{\pi} H(\theta_S, f)\,d\theta_S \\ &H_q^{(1)}(f) = \frac{1}{\pi}\int_{-\pi}^{\pi} H(\theta_S, f)\cos q\theta_S d\theta_S \quad H_q^{(2)}(f) = \frac{1}{\pi}\int_{-\pi}^{\pi} H(\theta_S, f)\sin q\theta_S d\theta_S \\ &H_0(f) = H_0^{(1)}(f) \quad H_q(f) = \frac{1}{2}\left[H_q^{(1)}(f) - jH_q^{(2)}(f)\right] \quad H_{-q}(f) = \frac{1}{2}\left[H_q^{(1)}(f) + jH_q^{(2)}(f)\right] \\ &q = 1, 2, 3\ldots \end{aligned} \tag{11.5.9}$$

In addition to frequency dependence, $\left\{H_q^{(1)}(f), H_q^{(2)}(f)\right\}$ or $\{H_q(f)\}$ are relevant to elevation, individuals, and ears. For simplicity, however, these variables are excluded from the following discussion. Moreover, the weights here are written as functions of frequency; they should not be confused with the notation of Hankel functions in Chapters 9 and 10.

If the HRTF is azimuthally bandlimited such that all azimuthal harmonics with the order $|q| > Q$ vanish [that is, $\tilde{H}_q^{(1)}(f) = \tilde{H}_q^{(2)}(f) = \tilde{H}_q(f) = 0$ for $|q| > Q$], Equation (11.5.8) becomes

$$\begin{aligned} H(\theta_S, f) &= H_0^{(1)}(f) + \sum_{q=1}^{Q}\left[H_q^{(1)}(f)\cos q\theta_S + H_q^{(2)}(f)\sin q\theta_S\right] \\ &= \sum_{q=-Q}^{+Q} H_q(f)\exp(jq\theta_S). \end{aligned} \tag{11.5.10}$$

In this case, $H(\theta_S, f)$ is composed of $(2Q + 1)$ azimuthal harmonics and determined by the $(2Q + 1)$ azimuthal Fourier coefficients $\left\{H_q^{(1)}(f), H_q^{(2)}(f)\right\}$ or $\{H_q(f)\}$. These $(2Q + 1)$ azimuthal Fourier coefficients can be evaluated from the HRTFs measured or sampled at

M uniform azimuths within $-\pi < \theta_S \leq \pi$ ($-180° < \theta_S \leq 180°$). Let $H(\theta_i, f)$ be the HRTFs measured at M uniform azimuths, so Equation (11.5.10) yields

$$\begin{aligned} H(\theta_i, f) &= H_0^{(1)}(f) + \sum_{q=1}^{Q}\left[H_q^{(1)}(f)\cos q\theta_i + H_q^{(2)}(f)\sin q\theta_i\right] \\ &= \sum_{q=-Q}^{+Q} H_q(f)\exp(jq\theta_i) \\ \theta_i &= \frac{2\pi i}{M}, \quad i = 0,1\ldots(M-1). \end{aligned} \tag{11.5.11}$$

If θ_i in Equation (11.5.11) exceeds π, it should be replaced with $\theta_i - 2\pi$ to keep the azimuthal variable within the range of $-\pi < \theta \leq \pi$ because of the periodic variation in azimuth θ_i. For

$$M \geq (2Q+1), \tag{11.5.12}$$

the $(2Q + 1)$ azimuthal Fourier coefficients can be solved using M linear equations expressed in Equation (11.5.11). Fourier coefficients are calculated using the discrete orthogonality of the trigonometric function from Equations (4.3.16) to (4.3.18):

$$\begin{aligned} & H_0^{(1)}(f) = \frac{1}{M}\sum_{i=0}^{M-1} H(\theta_i, f) \\ & H_q^{(1)}(f) = \frac{2}{M}\sum_{i=0}^{M-1} H(\theta_i, f)\cos q\theta_i \quad H_q^{(2)}(f) = \frac{2}{M}\sum_{i=0}^{M-1} H(\theta_i, f)\sin q\theta_i \quad 1 \leq q \leq Q \\ & H_q^{(1)}(f) = H_q^{(2)}(f) = 0 \quad Q < q \leq (M-1)/2. \end{aligned} \tag{11.5.13}$$

Substituting Equation (11.5.13) into Equation (11.5.11) leads to an interpolation equation for the azimuthal continuous HRTFs:

$$\hat{H}(\theta_S, f) = \frac{1}{M}\sum_{i=0}^{M-1} H(\theta_i, f)\frac{\sin\left[\left(Q+\frac{1}{2}\right)(\theta - \theta_i)\right]}{\sin\left(\frac{\theta - \theta_i}{2}\right)}. \tag{11.5.14}$$

The comparison of Equation (11.5.14) with Equation (11.5.1) yields the following weights for azimuthal interpolation:

$$A_i = \frac{1}{M}\frac{\sin[\left(Q+\frac{1}{2}\right)(\theta_S - \theta_i)}{\sin\left(\frac{\theta_S - \theta_i}{2}\right)} = \frac{1}{M}\left[1 + 2\sum_{q=1}^{Q}\left(\cos q\theta_i \cos q\theta_S + \sin q\theta_i \sin q\theta_S\right)\right]. \tag{11.5.15}$$

Overall, at each given elevation of ϕ_0, azimuthal HRTFs can be decomposed as a weighted sum of the azimuthal harmonics. If the azimuthal HRTF can be represented by the azimuthal harmonics up to order Q, azimuthal continuous HRTFs can be reconstructed from $M \geq (2Q + 1)$ azimuthal measurements uniformly distributed in the $-\pi < \theta \leq \pi$ region. In other words, the azimuthal sampling rate should be at least twice that of the azimuthal Fourier harmonic bandwidth of HRTFs; otherwise, spatial aliasing occurs in interpolated HRTFs. This statement is the azimuthal sampling theorem of HRTFs, which is similar to the Shannon–Nyquist theorem for time sampling.

The minimal number of azimuthal measurements for the recovery of azimuthal continuous HRTF is expressed in Equation (11.5.12). An analysis of the KEMAR–HRTFs in the horizontal plane indicates that the highest-order Q of the azimuthal harmonics increases with frequency, with $Q = 32$ within the frequency range of $f \leq 20$ kHz. In this case, the contributions of 32 preceding-order azimuthal harmonics to the mean relative energy of HRTFs are larger than 0.99 (relative energy error of less than 1%). The Shannon–Nyquist azimuthal sampling theorem requires the minimal azimuthal measurements of $M_{min} = (2Q + 1) = 65$ to recover the azimuthal continuous HRTFs in the horizontal plane. As source elevation deviates from the horizontal plane, the minimal azimuthal measurements required to recover the azimuthal continuous HRTFs decrease. The analyses of human HRTFs yield similar results. Similar to the case in Section 11.5.1, imposing arrival time correction on HRIRs or considering the HRTF magnitudes alone obviously reduces the minimal azimuthal measurements for recovering azimuthal continuous HRTFs.

The preceding discussion can be extended to a three-dimensional case. Far-field HRTFs can be decomposed by real- or complex-valued spherical harmonic functions (Evans et al., 1998). Similar to the case of spatial Ambisonics in Section 9.1.2, the source direction is denoted by the notation Ω_S, then

$$H(\Omega_S, \mathrm{f}) = \sum_{l=0}^{\infty}\sum_{m=0}^{1}\sum_{\sigma=l}^{2} H_{lm}^{(\sigma)}(f) Y_{lm}^{(\sigma)}(\Omega_S) = \sum_{l=0}^{\infty}\sum_{m=-l}^{l} H_{lm}(f) Y_{lm}(\Omega_S) \tag{11.5.16}$$

Therefore, $H(\Omega_S, f)$ is decomposed into a weighted sum of infinite orders of spherical harmonic functions. Spherical harmonic functions are the orthogonal basis functions of Ω_S. Frequency-dependent weights [spherical harmonic coefficients $H_{lm}^{(\sigma)}(f)$ or $H_{lm}(f)$], which represent the spherical harmonic spectrum of HRTFs, can be evaluated from a directional continuous $H(\Omega_S, f)$ by using the orthogonality of spherical harmonic functions, as given by Equation (A.12) in Appendix A.

If the HRTF is spatially bandlimited so that all spherical harmonic components with the order $l \geq L$ vanish, Equation (11.5.16) becomes

$$H(\Omega_S, f) = \sum_{l=0}^{L-1}\sum_{m=0}^{l}\sum_{\sigma=1}^{2} H_{lm}^{(\sigma)}(f) Y_{lm}^{(\sigma)}(\Omega_S) = \sum_{l=0}^{L-1}\sum_{m=-l}^{l} H_{lm}(f) Y_{lm}(\Omega_S). \tag{11.5.17}$$

In this case, the HRTF is determined by L^2 spherical harmonic coefficients. Given the measured HRTFs at M discrete-sampled directions, Equation (11.5.17) leads to M linear equations with regard to L^2 spherical harmonic coefficients:

$$H(\Omega_i, f) = \sum_{l=0}^{L-1}\sum_{m=0}^{l}\sum_{\sigma=1}^{2} H_{lm}^{(\sigma)}(f) Y_{lm}^{(\sigma)}(\Omega_i) = \sum_{l=0}^{L-1}\sum_{m=-l}^{l} H_{lm}(f) Y_{lm}(\Omega_i), \tag{11.5.18}$$

where $H(\Omega_i, f)$, $i = 0, 1\ldots(M - 1)$ are M-measured HRTFs. Similar to the case of spatial Ambisonics in Section 9.3.2, for $M \geq L^2$, Equation (11.5.18) can be solved by the pseudo-inverse method. In particular, if the measured directions (directional sampling on a spherical surface) are chosen such that spherical harmonic functions satisfy the discrete orthogonality up to the $(L - 1)$ order expressed in Equation (A.20) in Appendix A, an accurate solution of spherical harmonic coefficients can be found. Directional continuous HRTFs are obtained by substituting the resultant spherical harmonic coefficients into Equation (11.5.17). $M \geq L^2$ gives the low limit on the number of directional measurements required to recover the directional continuous HRTF, which is the consequence of the Shannon–Nyquist spatial sampling theorem. The practical required number of directional measurements is usually larger than the low limit, depending on the directional sampling scheme.

As stated in Section 11.2.1, near-field HRTF measurements are relatively difficult because of their source distance-dependent features. The spherical harmonic decomposition algorithm can be extended to the distance extrapolation of HRTFs, that is, to estimate the near-field HRTFs on the basis of far-field measurements (Duraiswami et al., 2004 ; Pollow et al., 2012; Zhang et al., 2010). According to the acoustic principle of reciprocity, the pressure is invariant after the positions of the sound source and the receiver are exchanged. Therefore, for a source located at the position of the ear, the pressure at the receiver position (r_S, Ω_S) and the HRTF can be decomposed by spherical harmonic functions as

$$\begin{aligned} H(r_S, \Omega_S, f) &= \sum_{l=0}^{\infty}\sum_{m=-l}^{l} H_{lm}(f)\Xi_l(kr_S)Y_{lm}(\Omega_S) \\ &= H_{00}^{(1)}(f)\Xi_0(kr_S)Y_{00}^{(1)}(\Omega_S) \\ &+\sum_{l=1}^{\infty}\sum_{m=0}^{l}\left[H_{lm}^{(1)}(f)\Xi_l(kr_S)Y_{lm}^{(1)}(\Omega_S)+H_{lm}^{(2)}(f)\Xi_l(kr_S)Y_{lm}^{(2)}(\Omega_S)\right], \end{aligned} \tag{11.5.19}$$

and

$$\Xi_l(kr_S) = (-j)^{l+1} kr_S \exp(jkr_S) h_l(kr_S), \tag{11.5.20}$$

where $h_l(kr_S)$ is the l-order spherical Hanker function of the secondary type. Similar to the case of far-field HRTFs, if all spherical harmonic components with the order $l \geq L$ vanish in Equation (11.5.20), then the spherical harmonic coefficients of decomposition can be evaluated from M directional measurements $H(r_0, \Omega_i, f)$, $i = 0, 1\ldots(M - 1)$ at a constant source distance $r_S = r_0$ (for example, at a far-field distance) provided that M satisfies the condition of the Shannon–Nyquist spatial sampling theorem. The distance- and directional continuous HRTFs can be obtained by substituting the resultant spherical harmonic coefficients into Equation (11.5.19).

The basic functions in spatial harmonic decomposition are continuous and predetermined by which spatial continuous HRTFs are recovered. However, the efficiency of spatial harmonic decomposition is not optimal from the point of data reduction because more basis functions are usually required. HRTFs can also be decomposed by other spatial basis functions although these spatial functions may not be continuous and predetermined. If HRTFs can be decomposed by a small set of spatial basis functions, the dimensionality of data is greatly reduced, and HRTFs in more directions can be recovered from a small set of directional measurements (fewer than the requirement of the Shannon–Nyquist spatial sampling theorem). This finding is observed because the Shannon–Nyquist spatial sampling theorem is a sufficient condition rather than a necessary one for recovering HRTFs in full directions. For example, a set of compacted spatial basis functions can be derived through spatial principal component analysis of a baseline HRTF database (SPCA, Xie, 2012). Individualized HRTF magnitudes for different subjects can be represented by 35 spatial basis functions, and the individualized HRTF magnitudes at 493 directions can be recovered from 73 directional measurements. This method is applied to simplify the HRTF measurement.

11.5.3 HRTF spatial interpolation and signal mixing for multichannel sound

The HRTF spatial interpolation and signal mixing of multichannel sounds are closely related to each other (Xie, 2006b). For example, the binaural pressures caused by a far-field target source (plane wave) in horizontal plane are continuous functions of the source azimuth θ_S and evaluated with Equation (1.4.2). For simplicity, the variable ϕ_S of the source elevation and the subscript for the left or right ear are omitted, and Equation (1.4.2) becomes

$$P(\theta_S, f) = H(\theta_S, f) P_{free}(f). \tag{11.5.21}$$

In multichannel horizontal surround sound, M loudspeakers are arranged in a circle around a listener. The azimuth of the ith loudspeaker is $\theta_i (i = 0, 1, \ldots M - 1)$, and the corresponding normalized signal amplitude is A_i. The HRTF of the ith loudspeaker to the concerned (left or right) ear is $H(\theta_i, f)$. Then, binaural pressures in reproduction are the superposition of these pressures caused by each loudspeaker:

$$P'(f) = \sum_{i=0}^{M-1} A_i H(\theta_i, f) E_A(f). \tag{11.5.22}$$

where $E_A(f)$ is the signal waveform in the frequency domain. A set of A_i (signal mixing or panning coefficients) is chosen so that binaural pressures in reproduction approximately match with those caused by the target source to recreate a virtual source at an arbitrary azimuth θ_S in a given loudspeaker configuration:

$$P'(f) \approx P(\theta_S, f). \tag{11.5.23}$$

If $E_A(f)$ is equal to the plane wave amplitude $P_A(f)$ caused by a far-field source, that is, $E_A(f) = P_A(f) = P_{free}(f)$, Equations (11.5.21), (11.5.22), and (11.5.23) yield

$$\hat{H}(\theta_S, f) \approx \sum_{i=0}^{M-1} A_i H(\theta_i, f). \tag{11.5.24}$$

Therefore, matching binaural pressures in multichannel sound reproduction with those created by a target source is analogous to HRTF directional interpolation in Equation (11.5.1). The different normalized amplitude A_i of loudspeaker signals is analogous to different weights in HRTF directional interpolation. In other words, different signal mixing or panning methods in multichannel sound are analogous to different HRTF directional interpolation schemes.

Conventional pair-wise amplitude panning in multichannel horizontal surround sound (for example, Section 4.1.2) is first analyzed. Pair-wise amplitude panning is analogous to the adjacent linear interpolation of HRTFs in Equation (11.5.5). Equation (11.5.5) is derived by retaining the first-order term in the Taylor expansion of θ_S on azimuthal continuous HRTFs.

Within the frequency range of $f < 3$ kHz, HRTFs with a rigid spherical head model in Equation (11.2.2) are valid, and HRTF is a function of $\sin\theta_S$. Therefore, replacing the azimuthal variable θ_S in the horizontal HRTF with $\xi = \sin\theta_S$ is a more reasonable approach. Expanding the horizontal HRTF as a Taylor series of $\xi = \sin\theta_S$ and retaining the first-order term yield

$$H(\xi, f) \approx H(\xi_i, f) + \frac{H(\xi_{i+1}, f) - H(\xi_i, f)}{\xi_{i+1} - \xi_i}(\xi - \xi_i). \tag{11.5.25}$$

where $\xi_i = \sin\theta_i$ and $\xi_{i+1} = \sin\theta_{i+1}$. Through the comparison of Equation (11.5.25) with Equation (11.5.5), the weights of the adjacent linear interpolation of HRTFs in Equation (11.5.6) are replaced with the following results:

$$\begin{aligned} A_{i+1} &= A_{i+1}(\theta_S) = \frac{\xi - \xi_i}{\xi_{i+1} - \xi_i} = \frac{\sin\theta_S - \sin\theta_i}{\sin\theta_{i+1} - \sin\theta_i} \\ A_i &= A_i(\theta_S) = 1 - \frac{\xi - \xi_i}{\xi_{i+1} - \xi_i} = 1 - \frac{\sin\theta_S - \sin\theta_i}{\sin\theta_{i+1} - \sin\theta_i}. \end{aligned} \tag{11.5.26}$$

Equation (11.5.26) also specifies the pair-wise amplitude panning functions (with constant-amplitude normalization) in multichannel horizontal surround sound signals.

The following observations are obtained from Equations (11.5.6) and (11.5.26):

1. If $\theta_i<\theta_S<\theta_{i+1}$ is close to the front ($\theta = 0°$), then $\sin\theta \approx \theta$, and Equations (11.5.6) and (11.5.26) yield the same results; otherwise, different results occur. In fact, the interaural phase delay difference ITD_p is a dominant localization cue at low frequencies. Multichannel sound reproduction with amplitude panning is designed to create the desired ITD_p at a low frequency and then recreate the virtual source in target directions. In addition, the low-frequency ITD_p in Equation (1.6.4) is directly proportional to $\sin\theta_S$. Therefore, replacing the weights A_i and A_{i+1} in the HRTF interpolation in Equation (11.5.6) with the results provided in Equation (11.5.26) is a more appropriate strategy for retaining accurate low-frequency localization cues. That is, the conventional adjacent linear interpolation formula of HRTFs should be revised at low frequencies.
2. In Equations (11.5.6) and (11.5.26), the smaller the interval between θ_i and θ_{i+1} and the smoother $H(\theta_S, f)$ azimuthal variation are, the more accurate the estimated HRTF from the adjacent linear interpolation will be. Correspondingly, a loudspeaker pair with an excessive span is usually unfavorable for stereophonic and multichannel sound reproduction in terms of virtual source quality, as stated in the analysis of Equation (3.2.30) and in Section 4.1.2.
3. When $\theta_i<\theta_S<\theta_{i+1}$, Equation (11.5.26) yields $0 < A_i, A_{i+1} < 1$. In this case, the weights of the adjacent linear interpolation of HRTFs are positive. Correspondingly, the signals of two adjacent loudspeakers for pair-wise amplitude panning are in phase.
4. Equations (11.5.6) and (11.5.26) can be extended to the case of $\theta_i<\theta_{i+1}<\theta_S$ or $\theta_S<\theta_i<\theta_{i+1}$, that is, the case of predicting $H(\theta_S, f)$ at a target θ_S outside the region of $[\theta_i, \theta_{i+1}]$ with a

pair of $H(\theta_i, f)$ and $H(\theta_{i+1}, f)$. In this case, the polarities of weights A_i and A_{i+1} are opposite with $A_{i+1} > 0$, $A_i < 0$ or $A_{i+1} < 0$, $A_i > 0$. Correspondingly, this case is analogous to creating an outside-boundary virtual source with out-of-phase loudspeaker signals in stereophonic or multichannel sound reproduction (Section 2.1.3).

5. For a pair of front–back symmetrical directions $\theta_{i+1} = 90° + \Delta\theta$ and $\theta_i = 90° - \Delta\theta$, the weights A_i and A_{i+1} evaluated from Equation (11.5.26) are infinite. Therefore, adjacent linear interpolation is unable to estimate the HRTF in the lateral direction from a pair of HRTFs in the front–back symmetrical directions. This conclusion can also be drawn from the symmetry of HRTFs. HRTFs with a rigid spherical head in Equation (11.2.2) are front–back symmetric. Practical HRTFs are approximately front–back symmetric with $H(90° - \Delta\theta, f) \approx H(90° + \Delta\theta, f)$ below the frequency of 1 kHz (Zhong and Xie, 2007; Zhong et al., 2013). In this case, whatever A_i and A_{i+1} are chosen, the adjacent linear interpolation always yields $H(90° - \Delta\theta, f)$ by multiplying constant A

$$A_i H(90° - \Delta\theta, f) + A_{i+1} H(90° + \Delta\theta, f) \approx AH(90° - \Delta\theta, f) \approx AH(90° + \Delta\theta, f)$$

In multichannel sound reproduction, a pair of lateral loudspeakers symmetrically arranged in left-front and left back directions is unable to recreate a stable lateral virtual source (Section 4.1.2).

6. In Equations (11.5.6) and (11.5.26), assuming that the azimuthal spans between all adjacent θ_i and θ_{i+1} are identical is unnecessary. Therefore, these equations are appropriate for nonuniform HRTF directional interpolation or irregular loudspeaker configurations in the horizontal plane (for example, 5.1-channel configuration).

As a special case, the loudspeaker configuration for a two-channel stereophonic sound is shown in Figure 2.1. The positions of the two loudspeakers are $\theta_{i+1} = \theta_L = \theta_0$, $\theta_i = \theta_R = -\theta_0$, and the signal amplitudes are $A_{i+1} = A_L$ and $A_i = A_R$. Replacing the target azimuth θ_S with the virtual source azimuth θ_I in Equation (11.5.26), the summing virtual source direction θ_I can be calculated as

$$\sin\theta_I = \frac{A_L - A_R}{A_L + A_R}\sin\theta_0. \tag{11.5.27}$$

Equation (11.5.27) is merely the stereophonic law of sine expressed in Equation (2.1.6). In contrast to Equation (2.1.6), Equation (11.5.27) is derived by the linear interpolation of the ear pressures created by a pair of stereophonic loudspeakers rather than by directly calculating ITD_p. Therefore, the analysis of Equation (11.5.26) is a more stringent derivation of the stereophonic law of sine.

The above analysis can be extended to a three-dimensional space. For example, the spherical triangular interpolation (Freeland et al., 2004) discussed in Section 11.5.1 is analogous to vector base amplitude panning in Section 6.3 (Pulkki, 1997).

The horizontal Ambisonic signal mixing is then discussed. A comparison of Equations (9.3.12) and (11.5.15) indicates that horizontal Ambisonic signal mixing for uniform (regular) loudspeaker configuration is analogous to the azimuthal interpolation of the HRTF expressed in Equation (11.5.14). Similarly, the spatial Ambisonic signal mixing discussed in Section 9.3.2 is analogous to the three-dimensional direction interpolation of HRTFs given in Equations (11.5.16) to (11.5.18). The near-field compensated higher-order Ambisonics in Section 9.3.4 is analogous to the distance extrapolation of HRTFs described in Equations (11.5.19) and (11.5.20).

In Equation (11.5.14), the HRTF at θ_S is estimated by a weighted sum of HRTFs at all M azimuthal measurements. Therefore, this condition is a global interpolation scheme, and the adjacent linear interpolation is a local interpolation method. Interestingly, Ambisonics is a global signal mixing method, and pair-wise amplitude panning is a local signal mixing method, as stated in Section 4.1.3.

HRTFs are consequences of the scattering and diffraction of incident sound waves by anatomical structures. The condition for the accurate reconstruction of a horizontal incident sound field in a local region is given in Section 9.3.1, which reveals the relationship of the radius of the region, the upper frequency limit, and the order of the azimuthal harmonics required for Ambisonic sound field reconstruction. This relationship is also applicable to HRTF directional interpolation. For example, letting $r = a = 0.0875$ m (average head radius), Equation (9.3.15) indicates that $Q = 32$-order horizontal Ambisonics with $M = (2Q + 1) = 65$ loudspeakers is required to reconstruct the target sound field within a region matching the head size and up to a frequency of 20 kHz. According to the analogy of Ambisonics and HRTF directional interpolation, 65 uniform-azimuthal measurements are needed to recover the azimuthal continuous HRTF in the horizontal plane up to 20 kHz by using Equation (11.5.14). This result is consistent with the analysis of the azimuthal harmonic components of practically measured HRTFs in Section 11.5.2. Correspondingly, the 32-order horizontal Ambisonics with $M \geq (2Q + 1) = 65$ loudspeakers can reconstruct binaural pressures accurately up to 20 kHz, with a mean relative energy error of less than 1%. However, such high-order Ambisonics is complicated and infeasible in most cases. Ambisonics with a moderate order is usually used in practical reproduction, which inevitably leads to errors in reconstructed binaural pressures (spatial aliasing) at high frequencies.

Similarly, Equation (9.3.30) indicates that up to $(L - 1) = 32$ order spherical harmonic components or $L^2 = 1089$ directional measurements are at least needed to recover the directional continuous HRTF in a three-dimensional space up to 20 kHz. More directional measurements than the lower limit are needed in practice, depending on the directional sampling scheme used in measurements.

In addition to various multichannel signal mixing methods, the focused source and local wave field synthesis methods in Section 10.2.4 are applicable to the distance extrapolation of HRTFs (Spors et al., 2011). This phenomenon is an example of an analogy between spatial sound reproduction and HRTF spatial interpolation.

In summary, some signal mixing methods in multichannel sound reproduction are analogous to certain interpolation and recovery schemes for HRTFs from the perspective of spatial function sampling, interpolation, and reconstruction. Different signal mixing methods are analogous to various interpolation schemes. In multichannel sound reproduction, after being scattered and diffracted by anatomical structures, sound waves created by multiple loudspeakers are superposed at the two ears. The binaural pressures of a target source are reconstructed by an "inherent interpolation scheme." However, limited by the Shannon–Nyquist spatial sampling theorem, practical multichannel sounds with an appropriate number of loudspeakers are only able to reconstruct binaural pressures below a certain frequency rather than within the full audible range. In this case, the psychoacoustic principle is incorporated to recreate the perceived virtual source.

One major purpose of this section is to reveal the close relationship between the signal mixing of multichannel sound and the spatial interpolation of HRTFs. As stated in Section 1.9.1, all existing spatial sound techniques are based on three methods and principles: sound field-based methods(such as higher-order Ambisonics and wave field synthesis),sound field approximation, and psychoacoustic-based methods (such as various stereophonic and multichannel sounds, including low-order Ambisonics), and binaural-based methods. The analyses in Chapters 9 and 10 (especially in Section 10.4) indicate that various sound field-based techniques are physically related, and sound field approximation and psychoacoustic-based methods are closely related to the sound field-based method. The analyses in this section demonstrate that binaural-based methods are closely related to two other methods. Various loudspeaker configurations in multichannel sound are analogous to different HRTF directional sampling schemes; some signal mixing methods in multichannel sound are analogous to different directional interpolation schemes. The combination of loudspeaker configuration and signal mixing method in multichannel sound is analogous to the combination of the HRTF directional sampling and interpolation scheme. However, this analogy was rarely explored in previous studies. By contrast, multichannel sound and VAD have been studied separately in many cases. On the basis of the analyses in this section, Section 3.1, Chapter 9, and Chapter 10, our conclusion is that ***various spatial sound techniques (including multichannel sound, VAD, Ambisonics, and even wave field synthesis) are unified under the theoretical framework of spatial function sampling, interpolation, and reconstruction.*** Addressing this unity is also ***a major purpose of this book*** because unity is vital to gaining insights into the physical nature of spatial sound reproduction.

The analogy between multichannel sound reproduction and HRTF directional interpolation enables the interchanging of some of the methods used for the two fields. This interchange is another major purpose of this section and convenient for research on spatial sounds. According to the preceding discussion, methods for analyzing the reconstructed sound field of Ambisonics are applied to examine the spatial interpolation of HRTFs [for example, Equation (11.5.15)]. Conversely, various methods are used to evaluate errors in HRTF interpolation and binaural reproduction is applicable to multichannel sounds. That is, the same method and results may be appropriate for HRTF and multichannel sound analyses. This problem is addressed in Chapter 12. Moreover, some methods in multichannel sound, especially psychoacoustic methods of signal mixing and summing localization, are utilized to simplify binaural synthesis processing in VAD (Section 11.6).

11.5.4 Spectral shape basis function decomposition of HRTFs

HRTF can be decomposed using spectral shape basis functions. In this case, the HRTF of a given ear is generally represented as

$$H(\Omega_S, f) = \sum_q W_q(\Omega_S) D_q(f), \tag{11.5.28}$$

where $D_q(f)$ is a series of spectral shape basis functions that depend only on frequency, and $W_q(\Omega_S)$ is the source-direction-dependent weight. When the basis function $D_q(f)$ is selected, $H(\Omega_S, f)$ is completely determined by $W_q(\Omega_S)$.

Various methods for deriving the basic functions, $D_q(f)$, are available. A set of orthonormal basis functions is often preferred. ***Principal component analysis (PCA)*** is an effective statistical algorithm for deriving basis functions. It eliminates the correlations among HRTFs so that HRTFs can be simply represented by the weighted sum of a small set of spectral shape basis functions (Martens, 1987; Kistler and Wightman, 1992; Middlebrooks and Green, 1992; Chen et al., 1995; Wu et al., 1997). Here, PCA is conducted in the frequency or time domain, and the spatial PCA mentioned at the end of Section 11.5.2 is performed in the spatial domain.

The measured HRTFs are usually represented by their samples at discrete frequencies and source directions. The far-field HRTFs in the M directions for a concerned ear are provided. The HRTF in each direction is represented by its samples at N discrete frequencies. HRTFs are simply denoted by $H(\Omega_i, f_k) = H(i, k)$, where $i = 0, 1\ldots (M-1)$ is the direction index, and $k = 0, 1\ldots (N-1)$ is the frequency index. Equation (11.5.28) is then written in the form of discrete variables:

$$H(i,k) = \sum_q W_q(i) D_q(k) + H_{av}(k) \quad i = 0, 1\ldots(M-1), \quad k = 0, 1\ldots(N-1), \tag{11.5.29}$$

where $H_{av}(k)$ is the mean of the HRTF across the source directions at frequency k,

$$H_{av}(k) = \frac{1}{M}\sum_{i=0}^{M-1} H(i,k). \tag{11.5.30}$$

The mean HRTF across the source directions is subtracted from each HRTF to eliminate the correlation of HRTFs and derive $D_q(k)$ effectively. The resultant data constitute an $N \times M$ matrix $[H_\Delta]$, whose entries are

$$H_{\Delta,k,i} = H(i,k) - H_{av}(k) \quad i = 0, 1\ldots(M-1), \quad k = 0, 1\ldots(N-1). \tag{11.5.31}$$

Thus, each row and column of the matrix corresponds to a specified frequency and direction, respectively. An $N \times N$ covariance matrix $[R]$ can be constructed from $[H_\Delta]$, whose entries describe the similarity between HRTFs in each pair of frequencies:

$$[\mathrm{COV}] = \frac{1}{M}[H_\Delta][H_\Delta]^{+}, \tag{11.5.32}$$

where + is the transpose and conjugation of the matrix.

[COV] is an $N \times N$ Hermitian matrix with real and non-negative eigenvalues. The spectral shape basis vector $\mathbf{D}_q = [D_q(0), D_q(1), \ldots D_q(N-1)]^{\mathrm{T}}$, which represents the samples of the spectral shape basis function at N discrete frequencies, can be obtained from the eigenvectors of the matrix [COV] of all $Q' \leq N$ positive eigenvalues. γ_q^2 is expressed as follows:

$$[\mathrm{COV}]\mathbf{D}_q = \gamma_q^2\ \mathbf{D}_q \qquad q = 1, 2\ldots Q' \qquad \gamma_1^2 > \gamma_2^2 > \ldots > \gamma_{Q'}^2 > 0. \tag{11.5.33}$$

The resultant spectral shape basis vectors are orthonormal to one another:

$$\mathbf{D}_{q'}^{+}\,\mathbf{D}_q = \begin{cases} 1 & q = q' \\ 0 & q \neq q' \end{cases}. \tag{11.5.34}$$

When the spectral shape basis vector $\mathbf{D}_q$ is obtained, direction-dependent weights can be evaluated using orthonormality in Equation (11.5.34):

$$W_q(i) = \sum_{k=0}^{N-1}\left[H(i,k) - H_{av}(k)\right] D_q^*(k), \tag{11.5.35}$$

where * denotes complex conjugation. Lastly, the spectral shape basis function decomposition of HRTFs is obtained by substituting $H_{av}(k)$ in Equation (11.5.30), $D_q(k)$ in Equation (11.5.33), and $W_q(i)$ in Equation (11.5.35) into Equation (11.5.29).

Each basis vector associated with its weight is termed a ***principal component*** (**PC**). When HRTFs are reconstructed by all Q' PCs, Equation (11.5.29) is the exact representation of the original HRTFs. If HRTFs are reconstructed by $Q < Q'$PCs associated with the preceding largest positive Q eigenvalues γ_q^2, $q = 1, 2, \ldots Q$, Equation (11.5.29) is an approximate representation of original HRTFs:

$$\hat{H}(i,k) \approx \sum_{q=1}^{Q} W_q(i) D_q(k) + H_{av}(k) \quad i = 0,1 \ldots (M-1), \; k = 0,1 \ldots (N-1). \qquad (11.5.36)$$

The preceding $Q<Q'$spectral shape basis vectors $\boldsymbol{D}_1, \boldsymbol{D}_2, \ldots \boldsymbol{D}_Q$ are orthonormal but incomplete. Accordingly, the more PCs included in Equation (11.5.36) are, the more exact and larger the dimensionality of the reconstructed HRTFs will be. The first PC contributes the most to the HRTF, while the contributions of the other PCs decline. The preceding Q PCs are the most important or representative, and they minimize the square errors in reconstruction. In this sense, PCA can also be regarded as a smoothing algorithm for HRTFs, retaining primary features while discarding minor details. These features constitute the basic principle of PCA decomposition and reconstruction of HRTFs. The cumulative percentage variance of the energy represented by the preceding Q PCs in Equation (11.5.29) is evaluated as

$$\eta = \frac{\sum_{i=0}^{M-1} \sum_{k=0}^{N-1} \left| \hat{H}(i,k) - H_{av}(k) \right|^2}{\sum_{i=0}^{M-1} \sum_{k=0}^{N-1} \left| H(i,k) - H_{av}(k) \right|^2} \times 100\% = \frac{\sum_{q=1}^{Q} \gamma_q^2}{\sum_{q=1}^{Q'} \gamma_q^2} \times 100\%. \qquad (11.5.37)$$

In addition, the dimensionality of the original HRTFs is N (frequencies) × M (directions). As indicated by Equation (11.5.36), the dimensionality of PCA-reconstructed HRTFs is $N \times Q + Q \times M + N = (N + M) \times Q + N$. With sufficient reconstruction accuracy, data dimensionality is reduced if Q satisfies the following condition:

$$Q < \frac{N(M-1)}{N+M}. \qquad (11.5.38)$$

The above discussion is suitable for complex-valued HRTFs and linear or logarithmic HRTF magnitude spectra in the frequency domain. The analysis is also applicable to complete or minimum-phase HRIRs in the time domain provided that discrete frequencies in the aforementioned analysis are replaced with discrete times.

If the arrival times of HRIRs are corrected prior to PCA, the number of required PCs is reduced for the same reconstruction accuracy. In addition, PCA can be extended to near-field HRTFs or HRIRs by using the discrete variable i in Equations (11.5.29) to (11.5.36) to represent the source position (including direction and distance).

PCA is applied to HRTFs for a specified individual and ear, and the resultant spectral shape basis functions and directional weights depend on the individual and concerned ear. PCA can be further extended to the binaural HRTFs of multiple individuals. The resultant spectral shape basis functions or vectors are independent of the subject and the concerned ear. The left–right and individualized differences in HRTFs are represented by different directional weights.

Many researchers applied PCA to decompose and reconstruct HRTFs and indicated that a few spectral shape basis functions or PCs are enough to represent HRTFs with sufficient accuracy. For example, Kistler and Wightman (1992) indicated that the preceding five PCs account for approximately 90% of the cumulative variation in the logarithmic HRTF magnitudes of 10 human subjects and 265 directions for each subject. Chen (1995) showed that 12 PCs are sufficient to represent the KEMAR–HRTFs in 2188 directions accurately, and the cumulative percentage variance of the energy [Equation (11.5.37)] reaches 99.9%.

Xie and Zhang (2012) applied PCA to the near-field HRIRs of KEMAR at 493 source directions and 9 distances from 0.2 m to 1.0 m at an equal interval of 0.1 m (Section 11.2.1). They indicated that minimum-phase HRIRs can be decomposed by 15 PCs, along with a time domain mean function, and the cumulative percentage variance of energy [Equation (11.5.37)] is 97.4%.

11.6 SIMPLIFICATION OF SIGNAL PROCESSING FOR BINAURAL SYNTHESIS

According to the discussion in Section 11.1.2, the binaural synthesis of a virtual source can be implemented by multiplying a mono stimulus with a pair of HRTFs in the target source direction (θ_S, ϕ_S) [Equation (11.1.1)] in the frequency domain or equivalently by convoluting the monostimulus with a pair of HRIRs in the time domain. Related signal processing is usually implemented using various practical HRTF-based filters.

The conventional implementation of binaural synthesis in Equation (11.1.1) necessitates two HRTF-based filters for each virtual source. In some practical applications (such as simulation of environment reflections to be presented in Section 11.10), U independent virtual sources should be simultaneously synthesized with $2U$ HRTF-based filters. Therefore, the computational cost of binaural synthesis increases linearly with U.

In the conventional implementation of binaural synthesis, directional continuous HRTFs are theoretically required to recreate a virtual source in an arbitrary direction. In practice, HRTFs are available in discrete directions. The directional resolution of HRTFs should satisfy the criterion of the Shannon–Nyquist spatial sampling theorem or at least satisfy certain psychoacoustic criteria so that HRTFs in arbitrary directions can be estimated by interpolation. Consequently, a large HRTF data set with sufficient directional resolution is required; consequently, the storage and loading of the data become difficult.

H_L and H_R in Equation (11.1.1) depend on the target source position. An HRTF-based filter pair should also be continually updated to simulate a moving virtual source via the conventional scheme, but doing so may cause some audible (commutation) artifacts. If mere the HRTFs in discrete directions are known, the pre-interpolation scheme described in the previous section is required. With all these factors, designing the hardware and software for signal processing is difficult, especially in real-time processing. The analogy between the signal mixing of multichannel sound and spatial interpolation of HRTFs is applied to simplify the signal processing of binaural synthesis for multiple and moving virtual sources.

11.6.1 Virtual loudspeaker-based algorithms

In multichannel sound reproduction, a real loudspeaker in the target virtual source direction is not always required. Instead, a virtual source can be created by the summing localization of loudspeakers located in other directions with appropriate signal mixing. Analogous to the case of multichannel sound, a virtual source in headphone-based binaural reproduction can be created by virtual loudspeakers. In other words, M virtual loudspeakers in appropriate directions are first created by the conventional HRTF-based algorithm for virtual source synthesis. Virtual sources in target directions are subsequently created by appropriate signal mixing or panning for M virtual loudspeakers according to the principle of the summing localization of multiple sound sources. This is the basic idea of ***virtual loudspeaker-based algorithm*** for binaural synthesis (Jot et al., 1998,1999). Although this algorithm usually creates a rough approximation of target binaural sound pressures in reproduction, it can recreate the desired auditory perception in accordance with the psychoacoustic principle.

The direction and signal of the ith virtual loudspeaker are Ω_i and $E_i(f)$, respectively. The HRTFs from the ith virtual loudspeaker to the left and right ears are $H_L(\Omega_i, f)$ and $H_R(\Omega_i, f)$, respectively. Then, the superposed binaural pressures or signals caused by M virtual loudspeakers are expressed as

$$E_{L,ear}(f) = \sum_{i=0}^{M-1} H_L(\Omega_i, f) E_i(f) \qquad E_{R,ear}(f) = \sum_{i=0}^{M-1} H_R(\Omega_i, f) E_i(f). \tag{11.6.1}$$

The normalized signal amplitudes $A_i = A_i(\Omega_S)$ of virtual loudspeakers are chosen according to a specified signal mixing method such that the signals of virtual loudspeakers are given as follows to synthesize a single virtual source at a target direction Ω_S:

$$E_i(f) = A_i(\Omega_S) E_A(f), \tag{11.6.2}$$

where $E_A(f)$ is complex-valued signal waveform in the frequency domain.

A similar method is applied to synthesize U virtual sources simultaneously. The direction of the uth target virtual source is Ω_u, corresponding signal waveform is $E_{A,u}(f)$, and the normalized amplitude or gain for the signal mixed with the ith virtual loudspeaker is $A_i(\Omega_u)$. Then, the overall signal of the ith virtual loudspeaker is the sum of those of U target virtual sources:

$$E_i(f) = \sum_{u=0}^{U-1} A_i(\Omega_u) E_{A,u}(f). \tag{11.6.3}$$

The synthesized binaural signals of U target virtual sources are obtained by substituting Equation (11.6.3) into Equation (11.6.1).

The virtual loudspeaker-based algorithm has two advantages over the conventional implementation. One is the nearly constant computational cost for multiple virtual source syntheses because of the dependence of the required number of HRTF-based filters on only the number of virtual loudspeakers rather than on the number of virtual sources. Therefore, computational cost is reduced when the number of virtual sources is larger than that of virtual loudspeakers. The other advantage is that the moving virtual source is created by changing the signal mixing $A_i(\Omega_u)$ to virtual loudspeakers rather than by updating HRTF-based filters. Therefore, audible artifacts can be easily avoided.

As an illustrative case (Xie et al., 2001),Figure 11.9 shows a configuration of $M = 8$ virtual loudspeakers, including six in the horizontal plane with $\phi = 0°$, $\theta_{LF}= 30°$, $\theta_L = 90°$, $\theta_{LB} = 150°$, $\theta_{RB}= -150°$, $\theta_R = -90°$,$\theta_{RF} = -30°$, and two in the lateral plane with $\theta_{LU} = 90°$, $\phi_{LU} = 30°$, $\theta_{RU} = -90°$, and $\phi_{RU}= 30°$. In Section 4.2.1, the horizontal 360° virtual source can be created through multichannel sound reproduction by using the six horizontal loudspeaker configuration in Figure 11.10 and pair-wise amplitude panning. With extension to three-dimensional directions, the eight virtual loudspeakers in Figure 11.10 and pair-wise amplitude panning create virtual sources in horizontal and upper lateral planes.

According to Equation (11.6.1), the configuration shown in Figure 11.10 requires 2 M = 16 HRTF-based filters to create $M = 8$ virtual loudspeakers. The algorithm is further simplified by taking advantage of the left–right symmetrical configuration of virtual loudspeakers and the assumption of roughly left–right symmetrical HRTFs (Zhong et al., 2013); thus, the shuffler implementation of signal processing is carried out (Cooper and Bauck, 1989; Bauck and Cooper, 1996).

The virtual loudspeaker pair at the left–right mirror positions is considered together. The horizontal left and right virtual loudspeaker pairs are taken as an example. H_{LL} and H_{RL} denote the HRTFs from the left virtual loudspeaker to two ears; H_{LR} and H_{RR} represent the HRTFs from the right virtual loudspeaker to two ears; and $E_L = E_L(f)$ and $E_R = E_R(f)$ indicate the signals of the left and right virtual loudspeakers, respectively. The binaural signals fed to a pair of headphones should be obtained as follows to synthesize the binaural pressures caused by the left and right virtual loudspeaker pairs:

$$\begin{bmatrix} E_{L,\,ear} \\ E_{R,\,ear} \end{bmatrix} = \begin{bmatrix} H_{LL} & H_{LR} \\ H_{RL} & H_{RR} \end{bmatrix} \begin{bmatrix} E_L \\ E_R \end{bmatrix}. \tag{11.6.4}$$

Four HRTF-based filters are needed to implement the algorithm given in Equation (11.6.4) directly. The acoustic transfer matrix in Equation (11.6.4) is symmetric by taking advantage of left–right symmetry, with $H_{LL} = H_{RR} = H_\alpha$ and $H_{LR} = H_{RL} = H_\beta$. Using the diagonalizing procedure for a symmetrical matrix confirms that Equation (11.6.4) is equivalent to

$$\begin{bmatrix} E_{L,\,ear} \\ E_{R,\,ear} \end{bmatrix} = \begin{bmatrix} 0.707 & 0.707 \\ 0.707 & -0.707 \end{bmatrix} \begin{bmatrix} H_\alpha + H_\beta & 0 \\ 0 & H_\alpha - H_\beta \end{bmatrix} \begin{bmatrix} 0.707 & 0.707 \\ 0.707 & -0.707 \end{bmatrix} \begin{bmatrix} E_L \\ E_R \end{bmatrix}. \tag{11.6.5}$$

As shown in Equation (11.6.5), E_L and E_R signals are first mixed by a 2 × 2 MS (mid-side or sum-subtract) matrix. Then, the resultant signals 0.707 (E_L+ E_R) and 0.707 (E_L− E_R) were

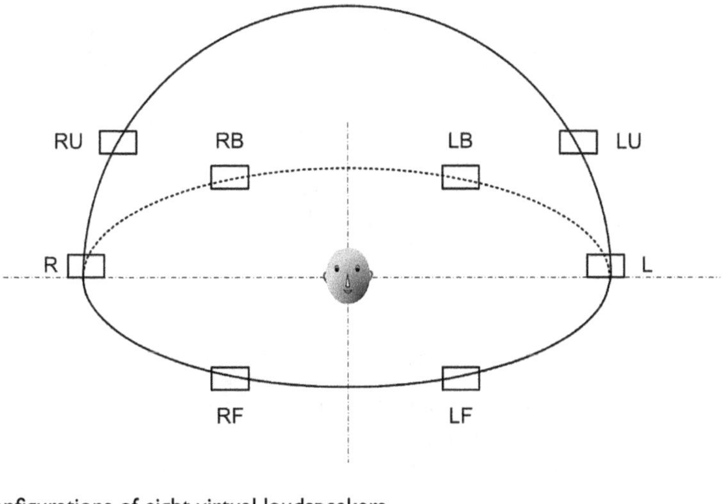

Figure 11.10 Configurations of eight virtual loudspeakers.

filtered by two filters $\Sigma = (H_\alpha + H_\beta)$ and $\Delta = (H_\alpha - H_\beta)$, yielding two signals of $0.707(H_\alpha + H_\beta)(E_L + E_R)$ and $0.707(H_\alpha - H_\beta)(E_L - E_R)$. Additional MS mixing is applied to form the binaural signals $E_{L,ear}$ and $E_{R,ear}$. Two filters are required for shuffler structure implementation in Equation (11.6.5). In this manner, synthesizing all four pairs of virtual loudspeakers in Figure 11.10 requires $2 \times 4 = 8$ filters, which is half the requirement for conventional implementation. Therefore, shuffler structure implementation is more efficient in simultaneously synthesizing four or more virtual sources. A virtual source localization experiment confirms the validity of the proposed algorithm.

In addition to pair-wise amplitude panning, other signal panning or mixing methods can be incorporated into the virtual loudspeaker-based algorithm. A virtual loudspeaker-based algorithm incorporated with Ambisonic signal mixing is termed ***binaural*** or ***virtual Ambisonics*** (Jot et al., 1998; Leitner et al., 2000; Noisternig et al., 2003). In binaural spatial Ambisonics, M virtual loudspeakers are arranged on the surface of a sphere with a far-field radius, according to the scheme in Section 9.3.2. The direction of the ith virtual loudspeaker is Ω_i, $i = 0, 1 \ldots (M - 1)$. The direction of a far-field target source is Ω_S, and the normalized amplitudes of virtual loudspeaker signals are expressed in Equation (9.3.27a). If multiple target virtual signals are synthesized simultaneously, the signals of the virtual loudspeaker are the sum of those for each virtual source. The synthesized binaural signals are obtained by substituting the resultant virtual loudspeaker signals into Equation (11.6.1). For the left–right symmetrical configuration of virtual loudspeakers, a shuffler structure implementation similar to Equation (11.6.5) simplifies the binaural Ambisonic signal processing. The near-field compensated higher-order Ambisonics in Section 9.3.4 can also be incorporated into a virtual loudspeaker-based algorithm to recreate a virtual source at various near-field distances in binaural reproduction (Menzies and Marwan, 2007). The advantage of this method is that far-field HRTFs at virtual loudspeaker positions are needed for synthesizing the virtual source at different directions and distances, and near-field HRTFs are not needed. Our experiment indicated that combined with the level cue, fifth-order dynamic binaural Ambisonics is able to recreate the auditory perception of a virtual source in various directions and distances closer than 1.0 m (Xie et al., 2021b).

The binaural Ambisonic method is also applied to convert the circular or spherical microphone array output for binaural reproduction. Ambisonic reproduction signals can be derived from the microphone array outputs according to Equations (9.8.7) and (9.8.14). Binaural signals are obtained by filtering Ambisonic reproduction signals with HRTFs in virtual loudspeaker directions (Duraiswami et al., 2005; Song et al., 2008). In comparison with binaural recording with an artificial head or human subject, the individualized characteristics of different subjects do not need to be considered in microphone array recording. Individualized binaural signals can be obtained using individualized HRTFs for signal conversion. Moreover, dynamic localization information is retained in microphone recording, which is addressed at the end of Section 11.10.2. Therefore, the microphone array is a flexible and universal recording method, as mentioned at the beginning of Section 9.8. However, a larger number of microphones and signal channels are needed in microphone array recording because of the Shannon–Nyquist spatial sampling theorem.

11.6.2 Basis function decomposition-based algorithms

The various basis function decompositions of HRTFs discussed in Sections 11.5.2 and 11.5.4 can also be used to simplify virtual source synthesis (Larcher et al., 2000). For the spatial basis function decomposition and spectral shape basis function decomposition, HRTFs for a specified ear and the arbitrary direction can be written in the form of Equation (11.5.28):

$$H(\Omega_S, f) = \sum_{q=0}^{Q} W_q(\Omega_S) D_q(f). \tag{11.6.6}$$

For spatial basis function decomposition, $W_q(\Omega_S)$ is the spatial basis function, and $D_q(f)$ is the frequency-dependent weight. For spectral shape basis function decomposition, $D_q(f)$ is spectral shape basis functions, and $W_q(\Omega_S)$ is the direction-dependent weight. Either $W_q(\Omega_S)$ or $D_q(f)$ is different for left or right HRTFs because HRTFs depend on the concerned ear.

According to Equation (11.6.6),Figure 11.11 illustrates the blocked diagram of the basis function decomposition-based algorithm (for the left ear only). The mono input stimulus $E_A(f)$ is weighted by $(Q + 1)$ directional functions $W_q(\Omega_S)$ and then filtered by a parallel bank of $(Q + 1)$ filters $D_q(f)$. The $(Q + 1)$ filters are invariable, and the virtual source direction is controlled by the $(Q + 1)$ directional functions or gains. Lastly, the outputs of all the $(Q + 1)$ filters are mixed to obtain the left ear signal.

In conventional binaural synthesis, each mono input signal $E_{A,u}(f)$ is filtered by a pair of $H(\Omega_u, f)$ (a pair of HRTFs at Ω_u) to synthesize multiple virtual sources in the U directions Ω_u with $u = 1, 2\ldots U$ simultaneously, and the outputs of all HRTF-based filters for each ear are mixed to obtain the overall signal for each ear.

$$E_{ear}(f) = \sum_{u=0}^{U-1} H(\Omega_u, f) E_{A,u}(f). \tag{11.6.7}$$

Decomposing each HRTF in Equation (11.6.7) according to Equation (11.6.6) and rearranging the equation yield

$$E_{ear}(f) = \left\{ \sum_{q=0}^{Q} D_q(f) \left[\sum_{u=0}^{U-1} W_q(\Omega_u) E_{A,u}(f) \right] \right\}. \tag{11.6.8}$$

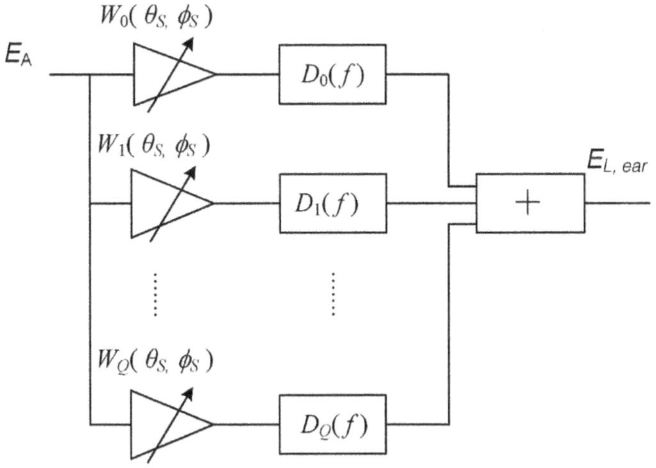

Figure 11.11 Block diagram of the basis function decomposition-based algorithm (for left ear only).

In Equation (11.6.8), the waveform signals of U target sources are mixed with frequency-independent weights $W_q(\Omega_u)$ and then filtered using $(Q + 1)$ parallel filters $D_q(f)$. The outputs of the $(Q + 1)$ filters are mixed to form the signal for a concerned ear. The filters are independent of the number and direction of the target virtual sources, and all the target virtual sources share a common set of $(Q + 1)$ filters. The direction of each target virtual source is controlled by source-direction-dependent weights or gains $W_q(\Omega_u)$.

Basis function decomposition-based algorithms can be regarded as encoding the spatial information of each target source into a set of non-individual "independent" signals $W_q(\Omega_u)E_{A,u}(f)$. Binaural signals are obtained by decoding the mix of independent signals for all target virtual sources by using a set of common filters $D_q(f)$. $D_q(f)$ is derived from the basis function decomposition of HRTFs (Jot et al., 1998). Individualized binaural signals are obtained if $D_q(f)$ is derived from a set of individualized HRTFs.

The basis function decomposition-based algorithm has two advantages similar to those of the virtual loudspeaker-based algorithm. That is, the computational efficiency for synthesizing multiple virtual sources is improved when the number U of virtual sources is larger than the number $(Q + 1)$ of the basis functions; the audible artifacts in creating a moving virtual source are avoided. Moreover, efficient decomposition methods enable HRTFs to be reconstructed with a small set of basis functions within a full audible bandwidth with sufficient accuracy. In other words, a basis function decomposition-based algorithm may enable the accurate and efficient synthesis of binaural signals with a small set of common filters. This feature is a prominent advantage of some basis function decomposition-based algorithms. By contrast, virtual loudspeaker-based algorithms in Section 11.6.1 are usually only able to synthesize binaural signals at low frequencies accurately and thus depend on the psychoacoustic principles of summing localization to create a virtual source unless a large number of virtual loudspeakers and their corresponding HRTF-based filters are used.

According to the discussion in Section 11.5.4, the PCA-based algorithm decomposes HRTFs by using a small set of spectral shape basis functions; thus, it is an effective means to simplify binaural synthesis. However, the PCA algorithm yields $W_q(\Omega_u)$ in discrete directions. An additional directional interpolation on $W_q(\Omega_u)$ is needed to synthesize virtual sources in arbitrary directions, and various directional interpolation schemes presented in Section 11.5 are applied to interpolate the weights. The PCA-based scheme can be extended to simplify binaural synthesis for near-field virtual sources at various distances. At the end of Section 11.5.4, the minimum-phase HRIRs at nine source distances and 493 directions at each distance can be decomposed by 15 PCs, along with a time domain mean function. Accordingly, binaural synthesis for near-field virtual sources can be implemented using $Q + 1 = 16$ common filters. This method improves the computational efficiency of dynamic binaural synthesis with multiple target sources (Section 11.10.2; Xie and Zhang, 2012). The spatial PCA algorithm at the end of Section 11.5.2 is also applied to simplify binaural synthesis with multiple target sources.

The spatial harmonic decomposition of HRTFs in Section 11.5.2, including azimuthal harmonic decomposition and spherical harmonic decomposition, can also be applied to binaural synthesis with multiple target sources. In fact, the spatial harmonic decomposition-based algorithm is closely related to binaural Ambisonics in Section 11.6.1 (Jot et al., 1998; Menzies and Marwan, 2007). Spatial harmonic decomposition yields directional continuous weights $W_q(\Omega_u)$ and thus can synthesize a virtual source in arbitrary directions. However, as indicated in Sections 11.5.2 and 11.5.3, the reconstruction of HRTFs up to 20 kHz requires a large number of spatial harmonics. Accordingly, a large number of common filters, $D_q(f)$, are required in binaural synthesis. Therefore, the efficiency of the spatial harmonic decomposition-based algorithm for binaural synthesis is low. For the lower-order binaural Ambisonics in Section 11.6.1, the spatial harmonic representation of the Ambisonic sound field is truncated to a lower order, but HRTFs for synthesizing virtual loudspeakers are intact. This procedure is analogous to Ambisonic reproduction with a few loudspeakers. By contrast, in the spatial harmonic decomposition-based algorithm, a truncation of spatial harmonics up to a lower order is equivalent to the simplification of the spatial harmonic representation of the Ambisonic sound field and that of HRTFs. The spatial harmonic decomposition-based algorithm is completely equivalent to binaural Ambisonics in Section 11.6.1 only in the case of a high order.

11.7 EQUALIZATION OF THE CHARACTERISTICS OF HEADPHONE-TO-EAR CANAL TRANSMISSION

11.7.1 Principle of headphone equalization

As stated in Section 11.1, the equalization of the characteristics of headphone-to-ear canal transmission is needed to ensure an accurate replication of target pressures at the eardrums in binaural reproduction. Mϕller (1992) comprehensively analyzed the characteristics of headphone-to-ear canal transmission and relevant equalization.

Figure 11.12(a) illustrates the anatomy of the human external ear, and Figure 11.12(b) shows the analog model for one-dimensional sound transmission through the ear canal (Mϕller, 1992). In the model, the complete sound field outside the ear canal is described by two variables from Thevenin's theorem: open-circuit pressure P_1 and generator impedance Z_1.Z_1 is the radiation impedance, as seen from the ear canal, into free air. P_1 does not physically exist in natural listening situations. However, if the ear canal is blocked to make the volume velocity zero (by analogy to electric current), P_1 can be measured just outside the blockage. The natural sound pressure P_2 at the open entrance to the ear canal is related to P_1 as follows:

$$\frac{P_2}{P_1} = \frac{Z_2}{Z_1 + Z_2}, \tag{11.7.1}$$

where Z_2 is the input impedance of the ear canal. Moreover, the sound pressure P_3 at the eardrum is related to P_1 and P_2:

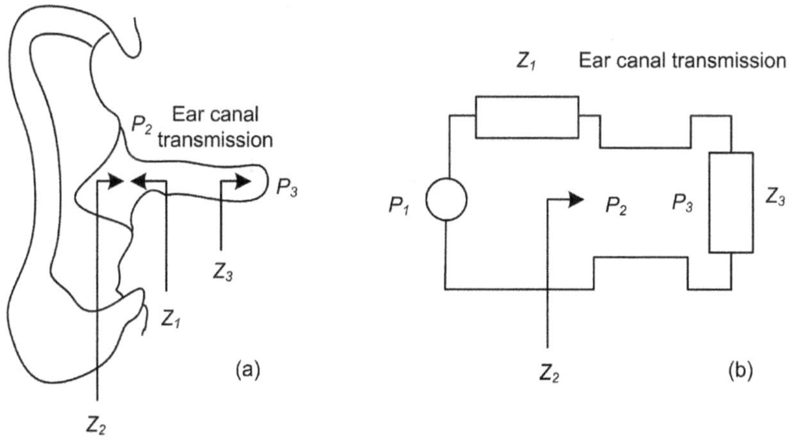

Figure 11.12 Models of sound transmission through the human external ear: (a) anatomical sketch and (b) analog model (adapted from Mϕller 1992).

$$\frac{P_3}{P_2} = \frac{Z_3}{Z_2} \qquad \frac{P_3}{P_1} = \frac{Z_3}{Z_1 + Z_2}. \tag{11.7.2}$$

where Z_3 is the eardrum impedance.

The entire sound transmission from a sound source to the eardrum is then divided into two consecutive parts: a direction-dependent part, describing the transmission from a sound source to the (blocked) ear canal entrance, and a direction-independent part, presenting the transmission along the ear canal to the eardrum. This finding suggests that sound pressures P_1, P_2, and P_3, or even those measured at other points in the ear canal, contain the same spatial information of the sound source. Therefore, they can all be used as reference points for binaural pressures and HRTF measurements. The entrance of the blocked ear canal is used as the reference point in the blocked ear canal technique for HRTF measurements.

Acoustical transmission processes, from a headphone to a listener's eardrum, are characterized by the headphone response and by the acoustic coupling between the headphone and the external ear. The corresponding analog model is shown in Figure 11.13. Similar to Figure 11.12, according to Thevenin's theorem, the acoustical characteristics in headphone reproduction are specified by the radiation impedance Z_4 and the "open-circuit" pressure P_4.Z_4 is the impedance observed outward at the entrance of the ear canal, and it includes a possible influence of volume enclosed by a circumaural headphone and by the contribution of the electrical, mechanical, and acoustic transfer characteristics of the headphone (all have been transferred to the acoustic side); P_4 is the "open-circuit" pressure at the entrance to the ear canal with zero volume velocity ("current"), which does not exist in actual cases but can be determined if the ear canal is physically blocked (Mϕller, 1992). The open-circuit pressure P_4 is related to the actual pressure P_5 at the entrance of the ear canal:

$$\frac{P_5}{P_4} = \frac{Z_2}{Z_2 + Z_4}, \tag{11.7.3}$$

where Z_2 is the impedance observed inward into the ear canal. The relationship of P_6 (the pressure at the eardrum) with P_4 and P_5 is expressed as

$$\frac{P_6}{P_5} = \frac{Z_3}{Z_2} \qquad \frac{P_6}{P_4} = \frac{Z_3}{Z_2 + Z_4}, \tag{11.7.4}$$

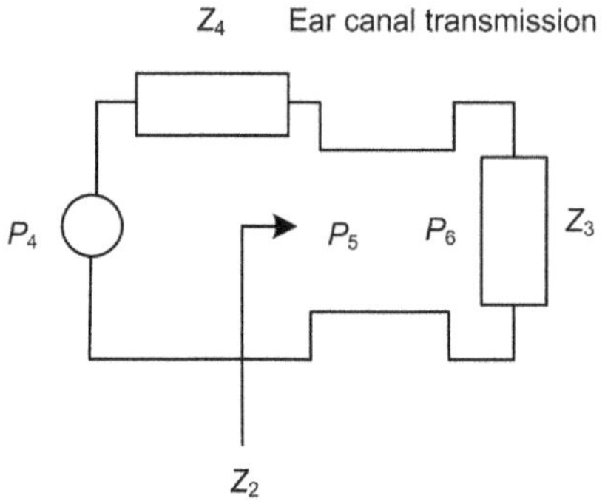

Figure 11.13 Analog model of headphone-external ear coupling and sound transmission in the ear canal (adapted from Mϕller, 1992).

where Z_3 is the impedance of the eardrum.

The binaural signal $E_{ear}(\xi) = E_\xi$ can be obtained by recording at a specific reference point ξ along the ear canal entrance to the eardrum or by filtering with the HRTFs obtained at the same reference point. It is proportional to the pressure P_ξ at the reference point generated by a sound source:

$$E_\xi = M_1 P_\xi, \tag{11.7.5}$$

where M_1 is the response of the recording microphone. P_ξ refers to P_1, P_2, and P_3 (Figure 11.12), denoting the entrance of the blocked ear canal, the entrance of the open ear canal, and the eardrum, respectively, which are chosen as the measurement reference points.

P_6 at the listener's eardrum is generated by the binaural signal E_x with the use of headphones. In the desired sound reproduction, this pressure should be identical to P_3 (which is caused by a real sound source) at the eardrum (Figure 11.12), that is,

$$P_6 = P_3. \tag{11.7.6}$$

Even when the anatomical difference between the listener and the head employed in binaural recording or synthesis is neglected, directly reproducing E_ξ by using headphones cannot guarantee Equation (11.7.6) because of the nonideal acoustical transmission characteristics from headphones to the eardrums and the nonideal response of the recording microphone. Before being sent to a headphone, E_ξ should be equalized. If $F = F(f)$ is the transfer function of the equalization filter, the actual signal sent to the headphone is

$$E = FE_\xi. \tag{11.7.7}$$

According to Equations (11.7.5), (11.7.6), and (11.7.7), P_6 at the listener's eardrum can be expressed as

$$P_6 = \frac{P_6}{E} E = \frac{P_6}{E} FM_1 P_\xi = \frac{P_6}{P'_\xi} \frac{P'_\xi}{E} FM_1 P_\xi, \tag{11.7.8}$$

where P'_ξ is the pressure at the reference point ξ in headphone reproduction. Ideally, P_6 in this equation should satisfy Equation (11.7.6):

$$\frac{P_6}{P'_\xi} \frac{P'_\xi}{E} FM_1 P_\xi = P_3 = \frac{P_3}{P_\xi} P_\xi. \tag{11.7.9}$$

When a point from the open ear canal to the eardrum is selected as the reference point ξ, we obtained $P_6/P'_\xi = P_3/P_\xi$ because of one-dimensional sound transmission in the ear canal (Figures 11.12 and 11.13). Then,

$$F = F(f) = \frac{1}{M_1 (P'_\xi / E)}. \tag{11.7.10}$$

The situation becomes complex when the entrance of the blocked ear canal is selected as the reference point. When $P_\xi = P_1$ in Figure 11.12 and $P'_\xi = P_4$ in Figure 11.13, substituting Equations (11.7.1) and (11.7.4) into Equation (11.7.9) yields

$$F = F(f) = \frac{Z_2 + Z_4}{Z_1 + Z_2} \frac{1}{M_1\left(P'_\xi/E\right)}. \tag{11.7.11}$$

Equation (11.7.11) involves the radiation impedance Z_1 observed outward from the entrance of the ear canal to the free air, the impedance Z_2 of the ear canal, and the radiation impedance Z_4 observed outward at the entrance of the ear canal in the case of headphone reproduction. Measuring these impedances is complicated and difficult. Fortunately, if the impedances satisfy either of these conditions: (1) $Z_1 << Z_2$, $Z_4 << Z_2$ and (2) $Z_1 \approx Z_4$. Equation (11.7.1) can be simplified. However, actual measurements reveal that condition (1) is invalid at some frequencies, whereas Z_1 is small below 1 kHz, satisfying $Z_1 << Z_2$. If the following condition is satisfied,

$$Z_1 \approx Z_4 \qquad Z_4 << Z_2 \quad (f < 1\ kHz), \tag{11.7.12}$$

then Equation (11.7.11) is simplified as

$$F = F(f) = \frac{1}{M_1\left(P'_\xi/E\right)} = \frac{1}{M_1\left(P_4/E\right)}. \tag{11.7.13}$$

The headphone that satisfies Equation (11.7.12) is called a headphone with ***free-air equivalent coupling to* the *ear* (*FEC*)**. Mϕller et al. (1995b) measured 14 headphones and indicated that all 14 headphones approximately satisfy the condition of the FEC up to 2 kHz. At above 2 kHz, all the headphones, except one, begin to deviate from the condition of the FEC. Measurements above 7 kHz are unreliable. In practical applications, whether a headphone can be considered an FEC-headphone depends on an acceptable error.

As indicated in Equations (11.7.10), (11.7.11), and (11.7.13), the characteristics of the headphone equalization filter are related to P'_ξ/E, which refers to the characteristics of transmission from the electric input signal of the headphone to the pressure at the reference point ξ in the ear canal. Hence, P'_ξ/E is called the ***headphone-to-ear-canal transfer function* (*HpTF*)** and is denoted by *Hp(f)*:

$$Hp(f) = \frac{P'_\xi}{E}. \tag{11.7.14}$$

Given that *Hp(f)* and the response M_1 of a microphone in binaural recording (or a microphone in HRTF measurements) are known, headphone equalization can be implemented according to Equations (11.7.10), (11.7.11), and (11.7.13). If the microphone has an ideal transmission response, M_1 in the aforementioned equations can be disregarded. In the practical measurement of *Hp(f)*, the same microphone as that employed for binaural recording or HRTF measurements can be used. In this case, substituting the microphone output $E'_\xi = M_1 P'_\xi$ into Equation (11.7.10) or Equation (11.7.13) yields

$$F = F(f) = \frac{1}{E'_\xi/E}. \tag{11.7.15}$$

Here, the final result is irrelevant to the transmission response of the microphone. Therefore, the influence of M_1 can be cancelled using either a microphone with an ideal transmission response or the same microphone used for binaural recording (or HRTF measurements). Then, Equations (11.7.10) and (11.7.13) become

$$F = F(f) = \frac{1}{Hp(f)}. \qquad (11.7.16)$$

HpTF is vital for headphone equalization. Numerous studies have measured the HpTFs (Mϕller et al., 1995b; Pralong and Carlile, 1996; Kulkarni and Colburn, 2000; Rao and Xie, 2006). Results vary across different studies. Some underlying principles can be determined by comparing headphone structures. A circumaural headphone with a larger volume of the cavity causes less compressive deformation of the pinnae; therefore, HpTFs among different headphone placements slightly differ. This type of headphone also exhibits reasonable repeatability in individualized HpTF measurements. A supra-aural headphone or a circumaural headphone with a smaller cavity volume makes the pinnae prone to compression deformation. The deformation varies with each measurement. In such a situation, the measured HpTF differs with headphone placement, especially at high Q-value spectral peaks and notches. An in-ear headphone is inserted into the ear canal to form a one-dimensional transmission system, and the influence of the compressive deformation of the pinnae is avoided. This problem should be noted in the choice of headphones.

Moreover, because of individual differences in the external ear, the HpTFs of circumaural headphones differ among individuals (Pralong and Carlile, 1996). Individualized differences in HpTFs mainly occur at frequencies above 6 kHz, similar to that of the spectral features of individualized HRTFs (Section 11.3.2). In fact, with circumaural headphones, the ear is entirely surrounded by a cushion. Therefore, the response of this type of headphone captures many of the external ear filtering effects similar to those in HRTFs is not surprising. Ideally, individualized HpTFs should be employed in headphone equalization to replicate binaural signals at the eardrum accurately. Otherwise, nonindividualized HpTF equalization likely impairs the localization cues in individualized HRTFs.

11.7.2 Some problems with binaural reproduction and VAD

Binaural reproduction can theoretically replicate the pressures at eardrums caused by a target source and thus recreate a perceived virtual source in a three-dimensional space. However, numerous experimental results indicate that subject-dependent directional errors or distortions generally exist. Examples of such errors include the following:

1. ***Reversal error*** (***front-back*** or ***back-front confusion***) A virtual source intended for the front hemisphere is perceived at a mirror position in the rear hemisphere or less frequently the reverse. In some instances, confusion arises regarding up- and down-source positions. Such confusion is termed ***up-down*** or ***down-up confusion***.
2. For ***elevation errors***, the angle of a virtual source in the front median plane is typically elevated to higher positions.
3. ***In in-the-head localization*** or ***intracranial lateralization*** during headphone presentation, perceived virtual sources or auditory events are often located on the surface of the head or even inside the head rather than outside the head although the cases of inside the head localization do not occur as frequently as those in conventional stereophonic or multichannel sound reproduction over headphones. Lateralization often exists in frontal target sources and leads to an unnatural hearing experience.

In Section 1.6, binaural cues (ITD and ILD) cannot determine the source direction completely, and they identify the cone of confusion where the source is located. The high-frequency spectral cue caused by the pinna and dynamic cue caused by head turning contributes greatly to front–back and vertical localization. In static binaural reproduction, the dynamic cue is omitted. In this case, high-frequency spectral cues are important. However, high-frequency spectral cues vary among individuals. Recording with an unmatched artificial head or binaural synthesis with unmatched HRTFs leads to incorrect high-frequency spectral cues in reproduction (Wightman and Kistler, 1989b; Mϕller et al., 1996). For example, Wenzel et al. (1993) showed that the front-back and up-down confusion rates of headphone reproduction, where representative (nonindividualized) HRTFs are used, increase from 19% to 31% and 6% to 18% compared with the those in free-field real source localization, respectively. Moreover, some other studies have indicated that the absence of headphone equalization, headphone equalization with nonindividualized HpTFs, and some types of headphones (e.g., supra-aural headphones) may impair the high-frequency spectral cue in reproduction (Pralong and Carlile, 1996). Errors at each stage of HRTF and HpTF measurements and signal processing cause similar problems. All of the above factors are possible reasons for the directional distortion of virtual sources in static binaural reproduction.

The errors introduced in binaural signal recording/synthesis and reproduction stages, including individualized (or matched) binaural recording, HRTF-based synthesis, and HpTF equalization, should be minimized to eliminate or reduce the reversal and elevation errors of the perceived virtual sources in headphone reproduction. However, ensuring accuracy at each stage is difficult because high-frequency spectral information is sensitive to various errors. In practice, all methods can improve the performance of static binaural reproduction partly rather than completely. A final and efficient method is to use dynamic binaural synthesis and reproduction. Introducing dynamic localization cues in binaural reproduction alleviates dependence on spectral cues for front–back and vertical localization because of redundancy in localization cues.

Many researchers agreed that lateralization is caused by incorrect spatial information on both ears during sound reproduction (Plenge, 1972, 1974). Binaural pressures and spatial information errors originate from numerous sources. Similar to the case of reversal and elevation errors, errors introduced in binaural signal recording/synthesis and reproduction stages may cause lateralization in binaural reproduction (Durlach et al., 1992). Therefore, an accurate replication of pressure at the eardrum is important for ***externalization*** (Hartmann and Wittenberg, 1996). The absence of dynamic cues is also an important reason for lateralization in static binaural reproduction (Loomis et al., 1990; Durlach et al., 1992; Wenzel, 1996; Zhang and Xie, 2013).

Many studies have emphasized that environmental reflection is essential for externalization (Durlach et al., 1992). In addition to direct sound, reflection is crucial for spatial hearing. As mentioned in Sections 1.6.6, environmental reflection is the key to distance perception. Free-field HRTFs (or HRIRs) are used in the preceding discussion on binaural synthesis; therefore, the resultant binaural signals only contain direct sound without environmental reflection. HRIRs can be replaced by binaural room impulse responses (BRIRs, Section 1.8.3) in binaural synthesis to eliminate the effect of lateralization. BRIRs can be obtained not only from measurements through an artificial head or a human subject but also from binaural room acoustic modeling or artificial reverberation algorithms (Section 7.5).Binaural signals with environmental reflections can also be directly recorded using an artificial head or a human subject. The results of the psychoacoustic experiment indicate that binaural synthesis with several preceding-order early reflections is sufficient to externalize auditory events in reproduction (Begault et al., 2001).

Recreating a virtual source or auditory event at various perceived distances is another aim of a VAD. The control of perceived distance is relatively difficult compared with the control of perceived virtual source directions because of multiple reasons.

1. The control of perceived distances should be based on a complete externalization in binaural reproduction.
2. Auditory distance perception is biased even for an actual sound source (Section 1.6.6).
3. Auditory distance perception depends on the combination of multiple cues, and the mechanisms behind auditory distance perception are less well understood than those underlying direction localization.

The control of perceived virtual source distances in VAD has been investigated by some researchers. Controlling environmental reflection is a common and effective way to manipulate the perceived distance of virtual sources, especially under a direct-to-reverberant energy ratio. This control is realized through artificial reflection. Reverberation algorithms are often used in practical cases to allow for certain errors.

The discussion in Section 11.3.3 indicates that the distance-dependent near-field ILD is a cue for absolute distance perception for a lateral sound source with a distance r_S of less than 1.0 m. As such, some studies have used near-field HRTFs to synthesize virtual sound sources at various distances (Brungart, 1999, Brungart and Rabinowitz, 1999, Brungart et al., 1999). This method is valid only within 1 m, and its accuracy diminishes when the virtual source approaches the median plane with a small ILD value.

11.8 BINAURAL REPRODUCTION THROUGH LOUDSPEAKERS

11.8.1 Basic principle of binaural reproduction through loudspeakers

Binaural signals from either binaural recordings or syntheses are originally intended for headphone presentation. When binaural signals are reproduced through a pair of left and right loudspeakers arranged in front of a listener, an unwanted ***crosstalk* occurs** from each loudspeaker to the opposite ear. Crosstalk impairs the directional information encoded in binaural signals. Prior to loudspeaker reproduction, binaural signals should undergo crosstalk cancellation to pre-cancel transmission from each loudspeaker to its opposite ear (Schroeder and Atal, 1963).

Let $E_{L,ear} = E_{L,ear}(f)$ and $E_{R,ear} = E_{R,ear}(f)$ denote the binaural signals in the frequency domain, and $E'_L = E'_L(f)$ and $E'_R = E'_R(f)$ represent the left and right loudspeaker signals, respectively. In Figure 11.14, binaural signals are pre-filtered by a 2 × 2 crosstalk cancellation matrix (crosstalk canceller) and then reproduced through loudspeakers. Loudspeaker signals are expressed as

$$\begin{bmatrix} E'_L \\ E'_R \end{bmatrix} = \begin{bmatrix} A_{11} & A_{12} \\ A_{21} & A_{22} \end{bmatrix} \begin{bmatrix} E_{L,ear} \\ E_{R,ear} \end{bmatrix}, \qquad (11.8.1)$$

where A_{11}, A_{12}, A_{21}, and A_{22} are the four transfer functions or filters that form the crosstalk cancellation matrix.

If H_{LL}, H_{RL}, H_{LR}, and H_{RR} denote the four acoustic transfer functions (HRTFs) from two loudspeakers to the two ears, these four transfer functions are determined by loudspeaker

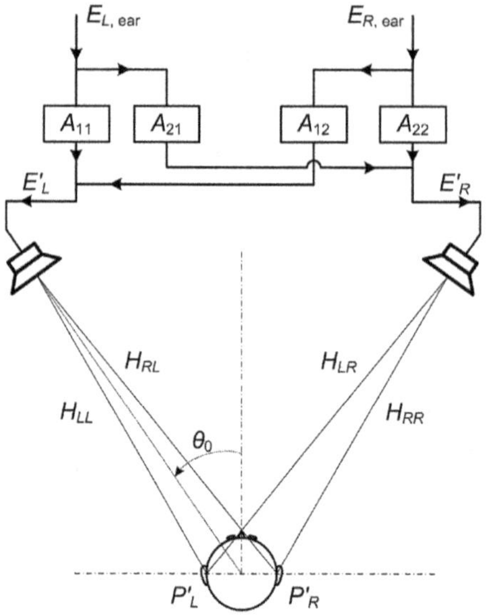

Figure 11.14 Binaural reproduction through a pair of frontal loudspeakers with crosstalk cancellation.

configuration and listener location. According to Equation (11.8.1), the reproduced pressures at the two ears are expressed as follows:

$$\begin{bmatrix} P'_L \\ P'_R \end{bmatrix} = \begin{bmatrix} H_{LL} & H_{LR} \\ H_{RL} & H_{RR} \end{bmatrix} \begin{bmatrix} E'_L \\ E'_R \end{bmatrix} = \begin{bmatrix} H_{LL} & H_{LR} \\ H_{RL} & H_{RR} \end{bmatrix} \begin{bmatrix} A_{11} & A_{12} \\ A_{21} & A_{22} \end{bmatrix} \begin{bmatrix} E_{L,ear} \\ E_{R,ear} \end{bmatrix}. \tag{11.8.2}$$

The transfer characteristics of the crosstalk cancellation matrix are appropriately selected such that the product of two 2 × 2 matrices in Equation (11.8.2) is equal to an identity matrix.

$$P'_L = E_{L,ear} \qquad P'_R = E_{R,ear}. \tag{11.8.3}$$

In this case, crosstalk is completely cancelled out, and the desired binaural signals are precisely transmitted to the listener's two ears.

If the acoustic transfer matrix $[H]$ in Equation (11.8.2) is nonsingular and therefore invertible, the crosstalk cancellation matrix is calculated as follows:

$$\begin{bmatrix} A_{11} & A_{12} \\ A_{21} & A_{22} \end{bmatrix} = \begin{bmatrix} H_{LL} & H_{LR} \\ H_{RL} & H_{RR} \end{bmatrix}^{-1} = \frac{1}{H_{LL}H_{RR} - H_{LR}H_{RL}} \begin{bmatrix} H_{RR} & -H_{LR} \\ -H_{RL} & H_{LL} \end{bmatrix}. \tag{11.8.4}$$

Equation (11.8.4) depicts the basic principle of crosstalk cancellation for loudspeaker reproduction.

If binaural signals are obtained through synthesis according to Equation (11.1.1). Thus,

$$E_{L,ear}(f) = H_L(\theta_S, f)E_A(f) \qquad E_{R,ear}(f) = H_R(\theta_S, f)E_A(f). \tag{11.8.5}$$

The source distance dependence r_S is excluded, and the target virtual source direction is simply denoted by azimuth θ_S in Equation (11.8.5). Substituting Equations (11.8.5) and (11.8.4) into Equation (11.8.1) yields

$$\begin{bmatrix} E_L' \\ E_R' \end{bmatrix} = \frac{1}{H_{LL}H_{RR} - H_{LR}H_{RL}} \begin{bmatrix} H_{RR} & -H_{LR} \\ -H_{RL} & H_{LL} \end{bmatrix} \begin{bmatrix} H_L(\theta_S, f) \\ H_R(\theta_S, f) \end{bmatrix} E_A(f). \tag{11.8.6}$$

Signal processing in Equation (11.8.6) involves two stages. The monophonic (mono) input stimulus $E_A(f)$ is initially filtered by a pair of HRTFs in the target source direction according to Equation (11.1.1). Then, the resultant binaural signals $E_{L,ear}$ and $E_{R,ear}$ are pre-filtered by a 2 × 2 crosstalk cancellation matrix to obtain signals E'_L and E'_R.

In most practical cases, the loudspeaker configuration is left–right symmetric with respect to listeners. If HRTFs are assumed to be left–right symmetric, the transfer functions from each loudspeaker to the ipsilateral and contralateral ears satisfy $H_{LL} = H_{RR} = H_\alpha$ and $H_{LR} = H_{RL} = H_\beta$, respectively. Then, the crosstalk cancellation matrix in Equation (11.8.4) is simplified as a symmetric solution:

$$A_{11} = A_{22} = \frac{H_\alpha}{H_\alpha^2 - H_\beta^2} \qquad A_{12} = A_{21} = \frac{-H_\beta}{H_\alpha^2 - H_\beta^2}. \tag{11.8.7}$$

Equation (11.8.6) is simplified as

$$\begin{bmatrix} E_L' \\ E_R' \end{bmatrix} = \frac{1}{H_\alpha^2 - H_\beta^2} \begin{bmatrix} H_\alpha & -H_\beta \\ -H_\beta & H_\alpha \end{bmatrix} \begin{bmatrix} H_L(\theta_S, f) \\ H_R(\theta_S, f) \end{bmatrix} E_A(f). \tag{11.8.8}$$

If signal processing is initially designed to create appropriate loudspeaker signals, the two stages of binaural synthesis and crosstalk cancellation can be merged. The matrix product in Equation (11.8.6) yields

$$E_L' = A_L(\theta_S, f) E_A(f) \qquad E_R' = A_R(\theta_S, f) E_A(f), \tag{11.8.9}$$

where

$$\begin{aligned} A_L(\theta_S, f) &= \frac{H_{RR}H_L(\theta_S, f) - H_{LR}H_R(\theta_S, f)}{H_{LL}H_{RR} - H_{LR}H_{RL}} \\ A_R(\theta_S, f) &= \frac{-H_{RL}H_L(\theta_S, f) + H_{LL}H_R(\theta_S, f)}{H_{LL}H_{RR} - H_{LR}H_{RL}}. \end{aligned} \tag{11.8.10}$$

In the symmetric case, we have

$$\begin{aligned} A_L(\theta_S, f) &= \frac{H_\alpha H_L(\theta_S, f) - H_\beta H_R(\theta_S, f)}{H_\alpha^2 - H_\beta^2} \\ A_R(\theta_S, f) &= \frac{-H_\beta H_L(\theta_S, f) + H_\alpha H_R(\theta_S, f)}{H_\alpha^2 - H_\beta^2}. \end{aligned} \tag{11.8.11}$$

Equation (11.8.9) demonstrates that the loudspeaker signals E'_L and E'_R for the target virtual sources in the direction θ_S can be directly synthesized by filtering a mono stimulus $E_A(f)$ with a pair of filters $A_L(\theta_S, f)$ and $A_R(\theta_S, f)$. This condition is the basic principle of binaural virtual source synthesis of sound reproduction through a pair of loudspeakers. Similar to conventional stereophonic sound, Equation (11.8.9) can be regarded as a special signal mixing method for the reproduction with two loudspeakers, which is called a ***binaural pan-pot*** in some instances. In addition, any factor common to HRTFs is cancelled out in the final responses $A_L(\theta_S, f)$ and $A_R(\theta_S, f)$ of a transaural synthesis filter because all the items in the numerator and denominator of Equation (11.8.10) contain a product of two HRTFs. In other words, $A_L(\theta_S, f)$ and $A_R(\theta_S, f)$ are independent of the reference point of the HRTFs used.

Equation (11.8.9) can also be derived using an alternative method. The frequency-domain binaural pressures caused by a source in the θ_S direction are given as

$$P_L = H_L(\theta_S, f)P_{free}(f) \qquad P_R = H_R(\theta_S, f)P_{free}(f). \tag{11.8.12}$$

If E'_L and E'_R are loudspeaker signals, then binaural pressures in reproduction are

$$\begin{bmatrix} P'_L \\ P'_R \end{bmatrix} = \begin{bmatrix} H_{LL} & H_{LR} \\ H_{RL} & H_{RR} \end{bmatrix} \begin{bmatrix} E'_L \\ E'_R \end{bmatrix}. \tag{11.8.13}$$

If signals E'_L and E'_R are appropriately chosen to guarantee that binaural pressures in reproduction are identical or directly proportional to those caused by a real source in the θ_S direction, a perceived virtual source at θ_S is then created. Accordingly, Equations (11.8.9) and (11.8.10) can be directly derived by letting Equation (11.8.13) be equal to Equation (11.8.12): $P_L = P'_L$, and $P_R = P'_R$.

Through the control of loudspeaker signals, the second method of derivation can be interpreted as a technique for controlling binaural pressures and synthesized virtual sources (Sakamoto et al., 1981, 1982); which is called ***transaural technique*** or ***transaural stereo*** (Cooper and Bauck, 1989; Bauck and Cooper, 1996). Correspondingly, the scheme of loudspeaker signal synthesis is called ***transaural synthesis***, and the filters specified by A_L and A_R in Equation (11.8.10) are called ***transaural (synthesis) filters***.

Crosstalk cancellation and transaural synthesis can be generalized to cases involving multiple loudspeakers and listeners (Bauck and Cooper, 1996). Therefore, M loudspeakers can accurately control the reproduced sound pressures at M receiver positions, allowing $M/2$ listeners to have an even number M at the same time.

The practical acoustic transfer matrix $[H]$ in Equation (11.8.4) may be singular and thus noninvertible at some frequencies. Some methods for solving the crosstalk cancellation matrix have been developed to address this problem, such as regularization method (Kirkeby and Nelson, 1999).

In practice, crosstalk cancellation and transaural synthesis can be implemented using various signal-processing structures (Mϕller, 1992; Iwahara and Mori, 1978). In the left–right symmetric case, a shuffler structure similar to Equation (11.6.5) simplifies the signal processing of crosstalk cancellation (Bauck and Cooper, 1996). Considering that transaural reproduction fails to provide stable spectral cues at high frequencies in most practical cases, Gardner (1997) proposed a bandlimited implementation of crosstalk cancellation below the frequency of 6 kHz. Causality should be considered in the implementation of the signal processing of crosstalk cancellation and transaural synthesis. The details of signal processing are referred to another book (Xie, 2008a, 2013).

In comparison with conventional stereophonic and multichannel sound, a remarkable advantage of transaural reproduction is that it is simple; only two independent signals and loudspeakers are sufficient to recreate the spatial information of sound within a certain directional region. However, transaural reproduction exhibits intrinsic problems related to the physical and psychoacoustic principles of transaural reproduction. These problems can be alleviated by appropriate measures, but they usually cannot be completely deleted. These problems are addressed in the succeeding sections.

11.8.2 Virtual source distribution in two-front loudspeaker reproduction

Binaural pressures or signals contain primary localization information. Provided that binaural signals are precisely replicated, headphone-based binaural reproduction can recreate perceived virtual sources at arbitrary horizontal, or even three-dimensional, directions. As discussed in Section 11.8.1, loudspeaker-based binaural reproduction by which crosstalk cancellation is incorporated can also generate the same binaural pressures as those produced by a real source. Theoretically, this reproduction technique can also recreate the perceived virtual sources in arbitrary horizontal or three-dimensional directions. In practical situations, however, such recreation is not implemented. Some authors reported that a pair of frontal loudspeakers can recreate perceived virtual sources in all horizontal or three-dimensional directions for listeners under a series of critical conditions (e.g., individualized HRTF processing, restriction of head movement, reproduction in anechoic rooms, etc.; Takeuchi et al., 1998). The experimental results indicate that the perceived virtual source positions are restricted to the region of frontal-horizontal quadrants under less critical conditions of reproduction (Nelson et al., 1996; Gardner, 1997). The virtual sources intended for rear-horizontal quadrants are often perceived at the mirror position in the frontal-horizontal quadrants. The virtual sources intended outside the horizontal plane are often perceived in frontal-horizontal quadrants in the same cone of confusion (Section 1.6.3).

In Sections 1.6 and 11.7.2, the dynamic cue caused by head turning and the spectral cues introduced by pinna reflection and diffraction are essential for front–back and vertical localization. The pinna effect is effective only for high-frequency sound waves with wavelengths comparable with the pinna dimension. HRTFs are approximately front–back symmetric below 1 kHz. Therefore, the spectral cue mainly works in the high-frequency range above 5–6 kHz. In addition, spectral cues are highly individual dependent. Even if pinna-induced spectral cues are carefully manipulated in loudspeaker reproduction, such cues work within a very limited listening region because high frequencies have short wavelengths. A distance deviation of 1/4 to 1/2 a wavelength from the optimal (default) listening position, or "sweet point," likely causes a radical change in binaural pressure. Therefore, high-frequency spectral cues caused by pinnae cannot be stably replicated in loudspeaker reproduction. On the other hand, the dynamic cue of the target source has been disregarded in conventional static binaural reproduction through loudspeakers. If a slight head rotation is allowed in reproduction with a pair of frontal loudspeakers, the virtual source intended for rear-horizontal quadrants can be perceived at the mirror azimuth in frontal-horizontal quadrants with the incorrect dynamic cue provided by the front loudspeakers, and the perceived azimuth is determined by binaural cues (especially ITD) in reproduction. The above analysis has been validated through an experiment involving a dynamic VAD (Liu and Xie, 2021).

Therefore, static transaural reproduction with two (or three) frontal loudspeakers cannot recreate stable perceived virtual sources in rear-horizontal quadrants or full three-dimensional directions, and it can recreate stable virtual sources in the frontal-horizontal quadrants only unless under critical reproduction conditions. This phenomenon is a defect in static transaural reproduction with two front loudspeakers. Adding a pair of rear loudspeakers can recreate a virtual source in the entire horizontal plane, but the loudspeaker configuration becomes complicated.

The virtual sources in the horizontal-lateral directions are unstable for static transaural reproduction with two frontal loudspeakers. The perceived azimuth of the lateral virtual source tends to move toward the front. For example, the target virtual source at $\theta_S = 90°$ is often perceived at $\theta_I = 70°$. The possible reasons for this instability include the following:

1. A mismatch occurs between the HRTFs (head radius) of binaural synthesis and those of practical listeners (Xie, 2002b).
2. The rotation of the listener's head around the vertical axis during reproduction causes serious directional distortion of the lateral virtual source (Xie, 2005).
3. A mismatch exists between the characteristics of the loudspeaker pair (Chi et al., 2009).

Such defects can be reduced by appropriate measures (Xie, 2008a, 2013).

11.8.3 Head movement and stability of virtual sources in Transaural reproduction

The performance of crosstalk cancellation and transaural reproduction depends greatly on the position of a listener with respect to loudspeakers. Theoretically, a given crosstalk cancellation and transaural processing is only effective for a specified listening position and head orientation. When the listener's head deviates from the specified listening position or rotates to another orientation, the condition of crosstalk cancellation is destroyed, and binaural pressures are changed. This observation is a general defect related to the physical principle of transaural reproduction, which can be partly alleviated rather than completely deleted by appropriate methods.

A listener's head can move in six degrees of freedom. For simplicity, the movement of the head is limited to cases of rotation around the vertical axis and translation in the horizontal plane. After head rotation, the directions of the loudspeakers with respect to the head change. Consequently, the superposed pressures at the two ears and the related localization cues change. Hill et al. (2000) analyzed the variations in ITD with head rotation in two frontal loudspeaker reproductions. The results indicate that variations in ITD exhibit a similar pattern to that of the real source in front quadrants and thus provide correct dynamic localization cues only for target virtual sources in front quadrants. However, it provides inconsistent or conflicting dynamic cues for target virtual sources at the rear quadrants. In fact, the ITD also varies with head rotation even for a real source. Xie (2005) further analyzed the stability of the perceived virtual source azimuth with head rotation in reproduction with two frontal loudspeakers. The results indicated that head rotation around the vertical axis mainly causes the perceived azimuthal distortion of the lateral virtual source.

Numerous studies have been performed on the stability of crosstalk cancellation and transaural reproduction against head translation, which is a hot topic in VAD, and details are referred to the book (Xie, 2008a, 2013). In the following, this stability is analyzed using a simple model (Xie et al., 2005c).

A pair of loudspeakers with a span angle of $2\theta_0$ is arranged at azimuths $\theta_L = \theta_0$ and $\theta_R = -\theta_0$ (Figure 2.21). If a listener is located at the default position (central line), the distances between the two loudspeakers relative to the head center (origin of coordinates) are identical. When the head center laterally deviates from the default position to the left and is located at the $(0, y_1)$ coordinate, the distances of the two loudspeakers relative to the head center become unequal with a difference in path length or arrival time:

$$\Delta t = \frac{r'_R - r'_L}{c}, \quad (11.8.14)$$

where c is the speed of sound. This arrival time difference is equivalent to an additional phase difference between loudspeaker signals:

$$\eta = 2\pi f \, \Delta t \approx \frac{4\pi f \, y_1 \sin\theta_0}{c}, \quad (11.8.15)$$

where $y_1 << r$ is assumed for the second approximate equality. When the head translation exceeds a certain distance such that $|\eta| \geq \pi/2$, that is, when the difference in the path length exceeds a quarter of a wavelength, the localization information encoded in the relative phase between the left and right loudspeaker signals is destroyed. Accordingly, the size of the listening region can be estimated by

$$|y_1| \leq y_{\max} = \frac{c}{8f \sin\theta_0}, \quad (11.8.16)$$

where y_{max}, which is inversely proportional to the frequency, is regarded as the limited distance for tolerable lateral head translation. As a result, the listening region is small at high frequencies. y_{max} is inversely proportional to $\sin\theta_0$. Thus, a narrow span angle between the left and right loudspeaker pairs reduces the variation in binaural pressures with head translation, thereby improving the stability of the virtual source and expanding the listening region. For example, y_{max} of the loudspeaker configuration with a 30° span angle is approximately twice that of a conventional loudspeaker configuration with a 60° span angle. At a frequency of 1.5 kHz, y_{max}= 0.11 m for a 30° span angle and y_{max}= 0.06 m for a 60° span angle.

To expand the listening region, Kirkeby et al. (1998a,1998b) proposed a pair of frontal loudspeakers at $\theta_L = 5°$ and $\theta_R = -5°$ with a span of $2\theta_0 = 10°$ for transaural reproduction. This concept is called a ***stereo dipole***. The crosstalk cancellation processing for a stereo dipole is identical to that described in Section 11.8.1. However, the difference between the HRTFs from a pair of loudspeakers with a narrow span angle to the ipsilateral and contralateral ears is small at low frequencies such that the acoustical transfer matrix in Equation (11.8.4) is close to the ill condition. Accordingly, a considerable increase in the low-frequency responses of transaural filters is required for lateral virtual sources.

For a lateral target virtual source at $\theta_S = 90°$, the two signals for a stereo dipole given in Equation (11.8.9) are out of phase. This phenomenon is similar to the case of an outside-boundary virtual source caused by out-of-phase loudspeaker signals in conventional stereophonic sound. In addition, a closely spaced loudspeaker pair with out-of-phase signals serves as an acoustical dipole, which is the origin of the term "stereo dipole." The close space between the two loudspeakers results in a slight difference in path lengths from the two loudspeakers to either of the ears. As a result, the out-of-phase loudspeaker signals, small differences in path length, and large wavelengths generate small summing low-frequency pressures in either ear. Thus, reasonable ear pressures at low frequencies can only be generated by considerably increasing the low-frequency component of the input stimulus.

Therefore, a stereo dipole improves the stability of virtual sources against lateral head translation and expands the listening region but at the cost of difficult signal processing. In practice, all these aspects should be comprehensively considered in choosing the span of loudspeakers.

Some studies have argued that a wide-span configuration enables large and stable channel separation within a wide frequency range because of the natural high-frequency crosstalk attenuation caused by the head shadow effect on contralateral loudspeakers (Bai and Lee, 2006). This phenomenon is mentioned in Section 7.2.4, where artificial head-recording signals are reproduced by a pair of surround loudspeakers in a 5.1-channel configuration. However, the performance of transaural reproduction is not uniquely determined by channel separation. Lateral head translation also causes a difference in the path between the left and right loudspeakers to the head center, especially for wide-span angle configurations. In turn, this difference introduces a variation in the relative arrival time between loudspeaker signals.

The optimal source distribution reproduction is also available (Takeuchi and Nelson, 2002) by which several loudspeaker pairs with different spans can be used. Each loudspeaker pair reproduces signals at specific frequency ranges to stabilize the crosstalk cancellation in a full audible bandwidth range.

11.8.4 Timbre coloration and equalization in transaural reproduction

Ideally, complete crosstalk cancellation yields the same binaural pressures as those of a real source. Nevertheless, the analysis in Sections 11.8.2 and 11.8.3 indicates that completely cancelling out crosstalk is difficult within a full audible frequency range. In practice, factors such as slight head translation or rotation, unmatched HRTFs, and room reflection inevitably lead to incomplete crosstalk cancellation. Thus, binaural pressures in reproduction deviate from those of a real source. The magnitude responses of the transaural synthesis filters $A_L(\theta_S, f)$ and $A_R(\theta_S, f)$ specified by Equations (11.8.9) and (11.8.10) are visibly frequency dependent. This dependence modifies the overall power spectra of the input signal $E_A(f)$, thereby altering the balance among the spectral components of the resultant loudspeaker signals. Consequently, the incomplete crosstalk cancellation caused by perturbations leads to virtual source localization errors and perceived timbre coloration, especially at high frequencies and for off-center listeners. These defects commonly occur in binaural reproduction through loudspeakers, necessitating additional timbre equalization.

The principle of timbre equalization in two-loudspeaker reproduction is explained as follows. Given the difficulty in robustly rendering fine high-frequency spectral cues to listener's ears in loudspeaker reproduction, the perceived virtual source direction is dominated by interaural cues (especially ITD) and limited to frontal-horizontal quadrants. The interaural cues are controlled by the relative, rather than the absolute, magnitude, and phase of the left and right loudspeaker signals. Scaling both loudspeaker signals with identical frequency-dependent coefficients does not alter their relative magnitude and phase or the perceived virtual source direction. However, this manipulation alters the overall power spectra of loudspeaker signals and therefore equalizes the timbre.

Timbre equalization has various algorithms. In the ***constant-power equalization algorithm*** for the left–right symmetrical case (Xie et al., 2005c, 2013b), the responses of the transaural synthesis filters $A_L(\theta_S, f)$ and $A_R(\theta_S, f)$ in Equations (11.8.9) and (11.8.11) are equalized by their root mean square (RMS):

$$E'_L = A'_L(\theta_S, f)E_A(f) \qquad E'_R = A'_R(\theta_S, f)E_A(f), \tag{11.8.17}$$

$$
\begin{aligned}
A'_L(\theta_S, f) &= \frac{A_L(\theta_S, f)}{\sqrt{|A_L(\theta_S, f)|^2 + |A_R(\theta_S, f)|^2}} \\
&= \frac{H_\alpha H_L - H_\beta H_R}{\sqrt{|H_\alpha H_L - H_\beta H_R|^2 + |-H_\beta H_L + H_\alpha H_R|^2}} \frac{|H_\alpha^2 - H_\beta^2|}{H_\alpha^2 - H_\beta^2}, \\
A'_R(\theta_S, f) &= \frac{A_R(\theta_S, f)}{\sqrt{|A_L(\theta_S, f)|^2 + |A_R(\theta_S, f)|^2}} \\
&= \frac{-H_\beta H_L + H_\alpha H_R}{\sqrt{|H_\alpha H_L - H_\beta H_R|^2 + |-H_\beta H_L + H_\alpha H_R|^2}} \frac{|H_\alpha^2 - H_\beta^2|}{H_\alpha^2 - H_\beta^2}.
\end{aligned}
\tag{11.8.18}
$$

The loudspeaker signals in Equation (11.8.18) satisfy the constant-power spectral relationship:

$$
|E'_L|^2 + |E'_R|^2 = E_A^2(f). \tag{11.8.19}
$$

Therefore, the overall power spectra of loudspeaker signals are equal to those of the input stimulus, reducing reproduction coloration.

He et al. (2006) indicated that timbre in transaural reproduction depends on the distribution of poles and zeros of transaural filters. The constant-power equalization algorithm cancels the poles and zeros close to the unit circle in the Z-plane, thereby smoothing or eliminating the peaks and notches in the magnitude responses of transaural filters. This smoothing also enables the easy implementation of signal processing. Using blocked ear canal HRTFs without pinnae for transaural synthesis with the constant-power equalization algorithm further reduces the peaks and notches in the magnitude responses of transaural filters and then decreases timbre coloration (He et al., 2007).

A high-frequency band equalization method can further reduce timbre coloration in transaural reproduction with two frontal loudspeakers (Liu and Xie, 2022), where the high-frequency responses of a pair of transaural filters are equalized by a frequency-dependent factor so that the overall power spectra of the responses remain constant, and the low-frequency responses of transaural filters are kept intact. The crossover frequency is 1.5 kHz.

11.9 VIRTUAL REPRODUCTION OF STEREOPHONIC AND MULTICHANNEL SURROUND SOUND

Conventional stereophonic (and multichannel surround) sound and VADs are two different categories of spatial sound systems. As stated in Section 11.5.3, VADs and multichannel surround sounds are closely related to each other. A type of special application of binaural and transaural reproduction is discussed in this section in which the principles of binaural and transaural synthesis are applied to reproduce stereophonic and multichannel surround sound to accommodate headphone presentation or various loudspeaker configurations.

11.9.1 Binaural reproduction of stereophonic and multichannel sound through headphones

Binaural signals are originally intended for headphone presentations. In Section 11.8.1, after appropriate crosstalk cancellation processing, binaural signals can be converted for loudspeaker reproduction. Amplitude stereophonic sound and multichannel sound signals are originally intended for loudspeaker reproduction. When they are directly presented by a pair of headphones, the spatial information of signals is spoiled, leading to an unnatural auditory perception of lateralization. Binaural synthesis is applied to convert stereophonic and multichannel sound signals for headphone presentation. The basic principle of this conversion is identical to that of the virtual loudspeaker-based algorithms described in Section 11.6.1. However, virtual loudspeaker-based algorithms are intended for different purposes, that is, they can be used to simplify binaural synthesis in VAD.

First, the binaural reproduction of stereophonic signals is discussed.

1. In Section 2.1, after being scattered/diffracted by the head and pinnae, signals are received by both ears when two-channel stereophonic signals are reproduced through a pair of loudspeakers. The pressure signal at each ear is a combination of these from two loudspeakers. The interchannel level difference (ICLD) between the two loudspeaker signals is then converted into an ITD in binaural pressure signals, resulting in the summing localization of virtual sources. By contrast, when presented by headphones, stereophonic signals are directly fed into two ears, generating ILD rather than ITD. This phenomenon differs from the case of a real source in which ITD is generated at low frequencies, and ITD and ILD are produced at high frequencies. Unnatural binaural pressure signals generate lateralization. In Section 2.2.4, recording with a near-coincident microphone pair introduces interchannel time difference (ICTD) with the same order as the ITD and directly improves the perceived performance for headphone presentation.
2. In contrast to loudspeaker reproduction, headphone reproduction does not cause listening room reflections. The presence of such reflections is important for externalization.
3. Static headphone reproduction lacks dynamic information caused by head movements.

Binaural synthesis simulates the transmissions from two loudspeakers to two ears so that stereophonic signals are reproduced by two virtual loudspeakers. Such processing presents correct stereophonic information for the two ears. Early in 1961, Bauer (1961b) introduced the basic idea of binaural processing to present stereophonic signals through headphones. Currently, processing is implemented using digital techniques (Zhang et al., 2000; Kirkeby, 2002).

H_{LL}, H_{RL}, H_{LR}, and H_{RR} are the transfer functions (HRTFs) from two frontal loudspeakers to two ears in stereophonic reproduction, and E_L and E_R are stereophonic or loudspeaker signals (Figure 11.15). The binaural pressures in reproduction are given as

$$P_L = H_{LL}E_L + H_{LR}E_R \quad P_R = H_{RL}E_L + H_{RR}E_R, \tag{11.9.1}$$

In headphone presentation, if E_L and E_R signals are preprocessed according to Equation (11.9.1), the resultant binaural pressures are equal or directly proportional to those in loudspeaker reproduction:

$$E_{L,ear} = H_{LL}E_L + H_{LR}E_R \quad E_{R,ear} = H_{RL}E_L + H_{RR}E_R. \tag{11.9.2}$$

The above scheme can be extended to the binaural reproduction of multichannel sounds. The binaural reproduction of the 5.1-channel sound is taken as an example. The principles of binaural reproduction of other multichannel sounds, such as the 22.2-channel sound in Section 6.5.1, are similar (ITU-R Report, BS 2159-7, 2015). This type of reproduction has various techniques and patents, such as Dolby Headphone, developed by Dolby Laboratories

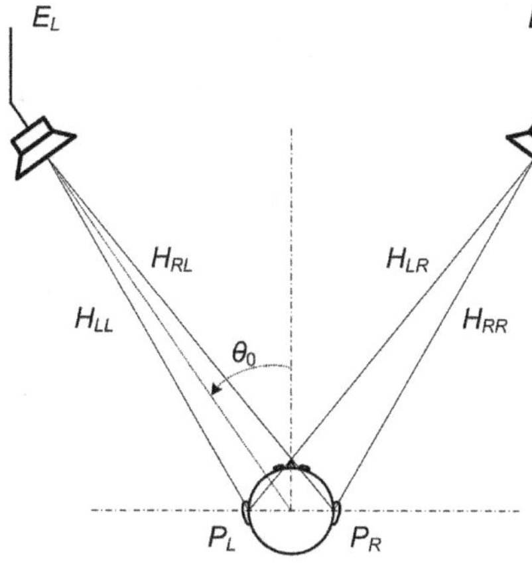

Figure 11.15 Transmission from two front loudspeakers to two ears.

(http://www.dolby.com). Although implementation details vary among different techniques and patents, their basic principles are similar.

In Section 5.2, the 5.1-channel sound has five independent channels with full audio bandwidths. The signals are denoted by E_L, E_C, E_R, E_{LS}, and E_{RS}, where the low-frequency effect channel is neglected.

When directly presented by headphones, 5.1-channel signals are downmixed into two-channel headphone signals E_{L0} and E_{R0}, according to Equation (8.2.3):

$$E_{L0} = E_L + \kappa_C E_C + \kappa_S E_{LS} \quad E_{R0} = E_R + \kappa_C E_C + \kappa_S E_{RS}, \tag{10.9.3}$$

In actual 5.1-channel loudspeaker reproduction, H_{LL}, H_{RL}, H_{LR}, and H_{RR} denote the HRTFs from the left and right loudspeakers to two ears; H_{LC} and H_{RC} refer to the HRTFs from the center loudspeaker to two ears; and H_{LLS}, H_{RLS}, H_{LRS}, and H_{RRS} correspond to the HRTFs from the left and right loudspeakers to the two ears (Figure 11.16). The binaural pressures P_L and P_R are the sum of these pressures caused by each loudspeaker expressed as

$$\begin{aligned} P_L &= H_{LL}E_L + H_{LC}E_C + H_{LR}E_R + H_{LLS}E_{LS} + H_{LRS}E_{RS} \\ P_R &= H_{RL}E_L + H_{RC}E_C + H_{RR}E_R + H_{RLS}E_{LS} + H_{RRS}E_{RS}. \end{aligned} \tag{11.9.4}$$

Similar to the case in Equation (11.9.2), 5.1-channel signals are filtered with five pairs of HRTFs that correspond to the locations of five loudspeakers. The signals are then mixed and reproduced through headphones:

$$\begin{aligned} E_{L,ear} &= H_{LL}E_L + H_{LC}E_C + H_{LR}E_R + H_{LLS}E_{LS} + H_{LRS}E_{RS} \\ E_{R,ear} &= H_{RL}E_L + H_{RC}E_C + H_{RR}E_R + H_{RLS}E_{LS} + H_{RRS}E_{RS}. \end{aligned} \tag{11.9.5}$$

Through the simulation of acoustic transmission from five loudspeakers to two ears, the binaural pressures generated in headphone reproduction are equal or directly proportional to those in actual loudspeaker reproduction, resulting in correct spatial information.

The algorithm expressed in Equation (11.9.5) requires 10 HRTF-based filters. With the advantage of the left–right symmetry, the algorithm can be simplified by a shuffler structure implementation with five or even four filters (Cooper and Bauck, 1989; Xie et al., 2005a).

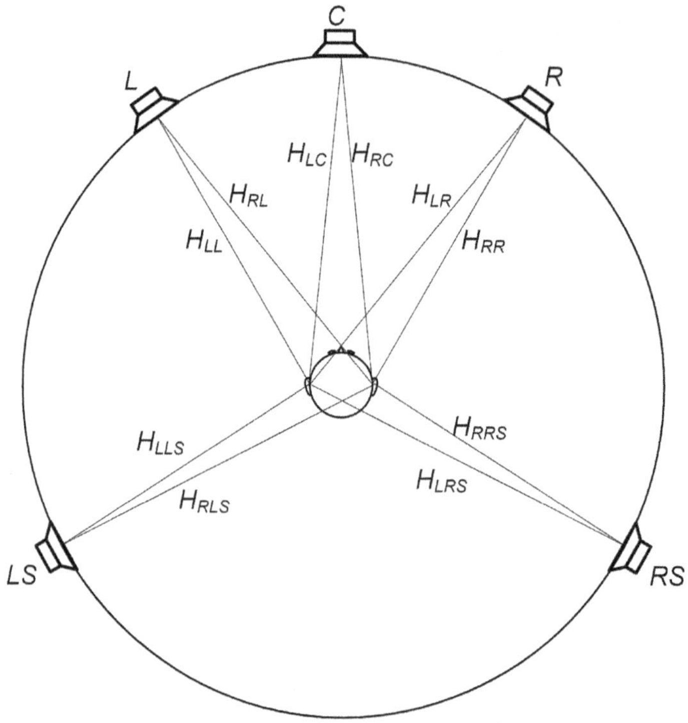

Figure 11.16 Transmission from five loudspeakers in 5.1-channel sound to two ears.

Various errors in conventional binaural reproduction, including reversal error, elevation error, and lateralization, may occur in the binaural reproduction of stereophonic and multichannel sound signals, especially for frontal virtual sources. The reasons for these defects are analyzed in Section 11.7.2. Dynamic binaural synthesis and reproduction (Section 11.10) greatly deletes these defects, but signal processing is complicated. Binaural synthesis with individualized or customized HRTFs partly improves localization performance in static reproduction (Xie et al., 2013b; Xie and Tian, 2014). Individualized HpTF equalization is also beneficial, but it is difficult in practice, especially in most consumer applications.

Incorporating reflections and reverberations to binaural signals improve externalization to a certain extent. Acoustic models of listening rooms have been introduced into some algorithms of the binaural reproduction of multichannel sounds, such as Dolby headphones. However, reflections should be carefully introduced for processing. Otherwise, an unnatural auditory perception may occur. Moreover, introducing reflections complicates signal processing.

1. In loudspeaker reproduction, strong reflections from a listening room degrade the spatial information in the original recording of a multichannel sound program. According to the recommendations of the ITU (ITU-R BS 1116-1, 1997), controlling the reflections from a listening room necessitates the appropriate treatment of sound absorption. Appropriately regulating the simulated reflections in the binaural reproduction of the 5.1-channel sound through headphones is difficult to accomplish. The contribution of weakly simulated reflections is insufficient for the externalization of virtual sources. However, strongly simulated reflections degrade the spatial information in the original program recording or cause unnatural perceived effects. In particular, some reflection

information is often mixed with the original 5.1-channel signals. In practical programs, the center signal C often records speech (such as a dialog in film). Adding superfluous reflections to speech signals may diminish intelligibility.

2. The length of the room impulse response (RIR) is in the order of reverberation time. Input signals should be convoluted with BRIRs to accurately simulating the reflections in a listening room. For a reverberation time of 0.3 s and a sampling frequency of 48 kHz, for example, the length of the BRIRs is on the order of 48000 × 0.3 = 14400 points. The real-time convolution of such a long impulse response is difficult. In practice, only a few early reflections are precisely simulated, possibly degrading the perceived performance in reproduction.

In many 5.1-channel surround sound programs, especially 5.1-channel music programs, some reflections are recorded in two surround channel signals. Externalization can also be achieved without incorporating reflections from a listening room if reflection information in surround channels is enhanced in binaural reproduction through headphones. For most 5.1-channel programs with pictures [such as films and digital television (DTV) programs], the center channel signal usually records speech (dialog), which is unsuitable for incorporating superfluous reflections for externalization. In this case, reintroducing room reflections for the externalization of the center channel is not always necessary because the picture captures the listener's attention.

Based on these considerations, Xie et al. (2005a, 2005b) suggested an improved algorithm to binaurally reproduce a 5.1-channel surround sound through headphones. The main points are as follows:

1. E_L and E_R are mixed with a −3 dB-attenuated center signal E_C and then filtered with free-field HRTFs at the corresponding virtual loudspeakers. Reflections from the listening room are not simulated.
2. The surround signals E_{LS} and E_{RS} are decorrelated and then reproduced by multiple virtual surround loudspeakers to simulate the 5.1-channel sound reproduction in commercial cinema.

In Section 7.5.4, some algorithms for audio signal decorrelation are available. The simplest one involves applying an appropriate delay to the original signal to reduce correlation to a certain extent. This algorithm tends to create timbre coloration during reproduction. Decorrelation can be implemented to minimize coloration by filtering signals with special all-pass filters, including all-pass filters with random phases and reciprocal maximal-length sequence filters (Xie et al., 2012). A shuffler structure implementation also simplifies the aforementioned signal processing (Xie et al., 2005a, Xie, 2013).

Timbre coloration may occur in the binaural reproduction of stereophonic and multichannel sound signals (Lorho et al., 2002). Some algorithms have been proposed to reduce timbre coloration (Merimaa, 2009, 2010), and their principles are similar to the constant-power equalization of transaural processing in Section 11.8.4.

11.9.2 Stereophonic expansion and enhancement

In Section 2.1, the stereophonic loudspeakers of a standard configuration are symmetrically arranged with a 60° span angle relative to a given listener. Stereophonic programs are designed for standard loudspeaker configuration. In some applications, however, stereophonic loudspeakers may not be arranged in compliance with standard configuration because of the limitations imposed by practical conditions. For example, stereophonic loudspeakers in TV sets, multimedia computers, and mobile devices may be arranged with a narrow span. However, the perceived performance is degraded using loudspeaker configuration with a narrow span in stereophonic reproduction. ***Stereophonic expansion*** (***stereo-base widening***) algorithms aim to compensate for the loudspeaker configuration with a narrow span. In other words, a pair of actual loudspeakers with a narrow span angle is used to simulate a pair of virtual loudspeakers with a standard configuration through which stereophonic signals are reproduced.

E_L and E_R are loudspeaker signals originally intended for a stereophonic loudspeaker configuration with a standard 60° span angle. For the left–right symmetric configuration, the transfer functions from two loudspeakers in the standard configuration to the two ears are denoted by $H^1_{LL} = H^1_{RR} = H_{\alpha 1}$, $H^1_{LR} = H^1_{RL} = H_{\beta 1}$. Then, the reproduced binaural pressures are identified by

$$\begin{bmatrix} P_L \\ P_R \end{bmatrix} = \begin{bmatrix} H_{\alpha 1} & H_{\beta 1} \\ H_{\beta 1} & H_{\alpha 1} \end{bmatrix} \begin{bmatrix} E_L \\ E_R \end{bmatrix}. \tag{11.9.6}$$

For a loudspeaker configuration with a narrow span angle, $H_{LL} = H_{RR} = H_\alpha$ and $H_{LR} = H_{RL} = H_\beta$ denote the transfer functions from two loudspeakers to the two ears, and E'_L and E'_R correspond to the actual loudspeaker signals. The reproduced binaural pressures are

$$\begin{bmatrix} P'_L \\ P'_R \end{bmatrix} = \begin{bmatrix} H_{LL} & H_{LR} \\ H_{RL} & H_{RR} \end{bmatrix} \begin{bmatrix} E'_L \\ E'_R \end{bmatrix}. \tag{11.9.7}$$

Binaural pressures in Equation (11.9.7) are equal to those in Equation (11.9.6), yielding

$$E'_L = A_L(\theta_L, f)E_L + A_L(\theta_R, f)E_R \qquad E'_R = A_R(\theta_L, f)E_L + A_R(\theta_R, f)E_R, \tag{11.9.8}$$

where θ_L= 30° and θ_R= −30° are the azimuths of the left and right loudspeakers, respectively, in the standard configuration, and

$$\begin{aligned} A_L(\theta_L, f) = A_R(\theta_R, f) &= \frac{H_\alpha H_{\alpha 1} - H_\beta H_{\beta 1}}{H_\alpha^2 - H_\beta^2} \\ A_R(\theta_L, f) = A_L(\theta_R, f) &= \frac{H_\alpha H_{\beta 1} - H_\beta H_{\alpha 1}}{H_\alpha^2 - H_\beta^2}. \end{aligned} \tag{11.9.9}$$

Therefore, rendering the precorrected signals E'_L and E'_R through a pair of loudspeakers with a narrow span can generate the same binaural pressures and consequently the same effect as in standard stereophonic reproduction.

The signal processing provided by Equation (11.9.8) can be regarded as an improvement or extension of the out-of-phase loudspeaker signals for recreating the outside-boundary virtual source in stereophonic reproduction (Section 2.4.1). They are equivalent at low frequencies. Above the frequency of 1.5 kHz, the summing localization principle in amplitude stereophonic reproduction gradually becomes invalid. Moreover, for most practical stereophonic signals (such as music, a considerable amount of energy comes from a low-frequency below 2–4 kHz. Therefore, the HRTFs of an artificial head without a pinna or spherical head model can be used to simplify stereophonic expansion processing. A bandlimited implementation of stereophonic expansion below 1.5 kHz can also result in satisfactory subjective performance (Xie and Zhang, 1999). Aarts (2000) used the simplest shadowless head model for stereophonic expansion processing. In addition, a shuffler structure implementation similar to Equation (11.6.5) simplifies signal processing by taking advantage of the left–right symmetry of loudspeaker configurations. The constant-power equalization scheme in Equation (11.8.18) can be incorporated to reduce timbre coloration.

Stereophonic enhancement algorithms are a class of techniques for consumer applications that manipulate conventional stereophonic signals and then reproduce them through a pair of frontal loudspeakers to enhance certain spatial perception effects (Maher, 1997). Various stereophonic enhancement algorithms, including SRS 3D, Q sound, and Spatialize, are available. These algorithms may differ in terms of implementation, but their principles are similar to those of stereophonic expansion. Some of these algorithms have been applied to commercial products or software-based plugins in multimedia computers. The algorithm is not always valid for all stereophonic signals. It is invalid for stereophonic signals with interchannel time differences, such as those recorded by the spaced microphone technique. Certain stereophonic enhancement algorithms also alter timbre, thereby causing perceivable coloration in reproduction. Olive (2001) conducted a subjective experiment to evaluate five software-based plugins of stereophonic enhancement algorithms. The results indicated that three of the five algorithms are inferior to conventional stereophonic reproduction.

11.9.3 Virtual reproduction of multichannel sound through loudspeakers

Multichannel sound reproduction requires multiple loudspeakers, which are complex and inconvenient for some practical applications, such as those for TV or multimedia computers. Various downmixing schemes in Section 8.2 convert multichannel sound signals for reproduction with a smaller number of loudspeakers (such as a pair of stereophonic loudspeakers), but this conversion loses some spatial information of multichannel sound. To solve these problems, researchers introduced some transaural and virtual loudspeaker-based approaches for multichannel sound reproduction, which can be regarded as a special downmixing technique called ***virtual surround sound***. A typical example is the virtual 5.1-channel surround sound (Bauck and Cooper, 1996; Davis and Fellers, 1997; Kawano et al., 1998; Toh and Gan, 1999; Hawksford, 2002; Bai and Shih, 2007). Through transaural processing, the center and two surround loudspeakers in the 5.1-channel surround sound can be simulated through a pair of actual stereophonic loudspeakers. During the past decades, this approach has been patented, and some commercial products have been introduced (e.g., TruSurround by SRS Labs, Qsurround by Qsound Labs, Dolby Virtual Speaker, Virtual Surround by Dolby Laboratories, etc.). Some of these techniques are outlined on web pages (http://www.dolby.com; http://www.dts.com/; http://www.qsound.com).

Existing virtual 5.1-channel surround sounds may differ in terms of implementation but have similar basic principles. Figure 11.17 shows a block diagram of the system (i.e., the low frequency effect (LFE) channel, which is processed similar to the central channel, is excluded). The five channel signals are processed and then mixed into signals E'_L and E'_R, which are reproduced through a pair of frontal loudspeakers at $\theta_L = 30°$ and $\theta_R = -30°$, respectively. The two loudspeaker signals can be written in the frequency domain as

$$\begin{aligned} E_L' &= E_L + 0.707E_C + A_L(\theta_{LS}, f)E_{LS} + A_L(\theta_{RS}, f)E_{RS} \\ E_R' &= E_R + 0.707E_C + A_R(\theta_{LS}, f)E_{LS} + A_R(\theta_{RS}, f)E_{RS}, \end{aligned} \tag{11.9.10}$$

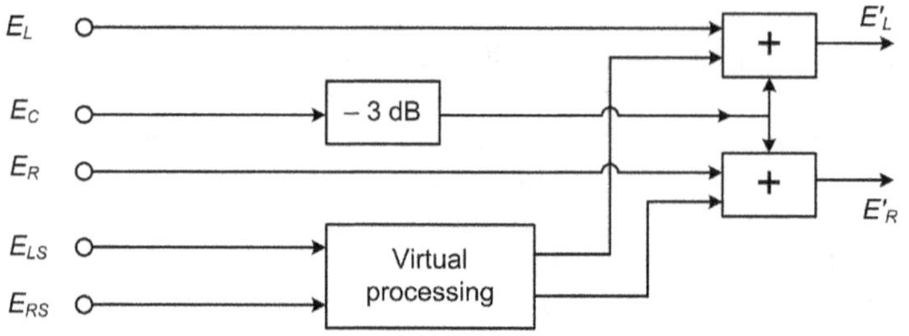

Figure 11.17 Block diagram of virtual 5.1-channel surround sound.

where θ_{LS} and θ_{RS} are the target azimuths for the virtual left and right surround loudspeakers, respectively. In Equation (11.9.10), E_L and E_R signals are directly fed into the left and right loudspeakers, respectively, to create a summing virtual source within the span of the two loudspeakers. E_C is attenuated by −3 dB and then simultaneously fed into the left and right loudspeakers. According to Equation (11.9.10), when other signals satisfy $E_L = E_R = E_{LS} = E_{RS} = 0$, we derive $E'_L = E'_R = 0.707E_C$. In this case, E_C is reproduced by the phantom center channel method, leading to a summing virtual source at the front $\theta_I = 0°$. The surround signals E_{LS} and E_{RS} are filtered by transaural synthesis filters and then fed into the loudspeakers. For the left surround signal E_{LS}, when other signals satisfy $E_L = E_R = E_C = E_{RS} = 0$, Equation (11.9.10) becomes

$$E'_L = A_L(\theta_{LS}, f)E_{LS} \qquad E'_R = A_R(\theta_{LS}, f)E_{LS}. \tag{11.9.11}$$

Similarly, for the right surround signal E_{RS}, when other signals satisfy $E_L = E_R = E_C = E_{LS} = 0$, Equation (11.9.10) becomes

$$E'_L = A_L(\theta_{RS}, f)E_{RS} \qquad E'_R = A_R(\theta_{RS}, f)E_{RS}. \tag{11.9.12}$$

HRTFs from the left or right virtual surround loudspeakers to the ipsilateral and contralateral ears indicate $H_L(\theta_{LS}, f) = H_R(\theta_{RS}, f) = H_{\alpha 2}$ and $H_R(\theta_{LS}, f) = H_L(\theta_{RS}, f) = H_{\beta 2}$ because of the left–right symmetry. The HRTFs from the actual left and right loudspeakers to the ipsilateral and contralateral ears are H_α and H_β, respectively. Equation (11.8.11) yields

$$\begin{aligned} A_L(\theta_{LS}, f) = A_R(\theta_{RS}, f) &= \frac{H_\alpha H_{\alpha 2} - H_\beta H_{\beta 2}}{H_\alpha^2 - H_\beta^2} \\ A_R(\theta_{LS}, f) = A_L(\theta_{RS}, f) &= \frac{H_\alpha H_{\beta 2} - H_\beta H_{\alpha 2}}{H_\alpha^2 - H_\beta^2} \end{aligned}. \tag{11.9.13}$$

Through a pair of actual front loudspeakers, transaural synthesis in Equation (11.9.13) creates a pair of virtual surround loudspeakers by which the surround signals E_{LS} and E_{RS} are reproduced. This phenomenon is the basic principle of a virtual 5.1-channel surround sound.

The virtual 5.1-channel surround sound is based on transaural synthesis. All the problems encountered in transaural reproduction, including narrow listening regions, localization errors, and coloration in reproduction, occur in virtual 5.1-channel surround sounds. The factors that cause such problems are analyzed in Sections 11.8.2, 11.8.3, and 11.8.4.

The following schemes for improving the performance of the virtual 5.1-channel surround sound are suggested (Xie et al., 2005c, 2005d). The basic considerations include the following:

1. Enlarging the listening region
   In the virtual 5.1-channel surround sound system shown in Figure 11.7, a pair of actual stereophonic loudspeakers is arranged with a conventional span angle of 60°. In Section 11.8.3, a narrow span angle between the left and right loudspeaker pairs improves the stability of virtual sources against lateral head translation, thereby expanding the listening region. However, a loudspeaker pair with a narrow span angle (such as a stereo dipole) requires a boost of the low-frequency components of the signals, leading to instability at low frequencies. As a trade-off between robustness and better low-frequency reproduction performance, a 20°–30°span angle (that is, the left and right loudspeakers are arranged at azimuths of ±10° to ±15°) is a better option. Unlike the conventional 60° span angle, such a loudspeaker configuration improves the stability of virtual sources and simultaneously avoids a large increase in the low-frequency components of the signals. It also facilitates loudspeaker arrangements on TV and multimedia computers. However, this loudspeaker configuration requires additional transaural synthesis filters other than those for virtual surround loudspeakers to retain the baseline width for frontal virtual sources.
2. Improving the distribution of virtual source directions
   In Section 11.8.2, the perceived virtual source positions in two-front loudspeaker reproduction are usually restricted to the region of frontal-horizontal quadrants. The virtual source intended for rear-horizontal quadrants is often perceived at the mirror position in the frontal-horizontal quadrants. In some virtual 5.1-channel surround sound algorithms, the target azimuths for virtual surround loudspeakers are ±110°, as specified by the ITU-recommended standard. Therefore, the actual perceived azimuths for virtual surround loudspeakers are often located at mirror azimuths of ±70°. This problem narrows the distribution region of perceived virtual sources. As a trade-off, the target azimuth of ±90° can be selected for two virtual surround loudspeakers. This surround loudspeaker configuration is a compromise in actual 5.1-channel sound reproduction when a listening room is unsuitable for the ITU-recommended configuration.
3. Timbre equalization
   Rumsey et al. (2005) demonstrated that a degradation of the timbre generally has a greater impact on the overall perception of the audio quality, than the degradation of the spatialization. The constant-power equalization algorithm discussed in Section 11.8.4 should be incorporated into signal processing to reduce timbre coloration in reproduction. Loudspeaker signals then become

$$\begin{aligned} E'_L &= A'_L(\theta_L, f)E_L + A'_L(\theta_R, f)E_R + 0.707E_C + A'_L(\theta_{LS}, f)E_{LS} + A'_L(\theta_{RS}, f)E_{RS} \\ E'_R &= A'_R(\theta_L, f)E_L + A'_R(\theta_R, f)E_R + 0.707E_C + A'_R(\theta_{LS}, f)E_{LS} + A'_R(\theta_{RS}, f)E_{RS}, \end{aligned} \tag{11.9.14}$$

where E_C is processed in the same manner as in Equation (11.9.10). It is attenuated by −3 dB and then simultaneously fed into the left and right loudspeakers to create the summing virtual source in the frontal direction $\theta = 0°$.

After the equalization algorithm is incorporated, $A_L(\theta_{LS}, f)$, $A_R(\theta_{LS}, f)$, $A_L(\theta_{RS}, f)$, and $A_R(\theta_{RS}, f)$ in Equation (11.9.13) are replaced with $A'_L(\theta_{LS}, f)$, $A'_R(\theta_{LS}, f)$, $A'_L(\theta_{RS}, f)$, and $A'_R(\theta_{RS}, f)$. For left–right symmetrical configuration, the transfer function from two virtual surround loudspeakers to two ears is denoted by $H_L(\theta_{LS}, f) = H_R(\theta_{RS}, f) = H_{\alpha 2}$ and $H_R(\theta_{LS}, f) = H_L(\theta_{RS}, f) = H_{\beta 2}$, Equation (11.8.18) yields

$$
\begin{aligned}
A'_L(\theta_{LS}, f) = A'_R(\theta_{RS}, f) &= \frac{H_\alpha H_{\alpha 2} - H_\beta H_{\beta 2}}{\sqrt{|H_\alpha H_{\alpha 2} - H_\beta H_{\beta 2}|^2 + |H_\alpha H_{\beta 2} - H_\beta H_{\alpha 2}|^2}} \frac{|H_\alpha^2 - H_\beta^2|}{H_\alpha^2 - H_\beta^2} \\
A'_R(\theta_{LS}, f) = A'_L(\theta_{RS}, f) &= \frac{H_\alpha H_{\beta 2} - H_\beta H_{\alpha 2}}{\sqrt{|H_\alpha H_{\alpha 2} - H_\beta H_{\beta 2}|^2 + |H_\alpha H_{\beta 2} - H_\beta H_{\alpha 2}|^2}} \frac{|H_\alpha^2 - H_\beta^2|}{H_\alpha^2 - H_\beta^2}
\end{aligned}
\tag{11.9.15}
$$

In contrast to H_α and H_β in Equation (11.9.13), those in Equation (11.9.15) denote the HRTFs from two loudspeakers at azimuths ±15° (rather than ±30°) to the ipsilateral and contralateral ears, respectively.

In contrast to the implementation in Equation (11.9.10), the original E_L and E_R should be processed to retain the baseline width of the frontal virtual source. That is, a pair of front loudspeakers with a 30°span angle is used to create a pair of virtual front loudspeakers with a 60°span angle. Here, $\theta_L = 30°$ and $\theta_R = -30°$ are the azimuths of the virtual left and right loudspeakers, respectively. For the left–right symmetrical configuration, the transfer functions from two virtual front loudspeakers to two ears satisfy $H_L(\theta_L, f) = H_R(\theta_R, f) = H_{\alpha 1}$ and $H_R(\theta_L, f) = H_L(\theta_R, f) = H_{\beta 1}$, similar to Equation (11.9.15), yielding

$$
\begin{aligned}
A'_L(\theta_L, f) = A'_R(\theta_R, f) &= \frac{H_\alpha H_{\alpha 1} - H_\beta H_{\beta 1}}{\sqrt{|H_\alpha H_{\alpha 1} - H_\beta H_{\beta 1}|^2 + |H_\alpha H_{\beta 1} - H_\beta H_{\alpha 1}|^2}} \frac{|H_\alpha^2 - H_\beta^2|}{H_\alpha^2 - H_\beta^2} \\
A'_R(\theta_L, f) = A'_L(\theta_R, f) &= \frac{H_\alpha H_{\beta 1} - H_\beta H_{\alpha 1}}{\sqrt{|H_\alpha H_{\alpha 1} - H_\beta H_{\beta 1}|^2 + |H_\alpha H_{\beta 1} - H_\beta H_{\alpha 1}|^2}} \frac{|H_\alpha^2 - H_\beta^2|}{H_\alpha^2 - H_\beta^2}
\end{aligned}.
\tag{11.9.16}
$$

A direct implementation of the algorithm in Equation (11.9.14) necessitates eight transaural synthesis filters. Similar to the case of Equation (11.6.5), left–right symmetry yields a simplified shuffler structure implementation. Equation (11.9.14) is equivalent to

$$
\begin{aligned}
\begin{bmatrix} E'_L \\ E'_R \end{bmatrix} = \begin{bmatrix} 0.707 & 0.707 \\ 0.707 & -0.707 \end{bmatrix} \Bigg\{ & \begin{bmatrix} 1 \\ 0 \end{bmatrix} E_C \\
& + \begin{bmatrix} \Sigma_1 & 0 \\ 0 & \Delta_1 \end{bmatrix} \begin{bmatrix} 1 & 1 \\ 1 & -1 \end{bmatrix} \begin{bmatrix} E_L \\ E_R \end{bmatrix} + \begin{bmatrix} \Sigma_2 & 0 \\ 0 & \Delta_2 \end{bmatrix} \begin{bmatrix} 1 & 1 \\ 1 & -1 \end{bmatrix} \\
& \times \begin{bmatrix} E_{LS} \\ E_{RS} \end{bmatrix} \Bigg\},
\end{aligned}
\tag{11.9.17}
$$

where

$$
\begin{aligned}
\Sigma_1 &= 0.707\left[A'_L(\theta_L, f) + A'_L(\theta_R, f)\right] & \Delta_1 &= 0.707\left[A'_L(\theta_L, f) - A'_L(\theta_R, f)\right] \\
\Sigma_2 &= 0.707\left[A'_L(\theta_{LS}, f) + A'_L(\theta_{RS}, f)\right] & \Delta_2 &= 0.707\left[A'_L(\theta_{LS}, f) - A'_L(\theta_{RS}, f)\right]
\end{aligned}.
\tag{11.9.18}
$$

Four filters are required to implement the algorithm given in Equation (11.9.17), thereby simplifying the signal processing. Figure 11.18 shows a block diagram for the signal processing designed according to Equation (11.9.17), where the LFE channel is supplemented. The filters in the block diagram are implemented by a FIR structure with a length of 1.3–2.7 ms; the corresponding perceived performance is validated by psychoacoustic experiments

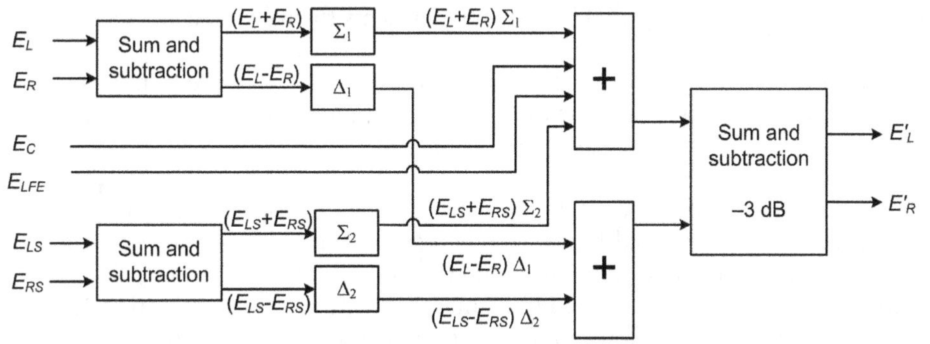

Figure 11.18 Block diagram for the shuffler structure implementation of signal processing in Equation (11.9.17).

(Xie et al., 2006). As expected, the algorithm can recreate the summing virtual source in frontal-horizontal quadrants, but the virtual source intended for rear-horizontal quadrants is often perceived at the mirror position in frontal-horizontal quadrants.

Similar algorithms are applicable to the virtual reproduction of more channel sounds. For example, Matsui and Ando (2010) proposed an algorithm for the virtual reproduction of a 22.2-channel sound (Section 6.5.1) with three actual loudspeakers. The original center channel signal is directly reproduced by an actual center loudspeaker to stabilize the frontal virtual source. After transaural processing, other original channel signals are reproduced by a pair of left- and right-front loudspeakers. Other researchers have suggested using horizontal 5.1- or 7.1-channel loudspeaker configurations to create vertical localization information (virtual loudspeakers in vertical directions, Lee et al., 2011; Kim et al., 2014). For example, Lee et al. suggested using a horizontal 7.1-channel loudspeaker configuration to reproduce the signals of Samsung 10.2-channel surround sound (Section 6.5.1). The vertical localization information is created by four loudspeakers in the 7.1-channel configuration and transaural synthesis. However, as discussed in Section 11.8.2, creating a vertical virtual source with a horizontal loudspeaker configuration and transaural processing algorithm is difficult. If all actual loudspeakers are arranged in the frontal horizontal plane, creating a virtual source in the rear-horizontal quadrants is also difficult. Therefore, using a few actual horizontal loudspeakers for transaural reproduction of multichannel spatial surround sounds inevitably leads to virtual source localization errors. Four actual loudspeakers can be used for the virtual reproduction of spatial surround sound (such as the 9.1-channel sound in Section 6.5.1) to improve the performance of vertical information reproduction in the front space. Four actual loudspeakers are arranged in the left-front and right-front directions in the horizontal plane, as well as in the left-front-up and right-front-up directions (Xie et al., 2021c).

11.10 RENDERING SYSTEM FOR DYNAMIC AND REAL-TIME VIRTUAL AUDITORY ENVIRONMENTS

11.10.1 Binaural room modeling

Room or environment reflection has been disregarded in the binaural synthesis of free-field virtual sources. However, as stated in Section 1.8, reflections exist in most actual rooms and are crucial to spatial auditory perception. Therefore, a complete VAD should include a reflection-modeling component hereafter called a ***virtual auditory or acoustic environment*** (*VAE*). The simulation of binaural information created by room reflections is called binaural room modeling.

Incorporating reflections into VAE processing presents the following advantages:

1. It recreates spatial auditory perception in a room or reflective environment.
2. It eliminates or reduces lateralization during headphone presentation.
3. It enables the control of the auditory distance perception.

Binaural reflection information can be obtained through measurement. A pair of BRIRs is first measured from an artificial head or human subject. Then, binaural signals are synthesized by convoluting the mono stimulus with HRIRs (Section 11.1.2). The principle of BRIR measurement is similar to that of the HRTF measurement described in Section 11.2.1.

Binaural reflection information can also be obtained through simulations or calculations. The simulation methods can be subdivided into two categories: ***physics-based methods*** and ***perception-based methods.***

Physics-based methods simulate the physical propagation of sound from the source to the receiver inside a room and then convert the sound field information into binaural information (BRIRs). Similar to the simulation of spatial room impulse responses, physical simulation involves source simulation and transmission or room acoustic simulation. A listener simulation (simulating the scattering and diffraction caused by a listener) should be supplemented to simulate BRIRs. Geometrical acoustic-based methods and wave acoustic-based methods are applicable to binaural reflection simulations. Geometrical acoustic-based methods usually yield the arrival times, directions, energies, and frequency spectra of direct and reflected sounds. The direct sound signal and each reflected sound signal are filtered with the HRTFs at the corresponding directions and then summed to form the complete BRIRs. Binaural reflection synthesis is often implemented using parametric and decomposed structures to improve computational efficiency. That is, source radiation patterns, sound propagation, surface materials, and air absorption, as well as HRTF-based filtering, are modeled in real time and then separately realized by corresponding low-order or simplified filters and delay lines. This approach indicates that BRIRs are approximated by appropriate series-parallel filters and delay line connections.

Perception-based methods simulate room reflections according to some general (statistical) rules and predetermined room acoustic parameters (such as reverberation time) from measurement or calculation and then recreate the desired auditory perception of reflections by signal processing algorithms from a perceptual rather than a physical viewpoint. Various artificial delay and reverberation algorithms described in Sections 7.5.1 and 7.5.2 are perception-based methods.

The binaural simulation of room reflections with complete BRIRs is complicated. A combined method is often used in practice. That is, early discrete reflections are simulated by BRIRs obtained from physical-based methods, and the late diffused reverberation is simulated using perception-based methods.

11.10.2 Dynamic virtual auditory environments system

The static VAD or VAE has been addressed in previous sections, where both virtual sources and listeners are assumed to be in fixed locations, and real-time processing is not always required. In a real acoustic environment, however, either source or listener movements alter binaural pressures and provide dynamic acoustic information. This dynamic information should be incorporated into VAD or VAE processing because it is significant for localizing source and recreating convincing auditory perceptions of acoustic environments. Therefore, in addition to modeling sound sources, room (environment), and listener, a sophisticated VAE should be able to constantly detect the position and orientation of a listener's head on

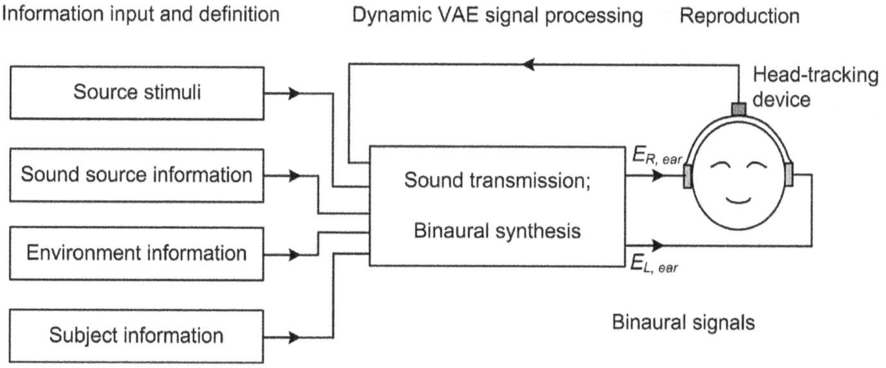

Figure 11.19 Structure of a typical dynamic VAE system.

whose basis signal processing is updated in real time. In other words, a realistic VAE should be an interactive, dynamic, and real-time rendering system hereafter called a ***dynamic and real-time VAE system*** or ***dynamic VAE system***.

Figure 11.19 shows the basic structure of a dynamic VAE system, which consists of three parts.

1. Information input and definition
   This part inputs prior information and data for the dynamic VAE through a user interface. This information involves source signals and characteristics (such as source position and frequency-dependent directivities), geometrical and physical characteristics of the environment (such as shape and size of room and absorption coefficients of surface materials and air), and physical data of listeners (HRTFs). A ***head tracke***r detects the temporary position and orientation of the listener's head and then provides information for the system.
2. Dynamic VAE signal processing
   According to prior information and data, this part simulates sound sources, the direct and reflected/scattered transmission in a room, and the scattering/diffraction of the listener in real time. The HRTFs for binaural synthesis are constantly updated according to the information provided by the head tracker to obtain dynamic binaural signals.
3. Reproduction
   The resultant binaural signals are reproduced through headphones after headphone-to-ear canal transmission equalization or through loudspeakers after crosstalk cancellation.

Ideally, binaural signals or auditory scenarios created by a dynamic VAE should synchronously vary with head movement and the target environment, as in real environments. Therefore, an ideal dynamic VAE should be a linear time-variable system. However, signal processing schemes in dynamic VAEs are deduced from the static scheme, where a series of short "static states" is used to approximate transient states. Thus, the dynamic behavior of VAEs should be considered.

The first dynamic behavior is the ***scenario update rate*** of a VAE. A dynamic VAE updates binaural signals and auditory scenarios at certain time intervals. The scenario update rate of a VAE refers to the number of update scenario manipulations per second. Another dynamic behavior is the ***system latency time***, which is defined as the time from which a listener's head moves to the time at which the corresponding change in the synthesized binaural signal output occurs. System latency is contributed by a series of factors, including latency in the response of the head tracker, data transmission and communication, time required for update and signal processing, and data buffer.

A dynamic VAE should possess a high scenario update rate and short system latency time to enhance the perceived performance. Limited by the hardware and software resources, the dynamic behavior of a practical AVE system is usually a compromise based on psychoacoustic results. Many psychoacoustic experiments have been performed on the acceptable scenario update rate and system latency time. The results vary in different experiments. Sandvad (1996) indicated that a scenario update rate of 20 Hz does not influence a listener's localization response time. Brungart et al. (2006) suggested that a system latency time of less than 60 ms is adequate for most applications.

Updating the scenario in a dynamic VAE may cause audible artifacts. For a smooth transition from one auditory scenario to the next, some crossfading methods should be incorporated into signal processing. Updating the HRTF-based filters is not required, and a smooth transition can be realized if virtual loudspeaker-based algorithms or basis function decomposition algorithms in Section 11.6 are incorporated into dynamic binaural synthesis. In particular, the PCA-based algorithm greatly improves the computational efficiency of simultaneously synthesizing multiple virtual sources (including real sources for direct sound and image sources for reflections), thus increasing the maximum allowable number of virtual sources to be synthesized. Conversely, the spherical harmonic function decomposition-based algorithm is convenient for dynamic information simulation. The transformation of spherical harmonic functions under sound field rotation (Section 9.4.2) is applicable to dynamic processing because the listener's head turning around a certain axis with an angle is equivalent to the whole sound field rotating around the same axis with an opposite angle.

Various VAE systems have been developed for scientific studies (such as hearing study and virtual reality), such as the SLAB (sound laboratory) by NASA (Wenzel et al., 2000), the digital interactive virtual acoustics (DIVA) by the Helsinki University of Technology (which is now merged as Aalto University, Saviojia et al., 1999), and the first- and second-generation systems by Ruhr-Universitat Bochum (Blauert et al., 2000; Djelani et al., 2000; Silzle et al., 2004). Some early systems were developed in the DSP platform, and late systems were mostly obtained in the PC platform and software.

The AVE system developed by the Acoustic Lab at the South China University of Technology is intended for psychoacoustic research. It is also a PC and C++ language-based system (Zhang and Xie, 2013). The system detects and simulates dynamic information for six degrees of freedom of head movement and can synthesize far- and near-field virtual sources at various distances and directions, as well as early reflections. Two alternative algorithms are used for dynamic binaural synthesis. The first is the conventional HRTF-based algorithm, and the second is the PCA-based algorithm, where HRTFs in various directions and distances are decomposed by PCA, and binaural synthesis is implemented by a parallel bank of 16 filters (Sections 11.5.4 and 11.6.2). The system can simultaneously render up to 280 virtual sources (direct or image) by using a conventional algorithm or 4500 virtual sources using a PCA-based algorithm. The scenario update rate is 120 Hz, and the system latency time is 25.4 ms. The improved system performance is attributed to improvements in PC performance and signal-processing algorithms. A psychoacoustic experiment demonstrates that dynamic binaural synthesis effectively eliminates lateralization and reversal errors even if nonindividualized HRTFs are used.

Some authors studied dynamic transaural reproduction through loudspeakers, where dynamic binaural synthesis and dynamic crosstalk cancellation are involved (Gardner, 1997;Lentz and Schmitz, 2002, Lentz et al., 2005). The signals recorded by a spherical microphone array can also be converted into dynamic binaural reproduction (Duraiswami et al., 2005). Based on the method discussed in the last paragraph of Section 11.6.1, the

HRTF-based filters are constantly updated according to the listener's head position and orientation information detected by the head tracker. Spherical microphone array recording retains dynamic information and overcomes the shortcomings of conventional artificial head recordings. In practical implementation, the head turning to a specified direction is equivalent to turning the sound field in opposite directions; therefore, dynamic information can be incorporated by applying a rotation transformation to the independent signals of virtual Ambisonics (Section 9.4.2, Enzner et al., 2013). In this case, HRTF-based filters corresponding to M virtual loudspeaker directions are sufficient, but directional continuous HRTFs and updating HRTFs are not required.

11.11 SUMMARY

Binaural reproduction and VAD aim to reconstruct binaural sound pressures. The related binaural signals can be obtained by binaural recording from an artificial head or human subject, and they can also be obtained by binaural synthesis.

HRTFs, which are essential for binaural synthesis, can be acquired using three methods: measurement, calculation, and customization. The technique for far-field HRTF measurement is mature, and many far-field HRTF databases have been established. Near-field HRTF measurements are relatively difficult. Rare near-field HRTF databases for artificial head and human subjects are available. The analytical solutions of HRTFs can only be calculated for rare simplified head/torso models. Based on the scanning of the anatomical surfaces of subjects, numerical calculations, such as the BEM, yield HRTFs within the full audible frequency range with an appropriate accuracy. HRTF customization has various methods. HRTF customization is simpler than measurement or calculation and usually yields a modest perceived performance. However, the accuracy of customization is inferior to that of measurement or calculation. HRTF customization should be improved in many aspects. HRTFs and HRIRs exhibit various features in time and frequency domains. These features are closely related to auditory localization.

Binaural synthesis is often implemented using digital filters. HRTF-based filters can be implemented using different filter models and designed using different methods. The minimum-phase approximation and spectral smoothing of HRTFs simplify filters.

Measurement yields HRTFs at discrete and finite directions. HRTFs in unmeasured directions can be reconstructed or estimated from the measured data by using various interpolation schemes.

A directional continuous HRTF can be decomposed via spatial basis functions. Spatial harmonics are a common type of spatial basis function. The spatial harmonic decomposition of HRTFs is closely related to the directional interpolation of HRTFs from which the Shannon–Nyquist spatial sampling theorem of HRTFs can be derived. The spatial interpolation of HRTFs and the signal mixing of multichannel surround sounds are closely related to each other. Some signal mixing methods in multichannel sound reproduction are analogous to certain interpolation and recovery schemes for HRTFs. Under the theoretical framework of spatial function sampling, interpolation, and reconstruction, various spatial sound techniques are unified. The analogy between multichannel sound reproduction and HRTF interpolation enables the interchanging of some of the methods used for the two fields.

HRTFs can also be decomposed using spectral shape basis functions. PCA is an effective statistical algorithm for deriving basis functions. It eliminates the correlations among HRTFs so that HRTFs can be simply represented by the weighted sum of a small set of spectral shape basis functions.

The analogy between the signal mixing of multichannel sound and spatial interpolation of HRTFs or the basis function decomposition of HRTFs are applied to simplify the signal processing of binaural synthesis for multiple and moving virtual sources, resulting in virtual loudspeaker-based algorithms and basis function decomposition-based algorithms.

The equalization of the transfer characteristics of the headphone-to-ear canal is needed in binaural reproduction, which is realized by inverse HpTF filters. Some types of headphones exhibit poor repeatability in HpTF measurements, which are related to the compression deformation of pinnae by headphones. Ideally, individualized HpTFs should be used for headphone equalization. Reversal error, elevation error, and lateralization often occur in the headphone presentation of binaural signals. These defects are caused by the absence of dynamic cues and errors in the high-frequency spectral cue in static binaural reproduction. Reflections are vital to the externalization of a virtual source. Based on externalization, auditory distance perception in binaural reproduction can be controlled by reflection and binaural synthesis with near-field HRTFs.

Crosstalk cancellation is required when binaural signals are reproduced through loudspeakers. Binaural synthesis and crosstalk cancellation can be merged as transaural processing. Transaural reproduction with two frontal loudspeakers can recreate stable virtual sources in frontal-horizontal directions. A given crosstalk cancellation and transaural processing are only effective for a specified listening position and head orientation. Therefore, the listening region for transaural reproduction is narrow, and the perceived timbre coloration often occurs in binaural reproduction through loudspeakers. Nevertheless, timbre coloration can be reduced by the constant-power equalization algorithm.

Lateralization occurs when stereophonic sound and multichannel sound signals are directly presented by a pair of headphones. Binaural synthesis is applied to convert stereophonic sound and multichannel sound signals for headphone presentation. The transaural method is used for the stereophonic expansion and virtual reproduction of multichannel sound.

A complete VAD should include a reflection-modeling component hereafter called a VAE. Therefore, various methods exist for binaurally simulating reflections. A dynamic and real-time virtual auditory environment system simulates the dynamic auditory information caused by head movement and thus accurately recreates various auditory events.

Chapter 12

Binaural pressures and auditory model analysis of spatial sound reproduction

In Chapters 9 and 10, the physical performance of loudspeaker-based spatial sounds, including Ambisonics, wave field synthesis, and conventional multichannel sounds, is evaluated by analyzing their reconstructed sound field. However, in an arbitrary sound field, a listener uses two ears to receive a sound wave. When the listener enters the sound field, his/her anatomical structures, such as the head, pinnae, and torso, scatter and diffract the sound wave. The course of these scattering and diffraction encodes the temporal and spatial information of sound into binaural pressures or signals. Therefore, the physical and perceptual characteristics of the sound field can be evaluated from binaural pressures incorporated with an appropriate model of auditory information processing. The summing localization equations for stereophonic and multichannel sounds in Sections 2.1, 3.2.1, and 6.1.1 are derived by analyzing binaural pressures and interaural localization cues. However, these analyses are based on a simplified head model in which the head shadow is ignored. Results are valid only at a low frequency.

Strictly, the scattering and diffraction of anatomical structures should be considered in the analysis of binaural pressures. In Sections 11.6.1 and 11.9.1, HRTF-based filtering or binaural synthesis is used to convert stereophonic and multichannel sound signals to binaural signals for headphone presentation. A similar method is applicable to convert stereophonic and multichannel sound signals for binaural analysis. Some localization-related physical cues, such as ITD, ILD, monoaural/binaural spectra, and dynamic variation in ITD, can be evaluated on the basis of binaural pressures. For further analysis of the perceptual performance of stereophonic and multichannel sounds, the course and mechanism of the comprehensive processing of binaural information by the human auditory system (including a high-level system) should be considered, and various auditory models are needed.

This chapter analyzes binaural pressures, summing virtual source, and some other auditory perceptions in stereophonic and multichannel sounds. In Section 12.1, the methods for binaural pressure analysis and physical cues for localization are presented, and summing localization and spatial auditory sensation in stereophonic and multichannel sound are further analyzed. In Section 12.2, the analysis of spatial sound by using binaural auditory modes is discussed. In Section 12.3, the binaural measurement system for assessing the performance of spatial sound reproduction is described.

12.1 PHYSICAL ANALYSIS OF BINAURAL PRESSURES IN SUMMING VIRTUAL SOURCE AND AUDITORY EVENTS

12.1.1 Evaluation of binaural pressures and localization cues

A strict binaural pressure analysis should consider the scattering and diffraction (comprehensive filtering) effect of anatomical structures (such as the head). In Section 1.4.2, this effect is described by HRTFs from which binaural pressures caused by a point source or plane wave

DOI: 10.1201/9781003081500-12

source in a free field are evaluated. For a far-field plane wave incident from (θ_S, ϕ_S), binaural pressures in the frequency domain are calculated as

$$P_L(\theta_S, \phi_S, f) = H_L(\theta_S, \phi_S, f)P_{free}(f) \quad P_R(\theta_S, \phi_S, f) = H_R(\theta_S, \phi_S, f)P_{free}(f), \quad (12.1.1)$$

where $P_{free}(f)$ is a complex-valued free-field sound pressure at the position of the head center with the head absent, e.g., the pressure of the incident plane wave. The distance variable r_S is dropped in Equation (12.2.1) because far-field HRTFs are nearly distance independent. For simplicity, the parameter a for anatomical structure and the dimension of each subject are also omitted. In the time domain, Equation (12.1.1) becomes

$$p_L(\theta_S, \phi_S, t) = h_L(\theta_S, \phi_S, t) \otimes_t p_{free}(t) \quad p_R(\theta_S, \phi_S, t) = h_R(\theta_S, \phi_S, t) \otimes_t p_{free}(t). \quad (12.1.2)$$

The time domain functions denoted by a lowercase letter in Equation (12.1.2) are related to the corresponding frequency domain functions by inverse Fourier transform. $h_L(\theta_S, \phi_S, t)$ and $h_R(\theta_S, \phi_S, t)$ are HRIRs in Equation (1.4.3). Notation "$\otimes_t$" denotes convolution in the time domain.

For a far-field incident plane wave, its magnitude is independent from the receiver position, and its phase is chosen to be zero at the origin, i.e., the head center, as shown in Equation (1.2.1). For a point source at the finite distance r_S, the distance dependence of HRTFs and free-field pressure at the origin should be considered. In this case, the propagation delay and attenuation in the free field should be supplemented. Distance dependence should also be supplemented into HRTFs. Then, binaural pressures are calculated as

$$\begin{aligned} P_L(r_S, \theta_S, \phi_S, f) &= \frac{Q_p(f)}{4\pi r_S} H_L(r_S, \theta_S, \phi_S, f)\exp\left(-j2\pi f\frac{r_S}{c}\right), \\ P_R(r_S, \theta_S, \phi_S, f) &= \frac{Q_p(f)}{4\pi r_S} H_R(r_S, \theta_S, \phi_S, f)\exp\left(-j2\pi f\frac{r_S}{c}\right). \end{aligned} \quad (12.1.3)$$

where $Q_p(f)$ is related to source strength, and c is the speed of sound.

In the time domain, Equation (12.1.3) becomes

$$\begin{aligned} p_L(r_S, \theta_S, \phi_S, t) &= \frac{1}{4\pi r_S} h_L(r_S, \theta_S, \phi_S, t) \otimes_t q_p\left(t - \frac{r_S}{c}\right). \\ p_R(r_S, \theta_S, \phi_S, t) &= \frac{1}{4\pi r_S} h_R(r_S, \theta_S, \phi_S, t) \otimes_t q_p\left(t - \frac{r_S}{c}\right). \end{aligned} \quad (12.1.4)$$

The time domain functions in Equation (12.1.4) are related to the corresponding frequency domain functions in Equation (12.1.3) via inverse Fourier transform.

In the case of reproduction with M loudspeakers, the loudspeakers are arranged at a constant far-field distance with respect to the origin. The direction of the ith loudspeaker is (θ_i, ϕ_i), and the corresponding HRTFs are $H_L(\theta_i, \phi_i, f)$ and $H_R(\theta_i, \phi_i, f)$. The frequency domain signal of the ith loudspeaker is $E_i(f)$. All loudspeaker signals have identical waveform but different amplitude and delay; loudspeaker signals can be written as

$$E_i(f) = A_i E_A(f), \quad (12.1.5)$$

where A_i is a normalized complex-valued amplitude or gain, which includes information on magnitude, phase, and linear delay. $E_A(f)$ is the signal waveform in the frequency domain. The superposed pressures caused by M loudspeakers in the two ears are calculated as

$$P_L'(f)=\sum_{i=0}^{M-1}H_L(\theta_i,\phi_i,f)E_i(f) \qquad P_R'(f)=\sum_{i=0}^{M-1}H_R(\theta_i,\phi_i,f)E_i(f). \tag{12.1.6}$$

In the time domain, Equation (12.1.6) becomes

$$p_L'(t)=\sum_{i=0}^{M-1}h_L(\theta_i,\phi_i,t)\otimes_t e_i(t) \qquad p_R'(t)=\sum_{i=0}^{M-1}h_R(\theta_i,\phi_i,t)\otimes_t e_i(t). \tag{12.1.7}$$

The time domain functions in Equation (12.1.7) are related to the corresponding frequency domain functions in Equation (12.1.6) by inverse Fourier transform.

When loudspeakers are arranged at finite distances, the position of the ith loudspeaker is (r_i, θ_i, ϕ_i), the corresponding signal is $E_i(f)$. Here, the distances of the loudspeakers to the origin may be identical or different. The superposed pressures, including the effects of propagation delay and attenuation caused by the finite loudspeaker distance and supplementing distance dependence on HRTFs, in the two ears are calculated as

$$\begin{aligned}P_L'(f)&=\sum_{i=0}^{M-1}\frac{1}{4\pi r_i}H_L(r_i,\theta_i,\phi_i,f)E_i(f)\exp\left(-j2\pi f\frac{r_i}{c}\right)\\ P_R'(f)&=\sum_{i=0}^{M-1}\frac{1}{4\pi r_i}H_R(r_i,\theta_i,\phi_i,f)E_i(f)\exp\left(-j2\pi f\frac{r_i}{c}\right).\end{aligned} \tag{12.1.8}$$

In the time domain, Equation (12.1.8) becomes

$$\begin{aligned}p_L'(t)&=\sum_{i=0}^{M-1}\frac{1}{4\pi r_i}h_L(r_i,\theta_i,\phi_i,t)\otimes_t e_i\left(t-\frac{r_i}{c}\right)\\ p_R'(t)&=\sum_{i=0}^{M-1}\frac{1}{4\pi r_i}h_R(r_i,\theta_i,\phi_i,t)\otimes_t e_i\left(t-\frac{r_i}{c}\right).\end{aligned} \tag{12.1.9}$$

The time domain functions in Equation (12.1.9) are related to the corresponding frequency domain functions in Equation (12.1.8) by inverse Fourier transform.

Some physical cues related to spatial auditory perception, such as ITD, ILD, and their dynamic variation with head turning and monaural/binaural spectra, can be analyzed on the basis of the resultant binaural pressures.

In addition to recreating a virtual source, recreating other spatial auditory perceptions or sensations in practical reproduction, such as auditory source width and listener envelopment similar to those in a concert hall, is necessary. The interaural cross-correlation in Section 1.7.3 is applicable to the comprehensive analysis of virtual source localization and some other spatial auditory perceptions in reproduction. The normalized interaural cross-correlation function in sound reproduction is calculated similar to that in Equation (1.7.1):

$$\Psi_{LR}(\tau)=\frac{\int_{t1}^{t2}P'_L(t)P'_R(t+\tau)dt}{\left\{\left[\int_{t1}^{t2}P'^2_L(t)dt\right]\left[\int_{t1}^{t2}P'^2_R(t)dt\right]\right\}^{1/2}} \qquad |\tau|\leq 1.0\ ms. \tag{12.1.10}$$

As stated in Section 1.8.3, the time region $[t_1, t_2]$ for interaural cross-correlation calculation depends on the time window for auditory processing and the problem to be investigated. For stationary stimuli or stationary random stimuli, the time region for calculation is chosen as $[-\infty, +\infty]$ for convenience.

By definition, $0 \leq \Psi_{LR}(\tau) \leq 1$. The auditory event in loudspeaker reproduction can be classified and analyzed according to the characteristics of $\Psi_{LR}(\tau)$.

1. The obvious and positive maximum value in the curve of $\Psi_{LR}(\tau)$ indicates a high interaural correlation and corresponds to a definite virtual source at a certain position. The parameter $\tau = \tau_{max}$ that maximizes $\Psi_{LR}(\tau)$ is defined as the interaural time difference ITD_{corre} calculated through interaural cross-correlation:

$$ITD_{corre}(\theta, \phi) = \tau_{max}. \tag{12.1.11}$$

   In this case, the interaural cross-correlation (IACC) defined in Equation (1.7.2) is identical to $IACC_{sign}$ defined in Equation (1.7.3). The closer the IACC to a unit is, the more definite the virtual source will be.
2. A reduction in the positive maximum value or a smooth curve of $\Psi_{LR}(\tau)$ indicates a decrease in the interaural correlation and leads to an extending and blur virtual source. A further reduction of interaural correlation leads to widening auditory events or the sensation of envelopment. An interaural correlation close to zero may cause two or more splitting auditory events.
3. An obvious and negative minimum value in the curve of $\Psi_{LR}(\tau)$ [or positive maximum value in $|\Psi_{LR}(\tau)|$] corresponds to a negative correlation between binaural pressures, e.g., a negative $IACC_{sign}$ in Equation (1.7.3). This negative correlation leads to unnatural auditory events, such as lateralization.

Various definitions and methods are used for ITD calculation. The ITDs calculated from interaural phase delay difference, interaural envelope delay difference, and interaural group delay difference are addressed in Section 1.6.1. The results and physical significance of ITDs calculated from various methods are different. Some explanations for ITD_{corre} are given as follows:

1. ITD_{corre} is consistent with the auditory signal processing of humans, especially at low frequencies below 1.0 kHz–1.5 kHz.
2. Equation (12.1.11) can be regarded as a type of weighted mean ITD_{corre} over an entire frequency range, and the result is independent from frequency. In practice, the weighted mean ITD_{corre} over a certain frequency range can be calculated by filtering (such as low-pass or band-pass filtering) $p'_L(t)$ and $p'_R(t)$ prior to calculating the cross-correlation function. For a tone stimulus, the resultant ITD_{corre} is identical to the interaural phase delay difference.
3. In practice, only the discrete time samples of binaural pressures are available. Hence, the integral on the continuous time t in Equation (12.1.10) is replaced by the summation

on discrete time. Binaural pressures are often upsampled before calculation to improve the resolution of the resultant ITD. For example, the time resolution of binaural pressures is only 23 μs at a sampling frequency of 44.1 kHz. The time resolution in calculation can be improved to 2.3 μs by 10-time upsampling.

4. Equation (12.1.10) can be calculated in the frequency domain by Fourier transform as

$$\Psi_{LR}(\tau) = \frac{\int P_L'^*(f) P_R'(f) \exp(j2\pi f\tau)\, df}{\left\{\left[\int |P_L'(f)|^2 df\right]\left[\int |P_R'(f)|^2 df\right]\right\}^{1/2}}, \tag{12.1.12}$$

where the superscript "*" denotes the complex conjugation. Calculating the integral in Equation (12.1.12) over different frequency ranges leads to $\Psi_{LR}(\tau)$ and ITD_{corre} in different frequency bands.

The ILD can be evaluated from binaural pressures. The frequency-dependent ILD caused by a real source is calculated from Equation (1.6.7). The mean ILD over certain frequency range of $f_L \le f \le f_H$ can be evaluated as

$$ILD(\theta_S, \phi_S) = 10 \log_{10}\left[\frac{\int_{f_L}^{f_H} |P_L(\theta_S, \phi_S, f)|^2 df}{\int_{f_L}^{f_H} |P_R(\theta_S, \phi_S, f)|^2 df}\right]. \tag{12.1.13}$$

In practice, the frequency range in ILD calculation can be selected as each 1/3 Oct or equivalent rectangular bandwidth (ERB). For reproduction with multiple loudspeakers, ILDs are evaluated by substituting binaural pressures in Equation (12.1.13) with the superposed binaural pressures.

The dynamic cue for localization, such as ITD and ILD variations caused by head turning, can also be evaluated from binaural pressures. For a real source at (θ_S, ϕ_S), the direction of the source with respect to the head becomes $(\theta_S - \delta\theta, \phi_S)$ after the head rotates around the vertical (z) axis anticlockwise with a small azimuth $\delta\theta$ (Figure 6.1a). After the head turns around the front–back (x) axis to the left with a small angle $\delta\gamma$, the new direction (θ''_S, ϕ''_S) of the source with respect to the head is related to (θ_S, ϕ_S) in Figure 6.1b:

$$\begin{aligned} \cos\theta''_S \cos\phi''_S &= \cos\theta_S \cos\phi_S \\ \sin\theta''_S \cos\phi''_S &= \sin\theta_S \cos\phi_S \cos\delta\gamma - \sin\phi_S \sin\delta\gamma. \\ \sin\phi''_S &= \sin\theta_S \cos\phi_S \sin\delta\gamma + \sin\phi_S \cos\delta\gamma \end{aligned} \tag{12.1.14}$$

The binaural pressures for a real source after head turning are calculated by substituting HRTFs with respect to the new orientation of the head into Equation (12.1.1). Similarly, for reproduction with multiple loudspeakers, the direction of each loudspeaker with respect to the head after head turning can be evaluated, and the superposed binaural pressures can be calculated by substituting the corresponding HRTFs into Equation (12.1.6). ITD and ILD after head turning are calculated from the corresponding binaural pressures, and the dynamic variations in ITD and ILD are evaluated by comparing the results before and after head turning.

Monoaural/binaural spectra can be evaluated from Equation (12.1.1) or (12.1.3) and Equation (12.1.6) or (12.1.8).

12.1.2 Method for summing localization analysis

When correlated signals are reproduced by two or more loudspeakers, the perceived direction and quality of a virtual source can be analyzed by the following procedures. These procedures are similar to the course of spatial auditory information processing.

1. For a given stimulus, binaural pressures caused by a target (real) source at various positions are calculated. Then, directional localization cues, such as ITD, ILD, and ITD and ILD variations caused by head turning, and monoaural/binaural spectra, are evaluated.
2. For a given reproduction method, stimulus, and loudspeaker signals, the superposed binaural pressures are calculated, and directional localization cues are evaluated.
3. The perceived virtual source direction is estimated by comparing the localization cues in reproduction with localization cues of the target source.
4. During analysis, the stimuli of the target source and the reproduced one should be identical, and the same set of HRTFs (from the same individual) should be used. In other words, comparison should be conducted under the same condition.

A comprehensive model of auditory spatial information processing is needed in Step (3) by which physically binaural pressures are mapped to the auditory space. As stated in Section 1.6.5, auditory localization is the comprehensive consequence of multiple localization cues in different frequency ranges. If all localization cues that match with those of the target source are created in reproduction, a listener perceives a definite virtual source at the target position. If inconsistent or even conflicting localization cues in different frequency ranges are created in reproduction, the auditory system appears to identify the virtual source position according to the dominant or more consistent cues. However, if too many conflicts or losses exist in localization cues, accuracy, and quality in localization are likely to be degraded, splitting virtual sources are perceived, or localization is even impossible. The course and mechanism of auditory spatial information processing are complicated, a complete and comprehensive model of this processing remains unavailable. In the current stage of investigation, various localization cues in different frequency bands are evaluated individually. Then, the localization information provided by these cues and its consistency are analyzed. This section focuses on the physical analysis of localization cues. Section 12.2 further analyzes the localization cues by using some preliminary binaural auditory models.

At low frequencies below 1.5 kHz, ITD is the dominant cue for lateral localization. An analysis of ITD determines the cone of confusion in which the virtual source is located. Additional analysis of the variation in ITD caused by head turning further determines the position of the virtual source on the cone of confusion. For the virtual source in the horizontal plane, ITD determines the lateral departure of the virtual source and the ITD variation caused by head rotation enables front–back discrimination. For tone and narrow-band stimuli, the interaural correlation-based method yields an ITD that is consistent or basically consistent with interaural phase delay difference. If the interaural cross-correlation in Equation (12.1.12) is calculated over a frequency range (for example, 0.1 kHz–1.5 kHz), the resultant ITD can be regarded as a kind of weighted mean ITD across the frequency range. Accordingly, the virtual source direction evaluated from the weighted mean ITD can be regarded as an approximate estimation of the weighted mean-perceived direction across the frequency range if the perceived virtual source direction in reproduction depends on frequency. However, the virtual source may widen, become blurry, or even fail to be localized because the dispersion of the virtual source direction at different frequencies. The quality or definiteness of the virtual source can also be evaluated by comparing $IACC_{sig}$ in sound reproduction with that of the target (actual) source.

Nakabayashi (1975) suggested the use of naturalness as a criterion to assess the quality of a low-frequency virtual source. Naturalness is assessed with a five-grade scale, e.g., scale of A, B, C, D, and E. It (subjective criterion) is related to the difference ΔILD between the ILD of sound reproduction and that of a real sound source (physical criterion). ΔILD is calculated from following unsigned difference:

$$\Delta ILD = | ILD_{SUM} - ILD |, \tag{12.1.15}$$

where ILD_{SUM} is the interaural level difference of sound reproduction or summing virtual source to distinguish from interaural level difference of a real source. For simplicity, both ILD_{SUM} of sound reproduction and ILD of a real source are represented by ILD if confusion does not occur.

The naturalness of a low-frequency virtual source is approximately related to ΔILD as follows:

1. Grade A, not different from a real source, for $\Delta ILD \approx 0$ dB (or approximately $\Delta ILD \leq$ 1 dB)
2. Grade B, slightly unnatural, for 1 dB < ΔILD < 6–7 dB
3. Grade C, unnatural, for 6–7 dB < ΔILD < 16–18 dB
4. Grade D, very unnatural, for ΔILD > 16–18 dB.
5. Grade E, extremely unnatural (no localization obtained)

Therefore, the naturalness of a low-frequency virtual source in sound reproduction can be approximately or semiquantitatively assessed by comparing the ILD in reproduction with that of a real source.

ILD at High-frequency above 4-5 kHz is also a lateral localization cue. The consistency of high-frequency ILD in sound reproduction and that of a real source can also be examined. However, as stated in Section 1.6.2, high-frequency ILD varies with direction and frequency in a complicated manner. Even within the horizontal azimuth of $0° \leq \theta_S \leq 90°$, high-frequency ILD for a real source with a narrow-band stimulus does not monotonically vary with θ_S. A mapping between ILD and source azimuth can only be obtained by comprehensively analyzing the ILD within an efficient bandwidth.

Errors in monaural/binaural pressure spectra in sound reproduction can also be evaluated by comparing the results of Equations (12.1.6) and (12.1.1). The results that are better related to perception can be obtained by introducing a binaural auditory model in Section 12.2 to analysis.

In addition, HRTFs measured from artificial heads or human subjects can be used in this analysis. The calculated HRTFs are also available (Sections 11.2.1 and 11.2.2). The HRTFs used in this analysis should represent the mean or typical data from a certain population (Xie, 2014). For simplicity, the HRTFs of a rigid-spherical head model in Section 11.2.2 are also used. HRTFs are continuous functions of a source direction, but usually HRTFs sampled at discrete and finite directions are available. HRTFs with efficient directional resolution are required in this analysis. Calculation yields HRTFs with high directional resolution, but measuring HRTFs with very high directional resolution is difficult. If the directional resolution of the measured HRTFs satisfies the condition of the Shannon–Nyquist spatial sampling theorem, HRTFs at unmeasured directions can be obtained through spatial interpolation. The scheme of HRTF spatial interpolation is usually included in the analysis.

12.1.3 Binaural pressure analysis of stereophonic and multichannel sound with amplitude panning

In this section, the method in Section 12.1.2 is applied to analyze some typical examples in stereophonic and multichannel sound reproduction with amplitude panning (Damaske, 1969/1970; Damaske and Ando, 1972; Xie, 1999c).

The first example is stereophonic reproduction with amplitude panning. A pair of stereophonic loudspeakers is arranged at a horizontal azimuth of ±30°. The normalized amplitudes of loudspeaker signals are $A_0 = A_L$ and $A_1 = A_R$. The HRTFs of a rigid-spherical head model with a radius of $a = 0.0875$ m [Equation (11.2.2)] are used in the analysis. Indeed, results may be more accurate if the measured HRTFs of an artificial head are used in the analysis. Interaural correlation-based ITD is first calculated, and the virtual source azimuth is evaluated. Figure 12.1 illustrates the curve of $\Psi_{LR}(\tau)$ for a sinusoidal stimulus with $f = 0.25$ kHz and ICLD of $d = 20\log_{10}(A_L/A_R) = 9$ dB. $\Psi_{LR}(\tau)$ exhibits a maximal value of the unit at $\tau = 0.195$ ms. Therefore, the ITD is 0.195 ms, corresponding to a virtual source at $\theta_I = 14°$. Figure 12.2 shows the virtual source azimuth versus the ICLD, e.g., $d = 20\log_{10}(A_L/A_R)$ at three different frequencies. For a given ICLD except $d = 0$ dB, the virtual source moves toward the direction of the loudspeaker (30°) as frequency exceeds 0.7 kHz, resulting in an unstable virtual source. This phenomenon is consistent with the result of Equation (2.1.5) and Figure 2.3. Equation (2.1.5) and Figure 2.3 are obtained from a shadowless head model with an equivalent or pre-corrected radius $a' = \kappa a$. The results in Figure 12.2 verify the validity of Equation (2.1.5). With Equation (12.1.13), the ILD in stereophonic reproduction does not exceed 0.2 dB at $f = 0.2$ kHz. This result is consistent with that of a real sound source. As frequency increases (especially above 1.5 kHz), the ILD caused by ICLD in stereophonic reproduction increases. The high-frequency ILD in stereophonic reproduction may not quantitatively match with that of a real source, and the difference may depend on individuals (Breebaart, 2013). However, the qualitative features of ILD in stereophonic reproduction are similar to those of a real source. Therefore, stereophonic sound with amplitude panning does not lead to a conflicting ILD cue and promises a better perceived quality (especially naturalness) of the virtual source.

The second example is the first-order horizontal Ambisonics with four loudspeakers. The loudspeaker configuration is shown in Figure 4.1, and the conventional solution of loudspeaker signals is given in Equation (9.3.12). The HRTFs of a rigid-spherical head model with a radius of $a = 0.0875$ m are used in analysis. Interaural correlation-based ITDs are first

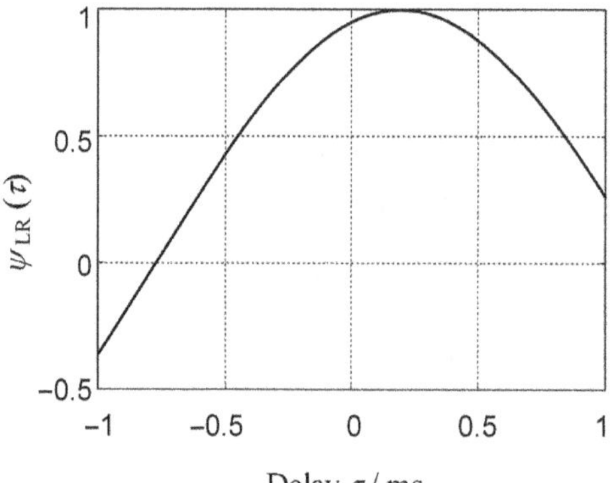

Figure 12.1 Curve of $\Psi_{LR}(\tau)$ for two-channel stereophonic reproduction in which a sinusoidal stimulus with a frequency of $f = 0.25$ kHz and ICLD of $d = 9$ dB.

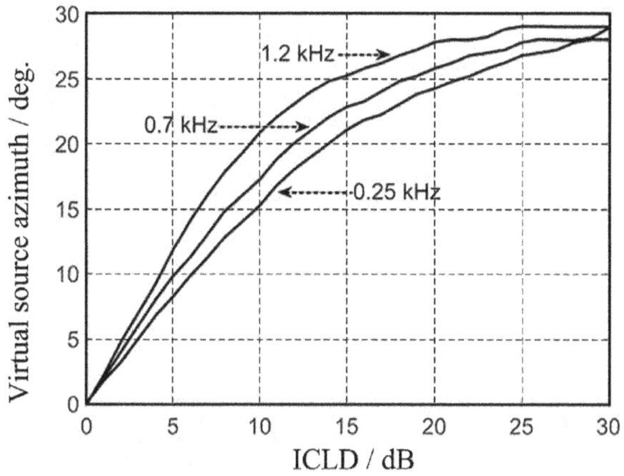

Figure 12.2 Curves of the virtual source azimuth versus the ICLD at three different frequencies.

calculated, and the virtual source azimuth is evaluated. During the evaluation, the front–back discrimination of a virtual source is implemented by considering the dynamic ITD variation caused by head rotation around the vertical axis. The perceived virtual source azimuth θ_I versus the target azimuth θ_S for sinusoidal stimuli with $f = 0.25$, 0.5, and 0.7 kHz is illustrated in Figure 12.3. For comparison, $\theta_I = \theta_S$ of Equation (4.1.16) for low-frequency localization is also presented in the Figure 12.3. The results within the target azimuth range of $0° \leq \theta_S \leq 90°$ are given only because of symmetry. As frequency increases, the virtual source moves toward the direction of the left-frontal (LF) loudspeaker (45°) except for the case of $\theta_S = 45°$. The results in Figure 12.3 are consistent with those in Section 4.1.3 (Figure 4.6), and they further verify the validity of the virtual source localization theorem used in Chapters 4 and 5.

The third example is the first-, second-, and third-order horizontal Ambisonics with eight loudspeakers. The eight loudspeakers are arranged at 0°, ±45°, ±90°, ±135°, and 180°, respectively. The stimulus is a 0.1–1.5 kHz band-pass filtered pink noise. The method and condition for analysis are similar to those in the second example. The perceived virtual source azimuth θ_I versus the target azimuth θ_S is illustrated in Figure 12.4. For comparison, $\theta_I = \theta_S$ of Equation (4.1.16) for low-frequency localization is also presented in Figure 12.4. For a stimulus with

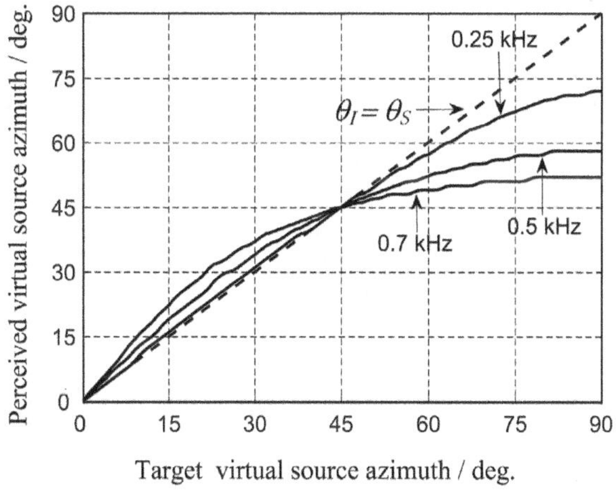

Figure 12.3 Perceived virtual source azimuth for the first-order horizontal Ambisonics with four loudspeakers and at three frequencies.

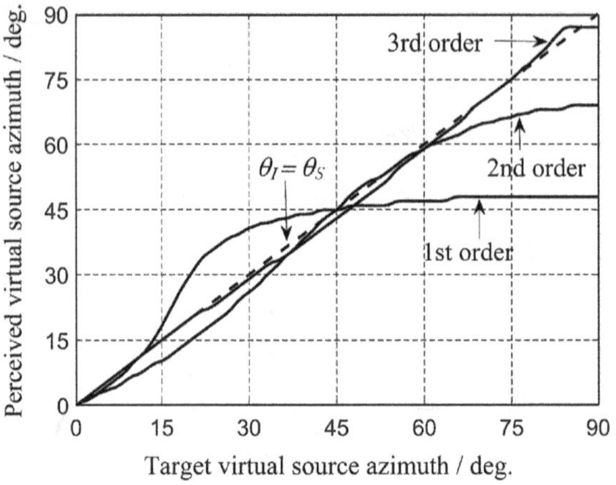

Figure 12.4 **Perceived virtual source azimuth for the first-, second-, and third-order horizontal Ambisonics with eight loudspeakers and a stimulus of 0.1–1.5 kHz band-pass filtered pink noise.**

a certain bandwidth, the resultant virtual source azimuth can be regarded as an approximate estimation of the weighted mean-perceived azimuth across the bandwidth. The results within the target azimuth range of $0° \leq \theta_S \leq 90°$ are given only because of symmetry. For the first-order reproduction, the virtual source, especially the lateral virtual source, moves toward the direction of the LF loudspeaker (45°). As the order increases, the movement of the lateral virtual source reduces. For the third-order reproduction, the perceived virtual source direction is basically consistent with the target direction, e.g., $\theta_I \approx \theta_S$ within the range of $0° \leq \theta_S \leq 90°$. Therefore, the third-order reproduction yields an ideal localization performance.

The definiteness of the virtual source can be evaluated qualitatively from the interaural correlation of binaural pressures. For example, for the stimulus of the 0.1-1.5 kHz band-pass filtered pink noise and the target azimuth of $\theta_S = 75°$, the curve of $\Psi_{LR}(\tau)$ for a real source and the first-, second-, and third-order horizontal Ambisonics is shown in Figure 12.5. For the real source, $\Psi_{LR}(\tau)$ maximizes at $\tau = 0.686$ ms with $IACC_{sig} = 0.980$. For the first-, second-, and third-order Ambisonics, $\Psi_{LR}(\tau)$ maximizes at $\tau = 0.520$, 0.653, and 0.687 ms and

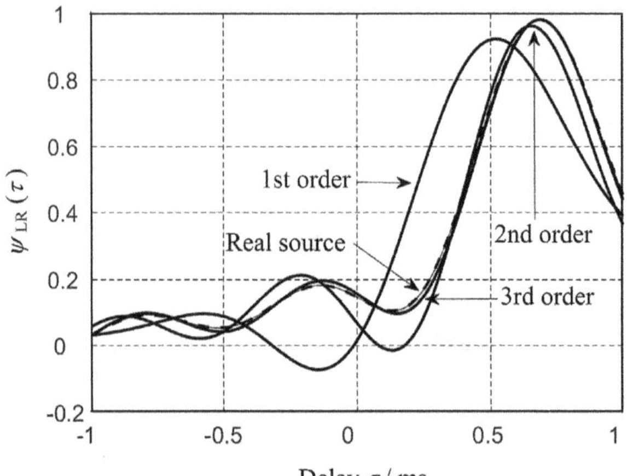

Figure 12.5 **Curves of $\Psi_{LR}(\tau)$ of the preceding three-order horizontal Ambisonics with a stimulus of 0.1–1.5 kHz band-pass filtered pink noise and $\theta_S = 75°$.**

with $IACC_{sig}$ = 0.922, 0.961, and 0.981, respectively. In comparison with the case of the real source, the first-order reproduction exhibits a deviated ITD and a reduced $IACC_{sig}$, thus leading to the movement of the virtual source direction and the reduction in the definiteness of the virtual source. The reduction of definiteness is due to the frequency-dependent perceived virtual source direction in lower-order Ambisonic reproduction; consequently, the virtual source of stimuli with some bandwidths broadens and becomes blurry. A blurred virtual source is not desired in terms of the quality of the virtual source. As the order of Ambisonics increases, the movement of the weighted mean-perceived direction of the virtual source reduces, and the definiteness of the virtual source improves. This condition is proven in Section 9.3.1, which shows that the reconstructed sound field of the higher-order Ambisonics approaches the target sound field.

The ILD in reproduction can also be analyzed. Figure 12.6 illustrates the curves of the ILD versus the frequency for the preceding three-order horizontal Ambisonics with a target azimuth of $\theta_S = 75°$. Figure 12.6 also illustrates the ILD of a real source. The ILD of the first-, second-, and third-order Ambisonics matches with that of the real source below 0.7, 1.2, and 1.8 kHz, respectively. Therefore, the upper frequency limit of a matching ILD increases as the order increases. The ILD for Ambisonics deviates from that of the real source above the upper limit. Some narrow peaks may form in the ILD for Ambisonics. The results are evaluated using the HRTFs of a rigid-spherical head model. Using measured HRTFs from an artificial head may lead to more accurate results, but the conclusion is similar. Equation (9.3.15) indicates that the first-, second-, and third-order-horizontal Ambisonics can reconstruct the target sound field accurately up to the frequencies of 0.62, 1.25, and 1.87 kHz within a region with a head radius of $r = a = 0.0875$ m, respectively. Therefore, the results of ILD analysis are basically consistent with those evaluated from Equation (9.3.15). Benjamin (2010) further analyzed the ILD of the horizontal Ambisonics with eight loudspeaker and signals of maximizing the energy vector magnitude at mid and high frequency. The results also deviate from those of a real source. Therefore, optimizing the Ambisonic signals with maximizing the energy vector magnitude at mid- and high-frequency does not reduce the deviation of mid- and high-frequency ILD in reproduction.

The fourth example is to analyze the dynamic localization cue in spatial Ambisonic reproduction (Xie et al., 2017a). The 28+1 loudspeaker configuration in Figure 9.6 is used. As stated in Section 9.4.1, this loudspeaker configuration is applicable to spatial Ambisonic

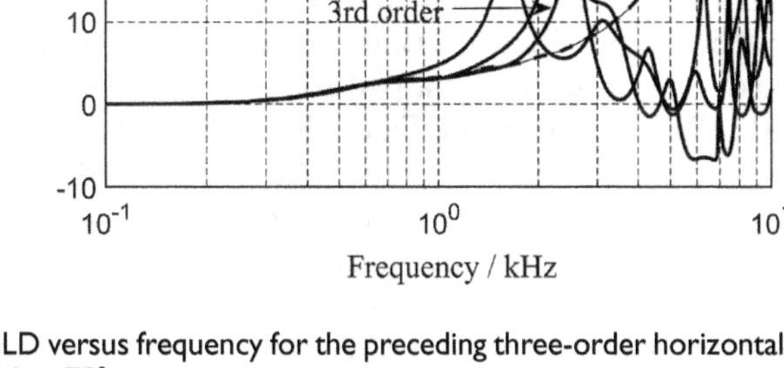

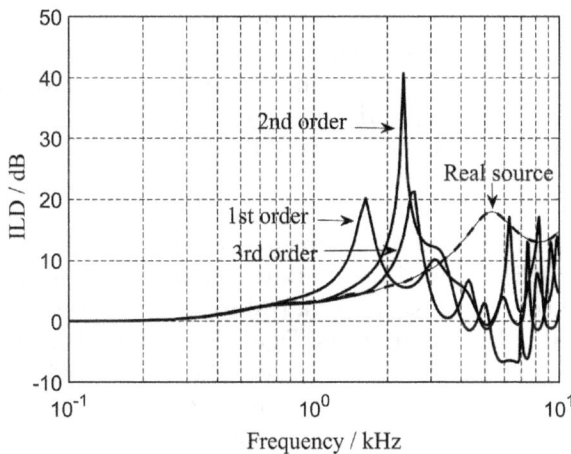

Figure 12.6 Curves of ILD versus frequency for the preceding three-order horizontal Ambisonics with a target azimuth of $\theta_S = 75°$.

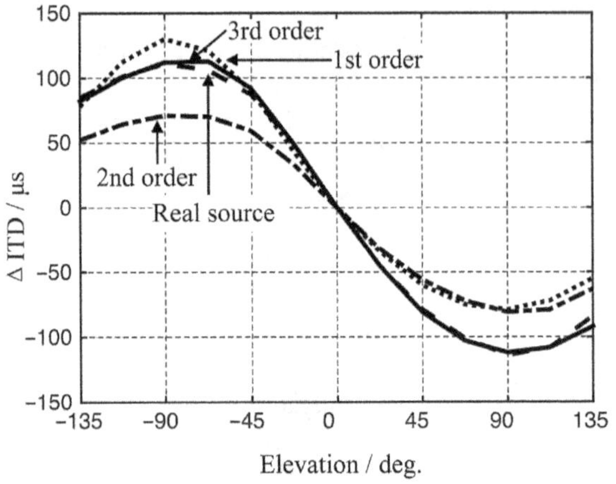

Figure 12.7 For a real source and the preceding three-order Ambisonics, various target directions in the median plane, the variation in ITDs after the head rotates anticlockwise around the vertical axis with an angle of 10°.

reproduction up to the third order. The ITDs and their variations with head turning for a real source and the preceding three-order Ambisonics are calculated. The loudspeaker signals are given in the conventional solution of Equation (9.3.23). The BEM-based calculated HRTFs of a KEMAR artificial head with DB60/61 small pinnae are used in the analysis (Section 11.2.2; Liu et al., 2016). The directional resolution of HRTFs is 1°. The calculation of interaural cross-correlation in Equation (12.1.12) is up to a frequency of 1.5 kHz because ITD is a dominant localization cue at low frequencies below 1.5 kHz.

The calculation results indicate that the ITDs for a real source and the preceding three-order Ambisonics are approximately zero (does not exceed 10 μs) for various target directions in the median plane. Figure 12.7 illustrates the variation in ITD after head rotating anticlockwise around the vertical (z) axis with an angle of 10°. Similar to the case of Figure 6.5, selecting a new elevation angle of $-180° < \varphi \leq 180°$ in the median plane is convenient. The new elevation angle is related to the default angle when $\theta = 0°$ and $\varphi = 90° - \phi$ and when $\theta = 180°$ and $\varphi = -90° + \phi$. Therefore, the top, front, and back directions are denoted by $\varphi = 0°$, 90°, and −90°, respectively. Figure 12.7 also shows that the dynamic ITD variation in Ambisonics exhibits a tendency similar to that of the real source. The variation in the ITD of the third-order Ambisonics is consistent with that of the real source. The results of head tilting around the front–back (x) axis are similar. Therefore, the third-order Ambisonics can create correct ITD, and its dynamic variation below 1.5 kHz.

The fifth example is the magnitude spectra of the binaural pressures in spatial Ambisonic reproduction. Gorzel et al. (2014) analyzed the binaural pressures of the third-order spatial Ambisonics. The results indicated that obvious spectral distortion at a high frequency occurs. A similar problem occurs in lower-order binaural Ambisonic reproduction (Section 11.6.1; Kearney and Doyle, 2015). Figure 12.8 illustrates the magnitude spectra of pressure at the entrance of the (blocked) left ear canal for a real source and for the preceding three-order spatial Ambisonics with 28+1 loudspeakers (Figure 9.6; Xie et al., 2017a). The target direction is $(\theta_S, \phi_S) = (0°, 67.5°)$ in the median plane. The BEM-based calculated HRTFs of a KEMAR artificial head with DB60/61 small pinnae are used in analysis. In Figure 12.8, the magnitude spectra of the first-, second-, and third-order reproduction match with that of the real source up to 0.6, 1.3, and 1.9 kHz, respectively. The upper frequency limit increases as the order increases. The magnitude spectra of Ambisonic reproduction deviate from that of the real source above the upper frequency limit. The results observed in the right ear are

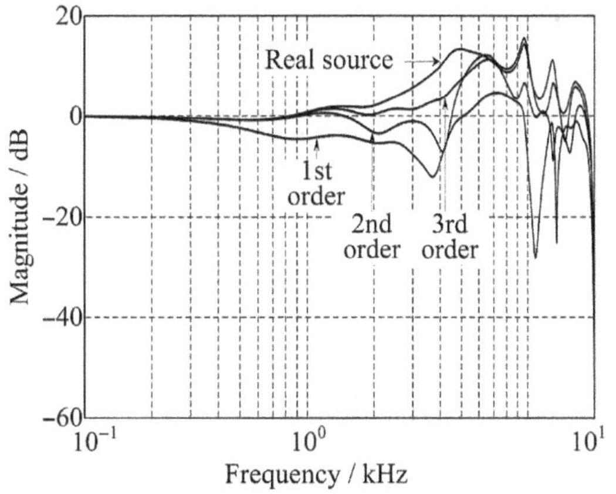

Figure 12.8 Magnitude spectra of pressure at the entrance of the (blocked) left ear canal for a real source and for the preceding three-order spatial Ambisonics with 28 + 1 loudspeakers. The target direction is $(\theta_S, \phi_S) = (0°, 67.5°)$.

similar because of symmetry. Equation (9.3.15) indicates that the first-, second-, and third-order-horizontal Ambisonics can reconstruct a target sound field accurately up to the frequency of 0.62, 1.25, and 1.87 kHz within a region with a head radius of $r = a = 0.0875$ m, respectively. Therefore, the results of binaural pressure analysis are consistent with those of the sound field analysis.

The sixth example involves exploring the influence of the number of loudspeakers on the binaural pressure error for Ambisonic reproduction (Jiang et al., 2018). For the $(L - 1)$-order reproduction, $M \geq L^2$ loudspeakers are arranged on a spherical surface with a far-field distance. The analysis indicates that error in binaural pressures is small for a head radius of $r = a = 0.0875$ m and below the upper frequency limit given in Equation (9.3.30). Near the upper frequency limit, increasing the number of loudspeakers from the low limit L^2 to $(L + 1)^2$ or $(L + 2)^2$ effectively reduces the error in binaural pressures. However, the relationship between the binaural pressure error and the number of loudspeakers becomes complicated when the number of loudspeakers increases to more than $(L + 2)^2$ or the frequency increases to a higher range. Based on the results of this work, the conclusion of some previous studies should change (Solvang, 2008). Therefore, the results of binaural pressure analysis are consistent with those of sound field analysis in Section 9.5.1.

The method presented in this section is also applicable to the analysis of the virtual source localization in other multichannel sound reproductions, such as 5.1- and 7.1-channel sound in Chapter 5 and multichannel spatial surround sounds with different loudspeaker configurations and amplitude panning techniques (such as VBAP) in Chapter 6. The results are similar to those of the simplified analyses in Chapters 5 and 6 and therefore omitted here.

12.1.4 Analysis of summing localization with interchannel time difference

Some psychoacoustic experimental results of the summing localization of two stereophonic loudspeakers with interchannel time difference (ICTD) are presented in Sections 1.7.1 and 2.1.4, in which the discussions are focused on wideband stimuli with transient characteristics. In this section, binaural pressures, localization cues, and summing localization of the stereophonic reproduction of steady and narrow-band stimuli at a low frequency with ICTD are further analyzed (Xie, 2002a).

A pair of stereophonic loudspeakers is arranged at azimuths of ± 30°. The magnitudes of two loudspeaker signals are identical, but the signal of the left loudspeaker is leading in time with Δt. In this case, the ICLD vanishes, and an ICTD = Δt exists in the loudspeaker signals. For a sinusoidal stimulus, the normalized complex-valued amplitudes of loudspeaker signals are given as

$$A_L = A_0 = |A_L| \exp(j2\pi f \Delta t) \qquad A_R = A_1 = |A_R| \qquad |A_L| = |A_R|. \tag{12.1.16}$$

The binaural pressure in reproduction is calculated using the method in Section 12.1.1. The HRTFs of a rigid-spherical head model with a radius of $a = 0.0875$ m [Equation (11.2.2)] are used in the analysis. The correlation-based ITD in reproduction is calculated. For a sinusoidal or narrow-band stimulus, the correlation-based ITD is equivalent to the interaural phase delay difference, and the latter is a dominant lateral localization cue at a low frequency of $f < 1.5$ kHz. θ_I of the virtual source can be evaluated by comparing the ITD in reproduction with that of a real source. Moreover, ILD can be calculated to analyze the naturalness of a summing virtual source.

Figure 12.9 illustrates the curves of the virtual source azimuth θ_I versus the ICTD for a sinusoidal stimulus of $f = 0.25$, 0.5, and 0.7 kHz. Figure 12.10 shows the curves of ILD versus ICTD. Figures 12.9 and 12.10 reveal the following:

1. For a stimulus with $f = 0.25$ kHz, the virtual source is located near the front direction of $\theta_I \approx 0°$ when $0 \leq \text{ICTD} \leq 1.0$ ms. However, |ILD| in reproduction increases as ICTD increases. |ILD| = 6.4 dB when ICTD = 1.0 ms. The comparison with the case of a real source at $\theta_S = 0°$ with ILD = 0 dB yields ΔILD = 6.4 dB. Therefore, the naturalness of the summing virtual source lies at the border between slightly unnatural and unnatural (Grades B and C).
2. For a stimulus with $f = 0.5$ kHz, the virtual source azimuth increases slowly as the ICDT increases from 0 to 0.4 ms. However, the virtual source azimuth increases quickly as the ICDT increases from 0.4 to 0.8 ms and may exceed the boundary bounded by two loudspeakers ($\theta_I > 30°$). When ICTD > 0.8 ms, the resultant ITD exceeds the maximum possible ITD caused by a horizontal real source at $\theta_S = 90°$. In this case, the

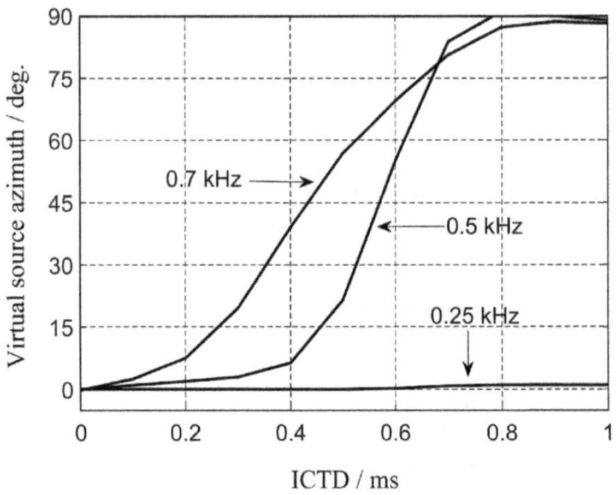

Figure 12.9 Variation in the virtual source azimuth versus interchannel time difference (ICTD) in stereophonic reproduction for a low-frequency sinusoidal stimulus at three different frequencies.

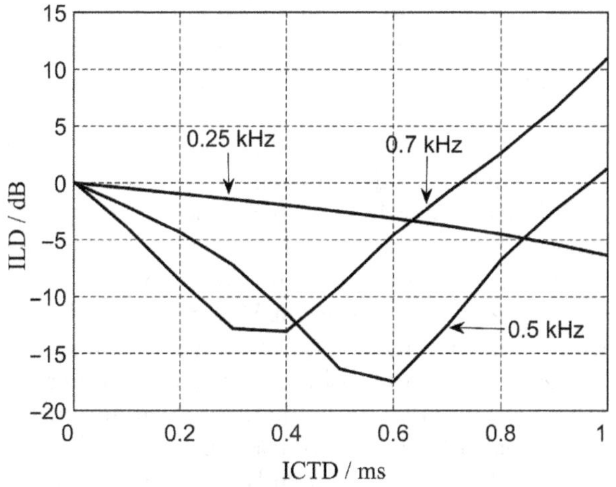

Figure 12.10 Variation in ILD versus ICTD in stereophonic reproduction for a low-frequency sinusoidal stimulus at three different frequencies.

virtual source azimuth cannot be evaluated from the resultant ITD. In practice, a slight lateral translation of a listener's head causes a difference in path lengths between the two loudspeakers to the head center. This difference in path lengths is equivalent to an additional ICTD between loudspeaker signals. A quick variation in the virtual source azimuth with ICTD indicates an unstable virtual source against head translation. The ICTD may also cause a larger |ILD|. Moreover, a conflicting ITD > 0 and ILD < 0 may occur. For example, when ICTD = 0.6 ms, ILD ≈ −17.5 dB. In this case, a virtual source is unnatural (Grade D) even though it exists (Grade D).

3. For a stimulus with f = 0.7 kHz, ICTD results in a different behavior in summing localization. The virtual source azimuth θ_I increases as the ICTD increases. When ICTD > 0.35 ms, the virtual source exceeds the boundary bounded by two loudspeakers ($\theta_I > 30°$). In addition, conflicting ITD and ILD occur.

Therefore, for a sinusoidal or narrow-band stimulus at a low frequency, ICTD leads to a complicated behavior in summing localization. At very low frequencies, ICTD cannot effectively create ITD for the lateral displacement of the virtual source. The function of ICTD in the lateral displacement of the virtual source becomes obvious only when f > 0.5–0.7 kHz. The direction of the virtual source created by ICTD only depends greatly on frequency, and the outside-boundary virtual source may occur.

For a sinusoidal or narrow-band stimulus at low frequencies, ICTD is closely related to interchannel phase difference, e.g., $\eta = 2\pi f \times$ ICTD. For a stimulus with f = 0.25 kHz, we have $\eta \leq \pi/2$ (90°) when 0 ≤ ICTD ≤ 1.0 ms. According to the analysis in Section 2.1.3, for stereophonic signals with equal magnitudes and a phase difference of less than 90°, the virtual source is located at the front of $\theta_I = 0°$. For a stimulus with f = 0.5 kHz, an increase in ICTD may cause $\eta > \pi/2$ and then lead to the case of an "outside-boundary virtual source" or an auditory event with an uncertain position. For example, ICTD = 1 ms corresponds to $\eta = \pi(180°)$. According to Section 2.1.3, stereophonic signals with equal magnitudes but opposite phase create auditory events with uncertain positions. Therefore, the results are basically consistent with the qualitative analysis in Section 2.1.3. As frequency increases further, the situation becomes more complicated and confused because of the periodic characteristic of interchannel phase difference.

The analysis in summing localization is based on low-frequency ITD only. If the low-frequency ILD caused by ICTD is considered, the situation becomes more complicated. A conflicting ILD degrades the naturalness of the summing virtual source (even if it exists) or even leads to confused spatial auditory events.

Overall, at a low frequency below 1.5 kHz, ICTD only creates a frequency-dependent or inconsistent ITD cue and simultaneously causes a conflicting ILD cue. For a sinusoidal or narrow-band stimulus, ICTD may develop a virtual source, but the naturalness or perceived quality of the virtual source is poor. This result has been proven by a virtual source localization experiment (Xie, 2002a). For steady stimuli with certain bandwidths, such as 0.1–1.5 kHz low-pass filtered pink noise, an inconsistent ITD cue across frequency cannot create a virtual source at a definite position. Instead, it produces a widening and blur auditory event or an auditory event with an uncertain position. This conclusion can also be derived from the IACC calculation in Equations (12.1.12) and (1.7.2). In addition, if the HRTFs of a rigid-spherical head model in Equation (11.2.2) are used in analysis, binaural pressures and the corresponding interaural localization cues depends on ka. The frequency-dependent virtual source position implies a head radius-dependent virtual source position. In this case, the perceived virtual source position depends on individuals.

The above method is also applicable to the analysis of the summing localization of low-frequency stimuli with a combination of ICLD and ICTD. Overall, when the ICTD is small or when the frequency is low enough so that the absolute value of interaural phase difference does not exceed $\pi/2$, different ICLDs create a virtual source between two loudspeakers. The virtual source direction depends on frequency and varies with ICLD in a similar manner at different frequencies. For stimuli with different frequency components, an integrated but widening and blur virtual source may be recreated. When the absolute value of interaural phase difference exceeds $\pi/2$ because of the combination of frequency and ICTD, inconsistent localization cues and confused spatial auditory events may occur. For various near-coincident microphone techniques in Section 2.2.4 (such as the ORTF pair), the smaller distance between two microphones leads to an interaural phase difference less than $\pi/2$ below certain frequencies (such as 0.7 kHz). Therefore, a near-coincident microphone technique often yields an appropriate localization performance in stereophonic reproduction.

The aforementioned analysis is valid only for a sinusoidal or narrow-band stimulus at low frequencies. In Section 1.7.1, for wideband stimuli with transient characteristics, ICTD only causes the lateral displacement of the virtual source, but the perceived quality of the virtual source is inferior to that created via the ICLD-based method. ICTD-based summing localization for wideband stimuli with transient characteristics has only been investigated through a psychoacoustic experiment. In the future, further analysis should be based on the auditory mechanism and model of high-level systems. For wideband stimuli with transient characteristics, when the ICTD exceeds the lower limit of the precedence effect, a different auditory model should be used in the analysis.

12.1.5 Analysis of summing localization at the off-central listening position

Low-frequency summing localization at a central listening position is preliminarily analyzed in Sections 2.1 and 3.2. A listener's head may deviate from the central listening position in practical reproduction. For example, in stereophonic reproduction (Section 2.4.3), a lateral translation of the head away from the central line (at off-central line) not only makes the two loudspeakers left–right asymmetric with respect to the head but also leads to differences in path length between the two loudspeakers and the head center. The difference in path length results in equivalent ICTD and ICLD. For some wideband stimuli with transient characteristics, if the lateral translation of the head is large so that the combination of equivalent ICTD and ICLD reaches the lower limit of the precedence effect, the perceived virtual source appears at the position of the nearer loudspeaker. If the lateral translation of the head is appropriate so that the combination of equivalent ICTD and ICLD does not reach the lower limit of precedence effect, the summing virtual source is located between two loudspeakers. Summing localization may be based on more consistent cues or some unknown mechanisms. However, the rule of summing localization with a combination of ICTD and ICLD for narrow-band stimuli at a low frequency is different from that for bandwidth (transient) stimuli. In the following, the summing localization at the off-central listening position with narrow-band stimuli at a low frequency is analyzed (Xie and Guo, 2004).

For finite loudspeaker distances with respect to the head center, the binaural superposed pressures are calculated with Equation (12.1.8). The distances and azimuths of left and right loudspeakers with respect to the head center are evaluated with Equations (2.4.6) and (2.4.7). For a rigid-spherical head model, distance-dependent HRTFs are calculated with Equation (11.2.3). Similar to the procedures in Section 12.1.1, the correlation-based ITD is examined from the binaural superposed pressures, and the direction of the summing virtual source is assessed. The ILD can also be analyzed.

When the head deviates from the center position with a distance y_1 (with $y_1 > 0$ representing a translation to the left and $y_1 < 0$ the right), the virtual source azimuth θ'_I with respect to the new position of the head center is evaluated by comparing the resultant ITD with that of a real source. The azimuth θ'_I can be converted to the azimuth θ_I with respect to origin in terms of the geometrical relationship in Figure 12.11 to evaluate the variation in the perceived virtual source azimuth caused by head translation:

$$\theta_I = \theta'_I + \arcsin\left[\frac{y_1 \cos(\theta'_I)}{r_S}\right]. \qquad (12.1.17)$$

For sinusoidal stimuli at a low frequency, the virtual source azimuth at the off-central position can be evaluated from the summing localization with a combination of ICTD and ICLD similar to that in Section 12.1.4. The results can also be used for low-frequency stimuli with a bandwidth of less than 1/3 Oct. A standard stereophonic loudspeaker configuration with $2\theta_0 = 60°$ and a distance of 2.0 m with respect to the origin is assumed. The radius of the spherical head model is $a = 0.0875$ m.

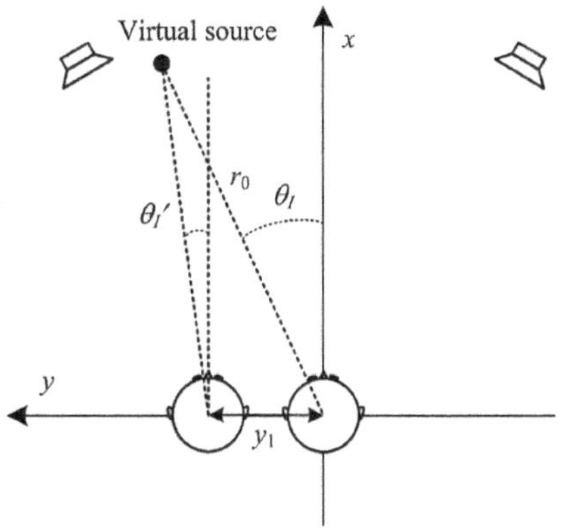

Figure 12.11 Virtual source position with respect to the central and off-central listening positions.

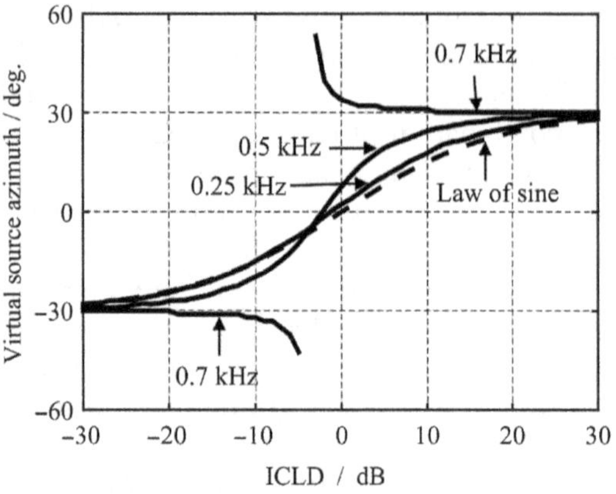

Figure 12.12 Virtual source azimuth versus ICLD = 20 $\log_{10}(A_L/A_R)$ (dB) in stereophonic reproduction and a listening position of 0.125 m translation to the left.

Figure 12.12 illustrates the perceived virtual source azimuth versus ICLD for the sinusoidal stimuli of 0.25, 0.5, and 0.7 kHz at a listening position of y_1 = 0.125 m to the left. For comparison, Figure 12.12 also presents the result of the central listening position evaluated from the law of sine in Equation (2.1.6). The following observations are detected:

1. For the stimuli of f = 0.25 and 0.5 kHz, the virtual source azimuth varies continuously from the direction of the right loudspeaker (−30°) to that of the left loudspeaker (30°) when ICLD changes from −30 dB to 30 dB. The pattern of variation is similar to that given by the law of sine. However, after a head translation to the left, the virtual source direction moves toward the direction of left loudspeakers when ICLD >−3 dB. The movement of the virtual source for the 0.5 kHz stimulus is larger than that for the 0.25 kHz stimulus. When ICLD <−3 dB, the virtual source direction for the 0.25 kHz stimulus is basically unchanged after the head translation to the left, but the virtual source direction for the 0.5 kHz stimulus moves toward the direction of the right loudspeaker (−30°).
2. For the stimuli of f = 0.7 kHz, summing localization exhibits a different feature and outside-boundary virtual source may occur. For ICLD ≥ −3 dB, the virtual source is outside the left loudspeaker with $\theta_I > 30°$; for ICLD ≤ −5 dB, the virtual source is outside the right loudspeaker with $\theta_I < -30°$. An abrupt variation or discontinuity in virtual source direction occurs near ICLD = −4 dB.

At an off-central listening position, the equivalent ICTD caused by the path difference between the left and right loudspeakers to the head center is evaluated with Equation (2.4.8). For a sinusoidal stimulus, the ICTD corresponds to an interaural phase difference. According to the analysis in Section 2.1.3, when the path difference between the left and right loudspeakers to the head center is less than a quarter of the wavelength so that the equivalent interaural phase difference $|\eta| \leq \pi/2$ (90°), the virtual source is located between two loudspeakers. Otherwise, when the equivalent interaural phase difference $|\eta| \geq \pi/2$, an outside-boundary virtual source or an auditory event with an uncertain position may appear.

According to Equation (2.4.8), when a listener's head deviates from the central listening positions, the equivalent phase difference of the sound wave from the left and right loudspeakers are evaluated by

$$\eta = 2\pi f \Delta t \approx \frac{4\pi f y_1 \sin\theta_0}{c}, \tag{12.1.18}$$

where c is the speed of sound, and the second approximate equality in Equation (12.1.18) is valid for $y_1 << r_0$. In the above example of $y_1 = 0.125$ m, Equation (12.1.8) yields $\eta = 0.18\pi$, 0.36π, and 0.51π for $f = 0.25$, 0.5, and 0.7 kHz, respectively. Therefore, for the stimuli of $f = 0.25$ and 0.5 kHz, the virtual source is between two loudspeakers. For the stimulus of $f = 0.7$ kHz, the virtual source may be outside the loudspeakers.

Under the condition of $|\eta| \leq \pi/2$, Equation (12.1.18) leads to

$$|y_1| \leq y_{\max} = \frac{c}{8f \sin\theta_0} \tag{12.1.19}$$

or

$$f \leq f_{\max} = \frac{c}{8\pi\, |y_1|\, \sin\theta_0}. \tag{12.1.20}$$

According to Equation (12.1.19), for a stimulus with f, when a listener's head deviates from the central position with a distance $|y_1| \leq y_{max}$, the virtual source is located between two loudspeakers. Otherwise, when $|y_1| > y_{max}$, the virtual source may be outside the loudspeakers. y_{max} can be regarded as the maximal allowable distance of head translation. It is inversely proportional to the frequency. That is, the higher the frequency is, the smaller y_{max} will be.

According to Equation (12.1.20), for a given distance $|y_1|$ of head translation, when $f \leq f_{max}$, the virtual source is located between two loudspeakers. Otherwise, when $f > f_{max}$, the virtual source may be outside the loudspeakers. f_{max} can be regarded as an upper frequency limit of the desired spatial information reproduction. It is inversely proportional to $|y_1|$. That is, the larger the head translation is, the smaller f_{max} will be.

If $|\eta| \leq \pi/2$ is required up to $f = 1.5$ kHz, Equation (2.1.19) yields $|y_1| < 0.057$ m. Therefore, the listening region for stereophonic reproduction is not wide. Similar to the case in Section 11.8.3, a narrow span angle between the left and right loudspeaker pairs widens the listening region.

According to the discussion in Sections 12.1.3 and 12.1.4, the equivalent ICTD or interchannel phase difference in Equation (12.1.18) results in ILD at low-frequency binaural pressures. Figure 12.13 illustrates the ILD of stereophonic reproduction at a listening position of $y_1 = 0.125$ m to the left, and the calculation conditions are identical to the abovementioned conditions. For $f = 0.25$ kHz, the magnitude of ILD is small. A large magnitude of ILD is observed at $f = 0.5$ and 0.7 kHz. When ICLD > 0 dB, conflicting ITD > 0 and ILD < 0 occurs. The ITD at a low frequency is considered in the above analysis of summing localization. The situation becomes more complicated if the influence of ILD on summing localization is further considered. A conflicting ILD degrades the naturalness of the perceived virtual source (if it exists) at least or even creates confusing auditory events.

In summary, at the off-central listening position, the perceived virtual source position for narrow-band stimuli at a low frequency exhibits the following features:

1. For a small head translation or at a low frequencies so that the absolute value of the equivalent interchannel phase difference does not exceed $\pi/2$, the summing virtual source is between two loudspeakers. The position of the summing virtual source varies with ICLD and frequency. Otherwise, an outside-boundary virtual source or even an auditory event with an uncertain position may occur.
2. Conflicting ITD and ILD degrades the naturalness of the virtual source or even leads to a confusion auditory event.

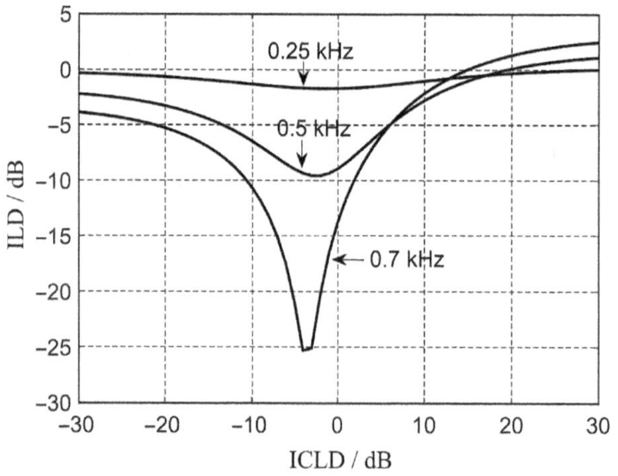

Figure 12.13 ILD for stereophonic reproduction at a listening position of y_l = 0.125 m to the left.

3. Head translation may result in confused auditory events because the rule of summing localization depends greatly on frequency, for low-frequency stimuli with a certain bandwidth (such as low-pass filtered noise).

A virtual source localization experiment validated the above analysis (Xie and Guo, 2004). The above analysis and conclusion are only valid for narrow-band stimuli at low frequencies. For wideband stimuli with transient characteristics, the auditory events of summing localization or the ones caused by the precedence effect may occur in the off-central listening position, depending on the path difference between two loudspeakers to the listening position and ICLD of loudspeaker signals. This case has been explored via psychoacoustic experiments, and further analysis may rely on the binaural auditory mechanism and model of high-level processing.

A similar method is applicable to the analysis of the virtual source localization at the off-central listening position in other multichannel sound reproduction processes, such as the frontal virtual source in 5.1-channel reproduction. The results indicate that adding a center loudspeaker improves the stability of the frontal virtual source and widens the listening region (Xie and Guo, 2004). Moreover, Liu and Xie (2013b) analyzed the error of binaural pressures caused by head translation in the second- or third-order horizontal Ambisonics with six or eight loudspeakers. The results indicated that the Ambisonics exhibits less error in binaural pressures compared with conventional pair-wise amplitude panning. With the advantage of the analog relationship between the HRTF directional interpolation and signal mixing methods of multichannel sound in Section 11.5.3, this analysis can be extended to evaluate the error in interpolated HRTF caused by a slight head movement in original HRTF measurement.

12.1.6 Analysis of interchannel correlation and spatial auditory sensations

The summing localization with two or more loudspeakers with correlated stimuli is analyzed in Sections 12.1.2 to 12.1.4. In Sections 1.7.3 and 3.3, two or more loudspeaker reproduction with partially or low-correlated stimuli create widening and blurred auditory events. Therefore, decorrelated stimuli are often used in stereophonic and multichannel sound reproduction to recreate spatial auditory sensations similar to those in a concert hall, such as auditory spatial impression. This section analyzes the relationship between interchannel correlation and spatial auditory sensations.

In two or more loudspeaker reproduction, binaural pressures are the superposition of those caused by all loudspeakers. In Section 12.1.1, the interaural cross-correlation of binaural pressures is closely related to spatial auditory perception. Even for a real concert hall, interaural correlations within different periods of reflections are closely related to but do not completely determine auditory spatial impression (including auditory source width and listener envelopment). Moreover, in Section 1.8.3, the fused auditory perceptions or scenes caused by direct, early, and late reflected sounds are consequences of the comprehensive and ensemble courses of auditory information processing. Analyzing the spatial auditory perceptions caused by these sounds individually is convenient but not always reasonable. The problem becomes more complicated when spatial auditory perceptions similar to those in a concert hall are simulated by two or more loudspeakers with decorrelated stimuli. A strict analysis should be based on the mechanism and model of spatial auditory information processing of the high-level neural system. However, such a model is still unavailable. As preliminary studies, some simple analyses on the binaural superposed pressures and interaural correlation may yield qualitative and interesting results.

Damaske (1969/1970; Damaske and Ando, 1972) analyzed the interaural correlation in the loudspeaker reproduction of decorrelated stimuli. A pair of decorrelated stimuli (0.25–2.0 kHz band-pass filtered white noise with ICCC of 0.13) is reproduced by a pair of loudspeakers arranged symmetrically in the frontal or rear horizontal plane. The interaural correlations for different loudspeaker arrangements are analyzed. Two physical criteria related to interaural correlation, e.g., IACC in Equation (1.7.2) and $\Psi_{LR}(\tau)$ at $\tau = 0$ in Equation (12.1.10) [denoted by $\Psi_{LR}(0)$] are calculated. A method equivalent to Equation (12.1.10) is used in the calculation. The cross-correlation functions of binaural pressures caused by a real source at different directions are first measured using an artificial head. Then, the cross-correlation functions of binaural pressures caused by a pair of loudspeakers are calculated according to the principle of linear superposition. The results indicate that the loudspeaker pair arranged at ±23°, ±67°, ±126°, or ±158° leads to $\Psi_{LR}(0) = 0$. The IACC of these loudspeaker arrangements is small (less than 0.4). These positions of the loudspeaker pair are front–back symmetry because of the approximately front–back symmetry of HRTFs below 2.0 kHz. Moreover, analysis indicates that $\Psi_{LR}(\tau)$ exhibits a maximum of 0.95 at $\tau = 0.3$ ms if decorrelated stimuli are reproduced by a pair of lateral loudspeakers in quadraphonic configuration (such as LF and LB loudspeakers in Figure 4.1). Therefore, reproducing decorrelated stimuli by a pair of lateral loudspeakers cannot create a low IACC. Using a similar method, Damaske et al. further analyzed the interaural correlation in the reproduction of decorrelated stimuli with various numbers and configurations of loudspeakers. The results are illustrated in Table 12.1. The five, six, and seven loudspeaker configurations probably have low IACCs. By

Table 12.1 IACC for different numbers and configurations of loudspeakers

Number of loudspeakers	*Loudspeaker configuration*	*IACC*
3	±54°, 180°	0.16
4	±54°, ±126°	0.30
5	±36°, ±108°, 180°	0.17
6	±36°, ±90°, ±144°	0.11
7	±36°, ±90°, ±144°, 180°	0.08
8	±18°, ±72°, ±108°, ±162°	0.29

contrast, the eight-loudspeaker configuration exhibits an increased IACC of 0.29. Therefore, an appropriate number and configuration with decorrelated stimuli can create low IACC in multichannel sound reproduction.

Shi and Xie (2010) further analyzed the interaural correlation in 5.1-channel loudspeaker configuration (Figure 5.2) with various interchannel correlations. The method in Section 12.1.1 is used in the analysis. The MIT-HRTFs of a KEMAR artificial head with DB-061 small pinna and DB-100 occluded ear simulator (Section 11.2.1) are used. The relationship between the interchannel correlation ($ICCC_{sign}$) and IACC or $\Psi_{LR}(0)$ for a pair of left and right loudspeakers, a pair of lateral loudspeakers and a pair of surround loudspeakers are preliminary analyzed. For a pair of left and right loudspeakers, IACC increases as the $ICCC_{sign}$ defined in Equation (1.7.6) increases when interaural correlation is positive or $\Psi_{LR}(0) > 0$. When interaural correlation is negative [or $\Psi_{LR}(0) < 0$], IACC decreases as $ICCC_{sign}$ increases because an absolute value of $\Psi_{LR}(\tau)$ is calculated in Equation (1.7.2). To avoid this problem, $IACC_{sign}$ defined by Equation (1.7.3) is used to substitute IACC. When binaural pressures are positively correlated, the results are consistent with the analysis using IACC. When binaural pressures are negatively correlated, the results are somewhat different from those of Kurozumi and Ohgushi in Section 1.7.3.

Figure 12.14 illustrates the variation in $IACC_{sign}$ versus $ICCC_{sign}$ of the left and right loudspeakers of the stimulus of pink noise with a fully audible bandwidth. $IACC_{sign}$ increases as $ICCC_{sign}$ increases monotonically. When $ICCC_{sign}$ changes from -1.0 to $+1.0$, $IACC_{sign}$ varies from -1.0 to $+1.0$. $IACC_{sign}$ vanishes when $ICCC_{sign} = 0.3$. For band-pass filtered stimuli with different center frequencies (such as 1 Oct band-pass filtered pink noise), the results are similar except for some differences in $ICCC_{sign}$ at which $IACC_{sign}$ vanishes. The experimental results between spatial auditory events and interchannel $ICCC_{sign}$ can be explained by the analysis here.

For a pair of surround loudspeakers in 5.1-channel configuration, the relationship between $IACC_{sign}$ and $ICCC_{sign}$ is similar to that of the left and right loudspeakers. However, the result for a pair of lateral loudspeakers is different. Changing $ICCC_{sign}$ between a pair of lateral loudspeakers cannot effectively control $IACC_{sign}$. The details are omitted here.

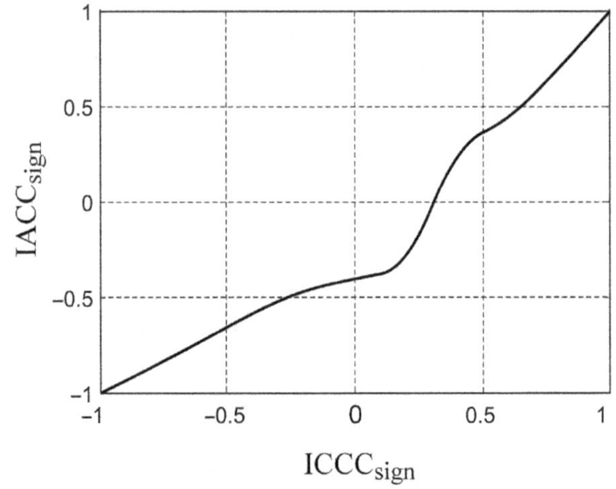

Figure 12.14 Variation in $IACC_{sign}$ versus $ICCC_{sign}$ of the left and right loudspeakers for the stimulus of pink noise with a fully audible bandwidth.

The interaural correlation caused by five or four (the left, right, left surround, and right surround) loudspeakers in 5.1-channel configuration with decorrelated stimuli can also be analyzed. In Section 3.3, in addition to low IACC, nearly equal root-mean-square magnitudes of binaural pressures should be created in multichannel sound reproduction to recreate the listener envelopment related to diffused reverberation. Accordingly, the ILD in reproduction should also be analyzed.

In the preceding analysis, a listener's head is fixed (oriented to the front). In practice, the listener may turn his/her head during listening. In an ideal diffused sound field, interaural correlation is invariable against head turning. This invariance should be persevered in multichannel sound reproduction. Walther and Faller (2011) further investigated the interaural correlation in horizontal multiple loudspeaker reproduction with decorrelated stimuli. Interaural correlation is analyzed in each ERB and compared with the just noticeable difference (JND) in the interaural correlation obtained from a psychoacoustic experiment. When the number of loudspeakers increases from four (the left, right, left surround, and right surround loudspeakers in 5.1-channel configuration) to eight with uniform configuration, the variation in interaural correlation with head turning reduces. In each ERB bandwidth, the variation of interaural correlation with head turning is less than the JND. In addition, the resultant ILD is less than the JND. Therefore, increasing the number of loudspeakers improves the perceived performance of the simulated diffused sound field.

Some studies have conducted a regressive analysis between the results of subjective experiments and physical parameters of sound field or binaural pressures and established the predictive models of the spatial perceived performance in multichannel sound (Rumsey et al., 2008; Conetta et al., 2008; Jackson et al., 2008; George et al., 2006, 2010). The spatial perceived performance in reproduction can be predicted on the basis of predictive model and measured parameters. Multichannel sound reproduction has various spatial perceived attributes. These models can predict some of these attributes, such as envelopment. Subjective envelopment may depend on multiple acoustic parameters, such as IACC, ICCC, directional distribution of reflections, ratio of average energy in rear channels to front channels, and the spectral and temporal characteristics of stimuli. These parameters can be evaluated when loudspeaker configuration and signals are given.

Similar methods are applicable to the assessment of the perceived quality of multichannel audio signal compression, coding, and decoding in Chapter 13. The perceived audio quality of coding and decoding is assessed by first inputting the signals after coding and decoding into the model, evaluating acoustic parameters, and comparing with the results of reference (original) signals. ITUs have recommended the model for assessment (ITU-R BS 1387-1, 1999), but the model is mainly for non-spatial perceived quality (such as timbre) assessment. In some succeeding works, spatial or binaural parameters and even the binaural auditory model in Section 12.2 have been incorporated into the analysis to assess the overall perceived performance (including the spatial performance) of audio coding and decoding (Choi et al., 2008; Seo et al., 2013). Some advanced mathematical tools, such as artificial neural network, are applicable to model training. The works on this issue are continuing

The analyses in this section are focused on horizontal loudspeaker configurations. In Sections 6.2 and 7.3.1, the cues and mechanism of the summing spatial auditory event and perception in vertical directions are different from those in horizontal directions. Different physical and auditory models, at least revised models, are needed to analyze the summing spatial auditory event and perception in vertical directions. Further work on this aspect is required.

12.2 BINAURAL AUDITORY MODELS AND ANALYSIS OF SPATIAL SOUND REPRODUCTION

Errors in binaural pressures and localization cues in spatial sound reproduction are analyzed in Section 12.1. However, the psychoacoustic cues and mechanisms of spatial auditory perception are not fully considered in these physical analyses. Physical or mathematical errors do not always reflect the perceived errors. Subjective assessment via a psychoacoustic experiment is a major method used to evaluate the perceived errors, and the results appropriately reflect the perceived performance in sound reproduction. However, a subjective assessment experiment is usually complicated and time consuming. An elaborate experimental design, a large number of experimental samples, and appropriate statistics on data are required to obtain significant results. Objective and subjective assessments are currently two complementary methods for assessing an electroacoustic system. The binaural auditory and high-level models of psychoacoustic processing should be further incorporated into the objective analyses of spatial sound reproduction.

A binaural auditory model describes the physical, physiological, and psychoacoustic processing of the hearing system on received sound waves. Analyzing the perceived attributes of sound by using various binaural auditory models is currently an important tendency in psychoacoustics (Pulkki and Karjalainen, 2015). That is, some functional models of auditory processing and perception are established on the basis of the results of psychoacoustic and physiological acoustic experiment, and the perceived attributes of a sound are assessed by inputting the sound signal into the model. Analysis and assessment with auditory models are usually simpler than and complementary with psychoacoustic experiments. Various binaural auditory models with different functions have been proposed. Some software tool boxes of binaural auditory models are available (Søndergaardand and Majdak, 2013). These models are applicable to the analysis of various binaural auditory perceptions, the perceived sound quality, and the performance of spatial sound. More accurate and elaborate analyses should involve models based on the neurophysiology of hearing, as stated in the discussion of ICTD-based summing localization of transient stimuli in Section 2.1.4. This kind of models is yet to be enhanced, but it is a promising direction. Studies on the neurophysiology of hearing should be performed to reveal the mechanism of spatial auditory perception and improve the perception performance in sound reproduction.

Spatial sound reproduction has various perceived attributes. Virtual source localization and timbre are two important attributes, which are analyzed using binaural auditory models in this section.

The traditional binaural auditory models used in this section consist of a bottom-up (signal-driven) architecture. Blauert (2012) pointed out and suggested that these traditional models are sufficient for a number of applications, including sound source localization. Auditory models that have inherent knowledge bases and employ top-down (hypothesis-driven) strategies are needed to evaluate some other perceived attributes in sound reproduction concerning auditory cognition. The model of auditory scene analysis (Section 1.7.4) may be applicable to the analysis of more complicated perceived attributes. These advanced models are an important direction in future research.

12.2.1 Analysis of lateral localization by using auditory models

In Section 1.6, when a listener's head is fixed, ITD, ILD, and spectral cues provide major information on localization. ITD and ILD are cues for lateral localization to determine the cone of confusion in which the source is located. The spectral cue contributes to front–back and vertical localization. Since Jeffress (1948) proposed the basic principle of a binaural auditory model, many researchers have improved this model and applied it to the objective assessment of perceived performance in spatial sound reproduction (Macpherson, 1991; Pulkki et al., 1999; Pulkki 2001a, Pulkki and Karjalainen, 2001; Takanen and Lorho, 2012).

According to considerations by Huopaniemi et al. (1999) and Pulkki et al. (1999), Figure 12.15 illustrates the blocked diagram of a binaural auditory model. This model is applicable to the analysis of summing virtual source localization and timbre coloration in spatial sound reproduction. However, the model does not involve the high-level processing of spatial information and is inappropriate for analyzing complicated spatial auditory perceptions, such as the precedence and cocktail party effects.

Given the input stimulus (such as pink noise or band-pass filtered pink noise), the binaural pressures of a real source or sound reproduction are calculated. The case of a real source is illustrated in Figure 12.15. For multiple loudspeaker reproduction, binaural pressures are calculated with Equations (12.1.6) to (12.1.9). In Sections 1.4.2 and 11.7.1, binaural pressures in the eardrums are obtained by filtering the input stimulus with a pair of HRTFs with the eardrums as the reference point. If a pair of HRTFs with the entrances of ear canals as a reference point is used, a pair of ear canal filters should be supplemented to obtain binaural pressures in the eardrums. Strictly speaking, a complete auditory model should include the transfer responses of the middle ear, which can be approximated by the inverse of the standard hearing threshold contour. In a left–right symmetric model, however, the middle ear has no effect on interaural cues, such as ITD and ILD. For simplicity, therefore, the early binaural auditory model (such as in Figure. 12.15) disregards the effect of the middle ear. However, some later models involve the effect of the middle ear.

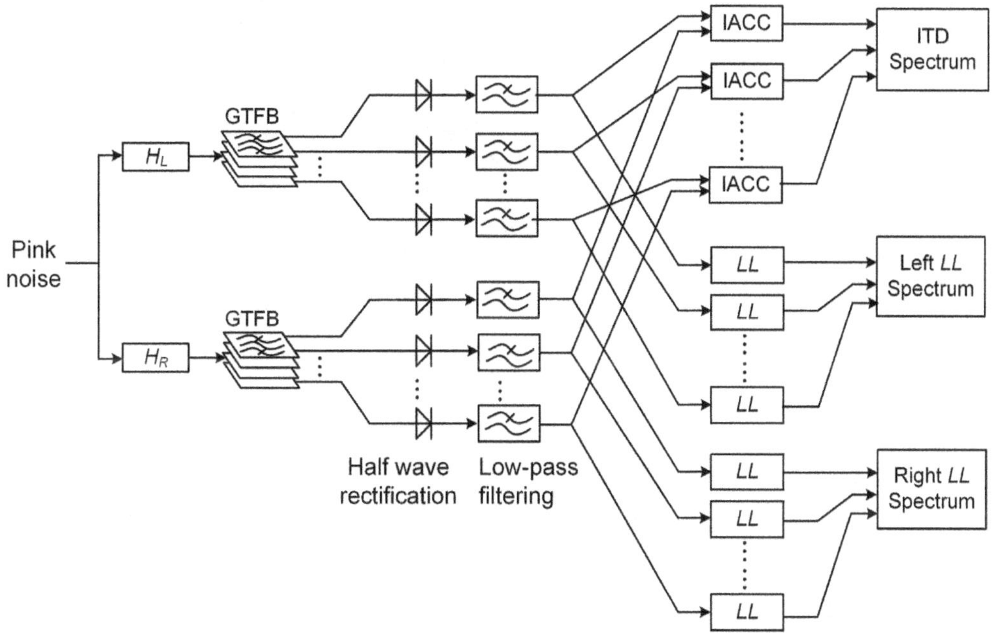

Figure 12.15 Block diagram of a binaural auditory model(adapted from Huopaniemi et al., 1999).

A series of parallel band-pass filter banks is analogous to the frequency analysis function of the inner ear, that is, auditory filter simulation. When a band-pass filter is represented by a GammaTone filter, its impulse response is

$$g(t) = \frac{At^{N-1}\cos(2\pi f_c t + \varphi)}{\exp(Bt)} \qquad t \geq 0, \tag{12.2.1}$$

where A is a parameter for filter gain, f_c is the central frequency (Hz) of the filter, φ is the initial phase, N is the order of the filter, and B is a parameter that characterizes the filter bandwidth. When $N = 4$ and $B = 2\pi ERB$, the GammaTone filter banks are an approximation of auditory filters. In Equation (1.3.3), the ERB of the auditory filter is given in Hertz, whereas the unit of central frequency is kilohertz. Huopaniemi et al. (1999) used 32 parallel GammaTone filters.

Following each ERB band-pass filtering, half-wave rectification and low-pass filtering with a cut-off frequency of 1 kHz are used to simulate the behavior of hair cells and auditory nerves in each band-pass channel. For simplicity, the adaptive process in the auditory system is excluded.

For the outputs of half-wave rectification and low-pass filters of the left and right ears, the normalized interaural cross-correlation function $\Psi_{LR}(\tau)$ in each ERB channel is calculated using the method similar to that in Section 12.1.1. Then, correlation-based ITD is evaluated as a function of frequency in the unit of ERBN given in Equation (1.3.4). For steady and stationary random signals, the integral in Equation (12.1.10) is calculated over a sufficient long temporal period for convenience in calculation. For nonstationary random signals, the integral in Equation (12.1.10) should be calculated over an appropriate time window.

Here, the frequency-dependent ITD is calculated at a frequency resolution in accordance with the resolution of the auditory system. Half-wave rectification and low-pass filtering are performed before $\Psi_{LR}(\tau)$ is calculated. Below the cut-off frequency, low-pass filters slightly influence the signal; as such, $\Psi_{LR}(\tau)$ and ITD depend on the final structure (phase) of binaural pressures (signals). Above the cut-off frequency, the signals are smoothened by low-pass filters, so $\Psi_{LR}(\tau)$ and ITD gradually depend on the envelop rather than the fine structure of binaural pressures. Therefore, the interaural phase delay is a cue for localization at low frequencies, and interaural envelop delay is a localization cue at mid and high frequencies.

The monoaural loudness spectra in units of Sones/ERB for each ERB bank can be calculated for either of the two ears by using the following formula:

$$L_L = \left(\overline{E_{L,ear}^2}\right)^{1/4} \qquad L_R = \left(\overline{E_{R,ear}^2}\right)^{1/4}, \tag{12.2.2}$$

where $\left(\overline{E_{L,ear}^2}\right)$ and $\left(\overline{E_{R,ear}^2}\right)$ are the average power over time for the left and right ears, respectively. Equation (11.2.2) is an approximation of the loudness formula provided by Zwicker. In original Zwicker's formula, however, the exponential factor is 0.23 rather than 1/4, as in Equation (12.2.2; Zwicker and Fastl, 1999).

The monaural loudness level spectra (LL) in units of Phons/ERB for each ERB bank are

$$LL_L = 40 + 10\log_2 L_L \qquad LL_R = 40 + 10\log_2 L_R. \tag{12.2.3}$$

In Equation (12.2.3), ILD as a function of frequency (in units of ERBN) can be evaluated on the basis of the difference in the loudness level between the left and right ears as follows:

$$\Delta LL = LL_L - LL_R. \qquad (12.2.4)$$

Binaural pressures (signals) in reproduction are inputted into the above model to evaluate the virtual source azimuth in spatial sound reproduction, and the loudness level spectra of the left and right ears and ITD are obtained as functions of frequency in the unit of ERBN. Lastly, the results are compared with those of a real source. More than one peak may be found in the curve of $\Psi_{LR}(\tau)$ in each ERB bandwidth, which prohibits the evaluation of ITD from directly maximizing $\Psi_{LR}(\tau)$. In this case, $\Psi_{LR}(\tau)$ at each ERB channel is first multiplied with the two $\Psi_{LR}(\tau)$ at two neighboring (above and below) ERB channels, and the ITD is evaluated by maximizing the multiplication. This manipulation is consistent with auditory signal processing. When the positions of the peak of $\Psi_{LR}(\tau)$ in two neighboring ERB channels are coincident, the ITD corresponding to this peak is more related to localization.

If all interaural localization cues (ITD and ILD in all frequency bands) in sound reproduction match with those of the target or real source, the virtual source direction in reproduction can be determined, or considering the symmetry of ITD and ΔLL, the cone of confusion in which the virtual source is located can be determined at least. If the localization information provided by an interaural cue (such as ITD) is inconsistent at different frequencies, or the localization information provided by two localization cues (ITD and ILD) are unmatched, the perceived direction of the virtual source is determined in terms of the relative importance of different cues in various frequency bands and more consistent information provided by these cues. However, excessively unmatched or conflicting information degrades the accuracy and quality of localization. Analysis of the localization of bandwidth stimuli with conflicting information relies on the mechanism of high-level processing and is therefore complicated. Existing models are still unable to do these complicated analyses. They can only apply to the localization of some stimuli with a certain bandwidth based on the known psychoacoustic rules.

Pulkki and Karjalainen (2001) used an aforementioned model to analyze the two-channel stereophonic sound with amplitude panning. Stereophonic reproduction creates an ITD matched with that of the target source up to the frequency of 1.1 kHz. The resultant ITD deviates from that of the target source above 1.1 kHz. As proven in Section 2.1.2, above 0.7 kHz, especially above 1.0 kHz, the virtual source azimuth evaluated from interaural phase difference varies with frequency. The first example in Section 12.1.3 yields similar results. Therefore, the results of previous analyses are basically consistent with those from the binaural auditory model. Analysis based on the binaural auditory model also indicates that the ILD in stereophonic reproduction is basically consistent with that of a target source at a low frequency below 0.5 kHz and a high frequency above 2.6 kHz. The ILD in stereophonic reproduction deviates from but at least does not conflict with that of a target source within the frequency range of 0.5–2.6 kHz. Therefore, for wideband stimuli, human hearing can localize the virtual source according to the more consistent information provided by ITD and ILD. Pulkki et al. (1999) also analyzed the virtual source localization in stereophonic reproduction with out-off phase signals and time panning. The results are similar to these in Sections 2.1.3 and 12.1.4.

Similar method is applicable to analyze the performance of virtual source localization in stereophonic and multichannel sound with various microphone techniques (Pulkki, 2002). Pulkki and Hirvonen (2005) further analyzed the virtual source localization of two kinds of horizontal loudspeaker configurations with different signal mixing types, including 5.1-channel configuration with the first-order Ambisonic signals, pair-wise amplitude panning signals, near-coincident microphone signals, regular eight-loudspeaker configuration with the first- and second-order Ambisonic signals, and pair-wise amplitude panning signals. The results are consistent with those in Chapter 4, Chapters 5, and Section 12.1.

12.2.2 Analysis of front-back and vertical localization by using a binaural auditory model

ITD and ILD specify the lateral displacement of the virtual source and are sufficient for analyzing two-channel stereophonic sound. A complete spatial sound deals with front–back and vertical localization. In Section 1.6, the monoaural/binaural spectra at high frequencies are one of the front–back and vertical localization cues. Therefore, monoaural/binaural spectra should be analyzed in sound reproduction. A simple method presented in Section 12.1.1 involves the direct comparison of monoaural/binaural spectra in reproduction with those of the target sources. An example of spatial Ambisonics is given in Section 12.1.3. Considering auditory resolution, the analysis of monoaural/binaural spectra by using auditory models yields results that are more consistent with practical auditory perceptions.

Pulkki (2001a) used the binaural auditory model in Section 12.2.1 to analyze the summing localization of a pair of loudspeakers at $\phi = -15°$ and $30°$ in the median plane with amplitude panning. The mean-abstracted monaural loudness level spectrum indicates that the resultant peak and notch in reproduction do not match with those of a real source.

Baumgartner et al. (2013, 2015) developed an auditory model to analyze vertical localization. Similar to the model in Figure 12.15, their model involves the transmission from a source to the entrance of the ear canal, the transmissions of the ear canal and middle ear, auditory filters, half-wave rectification, and low-pass filter to simulate the behavior of inner hair cells and auditory nerve in the inner ears. The model yields the internal representation of binaural signals. The probability distribution of the perceived virtual source direction is obtained by comparing the internal representation in reproduction with that of a special pattern. Individualized localization can be analyzed by an individualized calibration of the model. This model is applicable to the analysis of multichannel spatial surround sound (such as Auro 9.1 and USC 10.2 in Section 6.5.1). The results are similar to those obtained by Pulkki et al., e.g., high-frequency spectral cues in reproduction do not match with that of a real source.

The monaural loudness level spectrum in spatial Ambisonic reproduction can be analyzed using the loudness model in Section 12.2.1 or Moore's loudness model in Section 12.2.3. The results and the analysis by Pulkki and Baumgartner et al. indicated that errors in high-frequency monaural pressure spectra reproduction exist even after the smooth of auditory filters.

A similar problem occurs in multichannel horizontal surround sound reproduction. Multichannel horizontal surround sound with practical loudspeaker configuration and signal mixing cannot create an appropriate high-frequency spectral cue for front–back localization. For example, at high frequencies of above 5–6 kHz and even at the central listening position, the monoaural/binaural pressure spectra created by the third-order horizontal Ambisonics with eight or twelve loudspeakers deviate obviously from those of a real source.

Overall, the above examples indicate that various multichannel sounds with practical loudspeaker configurations and signal mixing cannot provide appropriate spectral cues for front–back and vertical summing localization. In Section 11.5.3, summing a virtual source via two or more loudspeakers with amplitude panning is analogous to the directional interpolation of HRTFs. Loudspeaker configurations and signal mixing in practical multichannel sound do not satisfy the condition of Shannon–Nyquist directional sampling theorem at high frequencies; as such, they cannot provide appropriate high-frequency spectral cues above

5–6 kHz. Practical multichannel sound reproduction may create a correct dynamic cue at low frequencies. For example, the fourth and fifth examples in Section 12.1.3 indicate that the third-order Ambisonics can create correct binaural pressures up to 1.9 kHz and form correct low-frequency ITD and its dynamic variation below 1.5 kHz. When the spectral cue is conflicting with a dynamic cue, the dynamic cue dominates front–back localization at least at low frequencies. Multichannel horizontal surround sound utilizes the dynamic cue for front–back localization at low frequencies. Dynamic ITD variation caused by head rotation has been included in the analysis of front–back localization in multichannel horizontal surround sound in Chapters 3 to 5 [such as Equation (3.2.9)]. Therefore, if the analysis of dynamic ITD variation is supplemented, a model similar to that in Section 12.2.1 is applicable to the analysis of the summing localization in a horizontal plane. Braasch et al. (2013) developed such a model and applied it examine surround sound reproduction.

For a real source in the free field, vertical localization is the comprehensive consequence of high-frequency spectral and dynamic cues. Multichannel sound reproduction cannot create an appropriate spectral cue, and vertical summing localization relies more on the dynamic cue. This result is described in Chapter 6. In Section 6.2, summing the virtual source in a vertical direction (such as median plane) may be blurry and unstable because of the error in summing a high-frequency spectral cue. The aforementioned model of Baumgartner et al. is appropriate for the case of a fixed head. In future studies, a further vertical localization model should comprehensively consider spectral and dynamic cues.

12.2.3 Binaural loudness models and analysis of the timbre of spatial sound reproduction

Timbre is one of the perceived attributes of sound. The American Standard Association (1960) defined the timbre as that "attribute of auditory sensation in terms of which a listener can judge that two sounds similarly presented and having the same loudness and pitch are dissimilar". Rumsey et al. (2005)have shown that in multichannel sound reproduction, a degradation of the timbre generally has a greater impact on the overall perception of the audio quality compared with the degradation of the spatialization. Therefore, timbre analysis is an important part of the assessment of the performance in spatial sound.

Perceived timbre is closely related to binaural loudness level spectra (BLLS), whereas the BLLS depends on various factors, including source direction. In spatial sound reproduction, the BLLS at different target source directions, e.g., directional loudness in Section 1.3.2, should be considered. The binaural auditory model in Figure 12.15 is applicable to the analysis of timbre after a pair of transmission filters of the middle ears is supplemented. The BLLS as a function of frequency (in unit of ERBN) is calculated as the sum of the monoaural loudness level spectra (MLLS) of the left and right ears given in Equation (12.2.2). Timbre depends on other factors, in addition to BLLS. Perceivable timbre coloration occurs when the variation in BLLS in reproduction exceeds the JND. Pulkki (2001b) used this model to analyze the timbre in two-channel stereophonic reproduction. The results indicated that considerable differences are observed in BLLSs in stereophonic reproduction and those of a real source in an anechoic environment. These differences are due to the comb-filtering effect caused by the coherent interference of the sound from two loudspeakers. However, the perceived differences become less obvious in a usual listening room because the reverberation of a listening room reduces the comb-filtering effect caused by the coherent interference.

In addition to the loudness formula by Zwicker in Equation (12.2.2), BLLS can be evaluated by Moore's loudness model and formula (Moore et al., 1997). Moore's model was included in the American national standard for loudness calculation (ANSI S3.4, 2007). Subsequently, Moore and Glasberg (2007) proposed a modified binaural loudness model by considering the inhibitory interactions between the two ears. The modified model simulates the transfer of sound through the external ear and middle ear to the cochlea, the frequency analysis and excitation patterns in the inner ear, the transform of excitation patterns to monaural specific loudness, and the transform of monoaural loudness to binaural loudness. In 2017, Zwicker's and Moore's models were recommended by ISO in the standard for loudness calculation (ISO 532-1, 2017; ISO 532-2, 2017). The details of Moore's model are referred to above literatures. Figure 12.6 illustrates the blocked diagram of Moore's modified binaural loudness model. The model involves following steps:

1. Scale of the input stimulus
   A mono stimulus is scaled such that its root-mean square (RMS) value of 1 corresponds to a pressure level of 0 dB in the free field (measured at the center of the head with the head absent).
2. Transfer through the external and middle ears
   The scaled stimulus is filtered by a pair of HRTFs with a reference point in the eardrum to simulate the transfer through the external ear, yielding binaural pressures (signals) in the eardrums. If the HRTFs measured at the blocked entrance of the ear canals are used, a pair of ear canal filters should be appended to obtain the pressures in the eardrums. Then, the pressures in the eardrums are filtered by the middle ear transfer function specified by the ANSI S3.4 (2007) standard, resulting in the physical spectra at the cochlea.
3. Calculation of the excitation patterns at the cochlea
   The frequency selectivity of the cochlea is modeled by a bank of band-pass filters with different center frequencies. The bandwidth and shape of the filters depend on the center frequency and level of input stimuli, as specified by ANSI S3.4 (2007). Cochlea filters transform the physical spectra to excitation patterns. These patterns, which are defined as the output power of auditory filters as functions of frequency (in unit of ERB) for a given input stimulus, represent the distribution of excitation in the cochlea caused by a sound wave. Excitation patterns are calculated at a frequency resolution of 0.1 ERBN to obtain sufficient accuracy in calculation.
4. Transformation to monaural specific loudness
   Excitation patterns are transformed to monaural specific loudness (loudness spectra) in units of Sones/ERB by the loudness formula of Moore.
5. Calculation of binaural specific loudness and conversion to BLLS
   The signal at the left ear inhibits loudness at the right ear and vice versa. Moore and Glasberg (2007) considered the inhibitory interactions between the two ears in the modified binaural loudness model. The original monaural specific loudness of the left and right ears obtained in Step (5) is used to calculate the inhibition factor of the opposite ear. Subsequently, the original monaural specific loudness in the left and right ears is divided by the inhibition factors and summed to obtain the binaural specific loudness (loudness spectra) in the unit of Sones/ERB. The binaural specific loudness is converted to BLLS in the unit of Phons/ERB.
6. Calculation of the overall loudness level
   The binaural specific loudness is summed across frequencies to obtain the overall binaural loudness in units of Sones and converted into overall binaural loudness level in units of Phons

Similar to the case in Section 12.1.1, the binaural pressures in sound reproduction are the superposition of those caused by all loudspeakers and can be calculated using HRTFs. At the off-central listening position, binaural pressures are calculated similar to those in Section 12.1.5. In addition, half-wave rectification and low-pass filtering for simulating

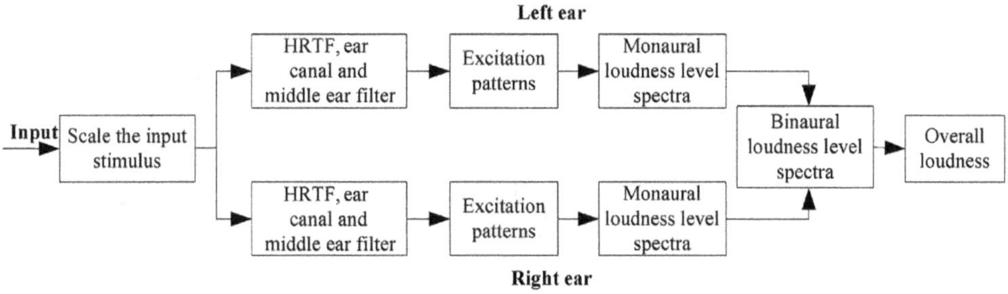

Figure 12.16 Blocked diagram of Moore's modified binaural loudness model.

the behavior of hair cells and auditory nerves in each band-pass channel are omitted in Figure 12.16. This omission is reasonable because the loudness spectra depend on the output power of auditory filters.

The BLLS in reproduction are calculated using the binaural loudness model and compared with those of the target sound field to analyze the timbre in spatial sound reproduction. BLLS represent the dependence of binaural loudness on frequency. The perceived timbre coloration occurs when the difference between the BLLS in reproduction and that of a target sound field exceeds the JND. The JND of the BLLS is related to the sound level discrimination. The level discrimination for 1 kHz pure tone is about 1 dB (Florentine et al., 1987), indicating that the JND of BLLS for a stimulus around 1 kHz is in the order of 1 Phon /ERB. Therefore, a rough estimation of JND = 1 Phon/ERB is used in the following analysis although this estimation is inaccurate.

The above method is applicable to the analysis of the timbre coloration in Ambisonic recording and reproduction (Liu and Xie, 2015). A horizontal Ambisonics with 12 loudspeakers is analyzed. The loudspeakers are arranged at a far-field distance and uniformly around the listener. The azimuths of loudspeakers are $\theta_i = 0°, \pm 30°, \pm 60°, \pm 90°, \pm 120°, \pm 150°$, and 180°, respectively. According to Equation (9.3.13), this loudspeaker configuration is appropriate for horizontal Ambisonics up to the order of $Q = 5$. The target sound field is a plane wave incident from the horizontal azimuth θ_S. The normalized amplitudes of loudspeaker signals are given by the conventional solution in Equation (9.3.12). The input stimulus is a pink noise and scaled to a free-field pressure level of 70 dB. The far-field HRTFs of KEMAR (with DB-060/061 small pinnae but without torso) calculated by the boundary element method with an angle interval of 1° are used for analysis (Liu et al., 2016).

Figure 12.17 illustrates the BLLS of the preceding five-order horizontal Ambisonic reproduction with the target plane wave incident from $\theta_S = -45°$. Figure 12.17(a) presents the case of the central listening position. As the order of Ambisonics increases, the BLLS in reproduction matches with that of the target field in an incremental frequency range. That is, the high-frequency limit without timbre coloration increases gradually. Figures 12.17(b) and 12.17(c) show the cases of 0.10 and 0.20 m deviations to the right from the central position, respectively. The high-frequency limit of reproduction also increases as the order of Ambisonics increases. However, in the cases of deviation from the central position, the difference in BLLS between Ambisonic reproduction and target source exceeds 1 Phon/ERB at lower frequencies. In other words, timbre coloration appears in a lower frequency range as compared with central position. For the third-order reproduction, the high-frequency limits of reproduction without perceivable timbre coloration are estimated to be 21, 15, and 11 ERBN for the central, 0.10 m, and 0.20 m right deviations from the center position, respectively, corresponding to 1.96, 0.92, and 0.52 kHz. The results of the target plane wave incident from other azimuths are similar.

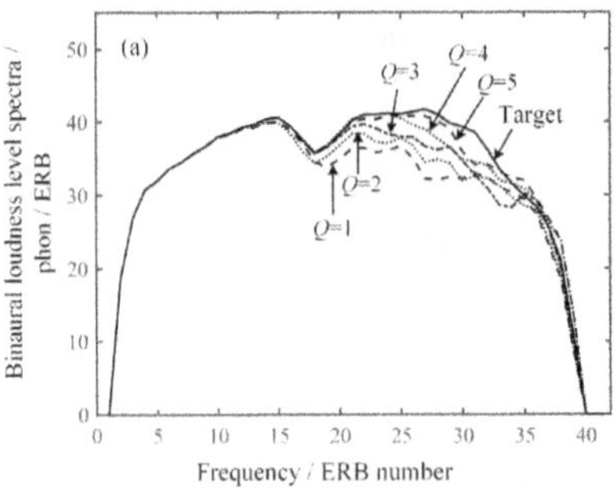

(a) Central listening position

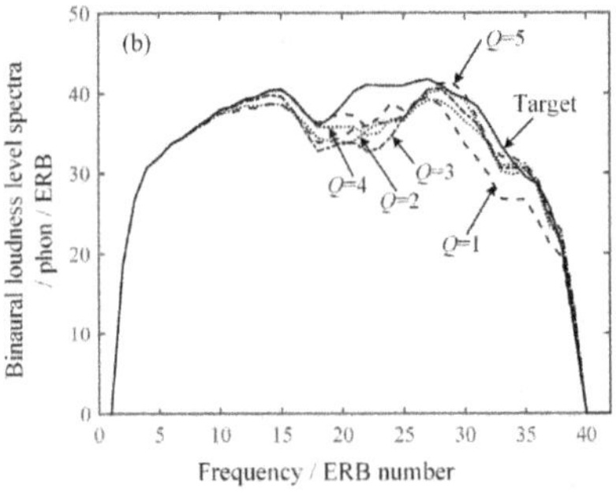

(b) 0.10 m deviation to the right

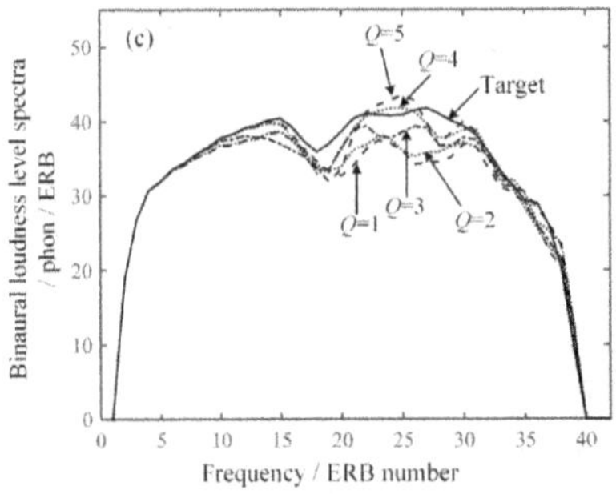

(c) 0.20 m deviation to the right

Figure 12.17 BLLS of horizontal Ambisonic reproduction with $\theta_S = -45°$ (a) Central listening position; (b) 0.10 m deviation to the right; (c) 0.20 m deviation to the right.

Given the order of Ambisonics and $r = r_H$ of a circular region, the upper frequency limit of accurate sound field reconstruction can be evaluated from Equation (9.3.15) as

$$f < f_{\max,H} = \frac{cQ}{2\pi r_H}, \tag{12.2.5}$$

where c is the speed of sound. For the mean head radius $a = 0.0875$ m and a deviation distance y_1 from the central position, the required radius of the region for accurate sound field reconstruction is $r_H = a + y_1$. Applying this condition to Equation (12.2.5) yields the upper frequency limit for various listening positions below which no perceivable timbre coloration occurs:

$$f_{\max,H} = \frac{cQ}{2\pi\left(a + y_1\right)}. \tag{12.2.6}$$

For instance, in the third-order horizontal Ambisonics, the upper frequency limits evaluated from Equation (12.2.6) are $f_{max.H} = 1.87$, 0.87, and 0.57 kHz for central, 0.1 m and 0.2 m right deviation from the central position, respectively. Therefore, the upper frequency estimated from difference in BLLS are basically consistent with those evaluated from Equation (12.2.6). If an accurate reconstruction of the target sound field within a radius of $a = 0.0875$ m and up to a frequency limit of f_{max}, $_H$ = 20 kHz is desired, Q = 32-order horizontal Ambisonics with $M = (2Q + 1) = 65$ loudspeakers at least is required. This result is consistent with that derived from the azimuthal sampling theorem of HRTFs in Section 11.5.2. In practice, such a high-order Ambisonics is difficult to be implemented. Therefore, timbre coloration occurs in practical Ambisonic reproduction to different extents.

The above discussion implies that the independent signals of Ambisonics are accurate. Errors in the final reconstructed sound field are caused by the error in reproduction only. However, the independent signals of Ambisonics may be obtained from a circular microphone array recording, e.g., from Equation (9.8.6). Given the radius r_M of the microphone array and the number M' of microphones, the independent signals in Equation (9.8.6) are accurate below the upper frequency limit $f_{max} \cdot _M$ imposed by Shannon–Nyquist directional sampling theorem. According to the discussion in Section 9.8.1, for a uniform microphone array with $M' \geq (2Q + 1)$, substituting r in Equation (9.3.15) with r_M yields

$$f < f_{\max, M} = \frac{(M'-1)c}{4\pi r_M} \approx \frac{M'c}{4\pi r_M}. \tag{12.2.7}$$

If the condition in Equation (12.2.7) is not satisfied, spatial aliasing occur in the independent signals of Equation (9.8.6). Therefore, the overall errors in the final reconstructed sound field are contributed by the errors in microphone array recording and loudspeaker reproduction. A further analysis of BLLS indicates that the influence of spatial aliasing errors of a microphone array on the final reconstructed sound field is negligible in an arbitrary receiver region, providing that the upper frequency limit of the microphone array is higher than that of reproduction. Equations (12.2.5), Equation (12.2.7), and the condition of $f_{max},_M > f_{max},_H$ yield

$$\frac{M'}{r_M} > \frac{2Q}{r_H}. \tag{12.2.8}$$

Based on the above analysis, a scheme of the comprehensive and optimized design of horizontal Ambisonic recording and reproduction is suggested.

1. r_H and $f_{max,\, H}$ are selected for reproduction according to application requirements. The order Q of Ambisonics and the number of loudspeaker $M \geq (2Q + 1)$ for reproduction are calculated using Equation (12.2.5).
2. M'/r_M is calculated with Equation (12.2.8), and the low-limit number M' of the microphones is determined if r_M can be selected in terms of the required signal-to-noise ratio at low frequencies (Rafaely, 2005).

A similar method is applied to analyze the timbre in horizontal Ambisonic reproduction with different decoding methods. A conventional decoding method leads to less timbre coloration than that of the in-phase method and method of maximizing the energy vector magnitude (Liu and Xie, 2016).

The results of analysis of the BLLS on spatial Ambisonic reproduction are similar to those on horizontal Ambisonics. In addition, the above analysis of the timbre of horizontal Ambisonics is valid for reproduction in an anechoic environment. Similar to the case of stereophonic reproduction (Pulkki, 2001b), the perceived timbre coloration for Ambisonic reproduction in a usual listening room may become less obvious.

The aforementioned method for analyzing the timbre is general and not limited to stereophonic and Ambisonics. It is applicable to the analysis of the timbre of other spatial sounds, such as various multichannel sounds, WFS, and VAD. For example, analyses on 2.5-dimensional WFS with the horizontal linear or circular array of secondary sources indicate that obvious timbre coloration exists above a certain upper frequency limit (Xie et al., 2015b). The correction of the frequency response of driving signals (choosing a flat response; Section 10.1.5) above the upper frequency limit of anti-spatial aliasing reduces the timbre coloration partly at most. Huopaniemi et al. (1999) first suggested using Zwicker's loudness formula to analyze an HRTF-based filter design and timbre. Moore's modified binaural loudness model has also been used to examine the audibility of the spectral detail of HRTF at high frequencies in a VAD (Xie and Zhang, 2010).

12.3 BINAURAL MEASUREMENT SYSTEM FOR ASSESSING SPATIAL SOUND REPRODUCTION

Binaural pressures for analysis can also be obtained by measurement. An artificial head is placed at the listening position in reproduction. Binaural pressures are measured by a pair of microphones in the two ears of the artificial head. The output signals of microphones are amplified and converted into digital signals by using an A/D converter and then examined by using a computer via the methods in Sections 12.1 and 12.2. Stereophonic and multichannel sound is evaluated in a free-field environment to achieve simplicity or avoid the uncertainty caused by the reflections in a listening room. Correspondingly, binaural pressure measurement for reproduction is often conducted in an anechoic chamber. In some instances, binaural pressure measurement is conducted in a listening room to evaluate the perceived performance of practical reproduction.

Tohyama and Suzuki (1989) used a KEMAR artificial head to measure binaural pressures and compare the IACC in each 1/3 Oct frequency band for an ideal diffused field with those for two- or four-loudspeaker reproduction. At the central listening position, the frequency dependence of IACCs for two and four front–back symmetric loudspeakers are greatly different from those of an ideal diffused field. However, four loudspeakers in a frontal horizontal plane create a frequency dependence of IACCs similar to that of an ideal diffused field. The position of four front loudspeakers in optional 7.1-channel configuration by ITU is consistent with the above results.

Muraoka and Nakazato (2007) measured the variation in IACC with a frequency for different numbers and configurations of loudspeakers. The results of ITU 5.1-channel loudspeaker configuration are similar to those of a target sound filed.

A PC-based binaural measurement system for objectively assessing stereophonic and multichannel sound reproduction can be designed on the basis of above consideration. To evaluate the virtual source localization in practical reproduction, Mac Cabe and Furlong (1994) designed a binaural measurement system and then measured the ITD, ILD, and IACC of Ambisonics and some other reproduction techniques. The measured results are compared with those of a real source, and the conclusions are similar to those in Section 12.1.2. Binaural measurement with an artificial head is often used to measure some room acoustic parameters, such as IACC. It has been specified in the appendix of the related international standard (ISO 3382-1, 2009).

Macpherson (1991) incorporated binaural auditory models into a PC-based binaural measurement system and used this system to analyze the binaural pressures at the off-central position in stereophonic reproduction. Blanco-Martin (2011) also conducted a similar work.

The validity of various binaural measurement systems depends greatly on the physical and auditory models used in the system. This case is similar to those of the calculation and analysis described in Sections 12.1 and 12.2.

12.4 SUMMARY

Binaural pressure analysis is a method for the objective assessment of spatial sound. Various localization cues and some other binaural-related physical attributes, such as ITD, ILD, interaural cross-correlation, monoaural/binaural spectra, and dynamic cues, can be evaluated through binaural pressure calculation. Then, auditory events or perceptions in reproduction can be analyzed. The results for two-channel stereophonic sound with amplitude panning and Ambisonics are similar to those in previous chapters, but the analyses in this chapter are more accurate and stringent. For two-channel stereophonic sound reproduction with ICTD or at the off-central listening position, the results of the analysis of narrow-band stimuli at low frequencies differ from the previous experimental results of wideband stimuli with transient characteristics. For narrow-band stimuli at low frequencies, outside-boundary virtual source and unnatural auditory events may occur. Interaural correlation can also be applicable to the assessment of the definiteness and perceived width of the virtual source and envelopment in reproduction.

Subjective attributes in spatial sound reproduction can be analyzed at a deeper level by incorporating a binaural auditory model. Binaural auditory models have been used to examine virtual source localization and timbre in spatial sound reproduction. In particular, analyses on monoaural pressure spectra or loudness spectra indicate that practical loudspeaker configurations and signal mixing in stereophonic and multichannel sounds cannot provide appropriate spectral cues for front–back and vertical summing localization. Therefore, future studies should perform analysis with a binaural auditory model.

In addition to calculation, binaural pressure (signals) for analysis can be obtained through measurement.

Chapter 13

Storage and transmission of spatial sound signals

Storage and transmission are important stages in the system chain of spatial sound by which spatial sound signals are allocated to users. Through storage, signals are recorded into a permanent medium. In transmission, signals are delivered to terminals via physical means. The spatial information of a sound field is encoded in the magnitude and phase relation between a set (two or more) of spatial sound signals. Therefore, in addition to the time/frequency-domain information of each channel signal, the relative magnitude and phase/time relation among all channel signals should be maintained through ideal storage and transformation.

Before the end of the 1970s, spatial sound signals (mainly stereophonic sound signals) were primarily stored and transmitted by analog techniques. Since the 1980s, digital techniques for audio signal storage and transmission have been developed and laid the foundation for the innovation and application of various advanced spatial sound techniques. Currently, digital audio techniques are the major storage and transmission methods of spatial sound signals.

The digital storage and transmission of audio signals is a special and important field in communication and audio. Numerous related works have been performed, and some techniques have been widely used. Relevant techniques have also been rapidly developed. Digital audio involves extensive topics, but this chapter does not explore these topics in depth. For further details, readers can refer to other books and studies (Spanias et al., 2007; Pohlmann, 2011; Bosi and Goldberg, 2003). However, to discuss spatial sound comprehensively, this chapter outlines the basic principles and techniques in the storage and transmission of spatial sound signals. Because related techniques and standards are being developed quickly, readers may focus on the latest studies and standards.

In Section 13.1, some conventional and representative analog techniques of the storage and transmission of spatial sound signals (mainly stereophonic signals) are briefly outlined. In Section 13.2, the basic concepts of digital audio storage and transmission are introduced. In Section 13.3, quantization noise and shaping are discussed. In Section 13.4, the principle and key techniques of digital audio coding and compression related to spatial sound are described. In Sections 13.5–13.10, some typical audio coding techniques and standards developed since 1990, including serial techniques and standards of Moving Pictures Expert Group (MPEG), serial techniques of Dolby and DTS, Meridian Lossless Packing (MLP) lossless coding technique, Adaptive Transform Acoustic Coding (ATRAC) coding technique, and serial techniques of Audio Video Coding Standard (AVS), are presented. In Section 13.11, optical disk techniques for audio storage are introduced. In Section 13.12, digital radio and television broadcasting are discussed. In Section 13.13, audio storage and transmission via computers and the Internet are briefly outlined.

DOI: 10.1201/9781003081500-13

13.1 ANALOG AUDIO STORAGE AND TRANSMISSION

Spatial sound signals were previously stored and transmitted by analog techniques. These techniques were vital to the popularization of two-channel stereophonic sound, and some of them were still used until 2021. Numerous analog techniques for the storage and transmission of stereophonic and quadraphonic signals have been proposed and developed. However, this section does not provide a detailed review of these analog techniques, given that they have been described in detail in another book (Xie X.F., 1981). To review the history of the development of spatial sound, this section outlines some widely used analog techniques of storage and transmission.

13.1.1 45°/45° Disk recording system

A gramophone disk is a traditional method used to record sound. As stated in Section 1.9.3, the commercialization of 45°/45° disk record by the end of the 1950s popularized stereophonic sound in domestic reproduction (Goldmark et al., 1958).

In a 45°/45° disk (Figure 13.1), two independent signals are engraved in groove modulations at two orthogonal directions, e.g., at ±45° to the vertical direction. This fashion of groove modulations ensures the symmetry between two signals. A cutter head for recording is shown in Figure 13.2. Excited by two-channel stereophonic signals, two independent coils drive the vibration of the cutter head with two degrees of freedom at ±45° to the vertical direction. Therefore, two-channel stereophonic signals are engraved in a groove modulation in a disk. Conversely, during playback, a pickup stylus vibrates with two degrees of freedom at ±45° to the vertical direction because of the groove modulation in the disk; thus, the two coils in the pickup generate two-channel stereophonic signals.

In a monophonic disk, the signal is recorded as the groove modulation at lateral direction. To be compatible with a monophonic disk, the phase of left or right channel signal in a 45°/45° disk record is reversal. The difference between left and right signals is represented by the vertical vibration and the sum of the left and right signals is represented by lateral vibration. The summing vector of vibration direction is illustrated in Figure 13.3.

A three-step plating process is involved in the practical industrial production of disks. The original lacquer in which grooves have been engraved is spritzed with a fine mist of silver and electroplated with a coat of nickel to create a negative "father" or "master" plate. The father plate is separated from the lacquer, and a positive plate called a "mother plate"is made

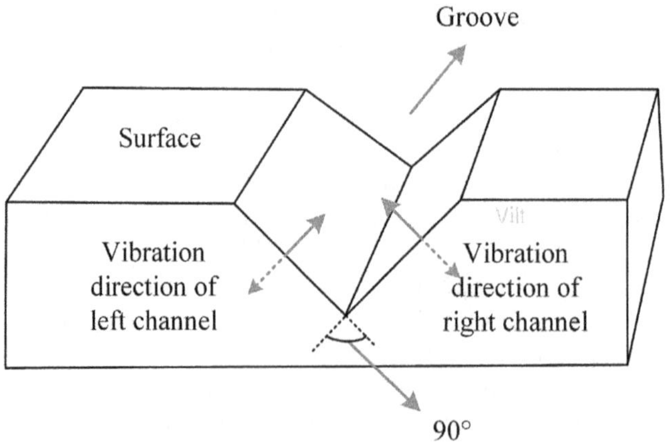

Figure 13.1 Groove and two independent vibration directions of a 45°/45° disk.

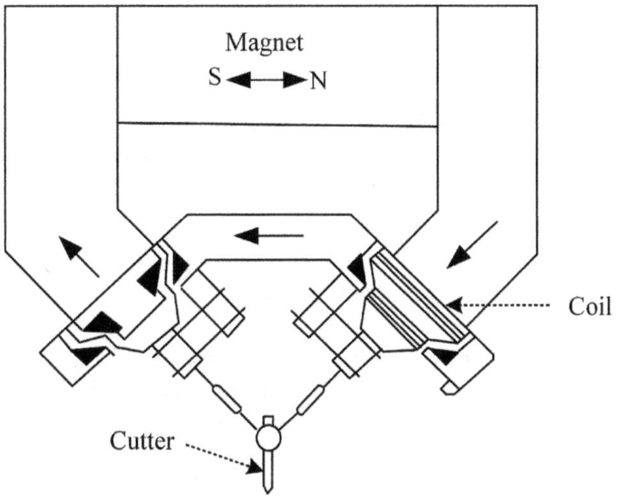

Figure 13.2 Cutter head of a 45°/45° disk record.

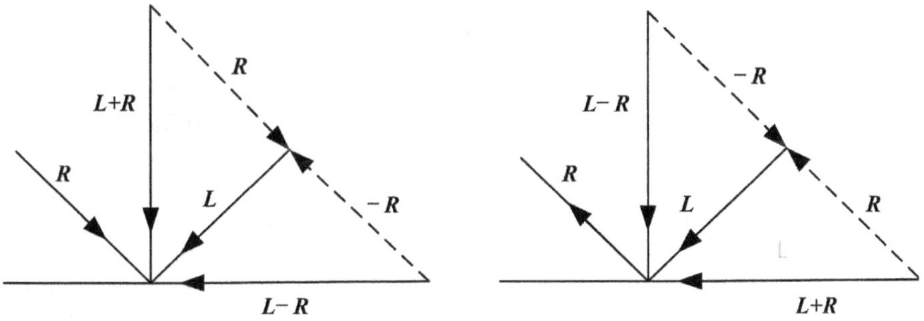

Figure 13.3 Summing vector of the vibration direction in the playback of 45°/45° disk recording.

by growing a second nickel coat on the father plate. The mother plate is another electroplate used to create a new negative plate called a "stamper," which is used to press disks.

Before the beginning of the 1980s, a 45°/45° disk record was used widely in domestic stereophonic reproduction. Since then, it has been substituted by a compact disk and is currently used by a small group of nostalgic hobbyists.

In the beginning of the 1970s, when quadraphones were developed, a method for recording the four-channel signals in the groove of a 45°/45° disk by using a modulated carrier called a CD-4 system was proposed (Inoue et al., 1971), but this method was seldom used in practice.

13.1.2 Analog magnetic tape audio recorder

In analog magnetic tape recording, analog audio signals are stored in a magnetic tape by converting signals into polarized magnetic domains. Tape recorders are often used for professional and consumer recording. Commercial tape recorders have two common types, i.e., reel-to-reel recorder and compact cassette recorder. The former is used for professional recording and partly consumer recording, and the latter is utilized for consumer recording. They have similar principles, but their structures are different. They are described in detail in other books (Guan, 1988). The widths of tape vary depending on the type of tape

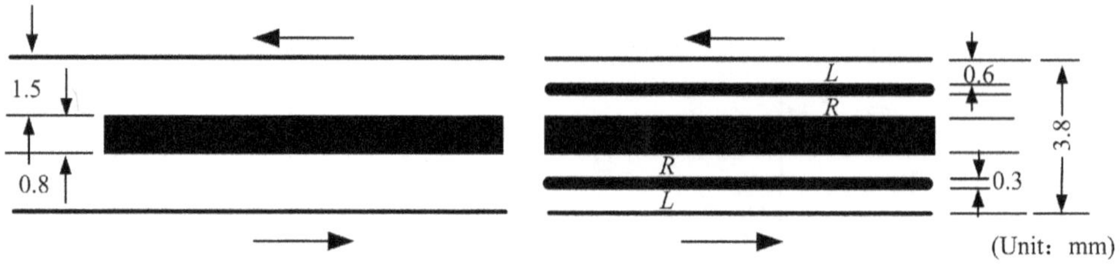

Figure 13.4 Soundtracks of a compact cassette tape. Left side: two sound tracks for monophonic sound; right side: four sound tracks for stereophonic sound.

recorder. For example, tapes for professional recorders have widths of 50.8 mm (2 inches), 25.4 mm (1 inch), and 12.7 mm (1/2 inch). For professional and consumer recorders, their width is 6.3 mm (1/4 inch). For compact cassette recorders, the width of tapes is 3.81 mm (0.15 inch). According to the standard of International Electrotechnical Commission (IEC) and International Radio Consultative Committee (CCIR), the tape speeds are 76.2 and 38.1 cm/s for professional recorders, 19.05 and 9.53 cm/s for consumer recorders, and 4.76 cm/s for compact cassette recorder.

A compact cassette tape is taken as an example to illustrate the method of two-channel signal recording (Ottens, 1967). The width of tape is 3.81 mm, and the tape speed is 4.76 cm/s. The position of four tracks (two on each side) for two-channel stereophonic recording with two sides is illustrated on the right side of Figure 13.4. The left and right channel signals are recorded into two adjacent tracks. Therefore, stereophonic is compatible with monophonic soundtrack shown on the left side of Figure 13.4. However, the position of two adjacent tracks is inclined to cause a crosstalk between left and right channel signals.

Reel-to-reel recorders are widely used in professional recording/playback. Since the 1970s, compact cassette recorders have been widely used in consumer recording/playback. With the development of various digital audio storage techniques, compact cassette recorders have been gradually removed.

Analog magnetic tape recorders had also been widely used in professional multichannel audio recording. The principle is similar to that of two-channel recording. However, as the number of channels increases, these recorders become more complicated and expensive.

13.1.3 Analog stereo broadcasting

Radio broadcasting is an important means to transmit audio signals to users. In the early experiment of two-channel stereophonic broadcasting, two separate transmission chains, including two radio transmitters and receivers, were used to transmit and receive left and right channel signals. However, this method caused serious problems, such as complicated system structure, incompatibility with monophonic broadcasting, and poor consistency between two channel signals. Since then, various systems and formats of analog stereophonic broadcasting have been suggested; for example, the GE Zenith system has been widely used (Eilers, 1961).

GE Zenith system is a subcarrier radio broadcasting system with frequency modulation (FM) on the main carrier at a very high frequency (VHF). Figure 13.5 illustrates the block diagram of its transmitters and receivers. The bandwidth of the left and right signals is restricted within 30 Hz to 15 kHz. After MS transformation (Section 2.2.2), left and right signals are converted into the sum signal $E_M = E_L + E_R$ and the difference signal $E_S = E_L - E_R$. The sum signal directly serves as the main signal. The difference signal is the amplitude modulated onto a subcarrier of 38 kHz by a suppressed carrier modulator, resulting in a signal within

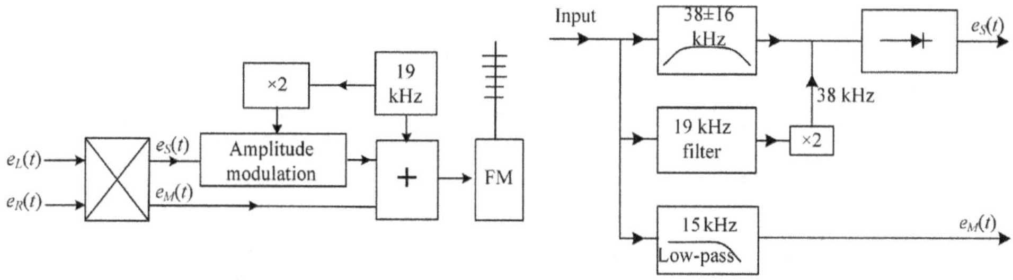

Figure 13.5 Block diagram of the transmitters and receivers of the GE Zenith system.

the bandwidth of 23 kHz to 53 kHz. A pilot signal with a frequency of 19 kHz (just half of that of the subcarrier) and at a very low level (10% of total modulation or permitted deviation) is also included to synchronize the receiver system. A mixed signal is created by a combination of the sum signal, the modulated subcarrier signal, and the pilot signal. Figure 13.5 illustrates the frequency band of the mixed signal. The total bandwidth of the mixed signal is 30 Hz–53 kHz, and the mixed signal is frequency modulated onto the main carrier at VHF with a maximum modulation of 90% of the permitted 75 kHz deviation.

$$u(t) = \left[0.9\frac{e_L(t)+e_R(t)}{2} + 0.9\frac{e_L(t)-e_R(t)}{2}\sin 2\omega t + 0.1\sin \omega t\right] \times 75 \text{ kHz}, \qquad (13.1.1)$$

where $\omega = 2\pi \times 19$ kHz, and $e_L(t)$ and $e_R(t)$ are the corresponding left and right channel signals in the time domain.

In the receiver, the main discriminator produces the mixed signal from which the sum signal, the modulated subcarrier signal, and the pilot signal are separated by filters. With the help of a frequency doubling version of the pilot signal, a second discriminator produces the difference signal from the modulated subcarrier. The left and right signals are recovered from the sum and difference signals after inverse MS transformation.

The GE Zenith system is downward and upward compatible with a monophonic FM broadcasting system. When the mixed signal of stereophonic broadcasting is obtained by demodulation in a monophonic receiver, the sum signal for reproduction is separated by low-pass filtering. A monophonic broadcasting system only transmits the sum signal, and MS transformation in a stereophonic receiver produces identical left and right signals.

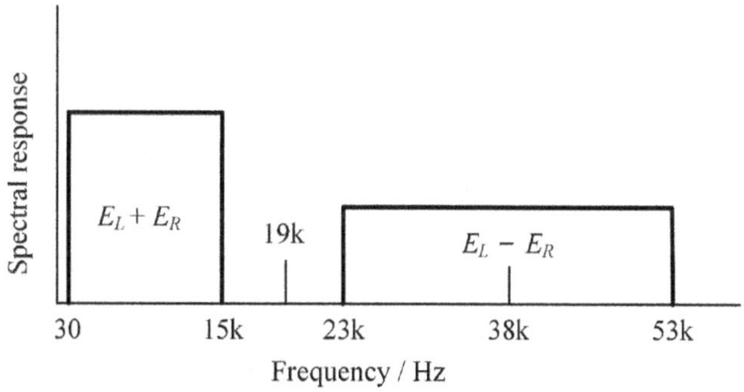

Figure 13.6 Frequency band of the mixed signal.

Since the early 1960s, FM stereo broadcasting has been used in USA, European countries, and Japan. In China, FM stereo broadcasting began in the early 1980s and is still being used (2021). In 1970s, some work suggested to add subcarriers to GE-Zennith for quadraphonic broadcasting (Gibson et al., 1972) Moreover, some amplitude modulation (AM) stereophonic radio broadcasting systems and formats were proposed (Mennie, 1978). For instance, the Motorola compatible-quadrature AM (C-QUAM) system has been used in the USA. AM stereophonic radio broadcasting can cover a large region, but its audio quality is usually inferior to that of FM stereophonic radio broadcasting. From the mid-1980s to early 1990s, AM stereophonic radio broadcasting was tested in China, but it has not been widely used because of the succedent development of digital audio broadcasting (DAB).

13.2 BASIC CONCEPTS OF DIGITAL AUDIO STORAGE AND TRANSMISSION

As stated in Section 13.1, various analog techniques were the primary means of storage and transmission of stereophonic signals. Some analog techniques and methods for the storage and transmission of quadraphonic signals were also proposed, but these techniques had seldom been widely used. Analog storage and transmission techniques usually exhibit some drawbacks, such as limited bandwidth, inclining to create various linear and nonlinear distortions in the course of transmission, storage, and copy. Therefore, the electroacoustic or perceptual performance of an analog storage/transmission system is often deficient. Moreover, the storage and transmission of multichannel audio signals with analog techniques are complicated and difficult. This difficulty was a bottleneck in the development of multichannel sound and one of the reasons for the development of 4-2-4 matrix quadraphones in the early 1970s.

Since the 1980s, especially since 1990, digital audio technique has been developed quickly. It has the advantages of a great bandwidth and a dynamic range, enhanced anti-noise, and improved anti-interference performance. It often promises a better electroacoustic or perceptual performance. Digital audio techniques also facilitate the storage, transmission, copy, and processing of multichannel audio signals. Digital audio signals can be manipulated simultaneously with other multimedia signals, such as image, video signals, and other data. They are adapted to modern information media, such as computers and Internet. Therefore, the development of digital audio technique enables the implementation and application of various spatial sound techniques. Currently, spatial sound or audio signals are usually stored and transmitted by digital techniques except some special cases described in Section 13.1.

An analog audio signal is represented by a continuous time function with a continuous amplitude. In digital audio, a signal is discretized in time and amplitude and approximately represented by a sequence of binary codes. Converting an analog signal into a digital signal includes three steps, e.g., sampling, quantization, and coding, which are implemented by an analog/digital (A/D) converter.

Sampling is performed to represent a continuous time signal by its values at discrete times, resulting in discrete time samples or signals (sequence). According to the Shannon–Nyquist sampling theorem, a continuous time signal with a bandwidth limited to f_m can be recovered from its discrete time samples by an ideal low-pass filter provided that the sampling frequency f_s is not less than twice of f_m. For example, a continuous time signal with a bandwidth limited to 20 kHz can be recovered from discrete time samples at a sampling frequency of not less than 40 kHz. However, an ideal low-pass filter with a rectangular bandwidth is difficult to be realized. The sampling frequency is usually chosen as 2.1 to 2.5 times of f_m to recover the continuous time signal efficiently. A higher sampling frequency is beneficial to reducing the error in the recovered signal. The sampling frequencies commonly used for audio signals are 32, 44.1, 48, 96, and 196 kHz.

Quantization is conducted to represent the amplitude of a discrete time signal with a finite number of discrete values. The full-scale range of the signal amplitude is divided into L discrete regions. For uniform quantization, the width of each region is equal and termed ***quantization width*** or ***step***. If the amplitude of a signal lies within a region, the output of a quantizer is the midvalue of that region. In other words, the continuous amplitude of a signal is mapped to L discrete values. L is termed ***quantization level***. For binary quantization, $L = 2^m$, and m is the ***quantizationbit***. Mapping the continuous amplitude of a signal to finite discreate values leads to a ***quantization error or noise***, which is addressed in Section 13.3.1. Generally, a quantization error reduces as the quantization level or bit increases. Spatial sound signals are usually quantized with m = 16, 20, 24, or even more bits to ensure audio quality.

Coding is carried out to convert quantized signals into digital pulses. Various coding methods are used. The simplest one is linear ***pulse-code modulation*** (***PCM***) through which each m-bit quantized sample is represented by m binary pulses. A PCM digital signal stream is formed by arranging the binary pulses of all samples in order of time. The ***bit rate*** of a digital signal is the number of binary pulses per unit time. For a mono signal, the bit rate is calculated as a product of sampling frequency and quantization bit. For two or more channel signals, the number of channels must be further multiplied. Therefore, the bit rate of digital audio signals increases as the sampling frequency, quantization bit, and number of channels increase. For example, at a sampling frequency of 44.1 kHz and 16-bit quantization, the bit rate of each channel signal is 705.6 kbit/s, and the bit rate of two-channel stereophonic signals is 1411.2 kbit/s = 1.4112 Mbit/s.

As the number of signal channels increases, the bit rate of PCM signals becomes very large. Accordingly, a very large bandwidth for transmission or very large capacity for storage is needed, but this requirement makes the practical transmission or storage difficult. Therefore, prior to transmission or storage, some compression and complicated coding methods are often applied to digital signals to reduce their bit rate and thus adapt to the bandwidth or capacity of practical media. Digital audio compression and coding are a kind of ***source coding***, which is addressed in Sections 13.4–13.8.

Physical defects and disruption may occur in storage media or transmission. These defects and disruptions cause bit errors and degrade audio quality or even make the storage or transmission invalid. Bit errors have various types depending on the cause of physical defects and disruptions. Errors that occur independently for each code or bit and have no relation to one another are called ***random errors***. Large errors in a number of codes or bits are called ***burst errors***. For example, a scratch in an optical disk causes a burst error. Random and burst errors should be corrected in reproduction, but dispersing errors are usually easier to be corrected than burst errors.

Source-coded signals are inappropriate for direct storage and transmission. Instead, ***channel coding*** should be applied to source-coded signals prior to storage and transmission to ensure the reliability of storage and transmission. Channel coding divides the audio bit stream into a series of frames with finite length, supplements correcting codes for error correction, codes for other information, and lastly arranges various codes and data according to a certain rule. Successive data in the stream should be interleaved prior to storage or transmission to correct burst errors effectively. When a burst error occurs in the successive codes of storage or transmission, de-interleaving processing at the receiver disperses the burst error and thereby enables the error to be corrected. In addition, frames in a binary bitstream are marked with a synchronization word to indicate the boundary of each frame and permit synchronization at the receiver. Even if the synchronization word in one or some frames is lost,

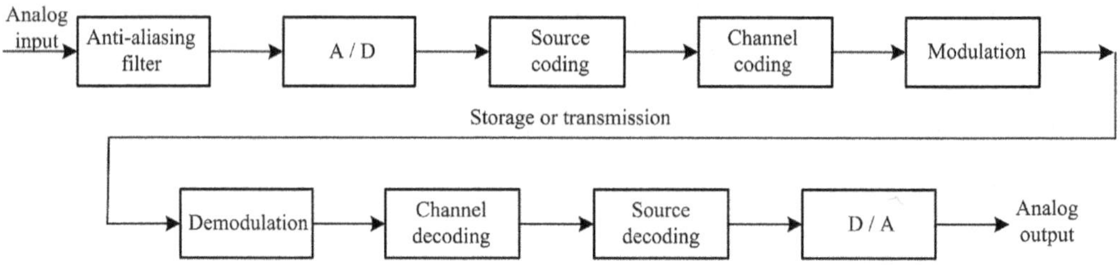

Figure 13.7 Block diagram of digital audio signal storage, transmission, and reproduction.

synchronization can be recovered at the subsequent frames. Various channel coding methods have been developed to meet the requirements of different digital storage and transmission techniques.

Channel-coded data should be modulated so that they are appropriate for storage or transmission. The chosen modulation scheme depends on the storage and transmission method. The basic criterion for choosing a modulation scheme is that it enables to extract synchronous information easily, occupies a narrow bandwidth, and possesses a strong anti-interference performance. It is uneasily influenced by the DC cutoff characteristic of the system.

Figure 13.7 illustrates the block diagram of digital audio signal storage and transmission, involving the courses of A/D conversion, source coding, channel coding, modulation, and corresponding inverse courses in reproduction. After decoding and error correction, PCM audio signals are converted back to analog audio signals by digital/analog (D/A) converts. The methods of source coding and compression related to spatial sound signals are discussed in Sections 13.4–13.8. This book does not describe channel coding and modulation in detail because it mainly deals with the problems of communication theory. Instead, some methods of channel coding and modulation are briefly mentioned when specific methods of storage and transmission are discussed.

13.3 QUANTIZATION NOISE AND SHAPING

13.3.1 Signal-to-quantization noise ratio

Mapping the continuous amplitude of a signal to finite discreate values leads to quantization error or noise (Pohlmann, 2011). For instance, if the amplitude of a bipolar signal is distributed in each quantization region with equal probability, within each quantization region with a width ΔQ, the probability distribution of quantization error is uniform, and the distribution function is given as

$$p_Q(x) = \frac{1}{\Delta_Q} \qquad -\frac{\Delta_Q}{2} \le x \le \frac{\Delta_Q}{2}. \tag{13.3.1}$$

Because the mean of x vanishes, the mean-square error of quantization is evaluated as

$$\mathrm{err}_Q = \int_{-\frac{\Delta_Q}{2}}^{\frac{\Delta_Q}{2}} x^2 p_Q(x)\,dx = \frac{\Delta_Q^2}{12}. \tag{13.3.2}$$

For a bipolar sinusoidal signal with amplitude E_0, the full-scale range of amplitude is $2E_0$. For m-bit quantization, we have

$$2E_0 = L\Delta_Q = 2^m \Delta_Q. \tag{13.3.3}$$

The mean power of a sinusoidal signal is calculated as

$$\text{Pow}_S = \frac{E_0^2}{2} = \frac{1}{2}\left(\frac{2^m \Delta_Q}{2}\right)^2. \tag{13.3.4}$$

The ***relative strength*** of quantization error or noise is evaluated by ***signal-to-quantization-noise ratio SNR$_Q$***, which is defined as the ratio between the mean power of a sinusoidal signal in Equation (13.3.4) and the mean-square error of quantization in the unit of decibel:

$$SNR_Q = 10\log_{10}\frac{Pow_S}{err_Q} = 10\log_{10}\left(\frac{3}{2}\times 2^{2m}\right) \approx 1.76 + 6m(dB). \tag{13.3.5}$$

SNR_Q increases as the quantization bit increases at a rate of 6 dB/bit. For m = 16-bit quantization, SNR_Q evaluated from Equation (13.3.5) is 98 dB.

SNR_Q of linear quantization is high when the signal amplitude is large. For a small signal amplitude, SNR_Q is low because the full-scale range of quantization is not fully utilized (only parts of bits are used for quantization). A nonlinear quantization scheme can be used to solve this problem through which a signal with a small amplitude is quantized with a small quantization width, and a signal with a large amplitude is quantized with a large quantization width.

13.3.2 Quantization noise shaping and 1-Bit DSD coding

The SNR_Q can be improved by an increase of quantization bit. However, limited by the precision of an A/D converter, the implementation of quantization with a very high bit is difficult. In addition, quantization with a very high bit increases the bit rate and makes storage and transmission difficult. Some other methods have been developed to improve SNR_Q and thus solve this problem (Pohlmann, 2011).

In oversampling, analog audio signals are sampled with a frequency much higher than the low limit of the Shannon–Nyquist sampling theorem, thereby improving SNR_Q. For a sampling frequency of f_{s0} and the given quantization bit, the power spectrum of quantization noise distributes uniformly within the frequency range from 0 to $f_{s0}/2$. For the same quantization bit, when the sampling frequency is K times f_{s0}, e.g., $f_s = K f_{s0}$, the overall power of quantization noise is unchanged, but the power spectrum of quantization noise is distributed uniformly within the frequency range from 0 to $K f_{s0}/2$. The power spectrum density of quantization noise is $1/K$ times of that of original case. Therefore, the power of quantization noise within the audible bandwidth is $1/K$ times of that of the original case, or equally SNR_Q is K times of the original case. A two-time oversampling improves SNR_Q by 3 dB, and a 4-time oversampling is equivalent to an increase in one quantization bit. The quantization noise out of the audible bandwidth can be removed by a low-pass filter. Figure 13.8 illustrates the power spectra of quantization noise for sampling with f_{s0}, as well as 4-time oversampling.

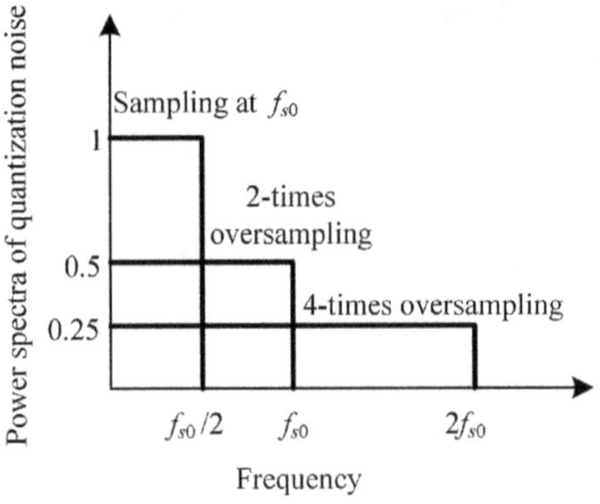

Figure 13.8 Power spectra of quantization noise.

Oversampling reduces the quantization bit at the cost of increasing the sampling frequency. The profit of oversampling alone on reducing quantization noise is not large. A combination of oversampling and noise shaping obviously improves SNR_Q. Noise shaping pushes the power distribution of quantization noise out of the audible frequency band while keeping the overall quantization noise power unchanged; therefore, SNR_Q increases in the audible frequency band. The quantization noise out of the audible bandwidth can be removed by a low-pass filter.

Noise shaping can be implemented by applying a frequency-dependent negative-feedback loop to quantization noise. Because the negative-feedback factor at low frequencies is larger than that at high frequencies, the quantization noise at a low (audible) frequency band is restrained. Figure 13.9 (a) illustrates the block diagram of the first-order noise shaping. For a discrete time input signal $e_x(n)$, the output of a quantizer is expressed as

$$e_y(n) = e_x(n) + q(n), \tag{13.3.6}$$

where $q(n)$ is the unshaped quantization noise. In Figure 13.9 (a), $q(n)$ is extracted by subtracting $e_x(n)$ from $e_y(n)$. Then, $q(n)$ is delayed by one sample and feedback to the input with an inverse phase. The state equation of the system is given as

$$e_y(n) = u(n) + q(n) \qquad u(n) = e_x(n) - q(n-1). \tag{13.3.7}$$

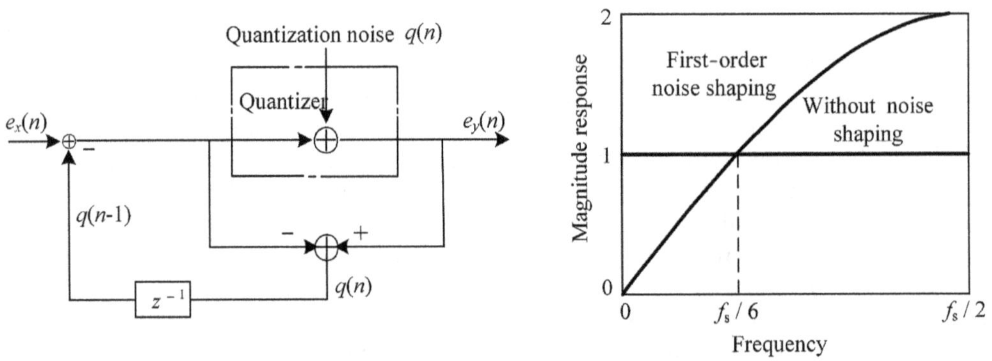

Figure 13.9 First-order noise shaping (a) Block diagram; (b) Magnitude response of filter.

The input/output equation of the system is expressed as

$$e_y(n) = e_x(n) + \left[q(n) - q(n-1)\right]. \tag{13.3.8}$$

Therefore, quantization noise is replaced by its differential value after noise shaping. Equation (13.3.8) can be converted into the Z-domain as

$$E_y(z) = E_x(z) + \left(1 - z^{-1}\right)Q(z) . \tag{13.3.9}$$

The functions denoted by capital letters in Equation (13.3.9) represent the signal and noise in the Z-domain. Therefore, quantization noise is shaped or filtered by the following filter:

$$H(z) = 1 - z^{-1}. \tag{13.3.10}$$

Let $z = \exp(j2\pi f/f_s)$, where f_s is the frequency of oversampling. The magnitude response of the filter is given as

$$|H(f)| = 2\left|\sin\frac{\pi f}{f_s}\right|. \tag{13.3.11}$$

Figure 13.9(b) illustrates the magnitude response of the first-order noise shaping filter given by Equation (13.3.11). Quantization noise is attenuated greatly within the frequency band $f/f_s < 1/6$.

Figure 13.10 is an alternative model to the noise shaping scheme in Figure 13.9 (a) and called the first-order Σ-Δ modulation. It involves an integrator (accumulator Σ), a differentiator (Δ), delay elements, and a quantizer. The integrator boosts the low-frequency component of signals. Therefore, the response of the model in Figure 13.10 is identical to that of the model in Figure 13.9(a).

After noise is shaped with the first-order Σ-Δ modulation, quantization even with a smaller number of bits also exhibits a high SNR_Q provided that the oversampling frequency is high enough. However, sampling with an excessively high frequency is also difficult. The schemes of higher-order Σ-Δ modulation or multistage noise shaping can be used to solve this problem. Higher-order Σ-Δ modulation utilizes a higher-order integrator to further boost the response at low frequencies. Multistage noise shaping is equivalent to using feedforward

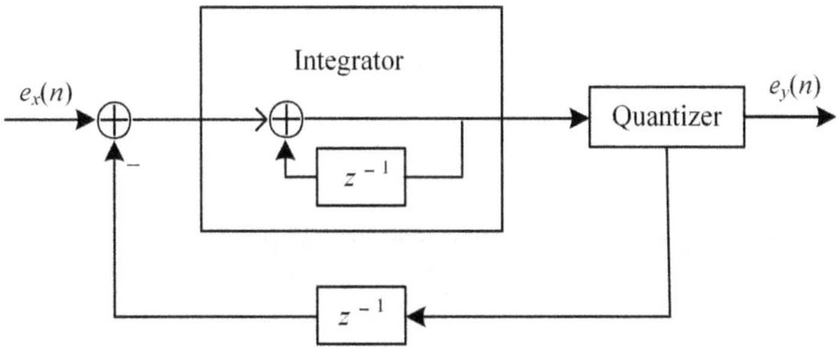

Figure 13.10 First-order Σ-Δ modulation.

rather than feedback to restrain quantization noise. This issue is described in detail in another book (Pohlmann, 2011).

At a sufficiently high sampling frequency, ***1-bit quantization*** can achieve precision that matches with that of 16-bit quantization. For example, if the sampling frequency increases to 44.1 kHz × 16 bits × 4, SNR_Q of 1-bit quantization within the audible bandwidth is identical to that of 16-bit quantization at a sampling frequency of 44.1 kHz. Actually, the amplitude of a signal is represented by the density of pulses in 1-bit quantization. Therefore, 1-bit quantization is also termed ***pulse density modulation*** (***PDM***). The D/A converter for 1-bit PDM codes is relatively simple, and a low-pass filter or integrator is enough to convert PDM codes into analog signals. The 1-bit A/D converter can be implemented by a 1-bit Σ-Δ modulator. Figure 13.11 illustrates the block diagram. Let $e_x(n)$ denote the input signal or time sequences and $e_y(n)$ indicate the output signal or time sequences after 1-bit quantization. In Figure 13.11, $e_u(n)$ corresponds to the difference between the current input $e_x(n)$ and $e_w(n)$ obtained by 1-bit D/A conversion to the output $e_y(n-1)$ at a previous instant; $e_v(n)$ is the sum of the current $e_u(n)$ and previous $e_u(n-1)$. In terms of $e_v(n)$ larger or less than 0, comparator outputs a code of 1 or 0. The state equation can be written as

$$e_u(n) = e_x(n) - e_w(n) \qquad e_v(n) = e_u(n) + e_u(n-1)$$
$$e_y(n) = \begin{cases} 1 & e_v(n) > 0 \\ 0 & e_v(n) < 0 \end{cases} \qquad e_w(n) = \begin{cases} 1 & e_y(n-1) = 1 \\ -1 & e_y(n-1) = 0 \end{cases}. \tag{13.3.12}$$

The ***direct stream digital*** (***DSD***) technique based on 1-bit quantization has two important applications. The first application is for the D/A conversion of a multibit digital signal. In Figure 13.12, the multibit input with f_s is first upsampled to a multibit signal with the sampling frequency Kf_s, converted to 1-bit samples at a sampling frequency of Kf_s by a 1-bit Σ-Δ modulator, and converted into an analog signal by a low-pass filter. Another important application of DSD technique is the direct adoption as a digital audio signal format as in the case of super audio compact disk (SACD) in Section 13.11.4.

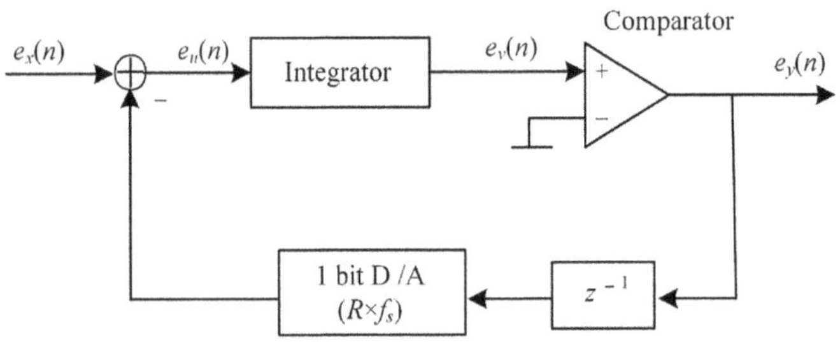

Figure 13.11 Block diagram of 1-bit A/D converter.

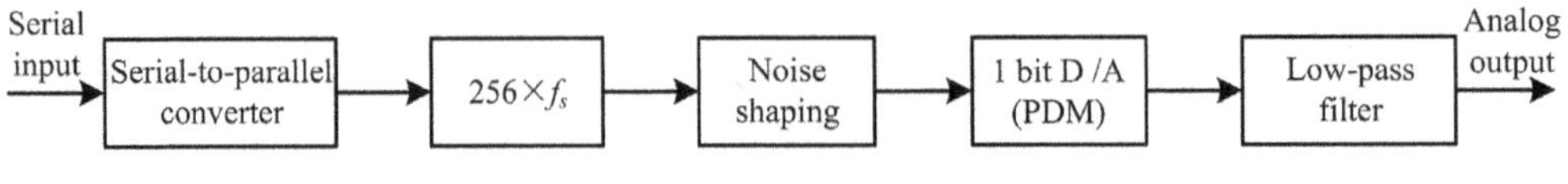

Figure 13.12 Application of DSD technique to the D/A conversion of multibit digital signal.

13.4 BASIC PRINCIPLE OF DIGITAL AUDIO COMPRESSION AND CODING

13.4.1 Outline of digital audio compression and coding

As stated in Section 13.2, digital audio signal is often compressed and coded to reduce its bit rate. According to communication theory, source coding is a mapping from a sequence of symbols from an information source to a sequence of alphabet symbols (usually bits) to improve the effectiveness of communications or reduce the redundancy of symbols from an information source.

Digital audio compression and coding are an active field in communication, signal processing, and audio technique. During the past decades, various coding techniques for speech and high-quality audio signals have been developed. The principles and requirements of coding techniques for different applications vary. Limited by the length, this chapter addresses only the coding techniques related to the storage and transmission of spatial sound signals and does not discuss various audio compression and coding techniques in detail. The signals of spatial sound have various components, such as music, speech, environment, and effect sounds. A high perceived quality can be achieved from coded spatial sound signals. By contrast, speech coding in communication often aims to reduce the channel resource for transmission while keeping speech intelligible.

From the point of recovery of original signals (data), audio signal compression and coding can be divided into two groups, e.g., lossless coding and lossy coding. Lossless coding allows original signals to be perfectly reconstructed from the compressed data. Because some less important information is omitted, lossy coding only permits original signals to be approximately recovered from the compressed data with some distortion. Generally, a lossless coding algorithm exhibits a high perceived audio quality at the cost of a low compression efficiency. By contrast, a lossy coding algorithm exhibits a high compression efficiency at the cost of reducing perceived audio quality. Investigations on audio coding aim to achieve a high compression efficiency as far as possible for a given perceived audio quality or a high perceived audio quality as far as possible for a given compression efficiency.

From the point of algorithm, audio coding can be divided into waveform coding, parameter coding, and hybrid coding.

1. *Waveform coding* directly manipulates the time or frequency samples of audio signals and compresses data by using the knowledge of the amplitude distribution of samples and the correlation between adjacent samples and other types of information about the samples. Waveform coding is theoretically applicable to arbitrary audio signals with a relatively higher perceived audio quality (applicable to spatial sound signals). The algorithms of waveform coding are relatively simple, but they have a disadvantage of a low coding efficiency.
2. *Parameter coding* represents signals with appropriate physical models and therefore converts signals into model parameters. Signals are resynthesized on the basis of models and parameters. For example, on the basis of the model of the human vocal system, speech signals can be represented and synthesized by an exciting stimulus and a time-variable filter model. Because signals are represented by the time-variable parameters of the model, the required bit rate for transmission is greatly reduced. However, the audio quality of parameter coding is usually inferior to that of waveform coding. Parameter coding is traditionally applied to speech communication and music synthesizer. Recently, it is also applied to coding of spatial sound signals, especially the coding of two or more channel signals.

3. ***Hybrid coding*** represents signals with appropriate physical models and converts signals into model parameters. It also codes the error between the reconstructed signals from parameter coding only and the original signals for the recovery of original signals.

From the point of physical and auditory mechanisms, spatial sound signal coding utilizes the redundancy among signals, including the redundancy in time, frequency, spatial, and perceptual domains.

1. ***Redundancy in the time domain.*** Generally, the probability distribution of the amplitude of a practical audio signal is nonuniform. The probability of a smaller amplitude is usually larger than that of a larger amplitude. Speech and music signals may be silent. The samples of signals may be correlated within a certain time length. An approximately periodic signal is also correlated between different periods. These features of audio signals are attributed to redundancy in the time domain.
2. ***Redundancy in the frequency domain.*** Generally, the long-term mean power spectra of audio signals are not uniform. Instead, they attenuate as frequency increases and exhibit a feature similar to that of pink noise. The short-term power spectra of speech signals exhibits the specific features of resonance peaks. These features of audio signals are attributed to redundancy in the frequency domain.
3. ***Redundancy in the spatial domain.*** Spatial sound utilizes the relationships among different channel signals, such as interchannel correlation, level difference, and time (phase) difference to represent the spatial information of sound. Therefore, different channel signals have redundancy. This feature of audio signals is attributed to redundancy in the spatial domain.
4. ***Redundancy in the perceptual domain.*** The resolutions of human hearing on frequency and amplitude are limited. Therefore, the details of audio signals are not always audible. Moreover, the quantization error (noise) is not always audible because of the masking effect. This perceptually irrelevant information can be regarded as redundancy in the perceptual domain and ignored. Various perceptual audio coding utilizes redundancy in the perceptual domain to compress audio signals. Perceptual audio coding is lossy coding. It has a high compression efficiency and retains a better perceived audio quality.

Overall, audio compression and coding remove redundancy within signals and reduce the bit rate. Based on the aforementioned consideration, various audio coding techniques have been developed, and some of these techniques have been applied to the storage and transmission of spatial sound signals. Some typical coding techniques related to spatial sound are addressed in the following sections. Practical spatial sound signal coding usually utilizes multiple methods to remove the redundancy of signals.

Compression and coding are technical means of the storage and transmission of spatial sound signals. ***Different compression and coding methods should not be confused with different spatial sound systems.*** The signals of a spatial sound system can be conveyed by using different compression and coding methods. For example, 5.1-channel sound signals can be conveyed by using Dolby Digital coding (Section 13.6.1) or DTS coherent acoustics coding (Section 13.7) method. Classifying spatial sound systems and techniques according to the methods of compression and coding is inappropriate. However, this problem often occurs in some commercial articles. As stated in Section 13.1, two-channel stereophonic signals can be delivered by different analog methods, such as 45°/45° disk recording and magnetic tape recording, but they are the signals of the same spatial sound system.

The performance of various audio compression and coding techniques is evaluated in terms of the audio quality, bit rate, complexity of algorithm, delay in coding, ability of anti-bit error, and other criteria. Audio quality is evaluated by objective and subjective methods. In addition to some physical indices such as signal-to-quantization-noise ratio, objective measurement can also obtain some indices in which the human auditory resolution has been considered. However, because redundancy in the perceptual domain is often used in spatial sound signal coding, objective measurement cannot reflect the practical perceived performance, so a subjective assessment experiment is needed. The International Telecommunications Union (ITU) has recommended some standards for the subjective assessment of the performance of audio coding, which is addressed in Chapter 15. Some international experiments on the subjective comparison and assessment of various audio coding methods have been conducted (e.g., EBU-Tech 3324, 2007).

13.4.2 Adaptive differential pulse-code modulation

Linear PCM directly quantizes and codes the amplitude of each sampled signal. As stated in Section 13.3.1, a large quantization bit is needed to ensure a sufficient SNR_Q or dynamic range, resulting in a high bit rate. Alternatively, a ***differential pulse-code modulation*** (***DPCM***) quantizes and codes the following differential value between two adjacent samples:

$$e_d(n) = e_x(n) - e_x(n-1). \tag{13.4.1}$$

If the adjacent samples of a discrete time signal are obviously correlated, the dynamic range and mean power of the differential value are smaller than those of the original signal. Accordingly, the required bit for the quantization of differential values is smaller than that for the original signal. Therefore, DPCM improves the efficiency of quantization.

The samples of a discrete time signal may be correlated within a certain time period. Therefore, the current sample can be predicted from previous samples within a certain period. ***Adaptive differential pulse-code modulation*** (***ADPCM***) quantizes and codes the difference between the unquantized input sample value and the predicted value (Gersho and Gray, 1992; Furui, 2000). Under an ideal condition, the correlation of samples in the time domain should be removed completely. In this case, the differential value (residual) is a random white noise. In comparison with DPCM, ADPCM utilizes a higher order and more accurate predictive model so that it removes the correlation in the time domain and improves the coding more efficiently.

The efficiency of ADPCM depends on the correlation of input signals in the time domain. ADPCM is effective only for signals with high correlation in the time domain. It is ineffective for random signals. For speech signals, DPCM and ADPCM increase the signal-to-noise ratio by 10 and about 14 dB compared with conventional PCM, respectively (Furui, 2000).

Adaptive prediction can be implemented by a finite impulse response (FIR) filter. Let $e_x(n)$ denote the input sample value and $\hat{e}_x(n)$ represent the predicted value. According to linear prediction theory, if the samples of a discrete time signal are correlated within a certain period, the current sample can be estimated from a linear combination of pass P sample values (forward prediction) as

$$\hat{e}_x(n) = \sum_{i=1}^{P} a_i e_x(n-i), \tag{13.4.2}$$

where a_i is P prediction coefficients. The difference between input and prediction, e.g., error of prediction is evaluated as

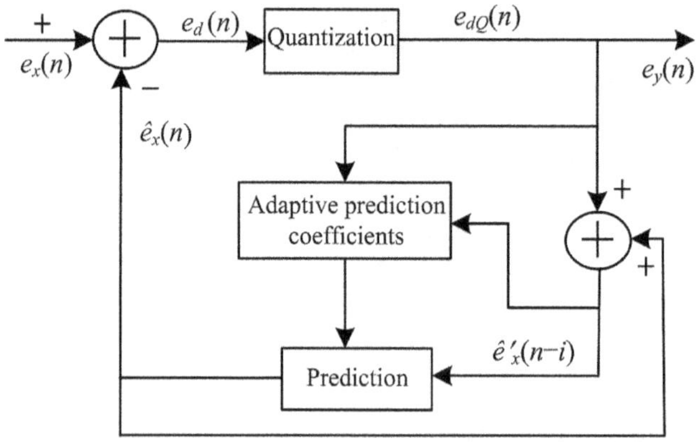

Figure 13.13 Block diagram of ADPCM.

$$e_d(n) = e_x(n) - \hat{e}_x(n) = e_x(n) - \sum_{i=1}^{p} a_i e_x(n-i). \tag{13.4.3}$$

Figure 13.13 is the block diagram of ADPCM. The differential signal between the input signal $e_x(n)$ and the predicted signal $\hat{e}_x(n)$ is $e_d(n)$. After quantization, it becomes $e_{dQ}(n)$. In contrast to Equation (13.4.2), the predicted signal $\hat{e}_x(n)$ is directly evaluated from the sum of the previous predicted signal and the quantized differential signal rather than from the input signal $e_x(n - i)$, e.g., the following $\hat{e}'_x(n-i)$ is used to replace $e_x(n - i)$ on the right side of Equation (13.4.2):

$$\hat{e}'_x(n-i) = \hat{e}_x(n-i) + e_{dQ}(n-i). \tag{13.4.4}$$

The prediction coefficients in Equation (13.4.2) should be updated constantly to adapt to the variation in the signal, e.g., $a_i = a_i(n)$. Various algorithms are used to calculate the prediction coefficients in linear prediction theorem. For example, using the stochastic gradient algorithm, the prediction coefficients at the next instant can be calculated from the current coefficients by using the following equation:

$$a_i(n+1) = a_i(n) + \Delta_i(n) e_d(n) e'_x(n-i) \quad i = 1, 2 \ldots N, \tag{13.4.5}$$

where $e_d(n)$ is obtained by substituting e_x with $\hat{e}'_x$ in Equation (13.4.3). $\Delta_i(n) > 0$ is the step lengths that control the stability and speed of the convergence of the algorithm.

A higher-order FIR filter model usually promises a better prediction performance, but it is complex. The requirement of accuracy in prediction usually conflicts with that of simplicity. Some schemes use an infinite impulse response (IIR) filter model for prediction. The system function of an IIR filter possesses poles and thus can preferably predict the signal peaks.

Prediction coefficients are calculated in the coder and transmitted as a part of the bit stream. They vary in each frame of signals. A decoder retrieves the complete signal from the coded differential signal and prediction coefficients. The predicted signal in a decoder is created by the course identical to that in an encoder. The decoder retrieves a complete signal by adding the predicted signal with an error signal.

13.4.3 Perceptual audio coding in the time-frequency domain

Perceptual audio coding utilizes the redundancy in the perceptual domain to compress audio signals and is a lossy coding method (Brandenburg and Bosi, 1997). The masking effect in Section 1.3.3 is the psychoacoustic basis for perceptual audio coding. Section 13.3.1 indicates that mapping the continuous amplitude of a signal to finite discreate values leads to quantization noise. In the absence of noise shaping, signal-to-quantization-noise ratio increases as the quantization bit increases. A larger quantization bit is needed to obtain high signal-to-quantization-noise ratio. However, quantization noise is not always audible because of the masking effect. If the level of quantization noise is below the threshold of a masking curve (pattern), the quantization noise is inaudible. If the level of a signal is below the hearing threshold or below the threshold of the masking curve, the signal is also inaudible. The above psychoacoustic principles are applicable to perceptual audio coding.

Because masking is related to the time-frequency resolution of human hearing, time domain signals should be transformed to time-frequency domain signals prior to perceptual coding. Transformation has two types. They are similar in nature but different in time-frequency resolution. In practical coding, appropriate transformations and related parameters should be chosen so that time-frequency resolution meets the requirement of auditory perception.

The first type of transformation uses an analysis filter bands to decompose time domain signals into subband components. The bandwidth of subband filters can be uniform or nonuniform. Filters with a uniform bandwidth are relatively simple, but filters with a nonuniform bandwidth can be adapted to the frequency resolution of human auditory. Generally, subband filters exhibit higher time resolution and lower frequency resolution. According to Shannon–Nyquist temporal sampling theorem (Oppenheim et al., 1999), each subband component can be downconverted into a baseband and then subsampled at a sampling frequency not less than twice the bandwidth of the subband.The original signal can be restored by first upsampling and upconverting the baseband representation of each subband component and combining the components of all subbands by using a synthesis filter bank. Ideally, analysis filters should have abrupt transition characteristics between the passband and the stopband to avoid the frequency domain overlap in the restored signal. In addition, the analysis filters should have linear-phase characteristics. Filters with such kind of characteristics are difficult to be implemented. In practice, the bandwidth of signals is often divided into K uniform subbands, where K is the power of 2. Then, analysis filtering is implemented by K ***quadrature mirror filters*** (*QMF*). Although the outputs of quadrature mirror filters overlap at the boundaries between the subbands, overlapping components are cancelled in synthesis filtering, and the original signal is restored. A polyphase quadrature mirror filter (PQMF) band is often used in practical subband coding. For MPEG-1 Layer I and Layer II coding in Section 13.5.1, a PQMF bands is used to divide the time domain input into 32 uniform subband components. An analysis filter bands with a critical bandwidth is used for MPEG-1 Layer III coding.

In the second type of transformation, the discrete time samples of the input signal are divided into block or frames with an appropriate length, and short-term discrete orthogonal transform is used to convert each block of time samples into spectral coefficients in the transform domain (such as in the frequency domain or more strictly in the time-frequency domain). Generally, various short-term discrete orthogonal transforms exhibit higher-frequency resolution and lower time resolution. A well-known short-term discrete orthogonal transform is the short-term Fourier transform (STFT) in Equation (8.3.15). However, ***modified discrete cosine transform*** (MDCT) is often used in spatial sound signal coding, such as Dolby Digital coding described in Section 13.6.1. The advantage of MDCT is that the power of a signal is dominated by the preceding spectral components, which are beneficial to signal compression. In addition to STFT and MDCT, other short-term discrete orthogonaltransforms, which yield the coefficients in the transform domain, are applicable to audio coding.

Similar to the case in Section 8.3, the discrete signal in the time domain is denoted by $e_x(n)$, and n is the discrete time. After subband filtering or short-term discrete orthogonal transform, the subband components or coefficients of the transform are denoted by $E_x(n', k)$. For subband filtering, n' is the variable of discrete time, and k is the index of the subband, e.g., $E_x(n', k)$ is the sample of the signal at the kth subband and time n'. For short-term discrete orthogonal transform, n' is the variable of time (block or frame), k is the variable in the transformed domain (such as frequency domain), and $E_x(n', k)$ is the coefficient of transform at time (block or frame) n'.

An N-point (even number) MDCT-modified discrete cosine transform can be written as follows by using the above notation (Bosi et al., 1997):

$$\begin{aligned} E_x(n', k) &= 2\sum_{n=NL}^{NH} e_x(n'+n)\cos\left[\frac{2\pi}{N}(n+n'+n_0)\left(k+\frac{1}{2}\right)\right] \\ n_0 &= \frac{N/2+1}{2} \qquad k = 0, 1 \ldots (N/2-1), \end{aligned} \tag{13.4.6}$$

where $NL \le 0$ and $NH > 0$ are the initial and end times for MDCT calculation, and $N = NH - NL + 1$ is the length of the block or frame for MDCT calculation. An N-point MDCT satisfies the following odd symmetry:

$$E_x(n', k) = -E_x(n', N-1-k). \tag{13.4.7}$$

Therefore, a signal can be completely described by the preceding $N/2$ MDCT coefficients with $k = 0, 1\ldots (N/2 - 1)$. An N-point inverse MDCT is given as

$$\begin{aligned} e_x(n'+n) &= \frac{2}{N}\sum_{k=0}^{N/2-1} E_x(n', k)\cos\left[\frac{2\pi}{N}(n+n'+n_0)\left(k+\frac{1}{2}\right)\right] \\ n &= NL, NL+1, \ldots . NH. \end{aligned} \tag{13.4.8}$$

Figure 13.14 presents the block diagram of the principle of perceptual audio coding. It involves the following stages.

1. The discrete time input signal is converted into the time-frequency domain by an analysis filter band or short-term discrete orthogonal transform, yielding subband components or spectral coefficients $E_x(n', k)$.
2. The short-term power spectra of input signals within a certain time window (sampling block) are evaluated by converting the input signal to the time-frequency domain (such

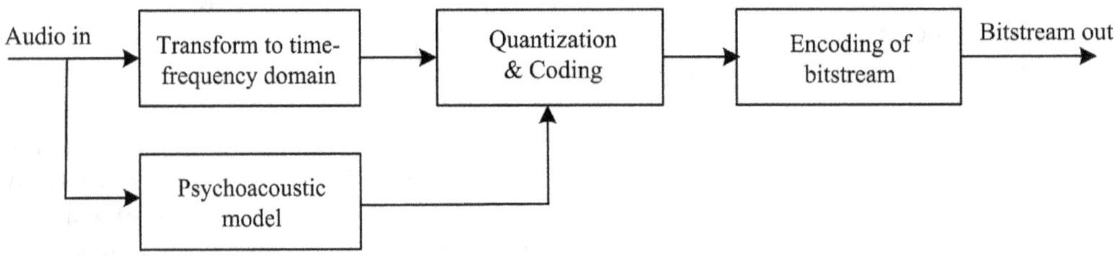

Figure 13.14 Block diagram of the principle of perceptual audio coding.

as using STFT) or directly evaluated from the result of stage (1). The resultant power spectra are analyzed by a psychoacoustic model inputted to the time-frequency domain.

3. The subband or spectral components of input signals are quantized and coded. Different subband components or spectral coefficients $E_x(n', k)$ are quantized with different bits. Given the available bit rate, the algorithms of dynamic bit allocation are often used. According to the short-term power spectra evaluated in stage (2) and certain psychoacoustic models, available bits are allocated to each subband component or spectral coefficients (set) to optimize the final perceived performance.
4. The quantized data are organized into frames and then assembled into the bit stream. In addition to the quantized and coded samples $E'_x(n', k)$, bit stream involves some side information, such as bit allocation information for reconstructing a signal in decoding.

A decoder reconstructs the signal from the bit stream of coded signals. Decoding is an inverse course of coding (Figure 13.14). The subband components or spectral coefficients of each frame are extracted from the bit stream from which PCM audio signals are reconstructed.

Quantization, dynamic bit allocation, and coding have various algorithms. The performance of these algorithms, such as computational cost, compression ratio, and perceived effect, varies. Dynamic bit allocation strategies in audio coding have two kinds. In forward-adaptive allocation, allocation is performed in the coder, and information is included in the bit stream. An advantage of forward-adaptive allocation is that the psychoacoustic model is only included in the coder. A revision of the psychoacoustic model does not influence the design of the decoder. A disadvantage is that some bit resources are needed to convey the allocation information to the decoder. In backward-adaptive allocation, bit allocation information in the decoder is derived from the coded audio data, and the transmission of bit allocation information is not needed. Backward-adaptive allocation has a higher transmission efficiency, but it consumes the computational resource of the decoder. In addition, the psychoacoustic model in the decoder cannot be easily improved when it is applied.

Psychoacoustic models simulate the perception of human hearing to sound. They are essential for perceptual audio coding, especially dynamic bit allocation. Various psychoacoustic models with different accuracies and complicities exist. The quantitative analysis and simulation of the masking effects are the cores of various psychoacoustic models related to audio coding. In many cases, SNR_Q in each subband (critical band) is calculated from the short-term power spectral level and given quantization bits for each frame (sampling block) of input signals. The larger the quantization bits are, the larger SNR_Q will be. According to the psychoacoustic pattern of masking, the signal-to-masking-ratio SMR in each subband, which is the difference between power spectral level of signal and minimum masking threshold, is calculated. The noise-masking ratio in each subband is determined using the following formula:

$$NMR = SMR - SNR_Q \quad (dB). \tag{13.4.9}$$

$NMR > 0$ means that quantization noise is audible. Figure 13.15 illustrates the relationship among NMR, SMR, and SNR_Q. For a given the overall bit rate, dynamic bit allocation, which is usually implemented by iterative algorithm, minimizes the overall NMR across all subbands. In addition, signal components with a level lower than the hearing threshold is inaudible. They are not coded or coded with lower bits.

Masking pattern models are needed to calculate SMR. These patterns depend on the components of stimuli (tonal or nontonal components). Many models in audio coding detect tonality in signals (such as local maximum in the spectrum or spectral flatness measure) to

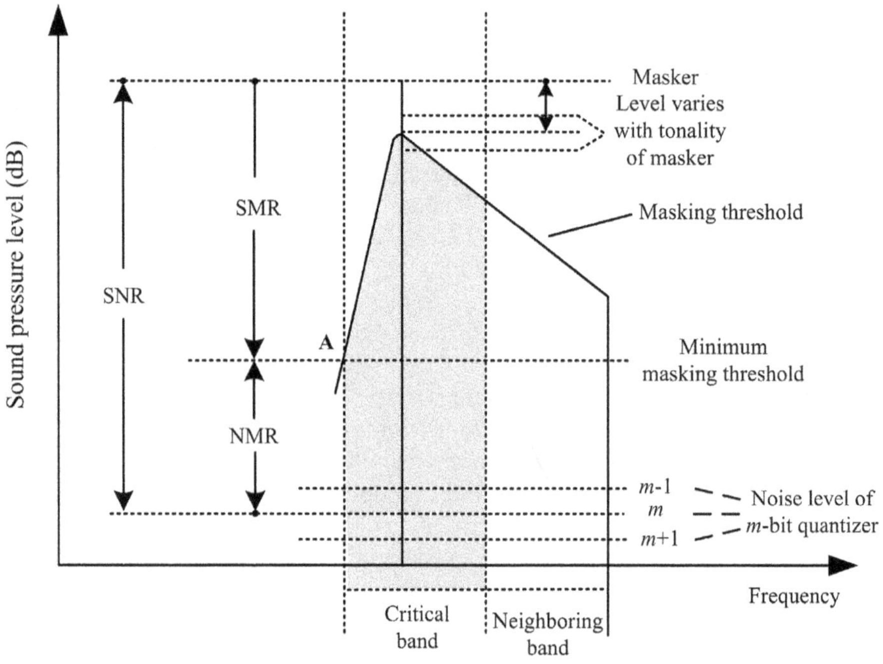

Figure 13.15 Relationship of *NMR*, *SMR*, and SNR_Q (Noll, 1997, with the permission of IEEE).

determine the range and amount of masking. In addition, the effect of the critical bandwidth component of a masker is not limited to a single critical band, but it is also distributed to other bands. The spreading function describes masking across several critical bands to simulate the masking response of the entire basilar membrane. The effects of maskers in several critical bands should be considered in the calculation of the overall masking threshold.

13.4.4 Vector quantization

Scalar quantization is discussed in the preceding sections in which each sample (or each differential sample) of the input signal is quantized and coded. ***Vector quantization*** (***VQ***) assembles some scalar data into a set according to a certain rule. Each set can be regarded as a vector and can be quantized jointly in a vector space (Gersho and Gray, 1992; Furui, 2000). VQ maximizes the statistical correlation between the components of the vector and compresses the data with less loss of information. As a lossy coding technique, VQ is widely used in speech coding and sometimes used in audio coding.

If the data of K scalar inputs $e_{x0}, e_{x1}\ldots e_{xK}$ constitute a K-dimensional vector $\boldsymbol{e}_x = [e_{x0}, e_{x1}\ldots e_{xK}]$, and the set $\{\boldsymbol{e}_x\}$ of all K-dimensional vectors constitutes a K-dimensional Euclid space. The Euclid space is divided into L subspaces that do not intersect. Each subspace is approximately represented by a vector $\boldsymbol{e}_{yl}$. The set $\{\boldsymbol{e}_{yl}\}$ of L representative vectors $\boldsymbol{e}_{yl}$ ($l = 0, 1\ldots L - 1$) is termed a ***code book***. The number L of representative vectors in a code book is called the length of a code book. Various divisions of subspaces or choices of the code book constitute different vector quantizers.

For an arbitrary input vector $\boldsymbol{e}_x$, a vector quantizer first determines the subspace in which $\boldsymbol{e}_x$ belongs to and then represents $\boldsymbol{e}_x$ by a corresponding vector $\boldsymbol{e}_{yl}$. Therefore, the nature of VQ is to map the arbitrary vector $\boldsymbol{e}_x$ in a K-dimensional Euclid space into a finite set $\{\boldsymbol{e}_{yl}\}$ of L representative vectors.

The input of a vector is the time sample. Samples are grounded according to their time order, and each group is an input vector e_x. The vector coder first matches the input vector to a representative vector in the code book. The match is conducted according to a certain rule of the least error. The vector coder only codes the index of the representative vector rather than the time sample of input, which improves the efficiency of coding and compresses the date. With a code book identical to that in the coder, a vector decoder finds the representative vector and then retrieves the input vector e_x approximately from the coded index of the representative vector. The key of VQ is to design an appropriate code book to represent the input vectors and improve the efficiency of coding.

13.4.5 Spatial audio coding

In Section 13.4.1, the signals of different channels in spatial sounds may be correlated, that is, the information represented in different channels is somewhat redundant. This redundancy can be utilized in the compression and coding of two or more channel signals. This principle is the basis for ***spatial audio coding*** (*SAC*; Herre et al., 2004; Breebaart and Faller, 2007). In comparison with the method of coding each channel signal separately, the efficiency of SAC in coding two and more channel signals is improved obviously. Various SAC-based schemes have been developed.

MS stereo coding (shorten for MS coding) is a typical SAC-based scheme through which the sum (M) and difference (S) of stereophonic signals in Equation (2.2.18) rather than the original left and right signals are coded. Therefore, this scheme is similar to GZ Zenith FM stereophonic broadcasting described in Section 13.1.3. When the original stereophonic signals are correlated or when the masked threshold for M/S coding is higher, the SAC improves the efficiency of coding. As a special case, when the left and right signals are identical, the difference signal vanishes and does not need to be coded. In a practical course of coding, the efficiency of MS coding is compared with that of original L/R coding. If the former is higher than the latter, M/S stereo coding is adopted. Otherwise, the original L/R coding is adopted. Because the MS transform is implemented in the time-frequency domain, MS coding can be selectively implemented in each time frame and frequency band. This selective implementation improves the efficiency of coding but complicates algorithm. The side information that describes the scheme of coding in each frequency band should be supplemented to the bit stream. The MPEG-1/2 Layer III selects the coding fashion in a full band, and the MPEG-2/4 AAC individually chooses the coding fashion in each band. MS coding is also applicable to each symmetric pair in multichannel sound signals, such as left and right signals in 5.1-channel sound. As an extension of MS coding, appropriate linear (matrix) transformation can be applied to multichannel signals to remove the redundancy among them and thereby improve the efficiency of coding. SU(n) transformation in Section 9.4.2 is an example. The PCA similar to that in Section 8.3.6 is also applicable (Briand et al., 2006).

Intensity stereo coding (shortened for intensity coding; Herre et al., 1994) is another typical SAC-based scheme. In Section 12.2, the energy-time envelops rather than the detailed structure of the high-frequency sound components mainly contributes to perceptions. Therefore, for certain types of signals, high-frequency components in two or multiple channels may be downmixed into a single channel and coded, then transmitted along with the scaling information of the energy-time envelops of each frequency band. For example, the high frequency of components of all channel signals is downmixed to a single signal to be coded. The decoder retrieves the high-frequency energy-time envelops of each channel signal approximately from the coded single signal by using scaling information in each frequency band. For signals with an interchannel time difference, downmixing to a single signal directly causes the comb-filtering effect. Similar to the case in Section 8.2, the adaptive equalization of the phases of signals prior to downmixing reduces the comb-filtering effect. Intensity coding is applicable to the high-frequency bands above 2 kHz and usually more appropriate for the bands above 4–6 kHz. Using intensity coding in low-frequency bands destroys the low-frequency-phase information among the channels and thereby spoils the virtual source localization.

MS coding and intensity coding belong to ***joint stereo coding***. However, the spatial information of a target sound field is not only represented by intensity relation but also by the phase and other relations between stereophonic signals. In ***parametric stereo coding***, stereophonic signals are downmixed into a single-channel signal and then coded. The side information in each time frame and frequency band obtained by time-frequency analysis, such as interchannel level difference, interchannel phase difference, and interchannel correlation, is also supplemented into the bit stream. The bit rate for transmitting the side information is much lower than that for transmitting the original signal. A decoder synthesizes stereophonic signals from the single coded signal and side information. A more exact filter band than that used in coding can be used to synthesize stereophonic signals and reduce audible artifacts.

Binaural cue coding can be regarded as a further extension of parametric stereo coding. (Baumgarte and Faller, 2003; Faller and Baumgarte, 2003; Breebaart and Faller, 2007). In headphone presentation, binaural cues of spatial perception, such as interaural time difference (ITD) and interaural level difference (ILD), are consistent with the interchannel relations between two channel signals. However, in loudspeaker reproduction, a hypothesis in binaural cue coding is that spatial cues and perceived effects are determined by the relation between signals. Inputs are downmixed into a single signal and then coded. Similar to the case in Section 8.2, appropriate signal processing prior to downmixing reduces the comb-filtering effect. The side information that describes relationships between input signals is evaluated from inputs and supplemented into bit streams. A decoder synthesizes two channel signals in the time-frequency domain from the single coded signal and side information by introducing an appropriate delay, level difference, and correlation between signals. The signals in the time-frequency domain are finally transformed into signals in the time domain.

As a general case in SAC, a larger number of channel signals are downmixed into a smaller number (not limited to one) of channel signals and coded. The side information describing the relations among the original input signals are also supplemented into the bit stream. A decoder synthesizes the larger number of channel signals from the coded signals and side information (Faller, 2004). For example, 5.1-channel signals can be downmixed into two channel signals and coded, and the downmixed signals are compatible with two-channel stereophonic reproduction. Downmixing into more than one channel signal preserves more spatial information of sound and thus promises a better perceived performance after decoding. This downmixing possesses downward compatibility. These advantages are obtained at the cost of an increase in the bit rate.

SAD is similar to the matrix surround sounds in Section 8.1 and the upmixing schemes of multichannel sound signals in Section 8.3 in some aspects. All these methods create a larger number of channel signals from a smaller number of channel signals by using information on the relations among different channel signals. However, SAD analyzes relations among different channel signals prior to coding and supplements the information of these relations into the bit stream. By contrast, the matrix surround and upmixing schemes of multichannel sound signals estimate the relations among different channel signals from the independent transmitted signals and implement "blind separation of information."

The DirAc in Section 7.6 can also be regarded as SAC. In this method, the sound intensity vector and diffuseness of sound field are used as the side information from which spatial sound signals are synthesized.

13.4.6 Spectral band replication

The resolution of human hearing at high frequencies is lower than that at low frequencies. Therefore, human hearing is more tolerant to the error caused by coding at high frequencies than that at low frequencies. Accordingly, some high-frequency components in signals can be omitted, and more bits are allocated to low-frequency components to reduce the perceivable quantization error at low frequencies. In the applications of DAB, mobile devices, and wireless audio transmission, the bit rate is limited. A direct solution to this problem is to restrict the high-frequency bandwidth of signals. However, omitting the high-frequency components of a signal degrades the overall perceived audio quality. ***Spectral band replication*** (***SBR***) aims to retrieve the high-frequency components of signals from the coded low-frequency components and thus improve the overall perceived audio quality (Groschel et al., 2003). SBR utilizes the correlation between the high- and low-frequency components of signals, that is, it uses the redundancy in the frequency domain. The low-frequency components of signals are coded, and some parameters describing the features of high-frequency components are extracted. A decoder retrieves the high-frequency components from the coded low-frequency components and high-frequency parameters. Therefore, SBR can be regarded as a hybrid of wavefront and parameter coding technique.

SBR is used in the MPEG-4 HE-AAC coding (Section 13.5.4). A quadrature mirror filter band decomposes the PCM input signal into subband signals in the time-frequency domain. Then, the envelope of signals is evaluated, and tone, noise components, and some other types of information in high-frequency components are analyzed. A decoder retrieves the high-frequency components from the coded low-frequency components and aforementioned information by controlling the envelope of high-frequency components so that it is close to the original signal as far as possible (Herre and Dietz, 2008).

13.4.7 Entropy coding

In PCM coding, all the samples of signals are coded into a bit string with a fixed length. Therefore, the bit rate of the bit stream is fixed. However, the probability distribution of the amplitude of a practical audio signal is nonuniform. The probability of a small amplitude is usually larger than that of a large amplitude. An entropy coder codes the samples of signals into a bit string with variable lengths in which the samples with a larger probability are coded into a bit string with a shorter length and vice versa. Accordingly, the instantaneous bit rate of the coded signal is variable, but the average bit rate over time is reduced. Therefore, entropy coding utilizes the statistical characteristics and redundancy in the time domain to compress signals and is especially appropriate for music signals with obvious peaks of amplitudes. It is lossless in which the information of signal is not lost, and information entropy is preserved.

A key to entropy coding is estimating the probability distribution of a signal amplitude as exactly as possible. On the basis of probability distribution, the set of bit strings with variable lengths is designed for coding. The efficiency of compression depends on the accuracy of estimated probability distribution. Some models and schemes of entropy coding, such as ***Huffman coding*** and arithmetic coding, have been designed. Different schemes of entropy coding may be more appropriate for different signals. Entropy coding is described in detail in other books (Gersho and Gray, 1992) and thus omitted here.

13.4.8 Object-based audio coding

As stated in Section 6.5.2, object-based spatial sound is a remarkable direction of development. In contrast to the case of channel-based spatial sound, one major feature of an object-based spatial sound is that signal mixing/rendering is moved to the reproduction stage (after transmission or storage). The advantages of object-based spatial sound include the following:

1. Flexibility in reproduction. The signals of an object-based spatial sound can be reproduced by different principles and methods, such as multichannel sounds, Ambisonics, WFS, and binaural reproduction, because transmitted or coded signals are independent from reproduction methods. For multichannel sound reproduction, signals are appropriate for different numbers and configurations of loudspeakers and various signal panning/mixing methods.
2. Convenience in the control of each audio object independent, such as loudness and direction of the object.
3. Interactivity, which enables listeners (users) to control the audio object interactively. Interactivity is vital to the applications of game and virtual reality. Even in conventional applications, interactivity enables listeners to achieve the individualized reproduced effect or enables dynamic rendering according to the temporary position and orientation of a listener's head.

Based on a new system structure, object-based spatial sound transmits the spatial information of sound with a manner different from that of channel-based spatial sound. Accordingly, object-based spatial sound uses a new coding structure. Moreover, the metadata depend on object sources and scenes. For a static target source, the metadata involve the level and position of a source at least. For a moving target source, its position should be described by time-variable parameters (parameter equation of the moving trajectory). For a target sound scene, the metadata involve the parameters of scenes, such as parameters of reflected sounds. The metadata are usually coded with a standard format and transmitted with the coded signals of objects.

Similar to the case of channel-based spatial sound, audio objects can be compressed and coded with various conventional coding methods. However, as the number of audio objects increases, coding each object individually leads to a very high bit rate. Multiple objects can be coded by the method of the parameterized spatial audio object coding to improve the coding efficiency. The coder downmixes a larger number of audio objects into a smaller number of signals and then codes the signals. The parameters that describe each object and the relation among objects are extracted and coded into the bit stream as side information. A decoder retrieves object signals from downmixing signals and side information.

The combinations of channel- and object-based techniques are often used in practical spatial sound systems, such as Dolby Atmos in Section 6.5.2. Therefore, the decoder may upmix or downmix channel-based beds to adapt the practical number and configuration of loudspeakers. The principle of upmixing/downmixing is addressed in Chapter 8.

13.5 MPEG SERIES OF AUDIO CODING TECHNIQUES AND STANDARDS

In1988, MPEG under the International Organization for Standardization (ISO) and the International Electrotechnical Commission (IEC) began the workon the standardization of compression and coding technique for video and associated audio. Since the early 1990s, a series of technical standard for stereophonic and multichannel sound signal coding have been constituted. This section outlines these techniques and standards

13.5.1 MPEG-1 audio coding technique

MPEG-1 Audio is the first audio coding standard in the MPEG series and was finished in 1992 (ISO/IEC 11172-3 1993; Brandenburg and Stoll, 1994; Brandenburg and Bosi, 1997; Noll, 1997). It is a perceptual audio coding technique based on subband decomposition in Section 13.4.3. The operating modes of MPEG-1 audio include single channel, dual channel (such as bilingual), stereo, and joint stereo. It supports PCM input with 16-bit quantization at a sampling frequency of 32, 44.1, and 48 kHz. The bit rate of coding output is 32–224 kbit/s per channel. MPEG-1 provides three levels for coding, e.g., Layer I, Layer II, and Layer III. The efficiency and complexity of coding increase as the number of layer increases. All three layers of MPEG-1 support intensity stereo coding, and Layer III supports MS stereo coding. MPEG-1 Layer I is used in a digital compact cassette (DCC); MPEG-1 Layer II is utilized for DAB and video CD (VCD); and MPEG-1 Layer III, e.g., the famous MP3, is employed in Internet audio transmission and mobile audio devices.

Figure 13.16 is the block diagram of the ***MPEG-1 Layer I audio coding***. The PCM input is converted to subband components (time-frequency domain) by analysis filter bands. Then, chunks are formed, and scale factors are evaluated. The PCM input is converted to the frequency domain by fast Fourier transform (FFT) at the same time. From the frequency domain signal and scale factor, the masking threshold at each subband is evaluated using the psychoacoustic model. Then, according to the output of the psychoacoustic model, available bits are allocated to each subband, and each subband component is quantized with a resolution of the allocated bit. Lastly, the quantized samples of all subband components, along with various auxiliary data and error correcting codes, are packed with a certain format, yielding the output of bit stream. Some details on Layer I are as follows

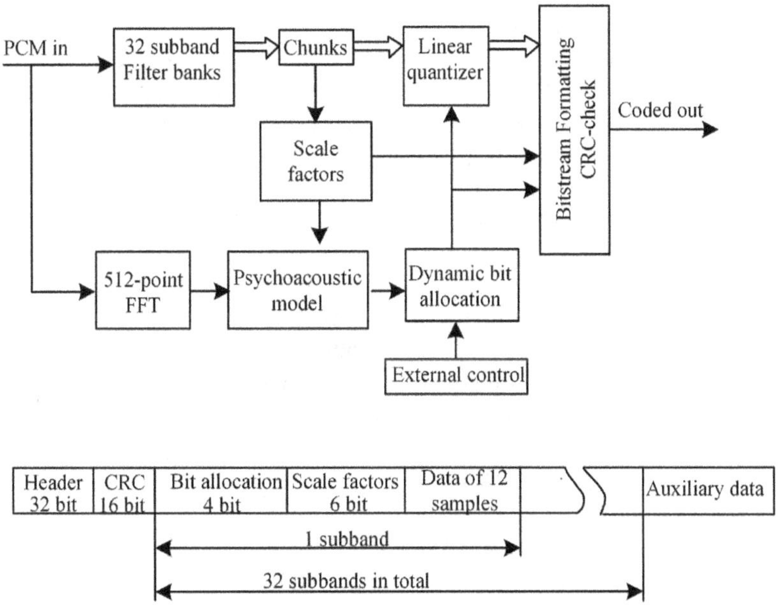

Figure 13.16 Block diagram and frame structure of the bit stream of MPEG-1 Layer I audio coding (adapted from Brandenburg and Stoll, 1994).

1. Input
   At a sampling frequency of 48 kHz, inputs are PCM signals at a bit rate of 48 kHz × 16 bits = 768 kbit/s per channel, with 384 samples in each frame.
2. Analysis filter band
   A PQMF band decomposes PCM signals into 32 subband components with a uniform bandwidth of 750 Hz. In addition, each subband component is subsampled at a sampling frequency of 48 kHz/32 = 1.5 kHz.
3. Chunks
   The scale factor of each subband is calculated for a block of 12 successive subband samples. The time length of each block is 12 (samples) × 32 (subbands)/48 (kHz) = 8 ms.
4. Scale factor
   In Layer I, quantization bits are dynamically allocated to each subband and block. The amplitude of each subband component in each block is normalized with the maximal amplitude of the 12 samples to maximize the dynamic range of quantization.
5. Fast Fourier transform (FFT)
   The PCM input is also converted to a frequency domain by a 512-point FFT or more strictly converted to a time-frequency domain by STFT. A 512-point FFT ensures a sufficient frequency resolution for the calculation of the signal-to-masking ratio in the psychoacoustic model. By contrast, the PQMF band ensures sufficient time resolution in the coding output.
6. Psychoacoustic model
   The signal-to-masking-ratio in each subband is calculated using a psychoacoustic model from the frequency domain signals given by the FFT. A relatively simple psychoacoustic model (Model 1) is used in Layer I. The model involves the following steps: FFT analysis of input, determination of the level and hearing threshold of signals, analysis/extraction of tonal and nontonal components, calculation of the masking threshold of individual components and global masking threshold, calculation of the minimum masking threshold, and signal-to-masking ratio of each subband and block.
7. Dynamic bit allocation and quantization
   According to the outputs of the psychoacoustic model and available bit resources, bits are dynamically allocated to each subband. The minimum bit for each subband is determined by the signal-masking ratio. If some surplus bits are available, they are added to subbands at which they are needed to improve the signal-to-noise ratio. Bit allocation is implemented by the iterative algorithm. When bit allocation is determined, the samples of each subband component are linearly quantized with the allocated bits.

Figure 13.16 also illustrates the frame structure of Layer I audio bit stream. It involves a synchrony head (32 bit), a cyclic redundancy code (CRC) code (cyclic redundancy check, 16 bits), bit allocation information (4 bits), scale factor (6 bits), the sample data of all subbands (the same length for each sample, 2–15 bits for each sample), and auxiliary data (with variable lengths).

Decoding is an inverse course of coding. After error correction, the coded data, bit allocation information, and scale factor are extracted from the packed bit stream. Then, each subband component of signals is restored. Lastly, the time-domain PCM signal is recovered from subband components.

The principle and structure of ***MPEG-1 Layer II*** audio coding are similar to those of Layer I. Some differences between them include the following:

1. At a sampling frequency of 48 kHz, each frame of input PCM signal has 1152 samples.

2. A 1024-point FFT is used to improve the frequency resolution of the data input to the psychoacoustic model.
3. Although a block in each subband consists of 12 successive subband samples, three successive blocks are coded in each frame of data. Therefore, a frame of data in each subband has 36 samples, corresponding to a signal length of 24 ms at a sampling frequency of 48 kHz.
4. The scale factor is calculated for each block with 12 subband samples; therefore, three scale factors for each frame of subband signals. Only one or two of the scale factors are transmitted for stationary signals that slowly vary with time to reduce the bit rate for the transmission of scale factors; and all three scale factors are transmitted for transient signals that quickly vary with time. Two-bit information that describes the strategy of scale factors in each frame of subband component is also included in the bit stream.

The efficiency of ***MPEG-1 Layer III*** audio coding is higher than that of Layers I and II, but the structure of Layer III is more complex than Layers I and II. Figure 13.17 illustrates the block diagram of MPEG-1 Layer III audio coding. The features of Layer III are as follows:

1. At a sampling frequency of 48 kHz, each frame of input PCM signal has 1,152 samples. A frame of data in each subband has 36 samples. A 1024-point FFT is used. All of them are identical to Layer II.
2. Thirty-two subband filters with acritical bandwidth rather than a uniform bandwidth are used to adapt to the nonuniform frequency resolution of human hearing.
3. The output samples of each of 32 subband filters are fed to an 18-channel MDCT filter bands to improve the frequency resolution and reduce time aliasing in subband filter outputs (Section 13.4.3). MDCT has two block lengths, e.g., 18 samples for the long block and 6 samples for the short block. Because a 50% overlap is used in MDCT, the corresponding window size of MDCT is 36 samples for the long window and 12 samples for the short windows. Because the length of three short blocks matches with that of a long block, a long block can be replaced with three short blocks. The block length of MDCT is chosen according to the characteristic of signals evaluated from the psychoacoustic model. A long block has high-frequency resolution and is appropriate for a stationary signal. A short block has high time resolution and thus appropriate for transient signal. In each time frame, MDCT with an identical window length is applied to all subbands, or MDCT with different window lengths is applied to different subbands; that is, a long window is applied to the two lowest subbands to improve the resolution

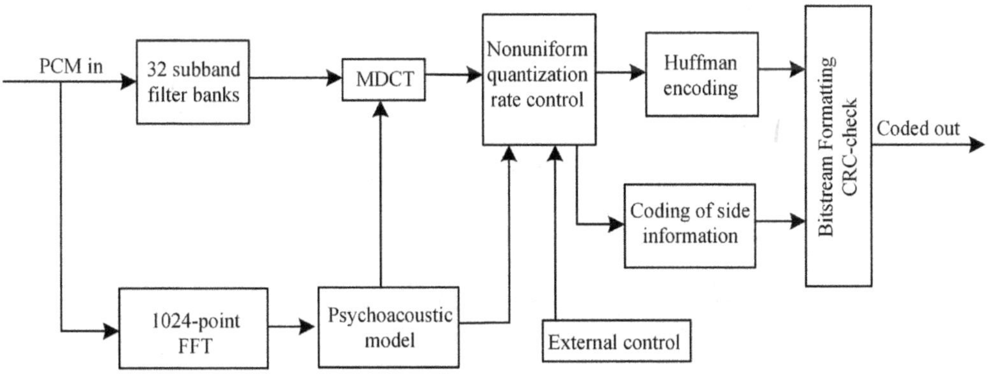

Figure 13.17 Block diagram of MPEG-1 Layer III audio coding (adapted from Brandenburg and Stoll, 1994).

at a low frequency, and a short window is used for other subbands. Switching between long and short blocks is implemented by a special time window.
4. A fine but complicated psychoacoustic model (Model 2) is used.
5. Nonuniform quantization is used.
6. Huffman (lossless) coding is introduced to further improve the coding efficiency.
7. Four modes of coding, such as traditional stereo coding, MS stereo coding, intensity stereo coding, and a combination of MS and intensity stereo coding, are supported.

13.5.2 MPEG-2 BC audio coding

MPEG-2 BCaudio coding is the abbreviation of ***MPEG-2 backward compatible audio coding***, which is also called ***MPEG-2 Audio***. It was developed in the 1990s for multichannel extension of MPEG-1 (ISO/IEC 13818-3, 1998; Brandenburg and Bosi, 1997; Noll, 1997). It supports the 5.1-channel signal coding and backward compatible with MPEG-1 Layer I and Layer 2.

Figure 13.18 illustrates the principle of downward compatible coding/decoding in MPEG-2 BC. For example, in 5.1-channel signal coding, the PCM inputs involve $e_L(n)$, $e_R(n)$, $e_C(n)$, $e_{LS}(n)$, $e_{RS}(n)$ [the input $e_{LFE}(n)$ of the left frequency effect channel is omitted in the figure]. A matrix similar to that in Equation (8.2.3) downmixes the five inputs into two-channel compatible stereophonic signals $e_{L0}(n)$ and $e_{R0}(n)$. The matrix also outputs three multichannel extension signals $e_{C0}(n)$, $e_{LS0}(n)$, and $e_{RS0}(n)$, which are identical to $e_C(n)$, $e_{LS}(n)$, and $e_{RS}(n)$ signals, to restore five channel signals in a decoder, respectively. Two-channel compatible stereophonic signals and multichannel extension signals are coded individually and then packed into a bit stream. In the decoder, two-channel compatible stereophonic signals and multichannel extension signals are extracted, and five channel signals are restored by an inverse matrix.

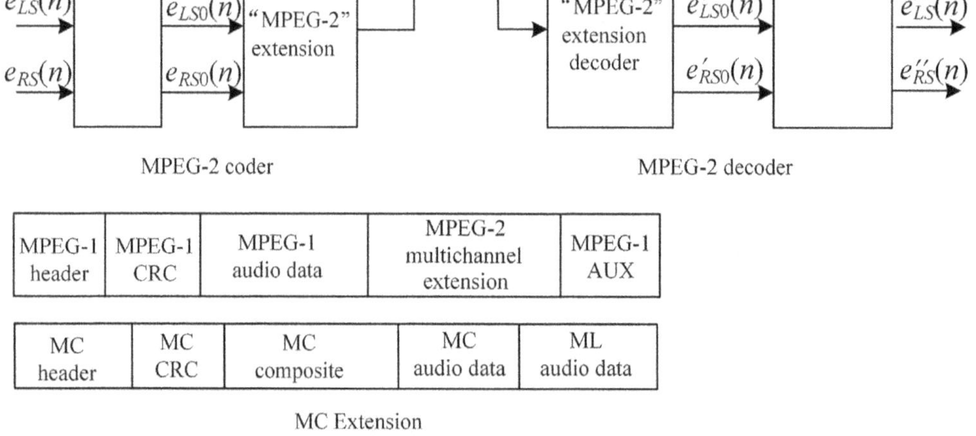

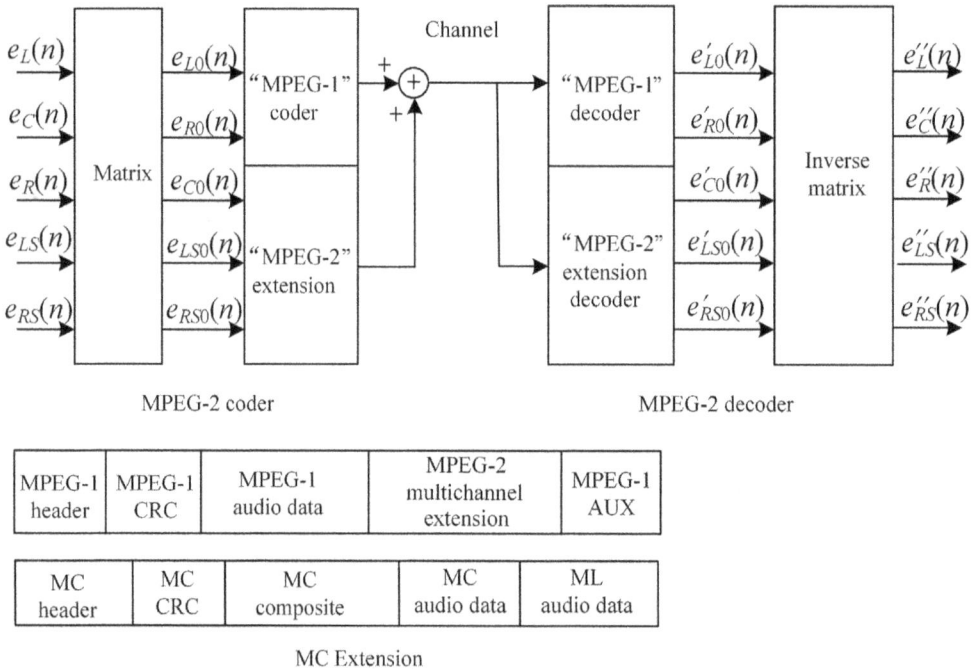

Figure 13.18 Principle of downward compatible coding/decoding in MPEG-2 BC (adapted from Pohlmann, 2011).

MPEG-2 BC utilizes most coding strategies in MPEG-1. Two-channel compatible stereophonic signals are coded identical to that in MPEG-1. Figure 13.18 also illustrates the data structure of the MPEG 2 BC bit stream. In comparison with MPEG-1, multichannel extension data are transmitted as auxiliary data in a MPEG-1 data frame. When the MPEG-2 BC data stream is decoded by a MPEG-1 decoder, multichannel extension data are ignored; therefore, data streams are compatible with MPEG-1 decoding.

In addition to multichannel extension, MPEG-2 BC supports multilingual and low sampling frequency extensions (single-channel or stereophonic signals at 16, 22.05, and 24 kHz). The details are omitted here.

MPEG-2 BC was intended for audio coding in high-definition television (HDTV) or digital television (DTV) in Europe. A remarkable feature of MPEG-2 BC is that it is compatible with MPEG-1 audio coding. However, the inverse matrix in the decoder may cancel most signals in a channel but does not cancel quantization noise. In other words, after the inverse matrix in the decoder, quantization noise may be audible because of unmasking. Therefore, a compatible design reduces the coding efficient of MPEG-2 BC. In the mid-1990s, some groups conducted a series of subjective experiments to assess the performance of MPEG-2 BC (Kirby, 1995; Kirby et al., 1996; Wüstenhagen et al., 1998). For 5.1-channel signals, MPEG-2 BC requires a bit rate of 640 kbit/s to achieve a good perceived quality. At the same bit rate, the perceived performance of MPEG-2 BC is inferior to that of Dolby Digital (Section 13.6.1). Therefore, MPEG-2 BC is not successful in practice.

13.5.3 MPEG-2 advanced audio coding

MPEG-2 advanced audio coding (*MPEG-2 AAC*) aims to improve the efficiency of multichannel coding while preserving the perceived audio quality. MPEG-2 AAC refers to some strategies in MPEG-2 Layer II, MPEG-I Layer III, and Dolby Digital but discards the consideration of downward compatibility. It becomes an international standard in 1997 (ISO/IEC 13818-7, 1997; Bosi et al., 1997; Brandenburg and Bosi, 1997). It supports the sampling frequency of 8–96 kHz. At the sampling frequency of 48 kHz, the maximal allowable bit rate is 288 kbit/s per channel. It supports the coding up to 48 main channels, 16 low-frequency effect channels, 16 multilingual channels, and 16 data streams. It is appropriate for single-channel, stereophonic, and multichannel coding.

Figure 13.19 illustrates the block diagram of the principle of MPEG-2 AAC. It mainly involves following modules, and some modules are optimal.

1. Gain control
   Gain control module aims to control the gain in different frequency bands and reduce the number of bits for coding. PQMF bands divide the PCM input into four uniform subbands. A gain detector provides the gain information for a gain modifier to control the gain in each PQMF band.
2. Analysis filter band
   Analysis filter bands divide the input into blocks and convert to time-frequency components. MDCT in Section 13.4.3 is used in AAC for time-frequency transform, and a time domain aliasing cancellation technique (TDAC) is used to remove the time domain aliasing in the restored time signal. The window function and its length for transform can be adaptively chosen according to the characteristic of input signals. Optional window includes sine window and Kaser–Bessel window. The former is appropriate for signals with dense spectral components, and the latter is appropriate for signals with sparse spectral components. AAC can seamlessly switch between two types of windows. The window length of MDCT includes 2,048 samples (long window) and 256 samples

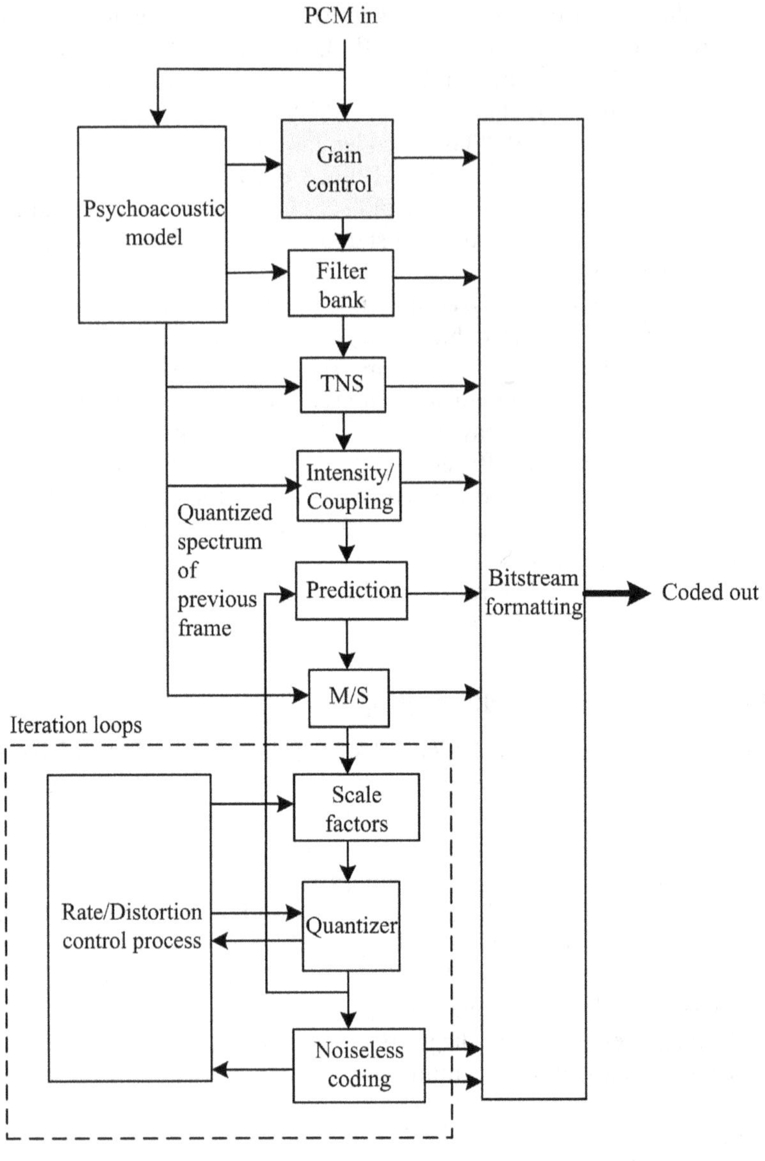

Figure 13.19 Block diagram of the principle of MPEG-2 AAC (adapted from ISO/IEC 13818-7 and Brandenburg and Bosi, 1997).

(short window). A 50% overlap is used in MDCT to avoid the abrupt change between the boundaries of two adjacent windows. Therefore, a long window involves 512 samples of the preceding block, 1,024 samples of the current block, and 512 samples of the following block. Because of the odd symmetry in Equation (13.4.7), MDCT with a long window yields 1,024 spectral coefficients, which are consistent with the number of samples in a block. Similarly, MDCT with a short window yields 128 spectral coefficients. MDCT with a long window has a higher coding efficiency and is appropriate for stationary signals. MDCT with a short window has a lower coding efficiency but possesses a higher time resolution. Window length is chosen adaptively according to the characteristic of input signal, and long and short windows are smoothly transited.

3. Psychoacoustic model
   A psychoacoustic model similar to model 2 in MPEG-1 Layer III is used. Given the sampling frequency, the samples in each time window (2,048 or 256 samples) are analyzed using the model and masking threshold in each subband is calculated, resulting in signal-to-masking ratios, scale factors, and type and length of MDCT block.
4. Temporal noise shaping
   For a time domain transient impulse after a silent or signal with a low amplitude, quantization noise temporally extendsafter decoding. The temporal mismatch between the masking threshold and quantization noise dissatisfies the condition of temporal masking (Section 1.3.3); therefore, a perceivable pre-echo of quantization noise occurs in the time region of silent or signal with a low amplitude. Using a short time window for a transit signal partly restrains the pre-echo. Temporal noise shaping (TNS) aims to further restrain the pre-echo according to the signal characteristic. It replaces the spectral coefficients with the residue of linear prediction. TNS improves the temporal resolution of signals and thus allows controlling the temporal structure of quantization errors in decoding the output so that it is masked by the signal. Thus, the pre-echo is restrained.
5. Joint stereo coding
   AAC supports joint stereo coding, including MS stereo coding and intensity stereo coding. The basic principle is outlined in Section 13.4.5 and omitted here.
6. Prediction
   As stated in Section 13.4.2, the samples of a discrete time signal may be correlated within a certain period. For stationary signals, prediction in coding reduces the redundancy of signals. Therefore, prediction is only used for long windows. In contrast to the case of ADPCM in Section 13.4.2, the predictor in AAC uses the spectral components of two preceding frames to estimate the spectral component of the current frame. The difference between the predicted and input signals is coded. Signal is restored from the predicted and differential signals similar to that in Section 13.4.2.
7. Quantization and coding
   Bits are dynamically allocated according to the results from psychoacoustic models. Nonuniform quantization is used. A strategy of two nested iteration loops determines the quantization step size and the application of the scale factor band. The spectrum is quantized per scale factor band. The step sizes of quantization are described by a set of scale factors and a global gain. The global and scale factors are quantized at a step size of 1.5 dB. The Huffman coding in Section 13.4.7 is also used to further reduce the average bit rate.
8. Bit stream output
   Coded audio data, along with the data of side information, are packed to some format, forming the bit stream output.

MPEG-2 AAC provides three profiles to enable a tradeoff between audio quality and complexity.

1. Main profile
   The main profile offers the best audio quality but is the most complicated. Its decoder can also decode a bit stream coded by a low complexity profile. This profile involves all the modules in Figure 13.15 except gain control modules.
2. Low complexity profile
   The low complexity profile is simpler than the main profile, but its audio quality is inferior to that of the main profile. This profile uses a limited order TNS and does not use the gain control and prediction modules.

3. Scalable sampling rate profile
   The scalable sampling rate profile is the simplest of the three profiles. This profile includes a gain control module and a limited order TNS. It does not involve prediction and intensity stereo coding. It can also provide a frequency-scalable signal.

AAC decoding is an inverse course of the aforementioned coding and omitted here.

Subjective experiments (Kirby et al., 1996) have indicated that at a compression ratio of 12:1 (at a sampling frequency of 48 kHz, with a bit rate of 64 kbit /s per channel or 320 kbit/s for five channels), the MPEG-2 AAC main profile provides a "indistinguishable" perceived quality. The overall perceived quality of MPEG-2 AAC at 320 kbit/s is better than that of MPEG-2 BC Layer II at 640 kbit/s. The average quality of the latter is not better than that of MPEG-2 AAC at 256 kbit/s. For two-channel stereophonic signals, AAC at 96 kbit/s exhibits an average quality comparable with that of MPEG-1 Layer II at 192 kbit/s or Dolby Digital at 160 kbit/s (Herre and Dietz, 2008). Some earlier subjective experiments (Soulodre et al., 1998) have demonstrated that AAC and Dolby Digital achieves the highest quality at 128 and 192 kbit/s for two-channel stereophonic signals, respectively. Therefore, MPEG-2 AAC is a highly efficient coding method.

13.5.4 MPEG-4 audio coding

***MPEG*-4 Audio** is a low-bit-rate coding standard for multimedia communication and entertainment applications. It was formulated in 1995, with the first and second editions released in 1999 and 2000, respectively (Brandenburg and Bosi, 1997; Väänänen and Huopaniemi, 2004). MPEG-4 Audio combines some previous techniques of high-quality audio coding, speech coding, and computer music with great flexibility and extensibility. It supports the synthetic audio coding (such as computer music), natural audio coding (such as music and speech) and synthetic-natural hybrid coding.

MPEG-4 natural audio coding provides three schemes, e.g., parametric audio coding (Section 13.4.1), code-excited linear prediction (CELP) and general audio (waveform) coding. The preceding two schemes are appropriate for speech or audio coding at a low bit rate. The parametric coding includes the tools of harmonic vector excitation coding (HVXC), as well as harmonic and individual line plus noise (HILN). For natural audio with a sampling frequency higher than 8 kHz and bit rate of 16–64 kbit/s (or higher), MPEG-4 Audio directly codes the wavefront. The core scheme of wavefront coding is the AAC in Section 13.5.3. The block diagram of MPEG-4 AAC is similar to that in Figure 13.18. In comparison with MPEG-2 AAC, MPEG-4 AAC increases the perceptual noise substitution (PNS) and long-term prediction (LTP) tools. PNS tools aim to improve the coding efficiency of noise-like signals. When PNS is used, a noise substitution flag and designation of the power of the coefficients are transmitted instead of quantized spectral components. The decoder inserts pseudo-random values scaled by the proper noise power level. Tone-like signals require a much higher coding resolution than that of noise-like signals. However, tone-like signals are predictable because of its long-term periodicity. The LTP tool uses forward-adaptive long-term prediction to remove the redundancy among the successive blocks.

The first version of MPEG-4 high-efficiency AAC (MPEG-4 HE-AAC v1) was developed in 2003 to improve the coding efficiency for low-bit-rate audio (Herre and Dietz, 2008). Based on the architecture MPEG-4 AAC, the SBR tool in Section 13.4.6 is used in MPEG-4 HE-AAC v1. When bit rates are 20, 32, and 48 kbit/s, the ranges of SBR are 4.5–15.4, 6.8–16.9, and 8.3–16.9 kHz, respectively. Afterward, a parametric stereo coding module is combined into MPEG-4 HE-AAC, resulting in MPEG-4 HE-AAC v2. The bit streams for the side information of SBR and parametric stereo coding are transmitted in the previously unused parts of the AAC bit stream, enabling the compatibility with existing AAC. The typical bit rate for this side information is a few kilobits per second. The typical bit rate of HE-AAC v2 is 32 kbit/s for stereophonic sound and 160 kbit/s for 5.1-channel sound to achieve near-transparent audio quality (which is obtained by AAC without extension at a bit rate of 320 kbit/s). At the bit rate of 24 kbit/s per channel, HE-AAC improves the coding efficiency by 25% compared with that of previous AAC. With the same quality, the bit rate of HE-AAC v1 is 33% higher than that of HE-AAC-v2. The bit stream of HE-AAC supports up to 48 channels. Although HE-AAC is a part of MPEG-4, it is not limited to be use in interactive multimedia video and audio. Because HE-AAC possesses a high coding efficient, it can be independently use to the cases of audio coding with a strictly limited bandwidth, such as DAB and wireless music download in mobile phone.

One feature of MPEG-4 is that it allows object-based synthetic and natural audio coding. It considers every sound source's signal (natural or synthesized sound) in the auditory scene as an independent transmitted object or element, then re-synthesizes it into a complete auditory (or more precisely audio-visual) scene in a user terminal. MPEG-4 adopts ***Audio Binary Format for Scene Description*** (audio BIFS) as a tool to describe sound scene parameters and achieve sound scene combination, while retaining flexibility in defining combination methods. Users can flexibly compile and combine these objects, and local interaction is allowable for synthesized scenes from different viewing (listening) positions and angles. MPEG-4 supports virtual auditory environment applications, which have actually become part of MPEG-4. Substantial research has been devoted to such applications (Scheirer et al., 1999; Väänänen and Huopaniemi, 2004; Jot and Trivi, 2006; Dantele et al., 2003; Seo et al., 2003).

The second edition of MPEG-4 provides parameters that describe three-dimensional acoustic environments in advanced audio BIFS, which includes the parameters of rectangular rooms (e.g., room size and frequency-dependent reverberation time), the parameters of sound source characteristics (e.g., frequency-dependent directivity, position, and intensity), and the acoustic parameters of surface materials (e.g., frequency-dependent reflection or absorption coefficients). Auditory scenes are synthesized at a user's terminal in terms of these parameters. Because MPEG-4 does not specify sound synthesis and reproduction methods, many types of sound synthesis and reproduction technologies can be adopted, depending on application requirements and hardware performance at a user's terminal. A real-time and dynamic virtual auditory environment system (Section 11.10) is usually an appropriate choice. In this case, a listener's movement in a virtual space causes changes in binaural signals. Interactive signal processing is supported to simulate the dynamic behavior of binaural signals in accordance with a listener's temporal head orientation.

13.5.5 MPEG parametric coding of multichannel sound and unified speech and audio coding

After MPEG-2 AAC, MPEG-4 AAC, MPEG-4 HE-AAC v1 and v2, MPEG provides the MPEG-D MPEG Surround (MPS) in 2007, a technique and standard of the generalized means for the parametric coding of channel-based multichannel sound signals with a high efficiency. As shown in Figure 13.20, in an MPS coder, multichannel inputs are downmixed into mono or stereophonic signals and then coded. The MPS spatial parameters that describe the relationship among multichannel inputs are extracted as side information to be transmitted. In addition, residual signals containing the error related to the parametric representation are calculated and coded by the low-complexity-profile MPEG-2 AAC. A decoder restores multichannel signals from coded signals, spatial parameters, and residual signals by re-upmixing (ISO/IEC 23003-1, 2007; Hilpert and Disch, 2009; Villemoes et al., 2006; Breebaart et al., 2007; Breebaart and Faller, 2007; Herre et al., 2008). MPS supports up to

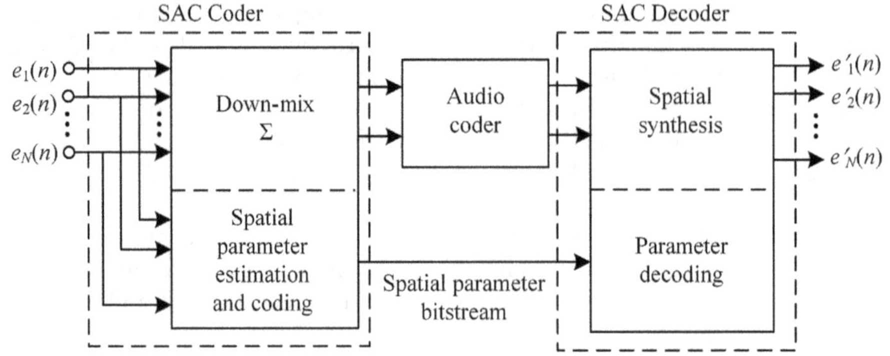

Figure 13.20 Block diagram of MPEG-D MPEG Surround coding and decoding (adapted from Hilpert and Disch 2009).

32 channel outputs. Signal downmixing in MPS coding enables a downward compatibility with stereophonic sound. In addition, a two-channel matrix-compatible downmixing similar to those in Section 8.1.4 can be chosen so that legacy receivers without MPS spatial parameter processing can still decode multichannel signals by conventional matrix decoding. MPS spatial parameters, such as level difference, the correlation between channels in the time-frequency domain can be evaluated from the output of QMF bands and transmitted at a bit rate of 3–32 kbit/s or higher. Existing techniques, such as MPEG-4 AAC, MPEG-4 HE-AAC, or MPEG-1 Layer II, are applicable to the core coding of downmixing mono and stereophonic signals in MPS. A subjective assessment experiment involving the MUSHRA method in Section 15.5 indicated that the average perceived quality reaches a good region at a bit rate of 64 kbit/s for the MPS with HE-AAC as core coding, crosses the border of the excellent region of 80 scales at 96 kbit/s, and achieves excellent quality at 160 kbit/s.

The MPS is a parameter coding technique for channel-based spatial sound, and the decoder yields signals for certain loudspeaker configuration. MPEG-D spatial audio object coding (SAOC), which was finalized in 2010, is a parametric coding technique for multiple objects [ISO/IEC23003-2 (2010); Herre et al., 2012]. As shown in Figure 13.21 (a), in a coder, multiple objects are downmixed into stereophonic or mono signals and then coded. At the same time, the parameters describing the relation among objects and each object are extracted and transmitted as SAOC parameters (side information). A SAOC decoder involves an object decoder and a mixer/render. An object decoder extracts the objects from the downmixing bit stream according to the SAOC parameters. In terms of the side information of each object and practical loudspeaker configuration, a mixer/render mixes object signals to loudspeaker signals by a rendering matrix. The object decoder and mixer can be integrated into one to improve the decoding efficiency, as shown in Figure 13.21 (b).

SACO-downmixed signals can be coded with existing coding schemes, such as HE-AAC. SAOC parameters include object-level differences, inter-object correlations, downmixing gains, and object energies. The SAOC object parameters are given in certain time-frequency resolution and transmitted as ancillary data with a bit rate of as low as 2–3 kbit/s per object or 3 kbit/s per scene. For some objects whose audio quality is needed to be enhanced, the residual signal (the differential signal between parametric reconstruction and original signal) is transmitted in the SAOC bit stream with AAC-based scheme so as to reconstruct the object signal exactly in the decoder.

SAOC has two decoding and rendering modes. The first mode is the SAOC decoder processing mode, which provides mono, stereophonic, and binaural outputs. As shown in Figure 13.22 (a), the SAOC bitstream, rendering matrix, and head-related transfer function (HRTF)

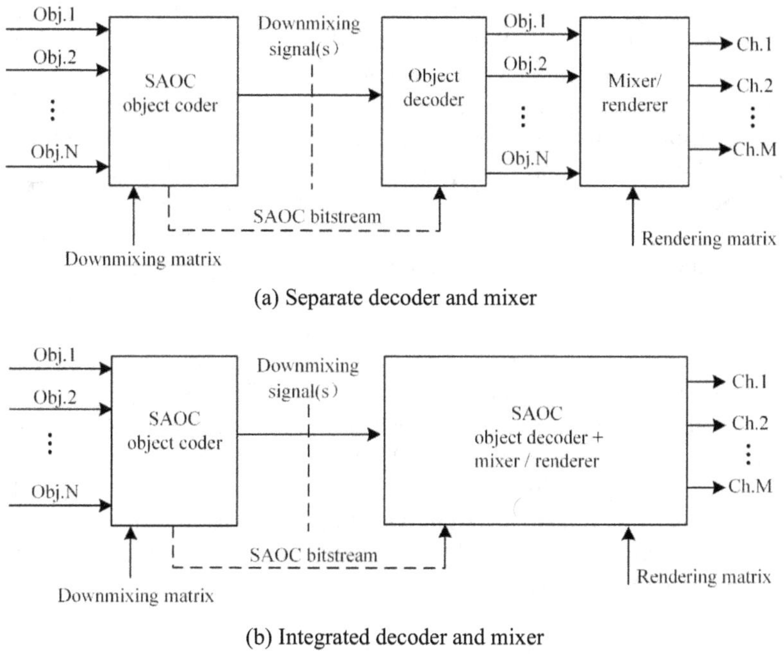

Figure 13.21 Block diagram of MPEG-D spatial audio object coding: (a) separate decoder and mixer; (b) integrated decoder and mixer (adapted from Herre et al., 2012).

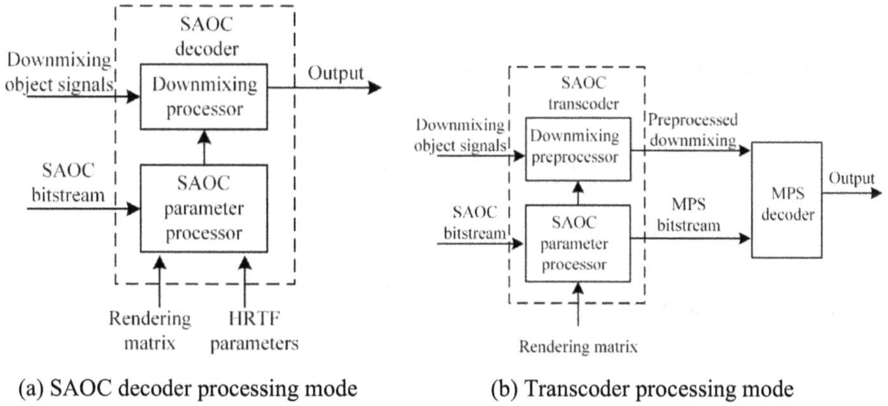

Figure 13.22 Two SAOC decoding and rendering modes (a) SAOC decoder processing mode; (b) Transcoder processing mode (adapted from Herre et al., 2012).

parameters (for binaural outputs) are sent to an SAOC processor. The downmixing processor directly generates output signals from downmixing signals and the output of the SAOC processor. In other words, object signal extraction, rendering, and even binaural synthesis are integrated into one stage to improve the efficiency of processing. An open SAOC interface is also included, which enables users to provide varying HRTF parameters. Dynamic binaural synthesis with the head tracker is also allowed. The second mode is the SAOC transcoder processing mode, which provides multichannel outputs. As shown in Figure 13.22 (b), SAOC downmixing object signals and parameters are converted to MPS bitstream and parameters and then decoded by an MPS decoder.

SAOC possesses a high coding efficiency. MPS and SAOC allow for a 5.1-channel signal coding at a low bit rate of 48 kbit/s. SAOC enables the independent manipulation of the gain, equalization, and effect of each object. Because its coding and transmission are independent from reproduction, SAOC is appropriate for different reproduction manners (different numbers and configurations of loudspeakers or headphones). SAOC is applicable to teleconferences, personalized music remixed in interactive reproduction, and games.

As stated in Section 13.4.1, general audio coding techniques are different from speech coding techniques. They were developed separately. Audio compression and coding are realized by removing the physical and perceptual redundancy in signals. Speech coding is usually based on the time-varying filter (linear prediction) model of the human vocal system. MPEG specified the standard of MPEG-D Unified Speech and Audio coding (USAC) [ISO/IEC 23003-1 (2012); Neuendorf et al., 2013] to adapt to the applications of digital radio, streaming multimedia, audio book, mobile devices. The USAC combines the enhanced HE-AAC v2 audio coding technique and speech coding technique AMR-WB+ into a unified system. According to the signal component, USAC flawlessly transits between two coding modes. HE-AAC v2 is improved in many aspects, including a time-warped MDCT, additional 50% window lengths of 512 and 1,024 samples (block lengths of 256 and 512 samples) for MDCT with 50% overlap, enhanced SBR bandwidth extension, unified stereo coding. USAC distinguishes from MPS in the following aspects:

1. In addition to interchannel level difference and interchannel correlation, interchannel phase difference is added as the side information of the USAC.
2. In contrast to MPS in which residual signals are coded independently from the downmixing signals, the USAC couples the coding of downmixing and residual signals closely to improve the transmission quality.
3. At higher bit rates, because the core coder can discretely handle a wider bandwidth and code multiple channels, the tools of SBR and parametric coding can be not used. In this case, complex prediction stereo coding is used to improve the efficiency of stereo coding.

The USAC can operate at a bit rate as low as 8 kbit/s for mono signal. Within the bit rate ranging from 8 kbit/s for mono to 64 kbit/s for stereophonic sound, the quality of the USAC exceeds that of HE-AAC v2.

13.5.6 MPEG-H 3D audio

Multichannel spatial surround sound with height in Chapter 6 is next-generation spatial sound technique and system. Various multichannel spatial surround sound systems have been proposed. These systems differ in performance and complexity. The prospect of these systems depends on their acceptability by consumers. The diversity of systems and formats causes serious problems of compatibility in program making, coding, and reproduction. Therefore, the standard for multichannel spatial sound should be developed.

MPEG-H 3D Audio is a coding technique and standard for new-generation spatial sounds. It has the advantages of high flexibility, quality, and efficiency [ISO/IEC 23008-3 (2015); Herre et al., 2014, 2015]. It was called for proposal in 2013 and published in 2015. The MPEG-H 3D Audio coding supports channel-based, objected-based, and higher-order Ambisonic signal inputs. After decoding is completed, it supports various reproduction methods, such as 2-channel, 5.1-channel to 22.2-channel, or even more channel reproduction. It also supports headphone presentation. The reproduction methods are basically independent from the input format, which avoids the problem of compatibility among different spatial sound systems. Therefore, MPEG-H 3D Audio is a universal standard that unites different spatial sound techniques and formats rather than a coding technique only. This feature distinguishes the MPEG-H 3D Audio from the previous MPEG standards such as MPEG-1 and MPEG-2 AAC audio coding.

The MPEG-H3D core coding uses many previous techniques, especially USAC in Section 13.5.5. Compared with USAC, the main enhancements of MPEG-H 3D core coding are as follows:

1. Joint coding of quadruple input channels
2. Enhanced noise filling

Therefore, the development of MPEG-H 3D Audio focuses on the conversion of various spatial sound signals and rendering/reproduction after coding and decoding. Figure 13.23 illustrates the block diagram of MPEG-H 3D Audio. After USAC-3D core decoding, the following signals and data are separated from the MPEG-H 3D Audio bit stream:

1. Channel-based signals
2. Object-based signals
3. Compressed object metadata
4. SAOC transmission channel signals and side information
5. Higher-order Ambisonics (scene-based) signals and side information

On the basis of the aforementioned signals and data, the decoder generates the reproduced signals that are mixed and then reproduced by loudspeakers or converted for headphone presentation.

The key strategies in MPEG-H 3D Audio include the following:

1. Channel-based reproduction and format conversion
   Channel-based signals are intended for a specific loudspeaker configuration or format, and each channel corresponds to the signal of a loudspeaker at a given position.

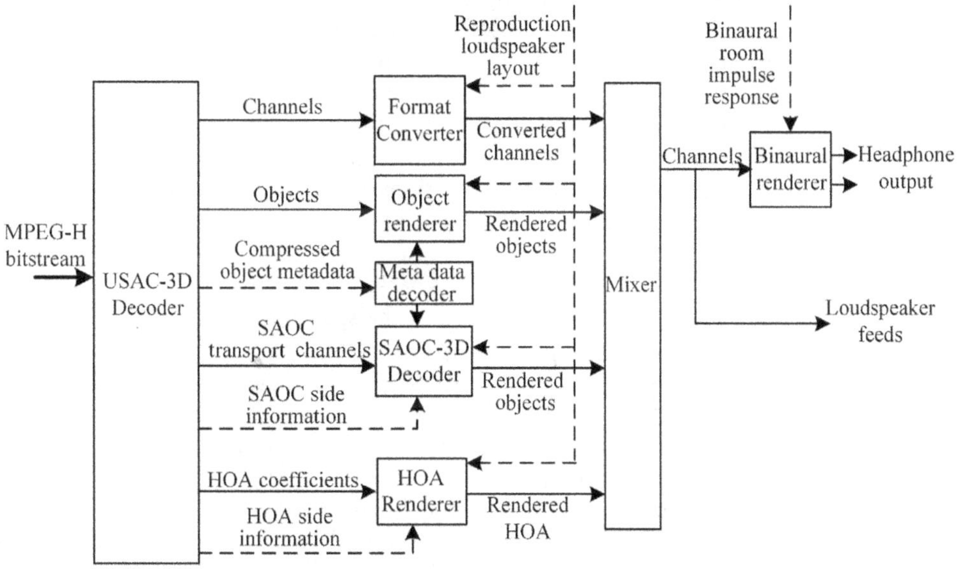

Figure 13.23 Block diagram of MPEG-H 3D Audio coding/decoding (adapted from Herre et al., 2014).

However, the configurations of loudspeakers are diversiform or even nonstandard. ***Format converter*** aims to convert a set of channel-based signals for reproduction with practical loudspeaker configuration, especially to convert the signals for reproduction with a larger number of channels (such as 22.2-channel) to signals for reproduction with a smaller number of channels (such as 5.1-channel). The principle is similar to that of downmixing in Section 8.2. To ensure the reproduction quality, MPEG-H 3D Audio generates optimized downmixing matrix automatically. It supports optionally transmitted downmixing matrix to preserve the artistic contents of original signals. In addition, it uses equalization filters to preserve the timbre and uses advanced active downmixing algorithm to avoid downmixing artifacts.

The format conversion module involves two parts: rule-based initialization and active downmix algorithm. In a decoder, rule-based initialization derives the optimized downmixing matrix that converts channel-based inputs to the signals of loudspeakers in practical configuration. Derivation is implemented by iteration according to a set of tuned mapping rules for each input channel. Each rule specifies rendering of one input channel to one or more output channels. When an input channel is just consistent with an output channel (loudspeaker), the input is mapped to that output channel to avoid using the phantom source method.

In Section 8.2, for multichannel-correlated inputs with interchannel time difference, or for inputs derived from different filtered versions of the same signals, downmixing directly may lead to comb-filtering and thus timbre coloration. An active downmixing algorithm analyzes the correlation between input signals and aligns the phases if necessary to address this problem. Frequency-dependent normalization is applied to the gain of downmixing to preserve the overall power spectra of signals after downmixing. Moreover, adaptive downmixing algorithm should not alter the uncorrelated input signals.

2. Object-based rendering

   USAC-3D core decoding yields object signals and object metadata. In addition to the spatial positions of objects (virtual sources), metadata of other characteristics of objects are included (Fug et al., 2014). Given the configuration of loudspeakers, the decoder feeds object signals to loudspeakers according to certain panning/mixing rules. Time-varying position data describe the spatial trajectory (or parameter equation) of a virtual source. Therefore, time-varying panning enables the simulation of a moving virtual source. The VBAP in Section 6.3 is used in MPEG-H 3D Audio for virtual source rendering. The three active loudspeakers in a spherical triangle are chosen according to the object metadata and practical loudspeaker configuration, and the gains of loudspeaker signals are calculated. However, most practical configurations lack loudspeakers below the horizontal plane. In these cases, imaginary (virtual) loudspeakers are needed for the VBAP.

3. Higher-order Ambisonics

   In Chapter 9, the independent signals of higher-order Ambisonics (HOA), which are the time-varying components or coefficients of the spatial harmonic decomposition of a sound field, represent the temporal and spatial information of sound field. In reproduction, independent signals are linearly decoded to loudspeaker signals. The linearly decoded matrix depends on the configuration of loudspeakers. To improve the performance of coding, MPEG-H 3D Audio does not directly transmit the independent signals of Ambisonics. Instead, it applies a two-stage coding process to HOA data, e.g., spatial coding of components and multichannel perceptual coding.

   The independent signals of HOA are decomposed into predominant and ambient components. Predominant components, which include directional sounds, can be coded as a set of plane wave incidence from certain directions. Therefore, each predominant component, along with the time-varying directional parameter, can be transmitted as audio stream. Ambient components mainly contain nondirectional information. Because the spatial resolution of human hearing to ambient components is relatively low, they can be handled by lower-order Ambisonics to improve the coding efficiency. However, because the Ambisonic representation or signals of ambient components may be highly correlated, the spatial unmasking of quantization noise may occur after decoding. Similar to the MS stereo coding in Section 13.4.5, the Ambisonic representation is decorrelated by transforming it to a different spatial domain for perceptual coding to avoid this problem.

The decoding of Ambisonic signals is an inverse course of coding. Based on Ambisonic data from the USAC 3D core decoding, decorrelated ambient components are first transformed to the HOA representation. The HOA representation of predominant components is also resynthesized from the coded data. The HOA representations of predominant and ambient components are then combined to form HOA-independent signals. According to the practical loudspeaker configuration, HOA-independent signals are linearly decoded into loudspeaker signals. A matrix that preserves the constant power (energy) is used in decoding.

4. SAOC-3D decoding and rendering (Murtaza et al., 2015)
MPEG-H 3D Audio also supports parametrically coded channel signals and audio objects, e.g., an extended spatial audio object coding called SAOC-3D. In comparison with the original SAOC in Section 13.5.5, the SAOC-3D is extended in the following aspects.
   - SAOC-3D in principle supports an arbitrary number of downmixing channels, while original SAOC supports a two-channel downmixing at most.
   - SAOC-3D supports the direct decoding/rendering to multichannel outputs for arbitrary loudspeaker configurations, including the enhanced decorrelation to output signals. SAOC only support by using a MPEG surround as a rendering engine.
   - Some SAOC tools, such as residual coding, are unnecessary and thus omitted in MPEG-H 3D Audio.
5. Binaural rendering
The mixed outputs in Figure 13.23 are for loudspeaker reproduction. They can be converted to signals for headphone presentation by binaural synthesis in Section 11.9.1. That is, the signal intended for each loudspeaker is convolved with a pair of binaural room impulse responses and then mixed to simulate transmission from loudspeakers to two ears in a listening room. Binaural rendering is vital in mobile devices.
6. Loudness and dynamic range control
7. Some loudness normalization and dynamic range information are embedded into the MPEG-H 3D Audio bit stream for loudness and dynamic range control in the decoder.

Subjective experiments indicated that MPEG-H 3D Audio exhibits an excellent quality at a bit rate of 1.2 Mbit/s or 512 kbit/s and shows good quality at a bit rate of 256 kbit/s. MPEG-H 3D Audio with a lower bit rate is being developed.

13.6 DOLBY SERIES OF CODING TECHNIQUES

Since the 1980s, Dolby Laboratories has developed a series of digital audio compression and coding techniques. These techniques have been widely used. However, the details of some of these techniques have not been published. This section outlines the basic principle of these techniques that have been published.

13.6.1 Dolby digital coding technique

The Dolby AC-1 developed in the early time is a stereophonic coding technique. It uses adaptive delta modulation (ADM) and combines with analog compounding. It is not a perceptual coding technique. The Dolby AC-2 developed in the 1980s is a coding technique for stereophonic and multichannel sound. It is a perceptual coding technique that consists of four single-channel coders/decoders (Fielder and Robinson, 1995; Brandenburg and Bosi, 1997).

Dolby Digital (AC-3) is a multichannel audio coding technique introduced in 1991. It was originally intended for 35 mm film soundtrack in a commercial cinema and subsequently specified as the audio coding standard of HDTV in USA. It has also been used widely for audio coding in DVD-Video (Davis, 1993; Davis and Todd, 1994; Todd et al., 1994; ETSI TS 102 366 V1.4.1, 2017; ATSC standard Doc.A52, 2012). Dolby Digital supports the sampling frequencies of 32, 44.1, and 48 kHz. It allows 5.1-channel coding (and mono, stereophonic, and three- and four-channel coding) with a bit rate ranging from 32 kbit/s to 640 kbit/s. A typical bit rate for 5.1-channel coding is 384 kbit/s. At this bit rate, Dolby Digital provides a good perceived audio quality (ITU-R Doc.10/51-E, 1995; Wüstenhagen et al., 1998; Gaston and Sanders, 2008).

Figure 13.24 (a) illustrates the block diagram of Dolby Digital coding. After a time window, PCM input samples are transformed into time-frequency coefficients by analyzing filter

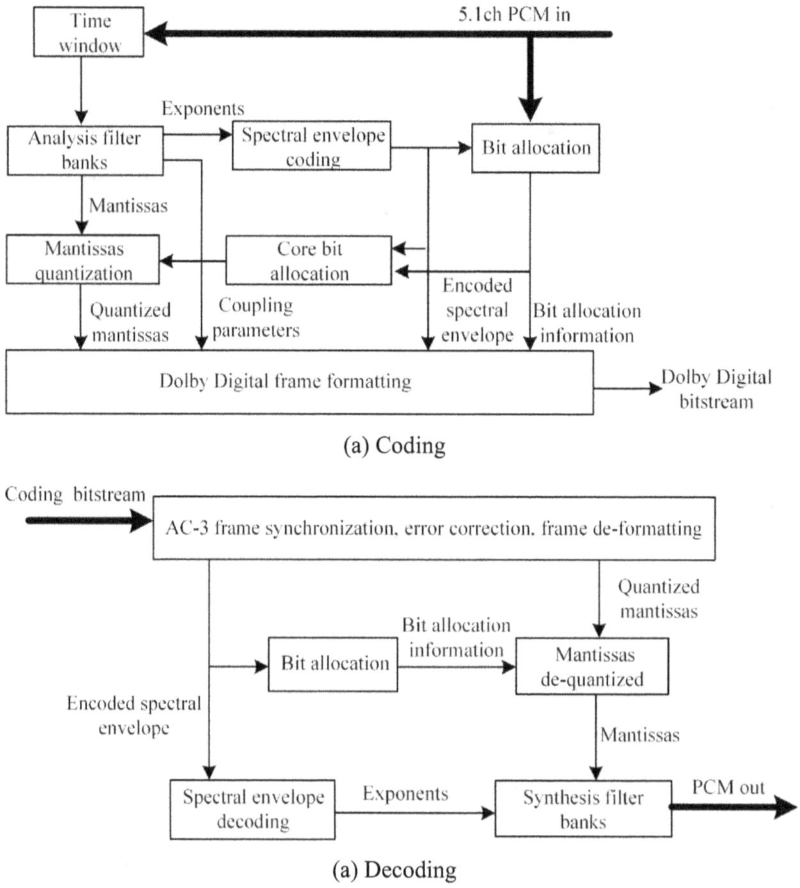

Figure 13.24 Block diagram of Dolby Digital coding/decoding (a) Coding; (b) Decoding (adapted from ETSI TS 102 366 V1.4.1, 2017).

bands. Coefficients are normalized so that their maximal absolute magnitudes do not exceed 1. Each normalized coefficient is represented by a binary exponent and a mantissa and subsequently coded. For example, the exponent for a 16-bit binary number 0.0010 1100 0011 0001, which represents the number of "0" after the decimal point, is 2 in the decimal system or 10 in the binary system; and the mantissa is 10 1100 0011 0001 in the binary system. Binary exponents represent a rough variation in the spectral envelop and the binary mantissas represent the detail variations in the spectra. In a coder, the core bit allocation for mantissas is determined by a spectral envelope and a psychoacoustic model. The final stream involves the coded audio data, synchronous data, bitstream information, and additional data.

Figure 13.24 (b) shows the block diagram of Dolby Digital decoding. Decoding is an inverse course of coding. Various data are extracted. Mantissas are de-quantized according to bit allocation information. Spectral coefficients are reconstructed from exponents and mantissas from which PCM signals are restored using synthesis filter bands.

Some technical details of Dolby Digital coding/decoding are outlined as follows:

1. Analysis filter bands
   Analysis filter bands are implemented by MDCT. The size of MDCT determines the time-frequency resolution. Dolby Digital dynamically chooses the size of MDCT. The PCM input is underground 8 kHz high-pass filters, and high-frequency energy is estimated. Stationary and transient signals are detected by comparing the resultant high-frequency energy with a pre-determined threshold.

   Stationary signals require a higher-frequency resolution; therefore, MDCT with a long window (512 samples) is used. Each PCM block is overlapped by 50% with its two neighbors to avoid theartifacts caused by the abrupt transition between the border of two adjacent blocks; that is, 512 audio samples for MDCT are constructed by taking 256 samples from the previous block and 256 samples from the current block. The 512-point MDCT yields 256 spectral coefficients because of the odd symmetric relation in Equation (13.4.7). A Kaiser–Bessel window is used to improve the frequency selectivity and reduce the influence of the block border. At a sampling frequency of 48 kHz, the frequency and time resolutions of MDCT with a long window are 187.5 Hz and 5.33 ms. Transient signals require a high time resolution; therefore, MDCT with a short window (256 samples) is used. Each PCM block is also overlapped by 50% with its two neighbors. The 256-point MDCT yields 128 spectral coefficients. At a sampling frequency of 48 kHz, the frequency and time resolutions of MDCT with a short window are 375 Hz and 2.67 ms.
2. Exponent coding strategy
   The exponents of spectral coefficients represent the rough variation in spectra. Dolby Digital coding allows a range of exponent values from 0 to 24 (in the decimal system). Spectral coefficients with an exponent value more than 24 (or the corresponding spectral coefficients less than 2^{-24}) are set to 24. A differential coding is used to code the exponent within a block. The first exponent of a full-bandwidth channel (or low-frequency effect, LFE channel) is coded with a 4-bit absolute value, corresponding to a variation from 0 to 15. Successive exponents at the ascending frequency are differentially coded and represented by one of five possible values ±2, ±1, and 0 corresponding to ±12, ±6, and 0 dB variation in magnitude, respectively. Differential exponents are combined into groups in the block. According to the bit rate and required frequency resolution, grouping is formed by one of three exponent coding strategies, namely, D_{15}, D_{25}, and D_{45} modes, where the index "5" denotes five quantization levels of differential exponents, and index "1," "2," or "3" denotes the number of spectral coefficients that

share the same differential exponent. For example, in the D_{15} mode, three spectral coefficients are combined into a group, each spectral coefficient requires a differential exponent, and each differential exponent can take one of the five possible values. Therefore, there are $5 \times 5 \times 5 = 125$ variations in differential exponents in D_{15} strategy. In this case, 7 bits is needed to code these variations, or 2.33 bit is required to code a differential exponent. Similar, in D_{25} and D_{45}strategies, 2.17 and 0.58 bits are respectively needed to a differential exponent. The bit rate and frequency resolution descend in the order of D_{15}, D_{25}, and D_{45}. The Dolby Digital coder chooses an optimal exponent coding strategy for each audio block. For stationary signals, a set of differential exponents can be shared by six MDCT blocks at most.

3. Mantissa quantization and adaptive bit allocation
   As stated, Dolby Digital uses a spectral envelop and a psychoacoustic model to determine the core bit allocation for mantissa coding. In contrast to other coding methods (such as MPEG-1 layers I/II), Dolby Digital coder employs a forward-backward-adaptive psychoacoustic model. The decoder also includes a core backward-adaptive model. Core bit allocation uses a psychoacoustic model based on certain assumptions on the masking properties of signals. Some parameters of the model are also transmitted by the data stream. Therefore, the actual psychoacoustic model in a decoder can be adjusted by the coder. The coder can perform an ideal bit allocation based on a complicated but accurate psychoacoustic model and compares the results with the core bit allocation. If core bit allocation can better match the ideal bit allocation by changing some parameters, the coder does so and finishes the bit allocation. Otherwise, the coder sends some information to the decoder.
4. Channel coupling and re-matrixing
   At a very low bit rate, the aforementioned compression and coding algorithms may not satisfy the bit rate requirement. In this case, spectral coefficients at high frequencies are combined into a single coupling channel to transmission. The coupling channel is formed by a vector summation of the spectral coefficients from all channels in coupling. This process is an extension of intensity stereo coding in Section 13.4.5. Channel coupling is based on the psychoacoustic principles that low-frequency ITD dominates lateral localization; at a high frequency, only the energy envelop contributes to localization. Dolby Digital decomposes signals into 18 subband components and applies the channel coupling above some subbands. Similar to original channels, spectral coefficients in the coupling channel are represented by binary exponents and a mantissa and coded. The coder calculates the powers of original signals and coupled signal. The resultant power ratio between the original signal and the coupled signal is evaluated for each input channel and each subband and transmitted as side information parameters. The decoder distributes the coupling channel signal to the output channels according to the side information parameters.

In re-matrixing, MS coding for a pair of channels with high correlation is used. It is applied to stereophonic signal coding. Its principle is outlined in Section 13.4.5.

In addition to the aforementioned technical details, Dolby Digital possesses some user features.

1. Loudness control
   The level of dialog varies in different programs. Switching between different programs directly causes a variation in perceived loudness. Dolby Digital stream includes a code for dialog level normalization, which enables users to set the gain in reproduction according to the required level.

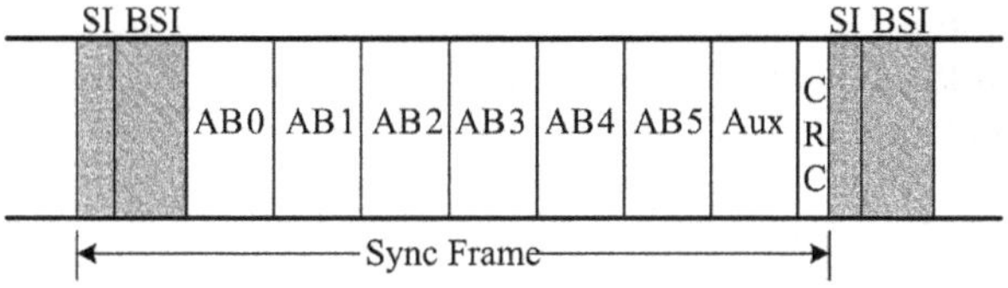

Figure 13.25 Dolby Digital synchronization frame.

2. Dynamic range control and compression
   Dolby Digital involves a code for dynamic range control and compression in the decoder to accommodate different applications.
3. Channel downmixing
   In this process, 5.1-channel signals can be downmixed to mono, stereophonic, and Dolby Pro-Logic signals. For optimal downmixing in all signal conditions, dynamic downmixing coefficients are provided in the Dolby Digital bitstream.

The Dolby Digital coding stream consists of a series of the successive synchronization frame. Figure 13.25 illustrates the structure of a synchronization frame. The synchronization frame consists of synchronization information (SI), bit stream information (BSI), six audio blocks AB0–AB5, auxiliary data field (AUX), and CRC. The size of SI is 40 bits, including 16-bit synchronization word, 16-bit cyclic redundancy core (applied to the first 5/8 of the frame), 8-bit sampling frequency, and frame size code. The size of BSI is variable, including the parameters of channel number and other audio data information. Each synchronization frame involves six coded audio blocks AB0–AB5, and each audio block usually contains 256 coded samples from each input channel. Therefore, a synchronization frame has 1,536 samples per channel. The relative length of six audio blocks within a 1536-sample synchronization frame is also variable. This feature is useful for a nonstationary audio signal over a synchronization frame. The total size of the coded audio blocks is also variable depending on various parameters, such as the number of channels. The AUX with a variable length is for additional data, and a 17-bit CRC word is for error detection.

13.6.2 Some advanced Dolby coding techniques

After Dolby Digital, Dolby Laboratories has introduced a series advanced coding techniques for multichannel sound. As stated in Section 8.1.4, Dolby Digital Surround EX extends the Dolby Digital 5.1-channel coding to 6.1-channel coding by adding a rear surround channel and using the matrix technique.

Dolby Digital Plus (Dolby Digital+), also called Enhanced AC-3 (E-AC-3), is an extension of Dolby Digital coding technique (Fielder et al., 2004; ETSI TS 102 366 V1.4.1, 2017; ATSC standard Doc.A52, 2012). Dolby Digital Plus is also a perceptual coding technique. It was originally intended for high-quality and high-bit-rate audio coding for digital cinema. However, to adapt to satellite and cable television transmission and streaming video service, Dolby Digital Plus also supports coding with low bit rates. Dolby Digital Plus supports bit rates ranging from 32 kbit/s–6.144 Mbit/s, which is much wider than that of Dolby Digital. Dolby Digital Plus ensures a good audio quality near the upper limit of bit rate. Even at low bit rate, Dolby Digital Plus also possesses an appropriate perceived quality

Dolby Digital Plus uses an extension structure. It starts from 7.1-channel and allows for at least 13.1-channel coding (it has been extended to 15.1-channel coding). It supports the sampling frequency of 32, 44.1, and 48 kHz and up to 24-bit quantization. To be convenient for converting to Dolby Digital 5.1-channel signals, Dolby Digital Plus preserves most filter bands, the basic framing structure of core and extension data in Dolby Digital. However, as an extension of Dolby Digital, Dolby Digital Plus adds some new coding tools and techniques. The main points are outlined as follows.

1. Flexible frame and data stream structure
   A frame of Dolby Digital core data stream involves six blocks. Each block has 256 samples of coefficients. The core stream of Dolby Digital Plus preserves the same six-block structure as that in Dolby Digital. It also supports a shorter frame of one to three blocks, and each block is 256 samples of coefficients. The core stream includes the signals of six channels and thus allows the transmission of 5.1-channel signals. Additional channel signals are transmitted by the additional substreams. Dolby Digital Plus allows up to eight additional substreams at least (for 13.1-channel coding).
2. Improved analysis filters and quantization
   Analysis filter bands in Dolby Digital apply an MDCT with 50% overlap (a window size of 512 samples) to each block of 256 new input samples, resulting in 256 MDCT coefficients for each block. To increase the frequency resolution of filter bands and improve the coding performance of stationary signals, Dolby Digital Plus utilizes an adaptive hybrid transfer (AHT) algorithm; that is, a non-window (rectangular window) and non-overlapped Type II discrete cosine transform (DCT) is supplemented to MDCT analysis in Dolby Digital. DCT is applied to the six successive blocks (a frame) of samples, and the resultant $256 \times 6 = 1536$ DCT coefficients for each channel form a hybrid transform block. According to the characteristics of input signals, MDCT and AHT are switched adaptively. AHT is applied only to the case in which the exponent strategy is shared over all six blocks in a frame.

   Dolby Digital analyzes the spectral envelope of the input signal to estimate the SMR in approximately 6 dB increments for bit allocation. By contrast, the quantization applied to AHT in Dolby Digital Plus allows for 1.5 dB increments in the SMR estimation for bit allocation at the lowest allocation level. At the same time, the six-dimensional VQ and gain adaptive quantization are used for AHT coefficients to improve the coding efficiency of some "difficult to code" signals.
3. Spectral band replication (SBR)
   Dolby Digital Plus uses SBR to extend the bandwidth at a low bit rate. The basic principle of SBR is stated in Section 13.4.6.
4. Enhanced channel coupling
   In Section 13.6.1, Dolby Digital optionally uses channel coupling at high frequencies to increase the coding efficiency. Dolby Digital Plus utilizes an enhanced channel coupling by analyzing and compensating the phase among channels to reduce the coupling cancellation in the coding and then preserve the phase relation after decoding. This method also extends the channel coupling to lower frequency than the previous method.
5. Transient pre-echo processing
   Dolby Digital Plus uses a new transient pre-echo processing tool that utilizes time scaling synthesis to reduce or remove the transient pre-echo caused by low-bit-rate coding.
6. Conversion to Dolby Digital signals
   Dolby Digital Plus coding signals should be converted to Dolby Digital signals to be downward compatible with the Dolby Digital decoder. Because Dolby Digital Plus and Dolby Digital employ the same analysis filter bands, bit allocation, and framing structure of the code stream, conversion can be conducted directly in the time-frequency domain common to Dolby Digital Plus and Dolby Digital. The courses of a full decoding Dolby Digital Plus stream to PCM samples and then re-coding into Dolby Digital stream are unnecessary. The direct method simplifies the conversion and avoids the quality loss caused by twice the coding.

Dolby Digital Plus supports more than 5.1-channel coding. The core data stream is 5.1-channel downmixing signals to be compatible with 5.1-channel reproduction. It can be decoded directly for 5.1-channel reproduction or converted to Dolby Digital 5.1-channel stream in the time-frequency domain at a bit rate of 640 kbit/s. More other replacement and supplemental channels are transmitted as substreams. They are used for substituting parts of downmixed 5.1-channels in more than 5.1-channel reproduction. The use of replacement and supplemental channels avoids audible quantization noise caused by the unmasking in the inverse matrix similar to that in MPEG-2 BC. For example, in 7.1-channel horizontal surround sound, seven channels with full audible bandwidth, namely, the left, center, right, left-surround, right-surround, left-back-surround, and right-back-surround channels, as well as an LFE channel, are involved. The steps of multichannel coding/decoding are as follows:

1. A 5.1-channel downmixing of 7.1-channel signals is created and coded as an independent (core) stream.
2. The replacement and supplemental channels, e.g., four surround channels in 7.1-channel signals are coded as a dependent substream.
3. Independent core stream and dependent substream are decoded, yielding 5.1-channel downmixing signals and four surround channel signals in 7.1-channel signals.
4. The complete 7.1-channel signals are obtained by substituting the two surround channel signals in 5.1-channel signals with the four surround channel signals in 7.1-channel signals.

The Dolby Ture HD introduced in 2005 is a lossless extension that uses Meridian Lossless Packing (MLP, Section 13.8) as the lossless code (Dressler, 2006). It can be used as a lossless audio coding format in a Blu-ray disk. Dolby Ture HD allows a maximal 192 kHz sampling frequency, 24-bit quantization, and 18 Mbit/s bit rate. Dolby Ture HD supports 7.1-channel and maximum up to 14-channel coding. At the largest bit rate, it allows 8-channel coding at 96 kHz sampling frequency and 24-bit quantization. The compression ratio of MLP in Dolby Ture HD ranges from 2:1 to 4:1, which is higher than 2:1 for MLP used in DVD-Audio. The low compression ratio for MLP in DVD-Audio is due to the continuous and rich-harmonic characteristics of music signals, which make the compression difficult.

For the compatibility and simplification of decoding, Dolby Ture HD remaps multichannel input signals into substreams (Dressler, 2006). For example, as shown in Figure 13.26, 7.1-channel inputs are downmixed into 5.1-channel signals, and the latter is further downmixed into two-channel stereophonic signals by another matrix. In this way, matrix downmixing remaps 7.1-channel inputs into three sets of signals, e.g., two-channel stereophonic signals, 3.1-channel extension signals A, and two-channel extension signals B. The

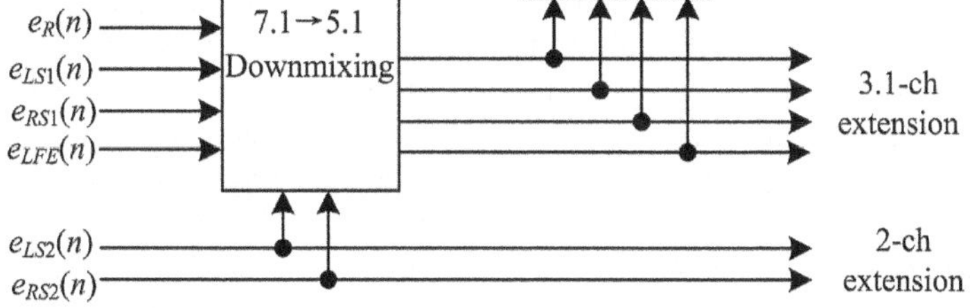

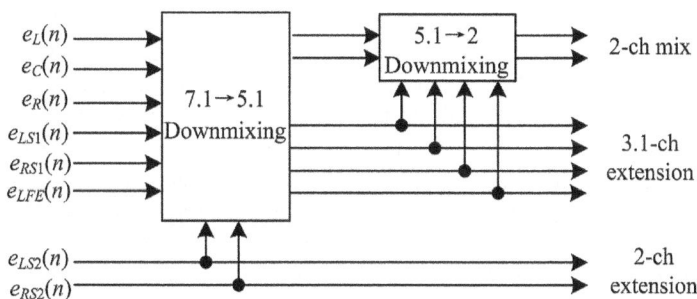

Figure 13.26 7.1-channel remapping in Dolby Ture HD.

three sets of signals are coded, yielding three streams. The decoder can choose parts or all streams. When only the core stream is chosen, decoding yields two-channel stereophonic signals. When the core stream and stream A are selected, decoding yields 5.1-channel signals. When all three streams are chosen, decoding yields 7.1-channel signals. Because MLP lossless coding is used in Dolby Ture HD, the unmasking of quantization noise after the matrix in the decoder does not occur. This phenomenon is an advantage of MLP coding, which distinguishes the MLP from some loss coding (such as MPEG-2 BC). In fact, this advantage of MLP has been utilized for 5.1-channel coding in a DVD-Audio. It is further developed in Dolby Ture HD.

In Section 6.5.2, Dolby Atmos is a new generation of spatial sound and coding technique. It involves two types of audio contents, namely, audio objects, and beds. Both expanded the Dolby Digital Plus and Dolby Ture HD support core coding of Dolby Atmos and support the downward compatibility of 5.1/7.1-channel signals (Dolby Laboratories, 2016). The expanded Dolby Digital Plus utilizes new bit stream metadata to extract Dolby Atmos object audio and outputs this information. The sampling frequency of Dolby Atmos contents is 48 kHz. The expanded Dolby Ture HD adds a fourth substream for Dolby Atmos, which is used for the lossless coding of fully object-based mixing. Dolby Atmos can be coded with the expanded Dolby Ture HD at different sampling frequencies (48, 96, and 192 kHz) and different quantization bits (16, 20, and 24 bits). Similar to MPEG-H 3D Audio, Dolby Atmos is a general spatial sound technique rather than an audio coding technique only.

Dolby AC-4 is a coding technique for video and audio entertainment services, such as broadcasting and Internet stream (ETSI TS 103 190-1 V1.3.1, 2018; ETSI TS 103 190-2 V1.2.1, 2018; Kjörling et al., 2016). The structure of AC-4 is somewhat similar to that of MPEG-H 3D Audio. AC-4 supports channel-based coding, such as stereophonic sound, 5.1-, 7.1- (with 5.1-horizontal and two height channels), 13.1- (with 9.1-horizontal and four height channels), and 22.2-channel coding. It also supports object-based coding and perceptually motivated sound field format. Therefore, AC-4 allows the coding of Dolby Atmos signals. In addition to the dynamic range and loudness control, AC-4 possesses some novel characteristics, such as supports immersive and personalized audio, advanced loudness management, video-frame synchronous coding, and dialog enhancement.

AC-4 is based on the waveform coding in the MDCT domain and parameter coding in the complex-pseudo QMF domain. In the MDCT domain, AC-4 provides two spectral front ends, e.g., the audio spectral front end (ASF) for arbitrary audio signals and the speech spectral front end (SSF) for speech signals. In ASF, audio signals are coded on the basis of the perceptual model and wavefront coding. The principle is similar to the aforementioned methods. Input signals are converted into MDCT coefficients by using one or several MDCT transforms for each coding frame. Five transform sizes can be chosen according to the characteristics of input signals. In SSF, speech is coded on the basis of the prediction model. The coder can check input signals and switch between ASF and SSF.

In the QMF domain, AC-4 introduces a companding tool to control the temporal distribution of the quantization noise. It also introduces advanced spectral extension, advanced coupling tool, advanced joint channel coding tool (A-JCC), and advanced joint object coding tool. A-JCC is an efficient tool for channel-based signal coding, which enables the parametric coding of more channel signals (such as 13.1-channel signals, with 9.1-horizontal channels and four height channels) by 5-channel downmixing, together with the side information of parameters (Lehtonen et al., 2017). The principle of parameter coding in the QMF domain is similar to that in the aforementioned MPEG series and Dolby series coding techniques and omitted here. Subjective assessment with MUSHRA method (Section 15.5) indicated that for channel-based stereophonic sound, 5.1- and 11.1-channel sound (with 7.1-horizontal channels and four height channels), AC-4 reaches an excellent range in a MUSHRA scale at bit rates of 96, 208, and 256 kbit/s, respectively. For immersive audio contents by using joint object coding in Dolby Atmos, AC-4 reaches an excellent range in the MUSHRA scale at bit rates of 384 kbit/s (Purnhagen et al., 2016).

13.7 DTS SERIES OF CODING TECHNIQUE

Since the mid-1990, DTS Inc. (originally Digital Theater, Inc.) has developed a series of audio coding techniques called coherent acoustic coding. These techniques have been applied to professional and consumer fields, such as cinema sound, DVD-Video, and multichannel music CD (Smyth et al., 1996). Coherent acoustic coding is an adaptive differential subband coding method. Original coherent acoustic coding possesses great flexibility. It supports various sampling frequencies from 8 kHz to 192 kHz and 16-bit quantization to 24-bit quantization from 1 to 8 channels, with bit rates from 32 kbit/s to 4.096 Mbit/s and compression from 1:1 to 40:1. The practical sampling frequency, quantization precision, the number of channels, and maximal bit rate should follow some restrictions, leading to multiple combinations.

Similar to MPEG-1 Layer I and Layer II and Dolby Digital, coherent acoustic coding is a perceptual coding technique. Figure 13.27 (a) illustrates the block diagram of coherent acoustics coding. The PCM input is divided into blocks and converted into subband components (time-frequency domain) by polyphase filter bands. Global bit allocation is performed according to the characteristics of input signals and psychoacoustic models. ADPCM is applied to each subband component. The coded signals are packed and outputted. Decoding is an inverse course of coding, as shown in Figure 13.27 (b).

Some details of coherent acoustic coding are outlined as follows:

1. Input PCM analysis frame
   The input PCM analysis frame (window length) is chosen by a comprehensive consideration of coding efficiency and audio quality. A long frame improves the coding

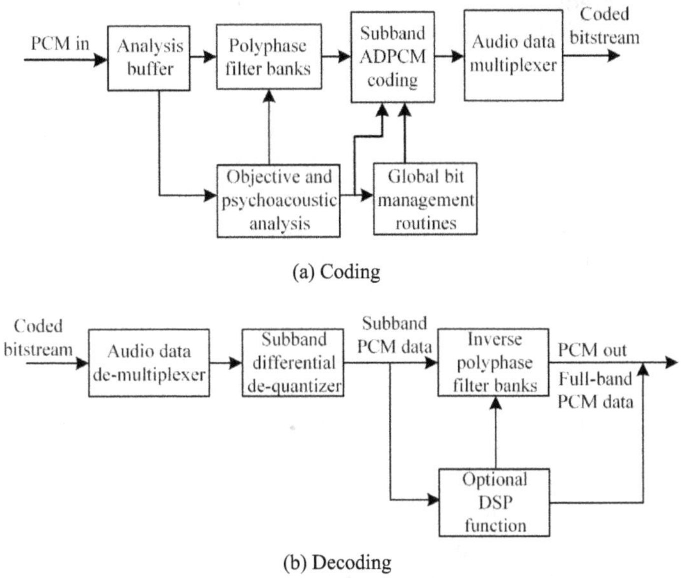

Figure 13.27 Block diagram of DTS coherent acoustics coding/decoding (a) Coding; (b) Decoding (adapted from Smyth, 1999).

efficiency at the cost of a reduction of the audio quality of transient signals and vice versa. In terms of sampling frequency and bit rate, coherent acoustics coding provides different frame sizes: 256, 512, 1024, 2048, and 4096 samples per channel. In fact, the available frame size is determined by the size of the input buffer in the decoder.

2. Subband filter bands
   Each frame of input PCM sample is transformed to subband components by using polyphase filter bands. For a sampling frequency of 48 kHz and below, each PCM channel input is divided into 32 uniform subbands. For example, for a frame of 1024 samples, polyphase filter bands yield 32 subband components in the time-frequency domain, and the size of each subband component is 32 samples. Polyphase filter bands have two choices, which are indicated by a flag embedded in the coded bit stream. Non-prefect-reconstruction (NPR) filter bands possess a narrow transient bandwidth and a high stop band rejection ratio; thus, it enhances the coding efficiency at low bit rates. The disadvantage is that signals cannot be reconstructed perfectly in the decoder. However, at a low bit rate, the influence of coding noise is more important, and NPR filter bands are usually used. Prefect reconstruction (PR) filter bands possess high reconstruction precision and are suitable for high bit rates.

   An alternative subband division is used for sampling frequency above 48 kHz. For example, at a sampling frequency of 96 kHz, the frequency range of 0–48 kHz of reconstruction is divided into two bands. The basic band between 0–24 kHz is subdivided into 32 uniform subbands; and the band between 24–48 kHz is subdivided into only 8 subbands.

3. Subband ADPCM
   Coherent acoustic coding uses the ADPCM to code 32 subbands independently in each channel. The principle of ADPCM is outlined in Section 13.4.2. Fourth-order forward-adaptive linear prediction is used to calculate the optimal prediction coefficients over a window for each subband. The sizes of window are 8, 16, and 32 subband PCM samples for the frame size of 256, 512, and 1024 (or greater) samples, respectively. Each set of prediction coefficients is quantized using a 4-element tree-search 12-bit vector code book and coded. The prediction coefficients are re-calculated and updated for different frames. For a frame size of 1024 or more samples, the prediction coefficients within a frame should also be updated. For example, when a frame of input of 4096 PCM samples is divided into 32 subband components, 4096/32 = 128 subband PCM samples are found in each subband for each frame of input. Therefore, the prediction coefficients in each subband are updated and transmitted 128/32 = 4 times in each frame of the PCM input.

   ADPCM improves the coding efficiency only for signals with high time domain correlation. Prior to ADPCM coding, the prediction gain is estimated. That is, the variances of differential signal and the subband component in each analysis window are compared. Signals with high time domain correlation led to a small differential signal and thus a sufficiently positive prediction gain. As a result, signals with a large prediction gain within an analysis window are coded with ADPCM; at the same time, "predictor mode" flags are embedded in the bit stream. Otherwise, signals with a negative prediction gain are not coded with ADPCM. Therefore, ADPCM coding is chosen dynamically within each subband and each analysis window.

   The position of transient within the analysis window is assigned to each of subband window and used for adjusting the coding to avoid the pre-echo of transient signals caused by ADPCM at a low bit rate.

   The coding of the LFE channel is different from that of the main channels. The LFE channel is derived by decimating a full-bandwidth input PCM stream. The resultant LFE channels are coded using ADPCM.

4. Psychoacoustic model and bit allocation
   The minimal SMR for each subband is calculated using a psychoacoustic model based on which available bits are allocated. At a low bit rate, bit allocation is determined either by the SMR values from the psychoacoustic model, SMR values modified using subband prediction gains, or a combination of both. At a high bit rate, bit allocation is determined by a combination of SMR and differential minimum mean-square error.
5. Entropy coding
   Entropy coding on the output of ADPCM coding further improves the efficiency up to 20%.
6. Digital signal processing (DSP) in the decoder
   The DSP in the decoder provides some post-processing functions of user programming. It can manipulate the signals of individual or all channels, individual or all subbands, such as downmixing, dynamic range control, and channel delay.

An early application of DTS coherent acoustics coding is the coding of 5.1-channel signals. At 48 kHz sampling frequency and 16-bit quantization, the typical bit rate is 768 kbit/s or 1.536 Mbit/s.

A structure of extension from the 5.1-channel core data stream is used in succedent DTS coding technique to be compatible with a legacy 5.1-channel decoder. The earlier extension has a structure of "core stream + core substream." The code stream contains original 5.1-channel data. The channel extension (XCH) adds a rear channel coding, yielding DTS-ES 6.1 coding. The sampling frequency extension (X96) extends the sampling frequency from 48 kHz to 96 kHz by secondarily encoding a residual signal following the baseband encoding.

The latter DTS-HD coding has a structure of "core stream + core substream + extension substream." The "core stream + core substream" of DTS-HD is identical to the aforementioned XCH and X96. The extension substream supports XXCH, XBR, and XLL extensions (Fejzo et al., 2005). The XXCH is the channel extension, which allows adding more discrete channels. XBR is a high-bit-rate extension, which allows an increase in bit rate to improve the audio quality.

XLL is a lossless extension known as the DTS-HD master audio, which utilizes a substream to accommodate lossless audio compression. The maximal bit rate of DTS-HD master audio is 24.5 Mbits/s. The DTS-HD master supports lossless coding up to 8 channels at a sampling frequency of 192 kHz and 24-bit quantization. In practical uses, XXCH, XBR, and XLL extensions are optional, some or all of them can be chosen. Because of the extension structure, the bit stream of DTS-HD is downward compatible with the original decoder. When downward compatibility is not required, XLL directly allows the lossless coding of signals without using the core stream to improve the efficiency.

In 2015, the DTS introduced the DTS:X (http://dts.com/), which is an object-based spatial sound system and coding technique. Its basic spatial sound part is based on the multi-dimensional audio (MDA) technique in Section 6.5.2. The technical details of DTS:X have not been published up to the end of 2021.

13.8 MLP LOSSLESS CODING TECHNIQUE

Meridian lossless packing (MLP) is a coding technique developed by Median Audio Ltd. (Gerzon et al., 2004). It aims to code the PCM audio signals losslessly for transmission and storage. MLP was first applied to the audio signal compression in DVD-Audio and then used in Dolby Ture HD. MLP does not limit the sampling frequency. It supports 16- to 24-bit quantization, and coding up to 63 channel signals, depending on the available bit rate of

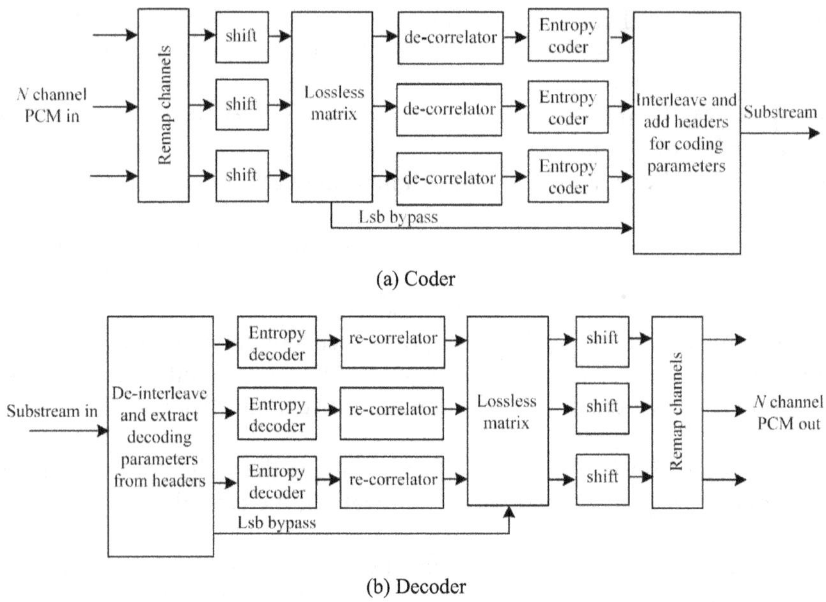

Figure 13.28 Block diagrams of MLP coder and decoder (a) Coder; (b) Decoder (adapted from Gerzon et al., 2004).

practical media. For application in DVD-Audio, MLP supports a sampling frequency of up to 192 kHz and exhibits a mean compression ratio of 2:1.

Figure 13.28 (a) illustrates the block diagram of the MLP coder. It involves the following steps. Multichannel inputs are mapped into two or more substreams; each channel is shifted and has undergone a lossless matrix. The output of the lossless matrix is decorrelated. The decorrelated signals are further optimized by entropy coding, and multiple data substreams are interleaved and packed for a carrier. The Lsb bypass in the figure represents the bypass of the least significant bit in the quantized data. Figure 13.28 (b) presents the block diagram of the MLP decoder. The inputs of substreams are de-interleaved, error corrected, and entropy decoded. Practical signals are restored by re-correlators, and multichannel PCM signals are obtained by lossless matrix, shift, and remapping.

Some technical components of MLP coding/decoding are outlined as follows:

1. Remapping of multichannel inputs
   For the compatibility and simplification of decoding, MLP uses a hierarchical stream structure, which involves multiple substreams re-mapped from multichannel inputs and hierarchical additional data. An MLP decoder decodes only the part of the required streams rather than all streams. For example, 5.1-channel inputs are re-mapped into two substreams. Decoding substream No. 0 yields two-channel stereophonic signals, and decoding both substreams No. 0 and No. 1 yields complete 5.1-channel signals.
2. Channel shifting
   Channel shifting is performed to recover unused capacity, such as less than 24-bit precision or less than full scale.
3. Lossless matrix transmission
   Different channel signals are correlated to some extent. Lossless matrix transmission aims to reduce these correlations. In Section 13.4.5, two or more channel signals can be linearly transformed by an appropriate matrix so that signals with a large amplitude are

concentrated on one or a few channels. Such kind of transformation optimizes signal coding. An example is MS transformation in stereophonic signals. However, conventional matrix transformation is not lossless because rounding errors are introduced into the inverse matrix. The MLP coder decomposes matrix transformation into a cascade of affine transformation. Each affine transformation modifies just one channel signal by adding a quantized linear combination of other channels. Lossless transformation is achieved by subtracting/adding a quantized linear combination of the other channels in the decoder.

4. Prediction and decorrelation in the time domain
   In Section 13.4.2, if each value of the future sample can be linearly predicted, the correlation of the signal in the time domain can be reduced. In this case, the prediction coefficients and differential values between the predicted and input samples are coded to improve efficiency. FIR filter models are used for prediction in many coding methods. However, the IIR filter model more effectively predicts the peaks of signals. Each coded channel in the MLP can be predicted independently with the FIR or IIR filter model up to the 8th order. The flexible choice of the filter model for prediction enables the effective compression of audio inputs with different characteristics. In addition, if the input of the low-frequency effect channel is included, it is regarded as a highly predictable signal and explored together with other channel signals in MLP. The low-frequency effect channel only requires a very low bit rate.
5. Entropy coding
   As stated in Section 13.4.7, entropy coding further improves the coding efficiency. Various optional entropy coding methods are employed in MLP.
6. Buffer
   For some transient or outburst signals that are difficult to be predicted, the peak of a bit rate may occur in MLP coding streams. First-in first-out shift registers (FIFO) are used in the MLP coder and decoder to reduce the variation in the bit rate. FIFO introduces a small overall delay, usually in the order of 75 ms. A FIFO management minimizes the delay in the decoder to allow start up or cuing. Therefore, the buffer in the decoder is usually almost empty except the cases of signals with high transient characteristics.
7. Interleave and packing
   Multiple data streams are interleaved and packed to code the output with fixed or variable bit rates.

13.9 ATRAC TECHNIQUE

Adaptive Transform Acoustic Coding (ATRAC) was a coding technique originally for SDDS cinema sound developed by Sony in 1993; afterward, it was applied to audio storage devices, such as MiniDisc (Tsutsui et al., 1992). ATRAC supports up to 8- (7.1-) channel audio coding. It achieves a 5:1 compression ratio at a sampling frequency of 44.1 kHz and 16-bit quantization.

ATRAC involves three steps, time-frequency analysis, bit allocation, and spectral quantization. It uses the principle of psychoacoustics to both bit allocation and time-frequency analysis. Figure 13.29 illustrates the block diagram of time-frequency analysis in ATRAC. Two-stage QMFs divide the input into three subband components, bandwidths of low-, mid-, and high-frequency subbands are 0–5.5, 5.5–11, and 11–22 kHz. QMFs effectively restore the original signals. Each subband component is converted to spectral coefficients by MDCT. The MDCT allows up to 50% overlap between time windows, which improves frequency resolution and maintains critical sampling. The block size of MDCT varies adaptively according

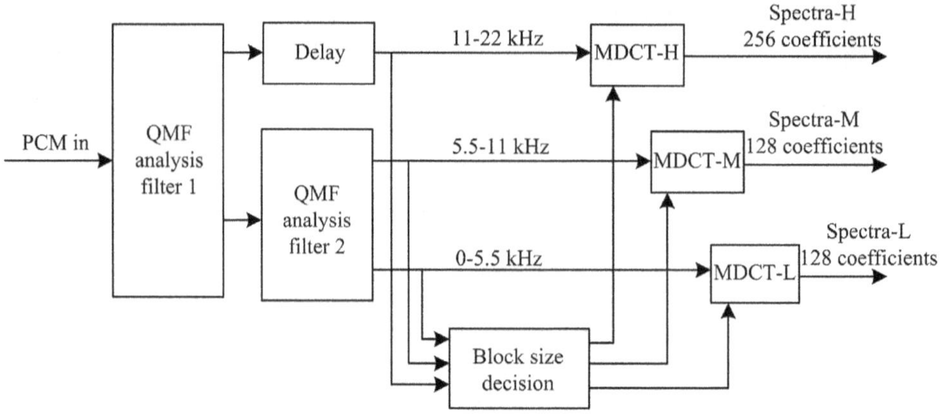

Figure 13.29 Block diagram of time-frequency analysis in ATRAC (adapted from Tsutsui et al., 1992).

to the characteristics of each subband components. For stationary signals, a long block of 11.6 ms improves frequency resolution. For transient signals, short blocks of 1.45 ms for high-frequency subband and 2.9 ms for mid-frequency subband ensure the time resolution. The block sizes of three subbands can be chosen independently. The outputs of three subband components are converted into 512 spectral coefficients, with 256 coefficients for high-frequency subband, 128 coefficients for each of mid- and low-frequency subbands. According to auditory frequency resolution (critical band), spectral coefficients are grouped into 52 block floating units (BFUs), and each unit involves a fixed number of spectral coefficients.

The spectral coefficients are quantized on the basis of scale factor and word length. The scale factor, which specifies the full-scale range of quantization, is chosen from a given list of possibilities. The word length, which specifies the quantization precision, is determined by bit allocation. For all spectral coefficients within each BFU, the scale factor and word length are identical. Similar to other psychoacoustic coding, bit allocation is calculated from the masking curve. However, ATRAC does not specify the algorithm of bit allocation and therefore possesses a large flexibility. Tsutsui et al. (1992) illustrated an algorithm of bit allocation. This algorithm uses a weighted combination of fixed and variable bits. The weights are determined by the characteristics of signals. Bit allocation is performed so that the final bits do not exceed the available bits.

ATRAC decoding is an inverse course of coding, the spectral components are reconstructed from the quantized spectral coefficients, and the output signal is obtained by time-frequency synthesis. In addition, some advanced ATRAC techniques, such as low-bit-rate coding techniques and lossless extension techniques, have also been developed.

13.10 AUDIO VIDEO CODING STANDARD

The Audio Video Coding Standard (AVS) is a series of techniques and standards developed by the AVS workgroup of China (Zhang et al., 2016; IEEE Computer Society, 2020). The first generation of AVS technique is termed AVS1 audio, including AVS1-P3 (advanced audio video coding), AVS1-P10 (mobile voice and audio), and AVS LS (lossless audio coding standard). The second generation of the AVS audio is termed AVS2 audio.

The AVS1-P3, which was finished in 2005, is for mono, stereophonic, and multichannel signal coding. It supports sampling frequencies ranging from 8 kHz to 96 kHz, and a bit rate ranging from 16 kbit/s to 96 kbit/s per channel. A bit rate of 64 kbit/s per channel can

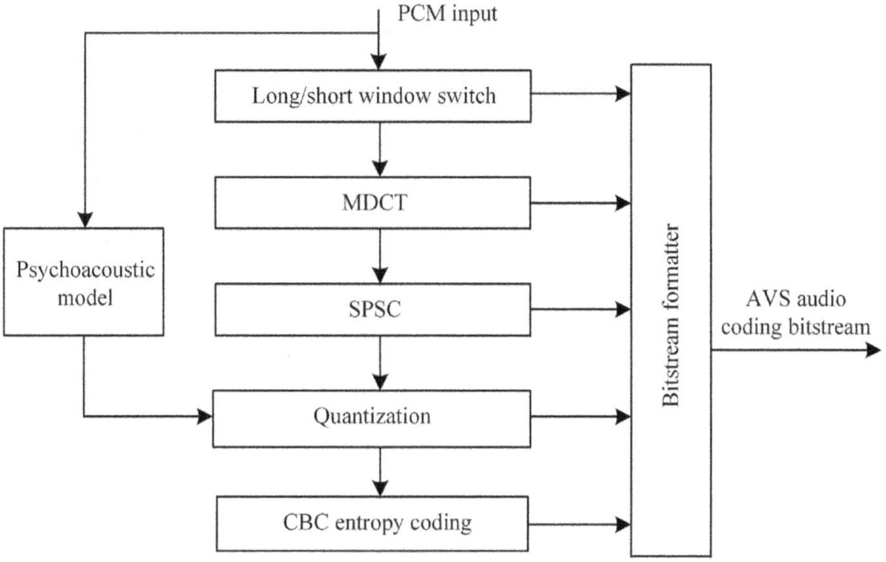

Figure 13.30 Block diagram of AVS1-P3 audio coding (adapted from Zhang et al., 2016).

provide transparent audio quality. Figure 13.30 illustrates the block diagram of AVS1-P3. The PCM input is analyzed by a psychoacoustic model from which a long/short window for analysis is chosen. The MDCT with a block overlap is used to convert the PCM input into spectral coefficients in the time-frequency domain and then coded using the psychoacoustic model. Square polar stereo coding (SPCS) may be used for stereophonic signals with high correlation. The principle of SPCS is similar to that of MS stereo coding in Section 13.4.5. In SPCS, either left or right signals with a larger amplitude and the difference in the left and right signals are coded. Quantization is similar to that of MPEG AAC. Context-dependent bit plane coding (CBC) is used for the entropy coding of spectral data.

The AVS LS is a technique of lossless audio coding, in through which a MS coding method is used for multichannel signal decorrelation. The input signal is decomposed into high- and low-frequency components by the integer lifting wavelet transform. Then, the two components are subjected to linear predictive coding (LPC). The residual signals of LPC are normalized and subjected to entropy coding.

The AVS2 audio is a 3D audio coding technique and standard similar to that of MPEG-H 3D Audio. It involves a basic channel profile for mono, stereophonic, multichannel, and bed for 3D audio coding and 3D profile for audio object coding. It also supports scene-based coding, such as Ambisonic coding. AVS2 is described according to the IEEE standard (IEEE Computer Society, 2020).

13.11 OPTICAL DISKS FOR AUDIO STORAGE

13.11.1 Structure, principle, and classification of optical disks

Optical disks are important media for storing various binary data based on the principle of optical storage. Since Philips and Sony introduced the compact disc-digital audio (CD-DA), optical disk technique has been greatly developed. It has been widely used for the storage of computer files and data, electronic publication, and video/audio programs (including various spatial sound programs).

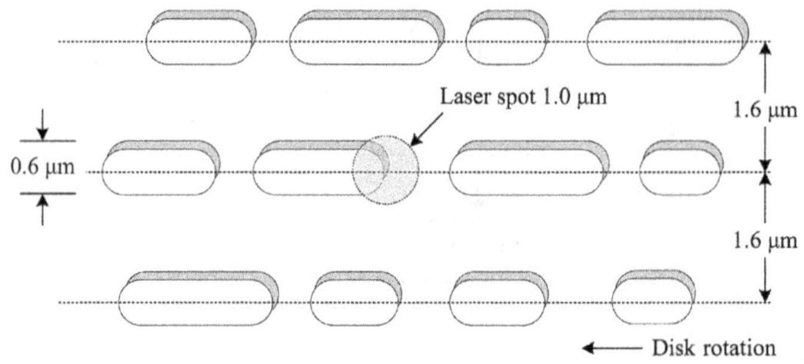

Figure 13.31 Pit track on the surface of an optical disk (take the compact disk as an example, adapted from Pohlmann, 2011 and referred to IEC 60906, 1999).

As shown in Figure 13.31, data are stored in a pit track on the surface of an optical disk. Pits are located in a spiral track that runs from inside to outside of the disk. Along the track, each pit edge represents a binary 1 to maximize the area of optical disk and increase the capacity of storage, and the flat lands between or within pits represent the binary 0. The reflection of pits and the lands between them are different. When a laser beam strikes to the land between pits, light is reflected almost completely. When a laser beam strikes a pit, light is cancelled because of the destructive interference between the reflections from the pit and surrounding land. The pit depth is chosen so that a completely destructive interference occurs. The binary data stored in an optical disk can be read by detecting the intensity of reflected light from the pit track.

Various methods are used to read the binary data from an optical disk. Figure 13.32 illustrates a three-beam laser pickup. A laser beam is created by a laser diode. It goes through a diffraction grating, a beamsplitter, and parallel optical lens, yielding a parallel laser beam.

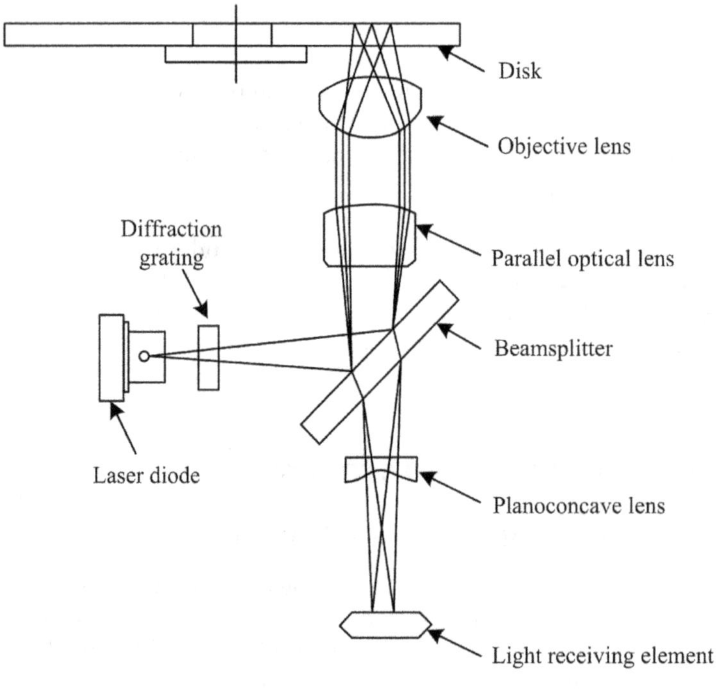

Figure 13.32 Three-beam laser pickup.

The parallel laser beam focuses on the disk surface by using an objective lens. The reflected laser beam from the surface of the disk propagates along a light path reversal to the incidence, goes through the beamsplitter in opposite and planoconcave lens, and finally reaches the light-receiving element. The data stored in the disk can be read from the electric output of the light-receiving element after de-modulation and error-correction processing.

The control system in a player is necessary to read the data from an optical disk. A servo system moves the objective lens up and down so that the laser beam focuses on the pit track with a size of about 1.0 μm. An auto-tracking system controls the laser beam in the radical direction to track the spiral pit sequence. A speed control system controls the angular speed of the rotation of the disk so that the spiral data track is read at a regular linear speed.

According to the characteristics of storage media, optical disks can be classified into three types: read-only, write-once, and erasable (rewritable) disks.

For a read-only disk, data are written into a disk when it is industrially manufactured. They are stored permanently and cannot be changed. Commercial read-only disks are manufactured by impression. A photoresist process similar to that used to manufacture integrated circuits is first used to produce a glass master plate. A glass plate is coated with photoresist and exposed to the laser beam that is modulated by the coded data "0" and "1". After the development, pits are left in the exposed areas of the photoresist. The master plate is finished by evaporating a silver coat onto the photoresist layer. Subsequent courses are similar to those in the industrial production of 45°/45° disk in Section 13.1.1. The metallized glass master plate is electroplated with a nickel coating, and the metal part is separated from the glass master plate to obtain the (negative) father. The resultant fathers are used to create some positive mothers. The process is repeated to create a number of (negative) stampers (sons) by which a great number of commercial read-only disks are impressed. Read-only disks are used for the storage of video/audio programs, computer software, and data.

For a write-once disk, data can be written once. Data are recorded in the disk permanently and can be read, but it cannot be erased or revised. Data are written into the spiral track on the surface of a pre-grooved disk. The coded data are inputted to an optical modulator to control the intensity of a laser beam. The modulated laser beam heats materials and changes its characteristics by focusing on media materials. Then, it creates pits in the spiral track. Because the intensity of the laser beam in a player is weak, it does not destroy the pit in a write-once disk.

For an erasable disk, data can be written, read, erased, and rewritten or reversed. Erasable disks include magneto-optical and phase-change disks. The principle of writing data in an erasable disk is similar to that of write-once disk, but the written data can be erased. For example, for phase-change disks, a high-intensity laser beam is used to change the crystal state of materials to write or erase data. When data are written, high temperature caused by a high-intensity laser beam changes materials from a crystalline phase (translucent and yielding high reflectivity from the metal layer) to an amorphous phase (absorptive and yielding low reflectivity). When data are erased, the course is reversed.

Up to the end of 2021, three families of optical disks, e.g., CD, DVD, and Blu-ray disk (BD) have been widely used. Disks have three types within each family, namely, read only, write once, and rewritable. The principle of three families is similar, but laser wavelength, data density, and other parameters are different for each family. International standards for various optical disks have been formulated.

13.11.2 CD family and its audio formats

CD (*compact disc*) stores the data in a single-side single layer on the disk surface. The CD consists of a transparent plastic substrate. A CD is actually made of three layers. The pit track of data is impressed along the top surface of the substrate layer and covered with a very

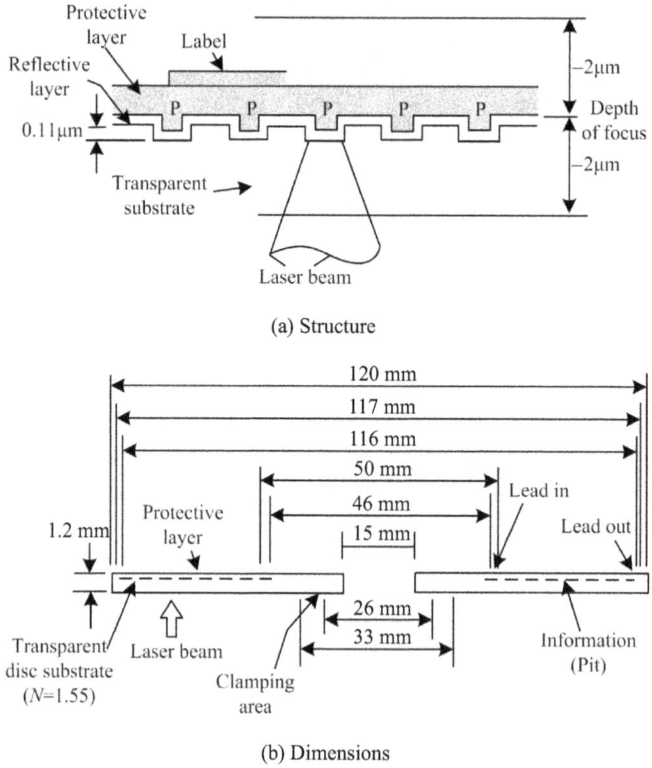

Figure 13.33 **Structure and dimensions of CD (a) Structure; (b) Dimensions (adapted from Pohlmann, 2011 and referred to IEC 60906, 1999).**

thin metal (reflecting) layer. The metallized pit surface is protected by another thin plastic layer above which the label is printed. Figure 13.33 (a) illustrates the structure of a CD and 13.33(b) is the dimensions of a common CD. The surface of a CD includes a lead-in area, a lead-out area, and an information area. Lead-in and lead-out areas contain the data for controlling the CD player. Some technical parameters of CD are listed in the second column of Table 13.1.

According to their functions, the CD family involves multiple members, such as CD-DA, Video CD (VCD), CD-ROM, CD-R, and CD-RW. The relation among these members is complex. The international standards for the CD family have been specified. Some contents related to the storage of spatial sound signals are outlined here, and the details are presented in literature book (IEC 60906, 1999; Pohlmann, 2011).

CD-DA is used to store music and other audio programs. Audio signal formats are as follows: two-channel stereophonic sound, linear PCM signals, 44.1 kHz sampling frequency, and 16-bit quantization. A CD-DA can store up to 74 min stereophonic programs. The bit rate is 1.41 Mbit/s. CD-DA possesses a spiral (physical) track that runs outward the disk. The spiral track can be divided into multiple sound tracks with variable lengths. A song or a movement of a symphony is usually stored in a sound track. A sound track involves multiple sectors, and each sector involves 98 frames. A frame is the smallest complete entity on a CD-DA. Figure 13.34 presents the following structure of a frame:

1. 24-bit (3 bytes) synchronous code, enabling the player to identify the beginning of each frame

Table 13.1 Some technical parameters of CD, DVD, and BD

	CD	*DVD (read-only)*	*BD*
Laser wavelength/nm	780 (infrared)	635/650 (red)	405 (blue)
Numerical aperture of objective lens	0.45	0.60	0.85
Laser spot diameter/μm	1.0	0.47	0.11
Disk diameter/mm	120*	120*	120*
Disk thickness/mm	1.2	1.2	1.2
Minimal pit length/μm	0.833	0.400 (single layer) 0.440 (dual layer)	0.149
Pit width/μm	0.6	0.3	
Pit depth/μm	0.11	0.16	
Track pitch/μm	1.6	0.74	0.32
Modulation code	EMF8/14+3	EMFPlus8/16	17PP
Error correction code	CIRC	RS-PC	Combination (Section 13.11.5)
User data rate/Mbit/s	1.41 (CD-DA) 1.23(CD-ROM, model 1)	10.08	48.0
Channel data rate/Mbit/s	4.3218	26.15625	
Capacity/GB	0.783 (CD-DA) 0.635 (CD-ROM, model 1)	4.7-17.0	25.0-54

* CD, DVD, and BD with a diameter of 80 mm are also available.

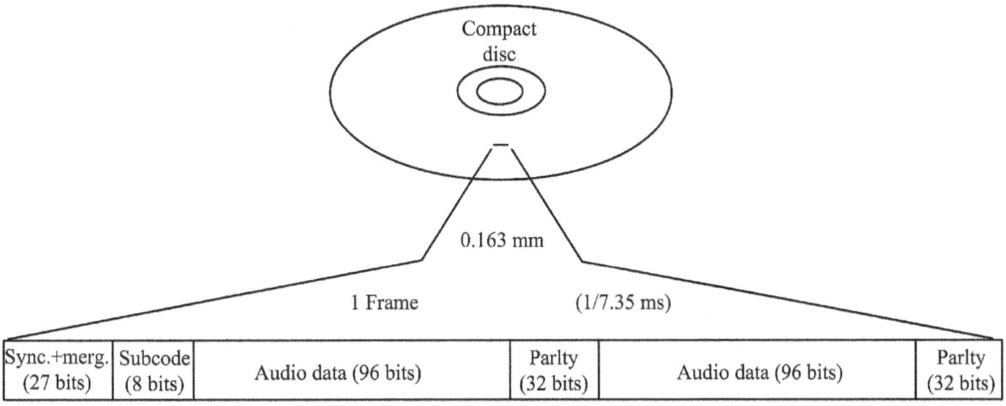

Figure 13.34 Structure of a frame in CD-DA(adapted from Pohlmann, 2011 and referred to IEC 60906, 1999).

2. 8-bit (1 byte) subcode, providing information of track numbers, beginning and ending of the track, index points, and other parameters
3. 192-bit (24 byte) audio data, including the data of two channels (left and right channel alternatively), 6 samples for each channel, and 16 bits for each sample. Successive audio data in the practical time frame are interleaved to scatter the burst error. Therefore, the samples in a data frame come from different time frames. The samples are de-interleaved in reproduction
4. 32-bit Q parity symbols and 32-bit P parity symbol are inserted in the middle and end of the frame. These parity symbols are used for error correction on the basis of cross-interleaved Reed–Solomon code (CIRC)

In addition, as stated in Section 13.2, coded data should be modulated so that they are appropriate for storage or transmission. An eight-to-fourteen (EMF) modulation is used in CD-DA, which maps each 8-bit subcode, audio data, andparity symbol into a 14-bit code. Therefore, EMF is an 8 to 14 modulation. For the reliable reading of data, three merging bits should be added between two adjacent codes. Therefore, each 8-bit data code is converted into a 17-bit channel code. Three merging bits should also be added after the synchronous code. Lastly, the information channel bit rate of CD-DA is 588 bits per frame or equally 4.32 Mbit/s, which is much higher than that of practical audio data.

Other members of the CD family are developed on the basis of CD-DA. They use different file formats, but their principle and data structure are similar to those of CD-DA. A video CD is used to store a video/audio program with a capacity of 650–740 MB. About 74 min MPEG-1 coding video/audio program data are stored in a video CD. The audio data are MPEG-1 Layer II coding stereophonic signals at a sampling frequency of 44.1 kHz. The video resolution is 240 × 352 pixels (NTSC format) or 288 × 352 pixels (PAL/SECAM format).

Compact disk read-only memory (CD-ROM) is used to store various electronic data, electronic publications, video/audio, and multimedia files. It has various formats. A mixed CD format can also be made in which some capacities of a CD are used for data with a CD-ROM format, and other capacities are used for data with a CD-DA format. CD-R and CD-RW are write-once and erasable disks, respectively. Data can be written into CD-R and CD-RW with CD-DA, video CD, or CD-ROM format.

13.11.3 DVD family and its audio formats

The principle of ***DVD*** (originally called digital video disk and renamed ***digital versatile disk***) is similar to that of CD (ISO/IEC 16488, 2002; Pohlmann, 2011). The appearance and dimensions of a DVD are identical to those of CD. A read-only DVD can be subdivided into the combinations of single/double-side and single/dual layer, e.g., (a) single-side single layer (DVD-5), (b) single-side duallayer (DVD-9), (c) double-side singlelayer (DVD-10), (d) double-side single/dual layer (DVD-14), and (e) double-side duallayer (DVD-18). Their capacities are 4.70,8.54, 9.40, 13.24, and 17.08 GB. All read-only DVD has a two-substrate structure, e.g., two 0.6 mm substrates are bonded together to form a 1.2 mm disk. Therein, a DVD-5 consists of a substrate with a single data layer and a blank substrate. A DVD-10 comprises two substrates, and each contains a single data layer. A substrate can also have two data layers. A DVD-9 is composed of a substrate with two data layers and a blank substrate. A DVD-18 consists of two substrates with each containing two data layers. Figure 13.35 illustrates the structure of a read-only DVD. In contrast to a CD, thedata layers in a DVD are placed near the internal interface of two substrates so that they are protected efficiently.

Data are recorded in a spiral data track on the data layer. For a substrate with two data layers, the layers are separated by a semi-reflective layer. The data in either layer can be read by focusing the laser beaming on it. Some technical parameters of read-only DVDs are listed in Table 13.1. A DVD player uses laser beam with a short wavelength and an objective lens with a large numerical aperture, which enables a smaller pit and a track pitch on the disk and thus an increase in data capacity in comparison with CD.

A DVD uses Reed–Solomon product code (RS-PC) for error correction through which two Reed–Solomon codes are combined as a product code. In addition, DVD uses EFM-Plus modulation. EFM-Plus is similar to EFM, but EMF-Plus uses two merging bitsand new lookup tables to convert each 8-bit data code into 16-bit channel code and thus improves the efficiency.

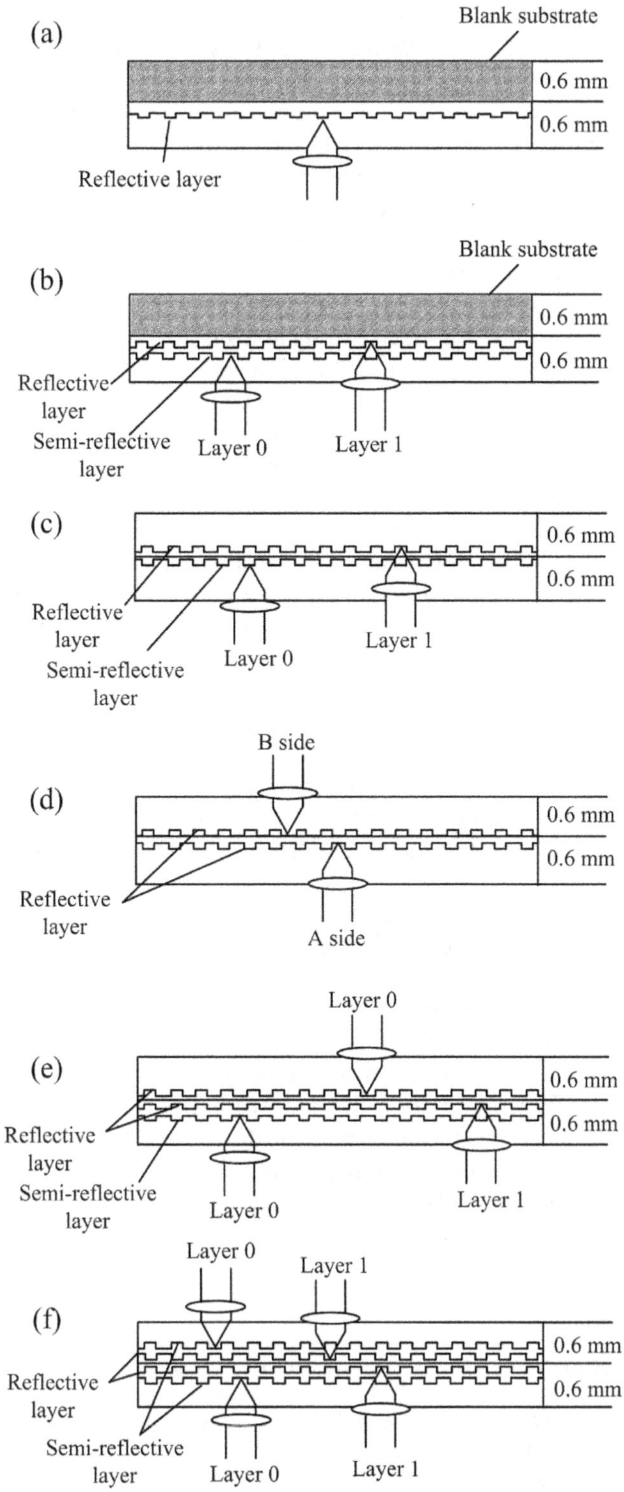

Figure 13.35 DVD disk types. (a) Single-side single layer, (b) single-side duallayer, (c) single-side duallayer (alternate format), (d) double-side singlelayer, (e) double-side single/dual layer, and (f) double-side duallayer (adapted from Pohlmann, 2011 and referred to ISO/IEC 16488, 2002).

According to their functions, a DVD family involves the members of DVD-ROM, DVD-Video, DVD-Audio, DVD-R, DVD-RW, and DVD-RAM. Therein, DVD-Video, DVD-Audio, DVD-ROM are read-only disks. They share the same specifications, physical formats, and file systems (the universal file format is termed universal disk` format (UDF) bridge). The file format and system of a DVD are different from those of CD. DVD-Video is utilized to store video/audio programs; DVD-Audio is used to store high-quality audio programs; DVD-ROM is used to store various electronic data, electronic publications, video/audio, and multimedia files. DVD-R and DVD-RW are write-once and erasable disks, respectively; DVD-RAM is random-access memory. The formats of DVD-R, DVD-RW, and DVD-RAM are different from read-only DVD.

DVD-Video utilizes MPEG-2 for video coding at a variable bit rate, with a maximal allowable bit rate of 9.8 Mbit/s (DVD-Forum, 1997). The video resolution is 720 × 480 pixels for NTSC format or 720 × 576 pixels for PAL format. DVD-Video supports multiple audio formats, including stereophonic and multichannel formats (Table 13.2). Moreover, DVD-Video allows up to eight independent audio signal streams and thus can involve at least two sound tracks with different audio formats. For the NTSC format, a Dolby Digital or a linear PCM sound track should be included. For a PAL format, a Dolby Digital, linear PCM, or MPEG-2 sound track should be included. DTS is an optional sound track. Other sound tracks, such as Sony SDDS, are also optional. For a linear PCM sound track, the maximal allowable audio bit rate is 6144 kbit/s = 6.144 Mbit/s. Therefore, the combinations of sampling frequency, quantization bit, and number of channels should be restricted so that the total bit rate does not exceed the allowable limit.

The maximal overall bit rate of video, audio, and subpicture (such as subtitles and explanation title) data for DVD-Video is 10.08 Mbit/s. Therefore, the combinations of the bit rate from all data should not exceed this limit. The playing time of a DVD-Video disk can be estimated from its capacity and overall bit rate. For example, at an average video/audio bit rate of 4.692 Mbit/s, the playing time of a DVD-5 disk with a capacity of 4.7 GB is about 133 min.

A DVD-Audio disk has two types (Fuchigami et al., 2000). One type only involves the main audio tracks, optional still pictures, real-time text, and visual menu. Another type involves the main audio tracks and optional video/audio. Video/audio tracks satisfy the specification of DVD-Video and therefore can be played by a DVD-Video player. DVD-Audio allows an audio bit rate up to 9.6 Mbit/s. DVD-Audio also supports multiple audio formats. The main formats are listed in Table 13.3. The linear PCM audio format is mandated, and other audio formats are optional. DVD-Audio also allows downmixing multichannel signals to stereophonic outputs. For a linear PCM sound track, the downmixed coefficients are provided by a content provider. For an MLP lossless coding soundtrack, two-channel stereophonic signals can be directly decoded from a substream (Section 13.8).

Table 13.2 Audio formats in DVD-Video

Audio format	*Sampling frequency/kHz*	*Quantization/bit*	*Number of channels*	*Bit rate/Mbit/s*
Linear PCM	48/96	16/20/24	1–8	Max 6.144
Dolby Digital	48	Max 24	2/5.1	Typical 0.384, Max 0.448
MPEG-1, Layer II	48	Max 20	2	Max 0.384
MPEG-2 BC	48	Max 20	Max 7.1	Max 0.912
DTS	48	Max 24	Max 6.1	Typical 0.768, Max 1.536

Table 13.3 Main audio formats and parameters of DVD-Audio

Audio format	*Sampling frequency/kHz*	*Quantization/bit*	*Number of channels*	*Bit rate/Mbit /s*
Linear PCM	192/176.4	16, 20, 24	2	Max 9.6
	96/88.2/48/44.1	16, 20, 24	1–6	
MLP	192/176.4	16, 20, 24	2	Max 9.6
	96/88.2/48/44.1	16, 20, 24	1–6	
Dolby Digital	48	16, 20, 24	1–6	
DTS	48/96	16, 20, 24	1–6	

The playing time of a DVD-Audio disk can be estimated from its capacity and bit rate. For a DVD-5 disk with a linear PCM, the following can be observed:

1. For two-channel signals with a sampling frequency of 192 kHz and 24-bit quantization, the playing time is 65 min.
2. For two-channel signals with a sampling frequency of 44.1 kHz and 16-bit quantization, the playing time is 422 min.
3. For six-channel signals with a sampling frequency of 48 kHz and 24-bit quantization, the playing time is 86 min.

13.11.4 SACD and its audio formats

Super audio CD (***SACD***) was developed by Sony and Philips in 1999. It is a ramification of DVD and used for the storage of high-quality and large-capacity music programs (Verbakel et al., 1998). The dimensions of SACD are identical to that of CD, with adiameter of 120 mm and a thickness of 1.2 mm. SACD has three types of structures, e.g., single-, double-, and hybrid-layer structures. A single-layer structure comprises a high-density layer only, with a data capacity of 4.7 GB. A double-layer structure is composed of two high-density layers, with a data capacity of 8.5 GB. A hybrid-layer structure contains a CD layer and a high-density layer. The capacity of the CD layer is 680 MB, which is used to store data in the CD-DA format. The capacity of the high-density layer is 4.7 GB. Audio data in the CD layer can be directly read by a CD-DA player, so it is compatible with CD-DA. The minimal pit length and track pitch in a high-density layer is 0.4 and 0.74 μm, respectively. Data in a high-density layer should be read by a laser beam with a wavelength of 650 nm (red), and the numerical aperture of the objective lens is 0.60. A practical SACD player is equipped with two laser pickups with wavelengths of 780 and 650 nm, respectively. The format and configuration of data in the high-density layer of a SACD are different from those of DVD.

SACD can store 6- (or 5.1-) channel signals and their two-channel downmixing. The following audio formats and techniques are used in SACD.

1. Audio signals are represented by 1-bit DSD samples (Section 13.3.2). The sampling frequency of DSD is 2.8224 MHz, which is 64 times of that for PCM sampling CD-DA (44.1 kHz).
2. A lossless coding algorithm termed direct stream transfer (DST) is used to compress the DSD data. A DST groups 37,632-bit data (corresponding to 1/75 s at a sampling frequency of 2.822 MHz) into a frame. The time domain correlation of signals is removed by linear prediction and entropy coding. Linear prediction is implemented by a FIR filter model, and entropy coding is implemented by an arithmetic coding scheme. For

different types of music signals, the DST yields a compression ratio from 2.4:1 to 2.7:1. A single high-density layer with a capacity of 4.7 G can store 74 min and eight-channel (six channels plus stereophonic downmixing) DSD coding signals. The practical data size in the high-density layer is much larger than that of DST data because data have been modulated, and error-correction codes have been added.

3. Reed–Solomon product code is used for error correction, and EFM-Plus modulation is utilized for channel coding.
4. For copyright protection, watermark is embedded into the substrate of SACD.

13.11.5 BD and its audio formats

***Blu-ray disk*(*BD*)** was developed to supersede the DVD. Its principle is similar to that of CD and DVD (ISO/IEC 30190, 2021; Pohlmann, 2011). The appearance and dimensions of a BD are identical to those of a CD or a DVD. BD can be classified into BD-ROM (read-only), BR-R (write-once), and BD-RE (rewritable). Their capacities are identical. The common BD structures include (a) single-side singlelayer (BD-25 or BD-27, with a capacity of 25.0 GB or 27.0 GB) and (b) single-side duallayer (BD-50 or BD-54, with a capacity of 50.0 GB or 54.0 GB). Multilayer BD has also been developed. Figure 13.36 illustrates the structure of a single-side BD. In contrast to CD and DVD, BD uses a substrate with a thickness of 1.1 mm, data layers are located on the side near the laser pickup. For a single-layer BD, a data layer is covered with a reflective layer and then a cover layer with a thickness of 0.1 mm. For a dual-layer BD, the inner data layer is covered with a reflective layer, and the outer data layer is covered with a semitransparent layer and a protective (transparent) layer with a thickness of 0.075 mm. Two data layers are separated by a 0.025 mm transparent separation layer. Because of the thin cover layer, the manufacture of BD is complicated.

In each data layer of BD, data are recorded in a spiral pit track and read outward from the inner radius. The fourth column in Table 13.3 lists some technical parameters of BD. In comparison with a DVD, a BD player uses a laser beam with a shorter wavelength and an

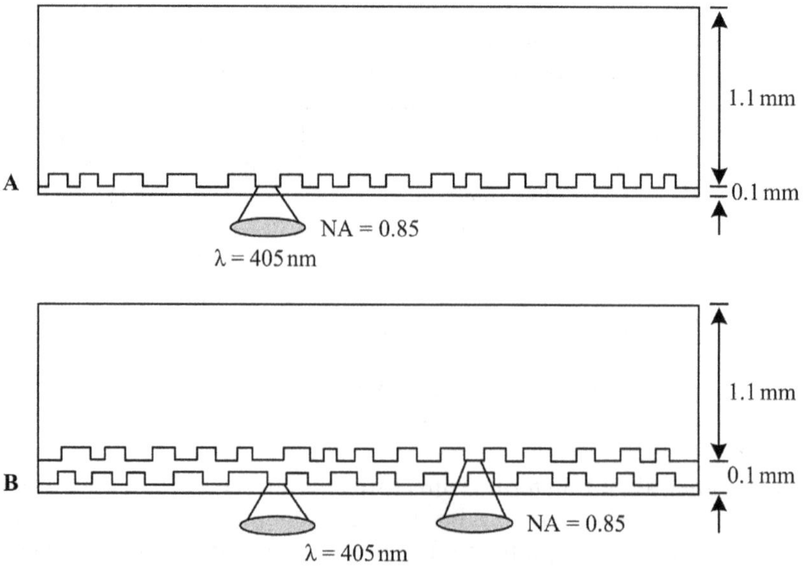

Figure 13.36 Structure of a single-side BD (A) Single layer and (B) dual layer(adapted from Pohlmann, 2011 and referred to ISO/IEC 30190, 2021).

Table 13.4 Main audio formats and parameters of BD

Audio format	*Sampling frequency/kHz*	*Quantization/bit*	*Number of channels*	*Bit rate/Mbit/s*
Linear PCM	48, 96 192	16, 20, 24 16, 20, 24	Max 8 Max 6	Max 27.648
Dolby Digital	48	16–24	Max 5.1	Max 0.640 (5.1-channel)
Dolby Digital Plus	48	16–24	7.1	Max 1.7
Dolby True HD	48, 96 192	16–24 16–24	Max 8 Max 6	Max 18.64
DTS	48	16, 20, 24	Max 5.1	Typical/max 1.524
DTS-HD	48, 96 192	16–24 16–24	8 6	Max 24.5

objective lens with a larger numerical aperture, the data density and capacity of a BD further increase.

Data on a BD, except for frame sync bits, are recorded using 1–7 parity preserve (PP/prohibit repeated minimum transition run length) modulation to convert data to channel codes. For error correction, it also combines a deep interleave with a long-distance code, a 64-kB Reed–Solomon code with a burst indicator.

BD allows a maximal bit rate of 48.0 Mbit/s. It is applicable to store a high-definition video program with a maximal resolution of 1920 × 1080 pixels, which are much higher than that of DVD-Video. BD supports multiple main video codes, such as MPEG-2 MP@ML, MP@HL/H1440L, MPEG-4 AVC (advanced video code), and SMPTE VC-1. The typical bit rates of high-definition video stream are 24.0 Mbit/s, 16.0 Mbit/s, and 18.0 Mbit/s, respectively. The maximal allowable bit rate of video stream is 40.0 Mbit/s.

BD supports up to 32 primary audio bitstreams and up to 32 secondary audio bitstreams. Primary audio is for the main sound tracks (such as sound tracks in a movie). Various coding formats are applicable to primary audio. Table 13.4 illustrates the parameters of these formats. PCM, Dolby Digital, and DTS formats are mandated, and other formats are optional. Secondary audio bitstreams are for commentary narration. Different coding formats are also available to secondary audio bitstreams, but the bit rate and audio quality are reduced in comparison with primary audio bitstreams.

The playing time of a BD depends on its capacity, content, and coding method. For example, BD-25 with 25 GB, playing time is about 2.3 h for an MPEG-2 video program with multiple sound tracks (bit rate of 24 Mbit/s); for a two-channel PCM program at 96 kHz sampling frequency and 24-bit quantization only (bit rate of 4.608 Mbit/s), playing time is about 12.1 h; for a Dolby Digital 5.1-channel program (bit rate of 0.640 Mbit/s), playing time is about 86.8 h. For a BD-50 with 50.0 GB, the playing time is about twice that of a BD-25.

13.12 DIGITAL RADIO AND TELEVISION BROADCASTING

13.12.1 Outline of digital radio and television broadcasting

Radio broadcasting is an important means for audio signal transmission. Limited by its performance (such as audio bandwidth), the first generation long-, medium-, and short-wave AM radio (analog) broadcasting is inappropriate for high-quality audio transmission. Second-generation FM radio broadcasting (Section 13.1.3) possesses better audio quality. It has been used as analog stereophonic broadcasting format for several decades. However, FM radio broadcasting is inefficient in anti-multipath interference, especially for receiving in a moving car, complicated buildings in city and montanic environment. Therefore, further improving the audio quality of a FM stereophonic broadcasting and extending it to multichannel sound broadcasting are difficult.

As a third-generation radio broadcasting technique, digital radio broadcasting combines the techniques of source compression and coding, channel coding, error correction, and digital modulation. Digital radio broadcasting efficiently utilizes the bandwidth of broadcasting and improves the audio quality. By using the principle of multiplexing, digital radio broadcasting can transmit multimedia information for various data and information services. Signals of digital radio (audio) broadcasting can be transmitted in different ways, including terrestrial, cable, and satellite transmissions. Various digital radio broadcasting formats and systems have been developed. Typical examples are DAB, digital AM broadcasting, and in-band on-channel DAB. In addition, some digital radio broadcasting techniques for satellite are used. Digital radio broadcasting usually supports a two-channel stereophonic sound format, and new-generation techniques may support multichannel sound formats.

Since the end of the 1980s, digital television has been developed greatly and has taken over the analog television. It can be implemented by terrestrial (wireless), cable, satellite, and mobile transmissions. It also supports stereophonic and multichannel audio.

Digital radio broadcasting and digital television have no universal standards. Various techniques and standards are used in different countries. Source coding, channel coding, and modulation used in these techniques and standards are different. However, these techniques are not described in detail in this paper. The basic principles of some typical digital radio broadcasting technique and audio techniques in digital television are briefly outlined (AES Staff Writer, 2004).

13.12.2 Eureka-147 digital audio broadcasting

Digital audio broadcasting (***DAB***) was a digital radio technique under the project of Eureka-147 in the 1980s. Since it became the European standard of digital radio broadcasting in 1995, it has been used in European and other countries (Kozamernik, 1995). DAB is appropriate for terrestrial, cable, and satellite transmissions and suitable for fixed and mobile receivers. For mobile receivers, DAB can operate at an arbitrary frequency of 30 MHz–3 GHz. It has an overall bandwidth of 1.536 MHz and allows an approximate bit rate of 1.5 Mbit/s. The bit rate of DAB ranges from 8 kbit/s to 192 kbit/s per channel. The bandwidth of DAB allows for five to six high-quality stereophonic data streams or up to 20 restricted quality mono data streams.

A DAB-transmitted signal consists of a multiplex of several independent audio and other data, which simultaneously allow several digital services. Figure 13.37 illustrates the block

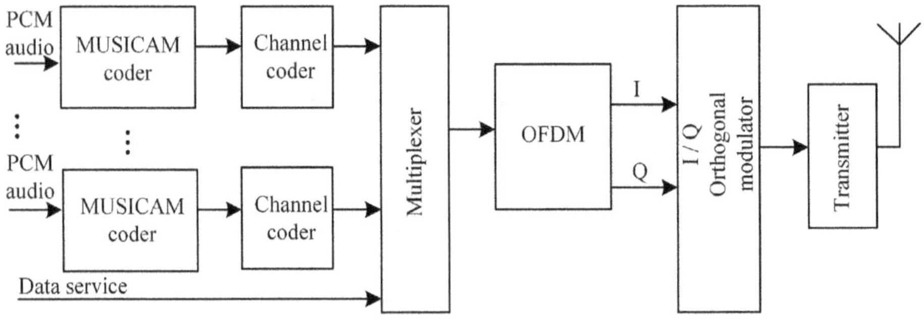

Figure 13.37 Block diagram of the transmission system of Eureka-147 DAB.

diagram of the transmission system of DAB. Each PCM audio input is coded independently in the source coding stage. The MUSICAM (e.g., the MPEG-1 Layer II in Section13.5.1) is used for audio source coding. Each source-coded audio data stream is further coded by a channel coder (and interleaved in the time domain). Multiple channel-coded audio data and other services are multiplexed (and interleaved in the frequency domain). An orthogonal frequency division multiplexing (OFDM) is applied through which the multiplexed bit stream is divided into a large number of bit streams with a low bit rate, and each low-bit-rate stream is used to modulate individual orthogonal subcarriers. The modulated carriers are transposed to appropriate radio frequency band and then transmitted.

OFDM in DAB involves a larger number of subcarriers. If the overall bandwidth of all subcarriers is Δ_W and the frequency interval of two adjacent subcarriers is Δf, the number of subcarriers is evaluated as

$$L = \frac{\Delta_W}{\Delta f}. \tag{13.12.1}$$

The effective symbol duration is given as

$$T_u = \frac{1}{\Delta f}. \tag{13.12.2}$$

A guard interval T_g is inserted between successive symbols to avoid the inter-symbol interference caused by multipath propagation. Itis usually chosen as one-fourth of effective symbol duration. If the time difference among multipath propagation does not exceed T_g, multipath signals are constructively interfered and enhanced in reception. Therefore, the practical duration of a symbol is given as

$$T_s = T_u + T_g. \tag{13.12.3}$$

According to the above analysis, a set of L subcarriers with uniform frequency interval can be expressed as

$$g_l(t) = \begin{cases} \exp(j2\pi f_l t) & 0 \le t \le T_s \\ 0 & \text{Other} \end{cases} \qquad f_l = f_0 + \frac{l}{T_s} \qquad l = 0, 1, \ldots (L-1). \tag{13.12.4}$$

Subcarriers satisfy the following orthogonality:

$$\int_0^{T_u} g_{l'}^*(t) g_l(t) dt = \begin{cases} T_s & l' = l \\ 0 & l' \neq l \end{cases}, \tag{13.12.5}$$

where the superscript "*" denotes complex conjugation. When L subcarriers are modulated by complex symbols $C_l(n) = \pm 1 \pm j$, $l = 0, 1\ldots (L-1)$, where n is discrete time, each subcarrier carries 2-bit information, and each symbols occupies 2 Lbits. The output of OFDM is given as

$$y(t) = \sum_{l=0}^{L-1} \sum_{n=-\infty}^{+\infty} C_l(n) g_l(t - nT_s). \tag{13.12.6}$$

The European standard of DAB specifies four transmission modes, with a maximal operating frequency of 375 MHz, 1.5 GHz, 3 GHz, and 1.5 GHz, respectively. The frequency intervals Δf of two adjacent subcarriers are 1, 4, 8, and 2 kHz, respectively. In all cases, the overall bandwidth Δ_W of all subcarriers is 1.536 MHz.

Practical DAB has been evolved to ***digital multimedia broadcasting*** (***DMB***). In addition to audio signals, DMB transmits various data, information, and real-time video signals. DMB signals can be received by various terminals, such as personal computers and mobile phones. The audio coding technique for DAB is further developed. The new-generation DAB+ uses MPEG-4 HE-AAC v2 (Section 13.5.4) for audio coding. In addition to improving audio quality, new-generation DAB may enable multichannel sound broadcasting.

13.12.3 Digital radio mondiale

Digital radio mondiale (DRM30) was a digital audio broadcasting technique below the frequency band of 30 MHz (ETSI ES 201 980 V3.2.1, 2012). DRM+ is extended to a frequency band from30 MHz to 300 MHz. The transmission part of DRM involves source coding, multiplexer, channel coding, and ODFM modulator. Receiving is an inverse course of transmission.

The bandwidth of DRM is 9 or 10 kHz identical to that of analog AM broadcasting. It can be extended to 18 or 20 kHz. Because the bandwidth is strictly restricted, high-efficient source coding is required to retain audio quality. DRM uses MPEG-4 (Section 13.5.4) as source coding. DRM offers three source coding methods. MPEG-4 AAC is mainly for mono and stereophonic music coding and achieves a relatively better audio quality. Sampling frequencies are 12 and 24 kHz for robustness modes A, B, C, and D and 24 and 48 kHz for robustness mode E. The granularity of the AAC bit rate is 20 bit/s for robustness modes A, B, C, and D and 80 bit/s for robustness mode E. MPEG-4 CELP is for mono speech coding with a relatively high quality. Sampling frequency is 8 or 16 kHz, and the bit rate ranges from 4 kbit/s to 20 kbit/s. MPEG-4 HVXC is for mono speech coding with a reasonable quality. The sampling frequency is 8 kHz, and the bit rate is 2 or 4 kbit/s. SBR is also used in coding.

DRM retains the existing bandwidth and frequency planning for analog AM broadcasting. After some revisions, the existing AM transmitter can be used for DRM, which facilitates a transition from analog to digital broadcasting.

13.12.4 In-band on-channel digital audio broadcasting

In-band on-channel (***IBOC***) is a hybrid method of simultaneously transmitting digital radio and analog radio broadcast signals on the same frequency band (AES Staff Technical Writer, 2006b). The HD Radio is an IBOC system developed byiBiquity Digital Corporation in the USA for the frequency range of medium-wave AM broadcasting and FM broadcasting. It has been standardized by The Federal Communications Commission (FCC) as the standard of digital radio and used in the USA.

HD Radio involves AM and FM HD Radios. AM and FM HD Radios use the bandwidth of 30 and 400 kHz, respectively. The source coding method is MPEG-4 HE-AAC v2. Viterbi-punctured convolutional coding is used for error-correction coding. Data are also interleaved in time and frequency domains. HD Radio also uses OFDM method for transmission, but the bandwidth and interval of carrier and the method of modulation are different. HD Radio does not alter the existing frequency planning for radio broadcasting and thus facilitates a transition from analog to digital broadcasting. However, it requires a larger transmission bandwidth.

13.12.5 Audio for digital television

Digital television (***DTV***) is a system that uses digital techniques to record, product, transmit, and receive video (and audio) signals. Digital television includes ***high-definition television*** (HDTV), ***standard definition television*** (***SDTV***), and ***interactive television*** (***ITV***). In particular, HDTV provides a video resolution of1280 × 720 pixels or 1920 × 1080pixels, and SDTV provides a video resolution similar to that of DVD. DTV supports digital stereophonic and multichannel sound formats to improve audio quality. As stated in Sections 1.9.3 and 5.1, the development of 5.1-channel sound for domestic use is closely related to HDTV sound. The recent multichannel spatial surround sound is also associated with an ultra-high-definition video system. The development of next-generations ultra-high-definition DTV and 3D DTV shown potential for the application of multichannel spatial surround sound.

DTV signals can be transmitted by different ways, including terrestrial, cable, satellite, and mobile communication transmission systems. However, no united standards are available for DTV or HDTV. Techniques for various transmission systems are different. Even for the same transmission system, different formats and standards are used in different countries. Therefore, transmission techniques for DTV are complicated. For terrestrial transmission, examples include the ATSC standard by Advanced Television System Committee in the USA, DVB in European countries, Digital terrestrial multimedia broadcast (DTMB) in Japan, and DTMB in China. These four standards differ in transmission techniques. They all support 5.1-channel sound, but they use various source coding schemes. ATSC, DVB, ISDB, and DTMB use Dolby Digital (AC-3), MPEG-2 Layer II, MPEG-2 AAC, and DRA (GB/T 26686, 2017), respectively.

13.13 AUDIO STORAGE AND TRANSMISSION BY PERSONAL COMPUTER

Since the mid-1980s and beginning of the 1990s, computer techniques have been developed quickly. Currently, a personal computer can comprehensively process the multimedia information of audio, video, still picture, text, and other data. It is a powerful tool for audio processing and production and provides an important means for audio storage and transmission. Computer-based audio processing, storage, and transmission are important tendencies.

In a Windows platform, most video/audio data are stored as a resource interchange file format (RIFF), such as WAV file for audio data. A RIFF file consists of chunks. In addition to core audio data, a RIFF audio data file includes the information of data, such as contents, sampling frequency, quantization bit, number of channels, and coding method of audio data. According to this information, the stream of audio data can be obtained in playback.

Magnetic hard disk, flash memory, and optical disk for a personal computer can be used to store audio files. A computer can be easily used to copy an audio file. Some measures may be taken to protect the copyright and prevent a copy of audio files. Various interfaces enable to exchange audio files between computers or other types of digital audio equipment. Flash memory is also the main means for storing video/audio files in various mobile devices.

Internet is an important means for various data and file transmission. Different video/audio files can be downloaded from the Internet and then played back. As the bandwidth of Internet increases, interactive streaming video/audio media for playing directly become a new means of transmission (Rumsey, 2017), various streaming video and audio services are popular on the internet. As a result, the boundary between media and broadcasting becomes blurred. However, the packet-switching transmission method of the internet may cause problems. Ideally, packets of audio data should be received in sequence. In practice, some packers may be missing and should be re-transmitted, resulting in a delay. This problem is not serious for downloading a file, but it may cause interruptions in direct playback signals. The high-efficient method of data compression and data buffer in receivers alleviate such kind of interruption.

13.14 SUMMARY

Storage and transmission are important stages in the system chain of spatial sound, by which spatial sound signals are delivered to the users. Before the end of the 1970s, spatial sound signals were primary storage and transmission by analog techniques. The 45°/45° disk recording system, analog magnetic tape audio recorder, and analog stereo broadcasting played important roles in the popularization of stereophonic sound. The GE Zenith FM stereophonic broadcasting system was still used by the end of 2021.

Digital audio techniques often promise better electroacoustic or perceptual performance. It facilitates the storage, transmission, copy, and processing of multichannel audio signals. It is adapted to modern information media, such as computers and Internet. Since the 1980s, especially after the 1990s, digital audio technique has been developing quickly. Digital audio storage and transmission techniques have been basically taken over traditional analog techniques.

In digital audio, a signal is discretized in time and amplitude and approximately represented by binary codes. Converting an analog signal into a digital signal includes three steps, e.g., sampling, quantization, and coding. Sampling is conducted to represent a continuous time signal by its values at discrete times. Sampling frequency is determined by Shannon–Nyquist sampling theorem. Quantization is performed to represent the amplitude of a discrete time signal with a finite number of discrete values. Mapping the continuous amplitude of a signal to finite discrete values leads to quantization noise. The signal-to-quantization noise ratio increases as the quantization bit increases. Quantization noise shapingimproves thesignal-to-quantization noise ratio.

The bit rate of digital audio signals increases as the sampling frequency, quantization bit, and number of channels increase. Prior to storage and transmission, digital audio signals are often compressed and coded to reduce the bit rate and adapt to the bandwidth or capacity of practical media. Various audio compression and coding methods have been developed. From the point of recovery of original signals (data), audio signal compression and coding can be divided into lossless coding and lossy coding. From the point of algorithm, audio coding can be divided into waveform coding, parameter coding, and hybrid coding. From the point of physical and auditory mechanisms, the spatial sound signal coding utilizes the redundancy among the signals, including the redundancy in time, frequency, spatial, and perceptual domains. Therefore, audio compression and coding remove redundancy within signals and reduce the bit rate. Various high-efficient coding techniques can be developed on the basis of masking effect and corresponding psychoacoustic models (redundancy in the perceptual domain).

Based on the principle of removing information redundancy, various audio coding techniques, including MPEG technique series and standards, Dolby technique series, DTS technique series, and MLP lossless coding technique, have been developed.

Optical disk is a common media for digital audio storage. According to their principle and development order, optical disk includes CD, DVD, and BD families. Various signal and recording formats within each optical disk family are available.

DAB is also a means for stereophonic or even multichannel signal transmission. It can improve the perceived audio quality. Some DAB techniques have been developed. Audio for digital television also supports stereophonic and multichannel sound formats.

Computers provide a new method for digital audio processing and storage. The Internet and stream media are new means for digital audio transmission.

Digital storage and transmission are being developed quickly. New studies on this field should be further explored.

Chapter 14

Acoustic conditions and requirements for the subjective assessment and monitoring of spatial sound

The basic principles of spatial sound are addressed in the preceding chapters. In practice, the acoustic conditions of reproduction, such as the characteristics of a listening room, the characteristics, and arrangement of loudspeakers, and their reproduction level, influence final perceived performance remarkably. To achieve the desired reproduction performance, acoustic conditions for reproduction should be appropriately designed according to physical and psychoacoustic principles. In particular, subjective assessment experiments or monitoring in the program production of spatial sound should be conducted under strict and standardized acoustic conditions to obtain consistent and reliable results. Therefore, acoustic conditions and designs for spatial sound reproduction should be specified.

Spatial sound can be reproduced by loudspeakers or headphones. Acoustic considerations and conditions for the subjective assessment and monitoring of spatial sound with loudspeakers are mainly addressed in this chapter. The case of spatial sounds intended for domestic reproduction is discussed in Sections 14.1–14.5. Therein, an outline and introduction are presented in Section 14.1. The acoustic considerations and design of a listening room are discussed in Section 14.2. The arrangements and characteristics of loudspeakers are discussed in Section 14.3. The problems of signal and listening level alignment are addressed in Section 14.4. Some related standards and guidance are introduced in Section 14.5. The methods of headphones and virtual monitoring are outlined in Section 14.6. Lastly, acoustic conditions and standards for spatial sound intended for cinema reproduction are presented in Section 14.7.

14.1 OUTLINE OF ACOUSTIC CONDITIONS AND REQUIREMENTS FOR SPATIAL SOUND INTENDED FOR DOMESTIC REPRODUCTION

Cinema and domestic reproduction are two traditional applications of spatial sound. Many spatial sound techniques and systems (such as a 5.1-channel system) developed after the 1980s are applicable to cinema and domestic reproduction. However, the number of listeners (and therefore the size of a listening region) and the required acoustic conditions for reproduction differ in two cases (Holman, 1991). Accordingly, the acoustic conditions for subjective assessment experiments and monitoring are different.

Acoustic conditions for the subjective assessment and monitoring of spatial sound intended for domestic reproduction are discussed here. Numerous works have been performed on this issue, and some standards and guidance have been specified. These standards and guidance are intended for the critical comparison of program materials and subjective experiments. The acoustic conditions in these standards and guidance allow the neutral and critical monitoring of sound reproduction to reveal the characteristics and deficiencies of program materials or systems. Moreover, these acoustic conditions not only lead to consistent and comparable

DOI: 10.1201/9781003081500-14

results in subjective experiments or monitoring but also positively affect the reproduction of high-quality program materials with an unimpaired system.

Usually, the acoustic conditions specified by the standards and guidance cannot be completely satisfactory for monitoring in practical program mixing and production. However, these standards and guidance provide a reference or baseline in practice. The practical conditions in domestic reproduction vary greatly and may differ considerably from those in the standard and guidance. Moreover, conditions and environments in domestic reproduction may be modified to enhance some perceived effects. Nevertheless, the standards and guidance still provide a reference for domestic reproduction at least in avoiding audible artifacts caused by inappropriate acoustic designs.

Most existing standards and guidance are originally intended for two-channel stereophonic and multichannel horizontal surround sound (such as 5.1-channel sound) because these two types of techniques are mature. The following sections in this chapter mainly focus on the acoustic conditions for two-channel and 5.1-channel reproduction. These acoustic conditions are incompletely appropriate for multichannel spatial surround sound although some basic principles and methods are similar. The acoustic conditions for multichannel spatial surround sound reproduction are considered in some updated standards, but further studies on this issue are needed.

The acoustic conditions and environment for cinema sound reproduction and monitoring differ considerably from those of domestic reproduction (Section 14.7).

14.2 ACOUSTIC CONSIDERATION AND DESIGN OF LISTENING ROOMS

Loudspeaker-based spatial sounds are usually reproduced in a room environment. The binaural pressures received by a listener are the superposition of those caused by the direct sound from all loudspeakers and reflections in the room. The resultant auditory events and sensations are determined by the superposed binaural pressures. As a stage of the system chain of spatial sound, the acoustic characteristics of a listening room greatly influence the final perceived effect. Therefore, the influences of listening rooms on the reproduced sound field and subjective perception should be considered (Rumsey, 2001).

For spatial sound reproduction intended to recreate the information of the original or target sound field (such as the information of a sound field in a concert hall), reflections in a listening room impair the target information to be reproduced. In this case, one of the purposes of the acoustic design of a listening room is to reduce the negative influences of its reflections. In practical use, however, listening rooms may sometimes serve as a part of the reproduction system, and the reflections of listening rooms are appropriately designed and utilized to create some desired perceived effects.

If reducing the influence of listening room reflections is considered only, an anechoic chamber is an ideal environment for spatial sound reproduction. Indeed, some scientific experiments on physical sound field reconstruction are conducted in anechoic chambers. In most cases, however, an anechoic chamber is inappropriate for the subjective assessment and monitoring of spatial sound reproduction because practical reproduction is conducted in a room with reflections rather than in an anechoic chamber. Therefore, the subjective results obtained in an anechoic chamber differ considerably from those of practical reproduction. Moreover, listening in an anechoic chamber for a long time may cause auditory fatigue and uneasiness to listeners. As a result, the subjective assessment and monitoring of spatial sound are usually conducted in a listening room (or studio) with some room reflections. The key is to design a listening room properly to reduce the influence of listening room reflections to some extent.

Various acoustic conditions are required for listening rooms with different purposes. The conditions of a ***reference listening room*** for subjective assessment experiments and program comparisons are the most critical. The conditions of a ***controlroom orastudio for mixing and monitoring***, especially for some small studios, can be appropriately compromised. For rooms in domestic reproduction, some simple acoustic treatments may improve the reproduction effect. Moreover, the considerations and designs of listening rooms for stereophonic and multichannel sound reproduction vary and depend on situation.

Despite the differences in the design of listening rooms for different purposes, a general rule is that early reflections arriving at the listening position within 15 or 20 ms after the direct sound from loudspeakers should be restrained. Although these early reflections may not create separate spatial auditory events due to the precedence effect (Sections 1.7.2 and 1.8), they may cause timbre coloration and change other spatial perceived attributes in reproduction. Therefore, the negative effect of early reflections in a listening room should be restrained. This result has also been validated by some psychoacoustic experiments on the perception of listening room reflections (Bech, 1995; Olive and Toole, 1989). The perceived threshold of listening room reflections depends on the characteristics of stimuli, arrival directions, and relative delay of direct and reflected sounds. Olive and Toole conducted a psychoacoustic experiment and indicated thatthe perceived thresholds of lateral reflections within 20 ms after the direct sound are about −15 and −20 dB for speech and music stimuli, respectively, with respect to the direct sound. Therefore, related standards and guidance specify the conditions of the early reflections and reverberation of a reference listening room.

Sound absorption treatment is a common method for restraining early reflections in a listening room. An appropriate sound absorption treatment effectively retrains early reflections in a reference listening room. By contrast, in a control room orstudio for mixing and production, reflections from the mixing console and the control room window are difficult to be avoided completely. In this case, the condition of restraining the early reflections can be properly relaxed.

A kind of live-end–dead-end (***LEDE***) design is popular among control roomsorstudios for stereophonic mixing and monitoring (Davis and Davis, 1980). In a LEDE design, the region around the front of a room is treated with strong sound absorption to restrain the early reflections from the front, side, and ceiling. The rear region of the room is treated with more reflection, e.g., some diffusers are used to create later diffuse reverberations around listening positions. Here, the reflections of a listening room areutilized to create some of the desired perceived effects.

Another method for restraining the early reflections caused by a wall behind a loudspeaker is mounting a loudspeaker flush with a wall. A rigid wall can be approximated as an infinitely extended and flat baffle. According to the acoustic principle of image sources in Section 1.2.2, the reflections from a wall and the direct sound from loudspeakers are superposed in phase at the receiver position, and the frequency-dependent interference between direct and reflected sounds vanishes. This method also increases the radiation efficiency, resulting in a 6 dB noticeable increase in the radiated pressure level at a low frequency. In control rooms and studios for stereophonic mixing and monitoring, two loudspeakers are often flush mounted on the two sides of the control room window. To restrain early reflections arriving at the listening region, some studies have also suggested a special reflective treatment of the areas around loudspeakers so that early reflections are directed away from the listening position and toward the rear of listening rooms (Walker, 1994).

A listening room for multichannel sound differs from that for stereophonic sound. Moreover, the conditions of listening rooms for different multichannel sound are different.

First, the shape of a listening room is considered. A long listening room is applicable to stereophonic reproduction. However, a wide listening room is more appropriate for horizontal 5.1-channel sound reproduction because two surround loudspeakers are arranged in side-rear positions. For horizontal 6.1- and 7.1-channel sound reproduction, a nearly square listening room may be more appropriate (if the distribution of a room mode is not considered) because rear surround loudspeakers are added. For spatial surround sound with more channels, situations are more complex. Second, the treatment of reflections in listening rooms for multichannel sound differs from that for stereophonic sound. Multichannel sound has multiple loudspeakers at different directions. Sounds from different loudspeakers are initially reflected by various walls. For surround loudspeakers at the side-rear position (such as those in 5.1-channel sound), the influence of reflections from an opposite wall becomes more obvious. For rear surround loudspeakers, the influence of reflections from the front wall or window of the control room becomes obvious. Lastly, flush-mounted loudspeakers with a rigid wall also cause a problem because the rigid wall reflects the sound from loudspeakers at the opposite direction.

Therefore, if all channels in a multichannel sound are regarded as equivalent, acoustic treatment (distribution of absorbing and diffusing treatments) in listening rooms should be even to provide a similar acoustic environment for all loudspeakers. Alternatively, if surround channels are only used for ambient sound, the sound absorption treatment is less critical for surround loudspeakers. Moreover, the diffused treatment of listening room reflections for radiation from surround loudspeakers (especially dipole surround loudspeakers) can recreate a late-diffused reverberation field in reproduction. The necessity of such a diffused treatment depends on the target content in surround channels.

Another problem is the interaction between the low-frequency radiations of loudspeakers and room modes. In Section 1.2.2, in enclosed spaces, such as rooms with multiple reflective surfaces, the interference of direct sound and reflections leads to a series of standing wave modes. In the case of low frequencies and small rooms, the pressure at a receiver position caused by a loudspeaker is closely related to room modes. According to wave acoustic theory (Morse and Ingrad, 1968; Kuttruff, 2009), if a loudspeaker is arranged at a position of the node (minimum pressure) of a room mode, it cannot effectively excite a particular mode. Conversely, if a loudspeaker is arranged at a position of the antinode (maximum pressure) of a room, it effectively excites this mode. The pressure at a receiver position depends on the position of nodes and antinodes. In particular, for a usual rectangular room, the center of a room coincides with the node of a series of asymmetric modes.

Overall, the interaction between loudspeaker radiation and room modes alters the low-frequency responses of loudspeakers, which depend on loudspeaker and receiver positions. The shape and dimensions of listening rooms should be appropriately chosen to alleviate the problems caused by room modes. An irrational ratio among the dimensions of a rectangular room can alleviate the uneven room response caused by the degeneracy of room modes. Some international standards and guidance have specified the radios among the dimensions of a listening room. The sound absorption treatment at low frequencies can also alleviate the influence of room modes. However, at a very low frequency, such a treatment is difficult or costly. In addition, the influence of room modes should be considered when choosing the arrangement of loudspeakers, especially subwoofers, in a listening room. This problem is addressed in Section 14.3.3.

The background noise in a listening room influences auditory perception in reproduction. Some measures should be taken to reduce background noise. An appropriate sound insulation treatment on listening rooms is needed to reduce the influence of exterior noise. Interior noises, such as those caused by the air-conditioning system and fan in a computer, should also be reduced.

14.3 ARRANGEMENT AND CHARACTERISTICS OF LOUDSPEAKERS

14.3.1 Arrangement of the main loudspeakers in listening rooms

Two common methods, namely, free-standing and flush-mounted arrangement, are used to arrange horizontal loudspeakers in a listening room. In ***free-standing arrangement***, loudspeakers are supported by stands in a listening room. However, this arrangement causes reflected notches in the low-frequency response of loudspeakers. At low frequencies, the radiation of a loudspeaker is diffracted to the back of the loudspeaker and then reflected by the wall behind the loudspeaker. Interference between the direct sound from the loudspeaker and the reflection from the wall occurs at the receiver position. If the straight line connecting the loudspeaker and the receiver position is perpendicular to the wall behind the loudspeaker, destructive interference between the direct and reflected sound maximizes and causes a notch when the distance between the loudspeaker and wall is equal to one-fourth of the wavelength. The depth of the reflected notch depends on the reflection coefficient of the wall, the directivity of loudspeakers, and the distance between a loudspeaker and a wall. The frequency of a reflected notch is related to the distance X_0 between a loudspeaker and a wall, as expressed in the following equation:

$$f_{notch} = \frac{c}{4X_0}, \tag{14.3.1}$$

where c is the speed of sound. For X_0 = 0.1, 0.5, 1.0, 2.0, and 3.0 m, Equation (14.3.1) yields f_{notch} = 858, 172, 86, 43, and 29 Hz, respectively.

An increase in the sound absorption of the wall reduces the depth of the reflected notch and thus alleviates the influence of reflections. However, applying a sound absorption treatment at low frequencies is practically difficult. Because f_{notch} is inversely proportional to X_0, an increase in X_0 can reduce f_{notch} and the depth of the reflected notch. However, a large X_0 and accordingly a large listening room, which are often infeasible, are required to reduce f_{notch} to an extent such that the influence of the reflectednotch on auditory perception is negligible. For example, a distance of X_0 = 1.7 m is necessary to reduce f_{notch} to 50 Hz. An opposite method can be used for a loudspeaker system with a large box, where a small distance X_0 and a relatively high f_{notch} are allowed. Because of the shadow effect of the loudspeaker box, the radiation in the rear of loudspeakers is attenuated above a certain frequency; therefore, the depth of the reflectednotch is reduced. However, when a loudspeaker is arranged near a wall, the mirror reflection of the wall increases the radiated pressure at a low frequency. In this case, equalization to the loudspeaker signal at low frequencies is needed. Bass management (Section 14.3.3) is also applicable to avoid the problem of reflected notch. In this method, low-frequency components below 80–120 Hz of all the main channels are reproduced by a subwoofer, and the main loudspeakers only reproduce the components above 80–120 Hz. Through this method, the low-frequency limit of the main loudspeakers is above the frequency of the reflected notch.

In ***flush-mounted arrangement***, the reflected notch caused by the wall behind loudspeakers can be completely avoided. In Section 14.2, the flush-mounted arrangement of loudspeakers is often used in control rooms and studios for stereophonic mixing and production, but it is somewhat difficult to be used for multichannel sound reproduction because the rigid wall for a flush-mounted loudspeaker reflects the sound from loudspeakers at opposite directions.

For the critical comparison of program materials and subjective experiments, loudspeakers should be arranged in a listening room in accordance with the related standards. For example, according to the ITU-R BS 775-3 (2012) standard, the acoustic center of frontal loudspeakers should be ideally at a height approximately equal to that of a listener's ears. The height of side/rear loudspeakers is less critical. Ideally, five loudspeakers with a full bandwidth should be arranged at an equal distance to the listener. If frontal loudspeakers are arranged in a straight-line base, a delay to the center channel signal is needed to compensate for the path difference of the center loudspeaker. The loudspeaker arrangement for domestic reproduction is not as critical as that for subjective experiments. The guidance of Dolby Laboratories allows a relatively flexible arrangement of loudspeakers in 5.1-channel reproduction (Dolby Laboratories, 1997, 2000). Strictly, a two-way loudspeaker system has better to be arranged vertically rather than horizontally to avoid the influence of the side lobe of loudspeaker radiation when a mixing engineer slightly deviates from the central listening position.

The arrangement of a center loudspeaker in reproduction with accompanying picture causes a problem. If an acoustically transparent projection screen is used for visual display, the center loudspeaker can be arranged behind the screen. However, if a flat panel display is used for visual display, the position of the flat panel display comes in conflict with that of the center loudspeaker. In this case, the center loudspeaker has to be arranged on top of the flat panel display, and the left and right loudspeakers are arranged on two sides of the flat panel display. Because the acoustic center of the center loudspeaker is not in line with those of the left and right loudspeakers, the summing localization with three frontal loudspeakers is influenced. For two-way loudspeaker systems, this problem can be partially remedied by arranging the left and right loudspeakers underneath the two sides of the flat panel display, with tweeters located below and a woofer on top. The center loudspeaker is arranged below the flat panel display.

The arrangement of the center loudspeaker should not obstruct the sight of the control window in a control room. In addition, the conventional elevated arrangement of frontal loudspeakers is inclined to cause early reflections from a mixing console. In practice, the lower arrangement of frontal loudspeakers to make their radiations at the grazing incidence over the mixing console is a better choice (Holman, 2008). Overall, the arrangement of frontal loudspeakers (especially center loudspeaker) may be conflicting with the positions of the control window, the flat panel display, and the mixing console in a control room. An appropriate design can alleviate this problem partly rather than completely.

Near-field monitoring is often used in a studio for program mixing and production through which small monitoring loudspeakers are arranged at a distance close to a listener. The original consideration of near-field monitors is that direct sound is dominant and that the influence of room reflections can be reduced at a close loudspeaker distance with respect to the listener. However, a loudspeaker creates full directional radiation when its size is shorter than the wavelength. Therefore, small loudspeakers are inclined to excite listening room reflections. Some authors argued that near-field monitoring may not completely achieve the intended purposes (Holman, 2008).

Although the above discussions focus on 5.1-channel sound reproduction, most conclusions are appropriate for horizontal surround sound with more channels. The situation for multichannel spatial surround sound is more complex than that for horizontal surround sound. Although investigations on these issues have been performed, no consensus on conclusions has been made. The above conclusions can be used as a reference only.

14.3.2 Characteristics of the main loudspeakers

The characteristics of loudspeakers (and a whole electroacoustic system) remarkably influence the sound field and the final perceived performance of spatial sound reproduction. The characteristics of the main loudspeakers should satisfy certain requirements, as specified by some international standards and guidance, to achieve accurate and consistent results in subjective assessment experiments and critical comparison of program materials. For example, the ITU standard (ITU-R BS 1116-3, 2015) and guidance by the AES Technical Council (2001) have specified some basic characteristics and permissible errors of the main loudspeakers and electroacoustic systems for the subjective assessment of multichannel sound, such as ***frequency response***, ***directivity index***, ***nor-linear distortion***, ***transient fidelity***, ***time delay***, and ***dynamic range***. A few characteristics specified by AES guidance are more critical than those in the ITU standard. These standard and guidance are discussed in Section 14.5.

In addition to the requirements on basic characteristics, the characteristics of all main loudspeakers should be matched. For two-channel stereophonic sound, two loudspeakers should be identical. For discrete multichannel sounds, such as 5.1-channel sound, all the main loudspeakers are desired to be identical or have similar acoustic characteristics at least in most standards and guidance (Dolby Laboratories, 2000). These characteristics and permissible errors are specified by related standards and guidance (ITU-R BS 1116-1, 1997; ITU-R BS 1116-3, 2015; AES Technical Council, 2001). In practical uses, however, the requirements of different sets of loudspeakers in multichannel sound reproduction may vary or even be considerably different.

For example, similar to the case of two-channel stereophonic sound, the left and right loudspeakers for 5.1-channel sound should be identical. Ideally, the center loudspeaker should also be identical to the left and right loudspeakers. However, as stated in Section 14.3.1, if a flat panel display is used for visual display, arranging the center loudspeaker may be difficult. In addition to the method presented in Section 14.3.1, a bar-shaped loudspeaker can be used for center loudspeaker and horizontally arranged above or below the flat panel display to reduce the difference between the center loudspeaker and left/right loudspeakers. The characteristics of the center loudspeaker should be matched with those of left/right loudspeakers. However, if a two-way loudspeaker system is used for center loudspeaker, the acoustic centers of the woofer and the tweeter of the center loudspeaker system in such an arrangement are not directly located at the front.

The sizes of the main loudspeakers can be reduced by using the method of bass management in Section 14.3.3, which is also beneficial to the arrangement of the center loudspeaker. A method similar to bass management is also allowed in the monitoring of Dolby Pro Logic reproduction (Dolby Laboratories, 1998). In this method, the low-frequency component of the center channel signal below 100 Hz is redirected to the left and right channels, and the component of the center channel signal above 100 Hz is reproduced by a center loudspeaker with characteristics similar to those of the left and right loudspeakers above 100 Hz. This method can also reduce the size of the center loudspeaker.

According to most standards and guidance, identical loudspeakers or loudspeakers with similar characteristics should be used for frontal and surround channels. However, surround loudspeakers with a reduced size are often used in practical domestic reproduction to save the cost and space. This method is appropriate for Dolby surround or Dolby Pro Logic reproduction because the bandwidth of their surround channel is inherently limited to 0.1–7 kHz. For other discrete multichannel sounds, such as 5.1-channel sound, surround loudspeakers with reduced sizes (and thus different characteristics) may influence the reproduced effects.

According to most standards and guidance, direct radiated loudspeakers with a certain directional pattern are used for the main loudspeakers in multichannel sound. The main axes of loudspeakers are pointed to the listening position. The main loudspeakers with such radiation characteristics create uniform radiation of the direct sound and reduce the reflections in the listening region to improve localization. The guidance of AES Technical Council suggests a directivity index of 8 dB ± 2 dB within the frequency range of 250–16 kHz for all the main loudspeakers in 5.1-channel reproduction. Alternatively, loudspeakers with a bipolar radiation pattern are sometimes used for surround loudspeakers in practical multichannel sound reproduction. The directional nulls of a bipolar loudspeaker are pointed to a listener. A combination of bipolar surround loudspeakers and reflections from the lateral and rear walls of a listening room recreates a diffused sound field and then enhances the envelopment in reproduction.

The optimal radiation pattern of surround loudspeakers has some controversies (Zacharov, 1998a; Holman, 2000). In fact, the optimal radiation pattern of surround loudspeakers depends on the characteristics of program materials. If surround channels are used for ambient sound (such as diffused reverberation), bipolar surround loudspeakers combined with diffusely reflective lateral and rear walls in listening rooms are appropriate. However, if surround channels are used to recreate directional information, bipolar surround loudspeakers, along with listening room reflections, degrade the localization performance. From this point of view, the acoustic design of a listening room also depends on the type of surround loudspeakers. Overall, the requirement of the radiation pattern for enhancing envelopment may come in conflict with that for improving localization. As such, the USC 10.2-channel system in Figure 6.14 involves two types of loudspeakers with different radiation patterns arranged at an azimuth of ±110° in the middle layer. In addition, to create reflections effectively, bipolar surround loudspeakers should be free-standing arranged with a certain distance to lateral and rear walls. Therefore, dipolar surround loudspeakers may be inappropriate for a small mixing studio with a limited space.

14.3.3 Bass management and arrangement of subwoofers

Arranging the main loudspeakers with lager size in practical multichannel sound reproduction is usually inconvenient. Bass management on main channel signals can solve this problem. Through bass management, low-frequency components in some main channel signals are separated and reproduced by subwoofers or other loudspeakers. Thus, this method increases the low-frequency limit and thus reduces the size of the main loudspeakers. Bass management is also applied to monitor in program mixing and production. The low-frequency limit of the main loudspeakers with "full bandwidth" lies in an order between 40 and 50 Hz at best, and the low-frequency limit of small monitor loudspeakers is usually higher than 50 Hz. Therefore, bass management is required to compensate for the deficiency of the main loudspeakers. It is based on psychoacoustic experiment and hypothesis results that the low-frequency component of a sound slightly contributes to directional localization. It is distinguished from a low-frequency effect channel, as indicated in Section 5.4.

Some authors investigated the influence of the crossover frequency of bass management on auditory perception. Borenius (1985) performed a psychoacoustic experiment and indicated that the directional deviation between a loudspeaker and a subwoofer is not perceptible or at least not disturbing for a crossover frequency below 200 Hz. Furthermore, the distance deviation between a loudspeaker and a subwoofer is perceptually disturbing, especially a speech stimulus. Kügler and Thiele (1992) further conducted a psychoacoustic experiment to investigate the audible difference between stereophonic reproduction with a pair of full-bandwidth loudspeakers and a pair of loudspeakers with low-frequency components coupling to a common subwoofer. They found that the difference is inaudible if the crossover frequency does not exceed 100–140 Hz. The upper limit of the crossover frequency depends on the location of a subwoofer in a listening room because of the interaction between the subwoofer and the modes of the listening room.

As stated in Section 14.3.2, Dolby Pro Logic allows bass management on a center channel. However, bass management on a surround channel is unnecessary because the bandwidth of a surround channel in Dolby Pro Logic is limited to 100–7 kHz.

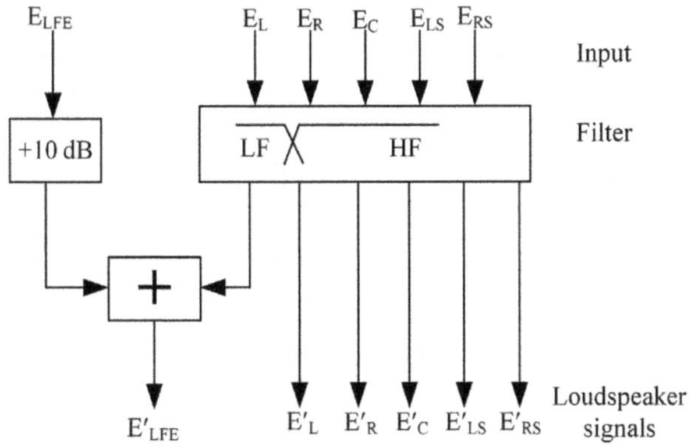

Figure 14.1 Block diagram of the bass management and reproduction of a low-frequency effects (LFE) channel signal for 5.1-channel sound. (adapted from AES Technical Council, 2001.)

For 5.1-channel sound, the low-frequency components of the main channel signals can be separately reproduced by several subwoofers (e.g., a subwoofer for frontal channels and another subwoofer for surround channels). However, a more common method is the reproduction of low-frequency components with a subwoofer. The crossover frequency of bass management usually lies between 80 and 160 Hz. A crossover frequency around 80 Hz yields a better effect. The low-frequency components of five main channel signals and the LFE channel signal can share a single subwoofer. Figure 14.1 illustrates the block diagram of the bass management and reproduction of LFE channel signals for 5.1-channel sound. The high-pass filtered components of five main channel signals are reproduced by five main loudspeakers; the low-pass filtered components of five main channel signals, along with the 10 dB-boosted LFE channel signal (Section 5.4), are reproduced by a subwoofer.

Bass management is also applied to the USC 10.2-channel sound in Section 6.5.1 (ITU-R Report BS 2159-7, 2015). The crossover frequency of bass management lies between 20 Hz and 50 Hz. The low-frequency components of all channels on the left sides of Figure 6.14 and those of the frontal center channel are fed to a subwoofer located at the left. Furthermore, the low-frequency components of all channels on the right sides of Figure 6.14 and those of the rear center channel are fed to a subwoofer located at the right.

No universal standard or specification on the arrangement of subwoofers in a listening room has been established. The location of subwoofers slightly influences virtual source localization provided that the crossover frequency is appropriately chosen. If localization is considered only, the arrangement of subwoofers is relatively flexible. However, some studies have suggested that a single subwoofer should be arranged in the central front; otherwise, low-frequency localization can be easily disturbed (Bell, 2000) possibly because the harmonic components above 120 Hz caused by nonlinear distortion in a subwoofer. In other words, the nonlinear distortion of a subwoofer may cause an audible directional difference.

The interaction between the radiation of subwoofers and room modes should be considered when subwoofers are arranged. According to waveacoustics-based calculation and experimental measurement, a subwoofer arranged on the floor at the corner of a room can effectively excite various room modes and yield a relatively smooth response at low frequencies (Nousaine 1997; Kuttruff, 2009). However, a subwoofer is often arranged on the floor of the frontal center of a listening room. For a left–right symmetric (such as a rectangular) listening room, the null of asymmetric lateral standing wave modes is consistent with the central line. A centrally located subwoofer cannot excite asymmetric lateral standing wave modes. Therefore, an offset arrangement of subwoofers is desired. In fact, the influence of room reflections on the radiation impedance and efficiency of a subwoofer depends on the location of subwoofers in a listening room. In particular, a subwoofer located at the corner of a room with rigid surfaces considerably increases the radiation pressures at low frequencies (Section 1.2.2). Some subwoofers are designed for arranging in particular locations, but others need to be moved around until the most subjectively satisfactory result is obtained (AES Technical Council, 2001). Phase shifts or delays are sometimes applied to subwoofer signals to correct the time/phase relationship among the subwoofers and main loudspeakers. However, simultaneously adapting the phase relationship of all the main loudspeakers is difficult when a single subwoofer is used.

Measurement and wave acoustic-based calculation indicate that the response of a subwoofer depends considerably on the subwoofer location and receiver position in a listening room. However, Zacharov et al. (1998a) conducted informal subjective listening and observed that for 5.1-channel reproduction with a crossover frequency of 85 Hz, the location of a subwoofer in a listening room appears to be noncritical, and a single subwoofer located at the boundary of a room likely suffices. Objective and subjective results may vary because objective measurement or calculation usually yields stationary room responses unless room impulse responses are further analyzed in the time–frequency domain; conversely, auditory processing is carried out within certain time windows.

The magnitude response of a subwoofer varies with the frequency and location of the subwoofer because of standing wave modes in a small room. Some studies have suggested using multiple subwoofers to reproduce decorrelated low-frequency signals. An appropriate choice of subwoofer locations and phase relationships among signals can excite various room modes evenly or cancel some of the standing wave modes, resulting in relatively even low-frequency responses. The sound field created by multiple subwoofers in a listening room can be evaluated by wave acoustic-based calculation or measurement. Generally, the optimized number and arrangement of subwoofers depend on the geometrical and acoustic characteristics of a listening room and the listening position (region). Various studies and results on this issue have been performed (Welti, 2002, 2012; Backman, 2009, 2010, 2011).

Pre-equalization on subwoofer signals can correct the low-frequency response at the receiver (listening) position. However, when a single subwoofer with a given location is used, pre-equalization can only correct the low-frequency response at a special receiver position at most. Multiple subwoofers can also simultaneously correct low-frequency responses at multiple receiver positions (Welti and Devantier, 2006). The principle of this correction method is similar to that of multiple receiver position matching described in Sections 9.6.3 and 9.7.

Although a low-frequency component below 100 or 80 Hz contributes slightly to localization, it may contribute to other spatial auditory perceptions. Griesinger (1997b, 1998) argued that interference between medial (front–back and up–down) and lateral (left–right) modes remarkably influences the low-frequency spaciousness in a small room. When asymmetric lateral modes with null in the center of a room are strongly reacted to medial modes at low frequencies, a high but possible unnatural spaciousness may be created. In practice, multiple subwoofers with decorrelated signals can be used to create natural spaciousness at low frequencies. Griesinger suggested utilizing a pair of subwoofers located at the two sides of a listening position to create natural spaciousness at low frequencies. The signals of two subwoofers exhibit a 90° phase shift to each other. The low-frequency components of the main channel signals separated by bass management can also be handled similarly.

Some advanced multichannel sounds (such as 10.2-channel and 22.2-channel sound in Section 6.5.1) have two independent LFE channels; therefore, the decorrelated signals in LFE channels can be created during program production.

14.4 SIGNAL AND LISTENING LEVEL ALIGNMENT

Many auditory perceived attributes and sensations in spatial sound reproduction depend on a listening level. The listening level in reproduction should be aligned to obtain consistent results in the critical comparison of program materials and subjective experiments. For compatibility and program exchange in broadcasting and recording, signal levels should also be aligned. Cinema sound reproduction also requires a strict level alignment. In different applications and fields, the procedures and methods of signal and level alignment have some differences or even controversies. The relevant details are in accordance with the guidance provided by AES Technical Council (2001) and described in a book by Rumsey (2001).In practical music program production, signal levels are not always aligned with the standard. In the following section, the problem of level alignment is addressed according to the ITU standard and document of AES Technical Council.

For the critical comparison of program materials and subjective experiments, the method of sound level alignment in early ITU standard and AES document is identical (ITU-R BS 1116-1, 1997; AES Technical Council, 2001). Pink noise with a root-mean-square (RMS) level of −18 dBFS is used for sound level alignment, where 0 dBFS is the clipping level of a digital tape recording. The overall ***reference listening level*** at the listening position is 85 dBA (IEC/A-weighted, slow). Then, the gain of each of M main channel should be adjusted to give the following reference sound pressure level:

$$L_{LISTref} = 85 - 10\log_{10} M \pm 0.25 \quad (dBA). \tag{14.4.1}$$

For example, in the case of $M = 5$ main channels, Equation (14.4.1) yields $L_{LISTref} = 78$ dBA. That is, when the signal for level alignment is individually reproduced by each channel, the sound level at the listening position should be adjusted to 78 dBA. Then, the overall sound pressure level created by five channels of noncorrelated alignment signals is 85 dBA. In practical reproduction, listeners may prefer different absolute listening levels although this option is not the preferred one in the ITU standard. For example, if a reproduced sound pressure level with −10 dBA reduction with respect to the reference level is preferred, 85 dBA on the right side of Equation (14.4.1) should be replaced by 75 dBA.

The late version of the ITU standard (ITU-R BS 1116-3, 2015) suggests a reference sound pressure level of 78 dBA per channel, e.g.,

$$L_{LISTref} = 78 \pm 0.25 \quad (dBA) \tag{14.4.2}$$

If the subjects do adjust the gain of the system, this fact should be noted in the test results.

In Section 5.4, the in-band gain of LFE channels should be 10 dB higher than that of the main channel. A low-pass filtered pink noise with the same bandwidth as LFE channels can be used to align the sound pressure level of LFE channels so that each 1/3 octave sound pressure level caused by LFE channels is 10 dB higher than that caused by the main channel at the listening position and within the LFE bandwidth.

The bandwidth of signals for sound level alignment has some controversies. Some scholars argued that full-band pink noise involves more low-frequency components; as such, measurement becomes sensitive to room modes. In addition, the polar pattern of a loudspeaker usually leads to a direction-dependent sound pressure level at high frequencies although the A-weighted level used in measurement gives less weights to high- and low-frequency components of sound. Therefore, band-pass filtered pink noise is also used for sound level alignment in some studies. The low-frequency limit of band-pass filtering may be as low as 200 Hz, and a high-frequency limit ranges from 1 kHz to 4 kHz. For example, Dolby Laboratories suggested the use of a 0.5–1 kHz band-pass pink noise as a signal for sound level alignment (Dolby Laboratories, 2000).

Different weighted sound levels are used for level alignment in various standards. For example, a C-weighted sound pressure level is used in some standards (such as that by a Japanese HDTV forum). The C-weighted curve is relatively even as compared with the A-weighted curve, giving relatively more weights to low-frequency components. For cinema applications, the International Organization for Standardization (ISO) and the Society of Motion Picture and Television Engineers (SMPTE) also formulated standards for level alignment (Section 14.7). Therefore, different levels of alignment yield various results, which cannot be compared directly.

Acoustic conditions and loudspeaker characteristics specified by various standards often cannot be satisfied in practical domestic reproduction. Under the project of Eureka 1653, Bech and Zacharov et al. investigated the influence of methods, signals, directivity, and loudspeaker arrangement on level alignment and explored the relationship between subjective loudness alignment and measurement (Bech and Zacharov, 1999; Suokuisma et al., 1998; Zacharov, 1998b; Zacharov et al., 1998b; Zacharov and Bech, 2000). The results indicated that the low-frequency components of signals can be ignored when the channel gain is aligned subjectively. Alignment with some signals may lead to relatively consistent subjective and objective results. However, further work on this issue is needed (AES Technical Council, 2001).

14.5 STANDARDS AND GUIDANCE FOR CONDITIONS OF SPATIAL SOUND REPRODUCTION

According to the technical requirements in spatial sound reproduction, some organizations, such as ITU, AES, European Broadcasting Union (EBU), SMPTE, German Surround Sound Forum, and Japanese HDTV Forum, have formulated standards or guidance related to stereophonic and multichannel sound reproduction. Some companies have also established standards for technical certification or guidance for technical operation. These standards and guidance are for different purposes, such as critical comparison of program materials and subjective experiments, monitoring in program production, cinema reproduction, and consumer applications. Generally, standards for the critical comparison of program materials and subjective experiments are the most critical. Although completely satisfying these standards in a practical control room or studio for program mixing and production is difficult, these critical standards provide a reference baseline for practical applications.

The ITU has recommended the characteristics of a reference listening room, a reference monitor loudspeaker, and a reference sound field in the standard related to the subjective assessment of multichannel sound systems (ITU-R BS 1116-1, 1997). The AES Technical Council (2001) has also formulated a related document. This document is a guidance (rather than a standard) with summarization of some standards. Most conditions in AES guidance are similar to these in ITU-R BS 1116-1, and EBU standards, but some conditions in AES guidance are more critical than the ITU standards. These standards and guidance are mainly for stereophonic and multichannel horizontal surround sound (such as 5.1-channel sound). ITU has updated its standards to ITU-R BS 1116-3 (2015) in which some considerations in advanced multichannel sound are supplemented.

Table 14.1 lists the conditions for a reference listening room and loudspeaker arrangement, as suggested by ITU-R BS 1116-3, and the AES guidance. For 5.1-channel sound, the

Table 14.1 Reference listening room and loudspeaker arrangement recommended by the ITU-R, BS 1116-3, and the technical document of AES Technical Council

Parameters	*Unit/Conditions*	*Reference values*	
		ITU	*AES*
Room size (floor area)	m^2	Mono, stereophonic 20–60 Multichannel 30–70	Mono, stereophonic >30 Multichannel >40
Room proportions	l = length w = width h = height	$1.1\ w/h \leq l/h \leq$ $4.5\ w/h - 4$ $l/h < 3$, $w/h < 3$	Identical to ITU (ratios within ±5% of integer values are considered unsatisfactory)
Base width	m	Stereophonic 2.0–3.0 Up to 4.0 m may be acceptable in a suitable designed room Multichannel 2.0–3.0 Up to 4.0 m may be acceptable and suitable in the designed room	Stereophonic and multichannel 2.0–4.0
Listening (spanned) angle between left and right loudspeakers	$2\theta_0/°$	Stereophonic and multichannel 60	Identical to ITU
Listening distance	m	Stereophonic 2 m to 1.7 × base width Multichannel Base width	Stereophonic and multichannel 2 m to 1.7 × base width
Listening region	m	Stereophonic Radius $\leq$ 0.7 m Multichannel Deviation to the front, back, left, and right should not exceed half of the base width	Stereophonic and multichannel 0.8
Height of the acoustical center of loudspeakers	m	Stereophonic and multichannel Seated listener's ear height	1.2
Distance to the wall	m	Stereophonic and multichannel $\geq$1	Identical to ITU

five main loudspeakers are arranged as illustrated in Figure 5.2. Most conditions in the ITU standard and AES guidance are consistent.

Table 14.2 lists the reference sound field conditions suggested by the ITU standard and AES guidance. In the ITU-R BS 1116-3 standard, the average reverberation time of a listening room, which is measured over the frequency range of 200 Hz to 4 kHz, is related to the volume V of the listening room and expressed as

$$T_m = 0.25\left(\frac{V}{V_0}\right)^{1/3}, \tag{14.5.1}$$

where V_0 = 100 m^3 is the reference volume. Figure 14.2 illustrates the tolerance limit of reverberation time over the frequency range from 63 Hz to 8 kHz. Within the frequency

Table 14.2 Reference sound field conditions recommended by the ITU-R BS 1116-3, and the technical document of AES Technical Council

		Reference values	
Parameters	*Unit/Conditions*	*ITU*	*AES*
Frequency response of direct sound	Measured under a free field	Tolerance limit is shown in Figure 14.3	Tolerance limit is shown in Figure 14.3
Early reflection	0–15 ms	<−10 dB in 1–8 kHz range with respect to the direct sound	Identical to ITU
Average and tolerance limit of reverberation time	s	Equation (14.5.1) and Figure 14.2	Equation (14.5.1) and Figure 14.3
Tolerance limits of an operational room response curve	Pink noise over 50–16 kHz, measured in 1/3 octave	Figure 14.4	Figure 14.5
Background noise		Preferably < NR 10, Never exceed NR15	Identical to ITU
Listening level	Input pink noise at −18 dBFS	Equation (14.4.2)	Equation (14.4.1), 78 dBA for five channels

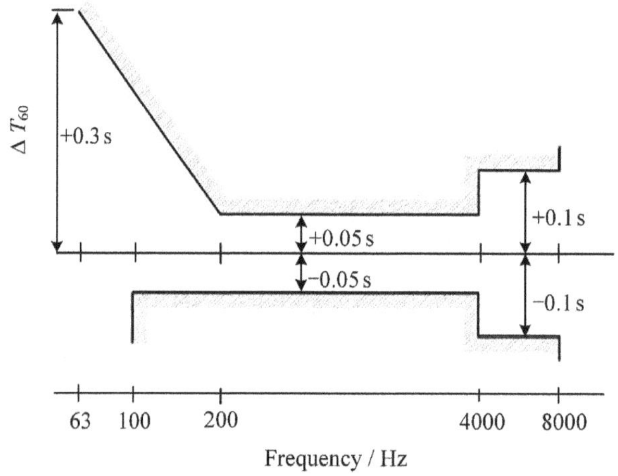

Figure 14.2 Average and tolerance limit of reverberation time recommended by the ITU-R BS 1116-3.

range of 200 Hz to 4 kHz, the tolerance limit of reverberation time is ±0.05 s; larger tolerance limits appear within the frequency ranges from 63 Hz to 200 Hz and from 4 kHz to 8 kHz. An increase in reverberation time at low frequencies is allowed because of difficulties in providing sufficient low-frequency sound absorption in a listening room. In the AES guidance, the average reverberation time of a listening room is calculated using Equation (14.5.1). Figure 14.3 shows the tolerance limit of reverberation time over the frequency in the AES guidance. Within the frequency range of 200 Hz to 4 kHz, the tolerance limits of reverberation time in AES guidance are identical to those of the ITU standard; however, within the frequency ranges of 63–200 Hz and 4–8 kHz, the tolerance limit of reverberation time in the AES guidance are slightly different from those of the ITU standard.

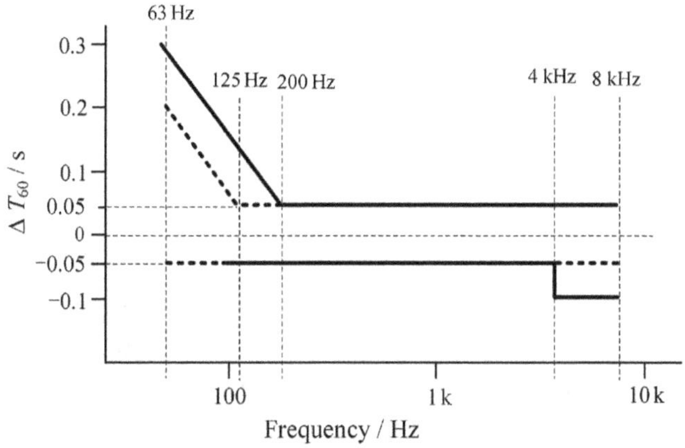

Figure 14.3 Average and tolerance limits of reverberation time recommended by the technical document of AES Technical Council.

The ITU standard and AES guidance also present the tolerance limits of an operational room response curve (Figures 14.4 and 14.5). The curve in the AES guidance is extended to a lower frequency and more critical than that in the ITU standard.

The ITU standard and the AES guidance also suggest the conditions and characteristics of reference monitor loudspeakers (Table 14.3). Some conditions in AES guidance are more critical than those in ITU standard. For example, the ITU standard suggests that nonlinear distortion should not exceed 3% below the frequency 250 Hz and should not exceed 1% above 250 Hz. While AES guidance suggests that the nonlinear distortion, should not exceed 3% below 100 Hz, and should not exceed 1% above 100 Hz.

Because the perceptual interaction between sound and pictures influences the assessment of sound quality, another standard by the ITU (ITU-R BS 2126-0, 2019) has further clarified relationships between distances from loudspeakers to the central listening position, the display sizes, and the viewing distances for the subjective assessment of sound systems with accompanying picture.

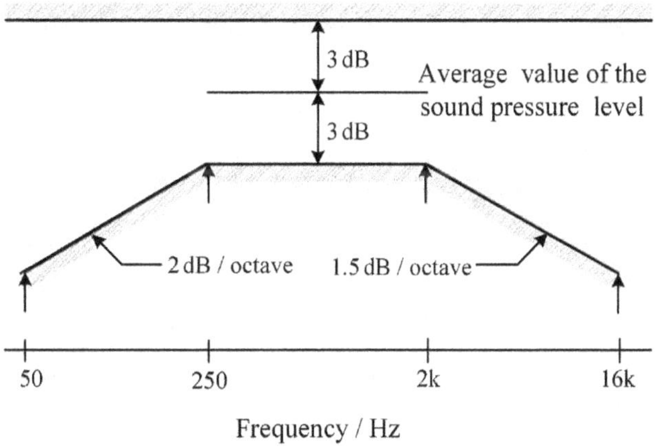

Figure 14.4 Tolerance limits of the operational room response curve recommended by the ITU-R BS 1116-3.

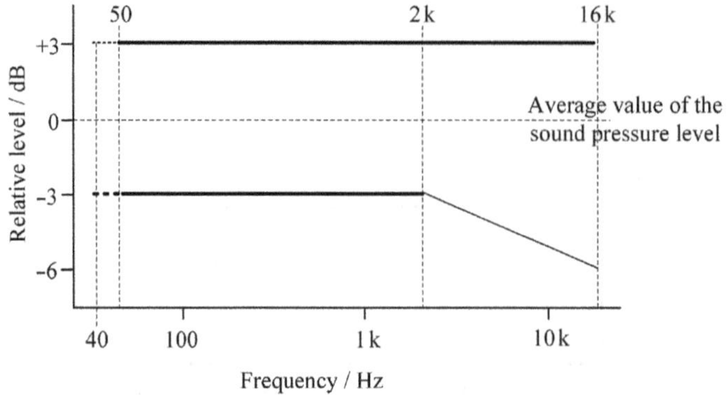

Figure 14.5 Tolerance limits of the operational room response curve recommended by the technical document of AES Technical Council.

Table 14.3 Reference monitor loudspeaker recommended by the ITU-R BS 1116-3, and the technical document of AES Technical Council

		Reference values	
Parameters	*Unit/Condition*	*ITU*	*AES*
Amplitude versus frequency response	40–16 kHz, measured in 1/3 octave	Directional angle of 0°, ≤ 4 dB. Directional angle ±10°, different from 0° ≤ 3 dB. Horizontal ±30°, different from 0° ≤ 4 dB. Difference within 250–2 kHz ≤1.0 dB	Identical to ITU Difference between front loudspeakers ≤0.5 dB
Directivity index		within 500Hz–10 kHz 6–12 dB	within 250Hz–16 kHz 8 ± 2 dB
Nonlinear distortion		< 250 Hz, −30 dB (=3%) > 250 Hz, −40 dB (=1%)	< 100 Hz, −30 dB (=3%) >100 Hz, −40 dB (=1%)
Transient fidelity (The decay time for an amplitude of 1/e or 0.37)	s	<5/*f* *f* is frequency	Identical ITU, but preferably 2.5/*f*
Time delay (between system channels)	μs	≤100	≤10
Dynamic range (maximum operating sound level measured according to IEC 20268)	dB	>108	>112
Noise level	dBA	<10	≤10

As stated, completely satisfying the conditions of the ITU standard or AES guidance for a control room or studio for program mixing and production is difficult. The Japanese HDTV Forum formulated a standard on the acoustic conditions of a control room or a studio and equipment for HDTV sound program mixing and production; in this standard, some conditions are loosened compared with those in the ITU standard and AES guidance (Rumsey, 2001; AES Technical Council, 2001).

14.6 HEADPHONES AND BINAURAL MONITORS OF SPATIAL SOUND REPRODUCTION

The acoustic conditions for monitoring loudspeaker-based spatial sound reproduction are discussed in the preceding sections. Headphones are used to monitor spatial sound reproduction in some cases. First, binaural reproduction and virtual auditory display in Chapter 11 are originally intended for headphone presentation. The problems of headphone presentation of binaural signals are addressed in Section 11.7, i.e., the equalization of headphone-to-external ear transmission is needed. Second, stereophonic and multichannel sound signals in studio are sometimes monitored using headphones. According to the ITU standard (ITU-R BS 1116-3, 2015), the diffuse-field frequency response of studio monitor headphones is recommended in ITU-R BS 708 (1990), and time delay differences between channels of a stereophonic system should not exceed 20 μs.

Headphone monitoring is not influenced by the reflections of a listening room, so conditions in an acoustic environment for a monitor can be loosened. However, as stated in Section 11.9.1, the headphone monitoring or presentation of stereophonic and multichannel sound signals directly leads to incorrect binaural information; as such, it is inappropriate for evaluating spatial attributes (such as virtual source localization). The method of binaural or virtual monitoring through headphones can be used to address this problem. Similar to the cases in Sections 11.6 and 11.9.1, HRTF-based binaural synthesis is used to create virtual loudspeakers in headphone presentation, and stereophonic or multichannel sound signals are monitored using virtual loudspeakers. If necessary, reflections of listening rooms can be involved in binaural synthesis to simulate the perceived effect on a practical listening room.

The reflections of a listening room can be simulated by various methods of binaural room modeling (Section 11.10.1;Vorländer, 2008; Lehnert and Blauert, 1992; Kleiner et al., 1993; Svensson and Kristiansen, 2002). Binaural signals with the reflections of listening rooms can also be synthesized by convolving multichannel sound signals with BRIRs (Section 1.8.3) of loudspeakers in a standard listening room. By rotating an artificial head, Karamustafaoglu et al. (1999) measured the BRIRs of various head orientations created by five loudspeakers located at different directions in a listening room. The method of dynamic VAE in Section 11.10.2 can be used for reproduction through which BRIRs for binaural synthesis are constantly updated according to a temporary head orientation of a listener. This method is termed ***binaural room scanning.***

Although the principle of binaural or virtual monitoring is identical to that of the binaural reproduction of stereophonic and multichannel sound through headphones, but the conditions of the former are more critical than those of the latter (consumer use). However, binaural monitoring should be further improved because of problems associated with binaural reproduction (Section 11.7.2).

14.7 ACOUSTIC CONDITIONS FOR CINEMA SOUND REPRODUCTION AND MONITORING

The reference acoustic conditions of the subjective assessment and monitoring of spatial sound for domestic reproduction (with a small-sized listening region) are addressed in preceding sections. This section outlines the acoustic conditions and designs for cinema sound reproduction and monitoring. The practical applications of spatial sound to commercial cinema and related problems are discussed in Section 16.1.1.

Commercial cinema includes traditional film cinema and current digital cinema. For a digital cinema technique, video/audio signals are recorded by digital techniques, stored with digital media (such as optical disk or hard disk), and reproduced by a high-definition projector and electroacoustic system. The size of a listening region in commercial cinema reproduction varies considerably from a small cinema with fewer than 100 seats to a medium-sized cinema with a few hundred seats and a large cinema with more than 1,000 seats. Nevertheless, the listening region for sound reproduction in a commercial cinema is much larger than that for domestic reproduction. Therefore, various spatial sound techniques and systems with a large listening region are required for cinema reproduction.

Acoustic conditions for commercial cinema have been specified by some international standards. Cinema sound tracks are mixed in dubbing theaters (or mixing rooms). The acoustic conditions and standards of a dubbing theater are identical to those of a practical cinema to ensure that the reproduced sound in a practical cinema is perceptually consistent with that in mixing. Variations in the acoustic design of different cinemas exist because of differences in cinema size and spatial sound techniques used, but some basic considerations are common.

Acoustic designs for cinemas include noise control and room acoustic designs. Noise in a cinema includes noise caused by the equipment inside cinemas, such as ventilating and air-conditioning systems, and noise caused by other sources outside the cinema, such as traffic noise. Therefore, some measurements should be taken to control internal noise and insulate external noise. For background noise, ISO 9568 (1993) requires that dubbing rooms should have a minimum rating of NC-20 and a maximum rating of NC-25. First-run cinemas should have a maximum rating of NC-30, and subsequent-run cinemas should have a maximum rating of NC-35.

A cinema is a large room, so the problem of room modes is not obvious. In the basic acoustic design of a cinema, sound focusing and echo caused by reflections must be avoided. A rear wall is usually treated with strong sound absorption to control reflections between the front and rear walls in a cinema; the front wall behind the screen where the front loudspeakers are mounted is also treated with sound absorption. In contrast to a concert hall, the auditory sensation of ambience in cinemas is created by surround channels and loudspeakers in a spatial sound reproduction system (rather than room reflections). Excessive reflections in cinemas degrade speech (dialog) intelligibility in films. Therefore, reflections in a cinema should be controlled to obtain an appropriate (relatively short) reverberation time. The technical guidance of some companies (such as Dolby, THX, and JBL) has suggested the reverberation time and tolerance limits for cinema with different volumes. Generally, the suggested reverberation time increases with the interior volume and decreases smoothly with the increase in frequency. For example, JBL suggested an acceptable reverberation time of 0.5–0.7 s at 500 Hz for a cinema with an interior volume of 2,700 m^3(about 500 seats, JBL Professional, 1998).

Various multichannel horizontal and spatial surround sound systems with different loudspeaker arrangements have been developed for cinema reproduction. Loudspeaker arrangements in these systems follow some basic rules. For example, in 5.1-channel loudspeaker arrangement in traditional cinemas, three screen loudspeakers are flush mounted with the front wall behind the screen; the acoustic centers of loudspeakers are located 2/3 the height of the screen. The main axes of loudspeakers are pointed to the height of ears of a sitting listener in a position at the 2/3 distance from the screen to the back wall. Left and right loudspeakers are arranged near two edges of the screen. Full-band loudspeaker systems composed of woofer elements and high-frequency horn elements are often used. Linear loudspeaker arrays are also utilized currently. The horizontal directivity of loudspeaker systems should sufficiently cover a wide listening region, while the vertical directivity of the loudspeaker system should be appropriately controlled to enhance the component of the direct sound arriving to the listening region. Each loudspeaker of channels should create a maximal undistorted sound level of 105 dBC in a position at the 2/3 distance from the screen to the back wall to ensure the dynamic range in reproduction. The left and right surround channel signals are reproduced by two symmetric sets of left and right surround loudspeakers, respectively (Figure 16.2), which are different from domestic 5.1-channel reproduction. Two- or three-way direct radiated loudspeaker systems are often employed for surround channel reproduction. The size and power of each surround loudspeaker are smaller than those of screen loudspeakers. Each set of loudspeakers for a surround channel should create a maximal undistorted sound level of 102 dBC in a position at the 2/3 distance from the screen to the back wall. Bipolar loudspeakers are sometimes used for surround channel. LFE channel signals are reproduced by subwoofers placed on the floor under the screen loudspeakers to create an in-band sound level larger than 110 dBC.

ISO and SMPTE have specified the standards of an acoustic response curve and its measurement method for cinema sound reproduction, e.g., ***curve-X* for *B-chain response*** (ISO 2969, 2015; SMPTE ST 202, 2010) to ensure the consistency of reproduction quality in different environments. The conditions in these two standards are basically identical. The B-chain refers to the system chain, including a frequency equalizer, an amplifier, a loudspeaker, and transmission from a loudspeaker to a receiver position. Therefore, the B-chain response is the transfer response of electroacoustic systems and rooms.

Since Allen (2006) conducted a study at Dolby Laboratories in the early 1970s, B-chain responses have been investigated. Near-field monitor loudspeakers with a flat response are initially arranged close to the console in a dubbing theater. Near-field monitoring aims to increase the direct-to-reverberation energy ratio in a listening position and then reduce the influence of reflections in dubbing theaters on the timbre. Numerous unequalized recordings of dialog and music are then reproduced with a near-field monitor system to explore the timbre difference with an original far-field reproduction system in a dubbing theater. Lastly, a far-field reproduction system is equalized to achieve the best timbre match with the near-field monitor system. The 1/3 octave response curve is measured by inputting wide-band pink noise to an equalized far-field reproduction system and recording with a microphone arranged close to the console in a dubbing theater.

The curve-X for B-chain responses in standards is obtained from the summarization of a number of measurements, which has been revised for several times. Figure 14.6 illustrates the curve-X for B-chain response and tolerance limits given by ISO 2969 (2015) and SMPTE ST 202 (2010). The response is flat within the frequency range of 50 Hz–2 kHz; it droops at rates of −3 dB/octave within 2–10 kHz and −6 dB/octave above 10 kHz. The response also droops at a rate of −3 dB/octave below 50 Hz. The response curve in Figure 14.6 is appropriate

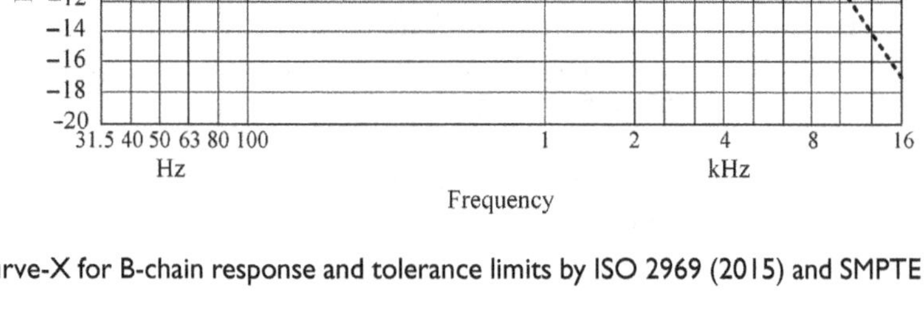

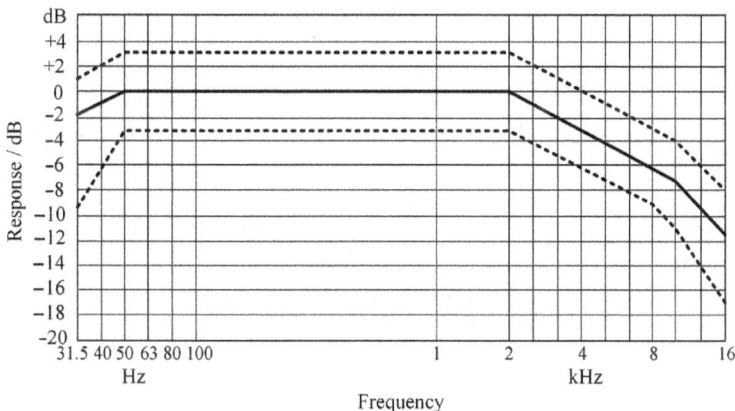

Figure 14.6 Curve-X for B-chain response and tolerance limits by ISO 2969 (2015) and SMPTE ST 202 (2010).

for medium-sized cinemas with about 500 seats. In the ISO standard, the response curve for cinemas is revised with different sizes. For small cinemas with fewer seats, the rate of drop within 2–10 kHz decreases. For large cinemas with more seats, the rate of drop within 2–10 kHz increases.

B-chain responses in dubbing theaters and practical cinemas should be equalized according to the X-curve or its revision. For multichannel sound reproduction, the B-chain response of each channel and its loudspeaker (or loudspeaker array) should be separately measured and equalized. The ISO/SMPTE standards have specified the method of B-chain response measurement. In particular, wide-band pink noise is used for measurement. For dubbing theaters, responses in each 1/3 octave are measured at each principal listening position with a normal ear height between 1.0 and 1.2 m. For practical cinemas, responses are measured at some positions within a square region around a center located at the 2/3 distance from the screen to the back wall.

The high-frequency timbre in a cinema is dominated by the direct sound generated by an electroacoustic system. The direct sound response of electroacoustic systems should be basically flat. An X-curve represents the stationary response of a combination of the direct sound and room reflections. Because of frequency-dependent sound absorption (by boundary, screen, and air) in a cinema, the decay of high-frequency energy is faster than that of low-frequency energy. Therefore, a high-frequency drop in the X-curve represents the reduction of reflected sound energy as frequency increases (Allen, 2006).

ISO and SMPTE have specified the method to align the reference sound pressure level for multichannel sound reproduction (ISO 22234, 2005; SMPTE RP 2096-1, 2017; SMPTE RP 2096-2, 2017; SMPTE RP 200, 2012). The signal used for alignment is band-pass pink noise with a bandwidth of 0.5–2 kHz. At a standard input signal level of −18 dBS, the reference sound level at the monitoring position created by each screen channel is aligned to 85 dBC (slow). Therefore, at an input signal level of −20 dBS, the reference sound level at the monitoring position is 83 dBC. The reference sound level at the monitoring position created by each surround channel is usually aligned to −3 dB below that of the screen channel. Such a reference sound level for each surround channel ensures that two surround channels with decorrelated signals create the same reference sound level at the monitoring position as that of a screen channel. The in-band reference sound level of LFE channels is aligned to 10 dB higher than that of screen channels. In addition, because the level alignment in cinema differs from that in domestic reproduction, a signal level should be re-aligned when cinema sound tracks are converted into DVD sound tracks.

Tomlinson Holman at Lucasfilm has developed a set of techniques and specifications to improve the effect of spatial sound reproduction, which is known as ***THX*** (derived from Tomlinson Holman's experiment). THX has formed a commercial certification process for spatial sound in cinema and domestic reproduction (www.thx.com). It originally aims to make the sound quality in a commercial cinema closer to that in sound mixing during production. The early THX specification was for Dolby Stereo and discrete multichannel (such as 5.1-channel) sound in cinemas. THX specification focuses on noise level control, optimal room acoustic design, characteristics of loudspeakers, and other types of electroacoustic equipment in a cinema. THX specification and certification have been extended to domestic theaters and known as Home THX. Home THX aims to convey the auditory experience in cinemas to homes. In addition to the specification of noise level, characteristics of loudspeakers, and other types of electroacoustic equipment, the following features characterize Home THX:

1. In accordance with the case of an optimal seat in a cinema, the frontal left and right loudspeakers are arranged at horizontal azimuths of ±22.5° rather than ±30°, as recommended by the ITU.

2. Signals are re-equalized to reduce the timbre difference caused by excessive high-frequency components when audio materials for cinemas are reproduced in small rooms.
3. Bipolar loudspeakers are used for surround channels, and decorrelation is applied to Dolby surround signal, to enhance the effect of ambient sound.

Existing methods and standards for the acoustic design and alignment of cinema sound reproduction are appropriate for multichannel (such as 5.1-channel) horizontal surround sound. Currently, cinema sound has been developed toward object-based spatial surround sound with height for which the system is more complex and the configuration of loudspeakers is more diversiform. Accordingly, the interaction between an electroacoustic system and a cinema environment becomes more complex. In addition, as the dimensionality of auditory information is extended from a horizontal plane to a three-dimensional space (with height), the concerned spatial auditory mechanism changes. Although some common rules under the acoustic conditions for various cinema sounds are established, acoustic conditions for advanced spatial sound in cinemas should be further investigated, and the corresponding standards should be supplemented. This issue has been explored in some studies. For example, HHK set up a dubbing studio for a 22.2-multichannel sound system (Sawaya et al., 2015).

14.8 SUMMARY

Acoustic conditions for spatial sound reproduction, which remarkably influences the final perceived performance, should be appropriately specified and designed according to physical and psychoacoustic principle. Subjective assessment experiments or monitoring in the program production of spatial sound should be conducted under strict and standardized acoustic conditions to obtain consistent and reliable results.

Acoustic conditions for domestic spatial sound reproduction differ from those for cinema reproduction. Numerous studies have been performed on acoustic conditions for the assessment and monitoring of domestic spatial sound reproduction, including the design and reflection control of listening rooms, the characteristics and arrangement of loudspeakers, the characteristics of sound field, and monitor level alignment. Some standards and guidance on this issue have been specified. These standards and guidance originally focused on stereophonic and 5.1-channel sound. Some new or updated standard and guidance have considered advanced multichannel sound. The acoustic conditions specified by the standards and guidance cannot be completely satisfied for monitoring in practical program mixing and production. However, these standards and guidance provide a reference or baseline in practice. Practical conditions in domestic reproduction may differ considerably from those in standards and guidance, but the standards and guidance still provide a reference for domestic reproduction.

Spatial sound program for cinema reproduction is produced in dubbing theaters. The acoustic environment of a dubbing theater is identical to those of a practical cinema and specified by the same standards. ISO and SMPTE have formulated the standards for an acoustic response and its measurement, as well as sound level alignment, in cinema sound reproduction. THX formulated a set of technical standards and commercial certification systems for commercial cinemas and domestic theaters.

Acoustic conditions and designs for multichannel spatial surround sound reproduction are more complex than those of multichannel horizontal surround sound although some basic considerations and methods are common. However, this issue should be further explored.

Chapter 15

Psychoacoustic and subjective assessment experiments on spatial sound

Validating the actual effects of spatial sounds necessitates the evaluation of their performance. Similar to the assessment of conventional electroacoustic systems, objective and subjective assessments are two major methods for evaluating the spatial sound. Reconstructed sound field and binaural analyses in the preceding chapters (especially Chapters 9–12) are objective methods. However, objective assessment based on purely physical and mathematical criteria cannot completely reflect the practical perceived performances, especially for spatial sound based on sound field approximation and psychoacoustic-based methods, because spatial sound reproduction deals with auditory psychology and physiology. In Section 12.2, binaural auditory and high-level models of psychoacoustic and physiological acoustic processing can be further incorporated into the objective analyses of spatial sound reproduction. However, advanced objective assessment is insufficient to reflect actual auditory perception accurately because the psychological and physiological aspects of hearing are highly complex issues influenced by many factors. At the current stage, therefore, the subjective assessment of spatial sound by psychoacoustic experiments is extraordinarily important. Various psychoacoustic experiment methods that correspond to different contexts have been proposed. The applicability of these methods depends on the aims of evaluation and the spatial sound systems and attributes to be evaluated.

In this chapter, psychoacoustic and subjective assessment experiments on spatial sounds are discussed. In Section 15.1, some basic considerations, conditions, and methods of psychoacoustic and subjective assessment experiments are outlined. In Section 15.2, the contents and attributes of subjective assessment are described. In Section 15.3, auditory comparison and discrimination experiments for qualitatively assessing perceivable differences are introduced. In Sections 15.4 and 15.5, methods for the subjective assessment of small impairments and intermediate quality levels of spatial sound reproduction are addressed, respectively. In Section 15.6, virtual source localization experiments and some important results are discussed.

15.1 OUTLINE OF PSYCHOACOUSTIC AND SUBJECTIVE ASSESSMENT EXPERIMENTS

Subjective assessment experiments for spatial sounds, or more generally, for electroacoustic systems, are closely related to but different from conventional psychoacoustic experiments. Conventional psychoacoustic experiments often aim to explore the auditory perceptions (ability) of humans, such as the perceptions of loudness, pitch, and source position. Subjective assessment experiments for electroacoustic systems aim to evaluate the performance or quality of electroacoustic systems.

DOI: 10.1201/9781003081500-15

In the subjective assessment of an electroacoustic system, the performance of a system is evaluated by a subject's auditory perception. Assessing the subjective auditory perception of sound is part of experimental psychology. Reliable results can only be obtained with strict psychoacoustic experimental methods and conditions, as well as appropriate statistical analyses. Commercial promotion and amateurish application result in the considerable misunderstanding of the subjective evaluation of sound reproduction. For example, conclusions are often drawn from informal listening tests, which cannot be applied to scientific experimental studies on spatial sounds.

Because the principles and reproduction methods (loudspeakers or headphone-based reproductions) differ in various spatial sound techniques, no consensus or standard for the assessment of spatial sounds—concerning evaluation contents, methods, experimental conditions, as well as data processing—has been reached. However, various subjective assessment experiments are common in some aspects. The International Telecommunication Union (ITU) developed standards for the subjective assessment of spatial sound (ITU-R BS 1116-1, 1997; ITU-R BS 1116-3, 2015; ITU-R BS 1534-3, 2015). These standards mainly focus on impairment that occurs in the low-bit-rate (compressive) coding of audio signals rather than on the quality of spatial sound. Nevertheless, standards can be referred to in some spatial sound evaluations, particularly the examination of the reproduction of stereophonic and multichannel sounds.

According to attributes and contents to be assessed, the subjective assessment of electroacoustic systems can be classified into two types (Rumsey, 2002): attribute judgments and preference rating. In attribute judgments, the performance or quality of systems is assessed in terms of some descriptive judgments on perceived attributes, such as perceived source direction, width, and envelopment. In preference rating, the performance or quality of systems is examined in terms of the extent of subjects' preference. Two types of assessment methods are related to each other because some characteristics of perceived attributes are preferred by most people. Searching the relationship between the perceived attributes and subjects' preferences is also a purpose of subjective assessment experiments.

Two schemes are typically used to assess the attributes or preferences of a tested system or object. In one scheme, an object is compared with a given or reminisced reference. For example, impairments in coded audio signals can be examined by comparing them with the original signals; the localization performance of a VAD can be evaluated by comparing the results of the VAD with those of a real source. In the other scheme, an object is directly analyzed, especially in the case of preference assessment without a reference. For instance, the immersive senses of a virtual auditory environment, which may not necessarily exist in natural environments, are assessed directly.

The design and content of subjective assessment experiments vary considerably, depending on the application and object to be assessed. As such, clear aims and objectives should be formulated in designing subjective experiments for spatial sound assessment. The contents and attributes to be assessed are first chosen according to the aim of experiments; then, experimental methods are selected, and paradigms are designed. Various experimental designs, contents, and attributes may be chosen depending on practical requirements. Some examples are given in the following sections.

Experiments should be conducted under strict conditions and environments. In Section 14.5, standards or guidance by ITU and AES Technical Council has specified the technical conditions of reference listening rooms, reference sound fields, and monitor loudspeakers for subjective assessment experiments (ITU-R BS 1116-1,1997; ITU-R BS 1116-3, 2015; AES Technical Council 2001). Although these standards or guidance is originally intended for stereophonic and multichannel horizontal surround sound (such as 5.1-channel sound), they can be used as a reference of the assessment experiments of other spatial sounds.

Similar to virtual monitors described in Section 14.6, virtual reproduction methods are sometimes used in the subjective assessment experiments of stereophonic and multichannel sound. In these methods, stereophonic and multichannel sound signals are reproduced by HRTF-based virtual loudspeakers in headphone presentation (Sections 11.6.1 and 11.9.1). If necessary, the dynamic virtual auditory environment system in Section 11.10 can be used to simulate the reflections in a listening room and the dynamic cue caused by head turning. Headphone-based virtual reproduction is a newly developed tool for psychoacoustic experiments and often used in scientific studies. It is simple and not influenced by the environment. Therefore, headphone-based virtual reproduction is a promising experimental tool although it should be further improved

The source materials and durations of signals for experiments depend on the objects, contents (attributes), and methods of assessments, which are discussed in the following sections. Although universal source materials for assessments under all conditions are unavailable, source materials that can reveal the characteristics or defects of the objects should be chosen. For example, to assess the basic audio quality of objects or systems, source materials that can reveal the differences in objects should be chosen. Speech materials should be included to assess the spatial sound used for speech communication. Speech, music, and wideband or narrowband noise stimuli are often used to evaluate the ability of a system for recreating a virtual source at various directions.

In psychoacoustic or subjective assessment experiments, the sound level presented by system should be modest to avoid excessive weakness that affects assessment or excessive strength that causes auditory fatigue and discomfort. For loudspeaker reproduction, sound levels are usually measured at the reference (central) listening position, and the scheme of sound level alignment is discussed in Section 14.4. For headphone presentation, sound pressures or levels are usually measured at the open or blocked entrances of ear canals (or occasionally eardrums). Because of the scattering and diffraction effects of anatomical structures (such as head and pinnae), pressures in the two ears differ from the pressure at the reference listening position with the head absent. Moreover, pressures in the left and right ears are different. For example, in the headphone presentation of binaural signals, differences between the pressures in the two ears (ILD) depend on target source direction, frequency, and individualized HRTFs. Binaural pressures in headphone presentation are sometimes converted into free-field pressures at the position of the head center with the head absent by using Equation (1.4.1) to compare with cases of a real source in a free-field or loudspeaker reproduction.

The early version of the ITU standard (ITU-R BS 1116-1, 1997) suggested aligning the gain of each channel equally so that for the test signal with a reference level, the total sound level at the reference listening position is 85 dBA (Section 14.4). The late version of the standard (ITU-R BS 1116-3, 2015) suggested a reference sound level of 78 dBA per channel. If subjects adjust the gain of a given system, this fact should be noted in the test results. The ITU standard of MUSHRA (ITU-R BS 1534-3, 2015) allows an individual adjustment of the listening level within a session, but it is limited to a range of 4 dB relative to the reference level (78 dBA per channel). Various reproduced levels may be selected for other experiments, such as virtual source localization experiments. A universal reproduced level for virtual source localization experiments has yet to be established. In some studies, the practical reproduced level for a localization experiment lies between 60 and 75 dB or (dBA) measured at the reference listening position or in the two ears.

The subjects recruited for experiments should have normal hearing. Young adults are typically chosen to avoid high-frequency hearing loss associated with increasing age. For the assessment of the audio quality of systems, experienced expert subjects should be selected for such experiments. As training progresses, some inexperienced subjects gradually become experienced subjects. More strictly, prior to experiments, each subject should be assessed by screening the hearing in accordance with the standard of ISO 389-1 (1998). Hearing losses for tones at 0.5, 1.0, and 2.0 kHz are often evaluated as in the medical diagnosis of hearing impairment for speech perception, and the average hearing loss of less than 25 dB at three frequencies is considered a normal range (ISO 1999, 1975). Some other researchers argued that an average hearing loss of less than 15 dB is a normal range. In practice, a more critical criterion than that for the medical diagnosis of hearing loss is used for subject selection. Screening the sense of hearing at high frequencies may also be supplemented. For example, some researchers used a hearing loss of no more than 10 dB within a frequency range of 250 Hz to 4 kHz and no more than 15 dB at 8 kHz as a criterion for subject selection (Hoffmann and Møller, 2006).

A sufficient number of subjects are needed to guarantee the validity of statistical results. In theory, the higher the number of subjects is, the more reliable the results will be. However, an increase in the number of subjects complicates experiments. The required number of subjects depends on experimental types and methods, which are discussed in the following sections. Repeated experiments on each condition and each subject are often conducted to expand experimental data samples. Experiments with more contents should be carried out at separate stages to avoid the hearing fatigue caused by lengthy experiments. Moreover, each trial should be less than half an hour with break intervals longer than the experimental period. Prior to formal experiments, training is necessary to familiarize subjects with evaluation methods and contents.

Raw data for experiments are collected from all subjects and repeated experiments. From the perspective of statistics, therefore, such experiments are subject to certain statistical rules on random variables. Experimental data should be further analyzed through appropriate statistical methods to yield the results of a certain confidence level, which is a crucial issue.

Consistencies in the raw data from each subject should be tested prior to statistical analysis. If the established experimental paradigm includes repeated judgments from each subject, the repeated data can be used for consistency tests. Otherwise, a few repeated judgments should be added to the pre-experiment stage. Consistencies can be evaluated in terms of differences or correlation between repeated data. For each subject, if the difference exceeds some thresholds or if the correlation is lower than some thresholds, data from this subject are inconsistent and should be eliminated in the following statistical analysis. Moreover, some invalid data should be identified using statistical method and excluded from mean calculation. A conventional method is pre-calculating the mean and confidence interval across all data. The data that lie beyond the confidence interval are invalid and thus excluded from the final mean calculation. However, this method is not always appropriate for pre-screening the results of all subjective/psychoacoustic experiments. Various methods for eliminating invalid data may be used in different experiments.

15.2 CONTENTS AND ATTRIBUTES FOR SPATIAL SOUND ASSESSMENT

Auditory attributes refer to the perceived attributes of sound stimuli. Spatial sound possesses multidimensional perceived attributes, and most of them can be classified into two categories, e.g., ***timbral attributes*** and ***spatial attributes***. The timbre is defined in Section 12.2.3. Spatial attributes have various definitions. Rumsey (2002) defined the spatial attributes as "*the three-dimensional nature of sources and their environment.*"

According to Section 1.7.4, the information provided by multiple sound sources and environments in source materials contributes to the auditory scene in reproduction. The final auditory scene in reproduction is determined by all stages in the system chain of a spatial sound, including source materials, systems, loudspeaker configuration, and characteristics of a listening room. Spatial sound assessment can be implemented by evaluating the overall auditory scene and the elements (or objects) that form a scene, e.g., by a ***scene-based paradigm***. A scene-based paradigm requires source materials that can reveal the comprehensive characteristics and defects of systems. This requirement is different from that in conventional psychoacoustic experiments, which aim to explore the auditory perceptions (ability) of humans. Stimuli with relatively simple elements are often used in conventional psychoacoustic experiments to avoid interference among different kinds of auditory information. However, some special and simple stimuli may sometimes be used to investigate some detailed characteristics or abilities of spatial sound systems. It can be regarded as an ***event or object-based paradigm***. For example, some simple stimuli (such as band-pass pink noise) are used in virtual source localization experiments (Section 15.6) to test the ability of a system to recreate virtual sources at different directions.

In a scene-based paradigm, a set of subjective attributes that can reveal the differences in the systems or objects to be assessed should be chosen. These attributes should be well defined to obtain stable and consistent results during assessment. Various attributes are often described by special terms, whose definitions should be completely understood by subjects. Conversely, Mason et al. (2001) pointed out that the nonverbal description (such as sketch) of spatial auditory perception also has an advantage, especially when the whole perceived spatial auditory scene is described. Figure 3.3 shows an example of describing a spatial auditory scene by using a sketch.

Various methods are utilized to present the subjective attributes to be assessed (Bagousse et al., 2010). For example, in one method, subjects can use a set of predefined (common) attributes and scales to assess. In another method, each subject employs an individual set of attributes and scales to assess, and statistical methods are applied to remove the redundancy among the data. Furthermore, differences in the verbal or literal expression of different subjects are considered to avoid the restriction caused by predefined attributes and scales. These two methods can be combined in practice, and a set of terms that describe common attributes can be derived from the second method. The second method has some examples (Berg and Rumsey, 2006; Francombe et al., 2017a, 2017b). For example, Berg and Rumsey applied the repertory grid technique (RGT) in psychology to elicit subjective attributes. It involves three steps.

1. Elicitation
   By comparing the three versions of stimuli, which are from the same source material and equal in duration, each subject judges which of the three versions is the most perceptually different from others and then describes its similarities and differences in a pair of opposite verbal descriptors. This course is repeated for all signals until no new verbal descriptors appear. The personal (bipolar) constructs of each subject are generated on the basis of bipolar verbal descriptors.
2. Scaling
   With the bipolar verbal descriptors as two ends of the scale, each subject rates each of his/her own personal constructs at a time with a five-point scale for each stimulus to be rated, yielding the results for each subject and each stimulus.
3. Data analysis
   Verbal protocol analysis is used to separate descriptive constructs from personal constructs, and cluster analysis is performed to extract the main descriptive attributes.

In addition, some other methods, such as perceptual structure analysis (Choisel and Wickelmaier, 2006), multidimensional scaling (Bech and Zacharov, 2006), and verbal transcript (Guastavino and Katz, 2004), are applicable to elicit subjective attributes.

Various attributes have been suggested for the assessment of spatial sound (systems), including timbral attributes, spatial attributes, and defects (Bagousse et al., 2014). These attributes vary in different standards and studies although they may be related to one another. Timbral attributes are important categories of perceived attributes. Rumsey et al. (2005) showed that the effect of the degradation of timbre on the overall perception of audio quality is greater than that of the degradation of spatialization in multichannel sound reproduction. Therefore, timbral attributes, such as timbral coloration, brightness, clearness, fullness, and naturalness, should be included in subjective assessment experiments (Bagousse et al., 2010).

Some review articles have described various spatial attributes (Zacharov and Koivuniemi, 2001a, 2001b, 2001c; Berg and Rumsey, 2001, 2003; Rumsey, 1998, 2002; Bagousse et al., 2010, 2014; Zacharov and Pedersen, 2015). The following spatial attributes are specified in the standard of the subjective assessment of small impairments in a spatial sound system according to the ITU (ITU-R BS 1116-3, 2015):

1. Stereophonic image quality (for stereophonic systems)
   "This attribute is related to differences between a reference and an object in terms of sound mage locations and sensations of the depth and reality of an audio event."
2. Front image quality (for multichannel sound systems)
   "This attribute is related to the localization of frontal sound sources. It includes stereophonic image quality and definition losses."
3. Impression of surround quality (for multichannel sound systems)
   "This attribute is related to spatial impression, ambience, or special directional surround effects."
4. Localization quality (for advanced sound systems)
   "This attribute is related to the localization of all directional sound sources. It includes stereophonic image quality and definition losses. This attribute can be separated into horizontal localization quality, vertical localization quality, and distant localization quality. In case of tests with an accompanying picture, these attributes can be also separated into localization quality on a display and localization quality around a listener."
5. Environment quality (extended the attribute of surround quality for advanced sound systems)
   "This attribute is related to spatial impression, envelopment, ambience, diffusivity, or spatial directional surround effects. It can be separated into horizontal environment quality, vertical environment quality, and distant environment quality."

ITU has also recommended a method for selecting and describing the attributes and terms in subjective assessment (ITU-R BS 2399-0, 2017). The suggested procedure includes: (1) attribute elicitation, (2) comprehension check, (3) suitability, (4) attribute descriptors, (5) redundancy check, (6) repeatability evaluation, and (7) final consensus. The ITU standard also defines some audio attributes and suggests adding some spatial audio attributes in future development.

In addition to the aforementioned ITU standards, other standards and studies deal with spatial attributes. Some of these standards and studies are established for spatial sound (ITU-R BS 1284-1, 2003), and others are created for the subjective assessment of loudspeaker systems (Toole, 1985; IEC 60268, 1998). However, Rumsey (2002) argued that the interpretations of spatial attributes in some standards and studies are unclear.

As stated in Section 1.8.1, auditory source width and listener envelopment are two important spatial attributes of a concert hall. They are related to early reflections and late reverberation in a concert hall. However, the cases of spatial sound reproduction are not completely identical to those of concert halls. Spatial sound is not always used to recreate the sensations of a concert hall, but it may be used to recreate auditory effects and sensations of different natural environments or even recreate the sensations of artificial environments. Although spatial sound assessment can refer to some methods in concert hall assessment, the method and attributes of concert hall assessment cannot be directly used for spatial sound assessment. Such difference is addressed when the "envelopment" of spatial sound reproduction is discussed in Section 3.1.

Rumsey (2002) proposed that the spatial attributes of an auditory scene in reproduction can be classified into four groups, e.g., individual source attributes, ensemble (such as the string instrument part in an orchestra) attributes, environment attributes, and scene attributes. Within each group, these attributes can be subdivided according to width, distance, depth, and immersion. That is,

1. Sub-attributes related to width
   - Individual source width
   - Ensemble width
   - Environment width
   - Scene width
2. Attributes related to distance and depth
   - Individual source distance
   - Ensemble distance
   - Individual source depth
   - Ensemble depth
   - Environment depth
   - Scene depth
3. Attributes related to immersion
   - Individual source envelopment
   - Ensemble source envelopment
   - Environmental envelopment
   - Presence

Berg and Rumsey (2002) suggested the following attributes for assessing five (5.1)-channel reproduction: naturalness, presence, preference, low-frequency content, ensemble width, individual source width, localization, source distance, source envelopment, room width, room size, room sound level, and room envelopment.

The appropriate selection of subjective attributes for spatial sound assessment depends on the system to be evaluated and its application. A few examples of subjective attributes for spatial sound assessment are outlined above, and most of them are mainly for horizontal surround sound. Some attributes should be supplemented when this assessment is extended to spatial surround sound. Spatial attributes that describe perceptual performance in a vertical dimension should be added. In fact, the aforementioned localization quality and environment quality specified by ITU-R BS 1116-3 (2015) involve the perceptual performance in the vertical dimension. In addition, with the development of various advanced multichannel sound and headphone-based spatial sound, some studies have explored attributes for the subjective assessment of a wide range of spatial sound techniques and systems (Zacharov and Pedersen, 2015; Francombe et al., 2017a, 2017b).

Even in the same standard or investigation, different subjective attributes are not always independent or orthogonal to one another. Therefore, some statistical analysis tools, such as principal component analysis and independent factor analysis, are often used to remove the correlation among subjective attributes and map independent perceptual components to objective physical attributes.

Various methods are used for subjective attribute assessment. The most direct method is discriminating the difference between an object (system) and a reference (system). Any perceivable difference can be used for discrimination. This method is simple, but it cannot reveal the details of various subjective attributes related to system quality. Because of multiplex subjective attributes for spatial sound, some attributes are chosen and assessed by an appropriate method depending on practical requirements and applications.

15.3 AUDITORY COMPARISON AND DISCRIMINATION EXPERIMENT

15.3.1 Paradigms of auditory comparison and discrimination experiment

Auditory comparison and discrimination experiments are typical approaches used to evaluate electroacoustic systems and audio processing. They are often utilized to examine spatial sound, especially headphone-based VAD. In this method, the performance of a target is assessed through an overall subjective comparison of a target and a reference (reproductions or signals) under controlled experimental conditions. Some commonly employed experimental design methods are described as follows.

In an ***A/B comparison experiment***, which is the simplest method, a subject evaluates whether a difference exists between samples A and B randomly rendered by turns. More strictly, one of the following auditory discrimination and forced choice paradigms are used:

1. The ***two-interval, two-alternative forced choice*** (*2I/2AFC*) paradigm includes two samples, namely, samples A and B, arranged randomly. Thus, two different combined sample sequences (AB and BA) are formed. Subjects determine which sample in the combined sequences contains the target (known) perception attributes. If they fail to identify the sample, randomly presented forced choices are needed. Each subject often performs repeated evaluation with equal repetitions of sequences AB and BA.
2. The ***three-interval two-alternative forced choice*** (*3I 2AFC*) paradigm includes three samples. The first sample is always labeled reference sample A, and it is randomly followed by sample A and target sample B. This order yields two different combined sample sequences, i.e., AAB and ABA. Subjects determine which of the second or third sample is different from (or identical to) the first sample. If they are unable to identify the sample, randomly presented forced choices are needed. Each subject conducts repeated evaluations with equal repetitions of AAB and ABA.
3. The ***three-interval three-alternative forced choice*** (*3I 3AFC*) paradigm includes three samples. Two of them are classified as sample A, and the remaining sample is sample B, thereby generating three different combined sample sequences, namely, AAB, ABA, and BAA. Subjects determine which sample in the combined sample sequence differs from the two other samples. If they fail to distinguish the required sample, randomly presented forced choices are needed. Frequently, each subject is asked to repeatedly evaluate the samples at equal repetitions of each combined sequence.
4. The ***four-interval, two-alternative forced choice*** (*4I/2AFC*) paradigm includes four samples. The first and fourth samples are always denoted as sample A, and the second and third samples are randomly named A or B; thus, two different combined sample sequences are obtained, i.e., AABA and ABAA. Subjects determine whether the second or third sample is different from (or identical to) the first (or fourth) sample. Forced choices are randomly presented upon failure to identify the required sample. Each subject is also often asked to provide repeated assessments at equal repetitions of each combined sequence.

5. The ***four-interval, three-alternative forced choice*** (4I/3AFC) paradigm includes four samples. The first sample is always A followed by one B and two repetitions of A in random order, producing three different combined sample sequences, namely, AAAB, AABA, and ABAA. Subjects determine which of the second, third, or fourth sample of the combined sample sequences differs from the first. If they cannot identify the required sample, forced choices are randomly presented. Each subject is asked to perform repeated evaluations at equal repetitions of each combined sequence.
6. The ***four-interval, one oddball, two-alternative forced choice*** (***4I/1O/2AFC***) paradigm is a deformation of the 4I/2AFC paradigm. It includes four samples with combined sequences, i.e., AABA, ABAA, BABB, and BBAB. Three out of the four samples are identical, and one (either the second or third) differs from the three other samples. Subjects determine which of the second or third sample in the combined sample sequences varies from the others. If they fail to identify the required sample, forced choices are randomly presented. Each subject performs repeated assessments at equal repetitions of each combined sequence.

Other methods, with principles similar to those underlying these six paradigms, are available. Given the similarity in principles, such methods can be regarded as variants.

The proportion of correct discriminations is usually calculated to analyze experimental data. Calculation is carried out on either the repetitions of all subjects or the repetitions of each subject, depending on experimental purpose and requirement.

Statistical methods are needed to further analyze and validate experimental data. The most commonly used approach is to test the hypothesis that the proportion of correct discriminations is equal to or greater than the expected value of random choice under certain significance levels. In the latter case, target and reference samples can be discriminated. The related statistical methods are outlined in Appendix B of this book, and details are described in another textbook (Marques de Sá, 2007).

Some issues concerning auditory comparison and discrimination experiments are worth noting.

1. The experiments aim to determine whether audible differences exist between reference and target samples, and any perceived difference can be used as a discrimination criterion. In this sense, the artificial artifacts in reproduction should be reduced as much as possible. For example, the overall level difference between reference and target samples should be avoided.
2. These types of experiments only qualitatively analyze the perceptual difference between reference and target samples. If further quantification analysis of specific perceived attributes is necessary, the methods to be discussed in Sections 15.4 and 15.5 would be helpful.
3. From the perspective of mathematical statistics, the number of samples (N) should be sufficiently large to enable the derivation of reliable statistical results. A value greater than 50 or 100 is usually preferred, with a low limit of $N \geq 30$. For a limited number of subjects, repetitions are necessary.

15.3.2 Examples of auditory comparison and discrimination experiment

Timbre coloration in Ambisonics is analyzed in Section 12.2.3. For a given order of Ambisonics and listening position, when frequency exceeds a certain upper limit, the difference in BLLS between reproduction and target source exceeds the JND of BLLS (with a rough estimated value of 1 Phon/ERB), and perceivable timbre coloration occurs. These results have been validated by a psychoacoustic experiment (Liu and Xie, 2015).

The workload is high because experiments should be carried out in a combination of different orders of Ambisonics, signals of different frequency ranges, and various listening positions. In the case of Ambisonic reproduction with real loudspeakers, controlling the listener's head position is difficult, especially if the accuracy of the head position over a long time must be ensured. Therefore, a scheme for the binaural reproduction of Ambisonics via headphones and the 3I 2AFC paradigm were used in the experiment. For a target plane wave at a given direction and a given listening position, reference binaural signals A were directly created by HRTF-based filtering. For a given order Ambisonics, loudspeaker configuration, and listening position, target binaural signal B was created by the binaural Ambisonic method in Section 11.6.1. The input stimuli were pink noise with a fully audible bandwidth and its low-pass versions with different cutoff frequencies. The 3I 2AFC paradigm was used to discriminate target B and reference A. Eight listeners participated in the experiment. Under each condition and stimulus order of AAB and ABA, each subject repeatedly examined the stimuli three times, so the following was obtained: 8 subjects × 2 stimulus orders × 3 repetitions = 48 samples. The statistical method in Appendix B was used to analyze the experimental results. At a significance level of 0.05, if the proportion of correct judgment exceeded 0.63 and thus was statistically larger than that of random judgment, the difference between the target and the reference was perceivable, and timbre coloration in reproduction occurred. Statistical analysis indicated that when the cutoff frequency of low-pass filtered pink noise was beyond a certain limit, the difference between the target and the reference was perceivable. The experimentally derived upper frequency limits were consistent with those from Moore's loudness model.

Auditory comparison and discrimination experiments are also used to evaluate various simplifications or approximations of signal processing in VAD. Accurate binaural signals are used as reference sample A, and simplified binaural signals are utilized as target sample B. If no perceivable difference is observed between a target and a reference, simplification is effective. These experiments are extensively used to evaluate the temporal windowing, smoothing, filter design, and spatial interpolation of HRTFs (Sections 11.4 and 11.5.1). Two examples are presented (Kulkarni and Colburn, 2004; Xie and Zhang, 2010), and relevant results are described (Xie, 2013).

15.4 SUBJECTIVE ASSESSMENT OF SMALL IMPAIRMENTS IN SPATIAL SOUND SYSTEMS

Auditory comparison and discrimination experiments (Section 15.3) can reveal the difference between target and reference reproduction or quality impairments in target reproduction, but such experiments cannot quantitatively assess differences or impairment. In 1997, the ITU recommended a standard for the subjective assessment of small impairments in multichannel sound (ITU-R BS 1116-1, 1997); in 2015, they revised it as ITU-R BS 1116-3. The technical conditions of reference listening rooms, reference sound fields, and monitor loudspeakers for subjective assessment experiments are described in Section 14.5.

A double-blind triple-stimulus with a hidden reference paradigm is recommended by the ITU standard. It is composed of three samples or signals (A, B, and C), and the known reference (such as an unimpaired reference) is always sample A. The hidden reference and the object are randomly assigned to B and C, respectively. Therefore, two kinds of combinations are generated: AAB and ABA. Subject rates the second and third stimuli and assigns the first stimulus as the reference stimulus. Repetitions are often employed, with the two combinations presented at equal numbers.

A five-grade scale ranging from 1.0 to 5.0 (with a precision of one decimal place) is used:

5.0: imperceptible
4.0: perceptible but not annoying
3.0: slightly annoying
2.0: annoying
1.0: very annoying

An experiment involves a training phase and a grading phase. In the training phase, subjects are trained to be familiar with experimental tasks. In the grading phase, subjects are free to switch among the stimuli (sample) in a triple-stimulus method.

The aforementioned method is applicable to assess the perceived attributes of sound reproduction. For mono sound, the main perceived attribute is basic audio quality. For stereophonic sound, the attributes of stereophonic image (virtual source) quality (Section 15.2) can be added. For multichannel sound, the spatial attributes of front image (virtual source) quality and the impression of surround quality can be supplemented. For advanced sound systems, the attributes of timbral quality, localization quality, and environment quality can be used. The aforementioned method is also utilized to assess other timbral and spatial attributes mentioned in Section 15.2 depending on the purposes of experiments (Bagousse et al., 2010). In an experimental design, subjects should completely understand the meaning of attributes to avoid confusion.

Although universally "suitable" source materials are unavailable for assessments under all conditions, critical materials that can reveal differences in objects should be chosen. In addition, the artistic content of a material should be neither so attractive nor so disagreeable or wearisome so that subjects are not distracted from detecting impairments. The duration of each stimulus is typically 10–25 s. The number of stimuli is about 1.5 times of the objects to be assessed and should not be fewer than five. It should also be equal among the objects to be assessed.

The ITU standard has not specified the number of listeners (subjects) for experiments but has emphasized that 20 listeners are usually sufficient. Because experiments aim to reveal slight impairments in reproduction rather than assess performance for average people, experienced expert subjects should be selected. In addition to the pre-screening methods in Section 15.1, a post-screening of listeners after the experiment should be included. Under each condition, each subject grades the hidden reference and the object. Ideally, if a subject can identify the object from the hidden reference, his/her mean grade difference between them should be statistically negative. Otherwise, his/her mean grade difference should be statistically zero. Therefore, data from each subject are subjected to a one-side t-test at a significant level of $\alpha = 0.05$ to test the hypothesis that the mean grade difference between the object and the hidden reference is statistically zero. If the hypothesis is rejected, the subject has sufficient expertise to justify the object. Otherwise, the data from that subject are excluded from the succeeding analysis.

Raw data from subjective assessment experiments should be subjected to preliminary statistical processing. Let x_n, $n = 1,2\ldots N$ denote the scores of N subjective evaluations, the mean and standard deviation are

$$\bar{x} = \frac{1}{N}\sum_{n=1}^{N} x_n \qquad \sigma_x = \sqrt{\frac{1}{N-1}\sum_{n=1}^{N}\left(x_n - \bar{x}\right)^2}, \tag{15.3.1}$$

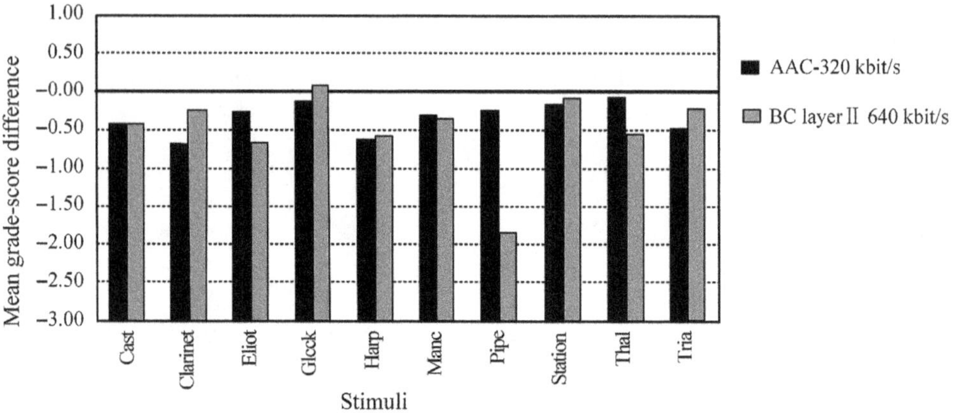

Figure 15.1 Perceived audio quality of the MPEG-2 AAC main profile at a bit rate of 320 kbit/s and MPEG-2 Layer II at a bit rate of 640 kbit/s given by BBC. The abscissa denotes different stimuli, and the ordinate corresponds to the mean score difference between the object and hidden reference among 23 subjects. (Brandenburg and Bosi, 1997; with the permission of the Audio Engineering Society.)

which represent the average score and dispersion of score data. The average score of all subjects is a measure of the audio quality impairment of an object. Further parametric and nonparametric statistical hypothesis tests (such as *t*-test and Wilcoxon signed-rank test) and ANOVA are needed to derive a definite conclusion from the data. Some statistical analysis methods are outlined in Appendix B and details are referred to relevant textbook (Marques de Sá, 2007).

The aforementioned method is often used to assess impairments in the perceived quality of multichannel sound. For the assessment of the quality of loss coding algorithms in Chapter 13, original (uncoded) multichannel sound signals are used as reference stimulus A, and coded signals are set as object signal B. For example, using the method specified by ITU-R BS 1116-1, BBC and NHK simultaneously assessed the perceived quality of MPEG-2 AAC coding scheme. Figure 15.1 shows the results given by BBC (Brandenburg and Bosi, 1997). The objects are MPEG-2 AAC main profile at a bit rate of 320 kbit/s and MPEG-2 Layer II at a bit rate of 640 kbit/s. Figure 15.1 illustrates the mean score difference between the object and hidden reference among 23 subjects, and Section 13.5.3 presents the conclusion.

15.5 SUBJECTIVE ASSESSMENT OF A SPATIAL SOUND SYSTEM WITH INTERMEDIATE QUALITY

The bit rate of many recent transmission media and playback devices, such as Internet, Digital Radio Mondiale, and digital satellite broadcasting, is restricted. These media and devices only provide intermediate-quality sound reproduction. A double-blind triple stimulus with a hidden reference paradigm (Section 15.4) is appropriate for evaluating minor impairments in spatial sound systems. However, it is not completely appropriate for examining spatial sound systems with lower quality because of its poor discrimination between slight differences in quality at the bottom of the scale. The ITU recommended a **multi-stimulus test with a hidden reference and anchor** (***MUSHRA***)for the subjective assessment of a spatial sound system with intermediate quality. The revised version of the MUSHRA standard in 2015 is referred to as ITU-R BS 1534-3 (2015).

The technical conditions of reference listening rooms, reference sound fields, and monitor loudspeakers for MUSHRA are identical to those of ITU-R BS 1116-3 in Section 15.4.

A set of MUSHRA-processed signals comprise a reference, a hidden reference, an object, and at least two hidden anchor signals. An original unprocessed signal is used as a reference and hidden reference signal. The low anchor and mid anchor are low-pass filtered versions of the original signal with cutoff frequencies of 3.5k and 7 kHz, respectively. The total number of signals should not exceed 12 (1 hidden reference + 9 test objects + 1 low anchor +1 mid anchor). Subjects simultaneously grade the hidden reference, multiple objects, and two anchors by comparing them with the reference. To compare differences in various objects, subjects can freely switch between the reference and any of the signals under testing. The grade scale varies continuously from 0 to 100. The following scores and their corresponding quality are used: 0–20, bad; 20–40, poor; 40–60, fair; 60–80, good; and 80–100, excellent. This scale is identical to that used for evaluating the quality of television pictures.

The MUSHRA method is applied to assess perceived quality in sound reproduction. For mono, stereophonic, and multichannel sounds, the main attributes to be evaluated are identical to that of the assessment of minor impairments in Section 15.3. Only one attribute should be examined in a trial. This method is also used to analyze other timbral attributes and spatial attributes in Section 15.2 (Bagousse et al., 2010).

Although universally "suitable" source materials are unavailable, materials representing a typical broadcasting program for desired applications should be chosen to reveal differences in the system to be tested. The duration of each stimulus is about 10 s and does not exceed 12s. The number of stimuli is about 1.5 times of the objects to be assessed and should not be fewer than five. It shall also be equal for each object to be assessed.

The MUSHRA method requires experienced subjects. Although the ITU standard has not specified the number of subjects for the MUSHRA test, it states that 20 subjects are usually sufficient. Subjects are chosen by a pre-screening paradigm similar to that in Section 15.1 and then trained. A post-screening of subjects after the experiment should also be included. The data from a subject are excluded from the succeeding analysis in either of the following cases: the subject grades the hidden reference for more than 15% of the test items lower than the score of 90, or the subject grades the mid anchor for more than 15% of the test items higher than the score of 90.

Prior to statistical analysis, the original scores for each test condition are converted linearly to normalized scores ranging from 0 to 100, with 0 being the bottom of the score. Statistical analysis is applied to the normalized scores. Some statistical analysis methods are outlined in Appendix B and described in detail in another textbook (Marques de Sá, 2007).

Using the NHK 22.2-channel sound as a reference, Kim et al. (2010) assessed the subjective directional quality and overall perceived quality of multichannel spatial sound reproduction with various numbers of loudspeakers in the upper (top) layer. They used a method similar to MUSHRA (Section 6.5.3). MPEG-D MPEG Surround coding (Section 13.5.5) and the ISO/MPEG-H 3D coding (Section 13.5.6; Herre et al., 2014, 2015) are evaluated with the MUSHRA method. Through this method, the perceived quality of various numbers of channels (loudspeakers), bit rates, and at-central or off-central listening position are examined.

In some cases, the known hidden reference and anchor in MUSHRA are unavailable. As such, the ITU further specified a standard for the subjective quality assessment of audio differences in sound systems by using multiple stimuli without a given reference (ITU-R BS 2132-0, 2019). Conditions and methods are similar to those in the MUSHRA standard in many aspects, but systems are assessed without a reference. Subjects are asked to grade each system under testing in terms of overall subjective quality and predefined sets of the selected attributes. The overall subjective quality rating is performed using a continuous quality scale, and attributes are rated on 100-point linear scales.

15.6 VIRTUAL SOURCE LOCALIZATION EXPERIMENT

15.6.1 Basic methods for virtual source localization experiments

The perceived sound source position is one of the subjective attributes of sound. One of the primary purposes of spatial sound is to generate virtual sources at different spatial positions in terms of direction and distance. Thus, the ability to generate differentially positioned virtual sources is an important measure of the performance of spatial sound. This ability is typically evaluated through virtual source localization experiments. Such experiments are commonly performed to examine the performance of spatial sound and have been conducted in many studies, although they are excluded from the ITU standard for the subjective assessment of stereophonic and multichannel sound.

A localization experiment involves an appropriate number of subjects determining the perceived spatial position (direction and distance) of headphone- or loudspeaker-rendered virtual sources under certain physical conditions. It is classified as absolute evaluation, which often requires further statistical analysis of raw results.

For different spatial sound systems, the type and duration of stimuli for localization experiments may vary. For multichannel sound, stationary random signals such as pink noise and white noise are often used. Speech, music, and impulse are also often used to investigate the localization for wideband stimuli with transient characteristics. Low-, high-, and band-pass filtered noise may be sometimes used to test the ability of recreating a virtual source in different frequency ranges. The length of stimuli usually varies from 1.0s to 10s.

For VADs, wideband noise signals such as white noise and pink noise are often used to examine the effects of various localization cues (especially spectral cues at high frequencies). In the case of static VADs, short-duration stimuli (e.g., hundreds of milliseconds) may be employed to avoid possible head movements during presentation. Conversely, in the case of dynamic VADs, long-duration stimuli (a few seconds or more) are frequently used to provide subjects enough time to move their heads through which subjects can fully utilize the resultant dynamic localization cues to localization.

In some experiments, a subject's head position should be fixed. This requirement is particularly demanding in static loudspeaker-based binaural reproduction through which the perceived virtual source position strongly depends on the head position. In this case, the head position can be monitored by various optic systems, cameras, or head-tracking systems. In some other experiments, such as Ambisonic and dynamic VAD reproduction, head turning is allowed or even encouraged to examine dynamic cues in reproduction.

In a horizontal plane, the position of a virtual source is specified by coordinates (r, θ), e.g., by the distance and azimuth of a source with respect to a subject's head center. In a three-dimensional space, the position of a virtual source is specified by spherical coordinates (r, θ, ϕ), as shown in Figure 1.1. An anticlockwise spherical coordinate system is often used in studies on stereophonic and multichannel sound, and a clockwise spherical coordinate system is utilized in studies on HRTF and VAD. For convenience in the analysis of lateral localization caused by interaural cues (ITD and ILD) and vertical localization caused by spectral cues in HRTF, an interaural polar coordinate system with respect to the head center is sometimes employed in localization experiments. In an interaural polar coordinate system, the source position is specified by (r, Θ, Φ), where the distance r is identical to that illustrated in Figure 1.1; the interaural polar elevation Φ is defined as the angle between the projection of the directional vector of the sound source to the median plane and the frontal axis with $-90° < \Phi \leq 270°$ (or $0° \leq \Phi < 360°$); and the interaural polar azimuth Θ is defined as the angle between the directional vector of the sound source and the median plane with $-90° \leq \Theta \leq +90°$.

Strictly, virtual source localization involves distance perception in terms of r and directional localization in terms of azimuth θ and elevation ϕ. However, a localization task is often simplified in practical experiments. For a horizontal virtual source, only the horizontal azimuth θ is reported in the directional localization. For virtual source in the median plane or upper lateral plane, using an interaural polar coordinate system is convenient. In this case, interaural polar elevation Φ or interaural polar azimuth Θ only is reported in the directional localization. Moreover, the ability of the human hearing to estimate the sound source distance is generally poorer than the ability to locate the sound source direction. The perceived distance control in spatial sound reproduction is also relatively difficult. Therefore, directional localization is included, and distance perception is omitted in many experiments. If source distance perception in headphone reproduction should be evaluated qualitatively, a judgment of virtual sources located within/on the surface or outside the head is often sufficient.

Various methods are used to identify virtual source positions in localization experiments. The most straightforward method is oral identification. In this case, a spatial coordinate is often set up in a listening room to help subjects determine spatial positions. Other methods include pointing by hand or with a laser pointer, clicking by mouse in a graphic interface representation of a three-dimensional space in a computer, touching a three-dimensional spherical model, and turning the head toward the perceived source direction in dynamic localization associated with a head-tracking system. Alternatively, some researchers predefine a certain number of spatial directions and ask subjects to select the single direction closest to the perceived direction.

15.6.2 Preliminary analysis of the results of virtual source localization experiments

Obtaining statistical results necessitates the preliminary statistical analysis of raw data. The simplest method is the calculation of the mean and standard deviation of a perceived source direction in terms of azimuth and elevation. Let us assume that, for a target virtual source direction (θ_S, ϕ_S), N experimental observations (samples) $[\theta_I(n), \phi_I(n)]$ exist; $n = 1, 2 \ldots N$, which may be from N different subjects, N repetitions of a subject, or K repetitions for each L subject with $N = L \times K$. Thus, the mean and standard deviation of these experimental samples are

$$\begin{aligned} \bar{\theta}_I &= \frac{1}{N}\sum_{n=1}^{N}\theta_I(n) \qquad \sigma_\theta = \sqrt{\frac{1}{N-1}\sum_{n=1}^{N}\left[\theta_I(n)-\bar{\theta}_I\right]^2} \\ \bar{\phi}_I &= \frac{1}{N}\sum_{n=1}^{N}\phi_I(n) \qquad \sigma_\phi = \sqrt{\frac{1}{N-1}\sum_{n=1}^{N}\left[\phi_I(n)-\bar{\phi}_I\right]^2} \end{aligned}, \tag{15.6.1}$$

where $\bar{\theta}_I$ and $\bar{\phi}_I$ represent the mean or centric-perceived azimuth and elevation, respectively; and σ_θ and σ_ϕ reflect the dispersion of the sample. Ideally, $\bar{\theta}_I$ and $\bar{\phi}_I$ should be equal to the target values of θ_S and ϕ_S, respectively. This equivalence means that the samples should be distributed in a diagonal straight line on the plot of function $\bar{\theta}_I$ (or $\bar{\phi}_I$) against θ_S (or ϕ_S). Otherwise, evaluation errors occur. The following equations are defined as the deviations (errors) between the mean perceived value (azimuth or elevation) and the target value:

$$\Delta\theta_1 = \frac{1}{N} \left| \sum_{n=1}^{N} \left[\theta_I(n) - \theta_S\right] \right| = \left| \bar{\theta}_I - \theta_S \right|$$
$$\Delta\phi_1 = \frac{1}{N} \left| \sum_{n=1}^{N} \left[\phi_I(n) - \phi_S\right] \right| = \left| \bar{\phi}_I - \phi_S \right|. \tag{15.6.2}$$

Furthermore, the following equation is defined as the mean unassigned (absolute) errors between the perceived value (azimuth or elevation) and target value across samples:

$$\Delta\theta_2 = \frac{1}{N} \sum_{n=1}^{N} \left| \theta_I(n) - \theta_S \right| \quad \Delta\phi_2 = \frac{1}{N} \sum_{n=1}^{N} \left| \phi_I(n) - \phi_S \right|. \tag{15.6.3}$$

For virtual sources within a horizontal plane, such as in the case of horizontal reproduction, a source direction is determined by azimuth θ_I alone so that only the error on θ_I is analyzed. Similarly, for sound sources within a lateral or median plane, only the error on the interaural polar azimuth Θ or interaural polar elevation Φ is analyzed.

For source distance perception, calculating the ratio or percentage of the perceived virtual sources located within/on the surface or outside the head is sufficient if merely qualitative evaluations are provided. If quantitative assessments are collected, a statistical analysis similar to that conducted for direction evaluations is needed. Given N samples $r_I(n)$, $n = 1, 2 \ldots N$ of the perceived source distances that correspond to a target distance r_S, the mean and standard deviation are

$$\bar{r}_I = \frac{1}{N} \sum_{n=1}^{N} r_I(n) \quad \sigma_r = \sqrt{\frac{1}{N-1} \sum_{n=1}^{N} \left[r_I(n) - \bar{r}_I\right]^2}. \tag{15.6.4}$$

When the perceived virtual source is close to the horizontal plane, these statistical methods for azimuth and elevation errors are basically reasonable. In a three-dimensional space, however, the arithmetic mean and standard deviation of $[\theta_I(n), \phi_I(n)]$, $n = 1, 2 \ldots N$ cannot be directly used to describe the directional dispersion of data because the source directions tested are distributed on the spherical surface whose center is consistent with that of the subject's head. For example, an actual deviation of $\Delta\theta_1$ or $\Delta\theta_2 = 30°$ in a horizontal plane of $\phi = 0°$ is considerably greater from the perspective of absolute direction than that at high elevations, such as $\phi = 60°$. This particular problem, coupled with the fact that azimuth and elevation errors are almost certainly not independent of each other, complicates the statistics of perceived source directions.

Wightman and Kistler (1989b) used spherical statistics to analyze three-dimensional perceived source directions. First, the target direction is represented by a unit vector $\boldsymbol{r}_S$ from the origin to (θ_S, ϕ_S); then, the result of the nth perceived sample is represented by a unit vector $\boldsymbol{r}_I(n)$ from the origin to the perceived direction $[\theta_I(n), \phi_I(n)]$. Similar to the case in Equation (15.6.3), the mean angular error is defined as the unassigned mean of the angular difference between target and perceived directions:

$$\Delta_2 = \frac{1}{N} \sum_{n=1}^{N} \left| \arccos\left[\boldsymbol{r}_I(n) \cdot \boldsymbol{r}_S\right] \right|, \tag{15.6.5}$$

where the dot denotes the scalar multiplication of two vectors. The mean perceived source direction $\overline{r}_I$ is obtained by the sum of all perceived direction vectors; thus,

$$\overline{r}_I = \sum_{n=1}^{N} r_I(n). \tag{15.6.6}$$

The angle between $\overline{r}_I$ and r_S, expressed as Δ_1, reflects the absolute directional deviation between the target and the mean perceived direction, with similar meanings to those indicated in Equation (15.6.2):

$$\Delta_1 = \arccos\left(\frac{\overline{r}_I \cdot r_S}{|\overline{r}_I|}\right). \tag{15.6.7}$$

The length of $\overline{r}_I$, that is, $R = |\overline{r}_I|$, indicates the dispersion of perceived source directions. Ideally, N samples of perceived directions should be identical to the target direction; thus, $R = N$. The shorter the length of R is, the greater the dispersion observed in the perceived directions will be. κ^{-1} can be defined to describe this dispersion. For small samples with $N < 16$, an approximate estimation of κ is given as

$$\kappa = \frac{(N-1)^2}{N(N-R)}. \tag{15.6.8}$$

Thus, the lower the value of κ^{-1} is, the less the dispersion will be. When $R = N$, $\kappa^{-1} = 0$.

Conversely, reversal errors (front-back and up-down confusion) may arise for some subjects in virtual source localization experiments, especially the localization experiment of headphone-based binaural reproduction (Section 11.7.2). In this case, the mean perceived source direction in Equation (15.6.1) or Equation (15.6.6) is meaningless, and the mean angular error provided by Equation (15.6.3) or Equation (15.6.5) may be considerable or potentially misleading. In the case of a low confusion rate, two approaches are commonly used to treat confusion:

1. Prior to processing, confusion is excluded from raw data.
2. Prior to processing, reversal is resolved (i.e., through spatial reflection, a response is coded as though it indicates the correct hemisphere).

After the application of these treatment approaches, the confusion rate must be determined to reflect the characteristics of experimental results. The spatial reflection for sound sources near the mirror plane imposes a low effect on the error in Equation (15.6.3) while increasing the confusion rate. For example, for a sound source with a target azimuth $\theta_S = 95°$ and perceived azimuth $\theta_I = 85°$, the angular error is 10°without spatial reflection. With spatial reflection, the angular error becomes zero, but the account of front-back confusion is added. In this case, spatial reflection is insignificant because an angular error of 10° may be attributed to a slight error in localization.

If the experimental results on different subjects considerably differ (such as in the case of binaural synthesis with nonindividualized HRTFs), the mean perceived source direction across subjects is insignificant. In this situation, many researchers simply calculate the mean perceived source direction across the repetitions for each subject and present results for each subject. As a simple alternative, the distribution of the perceived samples without the corresponding average is provided.

15.6.3 Some results of virtual source localization experiments

Many works on spatial sound involve virtual source localization experiments, and a few representative results are presented here. As mentioned in Sections 1.7.1, a summing localization experiment with two loudspeakers was conducted in the early age of stereophonic sound. Figures 1.26 and 1.27 illustrate the results of summing localization with two loudspeakers. The trading curve between ICLD and ICTD (Figure 1.28) was also obtained from an interpolation of data from a virtual source localization experiment. The results of summing localization with a pair of loudspeakers at horizontal azimuths of 45°and 135° and at 30° and 90° are illustrated in Figures 4.3 and 4.7, respectively (Thiele and Plenge, 1977).

A virtual source localization experiment in frontal, lateral, and rear regions was also conducted for ITU 5.1-channel loudspeaker configuration and pair-wise amplitude panning (Xie, 2001a). In this experiment, orchestral music was used as the stimulus, and eight subjects were included. The mean perceived azimuth and the corresponding standard deviation of the eight subjects were calculated according to Equation (15.6.1), and the analysis was validated by the results (Section 5.2.2). In addition, the results of summing localization with a pair of loudspeakers in a median plane for multichannel spatial sound are illustrated in Section 6.2.

Wightman and Kistler (1989b) conducted a three-dimensional directional localization experiment in an anechoic chamber with small loudspeakers as free-field real sources to compare the localization performance of free-field real sources and headphone-rendered virtual sources. They incorporated individualized HRTFs and HpTFs into binaural synthesis and equalization to precisely duplicate the at-eardrum sound pressures in headphone presentation as the pressures of free-field real sources.

Target source positions were uniformly distributed at 72 spatial directions, including six different elevation planes; that is, $\phi = -36°, -18°, 0°, 18°, 36°$, and $54°$. The stimulus in this experiment was a train of eight 250 ms bursts of band-pass Gaussian noise (200 Hz–14 kHz). To prevent subjects from becoming familiar with specific stimuli after several trials, the authors shaped the energy spectrum of noise according to an algorithm that divided the spectrum into critical bands. A random intensity (uniform distribution in a 20 dB range) was then assigned to the noise within each critical band.

Eight adult subjects (four males and four females) participated in the experiment. All had normal hearing, as verified by audiometric screening within 15 dB hearing loss (HL). Within the 72 target source directions under free-field real source conditions, each subject performed $K = 12$ repeated evaluations of each of the initial 36 target source directions and $K = 6$ repeated assessments of each of the remaining 36 target source directions. Within the 72 target source directions in headphone presentation, each subject carried out $K = 10$ repeated evaluations of each of the initial 36 target source directions and $K = 6$ repeated assessments of each of the remaining 36 target source directions. Lastly, the experimental results for each subject and for all subjects were separately analyzed.

Additional calculations yield the mean angular error Δ_2 in Equation (15.6.5), κ^{-1} in Equation (15.6.8), and the mirror confusion rate. Wightman showed the statistical results for each subject and the overall statistical results for eight subjects. The results indicated that the mean angular errors of the free-field real source and headphone presentation are similar, which is 20°in the low-elevation region and slightly larger in the high-elevation region. The averages of the confusion rate in all the spatial regions are about 11% for headphone presentation and 6% for the free-field real source.

Using a similar experimental method, Wenzel et al. (1993) further studied the effect of nonindividualized HRTFs on virtual source localization. Free-field real and virtual sources included 24 spatial directions, and headphone stimuli were synthesized using HRTFs and HpTFs from a subject with better localization in the experiment of Wightman and Kistler. Sixteen subjects (2 males and 14 females) participated in the experiment. In each of the 24 directions, every subject performed $K = 9$ repeated evaluations. Experimental data were statistically analyzed. The mean angular error Δ_2 calculated by Equation (15.6.5) and κ^{-1} calculated by Equation (15.6.8) indicated that after spatial reflection to resolve the reversal error in raw data, the localization of virtual sources is accurate and comparable with that of the free-field real sources for 12 of 16 subjects. The main problem caused by the use of nonindividualized HRTFs is the increasing confusion rate (especially front-back confusion). For virtual sources, the front-back and up-down confusion rates are 31% and 18%, respectively. For free-field real sources, the front-back and up-down confusion rates are 19% and 6%, respectively. These results have been frequently referred to in Sections 11.7.2.

Localization experimental results are naturally distributed in three-dimensional auditory space, which can be conveniently described using relevant spherical statistical and graphical methods (Leong and Carlile, 1998). Figure 15.2 is an example of the localization results for of dynamic binaural synthesis of free-field virtual source and headphone presentation (Section 11.10.2, Zhang and Xie, 2013). White noise was used as stimulus and MIT-KEMAR HRTFs were used in dynamic binaural synthesis. Twenty-eight target source directions, which were distributed in the right hemispherical space at four elevations ($\phi_S = -30°, 0°, 30°, 60°$) with seven azimuths ($\theta_S = 0°, -30°, -60°, -90°, -120°, -150°, 180°$) at each elevation, were chosen for the experiment. Six subjects were involved in the experiment and asked to evaluate the perceived virtual source positions in terms of direction and distance. Each subject assessed each target virtual source position thrice, yielding 18 evaluations of each target position. Figure 15.2 shows that the statistical results for the perceived virtual source directions after front-back and up-down confusion are resolved. The results are presented on the surface of a sphere and viewed from the front, right, and rear directions, respectively. The notation '+' represents the target virtual source direction. The black points at the center of ellipses are the average perceived directions across subjects and repetitions. These ellipses are the confidence regions at a significance level $\alpha = 0.05$. The results also indicate overall front-back and up-down confusion rates of 2.5% and 13.7%, respectively; the average perceived virtual source distances across subjects and repetitions range from 0.86 to 1.23 m. Therefore, dynamic binaural synthesis visibly reduces front-back and up-down confusion and improves externalization in headphone presentation even when nonindividualized HRTFs are used.

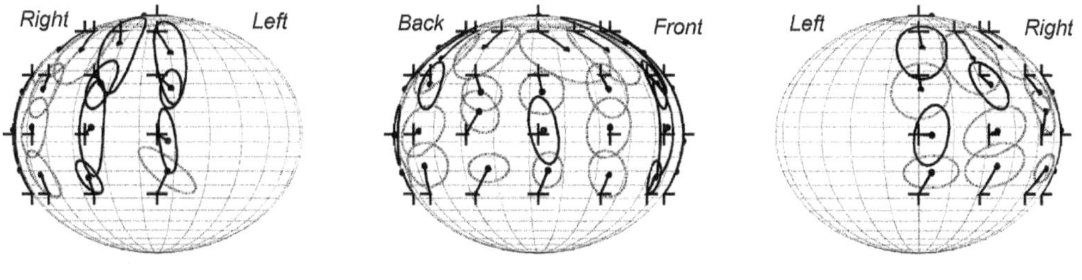

Figure 15.2 Graphical representation of the localization results of the dynamic binaural synthesis of a free-field virtual source: (a) view from the front; (b) view from the right; (c) view from the rear.

15.7 SUMMARY

Objective and subjective assessments are two major methods for evaluating spatial sound, but subjective assessment is more important. Subjective assessment experiments should be conducted with strict methods and conditions.

Assessing the subjective auditory perception of sound is a part of experimental psychology. Reliable results can only be obtained with strict psychoacoustic experimental methods and conditions, as well as appropriate statistical analyses.

The standard by ITU and guidance by AES Technical Council specify the technical conditions of reference listening rooms, reference sound fields, and monitor loudspeakers for the subjective assessment of stereophonic and multichannel sound. These standards and guidance can be used as a reference for some other spatial sound evaluations.

Spatial sound possesses multidimensional perceived attributes, and most of them can be classified into timbral attributes and spatial attributes. These attributes should be well defined to obtain stable and consistent results in the assessment. They vary across different standards and researches, although they may be related.

Various methods are used for the subjective assessment of spatial sound. They can be selected and designed according to the system, object, and attributes to be assessed. Source materials, stimulus duration, and the number of subjects should be carefully chosen, and invalid data should be eliminated from the final results. Final conclusions with a certain confidence level should be derived on the basis of appropriate statistical analysis of the raw experimental data, such as statistical hypothesis tests and ANOVA.

Auditory comparison and discrimination experiments are conducted to assess the perceived performance of an object by identifying the subjective difference between the object and a reference. Various paradigms are used for auditory comparison and discrimination experiments.

A paradigm of double-blind triple stimuli with a hidden reference has been recommended by the ITU for the subjective assessment of minor impairments in stereophonic and multichannel sound. Furthermore, MUSHRA has been recommended by the ITU for the subjective assessment of spatial sound systems with intermediate quality.

The virtual source localization experiment is a widely used method for evaluating the performance of spatial sound and has been used in numerous studies. Many of the experimental conclusions in the preceding and current chapters are derived from virtual source localization experiments.

Chapter 16

Applications of spatial sound and related problems

Cinema and domestic reproductions are two traditional and major applications of spatial sound. Since 1990, with the development of signal processing, communication, computer, and Internet technologies, spatial sound has been increasingly applied to various fields in scientific research, engineering, communication, consumer electronics, entertainment, and medicine, which are far beyond its traditional applications. In particular, sound reproduction in mobile devices is a new application of spatial sound. This chapter discusses the applications of spatial sound and related problems, including some recent applications since the 1990s. However, it does not intend to include all applications; instead, this chapter outlines some typical applications only. These typical examples are enough to demonstrate the wide applications of spatial sounds.

In Section 16.1, some typical applications of spatial sound, including those to commercial cinema, domestic reproduction, and recent automobile audio, are addressed. In Section 16.2, the applications of virtual reality, communication and information systems, and mobile devices, are discussed. In Section 16.3, the applications to scientific experiments on spatial hearing and psychoacoustics are addressed. In Section 16.4, applications to sound field auralization, especially for room sound fields, are discussed. In Section 16.5, the applications to clinical medicine are briefly outlined.

16.1 APPLICATIONS TO COMMERCIAL CINEMA, DOMESTIC REPRODUCTION, AND AUTOMOTIVE AUDIO

16.1.1 Application to commercial cinema and related problems

As stated in Section 14.7, commercial cinema includes traditional film cinema and current digital cinema. Cinema sound is a traditional application of spatial sound (Barbar, 2015).

The Disney film Fantasia in 1939 was the first practical application of multichannel sound techniques. The system involves three soundtracks and a special control track. The control track was used to pan the soundtracks to a number of loudspeakers (including the rear loudspeakers) automatically. This system can be regarded as an early form of object-based spatial sound. Soundtracks and control tracks were optically stored on a separate piece of film that operated synchronously with the picture.

Some multichannel sound formats for films were introduced in 1950 and used from 1950 to the 1960s. Warner Brothers introduced a four-channel format, including left, center, and right screen channels, and an effect or surround channel. The Cinemascope, a widescreen film introduced by 20th Century Fox, also used a four-channel sound format. Four-channel soundtracks were stored in the magnetic stripes in the 35 mm films, with one on each edge

DOI: 10.1201/9781003081500-16

of the film outside the perforations and one in each position between the picture and the perforations.

The Todd A-O six-channel format for 70 mm (widescreen) film sound involved five frontal channels and one rear channel. Six-channel soundtracks were stored in the magnetic stripes, with two on each edge of the film outside the perforations and one in each position between the picture and the perforations.

The magnetic soundtracks of films improved the audio quality and were popular in the 1950s. However, they were rejected because of a short lifetime and high cost; consequently, optical soundtracks became popular again in the 1970s. Dolby A and later Dolby-SR noise reduction technique improved the audio quality of optical soundtracks in films.

In Sections 1.9.3 and 8.1.2, Dolby Laboratories introduced Dolby Stereo in the mid-1970s. They encode four-channel signals into two independent signals by matrix and are stored in the optical soundtracks of a 35mm film. Dolby Stereo encoding technique, together with the adaptive decoding, Dolby A or Dolby-SR noise reduction techniques, has been widely used for film sound in cinema.

Since the 1990s, discrete multichannel sounds have been widely used in commercial cinema reproduction. Discrete multichannel sound for 35 mm film sound in commercial cinemas has three competitive formats. Dolby Digital was the first format, which utilized 5.1-channel sound and was coded with Dolby Digital (AC-3). DTS, the second format, similarly utilized 5.1-channel sound and was coded with coherent acoustics coding. Sony Dynamic Digital Sound (SDDS) was the third format, which employed 7.1-channel sound and was coded with ATRAC.

The data for the three formats are stored in different locations on a 35 mm film. Therefore, more than one soundtrack may be stored in a 35mm film. Figure 16.1 illustrates the positions of various data and soundtracks in a 35mm film. Data from Dolby Digital are stored optically in an area between the sprocket holes of a film. Analog optical soundtracks are stored in their original position for compatibility, e.g., along one side of the pictures. The SDDS data are stored on the other side of the picture. Moreover, the DTS soundtrack is recorded on a separate optical disk and played back by a laserdisc player. A special time code is recorded in the film to synchronize the audio with the picture. The sound format in films that combines

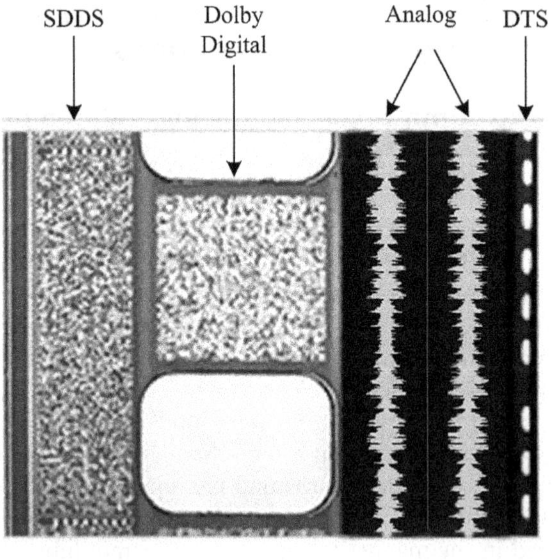

Figure 16.1 Positions of various sound/data tracks on a 35 mm film.

Dolby Digital data and analog optical soundtracks with Dolby-SR noise reduction is termed Dolby-SR-D.

Most multichannel sound systems after the 1980s can be used for both cinema and domestic reproduction, or more precisely, most multichannel sound systems for domestic reproduction are simplified versions of cinema sound systems. However, the requirement for a large-sized listening region makes the recording and production of cinema sound different from those of domestic sound programs. In particular, for almost all the listening positions in a cinema, the difference in propagation delays among different loudspeakers exceeds the upper limit of summing localization, and thus, recreating a virtual source between a pair of loudspeakers by summing localization is infeasible (Section 7.4.1).

Multichannel sounds for cinema aim to recreate a localization effect on the directions of loudspeakers and a subjective sensation of ambient sounds. Increasing the number of channels and loudspeakers is essential to improve the localization performance of cinema sound reproduction. As stated, the center channel and loudspeakers have been included in cinema sound since its early days to recreate auditory localization matching with the picture, especially matching with dialog in a film. The center channel and loudspeakers have also been introduced subsequently for sound reproduction in domestic theaters. In Section 14.7, for horizontal surround sound in cinemas, such as Dolby Stereo and 5.1-channel sound, the frontal channel signals are reproduced by frontal loudspeakers behind the screen. The surround channel signals are reproduced by a series of surround loudspeakers in the side and rear of the cinema. For Dolby Stereo, the mono surround channel signal is reproduced by all surround loudspeakers. For 5.1-channel sound, the left and right surround channel signals are reproduced by two symmetric sets of left and right surround loudspeakers, respectively. The frontmost surround loudspeakers are normally arranged at a 2/3 distance from the back wall to the screen. Figure 16.2 illustrates the loudspeaker configuration of 5.1-channel sound in a cinema. The difference in propagation delays leads to the decorrelation of sounds from different surround loudspeakers at a listening position because the distances from each loudspeaker to a listening position are different; as a result, a subjective sensation of envelopment by ambient sounds occurs. The subjective sensation of envelopment can be enhanced if surround channel signals are appropriately decorrelated and reproduced by a series of bipolar loudspeakers. This method is used in cinema Dolby Stereo reproduction by THX in Section 14.7.

To improve the reproduction of rear sound, 6.1-channel sound for cinema was introduced, such as Dolby Digital Surround EX in Section 5.3 and DTS-ES in Section 8.1.4. Introduced by Dolby Laboratories in 2010, Dolby Surround 7.1 (with 7.1 channels) has also been used for cinema sound.

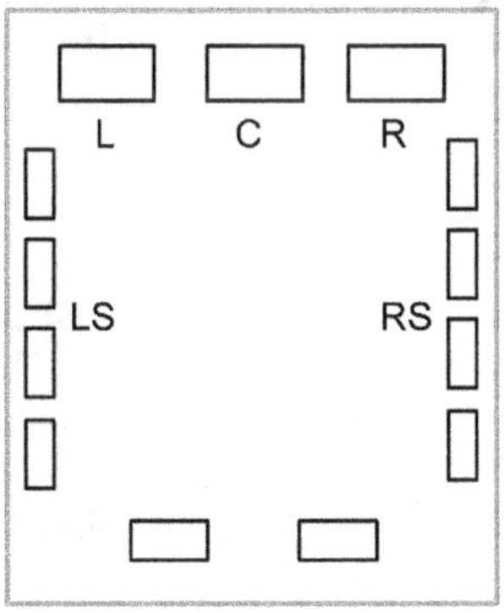

Figure 16.2 5.1-channel loudspeaker configuration in cinemas.

The IMAX film format, which was introduced in 1970, is a 70 mm film format with high image resolution. The IMAX film does not contain a soundtrack, which allows the picture to occupy more of the frame. The original IMAX had a 6-channel sound format, including left, center, right, left surround, right surround channels, and an upper center channel above the screen. The six-channel soundtracks were stored on a separate 35 mm magnetic film and played back on a film follower. The signal for the subwoofer was derived from the six-channel signals. In the 1990s, the magnetic film was replaced by a separate CD-ROM with a DTS-based soundtrack.

In Section 6.5, multichannel spatial surround sounds have emerged as new-generation sound reproduction techniques. Up to 2021, many cinemas have been equipped with multichannel spatial surround sound systems. Multichannel spatial surround sound has various (commercial and competitive) formats. Channel-based formats include the Auro-3D series and 22.2-channel sound by NHK. Object-based or a combination of object and channel-based formats include Dolby Atmos, MDA, and Auro MAX. For cinema use, the considerations and designs for different formats and systems may differ, but some basic methods are similar. As in the case of 5.1-channel reproduction in a cinema, each frontal (screen) channel of spatial surround sound is usually reproduced by an individual loudspeaker. Surround channels are reproduced by a series of surround loudspeakers arranged on the side, rear, or top. Surround loudspeakers can be divided into several groups that reproduce a surround channel signal. Bass management may be applicable to surround channels to reduce the size of surround loudspeakers. The low-frequency components of surround channels can be mixed and reproduced by surround subwoofers.

Dolby Atmos, as described in Section 6.5.2, is a hybrid of object- and channel-based systems with a variable number and configuration of loudspeakers. Dolby Laboratories (2015) has specified the characteristics and configurations of loudspeakers for Dolby Atmos in cinemas. Figure 16.3 illustrates a typical loudspeaker configuration comprising five screen loudspeakers and a screen subwoofer. A series of surround loudspeakers is arranged on each side of the wall and on two sides of the ceiling. The front-most surround loudspeakers are arranged nearer to the screen than those in 5.1-channel sound. Such a loudspeaker configuration is also appropriate for reproducing the beds in Dolby Atmos. The screen channels are reproduced by full-bandwidth loudspeakers, and bass management is not needed. However, Dolby Atmos allows for bass management of the surround channel. If a pair of surround subwoofers is used, the surround subwoofers are arranged in the back half of the cinema along the sidewalls, rear wall, or ceiling. Multiple pairs of subwoofers may also be used. The crossover frequency of bass management for surround channels should not exceed 100 Hz. To reduce the cost of the equipment in digital cinema, psychoacoustic experiments in Dolby Laboratories suggested that top surround loudspeakers with a higher low-frequency limit be used for Dolby Atmos. Accordingly, the crossover frequency of bass management for top surround channels is increased to 150 Hz (Hirvonen and Robinson, 2016). In addition, for digital cinema applications, the object data, metadata, and bed data of Dolby Atmos form a "print master" file, which is stored or transmitted after being packed with the industrial standard technology of material exchange format (MFX).

The former IOSONO (now Bacro Audio Techniques) pioneered an object-based spatial sound technique for films [ITU-R Report BS 2159-7 (2015)]. In this technique, a series of loudspeakers is arranged around and on top of the cinema. A combination of WFS, object-based, and channel-based methods is used to create a virtual point source at a practical loudspeaker distance, a virtual plane wave source behind the loudspeaker array (a virtual point source at a far-field distance), and a focused source inside the loudspeaker array. The IOSONO technique has been incorporated into Auro MAX. Various multichannel sound programs can be theoretically reproduced by virtual loudspeakers created by WFS. In other

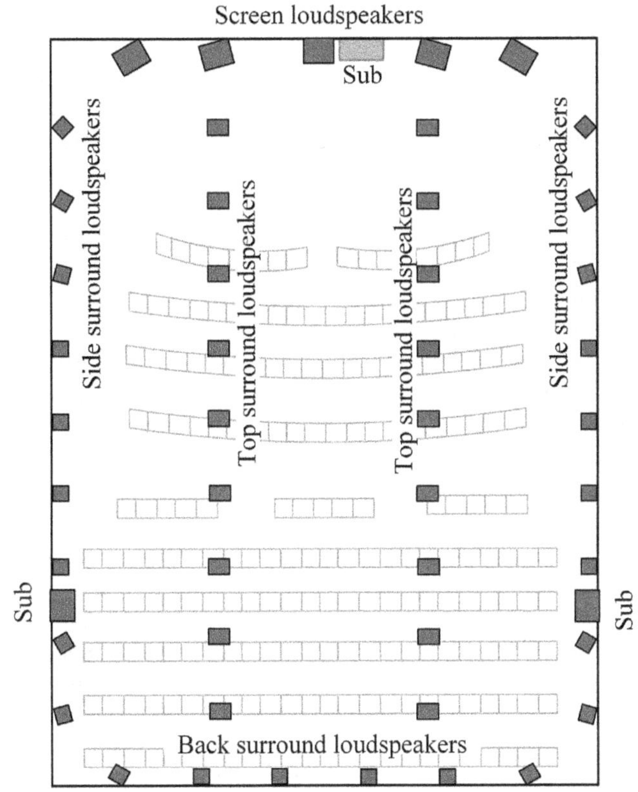

Figure 16.3 Typical loudspeaker configuration of Dolby Atmos for cinema. (adapted from Dolby Laboratories, 2015.)

words, the fixed loudspeaker array in WFS is theoretically appropriate for various multichannel sound reproductions. Boone et al. (1999) also suggested using virtual loudspeakers created by WFS for domestic multichannel sound reproduction. However, the interval between adjacent loudspeakers in WFS should satisfy the requirements of the Shannon–Nyquist spatial sampling theorem. Otherwise, spatial aliasing errors and audible artifacts may occur.

Overall, cinema sound has long been an important application of spatial sound. Since the 1990s, cinema sound has evolved from horizontal to three-dimensional spatial sound, from fewer channels to more channels, and from channel-based to object-based sound. Such evolutions are partly to improve sound effects in reproduction and partly due to commercial competition. Various competitive spatial sound techniques and formats are used in cinemas. The principles and considerations of these techniques are similar. The prospect of these formats depends partly on technical development and more on commercial and market requirements. This distinguishing feature is also observed in the development of spatial sound.

16.1.2 Applications to domestic reproduction and related problems

Domestic reproduction is another traditional application of spatial sound. In Section 1.9.3, stereophonic sound has been popular in domestic reproduction since the end of the 1950s or the beginning of the 1960s. Furthermore, 5.1-channel and some other multichannel horizontal surround sounds have been widely used in domestic reproduction since the 1990s. Some techniques for channel-based and a combination of object- and channel-based spatial surround sound have been developed quickly, and some of these techniques are used in domestic reproduction.

The reference acoustic conditions and standards for spatial sound reproduction are addressed in Chapter 14. These conditions and standards are specified for subjective experiment and monitoring and focused on 5.1-channel sound for great content. For domestic uses, if permitted, reproduction under reference acoustic conditions may result in a well-perceived performance. However, these reference acoustic conditions cannot be satisfied in most domestic reproductions. Therefore, some compromise should be made. For example, a compromise in loudspeaker arrangement is made if a practical domestic environment does not allow a standard arrangement of loudspeakers. The guidance by Dolby Laboratories allows a relatively flexible arrangement of surround loudspeakers in 5.1-channel sound reproduction. When some loudspeakers (especially center loudspeakers) are not arranged in a circle with an equal distance to the central listening position, a signal delay is applied to compensate for the arrival time difference among different loudspeakers.

The main loudspeaker with a full bandwidth usually has a larger size. In many cases, arranging multiple large main loudspeakers, especially in the case of multichannel sound reproduction, is inconvenient. Bass management can be used to reduce the size of the main loudspeakers. For example, multichannel sound can be reproduced by small-sized satellite loudspeakers and a subwoofer. However, a high crossover frequency in bass management (and thus small-sized main loudspeakers) often contradicts the requirement for localization. Surround loudspeakers with a smaller size than that of frontal loudspeakers are also applicable. For convenience, wireless and active surround loudspeakers may also be used.

Some methods to simplify the loudspeaker configuration for spatial surround sound in domestic reproduction have been suggested. A straightforward method is to downmix a larger number of signals to a smaller number of reproduced channels, for example, to downmix 22.2 channel signals to a smaller number of channel signals (Hamasaki, 2011). Other studies have suggested combining a pair of loudspeakers in the upper and middle layers of 22.2-channel reproduction into a single tallboy-type loudspeaker shown in Figure 16.4 [ITU-R Report BS 2159-7 (2015)].

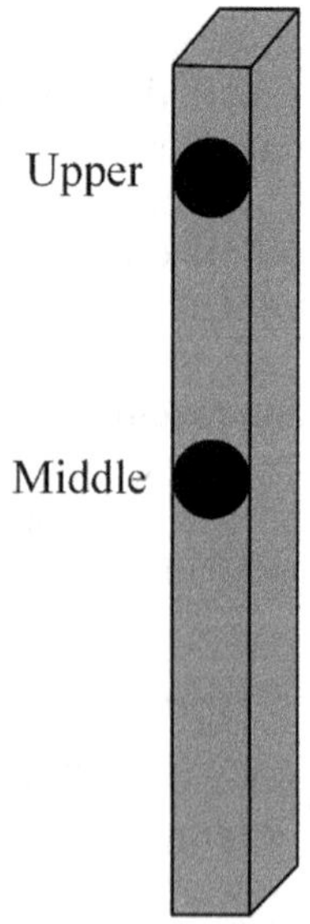

Figure 16.4 Tallboy-type loudspeaker used for 22.2-channel reproduction.

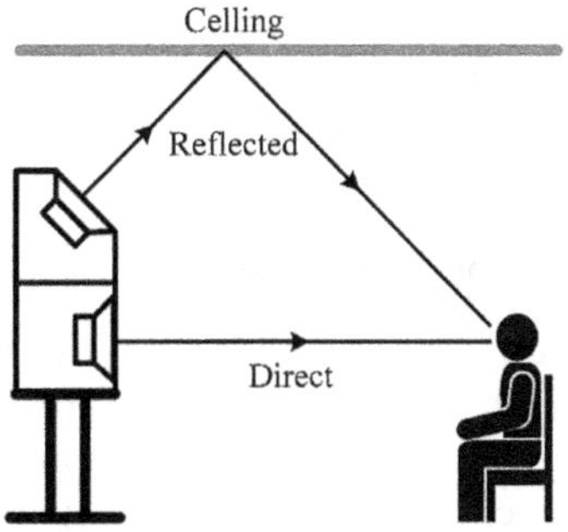

Figure 16.5 Dolby Atmos-enabled loudspeaker and ceiling reflection.

Figure 16.5 illustrates the loudspeaker configuration of Dolby Atmos for domestic reproduction (Dolby Laboratories, 2016). In practice, Dolby Atmos-enabled loudspeakers can be used when the top surround loudspeakers can be inconveniently arranged. The four Dolby Atmos-enabled loudspeakers are arranged on the tops of the horizontal left, right, left surround, and right surround loudspeakers, respectively. The radiation main axes of Dolby Atmos-enabled loudspeakers point to the ceiling. The reflections from the ceiling lead to an overhead perceptible effect. In this case, a reflective ceiling is needed. A pair of horizontal loudspeakers and a Dolby Atmos-enabled loudspeaker can also be integrated into a single loudspeaker with two sets of posts.

For a domestic theater system with a flat panel display, an array of small loudspeakers above, below, or around the display can be conveniently used for sound reproduction. Loudspeaker array reproduction can be designed on the basis of the principles of the multiple-receiver position-matching method in Section 9.6.3 or WFS in Chapter 10. In other words, frontal virtual loudspeakers for multichannel sound reproduction can be created by an array of real small loudspeakers. For example, Okubo et al. (2012) in NHK proposed using a loudspeaker array frame around a flat panel display to reproduce 22.2-channel signals. Reproduction with a loudspeaker array avoids the problem of conflict between the center loudspeaker and the visual display arrangement. However, limited by the principle of sound field reconstruction, a frontal loudspeaker array can only create virtual frontal loudspeakers. Beamforming can be used to reproduce side and rear surround channel signals with a frontal loudspeaker array. It creates several radiated beams with sharp directivity. These radiated beams are reflected to the listening position by the lateral walls of a listening room, thereby producing sound from lateral directions (Chung et al., 2012).

The virtual or transaural reproduction of multichannel sound in Section 11.9.3 is also applicable to simplify loudspeaker configuration in domestic reproduction. In particular, a pair of loudspeakers on the two sides of a flat panel display or a soundbar above or below the flat panel display can be used for virtual reproduction. Such loudspeaker configurations are convenient for domestic use. As stated in Section 11.9.3, virtual reproduction has some defects, such as narrow listening regions and localization errors. Methods of transaural reproduction with multiple loudspeakers (Section 11.8.1) and optimal source distribution (Section 11.8.3) can be used to enlarge the listening region. Multiple loudspeakers can be arranged above, below, or around the visual display. For example, Matsui and Ando (2013) in NHK proposed a method of transaural reproduction of 22.2-channel signals with a loudspeaker array frame around the flat panel display in a TV set.

Overall, various multichannel sound techniques and systems have been used in domestic reproduction. Ideal acoustic conditions and standards cannot be satisfied in practical domestic reproduction, and some compromises should be made. These compromises invariably degrade reproduction performance. Fortunately, domestic reproduction is not as critical as a scientific experiment or a professional monitor. A reasonable design and compromise may still yield an appropriate perceived performance.

16.1.3 Applications to automobile audio

Automobile audio has become popular in recent years. The acoustic environment in a car cabin is considerably different from that in a domestic room. Strictly speaking, the car cabin environment is inappropriate for high-quality sound reproduction. However, practical use requires sound reproduction in a car cabin. As a result, special acoustic design and processing in a car cabin are needed to reduce artifacts and improve perceived reproduction performance (Shively, 2000; Rumsey, 2016).

The background noise level in a car cabin is usually much higher than that in a domestic environment. The noise level varies quickly with the road condition and speed of the car, with a large part of the power spectra within the frequency range below 500 Hz. Background noise restricts the dynamic range of sound reproduction. The acoustic characteristics of an enclosed car cabin are determined by its interior shape and size and the sound absorption of interior surfaces. Because the dimensions of a car cabin are much smaller than those of an original domestic room, the frequency-dependent acoustic behavior in a car cabin shifts up in frequency. At about 50 Hz, at which the wavelength of sound is much longer than the usual dimensions of a car cabin, the car cabin acts as an acoustic system of lumped parameters. In this pressure region, a car cabin provides a large gain to the low-frequency radiation of the sound source (which may reach an order of 10–20 dB or more at the frequency of 20 Hz). Therefore, reproducing low-frequency components is relatively easier in a car cabin than in a domestic room. The reflections and interference caused by the surface inside the car cabin lead to various interior modes in a frequency range of dozens of Hertz to about 1 kHz, where the wavelength of sound is comparable with the dimensions of a car cabin. These modes result in frequency- and position-dependent radiated pressures with a level variation reaching the order of 8–12 dB. Absorptive and reflective regions appear above the frequency of 1 kHz, but reverberation fields do not occur because of the quick absorption of the surface in a small, enclosed space. The small, enclosed space in a car cabin also makes the statistical acoustic method for sound field analysis invalid. Some numerical methods based on wave acoustics, such as boundary element method, are needed for analyzing the sound field inside a car cabin with an irregular shape.

The loudspeaker-to-listener distance in a car cabin is usually small, ranging from the order of 2.0–2.5 m down to 0.15–0.2 m. The practical structures inside a car cabin are inappropriate for optimal loudspeaker arrangement. Moreover, the listener is usually not located in the central line of the car cabin. The aforementioned factors make the design of sound reproduction in a car cabin greatly different from that in a domestic environment. On the other hand, some parameters of a car cabin, such as the shape and dimensions, acoustic characteristics, and positions of loudspeakers, are known or predictable. On the basis of these parameters, the perceived artifacts in the listening position can be alleviated by appropriate acoustic design and signal processing. These designs and processing depend on the practical environment of the car cabin.

Loudspeakers can be symmetrically arranged with respect to the central line of the cabin. The reflective surfaces in a car cabin influence the sound field created by loudspeakers. The arrangement and high-frequency directivity of loudspeakers should be designed carefully to reduce the spectral peaks and notch in the listening position caused by the interference of reflection. Frontal loudspeakers are typically flush mounted on the dashboard, with the subwoofer located in the trunk.

The loudspeaker-to-receiver response can be optimized with a signal processing method. The path differences among different loudspeakers to the listening position can be compensated for by signal delay. Various signal processing techniques are implemented by DSP. A given equalization and compensation frequently improve the effect at one listening position while degrading it at another. In practice, equalization and compensation for each listening position can be designed, saved, and called for according to requirements. Some compromises should be made if the effects at more than one listening position need to be improved.

Automobile audio is mostly for music reproduction and thus is different from audio for a domestic theater, which is mostly for reproduction with an accompanying picture. Various spatial sound techniques and systems, including two-channel stereophonic sound, Dolby ProLogic, and 5.1- or 7.1-channel sound, have been used for automobile audio. The program materials come from various digital storage media (such as optical disks) or analog or digital radio broadcasting. Multichannel spatial surround sound may also be applicable for automobile audio reproduction. Currently, however, most program materials for multichannel spatial surround sound are intended for reproduction with accompanying picture. The situation may change in the future. The method of upmixing stereophonic or 5.1-channel program materials for multichannel spatial surround sound reproduction is also applicable. Bai and Lee (2010) suggested a combination of non-ideal transmission equalization and upmixing or downmixing of stereophonic or multichannel sound signals for automobile reproduction.

16.2 APPLICATIONS TO VIRTUAL REALITY, COMMUNICATIONS, MULTIMEDIA, AND MOBILE DEVICES

16.2.1 Applications to virtual reality

Virtual environments or virtual reality systems provide users with the feeling of being present in natural environments through computer-controlled artificial surroundings (Blauert et al., 2000). Virtual reality includes virtual visual, auditory, and tactile senses. The interaction and complementarity of multiple pieces of information on the aforementioned aspects strengthen the sense of reality and immersion. For virtual reality applications, a spatial sound system must recreate natural and immersive auditory senses rather than reconstruct a target sound field or binaural pressures accurately. From this point of view, various spatial sound techniques and systems may be applied to virtual reality, depending on application requirements (Hollier et al., 1997).

The hardware of headphone-based dynamic binaural reproduction is relatively simple. In early days, dynamic virtual auditory environment systems for a single user were implemented on a hardware platform composed of a personal computer, a flat panel display, and a head tracker. A system with multiple computer terminals, multiple visual displays, and multiple head trackers may be used for more than one user at a time. Static transaural reproduction with two frontal loudspeakers can only recreate the spatial information in the frontal-horizontal plane and allows for a narrow listening region. Therefore, it is suitable for a single user.

Since the 2010s, commercial three-dimensional visual displays have been developed quickly. Virtual reality can be effectively implemented through a combination of a head-mounted visual display and a dynamic virtual auditory environment system (Jin et al., 2005). Many products with head-mounted visual displays include a head tracker. When a user walks and turns in the virtual space, the head tracker detects the position and orientation of the user, and the system updates the virtual visual and auditory scenes dynamically, resulting in good immersive senses.

For large-region virtual reality systems with surround or 3D projection screens, sound reproduction within a large listening region is required to enable multiple users at the same time or allow a user to walk within the virtual space. In this case, spatial sound systems based on sound field reconstruction are relatively appropriate. Two examples include the system using warped B-format Ambisonics by Hollier et al. (1997) and the third-generation CAVE system using WFS by DeFanti et al. (2009).

An important application of virtual reality is virtual training. Unlike actual training, virtual reality offers a safe and low-cost task-training environment. An early example is driving training simulation (Krebber et al., 2000). A virtual car acoustic environment, which is part of the virtual driving environment, requires the following components:

1. An external moving sound source with respect to the driver (e.g., traffic flow with Doppler shift)
2. Fixed engine sound, which depends on engine speed and torque
3. Fixed tire sound, which depends on speed and road conditions
4. Fixed wind noise, which depends on speed
5. Background noises, commands to the driver, and other related elements

Virtual acoustic environment systems dynamically synthesize or call sound signals from pre-recorded sound databases according to a driver's control maneuvers, then reproduce the sound signals through headphones or loudspeakers with appropriate signal processing. Similar methods can be applied to special training environments, such as virtual aviation, aerospace, and submarine environments (Doerr et al., 2007).

Virtual auditory reality is also applied to various auditory scene displays (Hollier et al., 1997), exhibitions, entertainment (Kan et al., 2005), and the creation of special effects in video/audio program production.

Virtual reality was mainly applied to some professional fields. Since 2010, virtual reality with head-mounted visual display and dynamic virtual auditory environment systems has been applied to consumer fields, such as games, entertainment, media, education, and social intercourse. Therefore, auditory virtual reality has a wide application field.

16.2.2 Applications to communication and information systems

An important purpose of applying spatial sound to speech communication is to improve speech intelligibility. In real life, conversation usually occurs in environments with background noise and multiple speech sources concurrently competing. When target speech sources and other noise or competing speech sources are spatially separated, the hearing system can use the cocktail party effect (Section 1.7.5) to obtain expected information and guarantee speech intelligibility. This ability is attributed to binaural hearing.

However, mono signal transmission is dominantly used in currently available communication systems in which the inability to spatially separate targets and competing sources degrades speech intelligibility. Spatial sound methods can preserve the spatial information of sources or spatially separate the sources by signal processing and thus improve the quality of speech communication (Begault and Erbe, 1994; Drullman and Bronkhorst, 2000). Psychoacoustic experimental results indicate that spatially separating multiple speech sources by VADs enhances speech intelligibility for either full-bandwidth or 4 kHz low-pass (phone quality) speech signals (Begault, 1999).

From the point of auditory perception, various spatial sound techniques are theoretically applicable to speech communication, depending on practical requirements and costs. For speech communication with headphones, VAD is advantageous because its hardware is simple, and it requires only two independent signals (and therefore a low bandwidth for signal transmission). In addition, conventional headphone presentation is inclined to cause in-the-head localization and auditory fatigue for a long listening time. Incorporating VAD into speech communication can create natural auditory effects and easy auditory fatigue. For speech communication with loudspeakers, other spatial sound techniques may be needed.

Multiple talkers are present in a teleconferences at the same time (Kang and Kim, 1996; Evans et al., 1997). In addition to improvement in speech intelligibility in teleconferences, immersive and close-to-reality communication services are provided by spatial sound techniques. In remote conferencing, the direct approach to preserving spatial information and improving the intelligibility of transmitted speech is to combine and reproduce the binaural sound signals obtained by artificial-head recording in each meeting room if participants are distributed in two or more separate meeting rooms. Alternatively, the speech of each participant is captured by a microphone, then rendered by static or dynamic binaural synthesis according to a pre-defined spatial distribution and acoustic environment, and finally presented to all participants. Other techniques, such as discrete multichannel sounds, Ambisonics, WFS, and microphone arrays, are also applicable to teleconferencing (Boone and Bruijn, 2003) and create a virtual meeting environment. The DiRACc in Section 7.6 and SAOC in Section 13.5.5 are also applicable to teleconferencing (Herre et al., 2011). Similar applications include telepresence (Hollier et al., 1997), various emergency commands, and telephone systems in which multiple speech sources should be monitored simultaneously.

VADs also contribute to aeronautical communication, and numerous investigations on this application were undertaken by the NASA Ames Research Center (Begault, 1998). The projects are categorized as a combination of VAD applications for speech communication and information orientation. Given that civil aircraft cockpits are characterized by high environmental noise, headphones (aside from speech communication) are used to reproduce air traffic warnings based on which pilots determine target (e.g., other aircraft) directions or identify corresponding visual targets (e.g., radar display) and accordingly take appropriate measures. Applying VADs to aeronautical communication improves speech intelligibility and reduces the search or reaction time of pilots with the help of spatialized auditory warnings. The latter is important for flight safety. Additionally, headphone presentation may be combined with active noise control to reduce pilot exposure to binaural noise.

The mentioned VAD applications in aeronautical communication include auditory-based information display and orientation. In some cases, vision is often superior to hearing in terms of target identification and orientation. However, acoustic information becomes particularly important when a target is out of visual range (e.g., behind the human) or when visual information overload occurs (such as in the case of multiple visual targets). In real life, auditory information often guides visual orientation (Bolia et al., 1999), and goals can be localized through hearing even without visual help (Lokki and Gröhn, 2005). Therefore, revealing target information and orientation is another important application of spatial sound.

Audio navigation systems, which combine global positioning system (GPS) with VADs, reproduce sounds as they are emitted from target directions. These systems are applied primarily in civil or military rescue searches (Kan et al., 2004). A similar method can be used to present various types of spatial auditory information, such as that contained in guidance and information systems for the blind (Loomis et al., 1998; Bujacz et al., 2012) or tourism and museum applications (Gonot et al., 2006).

Monitoring multiple targets (such as different instruments and meters) is often necessary for practice; such targets cause visual overload. In this situation, VADs are used to alleviate the visual burden by transforming part of the visual presentation of spatial information into an auditory presentation (i.e., non-visual orientation). Various forms of sound design that provide useful information are called ***sonification*** (Barrass, 2012).

16.2.3 Applications to multimedia

The discussions in Sections 16.2.1 and 16.2.2 are based on some special applications of spatial sound. In professional applications, various functions, such as communication and virtual reality, may be separately implemented by corresponding equipment. In consumer applications, however, users may prefer multi-functional and integrated equipment.

Multimedia PCs, which are distinguished by integration and interaction, can handle a wide range of information, including audio, video, images, text, and data. Information exchange between computers is also possible through the Internet. Even standard PCs possess these functions, making them ideal platforms for communication, information processing, and virtual reality.

Since the 1990s, multimedia PCs have been an important application field for spatial sound, in addition to cinema, domestic, and automobile applications. Spatial sounds are widely incorporated into the entertainment functions of multimedia PCs. Currently, a multimedia PC is often used to play back various video and audio programs from optical disks and stream media. A common sound card in a PC supports two-channel stereophonic inputs and outputs. Some sound cards also support 5.1, 7.1, or even more channel outputs. Various video/audio playback software supports different audio-coded signals, such as MP3, AAC, Dolby Digital, and DTS. Some video/audio production software has powerful functions for multichannel signal editing, converting, and coding. Combined with an optical disk writer, a video/audio CD, DVD, or BD can be easily made on a multimedia PC. The development of hard disks, the Internet, and cloud computing facilitates the storage and transmission of spatial sound programs. Therefore, multimedia PCs provide an effective and convenient platform for video and audio program production and playback.

For a multimedia PC with a multichannel sound card, audio outputs can be directly reproduced with multichannel active loudspeakers. For a common multimedia PC, loudspeakers or headphones are often used for audio reproduction. For loudspeaker reproduction, two loudspeakers are often arranged on the two sides of the visual display with a small, spanned angle with respect to the listener. In this case, stereophonic expansion in Section 11.9.2 and the virtual reproduction of multichannel sound in Section 11.9.3 are applicable to improve the reproduced effect. For headphone presentation, binaural reproduction in Section 11.9.1 is also applicable.

Another entertainment function of a multimedia PC includes 3D games. VAD is often used in various 3D games on multimedia PCs to recreate spatial auditory effects. VAD has been incorporated into some 3D game software on the Windows platform. To create an authentic auditory effect, head trackers, as well as interactive and dynamic signal processing, can also be incorporated into the multimedia PC platform (López and González, 1999; Kyriakakis, 1998). A 3D game based on virtual reality (and VAD) is a promising application, as stated in Section 16.2.1. Other spatial sound techniques, such as Dolby Pro Logic IIz, may also be applied to 3D games (Tsingos et al., 2010).

Various applications of spatial sound to virtual reality, communication and information systems, and receivers of digital multimedia broadcasting can be implemented on multimedia PC platforms. A multimedia PC is also used as a teleconferencing terminal.

Multimedia applications also raise new requirements for the coding and transmission of spatial sound signals. The MPEG-4 coding standard in Section 13.5.4 is specified for multimedia video and audio.

16.2.4 Applications to mobile and handheld devices

Mobile communication and handheld sound reproduction devices, such as tablet computers, smartphones, and stream media players, have rapidly developed in recent years. From a practical perspective, using spatial sound for these types of products is a promising direction.

Since the 2010s, some corporations and research institutes have already launched relevant studies (AES Staff Technical Writer, 2006a; Yasuda et al., 2003; Paavola et al., 2005; Choi et al., 2006; Sander et al., 2012), and many commercial products have been introduced. Mobile products are characterized by a combination of functions, such as speech communication, interactive virtual auditory environments, teleconferencing, spatial auditory information presentation (e.g., traffic directions), and entertainment (e.g., video-audio reproduction, 3D games). Therefore, such products can be regarded as an application of multimedia technology. The increased speed and bandwidth of wireless communication networks favor the likelihood of carrying out the aforementioned functions. The application of spatial sound to mobile and handheld devices has been considered in the standard of MPEG-H 3D Audio (Section 13.5.6).

Compared with other uses, sound reproduction in mobile and handheld devices is restricted by the following two issues:

1. The limited processing and storage ability of the system, which requires simplification of algorithms and data,
2. The limited power supply by battery requires a reproduction method with low power consumption.

For sound reproduction in mobile devices, mini loudspeakers can be used, but this method may cause some problems. First, this method is unable to create a high-pressure level in reproduction due to the limited power supply in mobile devices and the restrictions imposed by the characteristics of mini loudspeakers. Second, the audio quality of mini loudspeakers is limited, especially at low frequencies. Third, the span between two loudspeakers in a mobile device is small (usually a few centimeters to a dozen centimeters). A mobile device is usually located 20–50 cm away from the listener. Accordingly, the spanned angle of two mini-loudspeakers with respect to the listener lies between 10° and 20°. Such a narrow span angle spoils the stereophonic sound effect.

The first and second problems mentioned have yet to be solved, but they may be changed with technical development. The third problem can be alleviated, and the effect can be improved by using the method of stereophonic expansion in Section 11.9.2 or the virtual reproduction of multichannel sound in Section 11.9.3 (Park et al., 2006; Breebaart et al., 2006). For 3D game use, the transaural method in Section 11.8 can also be utilized directly to create signals for two mini-loudspeakers from mono stimuli.

The above method deals with transaural reproduction via two loudspeakers with a narrow span angle. The analysis in Section 11.8.3 indicates that a loudspeaker configuration with a narrow-spanned angle requires a large boost at a low frequency in transaural processing; consequently, signal processing becomes difficult. Considering that the low-frequency limit of a mini-loudspeaker is 200–300 Hz at best, these low-frequency components can be filtered out in the design of transaural filters, and the difficulty in signal processing can be avoided. Moreover, near-field HRTFs can be used for transaural filters to adapt the practical distance in mobile device reproduction (Zhang et al., 2014).

Headphone presentation requires a relatively small power supply and usually provides better perceived audio quality. Therefore, it is appropriate for mobile devices. However, headphone presentation of stereophonic and multichannel sound signals directly may cause the problem of in-head-localization. To solve this problem, the binaural reproduction method in Section 11.9.1 can be used to convert the stereophonic and multichannel sound signals for headphone presentation on mobile devices. In particular, combining audio coding and binaural processing, the MPEG spatial audio coding and decoding (Section 13.4.5) reduces the bit rate of data and simplifies binaural synthesis processing (Breebaart et al., 2006). For

3D game use, the binaural signals can be synthesized by directly using the method in Section 11.1.2. In addition, with the development of low-cost head trackers, dynamic VAD or VAE in Section 11.10 can be used for mobile and handheld devices (Pörschmann, 2007). The accuracy of dynamic VAD for consumer uses is less critical and, therefore, a low-cost head tracker with moderate performance satisfies the requirement. For example, dynamic binaural synthesis can be used for stereophonic and multichannel sound reproduction with headphones in mobile devices (Zhang and Xie, 2014).

With the Cardboard introduced by Google, a smartphone can be used as a virtual reality platform. The screen of a smartphone can serve as a 3D visual display. The gyroscope and accelerometer in a smartphone can serve as head trackers for dynamic VAE. Google published an Android-based virtual reality (VR) development platform on which binaural Ambisonics was used for VAE. Dynamic binaural reproduction of multichannel (such as 22.2-channel) sound can also be implemented on the smartphone platform (Lin and Xie, 2018).

Spatial sound signal recording should also be specially designed to adapt to the size of a mobile device. In Section 2.2.4, a pair of microphones similar to Blumlein shuffling can be used for stereophonic recording on a mobile device. Some work suggested using an array of miniature microphones in a mobile device and beamforming to record the stereophonic and 5.1-channel signals (Bai et al., 2015).

16.3 APPLICATIONS TO THE SCIENTIFIC EXPERIMENTS OF SPATIAL HEARING AND PSYCHOACOUSTICS

In the scientific research of spatial hearing and psychoacoustics, the auditory perceptions of various types of information in the sound field should be explored. Spatial sound is applied to this research as a special application to virtual reality. As a tool of scientific research, the ability to achieve accurate and quantitative control of reconstructed sound fields or binaural pressures is vital. Therefore, sound field-based and binaural-based spatial sound techniques are appropriate for scientific purposes. However, it should be careful to use psychoacoustic-based spatial sound techniques for these purposes because the reconstructed sound field of this kind of technique has been simplified according to psychoacoustic principles and thus may not satisfy the requirement of accuracy.

The most direct method is to reconstruct a target sound field within an appropriate region. When a subject enters the sound field, his/her two ears receive the auditory information of the sound field exactly, which enables accurate research of the auditory perception in the target sound field. Higher-order Ambisonics and WFS, which are two representative sound field-based spatial sound techniques and systems, are theoretically appropriate for such research purposes. If the condition of the Shannon–Nyquist spatial sampling theorem is satisfied, high-order Ambisonics and WFS can reconstruct an arbitrary target sound field. However, high-order Ambisonics and WFS are often too complex to be used.

On the one hand, in some scientific experiments, auditory perception is explored only in some simple sound fields. Accordingly, a simple rather than complex target sound field needs to be reconstructed. In these cases, a simplified spatial sound system is enough. For example, to investigate the auditory perception caused by a direct sound and a few early reflections in a room, a system with several loudspeakers for spatially mapping the direct and reflected sounds is enough. Many of the experiments described in Section 1.8 were based on this method. Therefore, discrete multichannel sounds can serve as a tool for exploring auditory perception in a room or hall. On the other hand, the knowledge of auditory perception in a room is a foundation for the design and simplification of spatial sound. Similar experimental methods are also applicable to exploring the auditory sensation created by decorrelated source signals (Section 1.7.3) and the cocktail party effect (Section 1.7.5).

The Environment for Auditory Research (EAR) is a facility for spatial sound perception and speech communication research set up by the U.S. Army Research Laboratory (Ericson, 2011). The EAR consists of four indoor research spaces: a sphere room, a dome room, a distance room, a listening laboratory, and one outdoor research space. Dense loudspeakers are arranged in the research space for sound field and spatial perception experiments with various spatial sound techniques. For example, in the dome room, 180 loudspeakers are arranged within an elevation range of -20° to 40° with an azimuthal and elevation interval of 2° and 10°, respectively, which is close to the human auditory resolution. The facility in the dome room is appropriate for WFS and one-for-one spatial mapping and playback. In the sphere room, 57 loudspeakers are arranged on a spherical surface with a nearly uniform distribution, which is appropriate for higher-order spatial Ambisonics and vector base amplitude panning (VBAP) reproduction. Some other experimental facilities and spaces support stereophonic, 5.1-, 7.1-, 10.2-, 14.2-channel, and binaural reproduction. Therefore, various spatial sound techniques are incorporated into the EAR.

Generally, auditory experiments based on reconstructed sound fields have good stability and repeatability. These experiments should be theoretically conducted in an anechoic chamber to avoid the inference of environmental reflections; as such, they are relatively complicated. On the other hand, because binaural pressures include auditory information in an arbitrary sound field, HRTF-based binaural synthesis can be used to create various auditory events in a target sound field. Therefore, headphone-based VAD has become an important experimental tool for auditory scientific research.

Experimenting with headphone-based VAD is relatively simple and frees you from the inference of environmental reflections. More importantly, headphone-based VAD enables the exact manipulation of binaural signals. Some information may be added or deleted in binaural signals during binaural synthesis to explore the contribution of different types of information to auditory perception. However, headphone-based VAD has some defects, especially for static VAD (Section 11.7.2). To reduce the influence of these defects, each step in VAD should be designed carefully, such as using individual HRTF and HpTF processing. Practical measures depend on a given case.

Numerous binaural experimental studies have been conducted using VADs. Only a few examples are outlined here, but these examples are enough to illustrate the significance of VAD in auditory experiments. Wightman and Kistler (1992) used a VAD to investigate the relative salience of conflicting ITD and ILD as cues for localization. The conclusion is presented in Section 1.6.5. Wightman and Kistler (1999) also used dynamic VAD to investigate the contribution of dynamic cues caused by head rotation to front-back discrimination. The conclusion is presented in Section 1.6.3.

To explore the influence of high-frequency spectral cues caused by pinna on localization, a VAD can be used to manipulate the magnitude spectra in binaural signals. Langendijk and Bronkhorst (2002) employed this method to study the contribution of spectral cues to localization by eliminating the cues in some frequency bands at a frequency range above 4 kHz. The results show that removing the cues in 1/2-Octave bands does not affect localization, whereas removing the cues in 2-Octave bands makes correct localization virtually impossible.

Using VADs, Jin et al. (2004) investigated the relative contribution of monaural versus interaural spectral cues to resolving the directions within a cone of confusion. In their experiments, the natural values of the overall ITD and ILD were maintained. An artificial flat spectrum was presented at the left eardrum, and the right-ear sound spectrum was adjusted to preserve either the true right monaural spectrum or the true interaural spectrum. The localization experiments indicate that neither the preserved interaural spectral difference cue nor the preserved right monaural spectral cue adequately maintains accurate elevation in the presence of a flat monaural spectrum at the left eardrum.

VAD is also applicable to experiments on auditory distance perception (Zahorik, 2002a; Bronkhorst and Houtgast, 1999), multichannel speech intelligibility and talker recognition under different conditions (Drullman and Bronkhorst, 2000), the cocktail party effect (Crispien and Ehrenberg, 1995), spatial unmasking (Kopčo and Shinn-Cunningham, 2003), and auditory neurophysiology (Hartung et al., 1999).

16.4 APPLICATIONS TO SOUND FIELD AURALIZATION

16.4.1 Auralization in room acoustics

The sound field in a room consists of direct sounds and reflections (Sections 1.2.4 and 1.8). The subjective perception of room acoustic quality is closely related to the physical properties of the sound field in a room. The existing physical parameters or indexes used to describe the sound field, however, are currently insufficient to fully represent the subjective perception of the quality of room acoustics because of the incompleteness of these parameters or indexes. Thus, subjective assessment is necessary for evaluating the acoustic performance of rooms, such as concert halls, theaters, and multifunction halls.

Onsite listening is the most straightforward method of subjective assessment, but it is impractical in most cases. For example, employing onsite listening methods to compare the acoustic quality of different music halls located in different countries or cities is nearly impossible because of expensive travel costs and the short auditory memory of listeners. Therefore, various sound recording and reproduction technologies were used in research on room acoustic quality. However, the methods of onsite recording are for already-built rooms and are invalid for unbuilt rooms. Because resolving acoustic defects in an already-built room is difficult and costly, the objective and subjective properties of a room must be predictively evaluated in the design stage.

In the room design stage, various methods of room acoustic simulation can be used to predict the sound field, and the resultant sound field or binaural information can be reproduced by appropriate spatial sound technique. Such a method enables a predictive evaluation of the perceived performance of a room and resolves acoustic defects during the design stage. ***Auralization*** aims to record or simulate the sound field information by physical methods and then reproduce this information to re-generate or evoke auditory perceptions of the sound field. The principle of room auralization is similar to the cases of physical simulation of room reflections for multichannel sound signal production in Section 7.5.5 and binaural room modeling for dynamic VAE in Section 11.10.1. However, the purpose of auralization is different from the other two cases. In particular, auralization requires the exact simulation and reproduction of spatial information in the sound field, while consumer uses often omit part of the spatial information according to psychoacoustic principles.

Room auralization can be regarded as a special application of auditory virtual reality or auditory experiment based on a spatial sound platform in which special auditory environments are modeled. Various auralization methods have been proposed, as reviewed in a previous study by Kleiner et al. (1993) and in a book by Vorländer (2008). Similar to the case in Section 16.3, the accuracy of reconstructed sound fields or binaural pressures is vital for room auralization. For room auralization, sound field-based and binaural-based spatial sound techniques are appropriate. Psychoacoustic-based spatial sound techniques are usually inappropriate for room auralization because this technique cannot accurately reconstruct the complex room sound field.

Strictly speaking, auralization experiments with loudspeakers should be conducted in an anechoic chamber. In Section 3.1, researchers used an array with 65 loudspeakers to simulate a room sound field in the 1950s (Meyer and Thiele, 1956). According to its physical principle, higher-order Ambisonics can reconstruct an arbitrary sound field within a local region when the condition of the Shannon–Nyquist spatial sampling theorem is satisfied. Therefore, higher-order Ambisonics is appropriate for room auralization. Some studies have used lower-order Ambisonics for room auralization to simplify the reproduction system. This method is problematic. As indicated in Sections 9.3.1 and 9.3.2, the third-order Ambisonics can only reconstruct the target sound field up to 1.87 kHz within a region with an average human head size, which does not cover the frequency range important to room auditory perception. Lower-order Ambisonics reproduction causes timbre coloration above the upper-frequency limit (Section 12.2.3). A 32-order Ambisonics is needed to reconstruct the target sound field up to 20 kHz within a region with average human head size.

Ambisonic signals for room auralization can be obtained by an onsite recording with a circular or spherical microphone array and signal conversion as described in Section 9.8. Another method is to measure the multichannel spatial room impulse responses using a circular or spherical microphone array. These impulse responses include the temporal and spatial information of the sound field. Ambisonic impulse responses, which are equivalent to the impulse responses measured by a set of coincident microphones with different order directivities, can be derived from the impulse responses measured by the circular or spherical microphone array. Ambisonic impulse responses can be used for room acoustic analysis and auralization. In the latter case, the "dry" signal from an anechoic recording is convoluted with Ambisonic impulse responses and then decoded into loudspeaker signals. This method applies to room auralization and simulating room reflections in Ambisonic program production (Section 7.5.5). The physical nature of the two applications is identical. Microphone array recording is also restricted by the Shannon–Nyquist spatial sampling theorem. The reconstructed sound field is accurate only when both recording and reproduction satisfy the conditions of the spatial sampling theorem.

Ambisonic signals for room auralization can also be obtained by physical simulation. Various methods of physical simulation outlined in Section 7.5.5 apply to room auralization. The details of these methods are referred to the literature (Lehnert and Blauert, 1992; Kleiner et al., 1993; Svensson and Kristiansen, 2002; Vorländer, 2008). Ambisonic signals are derived from the simulated sound field. In particular, a virtual microphone array method for simulation is available. That is, the direct and reflected transmissions from a sound source to a virtual microphone array are first simulated. The outputs of the virtual microphone array are then converted to Ambisonic signals (Støfringsdal and Svensson, 2006).

The binaural method is another common tool for room auralization. Auralization based on the binaural method is called binaural auralization. Since the early developmental stages of room acoustics, binaural recording and reproduction have been used in research on room acoustic quality. Binaural signals for auralization can also be obtained by binaural room impulse response (BRIR)-based binaural synthesis. When both the sound source and the listener are fixed, the transmission from a source to a listener in a room can be regarded as a linear-time-invariant course. For an already-built room, BRIRs (Section 1.8.3) are measured from an artificial head or human subject. Binaural signals are created by convoluting a "dry" signal from an anechoic recording with BRIRs. The measured BRIRs can also be used for binaural-related room analysis, such as the interaural cross-correlation (IACC) in Section 1.8.3. Some professional software for room acoustic measurement, such as Dirac 7841, can be used for BRIR measurement and analysis. Binaural room impulse responses can also be derived from the spatial room impulse responses measured by a spherical microphone array. This method is flexible but restricted by the upper-frequency limit of the microphone array (Section 11.6). It should be emphasized that the physical characteristics (such as the directivity) of the sound source for BRIR or spatial room impulse response measurement influence the result of auralization greatly (Rao and Xie, 2007).

Dynamic binaural reproduction can also be introduced into auralization. The dynamic VAE in Section 11.10, microphone array recording, dynamic binaural reproduction in Section 11.6.1, and binaural room scanning in Section 14.6 can also be applied to auralization.

The University of Parma in Italy conducted numerous studies on BRIR measurement and auralization. In 2004, a large-scale research plan intended to facilitate research on famous cultural heritage was launched; the plan includes the room acoustic measurements (such as BRIR measurement) of 20 world-famous opera houses in Italy (Farina et al., 2004). Given the structures and purposes of opera houses, a dodecahedron loudspeaker with a subwoofer was used as the sound source for measurements in the orchestra pit, and a monitoring loudspeaker with directivity was used for measurements on the stage.

In conventional room acoustic design, an acoustic scale model is often used to evaluate the acoustic properties of the room being designed. In more detail, a modeled room with an N-times size reduction relative to a real room is first constructed. Then, the acoustic properties of the room being designed are predicted by re-scaling the acoustic properties of the modeled room 1/N times in terms of frequency. Xiang and Blauert (1991, 1993) incorporated binaural auralization into the acoustic scale model, in which the BRIRs for auralization were obtained by the acoustic scale.

With the advent of computer technology, binaural auralization with room acoustics modeling has become an important means of room acoustic research and design, especially as a useful tool for the predictive evaluation of room acoustic properties in the design stage. The principles of binaural room acoustics modeling are outlined in Section 11.10.1 and the details are referred to Lehnert and Blauert (1992) and Kleiner et al. (1993). For the application of auralization to room acoustic design, accurate room acoustic modeling is vital. From this point of view, physics-based methods of room acoustic modeling are appropriate for room auralization.

Some software with the function of binaural auralization has been developed for room acoustic design, such as ODEON, developed by the Technical University of Denmark (Naylor, 1993), and EASE/EARS, developed by SDA Software Design Ahnert GmbH, Germany (Ahnert and Feistel, 1993). The functions and methods of room acoustic modeling of these software products may be somewhat different, but the basic considerations and methods are similar. Such software products usually provide an interface for defining room geometry and the positions of sound sources and listeners. They also provide a large library of surface materials and physical data on sound sources for users. Some libraries are open to users, enabling later supplementation of materials and sound source data. On the basis of input information, these software products simulate acoustic transmission in a room, calculate various objective acoustic parameters and BRIRs, and realize auralization. These programs are constantly updated (http://www.odeon.dk, http://www.ada-acousticdesign.de/).

The binaural auralization is simple because an anechoic chamber is not needed in headphone presentation. Since the 1980s, binaural auralization has been developed and successfully used for room acoustic design and prediction, such as predicting the subjective attributes of ASW, LEV, and speech intelligibility and designing sound reinforcement systems. However, some problems with binaural auralization remain. In particular, perceptible differences exist between binaural auralization and onsite listening conditions. The differences are due to the errors or approximations in room acoustic modeling and the defects in binaural reproduction. Therefore, further improvement in the quality of binaural auralization is needed.

16.4.2 Other applications of auralization technique

Spatial sound technology is an extensively used method for sound event recording and reproduction. Aside from room acoustic research, this method has been generally used for sound archives and subjective assessment, such as noise evaluation (Gierlich, 1992; Song et al., 2008). Moreover, binaural recording and reproduction can be used to evaluate other loudspeaker-based reproduction systems (Toole, 1991). Auralization can also be used for other acoustic designs and evaluations.

Automobile sound quality designed to provide a comfortable sound environment in cars has recently received considerable attention from acousticians, automobile designers, and automobile manufacturers. Automobile sound quality is related to motor- and tire-relevant noise reduction, the speech intelligibility of interior colloquy, related communication equipment, and automobile audio reproduction quality. Spatial sound techniques, such as Ambisonics and the multiple-receiver position-matched method in Section 9.6.3, can be used to reconstruct the sound field and evaluate the sound quality in a car cabin. Information about the sound field can be obtained by microphone array recording or acoustic modeling with a computer. Because a car cabin is a small, enclosed space, numerical computation based on wave acoustics, rather than on geometrical acoustics, is required for acoustic simulation.

Binaural auralization can also be applied to the assessment of automobile sound quality. The signals for binaural auralization can be obtained by artificial-head recording or by convolving an anechoic stimulus with a pair of binaural impulse responses from a car cabin acquired from measurement or simulation. Researchers at Parama University, Italy, conducted numerous studies on this issue (Farina and Ugolotti, 1998).

Auralization is also used to simulate and assess the sound field of an aircraft cabin. For example, the sound field in an aircraft cabin can be recorded by a microphone array and reconstructed by the multiple-receiver position-matched method in Section 9.6.3 (Gauthier et al., 2015). Some researchers also applied binaural auralization to study the preference ratings for loudspeakers in different listening rooms (Hiekkanen et al., 2009).

16.5 APPLICATIONS TO CLINICAL MEDICINE

In Section 1.7.5, the auditory system takes advantage of the cocktail party effect to more effectively detect the target speech from an interfering background. The cocktail effect is attributed to binaural hearing and is closely related to spatial cues. Therefore, binaural hearing is vital for people with normal hearing to detect the target sound information in an interfering environment.

Hearing impairment is the most prevalent sensory deficit. According to the organs in which the lesion occurs, hearing impairment is classified into three basic categories: conductive hearing loss, sensorineural hearing loss, and mixed hearing loss. Conductive hearing loss is caused by external and middle ear diseases. Sensorineural hearing loss can be subdivided into three types: cochlear, neurological, and central loss, which are the results of cochlear lesions, spiral ganglion or auditory nerve conductive pathway lesions, and central system lesions (located in the brainstem and brain, involving the cochlear nucleus and its central auditory pathway, auditory cortex). Mixed hearing loss has elements of both conductive hearing loss and sensorineural hearing loss. Hearing-impaired people not only suffer from a partial or complete loss of unilateral or bilateral hearing but also often experience the loss of spatial auditory ability. Especially for some central (higher-level nervous system) disorders that lead to a decrease in the ability to process binaural information, their bilateral hearing loss may be slight, but their ability to identify binaural signals declines and even disappears, resulting in a phenomenon of "hearing without understanding."

A hearing aid is a device designed to improve the condition of hearing-impaired people, mainly those with conductive and non-severe sensorineural hearing impairments. Bilateral hearing aids may also improve the spatial auditory ability of some hearing-impaired people. Even for unilaterally hearing-impaired people, a combination of a unilateral hearing aid and normal hearing in another ear may improve their spatial auditory ability to some extent. Training is important to rehabilitate spatial auditory abilities for hearing aid users.

A bilateral hearing aid is essentially a binaural recording and reproduction system (Section 11.1) in which a pair of microphones placed at or close to the two ears is used to record binaural signals. After equalization and other processing to adapt to different conditions of hearing loss, binaural signals are presented through a headphone. Being fixed to the surface of the head, the microphones move with the turning of the head and therefore provide dynamic binaural signals for localization. According to the microphone position, hearing aids can be divided into three types: behind-the-ear (BTE), in-the-ear (ITE), and in-the-canal (ITC) hearing aids. The microphones of a BTE hearing aid are located behind the pinna. The interval between the microphones and headphones enables the hearing aid to have a high level of gain. However, the transmission from a sound source to two microphones BTE pinna differs from the transmission to the entrances of the ear canal. In other words, the binaural signals recorded by a pair of BTE microphones are incorrect and cannot provide the correct spectral cue for localization (Akeroyd and Whitmer, 2011). A similar problem occurs in ITE hearing aids because their microphones cover the auricular concha. For research on binaural hearing aids, Majdak et al. (2007) measured the HRTFs with microphones placed behind the pinna. Some problems associated with the bilateral hearing aid are still worth studying, such as (1) how to correct the spectral cue in binaural signals; (2) if a user can adapt to the new (incorrect) spectral cue provided by a hearing aid; and (3) the influence of spectral cue error on spatial hearing when the dynamic cue is preserved.

Many advanced hearing aids involve the processing of dynamic compression and adaptive beamforming with multiple microphones for target speech enhancement. If the signals of the left and right ears are operated independently, the possible differences in level compression and phase shift for the signals of the left and right ears may distort the resultant ITD and ILD. Some inappropriate speech enhancement algorithms may also spoil spatial auditory information and thus impair localization. Therefore, hearing aids with united left and right signal processing should be used to preserve spatial information (Bogaert et al., 2006, 2008).

A cochlear implant is a treatment for severe-to-profound sensorineural hearing loss caused by cochlear lesions. It captures sound signals through a microphone, converts them into certain encoded electrical signals by signal processing devices, and stimulates the auditory nerve directly through the implanted electrode array to restore or reconstruct hearing. Bilateral cochlear implants may also be thought of as a special "binaural recording and reproduction" system. The electrode system, not the headphone, serves as the transducer in "reproduction." Transducers display electrical signals rather than acoustic signals. Similar to the case of binaural hearing aids, bilateral cochlear implants (for patients with severe bilateral hearing loss) or unilateral cochlear implants combined with the residual hearing in the contralateral ear can restore or improve the ability of binaural hearing. However, bilateral cochlear implants also suffer from the problems caused by binaural hearing aids. More importantly, limited by technology, some simplification and approximation have been made in the signal processing strategy in artificial cochlea so that the temporal envelop information of sound is preserved and the final detail of sound is lost. Therefore, this signal processing strategy can preserve ILD information at most and lose low-frequency ITD information (Laback et al., 2015; Kan and Litovsky, 2015). However, further studies should be conducted to improve the performance of artificial cochlea in the rehabilitation of spatial auditory ability.

Some criterions related to spatial auditory perception, such as ability of localization and speech intelligibility in an interfering environment, can be used as a means for clinical auditory evaluations (Shinn-Cunningham, 1998). This method is used to evaluate not only the ability of natural hearing but also the clinical effects of hearing aids and cochlear implants. A multichannel loudspeaker system similar to that in Section 3.1 can be used for these purposes (Hoesel and Tyler, 2003; Seeber et al., 2004). A VAD test is also applicable to examine the ability of natural hearing. VAD is advantageous because its hardware and required environment are simple. Moreover, VAD can simulate various auditory environments, such as speech perception in different interfering conditions.

16.6 SUMMARY

Spatial sounds are applicable to a wide range of fields, and some representative applications are outlined in this section. Cinema and domestic reproductions are two traditional and major applications of spatial sound. Various spatial sound techniques and systems have been used in cinema and domestic reproductions. The technical requirements for these two applications are different. Automobile audio is a new application of spatial sound.

With the development of new technology, the applications of spatial sound have been extended to many new fields. For example, it can be applied to virtual reality, which is a new application, especially for special training. It can be used for communication and information systems for at least three purposes, e.g., enhancement of speech intelligibility, information display and orientation, and sonification to alleviate the visual burden. Multimedia PC is an ideal platform for the applications of spatial sound to communication, information processing, virtual reality, and entertainment. Spatial sound for mobile and handheld devices is a promising direction.

Spatial sound can serve as an experimental tool for scientific research on psychoacoustics and spatial hearing. By artificially manipulating the sound field or binaural signals, spatial sound enables us to investigate the contributions of various cues and information to auditory perception.

Auralization is an important tool for room acoustic design that enables a subjective evaluation and prediction of the sound quality of rooms. For room auralization, sound field- and binaural-based spatial sound techniques are appropriate. Some forms of software with the function of binaural auralization have been developed for room acoustic design. Auralization can also be used for other acoustic designs and evaluations. However, the performance of auralization should be further improved.

In clinical medicine, the methods of spatial sound are applicable to the evaluation of hearing and the improvement of the spatial auditory ability of people with hearing aids or cochlear implants. Further studies on this issue are needed.

Appendix A

Spherical harmonic functions

In the coordinate system A illustrated in Figure 1.1, spatial direction is specified by azimuth $-180° < \theta \leq 180°$ and elevation $-90° \leq \phi \leq 90°$, or equally, by two angles $0° \leq \alpha \leq 180°$ and $-180° < \beta \leq 180°$, where $\alpha = 90° - \phi$ and $\beta = \theta$. To simplify the notations, spatial direction is simply denoted by $\Omega = (\theta, \phi) = (\alpha, \beta)$ in the discussions related to spherical harmonic functions (SHFs).

Two equivalent forms of SHFs, i.e., real-valued and complex-valued SHFs, are available. Both are often applied to acoustical analysis. Normalized ***real-valued SHFs*** are defined as

$$\begin{aligned} Y_{lm}^{(1)}(\Omega) &= Y_{lm}^{(1)}(\alpha,\beta) = N_{lm}^{(1)} P_l^m(\cos\alpha)\cos(m\beta) \\ Y_{lm}^{(2)}(\Omega) &= Y_{lm}^{(2)}(\alpha,\beta) = N_{lm}^{(2)} P_l^m(\cos\alpha)\sin(m\beta) \\ l &= 0, 1, 2.....; \quad m = 0, 1, 2....l, \end{aligned} \tag{A.1}$$

where l is the order of SHFs; $P_l^m(x)$ is the associated Legendre polynomial and is defined by

$$P_l^m(x) = (1-x^2)^{\frac{m}{2}} \frac{d^m}{dx^m} P_l(x), \tag{A.2}$$

and $P_l(x)$ is the l-order Legendre polynomial, which is the solution to the following Legendre equation:

$$(1-x^2)\frac{d^2y}{dx^2} - 2x\frac{dy}{dx} + l(l+1)y = 0 \qquad -1 \leq x \leq 1 \tag{A.3}$$

N_{lm} is the normalized factor, expressed as

$$N_{lm}^{(1)} = N_{lm}^{(2)} = \sqrt{\frac{(l-m)!}{(l+m)!}\frac{(2l+1)}{2\pi\Delta_m}} \qquad \Delta_m = \begin{cases} 2 & m = 0 \\ 1 & m \neq 0 \end{cases} \tag{A.4}$$

The $l = 0$, 1, and 2 order real-valued SHFs are given by

$$
\begin{aligned}
&Y_{00}^{(1)}(\Omega)=\frac{1}{\sqrt{4\pi}}\\
&Y_{11}^{(1)}(\Omega)=\sqrt{\frac{3}{4\pi}}\sin\alpha\cos\beta \quad Y_{11}^{(2)}(\Omega)=\sqrt{\frac{3}{4\pi}}\sin\alpha\sin\beta \quad Y_{10}^{(1)}(\Omega)=\sqrt{\frac{3}{4\pi}}\cos\alpha\\
&Y_{22}^{(1)}(\Omega)=\sqrt{\frac{15}{16\pi}}\sin^2\alpha\cos 2\beta \quad Y_{22}^{(2)}(\Omega)=\sqrt{\frac{15}{16\pi}}\sin^2\alpha\sin 2\beta\\
&Y_{21}^{(1)}(\Omega)=\sqrt{\frac{15}{16\pi}}\sin 2\alpha\cos\beta \quad Y_{21}^{(2)}(\Omega)=\sqrt{\frac{15}{16\pi}}\sin 2\alpha\sin\beta\\
&Y_{20}^{(1)}(\Omega)=\sqrt{\frac{5}{16\pi}}\left(3\cos^2\alpha-1\right).
\end{aligned}
\tag{A.5}
$$

Figure A.1 illustrates the directional pattern of the $l = 0$, 1, and 2 order real-valued SHFs (the maximal magnitude of all SHFs are normalized to unit).

According to Equation (A.1) and Euler equation $\exp(\pm jm\beta) = \cos(m\beta) \pm j\sin(m\beta)$, normalized ***complex-valued SHFs*** are defined as

$$
\begin{gathered}
Y_{lm}(\Omega)=Y_{lm}(\alpha,\beta)=N_{lm}P_l^{|m|}(\cos\alpha)\exp(jm\beta)\\
l=0,1,2.....;\ m=0,\pm1,\pm2,....\pm l
\end{gathered}
\tag{A.6}
$$

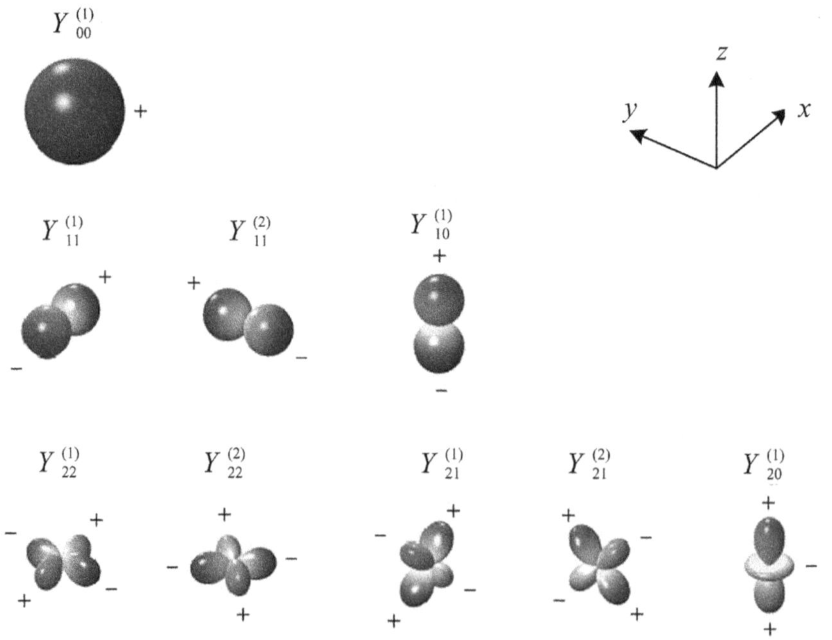

Figure A.1 The directional pattern of the $l = 0$, 1, and 2 order real-valued SHFs (the maximal magnitudes of all SHFs are normalized to unit).

The normalized factor is given by

$$N_{lm} = \sqrt{\frac{(l-|m|)!(2l+1)}{(l+|m|)\pi}}. \tag{A.7}$$

The $l = 0$, 1, and 2 order complex-valued SHFs are given by

$$\begin{aligned}
&Y_{00}(\Omega) = \frac{1}{\sqrt{4\pi}} \qquad Y_{1\pm1}(\Omega) = \sqrt{\frac{3}{8\pi}}\sin\alpha\ \exp(\pm j\beta) \qquad Y_{10}(\Omega) = \sqrt{\frac{3}{4\pi}}\cos\alpha \\
&Y_{2,\pm2}(\Omega) = \sqrt{\frac{15}{32\pi}}\sin^2\alpha\exp(\pm j2\beta) \qquad Y_{2,\pm1}(\Omega) = \sqrt{\frac{15}{32\pi}}\sin 2\alpha\exp(\pm j\beta) \\
&Y_{20}(\Omega) = \sqrt{\frac{5}{16\pi}}\left(3\cos^2\alpha - 1\right).
\end{aligned} \tag{A.8}$$

The complex and real-valued SHFs are related by the following equations:

$$\begin{aligned}
&Y_{l0}(\Omega) = Y_{l0}^{(1)}(\Omega) \\
&Y_{lm}(\Omega) = \begin{cases} \frac{\sqrt{2}}{2}\left[Y_{lm}^{(1)}(\Omega) + jY_{lm}^{(2)}(\Omega)\right] & m > 0 \\ \frac{\sqrt{2}}{2}\left[Y_{l,-m}^{(1)}(\Omega) - jY_{l,-m}^{(2)}(\Omega)\right] & m < 0 \end{cases}.
\end{aligned} \tag{A.9}$$

The real- and complex-valued SHFs exhibit some important characteristics. The SHFs satisfy orthonormality, thus:

$$\begin{aligned}
\int Y_{l'm'}^{(\sigma')}(\Omega)Y_{lm}^{(\sigma)}(\Omega)d\Omega &= \int_{\beta=0}^{2\pi}\int_{\alpha=0}^{\pi} Y_{l'm'}^{(\sigma')}(\alpha,\beta)Y_{lm}^{(\sigma)}(\alpha,\beta)\sin\alpha\ d\alpha\ d\beta \\
&= \delta_{ll'}\delta_{mm'}\delta_{\sigma\sigma'} \quad \sigma,\sigma' = 1, 2
\end{aligned}$$

$$\int Y_{l'm'}^{*}(\Omega)Y_{lm}(\Omega)d\Omega = \int_{\beta=0}^{2\pi}\int_{\alpha=0}^{\pi} Y_{l'm'}^{*}(\alpha,\beta)Y_{lm}(\alpha,\beta)\sin\alpha\ d\alpha\ d\beta = \delta_{ll'}\delta_{mm'}. \tag{A.10}$$

Because of the orthonormality and completeness of SHFs, any square integrable function $F(\Omega) = F(\alpha, \beta)$ can be decomposed by real-valued or complex-valued SHFs thus:

$$\begin{aligned}
F(\Omega) &= F_{00}^{(1)}Y_{00}^{(1)}(\Omega) + \sum_{l=1}^{\infty}\sum_{m=0}^{l}\left[F_{lm}^{(1)}Y_{lm}^{(1)}(\Omega) + F_{lm}^{(2)}Y_{lm}^{(2)}(\Omega)\right] \\
&= F_{00}Y_{00}(\Omega) + \sum_{l=1}^{\infty}\sum_{m=-l}^{l}F_{lm}Y_{lm}(\Omega).
\end{aligned} \tag{A.11}$$

Note that $Y_{l0}^{(2)}(\Omega) = 0$, which is retained in Eq. (A.11) only for convenience. With the orthonormality of SHFs given by Eqs. (A.10), the coefficients of SH decomposition are evaluated by

$$\begin{aligned}
F_{lm}^{(1)} &= \int F(\Omega) Y_{lm}^{(1)}(\Omega) d\Omega = \int_{\beta=0}^{2\pi} \int_{\alpha=0}^{\pi} F(\alpha, \beta) Y_{lm}^{(1)}(\alpha, \beta) \sin\alpha \, d\alpha \, d\beta \\
F_{lm}^{(2)} &= \int F(\Omega) Y_{lm}^{(2)}(\Omega) d\Omega = \int_{\beta=0}^{2\pi} \int_{\alpha=0}^{\pi} F(\alpha, \beta) Y_{lm}^{(2)}(\alpha, \beta) \sin\alpha \, d\alpha \, d\beta \\
F_{lm} &= \int F(\Omega) Y_{lm}^{*}(\Omega) d\Omega = \int_{\beta=0}^{2\pi} \int_{\alpha=0}^{\pi} F(\alpha, \beta) Y_{lm}^{*}(\alpha, \beta) \sin\alpha \, d\alpha \, d\beta \, .
\end{aligned} \tag{A.12}$$

The coefficients of real- and complex-valued decomposition are related by following equation:

$$F_{lm}^{(1)} = \frac{\sqrt{2}}{2}\left(F_{lm} + F_{l,-m}\right) \qquad F_{lm}^{(2)} = \frac{\sqrt{2}}{2} j\left(F_{lm} - F_{l,-m}\right) \qquad 1 \le m \le l \tag{A.13}$$

or

$$F_{l0} = F_{l0}^{(1)} \qquad F_{lm} = \begin{cases} \dfrac{\sqrt{2}}{2}\left[F_{lm}^{(1)} - jF_{lm}^{(2)}\right] & m > 0 \\ \dfrac{\sqrt{2}}{2}\left[F_{l,-m}^{(1)} + jF_{l,-m}^{(2)}\right] & m < 0 \end{cases}, \tag{A.14}$$

and

$$\sum_{m=0}^{l}\left[\left|F_{lm}^{(1)}\right|^2 + \left|F_{lm}^{(2)}\right|^2\right] = \sum_{m=-l}^{l} \left|F_{lm}\right|^2 . \tag{A.15}$$

As a special case, the Dirac delta function can be decomposed by SHFs thus:

$$\begin{aligned}
\delta(\Omega - \Omega') &= \frac{1}{\sin\alpha}\delta(\alpha - \alpha')\delta(\beta - \beta') = \delta(\cos\alpha - \cos\alpha')\delta(\beta - \beta') \\
&= \sum_{l=0}^{\infty}\sum_{m=0}^{l}\sum_{\sigma=1}^{2} Y_{lm}^{(\sigma)}(\Omega') Y_{lm}^{(\sigma)}(\Omega) = \sum_{l=0}^{\infty}\sum_{m=-l}^{l} Y_{lm}^{*}(\Omega') Y_{lm}(\Omega).
\end{aligned} \tag{A.16}$$

The SH decomposition yields the following addition equation:

$$\frac{2l+1}{4\pi} P_l(\cos\Delta\Omega') = \sum_{m=0}^{l}\sum_{\sigma=1}^{2} Y_{lm}^{(\sigma)}(\Omega') Y_{lm}^{(\sigma)}(\Omega) = \sum_{m=-l}^{l} Y_{lm}^{*}(\Omega') Y_{lm}(\Omega) \tag{A.17}$$

Where $\Delta\Omega'$ is the angle between directions Ω and Ω'.

A directional-continuous function $F(\Omega) = F(\alpha, \beta)$ is required for calculating the SH coefficients with Equation (A.12). Sometimes, however, only the discrete samples of the $F(\Omega)$ measured at M directions, i.e., $F(\Omega_i) = F(\alpha_i, \beta_i)$, $i = 0, 1...(M - 1)$, are available. Using appropriate directional sampling schemes, the integral in Equation (A.12) can be replaced by a weighted sum of M directional samples of $F(\Omega)$, so that SH coefficients up to a finite order $(L - 1)$ or L^2 SH coefficients are calculated. Accordingly, the $(L - 1)$ order truncated or spatial bandlimited SH representation of $F(\Omega)$ is obtained. Shannon–Nyquist spatial sampling theorem requires $L^2 \le M$. That is, SH coefficients up to the $\left(\sqrt{M}-1\right)$ order at most can be evaluated from the M directional samples of $F(\Omega)$.

The $(L - 1)$ order truncated SH representation of $F(\Omega)$, as well as the associated SH coefficients, are provided by

$$F(\Omega)=\sum_{l=0}^{L-1}\sum_{m=0}^{l}\sum_{\sigma=1}^{2}F_{lm}^{(\sigma)}Y_{lm}^{(\sigma)}(\Omega)=\sum_{l=0}^{L-1}\sum_{m=-l}^{l}F_{lm}Y_{lm}(\Omega), \tag{A.18}$$

$$F_{lm}^{(\sigma)}=\sum_{i=0}^{M-1}\lambda_i F(\Omega_i)Y_{lm}^{(\sigma)}(\Omega_i) \qquad F_{lm}=\sum_{i=0}^{M-1}\lambda_i F(\Omega_i)Y_{lm}^{*}(\Omega_i), \tag{A.19}$$

where λ_i is the quadrature weight that determines the relative contribution of the $F(\Omega)$ sample at direction Ω_i to the summation. Weight set $\{\lambda_i, i = 0, 1...(M - 1)\}$ depends on the directional sampling scheme. The directional sampling scheme and associated weight set should be appropriately selected to ensure that the M directional samples of SHFs satisfy the following discrete orthonormality:

$$\sum_{i=0}^{M-1}\lambda_i Y_{l'm'}^{(\sigma')}(\Omega_i)Y_{lm}^{(\sigma)}(\Omega_i)=\delta_{ll'}\delta_{mm'}\delta_{\sigma\sigma'} \qquad \sum_{i=0}^{M-1}\lambda_i Y_{l'm'}^{*}(\Omega_i)Y_{lm}(\Omega_i)=\delta_{ll'}\delta_{mm'}. \tag{A.20}$$

Some directional sampling schemes can be used to effectively calculate SH coefficients from the discrete directional samples of function $F(\Omega)$. Three common schemes are outlined here (Rafaely, 2005).

Scheme 1, *Equiangle sampling*. Elevation $\alpha = 90° - \phi$ and azimuth $\beta = \theta$ are respectively uniformly sampled at $2L$ angles, resulting in $M = 4L^2$ directional samples denoted by $(\alpha_q, \beta_{q'})$, $q, q' = 0, 1, ... (2L - 1)$. The SH coefficients in Eq. (A.19) are evaluated by

$$F_{lm}^{(\sigma)}=\sum_{q=0}^{2L-1}\sum_{q'=0}^{2L-1}\lambda_q F(\alpha_q,\beta_{q'})Y_{lm}^{(\sigma)}(\alpha_q,\beta_{q'}) \qquad F_{lm}=\sum_{q=0}^{2L-1}\sum_{q'=0}^{2L-1}\lambda_q F(\alpha_q,\beta_{q'})Y_{lm}^{*}(\alpha_q,\beta_{q'}). \tag{A.21}$$

Weights λ_q satisfy the following relationship:

$$\sum_{q=0}^{2L-1}\lambda_q P_l(\cos\alpha_q)=\frac{2\pi}{L}\delta_{l0} \qquad 0\le l\le(L-1). \tag{A.22}$$

The equiangle sampling scheme requires $M = 4L^2$ directional samples to calculate SH coefficients up to the $(L - 1)$ order, which is fourfold the lower bound imposed by the Shannon–Nyquist theorem.

Scheme 2, *Gauss-Legendre sampling*. Elevation is first sampled at L angles α_q ($q = 0, 1, \ldots L - 1$), so that $\{\cos(\alpha_q)\}$ are the Gauss-Legendre nodes (the zeros of the L-order Legendre polynomial). Then, the azimuth in each elevation are sampled at $2L$ angles $\beta_{q'} = \theta_{q'}$ ($q' = 0, 1, 2\ldots, 2L - 1$) with even intervals. Therefore, $M = 2L{*}L = 2L^2$ directional samples are derived and denoted by $(\alpha_q, \beta_{q'})$, $q = 0, 1,\ldots(L - 1)$, $q' = 0, 1, \ldots(2L - 1)$. The SH coefficients in Eq. (A.19) are calculated by

$$F_{lm}^{(\sigma)} = \sum_{q=0}^{L-1}\sum_{q'=0}^{2L-1} \lambda_q F(\alpha_q, \beta_{q'}) Y_{lm}^{(\sigma)}(\alpha_q, \beta_{q'}) \qquad F_{lm} = \sum_{q=0}^{L-1}\sum_{q'=0}^{2L-1} \lambda_q F(\alpha_q, \beta_{q'}) Y_{lm}^{*}(\alpha_q, \beta_{q'}). \tag{A.23}$$

The calculation of weights λ_q is referred to (Rafaely, 2015). The Gauss-Legendre sampling scheme requires $M = 2L^2$ directional samples to calculate SH coefficients up to the $(L - 1)$ order, which is twice the lower bound imposed by the Shannon–Nyquist spatial sampling theorem.

Scheme 3, *Nearly-uniform sampling*. Directional samples are uniformly or nearly-uniformly distributed on a spherical surface so that the distance between neighboring samples is constant or nearly constant. A uniform sampling scheme requires at least $M = L^2$ directional samples to calculate SH coefficients up to the $(L - 1)$ order, which is the lower bound imposed by the Shannon–Nyquist spatial sampling theorem. In practice, 1.3 to 1.5 times the lower bound is required, and the latter yields a set of equal weights $\lambda_i = 4\pi/M$ in the calculation of Equation (A.19). Nearly-uniform sampling is the most efficient in terms of the number of directional samples required, but it is also inconvenient for practical configuration.

There are only five kinds of directional sampling distributions that are strictly uniform and regular, including the tetrahedral, hexahedral (cubic), octahedral, dodecahedral, and icosahedral sampling. The maximal number of uniform directional sampling is twenty (Hollerweger, 2006). Usually, nearly-uniform sampling can be obtained. A direct method is to apply a tessellation to a regular polyhedron. For example, in the quaternary triangular mesh method, the mid-points at three sides of a triangular mesh surface of a polyhedron are found, and the mid-points are projected to the spherical surface to form four new spherical-triangular meshes. The tessellation method is simple and easy to implement. However, the number of directional sampling is fixed and cannot be adjusted when the initial polyhedron and the method of tessellation is given.

Another method to solve the nearly-uniform distribution of directional sampling is equivalent to the classical Thomson problem in electromagnetics. M unit charges are confined on the surface of a sphere. With exception of a scaling factor, the overall potential energy among the charges is calculated by:

$$U(M) = \sum_{i<i'} \frac{1}{R_{ii'}}, \tag{A.24}$$

where $R_{ii'}$ is the distance between the ith charge and the i'th charge. A stable distribution of M charges is solved by minimizing the overall potential energy in Equation (A.24). In this case, M charges are distributed nearly-uniformly on the spherical surface and the positions of charges are the nearly-uniform samples. There are some schemes to solve the nearly-uniform distribution for a given M, such as the relaxation scheme and the Monte-Carlo scheme

(Dragnev et al., 2002, Erber and Hockney, 2007; Cohn and Kumar, 2007). Liu (2015) illustrates the examples of nearly-uniform distribution for M = 100, 400, 900, 1600, 2500, and 3600.

Nearly-uniform distribution of directional sampling satisfies the discrete orthonormality in Equation (A.20) approximately. To evaluate the deviation from discrete orthonormality caused by the approximation, a $L^2 \times M$ matrix $[Y_{3D}]$ of real-valued SHFs is introduced (a matrix of complex-valued SHFs is also suitable). The entries of matrix are $Y_{lm}^{(\sigma)}(\Omega_i)$. The rows of this matrix are arranged in the order of $Y_{00}^{(1)}, Y_{11}^{(1)}, Y_{11}^{(2)}, Y_{10}^{(1)}\ldots.$, and the and columns are arranged in the order of $\Omega_0, \Omega_1 \ldots \Omega_{M-1}$. A equal weight of $\lambda_i = 4\pi/M$ is chose in Equation (A.20), the condition of discrete orthonormality can be written as

$$\frac{4\pi}{M}[Y_{3D}][Y_{3D}]^T = [I] \tag{A.25}$$

where $[I]$ is a $L^2 \times L^2$ identity matrix. The following $L^2 \times L^2$ matrix $[ERR]$ represents the deviation from orthonormality caused by nearly-uniform sampling

$$[ERR] = [I] - \frac{4\pi}{M}[Y_{3D}][Y_{3D}]^T \tag{A.26}$$

Appendix B

Some statistical methods for the data of psychoacoustic and subjective assessment experiments

The raw data of psychoacoustic and subjective experiments are collected from all subjects and repeated experiments. From the perspective of statistics, these data are subject to certain statistical rules on random variables. The data should be analyzed further by appropriate statistical methods to yield results at a certain confidence level. Statistical methods commonly used for data analysis include hypothesis test and analysis of variance. There are various methods of hypothesis test and analysis of variance. Some statistical methods related to experiments in Chapter 15 are outlined in this appendix. The details of data statistics are referred to relevant textbook (Marques de Sa, 2007). Some statistical software, such as SPSS and SAS, are also available.

1. Hypothesis test for the proportion of correct discrimination in the subjective comparison and discrimination experiment

For subjective comparison and discrimination experiments, following statistical hypothesis is tested under certain significance level: the mean proportion of correct discrimination is equal to that of random judgment. Accepting the hypothesis means that subjects cannot discriminate the object and reference. Or alternatively, the hypothesis that the mean proportion of correct discrimination is greater than some specific value is tested. Accepting the hypothesis means that subjects can discriminate the object and reference.

Let us assume that the result of discrimination is a random variable x, with $x = 1$ representing a correct evaluation and $x = 0$ representing failure. Thus, x obeys (0, 1) distribution, where the probability of $x = 0$ is $1 - p$ and the probability of $x = 1$ is p. The probability distribution function is

$$P(x = k) = p^k (1 - p)^{1-k} \quad k = 0, 1. \tag{B.1}$$

The corresponding expected value and variance are

$$\mu_0 = p \qquad \sigma^2 = p(1 - p). \tag{B.2}$$

In random selection, we have $p = p_0 = 0.5$ and 0.33 for the two- and three-alternative forced-choice paradigms, respectively.

Suppose that $(x_1, x_2, \ldots x_N)$ denotes N independent evaluations (samples) of the random variable x. Such evaluations may come from N different subjects, N repetitions of a specific subject, or K repetitions for each L subject with $N = L \times K$. The corresponding mean value and standard deviation of the samples are calculated by

$$\bar{x} = \frac{1}{N}\sum_{n=1}^{N} x_n \qquad \sigma_x = \sqrt{\frac{1}{N-1}\sum_{n=1}^{N}\left(x_n - \bar{x}\right)^2}. \tag{B.3}$$

Taking advantage of statistical hypothesis testing methods facilitates the evaluation of the discrimination between object and reference from actual samples (subjects' evaluations). Usually, hypothesis p as equal to a specific value p_0 is used as the criterion for categorizing evaluation as nondiscrimination, whereas p as larger than another specific value p_1 is the criterion used to classify assessment as discrimination. In the two-alternative forced-choice paradigm, $p_0 = 0.5$ and $p_1 = 0.75$ are usually selected. In the three-alternative forced-choice paradigm, $p_0 = 0.33$ and $p_1 = 0.67$ are typically used. Thus, two statistical hypotheses for the proportion of correct discrimination p are tested at a significance level of α:

1. Testing two-sided hypothesis H_0: $p = p_0$ and alternative hypothesis H_1: $p \neq p_0$. Accepting $p = p_0$ indicates that no audible difference (discrimination) between object and reference is detected.
2. Testing one-sided hypothesis H_0: $p \geq p_1$ and alternative hypothesis H_1: $p < p_1$. Accepting $p \geq p_1$ while rejecting $p = p_0$ indicates that audible difference (discrimination) between target B and reference A is detected.

For N independent samples of the random variable x, the probability of q correct evaluations is

$$P(N, q) = \frac{N!}{q!(N-q)!} p^q (1-p)^{N-q}. \tag{B.4}$$

For the two-alternative forced-choice paradigm with $p = p_0 = 0.5$ in two-sided hypothesis testing, we suppose that N is even for convenience. The overall probability of the correct evaluation number within $(N/2 - n) \leq q \leq (N/2 + n)$ is

$$P_{all} = \sum_{q=(N/2-n)}^{(N/2+n)} P(N, q). \tag{B.5}$$

Lower and upper bounds are determined by increasing n in Equation (B.5) from $n = 0$ in the range $0 \leq n \leq N/2$ by turns until $n = n_0$, where P_{all} is equal to or greater than $1-\alpha$. Then, the hypothesis of $p = p_0 = 0.5$ is accepted when the mean correct rate of the samples is within

$$\bar{x}_{low} < \bar{x} < \bar{x}_{upper} \qquad \bar{x}_{low} = \frac{1}{N}\left(\frac{N}{2} - n_0\right) \qquad \bar{x}_{upper} = \frac{1}{N}\left(\frac{N}{2} + n_0\right). \tag{B.6}$$

Failure occurs in the experiment when the mean correct rate of the samples is lower than $\bar{x}_{low}$.

As to the one-sided hypothesis testing of $p \geq p_1$, the overall probability of the correct evaluation number within $n \leq q \leq N$ is

$$P_{all} = \sum_{q=n}^{N} P(N, q). \tag{B.7}$$

The lower bound is determined by decreasing n in Equation (B.7) from round (p_1N) in the range $0 \le n \le$ round (p_1N) by turns until $n = n_0$, where P_{all} is equal to or greater than $1 - \alpha$. The one-sided hypothesis of $p \ge p_1$ is then rejected when the mean correct rate of the samples is within

$$\overline{x} \le \overline{x}'_{low} = \frac{n_0}{N}. \tag{B.8}$$

Moreover, $\overline{x}$ and

$$u = \frac{\overline{x} - p_0}{\sigma/\sqrt{N}} \tag{B.9}$$

are random variables. For a large N, variable u approximately obeys a normal (Gaussian) distribution of $N(0, 1)$ with zero mean and unit variance. Under this condition, the hypothesis test for the distribution of u at a significance level of α is more convenient.

Let $u_{1-\alpha}$ denote the quantile of $1 - \alpha$ of normal distribution $N(0, 1)$, which is defined by the integral on probability distribution function $f(\xi)$ $(-\infty < \xi < +\infty)$:

$$\int_{-\infty}^{u_{1-\alpha}} f(\xi)d\xi = 1-\alpha \quad or \quad \int_{u_{1-\alpha}}^{+\infty} f(\xi)d\xi = \alpha \qquad 0 \le \alpha \le 1\,. \tag{B.10}$$

For the two-sided hypothesis test H_0: $p = p_0$, when

$$u = \left|\frac{\overline{x} - p_0}{\sigma/\sqrt{N}}\right| < u_{1-\alpha/2} \tag{B.11}$$

hypothesis $p = p_0$ is accepted. For a normal distribution of $N(0, 1)$, $u_{1-\alpha/2} = 1.96$ at a significance level of $\alpha = 0.05$.

$$u = \frac{\overline{x} - p_0}{\sigma/\sqrt{N}} < u_{\alpha/2}. \tag{B.12}$$

means failure in the experiment

For the one-sided hypothesis test H_0: $p \ge p_1$, when

$$\frac{\overline{x} - p_1}{\sigma/\sqrt{N}} \le u_\alpha \tag{B.13}$$

hypothesis $p \ge p_1$ is rejected. For a normal distribution of $N(0, 1)$, $u_\alpha = -1.64$ at a significance level of $\alpha = 0.05$.

2. Mean value test on the data

Mean value test is used to verify whether the mean of a set of data is equal to, greater than, or less than a specific value μ_0. For example, in the double-blind triple-stimulus with hidden reference method recommended by the ITU (Section 15.4), mean value rest is applied to the

Table B.1 Mean value test on a single normal distribution

	Null hypothesis	*Alternative hypothesis*	*Reject the null hypothesis at significance level α if*
(1)	$\mu = \mu_0$	$\mu > \mu_0$	$\bar{x} \geq \mu_0 + \frac{\sigma_x}{\sqrt{N}} t_{1-\alpha}(N-1)$
(2)	$\mu = \mu_0$	$\mu < \mu_0$	$\bar{x} \leq \mu_0 - \frac{\sigma_x}{\sqrt{N}} t_{1-\alpha}(N-1)$
(3)	$\mu = \mu_0$	$\mu \neq \mu_0$	$\lvert \bar{x} - \mu_0 \rvert \geq \frac{\sigma_x}{\sqrt{N}} t_{1-\alpha/2}(N-1)$

data from each subject to test the hypothesis that the mean grade difference between object and hidden reference are statistically zero at significant level $\alpha = 0.05$.

The test for the mean value of N observed sample data is a hypothesis test on one population. Suppose that sample data are from a normally distributed population $N(\mu, \sigma^2)$ (this should not be confused with the number N of observed samples). μ and σ^2 are expectation and variance, respectively. The t-test in Table B.1 should be used in testing the hypothesis under a significance level α (usually, $\alpha = 0.05$). Here, $\bar{x}$ and σ_x are calculated from Equation (B.3); $t_{1-\alpha}(N - 1)$ and $t_{1-\alpha/2}(N - 1)$ are the $(1 - \alpha)$ and $(1 - \alpha/2)$ quantiles of t-distribution, respectively.

3. Mean value test for two sets of experimental data

Mean value test is also used to test if the means of two set of data are equal (or the mean of one set data is greater that of the other). For example, in the double-blind triple-stimulus with hidden reference method recommended by the ITU (Section 15.4), mean value test is applied to test difference between the mean scores of two objects; it is also applied to test the consistency in the raw data from each subject. In addition, to compare the localization accuracy of the free-field real source and virtual source, mean value test is used to test confusion rate and κ^{-1} [Equation (15.6.8)] of localization data for real and virtual source.

The mean value test for two sets of observed values from different sample data (two-mean test) is a hypothesis test on two populations. Suppose that the two sets of observed values x_n $(n = 1, 2,..., N_1)$ and y_n $(n = 1, 2,..., N_2)$ are from two normally distributed populations $N(\mu_1, \sigma_1^2)$ and $N(\mu_2, \sigma_2^2)$, respectively. μ_1, μ_2 and σ_1^2, σ_2^2 are corresponding expectations and variances. To verify the mean value, the homogeneity of variances should be tested beforehand; that is, testing hypothesis $\sigma_1^2 = \sigma_2^2$ at a significance level α. When

$$\frac{\sigma_x^2}{\sigma_y^2} \geq F_{1-\alpha/2}(N_1 - 1, N_2 - 1) \quad or \quad \frac{\sigma_y^2}{\sigma_x^2} \geq F_{1-\alpha/2}(N_2 - 1, N_1 - 1) \tag{B.14}$$

then the hypothesis should be rejected. $F_{1-\alpha/2}$ is the $(1 - \alpha/2)$ quantile of F distribution.

Only when hypothesis $\sigma_1^2 = \sigma_2^2$ is true can the mean values of the two sets of observed values be tested. Then, the t-test is used to validate the two sets of observed values $x_n(n = 1, 2,...,N_1)$ and $y_n(n = 1, 2,..., N_2)$ at a significance level α, as outlined in Table B.2. The definitions of the symbols in Table B.2 are the same as those in Table B.1 with

$$\sigma_w^2 = \frac{(N_1 - 1)\sigma_x^2 + (N_2 - 1)\sigma_y^2}{N_1 + N_2 - 2}. \tag{B.15}$$

Table B.2 Mean value test on two normally distributed populations

	Null hypothesis	*Alternative hypothesis*	*Reject the null hypothesis at significance level α if*
(1)	$\mu_1 = \mu_2$	$\mu_1 > \mu_2$	$\bar{x} - \bar{y} \geq t_{1-\alpha}(N_1 + N_2 - 2)\sigma_w \sqrt{\frac{1}{N_1} + \frac{1}{N_2}}$
(2)	$\mu_1 = \mu_2$	$\mu_1 < \mu_2$	$\bar{x} - \bar{y} \leq -t_{1-\alpha}(N_1 + N_2 - 2)\sigma_w \sqrt{\frac{1}{N_1} + \frac{1}{N_2}}$
(3)	$\mu_1 = \mu_2$	$\mu_1 \neq \mu_2$	$\lvert \bar{x} - \bar{y} \rvert \geq t_{1-\alpha/2}(N_1 + N_2 - 2)\sigma_w \sqrt{\frac{1}{N_1} + \frac{1}{N_2}}$

It is assumed in aforementioned mean value tests that the observed values of samples obey normal distribution. When the number of samples is greater enough, this assumption is valid (at least approximately valid). In this case, the mean value tests are parametric test of normal distribution. However, observed values of samples do not obey the normal distribution in some cases, such as in the case that the number of samples is fewer than 20. In these cases, some non-parametric statistical methods should be used to test the data. For example, Wilcoxon signed-rank test, whose function is similar to that of a *t*-test, can be used for non-parametric test for the difference of two sets of paired samples. Therefore, the applicability of a statistical analysis method should be carefully checked prior to using it.

4. Analysis of variance

The mean value test is for one set of data or two sets of data under two different conditions. As a more general case, in the localization experiments and experiment of quantitative assessment of quality impairments, validating the mean values of three or more sets of data under different controlled conditions is a common approach. Examples are: comparing the mean scores from three different objects, or comparing the confusion rate in localization of three systems. This is an issue of multi-population mean testing in statistics.

Many factors cause changes in experimental conditions. For example, in quantitative assessment of perceived audio quality, the change of object and source material may affect results. If only one factor changes, then the experiment is classified as a single-factor experiment. If more than one factor changes, then it is categorized as a multi-factor experiment. The different conditions of a factor are called levels. For example, in the case of four systems (objects) to be evaluated with six source materials, there are two factors, e.g., object factor and material factors. Object factor can be divided into four levels, namely $S_1, S_2,\ldots S_4$; and the material factors can be divided into six levels, namely $M_1, M_2\ldots M_6$.

As a most general case, mean value testing should be carried out on multiple populations under multiple factors. More than one factor and their interaction effects will likely influence results. This kind of problem, including evaluating statistically significant differences among multiple mean values and testing the significant effects of multiple factors, is addressed by analysis of variance (ANOVA) in statistics. ANOVA is important to analyzing the experimental data. Numerous conclusions on the psychoacoustic and subjective assessment experiments on spatial sound are drawn from ANOVA. The mathematical methods of ANOVA are discussed in relevant textbook (Marques de Sa, 2007). Details are not discussed here, given space limitations.

References

Aarts R.M. (1993): Enlarging the sweet spot for stereophony by time/intensity trading, *at the AES 94th Convention*, Berlin, Germany, Paper 3473.

Aarts R.M. (2000): Phantom sources applied to stereo-base widening, *J. Audio Eng. Soc.*, 48(3), 181–189.

Adami A., Haberts E.A.P., and Herre J. (2014): Perceptual evaluation of a coherence suppressing down-mix method, *at the AES 55th International Conference*, Helsinki, Finland.

AES Staff Technical Writer (2006a): Binaural technology for mobile applications, *J. AudioEng. Soc.*, 54(10), 990–995.

AES Staff Technical Writer (2006b): Digital radio broadcasting, *J. Audio Eng. Soc.*, 54(7/8), 771–774.

AES Staff Writer (2004): The world of digital radio, *J. Audio Eng. Soc.*, 52(12), 1272–1278.

AES Technical Council (2001): Multichannel surround sound systems and operations, AES Technical Council Document, AESTD1001.1.01-10.

Ahnert W., and Feistel R. (1993): EARS auralization software, *J. Audio Eng. Soc.*, 41(11), 894–904.

Ahrens J. (2012): *Analytic methods of sound field synthesis*, Springer-Verlag, Berlin, Germany.

Ahrens J., and Spors S. (2008a): Reproduction of moving virtual sound sources with special attention to the Doppler effect, *at the AES 124th Convention*, Amsterdam, The Netherlands, Paper 7363.

Ahrens J., and Spors S. (2008b): An analytical approach to sound field reproduction using circular and spherical loudspeaker distributions, *Acta Acust. United Ac.*, 94(6), 988–999

Ahrens J., and Spors S. (2008c): Focusing of virtual sound sources in higher order Ambisonics, *at the AES 124th Convention*, Amsterdam, The Netherlands, Paper 7378.

Ahrens J., and Spors S. (2009): Sound field reproduction employing non-omnidirectional loudspeakers, *at the AES 126th Convention*, Munich, Germany, Paper 7741.

Ahrens J. and Spors S. (2010): Sound field reproduction using planar and linear arrays of loudspeakers, *IEEE Trans. Audio, Speech, Language Process*, 18(8), 2038–2050.

Ahrens J., and Spors S. (2011): Wave field synthesis of moving virtual sound sources with complex radiation properties, *J. Acoust. Soc. Am.*, 130(5), 2807–2816.

Ahrens J., and Spors S. (2012): Wave field synthesis of a sound field described by spherical harmonics expansion coefficients, *J. Acoust. Soc. Am.*, 131(3), 2190–2199.

Ahveninen J., Kopco N., and Jaaskelainen, I.P. (2014): Psychophysics and neuronal bases of sound localization in humans, *Hearing Res.*, 307, 86–97.

Akeroyd M.A., and Whitmer W.M. (2011): Spatial hearing and hearing aids, *ENT Audiol News*, 20(5), 76–79.

Algazi V.R., Avendano C., and Duda R.O. (2001b): Elevation localization and head-related transfer function analysis at low frequencies, *J. Acoust. Soc. Am.*, 109(3), 1110–1122.

Algazi V.R., Duda R.O., Duraiswami R., et al. (2002): Approximating the head-related transfer function using simple geometric models of the head and torso, *J. Acoust. Soc. Am.*, 112(5), 2053–2064.

Algazi V.R., Duda R.O., Thompson D.M., et al. (2001a): The CIPIC HRTF database, in *Proceeding of 2001 IEEE Workshop on the Applications of Signal Processing to Audio and Acoustics*, New York, 99–102.

Allen I. (2006): The X-Curve, its origins and history, electro-acoustic characteristics in the cinema and the mix-room, the large room and the small, *SMPTE Motion Imag. J.*, 115(7/8), 264–275.

American Standards Association (1960): *Acoustical terminology SI*, New York, USA.
Ando Y. (1985): *Concert hall acoustics*, Springer-Verlag Press, Berlin, Germany.
Ando Y. (1998): *Architectural acoustics, blending sound sources, sound fields, and listeners*, Springer-Verlag, New York, USA.
Ando Y. (2009): *Auditory and visual sensations*, Springer-Verlag, New York, USA.
ANSI S3.25/ASA80 (1989): *Occluded ear simulator*, American National Standard, American National Standards Institute, New York, USA.
ANSI S3.36 (1985): *Manikin for simulated in-situ airborne acoustic measurements*, American National Standard, American National Standards Institute, New York, USA.
ANSI S3.4 (2007): *Procedure for the computation of loudness of steady sounds*, American National Standard, American National Standards Institute, New York, NY.
Arteaga D. (2013): An Ambisonics decoder for irregular 3D loudspeaker array, *at the AES 134th Convention*, Rome, Italy, Paper 8918.
Asano F., and Swanson D.C. (1995): Sound equalization in enclosures using modal reconstruction, *J. Acoust. Soc. Am.*, 98(4), 2062–2069.
Ashby T., Mason R., and Brookes T. (2013): Head movements in three-dimensional localisation, *at the AES 134th Convention*, Rome, Italy, Paper 8881.
Ashby T., Mason R., and Brookes T. (2014): Elevation localisation response accuracy on vertical planes of differing azimuth, *at the AES 136th Convention*, Berlin, Germany, Paper 9046.
ATSC Standard Doc A/342-1(2017): Audio common elements, Advanced Television system Committee, Washington, USA.
ATSC Standard Doc.A52(2012): Digital audio compression (AC-3, E-AC-3), Advanced Television System Committee, Washington, USA.
Auro Technologies, and Bacro Audio Technologies (2015): AuroMax, next-generation immersive sound system, www.auro-3D.com.
Avendano C., and Jot J.M. (2004): A frequency-domain approach to multichannel upmix, *J. Audio Eng. Soc.*, 52(7/8), 740–749.
Backman J. (2009): Subwoofers in symmetrical and asymmetrical rooms, *at the AES 126th Convention*, Munich, Germany, Paper 7748.
Backman J. (2010): Subwoofers in rooms: experimental modal analysis, *at the AES 128th Convention*, London, UK, Paper 7970.
Backman J. (2011): Subwoofers in rooms: modal analysis for loudspeaker placement, *at the AES 130th Convention*, London, UK, Paper 8323.
Baek Y.H., Jeon S.W., Park Y.C., et al. (2012): Efficient primary-ambient decomposition algorithm for audio upmix, *at the AES 133rd Convention*, San Francisco, CA, Paper 8754.
Bagousse S.L., Colomes C., and Paquier M. (2010): State of the art on subjective assessment of spatial sound quality, *at the AES 38th International Conference*, Piteå, Sweden.
Bagousse S.L., Paquier M., and Colomes C. (2014): Categorization of sound attributes for audio quality assessment—a lexical study, *J. Audio Eng. Soc.*, 62(11), 736–747.
Bai M.R., and Bai G. (2005): Optimal design and synthesis of reverberators with a fuzzy user interface for spatial audio, *J. Audio Eng. Soc.*, 53(9), 812–825.
Bai M.R., Hsu H., and Wen J.C. (2014): Spatial sound field synthesis and upmixing based on the equivalent source method, *J. Acoust. Soc. Am.*, 135(1), 269–282.
Bai M.R., Kuo M.C., and Hua Y.H. (2015): An application of miniature microphone array to stereophonic recording compatible to conventional practice, *J. Audio Eng. Soc.*, 63(4), 267–279.
Bai M.R., and Lee C.C. (2006): Objective and subjective analysis of effects of listening angle on crosstalk cancellation in spatial sound reproduction, *J. Acoust. Soc. Am.*, 120(4), 1976–1989.
Bai M.R., and Lee C.C. (2010): Comparative study of design and implementation strategies of automotive virtual surround audio systems, *J. Audio Eng. Soc.*, 58(3), 141–159.
Bai M.R., and Shih G.Y. (2007): Upmixing and downmixing two-channel stereo audio for consumer electronics, *IEEE Trans. Consumer Electron.*, 53(3), 1011–1019.
Balmages I., and Rafaely B. (2007): Open sphere designs for spherical microphone arrays, *IEEE Trans. Audio, Speech, Language Process.*, 15(2), 727–732.

Bamford J.S., and Vanderkooy J. (1995): Ambisonic sound for us, *at the AES 99th Convention*, New York, USA, Paper 4138.
Barbar S. (2015): Surround sound for cinema, in *Handbook for sound engineers* (5th edition, edited by Ballou G.), Focal Press, Burlinton, USA.
Barbour J.L. (2003): Elevation perception: phantom images in the vertical hemi-sphere, *at the AES 24th International Conference*, Banff, Canada.
Barrass S. (2012): Digital fabrication of acoustic sonifications, *J. Audio Eng. Soc.*, 60(9), 709–715.
Barron M. (2000): Measured early lateral energy fractions in concert halls and opera houses. *J. Sound Vib.*, 232(1), 79–100.
Barron M., and Marshall A. H. (1981): Spatial impression due to early lateral reflections in concert halls: the derivation of a physical measure. *J. Sound Vib.*, 77(2), 211–232.
Barry D., and Kearney G. (2009): Localisation quality assessment in source separation-based upmixing algorithms, *at the AES 35th International Conference*, London, UK.
Batke J.M., and Keiler F. (2010): Using VBAP-derived panning functions for 3D Ambisonics decoding, *at the 2nd International Symposium on Ambisonics and Spherical Acoustics*, Paris, France.
Batteau D.W. (1967): The role of the pinna in human localization, *Proc. Royal. Soc. London*, 168(Ser, B), 158–180.
Bauck J., and Cooper D.H. (1996): Generalized transaural stereo and applications, *J. Audio. Eng. Soc.*, 44(9), 683–705.
Bauer B.B. (1960): Broadening the area of stereophonic perception, *J. Audio Eng. Soc.*, 8(2), 91–94.
Bauer B.B. (1961a): Phasor analysis of some stereophonic phenomena, *J. Acoust. Soc. Am.*, 33(11), 1536–1539.
Bauer B.B. (1961b): Stereophonic earphones and binaural loudspeakers, *J. Audio. Eng. Soc.*, 9(2), 148–151.
Bauer B.B. (1979): A unified 4-4-4, 4-3-4, 4-2-4 SQ®-compatible system of recording and FM broadcasting (USQ™), *J. Audio Eng. Soc.*, 27(11), 866–880.
Bauer B.B., Allen R.G., and Budelman G.A. (1973b): Quadraphonic matrix perspective-advances in SQ encoding and decoding technology, *J. Audio Eng. Soc.*, 21(5), 342–350.
Bauer B.B., Budelman G.A., and Gravereaux D.W. (1973a): Recording techniques for SQ matrix quadraphonic discs, *J. Audio Eng. Soc.*, 21(1), 19–26.
Bauer B.B., Dimattia A.L., and Rosenheck A.J. (1965): Transmission of directional perception, *IEEE Trans. Audio*, 13(1), 5–8.
Bauer B.B., Gravereaux D.W., and Gust A.J. (1971): A compatible stereo-quadraphonic (SQ) record system, *J. Audio Eng. Soc.*, 19(8), 638–646.
Baumgarte F., and Faller C. (2003): Binaural cue coding part I, Phychoacoustic fundamentals and design principles, *IEEE Trans. Speech Audio Process*, 11(6), 509–519.
Baumgartner R., and Majdak P. (2015): Modeling localization of amplitude-panned virtual sources in sagittal planes, *J. Audio Eng. Soc.*, 63(7/8), 562–569.
Baumgartner R., Majdak P., and Laback B. (2013): Assessment of sagittal-plane sound-localization performance in spatial-audio applications, in *the Technology of binaural listening* (edited by Blauert J.), Springer-Verlag, Berlin Heidelberg.
Bech S. (1995): Perception of reproduced sound: audibility of individual reflections in a complete sound field, II, *at the AES 99th Convention*, New York, USA, Paper 4093.
Bech S., and Zacharov N. (1999): Multichannel level alignment, part III: the influence of loudspeaker directivity and reproduction bandwidth, *at the AES 106th Convention*, Munich, Germany, Paper 4909.
Bech S., and Zacharov N. (2006): *Perceptual audio evaluation – theory, method and application*, John Wiley & Sons, West Sussex, UK.
Begault D.R. (1994): *3-D Sound for virtual reality and multimedia*, Academic Press Professional, Cambridge, MA.
Begault D.R. (1998): Virtual acoustics, aeronautics, and communications, *J. Audio Eng. Soc.*, 46(6), 520–530.
Begault D.R. (1999): Virtual acoustic displaysfor teleconferencing: intelligibility advantagefor "telephone-grade audio", *J. Audio. Eng. Soc.*, 47(10), 824–828.

Begault D.R., and Erbe T. (1994): Multichannel spatial auditory display for speech communications, *J. Audio Eng. Soc.*, 42(10), 819–826.
Begault D.R., Wenzel E.M., and Anderson M.R. (2001): Direct comparison of the impact of head tracking, reverberation, and individualized head-related transfer functions on the spatial perception of a virtual speech source, *J. Audio. Eng. Soc.*, 49(10), 904–916.
Begault D.R., Wenzel E.M., Godfroy M., et al. (2010): Applying spatial audio to human interfaces: 25 years of NASA experience, *at the AES 40th International Conference*, Tokyo, Japan.
Bekesy G.V. (1960): *Experiments in hearing*, Mcgraw-Hill, New York, USA.
Bell D. (2000): Surround sound studio design, *Studio Sound*, 42(7), 55–58.
Benjamin E., Lee R., and Heller A. (2010): Why Ambisonics does work? *at the AES 129th Convention*, San Francisco, USA, Paper 8242.
Bennett J.C., Barker K., and Edeko F.O. (1985): A new approach for the assessment of stereophonic sound system performance, *J. Audio Eng. Soc.*, 33(5), 314–321.
Beranek L. (1996). *Concert halls and opera houses*, Acoustical Society of America, USA.
Berg J. (2009): The contrasting and conflicting definitions of envelopment, *at the AES 126th Convention*, Munich, Germany, Paper 7808.
Berg J., and Rumsey F. (2001): Verification and correlation of attributes used for describing the spatial quality of reproduced sound, *at the AES 19th International Conference*, Schloss Elmau, Germany.
Berg J., and Rumsey F. (2002): Validity of selected spatial attributes in the evaluation of 5-channel microphone techniques, *at the AES 112th Convention*, Munich, Germany, Paper 5593.
Berg J., and Rumsey F. (2003): Systematic evaluation of perceived spatial quality, *at the AES 24th International Conference*, Banff, Alberta, Canada.
Berg J., and Rumsey F. (2006): Identification of quality attributes of spatial audio by repertory grid technique, *J. Audio Eng. Soc.*, 54(5), 365–379.
Berkhout A.J. (1988): A holographic approach to acoustic control, *J. Audio Eng. Soc.*, 36(12), 977–995.
Berkhout A.J., Vries D.D., and Vogel P. (1993): Acoustic control by wave field synthesis, *J. Acoust. Soc. Am.*, 93(5), 2764–2778.
Bernfeld B. (1975): Simple equations for multichannel stereophonic sound localization, *J. Audio Eng. Soc.*, 23(7), 553–557.
Betlehem T., and Abhayapala T.D. (2005): Theory and design of sound field reproduction in reverberate rooms, *J. Acoust. Soc. Am*, 117(4), 2100–2111.
Betlehem T., and Poletti M.A. (2014): Two dimensional sound field reproduction using higher order sources to exploit room reflections, *J. Acoust. Soc. Am.*, 135(4), 1820–1833.
Blanco-Martin E., Casajús-Quirós F.J., Gómez-Alfageme J.J., et al. (2011): Objective measurement of sound event localization in horizontal and median planes, *J. Audio Eng. Soc.*, 59(3), 124–136.
Blauert J. (1997): *Spatial hearing: the psychophysics of human sound localization* (Revised edition), MIT Press, Cambridge, MA.
Blauert J. (2012): Modeling binaural processing: what next? (abstract), *J. Acoust. Soc. Am.* 132 (3, Pt2), 1911.
Blauert J., Brueggen M., Bronkhorst A.W., et al. (1998): The AUDIS catalog of human HRTFs, *J. Acoust. Soc. Am.*, 103(5), 3082.
Blauert J., Lehnert H., Sahrhage J., et al. (2000): An interactive virtual-environment generator for psychoacoustic research I: architecture and implementation, *Acta Acust. United Ac.*, 86(1), 94–102.
Blauert J., and Lindemann W. (1986): Auditory spaciousness: some further psychoacoustic analyses, *J. Acoust. Soc. Am.*, 80(2), 533–542.
Blauert J., and Rabenstein R. (2012): Providing surround sound with loudspeakers: a synopsis of current methods, *Arch. Acoust.*, 37(1), 5–18.
Blauert J., and Xiang N. (2008): *Acoustics for engineers*, Springer-Verlag, Berlin, Germany.
Blommer M.A., and Wakefield G.H. (1997): Pole-zero approximations for head-related transfer functions using a logarithmic error criterion, *IEEE Trans. Speech Audio Process.*, 5(3), 278–287.
Bloom P.J. (1977): Determination of monaural sensitivity changes due to the pinna by use of minimum-audible-field measurements in the lateral vertical plane, *J. Acoust. Soc. Am.*, 61(3), 820–828.
Blumlein A.D. (1931): Improvements in and relating to sound transmission, sound recording and sound reproducing systems, British Patent Specification 394, 325. *Reprint in J. Audio Eng. Soc.*, 6(2),91–98.

Boehm J. (2011): Decoding for 3D, *at the AES 130th Convention*, London, UK, Paper 8426.
Boer K.D. (1940): Stereophonic sound reproduction, *Philips Tech. Rev.*, 1940(5), 107–114.
Boer K.D. (1946): The formation of stereophonic image, *Philips Tech. Rev.*, 1946(8), 51–56.
Boer K.D. (1947): A remarkable phenomenon with stereophonic sound reproduction, *Philips Tech. Rev.*, 1947(9), 8–13.
Bogaert T.V.D., Doclo S., Wouters J., et al. (2008): The effect of multimicrophone noise reduction systems on sound localization by users of binaural hearing aids, *J. Acoust. Soc. Am.*, 124(1), 484–497.
Bogaert T.V.D., Klasen T.J., Moonen M., et al. (2006): Horizontal localization with bilateral hearing aids: without is better than with, *J. Acoust. Soc. Am.*, 119(1), 515–526.
Bolia R.S., D'Angelo W.R., and McKinley R.L. (1999): Aurally aided visual search in three-dimensional space, *Human Factors*, 41(4), 664–669.
Boone M.M. (2004): Multi-actuator panels (MAPs) as loudspeaker arrays for wave field synthesis, *J. Audio Eng. Soc.*, 52(7/8), 712–723.
Boone M.M., and Bruijn W.P.J.D. (2003): Improving speech intelligibility in teleconferencing by using wave field synthesis, *at the AES 114th Convention*, Amsterdam, The Netherlands, Paper 5800.
Boone M.M., Bruijn W.P.J.D., and Horbach U. (1999): Virtual surround speakers with wave field synthesis, *at the AES 106th Convention*, Munich, Germany, Paper 4928.
Boone M.M., and Verheijen E.N.G. (1998): Sound reproduction applications with wave-field synthesis, *at the AES 104th Convention*, Amsterdam, The Netherlands, Paper 4689.
Boone M.M., Verheijen E.N.G., and Van Tol P.F. (1995): Spatial sound field reproduction by wave field synthesis, *J. Audio Eng. Soc.*, 43(12), 1003–1012.
Borenius J. (1985): Perceptibility of direction and time delay errors in subwoofer reproduction, *at the AES 79th Convention*, New York, USA, Paper 2290.
Bosi M., Brandenburg K., Quackenbush S., et al. (1997): ISO/IEC MPEG-2 advanced audio coding, *J. Audio Eng. Soc.*, 45(10), 789–814.
Bosi M., and Goldberg R.E. (2003): *Introduction digital audio coding and standards*, Springer Science+Bussiness Media, New York, USA.
Bouéri M., and Kyirakakis C. (2004): Audio signal decorrelation based on a critical band approach, *at the AES 117th Convention*, San Francisco, USA, Paper 6291.
Bovbjerg B.P., Christensen F., Minnaar P., et al. (2000): Measuring the head-related transfer functions of an artificial head with high directional resolution, *at the AES 109th Convention*, Los Angeles, USA, Paper 5264.
Braasch J., Clapp S., Parks A., et al. (2013): A binaural model that analyses acoustic spaces and stereophonic reproduction systems by utilizing head rotations, in *the Technology of binaural listening* (edited by Blauert), Springer-Verlag, Berlin, Germany.
Bradley J.S., and Soulodre G.A. (1995): The influence of late arriving energy on spatial impression, *J. Acoust. Soc. Am.*, 97(4), 2263–2271.
Bradley J.S., and Soulodre G.A. (1996): Listener envelopment: an essential part of good concert hall acoustics, *J. Acoust. Soc. Am.*, 99(1), 22–23.
Brandenburg K., and Bosi M. (1997): Overview of MPEG Audio: current and future standards for low-bit-rate audio coding, *J. Audio Eng. Soc.*, 45(1/2), 4–21.
Brandenburg K., and Stoll G. (1994): ISO/MPEG-1 audio: a generic standard for coding of high-quality digital audio, *J. Audio Eng. Soc.*, 42(10), 780–792.
Breebaart J. (2013): Comparison of interaural intensity differences evoked by real and phantom sources, *J. Audio Eng. Soc.*, 61(11), 850–859.
Breebaart J., and Faller C. (2007): *Spatial audio processing: MPEG surround and other applications*, John Wiley & Sons, West Sussex, UK.
Breebaart J., Herre J., Villemoes L., et al. (2006): Multi-channel goes mobile: MPEG surround binaural rendering, *at the AES 29th International Conference*, Seoul, Korea.
Breebaart J., Hotho G., Koppens J., et al. (2007): Background, concept, and architecture for the recent MPEG surround standard on multichannel audio compression, *J. Audio Eng. Soc.*, 55(5), 331–351.
Breebaart J., Par S.V.D., Kohlrausch A., et al. (2005): Parametric coding of stereo audio, *EURASIP J. Appl. Signal Process.*, 2005(9), 1305–1322.

Bregman A.S. (1990): *Demonstrations of auditory scene analysis: the perceptual organization*, MIT Press, Cambridge, MA.

Briand M., Virette D., and Martin N. (2006): Parametric representation of multichannel audio based on principal component analysis, *at the AES 120th Convention*, Paris, France, Paper 6813.

Brimijoin W.O., and Akeroyd M.A. (2012): The role of head movements and signal spectrum in an auditory front/back illusion, *i-Perception*, 3(3), 179–182.

Brix S., Sporer T., and Plogsties J. (2001): CARROUSO-An European approach to 3D-audio, *at the AES 110th Convention*, Amsterdam, The Netherlands, Paper 5314.

Bronkhorst A.W. (2000): The cocktail party phenomenon: a review of research on speech intelligibility in multiple-talker conditions, *Acta Acust. United Ac.*, 86(1), 117–128.

Bronkhorst A.W., and Houtgast T. (1999): Auditory distance perception in rooms, *Nature*, 397, 517–520.

Brungart D.S. (1999): auditory localization of nearby sources. III. Stimulus effects, *J. Acoust. Soc. Am.*, 106(6), 3589–3602.

Brungart D.S., Durlach N.I., and Rabinowitz W.M. (1999): Auditory localization of nearby sources. II. Localization of abroadband source, *J. Acoust. Soc. Am.*, 106(4), 1956–1968.

Brungart D.S., Kordik A.J., and Simpson B.D. (2006): Effects of headtracker latency in virtual audio displays, *J. Audio Eng. Soc.*, 54(1/2), 32–44.

Brungart D.S., and Rabinowitz W.M. (1999): Auditory localization of nearby sources. Head-related transfer functions, *J. Acoust. Soc. Am.*, 106(3), 1465–1479.

Bujacz M., Skulimowski P., and Strumillo P. (2012): Naviton—a prototype mobility aid for auditory presentation of three-dimensional scenes to the visually impaired, *J. Audio Eng. Soc.* 60(9), 696–708.

Burkhard M.D., and Sachs R.M. (1975): Anthropometric manikin for acoustic research, *J. Acoust. Soc. Am.*, 58(1), 214–222.

Butler R.A., and Belendiuk K. (1977): Spectral cues utilized in the localization of sound in the median sagittal plane, *J. Acoust. Soc. Am.*, 61(5), 1264–1269.

Capra A., Fontana S., Adriaensen F., et al. (2007): Listening tests of the localization performance of stereodipole and Ambisonic systems, *at the AES 123rd Convention*, New York, USA, Paper 7187.

Cengarle G., Mateos T., and Bonsi D. (2011): A second-order Ambisonics device using velocity transducers, *J. Audio Eng. Soc.*, 59(9), 656–668.

Chang J.H., and Jacobsen F. (2012): Sound field control with acircular double-layer array of loudspeakers, *J. Acoust. Soc. Am.*, 131(6), 4518–4525.

Charpentier T. (2017): Normalization schemes in Ambisonic: does it matter? *at the AES 142nd Convention*, Berlin, Germany, Paper 9769.

Cheer J., Elliott S.J., and Gálvez M.F.S. (2013): Design and implementation of a car cabin personal audio system, *J. Audio Eng. Soc.*, 61(6), 412–424.

Chen J., Veen B.D.V., andHecox K.E. (1995): A spatial feature extraction and regularization model of the head-related transfer function, *J. Acoust. Soc. Am.*, 97(1), 439–452.

Chernyak R.I., and Dubrovsky N.A. (1968): Pattern of the noise image and binaural summation of loudness for different interaural correlation of noise, *Proceeding of the 6th International Congress on Acoustics*, Tokyo, Japan, 53–56.

Cherry E.C. (1953): Some experiments on the recognition of speech, with one and with two ears, *J. Acoust. Soc. Am.*, 25(5), 975–979.

Chétry N., Pallone G., Emerit M., et al. (2007): A discussion about subjective methods for evaluating blind upmix algorithms, *at the AES 31st International Conference*, London, UK.

Chi S., Xie B.S., and Rao D. (2009): Effect of mismatched loudspeaker pair on virtual sound image (in Chinese), *Appl. Acoust.*, 28(4), 291–299.

Choi I., Shinn-Cunningham B.G., Chon S.B., et al. (2008): Objective measurement of perceived auditory quality in multichannel audio compression coding systems, *J. Audio Eng. Soc.*, 56(1/2), 3–17.

Choi J., and Chang J.H. (2020): Exploiting deep neural networks for two-to-five channel surround decoder, *J. Audio Eng. Soc.*, 68(12), 938–949.

Choi J.W., and Kim Y.H. (2012): Integral approach for reproduction of virtual sound source surrounded by loudspeaker array, *IEEE Trans. Audio, Speech, Language Process.*, 20(7), 1976–1989.

Choi T., Park Y.C., and Youn D.H. (2006): Efficient out of head localization systemfor mobile applications, *at the AES 120th Convention*, Paris, France, Paper 6758.
Choisel S., and Wickelmaier F. (2006): Extraction of auditory features and elicitation of attributes for the assessment of multichannel reproduced sound, *J. Audio Eng. Soc.*, 54(9), 815–826.
Chowning J.M. (1971): The simulation of moving sound sources, *J. Audio Eng. Soc.*, 19(1), 2–6.
Chung H., Shim H., Nahn N., et al. (2012): Sound reproduction method by front loudspeaker array for home theater applications, *IEEE Trans. Consum. Electron.*, 58(2), 528–534.
Clack H.A.M., Dutton G.F., and Vanderlyn P.B. (1957): The "stereosonic" recording and reproduction system, *IRE Trans. Audio*, 5(4), 96–111.
Cobos M., and Lopez J.J. (2009): Resynthesis of sound scenes on wave-field synthesis from stereo mixtures using sound source separation algorithms, *J. Audio Eng. Soc.*, 57(3), 91–110.
Cobos M., and Lopez J.J. (2010): Interactive enhancement of stereo recordings using time-frequency selective panning, *at the AES 40th International Conference*, Tokyo, Japan.
Cohen E., and Eargle J. (1995): Audio in a 5.1 channel environment, *at the AES 99th Convention*, New York, USA, Paper 4071.
Cohn H., and Kumar A. (2007): Universally optimal distribution of points on spheres, *J. Amer. Math. Soc.*, 20(1), 99–148.
Conetta R., Rumsey F., Zielinski S., et al. (2008): QESTRAL (part 2): Calibrating the QESTRAL model using listening test data, *at the AES 125th Convention*, San Francisco, USA, Paper 7596.
Cook R.K., Waterhouse R.V., Berendt R.D., et al. (1955): Measurement of correlation coefficients in reverberant sound fields, *J. Acoust. Soc. Am.*, 27(6), 1072–1077.
Cooper D.H. (1982): Calculator program for head-related transfer function, *J. Audio. Eng. Soc.*, 30(1/2), 34–38.
Cooper D.H. (1987): Problems with shadowless stereo theory: asymptotic spectral status, *J. Audio. Eng. Soc.*, 35(9), 629–642.
Cooper D.H., and Bauck J.L. (1989): Prospects for transaural recording, *J. Audio. Eng. Soc.*, 37(1/2), 3–19.
Cooper D.H. and Shiga T. (1972): Discrete matrix multichannel stereo, *J. Audio Eng. Soc.*, 20(5), 346–360.
Copper D.H. (1974): QFMX-quadruplex FM transmission using the 4-4-4 QMX matrix system, *J. Audio Eng. Soc.*, 22(2), 82–87.
Cooper D.H., Shiga T., and Takagi T. (1973): QMX carrier channel disc, *J. Audio Eng. Soc.*, 21(8), 614–624.
Corteel E. (2006): Equalization in an extended area using multichannel inversion and wave field synthesis, *J. Audio Eng. Soc.*, 54(12), 1140–1161.
Corteel E., and Nicol R. (2003): Listening room compensation for wave field synthesis. What can be done? *at the AES 23rd International Conference*, Helsingφr, Denmark.
Craven P.G. (2003): Continuous surround panning for 5-speaker reproduction, *at the AES 24th International Conference*, Banff, Canada.
Crispien K., and Ehrenberg T. (1995): Evaluation of the "cocktail-party effect" for multiple speech stimuli within a spatial auditory display, *J. Audio Eng. Soc.*, 43(11), 932–941.
Damaske P. (1967/1968): Subjective investigation of sound fields, *Acta Acust. United Ac.*, 19(4), 199–213.
Damaske P. (1969/1970): Directional dependence of spectrum and correlation functions of the signals received at the ears, *Acta Acust. United Ac.*, 22(4), 191–204.
Damaske P., and Ando Y. (1972): Interaural crosscorrelation for multichannel loudspeaker reproduction, *Acta Acust. United Ac.*, 27(4), 232–238.
Daniel J. (2000): Acoustic field representation, application to the transmission and the reproduction of complex sound environments in a multimedia context (in French), PhD thesis, University of Paris 6, France.
Daniel J. (2003): Spatial sound encoding including near field effect: introducing distance coding filters and a viable, new Ambisonic format, *at the AES 23rd International Conference*, Copenhagen, Denmark.

Daniel J., and Moreau S. (2004): Further study of sound field coding with higher order Ambisonics, *at the AES 116th Convention*, Berlin, Germany, Paper 6017.

Daniel J., Nicol R., and Moreau S. (2003): Further investigations of high-order Ambisonics and wavefield synthesis for holophonic sound imaging, *at the AES 114th Convention*, Amsterdam, The Netherlands, Paper 5788.

Daniel J., Rault J.B., and Polack J.D. (1998): Ambisonics encoding of other audio formats for multiple listening conditions, *at the AES 105th Convention*, San Francisco, USA, Paper 4795.

Dantele A., Reiter U., Schuldt M., et al. (2003): Implementation of MPEG-4 audio nodes in an interactive virtual 3D environment, *at the AES 114th Convention*, Amsterdam, The Netherlands, Paper 5820.

Dattorro J. (1997): Effect design: part 1: Reverberator and other filters, *J. Audio Eng. Soc.*, 45(9), 660–684.

Davis D., and Davis C. (1980): The LEDE™ concept for the control of acoustic and psychoacoustic parameters in recording control rooms, *J. Audio Eng. Soc.*, 28(9), 585–595.

Davis M.F. (1987): Loudspeaker systems with optimized wide-listening-area imaging, *J. Audio Eng. Soc.*, 35(11), 888–896.

Davis M.F. (1993): The AC-3 multichannel coder, *at the AES 95th Convention*, San Francisco, USA, Paper 3774.

Davis M.F., and Fellers M.C. (1997): Virtual surround presentation of Dolby AC-3 and Pro Logic signal, *at the AES 103rd Convention*, New York, USA, Paper 4542.

Davis M.F., and Todd C.C. (1994): AC-3 operation, bitstream syntax, and features, *at the AES 97th Convention*, San Francisco, USA, Paper 3910.

DeFanti T.A., Dawe G., Sandin D.J., et al. (2009): The StarCAVE, a third-generation CAVE and virtual reality OptIPotal, *Future Gener. Comput. Syst.*, 25(2),169–178.

Djelani T., Porschmann C., Sahrhage J., et al. (2000): An interactive virtual-environment generator for psychoacousticresearch. II: collection of head-related impulse responsesand evaluation of auditory localization, *Acta Acust. United Ac.*, 86(6), 1046–1053.

Doerr K.U., Rademacher H., Huesgen S., et al. (2007): Evaluation of a low-cost 3D sound system for immersive virtual reality training systems, *IEEE Trans. Vis. Comput. Graph.*, 13(2), 204–212.

Dolby Laboratories (1997): Dolby professional encoding manual, http://www.dolby.com.

Dolby Laboratories (1998): Dolby surround mixing manual, http://www.dolby.com.

Dolby Laboratories (2000): 5.1 channel production guidelines, http://www.dolby.com

Dolby Laboratories (2002): Standards and practices for authoring Dolby digital and Dolby E bitstreams, http://www.dolby.com.

Dolby Laboratories (2012): Dolby Atmos, next-generation audio for cinema, http://www.dolby.com.

Dolby Laboratories (2015): Dolby Atmos specifications, http://www.dolby.com.

Dolby Laboratories (2016): Dolby Atmos for the home theater, http://www.dolby.com.

Dooley W.L., and Streicher R.D. (1982): M-S stereo: a powerful technique for working in stereo, *J. Audio Eng. Soc.*, 30(10), 707–718.

Dragnev P.D., Legg D.A., and Townsend D.W. (2002): Discrete logarithmic energy on the sphere, *Pacific J. Mat.*, 207(2), 345–358.

Dressler R. (1996): A step toward improved surround sound: making the 5.1 channel format reality, *at the AES 100th Convention*, Copenhagen, Denmark, Paper 4287.

Dressler R. (2000): Dolby surround Pro Logic II decoder principles of operation, http://www.dolby.com.

Dressler R. (2006): Audio coding for future entertainment formats, *at the AES 21st UK Conference*, Cambridge, UK.

Drullman R., and Bronkhorst A.W. (2000): Multichannel speech intelligibility and talker recognition using monaural, binaural, and three-dimensional auditory presentation, *J. Acoust. Soc. Am.*, 107(4), 2224–2235.

DTS Inc. (2006): DTS-HD Audio, consumer white paper for blu-ray disc and HD DVD applications, http://www.dts.com

Du G.H., Zhu Z.M., and Gong X.F. (2001): *Fundamental acoustics* (2nd edition, in Chinese), Nanjing University Press, Nanjing, China.

Duda R.O., and Martens W.L. (1998): Range dependence of the response of a spherical head model, *J. Acoust. Soc. Am.*, 104(5), 3048–3058.

Duraiswami R., Zotkin D.N., and Gumerov N.A. (2004): Interpolation and range extrapolation of HRTFs, *Proceedings of 2004 IEEE International Conference on Acoustics, Speech, and Signal Processing*, Montreal, Canada, Vol. 4, 45–48.

Duraiswami R., Zotkin D.N., Li Z.Y., et al. (2005): High order spatial audio capture and its binaural head-tracked playback over headphones with HRTF cues, *at the AES 119th Convention*, New York, USA, Paper 6540.

Durbin H.M. (1972): Playback effects from matrix recordings, *J. Audio Eng. Soc.*, 20(9), 729–733.

Durlach N.I., and Colburn H.S. (1978): Binaural phenomena, in *Handbook of perception*, Vol. IV, Academic Press, New York, USA.

Durlach N.I., Rigopulos A., Pang X.D., et al. (1992): On the externalization of auditory images, *Presence*, 1(2), 251–257.

DVD Forum (1997): *DVD specifications for read-only disc Part 3: video specifications*, Version 1.1, Tokyo, Japan.

Eargle J.M. (1971a): On the processing of two- and three-channel program material for four-channel playback, *J. Audio Eng. Soc.*, 19(4), 262–266.

Eargle J.M. (1971b): Multichannel stereo matrix systems: an overview, *J. Audio Eng. Soc.*, 19(7), 552–559.

Eargle J.M. (1972): 4-2-4 Matrix systems: standards, practice, and interchangeability, *J. Audio Eng. Soc.*, 20(10), 809–815.

Eargle J.M. (2006): *Handbook of recording engineering* (4th edition), Springer Science+Business Media Inc., New York, USA.

EBU-Tech 3324 (2007): *EBU evaluations of multichannel audio codecs*, European Broadcasting Union, Geneva, Switzerland.

Economou E.N. (2006): *Green's function in quantum physics* (3rd edition), Springer-Verlag, New York, USA.

Edwin P.C. (2002): In the light of 5.1 channel surround, "why A-B polycardiod centerfill" (AB-PC) is superior for symphony-orchestra recording, *at the AES 112th Convention*, Munich, Germany, Paper 5565.

Ehmer R.H. (1959a): Masking patterns of tones, *J. Acoust. Soc. Am.*, 31(8), 1115–1120.

Ehmer R.H. (1959b): Masking by tones vs noise bands, *J. Acoust. Soc. Am.*, 31(9), 1253–1256.

Ehret A., Groschel A., Purnhagen H., et al. (2007): Coding of "2+2+2" surround sound content using the MPEG surround standard, *at the AES 122nd Convention*, Vienna, Austria, Paper 6992.

Eilers C.G. (1961): Stereophonic FM broadcasting, *IRE Trans. Broadcasting TV Rec. BTR*, 7(2), 73–80.

Enzner G., Weinert M., Abeling S., et al. (2013): Advanced system options for binaural rendering of Ambisonic format, in *Proceeding of the 2013 IEEE International Conference on Acoustics, Speech and Signal Processing*, Vancouver, Canada, 251–255.

Epain N., Jin C.T., and Zotter F. (2014): Ambisonic decoding with constant angular spread, *Acta Acust. United Ac.*, 100(5), 928–936.

Erber T., and Hockney G.M. (2007): Complex systems: equilibrium configurations of N equal charges on a sphere ($2 \leq N \leq 112$), *Adv. Chem. Phys.*, 98, 495–594.

Ericson M.A. (2011): Multichannel sound reproduction in the environment for auditory research, *at the AES 131st Convention*, New York, USA, Paper 8513.

ETSI ES 201 980 V3.2.1 (2012): Digital radio mondiale (DRM); system specification, European Telecommunications Standards Institute, Sophia-Antipolis Cedex, France.

ETSI TS 102 366 V1.4.1 (2017): Digital audio compression (AC-3, Enhanced AC-3) standard, European Telecommunications Standards Institute, Sophia-Antipolis Cedex, France.

ETSI TS 103 190-1 V1.3.1 (2018): Digital audio compression (AC-4) standard, part 1: channel based coding, European Telecommunications Standards Institute, Sophia-Antipolis Cedex, France.

ETSI TS 103 190-2 V1.2.1 (2018): Digital audio compression (AC-4) standard, part 2: Immersive and personalized audio, European Telecommunications Standards Institute, Sophia-Antipolis Cedex, France.

ETSI TS 103 223 V1.1.1 (2015): MDA: object-based audio immersive sound metadata and bitstream, European Telecommunications Standards Institute, Sophia-Antipolis Cedex, France.

Evans M.J., Angus J.A.S., and Tew A.I. (1998): Analyzing head-related transfer function measurements using surface spherical harmonics, *J. Acoust. Soc. Am.*, 104(4), 2400–2411.

Evans M.J., Tew A.I., and Angus J.A.S. (1997): Spatial audio teleconferencing – which way is better?, in *Proceedings of the Fourth International Conference on Auditory Displays (ICAD 97)*, Palo Alto, California, USA, 29–37.

Evjen P., Bradley J.S., and Norcross S.G. (2001): The effect of late reflections from above and behind on listener envelopment, *Appl. Acoust.*, 62(2), 137–153.

Faller C. (2004): Coding of spatial audio compatible with different playback formats, *at the AES 117th Convention*, San Francisco, USA, Paper 6187.

Faller C. (2006): Multiple-loudspeakers playback of stereo signals, *J. Audio Eng. Soc.*, 54(11), 1051–1064.

Faller C. (2007): Matrix surround revised, *at the AES 30th International Conference*, Saariselka, Finland.

Faller C. (2010): Conversion of two closely spaced omnidirectional microphone signals to an XY stereo signal, *at the AES 129th Convention*, San Francisco, USA, Paper 8188.

Faller C., Altmann L., Levison J., et al. (2013): Multi-channel ring upmix, *at the AES 134th Convention*, Rome, Italy, Paper 8908.

Faller C., and Baumgarte F. (2003): Binaural cue coding part II, scheme and applications, *IEEE Trnas. Speech Audio Process.*, 11(6), 520–531.

Faller C., and Schillebeeckx P. (2011): Improved ITU and matrix surround downmixing, *at the AES 130th Convention*, London, UK, Paper 8339.

Farina A., Armelloni E., and Martignon P. (2004): An experimental comparative study of 20 Italian opera houses: measurement techniques, *J. Acoust. Soc. Am.*, 115(5), 2475.

Farina A., and Ayalon R. (2003): Recording concert acoustics for posterity, *at the AES 24th International Conference*, Banff, Canada.

Farina A., and Ugolotti E. (1998): Numerical model of the sound field inside cars for the creation of virtual audible reconstructions, *First COST-G6 Workshopon Digital Audio Effects (DAFX98)*, Barcelona, Spain.

Favrot S., and Buchholz J.M. (2012): Reproduction of nearby sound sources using higher-order Ambisonics with practical loudspeaker arrays, *Acta Acust. United Ac.*, 98(1), 48–60.

Favrot A., and Faller C. (2020): Wiener-based spatial B-format equalization, *J. Audio Eng. Soc.*, 68(7/8), 488–494.

Favrot S., Marschall M., and Kasbach J. et al. (2011): Mixed-order ambisonics recording and playback for improving horizontal directionality, *at the AES 131st Convention*, New York, USA, Paper 8528.

Fazi F.M., and Nelson P.A. (2010): The relation between sound field reproduction and near-field acoustical holography, *at the AES 129th Convention*, San Francisco, USA, Paper 8247.

Fazi F.M., and Nelson P.A. (2013): Sound field reproduction as an equivalent acoustical scattering problem, *J. Acoust. Soc. Am.*, 134(5), 3721–3729.

Feige F., and Kirby D.G. (1994): Report on the MPEG/Audio multichannel formal subjective listening tests, MPEG document ISO/IEC JTC1/SC29/WG11/N0685, International Organization for Standardization, Geneva, Switzerland.

Fejzo Z., Kramer L., McDowell K., et al. (2005): DTS-HD: technical overview of lossless mode of operation, *at the AES 118th Convention*, Barcelona, Spain, Paper 6445.

Fernando L.L. (2014): An architecture for reverberation in high order Ambisonics, *at the AES 137th Convention*, Los Angeles, USA, Paper 9109.

Fielder L.D., Andersen R.L., Crockett B.G., et al. (2004): Introduction to Dolby digital plus, an enhancement to the Dolby digital coding system, *at the AES 117th Convention*, San Francisco, USA, Paper 6196.

Fielder L.D., and Robinson D.P. (1995): AC-2 and AC-3: the technology and its application, *at the AES 5th Australian Regional Convention*, Sydney, Australian, Paper 4022.

Firtha G., and Fiala P. (2015a): Sound field synthesis of uniformly moving virtual monopoles, *J. Audio Eng. Soc.*, 63(1/2), 46–53.

Firtha G., and Fiala P. (2015b): Wave field synthesis of moving sources with retarded stationary phase approximation, *J. Audio Eng. Soc.*, 63(12), 958–965.
Fletcher H. (1940): Auditory patterns, *Rev. Mod. Psys.*, 12(1), 47–65.
Florentine M., Buus S., and Mason C.R. (1987): Level discrimination as a function of level for tones from 0.25 to 16 kHz, *J. Acoust. Soc. Am.*, 81(5), 1528–1541.
Franck A., Graefe A., Korn T., et al. (2007): Reproduction of moving sound sources by wave field synthesis: an analysis of artifacts, *at the AES 32nd International Conference*, Hillerød, Denmark.
Francombe J., Brookes T., and Mason R. (2017a): Evaluation of spatial audio reproduction methods (part 1): elicitation of perceptual differences, *J. Audio Eng. Soc.*, 65(3), 198–211.
Francombe J., Brookes T., Mason R., et al. (2017b): Evaluation of spatial audio reproduction methods (part 2): analysis of listener preference, *J. Audio Eng. Soc.*, 65(3), 212–225.
Freeland F.P., Biscainho L.W.P., and Diniz P.S.R. (2004): Interpositional transfer function for 3D-sound generation, *J. Audio Eng. Soc.*, 52(9), 915–930.
Fuchigami N., Kuroiwa T., Suzuki B.H. (2000): DVD-Audio specifications, *J. Audio Eng. Soc.*, 48(12), 1228–1240.
Fug S., Holzer A., and Borb C., et al. (2014): Design, coding and processing of metadata for 48(12), 1228–1240. Object-based interactive audio, *at the AES 137th Convention*, Los Angeles, Paper 9097.
Fukada A. (2001): A challenge in multichannel sound recording, *at the AES 19th International Conference*, Bavaria, Germany.
Fukada A., Tsujimoto K., and Akita S. (1997): Microphone techniques for ambient sound on a music recording, *at the AES 103rd Convention*, New York, USA, Paper 4540.
Furui S. (2000): *Digital speech processing, synthesis, and recognition* (2nd edition), Marcel Dekker, New York, USA.
Furuya H., Fujimoto K., Choi Y.J., and Higa N. (2001): Arrival direction of late sound and listener envelopment, *Appl. Acoust.*, 62(2), 125–136.
Furuya H., Fujimoto K., and Wakuda A. (2008): Psychological experiments on listener envelopment when both the early-to-late sound level and directional late energy ratios are varied, and consideration of calculated LEV in actual halls. *Appl. Acoust.*, 69(11), 1085–1095.
Furuya H., Fujimoto K., Wakuda A., et al. (2005): The influence of total and directional energy of late sound on listener envelopment, *Acoust. Sci. Tech*, 26(2), 208–211.
Fuster L., Lopez J.J., and Gonzalez A. (2005): Room compensation using multichannel inverse filters for wave field synthesis system, *at the AES 118th Convention*, Barcelona, Spain, Paper 6401.
Gardner W.G. (1995): Efficient convolution without input-output delay, *J. Audio Eng. Soc.* 43(3), 127–136.
Gardner W.G. (1997): 3-D audio using loudspeakers, Doctor thesis of Massachusetts Institute of Technology, Massachusetts, USA.
Gardner W.G. (2002): Reverberation algorithms, in *Applications of digital signal processing to audio and acoustics* (edited by Brandenburg K.), The International Series in Engineering and Computer Science, vol. 437, Springer, Boston, MA.
Gardner W.G., and Martin K.D. (1995): HRTF measurements of a KEMAR, *J. Acoust. Soc. Am.*, 97(6), 3907–3908.
Gaston L., and Sanders R. (2008): Evaluation of HE-AAC, AC-3 and E-AC-3 codecs, *J. Audio Eng. Soc.*, 56(3), 140–155.
Gauthier P.A., and Berry A. (2006): Adaptive wave field synthesis with independent radiation mode control for active sound field reproduction: theory, *J. Acoust. Soc. Am.*, 119(5), 2721–2737.
Gauthier P.A., and Berry A. (2007): Adaptive wave field synthesis for sound field reproduction, theory, experiment and future perspectives, *J. Audio Eng. Soc.*, 55(12), 1107–1124.
Gauthier P.A., and Berry A. (2008): Adaptive wave field synthesis with independent radiation mode control for active sound field reproduction: experimental results, *J. Acoust. Soc. Am.*, 123(4), 1991–2002.
Gauthier P.A., Berry A., and Wieslaw W. (2005): Sound field reproduction in-room using optimal control techniques: simulations in the frequency domain, *J. Acoust. Soc. Am.*, 117(2), 662–678.
Gauthier P.A., Camier C., Padois T., et al., (2015): Sound field reproduction of real flight recordings in aircraft cabin mock-up, *J. Audio Eng. Soc.*, 63(1/2), 6–20.

Gauthier P.A., Chambatte É., Camier C., et al. (2014a): Beamforming regularization, scaling matrices, and inverse problems for sound field extrapolation and characterization: part I– theory, *J. Audio Eng. Soc.*, 62(3), 77–98.

Gauthier P.A., Chambatte É.C., Camier C., et al. (2014b): Beamforming regularization, scaling matrices, and inverse problems for sound field extrapolation and characterization: part II– experiments, *J. Audio Eng. Soc.*, 62(4), 207–219.

GB/T22726-2008 (2008): Specification for multichannel digital audio coding technique (in Chinese), National Standard of the P.R China, National Institute of Standards of the People's Republic of China, Beijing, China.

GB/T26686-2017 (2017): General specification for digital terrestrial television receiver (in Chinese), National Standard of the P.R China, National Institute of Standards of the People's Republic of China, Beijing, China.

Geier M., Wierstorf H., and Ahrens J. (2010): Perceptual assessment of focused sources in wave field synthesis, *at the AES 128th Convention*, London, UK, Paper 8069.

Geisler C.D. (1998): *From sound to synapse: physiology of the mammalian ear*, Oxford University Press, New York, USA.

Gelfand S.A. (2010): *Hearing: An Introduction to psychological and physiological acoustics* (5th edition), Informa Healthcare, London, UK.

Geluso P. (2012): Capturing Height: the addition of Z microphones to stereo and surround microphone arrays, *at the AES 132nd Convention*, Budapest, Hungary, Paper 8595.

Genuit K., and Xiang N. (1995): Measurements of artificial head transfer functions for auralization and virtual auditory environment, in *Proceedings of 15th International Congress on Acoustics* (invited paper), Trondheim, Norway, II 469–472.

George S., Zielinski S., and Rumsey F. (2006): Feature extraction for prediction of multichannel spatial audio fidelity, *IEEE Trans. Audio, Speech, Language Process.*, 14(6), 1994–2005.

George S., Zielinski S., and Rumsey F. (2010): Development and validation of an unintrusive model for predicting the sensation of envelopment arising from surround sound recordings, *J. Audio Eng. Soc.*, 58(12), 1013–1031.

Germanenn A. (1998): The arrangements of microphones using three front channels, a systematic approach (in German), *in the Proceeding of Tonmeistertagung*, 518–542.

Gersho A., and Gray R.M. (1992): *Vector quantization and signal compression*, Springer, Boston, MA.

Gerzon M.A. (1973): Periphony: with height sound reproduction, *J. Audio Eng. Soc.*, 21(1), 2–10.

Gerzon M.A. (1975a): Recording concert hall acoustics for posterity, *J. Audio Eng. Soc.*, 23(7), 569–571.

Gerzon M.A. (1975b): A geometric model for two-channel four-speaker matrix stereo system, *J. Audio Eng. Soc.*, 23(2), 98–106.

Gerzon M.A. (1985): Ambisonics in multichannel broadcasting and video, *J. Audio Eng. Soc.*, 33(11), 859–871.

Gerzon M.A. (1986): Stereo shuffling: new approach-old technique, *Studio Sound*, 28(7), 122–130.

Gerzon M.A. (1990): Three channels, the future of stereo? *Studio Sound*, 32(6), 112–125.

Gerzon M.A. (1992a): General metatheory of auditory localisation, *at the AES the 92nd Convention*, Vienna, Austria, Paper 3306.

Gerzon M.A. (1992b): Optimum reproduction matrices for multispeaker stereo, *J. Audio Eng. Soc.*, 40(7/8), 571–589.

Gerzon M.A. (1992c): Panpot laws for multispeaker stereo, *at the AES 92nd Convention*, Vienna, Austria, Paper 3309.

Gerzon M.A. (1992d): Hierarchical transmission system for multispeaker stereo, *J. Audio Eng. Soc.*, 40(9), 692–705.

Gerzon M.A. (1992e): The design of distance panpots, *at the AES 92nd Convention*, Vienna, Austria, Paper 3308.

Gerzon M.A. (1992f): Compatibility of and conversion between multispeaker systems, *at the AES 93rd Convention*, San Francisco, USA, Paper 3405.

Gerzon M.A. (1994): Applications of Blumlein shuffling to stereo microphone techniques, *J. Audio Eng. Soc.*, 42(6), 435–453.

Gerzon M.A., and Barton G.J. (1992): Ambisonic decoder for HDTV, *at the AES 92nd Convention*, Vienna, Austria, Paper 3345.

Gerzon M.A., Craven P.G., Stuart J.R., et al. (2004): The MLP lossless compression system for PCM audio, *J. Audio Eng. Soc.*, 52(3), 243–260.

Gibson J.J., Christensen R.M., and Limberg A.L.R. (1972): Compatible FM broadcasting of Panoramic sound, *J. Audio Eng. Soc.*, 20(10), 816–822.

Gierlich H.W. (1992): The application of binaural technology, *Appl. Acoust.*, 36(3/4), 219–243.

Gnann V., and Spiertz M. (2008): Comb-filter free audio mixing using STFT magnitude spectra and phase estimation, *in the Proceeding of 11st International Conference of Digital Audio Effect (DAFx-08)*, Espoo, Finland.

Goldmark P.C., Bauer B.B., and Bachman W.S. (1958): The Columbia compatible stereophonic record, *IRE Trans. Audio*, 6(2), 25–28.

Goldstein H. (1980): *Classical mechanics* (2nd edition), Addison-Wesley Publishing Company Inc., Massachusetts, USA.

Gong M., Xiao Z., Qu T.S., et al. (2007): Measurement and analysis of near-field head-related transfer function, *Appl. Acoust.* (in Chinese), 26(6), 326–334.

Gonot A., Chateau N., Emerit M. (2006): Usability of 3D-sound for navigation in a constrained virtual environment, *at the AES 120th Convention*, Paris, France, Paper 6800.

Goodwin M.M. (2008a): Primary-ambient decomposition and dereverberation of two-channel and multi-channel audio, in *Proceeding of IEEE 42nd Asilomar Conference on Signals, Systems and Computers*, Pacific Grove, CA, 797–800.

Goodwin M.M. (2008b): Geometric signal decomposition for spatial audio enhancement, in *Proceeding of IEEE 2008 International Conference on Acoustics, Speech and Signal Processing*, Las Vegas, NV, 409–412.

Goodwin M.M., and Jot J.M. (2007): Primary-ambient decomposition and vector-based localization for spatial audio cording and enhancement, in *Proceeding of IEEE 2007 International Conference on Acoustics, Speech and Signal Processing*, Honolulu, HI, Vol. I, 9–12.

Gorzel M., Kearney G., and Boland F. (2014): Investigation of Ambisonic rendering of elevated sound source, *at the AES 55th International Conference*, Helsinki, Finland.

Grandjean P., Berry A., Gauthier P.A. (2021a): Sound field reproduction by combination of circular and spherical higher-order Ambisonics: part I–a new 2.5-D driving function for circular arrays, *J. Audio Eng. Soc.*, 69(3), 152–165.

Grandjean P., Berry A., and Gauthier P.A. (2021b): Sound field reproduction by combination of circular and spherical higher-order ambisonics: part II—hybrid system, *J. Audio Eng. Soc.*, 69(3), 166–181.

Grantham D.W., and Wightman F.L. (1978): Detectability of varying interaural temporal differences, *J. Acoust. Soc. Am.*, 63(2), 511–523.

Grassi E., Tulsi J., and Shamma S. (2003): Measurement of head-related transfer functions based on the empirical transfer function estimate, in *Proceedings of the 2003 International Conference on Auditory Display*, Boston, MA, 119–122.

Gribben C., and Lee H. (2014): The perceptual effects of horizontal and vertical interchannel decorrelation using the Lauridsen decorrelator, *at the AES 136th Convention*, Berlin, Germany, Paper 9027.

Gribben C., and Lee H. (2017): The perceptual effect of vertical interchannel decorrelation on vertical image spread at different azimuth positions, at the AES 142nd Convention, Berlin, Germany, Paper 9747.

Gribben C., and Lee H. (2018): The frequency and loudspeaker-azimuth dependencies of vertical interchannel decorrelation on the vertical spread of an auditory image, *J. Audio Emg. Soc.*, 66(7/8), 537–555.

Griesinger D. (1986): Spaciousness and localization in listening rooms and their effects on the recording technique, *J. Audio Eng. Soc.*, 34(4), 255–268.

Griesinger D. (1992a): IALF-binaural measures of spatial impression and running reverberance, *at the AES 92nd Convention*, Vienna, Austria, 1992, Paper 3292.

Griesinger D. (1992b): Measures of spatial impression and reverberance based on the physiology of human hearing, *at the AES 11th International Conference*, Portland, USA.

Griesinger D. (1996): Multichannel matrix surround decoder for two-eared listeners, *at the AES 101st Convention*, Los Angeles, USA, Paper 4402.
Griesinger D. (1997a): Progress in 5-2-5 matrix systems, *at the AES 103rd Convention*, New York, USA, Paper 4625.
Griesinger D. (1997b): Spatial impression and envelopment in small rooms, *at the AES 103rd Convention*, New York, USA, Paper 4638.
Griesinger D. (1998): Multichannel sound systems and their interaction with the room, *at the AES 15th International Conference*, Copenhagen, Denmark.
Grignon L.D. (1949): Experiments in stereophonic sound, *J. SMPTE*, 52(3), 280–292.
Groschel A., Schug M., Beer M., et al. (2003): Enhancing audio coding efficiency of MPEG Layer-2 with spectral band replication (SBR) for digital radio (EUREKA 147/DAB) in a backwards compatible way, *at the AES 114th Convention*, Amsterdam, The Netherlands, Paper 5850.
Guan S.Q. (1988): *Fundamental electroacoustic technology* (revised edition, in Chinese), Posts and Telecommunications Press, Beijing, China.
Guan S.Q. (1995): Some thoughts on Stereophonic (in Chinese), *Appl. Acoust.*, 14(6),6–11.
Guastavino C., and Katz B.F.G. (2004): Perceptual evaluation of multi-dimensional spatial audio reproduction, *J. Acoust. Soc. Am.*, 116(2), 1105–1115.
Gumerov N.A., O'Donovan A.E., and Duraiswami R., et al. (2010): Computation of the head-related transfer function via the fast multipole accelerated boundary element method and its spherical harmonic representation, *J. Acoust. Soc. Am.*, 127(1), 370–386.
Gundry K. (2001): A new active matrix decoder for surround sound, *at the AES 19th International Conference*, Schloss, Elmau, Germany.
Hahn N., Winter F., and Spors S. (2016): Local wave field synthesis by spatial band-limitation in the circular/spherical harmonics domain, *at the AES 140th Convention*, Paris, France, Paper 9596.
Hamasaki K. (2011): The 22.2 multichannel sounds and its reproduction at home and personal environment, *at the AES 43rd International Conference*, Pohang, Korea.
Hamasaki K., and Hiyama K. (2003): Reproduction spatial impression with multichannel audio, *at the AES 24th International Conference*, Banff, Canda.
Hamasaki K., Hiyama K., Nishiguchi T., et al. (2004): Advanced multichannel audio systems with superior impression of presence and reality, *at the AES 116th Convention*, Berlin, Germany, Paper 6053.
Hamasaki K., Nishiguchi T., Okumura R., et al. (2007): Wide listening area with exceptional spatial sound quality of a 22.2 multichannel sound system, *at the AES 122nd Convention*, Vienna, Austria, Paper 7037.
Hamdan E.C., and Fazi F.M. (2021): A modal analysis of multichannel crosstalk cancellation systems and their relationship to amplitude panning, *J. Sound Vib.*, 490, 115743.
Hammershφi D., and Mφller H. (1996): Sound transmission to and within the human ear canal, *J. Acoust. Soc. Am.*, 100(1), 408–427.
Han H.L. (1994): Measuring a dummy head in search of pinna cues, *J. Audio Eng. Soc.*, 42(1/2), 15–37.
Haneda Y., Makino S., Kaneda Y., et al. (1999): Common acoustical pole and zero modeling of room transfer functions, *IEEE Trans. Speech Audio Process.*, 7(2), 188–196.
Härmä A. (2010): Classification of time-frequency regions in stereo audio, *at the AES 128th Convention*, London, UK, Paper 7980.
Härmä A., Karjalainen M., Savioja L., et al. (2000): Frequency-warped signal processing for audio applications, *J. Audio Eng. Soc.*, 48(11), 1011–1031.
Hartmann W.M. and Wittenberg A. (1996): On the externalization of sound images, *J. Acoust. Soc. Am.*, 99(6), 3678–3688.
Hartung K., Sterbing S.J., Keller C.H., et al. (1999): Applications of virtual auditory space in psychoacoustics and neurophysiology, *J. Acoust. Soc. Am.*, 105(2), 1164.
Harvey F.K., and Uecke E.H. (1962): Compatibility problem in two-channel stereophonic recordings, *J. Audio Eng. Soc.*, 10(1), 8–12.
Harwood H.D. (1968): Stereophonic image sharpness, *Wireless World*, 74(July), 207–211.
Hawksford M.O.J. (2002): Scalable multichannel coding with HRTF enhancement for DVD and virtual sound systems, *J. AudioEng. Soc.*, 50(11), 894–913.

He J.J., Gan W.S., and Tan E.L. (2015): Time shifting based primary-ambient extraction for spatial audio reproduction, *IEEE Trans. Audio, Speech, Language Process.*, 23(10), 1576–1588.

He J.J., Tan E.L., and Gan W.S. (2014): Linear estimation based primary-ambient extraction for stereo audio signals, *IEEE Trans. Audio, Speech, Language Process.*, 22(2), 505–517.

He P., Xie B.S., and Rao D. (2006): Subjective and objective analyses of timbre equalized algorithms for virtual sound reproduction by loudspeakers (in Chinese), *Appl. Acoust.* (in Chinese) 25(1), 4–12.

He P., Xie B.S., and Zhong X.L. (2007): Virtual sound signal processing using HRTF without pinnae (in Chinese), *Appl. Acoust.*, 26(2), 100–106.

He Y.J., Xie B.S., and Liang S.J. (1993): Extension of localization equation for stereophonic sound image(in Chinese), *Audio Eng.*, 17(10), 2–4.

Hebrank J., and Wright D. (1974): Spectral cues used in the localization of sound sources on the median plane, *J. Acoust. Soc. Am.*, 56(6), 1829–1834.

Heller A.J., Benjamin E., and Lee R. (2010): Design of ambisonic decoders for irregular arrays of loudspeakers by non-linear optimization, *at the AES 129th Convention*, San Francisco, CA, Paper 8243.

Henning G.B. (1974): Detectability of interaural delay in high-frequency complex waveforms, *J. Acoust. Soc. Am.*, 55(1), 84–90.

Herre J., Brandenburg K., and Lederer D. (1994): Intensity stereo coding, *at the AES 96th Convention*, Amsterdam, The Netherlands, Paper 3799.

Herre J., and Dietz M. (2008): MPEG-4 high-efficiency AAC coding (Standards in a Nutshell), *IEEE Signal Process. Mag.*, 25(3),137–142.

Herre J., Falch C., Mahne D., et al. (2011): Interactive teleconferencing combining spatial audio object coding and DiRAC technology, *J. Audio Eng. Soc.*, 59(12), 924–935.

Herre J., Faller C., Disch S., et al. (2004): Spatial audio coding: next generation efficient and compatible coding of multichannel audio, *at the AES 117th Convention*, San Francisco, CA, USA, Paper 6186.

Herre J., Hilpert J., Kuntz A., et al. (2014): MPEG-H audio—The new standard for universal spatial/3D audio coding, *J. Audio Eng. Soc.*, 62(12), 821–830.

Herre J., Hilpert J., Kuntz A., et al. (2015): MPEG-H audio—The new standard for coding of immersive spatial audio, *IEEE J. Selected Topics Signal Process.*, 9(5), 770–779.

Herre J., Kjorling K., Breebaart H., et al. (2008): MPEG surround-The ISO/MPEG standard for efficient and compatible multichannel audio coding, *J. Audio Eng. Soc.*, 56(11), 932–955.

Herre J., Purnhagen H., Koppens J., et al. (2012): MPEG spatial audio object coding—The ISO/MPEG standard for efficient coding of interactive audio scenes, *J. Audio Eng. Soc.*, 60(9), 655–673.

Herrmann U., Henkels V., and Braun D. (1998): Comparison of 5 surround microphone method (in German), *in the Proceeding of Tonmeistertagung*, 508–517.

Hertz B.F. (1981): 100 years with stereo: the beginning, *J. Audio Eng. Soc.*, 29(5), 368–370.

Hibbing M. (1989): XY and MS microphone techniques in comparison, *J. Audio Eng. Soc.*, 37(10), 823–831.

Hidaka T., and Beranek L.L. (2000): Objective and subjective evaluations of twenty-three opera houses in Europe, Japan, and the Americas. *J. Acoust. Soc. Am.*, 107(1), 368–383.

Hidaka T., Beranek L.L., and Okano T. (1995): Interaural cross-correlation (IACC), lateral fraction (LF), and low- and high-frequency sound levels(G) as measures of acoustical quality in concert halls. *J. Acoust. Soc. Am.*, 98(2), 988–1007.

Hiekkanen T., Makivirta A., and Karjalainen M. (2009): Virtualized listening tests for loudspeakers, *J. Audio Eng. Soc.*, 57(4), 237–251.

Hill P.A., Nelson P.A., Kirkeby O., et al. (2000): Resolution of front–back confusion in virtual acoustic imaging systems, *J. Acoust. Soc. Am.*, 108(6), 2901–2910.

Hilpert J., and Disch S. (2009): The MPEG surround coding standard (Standards in a Nutshell), *IEEE Signal Process. Mag.*, 26(1),148–152.

Hirvonen T., and Robinson C.Q. (2016): Extended bass management methods for cost-efficient immersive audio reproduction in digital cinema, *at the AES 140th Convention*, Paris, France, Paper 9595.

Hiyama K., Komiyama S., and Hamasaki K. (2002): The minimum number of loudspeakers and its arrangement for reproducing the spatial impression of diffuse sound field, *at the AES 113rd Convention*, Los Angeles, USA, Paper 5674.

Hoang T.M.N., Ragot S., Kövesi B., et al. (2010): Parametric stereo extension of ITU-T G.722 based on a new downmixing scheme, in *Proceedings of the 2010 IEEE International Workshop on Multimedia Signal Processing*, Saint-Malo, France.

Hoesel R.J.M.V., and Tyler R.S. (2003): Speech perception, localization, and lateralization with bilateral cochlear implants, *J. Acoust. Soc. Am.*, 113(3), 1617–1630.

Hoffmann P.F., and Møller H. (2006): Audibility of spectral differences in head-related transfer functions, *at the AES 120th Convention*, Paris, France, Paper 6652.

Hollerweger F. (2006): Periphonic sound spatialization in multi-user virtual environment, Master's thesis at Graz University of Music and Dramatic art, Graz, Austria.

Hollier M.P., Rimell A.N., and Burraston D. (1997): Spatial audio technology for telepresence, *BT Technology J.*, 15(4), 33–41.

Holman T. (1991): New factors in sound for cinema and television, *J. Audio Eng. Soc.*, 39(7/8), 529–539.

Holman T. (1996): The number of audio channels, *at the AES 100th Convention*, Copenhagen, Denmark, Paper 4292.

Holman T. (2000): Comments on "subjective appraisal of loudspeaker directivity for multi-channel reproduction", and Zacharov N., Author's reply, *J. Audio. Eng. Soc.*, 48(4), 314–321.

Holman T. (2001): The number of loudspeaker channels, *at the AES 19th International Conference*, Schloss, Elmau, Germany.

Holman T. (2008): *Surround sound, up and running* (2ndedition), Focal Press, Burlington, MA.

Hosoe S., Nishino T., Itou K., et al. (2005): Measurement of head-related transfer functions in the proximal region, in *Proceeding of Forum Acusticum 2005*, Budapest, Hungary, 2539–2542.

Howie W., King R., and Martin D. (2016): A three-dimensional orchestral music recording technique, optimized for 22.2 multichannel sound, *at the AES 141st Convention*, Los Angeles, USA, Paper 9612.

Howie W., King R., and Martin D. (2017): Listener discrimination between common channel-based 3D audio reproduction formats, *J. Audio Eng. Soc.*, 65(10), 796–805.

Hull J. (1999): Surround sound past, present and future, Dolby Laboratories, www.dolby.com.

Hulsebos E., Schuurmans T., Vries D.D., et al. (2003): Circular microphone array for discrete multichannel audio recording, *at the AES 114th Convention*, Amsterdam, The Netherlands, Paper 5716.

Hulsebos E., and Vries D.D. (2002): Parameterization and reproduction of concert hall acoustics measured with a circular microphone array, *at the AES 112nd Convention*, Munich, Germany, Paper 5579.

Hulsebos E., Vries D.D., and Bourdillat E. (2002): Improved microphone array configurations for auralization of sound fields by wave-field synthesis, *J. Audio Eng. Soc.*, 50(10), 779–790.

Huopaniemi J., Zacharov N., and Karjalainen M. (1999): Objective and subjective evaluation of head-related transfer function filter design, *J. Audio. Eng. Soc.*, 47(4), 218–239.

IEC 60268 (1998): *Sound system equipment-part 13: listening tests on loudspeakers*, International Electrotechnical Commission, Geneva, Switzerland.

IEC 60906 (1999): *Audio recording –compact disc digital audio system*, International Electrotechnical Commission, Geneva, Switzerland.

IEC 60959 (1990): *Provisional head and torso simulator for acoustic measurement on air conduction hearing aids*, International Electrotechnical Commission, Geneva, Switzerland.

IEC 62574 (2011): *Audio, video and multimedia systems – general channel assignment of multichannel audio*, International Electrotechnical Commission, Geneva, Switzerland.

IEEE Computer Society (2020): *IEEE standard for second generation audio coding*, The Institute of Electrical and Electronics Engineers, New York, USA.

Inoue T., Takahashi N., and Owaki I. (1971): A discrete four-channel disc and its reproducing system (CD-4 system), *J. Audio Eng. Soc.*, 19(7), 576–583.

IRCAM Lab (2003): Listen HRTF database, http://recherche.ircam.fr/equipes/salles/listen/

Irwan R., and Aarts R.M. (2002): Two-to-five channel processing, *J. Audio Eng. Soc.*, 50(11), 914–926.

Ise S. (1999): A principle of sound field control based on the Kirchhof-Helmholtz integral equation and the theory of inverse systems, *Acta Acust. United Ac.*, 85(1), 78–87.

ISO 1999 (1975): *Acoustics-assessment of occupational noise exposure for hearing conservation purposes*, International Organization for Standardization, Geneva, Switzerland.

ISO 22234 (2005): *Cinematography – relative and absolute sound pressure levels for motion-picture multi-channel sound systems—measurement methods and levels applicable to analog photographic film audio, digital photographic film audio and D-cinema audio*, International Organization for Standardization, Geneva, Switzerland.

ISO 226 (2003): *Acoustics – normal equal-loudness-level contours*, International Organization for Standardization, Geneva, Switzerland.

ISO 2969(2015): *Cinematography – B-chain electroacoustic response of motion-picture control rooms and indoor theatres-specifications and measurements*, International Organization for Standardization, Geneva, Switzerland.

ISO 3382-1(2009): *Acoustics – measurement of room acoustic parameters, part 1: performance spaces*, International Organization for Standardization, Geneva, Switzerland.

ISO 389-1(1998): *Acoustics – reference zero for the calibration of audiometric equipment, part 1: reference equivalent threshold sound pressure levels for pure tones and supra-aural earphones*, International Organization for Standardization, Geneva, Switzerland.

ISO 532-1 (2017): *Acoustics – methods for calculating loudness – part 1: Zwicker method*, International Organization for Standardization, Geneva, Switzerland.

ISO 532-2 (2017): *Acoustics – methods for calculating loudness – part 2: Moore-Glasberg method*, International Organization for Standardization, Geneva, Switzerland.

ISO 9568 (1993): *Cinematography-background acoustic noise levels in theatres, review rooms and dubbing rooms*, International Organization for Standardization, Geneva, Switzerland.

ISO/IEC 11172-3 (1993): Information technology – coding of moving pictures and associated audio for digital storage media at up to about 1.5 Mbit/s, part 3: audio, International Organization for Standardization, Geneva, Switzerland.

ISO/IEC 13818-3 (1998): *Information technology – generic coding of moving pictures and associated audio, part 3: audio*, International Organization for Standardization, Geneva, Switzerland.

ISO/IEC 13818-7 (1997): *Information technology – generic coding of moving pictures and associated audio, advanced audio coding-part 7: advanced audio coding (AAC)*, International Organization for Standardization, Geneva, Switzerland.

ISO/IEC 16488 (2002): *Information technology – 120 mm DVD – read-only disk*, International Organization for Standardization, Geneva, Switzerland.

ISO/IEC 23001-8 (2015): *Information technology-MPEG systems technologies-part 8: coding-independent code points*, International Organization for Standardization, Geneva, Switzerland.

ISO/IEC 23003-1 (2007): *Information technology – MPEG audio technologies– part 1: MPEG surround*, International Organization for Standardization, Geneva, Switzerland.

ISO/IEC 23003-1 (2012): *Information technology – MPEG audio technologies – part 3: United speech and audio coding*, International Organization for Standardization, Geneva, Switzerland.

ISO/IEC 23003-2 (2010): *Information technology – MPEG audio technologies –part 2: Spatial audio object coding*, International Organization for Standardization, Geneva, Switzerland.

ISO/IEC 23008-3 (2015): *Information technology –high efficiency coding and media delivery in heterogeneous environments, part 3: 3D audio*, International Organization for Standardization, Geneva, Switzerland.

ISO/IEC 30190 (2021): *Information technology — digitally recorded media for information interchange and storage — 120 mm, single layer (25,0 Gbytes per disk)and dual layer (50,0 Gbytes per disk)BD recordable disk*, International Organization for Standardization, Geneva, Switzerland.

Itho R. (1972): Proposed universal encoding standards for compatible four-channel matrixing, *J. Audio Eng. Soc.*, 20(3), 167–173.

ITU-R BS 1116-1 (1997): *Methods for the subjective assessment of small impairments in audio systems including multichannel sound system*, International Telecommunication Union, Geneva, Switzerland.

ITU-R BS 1116-3 (2015): *Methods for the subjective assessment of small impairments in audio systems*, International Telecommunication Union, Geneva, Switzerland.

ITU-R BS 1284-1 (2003): *General methods for the subjective assessment of sound quality*, International Telecommunication Union, Geneva, Switzerland.

ITU-R BS 1387-1 (1999): *Method for objective measurement of perceived audio quality*, International Telecommunication Union, Geneva, Switzerland.

ITU-R BS 1534-3 (2015): *Method for the subjective assessment of intermediate quality level of audio systems*, International Telecommunication Union, Geneva, Switzerland.
ITU-R BS 1909 (2012): *Performance requirements for an advanced multichannel stereophonic sound system for use with or without accompanying picture*, International Telecommunication Union, Geneva, Switzerland.
ITU-R BS 2051-2 (2018): *Advanced sound system for programme production*, International Telecommunication Union, Geneva, Switzerland.
ITU-R BS 2126-0 (2019): *Methods for the subjective assessment of sound system with accompanying picture*, International Telecommunication Union, Geneva, Switzerland.
ITU-R BS 2132-0 (2019): *Method for subjective quality assessment of audio differences of sound systems using multiple stimuli without a given reference*, International Telecommunication Union, Geneva, Switzerland.
ITU-R Report BS 2159-7 (2015): *Multichannel sound technology in home and broadcasting applications*, International Telecommunication Union, Geneva, Switzerland.
ITU-R BS 2399-0 (2017): *Methods for selecting and describing attributes and terms, in the preparation of subjective tests*, Geneva, Switzerland.
ITU-R BS 708 (1990): *Determination of the electro-acoustical properties of studio monitor headphones*, International Telecommunication Union, Geneva, Switzerland.
ITU-R BS 775-1 (1994): *Multichannel stereophonic sound system with and without accompanying picture, Doc 10/63*, International Telecommunication Union, Geneva, Switzerland.
ITU-R BS 775-3 (2012): *Multichannel stereophonic sound system with and without accompanying picture*, International Telecommunication Union, Geneva, Switzerland.
ITU-R Doc.10/51-E (1995): *Low bit rate multichannel audio coder test results*, Geneva, Switzerland.
Iwahara M., and Mori T. (1978): Stereophonic sound reproduction system, United States Patent: 4, 118, 599.
Jackson J.D. (1999): *Classical electrodynamics* (3rd Edition), John Wiley & Sons, New York, USA.
Jackson P.J.B., Dewhirst M., Conetta R., et al., (2008): QESTRAL (part 3): system and metrics for spatial quality prediction, *at the AES 125th Convention*, San Francisco, USA, Paper 7597.
JBL Professional (1998): Cinema sound system design, https://www.jblpro.com/
Jecklin J. (1981): A different way to record classical music, *J. Audio Eng. Soc.*, 29(5), 329–332.
Jeffress L.A. (1948): A place theory of sound localization, *J. Comp. Physiol. Psych.*, 41(1), 35–39.
Jiang J.L., Xie B.S., and Mai H.M. (2018): The influence of the number of loudspeakers on the pressure error in ambisonics reproduction (in Chinese). *J. South China Univ. Technol.*, 46(3), 119–126.
Jiang J.L., Xie B.S., Mai H.M., et al. (2019): The role of dynamic cue in auditory vertical localization, *Appl. Acoust.*, 146, 398–408.
Jin C., Corderoy A., Carlile S., et al. (2004): Contrasting monaural and interaural spectral cues for human sound localization, *J. Acoust. Soc. Am.*, 115(6), 3124–3141.
Jin C., Epain N., and Parthy A. (2014): Design, optimization and evaluation of a dual-radius spherical microphone array, *IEEE Trans. Audio, Speech, Language Process.*, 22(1), 193–204.
Jin C., Leong P., Leung J., et al. (2000): Enabling individualized virtual auditory space using morphological measurements, in *Proceedings of the First IEEE Pacific-Rim Conference on Multimedia*, Sydney, Australia, 235–238.
Jin C., Tan T., and Kan A., et al. (2005): Real-time, head-tracked 3D audio with unlimited simultaneous sounds, in *Proceedings of Eleventh Meeting of the International Conference on Auditory Display (ICAD 05)*, Limerick, Ireland.
Joshi A.W. (1977): *Elements of group theory for physicist* (2nd edition), John Wiley & Sons, New York, USA.
Jot J.M., and Chaigne A. (1991): Digital delay networks for designing artificial reverberators, *at the AES 90th Convention*, Paris, France, Paper 3030.
Jot J.M., Larcher V., and Pernaux J.M. (1999): A comparative study of 3D audio encoding and rendering techniques, *at the AES 16th International Conference*, Rovaniemi, Finland.
Jot J.M., and Trivi J.M. (2006): Scene description model and rendering engine for interactive virtual acoustics, *at the AES 120th Convention*, Paris, France, Paper 6660.

Jot J.M., Wardle S., and Larcher V. (1998): Approaches to binaural synthesis, *at the AES 105th Convention*, San Francisco, California, USA, Paper 4861.

Juhasz G., and Piret E. (1980): Compatible correcting-matrix quadraphonic transmission system, *J. Audio Eng. Soc.*, 28(9), 596–600.

Julstrom S. (1987): A high-performance surround process for home video, *J. Audio Eng. Soc.*, 35(7/8), 536–549.

Julstrom S. (1991): An intuitive view of coincident stereo microphones, *J. Audio Eng. Soc.*, 39(9), 632–649.

Kahana Y., and Nelson P.A. (2007): Boundary element simulations of the transfer function of human heads and baffled pinnae using accurate geometric models, *J. Sound Vib.*, 300(3/5), 552–579.

Kan A., Jin C., Tan T., et al. (2005): 3DApe: a real-time 3D audio playback engine, *AES 118th Convention*, Barcelona, Spain, Preprint 6343.

Kan A., and Litovsky R.Y. (2015): Binaural hearing with electrical stimulation, *Hearing Res.*, 322, 127–137.

Kan A., Pope G., Jin C., and Schaik A.V. (2004): Mobile spatial audio communication system, in *Proceedings of Tenth Meeting of the International Conference on Auditory Display (ICAD 04)*, Sydney, Australia.

Kang S.H., and Kim S.H. (1996): Realistic audio teleconferencing using binaural and auralization techniques, *ETRI J.*, 18(1), 41–51.

Karamustafaoglu A., Horbach U., Pellegrin R., et al. (1999): Design and applications of a data-based auralization system for surround sound, *at the AES 106th Convention*, Munich, Germany, Paper 4976.

Karjalainen M., and Järveläinen H. (2007): Reverberation modeling using velvet noise, *at the AES 30th International Conference*, Saariselkä, Finland.

Kassier R., Lee H.K., Brookes T., et al. (2005): An informal comparison between surround sound microphone techniques, *at the AES 118th Convention*, Barcelona, Spain, Paper 6429.

Kates J.M. (1980): Optimum loudspeaker directional patterns, *J. Audio Eng. Soc.*, 28(11), 787–794.

Katz B.F.G. (2001): Boundary element method calculation of individual head-related transfer function. I. Rigid model calculation, *J. Acoust. Soc. Am.*, 110(5), 2440–2448.

Kawano S., Taira M., Matsudaira M., et al. (1998): Development of the virtual sound algorithm, *IEEE Trans. Consumer Electron.*, 44(3), 1189–1194.

Kearney G., and Doyle T. (2015): Height perception in Ambisonic based binaural decoding, *at the AES 139th Convention*, New York, USA, Paper 9423.

Keller A.C. (1981): Early Hi-Fi and stereo recordingat Bell Laboratories (1931–1932), *J. Audio Eng. Soc.*, 29(4), 274–280.

Kendall G.S. (1995): The decorrelation of audio signals and its impact on spatial imagery, *Comput. Music J.*, 19(4), 71–87.

Kessler R. (2005): An optimized method for capturing multidimensional "acoustic fingerprints", *at the AES 118th Convention*, Barcelona, Spain, Paper 6342.

Kim C., Mason R., and Brookes T. (2013): Head movements made by listeners in experimental and real-left listening activities, *J. Audio Eng. Soc.*, 61(6), 425–438.

Kim S. (Sungyoung), Ikeda M., and Martens W.L. (2014): Reproducing virtually elevated sound via a conventional home-theater audio system, *J. Audio Eng. Soc.*, 62(5), 337–344.

Kim S. (Sunmin), Lee Y.W., and Pulkki V. (2010): New 10.2-channel vertical surround system (10.2-VSS); comparison study of perceived audio quality in various multichannel sound systems with height loudspeakers, *at the AES 129th Convention*, San Francisco, USA, Paper 8296.

Kim Y.H., and Choi J.W. (2013): *Sound visualization and manipulation*, John Wiley & Sons, Singapore.

Kirby D.G. (1995): ISO/MPEG subjective tests on multichannel audio systems, *at the AES 99th Convention*, New York, USA, Paper 4066.

Kirby D.G., Cutmore N.A.F., and Fletcher J.A. (1998): Program origination of five-channel surround sound, *J. Audio Eng. Soc.*, 46(4), 323–330.

Kirby D.G., Warren K., and Watanabe K. (1996): Report on the formal subjective listening tests of MPEG-2 NBC multichannel audio coding, ISO/IEC JTC1/SC29/WG11 Nov.N1419, International Organization for Standardization, Geneva, Switzerland.

Kirkeby O. (2002): A balanced stereo widening network for headphones, *AES 22nd International Conference*, Espoo, Finland.

Kirkeby O., and Nelson P.A. (1993): Reproduction of plane wave sound fields, *J. Acoust. Soc. Am.*, 94(5), 2992–3000.

Kirkeby O., and Nelson P.A. (1999): Digital filter design for inversion problems in sound reproduction, *J. Audio Eng. Soc.*, 47(7/8), 583–595.

Kirkeby O., Nelson P.A., and Hamada H. (1998a): The "stereo dipole" – a virtual source imaging system using two closely spaced loudspeakers, *J. Audio Eng. Soc.*, 46(5), 387–395.

Kirkeby O., Nelson P.A., and Hamada H. (1998b): Local sound field reproduction using two closely spaced loudspeakers, *J. Acoust. Soc. Am.*, 104(4), 1973–1981.

Kirkeby O., Nelson P.A., and Orduna-Bustamante F. (1996): Local sound field reproduction using digital signal processing, *J. Acoust. Soc. Am.*, 100(3), 1584–1593.

Kistler D.J., and Wightman F.L. (1992): A model of head-related transfer functions based on principal components analysis and minimum-phase reconstruction, *J. Acoust. Soc. Am.*, 91(3), 1637–1647.

Kjörling K., Rödén J., Wolters M., et al. (2016): AC-4 –the next generation audio codec, *at the AES 140th Convention*, Paris, France, Paper 9491.

Kleczkowski P., Król A., and Malecki P. (2015): Multichannel sound reproduction quality improves with angular separation of direct and reflected sounds, *J. Audio Eng. Soc.*, 63(6), 427–442.

Kleijn W.B. (2018): Directional emphasis in Ambisonics, *IEEE Signal Process. Lett.*, 25(7), 1079–1083.

Kleiner M., Dalenbäck B.I., and Svensson P. (1993): Auralization-an overview, *J. Audio Eng. Soc.*, 41(11), 861–875.

Klepko J. (1997): 5-channel microphone array with binaural head for multichannel reproduction, *at the AES 103th Convention*, New York, USA, Paper 4541.

Klipsch P.W. (1958): Stereophonic sound with two tracks, three channels by means of a phantom circuit (2PH3), *J. Audio Eng. Soc.*, 6(2), 118–123.

Kohsaka O., Satoh E., and Nakayama T. (1972): Sound image localization in multichannel matrix reproduction, *J. Audio Eng. Soc.*, 20(7), 542–548.

Kolundžija M., Faller C., and Vetterli M. (2011): Reproducing sound fields using MIMO acoustic channel inversion, *J. Audio Eng. Soc.*, 59(10), 721–734.

Komiyama S. (1989): Subjective evaluation of angular displacement between picture and sound directions for HDTV sound systems, *J. Audio Eng. Soc.*, 37(4), 210–214.

Kopčo N., and Shinn-Cunningham B.G. (2003): Spatial unmasking of nearby pure-tone targets in a simulated anechoic environment, *J. Acoust. Soc. Am.*, 114(5), 2856–2870.

Koyama S., Furuya K., Wakayama K., et al. (2016): Analytical approach to transforming filter design for sound field recording and reproduction using circular arrays with a spherical baffle, *J. Acoust. Soc. Am.*, 139(3), 1024–1036.

Kozamernik F. (1995): Digital audio broadcasting – radio now and for the future, *EBU Tech. Rev.*, 1995(autumn), 2–27.

Kraft S., and Zölzer U. (2016): Low-complexity stereo signal decomposition and source separation for application in stereo to 3D upmixing, *at the AES 140th Convention*, Paris, France, Paper 9586.

Krebber W., Gierlich H.W., and Genuit K. (2000): Auditory virtual environments: basics and applications for interactive simulations, *Signal Process.*, 80(11), 2307–2322.

Kügler C., and Thiele G. (1992): Loudspeaker reproduction: study on the subwoofer concept, *at the AES 92nd Convention*, Vienna, Austria, Paper 3335.

Kuhn C., Pellegrini R., Leckschat D., et al. (2003): An approach to miking and mixing of music ensembles using wave field synthesis, *at the AES 115th Convention*, New York, Paper 5929.

Kuhn G.F. (1977): Model for the interaural time differences in the azimuthal plane, *J. Acoust. Soc. Am.*, 62(1), 157–167.

Kulkarni A. (1997): Sound localization in real and virtual acoustical environments, Doctor dissertation of Boston University, Boston, USA.

Kulkarni A., and Colburn H.S. (1998): Role of spectral detail in sound-source localization, *Nature*, 396, 747–749.

Kulkarni A., and Colburn H.S. (2000): Variability in the characterization of the headphone transfer-function, *J. Acoust. Soc. Am.*, 107(2), 1071–1074.

Kulkarni A., and Colburn H.S. (2004): Infinite-impulse-response models of the head-relatedtransfer function, *J. Acoust. Soc. Am.*, 115(4), 1714–1728.
Kulkarni A., Isabelle S.K., and Colburn H.S. (1999): Sensitivity of human subjects to head-related transfer-function phase spectra, *J. Acoust. Soc. Am.*, 105(5), 2821–2840.
Kuo S.M., and Morgan D.R. (1999): Active noise control: a tutorial review, *Proceedings of the IEEE*, 87(6), 943–973.
Kurozumi K., and Ohgushi K. (1983): The relationship between the cross-correlation coefficient of two-channel acoustic signals and sound image quality, *J. Acoust. Soc. Am.*, 74(6), 1726–1733.
Kuttruff H. (2009): *Room acoustics* (5th edition), Spon Press, Abingdon, UK.
Kyriakakis C. (1998): Fundamental and technological limitations of immersive audio systems, *Proc. IEEE*, 86(5), 941–951.
Kyriakakis C., Holman T., Lim J.S., et al. (1998): Signal processing, acoustics, and psychoacoustics for high quality desktop audio, *J. Vis. Commun. Image Represent*, 9(1), 51–61.
Laback B., Egger K., and Majdak P. (2015): Perception and coding of interaural time differences with bilateral cochlear implants, *Hearing Res.*, 322, 138–150.
Laitinen M.V., Kuech F., Disch S., et al. (2011): Reproducing applause-type signals with directional audio coding, *J. Audio Eng. Soc.*, 59(1/2), 29–43.
Laitinen M.V., Vilkamo J., Jussila K., et al. (2014): Gain normalization in amplitude panning as a function of frequency and room reverberance, *at the AES 55th International Conference*, Helsinki, Finland.
Langendijk E.H.A., and Bronkhorst A.W. (2002): Contribution of spectral cues to human sound localization, *J. Acoust. Soc. Am.*, 112(4), 1583–1596.
Larcher V., Jot J.M., Guyard J., et al. (2000): Study and comparison of efficient methods for 3D audio spatialization based on linear decomposition of HRTF data, *at the AES 108th Convention*, Paris, France, Paper 5097.
Leakey D.M. (1959): Some measurements on the effects of interchannel intensity and time differences in two channel sound systems, *J. Acoust. Soc. Am.*, 31(7), 977–986.
Leakey D.M. (1960): Further thoughts on stereophonic sound systems, *Wireless World*, 66, 154–160.
Lecomte P., Gauthier P.A., Langrenne C., et al. (2015): On the use of a Lebedev grid for ambisonics, *at the AES 139th Convention*, New York, USA, Paper 9433.
Lecomte P., Gauthier P.A., Langrenne C., et al. (2018): Cancellation of room reflections over an extended area using ambisonics, *J. Acoust. Soc. Am.*, 143(2), 811–828.
Lee H. (2010): A new time and intensity trade-off function for localisation of natural sound sources, *at the AES 128th Convention*, London, UK, Paper 8149.
Lee H. (2011): A new multichannel microphone technique for effective perspective control, *at the AES 130th Convention*, London, UK, Paper 8337.
Lee H. (2014): The relationship between interchannel time difference and level difference in vertical sound localization and masking, *at the AES 131st Convention*, New York, USA, Paper 8556.
Lee H. (2017): Sound source and loudspeaker base angle dependency of phantom image elevation effect, *J. Audio Eng. Soc.*, 65(9), 733–748.
Lee H. (2021): Multichannel 3D microphone arrays: a review, *J. Audio Eng. Soc.*, 69(1/2), 5–26.
Lee H., and Gribben C. (2014): Effect of vertical microphone layer spacing for a 3D microphone array, *J. Audio Eng. Soc.*, 62(12), 870–884.
Lee H., and Rumsey F. (2013): Level and time panning of phantom images for musical sources, *J. Audio Eng. Soc.*, 61(12), 978–988.
Lee J.M., Choi J.W., and Kim Y.H. (2013): Wave field synthesis of a virtual source located in proximity to a loudspeaker array, *J. Acoust. Soc. Am.*, 134(3), 2106–2117.
Lee K.S., Abel J.S., Välimäki V., et al. (2009): The switched convolution reverberator, *at the AES 127th Convention*, New York, USA, Paper 7927.
Lee Y.W., Kim S., Jo H., et al. (2011): Virtual height speaker rendering for Samsung 10.2-channel vertical surround system, *at the AES 131st Convention*, New York, USA, Paper 8523.
Lehnert H., and Blauert J. (1992): Principles of binaural room simulation, *Appl. Acoust.*, 36(3/4), 259–291.

Lehtonen H.M., Purnhagen H., Villemoes L., et al. (2017): Parametric joint channel coding of immersive audio, *at the AES 142nd Convention*, Berlin, Germany, Paper 9740.

Leitner S., Sontacchi A., and Höldrich R. (2000): Multichannel sound reproduction system for binaural signals – the Ambisonic approach, in *Proceedings of the COST G-6 Conference on Digital Audio Effects (DAFX-00)*, Verona, Italy.

Lentz T., Assenmacher I., Sokoll J., et al. (2005): Performance of spatial audio using dynamic cross-talk cancellation, *AES 119th Convention*, New York, USA, Paper 6541.

Lentz T., and Schmitz O. (2002): Realisation of an adaptive cross-talk cancellation system for a moving listener, *at the AES 21st International Conference*, St. Petersburg, Russia.

Leong P., and Carlile S. (1998): Methods for spherical data analysis and visualization, *J. Neurosci Met.*, 80(2), 191–200.

Li Z., and Duraiswami R. (2006): Headphone-based reproduction of 3D auditory scenes captured by spherical/hemispherical microphone arrays, in *Proceedings of IEEE 2006 International Conference on Acoustics, Speech and Signal Processing*, Toulouse, France, Vol 5, 337–340.

Lin H.B., and Xie B.S. (2018): Dynamic binaural reproduction of multichannel surround sound based on mobile phone (in Chinese), *Appl. Acoust.*, 37(2), 187–195.

Lipshitz S.P. (1986): Stereo microphone techniques, are the purists wrong?, *J. Audio Eng. Soc.*, 34(9), 716–744.

Litovsky R.Y., Colburn H.S., Yost W.A., et al. (1999): The precedence effect, *J. Acoust. Soc. Am.*, 106(4), 1633–1654.

Liu L.L., and Xie B.S. (2021): Analysis and experiment on the limitations of static and dynamic transaural reproduction with two frontal loudspeakers, *Arch. Acoust.*, 46(2), 213–228.

Liu L.L., and Xie B.S. (2022): A high-frequency–band timbre equalization method for transaural reproduction with two frontal loudspeakers, *J. Audio Eng. Soc.*, 70(1/2), 36–49.

Liu Y. (2014): Research on the stability and timbre of Ambisonics reproduction system (in Chinese), Dissertation of doctor degree, South China University of Technology, Guangzhou, China.

Liu Y. (2015): Research on spherical microphone array recording and binaural virtual rendering system (in Chinese), Dissertation of doctor degree, South China University of Technology, Guangzhou, China.

Liu Y., and Xie B.S. (2013a): Analysis on the stability of high-order Ambisonics system (in Chinese), *Tech. Acoust.*, 32(6), pt. 2, 247–248.

Liu Y., and Xie B.S. (2013b): Analysis on the stability of spatial interpolation of head-related transfer function and reproduction of multi-channel sound (in Chinese), *J. South China Univ. Technol.*, 41(8), 131–138.

Liu Y., and Xie B.S. (2015): Analysis with binaural auditory model and experiment on the timbre of Ambisonics recording and reproduction, *Chin. J. Acoustics*, 34(4), 337–356.

Liu Y., and Xie B.S. (2016): Analysis on the timbre of horizontal Ambisonics with different decoding methods, *at the AES 141st Convention*, Los Angeles, USA, Paper 9677.

Liu Y., Xie B.S., Yu G.Z., et al. (2016): Analysis on spatial discrimination threshold of head-related transfer function magnitude, *Chin J. Acoust.*, 35(1), 1–17.

Lokki T., and Gröhn M. (2005): Navigation with auditory cues ina virtual environment, *IEEE Multimedia*, 12(2), 80–86.

Loomis J.M., Golledge R.G., Klatzky R.L., et al. (1998): Navigation system for the blind: auditory display modes and guidance, *Presence*, 7(2), 193–203.

Loomis J.M., Hebert C., and Cicinelli J.G. (1990): Active localization of virtual sounds, *J. Acoust. Soc. Am.*, 88(4), 1757–1764.

López J.J., and González A. (1999): 3-D audio with dynamic tracking for multimedia environments, *at the 2nd COST-G6 Workshopon Digital Audio Effects(DAFx-1999)*, Trondheim, Norway.

Lopez-Poveda E.A., and Meddis R. (1996): A physical model of sound diffraction and reflections in the human concha, *J. Acoust. Soc. Am.*, 100(5), 3248–3259.

Lorho G., Isherwood D., Zacharov N., et al. (2002): Round robin subjective evaluation of stereo enhancement system for headphones, *at the AES 22nd International Conference*, Espoo, Finland.

Maa D.Y., and Shen H. (2004): *The handbook of acoustics* (Revised edition, in Chinese), Science Press, Beijing, China.

Mac Cabe C.J., and Furlong D.J. (1994): Virtual imaging capabilities of surround sound systems, *J. Audio Eng. Soc.*, 42(1/2), 38–49.
Mackenzie J., Huopaniemi J., Valimaki V., et al. (1997): Low-order modeling of head-related transfer functions using balanced model truncation, *IEEE Signal Process. Lett.*, 4(2), 39–41.
Macpherson E.A. (1991): A computer model of binaural localization for stereo imaging measurement, *J. Audio Eng. Soc.*, 39(9), 604–622.
Macpherson E.A. (2011): Head motion, spectral cues, and Wallach's "principle of least displacement" in sound localization, in *Principles and applications of spatial hearing* (Edited by SuzukiY., et al.), 103–120, World Scientific Publishing Co. Pte. Ltd., Singapore.
Macpherson E.A. (2013): Cue weighting and vestibular mediation of temporal dynamics in sound localization via head rotation, *at the 21st International Congress on Acoustics*, Montreal, Canada.
Maher R.C. (1997): Single-ended spatial enhancement using a cross-coupled lattice equalizer, *at the 1997 IEEE Workshop on Application of Signal Processing to Audio and Acoustics*, New Paltz, NY, USA.
Mai H.M., Xie B.S., and Jiang J.L. (2018): Analysis and experimental validation of the mixed-order Ambisonics Reproduction (in Chinese), *J. South China Univ. Technol.*, 46(3), 108–118.
Majdak P., Balazs P., and Laback B. (2007): Multiple exponential sweep method for fast measurement of head-related transfer functions, *at the AES 122nd Convention*, Vienna, Austria, Paper 7019.
Makita Y. (1962): On the directional localization of sound in the stereophonic sound filed, *EBU Rev. Pt. A*, 73(6), 102–108.
Malham D.G., and Myatt A. (1995): 3-D sound spatialization using Ambisonic technique, *Comput. Music J.*, 19(4), 58–70.
Marques de Sá J.P. (2007): *Applied statistics using SPSS, STATISTICA, MATLAB and R*, Springer-Verlag, Berlin, Heidelberg, New York.
Márschall M., Favrot S., and Buchholz J. (2012): Robustness of a mixed-order Ambisonics microphone array for sound field reproduction, *at the AES 132nd Convention*, Budapest, Hungary, Paper 8645.
Marshall A.H. (1967): A note on the importance of room cross-section in concert halls, *J. Sound Vib.*, 5(1), 100–112.
Marshall A.H., and Barron M. (2001): Spatial responsiveness in concert halls and the origins of spatial impression, *Appl. Acoust.*, 62(2):91–108.
Marston D. (2011): Assessment of stereo to surround upmixers for broadcasting, *at the AES 130th Convention*, London, UK, Paper 8448.
Martens W.L. (1987): Principal component analysis and resynthesisof spectral cues to perceived direction, in *Proceeding of the International computer Music Conference*, San Francisco, CA, USA, 274–281.
Martens W.L. (2001): Two-subwoofer reproduction enables increased variation inauditory spatial imagery, in *Proceedings of the 2nd International Workshop on Spatial Media*, Aizu-Wakamatsu, Japan, 86–97.
Martin G. (2005): A new microphone technique for five-channel recording, *at the AES 118th Convention*, Barcelona, Spain, Paper 6427.
Martin G., Woszczyk W., Corey J., et al. (1999): Sound source localization in a five-channel surround sound reproduction system, *at the AES 107th Convention*, New York, USA, Paper 4994.
Mason R. (2002): Elicitation and measurement of auditory spatial attributes in reproduced sound, Doctor dissertation of Philosophy, Surrey University, Guildford, UK.
Mason R., Ford N., Rumsey F., et al. (2001): Verbal and nonverbal elicitation techniques in the subjective assessment of spatial sound reproduction, *J. Audio Eng. Soc.*, 49(5), 366–384.
Matsudaira T.K., and Fukami T. (1973): Phase difference and sound image localization, *J. Audio Eng. Soc.*, 21(10), 792–797.
Matsui K., and Ando A. (2010): Binaural reproduction of 22.2 multichannel sound over loudspeakers, *at the AES 129th Convention*, San Francisco, CA, USA, Paper 8272.
Matsui K., and Ando A. (2013): Binaural reproduction of 22.2 multichannel sound with loudspeaker array frame, *at the AES 135th Convention*, New York, USA, Paper 8954.
Matsumoto M., Yamanaka S., and Tohyama M. (2004): Effect of arrival time correction on the accuracy of binaural impulse response interpolation, interpolation methods of binaural response, *J. Audio. Eng. Soc.*, 52(1/2), 56–61.

McKinnie D., and Rumsey F. (1997): Coincident microphone techniques for three-channel stereophonic reproduction, *at the AES 102nd Convention*, Munich, Germany, Paper 4429.

Meares D.J. (1991): Sound system for high definition television, *Appl. Acoust.*, 33(3), 229–243.

Meares D.J. (1992): Multichannel sound system for HDTV, *Appl. Acoust.*, 36(3/4), 245–257.

Meares D.J., and Ratliff P.A. (1976): The development of compatible 4-2-4 Quadraphonic Matrix system: B.B.C Matrix H, *EBU. Review-Tech., Pt.* 159(1976 Oct.), 208–217.

Melchior F., Thiergart O., Galdo G.D., et al. (2009): Dual radius spherical cardioid microphone arrays for binaural auralization, *at the AES 127th Convention*, New York, USA, Paper 7855.

Mennie D. (1978): AM stereo: five competing options, *IEEE J. Mag.*, 15(6), 24–31.

Menzies D. (2002): W-panning and O-format, tools for object spatialisation, *at the AES 22nd Conference*, Espoo, Finland.

Menzies D., and Al-Akaidi M. (2007): Ambisonic synthesis of complex sources, *J. Audio Eng. Soc.*, 55(10), 864–876.

Menzies D., and Marwan A.A. (2007): Nearfield binaural synthesis and ambisonics, *J. Acoust. Soc. Am.*, 121(3), 1559–1563.

Merchel S., and Groth S. (2010): Adaptively adjusting the stereophonic sweet spot to the listener's position, *J. Audio Eng. Soc.*, 58(10), 809–817.

Merimaa J. (2009): Modification of HRTF filters to reduce timbral effects in binaural synthesis, *at the AES 127th Convention*, New York, NY, USA, Paper 7912.

Merimaa J. (2010): Modification of HRTF filters to reduce timbral effects in binaural synthesis, part 2: individual HRTFs, *in AES 129th Convention*, San Francisco, CA, USA, Paper 8265.

Merimaa J., Goodwin M.M., and Jot J.M. (2007): Correlation-based Ambience extraction from stereo recordings, *at the AES 123rd Convention*, New York, USA, Paper 8265.

Merimaa J., and Pulkki V. (2005): Spatial impulse response rendering I: analysis and synthesis, *J. Audio Eng Soc.*, 53(12), 1115–1127.

Mertens H. (1965): Directional hearing in stereophony theory and experimental verification, *EBU Rev., Part A*, 92(Aug.), 146–158.

Meyer E., and Schodder G.R. (1952): On the influence of reflected sound on directional localization and loudness of speech (in German), *Nachr. Akad. Wiss, Göttingen, Math. Phys. Klasse IIa*, 6, 31–42.

Meyer E., and Thiele R. (1956): Room-acoustical investigations in numerous concert halls and radio studios by means of novel measuring technique (in German), *Acustica*, 6, 425–444.

Meyer J., and Elko G.W. (2004): Spherical microphone arrays for 3D sound recording, in *Audio signal processing for the next-generation multimedia communication systems* (edited by Huang Y. and Benesty J.), Kluwer Academic Publishers, Boston, USA, 67–89.

Middlebrooks J.C. (1992): Narrow-band sound localization related to external ear acoustics, *J. Acoust. Soc. Am.*, 92(5), 2607–2624.

Middlebrooks J.C. (1999a): Individual differences in external-ear transfer functions reduced by scaling in frequency, *J. Acoust. Soc. Am.*, 106(3), 1480–1492.

Middlebrooks J.C. (1999b): Virtual localization improved by scaling nonindividualized external-ear transfer functions in frequency, *J. Acoust. Soc. Am.*, 106(3), 1493–1510.

Middlebrooks J.C., and Green D.M. (1992): Observations on a principal components analysis of head-related transfer functions, *J. Acoust. Soc. Am.*, 92(1), 597–599.

Middlebrooks J.C., Makous J.C., and Green D.M. (1989): Directional sensitivity of sound-pressure levels in the human ear canal, *J. Acoust. Soc. Am.*, 86(1), 89–108.

Mills A.W. (1958): On the minimum audible angle, *J. Acoust. Soc. Am.*, 30(4), 237–246.

Miyasaka E. (1989): A sound reproduction system and transmission system for HDTV, *at the AES 7th Conference*, Toronto, Canada.

Momose T., Otani M., Hashimoto M., et al. (2015): Adaptive amplitude and delay control for stereophonic reproduction that is robust against listener position variations, *J. Audio Eng. Soc.*, 63(1/2), 90–98.

Monro G. (2000): In-phase corrections for Ambisonics, in *Proceedings of International Computer Music Conference*, Berlin, Germany, 292–295.

Moore B.C.J. (2012): *An introduction to the psychology of hearing* (6th edition), Emerald Group Publishing Limited, UK.

Moore B.C.J., and Glasberg B.R. (2007): Modeling binaural loudness, *J. Acoust. Soc. Am.*, 121(3), 1604–1612

Moore B.C.J., Glasberg B.R., and Bear T. (1997): A model for the prediction of thresholds, loudness, and partial loudness, *J. Audio Eng. Soc.*, 45(4), 224–240.

Moore B.C.J., Oldfield S.R., and Dooley G.J. (1989): Detection and discrimination of spectral peaks and notches at 1 and 8 kHz, *J. Acoust. Soc. Am.*, 85(2), 820–836.

Moore D., and Wakefield J. (2008): The design of Ambisonic decoders for the ITU 5.1 layout with even performance Characteristics, *at the AES 124th Convention*, Paper 7473.

Moorer J.A. (1979): About this reverberation business, *Comput. Music J.*, 3(2), 13–28.

Moreau S., Daniel J., and Bertet S. (2006): 3D sound field recording with higher order Ambisonics – objective measurements and validation of spherical microphone, *at the AES 120th Convention*, Paris, France, Paper 6857.

Morimoto M., Fujimori H., and Maekawa Z. (1990): Discrimination between auditory source width and envelopment, *J. Acoust. Soc. Japan*, 46(6), 448–457.

Morimoto M., and Iida K. (1993): A new physical measure for psychological evaluation of a sound field: front/back energy ratio as a measure for envelopment, *J. Acoust. Soc. Am.*, 93(4), 2282.

Morimoto M., and Iida K. (1995): A practical evaluation method of auditory source width in concert halls, *J. Acoust. Soc. Japan*, 16(2), 59–69.

Morimoto M., Iida K., and Sakagami K. (2001): The role of reflections from behind the listener in spatial impression, *Appl. Acoust.*, 62(2), 109–124.

Morrell M.J., and Reiss J.D. (2009): A comparative approach to sound localization within a 3-D sound field, *at the AES 126th Convention*, Munich, Germany, Paper 7663.

Morse P.M., and Ingrad K.U. (1968): *Theoretical acoustics*, McGraw-Hill, New York, USA.

Mourjopoulos J.N. (1994): Digital equalization of room acoustics, *J. Audio Eng. Soc.*, 42(11), 884–900.

Muraoka T., and Nakazato T. (2007): Examination of multichannel sound-field recomposition utilizing frequency-dependent interaural cross correlation (FIACC), *J. Audio Eng. Soc.*, 55(4), 236–256.

Murtaza A., Herre J., and Paulus J. (2015): ISO/MPEG-H 3D audio: SAOC-3D decoding and rendering, *at the AES 139th Convention*, New York, USA, Paper 9434.

Mϕller H. (1992): Fundamentals of binaural technology, *Appl. Acoust.*, 36(3/4), 171–218.

Mϕller H., Hammershϕi D., and Jensen C.B., et al. (1995b): Transfer characteristics of headphones measured on human ears, *J. Audio Eng. Soc.*, 43(4), 203–217.

Mϕller H., Hammershϕi D., Jensen C.B., et al. (1999): Evaluation of artificial heads in listening tests, *J. Audio Eng. Soc.*, 47(3), 83–100.

Mϕller H., Sϕrensen M.F., Hammershϕi D., et al. (1995a): Head-related transfer functions of human subjects, *J. Audio Eng. Soc.*, 43(5), 300–321.

Mϕller H., Sϕrensen M.F., Jensen C.B., et al. (1996): Binaural technique: do we need individual recordings? *J. Audio Eng. Soc.*, 44(6), 451–469.

Nakabayashi K. (1975): A method of analyzing the quadraphonic sound field, *J. Audio Eng. Soc.*, 23(3), 187–193.

Nakabayashi K., Kurozumi K., and Miyasaka E., et al. (1991): Three-one quadraphonic sound system for high definition television, *at the AES 10th International Conference*, London, UK.

Naylor G.M. (1993): ODEON – another hybrid room acoustical model, *Appl. Acoust.*, 38(2–4), 131–143.

Nelson P.A., and Elliott S.J. (1992): *Active control of sound*, Academic Press Inc., San Diego, USA.

Nelson P.A., and Kahana Y. (2001): Spherical harmonics, singular-value decomposition and the head-related transfer function, *J. Sound Vib.*, 239(4), 607–637.

Nelson P.A., Orduña-Bustamante F., and Engler E., et al. (1996): Experiments on a system for synthesis of virtual acoustic sources, *J. Audio Eng. Soc.*, 44(11), 990–1007.

Neuendorf M., Multrus M., and Rettelbach N., et al. (2013): The ISO/MPEG unified speech and audio coding standard—consistent high quality for all content types and at all bit rates, *J. Audio Eng. Soc.*, 61(12), 956–977.

Neukom M. (2006): Decoding second order Ambisonics to 5.1 surround systems, *at the AES 121st Convention*, San Francisco, CA, Paper 6980.

Neukom M. (2007): Ambisonic panning, *at the AES 123rd Convention*, New York, USA, Paper 7297.

Nicol R., and Emerit M. (1999): 3D-sound reproduction over an extensive listening area: a hybrid method derived from holophony and ambisonic, *at the AES 16th International Conference*, Rovaniemi, Finland.

Nielsen S.H. (1993): Auditory distance perception in different rooms, *J. Audio Eng. Soc.*, 41(10), 755–770.

Nikolic I. (2002): Improvements of artificial reverberation by use of subband feedback delay networks, *at the AES 112th Convention*, Munich, Germany, Paper 5630.

Nishino T., Inoue N., Takeda K., et al. (2007): Estimation of HRTFs on the horizontal plane using physical features, *Appl. Acoust.*, 68(8), 897–908.

Noisternig M., Sontacchi A., Musil T., et al. (2003): A 3D Ambisonic based binaural sound reproduction system, *at the AES 24th International Conference*, Banff, Canada.

Noll P. (1997): MPEG digital audio coding, *IEEE Signal Process. Mag.*, 14(5), 59–81.

Nousaine T. (1997): Multiple subwoofers for home theater, *at the AES 103rd Convention*, New York, USA, Paper 4558.

Nymand M. (2003): Introduction to microphone technique for 5.1 surround sound, at the DPA microphone workshop on mic techniques for multichannel audio, *the AES 24th International Conference*, Banff, Canada.

Ohgushi K., Komiyama S., Kurozumi K., et al. (1987): Subject evaluation of multi-channel stereophony for HDTV, *IEEE Trans. Broadcast.*, 33(4), 197–202.

Okano T., Beranek L.L., and Hidaka T. (1998): Relations among interaural cross-correlation coefficient ($IACC_E$), lateral fraction (LF_E), and apparent source width (ASW) in concert halls, *J. Acoust. Soc. Am.*, 104(1), 255–265.

Okubo H., Sugimoto T., Oishi S., et al. (2012): A method for reproducing frontal sound field of 22.2 multichannel sound utilizing a loudspeaker array frame, *at the AES 133rd Convention*, San Francisco, USA, Paper 8714.

Olive S. (2001): Evaluation of five commercial stereo enhancement 3D audio software plug-ins, *at the AES 110th Convention*, Amsterdam, The Netherlands, Paper 5386.

Olive S.E., and Toole F.E. (1989): The detection of reflections in typical rooms, *J. Audio Eng. Soc.*, 37(7/8), 539–553.

Olson H.F. (1969): Home entertainment: audio 1988, *J. Audio Eng. Soc.*, 17(4), 390–404.

Ono K., Nishiguchi T., Matsui K., et al. (2013): Portable spherical microphone for super hi-vision 22.2 multichannel audio, New York, USA, Paper 8922.

Oppenheim A.V., Schafer R.W., and Buck J.R. (1999): *Discrete-time signal processing* (2nd edition), Prentice-Hall, Upper Saddle River, NJ.

Orban R. (1970): A rational technique for synthesizing pseudo-stereo from monophonic sources, *J. Audio Eng. Soc.*, 18(2), 157–164.

Otani M., and Ise S. (2006): Fast calculation system specialized for head-related transfer function based on boundary element method, *J. Acoust. Soc. Am.*, 119(5), 2589–2598.

Ottens L.F. (1967): The compact-cassette system for audio tape recorders, *J. Audio Eng. Soc.*, 15(1), 26–28.

Paavola M., Karlsson E., and Page J. (2005): 3D audio for mobile devices via Java, *at the AES 118th Convention*, Barcelona, Spain, Paper 6472.

Park J.Y., Chang J.H., and Kim Y.H. (2010): Generation of independent bright zones for a two-channel private audio system, *J. Audio Eng. Soc.*, 58(5), 382–393.

Park Y.C., Chio T.S., and Jung J.W., et al. (2006): Low complexity 3D audio algorithms for handheld devices, *at the AES 29th International Conference*, Seoul, Korea.

Paul S. (2009): Binaural recording technology: a historical review and possible future developments, *Acta Acust. United Ac.*, 95(5), 767–788.

Perrett S., and Noble W. (1997): The effect of head rotations on vertical plane sound localization, *J. Acoust. Soc. Am.*, 102(4), 2325–2332.

Piere A.D. (2019): *Acoustics, an introduction to its physical principles and applications* (3rd edition), Springer, Cham, Switzerland,

Pihlajamaki T., Santala O., and Pulkki V. (2014): Synthesis of spatially extended virtual source with time-frequency decomposition of mono signals, *J. Audio Eng. Soc.*, 62(7/8), 467–484.

Plenge G. (1972): On the problem of inside-the-head locatedness, *Acustica*, 26(5), 241–252.
Plenge G. (1974): On the differences between localization and lateralization, *J. Acoust. Soc. Am.*, 56(3), 944–951.
Pohlmann K.C. (2011): *Principles of digital audio* (6th edition), McCraw-Hill Companies, Inc., New York, USA.
Poletti M.A. (1996): The design of encoding functions for stereophonic and polyphonic sound systems, *J. Audio Eng. Soc.*, 44(11), 948–963.
Poletti M.A. (2000): A unified theory of horizontal holographic sound systems, *J. Audio Eng. Soc.*, 48(12), 1155–1182.
Poletti M.A. (2005a): Effect of noise and transducer variability on the performance of circular microphone arrays, *J. Audio Eng. Soc.*, 53(5), 371–384.
Poletti M.A. (2005b): Three-dimensional surround sound systems based on spherical harmonics, *J. Audio Eng. Soc.*, 53(11), 1004–1025.
Poletti M.A. (2007): Robust two-dimensional surround sound reproduction for nonuniform loudspeaker layouts, *J. Audio Eng. Soc.*, 55(7/8), 598–610.
Poletti M.A. (2008): An investigation of 2D multizone surround sound systems, *at the AES 125th Convention*, San Francisco, USA, Paper 7551.
Poletti M.A., and Abhayapala T.D. (2011): Interior and exterior sound field control using general two-dimensional first order source, *J. Acoust. Soc. Am.*, 129(1), 234–244.
Poletti M.A., and Betlehem T. (2014): Creation of a single sound field for multiple listeners, in *Internoise 2014*, Melbourne, Australia.
Poletti M.A., Fazi F.M., and Nelson P.A. (2010a): Sound-field reproduction systems using fixed-directivity loudspeakers, *J. Acoust. Soc. Am.*, 127(6), 3590–3601.
Poletti M.A., Fazi F.M., and Nelson P.A. (2010b): Sound reproduction systems using variable -directivity loudspeakers, *J. Acoust. Soc. Am.*, 129(3), 1429–1438.
Politis A., Laitinen M.V., Ahonen J., et al. (2015): Parametric spatial audio processing of spaced microphone array recordings for multichannel reproduction, *J. Audio Eng. Soc.*, 63(4), 216–227.
Pollow M., Nguyen K.V., Warusfel O., et al. (2012): Calculation of head-related transfer functions for arbitrary field points using spherical harmonics decomposition, *Acta Acust. United Ac.*, 98(1), 72–82.
Pöntynen H., Santala O., and Pulkki H. (2016): Conflicting dynamic and spectral directional cues form separate auditory images, *at the AES 140th Convention*, Paris, France, Paper 9582.
Pörschmann C. (2007): 3-D audio in mobile communication devices: methods for mobile head-tracking, *J. Virtual Real. Broadcast.*, 4(13), 0009-6-11833.
Potard G., and Burnett I. (2004): Decorrelation techniques for the rendering of apparent sound source width in 3D audio display, in *Preceding of the 7th International Conference on Digital Audio Effect*, Naples, Italy, 280–284.
Power P., Davies W.J., Hirst J., et al. (2012): Localisation of elevated virtual sources in higher order Ambisonics sound fields, in *Proceedings Institute of Acoustics*, 34(Pt.4), Brighton, UK.
Pralong D., and Carlile S. (1996): The role of individualized headphone calibration for the generation of high fidelity virtual auditory space, *J. Acoust. Soc. Am.*, 100(6), 3785–3793.
Pueo B., López J., Escolano J., et al. (2010): Multiactuator panels for wave field synthesis: evolution and present developments, *J. Audio Eng. Soc.*, 58(12), 1045–1063.
Pulkki V. (1997): Virtual sound source positioning using vector base amplitude panning, *J. Audio Eng. Soc.*, 45(6), 456–466.
Pulkki V. (2001a): Localization of amplitude-panned virtual sources II: two- and three-dimensional panning, *J. Audio Eng. Soc.*, 49(9), 753–767.
Pulkki V. (2001b): Coloration of amplitude-panned virtual sources, *at the AES 110th Convention*, Amsterdam, The Netherlands, Paper 5402.
Pulkki V. (2002): Microphone techniques and directional quality of sound reproduction, *at the AES 112th Convention*, Munich, Germany, Paper 5500.
Pulkki V. (2007): Spatial sound reproduction with directional audio coding, *J. Audio Eng. Soc.*, 55(6), 503–516.
Pulkki V., and Hirvonen T. (2005): Localization of virtual sources in multichannel audio reproduction, *IEEE Trans. Speech, Audio Process.*, 13(1), 105–119.

Pulkki V., and Karjalainen M. (2001): Localization of amplitude-panned virtual sources I: stereophonic panning, *J. Audio Eng. Soc.*, 49(9), 739–752.

Pulkki V., and Karjalainen M. (2015): *Communication acoustics: an introduction to speech, audio and psychoacoustics*, John Wiley & Sons Ltd, West Sussex, UK.

Pulkki V., Karjalainen M., and Huopaniemi J. (1999): Analyzing virtual sound source attributes using a binaural auditory model, *J. Audio Eng Soc.*, 47(4), 203–217.

Pulkki V., and Merimaa J. (2006): Spatial impulse response rendering II: reproduction of diffuse sound and listening tests, *J. Audio Eng. Soc.*, 54(1/2), 3–20.

Pulkki V., Politis A., and Galdo G.D., et al. (2013): Parametric spatial audio reproduction with higher-order B-format microphone input, *at the AES 134th Convention*, Rome, Italy, Paper 8920.

Pulkki V., Pontynen H., and Santala O. (2019): Spatial perception of sound source distribution in the median plane, *J. Audio Eng. Soc.*, 67(11), 855–870.

Purnhagen H., Hirvonen T., and Villemoes L., et al. (2016): Immersive audio delivery using joint object coding, *at the AES 140th Convention*, Paris, France, Paper 9587.

Rafaely B. (2004): Plane-wave decomposition of the sound field on a spherical by convolution, *J. Acoust. Soc. Am.*, 116(4), 2149–2157.

Rafaely B. (2005): Analysis and design of spherical microphone arrays, *IEEE Trans. Speech, Audio Process.*, 13(1), 135–143.

Rafaely B. (2015): *Fundamentals of spherical array processing*, Springer-Verlag, Berlin Heidelberg.

Rao D., and Xie B.S. (2004): Multichannel spatial surround sound system, *Chin. J. Acoust.*, 23(2), 153–166.

Rao D., and Xie B.S. (2005): Head rotation and sound image localization in the median plane, *Chin. Sci. Bull.*, 50(5), 412–416.

Rao D., and Xie B.S. (2006): Repeatability analysis on headphone transfer function measurement (in Chinese), *Tech. Acoust.*, 25(supplement), 441–442.

Rao D., and Xie B.S. (2007): Influence of sound source directivity on binaural auralization quality (in Chinese), *Tech. Acoust.*, 26(5), 899–903.

Ratliff P.A. (1974): Properties of hearing related to quadraphonic reproduction, BBC RD38.

Riederer K.A.J. (1998): Head-related transfer function measurement, Master thesis of Helsinki University of Technology, Findland.

Rohr L., Corteel E., and Nguyen K.V., et al. (2013): Vertical localization performance in a practical 3-D WFS formulation, *J. Audio Eng. Soc.*, 61(12), 1001–1014.

Rubak P., and Johansen L.G. (1998): Artificial reverberation based on a pseudo-random impulse response, *at the AES 104th Convention*, Amsterdam, The Netherlands, Paper 4725.

Rubak P., and Johansen L.G. (1999): Artificial reverberation based on a pseudo-random impulse response II, *at the AES 106th Convention*, Munich, Germany, Paper 4900.

Rui Y.Q., Yu G.Z., and Xie B.S., et al. (2013): Calculation of individualized near-field head-related transfer function database using boundary element method, *at the AES 134th Convention*, Rome, Italy, Paper 8901.

Rumsey F. (1998): Subjective assessment of the spatial attributes of reproduced sound, *at the AES 15th International Conference*, Copenhagen, Denmark.

Rumsey F. (1999): Controlled subjective assessments of two-to-five channel surround sound processing algorithms, *J. Audio Eng. Soc.*, 47(7/8), 563–582.

Rumsey F. (2001): *Spatial audio*, Focal Press, Oxford, England.

Rumsey F. (2002): Spatial quality evaluation for reproduced sound: terminology, meaning, and a scene-based paradigm, *J. Audio Eng. Soc.*, 50(9), 651–666.

Rumsey F. (2013): Cinema sound in the 3D era, *J. Audio Eng. Soc.*, 61(5), 340–344.

Rumsey F. (2016): Automotive audio: they know where you sit, *J. Audio Eng. Soc.*, 64(9), 705–708.

Rumsey F. (2017): Broadcast and streaming: immersive audio, objects and OTT TV, *J. Audio Eng. Soc.*, 65(4), 338–341.

Rumsey F., Zielinski S., Jackson P., et al. (2008): QESTRAL (part 1): Quality evaluation of spatial transmission and reproduction using an artificial listener, *at the AES 125th Convention*, San Francisco, USA, Paper 7595.

Rumsey F., Zieliński S., and Kassier R. (2005): On the relative importance of spatial and timbral fidelities in judgments of degraded multichannel audio quality, *J. Acoust. Soc. Am.*, 118(2), 968–976.

Sakamoto N., Gotoh T., Kogure T., et al. (1981): Controlling sound-image localization in stereophonic sound reproduction, part 1, *J. Audio Eng. Soc.*, 29(11), 794–799.

Sakamoto N., Gotoh T., Kogure T., et al. (1982): Controlling sound-image localization in stereophonic sound reproduction, part 2, *J. Audio Eng. Soc.*, 30(10), 719–722.

Samsudin Kurniawati E., Ng B.H., et al. (2006): A stereoto mono downmixing scheme for MPEG-4 parametric stereo encoder, in *Proceeding of 2006 IEEE International Conference on Acoustics, Speech and Signal Processing*, Toulouse, France.

Sander C., Wefers F., and Leckschat D. (2012): Scalable binaural synthesis on mobile devices, *at the AES 133rd Convention*, San Francisco, USA, Paper 8783.

Sandvad J. (1996): Dynamic aspects of auditory virtual environments, *at the AES 100th Convention*, Copenhagen, Denmark, Paper 4226.

Saviojia L., Huopaniemi J., Lokki T., et al. (1999): Creating interactive virtual acoustic environments, *J. Audio. Eng. Soc.*, 47(9), 675–705.

Sawaguchi M. (editor) (2001): *Surround production handbook* (in Japanese), Kenrokukan Publishing, Japan.

Sawaya I., Sasaki K., Mikami S., et al. (2015): Dubbing studio for 22.2 multichannel sound system in NHK broadcasting center, *at the AES 138th Convention*, Warsaw, Poland, Paper 9327.

Scaini D., and Arteaga D. (2014): Decoding higher order Ambisonics to irregular periphonic loudspeaker arrays, *at the AES 55th International Conference*, Helsinki, Finland.

Scaini D., and Arteaga D. (2020): Wavelet-based spatial audio format, *J. Audio Eng. Soc.*, 68(9), 613–627.

Scheiber P. (1971): Four channels and compatibility, *J. Audio Eng. Soc.*, 19(4), 267–279.

Scheirer E.D., Väänänen R., and Huopaniemi J. (1999): AudioBIFS: describing audio scenes with the MPEG-4 multimedia standard, *IEEE Trans. Multimedia*, 1(3), 237–250.

Schoeffler M., Adami A., and Herre J. (2014): The influence of up- and down-mixes on the overall listening experience, *at the AES 137th Convention*, Los Angeles, USA, Paper 9140.

Schroeder M.R. (1958): An artificial stereophonic effect obtained from a single audio signal, *J. Audio Engrg. Soc.*, 6(2), 74–79.

Schroeder M.R. (1962): Natural sounding artificial reverberation, *J. Audio Eng. Soc.*, 10(3), 219–223.

Schroeder M.R. (1965): New method of measuring reverberation time, *J. Acoust. Soc. Am.*, 37(3), 409–412.

Schroeder M.R. (1987): Statistical parameters of the frequency response curves of large rooms, *J. Audio Eng. Soc.*, 35(5), 299–306.

Schroeder M.R. (1989): Self-similarity and fractals in science and art, *J. Audio Eng. Soc.*, 37(10), 795–808.

Schroeder M.R., and Atal B.S. (1963): Computer simulation of sound transmission in rooms, *Proceedings of the IEEE*, 51(3), 536–537.

Seeber B.U., Baumann U., and Fastl H. (2004): Localization ability with bimodal hearing aids and bilateral cochlear implants, *J. Acoust. Soc. Am.*, 116(3), 1698–1709.

Seo J., Park G.Y., Jang D.Y., et al. (2003): Implementation of interactive 3D audio using MPEG-4 multimedia standards, *at the AES 115th Convention*, New York, USA, Paper 5980.

Seo J.H., Chon S.B., Sung K.M., et al. (2013): Perceptual objective quality evaluation method for high-quality multichannel audio codecs, *J. Audio Eng. Soc.*, 61(7/8), 535–545.

Shaw E.A.G. (1974): Transformation of sound pressure level from the free field to the eardrum in the horizontal plane, *J. Acoust. Soc. Am.*, 56(6), 1848–1861.

Shaw E.A.G., and Teranishi R. (1968): Sound pressure generated in an external ear replica and real human ears by nearby point source, *J. Acoust. Soc. Am.*, 44(1), 240–249.

Shi B., and Xie B.S. (2008): Auditory spatial impression and some psychoacoustic problems in electro-acoustic reproduction (in Chinese), *Audio Eng.*, 32(9), 34–45.

Shi B., and Xie B.S. (2010): The cross-correlation of signals and spatial impression in surround sound reproduction, *Chin. J. Acoust.*, 29(3), 308–320.

Shinn-Cunningham B.G. (1998): Applications of virtual auditory displays, in *Proceedings of the 20th International Conference of the IEEE Engineering in Biology and Medicine Society*, Hong Kong, China, 20(3), 1105–1108.

Shinn-Cunningham B.G., Schickler J., Kopčo N., et al. (2001): Spatial unmasking of nearby speech sources in a simulated anechoic environment, *J. Acoust. Soc. Am.*, 110(2), 1118–1129.

Shively R. (2000): Automotive audio design (a tutorial), *at the AES 109th Convention*, Los Angeles, USA, Paper 5276.

Short K.M., Garcia R.A., and Daniels M.L. (2007): Multichannel audio processing using a unified-domain representation, *J. Audio Eng. Soc.*, 55(3), 156–165.

Silzle A., Novo P., and Strauss H. (2004): IKA-SIM: A system to generate auditory virtual environments, *at the AES 116th Convention*, Berlin, Germany, Paper 6016.

Simon L.S.R., and Mason R. (2010): Time and level localization curves for a regularly-spaced octagon loudspeaker array, *at the AES 128th Convention*, London, UK, Paper 8079.

Simonson G. (1984): Master's Thesis at the Technical University of Lyngby.

Sivonen V.P., and Ellermeier W. (2008): Binaural loudness for artificial-head measurements in directional sound fields, *J. Audio Eng. Soc.*, 56(6), 452–461.

SMPTE 320M (1999): *Television-channel assignments and levels on multichannel audio media, Proposed standard for Television, ITU information* doc.ITU-R 10C/11 and 10-11R/2, Society of Moving Picture and Television Engineers, NY, USA.

SMPTE RP 200 (2012): *Relative and absolute sound pressure levels for motion-picture multichannel sound systems – applicable for analog photographic film audio, digital photographic film audio and D-cinema*, Society of Moving Picture and Television Engineers, NY, USA.

SMPTE RP 2096-1(2017): *Cinema sound system baseline setup and calibration*, Society of Moving Picture and Television Engineers, NY, USA.

SMPTE RP 2096-2(2017): *Cinema Sound System Maintenance Calibration*, Society of Moving Picture and Television Engineers, NY, USA.

SMPTE ST 202 (2010): *Motion-pictures – dubbing theaters, review rooms and indoor theaters – B-Chain electroacoustic response*, Society of Moving Picture and Television Engineers, NY, USA.

SMPTE ST 2036-2-2008 (2008): Ultra high definition television–audio characteristics and audio channel mapping of program production, Society of Moving Picture and Television Engineers, NY, USA.

SMPTE ST 2098-5 (2018): *D-Cinema immersive audio channels and soundfield groups*, Society of Moving Picture and Television Engineers, NY, USA.

SMPTE428-3 (2006): D-Cinema distribution master, audio channel mapping and channel labeling, Society of Moving Picture and Television Engineers, NY, USA.

Smyth M. (1999): White paper, an overview of the coherent acoustics coding system, www.dts.com.

Smyth S.M.F., Smyth W.P., and Smyth M.H.C., et al. (1996): DTS coherent acoustics, delivering high quality multichannel sound the consumer, *at the AES 100th Convention*, Copenhagen, Denmark, Paper 4293.

Snow W. (1953): Basic principles of stereophonic sound, *J. SMPTE*, 61(5), 567–589.

So R.H.Y., Ngan B., Horner A., et al. (2010): Toward orthogonal non-individualised head-related transfer functions for forward and backward directional sound: cluster analysis and an experimental study, *Ergonomics*, 53(6), 767–781.

Solvang A. (2008): Spectral impairment of two-dimensional higher order ambisonics, *J. Audio Eng. Soc.*, 56(4), 267–279.

Søndergaardand P.L., and Majdak P. (2013): The Auditory modeling toolbox, in *the Technology of binaural listening* (edited by Blauert J.), Springer-Verlag, Berlin Heidelberg.

Song M.H., Choi J.W., and Kim Y.H. (2012): A selective array activation method for the generation of a focused source considering listening position, *J. Acoust. Soc. Am.*, 131(2), EL156–162.

Song W., Ellermeier W., and Hald J. (2008): Using beamforming and binaural synthesis for the psychoacoustical evaluation of target sources in noise, *J. Acoust. Soc. Am.*, 123(2), 910–924.

Sonke J.J., Labeeuw J. and Vries D.D.E. (1998): Variable acoustics by wavefield synthesis: a closer look at amplitude effects, *at the AES 104th Convention*, Amsterdam, The Netherlands, Paper 4712.

Sonke J.J., and Vries D.D.E. (1997): Generation of diffuse reverberation by plane wave synthesis, *at the AES 102th Convention*, Munich, Germany, Paper 4455.

Sontacchi A. (2003): *Dreidimensionale schallfeldreproduktion fuer lautsprecher-und kopfhoereranwendungen*, PhD thesis, Graz University of Technology, Styria Austria

Sontacchi A., and Hoeldrich R. (2000): Enhanced 3D sound field synthesis and reproduction using system by compensating interfering reflexions, in *Proceedings of DAFX-00*, Verona, Italy.

Sontacchi A., and Holdrich R. (2002): Distance coding in 3D sound fields, *at the AES 21st International Conference*, St. Petersburg, Russia.

Soulodre G.A., Grusec T., and Lavoie M., et al. (1998): Subjective evaluation of state-of-the-art two-channel audio codecs, *J. Audio Eng. Soc.*, 46(3), 164–177.

Spanias A., Painter T., and Atti V. (2007): *Audio signal processing and coding*, John Wiley & Sons, Hoboken, New Jersey.

Sporer T., Walther A., Liebetrau J., et al. (2006): Perceptual evaluation of algorithms for blind up-mix, *at the AES 121st Convention*, San Francisco, USA, Paper 6915.

Spors S., and Ahrens J. (2009): Spatial sampling artifacts of wave field synthesis for the reproduction of virtual point sources, *at the AES 126th Convention*, Munich, Germany, Paper 7744.

Spors S., and Ahrens J. (2010a): Local sound field synthesis by virtual secondary sources, *at the AES 40th International Conference*, Tokyo, Japan.

Spors S., and Ahrens J. (2010b): Reproduction of focused sources by spectral division method, in *Proceeding of the 4th IEEE International Symposium on Communications, Control and Signal Processing*, Limassol, Cyprus.

Spors S., and Ahrens J. (2010c): Analysis and improvement of pre-equalization in 2.5-dimensional wave field synthesis, *at the AES 128th Convention*, London, UK, Paper 8121.

Spors S., Buchner H., and Rabenstein R. (2004): Efficient active listening room compensation for wave field synthesis, *at the AES 116th Convention*, Berlin, Germany, Paper 6119.

Spors S., Buchner H., Rabenstein R., et al. (2007): Active listening room compensation for massive multichannel reproduction systems using wave-domain adaptive filtering, *J. Acoust. Soc. Am.*, 122(1), 354–369.

Spors S., Kuntz A., and Rabenstain R. (2003): An approach to listening room compensation with the wave synthesis, *at the AES 24th International Conference*, Banff, Canada.

Spors S., and Rabenstain R. (2006): Spatial aliasing artifacts produced by linear and circular loudspeaker arrays used for wave field synthesis, *at the AES 120th Convention*, Paris, France, Paper 6711.

Spors S., Rabenstain R., and Ahrens J. (2008): The theory of wave field synthesis revisited, *at the AES 124th Convention*, Amsterdam, The Netherlands, Paper 7358.

Spors S., and Wierstorf H. (2008): Comparison of higher order ambisonics and wave field synthesis with respect to spatial discretization artifacts properties and spatial sampling, *at the AES 125th Convention*, San Francisco, USA, Paper 7556.

Spors S., Wierstorf H., and Ahrens J. (2011): Interpolation and range extrapolation of head-related transfer functions using virtual local wave field synthesis, *at the AES 130th Convention*, London, UK, Paper 8392.

Spors S., Wierstorf H., Geier M., et al. (2009): Physical and perceptual properties of focused virtual sources in wave field synthesis, *at the AES 127th Convention*, New York, USA, Paper 7914.

Sports S., Renk M., and Rabenstein R. (2005): Limiting effect of active room compensation using wave field synthesis, *at the AES 118th Convention*, Barcelona, Spain, Paper 6400.

Stan G.B., Embrechts J.J., and Archambeau D. (2002): Comparison of different impulse response measurement techniques, *J. Audio. Eng. Soc.*, 50(4), 249–262.

Stanuter J., and Puckette M. (1982): Designing multi-channel reverberators, *Comput. Music J.*, 6(1), 52–65.

Start E.W. (1996): Application of curve array in wave field synthesis, *at the AES 100th Convention*, Copenhagen, Denmark, Paper 4143.

Start E.W., Valstar V.G., and Vries D.D. (1995): Application of spatial bandwidth reduction in wave field synthesis, *at the AES 98th Convention*, Paris, France, Paper 3972.

Steinberg J.C., and Snow W.B. (1934): Auditory perspective-physical factors, In Stereophonic Techniques, 3-7, Audio Engineering Society.

Steinke G. (1996): Surround sound – the new phase: an overview, *at the AES 100th Convention*, Copenhagen, Denmark, Paper 4286.

Støfringsdal B., and Svensson P. (2006): Conversion of discretely sampled sound field data to auralization formats, *J. Audio. Eng. Soc.*, 54(5), 380–400.
Streicher R., and Dooley W. (1985): Basic stereo microphone perspectives – a review, *J. Audio Eng. Soc.*, 33(7/8), 548–556.
Sugimoto T., Oode S., and Nakayama Y. (2015): Downmixing method for 22.2 channel sound signals in 8K super high-vision broadcasting, *J. Audio Eng. Soc.*, 63(7/8), 590–599.
Sun H., and Svensson U.P. (2011): Design 3-D high order Ambisonics encoding matrices using convex optimization, *at the AES 130th Convention*, London, UK, Paper 8402.
Suokuisma P., Zacharov N., and Bech S. (1998): Multichannel level alignment, part I: signals and methods, *at the AES 105th Convention*, San Francisco, USA, Paper 4815.
Suzuki H., Shinbara H., and Toyoshima S.M. (1993): Study on optimum rear loudspeaker height for 3-1 reproduction of HDTV audio, *at the AES 95th Convention*, New York, USA, Paper 3722.
Svensson U.P., Botts J., and Savioja L. (2017a): Computational modeling of room acoustics I: wave-based modeling, in *Architectural acoustics handbook* (edited by Xiang N.), J.Ross Publishing, USA.
Svensson U.P., Botts J., Savioja L., et al. (2017b): Computational modeling of room acoustics II: geometrical acoustics, in *Architectural acoustics handbook* (edited by Xiang N.), J. Ross Publishing, USA.
Svensson U.P., and Kristiansen U.R. (2002): Computational modeling and simulation of acoustic spaces, *at the AES 22nd International Conference*, Espoo, Finland.
Takane S., Arai D., Miyajima T., et al. (2002): A database of head-related transfer functions in whole directions on upper hemisphere, *Acoust. Sci. Tech.*, 23(3), 160–162.
Takanen M., and Lorho G. (2012): A binaural auditory model for the evaluation of reproduced stereophonic sound, *at the AES 45th International Conference*, Helsinki, Finland.
Takeuchi T., and Nelson P.A. (2002): Optimal source distribution for binaural synthesis over loudspeakers, *J. Acoust. Soc. Am.*, 112(6), 2786–2797.
Takeuchi T., Nelson P.A., Kirkeby O., et al. (1998): Influence of individual head-related transfer function on the performance of virtual acoustic imaging systems, *at the AES 104th Convention*, Amsterdam, The Netherlands, Preprint 4700.
Tervo S., Pätynen J., Kuusinen A., et al. (2013): Spatial decomposition method for room impulse responses, *J. Audio Eng. Soc.*, 61(1/2), 17–28.
Theile G. (1990): Further developments of loudspeaker stereophony, *at the AES 89 Convention*, Los Angeles, USA, Paper 2947.
Theile G. (1991a): HDTV sound systems: how many channels? *at the AES 9th International Conference*, Detroit, Michigan, USA.
Theile G. (1991b): On the naturalness of two-channel stereo sound, *J. Audio Eng. Soc.*, 39(10), 761–767.
Theile G. (1993): Trends and activities in the development of multichannel sound systems, *at the AES 12nd Conference*, Copenhagen, Denmark.
Theile G. (2001): Natural 5.1 channel recording based on psychoacoustic principles, *at the AES 19th International Conference*, Schloss Elmau, Germany.
Thiele G., and Plenge G. (1977): Localization of lateral phantom sources, *J. Audio Eng. Soc.*, 25(4), 196–200.
Theile G., and Steinke G. (1999): Surround sound guidelines for operational practice, *at the AES UK 14th Conference: Audio-The Second Century*, London, UK.
Theile G., and Wittek H. (2011): Principles in surround recordings with height, *at the AES 130th Convention*, London, UK, Paper 8403.
Thompson J., Smith B., Wamer A., et al. (2012): Direct-diffuse decomposition of multichannel signals using a system of pairwise correlations, *at the AES 133rd Convention*, San Francisco, USA, Paper 8807.
Thompson J., Wamer A., and Smith B. (2009): An active multichannel downmix enhancement for minimizing spatial and spectral distortions, *at the AES 127th Convention*, New York, USA, Paper 7913.
Todd C.C., Davidson G.A., and Davis M.F. (1994): AC-3, Flexible perceptual coding for audio transmission and storage, *at the AES 96th Convention*, Amsterdam, The Netherlands, Paper 3796.
Toh C.W., and Gan W.S. (1999): A real-time virtual surround sound system with bass enhancement, *at the AES 107th Convention*, New York, USA, Paper 5052.

Tohyama M., and Suzuki A. (1989): Interaural cross-correlation coefficients in stereo-reproduced sound field fields, *J. Acoust. Soc. Am.*, 85(2), 780–786.

Toole F.E. (1985): Subjective measurements of loudspeaker sound quality and listener performance, *J. Audio Eng. Soc.*, 33(1/2), 2–32.

Toole F.E. (1991): Binaural record/reproduction systems and their use in psychoacoustic investigations, *AES 91st Convention*, New York, USA, Preprint 3179.

Tregonning A., and Martin B. (2015): The vertical precedence effect: utilizing delay panning for height channel mixing in 3D audio, *at the AES 139th Convention*, New York, USA, Paper 9469.

Tsang P.W.M., and Cheung W.K. (2009): Development of a re-configurable ambisonic decoder for irregular loudspeaker configuration, *IET Circ. Dev. Syst.*, 3(4), 197–203.

Tsang P.W.M., Cheung W.K., and Leung C.S. (2009): Decoding Ambisonic signals to irregular loudspeaker configuration based on artificial neural networks, in *Proceedings of ICONIP*, part II, LNCS 5864, 273–280, Springer-Verlag, Berlin, Germany.

Tsingos N., Chabanne C., Robinson C., et al. (2010): Surround sound with height in games using Dolby Pro Logic IIz, *at the AES 129th Convention*, San Francisco, USA, Paper 8248.

Tsutsui K., Suzuki H., Shimoyoshi O., et al. (1992): ATRAC: adaptive transform acoustic coding for MiniDisc, *at the AES 93rd Convention*, San Francisco, USA, Paper 3456.

Uhle C., and Gampp P. (2016): Mono-to-stereo upmixing, *at the AES 140th Convention*, Paris, France, Paper 9528.

Uncini A. (2015): *Fundamentals of adaptive signal processing*, Springer International Publishing, Switzerland.

Usher J., and Benesty J. (2007): Enhancement of spatial sound quality: a new reverberation-extraction audio upmixer, *IEEE Trans. Audio, Speech, Language Process.*, 15(7), 2141–2150.

Väänänen R., and Huopaniemi J. (2004): Advanced audio BIFS: virtual acoustics modeling in MPEG-4 scene description, *IEEE Trans. Multimedia*, 6(5), 661–675.

Välimäki V., Parker J.D., and Saviojia L. (2012): Fifty years of artificial reverberation, *IEEE Trans. Audio, Speech, Language Process.*, 20(5), 1421–1447.

Vanderkooy J. (1994): Aspects of MLS measuring systems, *J. Audio Eng. Soc.*, 42(4), 219–231.

Vanderlyn P.B. (1954): British Patent Application. No.23989.

Verbakel J., Kerkhof L.Van De., Maeda M., et al. (1998): Super audio CD format, *at the AES 104th Convention*, Amsterdam, The Netherlands, Paper 4705.

Vilkamo J., Backstrom T., and Kuntz A. (2013): Optimized covariance domain framework for time-frequency processing of spatial audio, *J. Audio Eng. Soc.*, 61(6), 403–411.

Vilkamo J., Kuntz A., and Füg S. (2014): Reduction of spectral artifacts in multichannel downmixing with adaptive phase alignment, *J. Audio Eng. Soc.*, 62(7/8), 516–526.

Vilkamo J., Neugebauer B., and Plogsties J. (2011): Sparse frequency-domain reverberator, *J. Audio Eng. Soc.*, 59(12), 936–943.

Vilkamo J., and Pulkki V. (2014): Adaptive optimization of interchannel coherence with stereo and surround sound audio content, *J. Audio Eng. Soc.*, 62(12), 861–869.

Vilkamo J., Lokki T., and Pulkki V. (2009): Directional audio coding: virtual microphone-based synthesis and subjective evaluation, *J. Audio Eng. Soc.*, 57(9), 709–724.

Villemoes L., Herre J., Breebaart J., et al. (2006): MPEG surround: the forthcoming ISO standard for spatial audio coding, *at the AES 28th International Conference*, Pitea, Sweden.

Vinton M., McGrath D., Robinson C., et al. (2015): Next generation surround decoding and upmixing for consumer and professional applications, *at the AES 57th Convention*, Hollywood, CA, USA.

Vorlander M. (2004): Past, present and future of dummy head, *at the 2004 conference of the Federation of the Ibero-American acoustical societies*, Guimaraes, Portugal.

Vorländer M. (2008): *Auralization, fundamentals of acoustics, modelling, simulation, algorithms and acoustic virtual reality*, Springer-Verlag, Berlin, Germany.

Vries D.D.E. (1996): Sound reinforcement by wave field synthesis: adaptation of the synthesis operator to the loudspeaker directivity characteristics, *J. Audio Eng. Soc.*, 44(12), 1120–1131.

Vries D.D.E. (2009): *Wave field synthesis*, Audio Engineering Society, New York, USA.

Vries D.D.E., Reijnen A.J., and Schonewille M.A. (1994b): The wave-field synthesis concept applied to generation of reflections and reverberation, *at the AES 96th Convention*, Amsterdam, The Netherlands, Paper 3813.

Vries D.D.E., Start E.W., and Valstar V.G. (1994a): The wave-field synthesis concept applied to sound reinforcement restrictions and solutions, *at the AES 96th Convention*, Amsterdam, The Netherlands, Paper 3812.

Vries D.D.E., and Vogel P. (1993): Experience with a sound enhancement system based on wavefront synthesis, *at the AES 95th Convention*, New York, USA, Paper 3748.

Walker R. (1994): Early reflections in studio control rooms: the results from the first controlled image design installations, *at the AES 96th Convention*, Amsterdam, The Netherlands, Paper 3853.

Wallach H. (1940): The role of head movement and vestibular and visual cue in sound localization, *J. Exp. Psychol.*, 27(4), 339–368.

Waller J. K. (1996): The circle surround 5.2.5 5-channel surround system, Rocktron Corporation / RSP Technologies, White Paper.

Wallis R., and Lee H. (2014): Investigation into vertical stereophonic localisation in the presence of interchannel crosstalk, *at the AES 136th Convention*, Berlin, Germany, Paper 9026.

Walther A., and Faller C. (2011): Assessing diffuse sound field reproduction capabilities of multichannel playback systems, *at the AES 130th Convention*, London, UK, Paper8428.

Ward D.B., and Abhayapala T.D. (2001): Reproduction of a plane-wave sound field using an array of loudspeakers, *IEEE Trans. Speech, Audio Process.*, 9(6), 697–707.

Watkins A.J. (1978): Psychoacoustical aspects of synthesized vertical localecues, *J. Acoust. Soc. Am.*, 63(4), 1152–1165.

Weller T., Buchholz J.M., and Oreinos C. (2014): Frequency dependent regularization of a mixed-order ambisonics encoding system using psychoacoustically motivated metrics, *at the AES 55th Conference*, Helsinki, Finland.

Welti T. (2002): How many subwoofers are enough?, *at the AES 112nd Convention*, Munich, Germany, Paper 5602.

Welti T. (2012): Optimal configurations for subwoofers in rooms considering seat to seat variation and low frequency efficiency, *at the AES 133rd Convention*, San Francisco, USA, Paper 8748.

Welti T., and Devantier A. (2006): Low-frequency optimization using multiple subwoofers, *J. Audio Eng. Soc.*, 54(5), 347–364.

Wendt F., Frank M., and Zotter F. (2014): Panning with height on 2, 3, and 4 loudspeakers, *at the 2nd International Conference on Spatial Audio*, Erlangen, Germany.

Wenzel E.M. (1996): What perception implies about implementation of interactive virtual acoustic environments, *AES 101st Convention*, Los Angeles, CA, Paper 4353.

Wenzel E.M., Arruda M., Kistler D.J., et al. (1993): Localization using nonindividualized head-related transfer functions, *J. Acoust. Soc. Am.*, 94(1), 111–123.

Wenzel E.M., Miller D.J., and Abel J.S. (2000): Sound lab: a real-time, software-based system for the study of spatial hearing, *AES 108th Convention*, Paris, France, Paper 5140.

White J.V. (1976): Synthesis of 4-2-4 matrix recording systems, *J. Audio Eng. Soc.*, 24(4), 250–257.

Wierstorf H., Hohnerlein C., Spors S., et al. (2014): Coloration in wave field synthesis, *at the AES 55th International Conference*, Helsinki, Finland.

Wierstorf H., Raake A., Geier M., et al. (2013): Perception of focused sources in wave field synthesis, *J. Audio Eng. Soc.*, 61(1/2), 5–16.

Wiggins B. (2007): The generation of panning laws for irregular speaker arrays using heuristic methods, *at the AES 31st International Conference*, London, UK.

Wightman F.L., and Kistler D.J. (1989a): Headphone simulation of free-field listening, I: stimulus synthesis, *J. Acoust. Soc. Am.*, 85(2), 858–867.

Wightman F.L., and Kistler D.J. (1989b): Headphone simulation of free-field listening, II: psycho-physical validation, *J. Acoust. Soc. Am.*, 85(2), 868–878.

Wightman F.L., and Kistler D.J. (1992): The dominant role of low-frequency interaural time difference in sound localization, *J. Acoust. Soc. Am.*, 91(3), 1648–1661.

Wightman F.L., and Kistler D.J. (1997): Monaural sound localization revisited, *J. Acoust. Soc. Am.*, 101(2), 1050–1063.

Wightman F.L., and Kistler D.J. (1999): Resolution of front-back ambiguity in spatial hearing by listener and source movement, *J. Acoust. Soc. Am.*, 105(5), 2841–2853.

Wightman F.L., and Kistler D.J. (2005): Measurement and validation of human HRTFs for use in hearing research, *Acta Acust. United Ac.*, 91(3), 429–439.

Wightman F.L., Kistler D.J., and Arruda M. (1992): Perceptual consequences of engineering compromises in synthesis of virtual auditory objects, *J. Acoust. Soc. Am.*, 92(4), 2332.

Williams E.G. (1999): *Fourier acoustics, sound radiation and near-field acoustical holography*, Academic Press, London, UK.

Williams M. (1987): United theory of microphone systems for stereophonic and sound recording, *at the AES 82nd Convention*, London, UK, Paper 2466.

Williams M. (2002): Multichannel microphone array design, segment coverage analysis above and below the horizontal reference plane, *at the AES 112nd Convention*, Munich, Germany, Paper 5567.

Williams M. (2003): Multichannel sound recording practice using microphone arrays, *at the AES 24th International Conference*, Banff, Canada.

Williams M. (2004): Multichannel sound recording using 3, 4 and 5 channel arrays for front sound stage coverage, *at the AES 117th Convention*, San Francisco, USA, Paper 6230.

Williams M. (2007): Magic arrays, multichannel microphone array design applied to multi-format compatibility, *at the AES 122nd Convention*, Vienna, Austria, Paper 7057.

Williams M. (2008): Migration of 5.0 multichannel microphone array design to higher order MMAD (6.0, 7.0 & 8.0) with or without the inter-format compatibility criteria, *at the AES 124th Convention*, Amsterdam, The Netherlands, Paper 7480.

Williams M. (2012): Microphone array design for localization with elevation cues, *at the AES 132 and Convention*, Budapest, Hungary, Paper 8601.

Williams M. (2013): *Microphone arrays for stereo and multichannel sound recordings* (Vols I and II), Editrice II Rostro, Milano, Italy.

Williams M., and Du G.L. (1999): Microphone array analysis for multichannel sound recording, *at the AES 107th Convention*, New York, USA, Paper 4997.

Williams M., and Du G.L. (2000): Multichannel microphone array design, *at the AES 108th Convention*, Paris, France, Paper 5157.

Williams M., and Du G.L. (2001): The quick reference guide to multichannel microphone array, part I: using cardioid microphones, *at the AES 110th Convention*, Amsterdam, The Netherlands, Paper 5336.

Williams M., and Du G.L. (2004): The quick reference guide to multichannel microphone array, part II: using supercardioid and hypocardioid microphones, *at the AES 116th Convention*, Berlin, Germany, Paper 6059.

Wittek H. (2007): Perceptual differences between wavefield synthesis and stereophony, PhD thesis, University of Surrey, UK.

Wittek H., Rumsey F., and Theile G. (2007): Perceptual enhancement of wavefield synthesis by stereophonic means, *J. Audio. Eng. Soc.*, 55(9), 723–751.

Wittek H., and Theile G. (2002): The recording angle–based on localization curves, *at the AES 112nd Convention*, Munich, Germany, Paper 5568.

Wittek H., and Theile G. (2017): Development and application of a stereophonic multichannel recording technique for 3D audio and VR, *at the AES 143rd Convention*, New York, USA, Paper 9869.

Woodward J.G. (1975a): NQRC measurement of subjective aspects of quadraphonic sound reproduction, part i, *J. Audio. Eng. Soc.*, 23(1), 2–13.

Woodward J.G. (1975b): NQRC measurement of subjective aspects of quadraphonic sound reproduction, part II, *J. Audio. Eng. Soc.*, 23(2), 128–130.

Woodward J.G. (1977): Quadraphony – a review, *J. Audio Eng. Soc.*, 25(10/11), 843–854.

Woodworth R.S., and Schlosberg H. (1954): *Experimental psychology* (edited by Holt H.), New York, USA.

Woszczyk W., Beghin T., and De Francisco M., et al. (2009): Recording multichannel sound within virtual acoustics, *at the AES 127th Convention*, New York, USA, Paper 7856.

Woszczyk W., Leonard B., and Ko D. (2010): Space builder: an impulse response-based tool for immersive 22.2 ambiance design, *at the AES 40th International Conference*, Tokyo, Japan.

Wu S.X., and Zhao Y.Z. (2003): *Room and environmental acoustics* (in Chinese), Guangdong Science and Technology Press, Guangzhou, China.

Wu Y.J. and Abhayapala T.D. (2009): Theory and design of sound field reproduction using continuous loudspeaker concept, *IEEE Trans. Audio, Speech, Language Process.*, 17(1), 107–116.

Wu Y.J., and Abhayapala T.D. (2011): Spatial multizone soundfield reproduction, theory and design, *IEEE Trans. Audio, Speech, Language Process.*, 19(6), 1711–1720.

Wu Z., Chan F.H.Y., Lam F.K., et al. (1997): A time domain binaural model based on spatial feature extraction for the head-related transfer function, *J. Acoust. Soc. Am.*, 102(4), 2211–2218.

Wüstenhagen U., Feiten B., and Hoeg W. (1998): Subjective listening test of multichannel audio codecs, *at the AES 105th Conventiom*, San Francisco, USA, Paper 4813.

Xiang N., and Blauert J. (1991): A miniature dummy head for binaural evaluation of tenth-scale acoustic models, *Appl. Acoust.*, 33(2), 123–140.

Xiang N., and Blauert J. (1993): Binaural scale modeling for auralisation and prediction of acoustics in auditoria, *Appl. Acoust.*, 38(2/4), 267–290.

Xiang N., and Schroeder M.R. (2003): Reciprocal maximum-length sequence pairs for acoustical dual source measurements, *J. Acoust. Soc. Am.*, 113(5), 2754–2761.

Xiang N., Trivedi U., and Xie B.S. (2019): Artificial enveloping reverberation for binaural auralization using reciprocal maximum-length sequences, *J. Acoust. Soc. Am.*, 145(4), 2691–2702.

Xie B.S. (1992): Surround sound reproduction with N+1 channel cone array of loudspeakers (in Chinese), *Audio Eng.*, 16(6), 2–6.

Xie B.S. (1995): The advance in stereo technology in recent year (in Chinese), *Appl. Acoust.*, 14(4), 1–6.

Xie B.S. (1997): Analysis on a defect of the 5 channel 3/2 surround sound system (in Chinese), *Appl. Acoust.*, 16(5), 1–7.

Xie B.S. (1998): Interchannel phase difference and stereo sound image localization, *Chin J. Acoust.*, 17(1), 85–93.

Xie B.S. (1999a): Problem with multichannel and virtual surround sound (in Chinese), *Audio Eng.*, 23(5), 17–25.

Xie B.S. (1999b): Design consideration and quality assessment of multichannel surround sound (in Chinese), *Audio Eng.*, 23(8), 12–15.

Xie B.S. (1999c): Cross-correlation analysis on stereophonic and surround sound image (in Chinese), *J. Tongji Univ.*, 27(3), 361–365.

Xie B.S. (2001a): Signal mixing for a 5.1 channel surround sound system – analysis and experiment, *J. Audio Eng. Soc.*, 49(4), 263–274.

Xie B.S. (2001b): 6.1 channel general planar surround sound system, *Chin. J. Acoust.*, 20(2), 170–183.

Xie B.S. (2002a): Interchannel time difference and stereophonic sound image localization (in Chinese), *Acta Acust.*, 27(4), 332–338.

Xie B.S. (2002b): Effect of head size on virtual sound image localization (in Chinese), *Appl. Acoust.*, 21(5), 1–7.

Xie B.S. (2005): Rotation of head and stability of virtual sound image (in Chinese), *Audio Eng.*, 29(6), 56–59.

Xie B.S. (2006a): The meaning of the phase character of HRTF and interaural time difference (in Chinese), *Audio Eng.*, 30(11), 40–45.

Xie B.S. (2006b): Spatial interpolation of HRTFs and signal mixing for multichannel surround sound, *Chin. J. Acoust.*, 25(4), 330–341.

Xie B.S. (2008a): *Head-related transfer function and virtual auditory* (1st edition, in Chinese), National Defense Industry Press, Beijing, China.

Xie B.S. (2008b): The applications of virtual auditory to virtual reality, communications and information systems (in Chinese), *Audio Eng.*, 32(1), 70–75.

Xie B.S. (2008c): Principle, progress and problems of virtual auditory environment (in Chinese), *Audio Eng.*, 32(11), 39–44.

Xie B.S. (2009a): Head-related transfer function and virtual auditory display (in Chinese), *Sci. Sin.: Phys. Mech. Astron.*, 39(9), 1268–1285.

Xie B.S. (2009b): On the low frequency characteristics of head-related transfer function, *Chin J. Acoust.*, 28(2), 116–128.

Xie B.S. (2012): Recovery of individual head-related transfer functions from a small set of measurements, *J. Acoust. Soc. Am.*, 132(1), 282–294.
Xie B.S. (2013): *Head-related transfer function and virtual auditory display* (2nd edition), J. Ross Publishing, USA.
Xie B.S. (2014): Head-related transfer functions of typical subjects from Chinese-based database, *at the 21st International Congress on Sound and Vibration*, Beijing, China.
Xie B.S. (2020): Spatial sound –history, principle, progress and challenge, *Chin. J. Electronics*, 29(3), 397–416.
Xie B.S., and Guan S.Q. (2002): Development and psychoacoustic principle of multichannel surround sound (in Chinese), *Audio Eng.*, 26(2), 11–18.
Xie B.S., and Guan S.Q. (2004): Virtual sound and its application (in Chinese), *Appl. Acoust.*, 23(4), 43–47.
Xie B.S., Guan S.Q. (2012): Research and application of spatial sound – history, development and state of the art (in Chinese), *Appl. Acoust.*, 31(1), 18–27.
Xie B.S., and Guo T.K. (2004): Analysis on stereophonic sound image for an off center listener, (in Chinese), *Acta Acust.*, 29(5), 445–452.
Xie B.S., and Liang S.J. (1995): The effect of frequency on image localization in the surround sound systems (in Chinese), *Appl. Acoust.*, 14(3), 22–29.
Xie B.S., Liu L.L., and Jiang J.L. (2021b): Dynamic binaural Ambisonics scheme for rendering distance information of free-field virtual sources (in Chinese), *Acta Acust.*, 46(6), 1223–1233.
Xie B.S., Liu L.L., and Zhang C.Y. (2021c): Virtual reproduction of surround sound in frontal space using four loudspeakers, *Chin. J. Acoust.*, 40(2), 155–174.
Xie B.S., Mai H.M., Liu Y., et al. (2015b): Analysis on the timbre coloration of wave field synthesis using a binaural loudness model, *at the AES 138th Convention*, Warsaw, Poland, Paper 9320.
Xie B.S., Mai H.M., Rao D., et al. (2019): Analysis of and experiments on vertical summing localization of multichannel sound reproduction with amplitude panning, *J. Audio. Eng. Soc.*, 67(6), 382–399.
Xie B.S., Mai H.M., and Zhong X.L. (2017a): The median-plane summing localization in ambisonics reproduction, *at the AES 142nd Convention*, Berlin, Germany, Paper 9726.
Xie B.S., Mai H.M., and Zhong X.L. (2017b): Analysis on summing virtual source localization in different sagittal planes, *at the Inter-noise 2017*, Hong Kong, China.
Xie B.S., and Rao D. (2015): Analysis and experiment on summing localization of two loudspeakers in the median plane, *at the AES 139th Convention*, New York, USA, Paper 9452.
Xie B.S., Shi B., and Xiang N. (2012): Audio signal decorrelation based on reciprocal-maximal length sequence filters and its applications to spatial sound, *at the AES 133rd Convention*, San Francisco, USA, Paper 8805.
Xie B.S., Shi Y., Xie Z.W., et al. (2005c): Virtual reproduction system for 5.1 channel surround sound, *Chin. J. Acoust.*, 24(1), 76–88.
Xie B.S., Shi Y., Xie Z.W., et al. (2005d): Two-loudspeaker virtual 5.1 channel surround sound signal processing method, China patent No.ZL02134416.7.
Xie B.S., and Tian Z.J. (2014): Improving binaural reproduction of 5.1 channel surround sound using individualized HRTF cluster in the wavelet domain, *at the AES 55th International Conference*, Helsinki, Finland.
Xie B.S., Wang J., and Guan S.Q. (2001): A simplified way to simulate 3D virtual sound image (in Chinese), *Audio Eng.*, 25(7), 10–14.
Xie B.S., Wang J., Guan S.Q., et al. (2005a): Virtual reproduction of 5.1 channel surround sound by headphone, *Chin J. Acoust.*, 24(1), 63–75.
Xie B.S., Wang J., Guan S.Q., et al. (2005b): Headphone virtual 5.1 channel surround sound signal processing method, China patent No.ZL02134415.9.
Xie B.S., and Xie X.F. (1992b): The study of planar surround sound field (in Chinese), *Acta Acust.*, 17(3), 225–231.
Xie B.S., and Xie X.F. (1996): Analyse and sound image localization experiment study on multi-channel planar surround sound system, *Chin. J. Acoust.*, 15(1), 52–64.
Xie B.S., and Yu G.Z. (2021): Psychoacoustic principle, methods, and problems with perceived distance control in spatial audio, *Appl. Sci.*, 11(23), 111242.

Xie B.S., and Zhang C.Y. (1999): A simple method for stereophonic image stage extension (in Chinese), *Tech. Acoust.*, 18(supplement), 187–188.

Xie B.S., and Zhang C.Y. (2012): An algorithm for efficiently synthesizing multiple near-field virtual sources in dynamic virtual auditory display, *at the AES 132nd Convention*, Budapest, Hungary, Paper 8646.

Xie B.S., Zhang C.Y., and Zhong X.L. (2013c): A cluster and subjective selection-based HRTF customization scheme for improving binaural reproduction of 5.1 channel surround sound, *in the AES 134th Convention*, Rome, Italy, Paper 8879.

Xie B.S., Zhang L.S., and Guan S.Q., et al. (2006): Simplification and subjective evaluation of filters for virtual sound using loudspeakers (in Chinese), *Tech. Acoust.*, 25(6), 547–554.

Xie B.S., and Zhang T.T. (2010): The audibility of spectral detail of head-related transfer functions at high frequency, *Acta Acust. United Ac.*, 96(2), 328–339.

Xie B.S., and Zhong X.L. (2012): Similarity and cluster analysis on magnitudes of individual head-related transfer functions (abstract), *J. Acoust. Soc. Am.*, 131(4, Pt.2), 3305.

Xie B.S., Zhong X.L., and He N.N. (2015a): Typical data and cluster analysis on head-related transfer functions from Chinese subjects, *Appl. Acoust.*, 94(1), 1–13.

Xie B.S., Zhong X.L., Rao D., et al. (2007): Head-related transfer function database and its analyses, *Sci. China: Phys. Mech. Astron.*, 50(3), 267–280.

Xie B.S., Zhong X.L., Yu G.Z., et al. (2013b): Report on research projects on head-related transfer functions and virtual auditory displays in China, *J. Audio Eng. Soc.*, 61(5), 314–326.

Xie X.F. (1964a): A novel pseudo-stereophonic soundsystem (in Chinese), *at the 1st Acoustic meeting of China*, Paper C3.3, Bejing, China.

Xie X.F. (1964b): Serval simple circuits of pseudo-stereophonic sound, *at the 1st Acoustic meeting of China*, Paper C3.4, Bejing, China.

Xie X.F. (1964c): A pseudo-stereophonic soundsystem with single input and two outputs (in Chinese), *J. South China Inst. Technol.*, 2(2), 62–73.

Xie X.F. (1977): The 4-3-4 (four-channel) stereophonic system (in Chinese), *J. South China Inst. Techn.*, 5(1), 40–48.

Xie X.F. (1978a): Stereophonic sound, development and state of arts (in Chinese), *Audio Eng.*, 2(3), 1–11.

Xie X.F. (1978b): 4-3-4 transformation and N(≥3) channel reproduction of panoramic (stereophonic) sound (in Chinese), *J. South China Inst. Techn.*, 6(2), 54–70.

Xie X.F. (1981): *The principle of stereo* (in Chinese), Science Press, Beijing, China.

Xie X.F. (1982): The 4-3-N matrix multi-channel sound system, *Chin. J. Acoust.*, 1(2), 201–218.

Xie X.F. (1987): *The researches on stereophonic sound* (in Chinese), South China Institute of Technology Press, Guangzhou, China.

Xie X.F. (1988): A Mathematical analysis of three dimensional surrounding sound field (in Chinese), *Acta Acoust.*, 13(5), 321–328.

Xie X.F., and Xie B.S. (1992): Surround sound reproduction with folding loudspeaker arrangement (in Chinese), *Appl. Acoust.*, 11(5), 5–9.

Yamamoto T. (1973): Quadraphonic one point pickup microphone, *J. Audio Eng. Soc.*, 21(4), 256–261.

Yang J., and Gan W.S. (2008): Technique of creation of object-based radiation in sound field control and its applications, in *Innovation and harmony – Progress on acoustics in China* (edited by Cheng J.C. and Tian J.), Science Press, Beijing, China.

Yasuda Y., Ohya T., McGrath D., et al. (2003): 3-D audio communications services for future mobile networks, *at the AES 23rd International Conference*, Copenhagen, Denmark.

Yi K.L., and Xie B.S. (2020): Local Ambisonics panning method for creating virtual source in the vertical plane of the frontal hemisphere, *Appl. Acoust.*, 165, 107319.

Yin T.C.T. (1994): Physiological correlates of the precedence effect and summing localization in the inferior colliculus of the cat, *J. Neurosci.*, 14(9), 5170–5186.

Yoshikawa S., Noge S., and Funaki Y. (1993): Monitor Levels and Quality Evaluation of HDTV 3-1 Multichannel Sound, *at the AES 95th Convention*, New York, USA, Paper 3723.

Yost W.A., and Sheft S. (1993): Auditory perception, in *Human psychophysics* (edited by Yost W.A., Popper A.N., Fay R.R.), Springer-Verlag, New York, USA.

Yu G.Z., Liu Y., and Xie B.S. (2018a): Design and validation on a multiple sound source fast-measurement system of near-field head-related transfer functions, *Chin. J. Acoust.*, 37(2), 219–240.

Yu G.Z., Wu R.X., Liu Y., et al. (2018b): Near-field head-related transfer-function measurement and database of human subjects, *J. Acoust. Soc. Am.*, 143(3), EL194–198.

Yu G.Z., and Xie B.S. (2007): Head-related transfer function for nearby sources and its applications (in Chinese), *Audio Eng.*, 31(7), 45–50.

Yu G.Z., Xie B.S., Chen Z.W., et al. (2012a): Analysis on multiple scattering between the rigid-spherical microphone array and nearby surface in sound field recording, *at the AES 133rd Convention*, San Francisco, USA, Paper 8710.

Yu G.Z., Xie B.S., and Rao D. (2012b): Near-field head-related transfer functions of a artificial head and its characteristics (in Chinese), *Acta Acust.*, 37(4), 378–385.

Zacharov N. (1998a): Subjective appraisal of loudspeaker directivity for multi-channel reproduction, *J. Audio. Eng. Soc.*, 46(4), 288–303.

Zacharov N. (1998b): An overview of multichannel level alignment, *at the AES 15th International Conference*, Copenhagen, Denmark.

Zacharov N., and Bech S. (2000): Multichannel level alignment, part iv: the correlation between physical measures and subjective level calibration, *at the AES 109th Convention*, Los Angeles, Paper 5241.

Zacharov N., Bech S., and Meares D. (1998a): The use of subwoofers in the context of surround sound program reproduction, *J. Audio Eng. Soc.*, 46(4), 276–287.

Zacharov N., Bech S., and Suokuisma P. (1998b): Multichannel level alignment, part II: the influence of signals and loudspeaker placement, *at the AES 105th Convention*, San Francisco, USA, Paper 4816.

Zacharov N., and Koivuniemi K. (2001a): Unravelling the perception of spatial sound reproduction: technique and experimental design, *at the AES 19th International Conference*, Schloss, Elmau, Germany.

Zacharov N., and Koivuniemi K. (2001b): Unravelling the perception of spatial sound reproduction: analysis & external preference mapping, *at the AES 111st Convention*, New York, USA, Paper 5423.

Zacharov N., and Koivuniemi K. (2001c): Unravelling the perception of spatial sound reproduction: language development, verbal protocol analysis and listener training, *at the AES 111st Convention*, New York, USA, Paper 5424.

Zacharov N., and Pedersen T.H. (2015): Spatial sound attributes-development of a common lexicon, *at the AES 139th Convention*, New York, USA, Paper 9436.

Zahorik P. (2002a): Assessing auditory distance perception using virtual acoustics, *J. Acoust. Soc. Am.*, 111(4), 1832–1846.

Zahorik P. (2002b): Auditory display of sound source distance, in *Proceedings of the 2002 International Conference on Auditory Display*, Kyoto, Japan, 326–332.

Zahorik P., Brungart D.S., and Bronkhorst A.W. (2005): Auditory distance perception in humans: a summary of past and present research, *Acta Acust. United Ac.*, 91(3), 409–420.

Zeng J.Y. (2007): *Quantum mechanics* (4th edition, Vol. II, in Chinese), Science Press, Beijing, China.

Zhang C.Y., and Xie B.S. (2013): Platform for dynamic virtual auditory environment real-time rendering system, *Chin. Sci. Bull.*, 58(3), 316–327.

Zhang C.Y., and Xie B.S. (2014): Dynamic binaural reproduction of 5.1 channel surround sound with low cost head-tracking device, *at the AES 55th International Conference*, Helsinki, Finland.

Zhang C.Y., Xie B.S., and Xie Z.W. (2000): Elimination of effect of inside-the-head localization in sound reproduction by stereophonic earphone (in Chinese), *Audio Eng.*, 24(8), 4–6.

Zhang C.Y., Xie B.S., and Yu G.Z. (2014): A scheme of stereophonic expansion for handheld sound reproduction devices (in Chinese), *Appl. Acoust.*, 33(4), 324–329.

Zhang T., Zhang C.X., and Zhao X. (2016): Review of AVS audio coding standard, *ZTE Commun.*, 14(2), 56–62.

Zhang W., Abhayapala T.D., Kennedy R.A., et al. (2010): Insights into head-related transfer function: spatial dimensionality and continuous representation, *J. Acoust. Soc. Am.*, 127(4), 2347–2357.

Zhong X.L., and Xie B.S. (2004): Progress in the research of head-related transfer function, *Audio Eng.* (in Chinese) 28(12), 44–46.

Zhong X.L., and Xie B.S. (2005): Spatial characteristics of head related transfer function, *Chin. Phys. Lett.*, 22(5), 1166–1169.

Zhong X.L., and Xie B.S. (2007): Spatial symmetry of head-related transfer function, *Chin J. Acoust.*, 26(1), 73–84.

Zhong X.L., and Xie B.S. (2009): Maximal azimuthal resolution needed in measurements of head-related transfer functions, *J. Acoust. Soc. Am.*, 125(4), 2209–2220.

Zhong X.L., and Xie B.S. (2012): Approximation of individualized head-related transfer function – state of the art and problems (in Chinese), *Appl. Acoust.*, 31(6), 410–415.

Zhong X.L., Zhang F.C., and Xie B.S. (2013): On the spatial symmetry of head-related transfer functions, *Appl. Acoust.*, 74(6), 856–864.

Zhu Y. (2000): *Experimental psychology* (in Chinese), Peking University Press, Beijing, China.

Zielinski S.K., Rumsey F., and Bech S. (2003): Effects of the down-mix algorithms on quality of surround sound, *J. Audio Eng. Soc.*, 51(9), 780–798.

Zotkin D.N., Duraiswami R., and Davis L.S. (2004): Rendering localized spatial audio in a virtual auditory space, *IEEE Trans. Multimedia*, 6(4), 553–564.

Zotkin D.N., Duraiswami R., and Gumerov N.A. (2010): Plane-wave decomposition of acoustical scenes via spherical and cylindrical microphone arrays, *IEEE Trans. Audio, Speech, Language Process.*, 18(1), 2–16.

Zotter F., and Frank M. (2012): All-round Ambisonic panning and decoding, *J. Audio Eng. Soc.*, 60(10), 807–820.

Zotter F., and Frank M. (2019): *Ambisonics, A practical 3D audio theory for recording, studio production, sound reinforcement, and virtual reality*, Springer Open, Springer Nature, Cham, Switzerland.

Zotter F., Frank M., Kronlachner M., et al. (2014): Efficient phantom source widening and diffuseness in ambisonics, in *the Preceding of the EAA Joint Symposium on Auralization and Ambisonics*, Berlin, Germany.

Zotter F., Pomberger H., and Noisternig M. (2012): Energy-preserving ambisonic decoding, *Acta Acust. United Ac.*, 98(1), 37–47.

Zurek P.M. (1987): The precedence effect, in *Directional hearing* (edited by Yost W.A. and Gourevitch G.), Springer-Verlag, New York, USA.

Zwicker E., and Fastl H. (1999): *Psychoacoustics: facts and models* (2nd edition), Springer, Berlin, Germany.

Index

Pages in *italics* refers figures and **bold** refers tables.